OPERATIONS MANAGEMENT

OPERATIONS MANAGEMENT

TENTH EDITION

Nigel Slack

Alistair Brandon-Jones

Nicola Burgess

Harlow, England • London • New York • Boston • San Francisco • Toronto • Sydney • Dubai • Singapore • Hong Kong
Tokyo • Seoul • Taipei • New Delhi • Cape Town • São Paulo • Mexico City • Madrid • Amsterdam • Munich • Paris • Milan

PEARSON EDUCATION LIMITED
KAO Two
KAO Park
Harlow
CM17 9NA
United Kingdom
Tel: +44 (0)1279 623623
Web: www.pearson.com/uk

First published under the Pitman Publishing imprint 1995 (print)
Second edition (Pitman Publishing) 1998 (print)
Third edition 2001 (print)
Fourth edition 2004 (print)
Fifth edition 2007 (print)
Sixth edition 2010 (print)
Seventh edition 2013 (print and electronic)
Eighth edition 2016 (print and electronic)
Ninth edition 2019 (print and electronic)
Tenth edition 2022 (print and electronic)

ISBN: 978-1-292-40824-8 (print)
978-1-292-40821-7 (PDF)
978-1-292-40822-4 (ePub)

British Library Cataloguing-in-Publication Data
A catalogue record for the print edition is available from the British Library

Library of Congress Cataloging-in-Publication Data
Names: Slack, Nigel, author. | Brandon-Jones, Alistair, author. | Burgess, Nicola, author.
Title: Operations management / Nigel Slack, Alistair Brandon-Jones, Nicola Burgess.
Description: Tenth edition. | Harlow, England ; New York : Pearson, 2022. | Includes bibliographical references and index.
Identifiers: LCCN 2021056511 | ISBN 9781292408248 (print) | ISBN 9781292408217 (PDF) | ISBN 9781292408224 (ePub)
Subjects: LCSH: Production management. | Manufacturing processes. | Industrial management .
Classification: LCC TS155 .S562 2022 | DDC 658.5--dc23/eng/20220118
LC record available at https://lccn.loc.gov/2021056511

10 9 8 7 6 5 4 3 2 1
26 25 24 23 22

Front cover image: Andriy Onufriyenko/Moment/Getty Images
Cover design by Michelle Morgan, At The Pop Ltd.

Print edition typeset in 9.25/12 Sabon LT Pro by Straive
Printed in Slovakia by Neografia

NOTE THAT ANY PAGE CROSS REFERENCES REFER TO THE PRINT EDITION

BRIEF CONTENTS

CONTENTS

Companion Website

For open-access **student resources** specifically written to complement this textbook and support your learning, please visit **go.pearson.com/uk/he/resources**

Lecturer Resources

For password-protected online resources tailored to support the use of this textbook in teaching, please visit **go.pearson.com/uk/he/resources**

Pearson's Commitment to Diversity, Equity and Inclusion

Pearson is dedicated to creating bias-free content that reflects the diversity, depth and breadth of all learners' lived experiences. We embrace the many dimensions of diversity including, but not limited to, race, ethnicity, gender, sex, sexual orientation, socioeconomic status, ability, age and religious or political beliefs.

Education is a powerful force for equity and change in our world. It has the potential to deliver opportunities that improve lives and enable economic mobility. As we work with authors to create content for every product and service, we acknowledge our responsibility to demonstrate inclusivity and incorporate diverse scholarship so that everyone can achieve their potential through learning. As the world's leading learning company, we have a duty to help drive change and live up to our purpose to help more people create a better life for themselves and to create a better world.

Our ambition is to purposefully contribute to a world where:

- Everyone has an equitable and lifelong opportunity to succeed through learning.
- Our educational products and services are inclusive and represent the rich diversity of learners.
- Our educational content accurately reflects the histories and lived experiences of the learners we serve.
- Our educational content prompts deeper discussions with students and motivates them to expand their own learning and worldview.

We are also committed to providing products that are fully accessible to all learners. As per Pearson's guidelines for accessible educational Web media, we test and retest the capabilities of our products against the highest standards for every release, following the WCAG guidelines in developing new products for copyright year 2022 and beyond. You can learn more about Pearson's commitment to accessibility at:

https://www.pearson.com/us/accessibility.html

While we work hard to present unbiased, fully accessible content, we want to hear from you about any concerns or needs regarding this Pearson product so that we can investigate and address them.

- Please contact us with concerns about any potential bias at:

  https://www.pearson.com/report-bias.html

- For accessibility-related issues, such as using assistive technology with Pearson products, alternative text requests, or accessibility documentation, email the Pearson Disability Support team at:

  disability.support@pearson.com

Guide to 'Operations in practice' examples and case studies

Chapter	*Company example*	*Region*	*Sector/activity*	*Company size*
1 Operations management	LEGOLAND and LEGO both rely on their operations managers	Global	Manufacturing/ Leisure	Large
	MSF operations provide medical aid to people in danger	Global	Charity	Large
	Marina Bay Sands Hotel	Singapore	Hospitality	Large
	Servitisation and circular design at Philips Lighting	Europe	Lighting services	Large
	Two very different hospitality operations	Switzerland/ Global	Hospitality	Small/Medium/ Large
	Fjällräven products are voted the most sustainable in their field	Sweden	Manufacturing	Large
	Case study: Kaston-Trenton Service (KTS)	UK	Service	Small
2 Operations performance	Danone's path to B Corporation	Europe	Food manufacturing	Large
	Nutella shuts factory to preserve quality	France	Food manufacturing	Large
	Speeding response to save lives	UK	Emergency services	Various
	What does dependability mean when travelling by rail?	Global	Transport	Large
	566 quadrillion individual muesli mixes – now that's flexible	Germany	Internet retail	Small/Medium
	Everyday low prices at Aldi	Europe	Retail	Large
	Case study: IKEA looks to the future	Global	Retail	Large
3 Operations strategy	Operations is the basis of Ocado's strategy	Global	Retail	Large
	Dow Silicones' operations strategy	Global	Manufacturing	Large
	Tesco learns the hard way	UK	Retail	Large
	The rise of intangibles	All regions	Technology	Various
	Sustainability is high on Google's operations agenda	Global	Technology	Large
	Case study: McDonald's: half a century of growth	Global	Restaurants	Large

Chapter	*Company example*	*Region*	*Sector/activity*	*Company size*
4 Managing product and service innovation	The slow innovation progress of the Zip fastener	US/Sweden/Japan	Garment manufacturing	Various
	Gorilla Glass	USA	Technology	Medium/Large
	BT's open innovation ecosystem	UK/Global	Telecoms	Large
	Toyota's approach to organising innovation	Japan	Vehicle manufacturing	Large
	Product innovation for the circular economy	UK	Manufacturing	Small
	Case study: Widescale studios and the Fierybryde development	UK	Video game development	Small
5 The structure and scope of supply	Virtually like Hollywood	USA	Entertainment	Large
	Aalsmeer: a flower auction hub	Netherlands	Flower supply	Medium
	Adidas shuts its 'near market' factories	Germany/USA	Manufacturing	Large
	Aerospace in Singapore	Singapore	Aircraft servicing	
	Compass and Vodaphone – two ends of the outsourcing phenomenon	Global	Catering/Telecoms	Large
	Bangladesh disaster prompts reform – but is it enough?	Bangladesh	Manufacturing	Various
	Case study: Aarens Electronic	Netherlands	Manufacturing	Medium
6 Process design	Changi airport	Singapore	Transport	Large
	Fast (but not too fast) food drive-throughs	All regions	Restaurants	Various
	Legal & General's modular housing process	UK	Construction	Medium
	Ecover's ethical operation design	France and Belgium	Manufacturing	Large
	Dishang and Sands Film Studios – at opposite ends of the volume-variety spectrum	China/UK	Manufacturing	Small/Large
	London's underground tackles a bottleneck	UK	Transport	Large
	Case study: The Action Response Applications Processing Unit (ARAPU)	UK	Charity	Small
7 The layout and look of facilities	Ducati factory or Google office, they both have to look good	Italy/USA	Manufacturing/Technology	Large
	Reconciling quiet and interaction in laboratory layout	UK	Research	Small
	Supermarket layout	All regions	Retail	All
	Office layout and design	All regions	General	All
	Virtual reality brings layout to life	Switzerland	General	All
	Rolls-Royce factory is designed on environmental principles	UK	Manufacturing	Medium
	Case study: Misenwings SA	Switzerland	Catering	Small/Medium
8 Process technology	Go figure, or not	All regions	Technology	All
	Technology or people? The future of jobs	All regions	All	All
	Love it or hate it, Marmite's energy recycling technology	UK	Food manufacturing	Large
	Bionic duckweed	UK	Transport	Large
	'Wrong-shaped' parcels post a problem for UK Mail	UK	Delivery	Large
	Rampaging robots	USA	Entertainment	Large
	Case study: Logaltel Logistics	Europe	Logistics	Large

Chapter	*Company example*	*Region*	*Sector/activity*	*Company size*
9 People in operations	Do you want to own the company you work for?	UK	Software	Small
	Exoskeleton devices take the strain	USA	Manufacturing	Large
	Michelin calls it 'responsabilisation'	France	Manufacturing	Large
	Hybrid working divides opinions	Europe/USA	Several	Various
	The stress of high customer contact jobs	All regions	Service	Various
	Music while you work?	All regions	All	Various
	Technology and surveillance at work	All regions	All	Various
	Case study: Grace faces (three) problems	UK	Legal services	Small
10 Planning and control	Operations control at Air France	Global	Air transport	Large
	Can airline passengers be sequenced?	All regions	Air transport	Various
	The trials of triage	All regions	Healthcare	Various
	Sequencing and scheduling at London's Heathrow airport	UK	Air transport	Large
	Ryanair cancels flights after 'staff scheduling' errors	Europe	Air transport	Large
	Case study: Audall Auto Servicing	UK	Car servicing	Small
11 Capacity management	3M's COVID-19 surge capacity	Global	Manufacturing	Large
	How artificial intelligence helps with demand forecasting	All regions	Technology	Various
	Next-generation signal technology expands railway capacity	Global	Transport	Large
	Mass transport systems have limited options in coping with demand fluctuation	UK/ Singapore	Transport	Large
	Surge pricing helps manage demand for taxis and art galleries	UK	Service	Various
	Case study: FreshLunch	UK	Restaurants	Small
12 Supply chain Management	Zipline's drone-enabled supply network	Africa	Transport	Large
	Supply chain excellence at JD.com – the rise of an e-commerce titan	China	Retail	Large
	Twice around the world for Wimbledon's tennis balls	UK	Entertainment	Medium
	Considering the longer-term effects of COVID-19 on managing supply networks	Global	All	N/A
	Donkeys - the unsung heroes of supply networks	Global	Transport	Small
	TradeLens – blockchain revolutionises shipping	Denmark/ Global	Transport	Large
	Case study: Big or small? EDF's sourcing dilemma	UK	Energy	Large
13 Inventory management	An inventory of energy	All regions	Energy	Large
	Safety stocks for coffee and COVID	Switzerland/ All regions	Retail/ Healthcare	Large
	Mr Ruben's bakery	USA	Retail	Small
	Amazon's 'anticipatory inventory'	Global	Retail	Large
	France bans the dumping of unsold stock	France	Retail	Various
	Case study: supplies4medics.com	Europe	Healthcare	Medium

Chapter	Company example	Region	Sector/activity	Company size
14 Planning and control systems	Butcher's Pet Care coordinates its ERP	UK	Manufacturing	Large
	The computer never lies – really?	UK	Retail	Large
	The ERP for a chicken salad sandwich	UK	Manufacturing	Medium
	It's not that easy	Germany/ Australia/ Finland/USA	Various	Various
	Case study: Psycho Sports Ltd	N/A	Manufacturing	Small
15 Operations improvement	Kaizen at Amazon	Global	Retail	Large
	Disco balls and rice leads to Innovative Improvement	UK	Healthcare	Medium
	The checklist manifesto	USA	Healthcare	Various
	Six Sigma at Wipro	Global	Consultancy	Large
	Triumph motorcycles resurrected through benchmarking	UK	Manufacturing	Large
	Learning from Formula 1	UK	Retail	Large
	Schlumberger's InTouch technology for knowledge management	Global	Energy	Large
	Case study: Sales slump at Splendid Soup Co.	UK	Manufacturing	Large
16 Lean operations	Toyota: the lean pioneer	Global	Manufacturing	Large
	A very simple kanban	UK	Healthcare	Medium
	The rise of the personal kanban	All regions	N/A	Small
	Waste reduction in airline maintenance	All regions	Transport	Various
	Rapid changeover for Boeing and Airbus	Europe/USA	Transport	Large
	Jamie's 'lean' cooking	UK	Entertainment	Medium
	Autonomy at Amazon	All regions	Retail	Large
	Visual management at KONKEPT	Singapore	Retail	Medium
	Respect!	USA	Healthcare	Medium
	Case study: St Bridget's Hospital: seven years of lean	Sweden	Healthcare	Medium
17 Quality management	Quality at two operations: Victorinox and Four Seasons	Switzerland/ UK	Manufacturing/ Hospitality	Large
	Augmented reality technology adds to IKEA's service quality	Sweden	Retail	Large
	Virgin Atlantic offers a service guarantee for aviophobes	UK	Transport	Large
	Testing cars (close) to destruction	UK	Service	Medium
	Coin counting calculations	UK	Financial services	Various
	Keyboard errors – autofill and 'fat fingers'	UK/Germany	Government/ financial services	Large
	Quality systems only work if you stick to them	Japan	Manufacturing	Large
	Case study: Rapposcience Labs	Belgium	Mining	Small

Chapter	*Company example*	*Region*	*Sector/activity*	*Company size*
18 Managing risk and recovery	Time since last fatal crash. . . 12 years	USA	Transport	Various
	Volkswagen and the 'dieselgate' scandal	Germany	Manufacturing	Large
	Darktrace uses AI to guard against cyberattacks	UK	Technology	Large
	Pressing the passenger panic button	Global	Transport	Large
	Case study: Slagelse Industrial Services (SIS)	Denmark	Manufacturing	Medium
19 Project management	'For the benefit of all' – NASA's highs and lows	USA	Government	Large
	When every minute counts in a project – unblocking the Suez Canal	Egypt	Transport	Large
	McCormick's AI spice project	USA	Food	Large
	Berlin Brandenburg Airport opens at last	Germany	Transport	Large
	The risk of changing project scope – sinking the *Vasa*	Sweden	Military	Medium
	Ocado's robotics projects	UK	Retail	Large
	Case study: Kloud BV and Sakura Bank K.K.	Netherlands/ Japan	Financial services	Large

PREFACE

Operations may not run the world, but it makes the world run

This is our 10th edition

It's the 10th edition of this text, which means it's been around a long time! Since the first edition was published, a lot has happened to the subject of operations management. Supply networks, technologies, how people work and, above all, how the social responsibility of operations is viewed, have all changed radically. And so has this text. Over the years, we have changed the treatment and content to reflect key developments as (and often before) they fully emerge. Our philosophy has always been that we should keep pace with what is happening in the real world of operations management *as it is practised*.

One of the things that has affected the real world of operations management, is the COVID-19 pandemic. This edition was prepared as the Global pandemic was profoundly disrupting many established operations practices. Some pandemic-related changes will undoubtably endure, others will not. Some changes were simply accelerated versions of what was happening before the pandemic – for example, working from home. Others were relatively novel – workplace barriers, travel restrictions and socially distanced working. At the time of writing, it is not at all clear how widespread or long-lasting some of these changes will be. We have tried to use the COVID-19 pandemic to illustrate underlying principles of operations management and explain some of its effects, but without letting pandemic issues dominate the text.

It is adapting the content and coverage of the subject that has allowed us to maintain our market-leading position over the 10 editions. In 2021, this text was listed in the top 10 most highly cited business, marketing, accounting and economics textbooks worldwide, according to the *Financial Times* Teaching Power Rankings. It is our ambition to continue to include the many exciting developments in the subject well into the future. To help achieve this ambition, we are delighted to welcome a third author to the team. Our friend and colleague, Dr Nicola Burgess is a Reader at Warwick Business School. She has considerable teaching, research and administrative experience, and brings significant expertise to the team, particularly in the fields of 'lean' operations, operations improvement and healthcare management.

Why you need to study operations management

Because operations management is *everywhere*. Every time you experience a service and every time you buy a product, you are benefiting from the accomplishments of the operations managers who created them. Operations management is concerned with creating the services and products upon which we all depend. And all organisations produce some mixture of services and products, whether that organisation is large or small, manufacturing or service, for profit or not for profit, public or private. And, if you are a manager, remember that operations management is not confined to the operations function. All managers, whether they are called operations or marketing or human resources or finance, or whatever, manage processes and serve customers (internal or external). This makes at least part of their activities 'operations'.

Because operations management is *important*. Thankfully, most companies have now come to understand the importance of operations. This is because they have realised that, in the short-term, effective operations management gives the potential to improve both efficiency and customer service simultaneously. Even more important, operations management can provide the capabilities that ensure the survival and success of an enterprise in the long term.

Because operations management is *exciting*. It is at the centre of so many of the changes affecting the business world – changes in customer preference, changes in supply networks, changes in how we see the environmental and social responsibilities of enterprises, profound changes in technologies, changes in what we want to do at work, how we want to work, where we want to work and so on. There has rarely been a time when operations management was more topical or more at the heart of business and cultural shifts.

Because operations management is *challenging*. Promoting the creativity that will allow organisations to respond to so many changes is becoming the prime task of operations managers. It is they who must find the solutions to technological and environmental challenges, the pressures to be socially responsible, the increasing globalisation of markets and the difficult-to-define areas of knowledge management.

The aim of this text

This text provides a clear, authoritative, well-structured and interesting treatment of operations management as it applies to a variety of businesses and organisations. The text provides both a logical path through the activities of operations management and an understanding of their strategic context.

More specifically, this text is:

- *Strategic* in its perspective. It is unambiguous in treating the operations function as being central to competitiveness.
- *Conceptual* in the way it explains the reasons why operations managers need to take decisions.
- *Comprehensive* in its coverage of the significant ideas and issues that are relevant to most types of operation.
- *Practical* in that the issues and challenges of making operations management decisions *in practice* are discussed. The 'Operations in practice' boxes throughout each chapter and the case studies at the end of each chapter, all explore the approaches taken by operations managers in practice.
- *International* in the examples that are used. There are over 100 descriptions of operations practice from all over the world, over half of which are new for this edition.
- *Balanced* in its treatment. This means we reflect the balance of economic activity between service and manufacturing operations. Around 75 per cent of examples are from organisations that deal primarily in services and 25 per cent from those that are primarily manufacturing.

Who should use this text?

This text is for anyone who is interested in how services and products are created:

- *Undergraduate students* on business studies, technical or joint degrees should find it sufficiently structured to provide an understandable route through the subject (no prior knowledge of the area is assumed).
- *MBA students* should find that its practical discussions of operations management activities enhance their own experience.
- *Postgraduate students* on other specialist masters degrees should find that it provides them with a well-grounded and, at times, critical approach to the subject.

Distinctive features

Clear structure

The structure of the text uses the '4Ds' model of operations management that distinguishes between the strategic decisions that govern the ***direction*** of the operation, the ***design*** of the processes and operations that create products and services, planning and control of the ***delivery*** of products and services, and the ***development***, or improvement, of operations.

Illustrations-based

Operations management is a practical subject and cannot be taught satisfactorily in a purely theoretical manner. Because of this we have used short 'Operations in practice' examples that explain some of the issues faced by real operations.

Worked examples

Operations management is a subject that blends qualitative and quantitative perspectives; worked examples are used to demonstrate how both types of technique can be used.

Critical commentaries

Not everyone agrees about what is the best approach to the various topics and issues with operations management. This is why we have included 'critical commentaries' that pose alternative views to the ones being expressed in the main flow of the text.

Responsible operations

In every chapter, under the heading of 'Responsible operations', we summarise how the topic covered in the chapter touches upon important social, ethical and environmental issues.

Summary answers to key questions

Each chapter is summarised in the form of a list of bullet points. These extract the essential points that answer the key questions posed at the beginning of each chapter.

Case studies

Every chapter includes a case study suitable for class discussion. The cases are usually short enough to serve as illustrations, but have sufficient content also to serve as the basis of case sessions.

Problems and applications

Every chapter includes a set of problem-type exercises. These can be used to check your understanding of the concepts illustrated in the worked examples. There are also activities that support the learning objectives of the chapter that can be done individually or in groups.

Selected further reading

Every chapter ends with a short list of further reading that takes the topics covered in the chapter further, or treats some important related issues. The nature of each piece of further reading is also explained.

TO THE INSTRUCTOR. . .

Teaching and learning resources for the 10th edition

New for the 10th edition

In the 10th edition we have retained the extensive set of changes that we made in the 9th edition. In addition, with slight modification, we have retained the '4Ds' structure (direct, design, deliver and develop) that has proved to be exceptionally popular. Needless to say, as usual, we have tried to keep up to date with the (increasingly) rapid changes taking place in the wonderful world of operations.

Specifically, the 10th edition includes the following key changes:

- The coverage of 'lean operations', which was included in the 'Deliver' part in previous editions, has been moved to the 'Develop' part. This reflects the change in how 'lean' is seen in the subject. Its emphasis has shifted more towards a holistic approach to operations and improvement. And, while its role in planning and control remains relevant, lean is increasingly seen as an improvement approach.
- The 'Problems and applications' questions have been extended. Each chapter now has up to 10 questions that will help to practise analysing operations. They can be answered by reading the chapter. Model answers for the first two questions can be found on the companion website for this text. Answers to all questions are available to tutors adopting the text.
- Many totally new end-of-chapter case studies have been included. Of the 19 chapters, 10 cases are new to this text. We believe that these cases will add significantly to students' learning experience. However, several of the most popular cases have been retained.
- In every chapter we have included a new section called 'Responsible operations'. This summarises how the topic covered in the chapter touches upon important social, ethical and environmental issues. We have found that using this feature to develop the important issues of social, ethical and environmental responsibility through each session provides a useful learning thread that students respond to.
- We have extended and refreshed the popular 'Operations in practice' examples throughout the text. Of more than 100 examples, around 50 per cent are new to this text.
- We have further strengthened the emphasis on the idea that 'operations management' is relevant to every type of business and all functional areas of the organisation.
- We have placed greater stress on the worked examples in each chapter, so as to give students more help in analysing operations issues.
- Many new ideas in operations management have been incorporated. However, we have retained the emphasis on the foundations of the subject.
- A completely new instructor's manual is available to lecturers adopting this textbook, together with PowerPoint presentations for each chapter.

TO THE STUDENT. . .

Making the most of this text

All academic texts in business management are, to some extent, simplifications of the messy reality that is actual organisational life. Any text has to separate topics, in order to study them, which in reality are closely related. For example, technology choice impacts on job design that in turn impacts on quality management; yet, for simplicity, we are obliged to treat these topics individually. The first hint, therefore, in using this text effectively is to look out for all the links between the individual topics. Similarly with the sequence of topics, although the chapters follow a logical structure, they need not be studied in this order. Every chapter is, more or less, self-contained. Therefore, study the chapters in whatever sequence is appropriate to your course or your individual interests. But because each part has an introductory chapter, those students who wish to start with a brief 'overview' of the subject may wish first to study Chapters 1, 6, 10 and 15 and the chapter summaries of selected chapters. The same applies to revision – study the introductory chapters and summary answers to key questions.

The text makes full use of the many practical examples and illustrations that can be found in all operations. Many of these were provided by our contacts in companies, but many also come from journals, magazines and newsfeeds. So if you want to understand the importance of operations management in everyday business life, look for examples and illustrations of operations management decisions and activities in newsfeeds, social media and magazines. There are also examples that you can observe every day. Whenever you use a shop, eat a meal in a restaurant, download music, access online resources or ride on public transport, consider the operations management issues of all the operations of which you are a customer.

The end-of-chapter cases and problems are there to provide an opportunity for you to think further about the ideas discussed in the chapters. The problems can be used to test out your understanding of the specific points and issues discussed in the chapter and discuss them as a group, if you choose. If you cannot answer these you should revisit the relevant parts of the chapter. The cases at the end of each chapter will require some more thought. Use the questions at the end of each case study to guide you through the logic of analysing the issue treated in the case. When you have done this individually try to discuss your analysis with other course members. Most important of all, every time you analyse one of the case studies (or any other case or example in operations management) start off your analysis with the two fundamental questions:

- How is this organisation trying to compete (or satisfy its strategic objectives if a not-for-profit organisation)?
- What can the operation do to help the organisation compete more effectively?

Ten steps to getting a better grade in operations management

We could say that the best rule for getting a better grade is to be good. I mean really, really good! But there are plenty of us who, while fairly good, don't get the grade we really deserve. So, if you are studying operations management, and you want a really good grade, try following these simple steps:

Step 1 Practise, practise, practise. Use the 'Key questions' and the 'Problems and applications' to check your understanding.

Step 2 Remember a few **key models** and apply them wherever you can. Use the diagrams and models to describe some of the examples that are contained within the chapter.

Step 3 Remember to use both **quantitative and qualitative analysis**. You'll get more credit for appropriately mixing your methods: use a quantitative model to answer a quantitative question and vice versa but qualify this with a few well-chosen sentences. Each chapter incorporates qualitative and quantitative material.

Step 4 There's always a *strategic* **objective** behind any operational issue. Ask yourself, 'would a similar operation with a different strategy do things differently?' Look at the 'Operations in practice' examples in the text.

Step 5 Research widely around the topic. Use websites that you trust – don't automatically believe what you read. You'll get more credit for using references that come from genuine academic sources.

Step 6 Use **your own experience**. Every day, you're experiencing an opportunity to apply the principles of operations management. Why is the queue at the airport check-in desk so long? What goes on on in the kitchen of your favourite restaurant?'

Step 7 Always answer the question. Think 'what is really being asked here? What topic or topics does this question cover?' Find the relevant chapter or chapters, and search the key questions at the beginning of each chapter and the summary at the end of each chapter to get you started.

Step 8 Take account of the three tiers of accumulating marks for your answers:

(a) First, demonstrate your knowledge and understanding. Make full use of the text to find out where you need to improve.
(b) Second, show that you know how to illustrate and apply the topic. The case studies and 'Operations in practice' sections provide many different examples.
(c) Third, show that you can discuss and analyse the issues critically. Use the critical commentaries within the text to understand some of the alternative viewpoints.

Generally, if you can do (a) you will pass; if you can do (a) and (b) you will pass well; and if you can do all three, you will pass with flying colours!

Step 9 Remember not only **what** the issue is about, but also **understand why!** Try to understand why the concepts and techniques of operations management are important, and what they contribute to an organisation's success. Your new-found knowledge will stick in your memory, allow you to develop ideas and enable you to get better grades.

Step 10 Start now! Don't wait until two weeks before an assignment is due. Read on, and GOOD LUCK!

Nigel Slack, Alistair Brandon-Jones and Nicola Burgess

ABOUT THE AUTHORS

Nigel Slack is an Emeritus Professor of Operations Management and Strategy at Warwick University, and an Honorary Professor at Bath University. Previously he has been Professor of Service Engineering at Cambridge University, Professor of Manufacturing Strategy at Brunel University, a University Lecturer in Management Studies at Oxford University and Fellow in Operations Management at Templeton College, Oxford. He worked initially as an industrial apprentice in the hand-tool industry and then as a production engineer and production manager in light engineering. He holds a Bachelor's degree in Engineering and Master's and Doctor's degrees in Management, and is a Chartered Engineer. He is the author of many books and papers in the operations management area, including *The Manufacturing Advantage*, published by Mercury Business Books (1991), *Making Management Decisions* (1991) published by Prentice Hall, *Service Superiority* (with Robert Johnston, 1993), published by EUROMA, *The Blackwell Encyclopedic Dictionary of Operations Management* (with Michael Lewis) published by Blackwell, *Operations Strategy*, now in its 6th edition (with Michael Lewis, 2020) published by Pearson, *Perspectives in Operations Management* (*Volumes I to IV* with Michael Lewis, 2003) published by Routledge, *Operations and Process Management*, now in its 6th edition (with Alistair Brandon-Jones, 2021) published by Pearson, *Essentials of Operations Management*, now in its 2nd edition (with Alistair Brandon-Jones, 2018) also published by Pearson, and *The Operations Advantage*, published by Kogan Page (2017). He has authored numerous academic papers and chapters in books. He has also acted as a consultant to many international companies around the world in many sectors, especially financial services, transport, leisure, energy and manufacturing. His research is in the operations and manufacturing flexibility and operations strategy areas.

Alistair Brandon-Jones is a Full Chaired Professor of Operations and Supply Chain Management, and Head of the Information, Decisions and Operations Division in Bath University's School of Management. He is a Visiting Professor for Hult International Business School and Danish Technical University, and a non-executive director at Brevio (www.brevio.org) focused on smarter grant-making in the Third Sector. Between 2014 and 2017, he was Associate Dean for Post-Experience Education, responsible for the MBA, EMBA, DBA and EngDoc programmes. He was formerly a Reader at Manchester Business School, an Assistant and Associate Professor at Bath University, and a Teaching Fellow at Warwick Business School, where he also completed his PhD. His other books include *Operations and Process Management* (6th edition, 2021), *Essentials of Operations Management* (2nd edition, 2018) and *Quantitative Analysis in Operations Management* (2008). Alistair is an active empirical researcher focusing on digitisation of operations and supply chain management, professional service operations and healthcare operations. This research has been published extensively in world-elite journals including *Journal of Operations Management*, *International Journal of Operations & Production Management*, *International Journal of Production Economics* and *International Journal of Production Research*. Alistair has led Operations Management, Operations Strategy, Supply Chain Management, Project Management and Service Operations courses at all levels and has been invited to lecture at various international institutions, including the University of Cambridge, Hult International Business School, SDA Bocconi, Warwick Business School, NOVA University, Danish Technical University, Edinburgh Napier, Warwick Medical School and University College Dublin. In addition, he has extensive consulting and executive development experience with a range of organisations, including Maersk, Schroders Bank, Royal Bank of Scotland, Baker Tilly, Rowmarsh, QinetiQ Defence, Eni Oil and Gas, Crompton Greaves, Bahrain Olympic Committee, Qatar Leadership Centre, National Health Service and the Singapore Logistics Association. He has won a number of prizes for teaching excellence and contributions to pedagogy, including from *Times Higher Education*, Association of MBAs (AMBA), Production Operations Management Society (POMS), University of Bath, University of Manchester, University of Warwick and Hult International Business School.

Nicola Burgess is Reader in Operations Management at Warwick Business School. She has worked extensively with public sector organisations to understand operations management and improvement in a public sector context. Nicola's research has enabled her to work closely with policy makers as well as practitioners and she serves in an advisory capacity on healthcare programme boards. She also works closely with social enterprise in an advisory, research and teaching capacity. Her research has been published in world-leading journals including *Journal of Operations Management*, *European Journal of Operations Research*, *Human Resource Management* and the *British Medical Journal*. Nicola has taught operations management, operations strategy and supply chain management at all levels from undergraduate to postgraduate and contributes to the world-leading Distance Learning MBA at Warwick Business School. She is also Course Director for the innovative Foundation Year at Warwick Business School. Her teaching has been recognised by students as being 'passionate' and 'innovative', reflecting a desire to foster student engagement, enthusiasm and understanding of operations management, both inside and outside of the classroom.

AUTHORS' ACKNOWLEDGEMENTS

During the preparation of the 10th edition of this text (and previous editions) we have received an immense amount of help from friends and colleagues in the operations management community. In particular, everybody who has attended one of the regular 'faculty workshops' deserves thanks for the many useful comments. The generous sharing of ideas from these sessions has influenced this and all the other OM texts that we prepare. Our thanks go to everyone who attended these sessions and other colleagues who have helped us. It is, to some extent, invidious to single out individuals – but we are going to.

We thank Pär Åhlström of Stockholm School of Economics, James Aitken of University of Surrey,Eamonn Ambrose of University College Dublin, Erica Ballantyne of Sheffield University, Andrea Benn of University of Brighton, Yongmei Bentley of the University of Bedfordshire, Helen Benton of Anglia Ruskin University, Ran Bhamra of Loughborough University, Tony Birch of Birmingham City University, Briony Boydell of University of Portsmouth, Emma Brandon-Jones, John K Christiansen of Copenhagen Business School, Philippa Collins of Heriot-Watt University, Paul Coughlan of Trinity College Dublin, Doug Davies of University of Technology, Sydney, J.A.C. de Haan of Tilburg University, Ioannis Dermitzakis of Anglia Ruskin University, Stephen Disney of Cardiff University, Carsten Dittrich of the University of Southern Denmark, Tony Dromgoole of the Irish Management Institute, David Evans of Middlesex University, Ian Evans of Sunderland University, Margaret Farrell of Dublin Institute of Technology, Andrea Foley of Portsmouth University, Paul Forrester of Keele University, Abhijeet Ghadge of Heriot Watt University, Andrew Gough of Northampton University, Ian Graham of Edinburgh University, John Gray of The Ohio State University, Alan Harle of Sunderland University, Catherine Hart of Loughborough Business School, Susan Helper of Case Western Reserve University, Graeme Heron of Newcastle Business School, Steve Hickman of University of Exeter, Chris Hillam of Sunderland University, Ian Holden of Bristol Business School, Mickey Howard of Exeter University, Stavros Karamperidis of Heriot Watt University, Tom Kegan of Bell College of Technology, Hamilton, Benn Lawson of the University of Cambridge, Xiaohong Li of Sheffield Hallam University, John Maguire of the University of Sunderland, Charles Marais of the University of Pretoria, Lynne Marshall, Nottingham Trent University, Roger Maull of Exeter University, Bart McCarthy of Nottingham University, Peter McCullen of University of Brighton, John Meredith Smith of EAP, Oxford, Joe Miemczyk of ESCP Business School Europe, Michael Milgate of Macquarie University, Keith Millar of Ulster University, Keith Moreton of Staffordshire University, Phil Morgan of Oxford Brooks University, Adrian Morris of Sunderland University, Nana Nyarko of Sheffield Hallam University, Beverly Osborn of The Ohio State University, John Pal of Manchester Metropolitan University, Sofia Salgado Pinto of the Católica Porto Business School, Gary Priddis of University of Brighton, Carrie Queenan of University of South Carolina, Gary Ramsden of University of Lincoln, Steve Robinson of Southampton Solent University, Frank Rowbotham of University of Birmingham, James Rowell of University of Buckingham, Hamid Salimian of University of Brighton, Sarah Schiffling of University of Lincoln, Alex Skedd of Northumbria Business School, Andi Smart of Exeter University, Nigel Spinks of the University of Reading, Dr Ebrahim Soltani of the University of Kent, Rui Soucasaux Sousa of the Católica Porto Business School, Martin Spring of Lancaster University,James Stone of Aston University, R. Stratton of Nottingham Trent University, Ali Taghizadegan of University of Liverpool, Kim Hua Tan of the University of Nottingham, Dr Nelson Tang of the University of Leicester, Meinwen Taylor of South Wales University, Christos Tsinopoulos of Durham University, David Twigg of Sussex University, Arvind Upadhyay of University of Brighton, Helen Valentine of the University of the West of England, Andy Vassallo of University of East Anglia, Vessela Warren of University of Worcester, Linda Whicker of Hull University, John Whiteley of Greenwich University, Bill Wright of BPP Professional, Ying Xie of Anglia Ruskin University, Des Yarham of Warwick University, Maggie Zeng of Gloucestershire University and Li Zhou of University of Greenwich University.

In this edition we have received specific help with the new case studies. Our grateful thanks go to Vaggelis Giannikas, University of Bath, Jas Kalra, Newcastle

University, Jens Roehrich, University of Bath, Nigel Spinks, Henley and Brian Squire, University of Bath.

Our academic colleagues in the Operations Management Group at Warwick Business School and Bath University also helped, both by contributing ideas and by creating a lively and stimulating work environment.

At Warwick, thanks go to Vikki Abusidualghoul, Daniella Badu, Haley Beer, Mehmet Chakkol, Altricia Dawson, Mark Johnson, Anna Michalska, Pietro Micheli, Giovanni Radaelli, Ross Ritchie, Rhian Silvestro, and Chris Voss.

At Bath, thanks go to Meriem Bouazzaoui, Olivia Brown, Teslim Bukoye, Melih Celik, Soheil Davari, Brit Davidson, David Ellis, Jane Ellis-Brush, Malek El-Qallali, Güneş Erdogan, Vaggelis Giannikas, Elvan Gokalp, Andrew Graves, Gilbert Laporte, Michael Lewis, Sheik Meeran, Meng Meng, Zehra Onen Dumlu, Fotios Petropoulos, Lukasz Piwek, Jens Roehrich, Ozge Safak Aydiner, Ece Sanci, Mehrnoush Sarafan, Gamila Shoib, Michael Shulver, Brian Squire, Christos Vasilakis, Baris Yalabik.

We were lucky to receive continuing professional and friendly assistance from a great publishing team at Pearson. Special thanks to Rufus Curnow, Anita Atkinson, Felicity Baines and Diane Jones.

Finally, (and most importantly) to our families, who both supported and tolerated our nerdish obsession, thanks are inadequate, but thanks anyway to Angela and Kathy, Emma, Noah and George, and James, Maddy, Freya and Emily-Jane.

Nigel Slack, Alistair Brandon-Jones and Nicola Burgess

PART ONE

Directing the operation

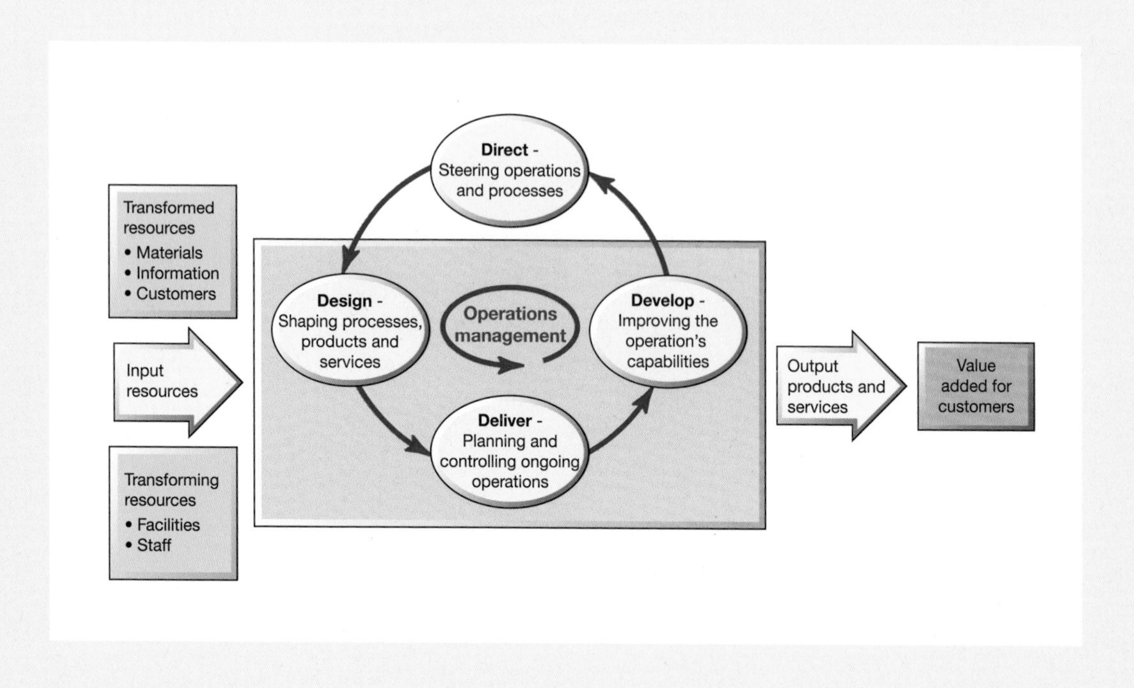

This part of the text introduces the idea of 'operations' and the operations function. It also examines the fundamental activities and decisions that shape the overall direction and strategy of the operations function. The chapters in this part are:

- **Chapter 1 Operations management**

  This introduces the common ideas that describe the nature and role of operations and processes in all types of organisation.

- **Chapter 2 Operations performance**

  This identifies how the performance of the operations function can be judged.

- **Chapter 3 Operations strategy**

  This examines how the activities of the operations function can have an important strategic impact.

- **Chapter 4 Managing product and service innovation**

  This looks at how innovation can be built into the product and service design process.

- **Chapter 5 The structure and scope of supply**

  This describes the major decisions that determine how, and the extent to which, an operation adds value through its own activities.

1 Operations management

KEY QUESTIONS

1.1 What is operations management?

1.2 Why is operations management important in all types of organisations?

1.3 What is the input–transformation–output process?

1.4 What is the process hierarchy?

1.5 How do operations (and processes) differ?

1.6 What do operations managers do?

INTRODUCTION

Operations management is about how organisations create and deliver services and products. Everything you wear, eat, sit on, use, read or knock about on the sports field comes to you courtesy of the operations managers who organised its creation and delivery. Everything you look up on a search engine, every treatment you receive at the hospital, every service you expect in the shops and every lecture you attend at university – all have been created by operations managers. While the people who supervised their creation and delivery may not always be called 'operations managers', that is what they really are. And that is what this text is concerned with – the tasks, issues and decisions of those operations managers who have made the services and products on which we all depend. This is an introductory chapter, so we will examine what we mean by 'operations management', how operations processes can be found everywhere, how they are all similar yet different, and what it is that operations managers do (see Figure 1.1).

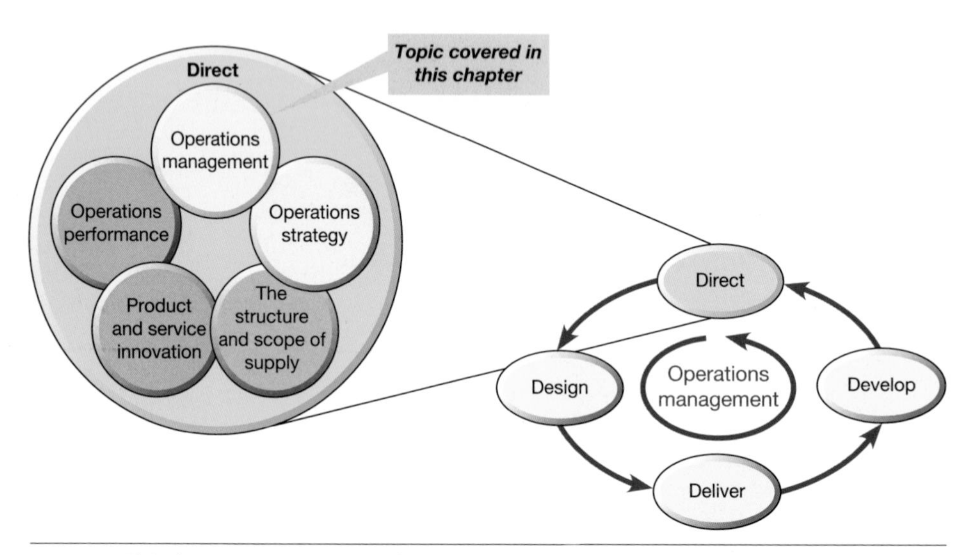

Figure 1.1 This chapter examines operations management

1.1 What is operations management?

Operations management is the activity of managing the resources that create and deliver services and products. The operations function is the part of the organisation that is responsible for this activity. Every organisation has an operations function because every organisation creates some type of service and/or product. However, not all types of organisation will necessarily call the operations function by this name. (Note in addition that we also use the shorter terms 'the operation' or 'operations' interchangeably with the 'operations function'.) Operations managers are the people who have particular responsibility for managing some, or all, of the resources that comprise the operations function. Again, in some organisations, the operations manager could be called by some other name. For example, they might be called the 'fleet manager' in a distribution company, the 'administrative manager' in a hospital, or the 'store manager' in a supermarket.

Operations principle
All organisations have 'operations' that produce some mix of services and products.

OPERATIONS IN PRACTICE — LEGOLAND® and LEGO® both rely on their operations managers

They may seem to be very different businesses, even though they partly share the same name. LEGOLAND is a world-renowned chain of location-based family leisure theme parks, and LEGO is one of the best-known makers of learning toys. But look in more detail and they share many common operations management activities. It is by looking at these activities that we can understand some of the similarities and difference between operations.

LEGOLAND[1]

Theme parks are a multi-billion-dollar industry. And one of the best-known brands in the industry is LEGOLAND®, whose LEGO-themed attractions hotels and accommodation are aimed primarily at families with children aged 3 to 12. LEGOLAND has parks in seven countries and across three continents. The first park opened over 60 years ago, near the LEGO factory in Billund, Denmark. Location is important. For example, LEGOLAND Deutschland is located in Bavaria close to Switzerland and Austria, all markets with a significant Lego following. All LEGOLAND parks are operated by the UK-based Merlin Entertainments, which also operates other branded attractions in the United Kingdom, Italy and Germany, such as Madame Tussauds, The London Eye, Warwick Castle and Alton Towers. What all of these have in common is that they provide their visitors with an 'experience'. Every stage of each attraction that customers (usually referred to as 'guests') move through has to be designed to create an intense or immersive experience centred on theming around movie or television characters, or in the case of LEGOLAND, LEGO intellectual property. The individual attractions in theme parks require considerable investment, often using sophisticated technology. Maintaining the utilisation of these attractions means trying to manage the flow of guests around the park so that they are queuing for as little time as possible. However, public holidays, seasons and weather will all impact on the number of guests wanting to visit each park. But however busy a park is, the quality of its guests' satisfaction with the experience is an important part of LEGOLAND'S operations management. What it calls its 'Guest Obsession' with creating smooth and memorable experiences for its guests includes regularly monitoring guest satisfaction scores and using 'net promoter' measurement (see Chapter 2 for a discussion of net promoter scores).

LEGO[2]

The LEGO Group, a privately held, family-owned company, with headquarters in Billund, Denmark, is one of the leading manufacturers of play materials. Lego bricks are manufactured at the Group's factories, located to be near its key markets in Europe and the United States. The company's success is founded on a deceptively simple idea. One LEGO brick is unremarkable but put one or two or more together and possibilities start to emerge. For example, there are more than 915 million possible ways of arranging six standard four-by-two bricks.[3] With all the elements, colours and decorations in the LEGO range, the total number of combinations becomes very large indeed. Yet however many bricks you assemble, and irrespective of what colour or set they are from, they will always fit together perfectly because they are made to very high levels of precision and quality. The company's motto is 'Only the best is good enough'. At the Billund operation, 60 tons of plastic is processed every 24 hours, with its moulding machines supplied by a complex arrangement of tubes. This stage is particularly important, because every LEGO piece must be made with tolerances as small as 10 micrometres. The moulds used by these machines are expensive, and each element requires its own mould. Robot trolleys travel between the machines, picking up boxes and leaving empty ones, an investment in automation that means that few people are required. In the packaging process the LEGO sets take their final form. The system knows exactly how much each packed box should weigh at any stage and any deviation sets off an alarm. Quality assurance staff perform frequent inspections and tests to make sure the toys are robust and safe. For every 1 million LEGO elements, only about 18 (that's 0.00002 per cent) fail to pass the tests. In addition, throughout the process, the company tries to achieve high levels of environmental sustainability. Plastic is extensively recycled in the factory.

Operations management is central to both businesses

Both LEGOLAND, which provides an entertainment service, and LEGO, which manufactures the famous LEGO bricks, depend on their operations managers to survive and prosper. It is they who design the stages that add value to the guests or the plastic that flows through the operation. They manage the activities that create services and products, they support the people whose skill and efforts contribute to adding value for both customers and the business itself. They attempt to match the operation's capacity with the demand placed upon it. They control quality throughout all the operation's processes. And they make whatever strategy each organisation has into practical reality. Without effective operations management, neither business would be as successful. Of course, there are differences between the two operations. One 'transforms' their guests, the other 'transforms' plastic. Yet they share a common set of operations management tasks and activities, even if the methods used to accomplish the tasks are different.

Table 1.1 Some activities of the operations function in various organisations

Internet service provider	Fast-food chain	International aid charity	Furniture manufacturer
▶ Maintain and update hardware ▶ Update software and content ▶ Respond to customer queries ▶ Implement new services ▶ Ensure security of customer data	▶ Locate potential sites for restaurants ▶ Provide processes and equipment to produce burgers, etc. ▶ Maintain service quality ▶ Develop, install and maintain equipment ▶ Reduce impact on local area ▶ Reduce packaging waste	▶ Provide aid and development projects for recipients ▶ Provide fast emergency response when needed ▶ Procure and store emergency supplies ▶ Be sensitive to local cultural norms	▶ Procure appropriate raw materials and components ▶ Make sub-assemblies ▶ Assemble finished products ▶ Deliver products to customers ▶ Reduce environmental impact of products and processes

If you want a flavour of some of the issues involved in managing a modern successful operation, look at the 'Operations in practice' example, 'LEGOLAND® and LEGO® both rely on their operations managers'. It illustrates how important the operations function is for any company whose reputation depends on creating high-quality, sustainable and profitable products and services. Their operations and their offerings are innovative, they focus very much on customer satisfaction, they invest in the development of their staff, and they play a positive role in fulfilling their social and environmental responsibilities. All of these issues are (or should be) high on the agenda of any operations manager in any operation. Continuing this idea, Table 1.1 shows just some of the activities of the operations function for various types of organisations.

Operations in the organisation

The operations function is central to the organisation because it creates and delivers services and products, which is its reason for existing. The operations function is one of the three core functions of any organisation. These are:

- the marketing (including sales) function – which is responsible for positioning and communicating the organisation's services and products to its markets in order to generate customer demand;
- the product/service development function – which is responsible for developing new and modified services and products in order to generate future customer demand;
- the operations function – which is responsible for the creation and delivery of services and products based on customer demand.

In addition, there are the support functions that enable the core functions to operate effectively. These include, for example, the accounting and finance function, the technical function, the human resources function and the information systems function. Remember that although different organisations may call their support functions by different names, almost all organisations will have the three core functions.

However, there is not always a clear division between functions. This leads to some confusion over where the boundaries of the operations function should be drawn. In this text, we use a relatively broad definition of operations. We treat much of the product/service development, technical and information systems activities and some of the human resource, marketing, and accounting and finance activities as coming within the sphere of operations management. We view the operations function as comprising all the activities necessary for the day-to-day fulfilment of customer requests within the constraints of social and environmental sustainability. This includes sourcing services and products from suppliers and delivering services and products to customers.

Operations principle
Operations managers need to cooperate with other functions to ensure effective organisational performance.

Figure 1.2 illustrates some of the relationships between operations and other functions in terms of the flow of information between them. Although not comprehensive, it gives an idea of the nature of each relationship. Note that the support functions have a different relationship with operations to the core functions. Operations management's responsibility to support functions is primarily to make sure that they understand operations' needs and help them to satisfy these needs. The relationship with the other two core functions is more equal – less of 'this is what we want' and more 'this is what we can do currently – how do we reconcile this with broader business needs?'

1.2 Why is operations management important in all types of organisations?

In some types of organisation, it is relatively easy to visualise the operations function and what it does, even if we have never seen it. For example, most people have seen images of a vehicle assembly line. But what about an advertising agency? We know vaguely what they do – they create the advertisements that we see online, in magazines and on television – but what is their operations function? The clue lies in the word 'create'. Any business that creates something must use resources to do so, and so must have an operations activity. Also, the vehicle plant and the advertising agency do have one important element in common; both have a higher objective – to make a profit from creating and delivering their products or services. Yet not-for-profit organisations also use their resources to create and deliver services, not to make a profit, but to serve society in some way. Look at the examples of what operations management does in five very different organisations in Figure 1.3 and some common themes emerge.

Start with the statement from the 'easy-to-visualise' vehicle plant. Its summary of what operations management does is: 'Operations management uses machines to efficiently assemble products that satisfy current customer demands'. The statements from the other organisations are similar but use slightly different language. Operations management uses not just machines but also 'knowledge', 'people', 'our and our partners' resources' and 'our staff's knowledge and experience', to efficiently (or effectively or creatively) assemble (or produce, change, sell, move, cure, shape, etc.)

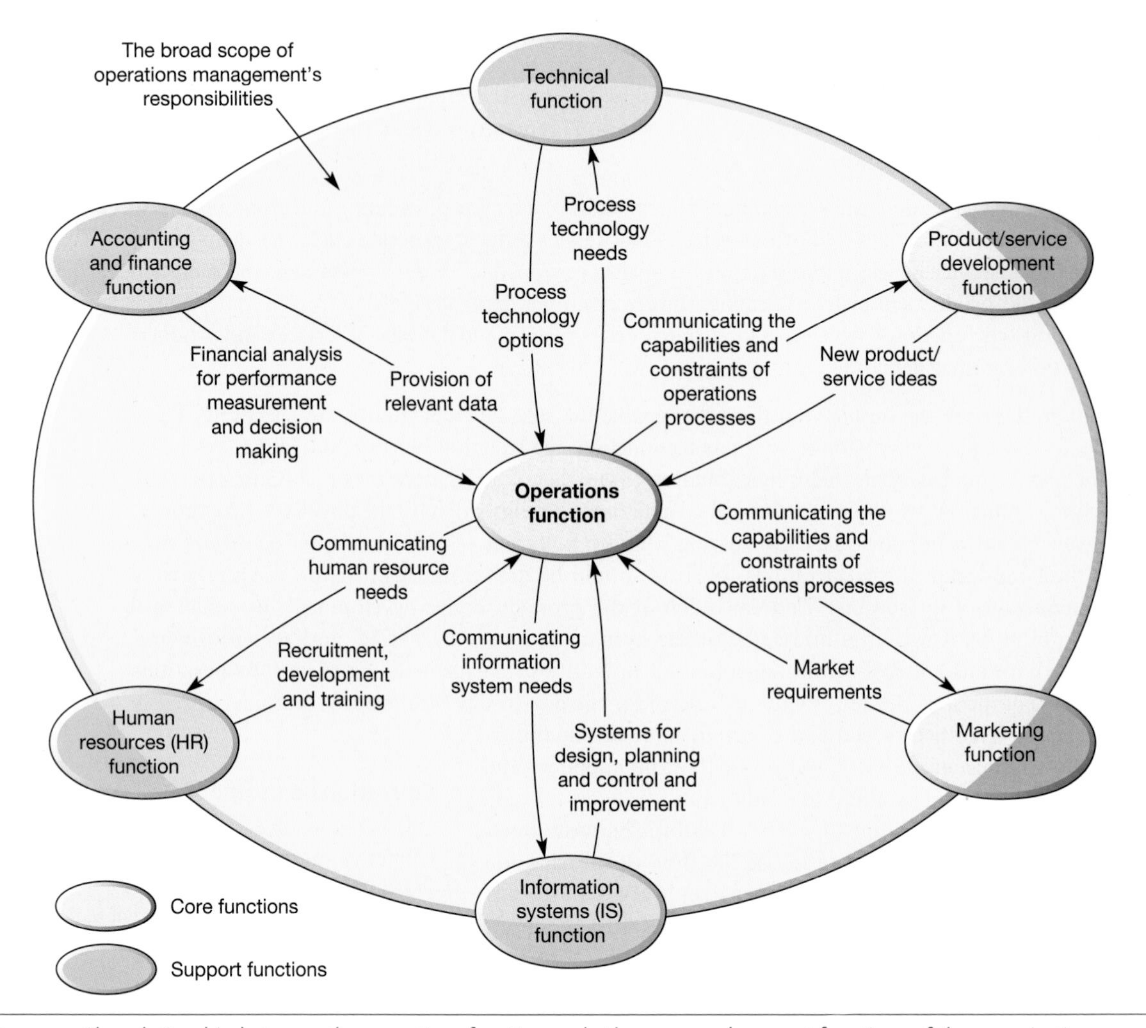

Figure 1.2 The relationship between the operations function and other core and support functions of the organisation

> **Operations principle**
> The economic sector of an operation is less important than its intrinsic characteristics in determining how it should be managed.

products (or services or ideas) that satisfy (or match or exceed or delight) customer (or clients' or citizens' or society's) demands (or needs or concerns or even dreams).

Whatever terminology is used there is a common theme and a common purpose to how we can visualise the operations activity in any type of organisation; small or large, service or manufacturing, public or private, profit or not-for-profit. Operations management uses 'resources to appropriately create outputs that fulfil defined market requirements' (see Figure 1.4). However, although the essential nature and purpose of operations management is the same in any type of organisation, there are some special issues to consider, particularly in smaller organisations and those whose purpose is to maximise something other than profit.

Operations management in the smaller organisation

Operations management is just as important in small organisations as it is in large ones. Irrespective of their size, all companies need to create and deliver their services and products efficiently and effectively. However, in practice, managing operations in a small or medium-size organisation has its own set of problems. Large companies may have the resources to dedicate individuals to specialised tasks, but smaller companies often cannot, so people may have to do different jobs as the need arises. Such an informal structure can allow the company to respond quickly as opportunities or problems present themselves. But decision making can also become confused as individuals' roles overlap. Small

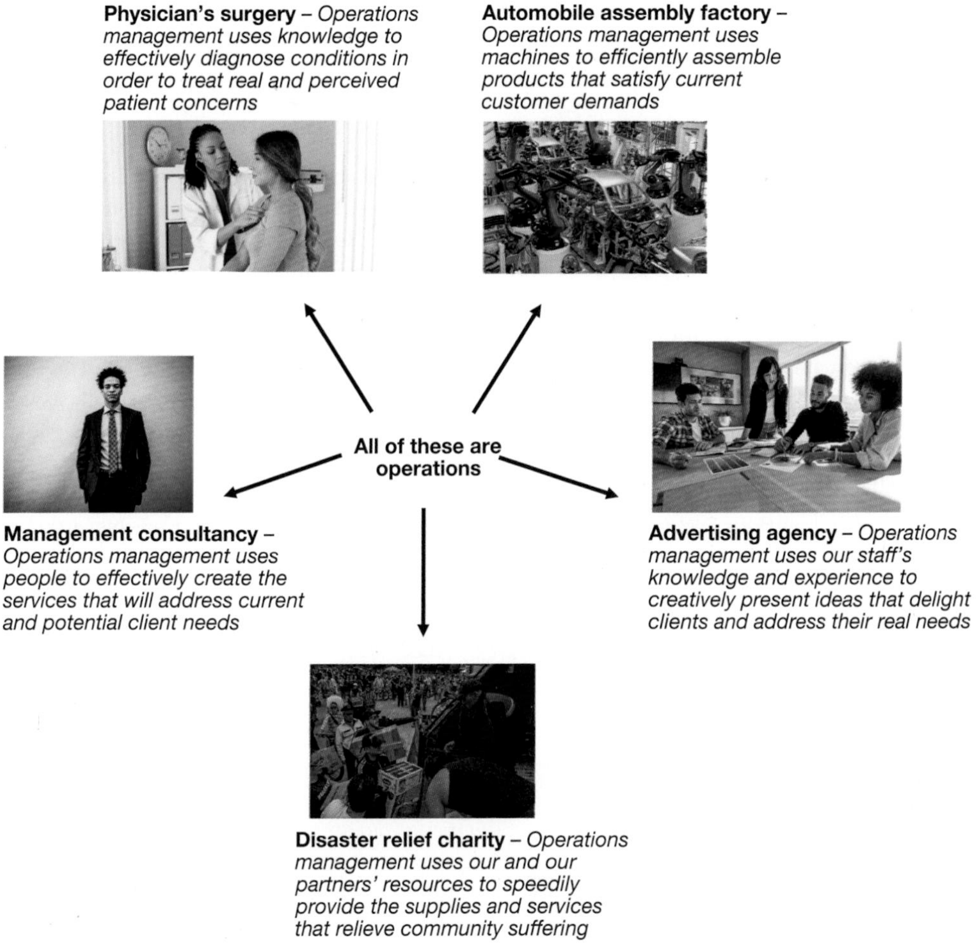

Physician's surgery – *Operations management uses knowledge to effectively diagnose conditions in order to treat real and perceived patient concerns*

Automobile assembly factory – *Operations management uses machines to efficiently assemble products that satisfy current customer demands*

All of these are operations

Management consultancy – *Operations management uses people to effectively create the services that will address current and potential client needs*

Advertising agency – *Operations management uses our staff's knowledge and experience to creatively present ideas that delight clients and address their real needs*

Disaster relief charity – *Operations management uses our and our partners' resources to speedily provide the supplies and services that relieve community suffering*

Figure 1.3 All of these are operations that produce some mix of products and services

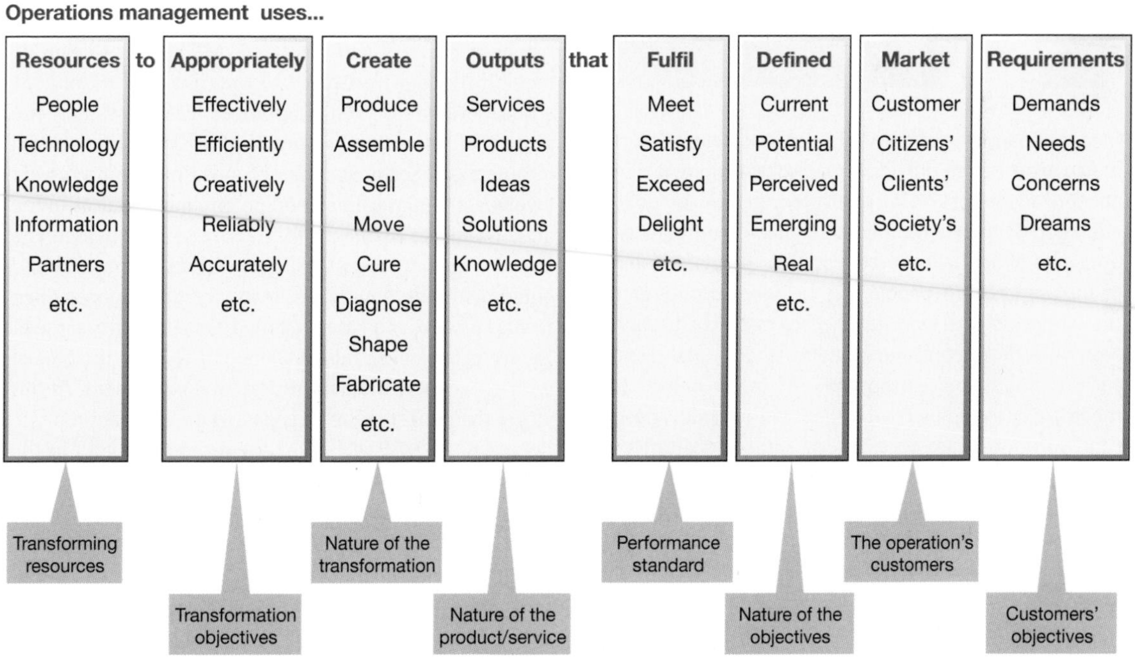

Operations management uses...

Resources	to	Appropriately	Create	Outputs	that	Fulfil	Defined	Market	Requirements
People		Effectively	Produce	Services		Meet	Current	Customer	Demands
Technology		Efficiently	Assemble	Products		Satisfy	Potential	Citizens'	Needs
Knowledge		Creatively	Sell	Ideas		Exceed	Perceived	Clients'	Concerns
Information		Reliably	Move	Solutions		Delight	Emerging	Society's	Dreams
Partners		Accurately	Cure	Knowledge		etc.	Real	etc.	etc.
etc.		etc.	Diagnose	etc.			etc.		
			Shape						
			Fabricate						
			etc.						
Transforming resources		Transformation objectives	Nature of the transformation	Nature of the product/service		Performance standard	Nature of the objectives	The operation's customers	Customers' objectives

Figure 1.4 Operations management uses *resources* to *appropriately create outputs* that *fulfil defined market requirements*

companies may have exactly the same operations management issues as large ones but they can be more difficult to separate from the mass of other issues in the organisation.

Operations management in not-for-profit organisations

Terms such as 'competitive advantage', 'markets' and 'business', which are used in this text, are usually associated with companies in the for-profit sector. Yet operations management is also relevant to organisations whose purpose is not primarily to earn profits. Managing the operations in an animal welfare charity, hospital, research organisation or government department is essentially the same as in commercial organisations. Operations have to take the same decisions – how to create and deliver services and products, invest in technology, contract out some of their activities, devise performance measures, improve their operations performance and so on. However, the strategic objectives of not-for-profit organisations may be more complex and involve a greater emphasis on political, economic, social or environmental objectives. Because of this there may be a greater chance of operations decisions being made under conditions of conflicting objectives. For example, it is the operations staff in a children's welfare department who have to face the conflict between the cost of providing extra social workers and the risk of a child not receiving adequate protection. Nevertheless, the vast

OPERATIONS IN PRACTICE

MSF operations provide medical aid to people in danger[4]

Médecins Sans Frontières (MSF) is an independent humanitarian organisation providing medical aid where it is most needed, regardless of 'race, religion, gender or political affiliation' with actions 'guided by medical ethics and the principles of neutrality and impartiality' to raise awareness of the plight of the people they help in countries around the world. Its core work takes place in crisis situations – armed conflicts, epidemics, famines and natural disasters such as floods and earthquakes. Its teams deliver both medical aid (including consultations with a doctor, hospital care, nutritional care, vaccinations, surgery, obstetrics and psychological care) and material aid (including food, shelter, blankets, etc.). Each year, MSF sends doctors, nurses, logisticians, water and sanitation experts, administrators and other professionals to work alongside locally hired staff. It is one of the most admired and effective relief organisations in the world. But no amount of fine intentions can translate into effective action without superior operations management. MSF must be able to react to any crisis with fast response, efficient logistics systems, and efficient project management.

Its response procedures are being developed continuously to ensure that it reaches those most in need as quickly as possible. The process has five phases: proposal, assessment, initiation, running the project and closing. The information that prompts a possible mission can come from governments, humanitarian organisations, or MSF teams already present in the region. Once the information has been checked and validated, MSF sends a team of medical and logistics experts to the crisis area to carry out a quick evaluation. When approved, MSF staff start the process of selecting personnel, organising materials and resources and securing project funds. Initiating a project involves sending technical equipment and resources to the area. Thanks to its pre-planned processes, specialised kits and emergency stores, MSF can distribute material and equipment within 48 hours, ready for the response team to start work as soon as they arrive. Once the critical medical needs have been met, MSF begins to close the project with a gradual withdrawal of staff and equipment. At this stage, the project closes or is passed on to an appropriate organisation. MSF will also close a project if risks in the area become too great to ensure staff safety. Whether it is dealing with urgent emergencies, or a long-running programme, everything MSF does on the ground depends on efficient logistics. Often, aircraft can be loaded and flown into crisis areas within 24 hours. But, if it is not a dire emergency, MSF reduces its costs by shipping the majority of material and drugs by sea.

majority of the topics covered in this text have relevance to all types of organisations, including non-profit, even if the context is different and some terms may have to be adapted.

The new operations agenda

Changes in the business environment have had a significant impact on the challenges faced by operations managers. Some of them are in response to changes in the nature of demand. Many (although not all) industries have experienced increased cost-based competition while their customers' expectations of quality and variety have increased simultaneously. What is possible technologically is also changing rapidly, as are customers' attitudes to social and environmental issues. At the same time, political, legal and regulatory structures have changed. In response, operations managers have had to adjust their activities to cope, especially in the following areas:

- *New technologies* – In both manufacturing and service industries, process technologies are changing so fast that it is difficult to predict exactly what their effect will be, only a few years in the future. Certainly, they are likely to have a dramatic effect, radically altering the operating practices of almost all types of operation.
- *Different supply arrangements* – Some markets have become more global, while others have been constrained by politically inspired trade restrictions. Some globalised supply markets are opening up new sourcing options, other supply chains have become increasingly risky. Often, opportunities for cost savings must be balanced against supply vulnerability and ethical issues.
- *Increased emphasis on social and environmental issues* – Generally, customers, staff, and even investors, have been developing an increased ethical and environmental sensitivity, leading to operations having to change the way they conceive and create their products and services. Similarly, there is a greater expectation about the ethical treatment of all an operation's stakeholders, including customers, the workforce, suppliers and society in general.

Figure 1.5 identifies just some of the operations responses in these three areas. (If you don't recognise some of the terms, don't worry, we will explain them throughout the text.) These responses form a major part of a new agenda for operations. The issues in Figure 1.5 are not comprehensive, nor are they universal. But very few operations functions will be unaffected by at least some of them. You will find 'Operations in practice' examples throughout this text that look at various aspects of these three areas and 'Responsible operations' sections in every chapter that look at social, environmental and ethical issues.

Operations principle
Operations management is at the forefront of coping with, and exploiting, developments in business and technology.

1.3 What is the input–transformation–output process?

All operations create and deliver services and products by changing inputs into outputs using an 'input–transformation–output' process. Figure 1.6 shows this general transformation process model that is the basis of all operations. Put simply, operations take in a set of input resources that are used to transform something, or are transformed themselves, into outputs of services and products. And although all operations conform to this general input–transformation–output model, they differ in the nature of their specific inputs and outputs. For example, if you stand far enough away from a hospital or a vehicle plant, they might look very similar, but move closer and clear differences do start to emerge. One is a service operation delivering 'services' that change the physiological or psychological condition of patients; the other is a manufacturing operation creating and delivering 'products'. What is inside each operation will also be different. The hospital contains diagnostic, care and therapeutic processes whereas the motor vehicle plant contains metal-forming machinery and assembly processes. Perhaps the most important difference between the two operations, however, is the nature of their inputs. The hospital transforms the customers themselves. The patients form part of the input to, and the output from, the operation. The vehicle plant transforms steel, plastic, cloth, tyres and other materials into vehicles.

Operations principle
All processes have inputs of transforming and transformed resources that they use to create products and services.

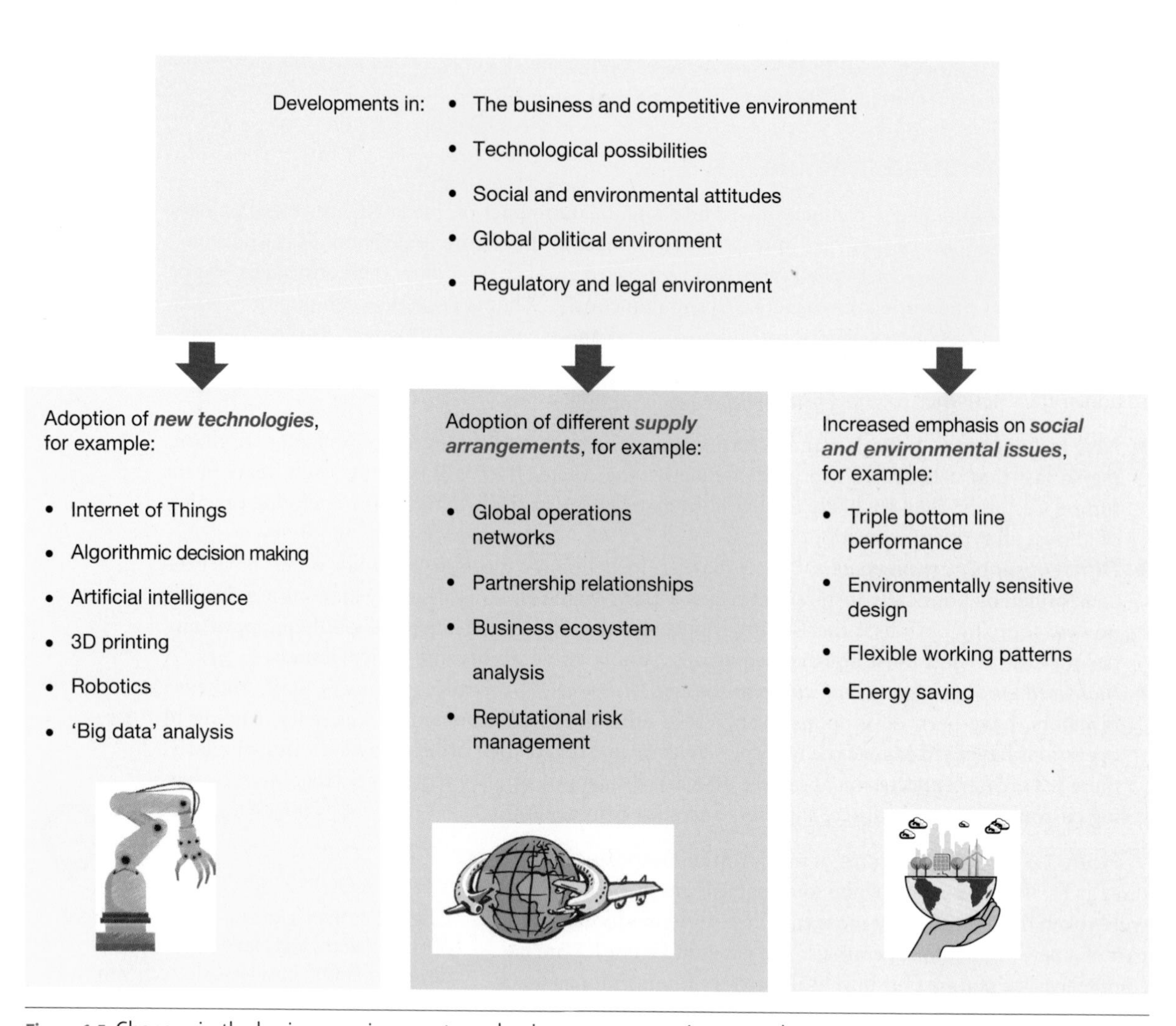

Developments in:
• The business and competitive environment
• Technological possibilities
• Social and environmental attitudes
• Global political environment
• Regulatory and legal environment
Adoption of new technologies, for example:
• Internet of Things
• Algorithmic decision making
• Artificial intelligence
• 3D printing
• Robotics
• 'Big data' analysis
Adoption of different supply arrangements, for example:
• Global operations networks
• Partnership relationships
• Business ecosystem analysis
• Reputational risk management
Increased emphasis on social and environmental issues, for example:
• Triple bottom line performance
• Environmentally sensitive design
• Flexible working patterns
• Energy saving

Figure 1.5 Changes in the business environment are shaping a new operations agenda

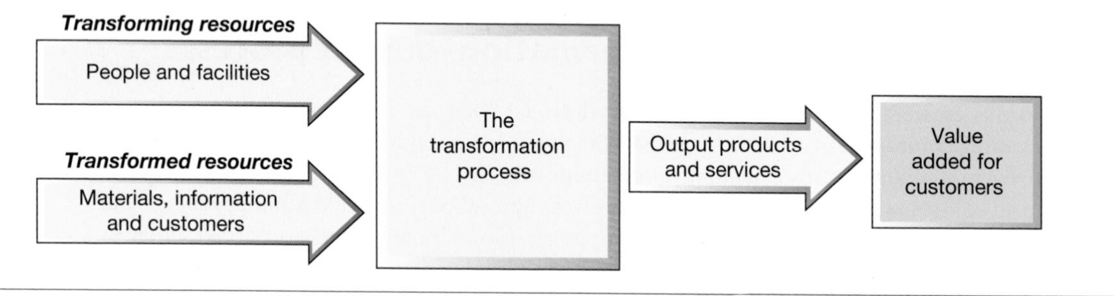

Figure 1.6 All operations are input–transformation–output processes

Inputs to the process – transformed resources

One set of inputs to any operation's processes are transformed resources. These are the resources that are treated, transformed or converted in the process. They are usually a mixture of the following:

- *Materials* – operations that process materials could do so to transform their physical properties (shape or composition, for example). Most manufacturing operations are like this. Other operations process materials to change their location (parcel delivery companies, for example). Some, like retail operations, do so to change the possession of the materials. Finally, some operations store materials, such as warehouses.

- ***Information*** – operations that process information could do so to transform their informational properties (that is, the purpose or form of the information); accountants do this. Some change the possession of the information: for example, market research and social media operations aggregate and sell information. Some store the information, such as archives and libraries. Finally, some operations, such as telecommunication companies, change the location of the information.
- ***Customers*** – operations that process customers might change their physical properties in a similar way to materials processors: for example, hairdressers or cosmetic surgeons. Some, like hotels, store (or more politely, accommodate) customers. Airlines and mass rapid transport transform the location of their customers, while hospitals transform their physiological state. Some are concerned with transforming their psychological state: for example, most entertainment services such as music, theatre, television, radio and theme parks. But customers are not always simple 'passive' items to be processed. They can also play a more active part: for example, they create the atmosphere in a restaurant; they provide the stimulating environment in learning groups in education, and so on.

Operations principle

Transformed resource inputs to a process are materials, information or customers.

Some operations have inputs of materials *and* information *and* customers, but usually one of these is dominant. For example, a bank devotes part of its energies to producing printed statements by processing inputs of material, but no one would claim that a bank is a printer. The bank also is concerned with processing inputs of customers at its branches and contact centres. However, most of the bank's activities are concerned with processing inputs of information about its customers' financial affairs. As customers, we may be unhappy with badly printed statements and we may be unhappy if we are not treated appropriately in the bank. But if the bank makes errors in our financial transactions, we suffer in a far more fundamental way. Table 1.2 gives examples of operations with their dominant transformed resources.

Inputs to the process – transforming resources

The other set of inputs to any operations process are transforming resources. These are the resources that act upon the transformed resources. There are two types, which form the 'building blocks' of all operations:

- facilities – the buildings, equipment, plant and process technology of the operation;
- staff – the people who operate, maintain, plan and manage the operation. (Note we use the term 'staff' to describe all the people in the operation, at any level.)

The exact nature of both facilities and staff will differ between operations. To a five-star hotel, its facilities consist mainly of 'low-tech' buildings, furniture and fittings. To a nuclear-powered aircraft carrier, its facilities are 'high-tech' nuclear generators and sophisticated electronic equipment. Staff will also differ between operations. Most staff employed in a factory assembling domestic

Table 1.2 Dominant transformed resource inputs of various operations

Predominantly processing inputs of materials	Predominantly processing inputs of information	Predominantly processing inputs of customer
▶ All manufacturing operations	▶ Accountants	▶ Hairdressers
▶ Mining companies	▶ Bank headquarters	▶ Hotels
▶ Retail operations	▶ Market research company	▶ Hospitals
▶ Warehouses	▶ Financial analysts	▶ Mass rapid transports
▶ Postal services	▶ News service	▶ Theatres
▶ Container shipping lines	▶ University research unit	▶ Theme parks
▶ Trucking companies	▶ Telecoms company	▶ Dentists

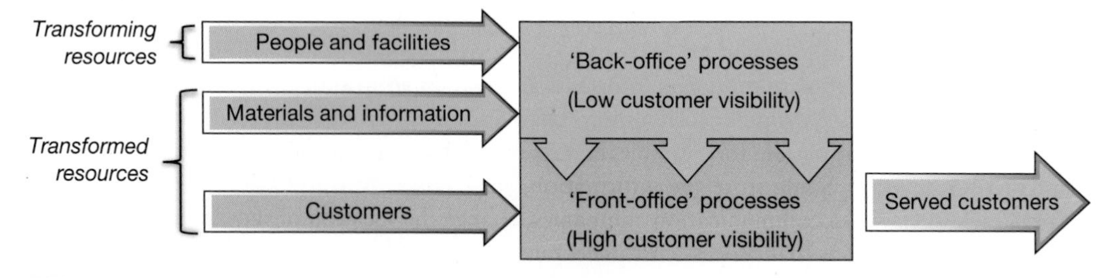

Figure 1.7 When the main transformed resource is the customers themselves, it is useful to distinguish between 'front-office' processes that act on customers directly and 'back-office' processes that provide indirect services

refrigerators may not need a very high level of technical skill. In contrast, most staff employed by an accounting company are, hopefully, highly skilled in their own particular 'technical' skill (accounting). Yet although skills vary, all staff can make a contribution. An assembly worker who consistently misassembles refrigerators will dissatisfy customers and increase costs just as surely as an accountant who cannot add up. The balance between facilities and staff also varies. A computer chip manufacturing company, such as Intel, will have significant investment in physical facilities. A single chip fabrication plant will cost billions of dollars, so operations managers will spend a lot of their time managing their facilities. Conversely, a management consultancy firm depends largely on the quality of its staff. Here operations management is largely concerned with the development and deployment of consultant skills and knowledge.

Operations principle

All processes have transforming resources of facilities (equipment, technology, etc.) and people.

Front- and back-office transformation

A distinction that is worth noting at this point, mainly because it has such an impact on how transforming resources are managed, is that between 'front-' and 'back-office' transformation. The '**front-office**' (or 'front-of-house') parts of an operation are those processes that interact with (transform) customers. '**Back-office**' (or 'back-of-house') operations are the processes that have little or no direct contact with customers, but perform the activities that support the front office in some way. The distinction is illustrated in Figure 1.7. But, as implied by the figure, the boundary between front and back offices is not clean. Different processes within an operation could have different degrees of exposure (what we refer to later as 'visibility') to customers.

OPERATIONS IN PRACTICE

Marina Bay Sands Hotel[5]

There are very few better examples of how back and front offices work together than the hotel industry. As customers, we naturally judge a hotel primarily on its front-office, client-facing, staff and facilities, but without effective back-office operations, customers would soon find that their front-office experience would be very much affected. This is certainly true for the Marina Bay Sands hotel in Singapore. Located in the heart of Singapore's Central Business District, Marina Bay Sands is an integrated, multi-award-winning, luxury resort owned by the Las Vegas Sands corporation, incorporating a hotel with over 2,500 rooms, a huge convention and exhibition centre, restaurants, a shopping mall, museum, two large theatres and the world's largest atrium casino. The hotel's three towers are crowned by the spectacular Sands SkyPark, which offers a 360-degree view of Singapore's skyline. It is home to lush gardens, an infinity edge swimming pool and an observation deck.

But the meticulous service provided by the hotel's highly trained front-of-house staff could not happen without the many back-of-house processes that customers do not always notice. Some of these processes are literally invisible to customers, for example those that keep the accounts, or those that maintain the hotel's air-conditioning systems, or the dim sum preparation (dim sum are steamed dumplings served in small, bite-sized portions – specialist chefs prepare 5,000 individual pieces every day). These processes are all important, and mass operations in their own right. Some back-of-house departments rely more on technology. The hotel's laundry must clean and press 4,000 pool towels every day, as well as thousands of items of room linen. Which is a problem for an organisation whose sustainability policy commits it to minimising its use of water. It took an investment of over £10 million in water-saving technology to reduce the hotel's usage by 70 per cent. Other back-of-house operations have a direct impact on how customers view the hotel. For example, the wardrobe department that keeps the hotel's over 9,000 staff looking smart is reputed to be the most high-tech in the world. Its 18 automated conveyors each have slots for 620 individual items of uniform, all of which have individual identification chips so that they can be tracked. Staff enter their number into a keypad, and, behind the scenes, the conveyor system automatically delivers the uniform. Some processes straddle the front-of-house/back-of-house divide. The valet parking operation parks up to 200 cars each hour in its 2,500 parking spaces, and retrieves them in a target retrieval time of seven minutes. Housekeeping cleans, tidies and stocks all the bedrooms. The hotel's 50 butlers serve the more exclusive suites and cater for a wide variety of demands (one guest asked them to arrange a wedding banquet at four hours' notice). It is a role that demands dedication and attention to detail.

Outputs from the process

Operations create products and services. Products and services are often seen as different. Products are physical things whereas services are activities or processes. A car or a newspaper or a restaurant meal is a product, whereas a service is the activity of the customer using or consuming that product. Yet, although some services do not involve many physical products, and some manufacturers do not give much service, most operations produce some mixture of products and services, even if one predominates. For example, services like consultancies produce reports, hairdressers sell hair gel and food manufacturers give advice on how to prepare their products.

Products or services, or does it matter?

The difference between a 'product' and a 'service' is not always obvious and has provoked a lot of (not always useful) academic debate. At an obvious, but simple, level, a product is a physical and tangible thing (you can touch a car, or television or phone). By contrast, a service is an activity that usually involves interaction with a customer (as with a doctor) or something representing the customer (as with a package delivery service). The resources that carry out these services may be tangible, but not the service they provide. For many years the accepted distinction between products and services was not confined to intangibility, but included other characteristics abbreviated to 'IHIP', standing for:

- Intangibility, in that they are not physical items.
- Heterogeneity, in that they are difficult to standardise because each time a service is delivered, it will be different because the needs and behaviour of customers will, to some extent, vary.
- Inseparability, in that their production and consumption are simultaneous. The service provider (who 'produces' the service) is often physically present when its consumption by a customer takes place.
- Perishability, in that they cannot be stored because they have a very short 'shelf life'. They may even perish in the very instant of their creation, like a theatre performance.

However, there are several problems with using these characteristics to define a 'service' – hence the academic debate. It is certainly not difficult to find examples of services that do not conform to them. Also, technology has had a significant effect; both on the extent to which the IHIP characteristics apply and how the limits that they place on service operations can be overcome. In particular, the development of information and communication technology has opened up many new types of service offerings. Yet, although they cannot totally define what is a 'service' and what is a 'product', each of the IHIP characteristics does have some validity.

Most operations produce outputs somewhere on a spectrum of the IHIP characteristics

Some operations produce just products. For example, mineral extraction operations (miners) are concerned almost exclusively with the product that comes from their mines. It is tangible, almost totally standardised, produced away from its consumption, and storable. Others produce just services. For example, a psychotherapy clinic provides personalised and close-contact therapeutic treatment for its customers with few, if any, tangible elements. However, most operations produce outputs that are somewhere in-between the two extremes, or a blend of the two. Figure 1.8 shows a number of the operations described in this chapter positioned in a spectrum using the IHIP characteristics, from almost 'pure' goods producers to almost 'pure' service producers. Both LEGO and Fjällräven are classic manufacturers on the left of the spectrum, making standard products. At the other extreme, LEGOLAND and the Marina Bay Sands hotel are (to slightly different degrees) producing intangible services. MSF and Philips lighting are somewhere in between.

Operations principle
Most operations produce a blend of tangible products and intangible services.

Using IHIP characteristics to distinguish between different types of output is of more than theoretical interest; they have real operational consequences. For example:

- Intangibility means it is difficult to define the 'boundary' of the less tangible elements of service. It therefore becomes particularly important to manage customers' expectations as to what the service comprises.
- Heterogeneity means that every service is different and difficult to standardise. Customers could ask for elements of service that are difficult to predict and may be outside the operation's capabilities. Cost efficiencies become difficult and staff must be trained to cope with a wide variety of requests.
- Inseparability means that production and consumption are simultaneous. So, to meet all demand, operations must have sufficient capacity in place to meet demand as it occurs. However, customer guidance can reduce the need for contact (e.g. the use of FAQs on a website).

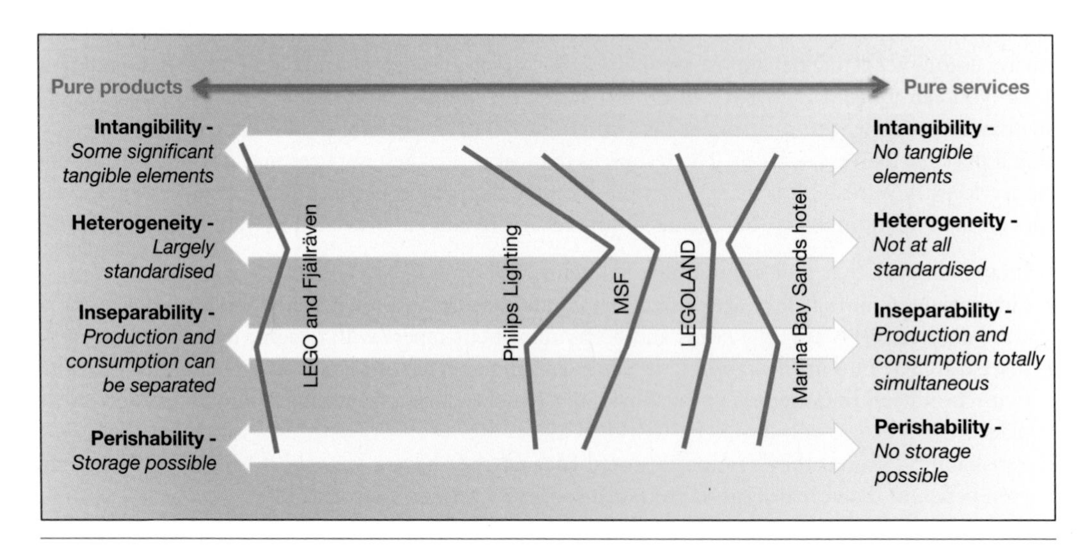

Figure 1.8 Relatively few operations produce either purely products or purely services. The output from most types of operations blend the characteristics of 'pure' goods and 'pure' services

- Perishability means that an operation's output is difficult to store and ceases to have value after a relatively short time, so matching capacity with demand (or vice versa) is important to avoid either underutilised resources or lost revenue.

Services and products are merging (and changing)

Increasingly the distinction between services and products is seen as not particularly useful. Some authorities see the essential purpose of all businesses, and therefore all operations, as being to 'serve customers'. Therefore, they argue, all operations are service providers who may (or may not) produce physical products as a means of serving their customers. This idea, that all operations should be seen as offering 'value propositions' through service, is called 'service-dominant logic'.[6] Among other things, it holds that service is the fundamental basis of exchange, that physical goods are simply the distribution mechanisms for the provision of service, and that the customer is always the co-creator of value. Our approach in this text is close to this in that we treat operations and process management as being important for all organisations. Whether they see themselves as manufacturers or service providers is very much a secondary issue.

Customers are part of the process – co-creation and co-production

If all operations can be seen as producing services, and services act on customers or their surrogates, then the role of customers in an operation's output should be considered. This is not a new idea, nor is it unusual for customers to play a central part in how they derive value from an operation's outputs (they take themselves around a supermarket, for example). Patients visiting the doctor with an ailment are required to describe their symptoms and discuss alternative treatments – the better they can do this, the better the value they derive. This idea of customer involvement is important because the distinction between the roles of 'producer' and 'consumer' are being blurred. The concept is usually known either as **co-creation** or **co-production** – there is some disagreement in what the two terms mean. Often co-creation implies customer involvement in the design of a product or service, and co-production is just the production of a pre-designed offering. The important point is that there is often a degree of customer involvement, engagement, participation or collaboration within an operation. The idea has important implications for all operations. Not only does it emphasise the importance of customers in shaping how an operation's outputs can create value, it establishes the importance of a full two-way interaction between an operation and its customers.

Servitisation

A term that is often used to indicate how operations, which once considered themselves exclusively producers of products, are becoming more service-conscious is '**servitisation**' (or servitization). Servitisation involves (often manufacturing) firms developing the capabilities they need to provide services and solutions that supplement their traditional product offerings. The best-known example of how servitisation works was when Rolls-Royce, the aero engine manufacturer, rather than selling individual engines, offered the option of customers being able to buy 'power-by-the-hour'. What this meant was that many of its customers in effect bought the power the aero engine delivers, with Rolls-Royce providing both the physical engines and all of the support (including maintenance, training, updates and so on) to ensure that they could continue to deliver power. This may sound like a small change, but the effects were important. First, Rolls-Royce became a provider of service (the power to make the aircraft fly) as opposed to a manufacturer of technically complex products. Second, it means that what customers really want (the reliable provision of power) and the objectives of the company are more closely aligned. Third, it provides an opportunity for companies to earn additional revenue from new services.

Operations principle

Servitisation involves firms developing the capabilities to provide services and solutions that supplement their traditional product offerings.

OPERATIONS IN PRACTICE

Servitisation and circular design at Philips lighting[7]

Operations managers are increasingly having to re-evaluate how they think about their products and services and how they produce them. Take, for example, Philips Lighting,[8] which responded to developments in its markets by combining and adopting two important changes to operations practice – servitisation and the **circular economy**.

The company's servitisation offering is called 'lighting-as-a-service' (LaaS), where it takes care of its customers' lighting needs from the initial design and installation of the lighting, to the operation and maintenance. By doing this, customers can save money because they pay only for the light they use, while at the same time avoiding the disturbance of having to replace and dispose of burnt-out bulbs or having to navigate system upgrades. The company originally became interested in LaaS when the architect Thomas Rau worked with Philips Lighting to supply a novel 'pay-per-lux' intelligent lighting system that was customised to fit the requirements of the Amsterdam office space of RAUArchitects, while also reducing price. When considering his lighting needs, Rau wanted to avoid buying an expensive over-engineered lighting system, only to eventually have to dispose of and replace it. Instead, he would rather purchase just the right amount of light 'as a service' that would suit the building. RAU and Philips developed a system that created a minimalist light plan making as much use as possible of the building's natural sunlight. It combined a sensor and controller system that helped keep energy use to an absolute minimum, by darkening or brightening the artificial lighting in response to motion within a space or the presence of daylight. From the customer's point of view, they not only save money by paying only for the light they use but also find it easier to optimise their use of energy, while avoiding the effort of managing the system. From the supplier's point of view, the agreement allowed Philips to retain control over how the lighting system worked, what products were supplied, how the system was maintained, how it was reconditioned and eventually how its products were recycled. The company struck a similar deal for the terminal buildings at Amsterdam Airport Schiphol. The airport pays only for the light it uses, while Philips remains the owner of all fixtures and installations, responsible for the performance and durability of the system and eventually its reuse and recycling at the end of its useful life. The collaboration between the supplier and user of the service resulted in reduced maintenance costs (because components could be individually replaced instead of the entire fixture being recycled) contributing to Schiphol's ambitious sustainability targets.

Customers

Any discussion about the nature of outputs from operations must involve consideration of the customers for whom they are intended. Remember that although customers may also be an input to many operations (see earlier), they are also the reason for their existence. Nor should 'customers' be seen as a homogeneous group. Marketing professionals spend much of their effort in trying to understand how customers can be usefully grouped together, the better to understand their various needs. This is called 'market segmentation', and is beyond the scope of this text. However, the implications of it are very important for operations managers. In essence, it means that different customer groups may want different things from an operation. We discuss this issue further in Chapter 3.

B2B and B2C

One distinction between different types of customers is worth describing at this point, because we shall be using the terminology at other points in the text. That is between business-to-business (B2B) and business-to-consumer (B2C) operations. B2B operations are those that provide their products or services to other businesses. B2C operations provide their products or services direct to the consumers who (generally) are the ultimate users of the outputs from the operation. Serving individual customers and serving other businesses are very different. This means that the operations serving these two types of customers will be faced with different kinds of concerns, and probably be organised in different

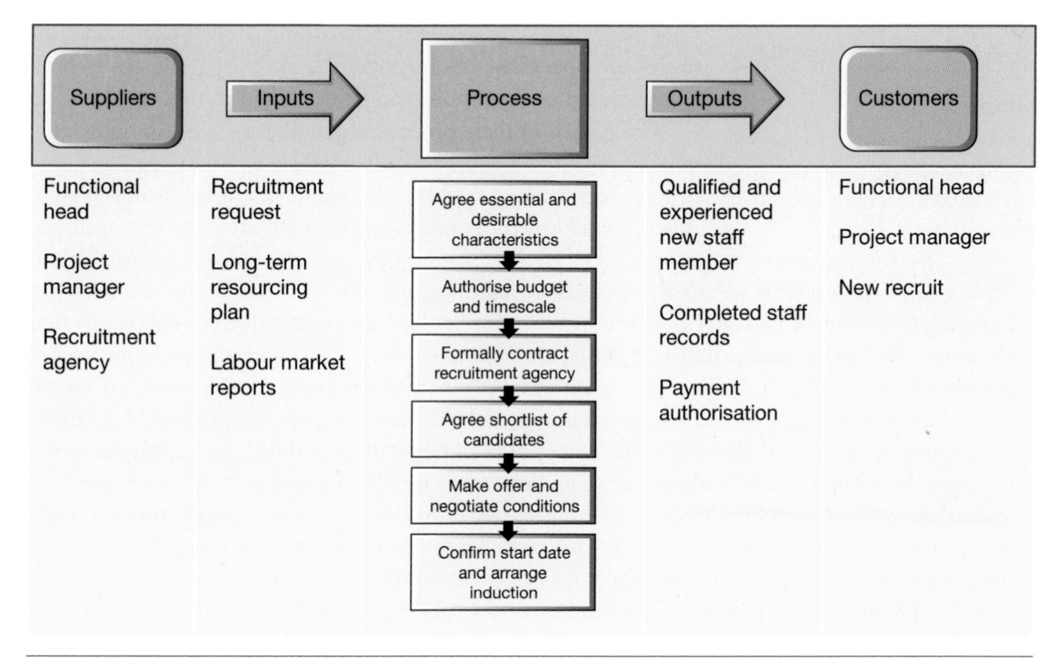

Figure 1.9 A simple SIPOC analysis for a recruitment process performed by the human resources function of a company

ways. Yet an understanding of customers is always important (whether business customers, or consumers). Without them, there would be no operation. It is critical that operations managers are aware of customers' needs, both current and potential.

> **Operations principle**
> An understanding of customer needs is always important, whether customers are individuals or businesses.

SIPOC analysis

Although the idea of the 'input–transformation–output' model is essentially simple, it can be the basis of a useful first step in understanding and improving processes. This is sometimes called **SIPOC analysis**. SIPOC stands for suppliers, inputs, process, outputs and customers. It is a method of formalising a process at a relatively general rather than a detailed level. Figure 1.9 shows an example that describes a recruitment process performed by the human resources function of a company. The advantage of such an analysis is that it helps all those involved in the process to understand (and, more important, agree) what it involves and where it fits within the business. More than this, it can prompt important questions that can sometimes be overlooked. For example, exactly what information should suppliers to the process provide? In what form should the information be given? What are the important steps in the process and who is responsible for them? And so on.

1.4 What is the process hierarchy?

So far, we have discussed operations management and the input–transformation–output model, at the level of 'the operation'. For example, we have described the toy manufacture, the theme park, the disaster relief operation and the hotel. But look inside any of these operations. One will see that all operations consist of a collection of **processes** (although these processes may be called 'units' or 'departments') interconnecting with each other to form an internal network. Each process acts as a smaller version of the whole operation of which it forms a part. Within any operation, the mechanisms that actually transform inputs into outputs are these processes. A 'process' is an arrangement of resources and activities that transform inputs into outputs that satisfy (internal or external) customer needs. They are the 'building blocks' of all operations, and they form an 'internal network' within an operation. Each process is, at the same time, an **internal supplier** and an **internal customer** for other processes. This 'internal customer' concept provides a model to analyse the internal activities of an operation. It is also a useful reminder that, by treating internal customers with the

Operations principle

A process perspective can be used at three levels: the level of the operation itself, the level of the supply network, and the level of individual processes.

same degree of care as external customers, the effectiveness of the whole operation can be improved. Table 1.3 illustrates how a wide range of operations can be described in this way.

Within each of these processes is another network of individual units of resource such as individual people and individual items of process technology (machines, computers, storage facilities, etc.). Again, transformed resources flow between each unit of transforming resource. Any business, or operation, is made up of a network of processes and any process is made up of a network of resources. But also, any business or operation can itself be viewed as part of a greater network of businesses or operations. It will have operations that supply it with the services and products it needs, and unless it deals directly with the end consumer, it will supply customers who themselves may go on to supply their own customers. Moreover, any operation could have several suppliers and several customers, and may be in competition with other operations creating similar services or products to itself. This network of operations is called the '**supply network**'. In this way, the input–transformation–output model can be used at a number of different 'levels of analysis'. Here we have used the idea to analyse businesses at three levels: the process, the operation and the supply network. But one could define many different 'levels of analysis', moving upwards from small to larger processes, right up to the huge supply network that describes a whole industry.

This idea is called the '**hierarchy of operations**' or the '**process hierarchy**', and is illustrated for a business that makes television programmes and videos in Figure 1.10. It has inputs of production, technical and administrative staff, cameras, lighting, sound and recording equipment, and so on. It transforms these into finished programmes, promotional videos, etc. At a more macro level, the business itself is part of a whole supply network, acquiring services from creative agencies, casting agencies and studios, liaising with promotion agencies, and serving its broadcasting company customers. At a more micro level within this overall operation there are many individual processes, manufacturing the sets, marketing its services, maintaining and repairing technical equipment, producing the videos and so on. Each of these individual processes can be represented as a network of yet smaller processes, or even individual units of resource. For example, the set manufacturing process could comprise four smaller processes – designing the sets, constructing them, acquiring the props and finishing the sets.

Table 1.3 Some operations described in terms of their processes

Operation	Some of the operation's processes
Airline	Passenger check-in assistance, baggage drop, security/seat check, board passengers, fly passengers and freight around the world, flight scheduling, in-flight passenger care, transfer assistance, baggage reclaim, etc.
Department store	Source merchandise, manage inventory, display products, give sales advice, sales, aftercare, complaint handling, delivery service, etc.
Police service	Crime prevention, crime detection, information gathering/collating, victim support, formally charging/detaining suspects, managing custody suites, liaising with court/justice system, etc.
Ice cream manufacturer	Source raw materials, input quality checks, prepare ingredients, assemble products, pack products, fast freeze products, quality checks, finished goods inventory, etc.

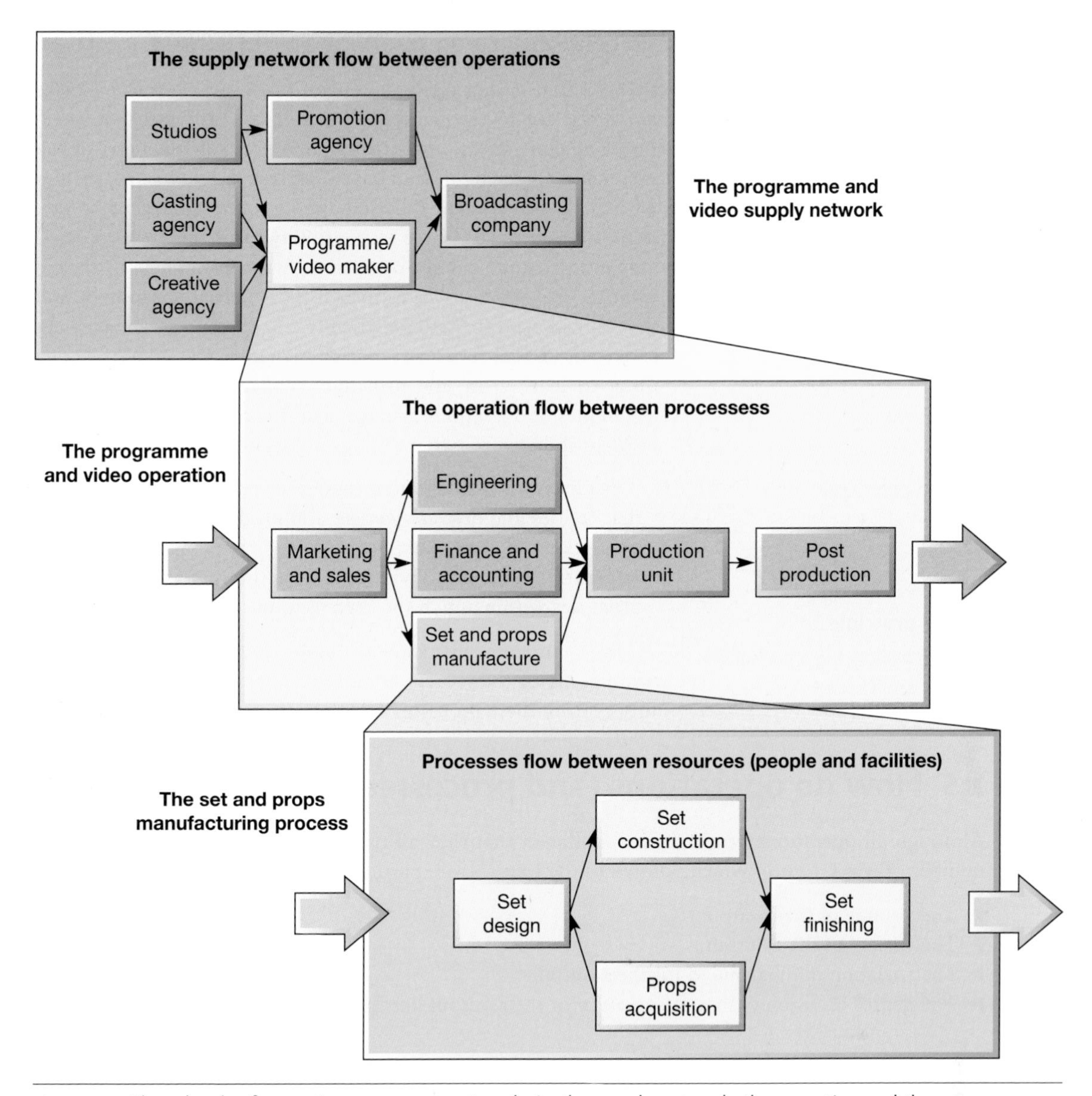

Figure 1.10 Three levels of operations management analysis, the supply network, the operation and the process

Critical commentary

The idea of the internal network of processes is seen by some as being over-simplistic. In reality, the relationship between groups and individuals is significantly more complex than that between commercial entities. One cannot treat internal customers and suppliers exactly as one does external customers and suppliers. External customers and suppliers usually operate in a free market. If an organisation believes that in the long run it can get a better deal by purchasing services and products from another supplier, it will do so. But internal customers and suppliers are not in a 'free market'. They cannot usually look outside either to purchase input resources or to sell their output services and products (although some organisations are moving this way). Rather than take the 'economic' perspective of external commercial relationships, models from organisational behaviour, it is argued, are more appropriate.

Operations management is relevant to all parts of the business

The example in Figure 1.10 demonstrates that it is not just the operations function that manages processes; all functions have processes. For example, the marketing function has processes that create demand forecasts, create advertising campaigns, create marketing plans, etc. All functions of the organisation have processes that need managing. Each function has its 'technical' knowledge, such as marketing expertise, finance expertise and so on. Yet each will also have a 'process management' role in producing its services. So, because all managers have some responsibility for managing processes, they are, to some extent, operations managers. They all should want to give good service to their (often internal) customers, and they all will want to do this efficiently. So, operations management is relevant for all functions, and all managers should have something to learn from the principles, concepts, approaches and techniques of operations management. It also means that we must distinguish between two meanings of 'operations':

> **Operations principle**
> All parts of the business manage processes, so all parts of the business have an operations role and need to understand operations management principles.

- 'Operations' as a function, meaning the part of the organisation that creates and delivers services and products for the organisation's external customers;
- 'Operations' as an activity, meaning the management of the processes within any of the organisation's functions.

> **Operations principle**
> Processes are defined by how the organisation chooses to draw process boundaries.

Table 1.4 illustrates just some of the processes that are contained within some of the more common non-operations functions, the outputs from these processes and their 'customers'.

1.5 How do operations (and processes) differ?

Although all operations processes are similar in that they all transform inputs, they do differ in a number of ways, four of which, known as the four Vs, are particularly important:

- The volume of their output.
- The variety of their output.
- The variation in the demand for their output.
- The degree of visibility that the creation of their output has for customers.

Table 1.4 Some examples of processes in non-operations functions

Organisational function	Some of its processes	Outputs from its processes	Customer(s) for its outputs
Marketing and sales	▶ Planning process ▶ Forecasting process ▶ Order-taking process	▶ Marketing plans ▶ Sales forecasts ▶ Confirmed orders	▶ Senior management ▶ Sales staff, planners, operations ▶ Operations, finance
Finance and accounting	▶ Budgeting processes ▶ Capital approval processes ▶ Invoicing processes	▶ Budgets ▶ Capital request evaluations ▶ Invoices	▶ Everyone ▶ Senior management, requesters ▶ External customers
Human resources management	▶ Payroll processes ▶ Recruitment processes ▶ Training processes	▶ Salary statements ▶ New hires ▶ Trained employees	▶ Employees ▶ All other processes
Information technology	▶ Systems review process ▶ Help desk process ▶ System implementation project processes	▶ System evaluation ▶ Systems advice ▶ Implemented working systems and aftercare	▶ All other processes in the business

The volume dimension

Take a familiar example of high-volume hamburger production. McDonald's serves millions of burgers around the world every day. Volume has important implications for the way McDonald's operations are organised. The first thing you notice is the repeatability of the tasks people are doing and the systemisation of the work, where standard procedures are set down specifying how each part of the job should be carried out. Also, because tasks are systematised and repeated, it is worthwhile developing specialised fryers and ovens. All this gives low unit costs. Now consider a small local cafeteria serving a few 'short order' dishes. The range of items on the menu may be similar to the larger operation, but the volume will be far lower, so the repetition will also be far lower, as will the number of staff (possibly only one person), so individual staff probably perform a wider range of tasks. This may be more rewarding for the staff, but less open to systemisation. Also, it is less feasible to invest in specialised equipment. So the cost per burger served is likely to be higher (even if the price is comparable).

The variety dimension

A taxi company offers a relatively high-variety service. It is prepared to pick you up from almost anywhere and drop you off almost anywhere. To do this it must be flexible. Drivers must have a good knowledge of the area, and communication between the base and the taxis must be effective. However, the cost per kilometre travelled will be higher for a taxi than for a less customised form of transport such as a bus service. Although both provide the same basic service (transportation), the taxi service has a higher variety of routes and times to offer its customers, while the bus service has a few well-defined routes, with a set schedule. Little, if any, flexibility is required from the bus operation. All is standardised and regular, which results in relatively low costs compared with using a taxi for the same journey.

The variation dimension

Consider the demand pattern for a summer holiday resort hotel. Not surprisingly, more customers want to stay in summer vacation times than in the middle of winter. At the height of 'the season' the hotel could be full to capacity, but off-season demand could be a small fraction of its capacity. Such a marked variation in demand means that the operation must change its capacity in some way: for example, by hiring extra staff for the summer. But, a hotel with high variation in demand will probably have high recruitment costs, overtime costs and underutilisation of its rooms, all of which increase the hotel's costs. By contrast, a hotel with level demand can plan its activities well in advance. Staff can be scheduled, food can be bought and rooms can be cleaned in a routine and predictable manner. This results in a high utilisation of resources and lower unit costs.

The visibility dimension

'Visibility' is slightly more difficult to envisage. It means how much of the operation's activities its customers experience, or how much the operation is exposed to its customers. Generally, customer-processing operations are more exposed to their customers than material- or information-processing operations. But even customer-processing operations have some choice as to how visible they wish to be. For example, a retailer could operate as a high-visibility 'bricks and mortar' shop or a lower-visibility web-based operation. A high-visibility 'bricks and mortar' operation will conform to most of the IHIP characteristics described previously. Customers will directly experience most of its 'value-adding' activities. They are likely to demand a relatively short waiting time. Their perceptions, rather than objective criteria, will also be important in how they judge the service. Customers could also request services or products that clearly would not be sold in such a shop, resulting in 'high received variety'. All of which make it difficult for high-visibility operations to keep costs down. Conversely, a web-based retailer, while not a pure low-contact operation, has far lower visibility. Behind its website, it can be more 'factory-like'. The time lag between the order being placed

> **Operations principle**
> The way in which processes need to be managed is influenced by volume, variety, variation and visibility.

and the items ordered by the customer being retrieved and dispatched does not have to be minutes, as in the shop, but can be hours or even days. Also, there can be relatively high staff utilisation. The web-based organisation can also centralise its operation on one (physical) site, whereas the 'bricks and mortar' shop needs many shops close to centres of demand. Therefore, the low-visibility web-based operation will have lower costs than the shop.

OPERATIONS IN PRACTICE

Two very different hospitality operations

Ski Verbier Exclusive[9]

It is the name of the company that gives it away; Ski Verbier Exclusive Ltd is a provider of 'upmarket' ski holidays in the Swiss winter sports resort of Verbier. With 23 years' experience of organising holidays, it looks after luxury properties in the resort that are rented from their owners for letting to Ski Verbier Exclusive's clients. The properties vary in size and the configuration of their rooms, but the flexibility to reconfigure the rooms to cater for the varying requirements of client groups is important. *'We are very careful to cultivate as good a relationship with the owners, as we are with our clients that use our holiday service'*, says Tom Avery, joint founder and director of the company. *'We have built the business on developing these personal relationships, which is why our clients come back to us year after year* [40 per cent to 50 per cent of clients are returners]. *We pride ourselves on the personal service that we give to every one of our clients; from the moment they begin planning their ski holiday, to the journey home. What counts is experience, expertise, obsessive eye for detail and the understated luxury of our chalets combined with our ability to customise client experience'*. And client requests can be anything from organising a special mountain picnic complete with igloos, to providing an ice sculpture of Kermit the Frog for a kids' party. The company's specialist staff have all lived and worked in Verbier and take care of all details of the trip well in advance, from organising airport transfers to booking a private ski instructor, from arranging private jet or helicopter flights to Verbier's local airport, to making lunch reservations in the best mountain restaurants. *'We cater for a small, but discerning market'*, says Tom. *'Other companies may be bigger, but with us it's our personal service that clients remember'*. However, snow does not last all the year round. The company's busiest period is mid-December to mid-April, when all the properties are full. The rest of the year is quieter, but the company does offer summer vacations in some of its properties. These can be either self-catering, or with the full concierge service that clients get in the ski season. *'We adapt to clients' requirements'*, says Tom. *'That is why the quality of our staff is so important. They have to be good at working with clients, be able to judge the type of relationship that is appropriate, and be committed to providing what makes a great holiday. That's why we put so much effort into recruiting, training and retaining our staff'*.

hotelF1[10]

Hotels are high-contact operations – they are staff-intensive and have to cope with a range of customers, each with a variety of needs and expectations. So, how can a highly successful chain of affordable hotels avoid the crippling costs of high customer contact? hotelF1, a subsidiary of the French Accor group, manages to offer outstanding value by adopting two principles not always associated with hotel operations – standardisation and an innovative use of technology. hotelF1 hotels are usually located close to the roads, junctions and cities that make them visible and accessible to prospective customers. The hotels themselves are built from state-of-the-art volumetric prefabrications. The prefabricated units are arranged in various configurations to suit the characteristics of each individual site. Rooms are 9 square metres in area, and are designed to be attractive, functional, comfortable and soundproof. Most important, they are designed to be easy to clean and maintain. All have the same fittings, including a double bed, an additional bunk-type bed, a wash basin, a storage area, a working table with seat, a wardrobe and a television set. The reception of a hotelF1 hotel is staffed only from 6.30 am to 10.00 am and from 5.00 pm to 10.00 pm. Outside these times an automatic machine sells rooms to credit card users, provides access to the hotel, dispenses a security code for the room and even prints a receipt. Technology is also evident in the washrooms. Showers and toilets are automatically cleaned after each use by using nozzles and heating elements to spray the room with a disinfectant solution and dry it before it is used again. To keep things even simpler, hotelF1 hotels do not include a conventional restaurant, as they are usually located near existing ones. However, a continental breakfast is available, usually between 6.30 am and 10.00 am, and of course on a 'self-service' basis!

The implications of the four Vs of operations processes

All four Vs have implications. Put simply, high volume, low variety, low variation and low customer contact all help to keep processing costs down. Conversely, low volume, high variety, high variation and high customer contact generally carry some kind of cost penalty. This is why the volume dimension is drawn with its 'low' end on the left, unlike the other dimensions, to keep all the 'low cost' implications on the right. The position of an operation on the four dimensions is determined by the demands of the market it is serving, although most operations have some discretion in moving themselves on the dimensions. Figure 1.11 summarises the implications of such positioning.

Operations principle

Operations and processes can (other things being equal) reduce their costs by increasing volume, reducing variety, reducing variation and reducing visibility.

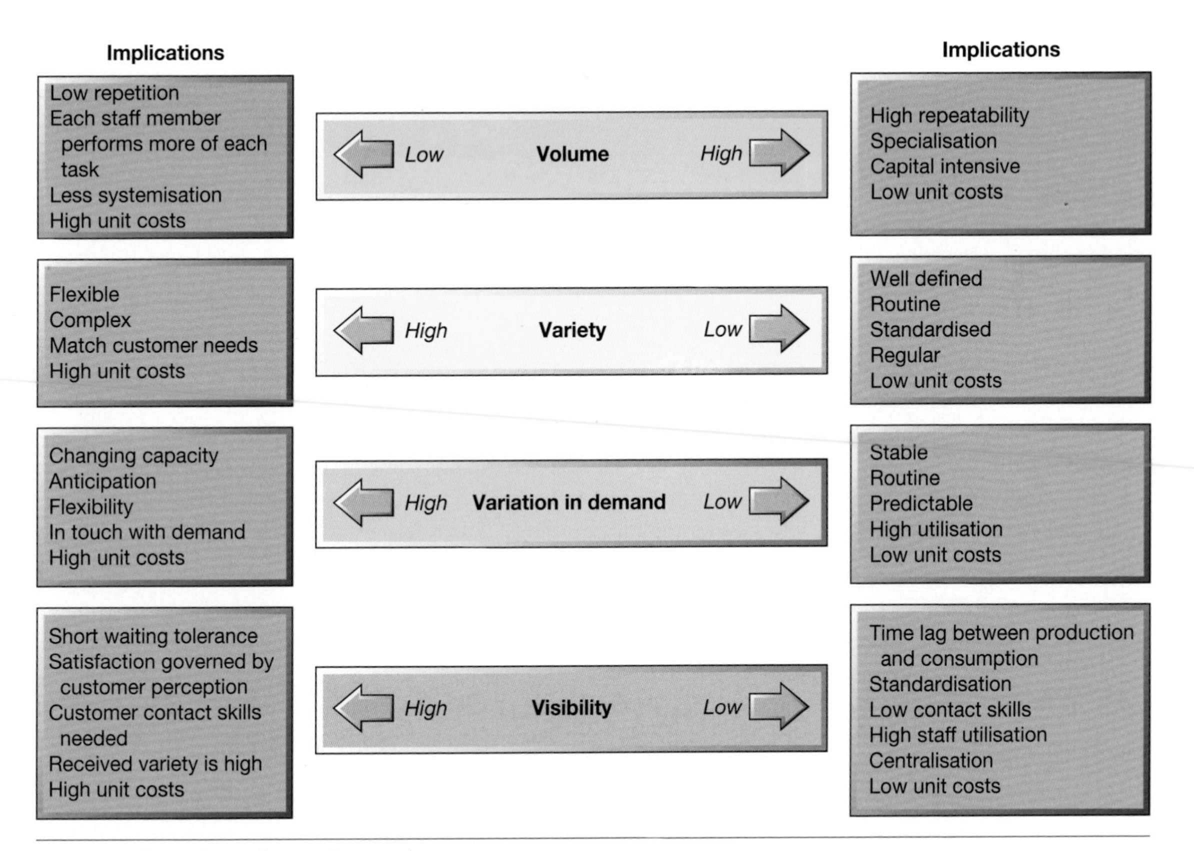

Figure 1.11 A typology of operations and processes

Worked example

Two very different hospitality operations positioned on the four Vs scales

Figure 1.12 illustrates the different positions on the dimensions of the Ski Verbier Exclusive operation and the hotelF1 hotel chain (see the 'Operations in practice' example on 'Two very different hospitality operations'). Although both provide the same basic service in that they accommodate people, they are very different. Ski Verbier Exclusive provides luxurious and bespoke vacations for a relatively small segment of the ski holiday market. Its variety of services is almost infinite in the sense that customers can make individual requests in terms of food and entertainment. Variation is high with four months of 100 per cent occupancy, followed by a far quieter period. Customer contact, and therefore visibility, is also very high. All of this is very different from the hotelF1 branded hotels, whose customers usually stay one night, where the variety of services is strictly limited, and business and holiday customers use the hotel at different times, which limits variation. Most notably, though, customer contact is kept to a minimum. Ski Verbier Exclusive has very high levels of service, which means it has relatively high costs. Its prices therefore are not cheap. Certainly not as cheap as hotelF1, which has arranged its operation in such a way as to provide a highly standardised service at minimal cost.

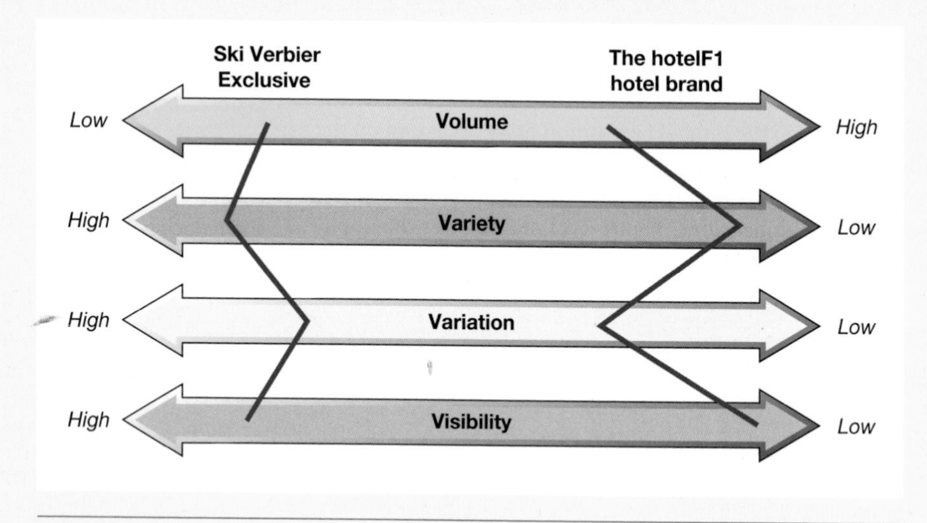

Figure 1.12 The four Vs profiles of two very different hospitality operations

1.6 What do operations managers do?

The exact details of what operations managers do will, to some extent, depend on the way an organisation defines the boundaries of the function. Yet there are some general classes of activities that apply to all types of operations no matter how the operations function is defined. We classify operations management activities under the four headings: direct, design, deliver and develop.

- Directing the overall strategy of the operation. A general understanding of operations and processes and their strategic purpose and performance, together with an appreciation of how strategic purpose is translated into reality, is a prerequisite to the detailed design of operations and process. This is treated in Chapters 1 to 5.
- Designing the operation's services, products and processes. Design is the activity of determining the physical form, shape and composition of operations and processes together with the services and products that they create. This is treated in Chapters 6 to 9.
- Planning and control process delivery. After being designed, the delivery of services and products from suppliers and through the total operation to customers must be planned and controlled. This is treated in Chapters 10 to 14.

- Developing process performance. Increasingly it is recognised that in operations, or any process, managers cannot simply deliver services and products routinely in the same way that they always have done. They have a responsibility to develop the capabilities of their processes to improve process performance. This is treated in Chapters 15 to 19.

OPERATIONS IN PRACTICE

Fjällräven products are voted the most sustainable in their field[11]

Developing a reputation for **sustainability** does not happen overnight. When Fjällräven's outdoor clothing and equipment products were voted the most sustainable in their field by Europe's largest brand study on sustainability, Sweden's Sustainable Brand Index, it was the result of many years dedication to sustainable-based decisions in design, testing, material choices, supply chain and production, right through to 'repairability' and what happens at the end of a product's life. Founded in 1960 by Åke Nordin in Örnsköldsvik, Sweden, the company was always committed to quality, functional and durable design, and in particular, acting responsibly towards people, animals and nature. So, Fjällräven prioritise the use of recycled, organic and renewable materials by applying the Higg Index criteria, an approach developed by the Sustainable Apparel Coalition that enables operations in the apparel industries to measure their sustainability performance. For example, the company produced a special edition of one of its most popular products, the Kånken backpack, which is made from 11 recycled plastic bottles. It is also dyed using the 'SpinDye' process, which uses much less water than traditional dying processes. Avoiding waste is important to Fjällräven. The type and amount of a material used is a key concern during the design process. The company's policy on material selection is 'Why use a raw material when a recycled one is available and offers the same quality?' It checks whether the amount of a material can be reduced by adapting the cut and fit of a garment or product to reduce waste. Moreover, looking to the future when garment recycling is more prevalent, the company tries to use just one or two materials in each product to make future recycling easier.

However, the company does recognise that achieving its sustainability goals is not always easy, and that compromises are sometimes necessary. 'We sometimes have to say "no" when we want to say "yes"', they say. All materials are evaluated for their efficiency, functional qualities, chemical composition and the amount needed. Its 'Preferred Materials And Fibres List' grades materials in terms of their impact on the environment, and is constantly updated to take account of new research and new materials. However, it doesn't matter how sustainable a material is if it does not do its job of keeping users warm and dry. Any material's functionality and efficiency has to be balanced with its environmental impact. Underlying the company's sustainability efforts is a long-term view of innovation and improvement. 'We have made, and will continue to make, mistakes' they say. 'But we try to learn from them [and] we aim to innovate and adapt. We're not ones to settle. We never sit back and relax thinking what we're doing now is good enough. At Fjällräven, the term "room for improvement" is ingrained in all of us'.

Operations management impacts social–environmental sustainability

Earlier, we identified the increasing importance of social–environmental sustainability on operations management practice. It is worth re-emphasising that many of the activities of operations managers have a huge impact on the natural environment, society broadly, and specific stakeholder groups such as the operation's staff, suppliers, investors and regulators (where relevant). **Social responsibility** is important to operations managers because of the profound impact operations practice can have on the environment and society at large, and conversely how operations practice is shaped by social–environmental considerations. Environmental sustainability means meeting the

Operations principle

Operations management activities will have a significant effect on the social, ethical and environmental performance of any type of enterprise.

needs of the present without compromising the ability of future generations to meet their own needs. Put more directly, it means the extent to which business activity negatively impacts on the natural environment. It is clearly an important issue, not only because of the obvious impact on the immediate environment of hazardous waste, air, and even noise, pollution, but also because of the less obvious, but potentially far more damaging issues around global warming.

Responsible operations

In every chapter, under the heading of 'Responsible operations', we summarise how the particular topic covered in the chapter touches upon important social, ethical and environmental issues.

There is a two-way relationship between operations management and **corporate social responsibility** (CSR, a term we will explain in the next chapter). Operations management practice can significantly affect such issues, and sensitivity to these issues has increasingly shaped what is regarded as good operations practice. One can think about this two-way relationship at different levels. Think about the pollution-causing disasters that make the headlines periodically. They seem to be the result of a whole variety of causes – oil tankers run aground, nuclear waste is misclassified, chemicals leak into a river, or polluting gas clouds drifting over industrial towns. But in fact they all have something in common. They were all the result of an operations-based failure. Somehow operations procedures were inadequate. Less dramatic in the short term, but perhaps more important in the long term, is the environmental impact of products that cannot be recycled and processes that consume large amounts of energy.

Just as important is the question of why organisations are increasingly careful to behave responsibly. One piece of research suggests that there are three reasons to engage in CSR activities:[12]

- The first is surprisingly altruistic. Some CSR focuses purely on philanthropy, where activities are not aimed explicitly at producing profits or specifically improve the operation's performance. For example, many operations donate funds or equipment to civic organisations, promote community enterprises and encourage employee volunteering.
- The second reason is more directly related to operations management. It involves activities that not only provide CSR benefits, but also support operations objectives by saving costs and/or enhancing revenue. Here CSR and the conventional concerns of operations management coincide. For example, such activities could include reducing waste or emissions (which may also reduce costs). In fact, many of operations management's environmental issues are concerned with waste. Operations management decisions in product and service design impact the utilisation of materials as well as long-term recyclability. Process design influences the proportion of energy, materials and labour that is wasted. Planning and control affects material wastage (packaging being wasted by mistakes in purchasing, for example) as well as energy and labour wastage. Improving working conditions for staff, or investing in training and education, may both enhance productivity and staff retention, as well as enhancing an organisation's reputation.
- The third reason is to explore new forms of business specifically to address social or environmental challenges, but at the same time provide business benefits. For example, in its Philippines operation, Unilever, the household and food brands company, supports women store-owners, while at the same time increasing its sales. Although these stores play an important role in many communities, they rarely have had the training or development necessary for growth, nor do they have access to business skills and information. The project, which helps the store-owning entrepreneurs gain the skills and knowledge to grow, both helps them boost their businesses and boosts the sales of Unilever brands.

The model of operations management

We can now combine two ideas to develop the model of operations and process management that will be used throughout this text. The first is the idea that operations and the processes that make up both the operations and other business functions are transformation systems that take in inputs and use process resources to transform them into outputs. The second idea is that the resources both in an organisation's operations as a whole and in its individual processes need to be managed in terms of how they are directed, how they are designed, how **delivery** is planned and controlled, and how they are developed. Figure 1.13 shows how these two ideas go together. This text will use this model to examine the more important decisions that should be of interest to all managers of operations and processes.

> **Operations principle**
> Operations management activities can be grouped into four broad categories: directing the overall strategy of the operation, designing the operation's resources and processes, planning and controlling delivery and developing performance.

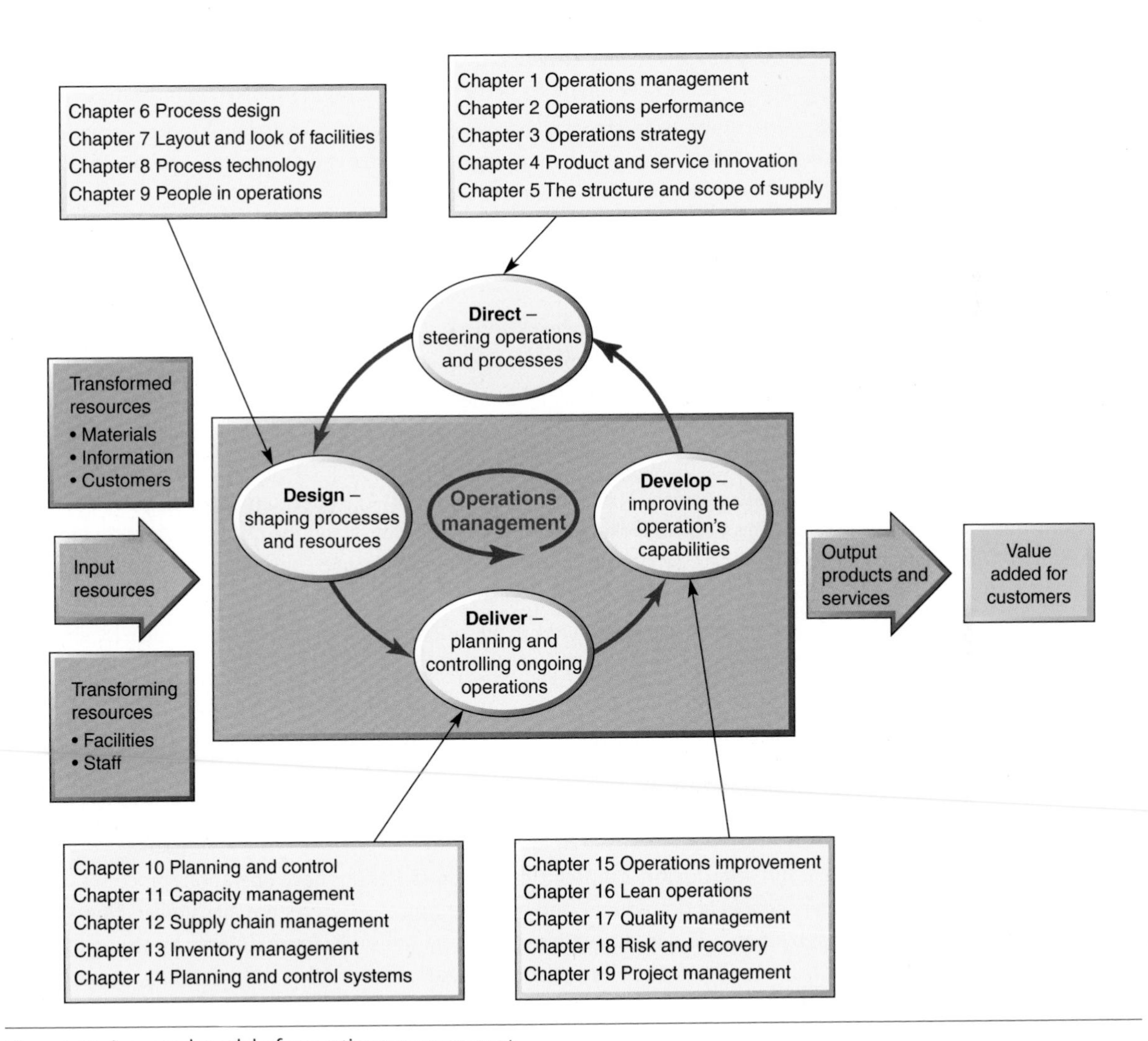

Figure 1.13 A general model of operations management

To be a great operations manager you need to. . .

So, you are considering a career in operations management, and you want to know, 'is it for you?' What skills and personal qualities will you need to make a success of the job as well as enjoying yourself? Well, the first thing to recognise is that there are many different roles encompassed within the general category of 'operations management'. Someone who makes a great risk control system designer in an investment bank may not thrive as a site manager in a copper mine. A video game project manager has a different set of day-to-day tasks when compared with a purchasing manager for a hospital. So, the first skill you need is to understand the range of operations-related responsibilities that exist in various industries; and there is no better way to do this than by reading this text! However, there are also some generic skills. Here are some of them:

- ***Enjoys getting things done*** – Operations management is about doing things and finishing tasks. It means hitting deadlines and not letting down customers, whether they are internal or external.
- ***Understands customer needs*** – Operations management is about fully understanding what 'value' means for customers. It means 'putting yourself in the customer's place'; knowing how to ensure that your services or products make the customer's life better.
- ***Places a high value on ethical, socially and environmentally sensitive decision-making*** – Given the potential impact of operations decisions, all operations practice needs to be set in the context of wider societal responsibilities.
- ***Communicates and motivates*** – Operations management is about directing resources to produce services or products in an efficient and effective manner. This means articulating what is required and encouraging people to do it. Interpersonal skills are vital. Operations managers must be 'people people'.
- ***Learns all the time*** – Every time an operations manager initiates an action (of any kind) there is an opportunity to learn from the result. Without learning there can be no improvement, and improvement is an imperative for all operations.
- ***Committed to innovation*** – Operations management is always seeking to do things better. This means creating new ways of doing things, being creative, imaginative and (sometimes) unconventional.
- ***Knows their contribution*** – Operations management may be the central function in any organisation, but it is not the only one. It is important that operations managers know how they can contribute to the effective working of other functions.
- ***Capable of analysis*** – Operations management is about making decisions. Each decision needs to be evaluated (sometimes with very little time). This involves looking at both the quantitative and the qualitative aspects of the decision. Operations managers do not necessarily have to be mathematical geniuses, but they should not be afraid of numbers.
- ***Keeps cool under pressure*** – Operations managers often work in pressured situations. They need to be able to remain calm no matter what problems occur.

Critical commentary

The central idea in this introductory chapter is that all organisations have operations processes that create and deliver services and products, and all these processes are essentially similar. However, some believe that by even trying to characterise processes in this way (perhaps even by calling them 'processes') one loses or distorts their nature and depersonalises or takes the 'humanity' out of the way in which we think of the organisation. This point is often raised in not-for-profit organisations, especially by 'professional' staff. For example, the head of one European 'medical association' (a doctors' trade union) criticised hospital authorities for expecting a 'sausage factory service based on productivity targets'.[13] No matter how similar they appear on paper, it is argued, a hospital can never be viewed in the same way as a factory. Even in commercial businesses, professionals, such as creative staff, often express discomfort at their expertise being described as a 'process'.

Summary answers to key questions

1.1 What is operations management?

- Operations management is the activity of managing the resources that are devoted to the creation and delivery of service and products. It is one of the core functions of any business, although it may not be called 'operations management' in some industries.
- Operations management is concerned with managing processes. And all processes have internal customers and suppliers. But all management functions also have processes. Therefore, operations management has relevance for all managers.

1.2 Why is operations management important in all types of organisations?

- Operations management uses the organisation's resources to create outputs that fulfil defined market requirements. This is the fundamental activity of any type of enterprise.
- Operations management is increasingly important because today's changing business environment requires new thinking from operations managers, especially in the areas of new technology, supply networks and environmental sustainability.

1.3 What is the input–transformation–output process?

- All operations can be modelled as input–transformation–output processes. They all have inputs of transforming resources, which are usually divided into 'facilities' and 'staff', and transformed resources, which are some mixture of materials, information and customers.
- Most operations create and deliver a combination of services and products, rather than being a 'pure' service or product operation.
- All operations can be positioned by their intangibility, heterogeneity, inseparabilit, and perishability characteristics.

1.4 What is the process hierarchy?

- All operations are part of a larger supply network which, through the individual contributions of each operation, satisfies end-customer requirements.
- All operations are made up of processes that form a network of internal customer–supplier relationships within the operation.

1.5 How do operations (and processes) differ?

- Operations and processes differ in terms of the volume of their outputs, the variety of outputs, the variation in demand for their outputs and the degree of 'visibility' they have.
- High volume, low variety, low variation and low customer 'visibility' are usually associated with low cost.

1.6 What do operations managers do?

- Responsibilities can be classed in four categories – direct, design, deliver and develop.
- Increasingly, operations managers have a responsibility for an operation's environmental performance.

CASE STUDY Kaston-Trenton Service (KTS)

Kaston-Trenton Service (KTS) is a domestic heating boiler maintenance company, based in the eastern part of the United Kingdom. Founded in the 1960s by plumber Christopher Trenton, it had grown substantially and was now run jointly by Christopher's two children, Ros, who looked after all marketing, sales and finance, and Mark, who looked after operations and supply issues. The company initially offered maintenance and repair services to domestic (household) customers with gas- or oil-burning boilers and expanded into offering similar services to business customers. Within the last two years KTS had also moved beyond simply servicing systems, to designing and installing HVAC (heating, ventilation and air conditioning) systems for business customers.

'Expanding into the design and installation business was something of a gamble', according to Ros. *'At the time, the B2B* [business to business] *part of our work was clearly showing more growth potential than our traditional domestic business and servicing business customers was also more profitable. So far, the installation venture has had mixed success. The jobs that we have done have been successful and our new customers very satisfied, but so far we have lost money on them. Partly, this is because we have had to invest in extra workshop space at our headquarters and employ a system designer, who is relatively expensive (but good) and only partly utilised at the moment. Hopefully, profitability will improve as the volume of installation jobs increases'*.

Table 1.5 shows the number of contracts and the revenue from domestic servicing, business servicing, and the design and installation businesses, both for the previous year and the forecast for the current year of operation (all figures as of end Qtr 3). The profitability of the three offerings was difficult to determine exactly, but Ros and Mark were satisfied with the contribution of domestic boiler servicing, and especially of the business boiler servicing activities.

KTS services

Domestic boiler servicing was seen by Ros and Mark as a 'cash cow', generating revenues at a fairly steady rate. There were many different makes of boiler installed, but KTS only contracted to service the most common that accounted for about 60 per cent of the installed base. Less common boilers were often serviced by the manufacturers that supplied them. Domestic servicing accounted for by far the most individual contracts for KTS, with customers spread over most of the East of England. Around 95 per cent of customers renewed their contracts each year, which was seen as a testament both to their quality of service and the company's keen pricing. *'It's a price sensitive market'*, said Ros. *'We have to be competitive, but that's not all that counts. Most visits by our technician are routine yearly services, but about 20 per cent of visits are 'call-outs' with varying degrees of urgency. If a home boiler stops working on a winter weekend, the householder obviously expects us to respond quickly, and we try our best to get a technician to them within 4 or 5 hours. If it's simply a non-urgent controller fault in summer, we would probably agree a mutually convenient time to visit within a couple of days. Actually, the idea of a "mutually convenient time" is important in this market. Householders often have to make special arrangements to be in, so we have to be flexible in arranging appointments and absolutely reliable in being there on time. Although call-outs are only 20 per cent of visits, they cause the majority of problems because both their timing and duration are unpredictable. Also, customers are sensitised to boiler performance following an emergency call-out. What we call the "robustness of the repair" has to be high. Once it's fixed, it should stay fixed, at least for a reasonable length of time'*.

Business boiler servicing was different. Most customers' systems had been, to some extent, customised, so the variety of technical faults that the technicians had to cope with was higher. Also, a somewhat higher proportion of visits were call-outs (between 25 and 30 per cent) so demand was slightly less predictable. The real difference between domestic and business customers, according to Mark, was the nature of the contact between KTS technicians and customers. *'Business customers want to be involved in knowing the best way to use their systems. They want advice, and they want to know what you are doing. So, for example, if you install an update to the system control software, they usually want to be informed.*

Table 1.5 The number of contracts and the revenue from the three activities

Activity	Previous year		Current year (forecast)	
	Number of contracts	Revenue (£000)	Number of contracts	Revenue (£000)
Domestic boiler servicing	7331	1408	9700	1930
Business boiler servicing	972	699	1354	1116
Design and installation	3	231	6	509
Total		2338		3555

They also either keep a servicing log themselves, or ask us to report on measures such as boiler efficiency, time between repairs, downtime due to failure or servicing (particularly important) and so on. Call-out response time is particularly important for them, but because there is usually someone always on their premises, it is easier to arrange a time to call for regular servicing'.

Both Ros and Mark were disappointed that the design and installation business had been slow to take off. The one system designer they had hired was proving an asset, and two of their technicians from the business servicing side of the operation had been moved over to installation work and were proving successful. *'It's a tight team of three at the moment'*, said Mark, *'and that should give us enough capacity for the remainder of the year. But we will eventually need to recruit more technicians as business (hopefully) builds up'*. The extra workshop space that the firm had rented (on the same site) and some new equipment had allowed the design and installation team to adapt and customise boiler and control systems to suit individual customers' requirements. *'Many installers are owned by boiler manufacturers and can be guilty of pushing a standard solution on customers'* said Mark. *'With us, every system is customised to each customer's needs'*.

KTS organisation

A small administrative office of four people reported directly to Ros and Mark and helped manage accounting, HR, invoicing, contract maintenance and purchasing activities. The office was adjacent to a workshop space shared by the domestic and business boiler technicians. KTS employed 42 technicians in total. Nominally 26 of these worked on domestic boiler servicing and repair, and 16 on business boiler servicing and repair, yet there was some flexibility between the two groups. *'We are lucky that our technicians are usually reasonable about helping each other out'*, said Mark. *'It is generally easier for the technicians used to serving business customers to serve domestic ones. They are not always as efficient as those used to domestic customers, but their customer-facing skills are usually better. Domestic boiler technicians do not always appreciate that business customers want more reassurance and information generally. Also, it is important for business customers to receive a full technical report within a couple of days of a visit. Domestic technicians are not used to doing that'*.

Improving service efficiency

Although both Ros and Mark were broadly happy with the way the business was developing, Mark in particular felt that they could be more efficient in how they organised themselves. *'Our costs have been increasing more or less in line with revenue growth, but we should really be starting to get some economies of scale. We need to improve our productivity, and I think we can achieve this by reducing waste. For example, we have found that our technicians can waste up to 30 per cent of their time on non-value-adding activities, such as form-filling or retrieving technical information'*.

Mark's solution was to tackle waste in a number of ways:

- ***Establish key performance measures (KPIs) and simple metrics*** – Performance measures must be clearly explained so that technicians understand the objectives that underlie their targets in terms of availability, utilisation and efficiency.
- ***Better forecasting*** – Demand was forecast only in the simplest terms. Historical data to account for seasonality had not been used, nor had obvious factors, such as weather, been monitored.
- ***Slicker processes*** – Administrative and other processes had been developed 'organically' with little consideration of efficiency.
- ***Better dispatching*** – Dispatching (the allocation of jobs to individual technicians) was usually done on a simple 'first come, first served' basis without taking the efficient use of technicians' time into account. It was believed that both travel time and 'time to uptime' could be improved by better allocation of jobs.
- ***Better training*** – In the previous two years, three technicians had retired, one had been dismissed and two had left for other jobs. Mark had experienced difficulty in replacing them with experienced people. It had become clear that it would become more important to hire inexperienced people and train them. In Mark's words, *'to get smart people with the right attitude and problem-solving skills, who don't mind get their hands dirty, and give them the technical skills'*.

In addition to thinking about how best to improve efficiency, future market growth was also a concern. Two developments were occupying Ros and Mark's thoughts, one in the short to medium term, the other in the longer term.

Future growth – short to medium term

Demand had been growing steadily, largely by KTS winning business from smaller competitors. But Mark wondered whether the nature of what customers would want was changing. An opportunity had been suggested by one of KTS's oldest business customers. They had been approached by another HVAC company that had asked if they would be interested in a 'total' service, where the company would both supply and operate a new heating system. In effect they were asking if KTS's customer would totally outsource their heating to them. It was an idea that Mark was intrigued by. *'I have heard about this type of deal before, but mainly for large businesses and offered by facilities management companies. It can involve companies like ours actually buying the heating system, installing it and taking responsibility for managing not just the system itself, but actually how much energy is used. Exactly how it might work will, I guess, depend on the terms of the contract. Does the customer pay an amount per unit of energy used (perhaps linked to the wholesale price of energy)? Or does the customer simply pay a fixed amount for agreed operating characteristics, such as maintaining a particular temperature range? We would have to think carefully about the implications for us before offering such a service. The customer who told us about the approach does not want to desert us, but who knows what they might do in the future'.*

The future – longer term

According to the Climate Change Committee (CCC), an independent advisory body that assisted the UK government in reaching required carbon levels, meeting the United Kingdom's target to reduce emissions would require reducing domestic emissions by at least 3 per cent per year – a challenging target. This would mean that within a few years it could become illegal to install gas boilers in new-build homes. One possible future that was discussed in the industry was a general move towards a hydrogen network (burning hydrogen produces no emissions and creates only water vapour and heat). However, a more likely future would probably involve combining different renewable technologies to provide low-carbon heat. The lowest-cost, long-term solution could be to replace gas and oil boilers with hydrogen alternatives alongside electric heating generated from renewable sources such as air source or ground source heat pumps, which use small amounts of electricity to draw natural heat from either the air or the ground. But, to make heat pumps effective, all existing and new-build homes would need to be made energy efficient by using far better levels of insulation.

Ros thought that these developments could prove far more challenging for KTS. *'Both Mark and I had assumed that we would be in this business for at least another 20 to 30 years. We both have families, so the long-term future of the business is obviously important to us. New heating technologies and fuels pose both opportunities and threats (yes, I've done an MBA!) for us. Reducing fossil fuel consumption will definitely mean that we have to change what we do. And some aspects of demand may reduce. For example, ground source systems require little maintenance. But if there is going to be an upswing in the installation market, we need to be on top of it'.*

QUESTIONS

1 **How would you position each of KTS's services on the four 'V' dimensions of volume, variety, variation and visibility?**

2 **What aspects of performance are important for KTS to win more servicing business?**

3 **How would you evaluate the potential of offering a 'total' service like the KTS customer had been offered?**

4 **What should KTS be doing to prepare for possible longer-term changes in their industry?**

Problems and applications

All chapters have 'Problems and applications' questions that will help you to practise analysing operations. They can be answered by reading the chapter. Model answers for the first two questions can be found on the companion website for this text.

1 Quentin Cakes make about 20,000 cakes per year in two sizes, both based on the same recipe. Sales peak at Christmas time, when demand is about 50 per cent higher than in the quieter summer period. Its customers (the stores that stock its products) order its cakes in advance through a simple internet-based ordering system. Knowing that Quentin Cakes have some surplus capacity, one of its customers has approached the company with two potential new orders.

The *Custom Cake* option – this would involve making cakes in different sizes where consumers could specify a message or greeting to be 'iced' on top of the cake. The consumer would give the inscription to the store, which would email it through to the factory. The customer thought that demand would be around 1,000 cakes per year, mostly at celebration times such as Valentine's Day and Christmas.

The *Individual Cake* option – this option involves Quentin Cakes introducing a new line of about 10–15 types of very small cakes intended for individual consumption. Demand for this individual-sized cake was forecast to be around 4,000 per year, with demand likely to be more evenly distributed throughout the year than its existing products.

The total revenue from both options is likely to be roughly the same and the company has capacity to adopt only one of the ideas. But which one should it be?

2 Re-read the 'Operations in practice' examples on LEGOLAND and LEGO. What kinds of operations management activities at each of these operations might come under the four headings of direct, design, deliver and develop?

3 Here are two examples of how operations try to reduce the negative effects of having to cope with high levels of variety. Research each of them (there is plenty of information on the web) and answer the following questions:

(a) What are the common features of these two examples?
(b) What other examples of standardisation in transport operations can you think of?

Example 1 – The Mumbai Tiffin Box Suppliers Association (search under dabbawallas) operates a service to transport home-cooked food from workers' homes to office locations in downtown Mumbai. Workers from residential districts must ride commuter trains to work. They can be conservative diners, who may also be constrained by cultural taboos on food handling. Their workers, known as dabbawallas, pick up the food in the morning in a regulation tin 'tiffin' box, deposit it at the office at lunchtime, and return it to the home in the afternoon. The dabbawallas take advantage of public transport to carry the tins, usually using otherwise underutilised capacity on commuter trains in the mid-morning and afternoon. Different colours and markings are used to indicate to the (sometimes illiterate) dabbawallas the process route for each tin.

Example 2 – Ports have had to handle an infinite variety of ships and cargoes with widely different contents, sizes and weights, and protect them from weather and pilferage, while in transit or in storage. Then the transportation industries, in conjunction with the International Organization for Standardization (ISO), developed a standard shipping container design. Almost overnight the problems of security and weather protection were solved. Anyone wanting to ship goods in volume only had to seal them into a container and they could be signed over to the shipping company. Ports could standardise handling equipment and dispense with warehouses (containers could be stacked in the rain if required). Railways and trucking companies could develop trailers to accommodate the new containers.

4 Figure 1.12 compares two hotel types on the four Vs dimensions. Where would the other 'Operations in practice' examples used in this chapter be positioned on these dimensions?

5 Not all surgery conforms to our preconceptions of the individual 'super-craftsperson', aided by their back-up team, performing the whole operation from first incision to final stitch. Many surgical procedures are fairly routine. An example is the process that was adopted by one Russian eye surgeon. The surgical procedure in which they specialise is a revolutionary treatment for myopia (short-sightedness) called 'radial keratotomy'. In the process, eight patients lie on moving tables arranged like the spokes of a wheel around its central axis, with only their eyes uncovered. Six surgeons, each with their own 'station', are positioned around the rim of the wheel so that they can access the patients' eyes. After the surgeons have completed their own particular portion of the whole procedure, the wheel indexes round to take patients to the next stage of their treatment. The surgeons check to make sure that the previous stage of the operation has been performed correctly and then go on to perform their own task. Each surgeon's activity is monitored on TV screens overhead and the surgeons talk to each other through miniature microphones and headsets.

(a) Compare this approach to eye surgery with a more conventional approach.
(b) What do you think are the advantages and disadvantages of this approach to eye surgery?

6 Write down five services that you have 'consumed' in the last week. Try and make these as varied as possible. Examples could include public transport, a bank, any shop or supermarket, attendance at an education course, a cinema, a restaurant, etc. Try to identify how these services are different and how they are similar.

7 The transforming resources of the input–transformation–output model of operations management are classified as 'facilities' and 'staff'. Should the information needed to make the transformation also be included?

8 What might be the 'back-office' processes in a theme park such as LEGOLAND?

9 Position pre-recorded lectures, non-interactive university lectures, small group tutorials and individual 'counselling' tutorials on the IHIP scales.

10 Why do some people think that analysing enterprises in terms of their processes 'takes the humanity out of the way in which we think of the organisation' as outlined in the final 'Critical commentary' in the chapter?

Selected further reading

Anupindi, R., Chopra, S., Deshmukh, S.D., Vam Mieghem, J.A. and Zemel, E. (2013) ***Managing Business Process Flows*****, 3rd edn, Pearson, Harlow.**
Takes a 'process' view of operations, it's mathematical but rewarding.

Barnes, D. (2018) ***Operations Management: An International Perspective*****, Palgrave, London.**
A text that is similar in outlook to this one, but with more of a (useful) international perspective.

Chase, R.B. and Jacobs, F.R. (2017) ***Operations and Supply Chain Management*****, McGraw-Hill, New York.**
There are many good general textbooks on operations management. This takes a supply chain view, although written very much for an American audience.

Hall, J.M. and Johnson, M.E. (2009) When should a process be art, not science?, ***Harvard Business Review*****, March.**
One of the few articles that looks at the boundaries of conventional process theory.

Hammer, M. and Stanton, S. (1999) How process enterprises really work, ***Harvard Business Review*****, November–December.**
Hammer is one of the gurus of process design. This paper is typical of his approach.

Holweg, M., Davies, J., De Meyer, A., Lawson, B. and Schmenner, R. (2018) ***Process Theory: The Principles of Operations Management*****, Oxford University Press.**
As the title implies, this is a book about theory. It is unapologetically academic but does contain some useful ideas.

Johnston, R. Shulver, M., Slack, N. and Clark, G. (2021) ***Service Operations Management*****, 5th edn, Pearson, Harlow.**
What can we say! A great treatment of service operations from the same stable as this text.

Slack, N. (2017) ***The Operations Advantage*****, Kogan Page, London.**
More of a practical treatment of how operations management can contribute to strategic success. Aimed at practising managers.

Slack, N. and Lewis, M.A. (2020) ***Operations Strategy*****, 6th edn, Pearson, Harlow.**
A more strategic coverage of operations management.

Notes on chapter

1. The information on LEGOLAND is based on the corporate websites of LEGOLAND www.legoland.com and Merlin Entertainment https://www.merlinentertainments.biz/ (accessed August 2021).
2. The information on LEGO is based on the corporate website of Lego System A/S https://www.lego.com/en-gb and Diaz, J. (2008) Exclusive look inside the Lego Factory, Gizmodo, 21 July, http://lego.gizmodo.com/exclusive-look-inside-the-lego-factory-5022769 (accessed August 2021).
3. Higgins, C. (2017) How many combinations are possible using 6 LEGO bricks?, Mental Floss, 12 February, https://www.mentalfloss.com/article/92127/how-many-combinations-are-possible-using-6-lego-bricks (accessed August 2021).
4. The information on which this example is based is taken from: www.msf.org and https://blogs.msf.org/about-us (accessed August 2021).
5. The information on which this example is based is taken from: the hotel's website, https://www.marinabaysands.com/ (accessed August 2021).
6. Vargo, S.L. and Lusch, R.F. (2008) Service-dominant logic: continuing the evolution, *Journal of the Academy of Marketing Science* 36 (Spring), 1–10.
7. The information on which this example is based is taken from: Phipps, L. (2018) How Philips became a pioneer of circularity-as-a-service, GreenBiz, 22 August; Philips website, The circular imperative, https://www.philips.com/a-w/about/environmental-social-governance/environmental/circular-economy.html (accessed August 2021). More information on the circular economy can be found at The Ellen MacArthur Foundation: Let's build a circular economy, https://ellenmacarthurfoundation.org (accessed August 2021).
8. In 2018 Philips Lighting changed its name to Signify, although it still uses the Philips brand for many of its products.
9. Based on a personal communication with Tom Avery CEO of Verbier Sky Exclusive.
10. Based on author's personal experience and the hotelF1 website, https://hotelf1.accor.com/home/index.en.shtml (accessed August 2021).
11. The information on which this example is based is taken from: the company website, https://www.fjallraven.com/ and Silven, R. (2020) Fjällräven voted most sustainable brand in its industry according to Sweden's Sustainable Brand Index, sgbonline.com, 29 April, https://sgbonline.com/press-release/fjallraven-voted-most-sustainable-brand-in-its-industry-according-to-swedens-sustainable-brand-index/ (accessed August 2021).
12. Rangan, V.K., Chase, L. and Karim, S. (2015) The truth about CSR, *Harvard Business Review*, January–February.
13. BBC News (2002) Politicians 'trample over' patient privacy, 1 July, http://news.bbc.co.uk/1/hi/in_depth/health/2002/bma_conference/2077391.stm (accessed August 2021).

2 Operations performance

KEY QUESTIONS

2.1 Why is operations performance vital in any organisation?

2.2 How is operations performance judged at a societal level?

2.3 How is operations performance judged at a strategic level?

2.4 How is operations performance judged at an operational level?

2.5 How can operations performance be measured?

2.6 How do operations performance objectives trade off against each other?

INTRODUCTION

Operations are judged by the way they perform. However, there are many ways of judging performance and there are many different individuals and groups doing the judging. Also, performance can be assessed at different levels. In this chapter, we start by describing a very broad approach to assessing operations performance at a societal level that uses the '**triple bottom line**' to judge an operation's social, environmental and economic impact. We then look at how operations performance can be judged in terms of how it affects an organisation's ability to achieve its overall strategy. The chapter then looks at the more directly operational-level aspects of performance – quality, speed, dependability, flexibility and cost. Finally, we examine how performance objectives trade off against each other. On our general model of operations management, the topics covered in this chapter are represented by the area marked on Figure 2.1.

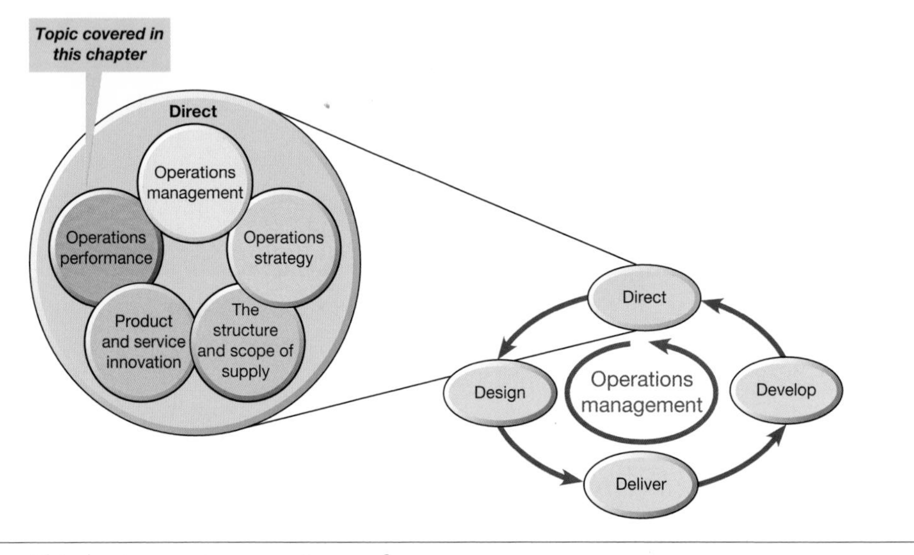

Figure 2.1 This chapter examines operations performance

2.1 Why is operations performance vital in any organisation?

It is no exaggeration to view operations management as being able to either 'make or break' any business – not just because the operations function is large and, in most businesses, represents the bulk of its assets and the majority of its people. Operations management can 'make' the organisation. First, operations management is concerned with doing things better, which can make operations the driver of improvement for the whole organisation. Second, it can build the 'difficult to imitate' capabilities that can have a significant strategic impact. (See the next chapter on operations strategy.) Third, operations management is very much concerned with 'process', with how things are done. And there is a relationship between process and outcome. But when things go wrong in operations, the reputational damage can last for years. We will deal with the risks of operations failures in Chapter 18, but the first point to make is that when operations do go wrong it is often the direct result of poor operations management. From irritatingly late parcel delivery to fatal air crashes, operations failures are both obvious and serious.

Performance at three levels

'Performance' is not a straightforward concept. First, it is multifaceted in the sense that a single measure can never fully communicate the success, or otherwise, of something as complex as an operation. Second, performance can be assessed at different levels, from the broad, long-term, societal level, to its more operational-level concerns over how it improves day-to-day efficiency and serves its individual customers. In the rest of this chapter we will look at how operations can judge its performance at each of these three levels, as illustrated in Figure 2.2:

Operations principle

Good operations performance is fundamental to the sustainable success of any organisation.

- The broad, societal level, using the idea of the 'triple bottom line'.
- The strategic level of how an operation can contribute to the organisation's overall strategy.
- The operational level, using the five operations 'performance objectives'.

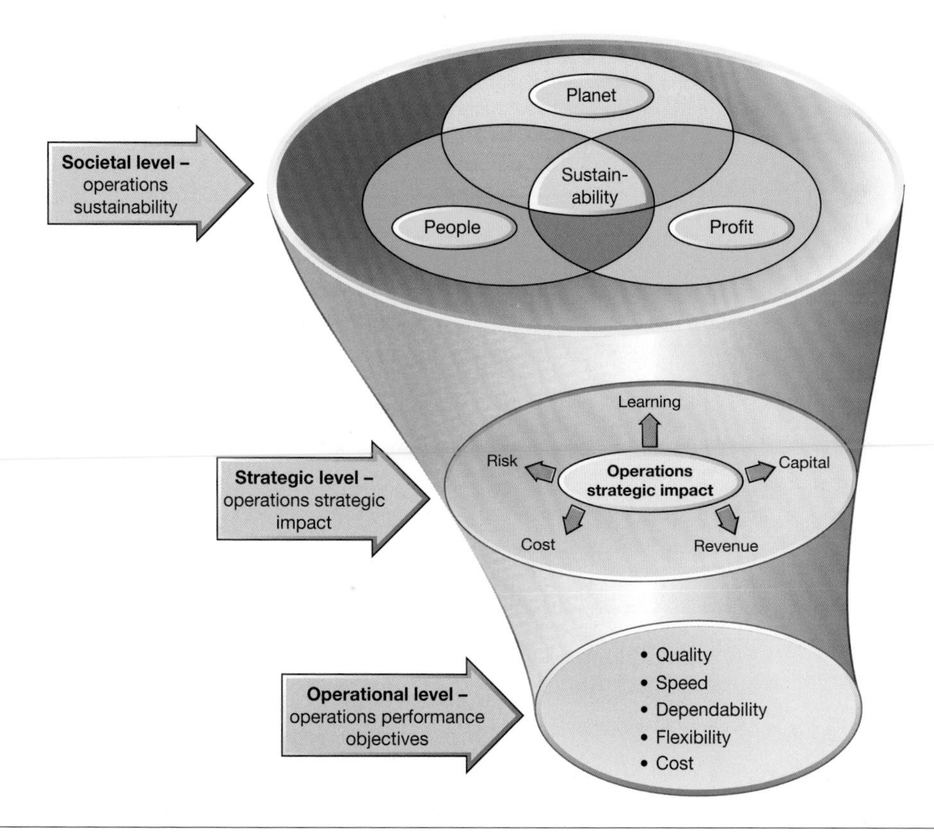

Figure 2.2 Three levels of operations performance

2.2 How is operations performance judged at a societal level?

The decisions that are made within any operation and the way it goes about its day-to-day activities will affect a whole variety of '**stakeholders**'. Stakeholders are the people and groups who have a legitimate interest in the operation's activities. Some stakeholders are internal, for example the operation's employees; others are external, for example customers, society or community groups and a company's shareholders. Some external stakeholders have a direct commercial relationship with the organisation, for example suppliers and customers; others do not, for example industry regulators. Sometimes stakeholder groups can overlap. So, voluntary workers in a charity may be employees, shareholders and customers all at once. It is always the responsibility of operations managers to understand the (sometimes conflicting) objectives of its stakeholders. Figure 2.3 illustrates just some of the stakeholder groups that would have an interest in how an organisation's operations function performs. But although each of these groups, to different extents, will be interested in operations performance, they are likely to have very different views of which aspect of performance is important.

> **Operations principle**
> All operations decisions should reflect the interests of stakeholder groups.

Corporate social responsibility (CSR)

This idea that operations should take into account their impact on a broad mix of stakeholders is often termed 'corporate social responsibility' (CSR). It is concerned with understanding, heeding and responding to the needs of all the company's stakeholders. CSR is fundamentally about how business takes account of its economic, ethical, social and environmental impacts in the way it operates. The issue of how CSR objectives can be included in operations management's activities is of increasing

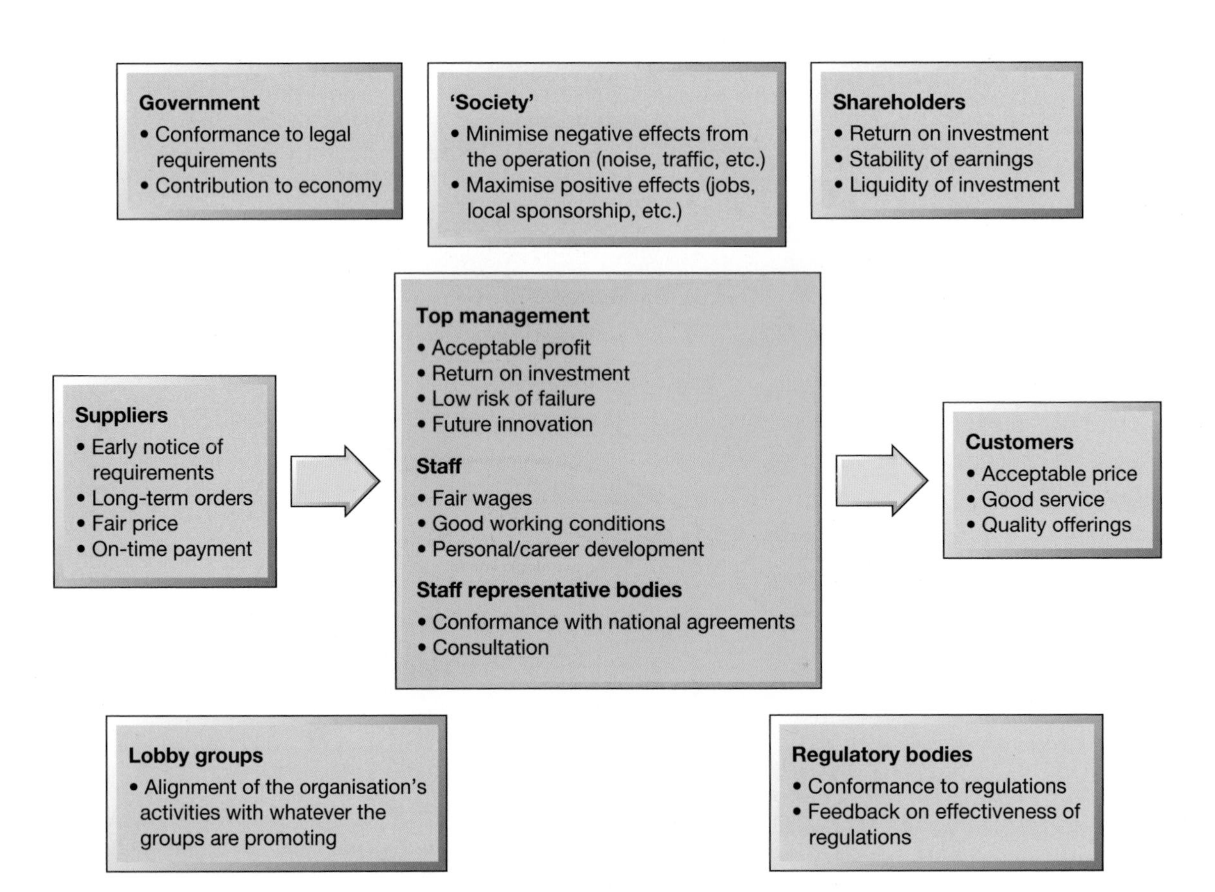

Figure 2.3 Stakeholder groups with typical operations objectives

importance, from both an ethical and a commercial point of view. It is treated several times at various points throughout this text and every chapter contains a 'Responsible operations' section that reflects how the topic covered in the chapter impacts on an operation's CSR.

The triple bottom line

One common term that tries to capture the idea of a broader approach to assessing an organisation's performance is the 'triple bottom line' (TBL, or 3BL), also known as 'people, planet and profit'.[1] The idea is that organisations should measure themselves not just on the traditional economic profit that they generate for their owners, but also on the impact their operations have on society (broadly, in the sense of communities, and individually, for example in terms of their employees) and the ecological impact on the environment. This is often summarised as 'sustainability'. A sustainable business is one that creates an acceptable profit for its owners but minimises the damage to the environment and enhances the existence of the people with whom it has contact. In other words, it balances economic, environmental and societal interests. The assumption underlying the triple bottom line (which is not universally accepted) is that a sustainable business is more likely to remain successful in the long term than one that focuses on economic goals alone.

> **Operations principle**
> Operations can judge themselves on the triple bottom line principle of people, planet and profit.

Environmental, social and governance (ESG)

As the importance of corporate social responsibility/triple bottom line performance has become recognised, the number of acronyms and terms to describe the idea have multiplied. One such is 'environmental, social and governance', or ESG. Derived from the ideas that we have already described, it treats CSR performance from an investor's perspective. It reflects the idea that any potential investor in an enterprise should focus on all ethical social and environmental factors rather than simply looking at the 'return' on that investment.

The social bottom line (people) – the social account, measured by the impact of the operation on the quality of people's lives

The idea behind the social bottom line performance is not just that there is a connection between businesses and the society in which they operate – that is self-evident. Rather it is that businesses should accept that they bear some responsibility for the impact they have on society and balance the external 'societal' consequences of their actions with the more direct internal consequences, such as profit. At the level of the individual, social bottom line performance means devising jobs and work patterns that allow individuals to contribute their talents without undue stress. At a group level, it means recognising and dealing honestly with employee representatives. In addition, businesses are also a part of the larger community and, it is argued, should be recognising their responsibility to local communities by helping to promote their economic and social well-being. Some ways that operations can impact the social bottom line performance include the following:

- Customer safety from products and services.
- Employment impact of an operation's location.
- Employment implications of outsourcing.
- Repetitive or alienating work.
- Staff safety and workplace stress.
- Non-exploitation of Global South suppliers by Global North companies.

The environmental bottom line (planet) – the environmental account, measured by the environmental impact of the operation

Environmental sustainability means making sure that the benefits from development actions more than compensate for any direct or indirect loss or degradation of the environment. Put more directly, it is generally taken to mean the extent to which business activity negatively impacts the natural environment. It is clearly an important issue, not only because of the obvious impact on the immediate environment of hazardous waste, air and even noise pollution, but also because of the less obvious, but potentially far more damaging, issues around global warming. Operations managers cannot avoid responsibility for environmental performance. It is often operational failures that are at the root of pollution disasters, and operations decisions (such as product design) that impact on longer-term environmental issues. Some ways that operations can impact the **environmental bottom line** performance include the following:

- Recyclability of materials, energy consumption, waste material generation.
- Reducing transport-related energy.
- Noise pollution, fume and emission pollution.
- Obsolescence and wastage.
- Environmental impact of process failures.
- Recovery to minimise impact of failures.

OPERATIONS IN PRACTICE

Danone's path to B Corporation[2]

The food industry is not always the most popular with some consumers who doubt the environmental sustainability of industrial-scale food and drink production. A common view among food bloggers is that the giants of the food industry, with their large factories and long supply chains, have disconnected consumers from the 'natural' source of their nourishment. Yet some are actively trying to balance financial, environmental and social objectives. Take Danone, for example, the French food company that employs over 100,000 people, and markets over €25 billion worth of its products, such as Activia yogurt and Evian mineral water, throughout 130 countries. It is one of Europe's largest and most well-known food companies, which built its reputation on a health-focused portfolio of food products. Danone believes that it should be looking beyond simply maximising profitability alone. It rejects the idea that a firm exists mainly to maximise the return to its owners, the shareholders. Its commitment to following a broader set of corporate objectives is exemplified by its ambition to become one of the first certified 'B Corp' multinationals. Certified B Corporations are businesses that meet the highest standards of verified social and environmental performance, public transparency and legal accountability to balance profit and purpose. The B Corp movement was launched in 2006 and has been gaining momentum around the world. **B Corps** use profits and growth as a means to a greater end: positive impact for their employees, communities and the environment. Danone decided to partner with B Lab to plan the most appropriate roadmap towards this goal. B Lab is a non-profit organisation that accredits B Corp certification to for-profit companies that demonstrate high standards of social and environmental performance. And Danone has actively translated these goals into action. It disposed of subsidiaries that produced biscuits, chocolate and beer, and is trying to become carbon neutral with its water brands. It invested in a way to make recycled plastic (which is often an unappealing grey colour) more attractive to consumers.

The economic bottom line (profit) – the economic account, measured by profitability, return on assets, etc., of the operation

The organisation's top management represent the interests of the owners (or trustees, or electorate, etc.) and therefore are the direct custodians of the organisation's economic performance. Broadly this means that operations managers must use the operation's resources effectively, and there are many ways of measuring this '**economic bottom line**'. Finance specialists have devised various measures (such as return on assets, etc.), which are beyond the scope of this text, to do this.

Some ways that operations can impact the financial bottom line performance include the following:

- Cost of producing products and services.
- Revenue from the effects of quality, speed, dependability and flexibility.
- Effectiveness of investment in operations resources.
- Risk and resilience of supply.
- Building capabilities for the future.

We will build on these 'economic bottom line' issues in the next section on judging operations performance at a strategic level.

Critical commentary

The dilemma with using this wide range of triple bottom line, stakeholders or CSR to judge operations performance is that organisations, particularly commercial companies, have to cope with the conflicting pressures of maximising profitability, on the one hand, and the expectation that they will manage in the interests of (all or part of) society in general with accountability and transparency, on the other. Even if a business wants to reflect aspects of performance beyond its own immediate interests, how is it to do it? At the economy-wide or social level, is it really possible to dictate what some call the single 'objective function' that should be maximised by firms? Or, to put the issue even more simply, how do we want the firms in our economy to measure their own performance? How do we want them to determine what constitutes 'good' or 'bad' decisions? Some economists argue that using a broad range of stakeholder perspectives gives undue weight to narrow special interests who want to use the organisation's resources for their own ends. In effect, the stakeholder perspective gives special interests a spurious legitimacy, which undermines the foundations of value-seeking behaviour.[3]

2.3 How is operations performance judged at a strategic level?

Many (although not all) of the activities of operations managers are operational in nature. That is, they deal with relatively immediate, detailed and local issues. However, it is a central idea in operations management that the types of decisions and activities that operations managers carry out can also have a significant strategic impact. Therefore, if one is assessing the performance of the operations function, it makes sense to ask how it impacts the organisation's strategic 'economic' position. We will examine operations' strategic role in the next chapter. But, at a strategic level, there are five aspects of operations performance that we identified as contributing to the 'economic' aspect of the triple bottom line that can have a significant impact (see Figure 2.4).

Operations management affects costs

It seems almost too obvious to state, but almost all the activities that operations managers regularly perform (and all the topics that are described in this text) will have an effect on the cost of producing products and services. Clearly the efficiency with which an operation purchases its transformed and transforming resources, and the efficiency with which it converts its transformed resources, will determine the cost of its products and services. And for many operations managers it is the most important aspect of how they judge their performance. Indeed, there cannot be many, if any, organisations that are indifferent to their costs.

Figure 2.4 Operations can contribute to financial success through low costs, increasing revenue, lowering risk, making efficient use of capital and building the capabilities for future innovation

Operations management affects revenue

Yet cost is not always necessarily the most important strategic objective for operations managers. Their activities also can have a huge effect on revenue. High-quality, error-free products and services, delivered fast and on time, where the operation has the flexibility to adapt to customers' needs, are likely to command a higher price and sell more than those with lower levels of quality, delivery and flexibility. And operations managers are directly responsible for issues such as quality, speed of delivery, dependability and flexibility, as we will discuss later in the chapter.

The main point here is that operations activities can have a significant effect on, and therefore should be judged on, the organisation's profitability. Moreover, even relatively small improvements in cost and revenue can have a proportionally even greater effect on profitability. For example, suppose a business has an annual revenue of €1,000,000 and annual costs of €900,000, and therefore a 'profit' of €100,000. Now suppose that, because of the excellence of its operations managers in enhancing quality and delivery, revenue increases by 5 per cent and costs reduce by 5 per cent. Revenue is now €1,050,000 and costs €855,000. So, profit is now €195,000. In other words, a 5 per cent change in cost and revenue has improved profitability by 95 per cent.

> **Operations principle**
> Profit is a small number made up of the difference between two very big numbers.

Net promoter score (NPS)

One popular method of measuring the underlying levels of customer satisfaction (an important factor in determining revenue) is the 'net promoter score' (NPS). This is computed by surveying customers and asking them how likely they are to recommend a company, service or product (on a scale of 1 to 10, where 1 = not at all likely, and 10 = extremely likely). Customers giving a score of 1 to 6 are called 'detractors', those giving a score of 9 or 10 are called 'promoters' and those giving a score of 7 or 8 are called 'passives'. The NPS is calculated by ignoring passives and subtracting the number of detractors from the number of promoters. So, if 200 customers are surveyed, and 60 are promoters, 110 passives and 30 detractors, the NPS is calculated as follows:

$$NPS = 60 - 30 = 30$$

What is considered an acceptable NPS varies, depending on the sector and nature of competition, but as a minimum target, a positive (>0) score might be regarded as (just) acceptable. NPS is a simple metric that is quick and easy to calculate, but some see that as its main weakness. It is without any sophisticated scientific basis, nor does it provide an accurate picture of customer behaviour. For example, if the survey mentioned above had resulted in 115 promoters, 0 passives and 85 detractors, it would still give an NPS of 30. Yet here, customers are far more polarised. However, notwithstanding its failings, using NPS over time can be used to detect possible changes in customer attitudes.

Operations management affects the required level of investment

How an operation manages the transforming resources that are necessary to produce the required type and quantity of its products and services will also have a strategic effect. If, for example, an operation increases its efficiency so that it can produce (say) 10 per cent more output, then it will not need to spend investment (sometimes called capital employed) to produce 10 per cent more output. Producing more output with the same resources (or sometimes producing the same output with fewer resources) affects the required level of investment.

Operations management affects the risk of operational failure

Well-designed and run operations should be less likely to fail. Not only are they more likely to operate at a predictable and acceptable rate without either letting customers down or incurring excess costs, they are less likely to intentionally or unintentionally cause environmental or social damage. Environmental and social performance (as encompassed by the triple bottom line concept, see earlier) is not confined to the very high-level perspective on performance that we described earlier; it is an important part of how operations strategies are formed, and how their success is measured. Also, if social, environmental or economic failures do occur, well-run operations should be able to recover faster and with less disruption. This is called resilience and is fully treated in Chapter 18.

Operations management affects the ability to build the capabilities on which future innovation is based

Operations managers have a unique opportunity to learn from their experience of operating their processes in order to understand more about those processes. This accumulation of process knowledge can build into the skills, knowledge and experience that allow the business to improve over time. But more than that, it can build into what are known as the 'capabilities' that allow the business to innovate in the future. We will examine this idea of operations capabilities in more detail in the next chapter.

Operations principle

All operations should be expected to contribute to their business at a strategic level by controlling costs, increasing revenue, making investment more effective, reducing risks and growing long-term capabilities.

2.4 How is operations performance judged at an operational level?

Assessing performance at a societal level and judging how well an operation contributes to strategic objectives are clearly important, particularly in the longer term, and form the backdrop to all operations decision-making. But running operations at an operational day-to-day level requires a more tightly defined set of objectives. These are called operations 'performance objectives'. There are five of them and they apply to all types of operation. Imagine that you are an operations manager in any kind of business – a hospital administrator, for example, or a production manager in a vehicle plant. What kinds of things are you likely to want to do in order to satisfy customers and contribute to competitiveness?

- You would want to do things right; that is, you would not want to make mistakes, and would want to satisfy your customers by providing error-free goods and services that are 'fit for their purpose'. This is giving a quality advantage.
- You would want to do things fast, minimising the time between a customer asking for goods or services and the customer receiving them in full, thus increasing the availability of your goods and services and giving a speed advantage.

▶ You would want to do things on time, so as to keep the delivery promises you have made. If the operation can do this, it is giving a **dependability** advantage.

▶ You would want to be able to change what you do; that is, being able to vary or adapt the operation's activities to cope with unexpected circumstances or to give customers individual treatment. Being able to change far enough and fast enough to meet customer requirements gives a **flexibility** advantage.

▶ You would want to do things cheaply; that is, produce goods and services at a cost that enables them to be priced appropriately for the market while still allowing for a return to the organisation; or, in a not-for-profit organisation, give good value to the taxpayers or whoever is funding the operation. When the organisation is managing to do this, it is giving a cost advantage.

> **Operations principle**
>
> Operations performance objectives can be grouped together as quality, speed, dependability, flexibility and cost.

The next part of this chapter examines these five performance objectives in more detail by looking at what they mean for four different operations: a general hospital, a vehicle factory, a city bus company and a supermarket chain.

Why is quality important?

Quality is consistent conformance to customers' expectations, in other words 'doing things right', but the things that the operation needs to do right will vary according to the kind of operation. All operations regard quality as a particularly important objective. In some ways quality is the most visible part of what an operation does. Furthermore, it is something that a customer finds relatively easy to judge about the operation. Is the product or service as it is supposed to be? Is it right or is it wrong? There is something fundamental about quality. Because of this, it is clearly a major influence on customer satisfaction or dissatisfaction. A customer perception of high-quality products and services means customer satisfaction and therefore the likelihood that the customer will return. Figure 2.5 illustrates how quality could be judged in four operations.

Quality inside the operation

When quality means consistently producing services and products to specification it not only leads to external customer satisfaction, but makes life easier inside the operation as well.

Quality could mean . . .

Hospital

- Patients receive the most appropriate treatment
- Treatment is carried out in the correct manner
- Patients are consulted and kept informed
- Staff or are courteous friendly and helpful

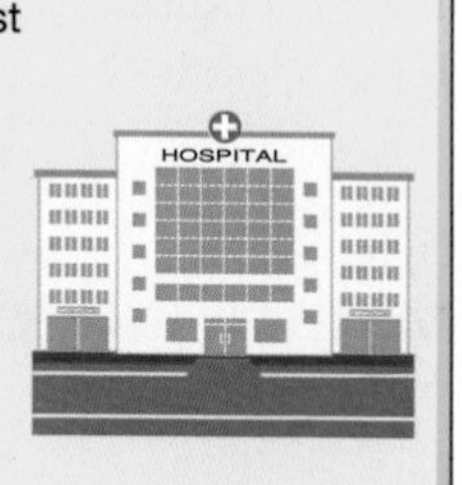

Automobile plant

- All parts are made to specification
- All assembly is to specification
- The product is reliable
- The product is attractive and blemish free

Bus company

- My buses are clean and tidy
- The buses are quiet and fume-free
- The timetable is accurate and user-friendly
- The staff are courteous, friendly and helpful

Supermarket

- Goods are in good condition
- The store is clean and tidy
- Decor is appropriate and attractive
- Staff are courteous, and friendly and helpful

Figure 2.5 Quality means different things in different operations

Quality reduces costs The fewer mistakes made by each process in the operation, the less time will be needed to correct the mistakes and the less confusion and irritation will be spread. For example, if a supermarket's regional warehouse sends the wrong goods to the supermarket, it will mean staff time, and therefore cost, being used to sort out the problem.

Quality increases dependability Increased costs are not the only consequence of poor quality. At the supermarket, it could also mean that goods run out on the supermarket shelves with a resulting loss of revenue to the operation and irritation to the external customers. Sorting out the problem could also distract the supermarket management from giving attention to the other parts of the supermarket operation. This in turn could result in further mistakes being made. So, quality (like the other performance objectives, as we will see) has both an external impact, which influences customer satisfaction, and an internal impact, which leads to stable and efficient processes.

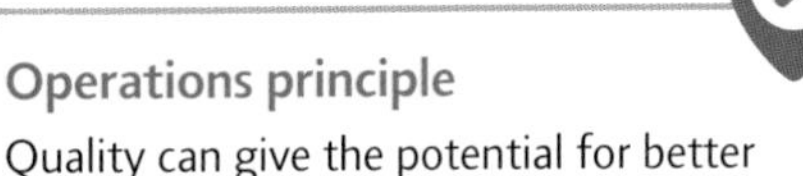

Operations principle

Quality can give the potential for better services and products and save costs.

OPERATIONS IN PRACTICE

Nutella shuts factory to preserve quality[4]

For well-known brands, particularly those associated with food products, quality has to be a priority. Any doubt that quality is not of the highest level can fundamentally damage a brand's reputation. Not surprising then, that Ferrero (the maker of Nutella, the hazelnut and chocolate spread) goes to extreme lengths to preserve the quality of its products. It even shut the biggest factory for its Nutella spread that produces one-quarter of the world's chocolate-hazelnut spread for five days. The company said that after reading the results of one of the quality checks at its Villers-Ecalles factory, operations managers spotted a quality defect in one of the semi-finished products that are used in the production of its Nutella and Kinder Bueno products. It said that temporary interruption to production was a precautionary measure that would allow further investigations to be carried out and countermeasures would be taken after a thorough quality investigation. It also reassured consumers that none of their products out in the market would be affected by the situation or the interruption in supply. When the problem was solved, which was very early in the process, during the grinding and roasting of the hazelnuts, the factory resumed production of its approximately 600,000 jars a day. Although any reputable food producer would treat any suspected problems of this type seriously, Nutella holds a special place in French consciousness (fights had broken out in French supermarkets when the product was discounted). Earlier the same year Nutella had advertised for 60 'sensorial judges' to taste its products. Applicants for the job must were required to undergo training to refine their taste buds and help them to express taste sensations. And of course, they must like hazelnuts.

Why is speed important?

Speed means the elapsed time between customers requesting products or services and their receiving them. Figure 2.6 illustrates what speed means for the four operations. The main benefit to the operation's (external) customers of speedy delivery of goods and services is that the faster they can have the product or service, the more likely they are to buy it, or the more they will pay for it, or the greater the benefit they receive (see the 'Operations in practice' example 'Speeding response to save lives').

Speed could mean . . .

Hospital

- The time between requiring treatment and receiving treatment kept to a minimum
- The time for test results, X-rays, etc. to be returned kept to a minimum

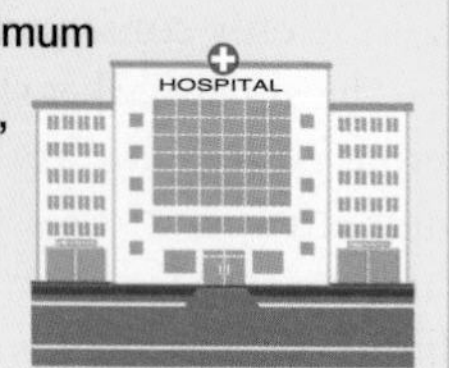

Automobile plant

- The time between dealers requesting a vehicle of a particular specification and receiving it kept to a minimum
- The time to deliver spares to service centres kept to a minimum

Bus company

- The time between a customer setting out on a journey and reaching their destination kept to a minimum

Supermarket

- The time taken for the total transaction of going to the supermarket, making the purchase, and returning kept to a minimum
- The immediate availability of goods

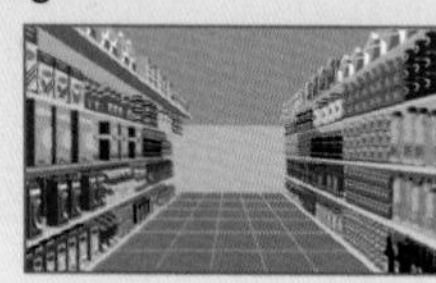

Figure 2.6 Speed means different things in different operations

Speed inside the operation

Inside the operation, speed is also important. Fast response to external customers is greatly helped by speedy decision-making and speedy movement of materials and information inside the operation. And there are other benefits.

Speed reduces inventories Take, for example, the vehicle plant. Steel for the vehicle's door panels is delivered to the press shop, pressed into shape, transported to the painting area, coated for colour and protection and moved to the assembly line where it is fitted to the vehicle. This is a simple three-stage process, but in practice material does not flow smoothly from one stage to the next. First, the steel is delivered as part of a far larger batch containing enough steel to make possibly several hundred products. Eventually it is taken to the press area, pressed into shape and again waits to be transported to the paint area. It then waits to be painted, only to wait once more until it is transported to the assembly line. Yet again it waits by the trackside until it is eventually fitted to the vehicle. The material's journey time is far longer than the time needed to make and fit the product. It actually spends most of its time waiting as stocks (inventories) of parts and products. The longer items take to move through a process, the more time they will be waiting and the higher inventory will be. This is an important idea, which will be explored in Chapter 16 on lean operations.

Speed reduces risks Forecasting tomorrow's events is far less of a risk than forecasting next year's. The further ahead companies forecast, the more likely they are to get it wrong. The faster the throughput time of a process, the later forecasting can be left. Consider the vehicle plant again. If the total throughput time for the door panel is six weeks, door panels are being processed through their first operation six weeks before they reach their final destination. The quantity of door panels being processed will be determined by the forecasts for demand six weeks ahead. If instead of six weeks, they take only one week to move through the plant, the door panels being processed through their first stage are intended to meet demand only one week ahead. Under these circumstances it is far more likely that the number and type of door panels being processed are the number and type that eventually will be needed.

Operations principle
Speed can give the potential for faster delivery of services and products and save costs.

OPERATIONS IN PRACTICE

Speeding response to save lives[5]

Few services have more need of speed than the emergency services. In responding to road accidents or acute illness, even seconds can be critical. Treating patients speedily means speeding up three elements of the total time to treatment: the time it takes for the emergency services to find out the details of the accident, the time it takes them to travel to the scene of the accident, and the time it takes to get the casualty to appropriate treatment. For example, for every 15 minutes a stroke is left untreated, three years are taken off the patient's life in the case of strokes, as a clot or a bleed cuts off blood supply to the brain. However, technology can help. In one trial, ambulance paramedics used video links to link to stroke specialists. Their advice helped make more accurate treatment decisions during the journey. Using this procedure, the number of patients receiving appropriate treatment within the critical time went from 5 per cent to 41 per cent. In another example, London's Air Ambulance service used an app that saved its emergency team two minutes in responding to emergencies. Rather than having to take all the details of an emergency before they rushed to their helicopter, the app, together with enhanced mobile communication, allows them to set off immediately and receive the details on their tablet when they are in the air. But is getting airborne two minutes sooner really significant? It is, when one considers that, if starved of oxygen, a million brain cells can die every minute. It allows the service's advanced trauma doctors and paramedics to perform procedures to relieve pain, straighten broken limbs, even perform open-chest surgery to restart the heart, often within minutes of injury. Including trauma medics in the team, in effect, brings the hospital to the patient, wherever that may be. When most rescues are only a couple of minutes' flying time back to the hospital, speed really can save lives. However, it is not always possible to land a helicopter safely at night (because of overhead wires and other potential hazards), so conventional ambulances will always be needed, both to get paramedics quickly to accident victims and to speed the patients to hospital.

Why is dependability important?

Dependability means doing things in time for customers to receive products or services exactly when they are needed, or at least when they were promised. Figure 2.7 illustrates what dependability means in the four operations. Customers might judge the dependability of an operation only after the product or service has been delivered. Initially this may not affect the likelihood that customers will select the service – they have already 'consumed' it. Over time, however, dependability can override all other criteria. No matter how cheap or fast a bus service is, if the service is always late (or unpredictably early) or the buses are always full, then potential passengers will be better off calling a taxi.

Dependability inside the operation

Inside the operation internal customers will judge each other's performance partly by how reliable the other processes are in delivering material or information on time. Operations where internal dependability is high are more effective than those which are not, for a number of reasons.

Dependability saves time Take, for example, the maintenance and repair centre for the city bus company. If the centre runs out of some crucial spare parts, the manager of the centre will need to spend time trying to arrange a special delivery of the required parts, and the resources allocated to service the buses will not be used as productively as they would have been without this disruption. More seriously, the fleet will be short of buses until they can be repaired and the fleet operations

Dependability could mean . . .

Hospital

- Proportion of appointments that are cancelled kept to a minimum
- Keeping to the appointment times
- Test results, X-rays, etc. returned as promised

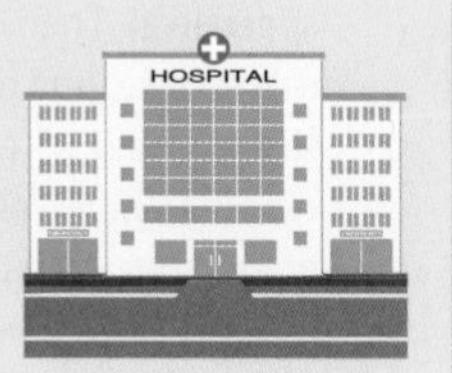

Automobile plant

- On-time delivery of vehicle to dealers
- On-time delivery of spares to service centres

Bus company

- Keeping to the published timetable at all points on the route
- Constant availability of seats for passengers

Supermarket

- Predictability of opening hours
- Proportion of goods out of stock kept to a minimum
- Keeping to reasonable queuing times
- Constant availability of parking

Figure 2.7 Dependability means different things in different operations

manager will have to spend time rescheduling services. So, entirely due to the one failure of dependability of supply, a significant part of the operation's time has been wasted coping with the disruption.

Dependability saves money Ineffective use of time will translate into extra cost. The spare parts might cost more to be delivered at short notice and maintenance staff will expect to be paid even when there is no bus to work on. Nor will the fixed costs of the operation, such as heating and rent, be reduced because the buses are not being serviced. The rescheduling of buses will probably mean that some routes have inappropriately sized buses and some services could have to be cancelled. This will result in empty bus seats (if too large a bus has to be used) or a loss of revenue (if potential passengers are not transported).

OPERATIONS IN PRACTICE

What does dependability mean when travelling by rail?[6]

When the operator of a private railway firm that serves the Tokyo suburbs issued an apology after one of its trains departed 20 seconds ahead of schedule, it made headlines around the world. Passengers on the 9.44.40 am Tsukuba Express from Minami-Nagareyama station, just north of Tokyo, were oblivious to the 20 second early departure when the train (which had arrived on time) pulled away at 9.44.20 am. In most parts of the world such a small error would not be any cause for comment (except perhaps to congratulate the company for being so close to the scheduled time), but the company that operates the Tokyo service clearly felt different. In a formal statement they said that they 'deeply apologise for the severe inconvenience imposed upon our customers'. They said that the conductor on the train had not properly checked the train's timetable. In future, they said, the crew had been instructed to 'strictly follow procedure to prevent a recurrence'. But what about any passengers who may have missed the train because it left 20 seconds early? Not to worry – another train arrived four minutes later (and left on time).

The story prompted news outlets to focus on how rail services in various countries performed in terms of their dependability (punctuality, as it would usually be called by rail operators). Japan is indeed one of the best on-time performing countries. The Japanese high-speed bullet train arrives a mere 54 seconds behind schedule on average. If a Japanese train is five minutes late or more, passengers are given a certificate that they can give to their boss or teacher as an excuse for being late. Perhaps predictably for a country famous for its clocks and watches, the Swiss operator SBB rates as among the most punctual in Europe with 88.8 per cent of all passengers arriving on time and customer satisfaction surpassing the company's target of 75 per cent. Other national rail networks with good dependability records include Sweden, Denmark, Germany and France. But comparisons are complicated by different definitions of exactly what 'on time' means. A Swiss train is late if it arrives more than three minutes after the advertised time. But in the United Kingdom it can be up to five minutes late (10 minutes if it's a longer journey) and still be recorded as 'on time'. In fact, there are a wide range of 'lateness' standards used throughout the world. In Ireland, a train is 'on time' if it's less than 10 minutes late (or 5 minutes for Dublin's Dart network). Trains in the United States are allowed 10 minutes' leeway for journeys up to 250 miles, and up to 30 minutes for journeys of more than 550 miles. In Australia, each rail company has its own definitions of punctuality. In Victoria, trains have between 5 and 11 minutes' flexibility. Queensland's trains have either 4 or 6 minutes, depending on the route. Also, countries measure different types of punctuality. In the United Kingdom, punctuality is measured when the train arrives at its final destination. In Switzerland, the punctuality of individuals is monitored (how late did each passenger arrive at whichever station they wanted to get off?).

Dependability gives stability The disruption caused to operations by a lack of dependability goes beyond time and cost. It affects the 'quality' of the operation's time. If everything in an operation is always perfectly dependable, a level of trust will have built up between the different parts of the operation. There will be no 'surprises' and everything will be predictable. Under such circumstances, each part of the operation can concentrate on improving its own area of responsibility without having its attention continually diverted by a lack of dependable service from the other parts.

> **Operations principle**
> Dependability can give the potential for more reliable delivery of services and products, and save costs.

Why is flexibility important?

Flexibility means being able to change the operation in some way. This may mean changing what the operation is doing, how it is doing it, or when it is doing it. Specifically, customers will need the operation to change so that it can provide four types of requirements:

- product/service flexibility – the operation's ability to introduce new or modified products and services;
- mix flexibility – the operation's ability to produce a wide range or mix of products and services;
- volume flexibility – the operation's ability to change its level of output or activity to produce different quantities or volumes of products and services over time;
- delivery flexibility – the operation's ability to change the timing of the delivery of its services or products.

Figure 2.8 gives examples of what these different types of flexibility mean to the four different operations.

Mass customisation

One of the beneficial external effects of flexibility is the increased ability of operations to do different things for different customers. So, high flexibility gives the ability to produce a high variety of products or services. Normally high variety means high cost (see Chapter 1). Furthermore, high-variety

Flexibility could mean . . .

Hospital

- Product/service flexibility – the introduction of new types of treatment
- Mix flexibility – a wide range of available treatments
- Volume flexibility – the ability to adjust the number of patients treated
- Delivery flexibility – the ability to reschedule appointments

Automobile plant

- Product/service flexibility – the introduction of new models
- Mix flexibility – a wide range of options available
- Volume flexibility – the ability ot adjust the number of vehicles manufactured
- Delivery flexibility – the ability to reschedule manufacturing priorities

Bus company

- Product/service flexibility – the introduction of new routes or excursions
- Mix flexibility – a large number of locations served
- Volume flexibility – the ability to adjust the frequency of services
- Delivery flexibility – the ability to reschedule trips

Supermarket

- Product/service flexibility – the introduction of new goods or promotions
- Mix flexibility – a wide range of goods stopped
- Volume flexibility – the ability to adjust the number of customers served
- Delivery flexibility – the ability to obtain out of stock items (very occasionally)

Figure 2.8 Flexibility means different things in different operations

operations do not usually produce in high volume. Some companies have developed their flexibility in such a way that products and services are customised for each individual customer. Yet they manage to produce them in a high-volume, mass-production manner, which keeps costs down. This approach is called **mass customisation**. Sometimes this is achieved through flexibility in design. For example, Dell is one of the largest-volume producers of personal computers in the world yet allows each customer to 'design' (albeit in a limited sense) their own configuration. Sometimes flexible technology is used to achieve the same effect. Another example is Paris Miki, an upmarket eyewear retailer, which has the largest number of eyewear stores in the world, and which uses its own 'Mikissimes Design System' to capture a digital image of the customer and analyse facial characteristics. Together with a list of customers' personal preferences, the system then recommends a particular design and displays it on the image of the customer's face. In consultation with the optician, the customer can adjust shapes and sizes until the final design is chosen. Within the store the frames are assembled from a range of pre-manufactured components and the lenses ground and fitted to the frames. The whole process takes around an hour. Another example is given in the mymuesli 'Operations in practice' example.

OPERATIONS IN PRACTICE

566 quadrillion individual muesli mixes – now that's flexible[7]

Three university students, Hubertus Bessau, Philipp Kraiss and Max Wittrock, in the small city of Passau, Germany, came up with the concept of mymuesli – the first web-based platform where you can mix your own organic muesli online, with a choice of 75 different ingredients. This makes it possible to create 566 quadrillion individual muesli mixes – and you can even name your own muesli mix. *'We wanted to provide customers with nothing else but the perfect muesli'*, they say. *'Of course, the idea of custom-mixing muesli online might sound wacky . . . but think about it – it's the breakfast you were always looking for'*. All muesli is mixed in the Passau production site according to strict quality standards and hygiene law requirements. Ingredients are strictly organic, without additional sugar, additives, preservatives or artificial colours. On visiting the website, customers first have to pick a muesli base. After this, they add other basics and ingredients such as fruit, nuts and seeds, and extras. And the company will deliver it direct by courier to your door! The chosen name is printed on the can to make it even more personal. One of mymuesli's great assets is the multitude of eccentric and unusual ingredients sourced from around the world. Philipp Kraiss, one of the company founders, is constantly on the lookout for 'new, crazy and tasty' muesli ingredients.

Flexibility inside the operation

Developing a flexible operation can also have advantages to the internal customers within the operation.

Flexibility speeds up response Fast service often depends on the operation being flexible. For example, if the hospital has to cope with a sudden influx of patients from a road accident, it clearly needs to deal with injuries quickly. Under such circumstances a flexible hospital that can speedily transfer extra skilled staff and equipment to the accident and emergency department will provide the fast service that the patients need.

Flexibility saves time In many parts of the hospital, staff have to treat a wide variety of complaints. Fractures, cuts or drug overdoses do not come in batches. Each patient is an individual with individual needs. The hospital staff cannot take time to 'get into the routine' of treating a particular complaint; they must have the flexibility to adapt quickly. They must also have sufficiently flexible facilities and equipment so that time is not wasted waiting for equipment to be brought to the patient. The hospital's time is being saved because it is flexible in 'changing over' from one task to the next.

Flexibility maintains dependability Internal flexibility can also help to keep the operation on schedule when unexpected events disrupt the operation's plans. For example, if the sudden influx of patients to the hospital requires emergency surgical procedures, routine operations will be disrupted. This is likely to cause distress and considerable inconvenience. A flexible hospital might be able to minimise the disruption by possibly having reserved operating theatres for such an emergency, and being able to bring in medical staff quickly who are 'on call'.

Operations principle

Flexibility can give the potential to create new, wider variety, differing volumes and differing delivery dates of services and products, and save costs.

Agility

In many competitive and dynamic markets, operations experience high rates of change in the environment in which they have to operate. It is for this reason that judging operations in terms of their agility has become popular. Agility means being able to sense changes in the (internal or external) environment of an operation and respond effectively, efficiently and in a timely manner. Moreover, it is often taken to imply that the agile operation can learn from its response to improve in some way. This last point is influential in shaping some definitions of agility, which see it as the capability of being able to survive and prosper in a competitive and/or turbulent environment of unpredictable, continuous change by reacting quickly and effectively to those changes. In many ways, agility is really a combination of all the five performance objectives but particularly cost, flexibility and speed.

Proponents of developing greater agility regularly refer to the propensity for organisations to start their life with high levels of agility (think business start-ups), only to become increasingly bureaucratic as they develop processes, policies and layers of management. The result is that when they grow beyond a certain point, they can struggle to retain those attributes that caused them to grow in the first place. To avoid this, proponents of agility recommend that operations focus on small, inventive, often multidisciplinary teams, who are capable of responding to change rather than 'sticking to the plan' regardless. Further, the culture and leadership of an operation needs to change, with organisational structures, budgets and staff incentives all adjusted to reflect the need for much shorter time frames. The term is particularly common in software development, where projects are frequently late and often run over budget. In this context, agility implies the ability to produce working-quality software in short, fast increments with development teams able to accept and implement fast-changing requirements. However, used more generally, agility means responding to market requirements by producing new and existing products and services fast and flexibly.

Why is cost important?

To the companies that compete directly on price, cost will clearly be their major operations objective. The lower the cost of producing their goods and services, the lower can be the price to their customers. Even those companies that do not compete on price will be interested in keeping costs low. Every euro or dollar removed from an operation's cost base is a further euro or dollar added to its profits. Not surprisingly, low cost is a universally attractive objective. The 'Operations in practice' example on 'Everyday low prices at Aldi' describes how one retailer keeps its costs down.

> **Operations principle**
> Cost is always an important objective for operations management, even if the organisation does not compete directly on price.

The ways in which operations management can influence cost will depend largely on where the operation's costs are incurred. The operation will spend its money on staff (the money spent on employing people), facilities, technology and equipment (the money spent on buying, caring for, operating and replacing the operation's 'hardware') and materials (the money spent on the 'bought-in' materials consumed or transformed in the operation). Figure 2.9 shows typical cost breakdowns for the hospital, car plant, supermarket and bus company.

Cost could mean . . .

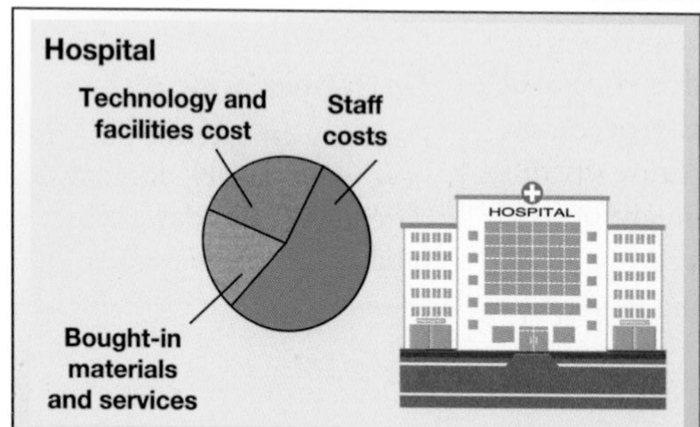

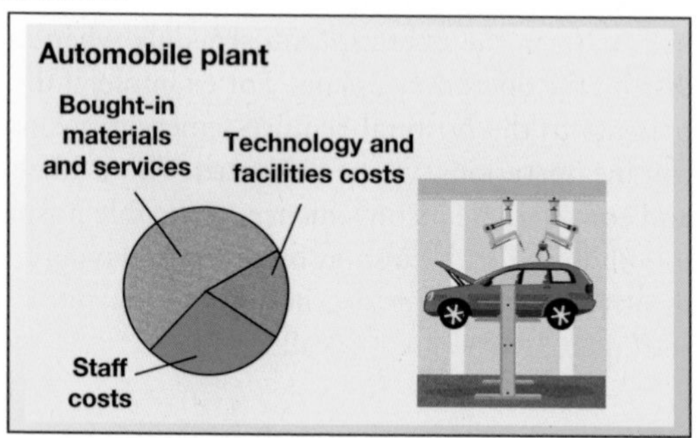

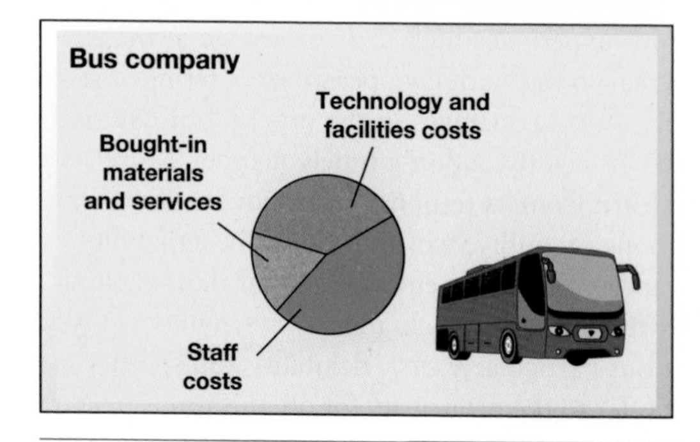

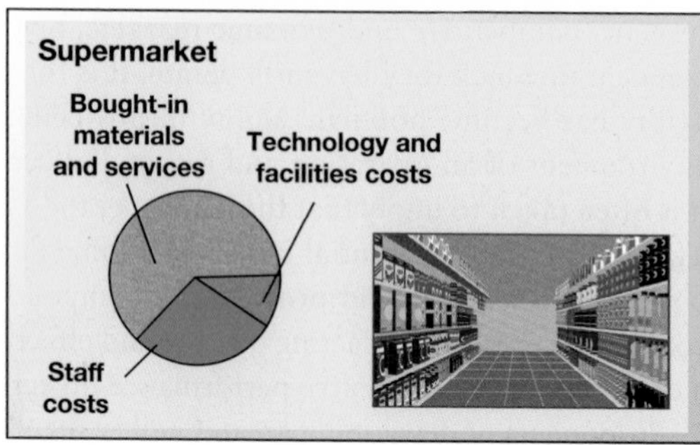

Figure 2.9 Cost means different things in different operations

Keeping operations costs down

All operations have an interest in keeping their costs as low as is compatible with the levels of quality, speed, dependability and flexibility that their customers require. The measure that is most frequently used to indicate this is productivity. Productivity is the ratio of what is produced by an operation (its output) to what is required to produce it (its input):

$$\text{Productivity} = \frac{\text{Output from the operation}}{\text{Input to the operation}}$$

Often partial measures of input or output are used so that comparisons can be made. So, for example, in the vehicle industry productivity is sometimes measured in terms of the number of cars produced per year per employee. This is called a single-factor measure of productivity:

$$\text{Single factor productivity} = \frac{\text{Output from the operation}}{\text{One input to the operation}}$$

OPERATIONS IN PRACTICE

Everyday low prices at Aldi[8]

Aldi is an international 'limited assortment' supermarket specialising in 'private label', mainly food products. It has carefully focused its service concept and delivery system to attract customers in a highly competitive market. The company believes that its unique approach to operations management makes it 'virtually impossible for competitors to match our combination of price and quality'. And it has proved especially successful in meeting the increasingly price-conscious behaviour of customers. How has it done this? By challenging the norms of retail operations. Its operations are kept deliberately simple, using basic facilities to keep down overheads. Most stores stock only a limited range of goods (typically around 700 compared with 25,000 to 30,000 stocked by conventional supermarket chains). The private label approach means that the products have been produced according to Aldi quality specifications and are sold only in Aldi stores. Without the high costs of brand marketing and advertising, and with Aldi's formidable purchasing power, prices can be 30 per cent below their branded equivalents. Other cost-saving practices include open-carton displays, which eliminate the need for special shelving, no grocery bags to encourage recycling as well as saving costs, multiple barcodes on packages (to speed up scanning) and using a 'cart rental' system, which requires customers to return the cart to the store to get their coin deposit back.

This allows different operations to be compared excluding the effects of input costs. One operation may have high total costs per car but high productivity in terms of number of cars per employee per year. The difference between the two measures is explained in terms of the distinction between the cost of the inputs to the operation and the way the operation is managed to convert inputs into outputs. Input costs may be high, but the operation itself is good at converting them to goods and services. Single-factor productivity can include the effects of input costs if the single input factor is expressed in cost terms, such as 'labour costs'. Total factor productivity is the measure that includes all input factors:

$$\text{Total factor productivity} = \frac{\text{Output from the operation}}{\text{All inputs to the operation}}$$

Improving productivity One obvious way of improving an operation's productivity is to reduce the cost of its inputs while maintaining the level of its outputs. This means reducing the costs of some or all of its transformed and transforming resource inputs. For example, a bank may choose to relocate its call centres to a place where its facility-related costs (for example, rent) are cheaper. A software developer based in Europe may relocate its entire operation to India or China, where skilled labour is available at significantly lower rates. A computer manufacturer may change the design of its products to allow the use of cheaper materials. Productivity can also be improved by making better use of the inputs to the operation. For example, garment manufacturers attempt to cut out the various pieces of material that make up the garment by positioning each part on the strip of cloth so that material wastage is minimised. All operations are increasingly concerned with cutting out waste, whether it is waste of materials, waste of staff time, or waste through the underutilisation of facilities.

Worked example

The health-check clinic

The health-check clinic has five employees and 'processes' 200 patients per week. Each employee works 35 hours per week. The clinic's total wage bill is £3,900 and its total overhead expenses are £2,000 per week. What is the clinic's single-factor labour productivity and its total factor productivity?

$$\text{Labour productivity} = \frac{200}{5} = 40 \text{ patients/employee/week}$$

$$\text{Labour productivity} = \frac{200}{(5 \times 35)} = 1.143 \text{ patients/labour hour}$$

$$\text{Total factor productivity} = \frac{200}{(3900 + 2000)} = 0.0339 \text{ patients/£}$$

Cost reduction through internal effectiveness Our previous discussion distinguished between the benefits of each performance objective externally and internally. Each of the various performance objectives has several internal effects, but all of them affect cost, so one important way to improve cost performance is to improve the performance of the other operations objectives (see Figure 2.10):

- High-quality operations do not waste time or effort having to redo things, nor are their internal customers inconvenienced by flawed service.
- Fast operations reduce the level of in-process inventory between processes, as well as reducing administrative overheads.
- Dependable operations do not spring any unwelcome surprises on their internal customers. They can be relied on to deliver exactly as planned. This eliminates wasteful disruption and allows the other processes to operate efficiently.
- Flexible operations adapt to changing circumstances quickly and without disrupting the rest of the operation. Flexible processes can also change over between tasks quickly and without wasting time and capacity.

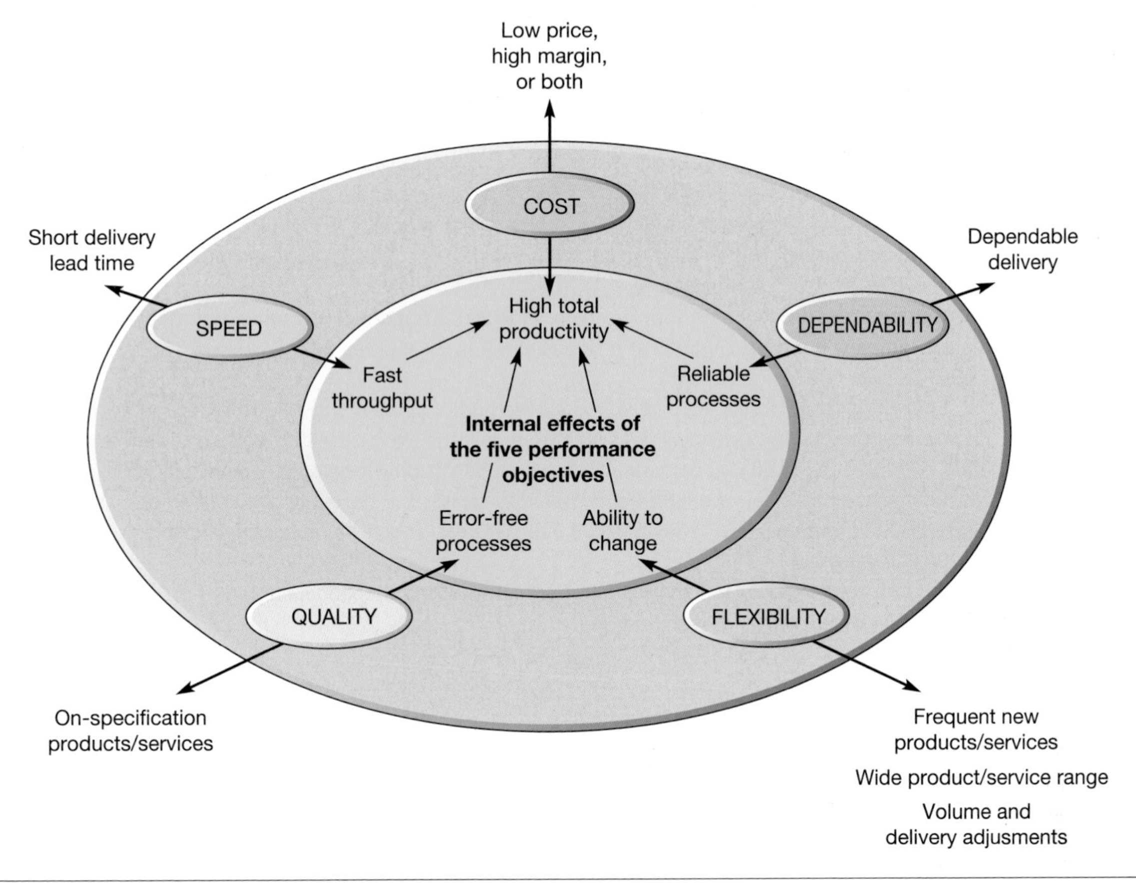

Figure 2.10 Performance objectives have both external and internal effects. Internally, cost is influenced by the other performance objectives

Worked example

Slap.com

Slap.com is an internet retailer of speciality cosmetics. It orders products from a number of suppliers, stores them, packs them to customers' orders, and then dispatches them using a distribution company. Although broadly successful, the business is very keen to reduce its operating costs. A number of suggestions have been made to do this. These are as follows:

- Make each packer responsible for their own quality. This could potentially reduce the percentage of mis-packed items from 0.25 per cent to near zero. Repacking an item that has been mis-packed costs €2 per item.
- Negotiate with suppliers to ensure that they respond to delivery requests faster. It is estimated that this would cut the value of inventories held by slap.com by €1,000,000.
- Institute a simple control system that would give early warning if the total number of orders that should be dispatched by the end of the day actually is dispatched in time. Currently 1 per cent of orders is not packed by the end of the day and therefore has to be sent by express courier the following day. This costs an extra €2 per item.

Because demand varies through the year, sometimes staff have to work overtime. Currently the overtime wage bill for the year is €150,000. The company's employees have indicated that they would be willing to adopt a flexible working scheme where extra hours could be worked when necessary, in exchange for having the hours off at a less busy time and receiving some kind of extra payment. This extra payment is likely to total €50,000 per year.

If the company dispatches 5 million items every year and if the cost of holding inventory is 10 per cent of its value, how much cost will each of these suggestions save the company?

Analysis

Eliminating mis-packing would result in an improvement in quality. Currently 0.25 per cent of 5 million items are mis-packed. This amounts to 12,500 items per year. At €2 repacking charge per item, this is a cost of €25,000 that would be saved.

Getting faster delivery from suppliers helps reduce the amount of inventory in stock by €1,000,000. If the company is paying 10 per cent of the value of stock for keeping it in storage, the saving will be €1,000,000 × 0.1 = €100,000.

Ensuring that all orders are dispatched by the end of the day increases the dependability of the company's operations. Currently, 1 per cent are late; in other words, 50,000 items per year. This is costing €2 × 50,000 = €100,000 per year, which would be saved by increasing dependability.

Changing to a flexible working hours system increases the flexibility of the operation and would cost €50,000 per year, but it saves €150,000 per year. Therefore, increasing flexibility could save €100,000 per year.

So, in total, by improving the operation's quality, speed, dependability and flexibility, a total of €325,000 can be saved.

The polar representation of performance objectives

A useful way of representing the relative importance of performance objectives for a product or service is shown in Figure 2.11(a). This is called the **polar representation** because the scales that represent the importance of each performance objective have the same origin. A line describes the relative importance of each performance objective. The closer the line is to the common origin, the less important is the performance objective to the operation. Two services are shown, a taxi and a bus service. Each essentially provides the same basic service, but with different objectives. The differences between the two services are clearly shown by the diagram. Of course, the polar diagram can be adapted to accommodate any number of different performance objectives. For example, Figure 2.11(b) shows how a **polar diagram** is used in a charity that promotes the growing and consumption of organically produced food.

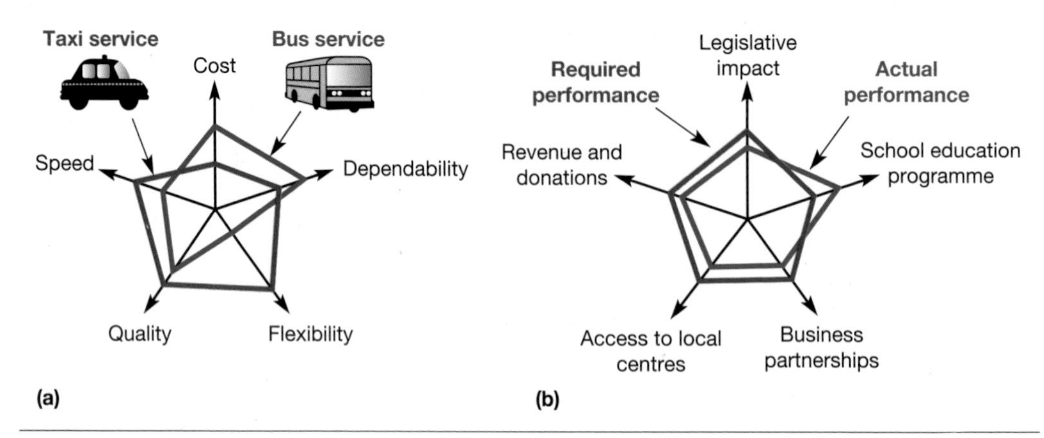

Figure 2.11 Polar diagrams for (a) a taxi service versus a bus service and (b) a charity promoting organically grown food

2.5 How can operations performance be measured?

Any operation needs to measure how well, or badly, it is doing. This is performance measurement. It is the process of quantifying action, where measurement means the process of quantification and the performance of the operation is assumed to derive from actions taken by its management. Some kind of performance measurement is a prerequisite for judging whether an operation is good, bad or indifferent. Without performance measurement, it is impossible to exert any control over an operation on an ongoing basis, or to judge whether any improvement is being made.

Performance measurement, as we are treating it here, concerns three generic issues:

- What factors should be included as performance measures?
- Which are the most important performance measures?
- What detailed measures should be used?

What factors should be included as performance measures?

Earlier in this chapter we explained how operations performance could be described and measured at three levels. There are two important points to make here. First, sometimes these measures are aggregated into 'composite' measures that combine several measures, such as 'customer satisfaction', 'overall service level' or 'operations agility'. These more aggregated 'composite' performance measures help to present a picture of the overall performance of a business, although they may include some influences outside those that operations performance improvement would normally address (customer satisfaction may partly be a function of how a service is advertised, for example). Second, all of the factors at each level can be broken down into more detailed measures. Figure 2.12 gives examples of this. These more detailed performance measures are usually monitored more closely and more often, and although individually they provide a limited view of an operation's performance, taken together they do provide a more descriptive and complete picture of what should be and what is happening within the operation. In practice, most organisations will choose to use performance measures from all three levels.

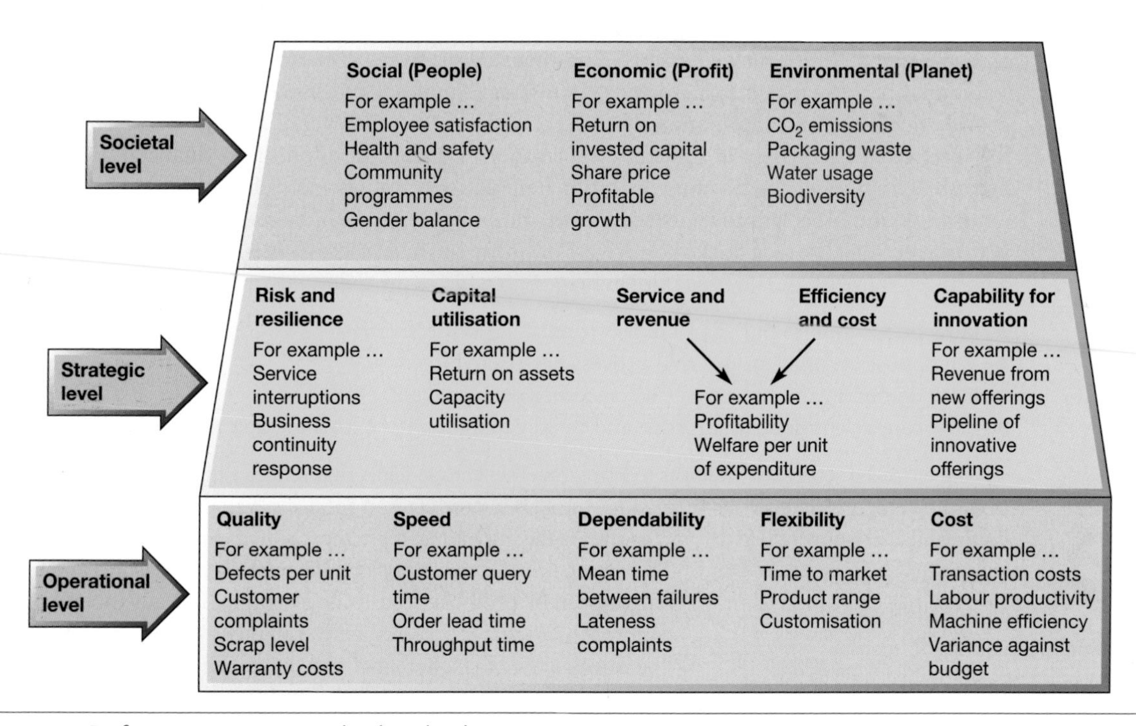

Figure 2.12 Performance measures at the three levels

Which are the most important performance measures?

One of the problems of devising a useful performance measurement system is trying to achieve some balance between having a few key measures on the one hand (straightforward and simple but may not reflect the full range of organisational objectives) or, on the other hand, having many detailed measures (complex and difficult to manage, but capable of conveying many nuances of performance). Broadly, a compromise is often reached by making sure that there is a clear link between the operation's overall strategy, the most important (or 'key') performance indicators (often called KPIs) that reflect strategic objectives, and the bundle of detailed measures that are used to 'flesh out' each key performance indicator. Obviously, unless strategy is well defined then it is difficult to 'target' a narrow range of key performance indicators.

What detailed measures should be used?

The five performance objectives – quality, speed, dependability, flexibility and cost – are really composites of many smaller measures. For example, an operation's cost is derived from many factors, which could include the purchasing efficiency of the operation, the efficiency with which it converts materials, the productivity of its staff, the ratio of direct to indirect staff, and so on. All of these measures individually give a partial view of the operation's cost performance, and many of them overlap in terms of the information they include. However, each of them does give a perspective on the cost performance of an operation that could be useful either to identify areas for improvement or to monitor the extent of improvement. If an organisation regards its 'cost' performance as unsatisfactory, disaggregating it into 'purchasing efficiency', 'operations efficiency', 'staff productivity', etc. might explain the root cause of the poor performance. The 'operational' level in Figure 2.12 shows some of the partial measures that can be used to judge an operation's performance.

The balanced scorecard approach

Arguably, the best-known performance measurement approach, and one used by many organisations, is the 'balanced scorecard' (BSC) devised by Kaplan and Norton[9] who, while retaining traditional financial measures, pointed out that they related to past events and were therefore inadequate for guiding companies in more competitive times. The framework they developed covers both strategic and operational levels. As well as including financial measures of performance, in the same way as traditional performance measurement systems, the balanced scorecard approach also attempts to provide the important information that is required to allow the overall strategy of an organisation to be reflected adequately in specific performance measures. In addition to financial measures of performance, it also includes more operational measures of customer satisfaction, internal processes, innovation and other improvement activities. In doing so it measures the factors behind financial performance that are seen as the key drivers of future financial success. In particular, it is argued that a balanced range of measures enables managers to address the following questions (see Figure 2.13):

- How do we look to our shareholders (financial perspective)?
- What must we excel at (internal process perspective)?
- How do our customers see us (the customer perspective)?
- How can we continue to improve and build capabilities (the learning and growth perspective)?

The balanced scorecard attempts to bring together the elements that reflect a business's strategic position, including product or service quality measures, product and service development times, customer complaints, labour productivity and so on. At the same time, it attempts to avoid performance reporting becoming unwieldy by restricting the number of measures and focusing especially on those seen to be essential. The advantages of the approach are that it presents an overall picture of the organisation's performance in a single report and, by being comprehensive in the measures of performance it uses, encourages companies to take decisions in the interests of the whole organisation rather than sub-optimising around narrow measures.

Operations principle
Multidimensional performance measurement approaches, such as the balanced scorecard, give a broader indication of overall performance.

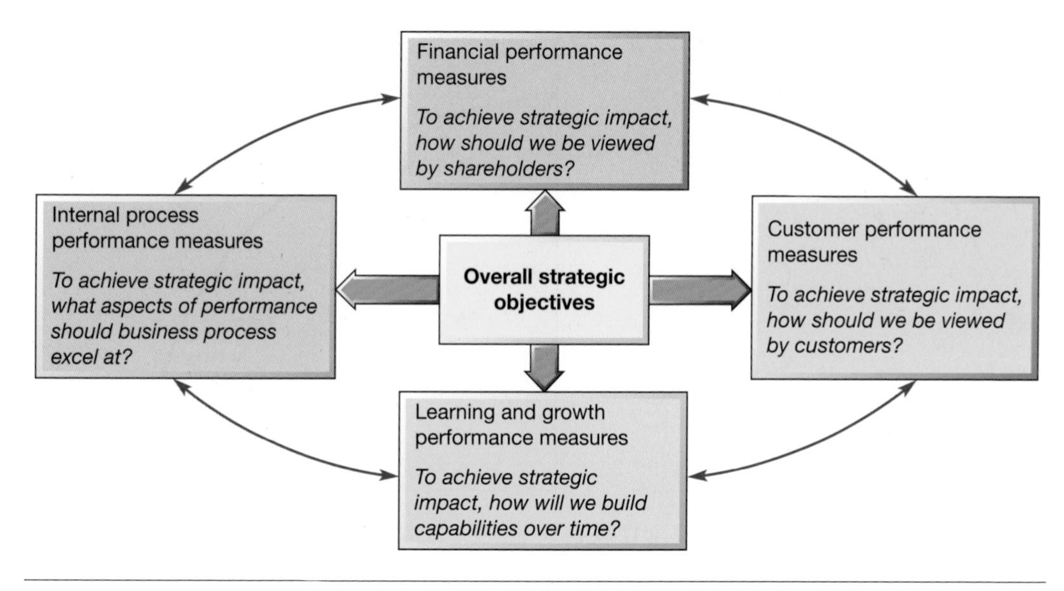

Figure 2.13 The measures used in the balanced scorecard

2.6 How do operations performance objectives trade off against each other?

Earlier we examined how improving the performance of one objective inside the operation could also improve other performance objectives. Most notably better quality, speed, dependability and flexibility can improve cost performance. But externally this is not always the case. In fact, there may be a 'trade-off' between performance objectives. In other words, improving the performance of one performance objective might only be achieved by sacrificing performance in another. So, for example, an operation might wish to improve its cost efficiencies by reducing the variety of products or services that it offers to its customers. 'There is no such thing as a free lunch' could be taken as a summary of this approach. Probably the best-known summary of the trade-off idea comes from Professor Wickham Skinner, who said: 'most managers will readily admit that there are compromises or trade-offs to be made in designing an airplane or truck. In the case of an airplane, trade-offs would involve matters such as cruising speed, take-off and landing distances, initial cost, maintenance, fuel consumption, passenger comfort and cargo or passenger capacity. For instance, no one today can design a 500-passenger plane that can land on an aircraft carrier and also break the sound barrier. Much the same thing is true in [operations]'.[10]

But there are two views of trade-offs. The first emphasises 'repositioning' performance objectives by trading off improvements in some objectives for a reduction in performance in others. The other emphasises increasing the 'effectiveness' of the operation by overcoming trade-offs so that improvements in one or more aspects of performance can be achieved without any reduction in the performance of others. Most businesses at some time or other will adopt both approaches. This is best illustrated through the concept of the 'efficient frontier' of operations performance.

Operations principle

In the short term, operations cannot achieve outstanding performance in all their operations objectives.

Trade-offs and the efficient frontier

Figure 2.14(a) shows the relative performance of several companies in the same industry in terms of their cost efficiency and the variety of products or services that they offer to their customers. Presumably all the operations would ideally like to be able to offer very high variety while still having very high levels of cost efficiency. However, the increased complexity that a high variety of product or service offerings brings will generally reduce the operation's ability to operate efficiently.

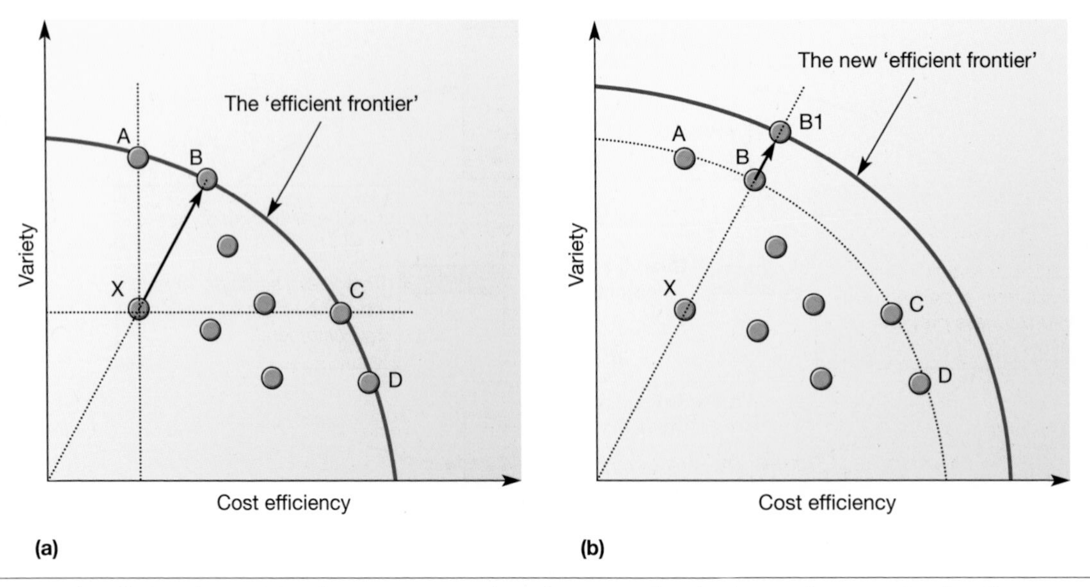

Figure 2.14 The efficient frontier identifies operations with performances that dominate other operations' performance

Conversely, one way of improving cost efficiency is to limit severely the variety on offer to customers. The spread of results in Figure 2.14(a) is typical of an exercise such as this. Operations A, B, C and D have all chosen a different balance between variety and cost efficiency. But none is dominated by any other operation in the sense that another operation necessarily has 'superior' performance. Operation X, however, has an inferior performance because operation A is able to offer higher variety at the same level of cost efficiency and operation C offers the same variety but with better cost efficiency. The convex line on which operations A, B, C and D lie is known as the 'efficient frontier'. They may choose to position themselves differently (presumably because of different market strategies) but they cannot be criticised for being ineffective. Of course, any of these operations that lie on the efficient frontier may come to believe that the balance they have chosen between variety and cost efficiency is inappropriate. In these circumstances, they may choose to reposition themselves at some other point along the efficient frontier. By contrast, operation X has also chosen to balance variety and cost efficiency in a particular way but is not doing so effectively. Operation B has the same ratio between the two performance objectives but is achieving them more effectively.

> **Operations principle**
>
> Operations that lie on the 'efficient frontier' have performance levels that dominate those which do not.

However, a strategy that emphasises increasing effectiveness is not confined to those operations that are dominated, such as operation X. Those with a position on the efficient frontier will generally also want to improve their operations effectiveness by overcoming the trade-off that is implicit in the efficient frontier curve. For example, suppose operation B in Figure 2.14(b) wants to improve both its variety and its cost efficiency simultaneously and move to position B1. It may be able to do this, but only if it adopts operations improvements that extend the efficient frontier. For example, one of the decisions that any supermarket manager has to make is how many checkout positions to open at any time. If too many checkouts are opened then there will be times when the checkout staff do not have any customers to serve and will be idle. The customers, however, will have excellent service in terms of little or no waiting time. Conversely, if too few checkouts are opened, the staff will be working all the time but customers will have to wait in long queues. There seems to be a direct trade-off between staff utilisation (and therefore cost) and customer waiting time (speed of service). Yet even the supermarket manager might, for example, allocate a number of 'core' staff to operate the checkouts but also arrange for those other staff who are performing other jobs in the supermarket to be trained and 'on call' should demand suddenly increase. If the manager on duty sees a build-up of customers at the checkouts, these other staff could quickly be used to staff checkouts. By devising a flexible system of staff allocation, the manager can both improve customer service and keep staff utilisation high.

This distinction between positioning on the efficient frontier and increasing operations effectiveness by extending the frontier is an important one. Any business must make clear the extent to which it is expecting the operation to reposition itself in terms of its performance objectives and the extent to which it is expecting the operation to improve its effectiveness in several ways simultaneously.

Operations principle

An operation's strategic improvement path can be described in terms of repositioning and/or overcoming its performance trade-offs.

Responsible operations

In every chapter, under the heading of 'Responsible operations', we summarise how the particular topic covered in the chapter touches upon important social, ethical and environmental issues.

Look at the corporate website of any large organisation. The strong possibility is that a significant part of it is devoted to the organisation trying to address the social–environmental impact of its activities. Some of the more ambitious (or defensive) even try to measure their corporate social responsibility (CSR) performance. There are several reasons for this. From a straightforward managerial perspective, measuring their impact supports better choices about which CSR initiatives to support since some will have a greater positive impact than others. It might also prompt managers to try to enhance the effectiveness of any CSR initiatives that are not achieving the expected impact. Just as important for most enterprises is the need to demonstrate their CSR credentials to a broad range of stakeholders. Customers (or at least, some customers) are taking an increasing interest in the CSR performance of who they buy from. Staff generally prefer to work for ethical enterprises, so it helps to recruit and retain talent. Even investors are examining organisations' CSR performance before they make investment decisions.

However, achieving a defensible method of accurately measuring CSR impact is not straightforward. In fact, some claim that there's no true set of universally understood or accepted criteria on how CSR initiatives are measured. Certainly, there is a wide variation in how effective organisations are in reporting on such on issues. One factor influencing reporting quality seems to be how important CSR issues are in the country where a firm is headquartered. Firms in countries with higher CSR standards, policies, regulations and common law structures produce significantly higher-quality CSR reports.[11] Yet the debate on how best to measure social–environmental performance remains unresolved. A practical approach is for operations to attempt to cover two aspects of CSR:

- Assess whether operations conform to the activities, procedures and policies that are designed to promote positive CSR behaviour. This could include such things as conformance to effective codes of conduct, number of meetings with stakeholders, extent of meeting diversity targets, health and safety statistics, number of social programmes provided to the community, process changes adopted to reduce waste emission and energy consumption, and so on.
- Assess outcome metrics that measure the impact of CSR efforts. This could include such things as the number of stakeholders perceiving the company as ethical, degree of reduction in the gap between highest-paid and lowest-paid employees, percentage increase in employee satisfaction scores, percentage reduction in workplace accidents, percentage of profits or income donated to community-based projects, number of employees who participate in company-sponsored volunteer activities, percentage reduction in waste and emissions, and so on.

Of course, much of this type of measurement is based on perceptions. But perceptions, as well as more objective reality, are important when assessing CSR. One can even use social media to judge the opinion of an operation's stakeholders. One study that used the opinions of companies' stakeholders expressed on Twitter,[12] showed that companies' CSR practices could be gauged using this method (and that there were often differences between the CSR opinions of stakeholders and what was reported by the companies themselves).

Summary answers to key questions

2.1 Why is operations performance vital in any organisation?

- Operations management can either 'make or break' any business. In most businesses, it represents the bulk of its assets.
- The positive effects of a well-run operation include a focus on improvement, the building of 'difficult to imitate' capabilities, and an understanding of the processes that are the building blocks of all operations.
- The negative effects of a poorly run operation include failures that are obvious to customers (and expensive for the organisation), a complacency that leads to the failure to exploit opportunities for improvement.

2.2 How is operations performance judged at a societal level?

- Operations decisions affect a variety of 'stakeholders'. Stakeholders are the people and groups who have a legitimate interest in the operation's activities.
- This idea that operations should take into account the impact on a broad mix of stakeholders is termed 'corporate social responsibility' (CSR).
- Performance at the societal level often uses the idea of the triple bottom line (TBL, or 3BL, also known as 'people, planet and profit'). It includes the social bottom line, the environmental bottom line and the economic bottom line.
- The social bottom line incorporates the idea that businesses should accept that they bear some responsibility for the impact they have on society and balance the external 'societal' consequences of their actions with the more direct internal consequences, such as profit.
- The environmental bottom line incorporates the idea that operations should accept that they bear some responsibility for the impact they have on the natural environment.
- The economic bottom line incorporates the conventional financial measures of performance derived from using the operation's resources effectively.

2.3 How is operations performance judged at a strategic level?

- The types of decisions and activities that operations managers carry out can have a significant strategic impact.
- In particular, operations can affect economic performance in five ways:
  - It can reduce the costs.
  - It can achieve customer satisfaction through service.
  - It can reduce the risk of operational failure.
  - It can reduce the amount of investment that is necessary.
  - It can provide the basis for future innovation.

2.4 How is operations performance judged at an operational level?

- The five 'performance objectives' that are used to assess the performance of operations at an operational level are quality, speed, dependability, flexibility and cost.
- Quality is important because: by 'doing things right', operations seek to influence the quality of the company's goods and services. Externally, quality is an important aspect of customer satisfaction or dissatisfaction. Internally, quality operations both reduce costs and increase dependability.
- Speed is important because: by 'doing things fast', operations seek to influence the speed with which goods and services are delivered. Externally, speed is an important aspect of customer service. Internally, speed both reduces inventories by decreasing internal throughput time and reduces risks by delaying the commitment of resources.
- Dependability is important because: by 'doing things on time', operations seek to influence the dependability of the delivery of goods and services. Externally, dependability is an important aspect of customer service. Internally, dependability within operations increases operational reliability, thus saving the time and money that would otherwise be taken up in solving reliability problems and also giving stability to the operation.
- Flexibility is important because: by 'changing what they do', operations seek to influence the flexibility with which the company produces goods and services. Externally, flexibility can produce new products and services (product/service flexibility), produce a wide range or mix of products and services (mix flexibility), produce different quantities or volumes of products and services (volume flexibility), produce products and services at different times (delivery flexibility). Internally, flexibility can help speed up response times, save time wasted in changeovers, and maintain dependability.
- Cost is important because: by 'doing things cheaply', operations seek to influence the cost of the company's goods and services. Externally, low costs allow organisations to reduce their price in order to gain higher volumes or, alternatively, increase their profitability on existing volume levels. Internally, cost performance is helped by good performance in the other performance objectives.

2.5 How can operations performance be measured?

- It is unlikely that for any operation a single measure of performance will adequately reflect the whole of a performance objective. Usually, operations have to collect a whole bundle of partial measures of performance.
- The balanced scorecard (BSC) is a commonly used approach to performance measurement and incorporates measures related to: How do we look to our shareholders (financial perspective)? What must we excel at (internal process perspective)? How do our customers see us (the customer perspective)? How can we continue to improve and build capabilities (the learning and growth perspective)?

2.6 How do operations performance objectives trade off against each other?

- Trade-offs are the extent to which improvements in one performance objective can be achieved by sacrificing performance in others. The 'efficient frontier' concept is a useful approach to articulating trade-offs and distinguishes between repositioning performance on the efficient frontier and improving performance by overcoming trade-offs.

CASE STUDY IKEA looks to the future[13]

For decades, IKEA has been one of the most successful retail operations in the world, with much of its success founded on how it organises its design, supply and retail service operations. With over 400 giant stores in 49 countries, IKEA has managed to develop its own standardised way of selling furniture. Its so-called 'big box' formula has driven IKEA to the global No. 1 position in furniture retailing. 'Big box' because the traditional IKEA store is a vast blue-and-yellow maze of a showroom (on average around 25,000 square metres) where customers often spend around two hours – far longer than in rival furniture retailers. This is because of the way it organises its store operations. IKEA's philosophy goes back to the original business, started in the 1950s in Sweden by the late Ingvar Kamprad. He was selling furniture, through a catalogue operation, and because customers wanted to see some of his furniture, he built a showroom on the outskirts of Stockholm and set the furniture out as it would be in a domestic setting. Also, instead of moving the furniture from the warehouse to the showroom area, he asked customers to pick the furniture up themselves from the warehouse, an approach that became fundamental to IKEA's ethos; what has been called the 'we do our part, you do yours' approach.

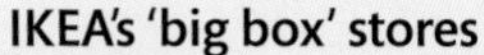

IKEA's 'big box' stores

IKEA offers a wide range of Scandinavian designs at affordable prices, usually stored and sold as a 'flat pack', which the customer assembles at home. *'It was an entirely new concept, and it drove the firm's success'*, says Patrick O'Brien, Retail Research Director at retail consultancy GlobalData. *'But it wasn't just what IKEA was selling that was different, but how it was selling it'*. The stores were located and designed around one simple idea – that finding the store, parking, moving through the store itself, and ordering and picking up goods should be simple, smooth and problem-free. Catalogues are available at the entrance to each store showing product details and illustrations. For young children, there is a supervised children's play area, a small cinema, a parent and baby room, and toilets, so parents can leave their children in the supervised play area for a time. Parents are recalled via the loudspeaker system if the child has any problems. Customers may also borrow pushchairs to keep their children with them.

Parts of the showroom are set out in 'room settings', while other parts show similar products together, so that customers can make comparisons. Given the volume of customers, there are relatively few staff in the stores. IKEA say it likes to allow customers to make up their own minds. If advice is needed, 'information points' have staff who can help. Every piece of furniture carries a ticket indicating its location in the warehouse from where it can be collected. Customers then pass into an area where smaller items are displayed that can be picked directly, after which they pass through the self-service warehouse where they can pick up the items they viewed in the showroom. Finally, customers pay at the checkouts, where a conveyor belt moves purchases up to the checkout staff. The exit area has service points and a large loading area allowing customers to bring their cars from the car park and load their purchases. Within the store, a restaurant serves, among other things, IKEA's famous Swedish meatballs. IKEA's fans say they can make a visit to the store a real 'day out'.

But not everyone is a fan

Yet not all customers (even those who come back time after time) are entirely happy with the traditional IKEA retail experience. Complaints include:

- It can be a long drive to reach one of their stores (unless you are 'lucky' enough to live near one).
- The long 'maze-like' journey that customers are 'encouraged' to take through the store is too prescriptive.
- There are too few customer-facing staff in the store.
- There are long queues at some points in the store, especially at checkouts and at busy times such as weekends.
- Customers have to locate, pick off the shelves and transport, sometimes heavy, products to the checkouts.
- IKEA designs can be 'bland' (or 'clean and aesthetically pleasing', depending on your taste).
- The furniture has to be assembled once you get it home, and the instructions are confusing.

Although many are

However, the impressive growth and success of IKEA over the years indicates that the company is doing many things right. The reasons customers give for shopping at IKEA include the following.

- Everything is available under one roof (albeit a very big roof).
- The range of furniture is far greater than at other stores.
- The products are displayed both by category (e.g. all chairs together) and in a room setting.
- Availability is immediate (competitors often quote several weeks for delivery).
- There is a kids' area and a restaurant so visiting the store is 'an event for all the family'.
- The design of furniture is 'modern, clean and inoffensive' – they fit anywhere.
- For the quality and design, the products are very good 'value for money'.

Was a new approach needed?

For decades, IKEA's unique retailing operations, combined with an excellent supply network and a customer-focused design philosophy, was an effective driver of healthy growth. However, there were indications that the company was starting to ask itself how it could solve some of the criticisms of its retail operations. '*We had to move away from conversations that began, "I love IKEA, but shopping at an IKEA store is not how I want to spend my time"*' (Gillian Drakeford, IKEA's UK boss). It needed to counter the complaints by some customers that its stores were understaffed, that the navigation of stores was too prescriptive, and that queues were too long. '*We have had a great proposition for 60 years, but the customer had to fit around it. But the world has changed and to remain relevant we need to have a proposition that fits around the customer*' (Gillian Drakeford).

IKEA was also realising that its 'big box' stores were under threat from a decline in car ownership (in 1994, 75 per cent of 21–29-year-olds held driving licences in the United Kingdom; by 2017 that had dropped to 66 per cent). Also, customers were increasingly wanting their flat-pack furniture to be delivered, rather than having to drive to an IKEA store to collect it. Ideally, they also wanted to order it online. IKEA did have an online presence, but compared with its competitors it was relatively underdeveloped. Not only that, but not all customers wanted to assemble their own furniture.

'*The entire premise that IKEA developed was that consumers would be willing to drive their cars 50 kilometres to save some money on something that looks amazing*', said Ray Gaul, a retail analyst at Kantar Retail. '*Young people like IKEA, but they can't or don't want to drive to IKEA*'. However, the traditional 'big box' strategy was still popular with many customers, and sales from its stores continued to grow. Yet, in most markets, there were plenty of potential customers who could not reach an IKEA store within a reasonable drive (assumed to be around two-and-a-half hours). And some degree of rethinking IKEA's operating model seemed to be required. Torbjörn Lööf, CEO of Inter IKEA (who manage the IKEA concept), summarised their commitment to a rethink. '*We have been successful on a long journey. But it is clear that one era is ending and another beginning*'.

Smaller stores to complement the larger ones

From 2015, IKEA opened several smaller-footprint stores in Europe, Canada, China and Japan. But not all were the same. As a deliberate strategy, each was slightly different. This allowed the company to test alternative ways of locating, designing and managing its new ventures. Should they have cafés? How big should they be? Should they carry a range of products, or focus on a single category? Should they be located in shopping malls or on the high street? So, a 'pop-up' IKEA store in central Madrid offered only bedroom furnishings. A store in Stockholm focused on kitchens. It allowed customers to cook in the store, and book a 90-minute consultation to plan their kitchen. A small store in London stocked a range of product categories, but had no café (only a coffee machine), and in place of a supervised kids' play area, computer games were provided. Other new stores were, in some ways, similar to traditional stores, but smaller, with fewer car parking spaces and less inventory, and acted as order-and-collection points. In some, customers could get expert advice on larger purchases, such as kitchens or bathrooms. Often in the smaller stores, only a few items could be purchased and taken home instantly. Rather, customers could use touch screens to order products and arrange for delivery or pickup at a convenient time. '*For me, it's a test lab for penetrating city centres*', said one senior IKEA executive. '*About 70 per cent of the people shopping there wouldn't go to a* [traditional IKEA] *store*'.

TaskRabbit

In 2017 IKEA bought TaskRabbit, whose app was one of the leaders in what was becoming known as the '**gig' economy**. Using its app, over 60,000 independent workers or 'taskers' (at the time of acquisition) offered their services to customers wanting to hire someone to do tasks such as moving or assembling furniture. '*In a fast-changing retail environment, we continuously strive to develop new and improved products and services to make our customers' lives a little bit easier. Entering the on-demand, sharing economy enables us to support that*', IKEA chief Jesper Brodin said in a statement. '*We will be able to learn from TaskRabbit's digital expertise, while also providing IKEA customers additional ways to access flexible and affordable service solutions to meet the needs of today's customer*'.

Web-based retailing

Arguably, the most significant retailing development in this period was the growth in online shopping. However, IKEA was slow to move online. Partly, this was because there was internal reluctance to interfere with its successful 'big box' retail operations that encouraged customers to spend a long time in store and make impulse purchases. However, it became clear that the company needed to become fully committed to 'multichannel' retail operations, including online sales. But it was also clear that there would not be a total shift to online sales. The idea was to offer both physical and digital options for customers who wanted to use both channels, and to win new customers online who would never make the journey to its 'big box' superstores. Some retail experts warned that the new strategy carried the same risks faced by all firms going online. According to Marc-André Kamel of consultants Bain & Company, *'customers are not shifting entirely to e-commerce, but wish to mix and match channels'*. And, although IKEA had little choice but to invest in online channels, the danger was that it could raise costs, especially as the company was also planning significant bricks-and-mortar expansion in new markets such as India, South America and South-East Asia.

Third-party sales

Another break with traditional IKEA practice came when it announced that it would consider selling its products through independent 'third-party' online retailers. Torbjörn Lööf, CEO of Inter IKEA, said the decision to supply online retailers was an important part of the broader overhaul of their operations. '[It] *is the biggest development in how consumers meet Ikea since the concept was founded*', he told the *Financial Times*.

Sustainability

IKEA was among the world's biggest users of wood (estimated as around 1 per cent of all wood used), and some environmental groups condemned what they saw as the 'disposable' nature of its furniture. Responding to this criticism, IKEA appointed a chief sustainability officer – the first time that sustainability was directly represented in the senior management team, and a recognition of the growing role of sustainability in determining how IKEA was perceived. It also recognised the ability of sustainability to drive business innovation. IKEA, like an increasing number of companies, accepted that it lived in a world of finite resources and recognised that consumption needed to reflect this. Because of this realisation, IKEA was seeking new ways to meet people's needs and aspirations while staying within the limits of our planet. It saw the emerging circular economy business model as a great opportunity to develop its business further. This was preferable to viewing sustainability as a risk to the business. Sooner or later, other companies would start creating business models that disrupted the IKEA way of selling home furnishings. In one initiative in Belgium IKEA offered its customers five options to give furniture a second life: selling old IKEA-furniture in the store (at the price paid to the customer who supplied it), renewing it by repainting or reassembling, repairing by offering replacement parts, returning old furniture through its transport service, and donating it to social organisations. Some commentators questioned the idea of selling longer-lasting products and trading pre-owned items without a mark-up as being bad for business. However, IKEA disagreed on the grounds that people may sometimes come to IKEA with a bit of a guilty conscience, wanting to buy stuff, but unable to completely forget the consequences. In fact, it believed that they actively welcomed the move. When it started buying back our furniture at Aalborg, it actually saw an increase in revenue.

QUESTIONS

1. **In the traditional IKEA 'big box' stores, what is the relative importance of the operational performance objectives (quality, speed, dependability, flexibility, cost), compared with a conventional high-street furniture store?**
2. **What trade-offs are customers who go to these big stores making?**
3. **How does the strategy of increasing IKEA's online presence impact on these trade-offs?**
4. **An IKEA executive was reported as saying that in some parts of the world *'we have reached the point of "peak stuff"'*. It was interpreted by some as a warning that consumer appetite for home furnishings had reached a crucial turning point. What are the implications of this for IKEA?**

Problems and applications

All chapters have 'Problems and applications' questions that will help you practise analysing operations. They can be answered by reading the chapter. Model answers for the first two questions can be found on the companion website for this text.

1 The environmental services department of a city has two recycling services – newspaper collection (NC) and general recycling (GR). The NC service is a door-to-door collection service that, at a fixed time every week, collects old newspapers that householders have placed in reusable plastic bags at their gates. An empty bag is left for the householders to use for the next collection. The value of the newspapers collected is relatively small; the service is offered mainly for reasons of environmental responsibility. By contrast the GR service is more commercial. Companies and private individuals can request a collection of materials to be disposed of, using either the telephone or the internet. The GR service guarantees to collect the material within 24 hours unless the customer prefers to specify a more convenient time. Any kind of material can be collected and a charge is made depending on the volume of material. This service makes a small profit because the revenue both from customer charges and from some of the more valuable recycled materials exceeds the operation's running costs. How would you describe the differences between the performance objectives of the two services?

2 Xexon7 is a specialist artificial intelligence (AI) development firm that develops algorithms for various online services. As part of its client service, it has a small (10-person) helpdesk call centre to answer client queries. Clients can contact them from anywhere in the world at any time of the day or night with a query. Demand at any point in time is fairly predictable, especially during the (European) daytime. Demand during the night hours (Asia and the Americas) is considerably lower than in the daytime and also less predictable. *'Most of the time we forecast demand pretty accurately and so we can schedule the correct number of employees to staff the work stations. There is still some risk, of course. Scheduling too many staff at any point in time will waste money and increase our costs, while scheduling too few will reduce the quality and response of the service we give'* (Binita Das, Help Desk Manager). Binita is, overall, pleased with the way in which her operation works. However, she feels that a more systematic approach could be taken to identifying improvement opportunities. *'I need to develop a logical approach to identifying how we can invest into improving things like sophisticated diagnostic systems. We need to both reduce our operating costs and maintain, and even improve, our customer service'*. What are the trade-offs that must be managed in this type of call centre?

3 The health-check clinic described in the worked example earlier in the chapter has expanded by hiring one extra employee and now has six employees. It has also leased some new health monitoring equipment, which allows patients to be processed faster. This means that its total output is now 280 patients per week. Its wage costs have increased to £4,680 per week and its overhead costs to £3,000 per week. What are its single-factor labour productivity and its multi-factor productivity now?

4 A publishing company plans to replace its four proofreaders, who look for errors in manuscripts, with a new scanning machine and one proofreader in case the machine breaks down. Currently the proof-readers check 15 manuscripts every week between them. Each is paid €80,000 per year. Hiring the new scanning machine will cost €5,000 each calendar month. How will this new system affect the proofreading department's productivity? (Proofreaders work 45 weeks per year.)

5 Look again at the figures in the chapter that illustrate the meaning of each performance objective for the four operations. Consider the bus company and the supermarket, and in particular consider their external customers. Draw the relative required performance for both operations on a polar diagram. Consider the internal effects of each performance objective. For both operations, identify how quality, speed, dependability and flexibility can help to reduce the cost of producing their services.

6 Visit the websites of two or three large oil companies such as Exxon, BP, Shell or Total. Examine how they describe their policies towards their customers, suppliers, shareholders, employees and society at large. Identify areas of the company's operations where there may be conflicts between the needs of these different stakeholder groups. Discuss or reflect on how (if at all) such companies try to reconcile these conflicts.

7 How should large supermarket companies measure their social and environmental performance?

8 Patagonia (the garment company) is also a 'B Corp', like Danone, which is described in the 'Operations in practice' example in the chapter. Why might it be more challenging for Danone to achieve this status than Patagonia?

9 The five performance objectives (quality, speed, dependability, flexibility and cost) measure the output from operations. Why might some operations also want to measure their internal processes?

10 What trade-offs are involved when airlines decide how many business-class seats to install in their aircraft?

Selected further reading

Blokdyk, G. (2019) ***Stakeholder Analysis: A Complete Guide*****, 5STARCooks.**
Very much a practical guide.

Bourne, M., Kennerley, M. and Franco-Santos, M. (2005) Managing through measures: a study of the impact on performance, ***Journal of Manufacturing Technology Management*****, 16 (4), 373–95.**
What it says on the tin.

Gray, D., Micheli, P. and Pavlov, A. (2015) ***Measurement Madness: Recognizing and Avoiding the Pitfalls of Performance Measurement*****, Wiley, New York, NY.**
Lots of examples of how companies can misuse performance measurement.

Kaplan, R.S. and Norton, D.P. (2015) ***Balanced Scorecard Success: The Kaplan-Norton Collection*****, Harvard Business Review Press, Boston, MA.**
A collection of the four books that trace these authors' work. The first is the most relevant here.

Neely, A. (2012) ***Business Performance Measurement: Unifying Theory and Integrating Practice*****, Cambridge University Press, Cambridge.**
A collection of papers on the details of measuring performance objectives.

Pine, B.J. (1993) ***Mass Customization: The New Frontier in Business Competition*****, Harvard Business Review Press, Boston, MA.**
The first substantial work on the idea of mass customization. Still a classic.

Redden, G, (2019) Questioning Performance Measurement: Metrics, Organizations and Power, SAGE Publications Ltd, London.
A critical look at the application of metrics that capture performance.

Savitz, A.W. and Weber, K. (2006) ***The Triple Bottom Line: How Today's Best-Run Companies Are Achieving Economic, Social and Environmental Success – and How You Can Too*****, Jossey-Bass, San Francisco, CA.**
Good on the triple bottom line.

Waddock, S. (2003) Stakeholder performance implications of corporate responsibility, ***International Journal of Business Performance Management*****, 5 (2/3), 114–24.**
An introduction to stakeholder analysis.

Notes on chapter

1. The phrase 'the triple bottom line' was first used in 1994 by John Elkington, the founder of a British consultancy called SustainAbility. Read Elkington, J. (1997) *Cannibals with Forks: The Triple Bottom Line of 21st Century Business*, Capstone. Also good is, Savitz, A.W. and Weber, K. (2006) *The Triple Bottom Line: How Today's Best-Run Companies Are Achieving Economic, Social and Environmental Success – and How You Can Too*, Jossey-Bass.
2. The information on which this example is based is taken from: B Corp website, https://bcorporation.net/about-b-corps (accessed August 2021); Economist (2018) Choosing plan B – Danone rethinks the idea of the firm, Business section, *Economist* print edition, 9 August.
3. See for example: Jensen, M.C. (2001) Value maximization, stakeholder theory, and the corporate objective function, *Journal of Applied Corporate Finance*, 14 (3), 8–21.
4. The information on which this example is based is taken from: Willan, P. (2018) Spread the word: dream job if you're nuts about chocolate, *The Times*, 28 July; Reuters Staff (2018) Ferrero stops production at biggest Nutella plant to assess quality issue, Reuters, 21 February, https://www.reuters.com/article/ferrero-nutella-stop-idUSL5N20G5NY (accessed August 2021); france 24 (2019) World's largest Nutella factory reopens after 'quality defect' france24, 25 February, https://www.france24.com/en/20190225-worlds-largest-nutella-factory-reopens-after-quality-defect (accessed August 2021); Sage, A. (2018) Nutella fistfights spread at Intermarché stores across France, *The Times*, 26 January.
5. The information on which this example is based is taken from: Palmer, M, (2020) Smart ambulances and wearables offer route to speedier treatments, *Financial Times*, 24 November. More on London's Air Ambulance Service can be found at https://www.londonsairambulance.org.uk (accessed August 2021).
6. The information on which this example is based is taken from: McCurry, J, (2017) Japanese rail company apologises after train leaves 20 seconds early, *Guardian*, 17 November; The Local (2017) SBB remains most punctual train company in Europe, news@thelocal.ch, 21 March.
7. The information on which this example is based is taken from: mymuesli website, http://uk.mymuesli.com/muesli/ (accessed August 2021).
8. The information on which this example is based is taken from: Aldi website, https://www.aldi.co.uk (accessed August 2021).
9. Kaplan, R.S. and Norton, D.P. (1993) *The Balanced Scorecard*, Harvard Business School Press, Boston, MA.
10. Skinner, W. (1985) *Manufacturing: The Formidable Competitive Weapon*, John Wiley & Sons, New York, NY.
11. Sethi, S.P., Martell, T.F. and Demir, M. (2017) An evaluation of the quality of corporate social responsibility reports by some of the world's largest financial institutions, *Journal of Business Ethics*, 140 (4), 787–805.
12. Barbeito-Caamaño, A. and Chalmeta, R. (2020) Using big data to evaluate corporate social responsibility and sustainable development practices, *Corporate Social Responsibility and Environmental Management*, 27 (6), 2831–48.
13. The information on which this case study is based is taken from: IKEA website, https://www.inter.ikea.com/ (accessed September 2021); Matlack, C. (2018) The tiny Ikea of the future, without meatballs or showroom mazes, *Bloomberg Businessweek*, 10 January; Milne, R. (2018) What will Ikea build next? *Financial Times*, 1 February; Economist (2017) Frictionless furnishing: IKEA undertakes some home improvements, *Economist* print edition, 2 November; Hipwell, D. (2017) This is no time to sit back and relax – we must deliver, says IKEA's UK boss, *The Times*, 10 February; Gerschel-Clarke, A. (2016) 'Peak Stuff': why IKEA is shifting towards new business models, Sustainablebrands.com, 17 February; Milne, R. (2017) Ikea turns to ecommerce sites in online sales push, *Financial Times*, 9 October; Hope, K. (2017) Ikea: why we have a love-hate relationship with the Swedish retailer, BBC News, 17 October; Armstrong, A. (2017) Revealed: how after 30 years, Ikea is undergoing a radical overhaul, *The Telegraph*, 15 October.

3 Operations strategy

KEY QUESTIONS

3.1 What is strategy and what is operations strategy?

3.2 How does operations strategy align with business strategy (top-down)?

3.3 How does operations strategy align with market requirements (outside-in)?

3.4 How does operations strategy align with operational experience (bottom-up)?

3.5 How does operations strategy align with operations resources (inside-out)?

3.6 How are the four perspectives of operations strategy reconciled?

3.7 How can the process of operations strategy be organised?

INTRODUCTION

In the long term, the major (and some would argue, only) objective of operations is to provide the organisation with some form of strategic advantage. That is why the management of processes, operations and supply networks must be aligned with overall strategy. While it will always be necessary to make adjustments to accommodate short-term events, simply reacting to current issues can lead to constant changes in direction with the operation becoming unstable. So all operations need the 'backdrop' of a well-understood strategy that indicates (at least, roughly) where they are heading and how they could get there. Once the operations function has understood its role in the business and after it has articulated its performance objectives, it needs to formulate a set of general principles that will guide its decision-making – this is its operations strategy. Many *enduringly* remarkable enterprises, from Apple to Zara, use their operations resources to gain long-term strategic success. It is the way they manage their operations that sets them apart from their competitors. This chapter considers four perspectives, each of which goes part way to illustrating the forces that shape operations strategy. It then examines how these perspectives can be reconciled and how the *process* of operations strategy can be organised effectively. Figure 3.1 shows the position of the ideas described in this chapter in the general model of operations management.

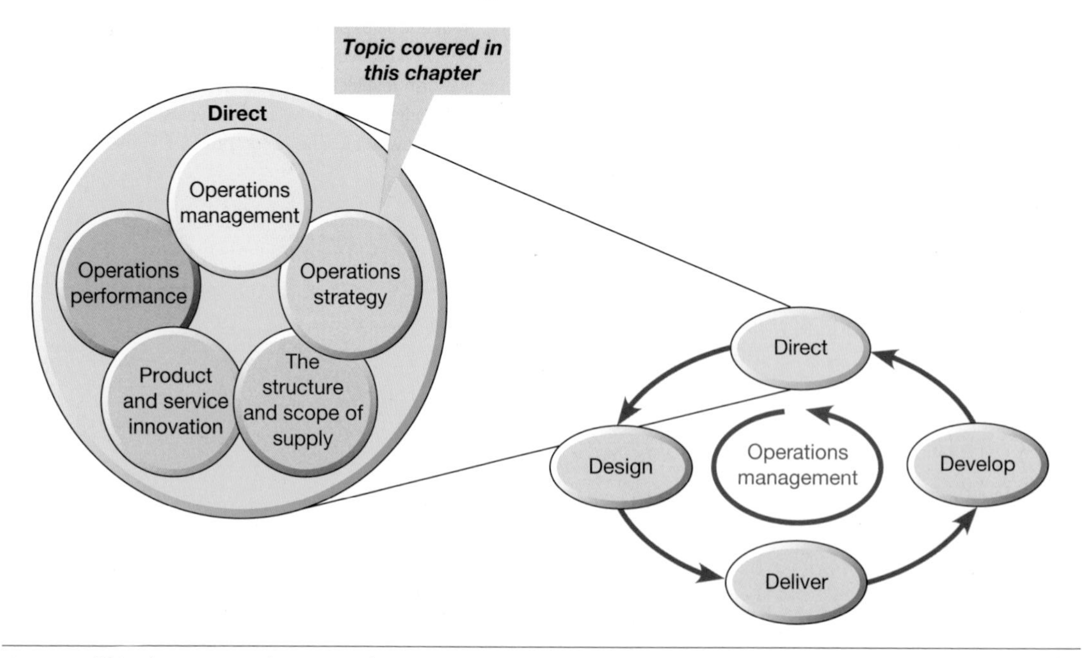

Figure 3.1 This chapter examines operations strategy

3.1 What is strategy and what is operations strategy?

Surprisingly, 'strategy' is not particularly easy to define. Linguistically the word derives from the Greek word '*strategos*' meaning 'leading an army'. And although there is no direct historical link between Greek military practice and modern ideas of strategy, the military metaphor is powerful. Both military and **business strategy** can be described in similar ways, and include some of the following:

- ▶ Setting broad objectives that direct an enterprise towards its overall goal.
- ▶ Planning the path (in general rather than specific terms) that will achieve these goals.
- ▶ Stressing long-term rather than short-term objectives.
- ▶ Dealing with the total picture rather than stressing individual activities.
- ▶ Being detached from, and above, the confusion and distractions of day-to-day activities.

From this perspective, **strategic decisions** are those that are widespread in their effect on the organisation, define the position of the organisation relative to its environment, and move the organisation closer to its long-term goals. But 'strategy' is more than a single decision; it is the *total pattern of the decisions* and actions that influence the long-term direction of the business. Thinking about strategy in this way helps us to discuss an organisation's strategy, even when it has not been explicitly stated. Observing the total pattern of decisions gives an indication of *actual* strategic behaviour.

The same points apply equally to ***operations strategy***. Operations strategy is defined as the 'pattern of decisions and actions that shape the long-term vision, objectives and capabilities of the operation and its contribution to the overall strategy of the business'.[1] At first, the term 'operations strategy' sounds like a contradiction. How can 'operations', a subject that is generally concerned with the day-to-day creation and delivery of products and services, be strategic? 'Strategy' is usually regarded as the opposite of those day-to-day routine activities. But '*operations*' is not the same as '*operational*'. 'Operations' are the resources that create products and services, and it is clear that they can have a real strategic impact. 'Operational' is the opposite of strategic, meaning day-to-day, detailed and often localised. So, one can examine both the operational *and* the strategic aspects of operations. It is also conventional to distinguish between the 'content' and the '**process' of operations strategy**. The ***content* of operations strategy** is the specific decisions and actions that set the operations role, objectives and activities. The *process* of operations strategy is the method that is used to make the specific 'content' decisions.

Operations principle

Operations is not the same as operational; it does have a strategic role.

OPERATIONS IN PRACTICE

Operations is the basis of Ocado's strategy[2]

Shopping is increasingly becoming an online activity. Even in food retailing, where shelf-life limits the ability to store many items for long periods, online grocery delivery is growing in many parts of the world. Making a success of operations strategy in this type of business is not easy, especially as profit margins can be painfully thin. Yet it can be done. Anyone buying groceries in the United Kingdom will be familiar with Ocado. It began by offering local grocery deliveries in the south of England, initially with a branding and sourcing arrangement with Waitrose, an upmarket UK supermarket. But it went on to become the world's largest dedicated online grocery retailer with over half a million active customers. Its objective, it says, is to provide its customers with the best shopping experience in terms of service, range and price, and build a strong business that delivers long-term value for its shareholders. But, just as important, it has developed an integrated high-tech supply process that provides both efficiency and high levels of service. Most online grocers fulfil web orders by ▶

gathering goods from the shelf of a local supermarket and then loading them into a truck for delivery. By contrast, Ocado has developed the Ocado Smart Platform (OSP), its exclusive solution for operating online retail businesses.

Ocado Smart Platform combines proprietary integrated end-to-end software and technology systems with physical fulfilment centre (warehouse) automation and delivery routing systems that manages the total user-input to delivery cycle. More to the point, it has allowed Ocado to offer its OSP as a service to other retailers. Ocado say that it has enabled them to replicate its unique capabilities for partners in other markets with a significantly lower cost than the alternative options available for these retailers. The company has long been known for its innovative investment in its own Customer Fulfilment Centre (CFC) technology. Rather than adopt the usual CFC process of transporting containers progressively using long conveyor belts, it uses a three-dimensional grid system to put together customers' orders. Robots move throughout the grid, picking up products and ferrying them to 'pick stations', where Ocado staff put the orders together. All movements are directed by an overarching control system that coordinates the robots, conveyors and orders. However, it is the total end-to-end system that underpins the company's own operations, and increasingly those of other retailers. Its technology allowed it to partner with Morrison's, a UK supermarket, and subsequently to sell its systems to Kroger (one of America's biggest supermarket chains) and to other supermarkets in Europe, Australia and Canada. For Ocado, the operations resources and processes that it uses to produce its services is more than simply a 'means to an end'. It has become, at least partially, the 'end' itself.

Using operations strategy to articulate a vision for the contribution of operations

Most businesses expect their operations strategy to improve operations performance over time, progressing from a state where they contribute little to the success of the business through to the point where they are directly responsible for its competitive success. The 'vision' of an operation is a clear statement of how operations intend to contribute value for the business. It is not a statement of what the operation wants to *achieve* (those are its objectives), but rather an idea of what it must *become* and what contribution it should make. A common approach to summarising operations contribution is the seminal Hayes and Wheelwright four-stage model.[3] The model traces the progression of the operations function from what is the largely negative role in 'stage 1' operations to a position where it is the central element of competitive strategy in 'stage 4'. Figure 3.2 illustrates the four steps involved in moving from stage 1 to stage 4.

Operations principle

Operations strategy should articulate a 'vision' for the operations function's contribution to overall strategy.

Stage 1: Internal neutrality

This is the very poorest level of contribution by the operations function and the effect is to harm the organisation's ability to compete effectively. In stage 1, the operations function is inward-looking and, at best, reactive with very little positive to contribute towards competitive success. Its vision is to be '**internally neutral**', so as to stop holding the organisation back in any way. It attempts to achieve this by 'avoiding making mistakes'.

Stage 2: External neutrality

The first step of breaking out of stage 1 is for the operations function to begin comparing itself with similar companies or organisations in the outside market. This may not immediately take it to the 'first division' of companies in the market, but at least it is measuring itself against its competitors' performance and trying to implement 'best practice'. In stage 2, the vision of the operations function is to become '**externally neutral**' with operations in the industry.

Stage 3: Internally supportive

Stage 3 operations have typically reached the 'first division' of their markets. For such operations, the vision becomes to be clearly and unambiguously the very best in the market. It achieves this by gaining a clear view of the company's competitive or strategic goals and supporting it by developing

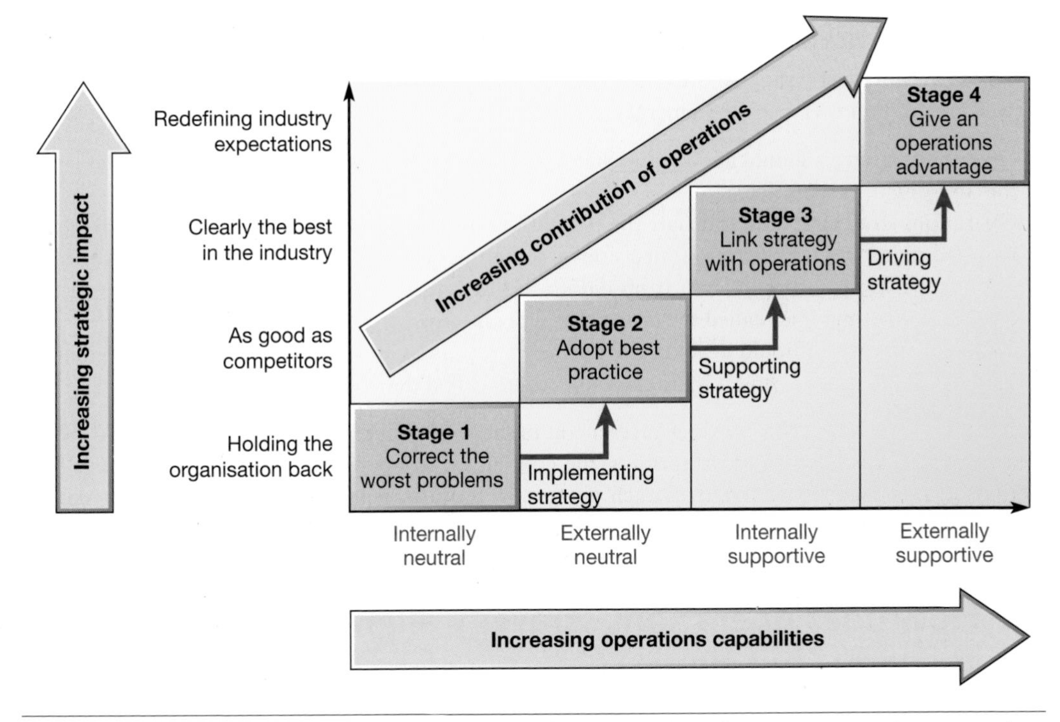

Figure 3.2 Hayes and Wheelwright's four-stage model of operations contribution sees operations as moving from implementation of strategy, through to supporting strategy, and finally to driving strategy

appropriate operations resources. The operation is trying to be '**internally supportive**' by providing a credible operations strategy.

Stage 4: Externally supportive

Stage 3 used to be viewed as the limit of the operations function's contribution. Yet the model captures the growing importance of operations management by suggesting a further stage. The difference between stages 3 and 4 is subtle, but important. A stage 4 operations function is one that is providing *the* foundation for an organisation's competitive success. It is forecasting likely changes in markets and supply, and it is developing the operations-based capabilities that will be required to compete in future market conditions. Stage 4 operations are innovative, creative and proactive, and are driving the company's strategy by being 'one step ahead' of competitors – what Hayes and Wheelwright call being '**externally supportive**'.

Critical commentary

The idea that operations can have a leading role in determining a company's strategic direction is not universally accepted. Both stage 4 of Hayes and Wheelwright's four-stage model and the concept of operations 'driving' strategy not only imply that it is possible for operations to take such a leading role, but are explicit in seeing this as a 'good thing'. A more traditional stance taken by some authorities is that the needs of the market will always be pre-eminent in shaping a company's strategy. Therefore, operations should devote all its time to understanding the **requirements of the market** (often defined by the marketing function within the organisation) and devote itself to its main job of ensuring that operations processes can actually deliver what the market requires. Companies can only be successful, they argue, by positioning themselves in the market (through a combination of price, promotion, product design, and managing how products and services are delivered to customers) with operations very much in a 'supporting' role. In effect, they say, Hayes and Wheelwright's model should stop at stage 3. The issue of an 'operations resource' perspective on operations strategy is discussed later in the chapter.

The four perspectives on operations strategy

Different authors have slightly differing views and definitions of operations strategy. However, between them, four 'perspectives' emerge, as shown in Figure 3.3:[4]

- Operations strategy should align with what the whole group or business wants – sometimes called the 'top-down' perspective.
- Operations strategy should translate the enterprise's intended market position so as to provide the required objectives for operations decisions – sometimes called the 'outside-in' perspective.
- Operations strategy should learn from day-to-day activities so as to cumulatively build strategic capabilities – sometimes called the 'bottom-up' perspective.
- Operations strategy should develop the business's resources and processes so that its capabilities can be exploited in its chosen markets – sometimes called the 'inside-out' perspective.

None of these four perspectives alone gives a comprehensive picture, but together they provide some idea of the pressures that go into forming the content of an operations strategy. In the next four sections of this chapter, we will treat each perspective in turn, before examining how these can be reconciled effectively.

3.2 How does operations strategy align with business strategy (top-down)?

A top-down perspective often identifies three related levels of strategy – corporate, business and functional:

- A *corporate strategy* should position the corporation in its global, economic, political and social environment. This will consist of decisions about what types of business the group wants to be in, what parts of the world it wants to operate in, how to allocate its cash between its various businesses, and so on.
- Each business unit within the corporate group will also need to put together its own *business strategy*, which sets out its individual mission and objectives. This business strategy guides the business in relation to its customers, markets and competitors, and also the strategy of the corporate group of which it is a part.

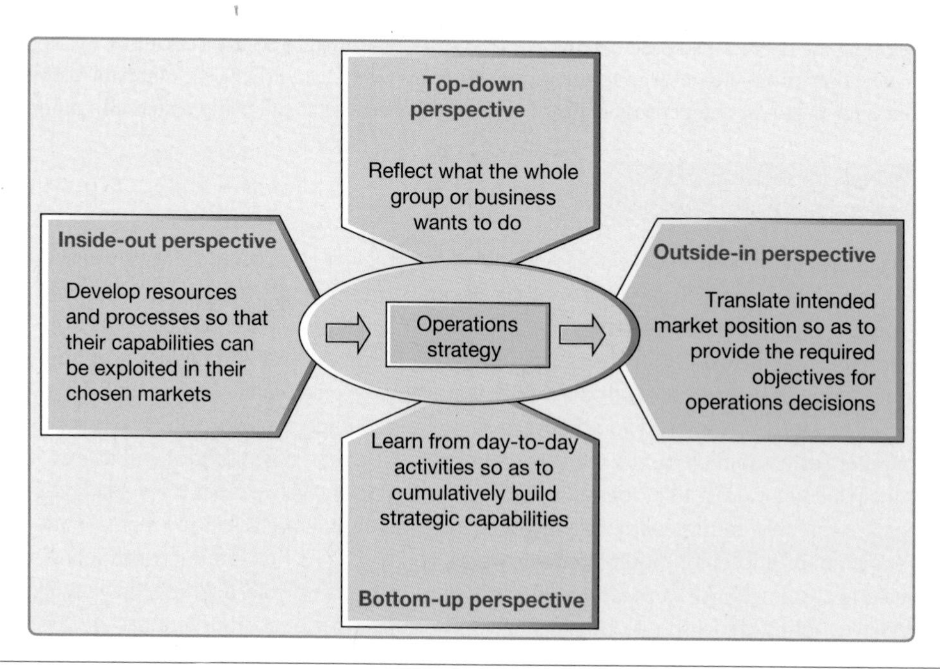

Figure 3.3 The four perspectives on operations strategy

- The operations, marketing, product/service development and other functions will then need to consider how best they should organise themselves to support the business's objectives. These *functional strategies* need to consider what part each function should play in contributing to the strategic objectives of the business.

So, the 'top-down' perspective on operations strategy is that it should take its place in this 'hierarchy of strategies'. As such, the role of operations is largely one of implementing or 'operationalising' higher-level strategy. For example, a printing services group has a company that prints packaging for consumer products. The group's management figures that, in the long term, only companies with significant market share will achieve substantial profitability. Its corporate objectives therefore stress market dominance. The consumer packaging company decides to prioritise volume growth, even above short-term profitability or return on investment. The implication for operations strategy is that it needs to expand rapidly, investing in extra capacity (factories, equipment and labour) even if it means excess capacity in some areas. It also needs to establish new factories in all parts of its market to offer relatively fast delivery. The important point here is that different business objectives would probably result in a very different operations strategy. Figure 3.4 illustrates this strategic hierarchy, with some of the decisions at each level and the main influences on the strategic decisions.

Operations principle
Operations strategies should reflect top-down corporate and/or business objectives.

Although this rather neat relationship between the levels of corporate, business and operations strategy may seem a little 'theoretical', it is still a powerful idea. What it is saying is that in order to understand strategy at any level, one has to place it in the context of what it is trying to do (the level above) and how it is trying to do it (the level below). At any level, a good top-down perspective should provide clarity and connection. It should clarify what an operations strategy should be prioritising, and give some guidance on the strategy to be achieved.

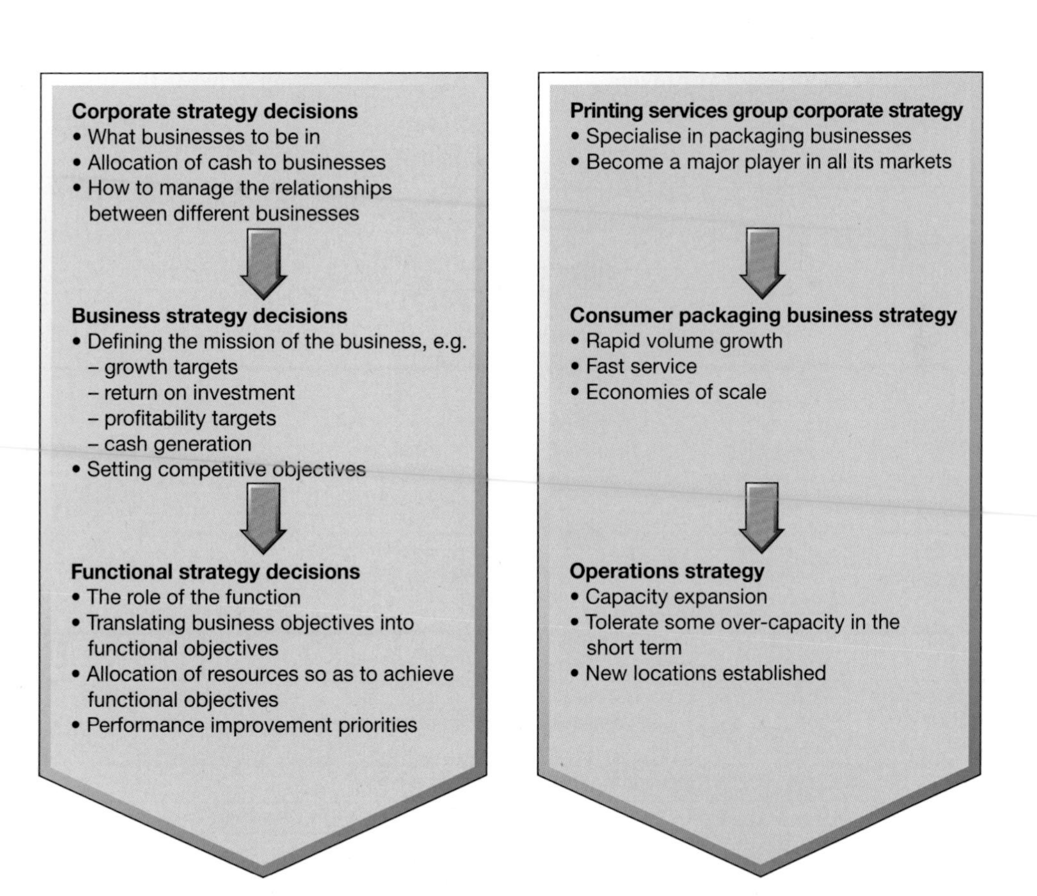

Figure 3.4 The top-down perspective of operations strategy and its application to a printing services group

Correspondence and coherence

Developing a functional strategy from a business strategy requires ambiguities to be clarified and conflicts to be reconciled. There should be a clear, explicit and logical connection between each functional strategy and the business strategy. Moreover, there should also be a clear, explicit and logical connection between a functional strategy and the decisions taken *within* the function. In other words, there should be *correspondence* between different levels of strategy. But *correspondence* is not all that is required. Operations strategy must also be *coherent,* both with other functional strategies and within itself. Coherence means that all decisions should complement and reinforce each other, not pull the operation in different directions. Figure 3.5 illustrates the two ideas of correspondence and coherence.

The concepts of the 'business model' and the 'operating model'

Two concepts have emerged over the last few years that are useful in understanding the top-down perspective on operations strategy – the 'business model' and the 'operating model'. The relationship between these two concepts is shown in Figure 3.6.

A *business model* is the plan that is implemented by a company to generate revenue and make a profit (or fulfil its social objectives if a not-for-profit enterprise). It includes the various parts and organisational functions of the business, as well as the revenues it generates and the expenses it incurs. It often includes such elements as:[5] the *value proposition* of what is offered to the market, the *target customer segments* addressed by the value proposition, the *distribution channels* to reach customers, the *core capabilities* needed to make the business model possible and the *revenue streams* generated by the business model. The idea of the business model is broadly analogous to the idea of a 'business strategy', but also adds an emphasis on *how* to achieve an intended strategy.

In contrast, the concept of an *operating model* is more operational in nature although there is no universally agreed definition. Here, we take it to mean a high-level design of the organisation that defines the structure and style that enables it to meet its business objectives.[6] Ideally, an operating model should provide a clear, 'big picture' description of what the organisation does

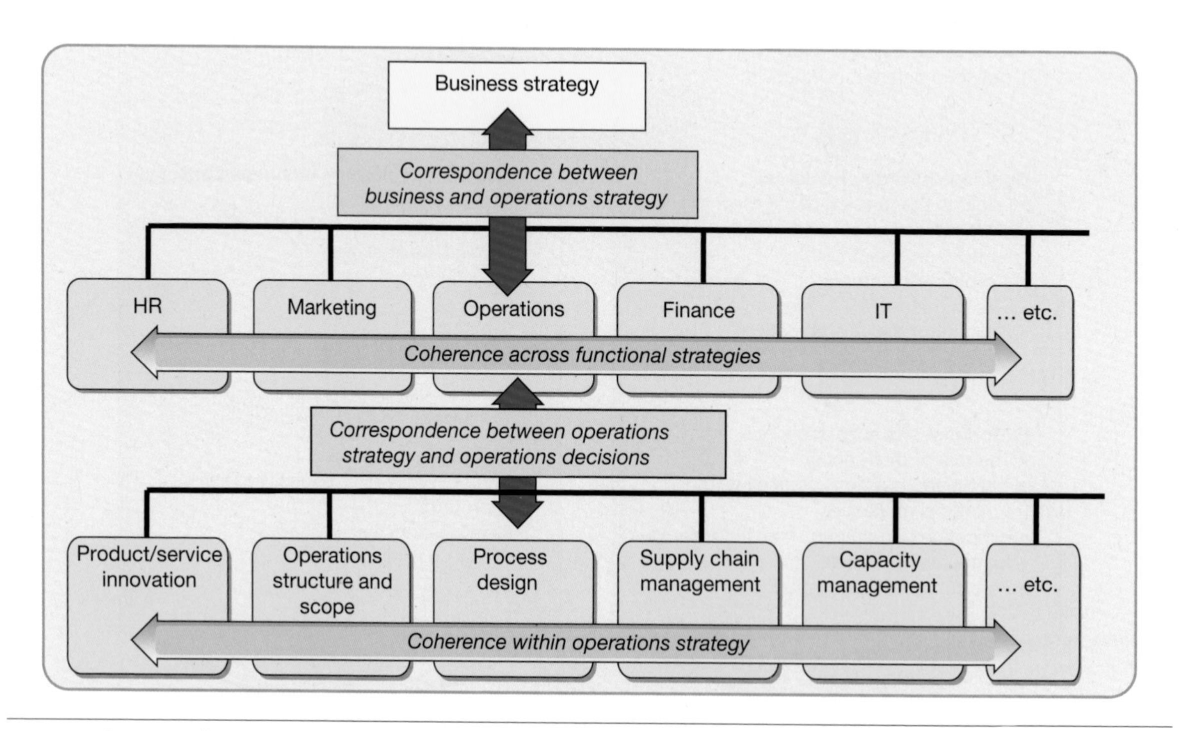

Figure 3.5 Correspondence and coherence are the two requirements of the top-down perspective of operations strategy

and how it does it. It defines how the critical work of an organisation is carried out. It should provide a way to examine the business in terms of the key relationships between business functions, processes and structures that are required for the organisation to fulfil its mission. It can include such elements as: key performance indicators (KPIs), with an indication of the relative importance of performance objectives, new investments and intended cash flows; who is responsible for products, geographies, assets, specific processes, systems and technologies etc.; and the structure of the organisation.

Note that an operating model reflects the idea that we proposed in Chapter 1 – that all managers are operations managers and all functions can be considered as operations because they comprise processes that deliver some kind of service. An operating model is like an operations strategy, but applied across all functions and domains of the organisation. Also, there are clear overlaps between the 'business model' and the 'operating model', although an operating model focuses more on how an overall business strategy is to be achieved.

3.3 How does operations strategy align with market requirements (outside-in)?

Any operations strategy should reflect the intended market position of the business. No operation that continually fails to serve its markets adequately is likely to survive in the long term. Organisations compete in different ways, so the operations function should therefore respond by providing the ability to perform in a manner that is appropriate for the intended market position. This is called a market (or outside-in) perspective on operations strategy.

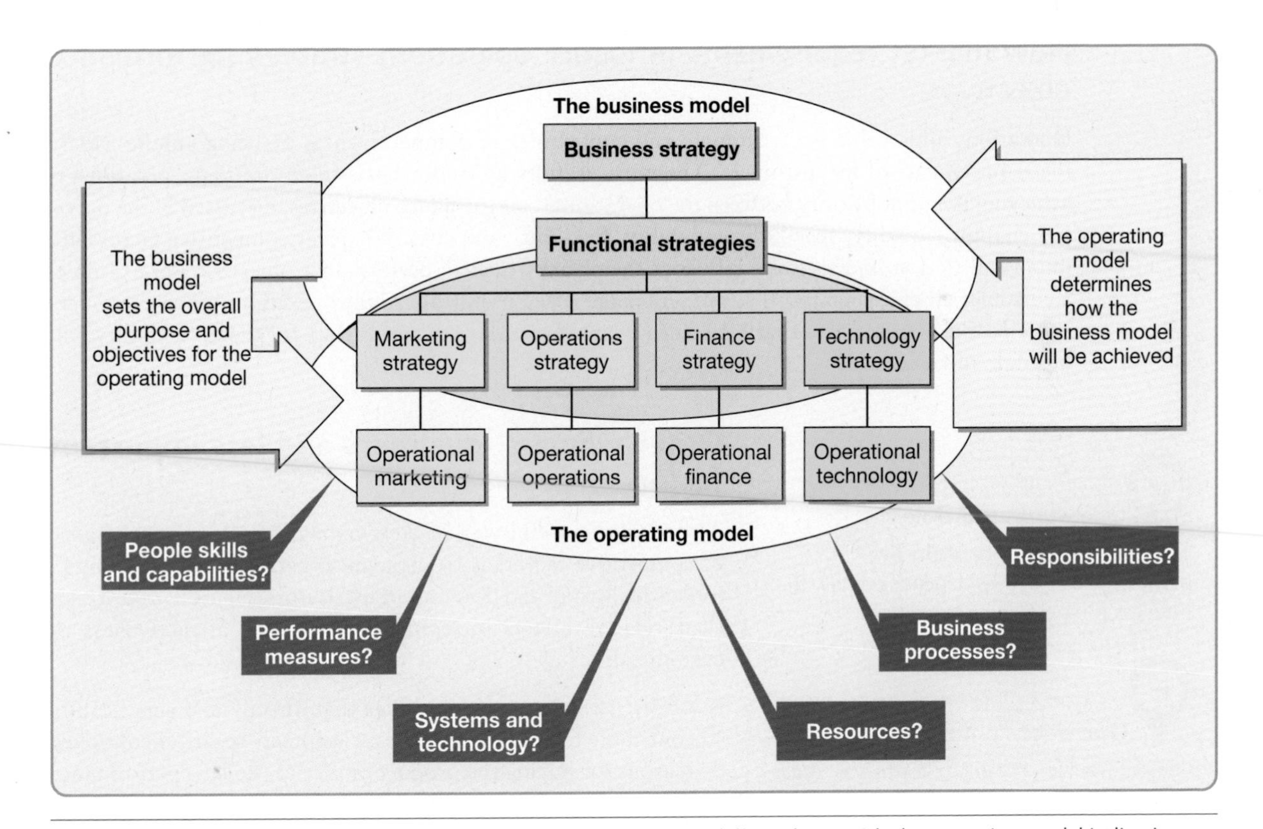

Figure 3.6 The concepts of the 'business model' and the 'operating model' overlap – with the operating model indicating how processes, resources, technology, people, measures and responsibilities are to be organised to support the business model

Worked example

Innovation at Micraytech (Part 1, top-down)[7]

Micraytech is a metrology systems company that develops integrated measurement systems for large international clients in several industries, and is part of the Micray Group, which includes several high-tech companies. It has grown through a strategy of providing products with a high degree of technical excellence and innovation together with an ability to customise its systems and offer technical advice to its clients. The group has set ambitious growth targets for the company over the next five years and has relaxed its normal 'return on sales' targets to help it achieve this. As part of this strategy, Micraytech attempted to be the first in the market with all appropriate new technical innovations. From a top-down perspective, its operations function, therefore, needed to be capable of coping with the changes that constant product innovation would bring. It developed processes that were flexible enough to develop and assemble novel components and systems while integrating them with software innovations.

The company's operations function realised that it needed to organise and train its staff to understand the way technology is developing so that it could put in place the necessary changes to the operation. It also needed to develop relationships with both existing and potentially new suppliers who could respond quickly when supplying new components. So, the top-down logic here is that everything about the operation – its processes, staff, systems and procedures – must, in the short term, do nothing to inhibit and, in the long term, actively develop the company's competitive strategy of growth through innovation.

How market requirements influence operations strategy performance objectives

Operations adds value for customers and contributes to competitiveness by being able to satisfy the requirements of its customers. The most useful way to do this is to ensure that operations is achieving the right priority between its 'operational' performance objectives, discussed in the previous chapter – quality, speed, dependability, flexibility and cost. Whatever competitive factors are important to customers should influence the priority of each performance objective. For example, a customer emphasis on fast delivery will make speed important to the operation. When customers value products or services that have been adapted or designed specifically for them, flexibility will be vital, and so on.

Order winners, qualifiers and less important factors

Operations principle
The relative importance of the five performance objectives depends on how the business competes in its market.

Operations principle
Competitive factors can be classified as 'order winners', 'qualifiers' or 'less important'.

A particularly useful way of determining the relative importance of competitive factors is to distinguish between 'order-winning', 'order-qualifying' and 'less important' factors. Figure 3.7 illustrates their relative effects on competitiveness (or attractiveness to customers):

- *Order winners* are those things that directly and significantly contribute to winning business. Customers regard them as key reasons for buying the product or service. Raising performance in an order-winning factor will either result in more business or at least improve the chances of gaining more business.
- *Order qualifiers* may not be the major competitive determinants of success, but are important in another way. They are factors where the operation's performance has to be above a particular level just to be considered by the customer. Performance below this 'qualifying' level of performance

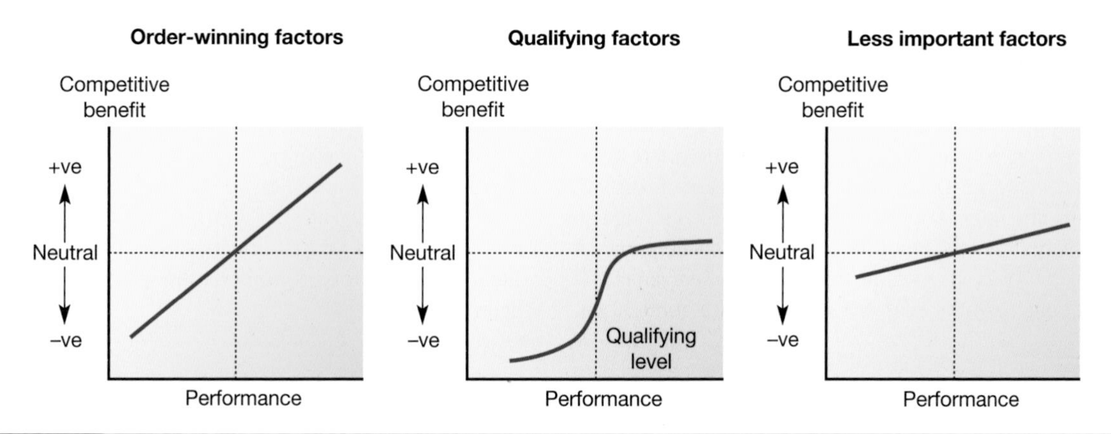

Figure 3.7 Order-winning, qualifying and less important competitive factors

often disqualifies the organisation from being considered by many customers. Conversely, further improvement above the qualifying level is unlikely to gain the company much competitive benefit.

- *Less important factors* are neither order winning nor qualifying. They do not influence customers in any significant way. They are worth including in any analysis only because they may be of importance in other parts of the operation's activities.

OPERATIONS IN PRACTICE

Dow Silicones' operations strategy[8]

For years, Dow Silicones (formerly Dow Corning) was a silicon business with a market position built on service and technical excellence. Customers had been willing to pay top prices for pioneering technology, premium products and customised service. Yet, as the market matured it became clear that some customers were becoming increasingly price sensitive. The premium price strategy was under attack both from large competitors that had driven down costs and from smaller competitors, with lower overheads. Dow Silicones was 'stuck in the middle'. In response, it decided to undertake detailed analysis of its market and segment customers based on the key factors motivating them to make purchases. Its work revealed four key groups of customers:

- *Innovative solution seekers* – customers who wanted innovative silicone-based products.
- *Proven solution seekers* – customers needing advice on existing proven products.
- *Cost-effective solution seekers* – customers who may even pay premium prices for a product, if it could take costs out of their business by improving their productivity.
- *Price seekers* – experienced purchasers of commonly used silicone materials wanting low prices and an easy way of doing business with their supplier.

Each of these segments held a distinct message for Dow Silicones' operations. For *innovative solution seekers,* there was a need to collaborate more closely with customers' R&D personnel in order to develop new products. To target *proven solution seekers,* the operations function took a more internal approach, working closely with the Dow Corning sales team. The aim was to help them understand its product range in more detail to improve conversion in the sales process. For *cost-effective solution seekers,* the focus was once again on working closely with the sales staff, but in this case the knowledge transfer was more bi-directional. The key was to build a stronger understanding of customers' processes and help better match their requirements with appropriate offerings. Finally, for the *price seeker* segment the focus was firmly on bringing down the costs of manufacturing and delivery. This last group was the most challenging for Dow Silicones. Its sales to this segment were small and declining, but the segment represented around 30 per cent of the total market for silicones and was expected to grow significantly. Dow Silicones' solution? Create a new offering, called Xiameter. This was a 'no-frills', low-price, restricted-range, minimum-order-quantity service, without any technical advice, that could only be accessed on the web (drastically cutting the costs of selling). Delivery times were sufficiently long to fit individual orders into the operation's existing manufacturing schedule.

The development of the Xiameter offering provides a good example of the sequence that organisations can follow in seeking to align the requirements of the market with the capabilities of its operations. The sequence looks something like this:

- *Segment the market:* This allowed Dow Silicones to identify the differing requirements of different customer groups.
- *Assess current performance:* Dow Silicones reviewed its performance for *each market segment* before making any decisions about how it might change direction.
- *Decide which segments to serve:* Dow Silicones decided that, while it was weak in the price seeker segment, it was worth pursuing ways in which it might compete.
- *Determine what is necessary for the business to compete:* For price seekers, Dow Silicones would need to supply at low cost, and abandon its technical advice service, because most customers in this segment were not willing to pay for it.
- *Determine what operations has to do:* For Xiameter to be successful, it would need to emphasise its 'no-frills' service (hence the new Xiameter brand), and reduce excess sales overheads (hence web-based sales). Most critically, for this high-volume, low-variety operation to work, customers would need to be prevented from asking for anything that would increase costs (hence limited product range, minimum order quantities and delivery times that do not disrupt production schedules).

The impact of product/service differentiation on market requirements

If an operation differentiates its services based on different customer segments, it will need to determine the performance objective for each segment. For example, Table 3.1 shows two customer segments in the banking industry. Here the distinction is drawn between the customers who are looking for banking services for their private and domestic needs (current accounts, overdraft facilities, savings accounts, mortgage loans, etc.) and those corporate customers who need banking services for their (often large) organisations. These latter services would include such things as letters of credit, cash transfer services and commercial loans.

The impact of the product/service life cycle on market requirements

One way of generalising the behaviour of both customers and competitors is to link it to the life cycle of the products or services that the operation is producing. The exact form of product/service life cycle will vary, but generally they are shown as the sales volume passing through four stages – introduction, growth, maturity and decline. The implication of this for operations management is that products and services will require different operations strategies in each stage of their life cycle, as shown in Figure 3.8.

Table 3.1 Different banking services require different performance objectives

	Retail banking	Corporate banking
Products	Personal financial services such as loans and credit cards	Special services for corporate customers
Customers	Individuals	Businesses
Range of services offered	Medium but standardised, little need for special terms	Very wide range, many need to be customised
Changes to service design	Occasional	Continual
Delivery	Fast decisions	Dependable service
Quality	Means error-free transactions	Means close relationships
Volume per service type	Most services are high volume	Most services are low volume
Profit margins	Most are low to medium, some high	Medium to high
Competitive factors		
Order winners	Price	Customisation
	Accessibility	Quality of service
	Ease of transaction	Reliability/trust
Qualifiers	Quality	Ease of transaction
	Range	Price
Less important		Accessibility
Internal performance objectives	Cost	Flexibility
	Speed	Quality
	Quality	Dependability

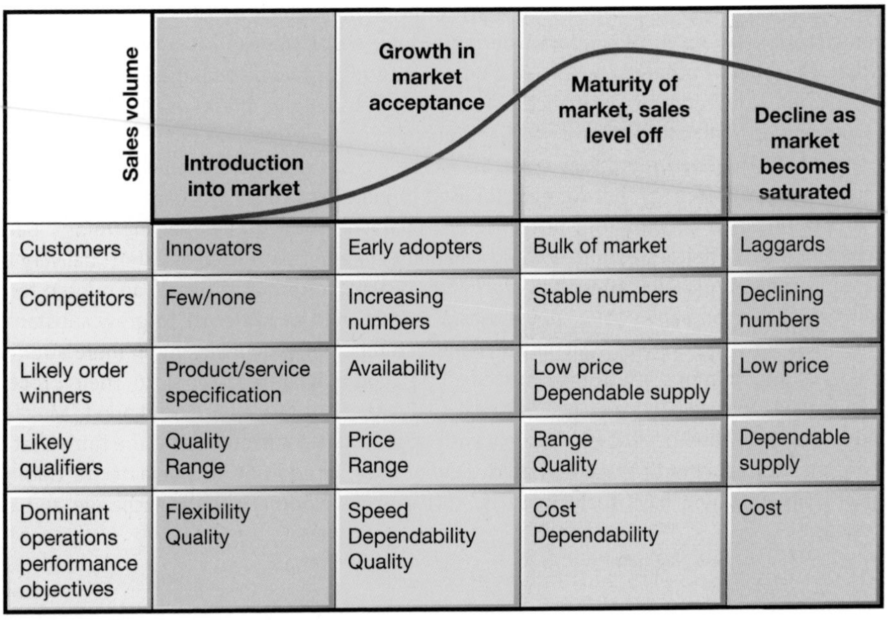

Sales volume	Introduction into market	Growth in market acceptance	Maturity of market, sales level off	Decline as market becomes saturated
Customers	Innovators	Early adopters	Bulk of market	Laggards
Competitors	Few/none	Increasing numbers	Stable numbers	Declining numbers
Likely order winners	Product/service specification	Availability	Low price Dependable supply	Low price
Likely qualifiers	Quality Range	Price Range	Range Quality	Dependable supply
Dominant operations performance objectives	Flexibility Quality	Speed Dependability Quality	Cost Dependability	Cost

Figure 3.8 The effects of the product/service life cycle on operations performance objectives

Introduction stage

When a product or service is first introduced, it is likely to be offering something new in terms of its design or performance, with few competitors offering the same thing. The needs of customers are unlikely to be well understood, so operations management needs to develop the flexibility to cope with any changes and be able to give the quality to maintain product/service performance.

Growth stage

As volume grows, competitors may enter the growing market. Keeping up with demand could prove to be the main operations preoccupation. Rapid and dependable response to demand will help to keep demand buoyant, while maintaining quality levels can ensure that the company keeps its share of the market as competition starts to increase.

Maturity stage

As demand starts to level off, some early competitors may have left the market and the industry may be dominated by a few larger companies. So, operations will be expected to get the costs down in order to maintain profits or to allow price cutting. Because of this, cost and productivity issues, together with dependable supply, are likely to be the operation's main concerns.

Operations principle

Operations strategy objectives will change depending on the stage of the business's services and products.

Decline stage

After time, sales will decline with more competitors dropping out of the market. There might be a residual market, but unless a shortage of capacity develops, the market will continue to be dominated by price competition. So, operations objectives continue to be dominated by cost.

Worked example

Innovation at Micraytech (Part 2, outside-in)

The Micray Group saw a major growth opportunity for Micraytech by continually incorporating technological innovations in its product offerings. However, Micraytech's marketing management knew that this could be achieved by focusing on one or both of two distinct markets. The first is the market for 'individual metrology devices'. These are 'stand-alone' pieces of equipment bought by all types of industrial customers and had traditionally been the company's main market. The second market was for 'integrated metrology systems'. These were larger, more complex, more expensive (and higher margin) offerings that were customised to individual customers' requirements. The two types of offering had overlapping, but different characteristics. 'Individual metrology devices' competed on their technical performance and reliability, together with relatively short delivery times compared to competitors. The 'integrated metrology systems' offerings currently accounted for only a small part of the company's sales, but it was a market that was forecast to grow substantially. The customers for these systems were larger manufacturers that were investing in more automated technologies and required metrology systems that could be integrated into their processes. From an 'outside-in' perspective, if it was to take advantage of this emerging market, Micraytech would have to learn how to work more closely with both its direct customers and the firms that were supplying its customers with the automated technologies. In addition to Micraytech's traditional technical skills, it would have to increase its software development, data exchange and client liaison skills.

OPERATIONS IN PRACTICE

Tesco learns the hard way[9]

When market conditions change, it usually means that operations strategy should change. But it can take time before changes become obvious, and even longer to react. This was a lesson that Tesco, the United Kingdom's biggest, and one of its most successful retailers, learned, when, by 2014, it had slumped to a £6.4 billion loss. Although it was still comfortably the market leader in grocery sales, its lead over rivals had worsened. Like-for-like sales (sales in its stores and online, while removing the effect of new stores opening) were down nearly 4 per cent; significant in the retail world. Tesco had not seen numbers this bad for 20 years. Why, asked its detractors, had the company not realised that its strategy was failing? One critic described Tesco as being 'like a juggernaut with a puncture and a worrying rattle in the engine'. Some of Tesco's problems at this time were beyond its control and a result of competitor activity. Waitrose (an upmarket supermarket, with a good reputation for quality) was serving the top end of the market, while German discount stores Aldi and Lidl were attracting more cost-conscious customers. However, some problems were of Tesco's own making, caused by its operations strategy failing to respond fast enough to market requirements. The strategy of building large out-of-town superstores was continued, even though a sharper monitoring of consumer behaviour would have revealed that such large-capacity units had lost their attraction as families cut down on weekly trips to the supermarket, and opted instead for home deliveries and topping up their groceries with trips to local stores. In fact, Philip Clarke, then Tesco's chief executive, admitted that he ought to have moved faster to cut back on planned superstore openings in response to clear radical changes in shopping habits. *'Hindsight is a wonderful thing. It's never really there when you need it'*, said Mr Clarke. *'I probably should have stopped more quickly that* [superstore] *expansion, I probably should have made the reallocation* [to small, local stores] *faster'.*

This episode in Tesco's (largely successful) history provided important lessons. Making significant changes in operations strategy can be extremely disruptive and costly in the short run, but necessary in the long run, even when the long-term consequences of a major change are unknowable (although better demand forecasting, can help – see Chapter 11). Understandable then, that in the face of such uncertainty, organisations, especially those with large inflexible investments, often delay change. However, having made major changes in its corporate, business and functional strategies, Tesco did recover in terms of like-for-like sales.

3.4 How does operations strategy align with operational experience (bottom-up)?

The 'top-down' perspective provides an orthodox view of how functional strategies *should* be formulated. However, while it is a convenient way of thinking about strategy, it does not represent the way strategies *are* formulated in most cases. When any group is reviewing its corporate strategy, it will take into account the circumstances, experiences and capabilities of the various businesses that form the group. Similarly, when reviewing their strategies, organisations will consult the individual functions about their capabilities and constraints and incorporate the ideas that come from each function's day-to-day experience. This is the *bottom-up* perspective, illustrated in Figure 3.9.

The bottom-up perspective accounts for the fact that in many cases organisations move in a particular strategic direction because their ongoing experience at an operational level convinces them that it is the right thing to do. The 'high-level' strategic decision-making, if it occurs at all, may simply confirm the general consensus around a given strategic direction and provide the resources to make

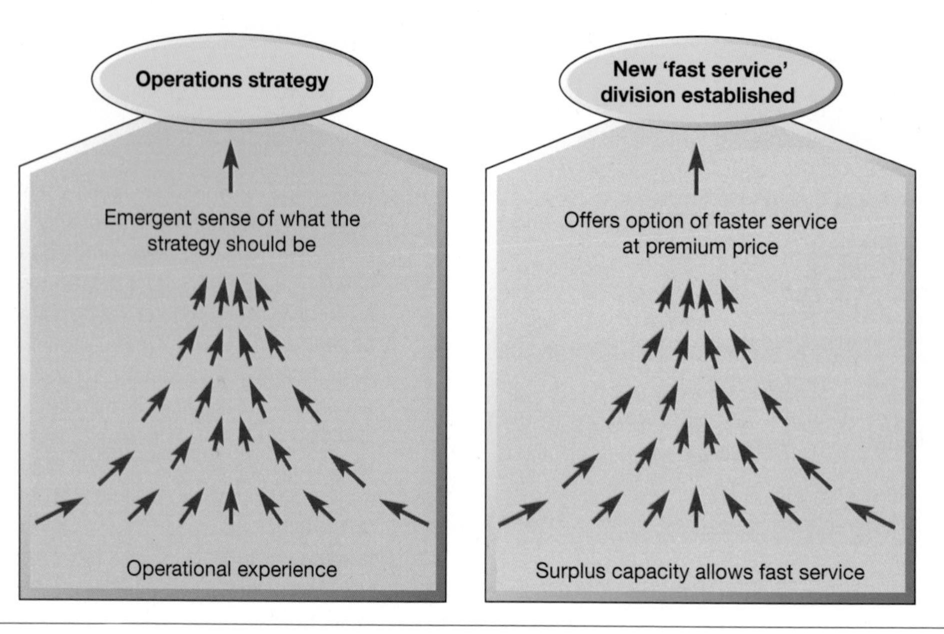

Figure 3.9 The 'bottom-up' perspective of operations strategy

it happen effectively. This is sometimes called the concept of '**emergent strategies**'.[10] It sees strategy making, at least partly, as a relatively unstructured and fragmented process to reflect the fact that the future is at least partially unknown and unpredictable.

For example, suppose the printing services company described previously succeeds in its expansion plans. However, in doing so it finds that having surplus capacity and a distributed network of factories allows it to offer an exceptionally fast service to customers, who are willing to pay considerably higher prices for such a responsive service. Its experiences lead the company to set up a separate division dedicated to providing fast, high-margin printing services to those customers willing to pay. The strategic objectives of this new division are not concerned with high-volume growth but with high profitability.

Operations principle

Operations strategy should reflect bottom-up experience of operational reality.

The reinforcing effect of top-down and bottom-up perspectives on operations strategy

The top-down and bottom-up perspectives are often seen as being diametrically opposite ways of looking at operations strategy, but they are not. In fact, the two perspectives can be mutually reinforcing.

Worked example

Innovation at Micraytech (Part 3, bottom-up)

Over time, as its operations strategy developed, Micraytech discovered that continual product and system innovation was having the effect of dramatically increasing its costs. Although it was not competing on low prices, and nor was it under pressure from the group to achieved high rates of return on sales, its rising costs were impacting profitability to an unacceptable degree. In addition, there was some evidence that continual updating of product and system specifications was confusing some customers. Partially in response to customer requests, the company's system designers started to work out a way of 'modularising' their system and product designs. This allowed one part of the system to be updated for those customers who valued the functionality that the innovation could bring, without interfering with the overall design of the main body of the system, of which the module was a part. Over time, this approach became standard design practice within the company. Customers appreciated the extra customisation, and modularisation reduced operations costs. Note that this strategy emerged from the company's experience and was therefore an example of a pure 'bottom-up' approach. Initially, no top-level board decision was taken to initiate this practice. Nevertheless, it emerged as the way in which the company's design engineers learned from their experience and used that learning to build their knowledge of how to lower some of the costs of innovation.

Figure 3.10 shows how this can work. The top-down perspective sets the overall direction and objectives for operations decisions and activities. In order to implement top-down strategy, the day-to-day activities of the operation must be aligned with the strategy. So, a way of judging operational day-to-day activities of an operation is to check that they fully reflect the overall top-down strategy of the organisation. But, as we have illustrated earlier, the experience gained from day-to-day activities can be accumulated and built into capabilities that an organisation could possibly exploit strategically. (We will expand this idea of 'capabilities' in the next section.)

3.5 How does operations strategy align with operations resources (inside-out)?

The final perspective of operations strategy is the operations resources (or 'inside-out') perspective. Its fundamental idea is that long-term competitive advantage can come from the capabilities of the operation's resources and processes, and these should be developed over the long term to provide the business with a set of capabilities or competences (we use the two words interchangeably).[11] So, the way an organisation inherits, acquires or develops its operations resources will, over the long term, have a significant impact on its strategic success. They can form the basis of the business's

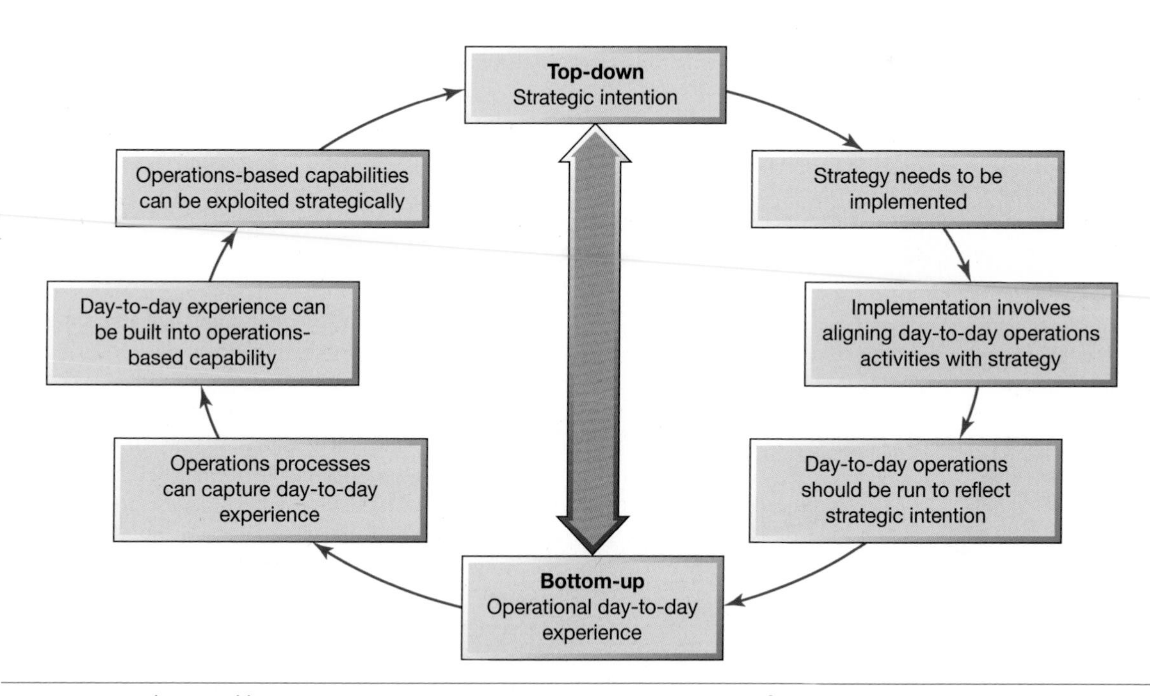

Figure 3.10 Top-down and bottom-up perspectives on operations strategy can reinforce each other

> **Operations principle**
> The long-term objective of operations strategy is to build operations-based capabilities.

ability to engage in unique and/or 'difficult to imitate' activities. Furthermore, the impact of an organisation's 'operations resource' capabilities will be at least as great as, if not greater than, that which it gets from its market position. So, understanding and developing the capabilities of operations resources, although often neglected, is a particularly important perspective on operations strategy.

Strategic resources and sustainable competitive advantage

The idea that building operations capabilities is a particularly important objective of operations strategy is closely linked with the 'resource-based view' (RBV) of the firm.[12] This holds that organisations with 'above-average' strategic performance are likely to have gained their sustainable competitive advantage because of their core competences (or capabilities). This means that the way an organisation inherits, or acquires, or develops its operations resources will, over the long term, have a significant impact on its strategic success. The RBV differs in its approach from the more traditional view of strategy that sees companies as seeking to protect their competitive advantage through their control of the market. For example, they may do this by creating *barriers to entry* through product differentiation, or making it difficult for customers to switch to competitors, or controlling the access to distribution channels (a major barrier to entry in gasoline retailing, for example, where oil companies own their own retail stations). By contrast, the RBV sees firms being able to protect their competitive advantage through *barriers to imitation*, that is, by building up 'difficult-to-imitate' resources.

Understanding existing capabilities and constraints

An operations resource perspective must start with an understanding of the resource capabilities and constraints within the operation. It must answer the simple questions: what do we have and what can we do? An obvious starting point here is to examine the transforming and transformed resource inputs to the operation. However, trying to understand an operation by listing its resources alone is like trying to understand a vehicle by listing its component parts. To understand the vehicle, we need to describe how the component parts form its internal mechanisms. Within the operation, the equivalents of these mechanisms are its *processes*. Yet even a technical explanation of the mechanisms does not convey its style or 'personality'. Something more is needed to describe these. In the same way, an operation is not just the sum of its processes; it also has *intangible* resources. See the 'Operations in practice' example on 'The rise of intangibles'.

OPERATIONS IN PRACTICE

The rise of intangibles

The (almost universal) assumption about operations strategy is that when businesses say they are investing in operations assets, they will be real, tangible items such as machines, computers and buildings. Yet, this was never universally true, and it is getting less true. Consider the following. The world's largest taxi company (Uber) does not own any vehicles. The world's largest provider of accommodation (Airbnb) does not own any real estate. The world's most popular media company (Facebook) does not create its content. One of the world's largest retailers (Alibaba) does not have any inventory. In fact, investment in tangible assets is becoming less and less important in many economies.[13] Instead, investment is

to the organisation is that its reputation (or brand) will suffer because market expectations exceed the operation's capability to perform at the appropriate level. At other times, the operation may make improvements before they can be fully exploited in the market. For example, the same online retailer may have improved its website so that it can offer extra services, such as the ability to customise products, before those products have been stocked in its distribution centre. This means that, although an improvement to its ordering processes has been made, problems elsewhere in the company prevent the improvement from giving value to the company. This is represented by point Y on Figure 3.13(b).

Using the importance–performance matrix to determine operations strategy improvement priorities

One can use the idea of comparing market and operations perspectives at a more focused and disaggregated level to provide direct guidance to operations managers. So, rather than ask generally, 'what are the market requirements for our products and/or services?' one asks, 'how important are the competitive factors that characterise a product or service?' The intention is to gain an understanding of the relative importance to customers of the various competitive factors. For example, do customers for a particular product or service prefer low prices to a wide range? The needs and preferences of customers shape the *importance* of operations objectives within the operation. Similarly, rather than ask generally, 'what are our operations capabilities?' one asks, 'how good is our operation at providing the required level of performance in each of the competitive factors?' But how good is our performance against what criteria? Strategically, the most revealing point of comparison is with competitors. Competitors are the points of comparison against which the operation can judge its performance. From a competitive viewpoint, as operations improves its performance, the improvement that matters most is that which takes the operation past the performance levels achieved by its competitors. The role of competitors then is in determining achieved *performance*. (In a not-for-profit context, 'other similar operations' can be substituted for 'competitors'.)

Both importance and performance have to be brought together before any judgement can be made as to the relative priorities for improvement. Just because something is particularly important to its customers does not mean that an operation should necessarily give it immediate priority for improvement. It may be that the operation is already considerably better than its competitors at serving customers in this respect. Similarly, just because an operation is not very good at something when compared with its competitors' performance, it does not necessarily mean that it should be immediately improved. Customers may not particularly value this aspect of performance.

Operations principle

Improvement priorities are determined by importance for customers and performance against competitors or similar operations.

- ▶ ***Judging importance to customers*** – Earlier, we introduced the idea of order-winning, qualifying and less important competitive factors, and one could take these three categories as an indication of the relative importance of each performance factor. But usually one needs to use a slightly more discriminating scale. One way to do this is to take our three broad categories of competitive factors – order winning, qualifying and less important – and to divide each category into three further points representing strong, medium and weak positions. Figure 3.14(a) illustrates such a scale.
- ▶ ***Judging performance against competitors*** – At its simplest, a competitive performance standard would consist merely of judging whether the achieved performance of an operation is better than, the same as, or worse than that of its competitors. However, in much the same way as the nine-point importance scale was derived, we can derive a more discriminating nine-point performance scale, as shown in Figure 3.14(b).

The priority for improvement that each competitive factor should be given can be assessed from a comparison of their importance and performance. This can be shown on an importance–performance matrix that, as its name implies, positions each competitive factor according to its scores or ratings on these criteria. Figure 3.15 shows an importance–performance matrix divided into zones of improvement priority (see later).

(a) Importance scale for competitive factors

Rating	Description	
1	Provides a crucial advantage to customers	High
2	Provides an important advantage to customers	
3	Provides a useful advantage to customers	
4	Needs to be up to good industry standard	
5	Needs to be up to median industry standard	
6	Needs to be within close range of rest of industry	
7	Not usually important but could become so	
8	Very rarely considered by customers	
9	Never considered by customers	Low

(b) Performance scale for competitive factors

Rating	Description	
1	Considerably better than similar organisations	Good
2	Clearly better than similar organisations	
3	Marginally better than similar organisations	
4	Sometimes marginally better than similar organisations	
5	About the same as similar organisations	
6	Slightly worse than the average of similar organisations	
7	Usually marginally worse than similar organisations	
8	Generally worse than most similar organisations	
9	Consistently worse than most similar organisations	Poor

Figure 3.14 Nine-point scales for judging importance and performance

The first zone boundary is the 'lower bound of acceptability', shown as line AB in Figure 3.15. This is the boundary between acceptable and unacceptable performance. When a competitive factor is rated as relatively unimportant (8 or 9 on the importance scale), this boundary will in practice be low. Most operations are prepared to tolerate performance levels that are 'in the same ball-park' as their competitors (even at the bottom end of the rating) for unimportant competitive factors.

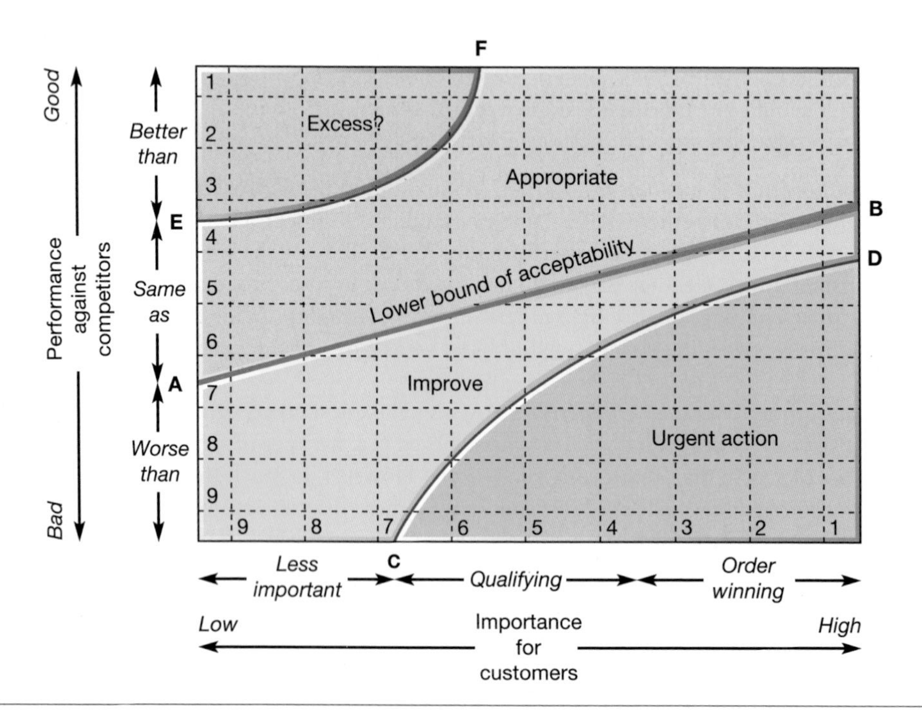

Figure 3.15 Priority zones in the importance–performance matrix

They only become concerned when performance levels are clearly below those of their competitors. Conversely, when judging competitive factors that are rated highly (1 or 2 on the importance scale) they will be markedly less sanguine at poor or mediocre levels of performance. Minimum levels of acceptability for these competitive factors will usually be at the lower end of the 'better than competitors' class. Below this minimum bound of acceptability, there is clearly a need for improvement; above this line there is no immediate urgency for any improvement. However, not all competitive factors falling below the minimum line will be seen as having the same degree of improvement priority. A boundary approximately represented by line CD represents a distinction between an urgent priority zone and a less urgent improvement zone. Similarly, above the line AB, not all competitive factors are regarded as having the same priority. The line EF can be seen as the approximate boundary between performance levels that are regarded as 'good' or 'appropriate' on the one hand and those regarded as 'too good' or 'excess' on the other. Segregating the matrix in this way results in four zones, which imply very different priorities:

- *The 'appropriate' zone* – competitive factors in this area lie above the lower bound of acceptability and so should be considered satisfactory.
- *The 'improve' zone* – lying below the lower bound of acceptability, any factors in this zone must be candidates for improvement.
- *The 'urgent-action' zone* – these factors are important to customers but performance is below that of competitors. They must be considered as candidates for immediate improvement.
- *The 'excess?' zone* – factors in this area are 'high performing', but not important to customers. The question must be asked, therefore, whether the resources devoted to achieving such performance could be used better elsewhere.

Worked example

YIR Laboratories

YIR Laboratories is a subsidiary of an electronics company. It carries out research and development as well as technical problem-solving work for a wide range of companies, including companies in its own group. It is particularly keen to improve the level of service that it gives to its customers. However, it needs to decide which aspect of its performance to improve first. It has devised a list of the most important aspects of its service:

- *The quality of its technical solutions* – the perceived appropriateness by customers.
- *The quality of its communications with customers* – the frequency and usefulness of information.
- *The quality of post-project documentation* – the usefulness of the documentation that goes with the final report.
- *Delivery speed* – the time between customer request and the delivery of the final report.
- *Delivery dependability* – the ability to deliver on the promised date.
- *Delivery flexibility* – the ability to deliver the report on a revised date.
- *Specification flexibility* – the ability to change the nature of the investigation.
- *Price* – the total charge to the customer.

YIR assigned a score to each of these factors using the 1–9 scale that we described in Figure 3.14. After this, YIR turned its attention to judging the laboratory's performance against competitor organisations. Although it has benchmarked information for some aspects of performance, it has to make estimates for the others. Both these importance and performance scores are shown in Figure 3.16.

YIR Laboratories plotted the importance and performance ratings it had given to each of its competitive factors on an importance–performance matrix. This is shown in Figure 3.17. It shows that the most important aspect of competitiveness – the ability to deliver sound technical solutions to its customers – falls comfortably within the appropriate zone. Specification flexibility and delivery flexibility are also in the appropriate zone, although only just. Both delivery speed and delivery ▶

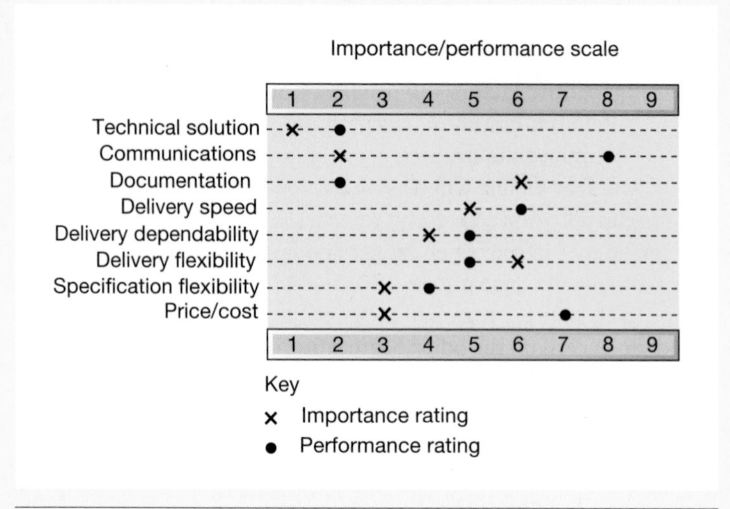

Figure 3.16 Rating 'importance to customers' and 'performance against competitors' on the nine-point scales for YIR Laboratories

dependability seem to be in need of improvement as each is below the minimum level of acceptability for its respective importance position. However, two competitive factors, communications and cost/price, are clearly in need of immediate improvement. These two factors should therefore be assigned the most urgent priority. The matrix also indicates that the company's documentation could almost be regarded as 'too good'.

The matrix may not reveal any total surprises. The competitive factors in the 'urgent-action' zone may be known to be in need of improvement already. However, the exercise is useful for two reasons:

- It helps to discriminate between many factors that may be in need of improvement.
- The exercise gives purpose and structure to the debate on improvement priorities.

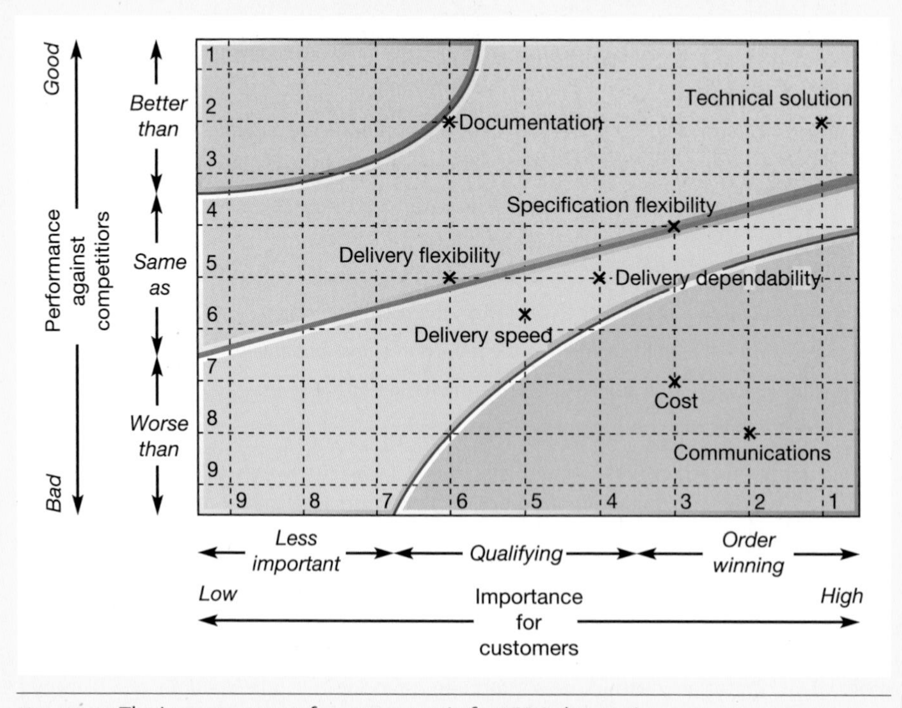

Figure 3.17 The importance–performance matrix for YIR Laboratories

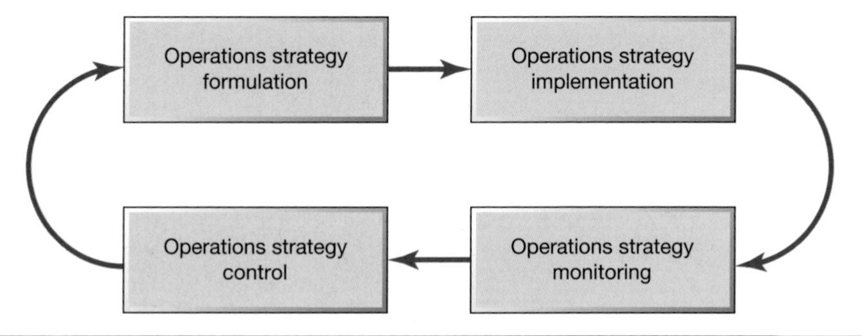

Figure 3.18 The stages of the process of operations strategy

3.7 How can the process of operations strategy be organised?

An operations strategy is the starting point for operations improvement. It sets the direction in which the operation will change over time. It is implicit that the business will want operations to change for the better. However, unless an operations strategy gives some idea as to *how* improvement will happen, it is not fulfilling its main purpose. This is where considering the *process* of operations strategy comes in. It is concerned with the method that is used to determine what an operations strategy should be and how it should be implemented. This is a complex and difficult thing to achieve in practice. And although any simple step-by-step model of how to 'do' operations strategy will inevitably be a simplification of a messy reality, we shall use a four-stage model to illustrate some of the elements of 'process'. This stage model is shown in Figure 3.18 and divides the process of operations strategy into *formulation*, *implementation*, *monitoring* and *control*. The four stages are shown as a cycle because strategies may be revisited depending on the experience gained from trying to make them happen.

Operations principle

The process of operations strategy involves formulation, implementation, monitoring and control.

Operations strategy formulation

The formulation of operations strategy is the process of clarifying the various objectives and decisions that make up the strategy, and the links between them. Unlike day-to-day operations management, formulating an operations strategy is likely to be only an occasional activity. Some firms will have a regular (e.g. annual) planning cycle, and operations strategy consideration may form part of this. However, the extent of any changes made in each annual cycle is likely to be limited. In other words, the 'complete' process of formulating an entirely new operations strategy will be a relatively infrequent event. There are many 'formulation processes', which are, or can be, used to formulate operations strategies. Most consultancy companies have developed their own frameworks, as have several academics.

Responsible operations

In every chapter, under the heading of 'Responsible operations', we summarise how the particular topic covered in the chapter touches upon important social, ethical and environmental issues.

If operations decisions are going to reflect social and environmental concerns, they must be reflected in how operations strategies are formulated. Indeed, corporate social responsibility (CSR) is increasingly seen not as a peripheral activity, but as a central and highly visible priority for most enterprises. (See the 'Operations in practice' example, 'Sustainability is high on Google's operations agenda'.) The media frequently carries reports of good corporate behaviour and most companies' corporate websites explain, often in great detail, their commitment to responsible operations ▶

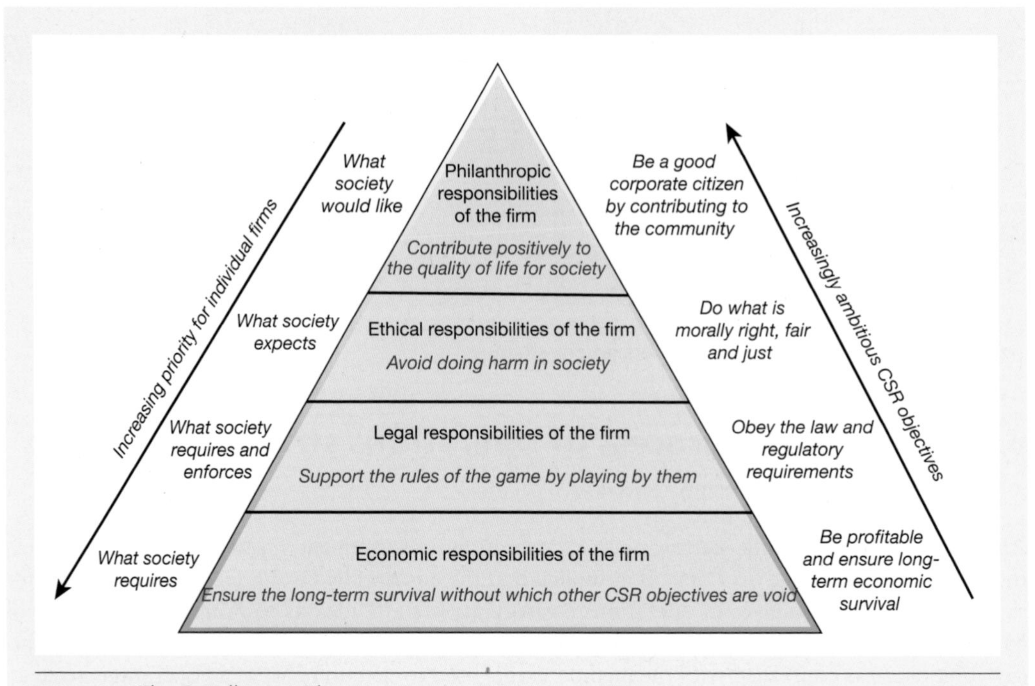

Figure 3.19 The Carroll pyramid

Source: Based on the work of Carroll, A.B. (1991) The pyramid of social responsibility: towards the moral management of organizational stakeholders, *Business Horizons*, July/August

practice. But how far should enterprises go in pursuing these objectives? Some would claim that the more philanthropic elements of CSR are 'nice to do', not 'must do'. One authority[16] sees corporate philanthropy at the peak of a pyramid comprising economic behaviours at its base, followed by legal and ethical behaviour (see Figure 3.19). Partly it is because it offers the potential to align economic and social objectives. Michael Porter and Mark Kramer, of Harvard Business School, suggest what they call a 'context-focused approach' to corporate philanthropy.[17] This overtly attempts to align social and economic strategic aims. They give as an example Cisco Systems, that created a 'win–win' by founding its Cisco Networking Academy in order to train computer network managers. In doing so, it filled a human resource gap in its foreign markets and simultaneously enhanced the job chances of the young people passing through the Academy.

Critical commentary

There is no shortage of individuals and institutions who do not accept the fundamental tenets of corporate social responsibility (CSR) and who believe that businesses get more from CSR than society does. Some even see CSR as a contradiction in terms because companies are legally bound to make money for their shareholders, so they can only be 'socially responsible' if they are being deliberately insincere. They argue that any social benefits from CSR (which they doubt) are more than overshadowed by losses to society from 'normal' corporate activity. CSR, they argue, is simply a means for corporations to boost their public image and avoid more stringent legal regulation. It is essentially a distraction that deliberately aims to deflect the responsibility for problems caused by corporate operations away from the business, while obstructing more legitimate efforts to tackle the social and environmental unfairness. But even less fundamentalist commentators on CSR question some aspects of how it is practised. One criticism is the lack of standards around CSR: it can be defined as whatever a business wants it to be. There is a lack of standards defining what counts as corporate responsibility and no agreed set of principles about what it means to be a responsible business. It is too easy for CSR managers to select, themselves, the areas of social benefit the company will address – often what is the most advantageous or convenient to the business. Another criticism is that, because of CSR, legislation has not been sufficiently developed to regulate business activity and therefore the pressure on businesses to be socially responsible is left to 'citizen (or media) oversight' rather than the law.

What should the formulation process be trying to achieve?

Before putting an operations strategy together, it is necessary to ask the question 'what should it be trying to achieve?' Clearly, it should provide a set of actions that, with hindsight, have provided the 'best' outcome for the organisation. But 'the best' is a judgement that can only be applied in hindsight. Yet, even if we cannot assess the 'goodness' of a strategy in advance, we can check it out for some attributes that could stop it being a success, as follows:

- *Is operations strategy comprehensive?* – In other words, does it include all important issues? Business history is littered with companies that simply failed to notice the potential impact of, for instance, new process technology or emerging changes in their supply network.
- *Is operations strategy coherent?* – As we discussed earlier, coherence is when the choices made in each decision area all direct the operation in the same strategic direction, with all strategic decisions complementing and reinforcing each other in the promotion of performance objectives.
- *Does operations strategy have correspondence?* – Again, as we discussed earlier, correspondence is when the strategies pursued correspond to the true priority of each performance objective.
- *Does operations strategy identify critical issues?* – The more critical the decision, the more attention it deserves. Although no strategic decision is unimportant, in practical terms some decisions are more critical than others. The judgement over exactly which decisions are particularly critical is very much a pragmatic one that must be based on the particular circumstances of an individual firm's operations strategy. However, the key is that they must be identified.

Operations principle

Operations strategies should be comprehensive, coherent, correspond to stated objectives and identify the critical issues.

Critical commentary

The argument has been put forward that strategy does not lend itself to a simple 'stage model' analysis that guides managers in a step-by-step manner through to the eventual 'answer' of a final strategy. Therefore, the models put forward by consultants and academics are of only limited value. In reality, strategies (even those that are made deliberately, as opposed to those that simply 'emerge') are the result of very complex organisational forces. Even descriptive models such as the four-stage model described above in Figure 3.19 can do little more than sensitise managers to some of the key issues that they should be taking into account when devising strategies. In fact, they argue, it is the articulation of the 'content' of operations strategy that is more useful than adhering to some over-simplistic description of a strategy process.

Operations strategy implementation

Operations strategy implementation is the way that strategies are operationalised or executed. It means attempting to make sure that intended strategies are actually achieved. It is important because no matter how sophisticated the intellectual and analytical underpinnings of a strategy, it remains only a document until it has been implemented. But the way one implements any strategy will very much depend on the specific nature of the changes implied by that strategy and the organisational and environmental conditions that apply during its implementation. However, strategy practitioners typically point to three key issues that are important in achieving successful implementation:

- *Clarity of strategic decisions* – There is a strong relationship between the formulation stage and the implementation stage of operations strategy. The crucial attribute of the formulation stage is clarity. If a strategy is ambiguous it is difficult to translate strategic intent into specific actions. With clarity, however, it should be easier to define the intent behind the strategy, the few important issues that need to be developed to deliver the intent, the way that projects should be led and resourced, who will be responsible for each task, and so on.
- *Motivational leadership* – Leadership that motivates, encourages and provides support is a huge advantage in dealing with the complexity of implementation. Leadership is needed to bring sense and meaning to strategic aspirations, maintain a sense of purpose over the implementation period and, when necessary, modify the implementation plan in the light of experience.
- *Project management* – Implementation means breaking a complex plan into a set of relatively distinct activities. Fortunately, there is a well-understood collection of ideas of how to do this. It is called 'project management' and a whole chapter is devoted to this subject (Chapter 19).

OPERATIONS IN PRACTICE

Sustainability is high on Google's operations agenda[18]

Large tech companies are not strangers to controversy. Their international spread, often complex tax arrangements, use of customer data, susceptibility to being used to promote questionable (or even illegal) content, and their sheer size and scope, put them at the centre of political and ethical debates. Not surprising then, that big tech businesses are very careful to place corporate social responsibility (CSR) at the heart of their operations strategies. Take Google for example, a company that has successfully integrated CSR into its strategies. At a time when the reputation of tech companies was at best stable, or declining, Google was positioned as having the best reputation in the world for corporate responsibility, according to the Reputation Institute.[19] Its study scored the corporate responsibility of companies according to their commitment to being a fair employer in the workplace, their citizenship role in society, and their ability to meet the financial obligations to their shareholders.

Google's commitment to energy saving is long-standing. Google is the world's largest corporate purchaser of clean energy and intends to run on carbon-free energy everywhere, at all times by 2030. Its design of 'living buildings' also contributes to its sustainability efforts. Living buildings (a term popularised by the International Living Future Institute, ILFI) are buildings that are regenerative – for example generating more energy than they use, harvesting and treating water on site, diverting waste from landfill and reusing materials. However, the CSR-related activities at Google (and its owner, Alphabet) range far wider than energy consumption. Creating sustainable workplaces, it says, is good both for people who work in them and for the environment. The company is well-known for its innovative design of office space (see Chapter 7). Google also stress that it sees its responsibilities as going beyond the boundaries of the firm and including its supply chain. It aspires, it says, to create a supply chain model for its more than 1,000 suppliers worldwide that creates a safer, fairer and equitable basis of supply. But maintaining high ethical standards in such a large company with such an extensive range of services is no easy task. One of the ways that Google puts its values into practice is through the 'Google Code of Conduct'.[20] This is a document that sets out the high standards of ethical business conduct that Google employees and Board members are expected to follow (failure to do so can mean disciplinary action).

Operations strategy monitoring

Especially in times when things are changing rapidly, as is the case during strategic change, organisations often want to track ongoing performance to make sure that the changes are proceeding as planned. Monitoring should be capable of providing early indications (or a 'warning bell' as some call it) by diagnosing data and triggering appropriate changes in how the operations strategy is being implemented. Having created a plan for the implementation, each part of it has to be monitored to ensure that planned activities are indeed happening. Any deviation from what should be happening (that is, its plans) can then be rectified through some kind of intervention in the operation.

Operations strategy control

Strategic control involves the evaluation of the results from monitoring the implementation. Activities, plans and performance are assessed with the intention of correcting future action if that is required. In some ways, this strategic view of control is similar to how it works operationally (which is discussed in Chapter 10), but there are differences. At a strategic level, control can be difficult because strategic objectives are not always clear and unambiguous. Ask any experienced manager and they will acknowledge that it is not always possible to articulate every aspect of a strategic decision in detail. Many strategies are just too complex for that. So, rather than adhering dogmatically to a predetermined plan, it may be better to adapt as circumstances change. And, the more uncertain the environment, the more an operation needs to emphasise this form of strategic flexibility and develop its ability to learn from events.

Summary answers to key questions

3.1 What is strategy and what is operations strategy?

- Strategy is the total pattern of decisions and actions that position the organisation in its environment and that are intended to achieve its long-term goals.
- Operations strategy concerns the pattern of strategic decisions and actions that set the role, objectives and activities of the operation. It can be used to articulate a vision for the (potential) contribution of operations to organisational success (i.e. moving from stage 1 to stage 4 of the Hayes and Wheelwright model of operations contribution).
- Operations strategy has content and process. The content concerns the specific decisions that are taken to achieve specific objectives. The process is the procedure that is used within a business to formulate its strategy.
- There are four key perspectives on operations strategy – the 'top-down', market requirements ('outside-in'), 'bottom-up' and operations resources ('inside-out').
- It is important to engage with a wide range of stakeholders, within and outside of the organisation, when developing operations strategy.

3.2 How does operations strategy align with business strategy (top-down)?

- The 'top-down' perspective views strategic decisions at a number of levels. Corporate strategy sets the objectives for the different businesses that make up a group of businesses. Business strategy sets the objectives for each individual business and how it positions itself in its marketplace. Functional strategies set the objectives for each function's contribution to its business strategy.
- It is important to consider *correspondence* between these different levels of strategy and *coherence* both with other functional strategies and within itself.
- The concepts of the 'business model' and 'operating model' are useful in understanding the top-down perspective on operations strategy.

3.3 How does operations strategy align with market requirements (outside-in)?

- A 'market requirements' (outside-in) perspective of operations strategy sees the main role of operations as satisfying markets. From this perspective, operations performance objectives and operations decisions should be primarily influenced by a combination of customers' needs and competitors' actions.
- Market requirements are influenced by product/service differentiation and the stage at which a product or service is within its lifecycle.

3.4 How does operations strategy align with operational experience (bottom-up)?

- The 'bottom-up' view of operations strategy emphasises the 'emergent' view of strategy development based on day-to-day operational experience. While the 'top-down' perspective may describe how operations strategy (and other function strategies) *should* be developed, it often doesn't describe how it *is* developed.
- Top-down and bottom-up perspectives are in fact complementary. In one direction, top-down perspectives can be used to judge the extent to which operational day-to-day activities reflect the higher-level strategies. In the other direction, experience gained from day-to-day activities can be accumulated and built into capabilities that can then be exploited strategically.

3.5 How does operations strategy align with operations resources (inside-out)?

- The 'operations resource' perspective (inside-out) of operations strategy is based on the resource-based view (RBV) of the firm and sees the operation's core competences (or capabilities) as being the main influence on operations strategy.
- An operations resource perspective should start by understanding existing capabilities and constraints within the operation.
- Identifying strategic decision areas can help support capability building for operations and their extended supply networks.
- Strategic resources (also called capabilities or competences) are critical in generating sustainable competitive advantage. These resources are valuable, rare, costly to imitate and organised in a way to allow the organisation to capture their value.

3.6 How are the four perspectives of operations strategy reconciled?

- Combined, the four perspectives give a good idea of how the content of operations strategy is developed and how operations excellence can act as a key source of competitive advantage. Among the models to support this activity are the operations strategy matrix, the 'line of fit' and the importance–performance matrix.

3.7 How can the process of operations strategy be organised?

- Formulating operations strategy is often called 'the process' of operations strategy and is made up of four stages – formulation, implementation, monitoring and control.
- Formulation is the process of clarifying the various objectives and decisions that make up the strategy, and the links between them. Implementation is the way that strategy is operationalised. Monitoring involves tracking ongoing performance and diagnosing data to make sure that the changes are proceeding as planned and providing early indications of any deviation from plan. Control involves the evaluation of the results from monitoring so that activities, plans and performance can be assessed with the intention of correcting future action if required.

CASE STUDY McDonald's: half a century of growth[21]

It's loved and it's hated. To some people, it is a shining example of how good-value food can be brought to a mass market. To others, it is a symbol of everything that is wrong with 'industrialised', capitalist, bland, high-calorie and environmentally unfriendly commercialism. To some, it is the best-known and most loved fast-food brand in the world with more than 39,000 restaurants, feeding 69 million customers per day (yes, per day!) or nearly 1 per cent of the entire world's population. But others consider it part of the homogenisation of individual national cultures, filling the world with Americanised, bland, identical, 'cookie cutter', Americanised and soulless operations that dehumanise its staff by forcing them to follow rigid and over-defined procedures. But whether you see it as friend, foe, or a bit of both, McDonald's has revolutionised the food industry, affecting the lives both of the people who produce food and of the people who eat it. It has also had its ups (mainly) and downs (occasionally) as markets, customers and economic circumstances have changed. Yet, even in the toughest times it has always displayed remarkable resilience. What follows is a brief (for such a large corporation) summary of its history.

Starting small

Central to the development of McDonald's is Ray Kroc, who by 1954 and at the age of 52 had been variously a piano player, a paper cup salesman and a multi-mixer salesman. He was surprised by a big order for eight multi-mixers from a restaurant in San Bernardino, California. When he visited the customer, he found a small but successful restaurant run by two brothers Dick and Mac McDonald. They had opened their 'Bar-B-Que' restaurant 14 years earlier, and by the time Ray Kroc visited the brothers' operation it had a self-service drive-in format with a limited menu of nine items. He was amazed by the effectiveness of their operation. Focusing on a limited menu including burgers, fries and beverages had allowed them to analyse every step of the process of producing and serving their food. Ray Kroc was so impressed that he persuaded the brothers to adopt his vision of creating McDonald's restaurants all over the United States, the first of which opened in Des Plaines, Illinois in June 1955. However, later, Kroc and the McDonald brothers quarrelled, and Kroc bought the brothers out. Now with exclusive rights to the McDonald's name, the restaurants spread, and in five years there were 200 restaurants through the United States. Yet through this, and later, expansions, Kroc insisted on maintaining the same principles that he had seen in the original operation. *'If I had a brick for every time I've repeated the phrase Quality, Service, Cleanliness and Value, I think I'd probably be able to bridge the Atlantic Ocean with them'* (Ray Kroc).

Priority to the process

Ray Kroc had been attracted by the cleanliness, simplicity, efficiency and profitability of the McDonald brothers' operation. They had stripped fast-food delivery down to its essence and eliminated needless effort, to make a swift assembly line for a meal at reasonable prices. Kroc wanted to build a process that would become famous for food of consistently high quality using uniform methods of preparation. His burgers, buns, fries and beverages should taste just the same in Alaska as they did in Alabama. The answer was the 'Speedee Service System'; a standardised process that prescribed exact preparation methods, specially designed equipment and strict product specifications. The emphasis on process standardisation meant that customers could be assured of identical levels of food and service quality every time they visited any store, anywhere. Operating procedures were specified in minute detail. Its first operations manual prescribed rigorous cooking instructions such as temperatures, cooking times and portions. Similarly, operating procedures were defined to ensure the required customer experience: for example, no food items were to be held more than 10 minutes in the transfer bin between being cooked and being served. Technology was also automated. Specially designed equipment helped to guarantee consistency using 'foolproof' devices. For example, the ketchup was dispensed through a metered pump. Specially designed 'clam shell' grills cooked both sides of each meat patty simultaneously for a pre-set time. And when it became clear that the metal tongs used by staff to fill French-fry containers were awkward to use efficiently, McDonald's engineers devised a simple aluminium scoop that made the job faster and easier.

For Kroc, the operating process was both his passion and the company's central philosophy. It was also the foundation of learning and improvement. The company's almost compulsive focus on process detail was not an end in itself. Rather it was to learn what contributed to consistent high-quality service in practice and what did not. McDonald's always saw learning as important. It founded 'Hamburger University', initially in the basement of a restaurant in Elk Grove Village, Illinois. It had a research and development laboratory to develop new cooking, freezing, storing and serving methods. Also, franchisees and operators were trained in the analytical techniques necessary to run a successful McDonald's. It awarded degrees in 'Hamburgerology'. But learning was not just for headquarters. The company also formed a 'field service' unit to appraise and help its restaurants by sending field service consultants to review their performance on a number of 'dimensions' including cleanliness, queuing, food quality and customer service. As Ray Kroc said, *'We take the hamburger business more seriously than anyone else. What sets McDonald's apart is the passion that we and our suppliers share around producing and delivering the highest-quality beef patties. Rigorous food safety and quality standards and practices are in place and executed at the highest levels every day'*.

No story illustrates the company's philosophy of learning and improvement better than its adoption of frozen fries. French-fried potatoes had always been important. Initially, the company tried observing the temperature levels and cooking methods that produced the best fries. The problem was that the temperature during the cooking process was very much influenced by the temperature of the potatoes when they were placed into the cooking vat. So, unless the temperature of the potatoes before they were cooked was also controlled (not very practical), it was difficult to specify the exact time and temperature that would produce perfect fries. But McDonald's researchers discovered that, irrespective of the temperature of the raw potatoes, fries were always at their best when the oil temperature in the cooking vat increased by three degrees above the low temperature point after they were put in the vat. So, by monitoring the temperature of the vat, perfect fries could be produced every time. But that was not the end of the story. The ideal potato for fries was the Idaho Russet, which was seasonal and not available in the summer months. At other times an alternative (inferior) potato was used. One grower, who, at the time, supplied a fifth of McDonald's potatoes, suggested that he could put Idaho Russets into cold storage for supplying during the summer period. Unfortunately, all the stored potatoes rotted. Not to be beaten, he offered another suggestion. Why don't McDonald's consider switching to frozen potatoes? But the company was initially cautious about meddling with such an important menu item. However, there were other advantages in using frozen potatoes. Supplying fresh potatoes in perfect condition to McDonald's rapidly expanding chain was increasingly difficult. Frozen potatoes could actually increase the quality of the company's fries if a method of cooking them satisfactorily could be found. Once again McDonald's developers came to the rescue. They developed a method of air-drying the raw fries, quick frying and then freezing them. The supplier, who was a relatively small and local suppler when he first suggested storing Idaho Russets, grew their business to supply around half of McDonald's US business.

Throughout its rapid expansion McDonald's focused on four areas – improving the product, establishing strong supplier relationships, creating (largely customised) equipment and developing franchise holders. But it was strict control of the menu that provided a platform of stability. Although its competitors offered a relatively wide variety of menu items, McDonald's limited itself to ten items. As one of McDonald's senior managers at the time stressed, *'It wasn't because we were smarter. The fact that we were selling just ten items* [and] *had a facility that was small, and used a limited number of suppliers created an ideal environment'*. Capacity growth (through additional stores) was also managed carefully. Well-utilised stores were important to franchise holders, so franchise opportunities were located only where they would not seriously undercut existing stores.

Securing supply

McDonald's says that it has been the strength of the alignment between the company, its franchisees and its suppliers (collectively referred to as the System) that has been the explanation for its success. But during the company's early years suppliers proved problematic. McDonald's approached the major food suppliers, such as Kraft and Heinz, but without much success. Large and established suppliers were reluctant to conform to McDonald's requirements, preferring to focus on retail sales. It was the relatively small companies that were willing to risk supplying what seemed then to be a risky venture. And as McDonald's grew, so did its suppliers, who also valued the company's less adversarial relationship. One supplier is quoted as saying, *'Other chains would walk away from you for half a cent. McDonald's was more concerned with getting quality. McDonald's always treated me with respect even when they became much bigger and didn't have to'*. Furthermore, suppliers were always seen as a source of innovation. For example, one of McDonald's meat suppliers, Keystone Foods, developed a novel quick-freezing process that captured the fresh taste and texture of beef patties. This meant that every patty could retain its consistent quality until it hit the grill. Keystone shared its technology with other McDonald's meat suppliers and today the process is an industry standard. Yet, supplier relationships were also rigorously controlled. For example, McDonald's routinely analysed its suppliers' products.

Fostering franchisees

McDonald's revenues consisted of sales by company-operated restaurants and fees from restaurants operated by franchisees. McDonald's views itself primarily as a franchisor and believes franchising is *'important to delivering great, locally – relevant customer experiences and driving profitability'*. However, it also believes that directly operating restaurants is essential to providing the company with real operations experience. Approximately 80 per cent were operated by franchisees. But where some restaurant chains concentrated on recruiting franchisees that were then left to themselves, McDonald's expected its franchisees to contribute their experiences for the benefit of all. Ray Kroc's original concept was that franchisees would make money before the company did, so he made sure that the revenues that went to McDonald's came from the success of the restaurants themselves rather from initial franchise fees.

Initiating innovation

Ideas for new menu items have often come from franchisees. For example, Lou Groen, a Cincinnati franchise holder, had noticed that in Lent (a 40-day period when some Christians fast and give up eating specific meats or choose to eat certain fish) some customers avoided the traditional hamburger. He went to Ray Kroc with his idea for a 'Filet-o-Fish': a steamed bun with a shot of tartar sauce, a fish fillet and cheese on the bottom bun. But Kroc wanted to push his own meatless sandwich, called the hula burger: a cold bun with a piece of pineapple and cheese. Groen and Kroc competed on a Lenten Friday to see whose sandwich would sell more. Kroc's hula burger failed, selling only six sandwiches all day, while Groen sold 350 Filet-o-Fish. Similarly, the Egg McMuffin was introduced by franchisee Herb Peterson, who wanted to attract customers into his McDonald's stores all through the day, not just at lunch and dinner. He came up with idea for the signature McDonald's breakfast item because he was reputedly *'very partial to eggs Benedict and wanted to create something similar'*.

Other innovations came from the company itself. When poultry became popular, Fred Turner, then the Chairman of McDonald's, had an idea for a new meal: a chicken finger-food without bones, about the size of a thumb. After six months of research, the food technicians and scientists managed to reconstitute shreds of white chicken meat into small portions that could be breaded, fried and frozen, then reheated. Test-marketing the new product was positive, and in 1983 they were launched under the name Chicken McNuggets. These were so successful that within a month McDonald's became the second largest purchaser of chicken in the United States. Some innovations came as a reaction to market conditions. Criticised by nutritionists who worried about calorie-rich burgers and shareholders who were alarmed by flattening sales, McDonald's launched its biggest menu revolution in 30 years in 2003 when it entered the prepared salad market. It offered a choice of dressings for its grilled chicken salad with Caesar dressing (and croutons) or the lighter option of a drizzle of balsamic dressing. Likewise, moves towards coffee sales were prompted by the ever-growing trend set by big coffee shops like Starbucks.

Problematic periods

Food, like almost everything else, is subject to swings in fashion. Not surprising then, that there have been periods when McDonald's has had to adapt. The period from the early 1990s to the mid-2000s was difficult for parts of the McDonald's empire. Growth in some parts of the world stalled. Partly this was due to changes in food fashion, nutritional concerns and demographic changes. Partly it was because competitors were learning to either emulate McDonald's operating system, or focus on one aspect of the traditional 'quick service' offering, such as speed of service, range of menu items, (perceived) quality of food, or price. Burger King promoted itself on its 'flame-grilled' quality. Wendy's offered a fuller service level. Taco Bell undercut McDonald's prices with its 'value pricing' promotions. Drive-through specialists sped up service times. Also, 'fast food' was developing a poor reputation in some quarters and, as its iconic brand, McDonald's was taking much of the heat. Similarly, the company became a lightning rod for other questionable aspects of modern life that it was held to promote, from cultural imperialism, low-skilled jobs (called 'McJobs' by some critics), abuse of animals and the use of hormone-enhanced beef, to an attack on traditional (French) values (in France). A French farmer called Jose Bové (who was briefly imprisoned) got other farmers to drive their tractors through, and wreck, a half-built McDonald's.

Similarly, in 2015 McDonald's closed more stores in its US home market than it opened – for the first time in its 60-year history. Partly this was a result of the increase in so-called 'fast casual' dining, a trend that combined the convenience of traditional McDonald's-style service with food that was seen as healthier, even if it was more expensive. Smaller rivals, such as Chipotle and Shake Shack, had started to take domestic market share.

Surviving strategies

Over recent years the company's strategy has been to become 'better, not just bigger', focusing on 'restaurant execution', with the goal of 'improving the overall experience for our customers'. In particular it has, according to some analysts, 'gone back to basics', a strategy used by McDonald's then Chief Executive Officer, Steve Easterbrook, when he was head of the company's British operation, where he redesigned the outlets to make them more modern, introduced coffee and cappuccinos, worked with farmers to raise standards and increased transparency about the supply chain. At the same time, he participated fully and forcefully with the company's critics in the debate over fast-food health concerns. But some analysts believe that the 'burger and fries' market is in terminal decline, and the McDonald's brand is so closely associated with that market that further growth will be difficult.

QUESTIONS

1. **How has competition to McDonald's changed over its existence?**
2. **What are the main operations performance objectives for McDonald's?**
3. **What are the most important structural and infrastructural decisions in McDonald's operations strategy, and how do they influence its main performance objectives?**

Problems and applications

All chapters have 'Problems and applications' questions that will help you practise analysing operations. They can be answered by reading the chapter. Model answers for the first two questions can be found on the companion website for this text.

1 ZNR Financial, a large accountancy corporation, is looking to assess the operations functions in three of its locations around the world. The ZNR Malaysia operations is marginally better than the operations of many of its competitors in the region, but still behind the very best players. The function is also viewed positively by other functions in the organisation and its 'voice is heard' when it comes to strategy conversations. Arguably, ZNR Japan operations continues to provide the basis on which ZNR Japan competes – it recently developed advanced AI software to enable the company to access new larger corporate clients who, in addition to basic accountancy services, value the customer intelligence that working with ZNR Japan can offer them. ZNR Hong Kong operations is now clearly better than most of its competitors and has an active voice in the strategic direction of the firm. Recently, the operations team worked closely with marketing to respond to a key client's request to develop more automated processing of high-volume, low-variety work on its behalf. The initiative has proved successful, so marketing is becoming increasingly keen to build on this internal 'win–win' relationship. Where would you position the three ZNR operations functions on the Hayes and Wheelwright model of operations contribution?

2 Giordano is one of the most widespread clothes retailers. It is based in Hong Kong and employs more than 8,000 staff in over 2,000 shops. But when it was founded, upmarket shops sold high-quality products and gave good service. Cheaper clothes were piled high and sold by sales assistants more concerned with taking the cash than smiling at customers. The company questioned why they could not offer value and service, together with low prices. To do this, they raised the wages of their salespeople by over 30 per cent, and gave all employees 60 hours of training. New staff were allocated a 'big brother' or 'big sister' from experienced staff to help them develop their service quality skills. Even more startling by the standards of their competitors, they brought in a 'no-questions asked' exchange policy irrespective of how long ago the garment had been purchased. Staff were trained to talk to customers and seek their opinion on products and the type of service they would like. This was fed back to the company's designers for incorporation into their new products. Their operating principles were summarised in its 'QKISS' list: Quality (do things right); Knowledge (keep experience up-to-date and share knowledge); Innovation (think 'outside of the box'); Simplicity (less is more); Service (exceed customers' expectations).

(a) In what way did an appreciation of competitors affect the market position of the Giordano operation?
(b) What are the advantages of sales staff talking to the customers?

3 Carry out an importance–performance analysis for an amusement park. In doing this, think about the competitive factors (i.e. the key ingredients) for this offering, their level of importance and their performance using the scale shown in Figure 3.14. Then map these onto an importance–performance matrix, as shown in Figure 3.15.

4 The Managing Partner of The Branding Partnership (TBP) was describing her business. *'It is about four years now since we specialised in the small to medium firms market. Before that we also used to provide brand consultancy services for anyone who walked in the door. However, within the firm, I think we could focus our activities even more. There seem to be two types of assignment that we are given. About 40 per cent is relatively routine. Typically, these assignments are conventional market research and focus group exercises. These activities involve a relatively standard set of steps that can be carried out by relatively junior staff. Of course, an experienced consultant is needed to make some decisions. Customers expect us to be relatively inexpensive and fast in delivering the service. Nor do they expect us to make simple errors; if we did this too often we would lose business. Fortunately,*

our customers know that they are buying a "standard package". However, specialist agencies have been emerging over the last few years and they are undercutting us on price. Yet I still feel that we can operate profitably in this market. The other 60 per cent of our work is for clients who require more specialist services, such as assignments involving major brand reshaping. These assignments are complex, large, take longer and require significant branding skill and judgement. It is vital that clients respect and trust the advice we give them in all "brand associated" areas such as product development, promotion, pricing and so on. Of course, they assume that we will not be slow or unreliable, but mainly it's trust in our judgement backed up by hard statistics that is important to the client. This is popular work with our staff. It is both interesting and very profitable'.

(a) How different are the two types of business described?

(b) It has been proposed that the firm is split into two separate businesses: one to deal with routine services and the other to deal with more complex services. What would be the advantages and disadvantages of doing this?

5 DSD designs, makes and supplies medical equipment to hospitals and clinics. Its success was based on its research and development culture. Around 50 per cent of manufacturing was done in-house. Its products were relatively highly priced, but customers were willing to pay for its technical excellence and willingness to customise equipment. Around 70 per cent of all orders involved some form of customisation from standard 'base models'. Manufacturing could take three months from receiving the specification to completing assembly, but customers were more interested in equipment being delivered on time rather than immediate availability. According to its CEO, *'manufacturing is really a large laboratory. The laboratory-like culture helps us to maintain our superiority in leading-edge product technology and customisation. It also means that we can call upon our technicians to pull out all the stops in order to maintain delivery promises. However, I'm not sure how manufacturing, or indeed the rest of the company, will deal with the new markets and products that we are getting into'*. The new products were 'small black box' products that the company had developed. These were devices that could be attached to patients, or implanted. They took advantage of sophisticated electronics and could be promoted directly to consumers as well as to hospitals and clinics. The CEO knew their significance. *'Although expensive, we have to persuade healthcare and insurance companies to encourage these new devices. More problematic is our ability to cope with these new products and new markets. We are moving towards being a consumer company, making and delivering a higher volume of more standardised products where the underlying technology is changing fast. We must become faster in our product development. Also, for the first time, we need some kind of logistics capability. I'm not sure whether we should deliver products ourselves or subcontract this. Manufacturing faces a similar dilemma. On one hand, it is important to maintain control over production to ensure high quality and reliability; on the other hand, investing in the process technology to make the products will be very expensive. There are subcontractors who could manufacture the products, they have experience in this kind of manufacturing but not in maintaining the levels of quality we will require. We will also have to develop a "demand fulfilment" capability to deliver products at short notice. It is unlikely that customers would be willing to wait the three months our current customers tolerate. Nor are we sure of how demand might grow. I'm confident that growth will be fast but we will have to have sufficient capacity in place not to disappoint our new customers. We must develop a clear understanding of the new capabilities that we will have to develop if we are to take advantage of this wonderful market opportunity'*. What advice would you give DSD? Consider the operational implications of entering this new market.

6 During manoeuvres in the Alps, a detachment of soldiers got lost. The weather was severe and the snow was deep. In these freezing conditions, after two days of wandering, the soldiers gave up hope and became reconciled to a frozen death on the mountains. Then, to their delight, one of the soldiers discovered a map in his pocket. Much cheered by this discovery, the soldiers were able to escape from the mountains. When they were safe back at their headquarters, they discovered that the map was not of the Alps at all, but of the Pyrenees. What is the relevance of this story to operations strategy?

7 Greenwashing is a derogatory term used to indicate that a business is exaggerating its environmental activities, or even deliberately conveying a false impression about how its activities are environmentally sound. Why might large technology companies be particularly vulnerable to this type of accusation?

8 Why can operations strategy never be exclusively concerned with 'strategic level' decisions?

9 Why might the 'order winner', 'qualifier', 'less important' classification underemphasise the importance of innovation?

10 Re-read the 'Operations in practice' example on Ocado. Why do you think has the company moved to sell its technology to other retailers?

Selected further reading

Braithwaite, A. and Christopher, M. (2015) ***Business Operations Models: Becoming a Disruptive Competitor,*** **Kogan Page, London.**
Aimed at practitioners, but authoritative and interesting.

Hayes, R.H., Pisano, G.P., Upton, D.M. and Wheelwright, S.C. (2005) ***Pursuing the Competitive Edge,*** **Wiley, Hoboken, NJ.**
The gospel according to the Harvard school of operations strategy. Articulate, interesting and informative.

Slack, N. (2017) ***The Operations Advantage,*** **Kogan Page, London.**
Apologies for self-referencing again! This book is written specifically for practitioners wanting to improve their own operations – short and to the point.

Slack, N. and Lewis, M. (2020) ***Operations Strategy,*** **6th edn, Pearson Education, Harlow.**
A book that takes a really deep dive into all aspects of operations strategy, with lots of cases and practical guidance.

Notes on chapter

1. For a more thorough explanation, see Slack, N. and Lewis, M. (2020) *Operations Strategy*, 6th edn, Pearson, Harlow.
2. The information on which this example is based was taken from: Braithwaite, T. (2020) How a UK supermarket nourished Silicon Valley's critics, *Financial Times*, 6 November; Chambers, S. (2019) Ocado the disruptor is being disrupted, *The Sunday Times*, 1 December.
3. Hayes, R.H. and Wheelwright, S.C. (1984) *Restoring our Competitive Edge: Competing Through Manufacturing*, John Wiley & Sons, Inc., New York, NY.
4. For a more thorough explanation, see Slack, N. and Lewis, M. (2020) *Operations Strategy*, 6th edn, Pearson, Harlow.
5. Alex Osterwalder (n.d.) What is a business model? https://www.strategyzer.com/expertise/business-models (accessed September 2021).
6. Based on the definitions developed by Capgemini.
7. The Micraytech examples have had names and some details changed to preserve commercial confidentiality.
8. Based on an example from Slack, N. (2017) *The Operations Advantage*, Kogan Page, London. Used by permission.
9. The information on which this example is based is taken from: Vandevelde, M. (2016) Tesco ditches global ambitions with retreat to UK, *Financial Times*, 21 June; Clark, A. and Ralph, A. (2014) Tesco boss defiant amid 4% plunge in sales, *The Times*, 5 June.
10. Mintzberg, H. and Waters, J.A. (1985) Of strategies: deliberate and emergent, *Strategic Management Journal*, 6, July/Sept, 257–72.
11. For a full explanation of this concept see Slack, N. and Lewis, M. (2020) *Operations Strategy*, 6th edn, Pearson, Harlow.
12. An idea proposed by Jay Barney. See Barney, J.B. (2001), Is the resource-based 'view' a useful perspective for strategic management research? Yes, *Academy of Management Review*, 26 (1), 41–56.
13. There are many economic publications making this point. Among the most accessible is Haskel, J. and Westlake, S. (2018) *Capitalism without Capital: The Rise of the Intangible Economy*, Princeton University Press, Princeton, NJ.

14. Again, from Haskel, J. and Westlake, S. op. cit.
15. Barney, J. (1991) The resource-based model of the firm: origins, implications and prospect, *Journal of Management*, 17 (1), 97–8.
16. Carroll, A.B. (1991) The pyramid of social responsibility: toward the moral management of organizational stakeholders, *Business Horizons*, 34 (4) July/August, 39–48.
17. Porter, M.E. and Kramer, M. (2002) The competitive advantage of corporate philanthropy, *Harvard Business Review*, 80 (12) 5–16.
18. The Google corporate website contains many examples of their commitment to sustainability. See https://sustainability.google/reports/ (accessed September 2021).
19. Czarnecki, S. (2018) Study: Google has the best reputation for corporate responsibility in the world, *PRWeek*, 11 October, https://www.prweek.com/article/1495753/study-google-best-reputation-corporate-responsibility-world (accessed August 2021).
20. See https://abc.xyz/investor/other/google-code-of-conduct/ (accessed September 2021).
21. The information on which this case study is based is taken from: Whipp, L. (2015) McDonald's to slim down in home market, *Financial Times,* 18 June: Smith, T. (2015) Where's the beef, *Financial Times,* 22 May; Whipp, L. (2015) McDonald's may struggle to replicate British success, *Financial Times,* 5 May; McDonald's Annual Report, 2017; Kroc, R.A. (1977) *Grinding it Out: The Making of McDonald's,* St. Martin's Press, New York; Cooper, L. (2015) At McDonald's the burgers have been left on the griddle too long, *The Times,* 24 August.

4 Managing product and service innovation

KEY QUESTIONS

4.1 What is product and service innovation?

4.2 What is the strategic role of product and service innovation?

4.3 What are the stages of product and service innovation?

4.4 How should product and service innovation be resourced?

INTRODUCTION

Customers value innovation. Companies such as Google, Amazon, Netflix, Nike, Airbnb, Apple and Dropbox have all been successful because they challenged the idea of what their markets wanted. Their products and services have been continually updated, altered and modified. Some changes are small, incremental adaptations to existing ways of doing things. Others are radical, major departures from anything that has gone before. The innovation activity is about successfully delivering change in its many different forms. Being good at innovation has always been important and it is also increasingly complex, with inputs from a wide variety of external sources. Organisationally, operations managers may not always have full responsibility for product and service innovation, yet they are always involved in some way and increasingly they are expected to take a greater and more active part. Unless a product, however well designed, can be produced to a high standard, and unless a service, however well conceived, can be implemented, they will never generate full benefits. In this chapter, we examine what is meant by product and service innovation; the strategic role of innovation; the key stages in the innovation process; and aspects of resources that must be considered to support innovation. Figure 4.1 shows where this chapter fits into the overall operations model.

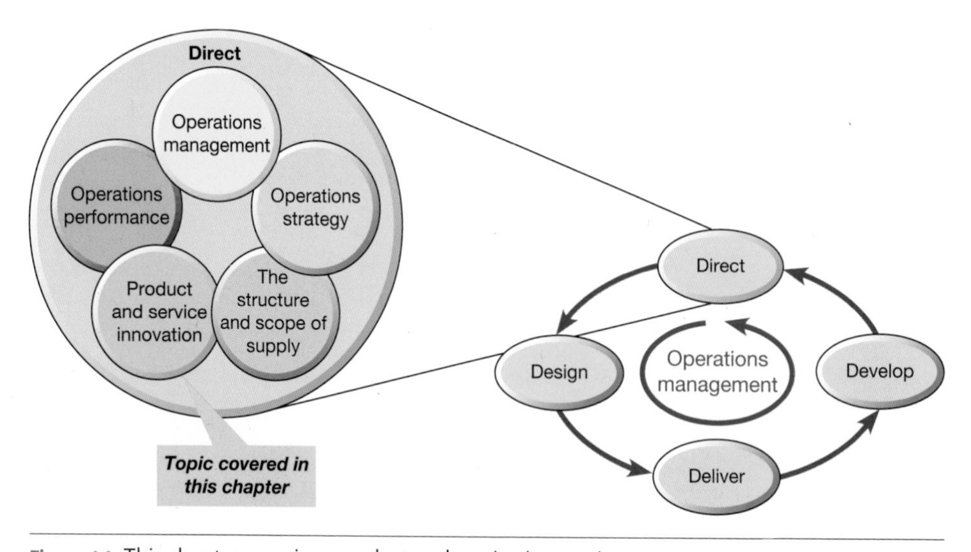

Figure 4.1 This chapter examines product and service innovation

4.1 What is product and service innovation?

There are a number of terms that we shall use in this chapter that have similar meanings, and are defined by different authorities in different ways, or overlap to some extent, and yet are related to each other. Specifically, we explore three related terms of creativity, innovation and design.

Creativity, innovation and design

The study of product and service innovation, what influences it and how to manage it, is a huge subject. However, a repeating theme in most innovation research is 'creativity'. Creativity is the ability to move beyond conventional ideas, rules or assumptions, in order to generate significant new ideas. It is a vital ingredient in innovation. It is seen as essential not just in product and service innovation, but also in the design and management of operations processes more generally. Partly because of the fast-changing nature of many industries, a lack of creativity (and consequently of innovation) is seen as a major risk.

'Innovation' is notoriously ambiguous and lacks either a single definition or measure. It is variously described as something that is new, a change that creates a new type of performance, the process of introducing something novel. The emphasis is always on novelty and change, yet innovation implies more than 'creativity' or 'invention' as it also suggests the process of transforming ideas into something that has the potential to be practical and provide a commercial return.

'Design' is the process that transforms innovative ideas into something more concrete. Innovation creates the novel idea; design makes it work in practice. Design is an activity that can be approached at different levels of detail. Figure 4.2 illustrates the relationship between creativity, innovation and design as we use the terms here. These concepts are intimately related, which is why we treat them in the same chapter. First we will look at some of the basic ideas that help to understand innovation.

Operations principle

The product and service innovation process must consider three related issues of creativity, innovation and design.

The innovation S-curve

When new ideas are introduced in services, products or processes, they rarely have an impact that increases uniformly over time. Usually performance follows an S-shaped progress, as illustrated in Figure 4.3. In the early stages of the introduction of new ideas, although often large amounts of resources, time and effort are needed to introduce the idea, relatively small performance improvements are experienced. However, with time, as experience and knowledge about the new idea grows, performance increases. As the idea becomes established, extending its performance further becomes increasingly difficult (see Figure 4.3(a)). When an idea reaches its mature, 'levelling-off' period, it is vulnerable to new ideas being introduced, which, in turn, move through their own S-shaped progression. This is how innovation works; the limits of one idea being reached prompt newer, better ideas, with each new S-curve requiring some degree of redesign (see Figure 4.3(b)).

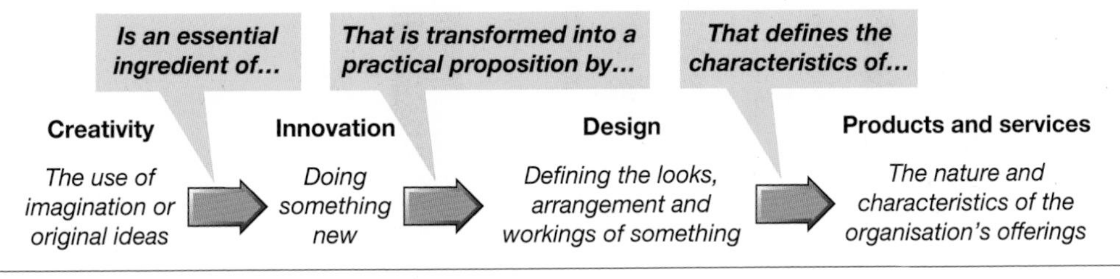

Figure 4.2 The relationship between creativity, innovation and design

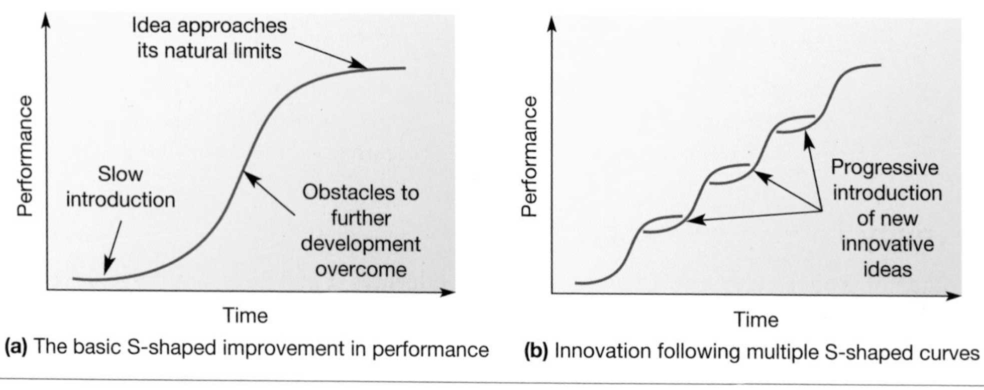

Figure 4.3 The S-shaped curve of innovation

OPERATIONS IN PRACTICE

The slow innovation progress of the zip fastener[1]

Some innovations take a long time to become successful. The zip (zipper, in the United States) is one such innovation. In 2017, the market for zip fasteners was around $11 billion, and is expected to grow to around $20 billion by 2024. This is based on the increasing global demand for clothing, luggage and other products that use zips, alongside the acceleration of fast fashion. Historically, it is a relatively recent innovation. For centuries, clothes were held together with loops and toggles, buckles, brooches, laces, or simply tied and wrapped. It was in fourteenth-century Britain that the hook and eye, what could be considered the earliest ancestor of the zip, began to be used. But they were both awkward to use and fragile. The first real zip-like device was patented in the United States by Whitcomb Judson's Universal Fastener Company of Chicago in 1893. His design specified a sliding guide to pull together a line of hooks and a line of eyes on a boot. Unfortunately, it proved rather unreliable and the company was taken over by Gideon Sundback, an engineer from Sweden. His innovation abandoned hooks and eyes and replaced them with rows of metal protuberances with a tooth on one side and a socket on the other, similar in principle to today's design. Around the same time, a Swiss inventor named Katharina Kuhn-Moos patented a similar design, but it was never manufactured. The device was still expensive compared to more conventional buttons.

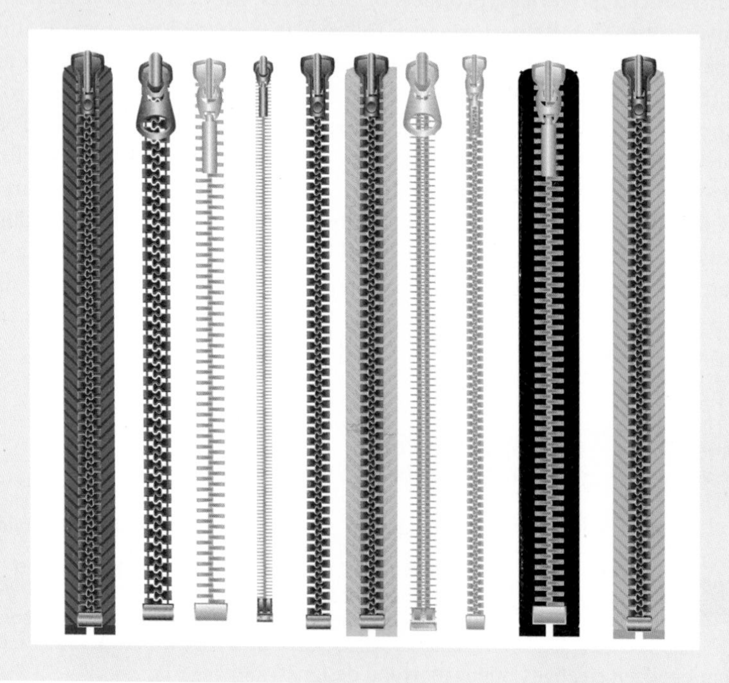

It was the Japanese company YKK that transformed the zip's prospects. Although there had been developments of the basic design, for example, plastic teeth to replace metal ones and continuous spirals of nylon used as teeth, the real breakthrough came from improving the quality in how zips were manufactured. Accuracy of manufacturing is important in how well a zip works. Very small misalignments can cause jams and breakages and, unlike losing a button, a broken zip can mean discarding a garment. Tadao Yoshida, known as 'the zipper king', had founded YKK in 1934, and had gained such a reputation for quality and reliability (guarantees that each of its zips will last for 10,000 uses) that in 1960, when Sundback's patents expired, YKK was able to move into the larger US market. It went on to gain around 40 per cent of the market, by value, and makes more zips every year than there are people on the planet.

Incremental or radical innovation

An obvious difference in how the pattern of new ideas emerges in different operations or industries is the rate and scale of innovation. Some industries, such as telecommunications, enjoy frequent and often significant innovations. Others, such as house building, do have innovations, but they are usually less dramatic. So, some innovation is radical, resulting in discontinuous, 'breakthrough' change, while other innovations are more incremental, leading to smaller, continuous changes. Radical innovation often includes large technological advancements that may require completely new knowledge and/or resources. Radical innovation is relatively rare – perhaps 5–10 per cent of all innovations could be classified as such – but creates major challenges for existing players within a market. This is because organisations are often unwilling to disrupt current modes of working in the face of a barely emerging market, but by the time the threat has emerged more fully it may be too late to respond. Clayton Christensen refers to this problem as the Innovator's Dilemma, which supports the ideas of the renowned economist Joseph Schumpeter that innovation should be a process of 'creative destruction'.[2] Incremental innovation, by contrast, is more likely to involve relatively modest technological changes and to build upon existing knowledge and/or resources, so existing services and products are not fundamentally changed. This is why established companies may favour incremental innovation: because they have the experience to have built up a significant pool of knowledge (on which incremental innovation is based). In addition, established companies are more likely to have a mindset that emphasises continuity, perhaps not even recognising potential innovative opportunities. New entrants to markets, however, have no established position to lose, nor do they have a vast pool of experience. As such, they may be more likely to try for more radical innovation.

Innovation is influenced by later stages in the value chain

At its simplest, a firm innovates in the form of a design for a product or service (or some blend of the two), produces or creates it through its core operations, and distributes it to its customers, who then use or experience it. Each of these stages is a transformation process. Innovation/design transforms ideas into workable designs. Production/creation transforms the design into a form that customers will find useful. Distribution disseminates it (physically or virtually). Finally, customers gain value by using it. But each of these stages are not independent of each other. Certainly, the innovation/design process is influenced by all the subsequent stages. Most of us are used to thinking about how product or service designs are judged primarily by how they add value for customers, but both production and distribution stages can also impact the design stage, see Figure 4.4.

Design for production/creation

Decisions taken during the design of a product or service can have a profound effect on how they can be created. For physical products, this has been well understood for decades and is usually termed 'design for production' (DFP), or 'design for manufacture' (DFM). But the same principle applies equally to services. How a service is designed and specified can make its execution easy or difficult in practice. The design of queuing areas in a theme park attraction could be easy or difficult to control. Technology can help. Virtual reality can help service engineers to (virtually) 'walk round' facilities. Designers can be (virtually) inside sports venues, aeroplanes, buildings and amusement parks, and so on.

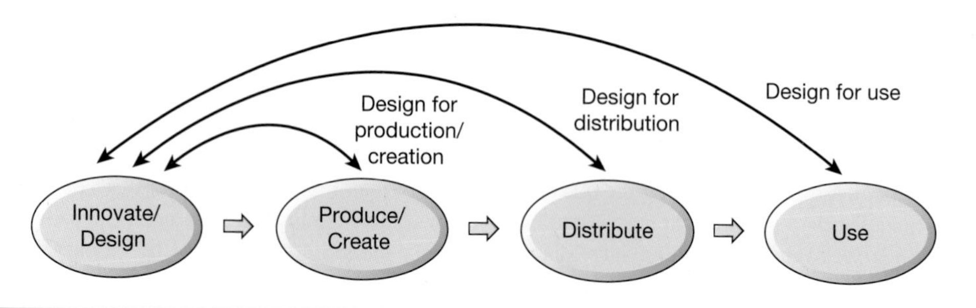

Figure 4.4 The influences on the design of a product or service are not limited to how it will be used, but include how it will be created and distributed

Design for distribution

The most obvious example of how the design of a product can be influenced by how it is distributed is 'flat pack' furniture. IKEA stores exemplify the extent to which clever 'flat pack' design can influence subsequent stages in the value chain. Designing their furniture to be sold in flat form allows for efficient transportation and efficient storage, which in turn allows customers in their stores to collect it. Similarly, some products are designed so that, when packaged, they fit conveniently onto transportation pallets, or containers. Again, the same idea applies to services. The design of online services can be influenced by how the service could be presented through its web page. Even purely 'artistic' offerings such as music are influenced by the way they are distributed. For example, the majority of revenue in the music industry comes from 'product' distributed through streaming services, where artists are paid per play. Provided, that is, that the listener plays the piece for at least 30 seconds. About one-third of all streams are played because a track has been included on a streaming company's playlist, usually selected by algorithms, the precise form of which are not always known. It is often assumed that the algorithms favour tracks that get to the 'catchy' part of the tune relatively fast. Some point out that this has led to the production of music tracks that are shorter, with truncated introductions and choruses that start sooner.[3]

4.2 What is the strategic role of product and service innovation?

Despite the obstacles to successful innovation, almost all firms strive to be innovative. The reason is that there is overwhelming evidence that innovation can generate significant payback for the organisations that manage the incorporation of innovative ideas in the design of their products and services. What matters is the ability to identify the innovations and manage their transformation into effective designs. Remember, an organisation's products and services are how markets judge it – they are its 'public face'. Effective product and service innovation processes add value to any organisation by:

- driving and operationalising innovation, increasing market share and opening up new markets;
- differentiating products and services, making them more attractive to customers, while increasing consistency in the company's range, and helping to ensure successful product launches;
- strengthening branding, so that products and services embody a company's values;
- reducing the overall costs associated with innovation, through more efficient use of resources, reduced project failure rate and faster time to market.

The process of design

The innovation activity is a process that involves many of the same design issues common to other operations processes. It conforms to the input—transformation–output model described in Chapter 1. Although organisations will have their own particular ways of managing innovation and design, the design process itself is essentially very similar across a whole range of industries. Moreover, the better

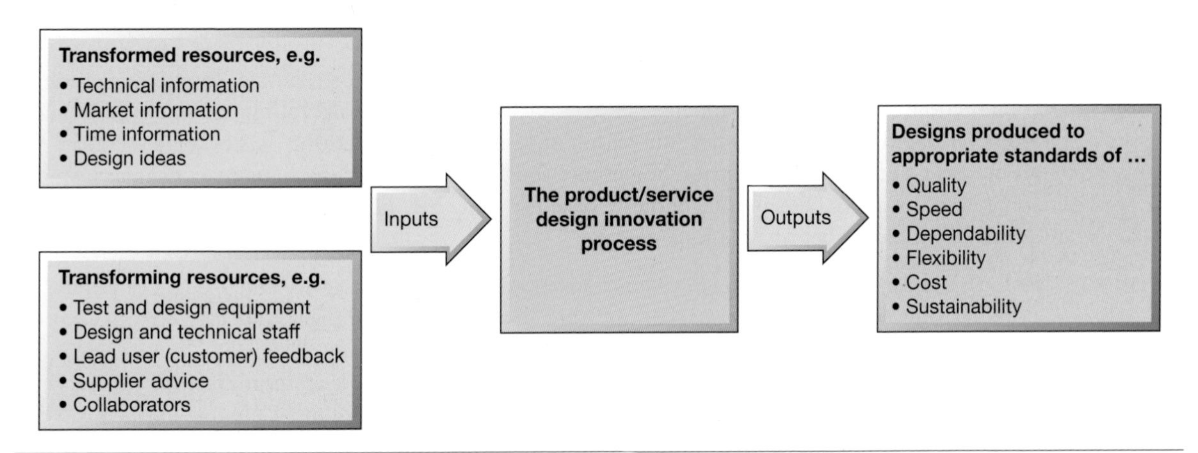

Figure 4.5 The product and service design innovation activity as a process

the process is managed, the better the service or product that is designed. Figure 4.5 illustrates the design activity as an input—transformation–output diagram. The transformed resource inputs will consist mainly of information in the form of market forecasts, market preferences, technical data and potential design ideas. It is these ideas and information that will be transformed in the design process into the final design. Transforming resources includes designs and those who manage them, specialist technical staff with the specific knowledge necessary to solve design problems, suppliers and interested customer groups if they have useful information, design technology such as simulation software, and so on.

Operations principle

The design activity is a process that can be managed using the same principles as other processes.

Performance objectives for the product and service innovation process

The performance of the design process can be assessed in much the same way as any other process, namely in terms of quality, speed, dependability, flexibility and cost. Because product and service design has such an influence on sustainability, we include it alongside our normal operational-level objectives.

What does quality mean for the innovation process?

Innovation quality is not always easy to define precisely, especially if customers are relatively satisfied with existing product and service offerings. Many software companies talk about the 'I don't know what I want, but I'll know when I see it' syndrome, meaning that only when customers use the software are they in a position to articulate what they do or don't require. Nevertheless, it is possible to distinguish high- and low-quality designs (although this is easier to do in hindsight) by judging them in terms of their ability to meet market requirements. In doing this, the distinction between the specification quality and the conformance quality of designs is important. No business would want a design process that was indifferent to 'errors' in its designs, yet some are more tolerant than others. For example, in pharmaceutical development the potential for harm is particularly high because drugs directly affect our health. This is why the authorities insist on such a prolonged and thorough 'design' process (more usually called 'development' in that industry). Although withdrawing a drug from the market is unusual, it does occasionally occur. Far more frequent are the 'product recalls' that are relatively common in, for example, the automotive industry. Many of these are design related and the result of 'conformance' failures in the design process. The 'specification' quality of design is different. It means the degree of functionality, or experience, or aesthetics, or whatever the product or service is primarily competing on.

Operations principle

Innovation processes can be judged in terms of their levels of quality, speed, dependability, flexibility, cost and sustainability.

What does speed mean for the innovation process?

The speed of design matters more to some industries than others. For example, design innovation in construction and aerospace happens at a much slower pace than in clothing or microelectronics. However, rapid product or service innovation or 'time-based competition' has become the norm for an increasing number of industries. Sometimes this is the result of fast-changing consumer fashion, sometimes a rapidly changing technology base forces it. Yet, no matter what the motivation, fast design brings a number of advantages:

- Early market launch – an ability to innovate speedily means that service and product offerings can be introduced to the market earlier and thus earn revenue for longer, and may command price premiums.
- Starting design late – fast design allows design decisions to be made closer to the time when service and product offerings are introduced to the market – important in fast-changing markets.
- Frequent market stimulation – rapid innovations allow frequent new or updated offerings to be introduced into the market.

What does dependability mean for the innovation process?

Design schedule slippage can extend design times, but worse, a lack of dependability adds to the uncertainty surrounding the innovation process. Professional project management (see Chapter 19) of the innovation process can help to reduce uncertainty and prevent (or give early warning of) missed deadlines, process bottlenecks and resource shortages. Disturbances to the innovation process may be minimised through close liaison with suppliers as well as market or environmental monitoring. Nevertheless, unexpected disruptions will always occur, especially for the most innovative product and service designs. This is why flexibility within the innovation process is one of the most important ways in which dependable delivery of new service and product offerings can be ensured.

What does flexibility mean for the innovation process?

Flexibility in the innovation process is the ability to cope with external or internal change. The most common reason for external change is that markets, or specific customers, change their requirements. Although flexibility may not be needed in relatively predictable markets, it is clearly valuable in more fast-moving and volatile markets, where one's own customers and markets change, or where the designs of competitors' offerings dictate a matching or leapfrogging move. Internal changes include the emergence of superior technical solutions. In addition, the increasing complexity and interconnectedness of service and product components in an offering may require flexibility. A bank, for example, may bundle together a number of separate services for one particular segment of its market. Privileged account holders may obtain special deposit rates, premium credit cards, insurance offers, travel facilities and so on together in the same package. Changing one aspect of this package may require changes to be made in other elements. So extending the credit card benefits to include extra travel insurance may also mean the redesign of the separate insurance element of the package.

What does cost mean for the innovation process?

The cost of innovation is usually analysed in a similar way to the ongoing cost of delivering offerings to customers. Often these cost factors are split up into three categories: the cost of buying the inputs to the process, the cost of providing the labour in the process and the other general overhead costs of running the process. In most in-house innovation processes the latter two costs outweigh the former. As we indicated earlier, delays in the delivery of an innovation can result in both more expenditure on the design and delayed (and probably reduced) revenue. The combination of these effects usually means that the financial break-even point for a new offering is delayed far more than the original delay in its launch (see the worked example 'Cyberdanss Software').

Worked example

Cyberdanss Software

'I have four of my best staff working on the CD08 project, the client is getting nervous and the costs are escalating. I can't understand why it's so far behind schedule. Soon we will become liable for late-delivery charges'. Lidiya Koval was the founder and owner of Cyberdanss, a software development firm, based in Kiev, that specialised in developing security software for larger firms that licenced products to financial service customers. The project, which had already taken four months, was due to be completed in two months' time, and was forecast to be at least two months late. The contract for this project specified that their client would pay a proportion of the licence revenue, which was estimated to be worth $50,000 per month to Cyberdanss; it also included a 'lateness charge' of $20,000 for every month the project was late. And, although the development project was initially seen by Lidiya as lucrative, even with her staff and other development costs totalling $15,000 per month, she knew that a two-month delay would push back the break-even for the development. But by how much would a delay affect the break-even time?

Analysis

If the development project had gone to plan, the income (payments from the client) would have started in month 7 and development costs finished in month 6. The resulting cash flow is shown in Table 4.1. The project would have broken even just before month 8.

The effect of a two-month delay is threefold. First, income is delayed. Second, development costs last for a further two months. Third, Cyberdanss is subject to two lateness payments. The results of this are shown in Table 4.2. The break-even point now occurs between months 11 and 12. In other words, a two-month delay has resulted in an almost four-month delay in break-even. Possibly just as significantly, the cash needed to fund the project has increased from a maximum of $90,000 to a maximum of $160,000.

Table 4.1 Income statement (in $'000) for development project CD08 (as planned)

	Month											
	1	2	3	4	5	6	7	8	9	10	11	12
Income							50	50	50	50	50	50
Development costs	15	15	15	15	15	15						
Cash position	(15)	(30)	(45)	(60)	(75)	(90)	(40)	10	60	110	160	210

Table 4.2 Income statement (in $'000) for development project CD08 (with two-month delay)

	Month											
	1	2	3	4	5	6	7	8	9	10	11	12
Income									50	50	50	50
Development costs	15	15	15	15	15	15	15	15				
Lateness penalty							20	20				
Cash position	(15)	(30)	(45)	(60)	(75)	(90)	(125)	(160)	(110)	(60)	(10)	40

Sustainability and the innovation process

The sustainability of a product or service is the extent to which it benefits the 'triple bottom line' – people, planet and profit. The design innovation process is particularly important in impacting ultimately on the ethical, environmental and economic well-being of stakeholders. Organisations increasingly consider sustainability in the design process. For example, some innovation activity is particularly focused on the ethical dimension of sustainability. Banks have moved to offer customers ethical investments that seek to maximise social benefit as well as financial returns. Such investments tend to avoid businesses involved in weaponry, gambling, alcohol and tobacco, for example, and favour those promoting worker education, environmental stewardship and consumer protection. Other examples of ethically focused innovations include the development of 'fair-trade' products. Similarly, garment manufacturers may establish ethical trading initiatives with suppliers; supermarkets may ensure animal welfare for meat and dairy products; and online companies may institute customer complaint charters. (This issue is taken further in the 'Responsible operations' feature later in this chapter.)

4.3 What are the stages of product and service innovation?

The design innovation process will generally have several stages. Even products and services that are normally considered purely 'artistic', with a large element of creativity, such as movies or theatrical productions, are actually progressed through clearly defined stages. For example, the process of developing video games follows three broad phases, pre-production, production and post-production, each of which also follows a sequence of steps, as shown in Figure 4.6. Like any innovation and design process, the video game design process will involve some blurring of the boundaries between stages, and often significant recycling and rework (see the case study at the end of this chapter). Furthermore, all such processes tend to move from a vaguely defined idea that is refined and made progressively more detailed until it contains sufficient information for turning into an actual service, product or process. The final design will not be evident until the very end of the process. Yet, many of the decisions that affect the eventual cost of delivery are made relatively early.

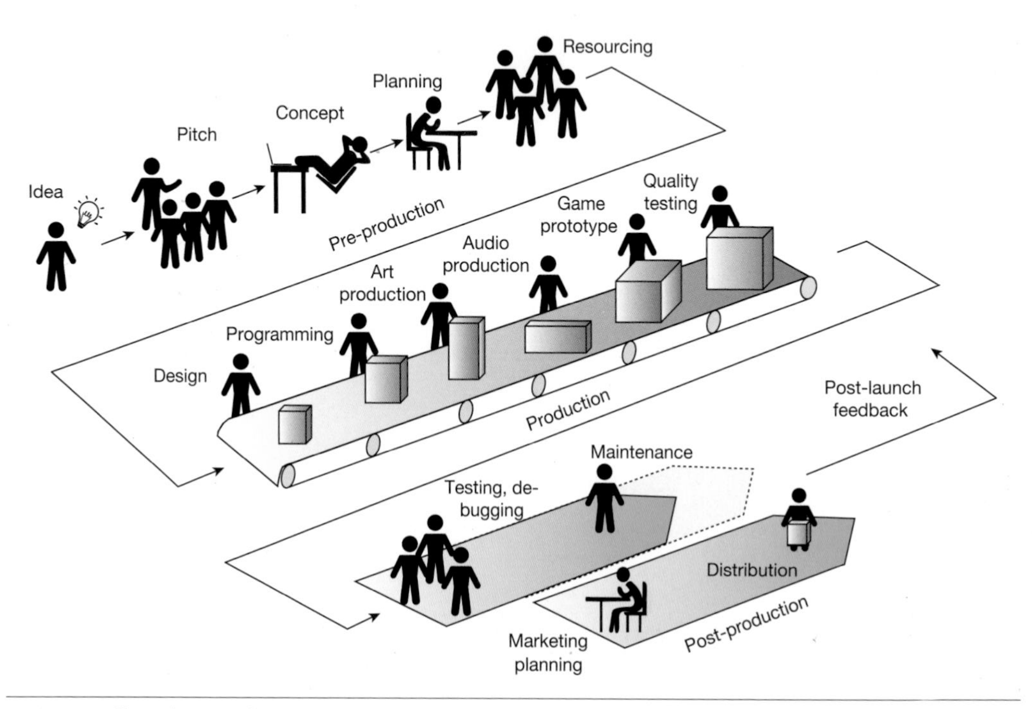

Figure 4.6 All products and services, even highly creative processes such as video game development, are created by operations processes

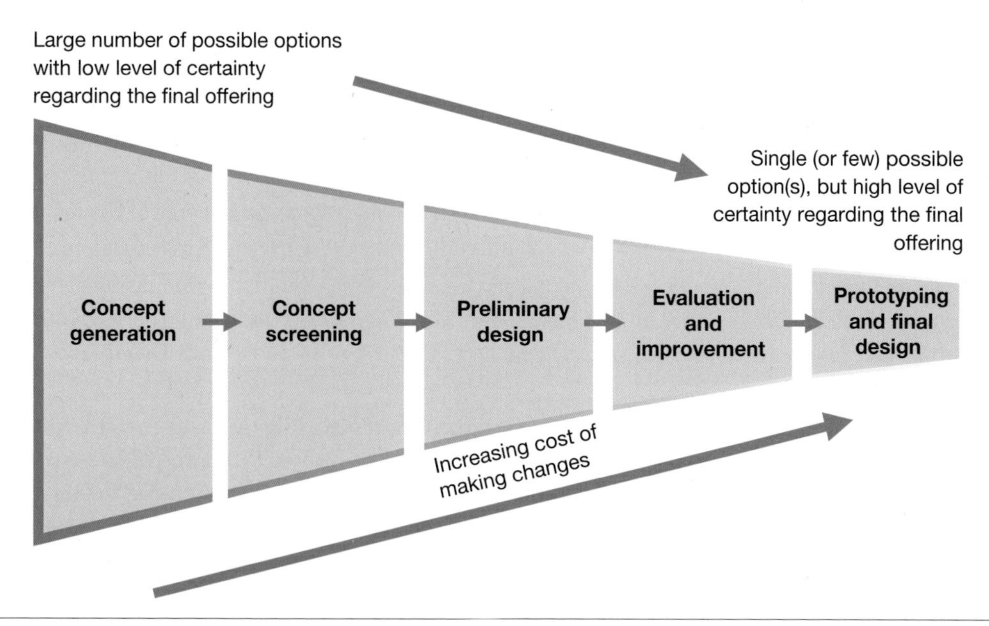

Figure 4.7 The stages of product/service innovation

The design funnel

Decisions taken during the innovation process progressively reduce the number of options that will be considered. For example, deciding to make the outside casing of a professional camera case from aluminium rather than plastic will constrain later decisions, such as the overall size and shape of the case. At each stage in the process, the level of certainty regarding the final design increases as design options are discarded and the number of options decreases. Figure 4.7 shows this idea, which is sometimes called the '**design funnel**'. But reducing design uncertainty also impacts on the cost of changing one's mind on some detail of the design. In most stages of design, changing a decision is bound to incur some sort of rethinking and recalculation of costs. Early on in the design activity, before too many fundamental decisions have been made, the costs of change are relatively low. However, as the design progresses, the interrelated and cumulative decisions already made become increasingly expensive to change. Although in practice the sequence of stages will differ to some extent, most product or service development will use a stage model similar to Figure 4.7. The innovation process starts with a general idea or 'concept' and progresses through to a fully defined specification. In-between these two states, the offering may pass through stages such as concept screening, preliminary design, evaluation and improvement, prototyping and final design.

Operations principle

Design processes involve a number of stages that move an innovation from a concept to a fully specified state.

Critical commentary

In this chapter we have treated product and service innovation together, implying that there is little difference between innovation in physical products and innovation in largely intangible services. It is not a totally misleading assumption. Yet, again, it is the tendency of many services to exhibit the IHIP characteristics (intangibility, heterogeneity, inseparability and perishability, see Chapter 1) that are put forward as adding distinctive features to service innovation. Some that have been suggested include the following:

- In service innovation it is more difficult to separate product, process and organisational innovations.

- Service innovation is more interactive in that, *by necessity*, it involves the engagement of both internal actors (marketing and technical staff, back-office and front-office staff, etc.) and external players (suppliers, regulators and especially customers). So, the process of introducing new services will be more complex and interdisciplinary.
- Developing this idea still further, some organisations employ a significant degree of co-creation (see Chapter 1) on the part of customers, both in the design and delivery of service.
- Service innovation is often more incremental and less radical than traditional product innovation. Innovations are often a mix of small, although not insignificant, innovations and changes introduced to solve single customers' problems (what is also sometimes called 'ad hoc' innovation).
- Service innovations are easier to copy and more difficult to protect. High-visibility services, where customers 'see' a significant part of the service are clearly more exposed to be inspected by anyone who wants to view them. It is common practice for service companies to 'sample' the service offered by their competitors. Moreover, whereas specific technologies embedded in a new manufactured product can be protected using patent law, it is not always as easy to do the same with service features.

Concept generation

Concept generation is where innovative ideas become the inspiration for new service or product concepts. And innovation can come from many different sources, both from within the organisation and from outside:

- ***Ideas from research and development*** – Many organisations have a formal research and development (R&D) function. As its name implies, its role is twofold. Research develops new knowledge and ideas in order to solve a particular problem or to grasp an opportunity. Development utilises and operationalises the ideas that come from research. And although 'development' may not sound as exciting as 'research', it often requires as much creativity and even more persistence. When the Rocket Chemical Company set out to create a rust-prevention solvent and degreaser to be used in the aerospace industry, it took them 40 attempts to get the water-displacing formula worked out. So that is what they called the product. WD-40 literally stands for Water Displacement, 40th attempt.
- ***Ideas from staff*** – The contact staff in a service organisation or the salesperson in a product-oriented organisation could meet customers every day. These staff may have good ideas about what customers like and do not like. They may have gathered suggestions from customers or have ideas of their own. One well-known example – which may be urban myth – is that an employee at Swan, the match maker, suggested having one instead of two sandpaper strips on the matchbox. It saved a fortune.
- ***Ideas from suppliers*** – Suppliers are often the experts in their field compared to their customers; a car seat manufacturer knows more about car seats (or should do) than the car manufacturer. They are also likely to be closer to many technological advances in their component; advances in respirator technology will probably picked up by respirator companies before the hospitals in which they are used. Finally, suppliers may supply several similar customers. They are in a good position to know what is going on in their industry.
- ***Ideas from competitor activity*** – Most organisations follow the activities of their competitors. A new idea from a competitor may be worth imitating, or better still, improving upon. Taking apart a competitor's service or product to explore potential new ideas is called 'reverse engineering'. Some aspects of services may be difficult to reverse engineer (especially 'back-office' services), as they are less transparent to competitors.
- ***Ideas from customers*** – Marketing may use market research tools for gathering data from customers in a structured way to test out ideas, or check services or products against predetermined

criteria. Ideas may also come from customers on a day-to-day basis; from complaints, or from every-day transactions. Organisations are increasingly developing mechanisms to facilitate the collection of this form of information. At a group level, crowdsourcing is the process of getting work or funding, or ideas (usually online) from a crowd of people.

Open sourcing – using a development community

Not all products or services are created by professional employed designers for commercial purposes. An open community, including the people who use the products, develops many of the software applications that we all use. If you use Google, Wikipedia or Amazon, you are using open-source software. The basic concept of open-source software is simple. Large communities of people around the world, who have the ability to write software code, come together and produce a software product. The finished product is not only available to be used by anyone or any organisation for free but is regularly updated to ensure it keeps pace with the necessary improvements. The production of open-source software is very well organised and, like its commercial equivalent, is continuously supported and maintained. However, unlike its commercial equivalent, it is absolutely free to use. Over the last decade, the growth of open source has been phenomenal with many organisations transitioning over to using this stable, robust and secure software.

Crowdsourcing

Closely related to the open sourcing idea is that of 'crowdsourcing'. Crowdsourcing is the process of getting work or funding, or ideas (usually online) from a crowd of people. Although in essence it is not a totally new idea, it has become a valuable source of ideas largely through the use of the internet and social networking. For example, Procter & Gamble, the consumer products company, asked amateur scientists to explore ideas for a detergent dye whose colour changes when enough has been added to dishwater. Other uses of the idea involve government agencies asking citizens to prioritise spending (or cutting spending) projects.

'Lead users' and 'harbingers of failure'

Ideas can come informally from customers, for example listening to customers during everyday transactions or from complaints. A particularly useful source of customer-inspired innovation, especially for products and services subject to rapid change, are so-called 'lead users'. These are users who are ahead of the majority of the market on a major market trend, and who also have a high incentive to innovate. They also are likely to have the real-world experience needed to problem solve and provide accurate data to enquiring market researchers. By contrast, another category of customer may be valuable because of their ability to consistently make bad purchase decisions. Some believe that this group of consumers has a tendency to purchase all kinds of failed products, time after time, flop after flop. These are the 'harbingers of failure'. These consumers keep on buying products that are subsequently taken from the shelves.

Ideas management

Several organisations, including the 3M Corporation, have successfully generated innovations by introducing formal incentives to encourage employee engagement. When employee-sourced ideas were processed using paper-based 'suggestion schemes' and a 'suggestion box', they were often only partly effective. Schemes had to be well resourced, and ideas could be difficult to evaluate consistently, so schemes could lose credibility. However, the advent of 'idea management' software systems helped. These systems can help operations to collect ideas from employees, assess them and, if appropriate, implement them quickly and efficiently. They can track ideas through from inception to implementation, making it much easier to understand important performance measures such as where ideas are being generated, how many ideas submitted are actually implemented, the estimated cost savings from submitted ideas and any new revenues generated by implemented ideas.

Table 4.3 Some typical evaluation questions for marketing, operations and finance

Evaluation criteria	Marketing	Operations	Finance
Feasibility	Is the market likely to be big enough?	Do we have the capabilities to deliver it?	Do we have access to finance to develop and launch it?
Acceptability	How much market share could it gain?	How much will we have to reorganise our activities to deliver it?	How much financial return will there be on our investment?
Vulnerability	What is the risk of it failing in the marketplace?	What is the risk of us being unable to deliver it acceptably?	How much money could we lose if things do not go to plan?

Concept screening

Concept screening is the stage where potential innovations are considered for further development against key criteria. Not every concept can be translated into viable product or service packages, so organisations need to be selective. For example, DuPont estimates that the ratio of concepts to marketable offerings is around 250 to 1. In the pharmaceuticals industry, the ratio is closer to 10,000 to 1. The purpose of concept screening is to take initial concepts and evaluate them for their feasibility (can we do it?), acceptability (do we want to do it?) and vulnerability (what are the risks of doing it?). Concepts may have to pass through many different screens, and several functions might be involved. Table 4.3 gives typical feasibility, acceptability and vulnerability questions for marketing, operations and finance functions.

Operations principle

The screening of designs should include feasibility, acceptability and vulnerability criteria.

OPERATIONS IN PRACTICE

Gorilla Glass[4]

Even if a new product or service idea does not make it to the end of a development process, the learning that derives from the development process can still prove useful. This is a lesson well understood by Corning, one of the world's biggest glassmakers. For nearly 170 years, it has combined its innovative ideas in glass science (and ceramics science and optical physics) with well-established manufacturing and engineering capabilities to develop its pioneering products and services. The scientists at Corning's research centre create thousands of new glass formulations and production innovations every year. Not all have immediate market potential. Those may have potential are sent to the research centre's small glassworks for trial production, but even these may not make it to market. But Corning knows that the results of its research, even if it does not result in a new product, still have value. Everything it learns from all its development projects is added to its 'knowledge bank'. Who knows, it could prove useful in the future.

This is exactly what happened during the development of Apple's revolutionary iPhone. Steve Jobs, the then boss (and founder) of Apple, had a request. Could Corning develop a faultlessly clear, strong and scratch-resistant glass that could be used to cover the screen of the newly designed iPhone, and please could it do it in six months?

Staff at Corning's research centre delved into their files and found the results of a project from back in the 1960s. It recounted a project to develop a toughened lightweight glass for industrial use. Small volumes of the tough glass had been made, but the product was abandoned when few customers were interested. Corning reworked the formula to create a glass that was able to be produced thin and strong enough to be suitable for touch screens. (If the screen on a mobile phone is too thick, finger movement on the screen is harder to detect.) It called the new glass Gorilla Glass. It went on to be progressively improved to retain its strength, even as screens became thinner as mobile phones became skinnier, and has also found markets in other touch-screen applications

Preliminary design

The first task in **preliminary design** is to define exactly what will go into the service or product. For service-dominant offerings this may involve documentation in the form of job instructions or 'service blueprints'. For product-dominant offerings, preliminary design involves defining product specifications (McDonald's has over 50 specifications for the potatoes used for its fries) and the component structure of the offering. This details all the components needed for a single product. For example, the components for a remote 'presentation' mouse may include the presentation mouse itself, a receiver unit and packaging. All of these three items are made up of components, which are, in turn, made up of other components, as shown in Figure 4.8.

There may be significant opportunities to reduce cost through design simplification at this stage. The most elegant product and service innovations are often the simplest. However, when an operation delivers a variety of services or products (as most do) the range can become complex, which in turn increases costs. Designers adopt a number of approaches to reduce design complexity. These include standardisation, commonality and modularisation.

Operations principle

A key objective in preliminary design should be simplification through standardisation, commonality and modularisation.

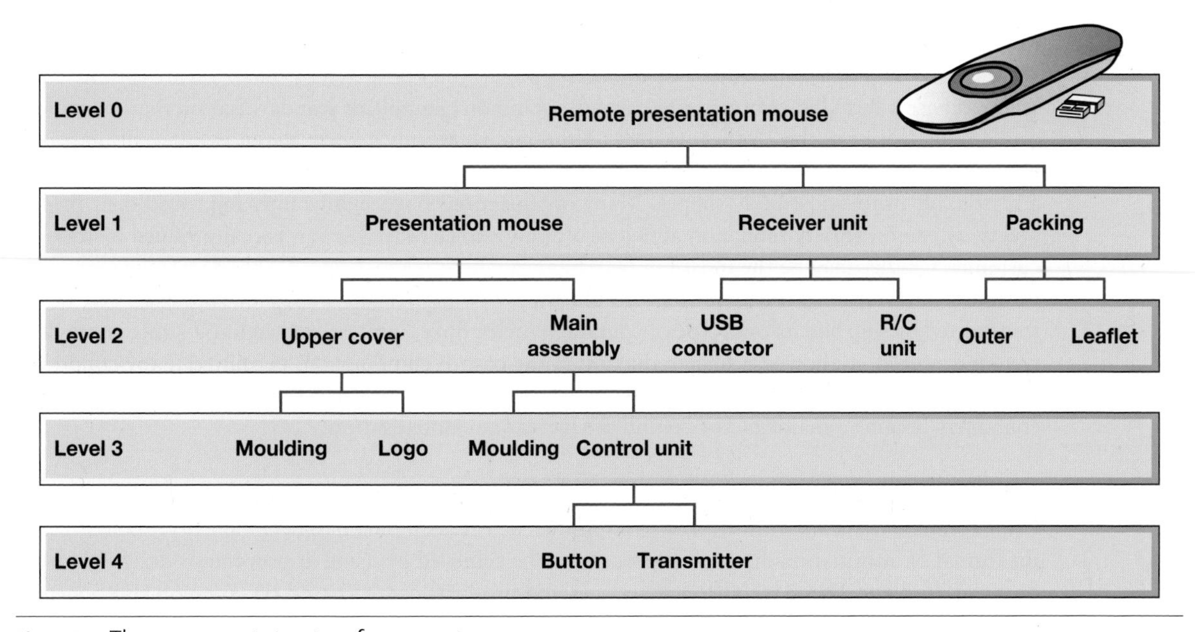

Figure 4.8 The component structure for a remote mouse

Critical commentary

The whole process-based approach to innovation could be interpreted as implying that all new offerings are created in response to a clear and articulated customer need. While this is usually the case, especially for services and products that are similar to their predecessors, more radical innovations are often brought about by the innovation itself creating demand. Nor does everyone agree with the dominant rational model, in which possible design options are progressively reduced stage by stage through the optimisation of known constraints and objectives. For some this neat model of the innovation process fails to reflect the creativity, arguments and chaos that sometimes characterise real innovation projects. First, managers do not start out with an infinite number of options. No one could process that amount of information – and anyway, designers often have some set solutions in their mind, looking for an opportunity to be used. Second, the number of options being considered often increases as time goes by. This may actually be a good thing, especially if the activity was unimaginatively specified in the first place. Third, the real process of innovation involves cycling back, often many times, as potential solutions raise fresh questions or become dead ends, and as requirements and constraints evolve. By contrast, the action-centric or co-evolution perspective of innovation proposes that offerings are designed through a combination of emotion and creativity; that the process by which this is done is generally improvised; and that the sequencing of innovation stages is in no way universal.

Standardisation

Operations sometimes attempt to overcome the cost penalties of high variety by standardising their products, services or processes. **Standardisation** is the application of commonality by using an agreed set of standards. This permits the individual components of a product or service to be created separately yet work or fit together. Standardisation is the process of making components, methods or processes uniform throughout an operation and between operations. It is the basis for the interchangeability of components and methods. There are many benefits of standardisation. It confers predictability on the design process, it allows for clear communication within and between enterprises, and it can have a dramatic effect on costs. Often it is the operation's outputs that are standardised. Examples of this are fast-food restaurants (you know exactly what you are getting when you purchase a Big Mac), but perhaps the most common example of standardisation that most of us are familiar with are the clothes we buy. Although everybody's body shape is different, garment manufacturers produce clothes in only a limited number of sizes. The range of sizes is chosen to give a reasonable fit for most body shapes. Many organisations have significantly improved their profitability by careful variety reduction aimed at offering choice only where it is really valued by the end customer. Clothes may be the most familiar, but arguably the most significant example of standardisation is the shipping container. Before the use of containers in shipping, whatever was being transported by ship had its own type of container: sacks, boxes, crates, or even loose. Once containers were introduced, it allowed a huge reduction in the cost of shipping and permitted many countries, for whom global was previously inaccessible, to trade on the world market. In effect, standard containers became a major factor behind the increase in globalisation.

Commonality

Using common elements within a service or product can also simplify design complexity. Standardising the format of information inputs to a process can be achieved by using appropriately designed forms or screen formats. The more different services and products can be based on common components, the less complex it is to produce them. For example, the European aircraft maker, Airbus, has designed its aircraft with a high degree of **commonality**. This means that 10 aircraft models ranging from the 100-seat A318 through to the world's largest aircraft, the A380, feature virtually identical flight decks, common systems and similar handling characteristics. In some cases, such as the entire A320 family, the aircraft even share the same 'pilot-type rating', which enables pilots with a single licence to fly any of them. The advantages of commonality for the airline operators include a much shorter training time for pilots and engineers when they move from one aircraft to another. In addition, when up to 90 per cent of all parts are common within a range of aircraft there is a reduced need to carry a wide range of spare parts.

Modularisation

The use of modular design principles, seen in computers for example, involves designing standardised 'sub-components' of a service or product, which can be put together in different ways. These standardised modules, or sub-assemblies, can be produced in higher volume, thereby reducing their cost. The package holiday industry can assemble holidays to meet a specific customer requirement, including pre-designed and purchased air travel, accommodation, insurance and so on. Similarly, in education there is an increasing use of modular courses, which allow 'customers' choice but ensure each module has economical student volumes.

Design evaluation and improvement

This stage in the innovation process takes the preliminary design and evaluates it to see if it can be improved. There are a number of techniques that can be employed at this stage to evaluate and improve the preliminary design. Perhaps the best known is quality function deployment (QFD).

Quality function deployment

The key purpose of QFD is to try to ensure that the eventual innovation actually meets the needs of its customers. It is a technique that was developed in Japan at Mitsubishi's Kobe shipyard and used extensively by Toyota and its suppliers. It is also known as the 'house of quality' (because of its shape) and the 'voice of the customer' (because of its purpose). The technique tries to capture what the customer needs and how it might be achieved. Figure 4.9 shows a simple QFD matrix used in the design of a promotional USB data storage stick. The QFD matrix is a formal articulation of how the company sees the relationship between the requirements of the customer (the whats) and the design characteristics of the new product (the hows):

- The whats, or 'customer requirements', are the list of competitive factors that customers find significant. Their relative importance is scored, in this case on a 10-point scale, with price scoring the highest.
- The competitive scores indicate the relative performance of the product, in this case on a 1 to 5 scale. Also indicated are the performance of two competitor products.
- The hows, or 'design characteristics' of the product, are the various 'dimensions' of the design, which will operationalise customer requirements within the service or product.
- The central matrix (sometimes called the relationship matrix) represents a view of the interrelationship between the whats and the hows. This is often based on value judgements made by the design team. The symbols indicate the strength of the relationship. All the relationships are studied, but in many cases, where the cell of the matrix is blank, there is none.
- The bottom box of the matrix is a technical assessment of the product. This contains the absolute importance of each design characteristic.
- The triangular 'roof' of the 'house' captures any information the team has about the correlations (positive or negative) between the various design characteristics.

Prototyping and final design

At around this stage in the design activity it is necessary to turn the improved design into a prototype so that it can be tested. It may be too risky to launch a product or service before testing it out, so it is usually more appropriate to create a 'prototype' (in the case of a product) or 'trial' (in the case of a service). Product prototypes include everything from clay models to computer simulations. Service

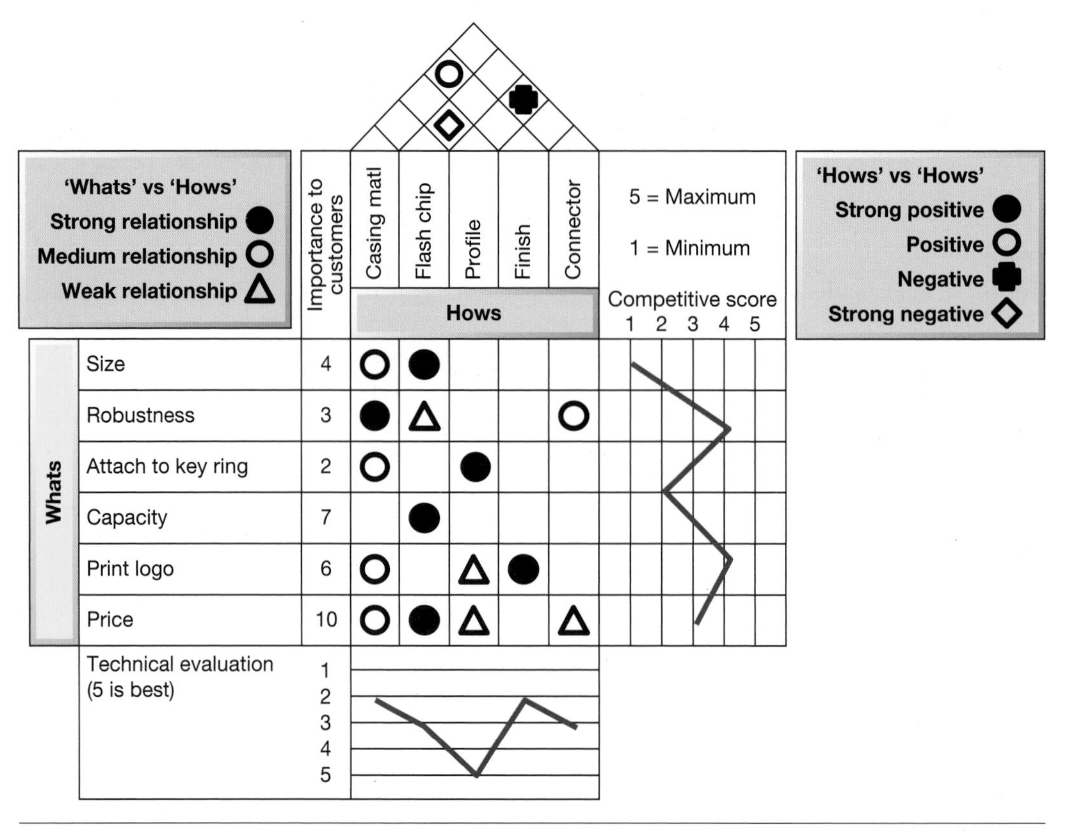

Figure 4.9 A QFD matrix for a promotional USB data storage stick

trials may include computer simulations but also implementing the service on a pilot basis. For example, retailers may pilot new services in a small number of stores to test customers' reaction. Virtual-reality-based simulations allow businesses to test new services and products as well as visualise and plan the processes that will produce them. Individual component parts can be positioned together virtually and tested for fit or interference. Even virtual workers can be introduced into the prototyping system to check for ease of assembly or operation.

Alpha and beta testing

A distinction that originated in the software development industry, but has spread into other areas, is that between the alpha and beta testing of a product or service. Most software products include both alpha and beta test phases, both of which are intended to uncover 'bugs' (errors) in the product. Not surprisingly **alpha testing** comes before beta testing. Alpha testing is essentially an internal process where the developers or manufacturers (or sometimes an outside agency that they have commissioned) examine the product for errors. Generally, it is also a private process, not open to the market or potential customers. Although it is intended to look for errors that otherwise would emerge when the product is in use, it is in effect performed in a virtual or simulated environment, rather than in 'the real world'. After alpha testing, the product is released for beta testing. **Beta testing** is when the product is released for testing by selected customers. It is an external 'pilot test' that takes place in 'the real world' (or near real world, because it is still a relatively short trial with a small sample) before commercial production. By the time a product gets to the beta stage, most of the worst defects should have been removed, but the product may still have some minor problems that may only become evident with user participation. This is why beta testing is almost always performed at the user's premises without any of the development team present. Beta testing is also sometimes called 'field testing', pre-release testing, customer validation, customer acceptance testing, or user acceptance testing.

4.4 How should product and service innovation be resourced?

As is the case with any type of process, for a product or service innovation process to function effectively, it must be appropriately designed and resourced. In this section, we examine five key questions that should be considered in resourcing the innovation process:

- What are the capacity requirements for innovation?
- Should innovation be carried out in-house or outsourced?
- What technology can be used to support the innovation process?
- What organisational structure is most suitable for the innovation process?
- How can the innovation process be compressed?

Operations principle

For innovation processes to be effective they must be adequately resourced.

Understanding capacity requirements for innovation activities

Capacity management involves deciding on the appropriate level of capacity and how it can be adjusted as demand changes. For innovation processes, demand is the number of new designs needed. The difficulty is that, even in large companies, the rate of new innovation is not constant. This means that innovation processes are subjected to uneven 'demand' for designs. Sometimes several new offerings may be introduced close together, while at other times little innovation is needed. This poses a resourcing problem because the capacity of an innovation activity is often difficult to flex. The expertise necessary for innovation is embedded within designers, technologists, market analysts and so on. Some expertise may be able to be hired in as and when it is needed, but much design resource is, in effect, fixed. This means that some organisations are reluctant to invest in innovation processes because they see them as an underutilised resource. A vicious cycle can develop in which companies fail to invest in innovation resources because many skilled design staff cannot simply be hired in the short term, resulting in innovation projects overrunning or failing to deliver appropriate solutions. This in turn may lead to the company losing business or otherwise suffering in the marketplace, which makes the company even less willing to invest in innovation resources.

OPERATIONS IN PRACTICE

BT's open innovation ecosystem[5]

BT is one of the world's leading providers of communications services and solutions, with customers in 180 countries. Its primary activities include the provision of networked IT services globally; local, national and international telecommunications services; TV and internet products and services; and converged fixed-mobile products and services. This is a business that thrives on innovation, and indeed BT has a world-renowned research and development organisation. But it is the way that the group has grown a broad ecosystem of innovation spanning both internal and external partners that has allowed BT to exploit innovation. Innovative ideas can come from many different sources. BT's 'open innovation' model is open in the broadest sense. It means working and collaborating with actual and potential suppliers, customers and its own staff. In fact, many of the novel ideas and technologies on which BT depend were created both within its own formal research and development facilities and from its staff more widely. The group's 'new ideas scheme' has an important role in the company's open innovation drive. It acknowledges and exploits the creativity of the group's 90,000-plus employees worldwide. They can use the scheme to propose ideas that could help with innovation in BT's processes, or improve its products and services. Staff can submit informal 'Eureka moment' ideas, or more carefully developed concepts. The ideas are then reviewed by experts in the business, with those ideas with the biggest potential taken further along the development process. Of the approximately 2,000 ideas submitted every year, around 50 make it through to implementation.

But it is the way that BT is 'open' to external innovation that has attracted much attention. In the world of telecoms, innovation is distributed around a number of innovation 'hot-spots' including Silicon Valley, Israel, Japan, Korea, Singapore, Hong Kong, India, China and throughout Europe. The central thrust of BT's global open innovation model comes from its technology scouting unit. This has been advancing the group's innovation efforts since 2000, when its then senior VP of technology and innovation was sent to Palo Alto to gain insights into emerging technologies and business models. Significantly, the head of BT External Innovation is still located not in the United Kingdom, but in Silicon Valley. At the heart of its network, the BT Infinity Lab in London allows it to keep track and often co-innovate with start-ups while innovation scouting teams scan the world of technology start-ups, venture capitalists and researchers for ideas. It is constantly looking for and evaluating ideas – everything from novel technologies, developments in how markets are developing, inventive operations processes, to pioneering business models. BT believes that its dedicated scouting capability amplifies its innovation-generating abilities by several orders of magnitude. BT's open innovation model also includes long-standing research partnerships with leading universities and business schools around the world.

Understanding whether innovation activities should be outsourced

Just as there are supply networks that deliver services and products, there are also networks that connect suppliers and customers in the innovation process. These networks are sometimes called 'design (or development) networks'. Innovation processes can adopt any position on a continuum of varying degrees of design engagement with suppliers, from retaining all the innovation capabilities in-house, to outsourcing all the innovation work. Between these extremes are varying degrees of internal and external capability. Figure 4.10 shows some of the more important factors that will vary depending on where an innovation process is on the continuum. Resources will be easily controlled if they are kept in-house because they are closely aligned with the company's normal organisational structures, but control should be relatively loose because of the extra trust present in working with familiar colleagues. Outsourced innovation work often involves greater control, with penalty clauses for delays often used in contracts.

From an open innovation perspective (discussed earlier in the chapter), firms should be willing to buy in innovations, rather than relying solely on those generated internally. Similarly, it may be beneficial to give access to underused proprietary innovations through joint ventures, licensing or spin-offs. However, a major inhibitor to open innovation is the fear of knowledge leakage. Firms become concerned that experience gained through collaboration with a supplier of design expertise may be transferred to competitors. There is a paradox here. Businesses usually outsource design primarily because of the supplier's capabilities, which are themselves an accumulation of specialist knowledge from working with a variety of customers. Without such knowledge 'leakage', the benefits of the supplier's accumulated innovation capabilities would not even exist.

Understanding what technologies to use in the innovation process

Technology has become increasingly important in innovation activities. Simulation software, for example, is now common in the design of everything from transportation services through to chemical factories. These allow developers to make design decisions in advance of the actual product or service being created by exploring possibilities, gaining insights and exploring the consequences of their

Responsible operations

In every chapter, under the heading of 'Responsible operations', we summarise how the particular topic covered in the chapter touches upon important social, ethical and environmental issues.

Increasingly, product or service innovation will include consideration of the environmental dimensions of sustainability. Critically examining the components of products with a view to a change of materials in the design could significantly reduce their environmental burden. Examples include the use of organic cotton or bamboo in clothing; wood or paper from managed forests used in garden furniture, stationery and flooring; recycled materials for carrier bags; and natural dyes in clothing, curtains and upholstery. Other innovations may be more focused on the use stage of an offering. The MacBook Air, for example, introduced an advanced power-management system that reduced its power requirements. In the detergent industry, Unilever and Procter & Gamble have developed products that allow clothes to be washed at much lower temperatures. Architecture firms are designing houses that can operate with minimal energy or use sustainable sources of energy such as solar panels. Some innovations focus on making product components within an offering easier to recycle or remanufacture once they have reached the end of their life. For example, some food packaging has been designed to break down easily when disposed of, allowing its conversion into high-quality compost. Mobile phones are often designed to be taken apart at the end of their life, so valuable raw materials can be reused. For example, Apple have introduced a robot (named Daisy) capable of disassembling 200 iPhones per hour in order to recover the various precious materials contained within. In the automotive industry, over 75 per cent of materials are now recycled.

Design innovation is not just confined to the initial conception of a product; it also applies to the end of its life. This idea is often called 'designing for the circular economy'. See the 'Operations in practice' example, 'Product innovation for the circular economy'. The circular economy (also called closed-loop or take-back economy) is proposed as an alternative to the traditional linear economy (or make–use–dispose, as it is termed). The idea is to keep products in use for as long as possible, extract the maximum value from them while in use and then recover and regenerate products and materials at the end of their service life. But the circular economy is much more than a concern for recycling as opposed to disposal. The circular economy examines what can be done right along the supply and use chain so that as few resources as possible are used, then products are recovered and regenerated at the end of their conventional life. This means designing products for longevity, reparability, ease of dismantling and recycling.

OPERATIONS IN PRACTICE

Product innovation for the circular economy[8]

Typical of the companies that have adopted the idea of the circular economy is Newlife Paints. It 'remanufactures' waste water-based paint back into a premium-grade emulsion. All products in its paint range guarantee a minimum 50 per cent recycled content, made up from waste paint diverted from landfill or incineration. The idea for the company began to take root in the mind of industrial chemist Keith Harrison. His garage was becoming a little unruly, after many years of do-it-yourself projects. Encouraged by his wife to clear out the mess, he realised that the stacked-up tins of paint represented a shocking waste. It was then that his search began for a sensible and environmentally responsible solution to waste paint. *'I kept thinking I could do something with it, the paint had an intrinsic value. It would have been a huge waste just to throw it away'.* Keith thought somebody must be recycling it, but no one was, and he set about finding a way to reprocess waste paint back to a superior-grade emulsion. After two years of research, Keith successfully developed his technology, which involves removing leftover paint from tins and blending and filtering them to produce colour-matched new paints. The company has also launched a premium brand, aimed at affluent customers with a green conscience, called Reborn Paints, the development of which was partly funded by Akzo Nobel, maker of Dulux Paints. Although Keith started small, he now licenses his technology to companies such as the giant French transnational waste company Veolia. *'By licensing we can have more impact and spread internationally'*, he says.

Simultaneous development

We described the design innovation process as a set of individual stages, each with a clear starting and ending point. Indeed, this sequential approach has traditionally been the typical form of product/service development. It has some advantages. The process is easy to manage and control because each stage is clearly defined, and each stage is completed before the next stage is begun, so each stage can focus its skills and expertise on a limited set of tasks. However, the sequential approach is both time-consuming and costly. When each stage is separate, any difficulties encountered during one stage might necessitate the design being halted while responsibility moves back to the previous stage. This sequential approach is shown in Figure 4.12(a).

Yet often there is really little need to wait until the absolute finalisation of one stage before starting the next. For example, perhaps while generating the concept, the evaluation activity of screening and selection could be started. It is likely that some concepts could be judged as 'non-starters' relatively early on in the process of idea generation. Similarly, during the screening stage, it is likely that some aspects of the design will become obvious before the phase is finally complete. Therefore, the preliminary work on these parts of the design could be commenced at that point. This principle can be taken right through all the stages, one stage commencing before the previous one has finished, so there is simultaneous or concurrent work on the stages (see Figure 4.12(b)).

> **Operations principle**
> Effective simultaneous development reduces time to market.

Early resolution of design conflict and uncertainty

Characterising product or service innovation as a whole series of decisions is a useful perspective. Importantly, a design decision, once made, need not irrevocably shape the final offering. In some cases, changing designs makes sense: for example, as new information emerges suggesting a better alternative. In addition, early decisions are often the most difficult to make because of the high level of uncertainty surrounding what may or may not work as a final design. This is why the level of debate, and even disagreement, over the characteristics of an offering can be at its most heated in the early stages of the process. One approach is to delay decision making in the hope that an obvious 'answer' will emerge. The problem with this is that, if decisions to change are made later in the innovation process, these changes will be more disruptive than if they are made earlier. Conversely, if the design team manages to resolve conflict early in the design activity, this will reduce the degree of uncertainty within the project and reduce the extra cost and, most significantly, time associated with either managing this uncertainty or changing decisions already made. There are two key implications of this. First, it is worth trying to reach consensus in the early stages of the innovation process even if this seems to be delaying the total process in the short term. Second, strategic intervention into the

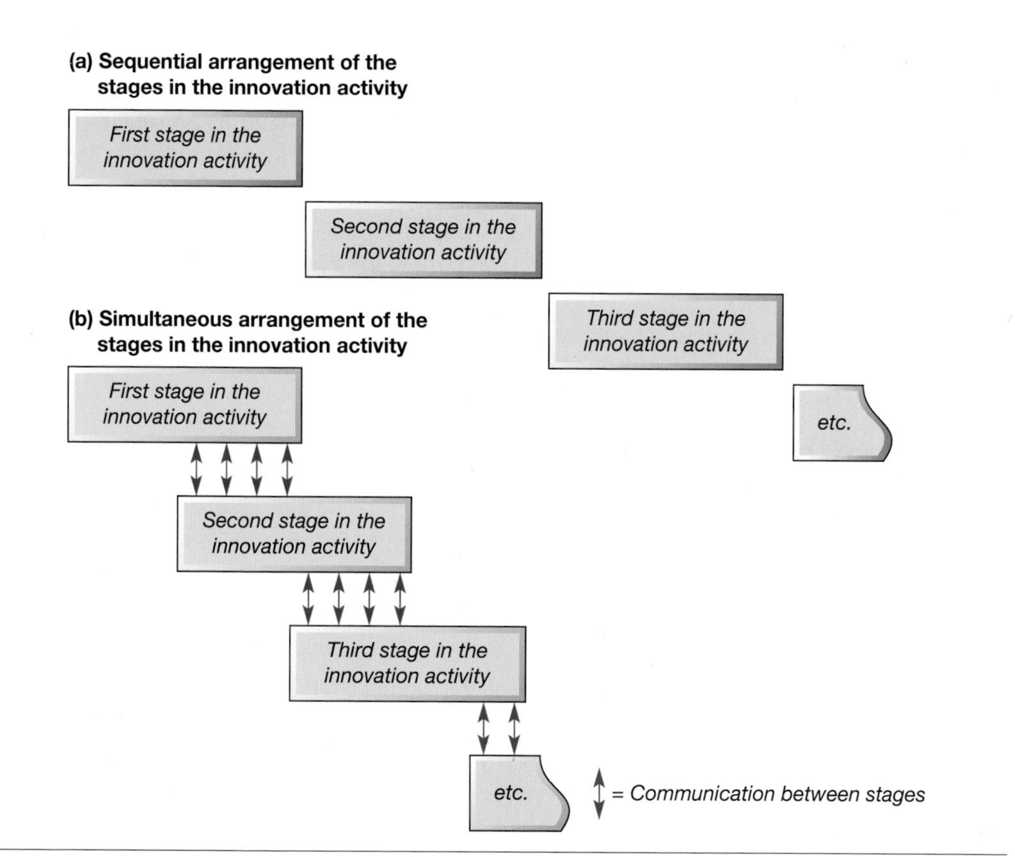

Figure 4.12 (a) Sequential arrangement of the stages in the design activity (b) Simultaneous arrangement of the stages in the design activity

innovation process by senior management is particularly needed during these early stages. Unfortunately, there is a tendency for senior managers, after setting the initial objectives of the innovation process, to 'leave the details' to the technical experts. They may only become engaged with the process again in the later stages as problems start to emerge that need reconciliation or extra resources.

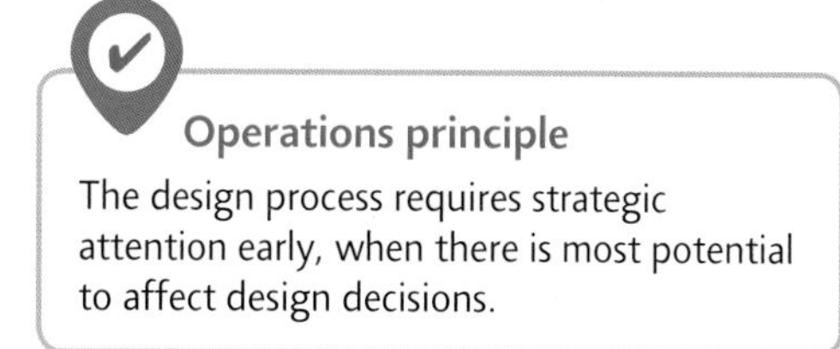

Operations principle

The design process requires strategic attention early, when there is most potential to affect design decisions.

Summary answers to key questions

4.1 What is product and service innovation?

- We explore three related terms of 'creativity', 'innovation' and 'design'. 'Creativity' is the ability to move beyond conventional ideas, rules or assumptions, in order to generate significant new ideas. 'Innovation' is the act of introducing something new with potential to be practical and give commercial return. 'Design' is the process that transforms innovative ideas into something more concrete. A design delivers a solution that will work in practice.
- The innovation S-curve describes the impact of an innovation over time, slow at first, increasing in impact, then slowing down before levelling off.
- Incremental and radical innovations differ in how they use knowledge. Radical innovation often requires completely new knowledge and/or resources, making existing services and products obsolete. Incremental innovation builds upon existing knowledge and/or resources.

4.2 What is the strategic role of product and service innovation?

- Despite the obstacles to successful innovation, almost all firms strive to be innovative given the significant payback for the organisations that manage to incorporate innovative ideas in the design of their products and services.
- The innovation activity is a process that involves many of the same design issues common to other operations processes. As such, it is a process that conforms to the input–transformation–output model described in Chapter 1.
- The performance of the innovation process can be assessed in the same way as any process, namely in terms of quality, speed, dependability, flexibility, cost and, more broadly, sustainability.

4.3 What are the stages of product and service innovation?

- Concept generation is where innovative ideas become the inspiration for new service or product concepts. Ideas come from inside R&D, staff, competitor activity, customers, open sourcing, crowdsourcing and lead users.
- Concept screening involves examining its feasibility, acceptability and vulnerability in broad terms to ensure that it is a sensible addition to the company's service or product portfolio.
- Preliminary design involves the identification of all the component parts of the service or product and the way they fit together. Designers adopt a number of approaches to reduce design complexity, including standardisation, commonality and modularisation.
- Design evaluation and improvement involves re-examining the design to see if it can be done in a better way, more cheaply or more easily. The best-known tool for this is quality function deployment (QFD).
- Prototyping and final design involves providing the final details that allow the service or product to be created or delivered.

4.4 How should product and service innovation be resourced?

- To be effective, the innovation process must be appropriately resourced. Issues to consider include understanding capacity requirements, deciding how much of the innovation process to outsource, and determining technology needs, organisational structure and methods to compress the innovation process.
- Organisations must understand the demand for innovation (i.e. number of new designs in a given period) and match this with their capacity for innovation.
- Organisations must determine whether all innovation activity should take place in-house or whether there are benefits in outsourcing some or all this activity to a third party.
- There are many technologies available to support the innovation process, including computer-aided design (CAD); digital twins; and knowledge management technologies.
- Selecting an appropriate organisational structure can support the innovation process. Typically, structures towards the project rather than functional end of the continuum (project matrix or project team) are seen as beneficial.
- The innovation process can often be compressed by integrating the design of the offering and the design of the process; adopting simultaneous development of the various stages in the overall process; and early resolution of design conflict and uncertainty.

CASE STUDY Widescale studios and the Fierybryde development

'Anyone who has been involved with designing and constructing video games, will tell you that game development never goes as planned. I sometimes think that it is a miracle that any game gets developed. Technical glitches, bottlenecks in production, conflicting creative egos, pressure from publishers, they will all throw you off course during the development cycle. It is a process that occupies the area on the borderline between art and technology. Yet, although video game development is an uncertain and complex process, it is how the development process is managed that is the key feature in whether a game will go on to be a success' (Izzy McNally, Co-Owner Widescale Studios).

Widescale Studios was a video game development studio, located in the midlands of the United Kingdom. It had been founded seven years ago by Izzy McNally and Oli Chambers, when they left a larger studio to gain 'some creative independence'. Video game software development studios are the organisations that actually create the games. There are many thousands of such studios worldwide, some large, but most employing less than 30 people.[9] Some studios are owned by video game publishers, some of which also produce gaming hardware, and some like Widescale, are independent. Publishers market and sell the games, manage relationships with distributers, platform providers and retailers, conduct market research and advertise games.

Originally from California, Izzy was, by background, an artist and writer, Oli started as a programmer, but had moved into becoming an executive producer. (In the industry, an executive producer is the person who is responsible for the overall coordination of the development.) Both Izzy and Oli admitted that their desire for more creative independence had not fully materialised. *'Since we started, we have been surviving as an independent studio by taking on contracts from the bigger studios, and we have built a good reputation. But if we don't have another contract ready to go when the last one finishes, we are in trouble. It can be dispiriting constantly looking for work to keep us afloat. That was why Fierybryde was so exciting'* (Oli Chambers).

The Fierybryde project

Fierybryde was an idea for a role-playing game (RPG) that had come out of a number of **brainstorming** sessions between Izzy, Oli and Hussein Malik in the middle of an unusually warm and pleasant summer. Hussein was a developer and self-confessed 'fanatical gamer' who had joined Widescale soon after it was founded. A role-playing game is a video game in which players assume the roles of characters who are protagonists in a fictional setting. The senior team at Widescale were excited at the concept of Fierybryde, and saw it as an opportunity to develop a game of their own that would (potentially) give them both creative and financial independence. The Fierybryde concept was intriguing, although not totally novel. The game's setting was a combination of space exploration and 'wild west' adventure (Fierybryde was the name of the spaceship in the story), with various characters who possessed different skills and psychological traits. The purpose of the game was to build an intergalactic trading empire while avoiding interference from political and commercial rivals.

Traditionally, independent studios that wanted to develop a game such as Fierybryde had four methods of raising funds. First, they could pitch the idea to a publisher. Most video game development was funded by big publishers. However, publishers almost always insisted on terms that were more favourable to them than the developers. Second, the studio could seek private investors who would put their own money into the company and share any subsequent profits. The downside to this for Izzy and Oli would be a certain loss of independence. Third, the studio could attempt to raise money by crowdfunding, asking for (relatively small) donations from thousands of potential future users of the game in return for preferential access to the finished game. It was an increasingly popular method of raising funding, but limited to relatively small sums in total, often less than £1 million (the typical budget for a RPG would be tens of millions or more). Finally, the studio could start the development from their own saved capital, then fund the ongoing costs from the profits from their other work. This was the approach chosen by Widescale, which had a retained cash pot of around £700,000. If successful, Fierybryde could provide a stable stream of income, without substantial rights and royalties going to some big publisher. In turn, this would let the studio pursue more interesting projects in the future. Table 4.4 shows Widescale's projected cash flow forecast as of the start of the project.

Table 4.4 Widescale's planned total cash flow (£000)

Year	1		2				3				4	
Qtr	3	4	1	2	3	4	1	2	3	4	1	2
General revenue	2500	2500	2300	2500	2000	2400	2000	2000	2000	2000	2000	2000
General costs	1600	1600	1600	1600	1600	1600	1600	1600	1600	1600	1600	1600
Fierybryde revenue								1500	4500	3500	1500	1500
Fierybryde costs	300	500	550	600	800	800	800	600	300	200	100	100
Cash	1300	1700	1850	2150	1750	1750	1350	2650	7250	10950	12750	14550

The Development process

Video game development is an uncertain and complex process, but a key feature in whether a game will go on to be a success is the way the development process is managed from concept through to launch. Although different studios use slightly different terms, game development is broken down into three stages: pre-production, production and post-production. Pre-production is the stage where the developers have to answer some fundamental questions about the game, including the market it is aimed at, the platform it will play on, the type of game it is going to be, the budget, the basic storyline and the timescale (at least nominally). The production phase is usually the most resource intensive phase, and is often the phase that is the most uncertain and difficult to plan. It involves programmers, character artists, graphic designers, audio designers, voice actors, quality testers and producers who, in the words of Oli, *'provide the glue that makes it all happen'*. The aim of all of them is to make a game that will be original, fun and involving. They do this by using new gameplays, gripping underlying stories, enhanced graphics and convincing characters. If a game fails to meet users' concepts of 'quality play', they could readily switch to other games. Post-production manages the transition of the game into the market. Often a publisher will become involved at this stage, if they weren't before. Even at this stage, quality assurance continues because bugs in the software always continue to emerge. A 'hype' video extract from the game with a mix of graphics and sample gameplay will probably be released for marketing purposes, and a spot at one of the major gaming conventions may be arranged.

Starting the Fierybryde development

Work started on the outline of the Fierybryde concept over the summer, with Izzy drafting an outline script and Oli working on some technical issues such as the number of 'levels' the game should have and how many maps it should contain. The project was formalised with its own budget in September when Hussein was asked to put together plans for how the development would progress. His first decision was to hire Ross Avery, who had been his boss at his previous studio. Ross had wide experience in the game development industry, largely in senior executive developer and producer roles.

Ross and Hussein formed the core of the Fierybryde team and were joined by planners and developers, both newly recruited and moved from Widescale's other work. However, Oli recognised that Fierybryde should not put any of the studio's regular projects at risk. *'They are our "bread and butter", each has a deadline and a budget that we must stick to. Fierybryde has more flexibility because it's directly under our control. Of course, we had a budget for it, but there was still considerable flexibility. So we looked at the budget and asked the question, do we want a team of 10 people working for 10 months or a team of 20 people working for five months? Theoretically, one could even have had a team of two people working for 50 months, but that would have been ridiculous. Also, at different times in the development process one will need different numbers of developers with specific skills. The balance was always between allocating the appropriate resources to Fierybryde without interfering with our other work'.*

By November it had become clear that Izzy would have to decide whether to take on responsibility for developing the games script herself or to hire in a script writer. In the end she hired a part-time script writer who had experience of television at work. Izzy admitted that it was a mistake. *'Script writing for a video game is totally different from writing for television or writing a novel. I underestimated this. In a script for television the narrative moves in a linear direction. With a video game the narrative is more like a tree. Each player can move along different branches of the tree depending on the decisions that they make. A script writer has to make up dialogue for many different scenarios, knowing that each individual player will see only one of them. It wasn't the fault of the writer we hired, it was my fault in underestimating the differences. In the end I had to take over far more of the script writing than I had intended'.*

The scripting and storyboarding of the game continued into the New Year, but by January tensions had begun to emerge between Oli, who was concerned about the rate that the project was burning though the budget, and Ross, who wanted the script, characterisations and overall architecture of the game settled before the production phase commenced. Oli wanted to get the production stage of development started as soon as possible. *'Widescale's strength was in the actual production stage of development. That's what we spend most of our time doing. If we weren't good at keeping to schedule, we couldn't have survived as a contract developer. Also, I thought that we had an outline script and the overall structure of the game more-or-less sorted from mine and Izzy's work over the summer. I do understand that when a new person like Ross first joins the team, the temptation is to try to sit down at the beginning of the development process and settle the whole script from start to finish. But it has to develop naturally; developing the script for a game is essentially an iterative process'.*

The production phase

Although there were still uncertainties, and some disagreement around the game's storyline, by the end of January Oli decided to formally move on to the production phase and allocated developers and artists to the project. He also started briefing the freelance graphics designers, sound designers and voice artists that they would need later in the process. In early February Ross resigned. He was philosophical about it. *'It's not unusual in this business. There will always be some tension between whoever is in charge of operationalising a concept and the studio owner. The important thing is who holds the budget. In this case the owners* [Oli and Izzy] *didn't want to give up control. It's their company and their money, so I guess they have the last word on any decisions, big or small. But, personally, I like to have more control than they wanted to give me'.*

From that point Oli acted as executive producer for the project, with Hussein overseeing technical issues and Izzy 'creative' ones. However, during the spring, the development fell increasingly behind schedule. Hussein admitted that many of the problems were the result of their own decisions. *'We started using a new 3D graphics package two months into the initial development. It allowed a new rendering approach that looked particularly exciting. It made the graphics better than we thought possible. It did give us some spectacular effects, but also gave us two problems. First, we totally underestimated the learning curve necessary to master the package. It took our developers a month or two to get used to the package and this delayed things more than we envisaged. Second, it became clear relatively quickly that the effect of the change was to knock the game's "frame rate" down to the point where it looked poor. We knew the choice would affect frame rate but we just didn't anticipate the impact this would have on what the game felt like to play. Both these things undermined our ability to estimate how long some key stages of the development might take. Without the ability to estimate the individual development tasks it became particularly difficult to schedule the development as a whole'.*

By June, the development team were overcoming the problems with the new graphics package when a new problem emerged. Hussein and Oli had decided to use a previously untried (by Widescale) game engine. (A game engine provides the software framework that allows developers to create video games.) Many commercial game engines are available to help game developers. Using one means that developers can focus solely on the logic of the game rather than getting bogged down in detail. A game engine allows code reuse, which usually means shorter development time and reduced cost. But not when a development team needs to learn it for the first time! Once again, progress slowed. Izzy believed that the problems were the result of trying to balance the eventual quality of the game with the costs of developing it. *'There is always something of a trade-off between the efficiency of the development process and the quality of the game that comes out of the process. Just to make things more complicated you have to wait until the game is almost fully developed before you can judge quality, in terms of how much fun it is to play. So you have to manage the development process in the best way you think will promote both creativity and efficiency'.*

Both Izzy and Ollie liked the idea of giving as much freedom as possible to individual developers within the team. Compared with many studios the atmosphere was relaxed. On the whole it was thought that this had led to a good creativity that would eventually show through in the final game. Oli also thought that a more relaxed attitude helped to develop and retain the best development talent. *'Developers value flexibility to innovate and more ownership over the content they are working on, and some degree of independence from micromanagement. All too easily they can find another studio where they feel their skills are more valued'.*

However, there were times when it proved less than fully efficient. For example, at one time two developers each designed their own different versions of the same scene because their work had not been coordinated, costing several days of wasted effort. There were also problems in managing the studio's regular contract work alongside the Fierybryde project. At times the studio's other work took development resource away from the project. As one frustrated developer put it, *'It could be frustrating to suddenly have a colleague taken off the project for a week to work on another job, but hey, it was the revenue from this other work that was funding the project so naturally it took priority. Nevertheless, there were times when the other jobs were vacuuming up resources, and the whole development process was like being in a pressure cooker'.*

Project crunch and financial crunch

By November it was becoming obvious that Fierybryde was in trouble. It had fallen well behind schedule and the studio's cash projections were looking bad. The studio's cash statement and projection at this time is shown in Table 4.5. It indicated that the company would need to draw further on

Table 4.5 Widescale's actual and projected cash flow (£000)

	Actual						Forecast					
Year	1		2				3				4	
Qtr	3	4	1	2	3	4	1	2	3	4	1	2
General revenue	2273	2332	2105	2117	2306	2308	2205	1886	2000	2000	2000	2000
General costs	1891	1764	1792	1898	1894	1869	1800	1800	1800	1800	1800	1800
Fierybryde revenue								0	800	4500	3500	2500
Fierybryde costs	302	550	499	614	855	842	850	700	400	250	200	100
End of Qtr. Cash	780	848	712	442	44	(386)	(831)	(1445)	(845)	3605	7105	9705

its overdraft facilities in the current quarter and would need more substantial funding in the new year. It was clear that even if all went well and it had no further problems, the soonest it could have got the game into a form to take to a publisher would be half-way through the following year. Even achieving this would probably involve what game developers call 'crunch': working extended hours of overtime (paid and unpaid) for periods of time in order to hit a particular deadline.

Oli was despondent, *'The real frustration is that the game is looking good. Everybody who has worked on it loves the storyline and are wowed by the graphics. We just need a final effort. It's tempting to see "crunch" as a failure of planning. But honestly I've never work on a development that has not involved some degree of crunch'.* It was Izzy who finally made the decision in the November. *'We have been working on this project for 18 months, that's not long for a game of this complexity. And it's really good, everyone agrees. But the potential of a game and the financial viability of its development process are different things. Basically, we have run out of credit and we have to accept that we need help. The most likely source of help is going to be a publisher. They could fund the remainder of this development from their small change. OK, they will demand a part of the company, and we would lose much of our independence. But it's either that, or abandon the development, let go probably a third of our staff, and try to get an emergency loan from our bank'.*

QUESTIONS

1 **Was it a mistake for Widescale to embark on the Fierybryde development?**

2 **List the reasons that could have contributed to the Fierybryde development falling behind schedule.**

3 **What would you have done differently?**

4 **What would you advise Izzy and Oli to do now, and why?**

Problems and applications

All chapters have 'Problems and applications' questions that will help you practise analysing operations. They can be answered by reading the chapter. Model answers for the first two questions can be found on the companion website for this text.

1 One product where customers value a very wide range of product types is domestic paint. Most people like to express their creativity in the choice of paints and other home decorating products that they use in their homes. Clearly, offering a wide range of paint must have serious cost implications for the companies that manufacture, distribute and sell the product. Visit a store that sells paint and get an idea of the range of products available on the market. How do you think paint manufacturers and retailers could innovate so as to increase variety but minimise costs?

2 '*We have to get this new product, and fast!*', said the operations director. '*Our competitors are close behind us and I believe their products will be almost as good as ours when they launch them*'. She was talking about a new product that the company hoped would establish it as the leader in the market. The company had put together a special development team together with its own development laboratory. It had spent £10,000 on equipping the laboratory and the cost of the development engineers would be £20,000 per quarter. It was expected that the new product would be fully developed and ready for launch within six quarters. It would be sold through a specialist agency that charged £10,000 per quarter and would need to be in place two quarters prior to the launch. If the company met its launch date, it was expected that it could charge a premium price that would result in profits of approximately £50,000 per quarter. Any delay in the launch would result in profits being reduced to £40,000 per quarter. If this development project were delayed by two quarters, how far would the break-even point for the project be pushed back?

3 Innovation becomes particularly important at the interface between offerings and the people who use them. Consider two types of websites:

(a) those that are trying to sell something, such as amazon.com; and
(b) those that are primarily concerned with giving information, for example reuters.com or nytimes.com

What constitutes good innovation for these two types of websites? Find examples of particularly good and particularly poor web design and explain the issues you've considered in making the distinction between them.

4 According to the Ellen MacArthur Foundation, a circular economy is 'a systemic approach to economic development designed to benefit businesses, society, and the environment'.[10] Also see the example earlier in this chapter. What do you see as the main barriers to a more widespread adoption of the idea?

5 A janitor called Murray Spangler invented the vacuum cleaner in 1907. One year later he sold his patented idea to William Hoover, whose company went on to dominate the market. Now, the Dyson vacuum cleaner has jumped from nothing to a position where it dominates the market. The Dyson product dates back to 1978 when James Dyson designed a cyclone-based cleaner. It took five years and 5,000 prototypes before he had a working design. However, existing vacuum cleaner manufacturers were not as impressed – two rejected the design outright. So Dyson started making his new design himself. Within a few years Dyson cleaners were outselling the rivals that had once rejected them. The aesthetics and functionality of the design help to keep sales growing in spite of a higher retail price.

(a) What was Spangler's mistake?
(b) What do you think makes 'good design' in markets such as the domestic appliance market?
(c) Why do you think the two major vacuum cleaner manufacturers rejected Dyson's ideas?
(d) How did design make Dyson a success?

6 It sounds like a joke, but it is a genuine product innovation. It's green, it's square and it comes originally from Japan. It's a square watermelon. Why square? Because Japanese grocery stores are not

large and space cannot be wasted. Similarly, a round watermelon does not fit into a refrigerator very conveniently. There is also the problem of trying to cut the fruit when it keeps rolling around. So an innovative Japanese farmer solved the problem with the idea of making a cube-shaped watermelon that could be packed and stored easily. There is no genetic modification or clever science involved in growing watermelons. It simply involves placing the young fruit into wooden boxes with clear sides. During its growth, the fruit naturally swells to fill the surrounding shape.

(a) Why is a square watermelon an advantage?
(b) What does this example tell us about product design?

7 Is there a fundamental conflict between encouraging creativity in the product and service innovation process and the very concept of a 'process'?

8 What do you think are the differences between innovation in product design and innovation in service design?

9 Standardisation is an important concept in design. In addition to the examples mentioned in the chapter, what other examples of standardisation can you think of?

10 How could conventional 'bricks-and-mortar' bookshops innovate their services in order to compete with online book retailers?

Selected further reading

Christensen, C.M. (1997) ***The Innovator's Dilemma: When New Technologies Cause Great Firms to Fail*****, Harvard Business School Press, Boston, MA.**
Ground-breaking book on disruptive innovation that has had a major influence on innovation theory over the last 20 years.

Goffin, K. and Mitchell, R. (2016) ***Innovation Management: Effective Strategy and Implementation*****, 3rd edn, Palgrave Macmillan, London.**
General advice from two experts in the subject.

Gutsche, J. (2020) ***Create the Future + The Innovation Handbook: Tactics for Disruptive Thinking*****, Fast Company Press, New York, NY.**
A popular and practical treatment, aimed at practitioners but all could gain from it.

Kahney, L. (2014) ***Jony Ive: The Genius Behind Apple's Greatest Products*****, Penguin, London.**
Inside the mind of one of the world's great designers.

Reason, B., Løvlie, L. and Brand Flu, M. (2016) ***Service Design for Business: A Practical Guide to Optimizing the Customer Experience*****, John Wiley & Sons, Chichester.**
A book that looks at the 'entire service experience' and is very readable.

Ridley, M. (2020) ***How Innovation Works: And Why It Flourishes in Freedom*****, Fourth Estate, New York, NY.**
One person's view, but with many fascinating examples.

Rose, D. (2015) ***Enchanted Objects: Innovation, Design, and the Future of Technology*****, Scribner Book Company, New York, NY.**
An interesting book examining how technology is and will continue to impact on design.

Tidd, J. and Bessant, J.R. (2018) ***Managing Innovation: Integrating Technological, Market and Organizational Change*****, 7th edn, John Wiley & Sons, Chichester.**
Well-established text that deserves its longevity.

Trott, P. (2016) ***Innovation Management and New Product Development*****, 6th edn, Pearson, Harlow.**
A standard text in the area.

Notes on chapter

1. The information on which this example is based is taken from: Economist (2018) The invention, slow adoption and near perfection of the zip, *Economist* print edition, 18 December.
2. Christensen, C.M. (1997) *The Innovator's Dilemma: When New Technologies Cause Great Firms to Fail*, Harvard Business School Press, Boston, MA.
3. Economist (2019) Don't stop me now: the economics of streaming is changing pop songs, *Economist* print edition, 5 October.
4. The information on which this example is based is taken from: Economist (2017) One of the world's oldest products faces the digital future, *Economist* print edition, 12 October.
5. The information on which this example is based is taken from: BT website, How BT innovates, https://www.bt.com/about/innovation/how-bt-innovates (accessed August 2021); BT News (2018) BT launches Better World Innovation Challenge for start-ups & SMEs, press release from BT, 8 May; BT Group plc Annual Report, Strategic Report, 2019; Fransman, M. (2014) *Models of Innovation in Global ICT Firms: The Emerging Global Innovation Ecosystems* (ed. M. Bogdanowicz), JRC Scientific and Policy Reports – EUR 26774 EN. Seville: JRC-IPTS.
6. See Grieves, M. and Vickers, J. (2017) Digital twin: mitigating unpredictable, undesirable emergent behavior in complex systems, in F.-J. Kahlen, S. Flumerfelt and A. Alves (eds), *Transdisciplinary Perspectives on Complex Systems: New Findings and Approaches*, Springer International Publishing.
7. The information on which this example is based is taken from: Morgan, J. and Liker, J.K. (2006) *The Toyota Product Development System: Integrating People, Process, and Technology*, Productivity Press, New York, NY; Sobek II, D.K., Liker, J. and Ward, A.C. (1998) Another look at how Toyota integrates product development, *Harvard Business Review* (July–August)
8. For further information, see: Goodwin, L. (2015) How to bust the biggest myths about the circular economy, *Guardian*, 12 March; Clegg, A. (2015) Sustainable innovation: shaped for the circular economy, *Financial Times*, 26 August; Company website, Newlife Paints, https://www.newlife-paints.com/about (accessed September 2021).
9. According to the techjury website, the global games market for video games is likely to be $200 billion by 2023, see https://techjury.net/blog/video-games-industry-statistics/ (accessed September 2021).
10. See https://archive.ellenmacarthurfoundation.org (accessed September 2021).

5 The structure and scope of supply

KEY QUESTIONS

5.1 What is the structure and scope of supply?

5.2 How should the supply network be configured?

5.3 How much capacity should operations have?

5.4 Where should operations be located?

5.5 How vertically integrated should an operation's supply network be?

5.6 What activities should be in-house and what should be outsourced?

INTRODUCTION

No operation exists in isolation – it is part of a larger and interconnected network of other operations, referred to as a supply network. An operation's supply network includes its suppliers and customers. It will also include suppliers' suppliers and customers' customers, and so on. At a strategic level, operations managers are involved in deciding both the structure and scope of these supply networks. Structure decisions involve deciding the overall configuration of the supply network, the amount of capacity needed within the network and where operations should be located. Scope decisions involve deciding the extent to which an operation performs activities itself, as opposed to requesting a supplier (or sometimes a customer) do them on its behalf. This chapter examines the fundamental and relatively strategic issues of supply networks. Later, in Chapter 12, we look at the more operational activities involved in managing the individual supply chains that run through networks. Figure 5.1 places the structure and scope of operations in the overall model of operations management.

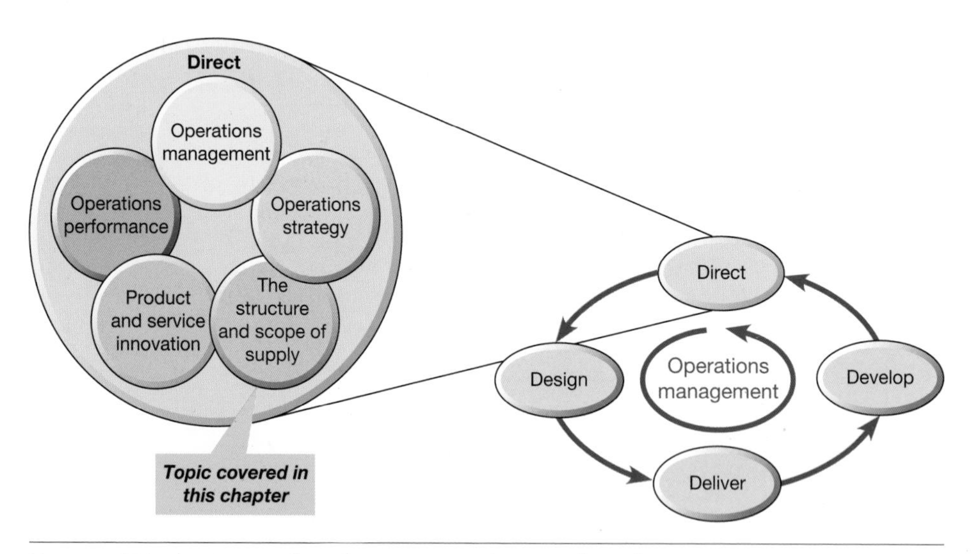

Figure 5.1 This chapter examines the structure and scope of supply

5.1 What is the structure and scope of supply?

A supply network is an interconnected set of operations. The structure of a supply network relates to its shape and form. The scope of an operation's supply network relates to the extent to which an operation decides to do the activities performed by the network itself, as opposed to requesting a supplier to do them. So, before we examine these two elements, first we need to establish what we mean by a 'supply network' and why it is important that an operation understands its position within it.

Supply networks

The supply network includes the chains of suppliers providing inputs to the operation, the chains of customers who receive outputs from the operation, and sometimes other operations that may at times compete and at other times cooperate. Materials, parts, information, ideas and sometimes people all flow through the network of customer–supplier relationships formed by all these operations. On its **supply side** an operation has its **first-tier** suppliers, who themselves have their own suppliers (second-tier suppliers) who in turn could also have suppliers, and so on. However, some **second-tier** suppliers may also supply an operation directly, thus missing out a link in the network. On the **demand side** of the network, 'first-tier' customers are the main customer group for the operation. These customers might not be the final consumers of the operation's products or services; they might have their own set of customers ('second-tier' customers). Again, the operation may at times supply second-tier customers directly. The suppliers and customers who have direct contact with an operation are called its **immediate supply network**, whereas all the operations that form the network of suppliers' suppliers and customers' customers, etc., are called the **total supply network**. Along with the forward flow of transformed resources (materials, information and customers) in the network, each customer–supplier linkage will feed back orders and information. For example, when stocks run low, retailers place orders with distributors, which likewise place orders with the manufacturer, which will in turn place orders with its suppliers, which will replenish their own stocks from their own suppliers. So, flow is a two-way process with products or services flowing one way and information flowing the other.

Figure 5.2 illustrates the total supply network for two different operations. First is a plastic homeware (kitchen bowls, etc.) manufacturer. On the demand side it supplies products to wholesalers that supply retail outlets. However, it also supplies some retailers directly, bypassing a stage in the network. As products flow from suppliers to customers, orders and information flow the other way from customers to suppliers. Yet it is not only manufacturers that are part of a supply network. The flow of physical materials may be easier to visualise, but service operations also sit within supply networks. One way to visualise the supply networks of some service operations is to consider the downstream flow of information that passes between operations. Most financial service supply networks can be thought about like this. However, not all service supply networks deal primarily in information. For example, the second illustration in Figure 5.2 shows the supply network centred on a shopping mall. It has suppliers that provide security services, cleaning services, maintenance services and so on. These first-tier suppliers will themselves receive services from recruitment agencies, consultants, etc. First-tier customers of the shopping mall are the retailers that lease retail space within the mall, who themselves serve retail customers. This is a supply network like any other. What is being exchanged between operations is the quality, speed, dependability, flexibility and cost of the services each operation supplies to its customers. In other words, there is a flow of 'operations performance' through the network. And although visualising the flow of 'performance' through supply networks is an abstract approach to visualising supply networks, it is a unifying concept. Broadly speaking, all types of supply network exist to facilitate the flow of 'operations performance'.

The importance of a supply network perspective

Understanding the nature of the supply network and the operations role within it is critical in understanding competitiveness, identifying significant links in the network, and shifting towards a longer-term perspective.

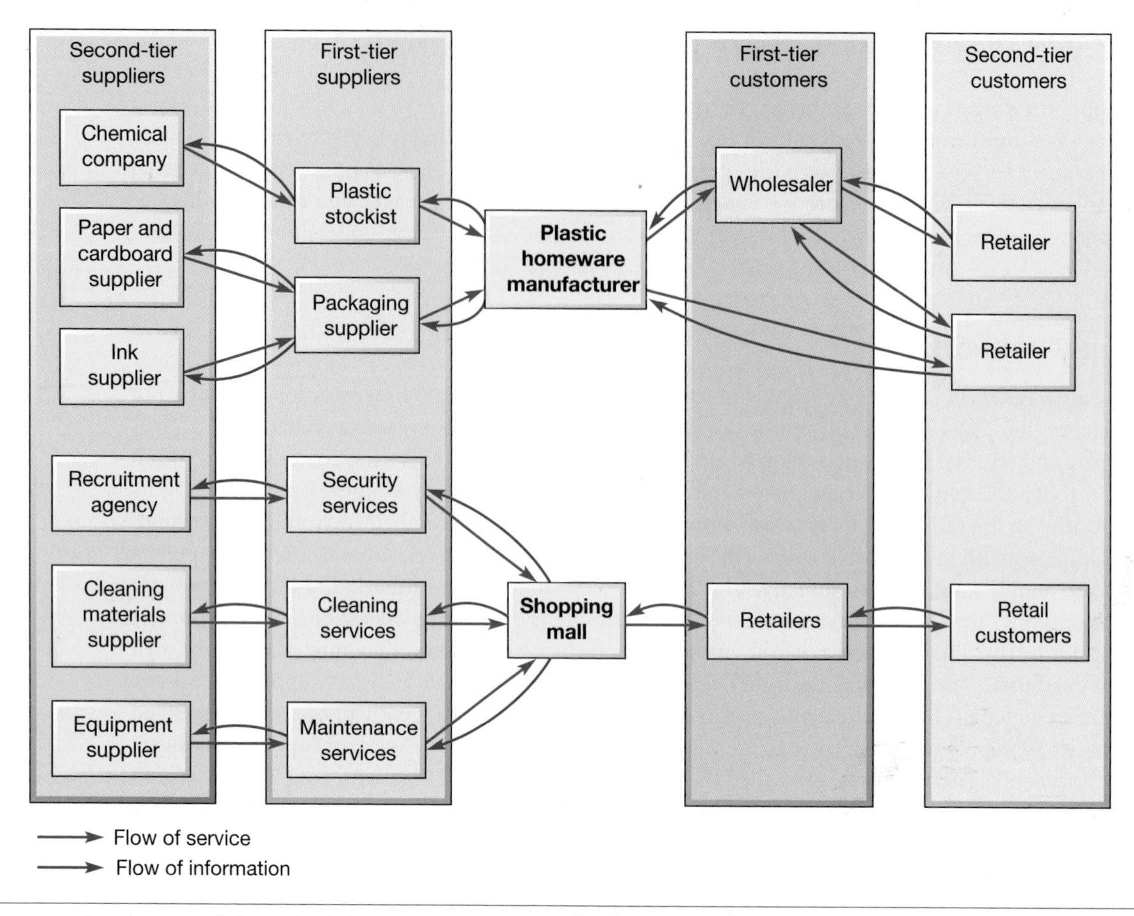

Figure 5.2 Supply network for a plastic homeware company and a shopping mall

Understanding competitiveness

Immediate customers and immediate suppliers, quite understandably, are the main concern for companies. Yet sometimes they need to look beyond these immediate contacts to understand why customers and suppliers act as they do. Any operation has only two options if it wants to understand its ultimate customers' needs at the end of the network. It can rely on all the intermediate customers and customers' customers, etc., which form the links in the network between the company and its end customers. Alternatively, it can look beyond its immediate customers and suppliers. Relying on one's immediate network is seen as putting too much faith in someone else's judgement of things that are central to an organisation's own competitive health.

Identifying significant links in the network

Not everyone in a supply network has the same degree of influence on the overall performance of the network; some contribute more than others. An analysis of networks needs to understand the downstream and the upstream operations that contribute most to end-customer service. For example, the end customers for domestic plumbing parts and appliances are the installers and service companies that deal directly with consumers. They are supplied by 'stockholders', who must have all parts in stock and deliver them fast. Suppliers of parts to the stockholders can best contribute to installers' competitiveness partly by offering fast delivery, but mainly through dependable delivery. The key players in this example are the stockholders. The best way of winning end-customer business in this case is to give the stockholder prompt delivery, which helps keep costs down while providing high availability of parts.

Taking a longer-term perspective

There are times when circumstances render parts of a supply network weaker than its adjacent links. High-street music stores, for example, have been largely displaced by music streaming services. A long-term supply network view would involve constantly examining technology and market changes to see how each operation in the supply network might be affected.

Operations principle

A supply network perspective helps to make sense of competitive, relationship and longer-term operations issues.

Structure and scope

What do we mean by the structure and scope of an operation's supply network? The first point to make is that structure and scope are strongly related (which is why we treat them together). For example, look again at the supply network for the shopping mall in Figure 5.2. Suppose that the company that runs the mall is dissatisfied with the service that it is receiving from the firm supplying security services. It may consider three alternatives. Option 1 is to switch suppliers and award the security contract to another security services supplier. Option 2 is to accept an offer from the company that supplies cleaning services to supply both security *and* cleaning services. Option 3 is to take over responsibility for security itself, hiring its own security staff. These options are illustrated in Figure 5.3. The first changes neither the structure nor the scope of this part of the supply network. The shopping mall still has three suppliers and is doing exactly what it did before. All that has changed is the (hopefully better) supplier. However, option 2 changes the structure of the supply network (the mall now has only two suppliers, the combined cleaning and security supplier, and the maintenance supplier), but not the scope of what the mall does, which is exactly what it did before. Option 3 changes both the structure of the network (again, the mall has only two suppliers: cleaning and maintenance services) and the scope of what the mall does (it now also takes on responsibility for security itself).

So, decisions relating to structure and scope are often interrelated. But, for simplicity, we will treat them separately. The second point to make is that both structure and scope decisions actually comprise a number of other 'constituent' decisions. These are shown in Figure 5.4. The structure of an operation's supply network is determined by three sets of decisions:

- How should the network be configured?
- What physical capacity should each part of the network have? (The long-term capacity decision.)
- Where should each part of the network be located? (The location decision.)

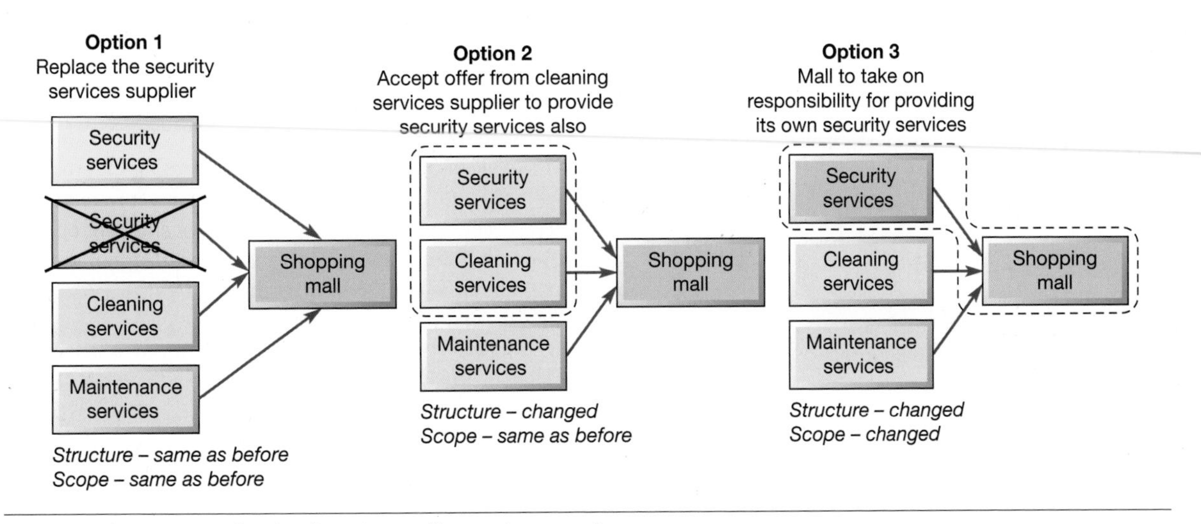

Figure 5.3 Three options for the shopping mall's supply network

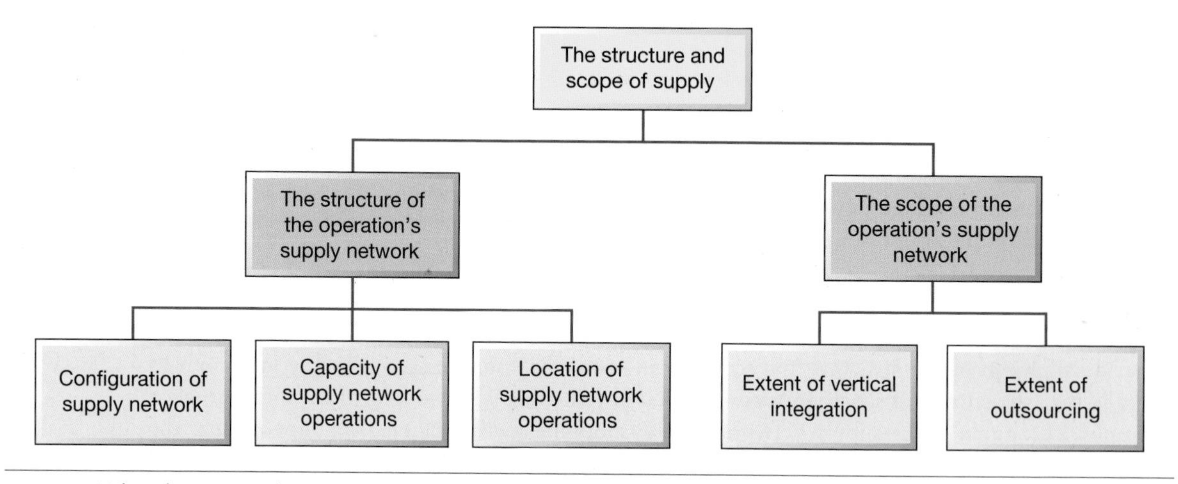

Figure 5.4 What determines the structure and scope of supply?

The scope of an operation's activities within the network is determined by two decisions:

- The extent and nature of the operation's vertical integration.
- The nature and degree of outsourcing it engages in.

The final point to make here is that structure and scope decisions are undeniably strategic. Different approaches to the structure and scope of operations define how different organisations do business even when they are in similar markets. There are few decisions that are more strategic than which other businesses you are going to trade with (structure) and how much of the total activities in the supply network you are going to take responsibility for (scope). However, both structure and scope also have a more operational aspect. As we illustrated in Figure 5.3, an operation such as the shopping mall can change its supply arrangements in a relatively short-term manner. We will treat the more operational day-to-day aspects of structure and scope in Chapter 12 on supply chain management.

OPERATIONS IN PRACTICE

Virtually like Hollywood

Could that most ephemeral of all industries, Hollywood's film-making business, hold messages about scope and structure for even the most sober of operations? It is an industry whose complexity most of us do not fully appreciate. In American writer F. Scott Fitzgerald's unfinished novel *The Last Tycoon*, the narrator of the story, Cecelia Brady, said, '*You can take Hollywood for granted like I did, or you can dismiss it with the contempt we reserve for what we don't understand . . . not half a dozen men have ever been able to keep the whole equation of* [making] *pictures in their heads*'. The 'equation' involves balancing the artistic creativity and fashion awareness necessary to create a market for its products, with the efficiency and tight operations practices that get films made and distributed on time. But although the form of the equation remains the same, the way its elements relate to each other has changed profoundly. The typical Hollywood studio once did everything itself. It employed everyone from the carpenters, who made the stage, through to the film stars. The film star Cary Grant (one of the biggest in his day) was as much of an employee as the chauffeur who drove him to the studio, though his contract was probably more restrictive. The finished products were rolls of film that had to be mass produced and physically distributed to the cinemas of the world. No longer. Studios now deal almost exclusively in ideas. They buy and sell concepts, they arrange finance, they cut marketing deals and, above all, they manage the virtual network of creative and not-so-creative talent that goes into a film's production. A key skill is the ability to put together teams of self-employed film stars and the small, technical specialist operations that provide technical support. It is a world that is less easy for the studios to control. The players in this virtual network, from film stars to electricians, have taken the opportunity to raise their fees to the point where, in spite of an increase in cinema attendance, returns are lower than at many times in the past. This opens opportunities for the smaller, independent studios. One way to keep costs low is by using inexpensive new talent. Technology could also help this process. Digital processes allow easier customisation of the 'product' and also mean that movies can be streamed direct to cinemas and direct to individual consumers' homes.

5.2 How should the supply network be configured?

Configuring a supply network means determining the overall pattern, shape or arrangement of the various operations that make up the supply network. Even when an operation does not directly own other operations in its network, it may still wish to change the shape of the network. A number of trends are reshaping networks in many industries. The most common example of network reconfiguration is the trend to reduce the number of direct suppliers that organisations work with. The complexity of dealing with many hundreds of suppliers is both expensive and can prevent operations from developing close relationships with its suppliers. Other configuration decisions include the **disintermediation** of some parts of the network, and a greater tolerance of other operations being both competitors and complementors at different times (**co-opetition**), the choice between using hubs or direct connections between operations, the development of **business ecosystems** and the increasing use of a triadic as opposed to dyadic perspective in supply networks.

Disintermediation

One trend in some supply networks is that of companies within a network bypassing customers or suppliers to make contact directly with customers' customers or suppliers' suppliers. 'Cutting out the middle men' in this way is called disintermediation. An obvious example of this is the way the internet has allowed some suppliers to 'disintermediate' traditional retailers in supplying goods and services to consumers. For example, many services in the travel industry that used to be sold through retail outlets (travel agents) are now also available direct from the suppliers. The option of purchasing the individual components of a vacation through the websites of the airline, hotel, car-hire operation, etc. became straightforward for consumers. They may still wish to purchase a service package from travel agents, which can have the advantage of convenience. Nevertheless, the process of disintermediation has developed new linkages in the supply network.

Co-opetition

One approach to thinking about supply networks sees any business as being surrounded by four types of players: suppliers, customers, competitors and complementors. Complementors enable one's products or services to be valued more by customers because they also can have the complementor's products or services, as opposed to when they have yours alone. Competitors are the opposite; they make customers value your product or service less when they can have their product or service, rather than yours alone. Competitors can also be complementors and vice versa. For example, adjacent restaurants may see themselves as competitors for customers' business. Yet in another way they are complementors. Would that customer have come to the area unless there was more than one restaurant to choose from? Restaurants, theatres, art galleries and tourist attractions generally all cluster together in a form of cooperation to increase the total size of their joint market. It is important to distinguish between the way companies cooperate in increasing the total size of a market and the way in which they then compete for a share of that market. Customers and suppliers, it is argued, should have 'symmetric' roles. Harnessing the value of suppliers is just as important as listening to the needs of customers. All the players in the network, whether they are customers, suppliers, competitors or complementors, can be both friends and enemies at different times. The term used to capture this idea is 'co-opetition'.

Network hubs versus direct connection

The more operations are involved in a supply network, the more complex it can become. One method of trying to reduce this complexity is to connect the operations within a network through a 'hub' operation. Rather that establish a connection between each operation in the network, they are routed through this common facility. In a logistics network the hub is usually a central warehouse that serves various other destinations such as smaller warehouses, retailers or customers. In passenger transportation a hub is a location, for example an airport, where passengers can transfer en route to their eventual destination. In air transport this idea is usually called the 'hub and spoke' model. Figure 5.5 illustrates a hub network structure as opposed to a 'direct' model.

The hub structure does have several advantages. The most obvious is that there are fewer routes to plan and maintain. If a network has n facilities, one of which is a hub, it will need $(n - 1)$ routes. If direct connections are used, $n(n - 1) / 2$ routes will be needed. It can also be easier to operate, especially if one business owns the whole network. For example, airlines often have a hub airport so that passengers can travel to a wide variety of destinations without switching airlines. Moreover, arrivals and departures from the hub can be coordinated so as to minimise waiting time or make loading at the hub more even. Also, a hub structure can be relatively easily expanded. An extra operation in the network simply means one more connection to the hub. Yet there are also advantages of direct connections. One of the biggest is reduced travel time, both by eradicating the transfer time at the hub and by avoiding convoluted routings. In an airline, for example, passengers could save time without having long layovers between connecting flights. Nor would they have to deal with any problems caused by a delayed flight. Direct connection structures can also be more robust because they do not rely on a single hub facility that may fail. For example, if a hub airport is closed by bad weather, the whole network will be affected.

Which structure is better will depend on a number of factors. The cost of movement is one factor. As aircraft became more efficient, airlines (especially domestic airlines) began to favour direct structures. However, the direct connection model does require a minimum level of traffic, otherwise infrequently used routes may become unviable. When airline traffic collapsed during the COVID-19 pandemic, several airlines reverted to the more efficient hub structure. The 'Operations in practice' example 'Aalsmeer: a flower auction hub' illustrates how a hub structure is used in the flower distribution sector.

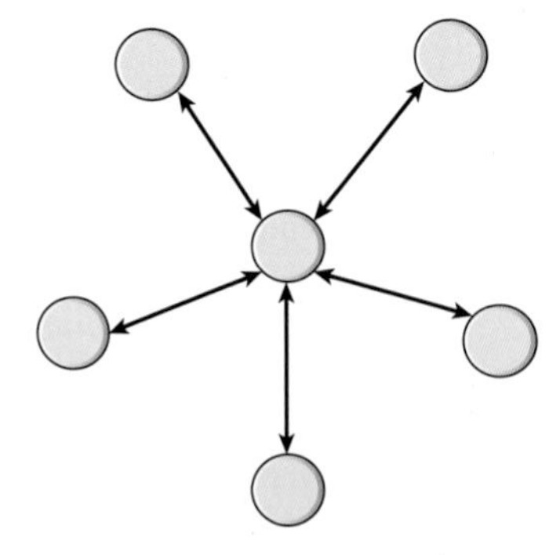

Hub network structure, 6
operations one of which acts
as a hub, requires 5 routes

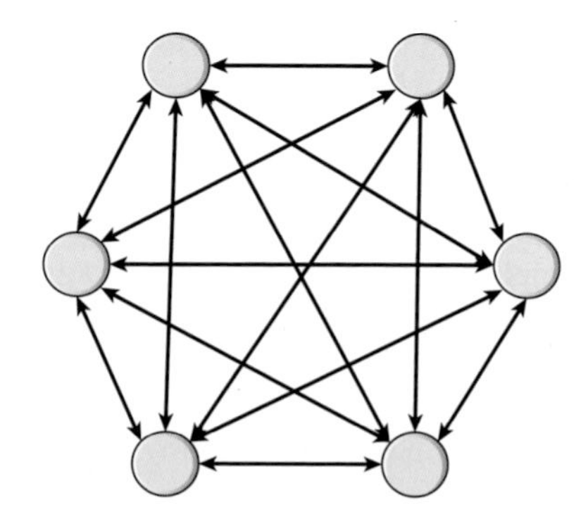

Direct network structure, 6
operations directly connected to
each other, requires 15 routes

Figure 5.5 Hub versus direct network structures

OPERATIONS IN PRACTICE

Aalsmeer: a flower auction hub[1]

A key feature in the configuration of any supply network is whether the separate supply chains that make up the network come together to form a 'hub'. A 'hub' is a point where several routes converge to share resources. They are often a key feature of a network because they can provide an opportunity to increase the utilisation of capacity, technology and expertise, as well as smooth the flow of goods through the network. This potentially leads to an increase in service levels, a reduction in throughput times and a reduction in costs. All these benefits can be seen in the Aalsmeer Flower Auction, the world's largest flower auction, operated by Royal Flora Holland and located near Amsterdam. Every day, from 4.00 am to 11.00 am, over 40 million flowers and plants from across the globe are traded in the second-largest building in the world, representing over 50 per cent of world trade in flowers. At first sight, the operations appear to be extraordinarily complex, with around 20,000 species of plants and flowers. Within the building, small 'trains' of crates containing these products are driven (or increasingly cycled for improved worker health) between quality check, sorting, auction rooms and despatch. Within the auction rooms, there is similar frenzied activity where 120,000 transactions take place in a matter of hours each day from when trading starts at 7.00 am. Products are then shipped via lorry for local markets and airfreight for markets worldwide within 24 hours, via the nearby Schiphol Airport. All processes, both within Aalsmeer Flower Auction and across its network of operations, are carefully designed and coordinated. For example, standardisation of containers that hold the wide variety of flowers and plants has dramatically reduced operational variety while still allowing for a huge range of products to flow through the giant floral hub. Beyond the auction itself, Royal Flora Holland plays an active role in coordinating its network of operations. Partnership agreements with suppliers, sharing of information (including more accurate demand forecasts), live tracking of products and their condition during transportation, training of farmers around new methods to improve product quality, shared R&D, and visits to both suppliers and buyers are all used to continually improve the competitiveness of the entire network.

Business ecosystems

An idea that is closely related to that of co-opetition in supply networks is that of the 'business ecosystem'. Like supply networks, business ecosystems include suppliers and customers. However, they also include stakeholders that may have little or no direct relationship with the main supply network yet interact with it by complementing or contributing significant components of the value proposition for customers. Many examples come from the technology industries where innovative products and services cannot evolve in a vacuum. They need to attract a whole range of resources, drawing in expertise, capital, suppliers and customers to create cooperative networks. For example, the app developers that develop applications for particular operating system platforms may not be 'suppliers' as such, but the relationship between them and the supply network that supplies the mobile device is mutually beneficial. Building an ecosystem of developers around a core product can increase its value to the end customer and by doing so increase the usage of the core product. Such an ecosystem of complementary products and services can also create significant barriers to entry for new competitors. Any possible competitors would not only have to compete with the core product, but also have to compete against the entire ecosystem of complementary products and services.

The triadic perspective on supply networks

Supply networks have many operations, all interacting in different ways. For simplicity, supply network academics and professionals often choose to focus on individual interaction between two specific operations in the network. This is called a 'dyadic' (simply meaning 'two') interaction, or

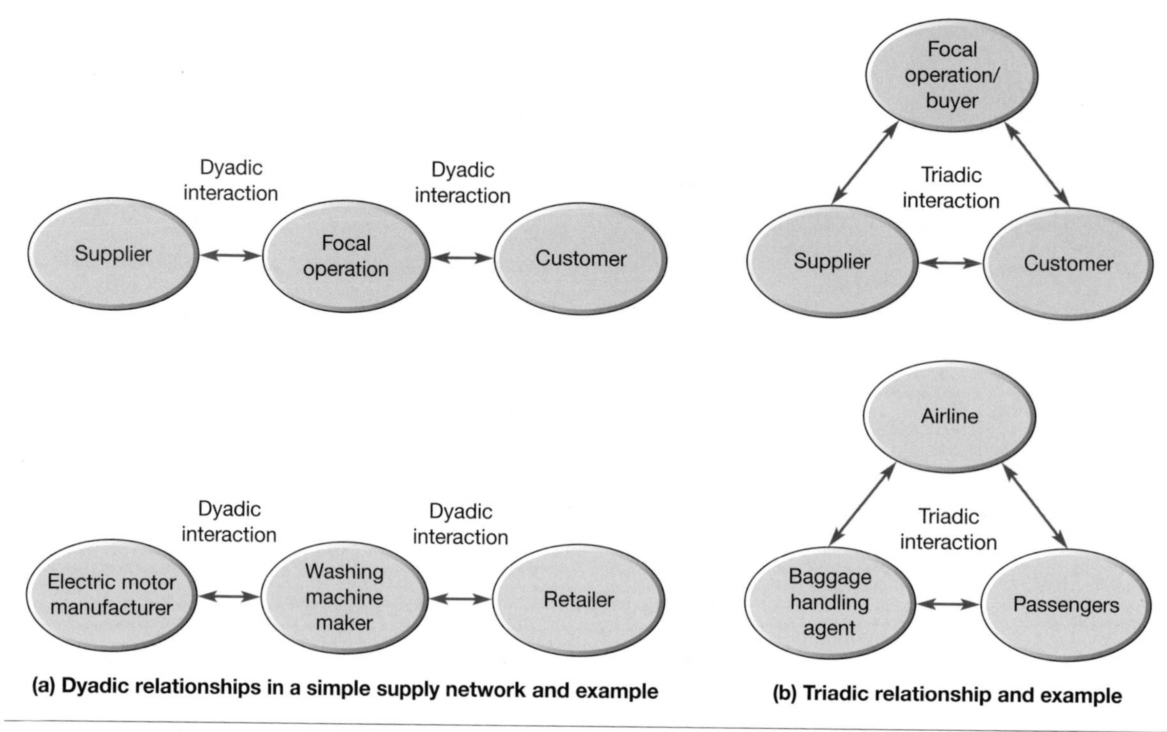

Figure 5.6 Dyadic and triadic relationships in two simple supply networks and examples

dyadic relationship, and the two operations are referred to as a 'dyad'. So if one wanted to examine the interactions that a focal operation had with one of its suppliers and one of its customers, one would examine the two dyads of 'supplier–focal operation' and 'focal operation–customer' (see Figure 5.6(a)). Much discussion (and research) is based on dyadic relationships because all relationships in a network are based on the simple dyad. However, more recently, and certainly when examining service supply networks, many authorities make the point that dyads do not reflect the real essence of a supply network. Rather, they say, it is triads, not dyads, that are the basic elements of a supply network (see Figure 5.6(b)). No matter how complex a network, it can be broken down into a collection of triadic interactions. The idea of triads is especially relevant in service supply networks. Operations are increasingly outsourcing the delivery of some aspects of their service to specialist providers, which deal directly with customers on behalf of the focal firm (more usually called the 'buying operation', or just 'buyer' in this context). For example, Figure 5.6(b) illustrates the common example of an airline contracting a specialist baggage-handling company to provide services to its customers on its behalf. Similarly, internal services are increasingly outsourced to form internal triadic relationships. For example, if a company outsources its IT operations, it is forming a triad between whoever is purchasing the service on behalf of the company, the IT service provider and the employees who use the IT services.

Critical commentary

The idea of widening the discussion of supply networks to include the 'business ecosystem' concept, described earlier, is also not without its critics. Some see it as simply another management 'buzzword', indistinguishable from the longer-established idea of the supply network. Other critics believe that the ecosystem metaphor is just a way for business to appear 'green'. They claim that the metaphor is used to suggest that the commercial relationships, on which almost all supply networks are based, have developed and are run using 'natural' values and therefore should be left to operate free from societal or government interference.

Thinking about supply networks as a collection of triads rather than dyads is strategically important. First, it emphasises the dependence of organisations on their suppliers' performance when they outsource service delivery. A supplier's service performance makes up an important part of how the buyer's performance is viewed. Second, the control that the buyer of the service has over service delivery to its customer is diminished in a triadic relationship. In a conventional supply chain, with a series of dyadic relationships, there is the opportunity to intervene before the customer receives the product or service. However, products or services in triadic relationships bypass the buying organisation and go directly from provider to customer. Third, and partially as a consequence of the previous point, in triadic relationships the direct link between service provider and customer can result in power gradually transferring over time from the buying organisation to the supplier that provides the service. Fourth, it becomes increasingly difficult for the buying organisation to understand what is happening between the supplier and customer at a day-to-day level. It may not even be in the supplier's interests to be totally honest in giving performance feedback to the buyer. This is referred to as the 'principal–agent' problem, where the principal here is the buyer and the agent is the supplier. Finally, this closeness between supplier and customer, if it excludes the buyer, could prevent the buyer from building important knowledge. For example, suppose a specialist equipment manufacturer has outsourced the maintenance of its equipment to a specialist provider of maintenance services. This reduces the ability of the equipment manufacturer to understand how its customers are using the equipment, how the equipment is performing under various conditions, how customers would like to see the equipment improved, and so on. The equipment manufacturer may have outsourced the cost and trouble of providing maintenance services, but it has also outsourced the benefits and learning that come from direct interaction with customers.

OPERATIONS IN PRACTICE

Adidas shuts its 'near market' factories[2]

It sounded like such a good idea. Like almost all of its rivals, Adidas had tended to concentrate on the design, marketing and distribution of its trainers (sneakers), subcontracting the 'making' part of the total process to a complex supply network, located largely in Asia. It had not run or owned its own manufacturing operations since the 1990s. The network of suppliers it employed spread over more than 1,000 facilities in 63 countries. Yet, like other similar companies, Adidas faced some problems with its Asian outsourcing model. Growing affluence in the area had increased costs, the longer and more complex supply networks were difficult to control, and such a globalised and complex supply chain meant a long lead time (around 18 months) between conceiving a new trainer and it arriving in the shops. And it is this last point that was the most problematic, particularly for fashionable trainers with a short 'fashion life'. Even orders to replenish stocks can take two to three months. But fashion cycles for trainers are getting shorter, with some designs lasting only one to three years. Faced with this, Adidas developed its 'Speedfactory' operation, the first one of which was located in Germany and the second one in the United States. The Speedfactory was totally automated, and designed to be able to accommodate new technologies, such as **3D printing** enabled by motion capture technology. And because almost all the stages of manufacturing were done on the same site, the intention was to make Adidas faster and more flexible, especially in producing small batches of fashionable products. It was hoped that the Speedfactories could produce shoes in days and replenish the fastest-selling products during the same season.

Yet within four years of the Speedfactories opening, Adidas announced it would cease production at the facilities. The company said it made more sense for the company to concentrate its Speedfactory production in Asia where the know-how and the vast majority of its suppliers were located, and where Adidas already makes more than 90 per cent of its products. Adidas said it would use its Speedfactory technology at two Asian supplier factories, and concentrate on modernising its other suppliers. One reason for the relative failure of the Speedfactories was the restricted range of models they could make. Adidas had set up Speedfactories to make trainers with a knit upper and Adidas's unique bouncy 'Boost' midsole, but it could not make leather shoes with a rubber sole because that used a different kind of joining process. So, as was pointed out by commentators, the effort was a failure not because its objective was flawed, but rather because it paid insufficient attention to the manufacturing process itself. Moreover, as Adidas pointed out, the learning that it gained from the Speedfactories would be used in its Asian supply base.

Structural complexity in retail supply networks

Some supply networks are relatively straightforward, both to describe and to manage. Others are more complex. At a simple level, the three-stage supply dyadic relationship shown in Figure 5.6(a) is less complex than the triadic supply network relationship shown in Figure 5.6(b). Put many triadic relationships together into an interrelated network and things become more complex. This is especially true when customers have some degree of choice over how and when they interact with a network. An example of such structural complexity is the move towards what have been called **omnichannel** retail networks. The original relationship between a retailer and its customers was straightforward – the retailer expected customers to collect goods from its stores. As online retailing increased, even when conventional retailers developed an online presence, they often kept their online operations separate from their high-street stores. This is the 'single-channel' model, as shown in Figure 5.7.

However, with the availability of more methods of contacting retailers (mobile phones, apps, social media, etc.) many retailers struggled to integrate the many alternative 'channels' of communication. In fact, they often treated them specifically as independent entities so that they could align each channel with specific targeted customer segments. This is the 'multichannel' model in Figure 5.7. Later came the first attempts to integrate high-street stores with (initially) web and other channels in order to

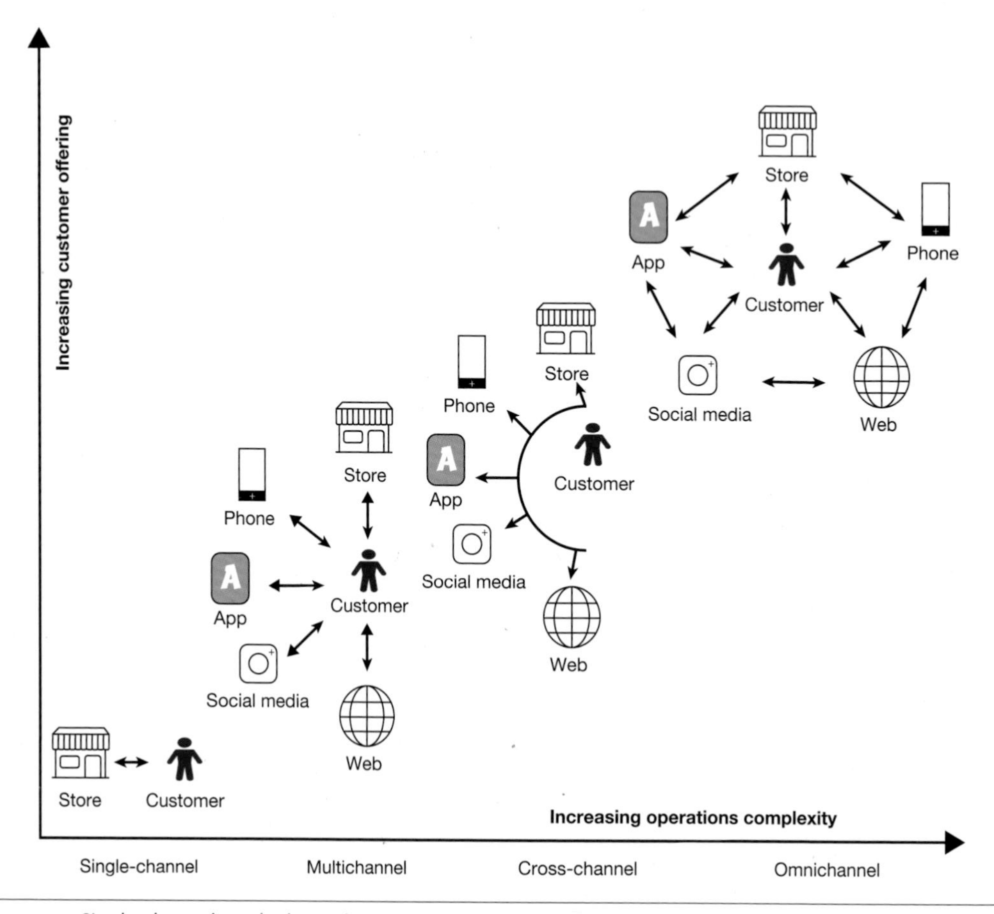

Figure 5.7 Single-channel, multichannel, cross-channel and omnichannel retail models

promote a better shopping experience for customers. Finally, the 'omnichannel' model is seen by many retail analysts as being one of the most important retail developments since the advent of online services. It seeks to provide a seamless all-inclusive customer experience by fully integrating all possible channels so customers can use whichever is the most convenient for them at whatever stage of the transaction. So, a customer could browse alternative products through social media, order their choice via an app, manage their account through a website, pay using their phone, and return the product to a physical store should they wish to. This requires a degree of technical sophistication and coordination between internal functions such as marketing, retailing operations, distribution and IT.

5.3 How much capacity should operations have?

The next set of 'structure' decisions concerns the size or capacity of each part of the supply network. Here we treat capacity in a general long-term sense. The specific issues involved in measuring and adjusting capacity in the medium and short terms are examined in Chapter 11.

The amount of capacity an organisation will have depends on its view of current and future demand. When an organisation has to cope with changing demand, a number of capacity decisions need to be taken. These include choosing the optimum capacity for each site and timing the changes in the capacity of each part of the network. Important influences on these decisions include the concepts of economy and diseconomy of scale.

The optimum capacity level

Most organisations need to decide on the size (in terms of capacity) of each of their facilities. A chain of truck service centres, for example, might operate centres that have various capacities. The effective cost of running each centre will depend on the average service bay occupancy. Low occupancy because of few customers will result in a high cost per customer served because the fixed costs of the operation are being shared between few customers. As demand, and therefore service bay occupancy, increases, the cost per customer will reduce.

The blue curves in Figure 5.8 show this effect for the service centres of 5-, 10- and 15-bay capacity. As the nominal capacity of the centres increases, the lowest cost point at first reduces. This is because the fixed costs of any operation do not increase proportionately as its capacity increases. A 10-bay centre has less than twice the fixed costs of a 5-bay centre. Also, the capital costs of constructing the operations do not increase proportionately to their capacity. A 10-bay centre costs less to build than twice the cost of a 5-bay centre. These two factors, taken together, are often referred to as economies of scale – a universal concept that applies (up to a point) to all types of operation. However, economies of scale do not go on forever. Above a certain size, the lowest cost point on curves such as that shown in Figure 5.8 may increase. This occurs because of what are called diseconomies of scale, two of which are particularly important. First, complexity costs increase as size increases. The communications and coordination effort necessary to manage an operation tends to increase faster than capacity. Although not seen as a direct cost, this can nevertheless be very significant. Second, a larger centre is more likely to be partially underutilised because demand within a fixed location will be limited. The equivalent in operations that process physical items is transportation costs. For example, if a manufacturer supplies the whole of its European market from one major plant in Denmark, all supplies may have to be brought in from several countries to the single plant and all products shipped from there throughout Europe.

Operations principle
Most operations exhibit economy of scale effects where operating costs reduce as the scale of capacity increases.

Operations principle
Diseconomies of scale increase operating costs above a certain level of capacity, resulting in a minimum cost level of capacity.

Operating at very high levels of capacity utilisation (occupancy levels close to capacity) can mean longer customer waiting times and reduced customer service. There may also be less obvious cost penalties of operating centres at levels close to their nominal capacity. For example, long periods of overtime may reduce productivity levels as well as costing more in extra payments to staff; utilising bays at very high utilisation reduces maintenance and cleaning time that may increase breakdowns, reduce effective life, and so on. This usually means that average costs start

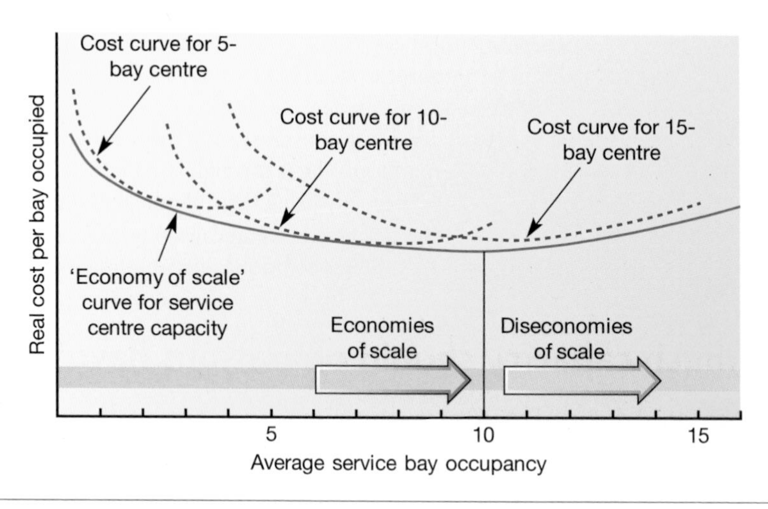

Figure 5.8 Unit cost curves for individual truck service centres of varying capacities

to increase after a point that is often lower than the theoretical capacity of the operation. In addition, while large-scale operations usually have a cost advantage over smaller units, there are potentially significant advantages that can be exploited by small-scale operations. These include:

- more responsive to change and an ability to act more like entrepreneurs in launching new products and services;
- more flexibility in decision making, with greater levels of autonomy granted to individuals than in most large-scale operations;
- greater market sensing given proximity to changing markets with more, but smaller, units of capacity.

Timing changes in capacity

Operations principle

Capacity-leading strategies increase opportunities to meet demand. Capacity-lagging strategies increase capacity utilisation.

Changing the capacity of any operation in a supply network is not just a matter of deciding on its optimum capacity. The operation also needs to decide when to bring new capacity 'on-stream'. For example, Figure 5.9 shows the forecast demand for a manufacturer's new product. In deciding when new capacity is to be introduced the company can mix the three strategies:

- Capacity is introduced to generally lead demand – timing the introduction of capacity in such a way that there is always sufficient capacity to meet forecast demand.
- Capacity is introduced to generally lag demand – timing the introduction of capacity so that demand is always equal to or greater than capacity.
- Capacity is introduced to sometimes lead and sometimes lag demand, but inventory built up during the 'lead' times is used to help meet demand during the 'lag' times. This is called 'smoothing with inventory'.

Each strategy has its own advantages and disadvantages. These are shown in Table 5.1. The actual approach taken by any company will depend on how it views these advantages and disadvantages. For example, if the company's access to funds for capital expenditure is limited, it is likely to find the delayed capital expenditure requirement of the capacity-lagging strategy relatively attractive. Of course, the third strategy, smoothing with inventory, is only appropriate for operations that produce products that can be stored. Customer-processing operations such as hotels cannot satisfy demand in one year by using rooms that were vacant the previous year.

Break-even analysis of capacity expansion

An alternative view of capacity expansion can be gained by examining the cost implications of adding increments of capacity on a break-even basis. Figure 5.10 shows how increasing capacity can move an operation from profitability to loss. Each additional unit of capacity results in a fixed-cost break

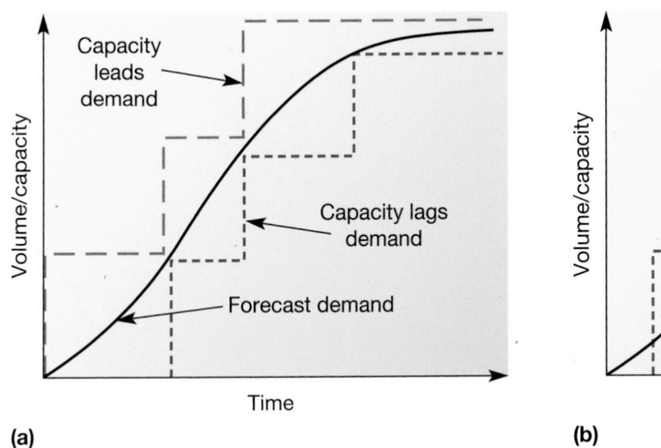

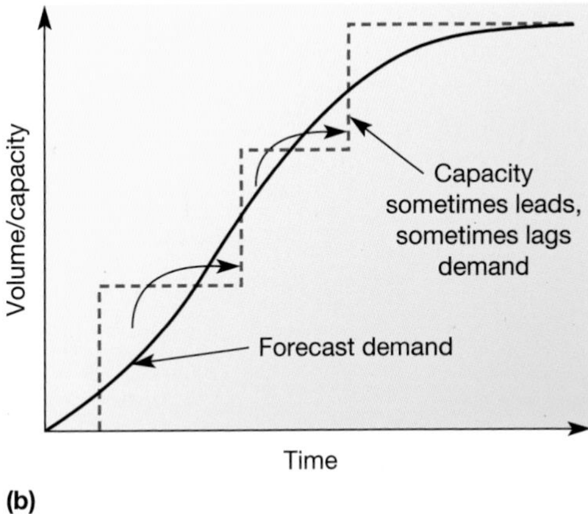

Figure 5.9 (a) Capacity leading and capacity lagging strategies, (b) Smoothing with inventories means using the excess capacity in one period to produce inventory that supplies the under-capacity period

Table 5.1 Arguments for and against leading, lagging and smoothing capacity strategies

Advantages	Disadvantages
Capacity-leading strategies	
Always sufficient capacity to meet demand, therefore revenue is maximised and customers satisfied	Utilisation of the plants is always relatively low, therefore costs will be high
Most of the time there is a 'capacity cushion' that can absorb extra demand if forecasts are pessimistic	Risks of even greater (or even permanent) over-capacity if demand does not reach forecast levels
Any critical start-up problems with new operations are less likely to affect supply	Capital spending on capacity will be early
Capacity-lagging strategies	
Always sufficient demand to keep the operation working at full capacity, therefore unit costs are minimised	Insufficient capacity to meet demand fully, therefore reduced revenue and dissatisfied customers
Over-capacity problems are minimised if forecasts prove optimistic	No ability to exploit short-term increases in demand
Capital spending on the operation is delayed	Under-supply position even worse if there are start-up problems with the new operations
Smoothing with inventory strategies	
All demand is satisfied, therefore customers are satisfied and revenue is maximised	The cost of inventories in terms of working capital requirements can be high. This is especially serious at a time when the company requires funds for its capital expansion
Utilisation of capacity is high and therefore costs are low	Risks of product deterioration and obsolescence
Very short-term surges in demand can be met from inventories	

that is a further lump of expenditure, which will have to be incurred before any further activity can be undertaken in the operation. The operation is unlikely to be profitable at very low levels of output. Eventually, assuming that prices are greater than marginal costs, revenue will exceed total costs. However, the level of profitability at the point where the output level is equal to the capacity of the operation may not be sufficient to absorb all the extra fixed costs of a further increment in capacity. This could make the operation unprofitable in some stages of its expansion.

Operations principle

Using inventories to overcome demand–capacity imbalance tends to increase working capital requirements.

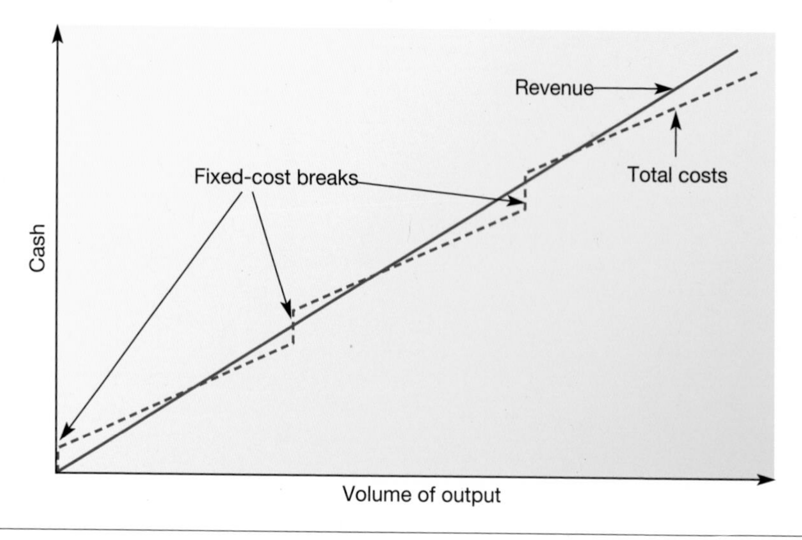

Figure 5.10 Repeated incurring of fixed costs can raise total costs above revenue

Worked example

De Vere Graphics

De Vere Graphics is investing in a new machine that enables it to make high-quality prints for its clients. Demand for these prints is forecast to be around 100,000 units in year 1 and 220,000 units in year 2. The maximum capacity of each machine the company will buy to process these prints is 100,000 units per year. They have a fixed cost of €200,000 per year and a variable cost of processing of €1 per unit. The company believe they will be able to charge €4 per unit for producing the prints. What profit are they likely to make in the first and second years?

Year 1 demand = 100,000 units; therefore the company will need one machine
Cost of producing prints = fixed cost for one machine + variable cost × 100,000
= €200,000 + (€1 × 100,000)
= €300,000
Revenue = demand × price
= 100,000 × €4
= €400,000
Therefore profit = €400,000 − €300,000
= €100,000
Year 2 demand = 220,000; therefore the company will need three machines
Cost of producung prints = fixed cost for three machines + variable cost × 220,000
= (3 × €200,000) + (€1 × 220,000)
= €820,000
Revenue = demand × price
= 220,000 × €4
= €880,000
Therefore profit = €880,000 − €820,000
= €60,000

Note: the profit in the second year will be lower because of the extra fixed costs associated with the investment in the two extra machines.

5.4 Where should operations be located?

The location of each operation in a supply network is both a key element in defining its structure, and will also have an impact on how the network operates in practice. If any operation in a supply network gets the location wrong, it can have a significant impact, not just on its own profits, but also on those of others in the network. For example, siting a data centre where potential staff with appropriate skills do not live will affect its performance and the service it gives its customers. Location decisions will usually have an effect on an operation's costs as well as its ability to serve its customers (and therefore its revenues). Also, location decisions, once taken, are difficult to undo. The costs of moving an operation can be hugely expensive and the risks of inconveniencing customers very high. No operation wants to move very often.

Reasons for location decisions

Not all operations can logically justify their location. Some are where they are for historical reasons. Yet even the operations that are 'there because they're there' are implicitly making a decision not to move. When operations do change their location, it is usually because of changes in demand and/or changes in supply, and an assumption that the potential benefits of a new location will outweigh any cost and disruption involved in the change.

Changes in demand

Customer demand shifting may prompt a change in location. For example, as garment manufacturers moved to Asia, suppliers of zips, threads, etc. started to follow. Changes in the volume of demand can also prompt relocation. To meet higher demand, an operation could expand its existing site, or choose a larger site in another location, or keep its existing location and find a second location for an additional operation; the last two options will involve a location decision. High-visibility operations may not have the choice of expanding on the same site to meet rising demand. A dry-cleaning service may attract only marginally more business by expanding an existing site because it offers a local, and therefore convenient, service. Finding a new location for an additional operation is probably its only option for expansion.

Changes in supply

The other stimulus for relocation is changes in the cost, or availability, of the supply of inputs to the operation. For example, a mining or oil company will need to relocate as the minerals it is extracting become depleted. The reason so many software companies located in India was the availability of talented, well-educated, but relatively lower-cost staff.

OPERATIONS IN PRACTICE

Aerospace in Singapore[3]

It is not immediously obvious why Singapore has been so successful at attracting a significant proportion of Asia's aerospace business. Unlike most states in the region, it has practically no internal air transport. But it does have one of the most respected airlines in the world in Singapore International Airlines (SIA). Yet it has attracted such global aerospace firms as Airbus, Rolls Royce, Pratt & Whitney, Thales, Bombardier and many more. There are a number of reasons for this. First, the location has access to the skills and infrastructure to support technically complex manufacturing. It was ranked number 1 globally in the 2019 World Economic Forum Global Competitiveness Report. Second, the country was trusted to provide an ethical framework for business. Again, it was ranked first in Asia for intellectual property rights protection in the 2019 World Economic Forum Global Competiveness Report.[4] Third, Asia is where the demand is. The world's fastest-growing airlines are in China, Singapore, Indonesia, India and in the Gulf. Fourth, the Singapore government offered significant help for companies wanting to invest in the sector, including generous tax incentives. Yet, although important, these incentives were not as beneficial as the 'soft' factors that make Singapore so attractive. In particular, the City State's universities and colleges, which produce the skilled scientists, engineers and staff who are vital to producing products that cannot be allowed to fail. The talent pipeline that the sector needed was seen as excellent and sustainable. Aerospace engineering courses are popular at Singapore's institutes of higher learning, and aerospace companies partner schools to develop courses and offer attachments and on-the-job training. Nor is government encouragement confined to large international aerospace firms. Many small and medium-sized enterprises in the sector built technical capabilities, particularly in areas such as aircraft and engine parts manufacturing, providing aerospace component maintenance, repair and overhaul (MRO) services.

Evaluating potential changes in location

Evaluating possible locations is almost always a complex task because the number of location options, the criteria against which they could be evaluated and the comparative rarity of a single location that clearly dominates all others, make the decision strategically sensitive. Furthermore, the decision often involves high levels of uncertainty. Neither the relocation activity itself nor the operating characteristics of the new site might be as assumed when the decision was originally made. Because of this, it is useful to be systematic in terms of (a) identifying alternative options, and (b) evaluating each option against a set of rational criteria.

Operations principle

An operation should change its location only if the benefits of moving outweigh the costs of operating in the new location plus the cost of the move itself.

Identify alternative location options

The first relocation option to consider is not to. Sometimes relocation is inevitable, but often staying put is a viable option. Even if seeking a new location seems the obvious way forward, it is worth evaluating the 'do nothing' option, if only to provide a 'base case' against which to compare other options. In addition to the 'do nothing' option, there should be a number of alternative location options. It is a mistake to consider only one location, but seeking out possible locations can be a time-consuming activity. Increasingly, for larger companies, the whole world offers possible locations. The implication of the globalisation of the location decision has been to increase both the number of options and the degree of uncertainty in their relative merits. The sheer number of possibilities makes the location decision impossible to 'optimise'. Rather, the process of identifying location options usually involves selecting a limited number of sites that represent different attributes. For example, a distribution centre, while always needing to be close to transport links, could be located in any of several regions and could be either close to population centres, or in a more rural location. The options may be chosen to reflect a range of both these factors. However, this assumes that the 'supply' of location options is relatively large, which is not always the case. In many retail location decisions, there are a limited number of high-street locations that become available at any point in time. Often, a retailer will wait until a feasible location becomes available and then decide whether to either take up that option or wait and take the chance that a better location becomes available soon. In effect, the location decision here is a sequence of 'take or wait' decisions.

Set location evaluation criteria

Although the criteria against which alternative locations can be evaluated will depend on circumstances, the following five broad categories are typical:

- *Capital requirements* – The capital or leasing cost of a site is usually a significant factor. This will probably be a function of the location of the site and its characteristics. In addition, the cost of the move itself may depend on which site is eventually chosen.
- *Market factors* – Location can affect how the market perceives an operation. Locating a general hospital in the middle of the countryside may have many advantages for its staff, but would be

very inconvenient for patients. Likewise, restaurants, stores, banks, petrol filling stations and many other high-visibility operations must all evaluate how alternative locations will determine their image and the level of service they can give. The same arguments apply to labour markets. Location can affect the attractiveness of the operation in terms of staff recruitment and retention. For example, 'science parks' are often located close to universities because they hope to attract companies that are interested in using the skills available at the university. But not all locations will necessarily have appropriate skills available. Staff at a remote call centre in the western islands of Scotland, used to a calm and tranquil life, were stunned by the aggressive nature of many callers, and needed assertiveness training.

- ***Cost factors*** – Two major categories of cost are affected by location. The first is the costs of producing products or services. For example, labour costs vary between different areas, especially internationally. They can exert a major influence on location, particularly in industries such as clothing, where labour costs as a proportion of total costs are relatively high. Other cost factors, known as community factors, derive from the social, political and economic environment of the site, for example local tax rates, government financial assistance, political stability, local attitudes to 'inward investment', language, local amenities (schools, theatres, shops, etc.), the availability of support services, the history of labour relations and behaviour, environmental restrictions and planning procedures. The second category of costs relate to both the cost of transporting inputs from their source to the location of the operation and the cost of transporting products and services from the location to customers. Whereas almost all operations are concerned to some extent with the former, not all operations are concerned with the latter, either because customers come to them (for example, hotels), or because their services can be 'transported' at virtually no cost (for example, some technology helpdesks). For supply networks that process physical items, however, transportation costs can be very significant.
- ***Future flexibility*** – Because operations rarely change their location, any new location must be capable of being acceptable, not only under current circumstances, but also under possible future circumstances, which is why in uncertain environments, any evaluation of alternative locations should include some kind of scenario planning that considers a range of possible futures.
- ***Risk factors*** – Closely related to the concept of future flexibility is the idea of evaluating the risk factors associated with possible locations. The risk criteria can be divided into 'transition risk' and 'long-term risk'. Transition risk is simply the risk that something goes wrong during the relocation process. For example, moving to an already congested location could pose higher risks to being able to move as planned than moving to a more accessible location. Long-term risks could again include damaging changes in input factors such as exchange rates or labour costs, but can also include more fundamental security risks to staff or property.

5.5 How vertically integrated should an operation's supply network be?

The scope of an operation's control of its supply network is the extent that it does things itself as opposed to relying on other operations to do things for it. This is often referred to as 'vertical integration' when it is the ownership of whole operations that is being decided, or 'outsourcing' when individual activities are being considered. We will look at the 'outsourcing' decision in the next section. The virtual integration decision involves an organisation assessing the wisdom of acquiring suppliers or customers, as well as the direction of integration, the extent of integration, and the balance among the vertically integrated stages. The decision as to whether to vertically integrate is largely a matter of a business balancing the advantages and disadvantages as they apply to that business.

In reality, different companies, even in the same industry, can make very different decisions over how much and where in the network they want to be. Figure 5.11 illustrates the (simplified) supply network for the wind turbine power generation industry. Original equipment manufacturers (OEMs) assemble the wind turbine nacelle (the nacelle houses the generator and gearbox). Towers and blades are often built to the OEM's specifications, either in-house or by outside suppliers. Installing wind turbines involves assembling the nacelle, tower and blades on site, erecting the tower and connecting

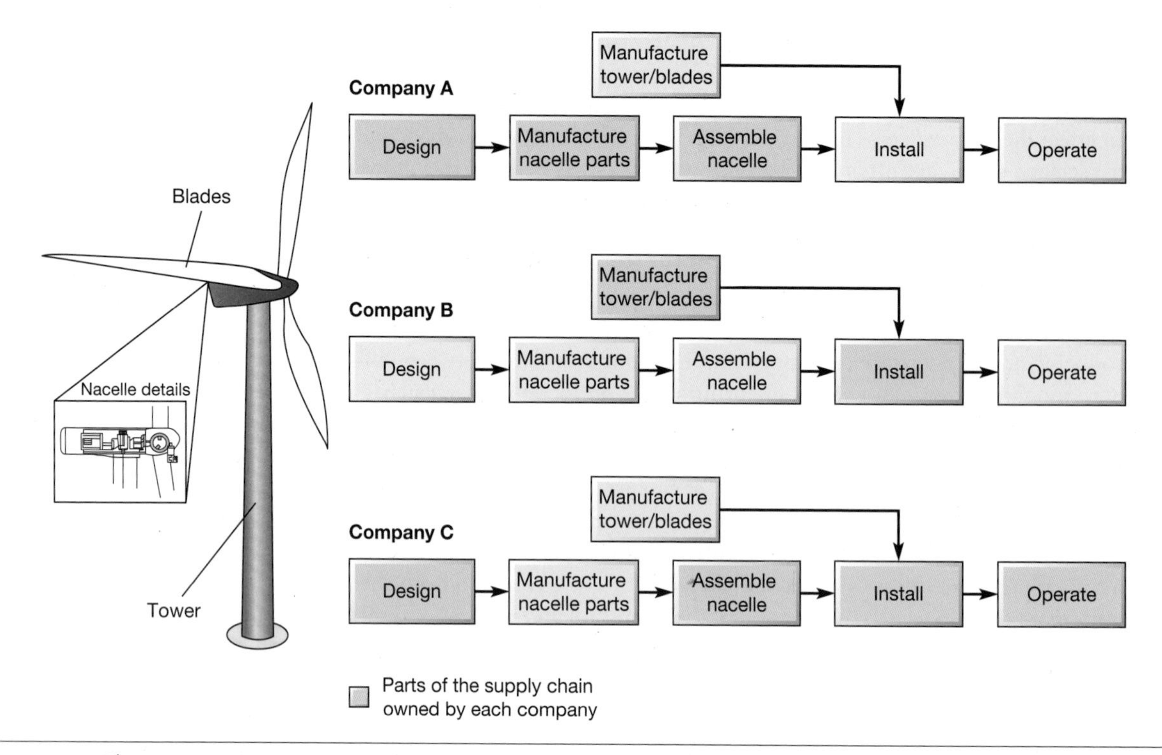

Figure 5.11 Three companies operating in the wind power generation industry with different vertical integration positions

to the electricity network. The extent of vertical integration varies by company and component. The three companies illustrated in Figure 5.11 have all chosen different vertical integration strategies. Company A is primarily a nacelle designer and manufacturer that also makes the parts. Company B is primarily an installer that also makes the tower and blades (but buys in the nacelle itself). Company C is primarily an operator that generates electricity and also designs and assembles the nacelles as well as installing the whole tower (but it outsources the manufacture of the nacelle parts, tower and blades).

An organisation's vertical integration strategy should consider three key elements – the direction of integration, the extent of integration and the balance among vertically integrated stages:

- ***The direction of integration*** – If a company decides that it should control more of its network, should it expand by buying one of its suppliers or should it expand by buying one of its customers? The strategy of expanding on the supply side of the network is called backward or '**upstream**' vertical integration and expanding on the demand side is called forward or '**downstream**' vertical integration. Backward vertical integration, by allowing an organisation to take control of its suppliers, is often used either to gain cost advantages or to prevent competitors gaining control of important suppliers. Forward vertical integration, on the other hand, takes an organisation closer to its markets and allows more freedom to make contact directly with its customers, and possibly sell complementary products and services.
- ***The extent of the process span of integration*** – Some organisations deliberately choose not to integrate far, if at all, from their original part of the network. Alternatively, some organisations choose to become very vertically integrated. Take many large international oil companies, such as Exxon, for example. Exxon is involved with exploration and extraction as well as the refining of the crude oil into a consumable product – gasoline. It also has operations that distribute and retail the gasoline (and many other products) to the final customer. This path (one of several for its different products) has moved the material through the total network of processes, all of which are owned (wholly or partly) by the one company.
- ***The balance among the vertically integrated stages*** – This is not strictly about the ownership of the network; it concerns the capacity and, to some extent, the operating behaviour of each stage in the network that is owned by the organisation. It refers to the amount of the capacity at each stage in the network that is devoted to supplying the next stage. So, a totally balanced network

relationship is one where one stage produces only for the next stage in the network and totally satisfies its requirements. Less than full balance in the stages allows each stage to sell its output to other companies or buy in some of its supplies from other companies.

Advantages of vertical integration

Although extensive vertical integration is no longer as popular as it once was, there are still companies that find it advantageous to own several sequential stages of their supply network. What then are the reasons why companies still choose to vertically integrate? Most justifications for vertical integration fall under four categories. These are:

- ***It secures dependable access to supply or markets*** – The most fundamental reasons for engaging in some vertical integration is that it can give more secure supply or bring a business closer to its customers. In some cases, there may not even be sufficient capacity in the supply market to satisfy the company. It therefore has little alternative but to supply itself. Downstream vertical integration can give a firm greater control over its market positioning. For example, Apple has always adopted a supply network model that integrates hardware and software, with both its hardware and software designed by Apple.
- ***It may reduce costs*** – The most common argument here is that 'We can do it cheaper than our supplier's price'. Such statements are often made by comparing the marginal direct cost incurred by a company in doing something itself against the price it is paying to buy the product or service from a supplier. But costs saving should also take into account start-up and learning costs. A more straightforward case can be made when there are technical advantages of integration. For example, producing aluminium kitchen foil involves rolling it to the required thickness and then 'slitting' it into the finished widths. Performing both activities in-house saves the loading and unloading activity and the transportation to another operation. Vertical integration also reduces the 'transaction costs' of dealing with suppliers and customers. Transaction costs are expenses, other than price, which are incurred in the process of buying and selling, such as searching for and selecting suppliers, setting up monitoring arrangements, negotiating contracts, and so on. However, if transaction costs can be lowered to the point where the purchase price plus transaction costs are less than the internal cost, there is little justification for the vertical integration of the activity.
- ***It may help to improve product or service quality*** – Sometimes vertical integration can be used to secure specialist or technological advantage by preventing knowledge getting into the hands of competitors. The exact specialist advantage may be anything from the 'secret ingredient' in fizzy drinks through to a complex technological process. For example, Dyson controls the majority of its value stream in developing, manufacturing and distributing its highly innovative vacuum cleaners, fans and, more recently, hairdryers. The main reason cited by its owner, Sir James Dyson, is to protect as much intellectual property as possible.
- ***It helps in understanding other activities in the supply network*** – Some companies, even those which are famous for their rejection of traditional vertical integration, do choose to own some parts of the supply network other than what they regard as core. For example, McDonald's, the restaurant chain, although largely franchising its retail operations, does own some retail outlets. How else, it argues, could it understand its retail operations so well?

Disadvantages of vertical integration

The arguments against vertical integration are often the following:

- ***It creates an internal monopoly*** – Operations, it is argued, will only change when they see a pressing need to do so. Internal supply is subject to reduced competitive forces that keep operations motivated to improve. If an external supplier serves its customers well, it will make higher profits; if not, it will suffer. Such incentives and sanctions do not apply to the same extent if the supplying operation is part of the same company.
- ***You cannot exploit economies of scale*** – Any activity that is vertically integrated within an organisation is probably also carried out elsewhere in the industry. But the effort it puts into the process

will be a relatively small part of the sum total of that activity within the industry. Specialist suppliers that can serve more than one customer are likely to have volumes larger than any of their customers could achieve doing things for themselves. This allows specialist suppliers to reap some of the cost benefits of economies of scale, which can be passed on in terms of lower prices to their customers.

- ***It results in loss of flexibility*** – Heavily vertically integrated companies by definition do most things themselves. This means that a high proportion of their costs will be fixed costs. They have, after all, invested heavily in the capacity that allows them to do most things in-house. A high level of fixed costs relative to variable costs means that any reduction in the total volume of activity can easily move the economics of the operation close to, or below, its break-even point.
- ***It cuts you off from innovation*** – Vertical integration means investing in the processes and technologies necessary to create products and services in-house. But as soon as that investment is made, the company has an inherent interest in maintaining it. Abandoning such investments can be both economically and emotionally difficult. The temptation is always to wait until any new technology is clearly established before admitting that one's own is obsolete. This may lead to a tendency to lag in the adoption of new technologies and ideas.
- ***It distracts you from core activities (loss of focus)*** – The final, and arguably most powerful, case against vertical integration concerns any organisation's ability to be technically competent at a very wide range of activities. All companies have things that they need to be good at. It is far easier to be exceptionally good at something if the company focuses exclusively on it rather than being distracted by many other things. Vertical integration, by definition, means doing more things, which can distract from the (few) particularly important things.

5.6 What activities should be in-house and what should be outsourced?

Outsourcing is the activity of taking activities that have been carried out in-house and moving them to outside suppliers. Theoretically, 'vertical integration' and 'outsourcing' are almost the same thing, with the difference between them one of scale. Vertical integration is a term that is usually applied to whole operations. 'Outsourcing' (also known as the 'do-or-buy' or 'make-or-buy' decision) usually applies to smaller sets of activities that have previously been performed in-house. For example, asking a specialist laboratory to perform quality tests previously performed by one's own quality department is an outsourcing decision. Although most companies have always outsourced some of their activities, a larger proportion of direct activities are now being bought from suppliers. In addition, many indirect and administrative processes are now being outsourced. This is often referred to as business process outsourcing (BPO). In a similar way, non-core processes such as providing catering services could be outsourced to a specialist. Such processes may still be physically located where they were before, but the outsourcing service provider manages the staff and technology. This is often done to reduce cost, but there can sometimes also be significant gains in the quality and flexibility of service offered.

Outsourcing and offshoring

Two supply network strategies that are often confused are those of outsourcing and offshoring. Outsourcing means deciding to buy in products or services rather than perform the activities in-house. Offshoring means obtaining products and services from operations that are based outside one's own country. Of course, one may both outsource and offshore. Offshoring is very closely related to outsourcing and the motives for each may be similar. Offshoring to a lower-cost region of the world is usually done to reduce an operation's overall costs, as is outsourcing to a supplier that has greater expertise or scale or both.

Making the outsourcing decision

Outsourcing is rarely a simple decision. Operations in different circumstances with different objectives are likely to take different decisions. Yet the question itself is relatively simple, even if the decision itself is not: 'Does in-house or outsourced supply in a particular set of circumstances give the appropriate performance objectives that the operation requires to compete more effectively in its markets?' For example, if the main performance objectives for an operation are dependable delivery and meeting short-term changes in customers' delivery requirements, the key question should be: 'How does in-house supply or outsourcing give better dependability and delivery flexibility performance?' This means judging two sets of opposing factors – those that give the potential to improve performance, and those that work against this potential being realised. Table 5.2 summarises some arguments for in-house supply and outsourcing in terms of each performance objective.

Operations principle

Assessing the advisability of outsourcing should include how it impacts on relevant performance objectives.

Incorporating strategic factors into the outsourcing decision

Although the effect of outsourcing on the operation's performance objectives is important, there are other factors that companies take into account when deciding if outsourcing an activity is a sensible option. For example, if an activity has long-term strategic importance to a company, it is unlikely to outsource it. A retailer might choose to keep the design and development of its website in-house, even though specialists could perform the activity at less cost, because it plans to shift to exclusively online operations at some point. Nor would a company usually outsource an activity where it had specialised skills or knowledge. For example, a company making laser printers may have built up specialised knowledge in the production of sophisticated laser drives. This capability may allow it to introduce product or process innovations in the future. It would be foolish to 'give away' such capability. After these two more strategic factors have been considered, the company's operations performance can be taken into account. Obviously if its operations performance is already superior to any potential supplier, it would be unlikely to outsource the activity. But even if its performance is currently below that of potential suppliers, it may not outsource the activity if it feels that it could significantly improve its performance. Figure 5.12 illustrates this decision logic.

Operations principle

Assessing the advisability of outsourcing should include consideration of the strategic importance of the activity and the operation's relative performance.

Table 5.2 How in-house and outsourced supply may affect an operation's performance objectives

Performance objective	'Do it yourself' In-house supply	'Buy it in' Outsourced supply
Quality	The origins of any quality problems are usually easier to trace in-house and improvement can be more immediate but there can be some risk of complacency	Supplier may have specialised knowledge and more experience, also may be motivated through market pressures, but communication more difficult
Speed	Can mean synchronised schedules, which speeds throughput of materials and information, but if the operation has external customers, internal customers may be low priority	Speed of response can be built into the supply contract where commercial pressures will encourage good performance, but there may be significant transport/delivery delays
Dependability	Easier communications can help dependability, but if the operation also has external customers, internal customers may receive low priority	Late delivery penalties in the supply contract can encourage good delivery performance, but organisational barriers may inhibit communication
Flexibility	Closeness to the real needs of a business can alert the in-house operation to required changes, but the ability to respond may be limited by the scale and scope of internal operations	Outsourced suppliers may be larger and have wider capabilities than in-house suppliers and more ability to respond to changes, but may have to balance conflicting needs of different customers
Cost	In-house operations do not have to make the margin required by outside suppliers, so the business can capture the profits that would otherwise be given to the supplier, but relatively low volumes may mean that it is difficult to gain economies of scale or the benefits of process innovation	Probably the main reason why outsourcing is so popular. Outsourced companies can achieve economies of scale and they are motivated to reduce their own costs because these directly impact on their profits, but transaction costs of working with a supplier need to be taken into account

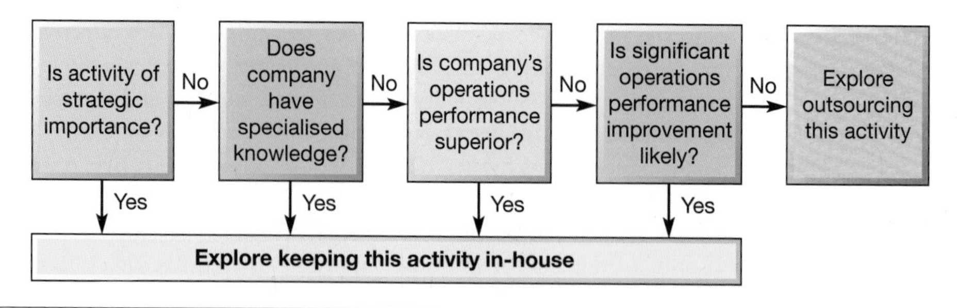

Figure 5.12 The decision logic of outsourcing

OPERATIONS IN PRACTICE

Compass and Vodafone – two ends of the outsourcing phenomenon[5]

Some companies have built their success on offering outsourcing services, and an obvious candidate for outsourcing is in-house catering, which is why there are many food service companies willing to take on that responsibility. The biggest in the world is the Compass Group, which provides services for a wide range of customers, from factories and office canteens, to schools and universities, oil rigs, the military, prisons and sporting events. Business customers that have outsourced their catering to Compass include Google, Microsoft, Nike, HSBC and Intel. Serving five and a half *billion* meals every year in 55,000 client locations, its 600,000 employees meet the catering needs of many different types of client. The group has gained its leading position in foodservice outsourcing partly because of its sector approach that distinguishes between the varying market needs of 'business and industry', 'healthcare and seniors', 'education', 'sports and leisure', and 'defence, offshore and remote'. But although Compass has been particularly successful in establishing itself as a leading player in the industry, there are very good reasons for any organisation to consider outsourcing its catering operations. For most businesses, catering is clearly not their core business. However good an in-house operation is, it will not be able to provide the variety, or have the expertise or the scale of a company like Compass. Nor is it likely to be able to stay ahead of food trends. Also, a company like Compass could help in providing non-routine catering such as events, reception services, office cleaning and some facilities management.

Getting outsourced catering wrong can upset staff, but getting outsourced customer servicing wrong can upset customers, which impacts revenues, and any operation's future. But the arguments for outsourcing customer servicing are similar to outsourcing catering. It holds the potential to give the same quality of customer support as in-house operations, but at a lower cost, particularly if it is outsourced to a country with a lower cost of living. It could provide a wider range of services and meet unexpectedly high demand. And specialist outsourced operations can invest in the latest technologies. Even so, customer care can be seen as so important that outsourcing it will always involve some risk. Customer frustration, especially if routed to overseas locations can prompt companies to consider bringing call-centre operations back to their home countries. One company to do this is Vodafone, the UK-based telecommunications company, who announced that it would be transferring over 2,000 jobs from its overseas call centres (mainly in South Africa, where it was using an external agency) to the United Kingdom. The new roles, it said, would be spread across the existing Vodafone call centres in various parts of the United Kingdom. They said that reshoring of the call centre jobs would bring the company's consumer mobile customer services in line with its UK-based call centres for UK business and broadband customers, and that the new roles would make a real difference to their customers as well as a real difference to the communities that would be the focus of their customer services investment. However, although Vodafone's move was directed at enhancing the quality of its UK customer services operations, in some parts of the world, legislation can be a factor. For example, the Italian government introduced a law that gave consumers calling a company the option of speaking to a call-centre worker in that country rather than someone overseas.

Globalisation, geopolitics, 'reshoring' and technology

Especially in the last few decades, the use of geographically dispersed suppliers to outsource at least some activities has become routine. This has been driven partly by labour cost differentials, partly by cheap and efficient IT connecting businesses, partly by trade deals, and partly by reduced transport costs. This is 'globalisation', where products, raw materials, money, technology and ideas move (relatively) smoothly across national boundaries. Globalisation is far from a recent phenomenon; it dates back to at least the 1st Century AD, with trade links between China and Europe. And there is a reason for its rise – globalised outsourcing is efficient. Apple and others can design their products in California and assemble them in China. A French aerospace company can direct the activities of its Brazilian suppliers almost as effectively as if they were in the next town. It pushes countries to specialise and swap, making them richer and the world smaller.

But it also introduces vulnerabilities. Trade differences and conflicts, human-rights boycotts, political nationalism and general geopolitical instability have slowed the progress of globalisation. In addition, the realisation that some supplies were concentrated in particular regions led to the consideration of supply vulnerabilities. For example, prior to 2020 China alone supplied about 42 per cent of the world's exports of personal protective equipment. Similarly, with other medical supplies, over 70 per cent of Italy's imported blood thinners came from China, as did 60 per cent of the ingredients for antibiotics imported by Japan. Not surprising then, that several countries put in policies that tried to reduce their dependence on any single supplier country.[6]

Increasingly some economists and business commentators question whether the boom in globalised operations is over. Some cite protectionist pressures in some countries in the Global North. Others see rising wages in (previously) Global South countries as reducing cost differentials. In addition, the operations-related advantages of sourcing from suppliers closer to home can be significant. Reducing reliance on complicated international supply chains can save transport and inventory costs, is less polluting, and is potentially less prone to reputational risk from any malpractice by remote suppliers. It can also increase supply flexibility. For example, the Spanish fast-fashion brand, Zara, manufactures some of its 'steady selling' items in low-cost factories in Asia, but makes the vast majority of its (less-predictable demand) garments closer to its markets so that it can respond quickly to changing fashions. Developments in technology could reinforce this so-called '**reshoring**' (also referred to as 'back-shoring', 'home-shoring' and 'on-shoring') process. Automation may encourage a trend towards 'radical insourcing' where countries in the Global North no longer need to outsource production to countries where wages are low.

Responsible operations

In every chapter, under the heading of 'Responsible operations', we summarise how the particular topic covered in the chapter touches upon important social, ethical and environmental issues.

Almost every decision described in this chapter has social, ethical or environmental implications. Changing the structure or scope of supply networks inevitably means some kind of profound change for the people and societies affected. And, although decisions regarding capacity undoubtedly have social and ethical implications, it is changes to the location and ownership of parts of a supply network that have the potential to be controversial, especially when considered in the context of 'globalisation'.

Any decision to outsource some degree of an operation's existing activities involves navigating a path through the tangle of incompatible priorities. Should customer (implicit) demands for cheaper products and services outweigh the loss of jobs suffered by current employees? If activities are outsourced to the Global South, how much responsibility should an enterprise take for conditions of employment in its suppliers? Working conditions acceptable, common and even welcomed in the Global South could be judged very differently in more prosperous countries. How should an operation draw the line between what is unacceptable in any location and under any circumstances (for example, slave or forced labour), and what is acceptable (for example, lower wage rates than would be acceptable in one's own country)?

Even where outsourcing does not involve any change in location (for example, outsourcing catering arrangements – see the 'Operations in practice' example 'Compass and Vodafone – two ends of the outsourcing phenomenon') there are ethical issues. Trade unions often point out that the only reason that outsourcing companies can do the job at lower cost is that they either reduce salaries, reduce working conditions, or both. Furthermore, they say, flexibility is achieved only by reducing job security. Employees who were once part of a large and secure corporation could find themselves as far less secure employees of a less benevolent employer with a philosophy of permanent cost cutting. Even some proponents of outsourcing are quick to point out the problems. There can be significant obstacles, including understandable resistance from staff who find themselves 'outsourced'. Some companies have also been guilty of 'outsourcing a problem'. In other words, having failed to manage a process well themselves, they ship it out rather than face up to why the process was problematic in the first place. There is also evidence that, although long-term costs can be brought down when a process is outsourced, there may be an initial period when costs rise as both sides learn how to manage the new arrangement.

OPERATIONS IN PRACTICE

Bangladesh disaster prompts reform – but is it enough?[7]

Outsourcing carries risks and responsibilities, often linked to a loss of control over suppliers. The Rana Plaza disaster provides an appalling example of the effect of this. On 24 April 2013 the Rana Plaza clothing factory near Dhaka in Bangladesh collapsed, killing a total of 1,134 people and injuring more than 2,500, most of whom were women and children. Many well-known clothing brands were sourcing products, either directly or indirectly, from the factory. It was claimed that local police and an industry association had issued a warning that the building was unsafe, but the owners had responded by threatening to dismiss people who refused to carry on working as usual. Understandably, there was an immediate call for tighter regulation and oversight by the Bangladeshi authorities. For years they had made only relatively weak attempts to enforce national building regulations, especially if the landlords involved were politically well connected. After the disaster, they promised to apply the laws more rigorously, but such promises had been made before.

There was also pressure for the predominantly American and European retailers that sourced from the Rana Plaza, and similar unsafe factories, to accept some of the responsibility for the disaster and change their buying policies. So, what are the options for these retailers? One option is to carry on as before and simply source garments from wherever is cheapest, although doing so is ethically questionable. Alternatively, retailers could stop sourcing from Bangladesh until its standards improve. But that may be difficult to enforce unless they took on the responsibility to police the whole supply chain right back to the cotton growers. It would also damage all Bangladeshi firms, even those that try to abide by safety rules. The third option is to stay and try to change how things are done in the country. Even before the Rana Plaza disaster some retailers had tried to improve safety in Bangladesh's 5,000 factories. Since the Rana Plaza disaster, some progress has been made in the country's garment industry. Campaigning organisations such as 'Fashion Revolution', 'Labour Behind the Label', 'War on Want' and 'Made in Europe' have urged retailers to be more transparent about their supply chains. Some brands did make moves to be more transparent about where their clothes were made and by whom. Other organisations focused on working conditions directly. The Fair Wear Foundation pioneered the establishment of anti-harassment committees and established programmes that provided training for managers, supervisors and workers in garment factories (85 per cent of female garment workers feared being sexually harassed at their workplace, with up to 60 per cent experiencing some form of sexual harassment). 'Fashion Revolution', the world's largest fashion activism movement founded a year after the disaster, published a manifesto outlining 10 clear demands for a better, responsible fashion industry. However, campaigning organisations admitted that progress had been limited. A report from the New York University Stern Centre for Business and Human Rights said that dangerous conditions continued to prevail at thousands of garment factories in Bangladesh.

Summary answers to key questions

5.1 What is the structure and scope of supply?

- A supply network includes the chains of suppliers providing inputs to the operation and the chains of customers who receive outputs from the operation.
- Understanding the nature of the supply network and the operations role within it is critical in understanding competitiveness, identifying significant links in the network, and shifting towards a longer-term perspective.
- The 'structure' of an operation's supply network relates to the shape and form of the network. It involves decisions around network configuration, capacity levels for each part of the network, and the location of each part of the network.
- The 'scope' of an operation's supply network relates to the extent that an operation decides to do the activities performed by the network itself, as opposed to requesting a supplier to do them. This involves deciding the extent of vertical integration and the degree of outsourcing.

5.2 How should the supply network be configured?

- Configuring a supply network means determining the overall pattern, shape or arrangement of the various operations that make up the supply network.
- Changing the shape of the supply network may involve reducing the number of suppliers to the operation so as to develop closer relationships, and bypassing or disintermediation of operations within the network.
- All the players in the network, whether they are customers, suppliers, competitors or complementors, can be both friends and enemies at different times. The term used to capture this idea is 'co-opetition'.
- An idea that is closely related to that of co-opetition is that of the 'business ecosystem'. Like supply networks, business ecosystems include suppliers and customers, but they also include stakeholders that have little direct relationship with the main network, yet interact by complementing or contributing significant components of value for end customers.
- Operations are increasingly outsourcing the delivery of some aspects of their service to specialist providers, which deal directly with customers on behalf of the focal firm. This marks a shift from a 'dyadic' perspective to a 'triadic' perspective.

5.3 How much capacity should operations have?

- The amount of capacity an organisation will have depends on its view of current and future demand. Key long-term capacity decisions include choosing the optimum capacity for each site and timing the changes in the capacity increase (or decrease) of each part of the network.
- When deciding the optimum capacity level, the concepts of economy and diseconomy of scale are critical.
- When deciding the timing of capacity change, organisations can consider a mix of three strategies – capacity introduced to lead demand, capacity introduced to lag demand and capacity smoothing, where inventory is used in lead periods to meet demand in lag periods.

5.4 Where should operations be located?

- The location of each operation in a supply network is both a key element in defining its structure and will also have an impact on how the network operates in practice.
- Key reasons for location decisions include changes in demand and/or changes in supply.
- Evaluating potential changes in location involves two key steps:
  - identifying alternative location options;
  - setting location evaluation criteria, including capital requirements, market factors, cost factors, future flexibility and risk factors.

5.5 How vertically integrated should an operation's supply network be?

- The scope of an operation's control of its supply network is the extent to which it does things itself as opposed to relying on other operations to do things for it. This is often referred to as 'vertical integration' when it is the ownership of whole operations that is being decided, or 'outsourcing' when individual activities are being considered.
- An organisation's vertical integration strategy can be defined in terms of the direction of integration, the extent of the process span of integration, and the balance among the vertically integrated stages.
- Advantages of vertical integration may include securing access to supply or markets; reducing costs; improving product or service quality; and improved understanding of supply network activities.
- Disadvantages of vertical integration may include the creation of an internal monopoly; lack of economies of scale; potential loss of flexibility; and isolation from innovation.

5.6 What activities should be in-house and what should be outsourced?

- Outsourcing is the activity of taking activities that could be or have been carried out in-house and moving them to outsourced suppliers. It is also known as the 'do-or-buy' or 'make-or-buy' decision.
- Making the outsourcing decision involves comparing the relative impact on key performance objectives of doing an activity in-house versus using an outsourced supplier. It also requires incorporation of other strategic factors, such as long-term competitive advantage and risk.
- There is a key difference between outsourcing and offshoring. Outsourcing means deciding to buy in products or services rather than perform the activities in-house. Off-shoring means obtaining products and services from operations that are based outside one's own country.
- While globalisation is a key trend that has led to geographically dispersed networks of operations, the last decade has seen some reversing of this trend, often referred to as reshoring.

CASE STUDY Aarens Electronic

Just outside Rotterdam in the Netherlands, Francine Jansen, the Chief Operating Officer of Aarens Electronic (AE) was justifiably proud of what she described as *'the most advanced machine of its type in the world, which will enable us to achieve new standards of excellence for our products requiring absolute cleanliness and precision'* and *'a quantum leap in harnessing economies of scale* [and] *new technology to provide the most advanced operation for years to come'*. The Rotterdam operation was joining AE's two existing operations in the Netherlands. The company offered precision custom coating and laminating services to a wide range of customers, among the most important being Phanchem, to which it supplied dry photoresist imaging films, a critical step in the manufacturing of microchips. Phanchem then processed the film further and sold it direct to microchip manufacturers.

The Rotterdam operation

The decision to build the Rotterdam operation had been taken because AE believed that a new low-cost operation using 'ultra-clean' controlled environment technology could secure a very large part of Phanchem's future business – perhaps even an exclusive agreement to supply 100 per cent of its needs. When planning the new operation three options were presented to AE's Executive Committee:

- **A** Expand an existing site by building a new machine within existing site boundaries. This would provide around 12 to 13 million square metres (MSM) per year of additional capacity and require around €19 million in capital expenditure.
- **B** Build a new facility alongside the existing plant. This new facility could accommodate additional capacity of around 15 MSM per year but, unlike option A, would also allow for future expansion. Initially this would require around €22 million in capital.
- **C** Set up a totally new site with a much larger increment of capacity, probably around 25 MSM per year. This option would be the most expensive at around €30 million.

Francine Jansen and her team initially favoured option B, but in discussion with the AE Executive Committee, opinion shifted towards the more radical option C. *'It may have been the highest-risk option but it held considerable potential and it fitted with the AE Group philosophy of getting into high-tech specialised areas of business. So we went for it'* (Francine Jansen). The option of a very large, ultra-clean, state-of-the-art facility also had a further advantage – it could change the economics of the photoresist imaging industry. In fact, global demand and capacity did not immediately justify investing in such a large increase in capacity. There was probably some overcapacity in the industry. But a large-capacity, ultra-clean type operation could provide a level of quality at such low costs that, if there were overcapacity in the industry, it would not be AE's capacity that would be lying idle.

Designing the new operation

During discussions on the design of the new operation, it became clear that there was one issue that was underlying all the team's discussions – how flexible should the process be? Should the team assume that they were designing an operation that would be dedicated exclusively to the manufacture of photoresist imaging film, and ruthlessly cut out any technological options that would enable it to manufacture other products, or should they design a more general-purpose operation that was suitable for photoresist imaging film, but could also make other products? It proved a difficult decision. The advantages of the more flexible option were obvious. *'At least it would mean that there was no chance of me being stuck with an operation and no market for it to serve in a couple of years' time'* (Francine Jansen). But the advantages of a totally dedicated operation were less obvious, although there was a general agreement that both costs and quality could be superior in an operation dedicated to one product.

Eventually the team decided to focus on a relatively non-flexible, focused and dedicated large machine. *'You can't imagine the agonies we went through when we decided not to make this a flexible machine. Many of us were not comfortable with saying, "This is going to be a photoresist machine exclusively, and if the market goes away we're in real trouble". We had a lot of debate about that. Eventually we more or less reached a consensus for focus but it was certainly one of the toughest decisions we ever made'* (Francine Jansen). The capital cost savings of a focused facility and operating costs savings of up to 25 per cent were powerful arguments, as was the philosophy of total process dedication. *'The key word for us was focus. We wanted to be quite clear about what was needed to satisfy our customer in making this single type of product. As well as providing significant cost savings to us it made it a lot easier to identify the root causes of any problems because we would not have to worry about how it might affect other products. It's all very clear. When the line was down we would not be generating revenue! It would also force us to understand our own performance. At our other operations, if a line goes down, the people can be shifted to other responsibilities. We don't have other responsibilities here – we're either making it or we're not'* (Francine Jansen).

When the Rotterdam operation started producing, the team had tweaked the design to bring the capacity at start-up to 32 MSM per year. Notwithstanding some initial teething troubles, it was, from the start, a technical and commercial success. Within six months a contract was signed with Phanchem to supply 100 per cent of its needs for the next ten years. Phanchem's decision was based on the combination of manufacturing and business focus that the Rotterdam team had achieved, a point stressed by Francine Jansen. *'Co-locating all necessary departments on the Rotterdam site was seen as particularly important. All the technical functions and the marketing and business functions are now on site'.*

Developing the supply relationship

At the time of the start-up, products produced in Rotterdam were shipped to Phanchem's facility near Frankfurt, Germany, almost 500 km away. This distance caused a number of problems, including some damage in transit and delays in delivery. However, the relationship between AE and Phanchem remained sound, helped by the two companies' cooperation during the Rotterdam start-up. *'We had worked closely with them during the design and construction of the new Rotterdam facility. More to the point, they saw that they would certainly achieve cost savings from the plant, with the promise of more savings to come as the plant moved down the learning curve'* (Francine Jansen). The closeness of the relationship between the two companies was a result of their staff working together. AE engineers were impressed by their customer's willingness to help out while they worked on overcoming the start-up problems. Similarly, AE had helped Phanchem when it needed extra supplies at short notice. As Francine Jansen said, *'partly because we worked together on various problems the relationship has grown stronger and stronger'.*

In particular the idea of a physically closer relationship between AE and Phanchem was explored. *'During the negotiations with Phanchem for our 100 per cent contract there had been some talk about co-location but I don't think anyone took it particularly seriously. Nevertheless, there was general agreement that it would be a good thing to do. After all, our success as Phanchem's sole supplier of coated photoresist was tied in to their success as a player in the global market – what was good for Phanchem was good for AE'* (Francine Jansen). Several options were discussed within and between the two companies. Phanchem had, in effect, to choose between four options:

- Stay where it was near Frankfurt.
- Relocate to the Netherlands (which would give easier access to port facilities) but not too close to AE (an appropriate site was available 30 km from Rotterdam).
- Locate to a currently vacant adjacent site across the road from AE's Rotterdam plant.
- Co-locate within an extension that could be specially built onto the AE plant at Rotterdam.

Evaluating the co-location options

Relatively early in the discussions between the two companies, the option of 'doing nothing' by staying in Frankfurt was discounted. Phanchem wanted to sell its valuable site near Frankfurt. The advantages of some kind of move were significant. The option of Phanchem moving to a site 30 km from Rotterdam was considered but rejected because it had no advantages over locating even closer to the Rotterdam plant. Phanchem also strongly considered building and operating a facility across the road from the Rotterdam plant. But eventually the option of locating in a building attached to AE's Rotterdam operation became the preferred option. Co-location would have a significant impact on Phanchem's competitiveness by reducing its operating costs, enabling it to gain market share by offering quality film at attractive prices, thus increasing volume for AE. The managers at the Rotterdam plant also looked forward to an even closer operational relationship with the customer. *'Initially, there was some resistance in the team to having a customer on the same site as ourselves. No one in AE had ever done it before. The step from imagining our customer across the road to imagining them on the same site took some thinking about. It was a matter of getting use to the idea, taking one step at a time'* (Francine Jansen).

The customer becomes a paying guest

However, when Francine and the Rotterdam managers presented their proposal for extending the plant to the AE board the proposal was not well received. *'Leasing factory space to our customer seemed a long way from our core business. As one Executive Committee member said, we are manufacturers; we aren't in the real estate business. But we felt that it would be beneficial for both companies'* (Francine Jansen). Even when the proposal was eventually accepted, there was still concern over sharing a facility. In fact the Executive Committee insisted that the door between the two companies' areas should be capable of being locked from both sides. Yet the construction and commissioning of the new facility for Phanchem was also a model of cooperation. Now, all visitors to the plant are shown the door that had to be 'capable of being locked from both sides' and asked how many times they think it has been locked. The answer, of course, is 'never'.

QUESTIONS

1 What were the key structure and scope decisions taken by Aarens Electronic?

2 What were the risks involved in adopting a process design that was 'totally dedicated' to the one customer's needs?

3 What were the advantages and disadvantages of each location option open to Phanchem, and why do you think it eventually chose to co-locate with AE?

This part of the text looks at how the resources and processes of operations are designed. By 'design' we mean how the overall form, arrangement and nature of transforming resources impact on the flow of transformed resources as they move through the operation. And that is the order in which we treat the four key issues that concern the design of operations. The chapters in this part are:

- **Chapter 6 Process design**

  This examines various types of processes, and how these 'building blocks' of operations are designed.

- **Chapter 7 The layout and look of facilities**

  This looks at how different ways of arranging physical facilities affect the appearance and the nature of flow through the operation.

- **Chapter 8 Process technology**

  This describes how the effectiveness of operations is influenced by the fast-moving developments in process technology.

- **Chapter 9 People in operations**

  This looks at the elements of human resource management that are traditionally seen as being directly within the sphere of operations management.

6 Process design

KEY QUESTIONS

6.1 What is process design?

6.2 What should be the objectives of process design?

6.3 How do volume and variety affect process design?

6.4 How are processes designed in detail?

INTRODUCTION

In Chapter 1 we described how all operations consist of a collection of processes that interconnect with each other to form an internal network. Each process acts as a smaller version of the whole operation of which they form a part, and transformed resources flow between them. We also defined a process as 'an arrangement of resources and activities that transform inputs into outputs that satisfy (internal or external) customer needs'. They are the 'building blocks' of all operations, and as such they play a vital role in how well operations operate. This is why process design is so important. Unless its individual processes are well designed, an operation as a whole will not perform as well as it could. And operations managers are at the forefront of how processes are designed. In fact, all operations managers are designers. When they purchase or rearrange the position of a piece of equipment, or when they change the way of working within a process, it is a design decision because it affects the physical shape and nature of their process, as well as its performance. This chapter examines the design of processes. Figure 6.1 shows where this topic fits within the overall model of operations management.

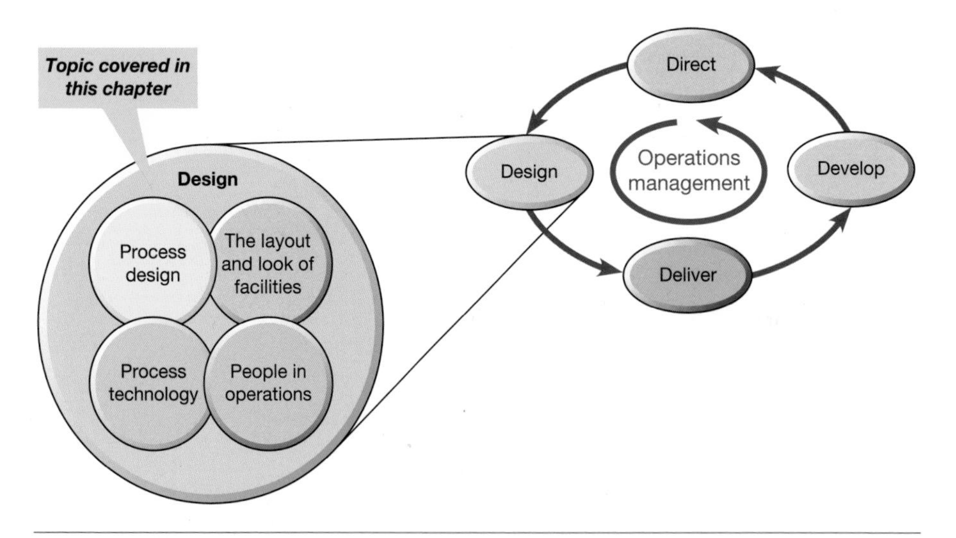

Figure 6.1 This chapter examines process design

6.1 What is process design?

To 'design' is to conceive the looks, arrangement and workings of something *before it is created*. In that sense, it is a conceptual exercise. Yet it is one that must deliver a solution that will work in practice. Design is also an activity that can be approached at different levels of detail. One may envisage the general shape and intention of something before getting down to defining its details. This is certainly true for **process design**. At the start of the process design activity it is important to understand the design objectives, especially when the overall shape and nature of the process is being decided. The most common way of doing this is by positioning it according to its volume and variety characteristics. Eventually the details of the process must be analysed to ensure that it fulfils its objectives effectively.

Process design and product/service design are interrelated

Often we will treat the design of services and products, on the one hand, and the design of the processes that make them, on the other, as though they were separate activities. Yet they are clearly interrelated. It would be foolish to commit to the detailed design of any product or service without some consideration of how it is to be produced. Small changes in the design of products and services can have profound implications for the way the operation eventually has to produce them. Similarly, the design of a process can constrain the freedom of product and service designers to operate as they would wish (see Figure 6.2). This holds good for all operations. However, the overlap between the two design activities is generally greater in operations that produce services. Because many services involve the customer in being part of the transformation process, the service, as far as the customer sees it, cannot be separated from the process to which the customer is subjected. Certainly, when product designers also have to make or use the things that they design, it can concentrate their minds on what is important. For example, in the early days of flight, the engineers who designed the aircraft were also the test pilots who took them out on their first flight.

> **Operations principle**
> The design of processes cannot be done independently of the services and/or products that are being created.

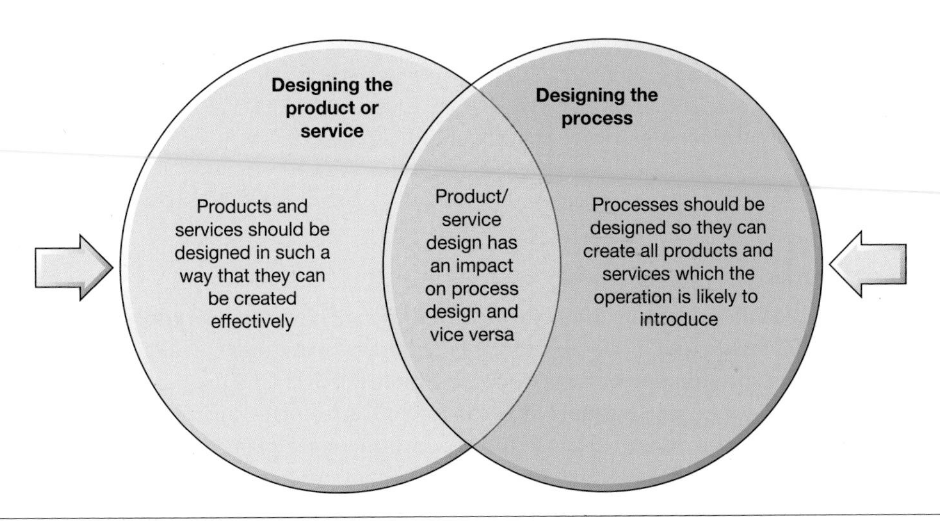

Figure 6.2 The design of products/services and the design of processes are interrelated and should be treated together

OPERATIONS IN PRACTICE

Changi airport[1]

Airports are complex operations – really complex. Their processes handle passengers, aircraft, crew, baggage, commercial cargo, food, security, restaurants and numerous customer services. Their operations managers must cope with aviation administration rules and regulations, a huge number of airport service contracts, usually thousands of staff with a wide variety of specialisms, airlines with competing claims to service priority, and customers, some of whom are experienced, others less so. Their processes are also vulnerable to disruptions from late arrivals, aircraft malfunction, weather, the industrial action of workers two continents away, conflicts and terrorism. Designing the processes that can operate under these conditions must be one of the most challenging operations tasks. So, to win prizes for 'Best Airport' customer service and operating efficiency year after year has to be something of an achievement, which is what the sixth-busiest international airport, Changi airport in Singapore, has done. As a major air hub in Asia, Changi serves more than 100 international airlines flying to some 300 cities in about 70 countries and territories worldwide. It handles almost 60 million passengers (that's roughly 10 times the size of Singapore's population). A flight takes off or lands at Changi roughly once every 90 seconds.

When Changi opened its new Terminal 4, it increased the airport's annual passenger handling capacity to around 82 million. Every stage of the customers' journey through the terminal was designed to be as smooth as possible. The aim of all the processes within and around the terminal was to provide fast, smooth and seamless flow for passengers. Each stage in the customer journey was provided with enough capacity to cope with anticipated demand. Once passengers arrive at the two-storey terminal building they pass through kiosks and automated options for self-check-in, self-bag tagging and self-bag-drops. Their bags are transported to the aircraft via an advanced and automated baggage handling system. Similarly, automated options, including face recognition technology, are used at immigration counters and departure gates. Biometric technology and 'fast and seamless travel' (FAST) services help to speed passenger throughput and increase efficiency. After security checks, passengers find themselves in 15,000 m^2 of shopping, dining and other retail spaces. The feelings of passengers were an important part of the design of T4. Architecturally, it aimed to be functional, and yet have its own aesthetic character, while ensuring that the design was passenger-centric and user-friendly. And with so many different companies involved in the day-to-day operation of the airport it was vital to include as many stakeholders as possible during the design. Workshops were conducted with various stakeholders, including airlines, ground handlers, immigration and security agencies, retail, and food and beverage operators as well as other users to ensure that the T4 design met the needs of each party.

Process networks

In Chapter 1 we used the 'hierarchy of operations' to illustrate how any operation is both made up of networks (of processes) and a part of networks (of other operations). This idea is essential in making all networks, including process networks, operate effectively. Figure 6.3 shows a simplified internal process network for one business. It has many processes that transform items and transfer them to other internal processes. Through this network there are many 'process chains', that is, threads of processes within the network. And thinking about processes as part of a network has a number of advantages. First, understanding how and where a process fits into the internal network helps to establish appropriate objectives for the process. Second, one can check to make sure that everyone in a process has a clear 'line of sight' forward through to end customers, so that the people working in each process have a better chance of seeing how they contribute to satisfying the operation's customers. Even more important, one can ask the question, 'how can each process help the intermediate processes that lie between them and the customer, to operate effectively?' Third, a

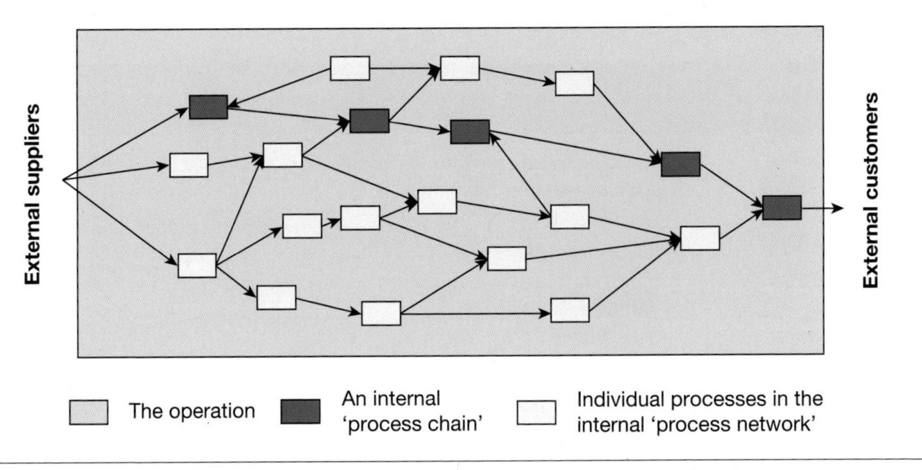

Figure 6.3 A process network within an operation showing an internal 'process chain'

clear 'line of sight' backwards through to the operation's suppliers makes the role and importance of suppliers easier to understand.

6.2 What should be the objectives of process design?

The whole point of process design is to make sure that the performance of the process is appropriate for whatever it is trying to achieve. For example, if an operation competes primarily on its ability to respond quickly to customer requests, many of its processes will need to be designed to give fast throughput times. Similarly, if an operation competes on low price, cost-related objectives are likely to dominate its process design. In other words, some kind of logic should link what the operation as a whole is attempting to achieve, and the performance objectives of its individual processes. As when we examined product and service design innovation in Chapter 4, we will include 'sustainability' as an operational objective of process design, even though it is really a far broader societal issue that is part of the organisation's 'triple bottom line' (see Chapter 2). This is illustrated in Table 6.1.

Operations principle

The design of any process should be judged on its quality, speed, dependability, flexibility, cost and sustainability performance.

Table 6.1 The impact of strategic performance objectives on process design objectives and performance

Operations performance objective	Typical process design objectives	Some benefits of good process design
Quality	▶ Provide appropriate resources, capable of achieving the specification of products or services ▶ Error-free processing	▶ Products and services produced 'on-specification' ▶ Less recycling and wasted effort within the process
Speed	▶ Minimum throughput time ▶ Output rate appropriate for demand	▶ Short customer waiting time ▶ Low in-process inventory
Dependability	▶ Provide dependable process resources ▶ Reliable process output timing and volume	▶ On-time deliveries of products and services ▶ Less disruption, confusion and rescheduling within the process

Table 6.1 (Continued)

Operations performance objective	Typical process design objectives	Some benefits of good process design
Flexibility	▶ Provide resources with an appropriate range of capabilities ▶ Change easily between processing states (what, how, or how much is being processed?)	▶ Ability to process a wide range of products and services ▶ Low cost/fast product and service change ▶ Low cost/fast volume and timing changes ▶ Ability to cope with unexpected events (e.g. a supply or processing failure)
Cost	▶ Appropriate capacity to meet demand ▶ Eliminate process waste in terms of: ▶ excess capacity ▶ excess process capability ▶ in-process delays ▶ in-process errors ▶ inappropriate process inputs	▶ Low processing costs ▶ Low resource costs (capital costs) ▶ Low delay/inventory costs (working capital costs)
Sustainability	▶ Minimise energy usage ▶ Reduce local impact on community ▶ Produce for easy disassembly	▶ Lower negative environmental and societal impact

'Micro' process objectives

Because processes are managed at a very operational level, process design also needs to consider a more 'micro'and detailed set of objectives. These are largely concerned with flow through the process. When whatever is being 'processed' enters a process, it will progress through a series of activities where it is 'transformed' in some way. Between these activities it may dwell for some time in inventories, waiting to be transformed by the next activity. This means that the time that a unit spends in the process (its throughput time) will be longer than the sum of all the transforming activities that it passes through. Also, the resources that perform the process's activities may not be used all the time because not all items will necessarily require the same activities and the capacity of each resource may not match the demand placed upon it. So, neither the items moving through the process, nor the resources performing the activities may be fully utilised. Because of this, the way that items leave the process is unlikely to be exactly the same as the way they arrive at the process. It is common for more 'micro' performance flow objectives to be used that describe process flow performance. For example:

- Throughput rate (or flow rate) is the rate at which items emerge from the process, i.e. the number of items passing through the process per unit of time.
- Cycle time is the reciprocal of throughput rate: it is the time between items emerging from the process. The term 'takt time' is the same, but is normally applied to 'paced' processes like moving-belt assembly lines. It is the 'beat' or tempo of working required to meet demand.[2]
- Throughput time is the average elapsed time taken for inputs to move through the process and become outputs.
- 'Work-in-progress' or process inventory is the number of items in the process, as an average over a period of time.
- The utilisation of process resources is the proportion of available time that the resources within the process are performing useful work.

Operations principle

Process flow objectives should include throughput rate, throughput time, work-in-progress and resource utilisation, all of which are interrelated.

OPERATIONS IN PRACTICE

Fast (but not too fast) food drive-throughs[3]

Some claim the first drive-through (or drive-thru, if you prefer) was the In-N-Out in California. Other claimants include the Pig Stand restaurant in Los Angeles, that allowed customers to drive by the back door where the chef would come out and deliver the restaurant's famous 'Barbequed Pig' sandwiches. What became apparent though, as the drive-through idea began to spread (and included other services such as banks), was that their design could have a huge impact on their efficiency and profitability. Today, drive-through processes are slicker, and far, far, faster, although most stick to a proven formula with orders generally placed by the customer using a microphone and picked up at a window. It is a system that allows drive-throughs to provide fast and dependable service. In fact, there is strong competition between drive-throughs to design the fastest and most reliable process. For example, some Starbucks drive-throughs have strategically placed cameras at the order boards so that servers can recognise regular customers and start making their order – even before it's placed. Other drive-throughs have experimented with simpler menu boards and see-through food bags to ensure greater accuracy. There is no point in being fast if you don't deliver what the customer ordered. These details matter. It has been estimated that sales increase 1 per cent for every six seconds saved at a drive-through. One experiment in making drive-through process times slicker was carried out by a group of McDonald's restaurants. On California's central coast, 150 miles from Los Angeles, a call centre took orders remotely from 40 McDonald's outlets around the country. The orders were then sent back to the restaurants and the food was assembled only a few metres from where the order was placed. Although saving only a few seconds on each order, it could add up to extra sales at busy times of the day. Another innovation is express lines for customers who place digital orders ahead of time. A good drive-through process should also help customers to contribute to speeding things up. So, for example, menu items must be easy to read and understand.

This is why what are often called 'combo meals' (burger, fries and a cola) can save time at the ordering stage. By contrast, complex individual items or meals that require customisation can slow down the process. This can become an issue for drive-through operators when fashion moves towards customised salads and sandwiches. Yet there are signs that above a certain speed of service, other aspects of process performance become more important. As one drive-through chief operations manager points out, *'you can get really fast but ruin the overall experience, because now you're not friendly'.*

Standardisation of processes

One of the most important process design objectives, especially in large organisations, concerns the extent to which process designs should be standardised. By standardisation in this context we mean 'doing things in the same way', or 'adopting a common sequence of activities, methods and use of equipment'. It is a significant issue in large organisations because, very often, different ways of carrying out similar or identical tasks emerge over time in the various parts of the organisation. But, why not allow many different ways of doing the same thing? That would give a degree of autonomy and freedom for individuals and teams to exercise their discretion. The problem is that allowing numerous ways of doing things causes confusion, misunderstandings and, eventually, inefficiency. In healthcare processes, it can even cause preventable deaths. For example, the Royal College of Physicians in the United Kingdom revealed that more than 100 different charts were used for monitoring patients' vital signs in UK hospitals.[4] Clinicians have to learn how to read new ones whenever they move, leading to confusion. Potentially, many hospital deaths could be prevented if clinicians used a standardised bed chart and process. The practical dilemma for most organisations is how to draw the line between processes that are required to be standardised, and those that are allowed to be different.

Operations principle

Standardising processes can give some significant advantages, but not every process can be standardised.

OPERATIONS IN PRACTICE

Legal & General's modular housing process[5]

Legal & General (L&G) is not the type of company one would expect to be building houses. One of the United Kingdom's leading financial services groups, which has invested over £19 billion in projects including homebuilding, urban regeneration and clean energy, it became involved in building modular homes. Modular construction of housing is more like the way you would expect a vehicle to be made. Modules are made 'off-site' in a factory then transported to the building site. As some modular construction proponents pointed out, once all cars were hand-built, but now cars are assembled in a factory. Rosie Toogood, the Chief Executive, Modular Homes, and champion of the modern manufacturing and construction approach, came from aerospace company Rolls-Royce, and Stuart Lord, the Manufacturing Operations Director, spent his prior career in the automotive industry.

Starting a modular homes business from scratch required the considerable investment that a major financial group like L&G could provide. By retaining quality construction but using modern processes L&G called its approach 'everything new, but nothing new'. What this meant was an assembly line that had four giant computer-operated cutting and milling machines, and four smaller ones, all of which were capable of cutting timber panels to higher levels of precision than could normally be achieved on a conventional building site. The finished modules, fitted with wiring and plumbing, decorated, carpeted and fitted out with kitchens and bathrooms, can then be loaded on to a lorry and delivered to sites. A home can consist of a single module or several combined. This approach meant digitally modelling every millimetre of every home before production started, standardising and simplifying processes for efficient, high-quality production. Just as important, it involved embracing continuous improvement – an approach that almost halved the time to deliver a completed house. By standardising and simplifying processes, it could drive up quality and productivity – driving down costs. Also, the adoption of mass processing was influenced by increasingly tough energy targets. The process used less water than conventional building methods and reduced site-generated waste. This, in turn, required fewer skips and provided a tidier, safer site. A report by the Ellen MacArthur Foundation highlighted several potential environmental benefits from off-site construction, including more energy-efficient homes.

Environmentally sensitive process design

With the issues of environmental protection becoming more important, process designers have to take account of 'green' (sustainability) issues. In many countries in the Global North, legislation has already provided some basic standards. Interest has focused on some fundamental issues:

- *The sources of inputs* to a product or service. (Will they damage rainforests? Will they use up scarce minerals? Will they exploit the poor or use child labour?)
- *Quantities and sources of energy* consumed in the process. (Do plastic beverage bottles use more energy than glass ones? Should waste heat be recovered and used in fish farming?)
- *The amounts and type of waste material* that are created in the manufacturing processes. (Can this waste be recycled efficiently, or must it be burnt or buried in landfill sites?)
- *The life of the product itself.* (If a product has a long useful life, will it consume fewer resources than a product with a shorter life?)
- *The end-of-life of the product.* (Will the redundant product be difficult to dispose of in an environmentally friendly way?)

Designers are faced with complex trade-offs between these factors, although it is not always easy to obtain all the information that is needed to make the 'best' choices. To help make more rational decisions in the design activity, some industries are experimenting with *life cycle analysis*. This technique analyses all the production inputs, the life cycle use of the product and its final disposal, in terms of total energy used and all emitted wastes. The inputs and wastes are evaluated at *every* stage of a service or product's creation, beginning with the extraction or farming of the basic raw materials. The 'Operations in practice' example 'Ecover's ethical operation design' demonstrates that it is possible to include ecological considerations in all aspects of product and process design.

> **Operations principle**
> The design of any process should include consideration of ethical and environmental issues.

Ecover's ethical operation design[6]

Ecover cleaning products, such as washing liquid, are famously ecological. In fact, it is the company's whole rationale. 'We clean with care', say Ecover, 'whether you're washing your sheets, your floors, your hands or your dishes, our products don't contain those man-made chemicals that can irritate your skin'. But it isn't just the company's products that are based on an ecologically sustainable foundation. Ecover's ecological factories in France and Belgium also embody the company's commitment to sustainability. Whether it is the factory roof, the use of energy or the way it treats the water used in the production processes, Ecover points out that it does its best to limit environmental impact. For example, the Ecover factory operates entirely on green electricity – the type produced by wind generators, tidal generators and other natural sources. What is more, it makes the most of the energy it does use by choosing energy-efficient lighting, and then using it only when needed. And, although the machinery it uses in the factories is standard for the industry, it keeps its energy and water consumption down by choosing low-speed appliances that can multi-task and don't require water to clean them. For example, the motors on its mixing machines can mix 25 tonnes of Ecover liquid while 'consuming no more electricity than a few flat irons'. And it has a 'squeezy gadget that's so efficient at getting every last drop of product out of the pipes, they don't need to be rinsed through'. Ecover say that they 'hate waste, so we're big on recycling. We keep the amount of packaging used in our products to a minimum, and make sure whatever cardboard or plastic we do use can be recycled, reused or re-filled. It's an ongoing process of improvement; in fact, we've recently developed a new kind of green plastic we like to call "Plant-astic" that's 100% renewable, reusable and recyclable – and made from sugarcane'.

Even the building is ecological. It is cleverly designed to follow the movement of the sun from east to west, so that production takes place with the maximum amount of natural daylight (good for saving power and good for working conditions). The factory's frame is built from pine rather than more precious timbers and the walls are constructed using bricks that are made from clay, wood pulp and mineral waste. They require less energy to bake, yet they're light, porous and insulate well. The factories' roofs are covered in thick, spongy Sedum (a flowering plant, often used for natural roofing) that gives insulation all year round. In fact, it's so effective, that they don't need heating or air conditioning – the temperature never drops too low or rises too high.

6.3 How do volume and variety affect process design?

In Chapter 1 we saw how processes range from those producing at high volume (for example, credit card transaction processing) to a low volume (for example, funding a large complex takeover deal). Processes can also range from producing a very low variety of products or services (for example, in an electricity utility) to a very high variety (for example, in an architects' practice). Usually, the two

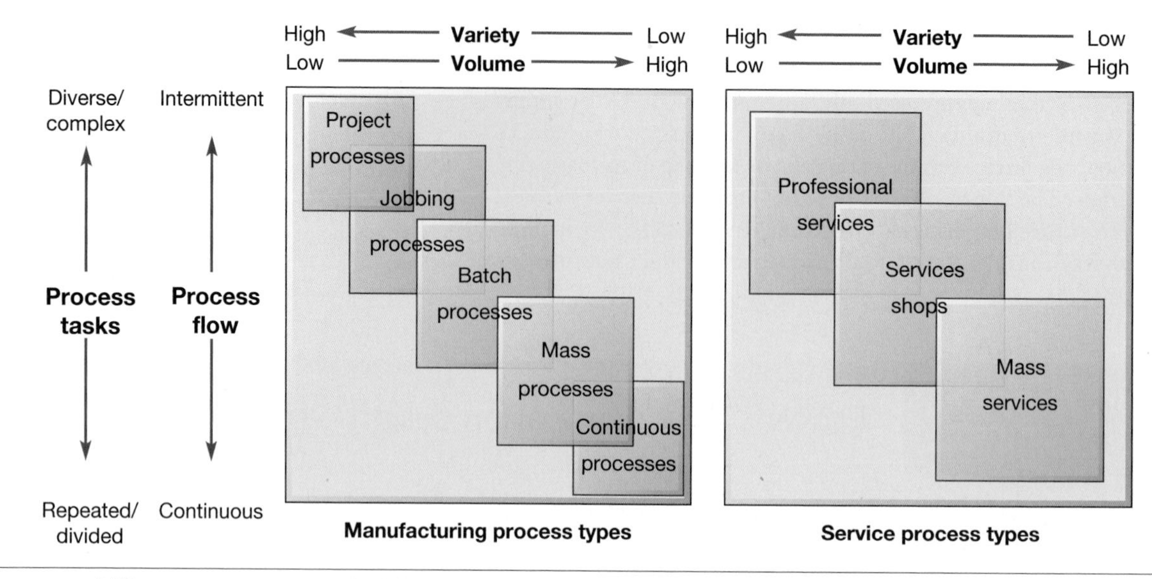

Figure 6.4 Different process types imply different volume–variety characteristics for the process

dimensions of volume and variety go together in a reversed way. Low-volume processes often produce a high variety of products and services, and high-volume processes, a narrow variety. Thus, there is a continuum from low-volume–high variety through to high-volume–low variety, on which we can position processes. And within a single operation there could be processes with very different positions on this volume–variety spectrum. For example, compare the approach taken in a medical service during mass medical treatments, such as large-scale immunisation programmes, with that taken in transplant surgery where the treatment is designed specifically to meet the needs of one person. In other words, no one type of process design is best for all types of requirements in all circumstances – different products or services with different volume–variety positions require different processes.

Operations principle

The design of any process should be governed by the volume and variety it is required to produce.

Process types

The position of a process on the volume–variety continuum shapes its overall design and the general approach to managing its activities. These 'general approaches' are called process types. Different terms are used to identify process types depending on whether they are predominantly manufacturing or service processes, and there is some variation in the terms used. For example, it is not uncommon to find the 'manufacturing' terms used in service industries. Figure 6.4 illustrates how these 'process types' are used to describe different positions on the volume–variety spectrum.

Project processes

Project processes deal with discrete, usually highly customised products, often with a relatively long timescale between the completion of each item, where each job has a well-defined start and finish. Project processes have low volume and high variety. Activities involved in the process can be ill-defined and uncertain. Transforming resources may have to be organised especially for each item (because each item is different). The process may be complex, partly because the activities in such processes often involve significant discretion to act according to professional judgement. Examples of project processes include software design, movie production, most construction work and large fabrication operations such as those manufacturing turbo generators.

The major construction site shown in the picture is a project process. Each 'item' (building) is different and poses different challenges to those running the process (civil engineers).

Jobbing processes

Jobbing processes also deal with high variety and low volumes. However, while in project processes each item has resources devoted more or less exclusively to it, in jobbing processes each product has to share the operation's resources with many others. Resources will process a series of items but, although each one will require similar attention, they may differ in their exact needs. Many jobs will probably be 'one-offs' that are never repeated. Again, jobbing processes could be relatively complex, however they usually produce physically smaller products and, although sometimes involving considerable skill, such processes often involve fewer unpredictable circumstances. Examples of jobbing processes include the work of made-to-measure tailors, many precision engineers such as specialist toolmakers, furniture restorers and the printer that produces tickets for the local social event.

This craftsperson is using general purpose wood-cutting technology to make a product for an individual customer. The next product made will be different (although maybe similar) for a different customer.

Batch processes

Batch processes may look like jobbing processes, but do not have the same degree of variety. As the name implies, batch processes produce more than one item at a time, so each part of the process has periods when it is repeating itself, at least while the 'batch' is being processed. If the size of the batch is just two or three items, it is little different to jobbing. Conversely, if the batches are large, and especially if the products are familiar to the operation, batch processes can be fairly repetitive. Because of this, the batch type of process can be found over a wide range of volume and variety levels. Examples of batch processes include machine tool manufacturing, the production of some special gourmet frozen foods and the manufacture of most of the component parts that go into mass-produced assemblies such as vehicles.

In this kitchen, food is being prepared in batches. All batches go through the same sequence (preparation, cooking and storage) but each batch is of a different dish.

Mass processes

Mass processes are those that produce items in high volume and relatively narrow variety (narrow in terms of its fundamentals – a vehicle assembly process might produce thousands of variants, yet essentially the variants do not affect the basic process of production). The activities of mass processes are usually repetitive and largely predictable. Examples of mass processes include frozen food production, automatic packing lines, vehicle production plants and television factories.

The vehicle production plant is everyone's idea of a mass process. Each product is almost (but not quite) the same, and made in large quantities.

Continuous processes

Continuous processes have even higher volume and usually lower variety than mass processes. They also usually operate for longer periods of time. Sometimes they are literally continuous in that their products are inseparable, being produced in an endless flow. They often have relatively inflexible, capital-intensive technologies with highly predictable flow, and although products may be stored during the process, their predominant characteristic is of smooth flow from one part of the process to another. Examples of continuous processes include water processing, petrochemical refineries, electricity utilities, steel making and some paper making.

This continuous water treatment plant almost never stops (it only stops for maintenance) and performs only one task (filtering impurities). Often, we only notice the process if it goes wrong.

Professional services

Professional services are high-contact processes where customers spend a considerable time in the service process. They can provide high levels of customisation (the process being highly adaptable in order to meet individual customer needs). Professional services tend to be people-based rather than equipment-based, and usually staff are given considerable discretion in servicing customers. Professional services include management consultants, lawyers' practices, architects, doctors' surgeries, auditors, health and safety inspectors, and some computer field service operations.

Here consultants are preparing to start a consultancy assignment. They are discussing how they might approach the various stages of the assignment, from understanding the real nature of the problem through to the implementation of their recommended solutions. This is a process map, although a very high-level one. It guides the nature and sequence of the consultants' activities.

Service shops

Service shops have levels of volume and variety (and customer contact, customisation and staff discretion) between the extremes of professional and mass services (see next paragraph). Service is provided via mixes of front- and back-office activities. Service shops include banks, high-street shops, holiday tour operators, car rental companies, schools, most restaurants, hotels and travel agents.

The health club shown in the picture has front-office staff who can give advice on exercise programmes and other treatments. Although every client has a unique fitness programme, certain activities (for example, safety issues) have to follow defined processes.

Mass services

Mass services have many customer transactions, involving limited contact time and little customisation. Staff are likely to have a relatively defined division of labour and have to follow set procedures. Mass services include supermarkets, a national rail network, an airport, telecommunications service, library, television station, the police service and the enquiry desk at a utility. For example, one of the most common types of mass service are the call centres used by almost all companies that deal directly with consumers. Coping with a very high volume of enquiries requires some kind of structuring of the process of communicating with customers. This is often achieved by using a carefully designed enquiry process (sometimes known as a script).

> **Operations principle**
> Process types indicate the position of processes on the volume–variety spectrum.

This is the 'back office' of part of a retail bank (the type that we all use). It is a call centre that deals with many thousands of calls every day. Staff are required to follow defined processes (scripts) to make sure customers receive a standard service.

The product–process matrix

The most common method of illustrating the relationship between a process's volume–variety position and its design characteristics is shown in Figure 6.5. Often called the 'product–process' matrix, it can in fact be used for any type of process, whether producing products or services.[7] The underlying idea of the product–process matrix is that many of the more important elements of process design are strongly related to the volume–variety position of the process. So, for any process, the tasks that it undertakes, the flow of items through the process, the layout of its resources, the technology it uses and the design of jobs, are all strongly influenced by its volume–variety position. This means that most processes should lie close to the diagonal of the matrix that represents the 'fit' between the process and its volume–variety position. This is called the 'natural' diagonal, or the 'line of fit'.

Critical commentary

Although the idea of process types can be useful, it is in many ways simplistic. In reality, there is no clear boundary between process types. For example, many processed foods are manufactured using mass-production processes but in batches. So, a 'batch' of one type of cake (say) can be followed by a 'batch' of a marginally different cake (perhaps with different packaging), followed by yet another, etc. Essentially this is still a mass process, but not quite as pure a version of mass processing as a manufacturing process that only makes one type of cake. Similarly, the categories of service processes are likewise blurred. For example, a specialist camera retailer would normally be categorised as a service shop, yet it also will give, sometimes very specialised, technical advice to customers. It is not a professional service like a consultancy, of course, but it does have elements of a professional service process within its design. This is why the volume and variety characteristics of a process are sometimes seen as being a more realistic way of describing processes. The product–process matrix adopts this approach.

Moving off the natural diagonal

A process lying on the natural diagonal of the matrix shown in Figure 6.5 will normally have lower operating costs than one with the same volume–variety position that lies off the diagonal. This is because the diagonal represents the most appropriate process design for any volume–variety position. Processes that are on the right of the 'natural' diagonal would normally be associated with lower volumes and higher variety. This means that they are more flexible than is warranted by their actual volume–variety position. They are not taking advantage of their ability to standardise their activities, so their costs are likely to be higher than if they were closer to the diagonal. Conversely, processes that are on the left of the diagonal have adopted a position that would normally be used for

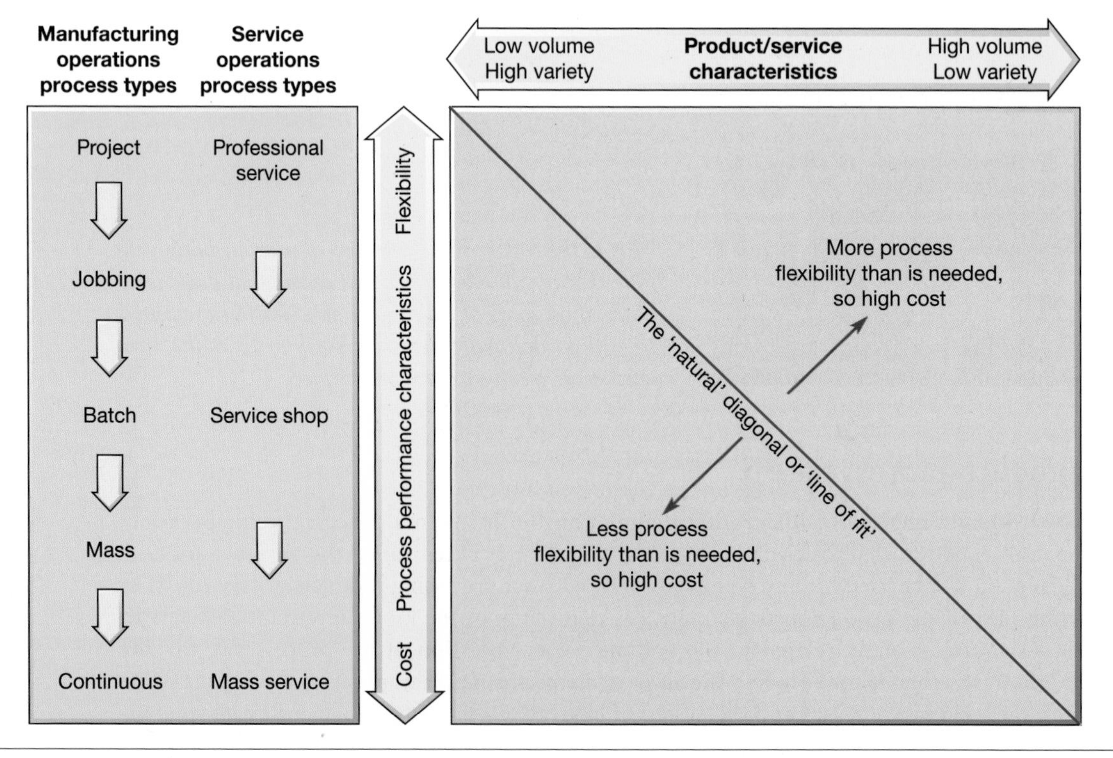

Figure 6.5 Deviating from the 'natural' diagonal on the product–process matrix has consequences for cost and flexibility

Source: Based on Hayes and Wheelright (1984)

higher-volume and lower-variety processes. These processes are 'over-standardised' and probably too inflexible for their volume–variety position. This lack of flexibility can also lead to high costs because the process will not be able to change from one activity to another as readily as a more flexible process.[8] So, a first step in examining the design of an existing process is to check if it is on the natural diagonal of the product–process matrix.

Operations principle
Moving off the 'natural diagonal' of the product–process matrix will incur excess cost.

OPERATIONS IN PRACTICE — Dishang and Sands Film Studios – at opposite ends of the volume-variety spectrum[9]

Making clothes must have been one of the very first 'production' tasks carried out by early humans, and it is still an important industrial sector. With the world garment market worth around €1.3 trillion, it is estimated as employing up to 75 million people.[10] Of all garment producing countries, China has long been seen as the master of clothing manufacturing, and its largest producer and exporter of apparel and textiles is the Dishang Group. Founded in 1993, Dishang has an annual turnover of $1.5 billion, producing garments for such well-known brands as Zara, Matalan and Adidas. But, although China is still a leader in terms of technical expertise and production efficiency, when its labour costs increased, Dishang, like other producers, expanded to set up operations in Cambodia, Myanmar and Bangladesh. It is a huge and sophisticated enterprise, manufacturing over 73 million garments per year from 80 wholly owned factories across 12 global locations. Its modern operations include automated technology, full online and end-of-line quality control systems and the use of acceptable quality limit (AQL) inspection levels according to the requirements of each customer. At Dishang's headquarters, customers can peruse 50,000 fabrics in the company's large digital library. Once customers have chosen a pattern they like, they can upload a picture into an internal system and be shown similar styles. Dishang's Chairman, Lihua Zhu believes that the Group's success is down to three factors. *'Firstly our volume strength is very important. Due to the size of the group we can secure the more competitive prices for our customers (when purchasing raw materials and components). Secondly we offer in-house design expertise and have factories and internal teams that specialise in different products, meaning brands and retailers can come to us for everything. And thirdly, we approach international markets differently with our own offices, which saves costs by cutting out the middle man'.*[11]

One certainly would not see such high volume at Sands Film's costume-making workshop. Every film or television programme that is set in any period, other than the present day, needs costumes for its actors. And most films have a lot of characters, so that means a lot of costumes. Sands Films in London has a well-established and permanent garment workshop. It is what we described earlier in the chapter as a typical 'jobbing' process. Sands Films provides a wide range of wardrobe and costume services. Its customers are the film, stage and TV production companies, each of which have different requirements and time constraints. And because each project is different and has different requirements, the workshop's jobs go from making a single simple outfit to providing a wide variety of specially designed costumes and facilities over an extended production period. The facilities include most normal tailoring processes such as cutting, dyeing and printing, and varied specialist services such as corset and crinoline making as well as millinery (hats). During the design and making process actors often visit the workshop, which has been called an 'Aladdin's cave' of theatrical costumes. The workshop is where actors come face to face with their character for the first time. Making a costume can only start once a project has been approved and a costume designer appointed, although discussions with the workshop may have started prior to that. When the budget and the timing have been agreed, the designer can start to present ideas and finished design to the workshop. And although the processes in the workshop are well established, each costume requires different skills and so has different routes through the stages.

Worked example

The 'meter installation' unit

The 'meter installation' unit of a water utility company installed and repaired water meters. Each installation job could vary significantly because meters had to be fitted into different water pipe systems. When a customer requested an installation, a supervisor would survey the customer's water system. An appointment would then be made for an installer to visit the customer and install the meter. Then the company decided to install a new 'standard' remote-reading meter to replace the wide range of existing meters. This new meter was designed to make installation easier by including universal quick-fit joints that reduced installation times. As a pilot, it was also decided to prioritise those customers with the oldest meters and conduct trials of how the new meter worked in practice. All other aspects of the installation process were left as they were. However, after the new meters were introduced the costs of installation were far higher than forecast, so the company decided to cut out the survey stage of the process because, using the new meter, 98 per cent of installations could be fitted in one visit. Just as significantly, fully qualified installers were often not needed, so installation could be performed by less-expensive staff.

This example is illustrated in Figure 6.6. The initial position of the installation process is at point A. The installation unit was required to install a wide variety of meters into a very wide variety of water systems. This needed a survey stage to assess the nature of the job and the use of skilled staff to cope with the complex tasks. The installation of the new type of meter changed the volume–variety position for the process, reducing the variety and increasing volume. However, the process was not changed so the design of the process was appropriate for its old volume–variety position, but not the new one. In effect, it had moved to point B in Figure 6.6. It was off the diagonal, with unnecessary flexibility and high operating costs. Redesigning the process to take advantage of the reduced variety and complexity of the job (position C on Figure 6.6) allowed installation to be performed more efficiently.

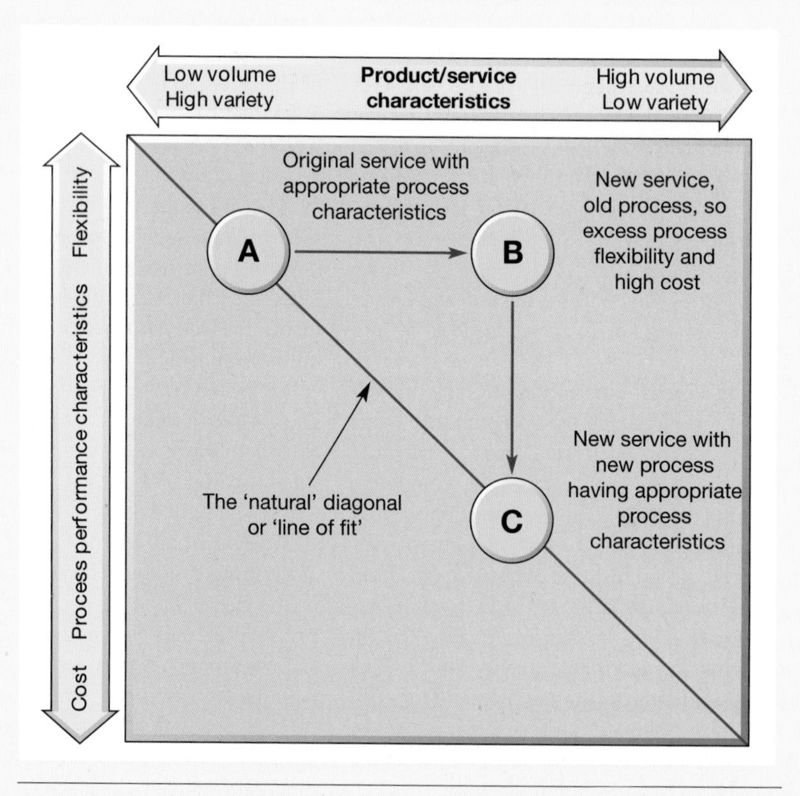

Figure 6.6 A product–process matrix with process positions from the water meter example

6.4 How are processes designed in detail?

After the overall design of a process has been determined, its individual activities must be configured. At its simplest this detailed design of a process involves identifying all the individual activities that are needed to meet the objectives of the process, and deciding on the sequence in which these activities are to be performed and who is going to do them. There will, of course, be some constraints to this. Some activities must be carried out before others and certain people or equipment can only do some activities. Nevertheless, for a process of any reasonable size, the number of alternative process designs is usually large. Because of this, process design is often done using some simple visual approach such as process mapping.

Process mapping

Process mapping simply involves describing processes in terms of how the activities within the process relate to each other. There are many techniques that can be used for *process mapping* (or process blueprinting, or process analysis, as it is sometimes called). They all identify the different *types of activity* and show the flow of materials or people or information through the process. They use process mapping symbols to classify different types of activity. And although there is no universal set of symbols, there are some that are commonly used. Most of these derive either from the early days of 'scientific' management around a century ago (see Chapter 9) or, more recently, from information system flowcharting. These symbols can be arranged in order, and in series or in parallel, to describe any process. For example, Figure 6.7 shows one of the processes used in a theatre lighting operation that hires out lighting and stage effects equipment to theatrical companies and event organisers. It deals with how customers' calls are processed by the technicians.

Operations principle
Process mapping is needed to expose the reality of process behaviour.

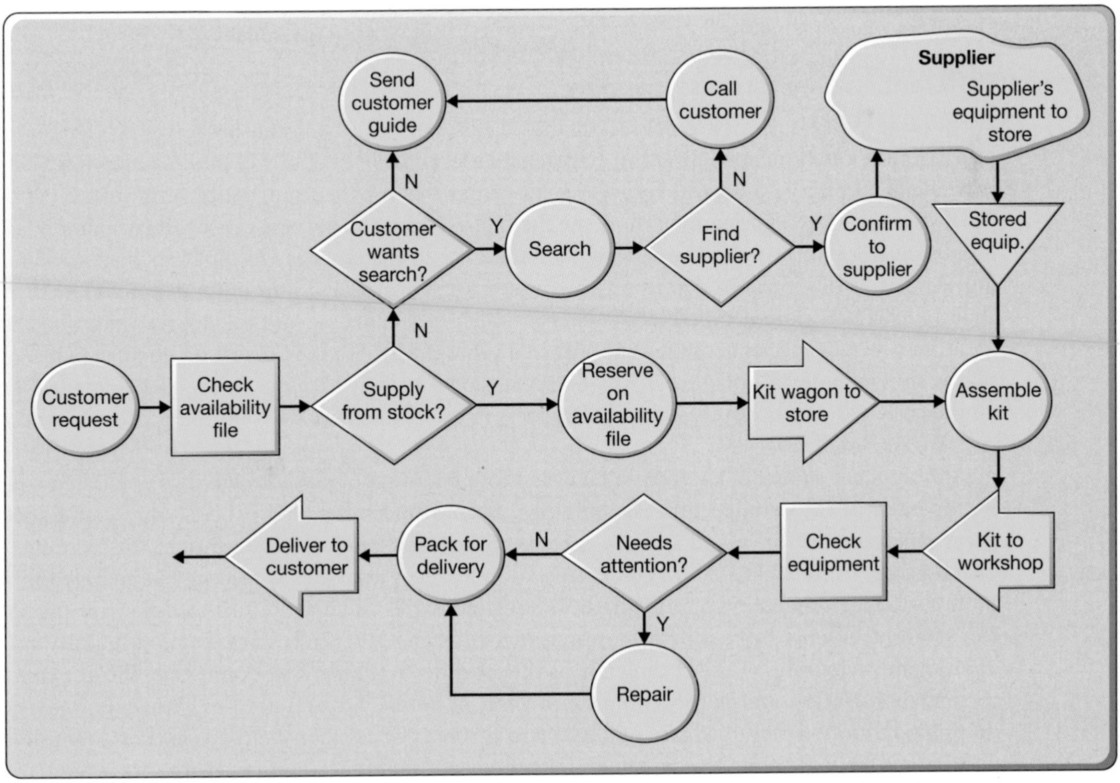

Figure 6.7 Process map for 'enquire to delivery' process at stage lighting operation

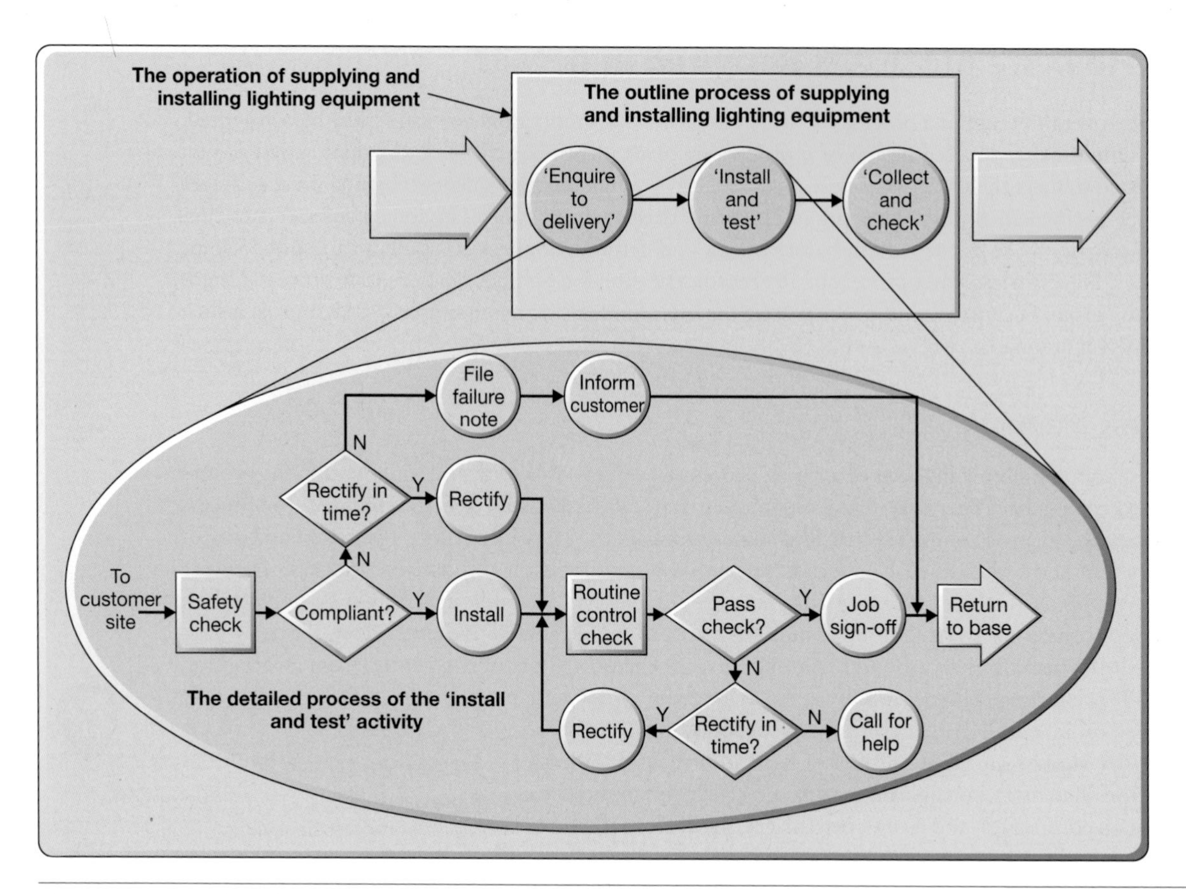

Figure 6.8 The 'supply and install' operations process mapped at three levels

Different levels of process mapping

For a large process, drawing process maps at this level of detail can be complex. This is why processes are often mapped at a more aggregated level, called **high-level process mapping**, before more detailed maps are drawn. Figure 6.8 illustrates this for the total *'supply and install lighting'* process in the stage lighting operation. At the highest level the process can be drawn simply as an input–transformation–output process with materials and customers as its input resources and lighting services as outputs. No details of how inputs are transformed into outputs are included. At a slightly lower or more detailed level, what is sometimes called an **outline process map** (or chart) identifies the sequence of activities but only in a general way. So, the process of *'enquire to delivery'* that is shown in detail in Figure 6.7 is here reduced to a single activity. At the more detailed level, all the activities are shown in a 'detailed process map' (the activities within the process 'install and test' are shown).

Although not shown in Figure 6.8, an even more micro set of process activities could be mapped within each of the detailed process activities. Such a micro detailed process map could specify every single motion involved in each activity. Some quick-service restaurants, for example, do exactly that. In the lighting hire company example, most activities would not be mapped in any more detail than that shown in Figure 6.8. Some activities, such as 'return to base', are probably too straightforward to be worth mapping any further. Other activities, such as 'rectify faulty equipment', may rely on the technician's skills and discretion to the extent that the activity has too much variation and is too complex to map in detail. Some activities, however, may need mapping in more detail to ensure quality or to protect the company's interests. For example, the activity of safety checking the customer's site to ensure that it is compliant with safety regulations will need specifying in some detail to ensure that the company can prove it exercised its legal responsibilities.

Mapping visibility in process design

'Processing' people is different. Processes with a high level of customer 'visibility' cannot be designed in the same way as processes that deal with inanimate materials or information. As we discussed in Chapter 1, operations and processes that primarily 'transform', people *experience* the process. When customers 'see' part of the process, it is useful to map them in a way that makes the degree of visibility of each part of the process obvious. Figure 6.9 shows yet another part of the lighting equipment company's operation: 'the collect and check' process. The process is mapped to show the visibility of each activity to the customer. Here four levels of visibility are used. There is no hard and fast rule about this; many processes simply distinguish between those activities that the customer *could* see and those that they couldn't. The boundary between these two categories is often called the 'line of visibility'. In Figure 6.9 three categories of visibility are shown. At the very highest level of visibility, above the 'line of interaction', are those activities that involve direct interaction between the lighting company's staff and the customer. Other activities take place at the customer's site or in the presence of the customer but involve less or no direct interaction. Yet further activities (the two transport activities in this case) have some degree of visibility because they take place away from the company's base and are visible to potential customers, but are not visible to the immediate customer.

Visibility, customer experience and emotional mapping

When customers experience a process, it results in the customer feeling emotions, not all of which are necessarily rational. Most of us have been made happy, angry, frustrated, surprised, reassured or furious as customers in a process. Nor is the idea of considering how processes affect customer

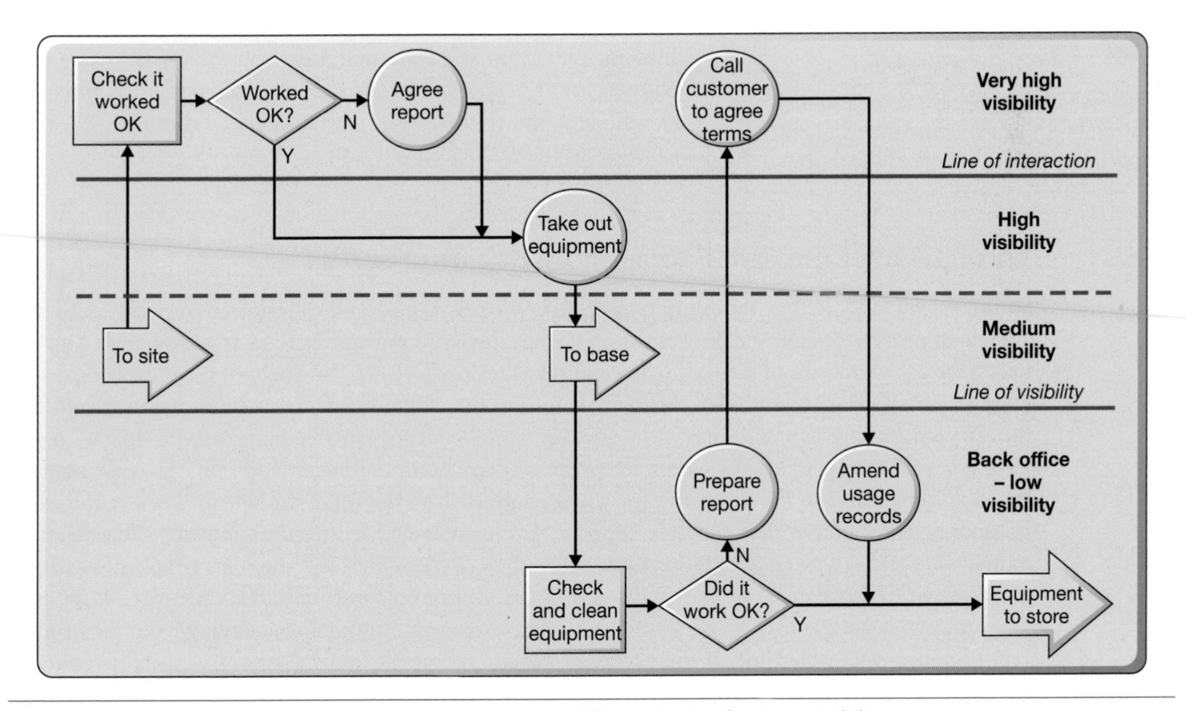

Figure 6.9 The 'collect and check' process mapped to show different levels of process visibility

emotions confined to those processes that are intended to engage the emotions: for example, entertainment-type organisations such as theme parks. Any high customer contact product (or more likely, service) always creates an experience for the customer. Moreover, customer experience will affect customer satisfaction, and therefore has the potential to produce customer loyalty, influence expectations and create emotional bonds with customers. This is why many service organisations are seeing how customers experience their processes (the so-called 'customer journey') at the core of their process design.

Designing the customer experience

Designing processes with a significant experience content requires the systematic consideration of how customers may react to the experiences that the process exposes them to. This will include the sights, sounds, smells, atmosphere and general 'feeling' of the service. The concept of a '**servicescape**', which is discussed in the next chapter, is strongly related to consideration of engaging customers so that they connect with the process in a personal way. One of the most common methods of designing such processes is to consider what are commonly called '**touchpoints**'. These have been described as 'Everything the consumer uses to verify their service's effectiveness'.[12] They are the points of contact between a process and customers, and there might be many different touchpoints during the customer journey. It is the accumulation of all the experiences from every touchpoint interaction that shapes customers' judgement of the process. The features of a process at the touchpoints are sometimes called 'clues' (or 'cues'): these are the messages that customers receive or experience as they progress through the process. The emotions that result from these clues contain the messages that the customer will receive and therefore influence how a customer will judge the process.

When designing processes, managers need to ensure that all the messages coming from the clues at each stage of the process are consistent with the emotions they want the customers to experience and do not give them wrong or misleading messages about the process. In the same way as process mapping indicates the sequence and relationship between activities, so **emotional mapping** can indicate the type of emotions engendered in the customer's mind as they experienced the process. Figure 6.10 shows how this might work for a visit to a clinic for a computerised tomography (CT) scan. There are many ways that emotions can be mapped and different diagrammatic representations can be used. In this case experiences are captured by asking what the patient is intended to, and actually, thinks, feels, says and acts. From this, a simple scoring system has been used ranging from +3 (very positive) to −3 (very negative).

> **Operations principle**
> The design of processes that deal with customers should consider the emotions engendered at each stage of the process.

Throughput time, cycle time and work-in-progress

So far, we have looked at the more conceptual (process types) and descriptive (process mapping) aspects of process design. We now move on to the equally important analytical perspective. And the first stage is to understand the nature of, and relationship between, throughput time, cycle time and work-in-progress. As a reminder, throughput time is the elapsed time between an item entering the process and leaving it, cycle time is the average time between items being processed and work-in-progress is the number of items within the process at any point in time. In addition, the work content for each item will also be important for some analysis. It is the total amount of work required to produce a unit of output. For example, suppose that in an assemble-to-order sandwich shop, the time to assemble and sell a sandwich (the work content) is two minutes and that two people are staffing the process. Each member of staff will serve a customer every two minutes; therefore, every two minutes, two customers are being served and so on average a customer is emerging from the process every minute (the cycle time of the process). When customers join the queue in the process they become work-in-progress (sometimes written as WIP). If the queue is ten people long (including that customer) when the customer joins it, they will have to wait ten minutes to emerge from the process. Or put more succinctly:

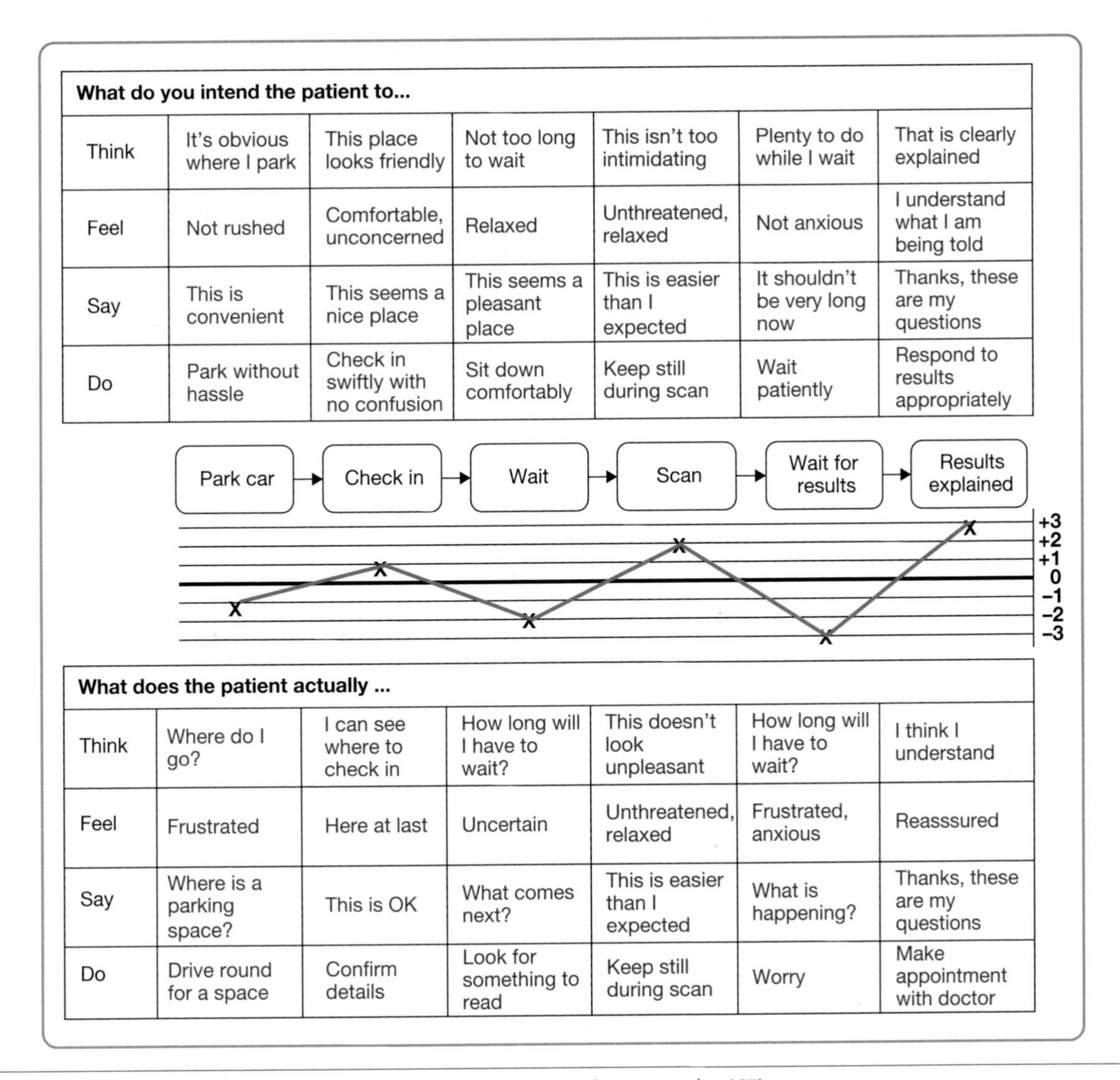

What do you intend the patient to...						
Think	It's obvious where I park	This place looks friendly	Not too long to wait	This isn't too intimidating	Plenty to do while I wait	That is clearly explained
Feel	Not rushed	Comfortable, unconcerned	Relaxed	Unthreatened, relaxed	Not anxious	I understand what I am being told
Say	This is convenient	This seems a nice place	This seems a pleasant place	This is easier than I expected	It shouldn't be very long now	Thanks, these are my questions
Do	Park without hassle	Check in swiftly with no confusion	Sit down comfortably	Keep still during scan	Wait patiently	Respond to results appropriately

What does the patient actually ...						
Think	Where do I go?	I can see where to check in	How long will I have to wait?	This doesn't look unpleasant	How long will I have to wait?	I think I understand
Feel	Frustrated	Here at last	Uncertain	Unthreatened, relaxed	Frustrated, anxious	Reasssured
Say	Where is a parking space?	This is OK	What comes next?	This is easier than I expected	What is happening?	Thanks, these are my questions
Do	Drive round for a space	Confirm details	Look for something to read	Keep still during scan	Worry	Make appointment with doctor

Figure 6.10 Customer experience map of a visit for a computerised tomography (CT) scan

$$\text{Throughput time} = \text{Work-in-progress} \times \text{Cycle time}$$

In this case: $\quad 10 \text{ minutes wait} = 10 \text{ people in the system} \times 1 \text{ minute per person}$

Little's law

This mathematical relationship (throughput time = work-in-progress × cycle time) is called Little's law. It is simple but very useful, and it works for any stable process. Little's law states that the average number of things in the system is the product of the average rate at which things leave the system and the average time each one spends in the system. Or, put another way, the average number of objects in a queue is the product of the entry rate and the average holding time. For example, suppose it is decided that in a new sandwich assembly and sales process, the average number of customers in the process should be limited to around ten and the maximum time a customer is in the process should be on average four minutes. If the time to assemble and sell a sandwich (from customer request to the customer leaving the process) in the new process has been reduced to 1.2 minutes, how many staff should be serving?

Operations principle

Process analysis derives from an understanding of the required process cycle time.

Putting this into Little's law:

$$\text{Throughput time} = 4 \text{ minutes}$$

$$\text{Work in progress, WIP} = 10$$

So, since: $$\text{Throughput time} = \text{WIP} \times \text{cycle time}$$

$$\text{Cycle time} = \frac{\text{Throughput time}}{\text{WIP}}$$

$$\text{The cycle time for the process} = \frac{4}{10} = 0.4 \text{ minutes}$$

That is, a customer should emerge from the process every 0.4 minutes, on average. Given that an individual can be served in 1.2 minutes:

$$\text{The number of servers required} = \frac{1.2}{0.4} = 3$$

In other words, three servers would serve three customers in 1.2 minutes: that is, one customer in 0.4 minutes.

Operations principle

Little's law states that throughput time = work-in-progress × cycle time.

Worked example

You'll never get them back in time

Mike was totally confident in his judgement: *'You'll never get them back in time'*, he said. *'They aren't just wasting time, the process won't allow them all to have their coffee and get back for 11 o'clock'.* Looking outside the lecture theatre, Mike and his colleague Dina were watching the 20 businesspeople who were attending the seminar queuing to be served coffee and biscuits. The time was 10.45 and Dina knew that unless they were all back in the lecture theatre at 11 o'clock there was no hope of finishing his presentation before lunch. *'I'm not sure why you're so pessimistic'*, said Dina. *'They seem to be interested in what I have to say and I think they will want to get back to hear how operations management will change their lives'.* Mike shook his head. *'I'm not questioning their motivation'*, he said, *'I'm questioning the ability of the process out there to get through them all in time. I have been timing how long it takes to serve the coffee and biscuits. Each coffee is being made fresh and the time between the server asking each customer what they want and them walking away with their coffee and biscuits is 48 seconds. Remember that, according to Little's law, throughput equals work-in-progress multiplied by cycle time. If the work-in-progress is the 20 managers in the queue and cycle time is 48 seconds, the total throughput time is going to be 20 multiplied by 0.8 minutes which equals 16 minutes. Add to that sufficient time for the last person to drink their coffee and you must expect a total throughput time of a bit over 20 minutes. You just haven't allowed long enough for the process'.* Dina was impressed. *'Err . . . what did you say that law was called again?'* *'Little's law'*, said Mike.

Worked example

Workstation renovation

Every year it was the same. All the workstations in the building had to be renovated (tested, new software installed, etc.) and there was only one week in which to do it. The one week fell in the middle of the August vacation period when the renovation process would cause minimum disruption to normal working. Last year the company's 500 workstations had all been renovated within one working week (40 hours). Each renovation last year took on average 2 hours and 25 technicians had completed the process within the week. This year there would be 530 workstations to renovate but the company's IT support unit had devised a faster testing and renovation routine that would only take on average

1.5 hours instead of 2 hours. How many technicians will be needed this year to complete the renovation processes within the week?

Last year:

$$\text{Work-in-progress (WIP)} = 500 \text{ workstations}$$

$$\text{Time available } (T_t) = 40 \text{ hours}$$

$$\text{Average time to renovate} = 2 \text{ hours}$$

$$\text{Therefore throughput rate } (T_r) = 0.5 \text{ hours per technician}$$

$$= 0.5N$$

$$\text{where } N = \text{number of technicians}$$

$$\text{Little's law WIP} = T_t \times T_r$$

$$500 = 40 \times 0.5N$$

$$N = \frac{500}{40 \times 0.5}$$

$$= 25 \text{ technicians}$$

This year:

$$\text{Work in progress (WIP)} = 530 \text{ workstations}$$

$$\text{Time available} = 40 \text{ hours}$$

$$\text{Average time to renovate} = 1.5 \text{ hours}$$

$$\text{Throughput rate } (T_r) = 1/1.5 \text{ per technician}$$

$$= 0.67N$$

$$\text{where } N = \text{number of technicians}$$

$$\text{Little's law WIP} = T_t \times T_r$$

$$530 = 40 \times 0.67N$$

$$N = \frac{530}{40 \times 0.67}$$

$$= 19.88 \text{ (say 20) technicians}$$

Throughput efficiency

This idea that the throughput time of a process is different from the work content of whatever it is processing has important implications. What it means is that for significant amounts of time no useful work is being done to the materials, information or customers that are progressing through the process. In the case of the simple example of the sandwich process described earlier, customer throughput time is restricted to 4 minutes, but the work content of the task (serving the customer) is only 1.2 minutes. So, the item being processed (the customer) is only being 'worked on' for 1.2/4 = 30 per cent of its time. This is called the **throughput efficiency** of the process:

$$\text{Percentage throughput efficiency} = \frac{\text{Work content}}{\text{Throughput time}} \times 100$$

In this case the throughput efficiency is very high, relative to most processes, perhaps because the 'items' being processed are customers who react badly to waiting. In most material and information transforming processes, throughput efficiency is far lower, usually in single percentage figures.

Worked example

The vehicle licensing centre

A vehicle licensing centre receives application documents, keys in details, checks the information provided on the application, classifies the application according to the type of licence required, confirms payment and then issues and mails the licence. It is currently processing an average of 5,000 licences every eight-hour day. A recent spot check found 15,000 applications that were 'in progress' or waiting to be processed. The sum of all activities that are required to process an application is 25 minutes. What is the throughput efficiency of the process?

$$\text{Work in progress} = 15{,}000 \text{ applications}$$

$$\text{Cycle time} = \text{time producing}$$

$$\frac{\text{Time producing}}{\text{Number produced}} = \frac{8 \text{ hours}}{5{,}000} = \frac{480 \text{ minutes}}{5{,}000} = 0.096 \text{ minutes}$$

$$\text{From Little's law, throughput time} = \text{WIP} \times \text{cycle time}$$

$$\text{Throughput time} = 15{,}000 \times 0.096$$

$$= 1{,}440 \text{ minutes} = 24 \text{ hours} = 3 \text{ days of working}$$

$$\text{Throughput efficiency} = \frac{\text{Work content}}{\text{Throughput time}} = \frac{25}{1{,}440} = 1.74 \text{ per cent}$$

Although the process is achieving a throughput time of three days (which seems reasonable for this kind of process) the applications are only being worked on for 1.7 per cent of the time they are in the process.

Value-added throughput efficiency

The approach to calculating throughput efficiency that is described above assumes that all the 'work content' is actually needed. Yet we have already seen from 'The vehicle licensing centre' worked example that changing a process can significantly reduce the time that is needed to complete the task. Therefore, work content is actually dependent upon the methods and technology used to perform the task. It may be also that individual elements of a task may not be considered 'value-added'. In 'The vehicle licensing centre' worked example, the new method eliminated some steps because they were 'not worth it': that is, they were not seen as adding value. So, **value-added throughput efficiency** restricts the concept of work content to only those tasks that are literally adding value to whatever is being processed. This often eliminates activities such as movement, delays and some inspections. For example, if in the worked example 'The vehicle licensing centre', of the 25 minutes of work content, only 20 minutes was actually adding value, then:

$$\text{Value-added throughput efficiency} = \frac{20}{1{,}440} = 1.39 \text{ per cent}$$

Workflow

When the transformed resource in a process is information (or documents containing information), and when information technology is used to move, store and manage the information, process design is sometimes called 'workflow' or 'workflow management'. It is defined as 'the automation of procedures where documents, information or tasks are passed between participants according to a defined set of rules to achieve, or contribute to, an overall business goal'. Although workflow may be managed manually, it is almost always managed using an IT system. The term is also often associated with business process reengineering (see Chapter 15). More specifically, workflow is concerned with the following:

- Analysis, modelling, definition and subsequent operational implementation of business processes.
- The technology that supports the processes.
- The procedural (decision) rules that move information/documents through processes.
- Defining the process in terms of the sequence of work activities, the human skills needed to perform each activity and the appropriate IT resources.

Process bottlenecks

A bottleneck in a process is the activity or stage where congestion occurs because the workload placed is greater than the capacity to cope with it. In other words, it is the most overloaded part of a process. And as such it will dictate the rate at which the whole process can operate. For example, look at the simple process illustrated in Figure 6.11. It has four stages and the total time to complete the work required for each item passing through the process is 10 minutes. In this simple case, each of the four stages has the same capacity. In the first case (a) the 10 minutes of work is equally allocated between the four stages, each having 2.5 minutes of work. This means that items will progress smoothly through the process without any stage holding up the flow, and the cycle time of the process is 2.5 minutes. In the second case (b) the work has not been allocated evenly. In fact, this is usually the case because usually it is difficult (actually close to impossible) to allocate work absolutely equally. In this case, stage 4 of the process has the greatest load (3 minutes). It is the bottleneck, and will constrain the cycle time of the process to 3 minutes. Bottlenecks reduce the efficiency of a process because, although the bottleneck stage will be fully occupied, the other stages will be underloaded. The activity of trying to allocate work equally between stages is called 'balancing'.

> **Operations principle**
> Allocating work equally to each stage in a process (balancing) smooths flow and avoids bottlenecks.

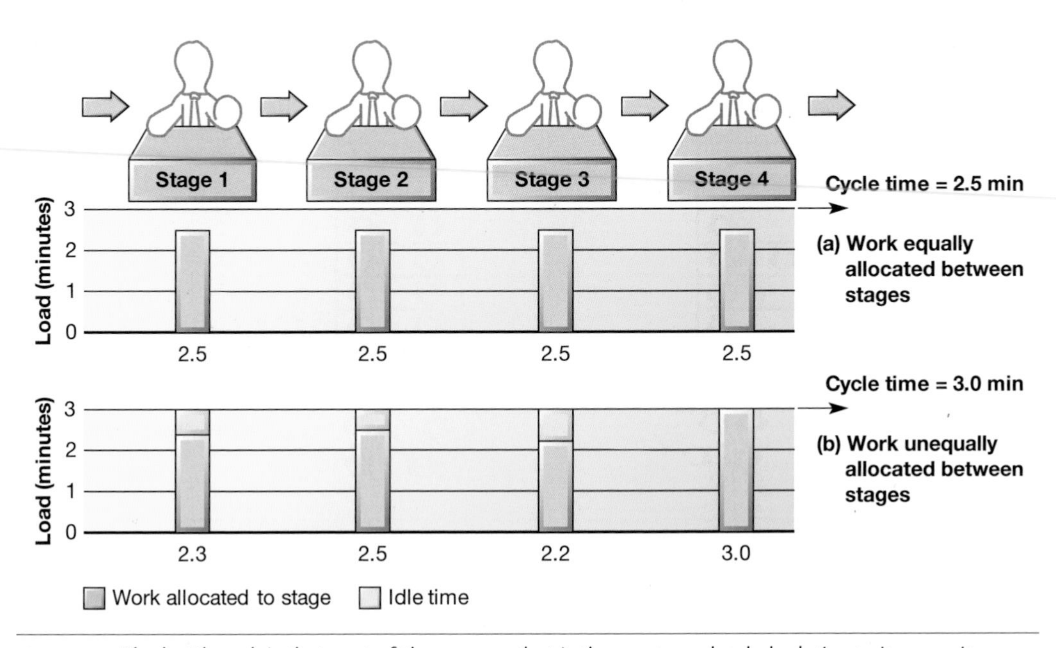

Figure 6.11 The bottleneck is that part of the process that is the most overloaded relative to its capacity

OPERATIONS IN PRACTICE

London's underground tackles a bottleneck[13]

Anyone who has travelled on a busy mass transport system like London Underground know how busy it can be, often with queues of passengers building up at various points as they move to or from their trains. One point that can become a bottleneck for passengers on London Underground is the escalators. Traditionally, in London, passengers stand on the right-hand side of the escalator, leaving the left-hand side free for those who want to walk up or down. But in an attempt to reduce the bottleneck at the escalators, Transport for London, who run the system, trialled a new arrangement that they believed would increase the capacity of its escalator at the Holborn station. Building new stations is expensive, so any way of increasing the capacity of existing ones is going to be attractive, and Holborn is a particularly busy station. The new (and radical, for Londoners) arrangement was to instruct passengers at peak times not to walk, but to stand on both sides of the escalator. The decision was also based on the fact that the escalators at Holborn are over 24 metres high. Apparently, height makes a big difference to the willingness of passengers to walk up escalators. When they are only a few metres high, most people will walk up them. At 30 metres, only the very energetic will. As shown in Figure 6.12, the trial was technically successful in that capacity increased significantly. However, the experiment was not made permanent. Why? Apparently it offended two aspects of human behaviour. First, it slowed the (vocal) minority of people who want to race up the escalator as their gym workout for the day. Second, it took away the feeling that travellers had at least some degree of choice (even if most chose not to exercise it).

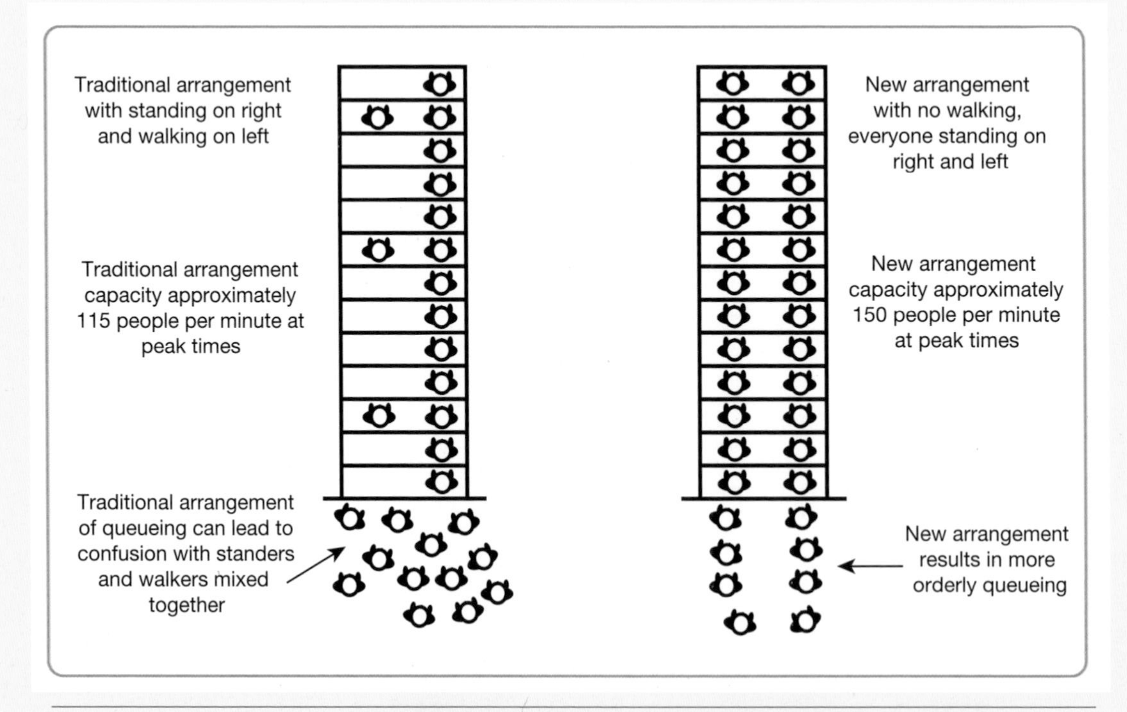

Figure 6.12 Requiring passengers to stand on both sides of the escalator makes it less of a bottleneck

Balancing work time allocation

Allocating work to process stages must respect the 'precedence' of the individual tasks that make up the total work content of the job that the process is performing. The most common way of showing task precedence is by using a 'precedence diagram'. This is a representation of the ordering of the elements, where individual tasks are represented by circles connected by arrows, which signify the ordering of the tasks. Figure 6.13 in the worked example 'Karlstad Kakes' illustrates how precedence diagrams can be used.

Operations principle
Process design must respect task precedence.

Worked example

Karlstad Kakes

Karlstad Kakes (KK) is a manufacturer of speciality cakes, which has recently obtained a contract to supply a major supermarket chain with a speciality cake in the shape of a space rocket. It has been decided that the volumes required by the supermarket warrant a special production process to perform the finishing, decorating and packing of the cake. This line would have to carry out the elements shown in Table 6.2.

Table 6.2 The individual tasks that make up the total job of the finishing, decorating and packing of the cake

Task a – De-tin and trim	Task d – Clad in top fondant	Task g – Apply blue icing
Task b – Reshape	Task e – Apply red icing	Task h – Fix transfers
Task c – Apply base fondant	Task f – Apply green icing	Task i – To base and pack

Figure 6.13 shows the precedence diagram for the total job. The initial order from the supermarket is for 5,000 cakes a week and the number of hours worked by the factory is 40 per week. From this:

$$\text{The required cycle time} = \frac{40 \text{ hrs} \times 60 \text{ mins}}{5{,}000} = 0.48 \text{ mins}$$

$$\text{The required number of stages} = \frac{1.68 \text{ mins (the total work content)}}{0.48 \text{ mins (the required cycle time)}} = 3.5 \text{ stages}$$

This means four stages.

Working from the left on the precedence diagram, tasks a and b can be allocated to stage 1. Allocating task c to stage 1 would exceed the cycle time. In fact, only task c can be allocated to stage 2 because including task d would again exceed the cycle time. Task d can be allocated to stage 3. Either task e or task f can also be allocated to stage 3, but not both or the cycle time would be exceeded. In this case, task e is chosen. The remaining tasks then are allocated to stage 4. The dotted lines in Figure 6.13 show the final allocation of tasks to each of the four stages.

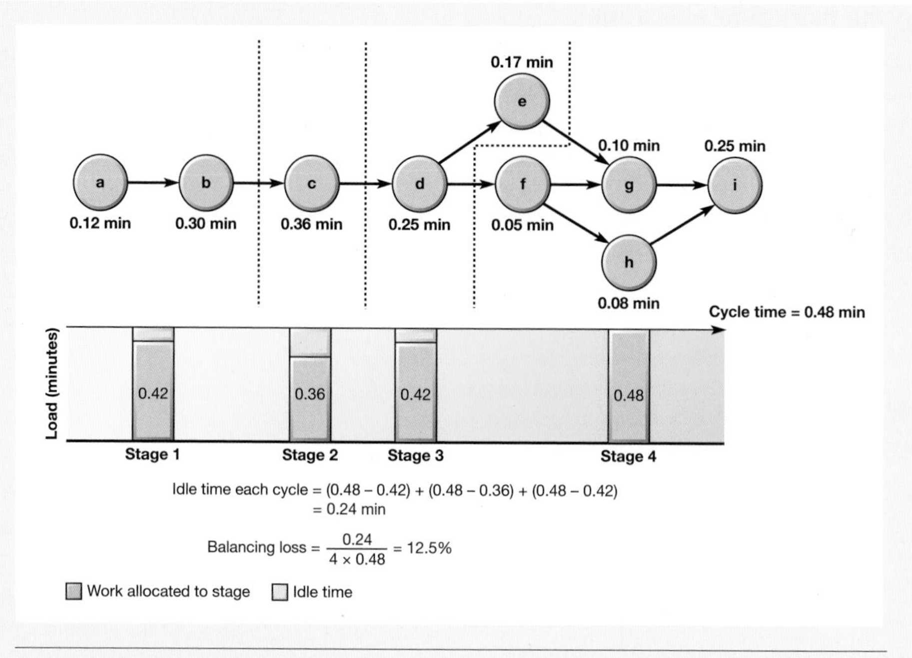

Figure 6.13 Precedence diagram for Karlstad Kakes with allocation of tasks to each stage

Arranging the stages

All the stages necessary to fulfil the requirements of the process may not be arranged in a sequential 'single line'. For example, suppose a mortgage application process requires four stages working on the task to maintain a cycle time of one application processed every 15 minutes. One possible arrangement of the four stages would be to arrange them sequentially, each stage having 15 minutes' worth of work. However, (theoretically) the same output rate could also be achieved by arranging the four stages as two shorter lines, each of two stages with 30 minutes' worth of work. Alternatively, following this logic to its ultimate conclusion, the stages could be arranged as four parallel stages, each responsible for the whole work content. Figure 6.14 shows these options.

This is a simplified example, but it represents a genuine issue. Should the process be organised as a single '**long-thin**', sequential arrangement, or as several '**short-fat**', parallel arrangements, or somewhere in-between? (Note that 'long' means the number of stages and 'fat' means the amount of work allocated to each stage.) In any particular situation, there are usually technical constraints, which limit either how 'long and thin' or how 'short and fat' the process can be, but there is usually a range of possible options within which a choice needs to be made. The advantages of each extreme of the long thin to short fat spectrum are very different and help to explain why different arrangements are adopted.

The advantages of the long-thin arrangement include:

- *Controlled flow of items* – This is easy to manage.
- *Simple handling* – This is especially true if the items being processed are heavy, large or difficult to move.
- *Lower capital requirements* – If a specialist piece of equipment is needed for one task in the job, only one piece of equipment would need to be purchased; on short-fat arrangements every stage would need one.
- *More efficient operation* – If each stage is performing only a small part of the total job, the person at the stage will have a higher proportion of direct productive work as opposed to the non-productive parts of the job, such as picking up tools and materials.

(This latter point is particularly important and is fully explained in Chapter 9 when we discuss job design.)

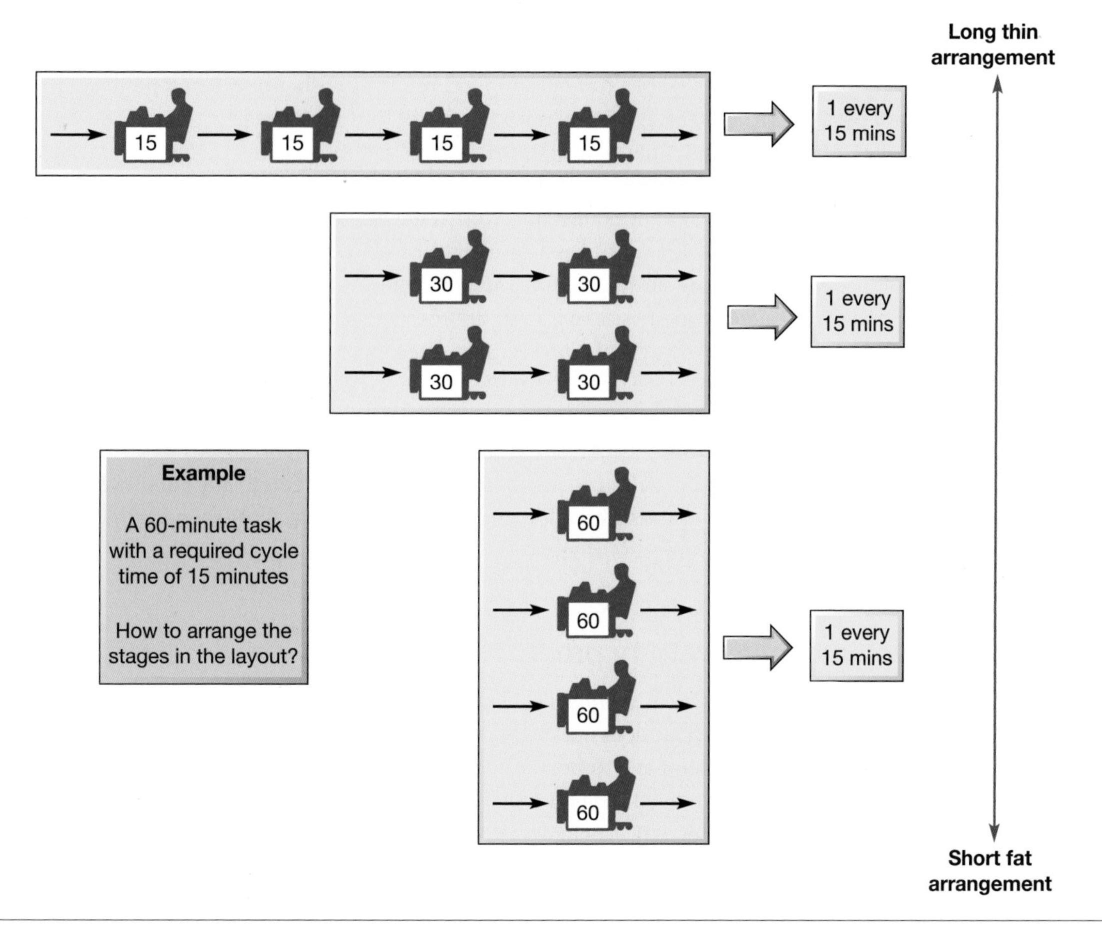

Figure 6.14 The arrangement of stages in a process can be described on a spectrum from 'long thin' to 'short fat'

The advantages of the short-fat arrangement include:

- *Higher mix flexibility* – If the process needs to work on several types of item, each stage or whole process could specialise in different types.
- *Higher volume flexibility* – As volume varies, stages can simply be closed down or started up as required; long-thin arrangements would need rebalancing each time the cycle time changed.
- *Higher robustness* – If one stage breaks down or ceases operation in some way, the other parallel stages are unaffected; a long-thin arrangement would cease operating completely.
- *Less monotonous work* – In the mortgage example, the staff in the short-fat arrangement are repeating their tasks only every hour; in the long-thin arrangement, it is every 15 minutes.

Responsible operations

In every chapter, under the heading of 'Responsible operations', we summarise how the particular topic covered in the chapter touches upon important social, ethical and environmental issues.

When we presented the advantages and disadvantages of designing processes to have a short cycle time (as in the 'long-thin' arrangement of process stages) we made no comment, or judgement, on the ethics of such highly repetitive work. Yet there clearly are issues of how repeating the same tasks over long periods, with little or no variety, affects both the mental and physical health of those asked to perform such work. That relatively few people actively prefer highly repetitive work with little variety is itself a telling indication of how unattractive most of us would find it. The general assumption is that task repetition leads to monotony, boredom, a lack of a feeling of meaningfulness and an increase in stress. In turn, this increases the chance of job dissatisfaction, absenteeism, burnout and labour turnover. In fact, the relationship between repetitive jobs and negative psychological effects is more nuanced than this, but nevertheless it would be difficult to argue that repetitive jobs are more fulfilling than those with more variety. Moreover, it is not only the potential psychological damage of repetitive work that needs to be considered. When the repeated task is physical, as in assembly-line work, there is also the possibility of damage such as hand pain and tendonitis. Again, studies are not always totally unambiguous, but repeated movements that involve awkward positions, and especially high levels of force and repetition have been shown to be associated with physical pain and disorders.[14]

Confronting the negative effect of repetitive jobs often involves one of two 'solutions' – redesign the job so that it is less repetitive (**job enrichment**) or automate it so that a human does not have to do it. (We will deal with the idea of job enrichment in Chapter 9 (People in operations).) The automation of repetitive jobs was, for many years, confined to some manufacturing tasks. However, the increase in the use of **robotic process automation** (RPA, treated later) and even **artificial intelligence (AI)** is allowing the automation of routine tasks, such as those found in the back-office processes found in many professional services.

Low-volume, high-variety processes

Many of the ideas and analytical approaches described in this chapter derive largely from high-volume, low-variety processes. This does not mean that they cannot be used in low-volume, high-variety process design, but they often have to be modified or adapted in some way. For example, splitting activities into very small increments so that work can be balanced between stages (see earlier) is often neither possible nor desirable when the variety of activities is very wide. This does not mean that trying to allocate work equally between work groups is not important, just that it will need to be done in a more approximate way. Even process mapping can be problematic. Some low-volume, high-variety processes are intrinsically difficult to describe as simple step-by-step sequential activities. There may be many alternative routes through a process that can be taken by whatever is being processed. Decisions about how to treat whatever is being processed may be a matter of judgement. The exact circumstances associated with processing something may not have occurred before. If it is information that is being processed, the information may be partial, uncertain or ambiguous.

Automating processes

The majority of processes used in this chapter to illustrate various aspects of process design are essentially manual in nature. That is, they involve a person or persons performing some kind of tasks. Go back far enough and almost all processes (although they would not have been called that) would have been manual. The history of operations management could be told as one of the progressive substitution of technology for people-based effort. First it was manufacturing processes that were automated in some way; and although there are still plenty of manual manufacturing processes, there are also factories that operate virtually 'dark', with very few humans involved. The equivalent automation in many service operations, especially those that primarily process information are specifically designed information technology (IT) systems. Anyone with a bank account is the recipient of the service provided by these IT systems. They are automated mass processes, usually designed from first principles, that may be sometimes rigid and impersonal but are remarkably efficient compared with performing such tasks manually.

Robotic process automation

Yet, although what earlier we called 'core' operations processes, especially high-volume ones, have increasingly become automated, there are many processes outside the operations function that could be automated. These processes are often lower volume than routine core operations processes, yet still follow a logical set of rules. They have been called 'swivel chair' processes – meaning that people take inputs of information from one set of systems (emails for example), process the information using rules, and then record the processed outputs into another system. Examples might include the 'onboarding' process for new employees, recording and entering invoices onto internal payment authorisation systems, entering details of client information onto customer relationship management systems, and so on. Typically, these processes are routine, predictable, rules-based and performed by professional employees whose time could be more profitably employed.

This is the area of application for what has become known (rather tautologically) as 'robotic process automation', or RPA. It is a general term for tools that function on the human interface of other computer systems. It does not, of course, use actual physical robots, rather it deploys software routines (often just called 'bots') to perform the most mundane and repetitive tasks previously done by people. Its aim is to enhance efficiency by automating the everyday processes that would otherwise require human effort. Admittedly, the same aim could be attributed to almost any IT-based automation. The difference between RPA and traditional IT systems is:

- RPA is best used away from the extremes of the volume–variety spectrum. High-volume, low-variety processes can be automated using conventional specifically designed IT systems. At the other extreme, very-high-variety, low-volume tasks are likely to need the flexible thought processes and decision-making of humans. RPA can be used in-between these extremes.
- RPA is relatively easy to develop compared with specifically designed IT systems. The latter require significant systems analysis and coding skills. RPA often uses simple 'drag and drop' instructions that can be used by people who understand the purpose of the processes being automated.
- RPA works around existing processes rather than trying to reengineer them. It is sometimes referred to as 'lightweight' IT because it tries not to disturb underlying computer systems.

The effects of process variability

So far in our treatment of process design we have assumed that there is no significant variability either in the demand to which the process is expected to respond, or in the time taken for the process to perform its various activities. Clearly, this is not the case in reality. So, it is important to look at the variability that can affect processes and take account of it.

There are many reasons why variability occurs in processes. These can include the late (or early) arrival of material, information or customers, a temporary malfunction or breakdown of process technology within a stage of the process, the recycling of 'mis-processed' materials, information or customers to an earlier stage in the process, variation in the requirements of items being processed, etc. All these sources of variation interact with each other, but result in two fundamental types of variability:

- Variability in the demand for processing at an individual stage within the process, usually expressed in terms of variation in the inter-arrival times of items to be processed.
- Variation in the time taken to perform the activities (i.e. process a unit) at each stage.

Critical commentary

Some commentators are critics of the very idea of thinking in terms of 'processes'. They claim that defining jobs as processes incites managers to look on all activities as a machine-like set of routine activities, verging on the mindless. At best, it encourages going through the stages in a process without thinking about what is really involved (what is known as 'box ticking'). At worst, defining all activities into the straitjacket of 'process' kills the essential humanity of working life. The counterargument is that this is a misunderstanding of what is (or should be) meant by a 'process'. A process is simply a framework, around which you can think about who should do what, and when. It simply means that one has thought about, and described, how to do something. Processes need not necessarily be formal, highly constrained or detailed – though they might be. When a process is seen as being too rigid, it is usually because it has been designed inappropriately for its volume–variety position.

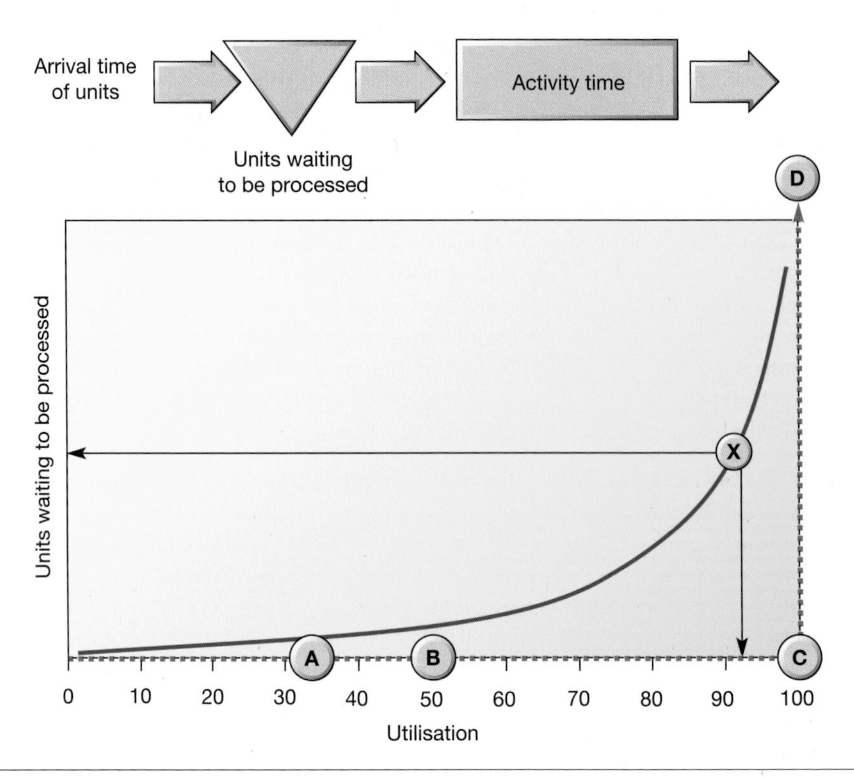

Figure 6.15 The relationship between process utilisation and number of items waiting to be processed for constant and variable arrival and process times

To understand the effect of arrival variability on process performance it is first useful to examine what happens to process performance in a very simple process as arrival time changes under conditions of no variability. For example, the simple process shown in Figure 6.15 comprises one stage that performs exactly 10 minutes of work. Items arrive at the process at a constant and predictable rate. If the arrival rate is one unit every 30 minutes, then the process will be utilised for only 33.33 percent of the time, and the items will never have to wait to be processed. This is shown as point A on Figure 6.15. If the arrival rate increases to one arrival every 20 minutes, the utilisation increases to 50 per cent, and again the items will not have to wait to be processed. This is point B on Figure 6.15. If the arrival rate increases to one arrival every 10 minutes, the process is now fully utilised but, because a unit arrives just as the previous one has finished being processed, no unit has to wait. This is point C on Figure 6.15. However, if the arrival rate ever exceeded one unit every 10 minutes, the waiting line in front of the process activity would build up indefinitely, as is shown as point D in Figure 6.15. So, in a perfectly constant and predictable world, the relationship between process waiting time and utilisation is a rectangular function as shown by the red line in Figure 6.15.

> **Operations principle**
> Variability in a process acts to reduce its efficiency.

However, when arrival and process times are variable, then sometimes the process will have items waiting to be processed, while at other times the process will be idle, waiting for items to arrive. Therefore, the process will have both a 'non-zero' average queue and be underutilised in the same period. So, a more realistic point is that shown as point X in Figure 6.15. If the average arrival time were to be changed with the same variability, the blue line in Figure 6.15 would show the relationship between average waiting time and process utilisation. As the process moves closer to 100 per cent utilisation, the average waiting time will become longer. Or, to put it another way, the only way to guarantee very low waiting times for the items is to suffer low process utilisation.

The greater the variability in the process, the more the waiting time–utilisation relationship deviates from the simple rectangular function of the 'no variability' conditions that was shown in

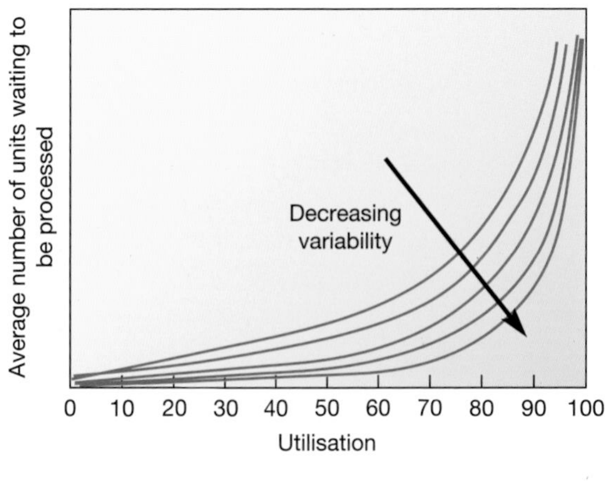

(a) Decreasing variability allows higher utilisation without long waiting times

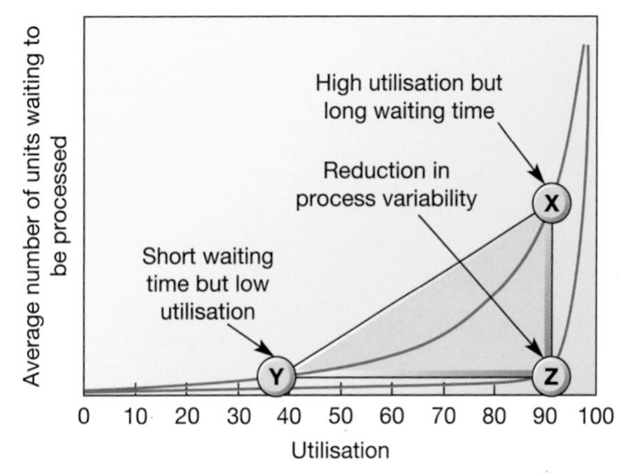

(b) Managing process capacity and/or variability

Figure 6.16 The relationship between process utilisation and number of items waiting to be processed for variable arrival and activity times

Figure 6.15. A set of curves for a typical process is shown in Figure 6.16(a). This phenomenon has important implications for the design of processes. In effect, it presents three options to process designers wishing to improve the waiting time or utilisation performance of their processes, as shown in Figure 6.16(b):

- accept long average waiting times and achieve high utilisation (point X);
- accept low utilisation and achieve short average waiting times (point Y); or
- reduce the variability in arrival times, activity times, or both, and achieve higher utilisation and short waiting times (point Z).

To analyse processes with both inter-arrival and activity time variability, queuing or 'waiting line' analysis can be used. This is treated in the supplement to Chapter 11. But do not dismiss the relationship shown in Figures 6.15 and 6.16 as some minor technical phenomenon. It is far more than this. It identifies an important choice in process design that could have strategic implications. Which is more important to a business, fast throughput time or high utilisation of its resources? The only way to have both of these simultaneously is to reduce variability in its processes, which may itself require strategic decisions such as limiting the degree of customisation of products or services, or imposing stricter limits on how products or services can be delivered to customers, and so on. It also demonstrates an important point concerned with the day-to-day management of process – the only way to absolutely guarantee 100 per cent utilisation of resources is to accept an infinite amount of work-in-progress and/or waiting time.

Operations principle

Process variability results in simultaneous waiting and resource underutilisation.

Summary answers to key questions

6.1 What is process design?

- Design is the activity that shapes the physical form and purpose of both products and services and the processes that produce them.
- The design activity is more likely to be successful if the complementary activities of product or service design and process design are coordinated.

6.2 What should be the objectives of process design?

- The overall purpose of process design is to meet the needs of customers through achieving appropriate levels of quality, speed, dependability, flexibility and cost.
- The design activity must also take account of environmental issues. These include examination of the source and suitability of materials, the sources and quantities of energy consumed, the amount and type of waste material, the life of the product itself and the end-of-life state of the product.

6.3 How do volume and variety affect process design?

- The overall nature of any process is strongly influenced by the volume and variety of what it has to process.
- The concept of process types summarises how volume and variety affect overall process design.
- In manufacturing, these process types are (in order of increasing volume and decreasing variety) project, jobbing, batch, mass and continuous processes. In service operations, although there is less consensus on the terminology, the terms often used (again in order of increasing volume and decreasing variety) are professional services, service shops and mass services.

6.4 How are processes designed in detail?

- Processes are designed initially by breaking them down into their individual activities. Often common symbols are used to represent types of activity. The sequence of activities in a process is then indicated by the sequence of symbols representing activities. This is called 'process mapping'. Alternative process designs can be compared using process maps and improved processes considered in terms of their operations performance objectives.
- The throughput time, work-in-progress and cycle time aspects of process performance are related by a formula known as Little's law: throughput time equals work-in-progress multiplied by cycle time.
- Variability has a significant effect on the performance of processes, particularly the relationship between waiting time and utilisation.

CASE STUDY The Action Response Applications Processing Unit (ARAPU)

Introduction

Action Response is a London-based charity dedicated to providing fast responses to critical situations throughout the world. It was founded by Susan N'tini, its Chief Executive, to provide relatively short-term aid for small projects until they could obtain funding from larger donors. The charity receives requests for cash aid, usually from an intermediary charity, and looks to process the request quickly, providing funds where and when they are needed. *'Give a man a fish and you feed him today, teach him to fish and you feed him for life. It's an old saying and it makes sense but, and this is where Action Response comes in, he might starve while he's training to catch fish'* (Susan N'tini).

Nevertheless, Susan does have some worries. She faces two issues in particular. First, she is receiving complaints that funds are not getting through quickly enough. Second, the costs of running the operation are starting to spiral. She explains: *'We are becoming a victim of our own success. We have striven to provide greater accessibility to our funds; people can access application forms via the internet, by post and by phone. But we are in danger of losing what we stand for. It is taking longer to get the money to where it is needed and our costs are going up. We are in danger of failing on one of our key objectives: to minimise the proportion of our turnover that is spent on administration. At the same time, we always need to be aware of the risk of bad publicity through making the wrong decisions. If we don't check applications thoroughly, funds may go to the "wrong" place and if the newspapers get hold of the story we would run a real risk of losing the goodwill, and therefore the funds, from our many supporters'.*

Susan held regular meetings with key stakeholders. One charity that handled a large number of applications for people in Nigeria told her of frequent complaints about the delays in the processing of the applications. A second charity representative complained that when they telephoned to find out the status of an application the ARAPU staff did not seem to know where it was or how long it might be before it was complete. Furthermore, they felt that this lack of information was eroding their relationship with their own clients, some of whom were losing faith in them as a result: *'trust is so important in the relationship',* they explained.

Some of Susan's colleagues, while broadly agreeing with her anxieties over the organisation's responsiveness and efficiency, took a slightly different perspective. *'One of the really good things about Action Response is that we are more flexible than most charities. If there a need and if they need support until one of the larger charities can step in, then we will always consider a request for aid. I would not like to see any move towards high process efficiency harming our ability to be open-minded and consider requests that might seem a little unusual at first'* (Jacqueline Horton, Applications Assessor).

Others saw the charity as performing an important counselling role. *'Remember that we have gained a lot of experience in this kind of short-term aid. We are often the first people that are in a position to advise on how to apply for larger and longer-term funding. If we developed this aspect of our work, we would again be fulfilling a need that is not adequately supplied at the moment'* (Stephen Nyquist, Applications Assessor).

The Action Response Applications Processing Unit (ARAPU)

Potential aid recipients, or the intermediary charities representing them, apply for funds using a standard form. These forms can be downloaded from the internet or requested via a special helpline. Sometimes the application will come directly from an individual community leader but more usually it will come via an intermediary charity that can help the applicant to complete the form. The application is sent to ARAPU, usually by fax or post (some are submitted online, but few communities have this facility).

ARAPU employs seven applications assessors with support staff who are responsible for data entry, coding, filing and 'completing' (staff who prepare payment, or explain why no aid can be given). In addition, a board of non-paid trustees meets every Thursday, to approve the assessors' decisions. The unit's IT system maintain records of all transactions, providing an update on the number of applications received, approved and declined, and payments allocated. These reports identified that the Unit received about 300 new applications per week and responded to about the same number (the Unit operates a 35-hour week). But while the Unit's financial targets were being met, the trend indicated that cost per application was increasing. The target for the turnaround of an application, from receipt of application to response, was 20 days, and although this was not measured formally, it was generally assumed that turnaround time was longer than this. Accuracy had never been an issue as all files were thoroughly assessed to ensure that all the relevant data were collected before the applications were processed. Productivity seemed high and there was always plenty of work waiting for processing at each section, with the exception that the 'completers' were sometimes waiting for work to come from the committee on a Thursday. Susan had conducted an inspection of all sections' in-trays that had revealed a rather shocking total of about 2,000 files waiting within the process, not counting those waiting for further information.

Processing applications

The processing of applications is a lengthy procedure requiring careful examination by applications assessors trained to make well-founded assessments in line with the charity's guidelines and values. Incoming applications are opened by one of the four 'receipt' clerks who check that all the necessary forms have been included in the application; the receipt clerks take about 10 minutes per application. These are then sent to the coding staff, in batches twice a day. The five coding clerks allocate a unique identifier to each application and key the information on the application into the system. The coding stage takes about 20 minutes for each application. Files are then sent to the senior applications assessors' secretary's desk. As assessors become available, the secretary provides the next job in the line to the assessor.

About 100 of the cases seen by the assessors each week are put aside after only 10 minutes' 'scanning' because the information is ambiguous, so further information is needed. The assessor returns these files to the secretaries, who write to the applicant (usually via the intermediate charity) requesting additional information and return the file to the 'receipt' clerks who 'store' the file until the further information eventually arrives (usually between one and eight weeks). When it does arrive, the file enters the process and progresses through the same stages again. Of the applications that require no further information, around half (150) are accepted and half (150) declined. On average, those applications that are not 'recycled' take around 60 minutes to assess.

All the applications, whether approved or declined, are stored prior to ratification. Every Thursday the Committee of Trustees meets to formally approve the applications assessors' decisions. The committee's role is to sample the decisions to ensure that the guidelines of the charity are upheld. In addition, they will review any particularly unusual cases highlighted by the applications assessors. Once approved by the committee the files are then taken to the completion officers. There are three 'decline' officers whose main responsibility is to compile a suitable response to the applicant pointing out why the application failed and offering, if possible, helpful advice. An experienced decline officer takes about 30 minutes to finalise the file and write a suitable letter. Successful files are passed to the four 'payment' officers where again the file is completed, letters (mainly standard letters) are created and payment instructions are given to the bank. This usually takes around 50 minutes, including dealing with any queries from the bank about payment details. Finally, the paperwork itself is sent, with the rest of the file, to two 'dispatch' clerks who complete the documents and mail them to the applicant. The dispatch activity takes, on average, 10 minutes for each application.

The feeling among the staff was generally good. When Susan consulted the team, they said their work was clear and routine, but their life was made difficult by charities that rang in expecting them to be able to tell them the status of an application they had submitted. It could take them hours, sometimes days, to find any individual file. Indeed, two of the 'receipt' clerks were now working almost full time on this activity. The team also said that charities frequently complained that decision making seemed slow.

QUESTIONS

1 What objectives should the ARAPU process be trying to achieve?

2 What is the main problem with the current ARAPU process?

3 How could the ARAPU process be improved?

Problems and applications

All chapters have 'Problems and applications' questions that will help you practice analysing operations. They can be answered by reading the chapter. Model answers for the first two questions can be found on the companion website for this text.

1 Visit a branch of a retail bank and consider the following questions:

   (a) What categories of service does the bank seem to offer?
   (b) To what extent does the bank design separate processes for each of its types of service?
   (c) What are the different process design objectives for each category of service?

2 Most of us are familiar with the 'drive-through' fast-food operations described in the chapter. Think about (or better still, visit) a drive-through service and try mapping what you can see (or remember) of the process (plus what you can infer from what may be happening 'behind the scenes').

3 The International Frozen Pizza Company (IFPC) operates in three markets globally. Market 1 is its largest market, where it sells 25,000 tons of pizza per year. In this market it trades under the name 'Aunt Bridget's Pizza' and positions itself as making pizza 'just as your Aunt Bridget used to make' (apparently she was good at it). It is also known for innovation, introducing new and seasonal pizza toppings on a regular basis. Typically, it sells around 20 varieties of pizza at any one time. Market 2 is smaller, selling around 20,000 tons per year under its 'Poppet's Pizza' brand. Although less innovative than Market 1, it still sells around 12 varieties of pizza. Market 3 is the smallest of the three, selling 10,000 tons per year of relatively high-quality pizzas under its 'Deluxe Pizza' brand. Like Aunt Bridget's Pizza, Deluxe Pizza also sells a relatively wide product range for the size of its market. Currently, both Markets 1 and 3 use relatively little automation and rely on high numbers of people, employed on a shift system, to assemble their products. Market 2 has always been keen to adopt more automated production processes and uses a mixture of automated assembly and manual assembly. Now, the management in Market 2 has developed an almost fully automated pizza assembly system (APAS). They claim that the APAS could reduce costs significantly and should be adopted by the other markets. Both Markets 1 and 3 are sceptical. *'It may be cheaper, but it can't cope with a high variety of products'*, is their response. Use the product–process matrix to explain the proposal by the management of Market 2.

4 A direct marketing company sells kitchen equipment through a network of local representatives working from home. Typically, individual orders usually contain 20 to 50 individual items. Much of the packing process is standardised and automatic. The Vice President of Distribution is proud of his distribution centre. *'We have a slick order fulfilment operation with lower costs per order, few packing errors, and fast throughput times. Our main problem is that the operation was designed for high volumes but the direct marketing business using representatives is, in general, on a slow but steady decline'*. Increasingly, customers are moving towards using the company's recently launched website or just buying from supermarkets and discount stores. Bowing to the inevitable, the company has started selling its products through discount stores. The problem is how to distribute their products through these new channels. Should they modify their existing fulfilment operation or subcontract the business to specialist carriers? *'Although our system is great at what it does, it would be difficult to cope with very different types of orders. Website orders will mean dealing with a far greater number of individual customers, each of whom will place relatively small orders for one or two items. We are not designed to cope with that kind of order. We would have the opposite problem delivering to discount stores. There, comparatively few customers would place large orders for a relatively narrow range of products. That is the type of job for a conventional distribution company. Another option would be to accept an offer from a non-competitor company that sell its products in a very similar way'*. What are the implications of the different sales channels for the existing distribution centre?

5 Revisit the case example that examines Legal and General's modular housing venture. Does their use of a factory to 'build' houses invalidate the idea that volume and variety govern the nature of operations processes?

6 A gourmet burger shop has a daily demand for 250 burgers and operates for 10 hours.

(a) What is the required cycle time in minutes?
(b) Assuming that each burger requires 7.2 minutes of work, how many servers are required?
(c) The burger shop has a three-stage process for making burgers. Stage 1 takes 2.0 minutes, stage 2 takes 3.0 minutes, and stage 3 takes 2.2 minutes. What is the balancing loss for the process?

7 At the theatre, the interval during a performance of *King Lear* lasts for 20 minutes and in that time 86 people need to use the toilet cubicles. On average, a person spends three minutes in the cubicle. There are 10 cubicles available.

(a) Does the theatre have enough toilets to deal with the demand?
(b) If there are not enough cubicles, how long should the interval be to cope with demand?

8 '*It is a real problem for us*', said Angnyeta Larson, '*We now have only ten working days between all the expense claims coming from the departmental coordinators and authorising payments on the next month's payroll. This really is not long enough and we are already having problems during peak times*'. Angnyeta was the department head of the internal financial control department of a metropolitan authority in southern Sweden. Part of her department's responsibilities included checking and processing expense claims from staff throughout the metropolitan authority and authorising payment to the salaries payroll section. She had 12 staff who were trained to check expense claims and all of them were devoted full time to processing the claims in the two weeks (10 working days) prior to the deadline for informing the salaries section. The number of claims submitted over the year averaged around 3,200, but this could vary between 1,000 during the quiet summer months and 4,300 in peak months. Processing claims involved checking receipts, checking that claims met with the strict financial allowances for different types of expenditure, checking all calculations, obtaining more data from the claimant if necessary and (eventually) sending an approval notification to salaries. The total processing time took on average 20 minutes per claim.

(a) How many staff does the process need on average, for the lowest demand, and for the highest demand?
(b) If a more automated process involving electronic submission of claims could reduce the average processing time to 15 minutes, what effect would this have on the required staffing levels?
(c) If department coordinators could be persuaded to submit their batched claims earlier (not always possible for all departments) so that the average time between submission of the claims to the finance department and the deadline for informing the salaries section was increased to 15 working days, what effect would this have?

9 The headquarters of a major creative agency offered a service to all its global subsidiaries that included the preparation of a budget estimate that was submitted to potential clients when making a 'pitch' for new work. This service had been offered previously only to a few of the group's subsidiary companies. Now that it was to be offered worldwide, it was deemed appropriate to organise the process of compiling budget estimates on a more systematic basis. It was estimated that the worldwide demand for this service would be around 20 budget estimates per week and that, on average, the staff who would put together these estimates would be working a 35-hour week. The elements within the total task of compiling a budget estimate are shown in Table 6.3.

Table 6.3 The elements within the total task of compiling a budget estimate

Element	Time (mins)	What element (s) must be done prior to this one?
A – obtain time estimate from creatives	20	None
B – obtain account handler's deadlines	15	None
C – obtain production artwork estimate	80	None
D – preliminary budget calculations	65	A, B and C
E – check on client budget	20	D
F – check on resource availability and adjust estimate	80	D
G – complete final budget estimate	80	E and F

(a) What is the required cycle time for this process?
(b) How many people will the process require to meet the anticipated demand of 20 estimates per week?
(c) Assuming that the process is to be designed on a 'long-thin' basis, what elements would each stage be responsible for completing? And what would be the balancing loss for this process?
(d) Assuming that instead of the long-thin design, two parallel processes are to be designed, each with half the number of stations of the long-thin design, what now would be the balancing loss?

10 Re-read the 'Operations in practice' example 'London's underground tackles a bottleneck'. What general lessons about designing processes for crowds of people would you draw from this example?

Selected further reading

Damelio, R. (2011) ***The Basics of Process Mapping,*** **2nd edn, Productivity Press, New York, NY.**
A practitioner book that is both very comprehensive and up-to-date.

Hammer, M. (1990) Reengineering work: don't automate, obliterate, ***Harvard Business Review,*** **July–August.**
This is the paper that launched the whole idea of business processes and process management in general to a wider managerial audience. Slightly dated but worth reading.

Harrington, H.J. (2011) ***Streamlined Process Improvement,*** **McGraw Hill Professional, New York, NY.**
Harvard Business Review (2011) ***Improving Business Processes*** **(Harvard Pocket Mentor), Harvard Business School Press, Boston, MA.**
A collection of HBR papers.

Hopp, W.J. and Spearman, M.L. (2011) ***Factory Physics,*** **3rd edn, Waveland Press Inc., Long Grove, IL.**
Very technical so don't bother with it if you aren't prepared to get into the maths. However, some fascinating analysis, especially concerning Little's law.

Ramaswamy, R. (1996) ***Design and Management of Service Processes,*** **Addison-Wesley Longman, Reading, MA.**
A relatively technical approach to process design in a service environment.

Slack, N. (2017) ***The Operations Advantage: A Practical Guide to Making Operations Work,*** **Kogan Page, London.**
The chapter on 'internal processes' expands on some of the issues discussed here.

Sparks, W. (2016) ***Process Mapping Road Trip: Improve Organizational Workflow in Five Steps,*** **Promptitude Publishing, Washington, DC.**
A practitioners' guide – straightforward and sensible.

Notes on chapter

1. The information on which this example is based is taken from: Zhang, S. (2016) 'How to fit the world's biggest indoor waterfall in an airport', *Wired*, 9 July; Airport Technology (2014) Terminal 4, Changi International Airport, https://www.airport-technology.com/projects/terminal-4-changi-international-airport-singapore/ (accessed September 2021); Driver, C. (2014) And the winners are . . . Singapore crowned the best airport in the world (and Heathrow scoops top terminal), Mailonline, 28 March, https://www.dailymail.co.uk/travel/article-2591405/Singapore-crowned-best-airport-world-Heathrow-scoops-terminal.html (accessed September 2021).
2. Takt time was originated by Toyota, the car company. It is their adaptation of the German word *'taktzeit'*, originally meaning 'clock cycle'.
3. The information on which this example is based is taken from: Oches, S. (2013) The drive-thru performance study, *QSR magazine*, September; Horovitz, A. (2002) Fast food world says drive-through is the way to go, *USA Today*, 3 April; Richtel, M. (2006) The long-distance journey of a fast-food order, *The New York Times*, 11 April, https://www.nytimes.com/2006/04/11/technology/the-longdistance-journey-of-a-fastfood-order.html (accessed September 2021).
4. Press Association (2012) Standardised bed chart 'could prevent hundreds of hospital deaths', *Guardian*, 27 July.
5. The information on which this example is based is taken from: Wilmore, J. (2019) We take a look around L&G's housing factory, *Inside Housing*, 14 February; Legal and General Modular Homes website, https://www.legalandgeneral.com/modular/a-modern-method/; The built environment: Achieving a resilient recovery with the circular economy', report by the Ellen MacArthur Foundation, https://www.ellenmacarthurfoundation.org/our-work/activities/covid-19/policy-and-investment-opportunities/the-built-environment (accessed August 2021).
6. The information on which this example is based is taken from: Qureshi, W. (2020) Ecover relaunches biodegradable detergents in PCR plastic, *Packaging News*, 21 January; Cornwall, S. (2013) Ecover announces world-first in plastic packaging, *Packaging Gazette*, 7 March; Ecover website, http://www.ecover.com (accessed August 2021).
7. The idea of the product-process matrix was originally presented in a different form in Hayes, R.H. and Wheelwright, S.C. (1984) *Restoring our Competitive Edge: Competing Through Manufacturing*, John Wiley & Sons, Inc., New York, NY.
8. One note of caution regarding this idea: although logically coherent, it is a conceptual model rather than something that can be 'scaled'. Although it is intuitively obvious that deviating from the diagonal increases costs, the precise amount by which costs will increase is very difficult to determine.
9. The information on which this example is based is taken from: Sutherland, E. (2017) Weihai and mighty, Drapersonline, 16 June; and www.sandsfilms.co.uk (accessed August 2021).
10. Global fashion industry statistics – international apparel', Fashion United, http://www.fashionunited.com/global-fashion-industry-statistics-international-apparel (accessed August 2021).
11. Quoted on Dishang Group's website, www.dishang-group.com (accessed August 2021).
12. Shostack, G.L. (1984) Designing services that deliver, *Harvard Business Review*, 62 (1), 133–9.
13. The information on which this example is based is taken from: Matthews, T. and Trim, L. (2019) London Underground: why it would be better if we stood on both sides of the escalators, MyLondon Local News, 13 August; Sleigh, S. (2017) TfL scraps standing only escalators – despite trial being deemed a 'success', *London Evening Standard*, 8 March.
14. There are many studies that investigate this issue: see, for example, Thomsen, J.F., Mikkelsen, S., Andersen, J.H., Fallentin, N., Loft, I.P., Frost, P., Kaergaard, A., Bonde, J.P. and Overgaard, E. (2007) Risk factors for hand-wrist disorders in repetitive work, *Occupational and Environmental Medicine*, 64 (8), 527–33.

7 The layout and look of facilities

KEY QUESTIONS

7.1 How can the layout and look of facilities influence performance?

7.2 What are the basic layout types and how do they affect performance?

7.3 How does the appearance of an operation's facilities affect its performance?

7.4 What information and analysis is needed to design the layout and look of facilities?

INTRODUCTION

The layout and look of an operation's facilities determines their physical positioning relative to each other and their aesthetic appearance. It involves deciding where to put all the facilities, desks, machines and equipment in the operation. Because people and facilities work together in most operations, this also impacts on how, and where, staff operate within the operation. It is also concerned with the physical appearance of an operation in a broader sense – an issue that is seen as increasingly important, given the effect it has on the people working in the operation and any customers who 'experience' the operation. Both the layout and look of facilities govern how safe, how attractive, how flexible and how efficient an operation is. They also determine how transformed resources – materials, information and customers – flow through an operation. For all these reasons, it is an important activity. It is it the first thing that most of us notice when we enter an operation. Also, relatively small changes – moving displays in a supermarket, the décor of a restaurant, the changing rooms in a sports centre, or the position of a machine in a factory – can affect flow through the operation, which, in turn, affects the costs and general effectiveness of the operation. Figure 7.1 shows the layout activity in the overall model of design in operations. In this chapter we will look briefly at what operations managers are trying to achieve when they change the layout or look of their facilities, we describe a number of recognised 'layout types', we look at how the physical appearance of operations influences their effectiveness and we look at the information needed to decide on facilities' layout and look.

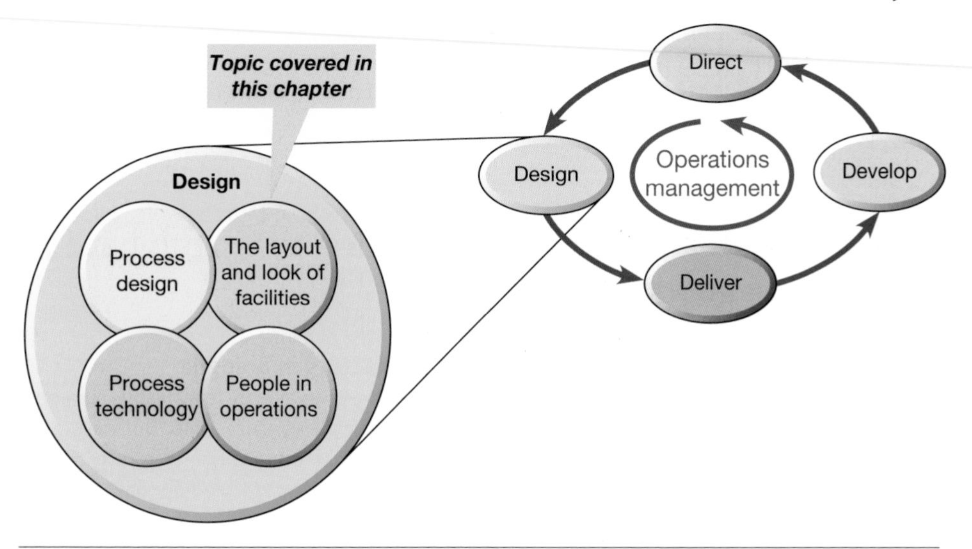

Figure 7.1 This chapter examines the layout and look of facilities

7.1 How can the layout and look of facilities influence performance?

The 'layout and look' of an operation or process means how its facilities are positioned relative to each other and how their general appearance is designed. These decisions will dictate the pattern and nature of how transformed resources progress through the operation or process. They also affect both how the people who staff the operation and, in high-visibility operations (where customers form part of the transformed resource), how customers judge their experience of being in the operation. Figure 7.2 shows how both layout and look of facilities affect some of the factors on which operations facilities are judged. Both are important decisions. If done badly, they can lead to over-long or confused flow patterns, long process times, inflexible operations, unpredictable flow, high costs, frustration for the people working in the operation and, in high-visibility operations, a poor customer experience.

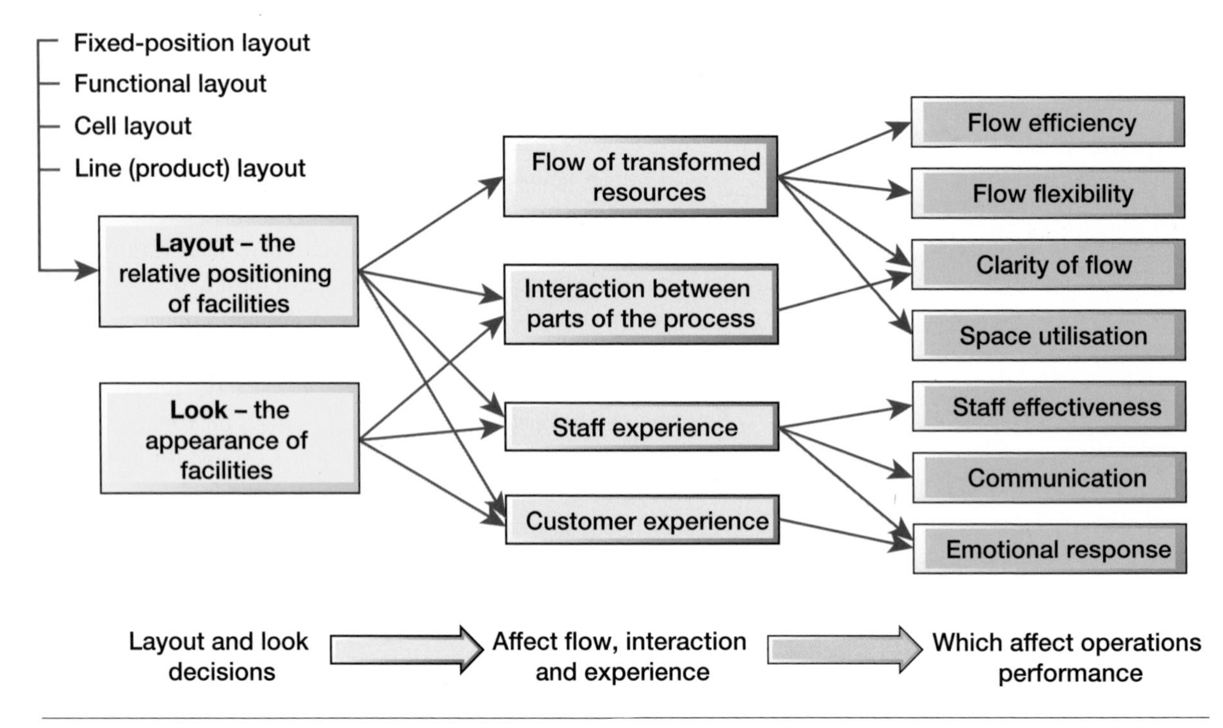

Figure 7.2 The layout and look of facilities involves the relative positioning of transforming resources within operations and processes, and their general appearance, which together dictate the nature and pattern of the flow of transformed resources and the experience of staff and, in high-visibility operations, customers

OPERATIONS IN PRACTICE

Ducati factory or Google office, they both have to look good[1]

Don't be tempted to think that the principles of design that govern the layout and look of an engineering factory and the head office of a global high-tech company should be very different. Where facilities are located within the operation, and what they look like are important for both. Here are two examples.

The Ducati factory in Bologna

Motorcycles have been built by Ducati at its current factory in Bologna continuously since 1946. But now, as well as producing 350 bikes every day in the busy season, the factory hosts visitors keen to see how the famous motorcycles are made. The assembly process starts in the

'supermarket' area of the factory. Each of the models made by Ducati has its own 'supermarket' zone. Here manufacturing trays are put together with exactly the right parts that will be needed for each of the subsequent stages in the assembly process. Because these 'kits' of parts exactly match the requirements for each product, any parts left in the tray at the end of assembly are an indication that something has gone wrong. Livio Lodi, who has worked in the Ducati factory for more than 26 years, is the official historian and curator of the Ducati Museum on the site. He explains: *'With this new style of just-in-time production, we have been able to reduce over 85 per cent of the defects in the final product. Porsche engineers came to us here and explained the way to set up the production in a just-in-time philosophy which they got from Toyota originally'.*

The factory has also become a tourist attraction thanks to the regular and popular factory tours. Partly it is a customer relations and marketing device as well as a production plant. So, the facilities have to be designed to accommodate visitors as well as motorcycle parts. Reviewers of the tour enthuse about the *'fantastic factory tour with wonderful friendly guides'*, and how it was *'very informative and knowledgeable'* and *'interesting to see all aspects of the assembly of the wonderful Ducati motorcycle'*; they mention how *'the museum was great and historic with some beautiful old bikes and some up-to-date winners'*. Many visitors are Ducati owners, but the company shows the factory and the museum to everybody. Ducati say that they don't care which kind of motorcycle you have. But, if you ride a Ducati there, you can park your bike inside the gates. If you don't ride a Ducati, you can leave it outside the entrance.

Google's revolutionary offices

Operations, and therefore operations layouts, are not confined to factories, warehouses, shops and other such workspaces. Many of us who work in operations actually work in offices. Financial services, government, call centres and all the creative industries, all work for the most part sitting at their desks. So, the layout of offices can affect operations performance for these industries just as much as layout can in a factory. Google, like many high-tech companies, is paying much more attention to its employees' work environment, the better to promote creativity and productivity. In fact, Google is famous for its innovative use of its workspaces. This is because it thrives on creativity and it believes that the designs of its offices will provide every employee with a space that will encourage creativity. The layouts of Google's offices are designed to promote creativity and collaboration. How people move about the space and who they meet and talk to are vital pieces of information that should contribute to any design. In addition to examining the formal needs of people's jobs, it is valuable to examine employee behaviour. For example, where do people actually spend the majority of their time? Where and when do the most productive meetings happen? Where and when do people make phone calls? When is the office at its emptiest? Elliot Felix, who led the team that wrote Google's global design guidelines for its offices, says that *'the ingredients of the offices should all fit together cohesively for a consistent employee experience. We're never just talking about space. We're talking about culture, etiquette, and rituals. What a lot of people forget is that we imbue space with our values'.*

What makes a good layout?

As with most operations design decisions, what constitutes a 'good' design will depend partly on the strategic objectives of the operation. But whatever 'good' is for any specific operation, it will usually be judged against a common set of criteria, as indicated in Figure 7.2. These are:

- ***The flow of transformed resources*** – The route taken by transformed resources as they progress through an operation or process is governed by how its transforming resources are positioned relative to each other. Often the objective is to achieve high flow efficiency that minimises distance travelled. But not always. For some customer-transforming operations, supermarkets for example, layout objectives can include encouraging customers to 'flow' in particular ways that maximise sales. However, sometimes high flow efficiency can be achieved only by sacrificing flow flexibility – the ability of transformed resources to take many different routes. Additional objectives can include the clarity of the flow of materials and/or customers and an effective use of the space available in the operation.
- ***The interaction between parts of the process*** – The individual facilities or parts of a process can suffer or benefit from being positioned close to each other. Dirty processes should not be located near to other parts of the process where their pollution could reduce its effectiveness. Noisy processes should not be located near processes that require concentration (see the 'Operations in practice' example, 'Reconciling quiet and interaction in laboratory layout'). Conversely there may be a positive effect of locating parts of an operation close to each other, for example to encourage communication between staff (see the 'Operations in practice' example of Google's office layout).
- ***Staff experience*** – An obvious prerequisite for any layout in any type of operation is that it should not constitute any physical or emotional danger to staff. So, 'fire exits should be clearly marked with uninhibited access', 'pathways should be clearly defined and not cluttered', etc. Unnecessary movement, caused by poor layout, will take productive time away from value-adding tasks. But just as important is the 'look, touch, taste, smell and feel' of an operation that will influence the 'employee experience' and hence staff productivity and morale.
- ***Customer experience*** – In high-visibility operations such as retail shops or bank branches, the layout and particularly the look of an operation can help to shape its image and the general experience of customers. Layout and look can be used as a deliberate attempt to establish a company's brand.

Operations principle

The layout and look of operations and processes is important because it affects their flow characteristics, the interaction between parts of the process and the experience of staff and customers.

Reconciling objectives

Some objectives, such as safety, security and staff welfare, are absolutely required. Others may have to be compromised, or traded off with other objectives. For example, two processes may have need of the same piece of equipment and could quite feasibly share it. This would mean good use of the capital used to acquire that equipment. But having both processes using it could mean longer and/or more confused process routes. Buying two pieces of equipment would underutilise them, but give shorter distance travelled. The 'Operations in practice' example 'Reconciling quiet and interaction in laboratory layout' is an example of how objectives have to be reconciled.

7.2 What are the basic layout types and how do they affect performance?

Most practical layouts are derived from only four basic layout types. These are:

- Fixed-position layout.
- Functional layout.

Table 7.1 Alternative layout types for each process type

Manufacturing process type	Potential layout types		Service process type
Project	Fixed position layout Functional layout	Fixed position layout Functional layout Cell layout	Professional service
Jobbing	Functional layout Cell layout	Functional layout Cell layouts	Service shop
Batch	Functional layout Cell layout		
Mass	Cell layout Product layout	Cell layout Product layout	Mass service
Continuous	Product layout		

- ▶ **Cell layout.**
- ▶ **Line** (sometimes called '**product**') layout.

These layout types are loosely related to the process types described in Chapter 6. As Table 7.1 indicates, a process type does not necessarily imply only one particular basic layout.

Fixed-position layout

Fixed-position layout is in some ways a contradiction in terms, since the transformed resources do not move between the transforming resources. Instead of transformed resources flowing through an operation, the recipient of the processing is stationary and the facilities and people who do the processing move as necessary. This could be because the transformed resources are too large, too delicate or too inconvenient to move; for example:

- ▶ *Motorway construction* – the product is too large to move.
- ▶ *Open-heart surgery* – patients are too delicate to move.
- ▶ *High-class service restaurant* – customers would object to being moved to where food is prepared.
- ▶ *Shipbuilding* – the product is too large to move.
- ▶ *Mainframe computer maintenance* – the product is too big and probably also too delicate to move, and the customer might object to bringing it in for repair.

OPERATIONS IN PRACTICE

Reconciling quiet and interaction in laboratory layout[2]

The layout of scientific laboratories is rarely straightforward. Not only can different areas of a laboratory require very different service needs (temperature, extraction, lack of vibration, etc.) but also two types of work in which all scientists engage can have different and opposing needs. On one hand, the development of new ideas is encouraged by free, and sometimes random, meetings between researchers. On the other hand, there are times when quiet reflection is vital to work through the implications of those same ideas. Moreover, different individuals have different preferred working patterns. The conversations, discussion and, sometimes noisy, debate between some researchers can both irritate and distract other staff who prefer quiet to think and write up their work. Even in prestigious and high-profile research operations, this conflict can be difficult to reconcile. For example, some of the researchers working at the Francis Crick Institute's laboratory in central ▶

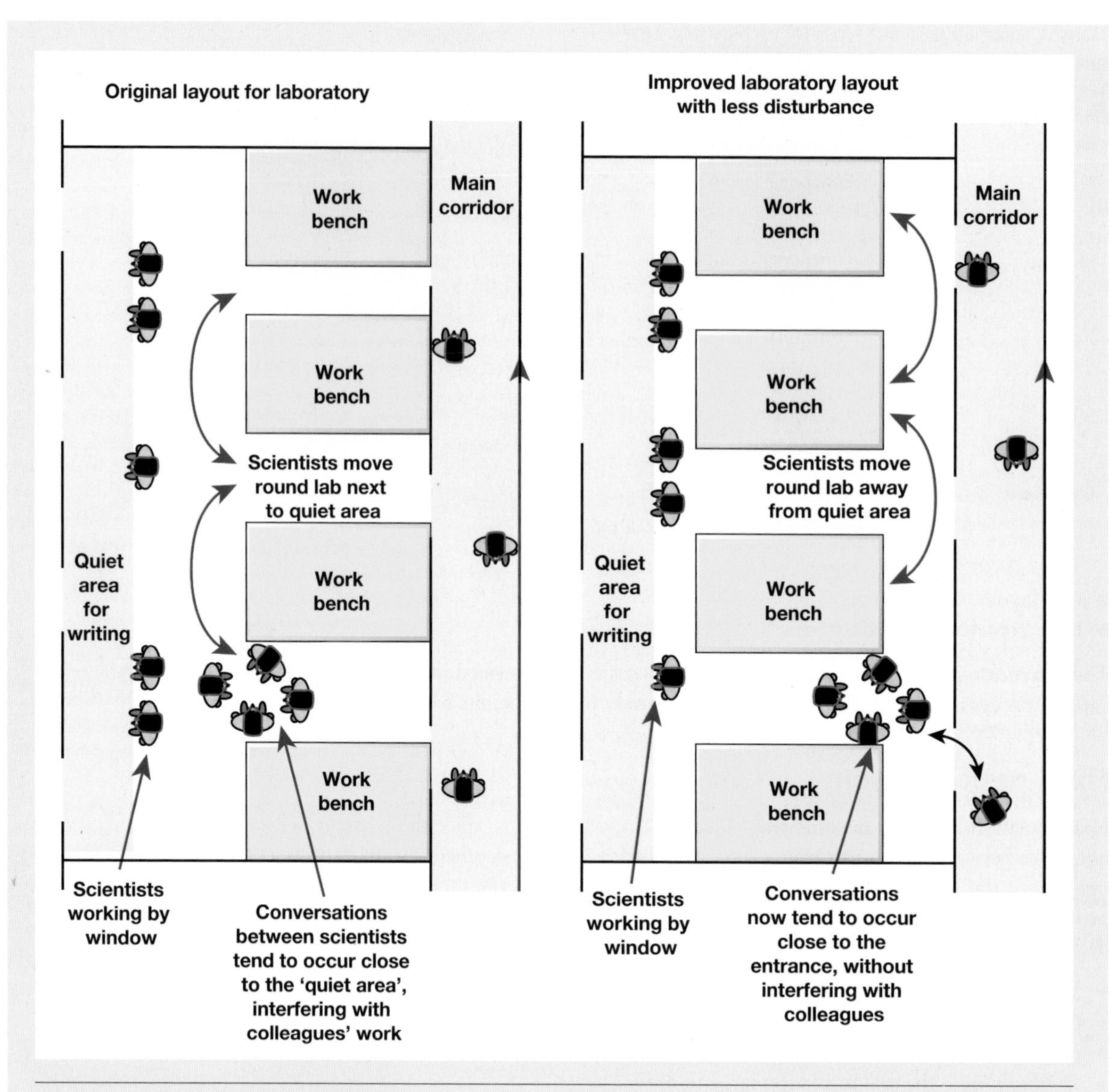

Figure 7.3 Example of an improved laboratory layout that reduces the degree of interference between different types of work (conversations and writing)

London complained that its open-plan layout, designed to encourage collaboration, made it difficult to concentrate on their work. Some people like the background noise, which can be similar to working in a café, while others prefer total silence, although many agree that the layout has been extremely successful in terms of promoting ad hoc meetings and has created new collaborations. Professor Alan Penn, who has been investigating how open-plan layouts (for example, those in advertising agencies or science laboratories) affect behaviour, points out how designing laboratories with busy circulation spaces allows scientists from different research groups to share ideas effectively. People walking around can stop and join a conversation in the door of a laboratory. Conversations inside the laboratory, when they are next to where the relatively high-flow movement along the corridor occurs, lead to discussions between research groups.

Figure 7.3 illustrates how laboratory design can, to some extent, reduce the conflict between the benefits of interaction and the need for quiet. The conventional layout on the left allows potentially disruptive conversations to interfere with quiet areas. The marginally modified layout on the right encourages conversations to happen closer to the entrance, without interfering with colleagues.

Functional layout

In functional layout, similar transforming resources are located together. This may be because it is convenient to group them together, or because their utilisation is improved. It means that when transforming resources flow through the operation, they will take a route from activity to activity according to their needs. Different products or customers will have different needs and therefore take different routes. Usually this makes the flow pattern in the operation very complex. Examples of functional layouts include:

- *Hospital* – some processes (e.g. X-ray machines and laboratories) are required by several types of patient; some processes (e.g. general wards) can achieve high staff and bed utilisation.
- *Supermarket* – some products, such as tinned goods, are convenient to restock if grouped together. Some areas, such as those holding frozen vegetables, need the common technology of freezer cabinets. Others, such as the areas holding fresh vegetables, might be together because that way they can be made to look attractive to customers.
- *Machining the parts that go into aircraft engines* – some processes (e.g. heat treatment) need specialist support (heat and fume extraction); some processes (e.g. machining centres) require the same technical support from specialist setter-operators, or need high utilisation.

Like most functional layouts, a library has different types of users with different traffic patterns. The college library in Figure 7.4 has put its users into three categories, as follows (in fact very similar to the categories used by retail customers):[3]

- *Browsers* – who seek interesting or useful materials by surfing the internet, browsing shelves and examining items, and moving around slowly while assessing the value of items.
- *Destination traffic* – who have a specific purpose or errand and are not deterred from it by surroundings or other library materials.
- *Beeline traffic* – who concentrate on goals unconnected with personal use of the library: for example, messengers, delivery staff or maintenance workers.

Based on studies tracking these different types of customers, the library derived the following guide rules for the layout of its library:

- Position displays and services that need to be brought to users' attention at the front of the facility.

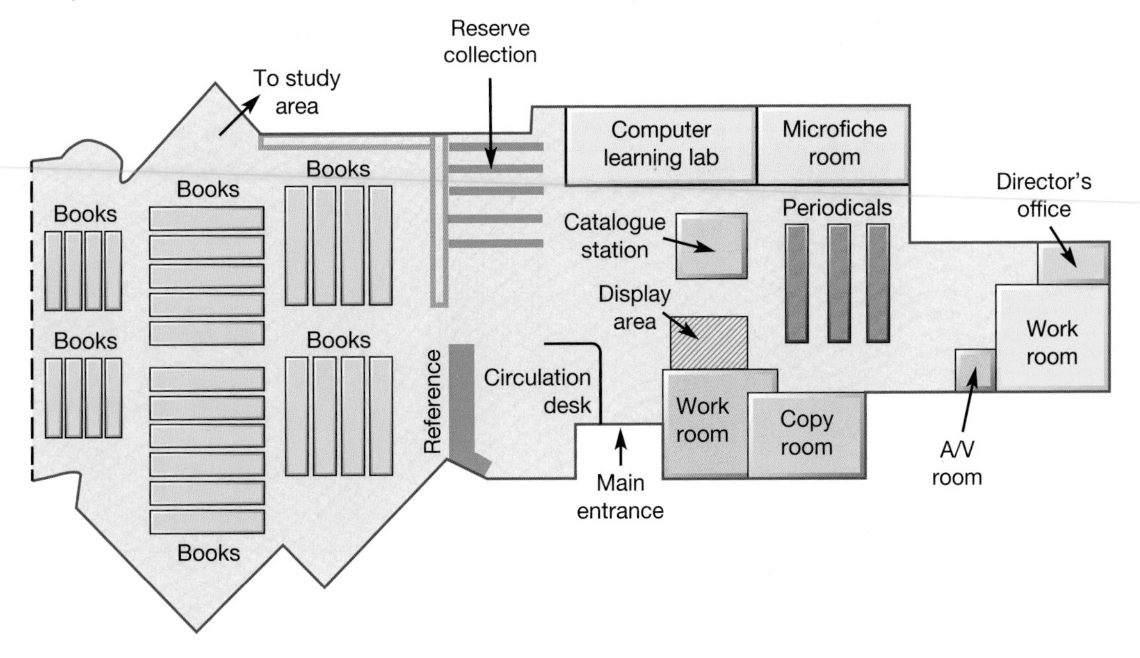

Figure 7.4 An example of a functional layout in a library

- To the right of the entrance should be: new acquisitions; items that might be selected on impulse and have no satisfactory substitutes; and items that require repeated exposure before users select them.
- On the left at the front should be items that probably will not be used unless there is maximum convenience for the user, such as reference books.
- The circulation desk should be on the left of the entrance, the last thing the user passes before leaving.
- The rear of the library should house items for which user motivation is strong, such as classroom-assigned materials and meeting rooms, or those that the user is willing to spend time and effort obtaining, such as microfiche printouts.

Cell layout

A cell layout is one where the transformed resources entering the operation are pre-selected (or pre-select themselves) to move to one part of the operation (or cell) in which all the transforming resources, to meet their immediate processing needs, are located. The cell itself may be arranged in either a functional or line (see next section) layout. After being processed in the cell, the transformed resources may go on to another cell. In effect, cell layout is an attempt to bring some order to the complexity of flow that characterises functional layout. Examples of cell layouts include:

- *'Lunch' products area in a supermarket* – some customers use the supermarket just to purchase sandwiches, savoury snacks, cool drinks, yoghurt, etc. for their lunch. These products are often located close together so that customers who are just buying lunch do not have to search around the store.
- *Maternity unit in a hospital* – customers needing maternity attention are a well-defined group who can be treated together and who are unlikely to need the other facilities of the hospital at the same time that they need the maternity unit.
- *Some computer component manufacture* – the processing and assembly of some types of computer parts may need a special area dedicated to the manufacturing of parts for one particular customer who has special requirements, such as particularly high quality levels.

Although the idea of cell layout is often associated with manufacturing, the same principle can be, and is, used in services. In Figure 7.5 the ground floor of a department store is shown, the

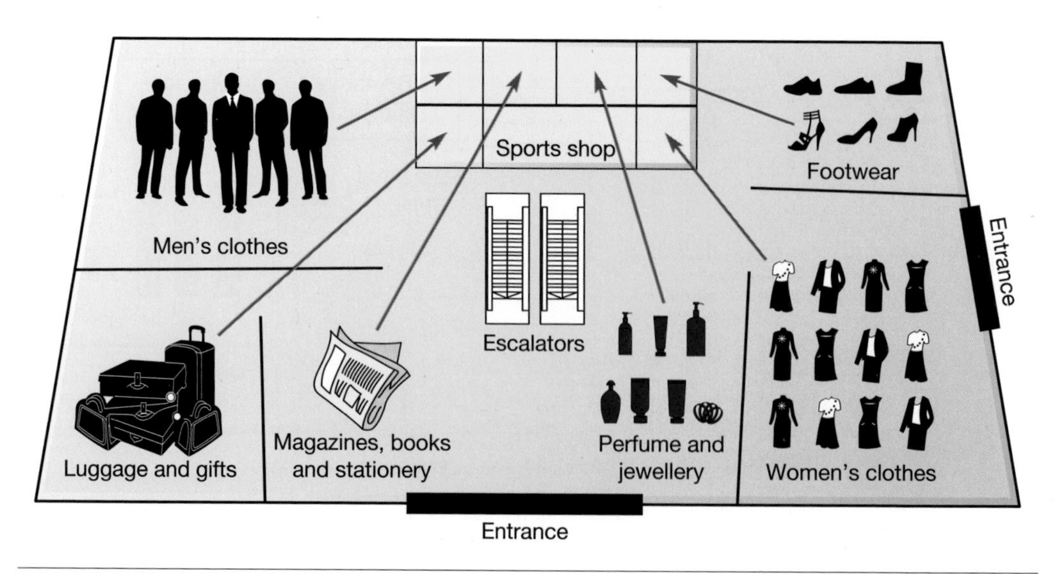

Figure 7.5 The floor plan of a department store showing the sports goods 'shop-within-a-shop' cell within the functional layout of the rest of the store

predominant layout of which is a functional layout, with separate areas devoted to selling each type of goods. The exception is the 'shop-within-a-shop' area that is devoted to many goods that have a common sporting theme (sports clothes, sports shoes, sports bags, sports magazines, sports books, sports equipment, etc). They have been located in the 'cell' not because they are similar goods (shoes, books and drinks would not usually be located together) but because they are needed to satisfy the needs of a particular type of customer. The store management calculates that enough customers come to the store to buy 'sports goods' in particular to devote an area specifically to them.

Line layout

Line layout involves locating the transforming resources entirely for the convenience of the transformed resources. Each product, piece of information or customer follows a prearranged route in which the sequence of activities required corresponds to the sequence in which facilities have been located. The transformed resources 'flow' along a 'line' according to their 'product' needs. This is why this type of layout is sometimes called flow or product layout. Flow is clear, predictable and therefore relatively easy to control. Usually, it is the standardised requirements of the product or service that lead to operations choosing line layouts. Examples of line layout include:

- *Mass-immunisation programme* – all customers require the same sequence of clerical, medical and counselling activities.
- *Self-service cafeteria* – generally the sequence of customer requirements (starter, main course, dessert and drink) is common to all customers, but layout also helps control customer flow.
- *Vehicle assembly* – almost all variants of the same model require the same sequence of processes.

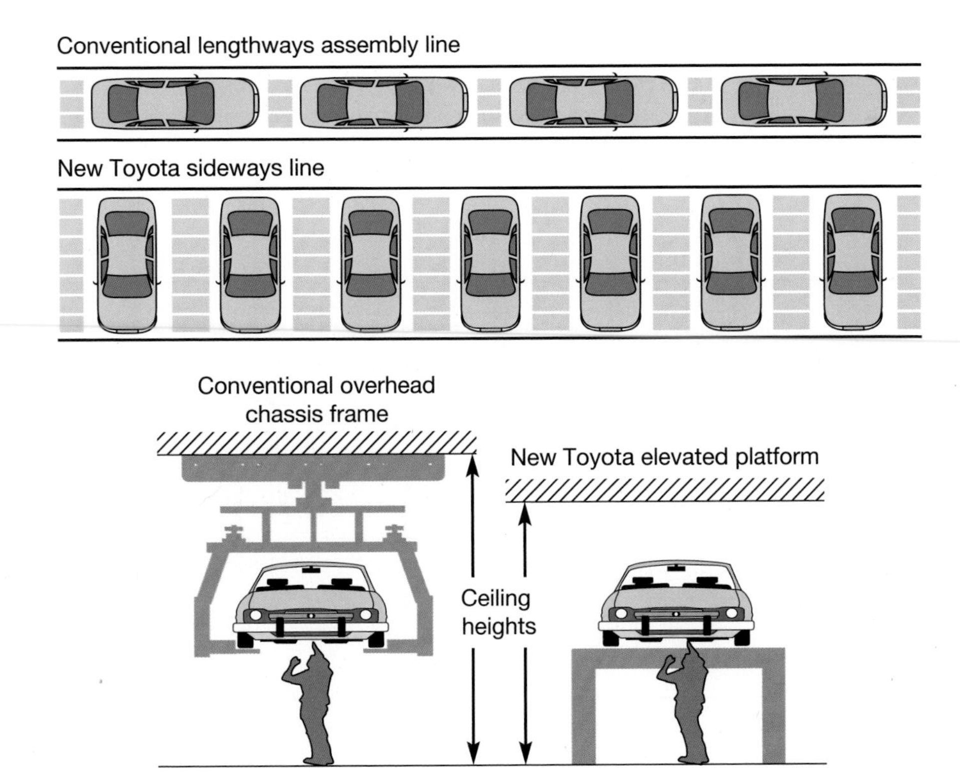

Figure 7.6 Contrasting arrangements in product layout for vehicle assembly plants

But don't think that line layouts are not changing. Toyota, like other Japanese firms, has built factories in other parts of the world, but still wants to manufacture in Japan, so cost savings have to be made. Figure 7.6 shows just two of the ideas that Toyota employed at its Miyagi factory in Japan to make assembly lines even more efficient. The top illustration shows how Toyota has positioned vehicles sideways rather than the conventional lengthways positioning. This shortens the line by 35 per cent (which increases space utilisation) and shortens the distance that workers have to walk between cars (which increases flow efficiency). The bottom illustration shows how, instead of the vehicle chassis hanging from overhead conveyor belts, they are positioned on raised platforms. This costs only half as much to construct and allows ceiling heights to be lowered, which is more space efficient and reduces heating and cooling costs by 40 per cent.

Mixed layouts

Many operations either design hybrid layouts that combine elements of some or all of the basic layout types, or use the 'pure' basic layout types in different parts of the operation. For example, a hospital would normally be arranged on functional layout principles – each department representing a particular type of function (the X-ray department, the operating theatres, the blood-processing laboratory and so on). Yet within each department, quite different layouts could be used. The X-ray department is probably arranged in a functional layout, the operating theatres in a fixed-position layout and the blood-processing laboratory in a line layout.

Another example is shown in Figure 7.7. Here a restaurant complex is shown with three different types of restaurant and the kitchen that serves them all. The kitchen is arranged in a functional layout, with the various processes (food storage, food preparation, cooking processes, etc.) grouped together. The traditional service restaurant is arranged in a fixed-position layout. The customers stay at their tables while the food is brought to (and sometimes cooked at) the tables. The buffet restaurant is arranged in a cell-type layout with each buffet area having all the processes (dishes) necessary to serve customers with their starter, main course or dessert. Finally, in the cafeteria restaurant, all customers take the same route when being served with their meal. They may not take the opportunity to be served with every dish but they move through the same sequence of processes.

> **Operations principle**
> There are four basic layout types, fixed-position, functional, cell and line, although layouts can combine elements of more than one of these.

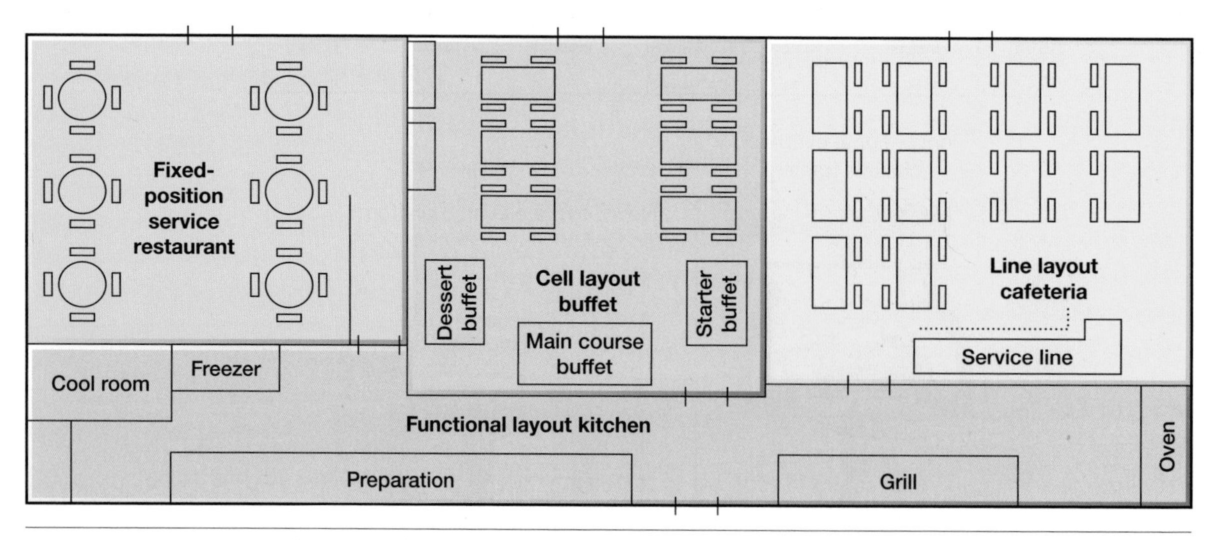

Figure 7.7 A restaurant complex with all four basic layout types

What type of layout should an operation choose?

The importance of flow to an operation will depend on its volume and variety characteristics (Figure 7.8). When volume is very low and variety is relatively high, 'flow' is not a major issue. For example, in telecommunications satellite manufacture, a fixed-position layout is likely to be appropriate because each product is different and because products 'flow' through the operation very infrequently, so it is just not worth arranging facilities to minimise the flow of parts through the operation. With higher volume and lower variety, flow becomes an issue. If the variety is still high, however, an entirely flow-dominated arrangement is difficult because there will be different flow patterns. For example, the library in Figure 7.4 will arrange its different categories of books and its other services partly to minimise the average distance its customers have to 'flow' through the operation. But, because its customers' needs vary, it will arrange its layout to satisfy the majority of its customers (but perhaps inconvenience a minority). When the variety of products or services reduces to the point where a distinct 'category' with similar requirements becomes evident but variety is still not small, cell layout could become appropriate, as in the sports goods cell in Figure 7.5. When variety is relatively small and volume is high, flow can become regularised and a line layout is likely to be appropriate, as in an assembly plant (see Figure 7.6).

Operations principle

Resources in low-volume, high-variety processes should be arranged to cope with irregular flow.

Although the volume–variety characteristics of the operation will narrow the choice down to one or two layout options, there are other associated advantages and disadvantages, some of which are shown in Figure 7.9. However, the type of operation will also influence the relative importance of these advantages and disadvantages. For example, a high-volume television manufacturer may find the low-cost characteristics of a product layout attractive, but an amusement theme park may adopt the same layout type primarily because of the way it 'controls' customer flow.

Operations principle

Resources in high-volume, low-variety processes should be arranged to cope with smooth, regular flow.

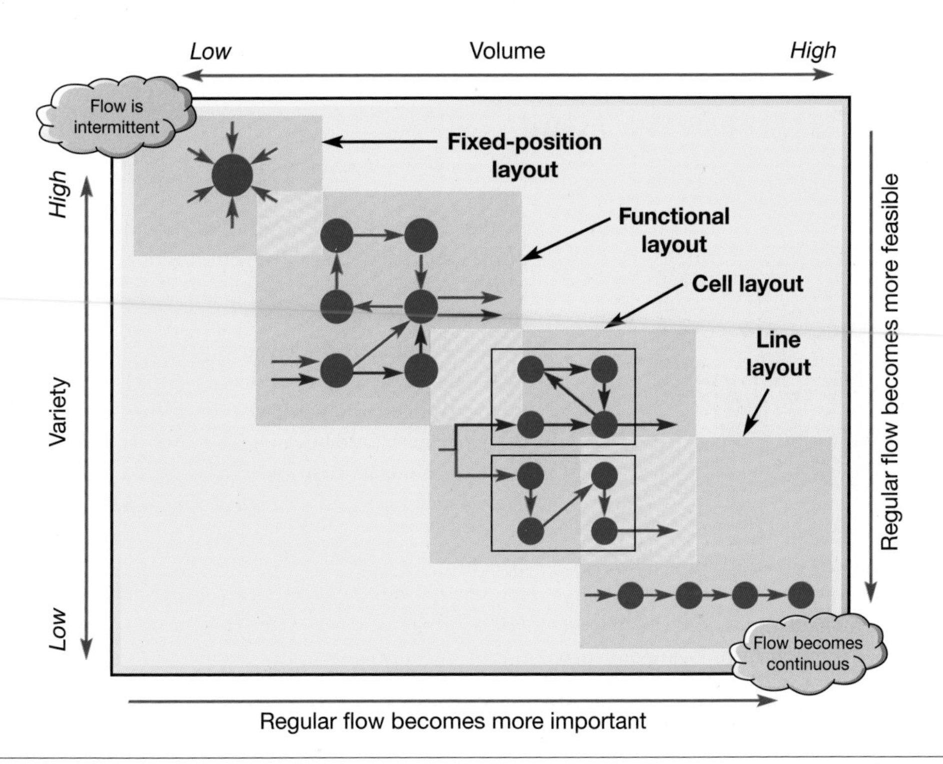

Figure 7.8 Different process layouts are appropriate for different volume–variety combinations

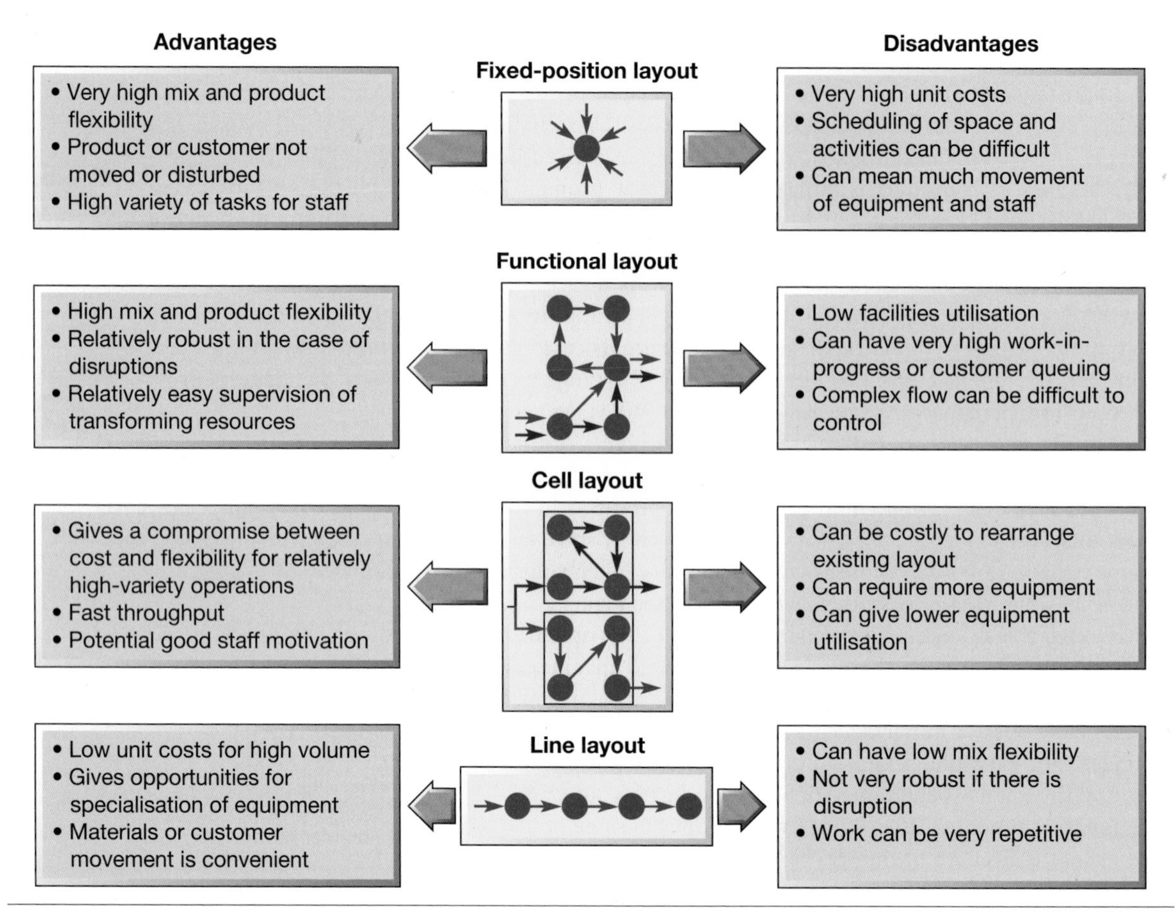

Advantages
Disadvantages
Fixed-position layout
• Very high mix and product flexibility
• Product or customer not moved or disturbed
• High variety of tasks for staff
• Very high unit costs
• Scheduling of space and activities can be difficult
• Can mean much movement of equipment and staff
Functional layout
• High mix and product flexibility
• Relatively robust in the case of disruptions
• Relatively easy supervision of transforming resources
• Low facilities utilisation
• Can have very high work-in-progress or customer queuing
• Complex flow can be difficult to control
Cell layout
• Gives a compromise between cost and flexibility for relatively high-variety operations
• Fast throughput
• Potential good staff motivation
• Can be costly to rearrange existing layout
• Can require more equipment
• Can give lower equipment utilisation
Line layout
• Low unit costs for high volume
• Gives opportunities for specialisation of equipment
• Materials or customer movement is convenient
• Can have low mix flexibility
• Not very robust if there is disruption
• Work can be very repetitive

Figure 7.9 Some advantages and disadvantages of layout types

OPERATIONS IN PRACTICE Supermarket layout

Successful supermarkets know that the design and layout of their stores has a huge impact on their profitability. This is why all big supermarket chains conduct extensive research into the most effective ways of laying out each part of their supermarkets and how best to position specific items. They must maximise their revenue per square metre and minimise the costs of operating the store, while keeping customers happy. At a basic level, supermarkets have to get the amount of space allocated to the different areas right, and provide appropriate display and storage facilities. But it is not just the needs of the products that are being sold that are important; the layout must also take into consideration the psychological effect that the layout has on their customers. Aisles are made wide to ensure a relatively slow flow of trolleys so that customers pay more attention to the products on display (and buy more). However, wide aisles can come at the expense of reduced shelf space that would allow a wider range of products to be stocked. The actual location of all the products is a critical decision, directly affecting the convenience to customers, their level of spontaneous purchase, and the cost of filling the shelves. Although the majority of supermarket sales are packaged tinned or frozen goods, the displays of fruit and vegetables are usually located adjacent to the main entrance, as a signal of freshness and wholesomeness, providing an attractive and welcoming point of entry. Basic products that figure on most people's shopping lists, such as flour, meat, sugar and bread, may be spread out towards the back of the store and apart from each other so that customers have to walk along more aisles, passing higher-margin items as they search. High-margin items are usually put at eye level on shelves (where they are more likely to be seen) and low-margin products lower down or higher up. Some customers also go a few paces up an aisle before they start looking for what they need. Some supermarkets call the shelves occupying the first metre of an aisle 'dead space', not a place to put impulse-bought goods. But the prime site in a supermarket is the 'gondola-end', the shelves at the end of the aisle. Moving products to this location can increase sales 200 or 300 per cent. It is not surprising that suppliers are willing to pay for their products to be located here. The circulation of customers through the store must also be right, but this can vary depending on which country you live in. In the United States shoppers like to work their way through a supermarket anti-clockwise. By contrast, shoppers in the United Kingdom like to shop clockwise. The reason for this is a bit of a mystery.

Cost analysis

Of all the characteristics of the various layout types, perhaps the most generally significant are the unit cost implications of layout choice. This is best understood by distinguishing between the fixed and variable cost elements of adopting each layout type. For any particular product or service, the fixed costs of physically constructing a fixed-position layout are relatively small compared with any other way of producing the same product or service. However, the variable costs of producing each individual product or service are relatively high compared to the alternative layout types. Fixed costs then tend to increase as one moves from fixed-position, through process and cell, to line layout. Variable costs per product or service tend to decrease, however. The total costs for each layout type will depend on the volume of products or services produced and are shown in Figure 7.10(a). This seems to show that for any volume there is a lowest-cost basic layout. However, in practice, the cost analysis of layout selection is rarely as clear as this. The exact cost of operating the layout is difficult

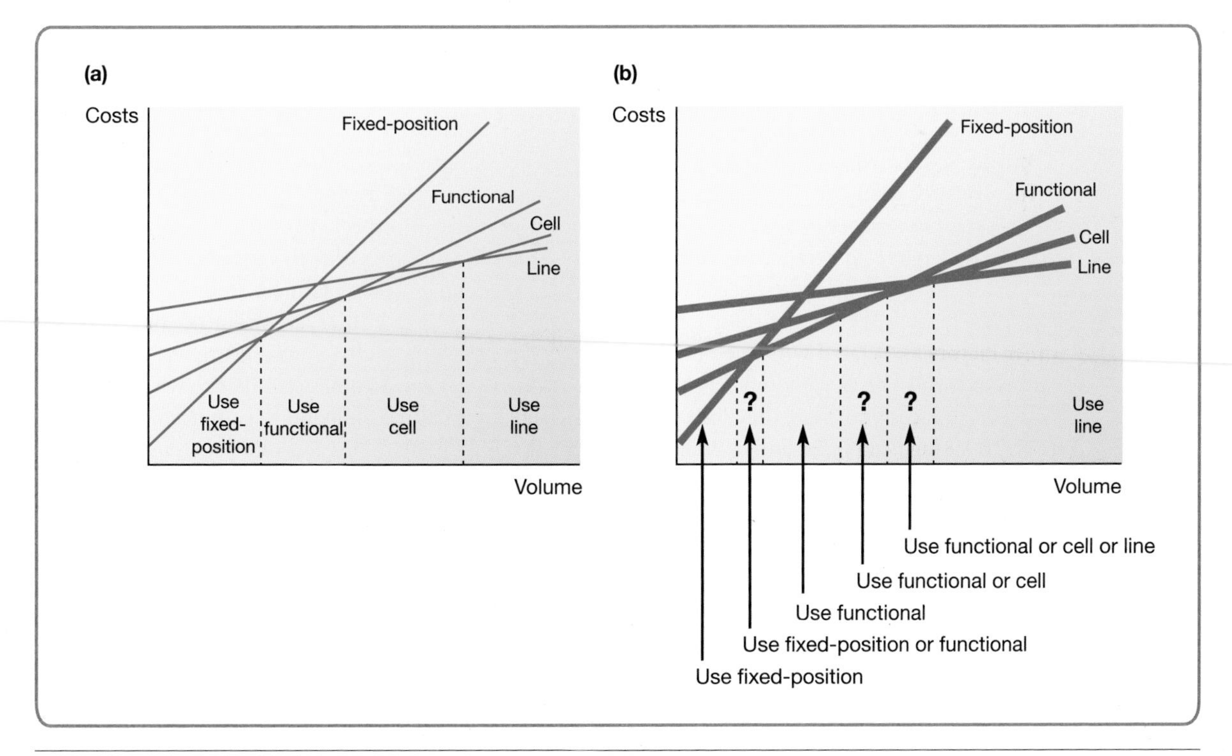

Figure 7.10 (a) The basic layout types have different fixed and variable cost characteristics which seem to determine which one to use. (b) In practice the uncertainty about the exact fixed and variable costs of each layout means the decision can rarely be made on cost alone

to forecast and will probably depend on many often difficult to predict factors. Rather than use thin lines to represent the cost of layout as volume increases, broad bands, within which the real cost is likely to lie, are probably more appropriate (see Figure 7.10(b)). The discrimination between the different layout types is now far less clear. There are ranges of volume for which any of two or three layout types might provide the lowest operating cost. The less certainty there is over the costs, the broader the cost 'bands' will be, and the less clear the choice will be. The probable costs of adopting a particular layout need to be set in the broader context of advantages and disadvantages shown in Figure 7.9.

Operations principle

Different layout types have different fixed and variable costs which determine the appropriateness of layout for varying volume–variety characteristics.

7.3 How does the appearance of an operation's facilities affect its performance?

Traditionally, operations managers have focused on the more evident 'pattern of flow' issues associated with facilities layout. Yet the aesthetics of a layout (in other words, what it looks and feels like) are also important and are increasingly seen as being within the scope of what operations management should be concerned with. Organisations with high-visibility operations such as retailers, hospitals or hotels have always understood that the look of their operations will affect customers' experience. Now, operations in most industries have come to understand the importance of aesthetics. Partly, this is because of the accumulating research evidence that the aesthetics of an operation can evoke positive or negative emotional response in people (whether they be customers or staff), and therefore affect their behaviour and well-being. Partly also, the approach of those technology companies that pioneered more relaxed and fluid working settings with the intention of encouraging casual encounters has started to influence other, less glamorous industries (see the 'Operations in practice' example on Ducati and Google earlier in the chapter).

The effect of workplace design on staff

There are some obvious and basic aspects of workplace design that will affect anyone working there. These are such things as: Is it warm enough? Too warm? Sufficiently well lit to see adequately? Not too noisy? These are all factors that deal with the physiological aspects of working – how we fit in with our physical working environment. Clearly, people who are cold, or irritated by their noisy environment, or straining to see what they are doing, will probably not be feeling, or working, particularly well. We look at these issues in Chapter 9 when we consider 'ergonomics'. But there are other factors associated with the design of a workplace that could affect staff attitudes, motivation and behaviour. This is why in recent years many companies have devoted resources to what goes into their workplaces and what they look like. Increasingly, special meeting zones, cappuccino bars, fish tanks, relaxing bean bags, games consoles, hammocks, ping pong tables and other such features have been integrated into workspaces. Why is this?

The core of the argument for using these design features is that a workplace is more than simply the arrangement of facilities and the pattern of flow that it creates. It is also the furniture; the way space is used and even the colour of the paint on the walls. Some workplace designers would go further. The aesthetics of the workplace also reflect the culture of the organisation. (There is no single authoritative definition of organisational culture, but generally it is taken to mean what it feels like to be part of an organisation, 'the organisation's climate'.)[4] Therefore, they argue, the appearance of a workplace should reflect the organisation's culture. The key questions are 'what does that workplace say about our culture?' and 'how can we create an environment that further promotes our culture?' What works for one company may be counter-cultural at another.

Operations principle

Layout should take into consideration the aesthetic appearance of the workplace and types of facilities available to staff.

Critical commentary

The question of how much difference the aesthetics and components of the working environment make is not uncontentious. Although many authorities agree that the look and feel of a workplace can have a positive influence on staff performance, not all studies agree. For example, one study that evaluated the importance of different productivity drivers among knowledge workers, while stressing the importance of their feelings of well-being at work, failed to confirm the role of the servicescape on productivity.[5] There is also some evidence that open plan workplaces, casual meeting rooms, and more 'living-room-like spaces' are not universally popular with staff. Gensler, a design firm, surveyed more than 90,000 people across ten industries about their views on 'fluid', 'open-plan' working. It found a surprising amount of opposition, with many staff claiming that open-plan offices, in particular, make it difficult to concentrate because of excessive noise. What they actually valued were as few distractions as possible so that they could focus on their jobs.[6]

The Allen curve

Arranging the facilities in any workplace will directly influence how physically close individuals are to each other. And this, in turn, influences the likelihood of communication between individuals. So, what effects does placing individuals close together or far apart have on how they interact? The work of Thomas J. Allen at the Massachusetts Institute of Technology first established how communication dropped off with increased distance. In 1984 his book, *Managing the Flow of Technology*, presented what has become known as the '**Allen curve**'. It showed a powerful negative correlation between the physical distance between colleagues and their frequency of communication. The 'Allen curve' estimated that we are four times as likely to communicate regularly with a colleague sitting two metres away from us as with someone 20 metres away (for example, separate floors mark a cut-off point for the regular exchange of certain types of technical information). As some experts have pointed out, the office is no longer just a physical place; email, remote conferencing and collaboration tools mean that colleagues can communicate without ever seeing each other. However, one study[7] showed that so-called distance-shrinking technology actually makes close proximity more important, with both face-to-face and digital communications following the Allen curve.

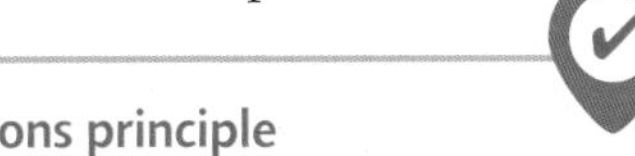

Operations principle

The likelihood of communication between people in their workplace falls off significantly as the distance between them increases. This is called the Allen curve.

OPERATIONS IN PRACTICE

Office layout and design[8]

Notwithstanding the severe reduction in people commuting daily to an office to perform their work, forced by the COVID-19 pandemic, offices are unlikely to disappear in the next few years. So-called 'hybrid working' where people spend some of their time working from home and some time in an office (a topic we look at in Chapter 9) has changed the need for offices, their location and their internal design. But they are still an important workplace. Before the pandemic, around 400 million people worked primarily in offices in 40 economies in the Global North. For example, in the United Kingdom, office workers were paid 55 per cent of all income. Changes to how offices were designed did come, but they had been changing for years before the pandemic.

One of the most significant was the move, towards the end of the twentieth century, from standardised rows of desks, a bit like a university examination room, to (equally standardised, some would say) individual cubicles. The man who invented the concept, Robert Propst, a designer working for the office furniture firm Herman Miller, hoped it would bring flexibility and independence to the office environment. He proposed what was the first modular office system, called the 'Action Office II'. Using his system, space could be divided up by wall-like vertical panels that could be slotted together in various ways. Propst believed that the best way to arrange the 'walls' would be to join the panels at 120° angles. However, to his disappointment, office designers realised that they could squash more people into the available space if they arranged the 'walls' at 90° to form the classic cubicle. Cubicles were not universally popular, possibly because they masked the social cues such as facial expressions and body language that influence social interactions. But cubicles are still used. One explanation for this is that privacy is so valued that office planners try to create the illusion of it. This seems to be borne out by the way people personalise their cubicles with pictures, flowers and rugs.

Open-plan, but less formal designs became more popular in the 2000s, often with moveable furniture, modular walls and a less formal décor (see the 'Operations in practice' example on Google's office design, earlier in this chapter). But the open-plan nature of some of these office designs is not always popular. The most common complaint is apparently the perceived noise level in open-plan offices, although this can be reduced to some extent by using acoustic panelling or soft furnishings to absorb noise. (Yet some office workers complain about their offices being too quiet.) Some open-plan offices also use an unassigned 'hot-desking' approach to allocating personal space. Although 'hot-desking' gives flexible capacity and saves money, it does take away a degree of 'personalisation'. Some also claim that it wastes employees' time when they have to search for a desk when the office is busy. It also takes time to perform ancillary tasks such as setting up a computer, adjusting a chair and finding out where the people one needs to talk to might be located that day.

The effect of workplace design on customers – servicescapes

If the appearance of an operation affects how its staff feel about working there, it certainly will also affect customers if they enter the workplace, as they do in 'high-visibility' operations. The term that is often used to describe the look and feel of the environment within an operation from a customer's perspective is its 'servicescape' (although it is sometimes also applied to how staff view their environment). There are many academic studies that have shown that the servicescape of an operation plays an important role, both positive and negative, in shaping customers' views.[9] The general idea is that ambient conditions, space factors and signs and symbols in a service operation will create an 'environmental experience' both for employees and for customers; and this environmental experience should support the service concept.

Operations principle

Layout should include consideration of the look and feel of the operation to customers and/or staff.

The individual factors that influence this experience will then lead to certain responses (again, in both employees and customers). These responses can be put into three main categories:

- cognitive (what people think);
- emotional (what they feel); and
- physiological (what their bodies experience).

However, remember that a servicescape will contain not only objective, measurable and controllable stimuli but also subjective, immeasurable and often uncontrollable stimuli, which will influence customer behaviour. The obvious example is other customers frequenting an operation. As well as controllable stimuli such as the colour, lighting, design, space and music, other factors such as the number, demographics and appearance of one's fellow customers will also shape the impression of the operation.

7.4 What information and analysis is needed to design the layout and look of facilities?

Designing the layout and look of any operation's facilities will eventually move on to considering the details of the design. This means operationalising the broad principles that governed the choice of whichever basic layout type was chosen and whatever aesthetic effect is wanted. But any detailed design should be based on the collection and manipulation of information regarding the nature and volume of the flow that the layout must accommodate, and the behaviour and preferences of staff and (if appropriate) customers.

Information for flow analysis of layouts

At a detailed level, layout can be complex. Often there are very many alternative ways to position facilities relative to each other. For example, in the very simplest case of just two work centres, there are only two ways of arranging these relative to each other. But there are six ways of arranging three centres and 120 ways of arranging five centres. This relationship is a factorial one. For N centres, there are factorial N ($N!$) different ways of arranging the centres, where:

$$N! = N \times (N - 1) \times (N - 2) \times \cdots . \tag{1}$$

So, for a relatively simple layout with, say, 20 work centres, there are $20! = 2.433 \times 10^{18}$ ways of arranging the operation. This is called 'combinatorial complexity' and it makes optimal layout solutions difficult to achieve in practice. Most layouts are designed by a combination of intuition, common sense, and systematic trial and error, a process that can be supported by using computer-aided design (CAD) software. Some of these treat the combinatorial complexity issue by using heuristic procedures. Heuristic procedures use what have been described as 'shortcuts in the reasoning process' and 'rules of thumb' in the search for a reasonable solution. They do not search for an optimal solution (though they might find one by chance) but rather attempt to derive a good suboptimal solution. However, both the information that is required, and how it is used, depends on the basic layout type that has been chosen.

Operations principle

Layouts are often combinatorially complex; there are many alternative layouts.

Information and analysis for the design of fixed-position layouts

In fixed-position arrangements the location of resources will be determined not on the basis of the flow of transformed resources, but on the convenience of transforming resources themselves. The objective of the detailed design of fixed-position layouts is to achieve a layout for the operation that allows all the transforming resources to maximise their contribution to the transformation process by allowing them to provide an effective 'service' to the transformed resources. The detailed layout of some fixed-position layouts, such as building sites, can become very complicated, especially if the planned schedule of activities is changed frequently. Imagine the chaos on a construction site if heavy trucks continually (and noisily) drove past the site office, delivery trucks for one contractor had to cross other contractors' areas to get to where they were storing their own materials, and the staff who spent most time at the building itself were located furthest away from it. Although there are techniques that help to locate resources on fixed-position layouts, they are not widely used.

OPERATIONS IN PRACTICE

Virtual reality brings layout to life[10]

You will have noticed that all the layouts illustrated in this chapter are 2D drawings. The same limitation was shared by layout practitioners. 3D models could be constructed, which would give a more realistic impression of how a layout would look once implemented, but these were expensive. Then, with the advent of more powerful computer-aided design, 3D simulations helped to visualise both the look and layout of facilities. Now, by wearing virtual reality (VR) headsets and getting a 360-degree view, designers can actually feel what it would be like to be right inside a proposed layout. More than that, groups of designers also wearing headsets (even if they are on

a space that has yet to be developed. One of the systems offering VR capabilities for designing layouts is Revizto, which was created by Vizerra SA based in Lausanne, Switzerland. Revizto (derived from the Latin for 'visual check') is cloud-based 'visual collaboration' software intended mainly for architects, engineers and contractors to visualise their designs and communicate information within the team responsible for the project. The software's interactive 3D environments allow designers to navigate their planned layout as if in a video game, highlighting structural and operational issues as they investigate. One of the more obvious advantages of using VR software such as Revizto is that designers can collaborate by sharing a realistic impression of their layouts based on the assumptions underlying their actual geometry. But just as important is the ability of such software to integrate the details of a layout with other information that will be used in its implementation, such as the quantity of materials needed and building schedules.

the other side of the world), can share the same lifelike experience, discuss design options and recommend changes to the design, all in real time. But it is not only designers who can collaborate using VR; other stakeholders such as clients (if architects are designing a building), or customers and staff who will use a facility, can visualise

Information and analysis for the design of functional layout

Before starting the process of detailed design in functional layouts there are some essential pieces of information that the designer needs:

- the area required by each work centre;
- the constraints on the shape of the area allocated to each work centre;
- the degree and direction of flow between each work centre (for example, number of journeys, number of loads or cost of flow per distance travelled);
- the desirability of work centres being close together or close to some fixed point in the layout.

The degree and direction of flow are usually shown on a flow record chart like that shown in Figure 7.11 in the worked example. This information could be gathered from routing information, or where flow is more random, as in a library for example, where the information could be collected by observing the routes taken by customers over a typical period of time.

Minimising distance travelled

In most examples of functional layout, the prime objective is to minimise the costs to the operation that are associated with flow through the operation. This usually means minimising the total distance travelled in the operation, as in Figure 7.12 in the worked example. The effectiveness of the layout, at this simple level, can be calculated from:

$$\text{Effectiveness of layout} = \sum F_{ij} D_{ij} \text{ for all } i \neq j$$

where

F_{ij} = the flow in loads or journeys per period of time from work centre i to work centre j

D_{ij} = the distance between work centre i and work centre j.

The lower the effectiveness score, the better the layout.

The steps in determining the location of work centres in a functional layout is illustrated in the worked example, 'Rotterdam Educational Group'.

Worked example

Rotterdam Educational Group

Rotterdam Educational Group (REG) is a company that commissions, designs and manufactures education packs for distance-learning courses and training. It has leased a new building with an area of 1,800 square metres, into which it needs to fit 11 'departments'. Prior to moving into the new building, it has conducted an exercise to find the average number of trips taken by its staff between the 11 departments. Although some trips are a little more significant than others (because of the loads carried by staff), it has been decided that all trips will be treated as being of equal value.

Step 1 – Collect information

The areas required by each department, together with the average daily number of trips between departments, are shown in the flow chart in Figure 7.11. In this example the direction of flow is not relevant and very low flow rates (less than five trips per day) have not been included.

Step 2 – Draw schematic layout

Figure 7.12 shows the first schematic arrangement of departments. The thickest lines represent high flow rates between 70 and 120 trips per day; the medium lines are used for flow rates between 20 and 69 trips per day; and the thinnest lines for flow rates between 5 and 19 trips per day. The objective here is to arrange the work centres so that those with the thick lines are closest together. The higher the flow rate, the shorter the line should be.

Step 3 – Adjust the schematic layout

If departments were arranged exactly as shown in Figure 7.12(a), the building that housed them would be of an irregular, and therefore high-cost, shape. The layout needs adjusting to take into account the shape of the building. Figure 7.12(b) shows the departments arranged in a more ordered fashion, which corresponds to the dimensions of the building.

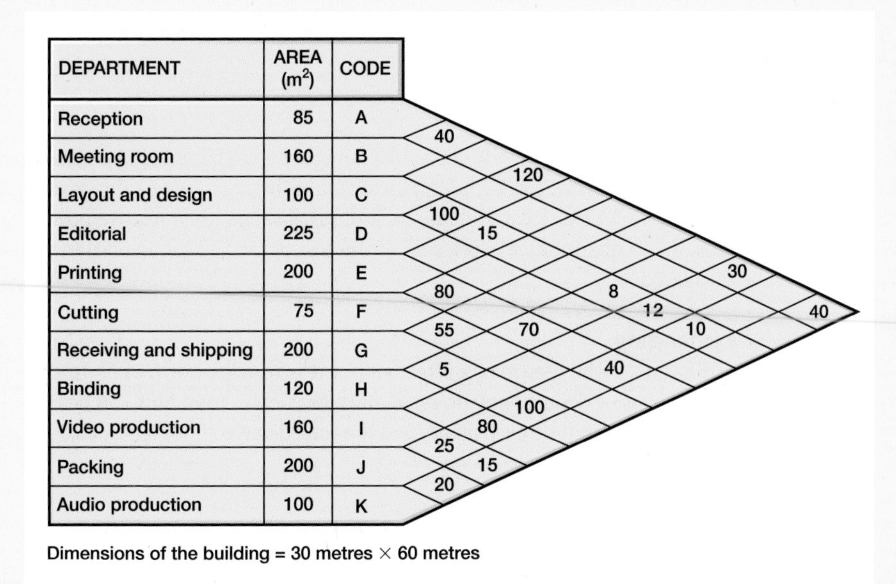

DEPARTMENT	AREA (m^2)	CODE
Reception	85	A
Meeting room	160	B
Layout and design	100	C
Editorial	225	D
Printing	200	E
Cutting	75	F
Receiving and shipping	200	G
Binding	120	H
Video production	160	I
Packing	200	J
Audio production	100	K

Figure 7.11 Flow information for Rotterdam Educational Group

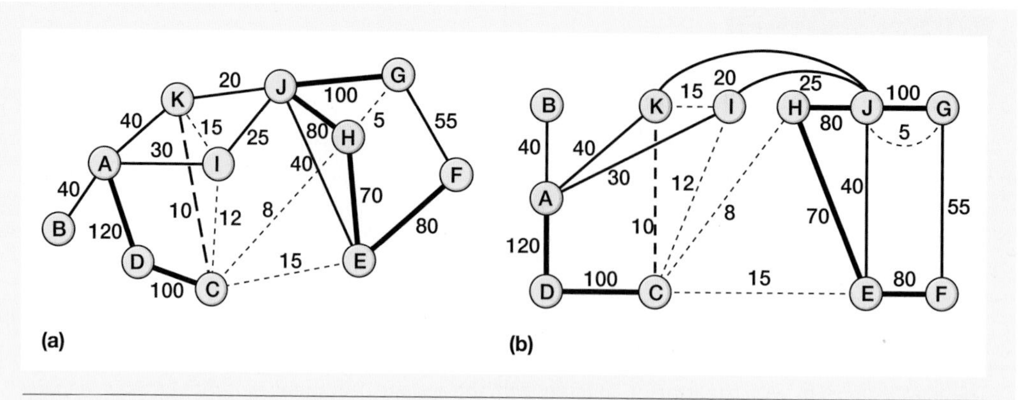

Figure 7.12 (a) Schematic layout placing centres with high traffic levels close to each other, (b) Schematic layout adjusted to fit building geometry

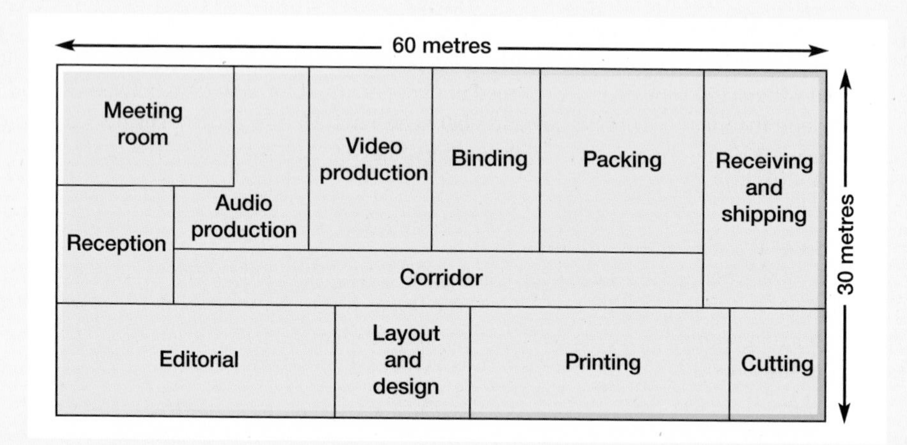

Figure 7.13 Final layout of building

Step 4 – Draw the layout

Figure 7.13 shows the departments arranged with the actual dimensions of the building and occupying areas that approximate to their required areas. Although the distances between the centroids of departments have changed from Figure 7.12 to accommodate their physical shape, their relative positions are the same. It is at this stage that a quantitative expression of the cost of movement associated with this relative layout can be calculated.

Step 5 – Check by exchanging

The layout in Figure 7.13 seems to be reasonably effective but it is usually worthwhile to check for improvement by exchanging pairs of departments to see if any reduction in total flow can be obtained. For example, departments H and J might be exchanged, and the total distance travelled calculated again to see if any reduction has been achieved.

Information and analysis for the design of cell layout

Figure 7.14 shows how a functional layout has been divided into four cells, each of which has the resources to process a 'family' of parts. In doing this the operations management has implicitly taken two interrelated decisions regarding:

- the extent and nature of the cells it has chosen to adopt;
- which resources to allocate to which cells.

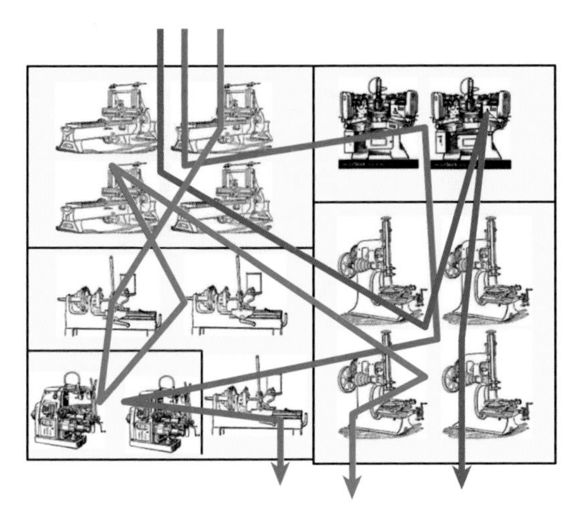

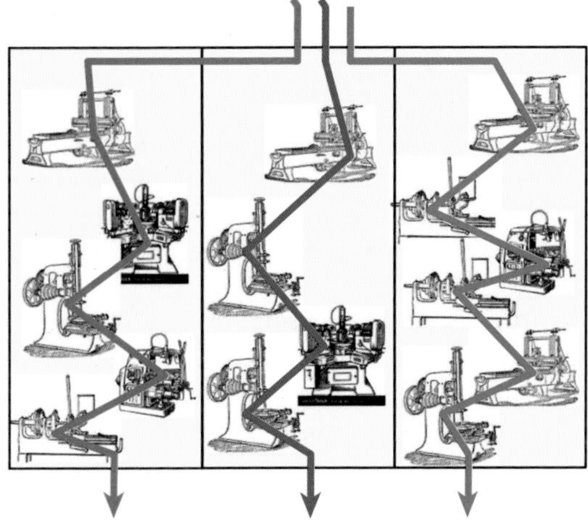

Figure 7.14 Cell layout groups the processes together that are necessary for a family of products/services

OPERATIONS IN PRACTICE

Rolls-Royce factory is designed on environmental principles[11]

Rolls-Royce Motor Cars has always been associated with top-level engineering and the very highest of quality. So it is not surprising that its headquarters and assembly plant at Goodwood, in the south of the United Kingdom, is also built to the same exacting standards. And, although the Rolls-Royce brand dates from 1906, its innovative and environmentally friendly building close to the famous Goodwood racing circuit is far newer. It was opened in 2003 after the company became a wholly owned subsidiary of the BMW Group, although planning had started four and a half years earlier. Determined to follow the advice of its founder, Sir Henry Royce, at every stage of planning and construction, the company say it was inspired to 'Strive for perfection in everything you do'. Which partly explains the £65 million cost of the original building, and the further £10 million invested in the extension that was added in 2013.

Environmental objectives were always fundamental to the building, and for almost 20 years it has been accredited to the international standard ISO 14001 for its environmental management and pollution prevention systems. The Goodwood site is in what the UK Government specifies as an 'Area of Outstanding Natural Beauty', so the building itself had to fit into the landscape using an innovative design and careful choice of materials. Just as important, the building was designed to have the smallest possible environmental impact. Sustainability informed all aspects of the building. The exterior is clad with a mixture of limestone and cedar wood, both from sustainable sources. Recycled materials are used wherever possible, including gravel extracted from the site prior to construction. Timber louvres activated by a weather station on the roof control the light entering the building, reducing the need for artificial lighting and helping to regulate the internal temperature. The walls are 25 per cent more thermally efficient than required by local building regulations. The company has established over 400,000 plants and trees of more than 120 species across the site, many of which occur naturally in the local area. Of course, the company composts all the green waste produced from its 42 acres of grounds.

Also included in the site was a large central lake, which not only attracted many wild birds but also acted as a heat sink for the plant's climate-control systems, thereby saving cost and energy compared with conventional air conditioning. The lake was also designed to form part of the sustainable water management and flood prevention system that stored filtered runoff from the roof and car parks before allowing it to drain naturally into the ground. A particular feature of the design for the main building was the 32,000 square metre living roof, which was planted with thousands of resilient, low-maintenance sedum plants, making it the largest living roof in the United Kingdom. As well as providing extremely effective natural concealment, the roof improved insulation, reduced rainwater runoff and became a haven for wildlife. But it is not just the building itself that is environmentally innovative. Inside, all the company's production processes have been designed to minimise waste, energy usage and water consumption. In excess of 60 per cent of waste, including cardboard, paper, plastic, tyres and polystyrene, is recycled. Leather offcuts from the vehicles' luxury upholstery are recycled to be used in the fashion and footwear industries. Spare pieces and offcuts of wood and veneer are given to a local charity, which uses them to make furniture and other fund-raising products.

Production flow analysis

The detailed design of cellular layouts is difficult, partly because the idea of a cell is itself a compromise between process and product layout. To simplify the task, it is useful to concentrate on either the process or product aspects of cell layout. If cell designers choose to concentrate on processes, they could use cluster analysis to find which processes group naturally together. This involves examining each type of process and asking which other types of processes a product or part using that process is also likely to need.

One approach to allocating tasks and machines to cells is production flow analysis (PFA), which examines both product requirements and process grouping simultaneously. In Figure 7.15(a) a manufacturing operation has grouped the components it makes into eight families – for example, the components in family 1 require machines 2 and 5. In this state the matrix does not seem to exhibit any natural groupings. If the order of the rows and columns is changed, however, to move the crosses as close as possible to the diagonal of the matrix, which goes from top left to bottom right, then a clearer pattern emerges. This is illustrated in Figure 7.15(b) and shows that the machines could be grouped together conveniently in three cells, indicated on the diagram as cells A, B and C.

Although this procedure is a particularly useful way to allocate machines to cells, the analysis is rarely totally clean. This is the case here where component family 8 needs processing by machines 3 and 8, which have been allocated to cell B. There are some partial solutions for this. More machines could be purchased and put into cell A. This would clearly solve the problem but requires investing capital in a new machine that might be underutilised. Or, components in family 8 could be sent to cell B after they have been processed in cell A (or even in the middle of their processing route if necessary). This solution avoids the need to purchase another machine but it conflicts partly with the basic idea of cell layout – to achieve a simplification of a previously complex flow. Or, if there are several components like this, it

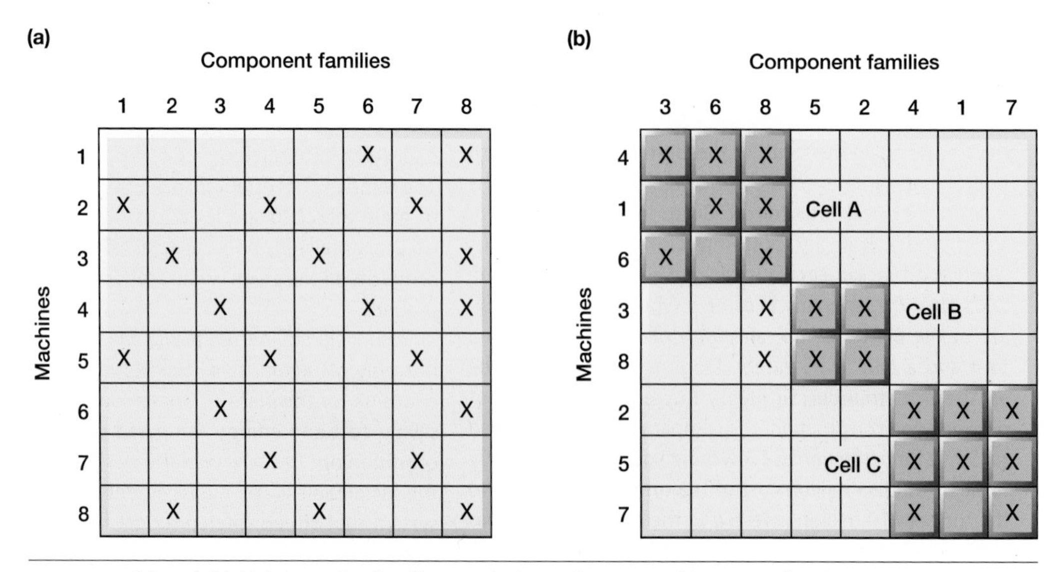

Figure 7.15 (a) and (b) Using production flow analysis to allocate machines to cells

might be necessary to devise a special cell for them (usually called a remainder cell) that will almost be like a mini functional layout. This remainder cell does remove the 'inconvenient' components from the rest of the operation, however, leaving it with a more ordered and predictable flow.

Information and analysis for the design of line layout

The nature of the line layout design decision is a little different to the other layout types. Rather than 'where to place what', it is concerned more with 'what to place where'. Locations are frequently decided upon and then work tasks are allocated to each location. So, the 'layout' activity is very similar to aspects of process design, which we discussed in Chapter 6. The main product layout decisions are as follows:

- What cycle time is needed?
- How many stages are needed?
- How should the task-time variation be dealt with?
- How should the layout be balanced?
- How should the stages be arranged ('long thin' layout to 'short fat' layout)?

Information and analysis for the design of the appearance of facilities

Traditionally, survey questionnaires and interviews have been used to assess both staff and customers' reaction to current and potential future workplace designs. Generally, indicators are developed for the questionnaires or interviews about both the level of satisfaction and the level of dissatisfaction. This is necessary because high levels of satisfaction do not automatically mean a low level of dissatisfaction. Different staff and/or customers can hold very different views about the positioning and appearance of layouts. More recently, some organisations have used remote recording to understand customers' and (sometimes) their staff's reaction to the layout and look of their operations.

Responsible operations

In every chapter, under the heading of 'Responsible operations', we summarise how the particular topic covered in the chapter touches upon important social, ethical and environmental issues.

A first thought on how facilities layout affects sustainability objectives might conclude that this is one of those issues where normal financial objectives perfectly align with environmental ones. After all, the common layout objective of minimising the distance travelled by physical items not only saves costs and time, it also minimises whatever energy is used to transport the items. Yet widen the scope of the issue to include a more holistic consideration of the total look and layout of working spaces, and several issues become evident. Think about any work site within which an operation is located, from choosing a site, through sourcing construction materials, building the structure itself and designing its interior, to its ongoing operation. Within this whole cycle, there are many opportunities to create a responsible operation as well as many potential problems that could undermine sustainability objectives. Start with choosing a site. Does it complement its surroundings? Can building avoid damaging the natural environment? Can the construction of a building be completed without excessive disruption? And so on. In recent years, the discipline of sustainable (or 'green' or 'environmental') architecture has emerged to promote designs that create sustainable working (and living) environments while minimising negative environmental impacts and energy consumption. (According to the UN Environment Programme Global Status Report 2017, buildings and construction account for more than 35 per cent of global final energy use and nearly 40 per cent of energy-related CO_2 emissions.)[12] It is a complex subject and beyond the scope of this text, but most research in the area is concerned with saving energy and water and making buildings more environmentally friendly by, for example, reducing carbon emissions. However, most operations managers will be more concerned with the sustainability of ▶

their ongoing operations (while accepting that these will be partly determined by its architecture). Figure 7.16 illustrates some of the energy-related factors that are likely to be relevant. These fall into four groups:

- *The environmental effectiveness of the workspace or building itself* – heat lost through its perimeter, amount of natural light utilised, filtration to preserve air purity, etc.
- *The effectiveness of the building services systems* – energy consumption of its heating, ventilation, air-conditioning (HVAC) technology, efficient use of other services such as water, gases, etc.
- *The needs of the individual facilities (e.g. machines) in the operation* – energy and other services requirements.
- *Energy consumed because of the layout* – energy required to maintain flow of items through the operation.

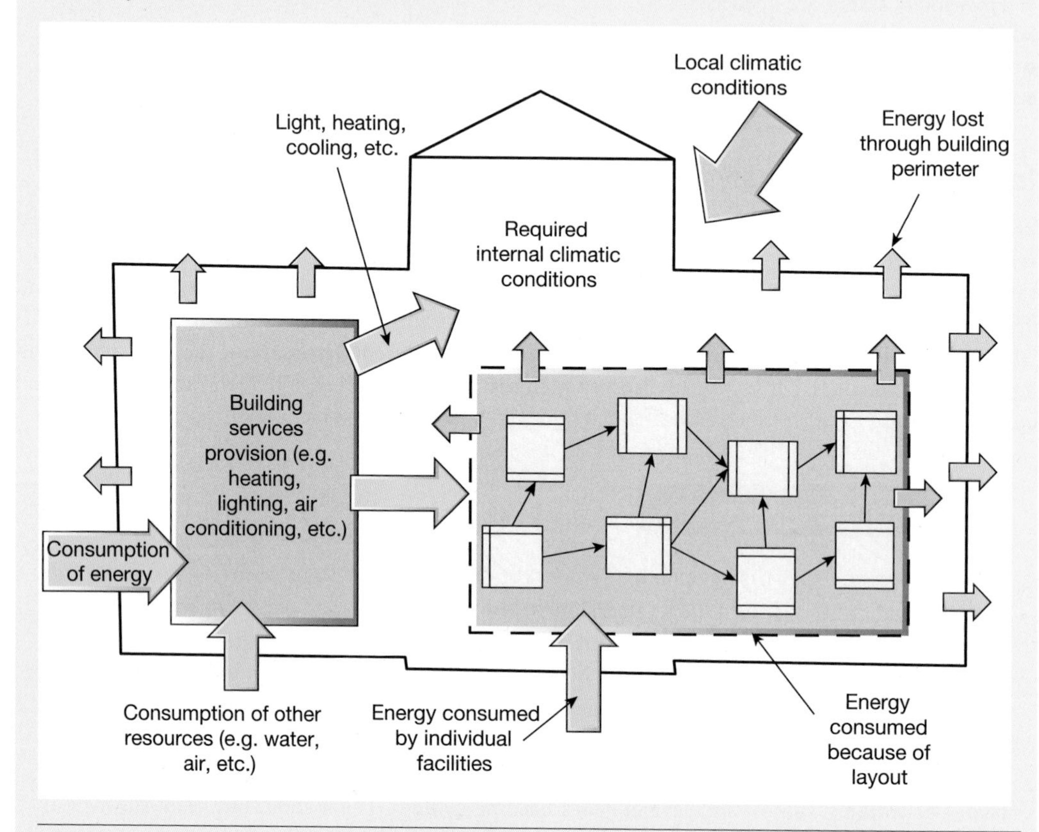

Figure 7.16 Some of the energy-related factors relevant in the design of workplace buildings

Summary answers to key questions

7.1 How can the layout and look of facilities influence performance?

- The 'layout and look' of an operation or process means how its facilities are positioned relative to each other and how their general appearance is designed.
- These decisions will dictate the pattern and nature of the flow of transformed resources as they progress through the operation or process. They also affect how both the people who staff the operation and, in high-visibility operations (where customers form part of the transformed resource), how customers judge their experience of being in the operation.

- The objectives of layout include: minimising (or sometimes maximising) the flow of transformed resources, minimising or maximising aspects of the interaction between parts of the process, and enhancing the experience of both staff and, where appropriate, customers.

7.2 What are the basic layout types and how do they affect performance?

- There are four basic layout types. They are fixed-position layout, functional layout, cell layout and line layout.
- The type of layout an operation chooses is influenced partly by the nature of the process type, which in turn depends on the volume–variety characteristics of the operation. The decision will also depend partly on the objectives of the operation. Cost and flexibility are particularly affected by the layout decision.
- The fixed and variable costs implied by each layout differ such that, in theory, one particular layout will have the minimum costs for a particular volume level. However, in practice, uncertainty over the real costs involved in layout design make it difficult to be precise about which is the minimum-cost layout.

7.3 How does the appearance of an operation's facilities affect its performance?

- The general appearance and aesthetics of a layout affects how staff view the operation in which they work, and how customers behave.
- The communication between people reduces with the distance between them. This is called the 'Allen curve'.
- In addition to the conventional operations objectives that will be influenced by the layout design, it will also influence the feel and general impression of the operation. This is often called the 'servicescape' of the operation.

7.4 What information and analysis are needed to design the layout and look of facilities?

- In fixed-position layout, formal layout techniques are rarely used but some, such as resource location analysis, bring a systematic approach to minimising the costs and inconvenience of flow at a fixed-position location.
- In functional layout, the detailed design task is usually (although not always) to minimise the distance travelled by the transformed resources through the operation. Either manual or computer-aided methods can be used to devise the detailed design.
- In cell layout, the detailed design task is to group the products or customer types such that convenient cells can be designed around their needs. Techniques such as production flow analysis can be used to allocate products to cells.
- In line layout, the detailed design of product layouts includes a number of decisions, such as the cycle time to which the design must conform, the number of stages in the operation, the way tasks are allocated to the stages in the line and the arrangement of the stages in the line.
- Gathering information as the basis for the analysis of the appearance of facilities usually uses survey questionnaires and interviews to assess both staff and customers' reaction to current and potential future workplace designs. More recently, some operations have used remote recording to understand customers' and (sometimes) their staff's reaction to the layout and look of their operations.

CASE STUDY Misenwings SA

'Our industry [in-flight catering] *is dominated by two giant operations, LSG Sky Chefs and Gate Gourmet. Between them, they provided over 800 million meals a year for over 500 airlines at hundreds of airports. We are very much at the other end of the scale compared with those companies. Last year we provided less than 200,000 meals to a mixture of small regional airlines and private jet hire companies. But we have operated at the quality end of the market and our margins have been healthy. Currently, we are in the middle of a slump in the airline industry, but when things get back to normal, we see ourselves growing and moving more into the mainstream'* (Annette Müller, CEO and founder, Misenwings).

Misenwings was founded when Annette graduated from the Swiss hospitality college that she had attended after giving up her previous job in the advertising industry. Originally, the company had supplied meals for private hire and charter firms. Switzerland has a surprisingly large business aviation market. Geneva is one of the Europe's busiest business aviation airports (the busiest European business aviation airport is Paris Le Bourget). It was a profitable business to be in. Most customers wanted high-quality 'gourmet' food that commanded a high price, and relatively high margin. However, after five successful years of operation, Annette had noticed that demand in this market was slowing. *'Increasingly, the private hire business was being promoted as a fast and convenient method of getting around Europe, rather than a luxury experience. Some customers of the charter companies did not even want a meal on the flight. That was when we expanded into supplying meals to conventional commercial airlines'* (Annette Müller). However, rather than trying to move into the market for supplying large airlines with a full range of meals, from economy snacks through to first-class meals, Misenwings offered what it called an 'Exclusive Meal' service to airlines that only served a hot meal to their business and first-class passengers while offering (or selling) prepared sandwiches to economy class passengers (which would be supplied by a separate company). It proved to be a wise move and within three years, the company grew from providing under 20,000 meals a year to around 200,000. At the same time, its revenue grew by 600 per cent.

Airline meals are usually prepared on the ground in kitchens close to the airport. They are then transported to the aircraft, placed in refrigerators, and heated by flight attendants before being served on board. Airline meals, and the aircraft on-board ovens used to heat them, are designed so that the food is not adversely affected by the change in altitude and pressure. A meal supplier will usually provide instructions for the cabin crew on how to heat and plate the food. The design of a menu is often a shared responsibility between the airline, which will employ specialist staff, and the meal supplier. In-flight meals have to be timed perfectly to match the flight for which the meal is intended, and Misenwings had developed a good reputation for this. Before an aircraft takes off, thousands of individual items must be delivered in exact quantities and on time. Making sure that the right items are on the right flight at the right time is a complex challenge, which requires exact coordination for delivering, loading and unloading trucks at the airport. Schedule disruptions mean that flexibility in the kitchen is important to meet turn-around times. However, Annette was keen to stress that it was the quality of its meals that was the foundation of its reputation. *'Of course, the dependability and flexibility of our kitchen operation is important. But that would mean nothing without our expertise in producing meals of the very highest quality'*.

The slump in the airline industry of the early 2020s affected Misenwings like every other airline caterer, although not as much as those serving the larger airlines. Nevertheless, Annette saw the reduction in business as an opportunity to reshape her operations. In particular, the Geneva preparation kitchen had not been changed since the company had served only the private hire business. *'What worked when we prepared relatively few meals, which were different every day, is not appropriate now that we are preparing far more meals to menus agreed with customer weeks in advance. We need to be more "industrial" in our approach. Our preparation kitchen needs totally rethinking'*.

Possible kitchen designs

Annette knew that, ultimately, every kitchen had its own specific requirements; what works for one business might not necessarily work in another, but she was convinced that the current kitchen configuration could be improved. The design would have to consider the size and style of the building, local planning issues, the budget available and, above all, the variety of menus and volume of meals to be prepared. But whatever the design of the new kitchen, it would have to contain five basic elements:

- Food preparation areas that segregate different types of food during preparation.
- Cooking areas that cook hot meals. This is the heart of the kitchen where the design of the area is vital to a smooth operation.
- Storage areas for various types of ingredients that must be kept at the required temperature and free from contamination.
- Packing area, where meals are prepared for transportation from the kitchen to the aircraft.
- Cleaning and washing area that receives and washes crockery, glasses, cutlery and so on, for reuse.

Annette set out her principles for a new kitchen design. *'Facilities and equipment will need to be positioned so that staff take as few steps as possible, they are on their feet all day as it is. There should also be minimal bending, reaching, walking or turning. It is also good if hot and cold areas are not too close. For example, the main oven range should not be right next to the fridge, freezer and blast chiller, which should themselves be close together. In addition, any new design should be energy-efficient. It could save us a lot of money in the long term, as well as being ethically the right thing to do'.*

There are three basic designs used in commercial kitchens:

- Island style – usually features a main 'island' block in the middle of the kitchen. Often the island contains the cooking equipment, with food preparation, storage and serving or 'packing for transport' areas on the outer walls.
- Zone style – involves dividing the kitchen into different sections, such as cooking, food preparation, storage, washing and so on. Food and staff move between the zones as required.
- Assembly line – usually used where a kitchen produces relatively large quantities of similar types of meal food. The kitchen is laid out in a line, in the order of how its equipment is used, which creates a conveyor-belt type of process.

The current Geneva kitchen was laid out in what would be classed as an 'island style' arrangement, but in a reversed manner, with preparation areas on the centre island and cooking equipment at the perimeter of the area. Annette admitted that it was not ideal. *'We just didn't think about layout when we built our current kitchen in Geneva. Frankly it was never ideal, even for the low volumes and wide range of meals that we started producing. And as volumes have grown the inefficiencies of the design have become very obvious. Food has to cross the kitchen area both when it comes from the stage areas and when it moves to the cooking areas on the outside walls. This is causing congestion as well as increasing the risk of contamination. Given that we are likely to be increasing volumes, we need to rethink'.*

The layout decision

Fortunately, the company had managed to acquire a site adjacent to their current site. This meant that they could both build a larger facility and continue working during its construction. The main dilemma was the layout of the new facility. There was general agreement that an island layout was not appropriate. Where there was less agreement was whether they should move to a zone-style layout alone, or combine a zone layout with an assembly-line layout in the assembly/packing area. The proposed zone layout of the whole kitchen is shown in Figure 7.17. The inclusion of a separate conference/meeting space was something of an afterthought. Many new airline catering kitchens had incorporated a conference space offering a comfortable area for collaborative meetings with customers or regulators for presentations and other affairs that required a pleasant showroom area. Misenwings thought that such a space would be useful.

Annette and her team considered how the assembly/packing area could be designed. Some of the suggested alternative layouts for this area are shown in Figure 7.18. Each is shown with six staff working in the area. This was thought to be the maximum number that would be needed as business increased over the next two or three years. Often fewer than six staff would be needed. Alternative (a) is the closest to how meals are assembled currently, with staff sharing two 'island' spaces to assemble meals (currently there is one slightly larger island used for this). Alternative (b) is very different. It uses a single 'paced' moving-belt assembly-line arrangement. This type of layout is used in some of the larger airline meals suppliers. Alternative (c) adopts 'roller conveyor', unpaced lines that allow assembly staff slightly more flexibility and some ability for staff to share work. The option of staff sharing work is also evident in the (unpaced) assembly line of alternative (d). By contrast, alternative (e) allows each of the assembly staff to work individually putting the whole meal together.

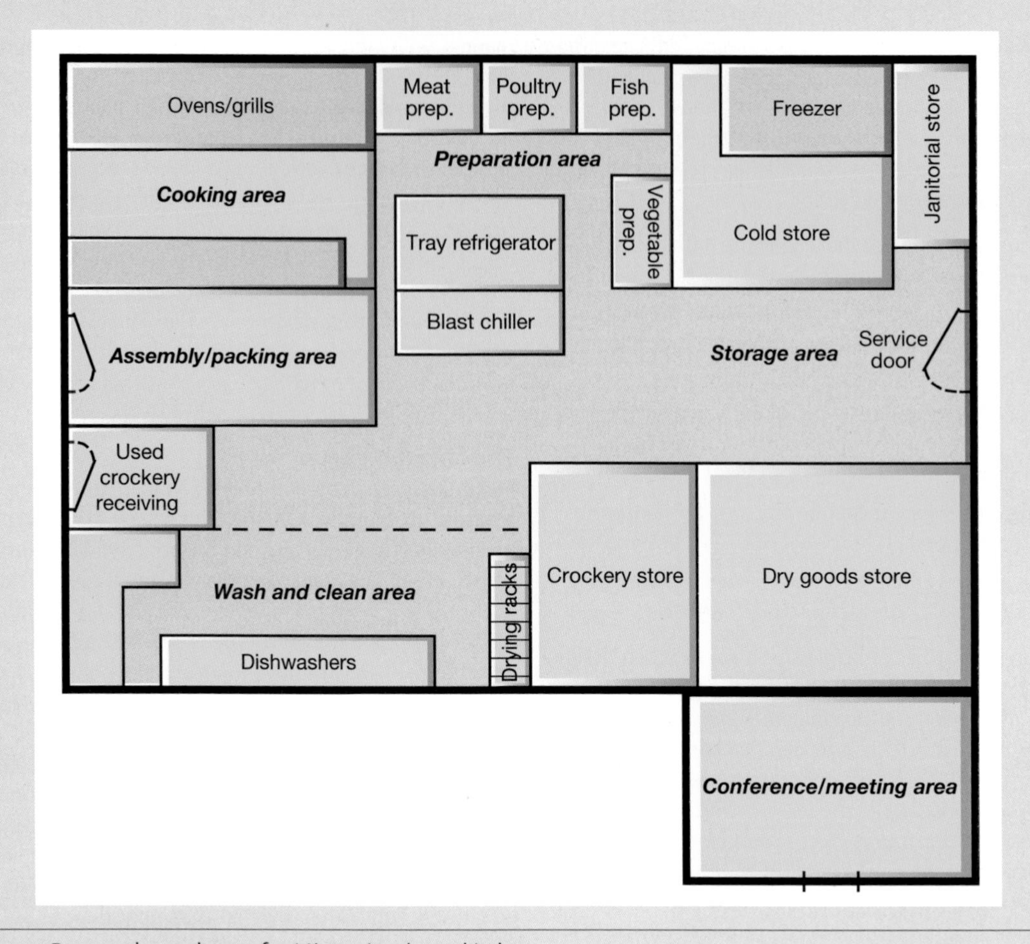

Figure 7.17 Proposed zone layout for Misenwings' new kitchen

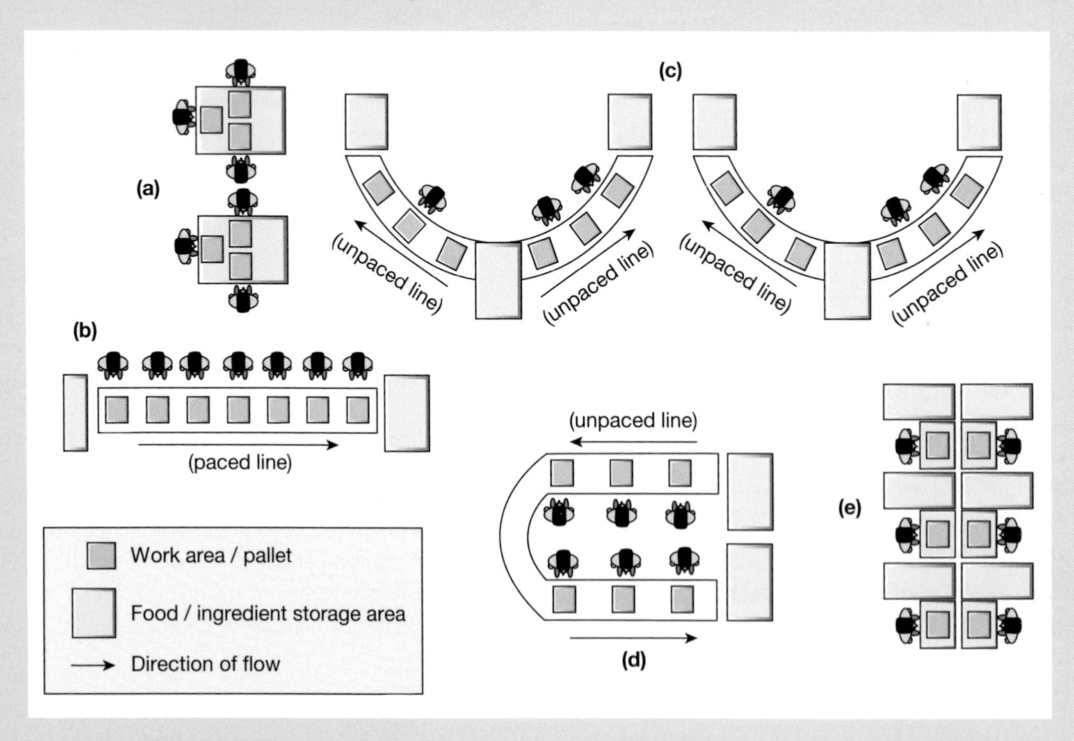

Figure 7.18 Alternative suggestions for the layout of the assembly/packing area (not to scale). All shown with the likely maximum of six staff

QUESTIONS

1 How do the three kitchen layouts of 'island', 'zone' and 'assembly line' correspond with the more conventional layout types used in the chapter?

2 In suggesting the layout in Figure 7.17, what layout objectives do you think were used?

3 What do you think are the advantages and disadvantages of the alternative layouts for the assembly/packing area illustrated in Figure 7.18? What other layouts should be considered for this area?

4 What is important in the design of the conference/ meeting area?

Problems and applications

All chapters have 'Problems and applications' questions that will help you practise analysing operations. They can be answered by reading the chapter. Model answers for the first two questions can be found on the companion website for this text.

1 Revisit the 'Operations in practice' example 'Supermarket layout'. Then visit a supermarket and observe people's behaviour. You may wish to try to observe which areas they move past slowly and which areas they seem to move past without paying attention to the products. (You may have to exercise some discretion when doing this; people generally don't like to be stalked round the supermarket too obviously.) Try to verify, as far as you can, some of the principles that were outlined in the example. What layout type is a conventional supermarket and how does it differ from a manufacturing operation using the same layout type?

2 Humans (known as 'Trackers') are still used by some retailers to follow customers (discreetly) through stores to see the flow between the various parts of a store, but technology is replacing them. These include video surveillance, thermal sensing, lasers, face recognition and so-called device-based solutions. Some of these technologies are sophisticated, but still being developed, such as face recognition linked with advanced artificial intelligence (AI) analysis. Others are more common with a proven track record, such as thermal imaging, which detects emissions from moving objects and, because it is not sensitive to light, can work in any space, no matter how well lit. But the technology that has, arguably, the most potential for capturing customer-movement data, is the use of sensors that pick up the unique identifier signals emitted by mobile phones as they automatically seek wi-fi networks that they can join. What do you believe are the advantages and disadvantages of using such technology?

3 In an assembly operation for customised laboratory equipment, the flow of materials through eight departments is as shown in Table 7.2. Assuming that the direction of the flow of materials is not important, construct a relationship chart, a schematic layout and a suggested layout, given that each department is the same size and the eight departments should be arranged four along each side of a corridor.

4 The assembler of customised laboratory equipment negotiates a long-term arrangement to supply a simplified standard product to be sold to forensic laboratories worldwide. This product requires an assembly sequence that takes it, in order, from Departments 2 to 4 to 8 to 5. Estimates of demand indicate that the new product would account for 30 standard container loads per day. If this new product came to be seen as a permanent addition to the operation's work, how might the layout need to be changed?

5 A company that produces a wide range of specialist educational kits for 5–10-year-olds is based in an industrial unit arranged in a simple layout of six departments, each performing a separate task. The layout is shown in Figure 7.19, together with the results of an investigation of the flow of parts and products between each department. However, the company plans to revamp its product range.

Table 7.2 Flow of materials between departments in standard container loads per day

	From . . .							
	Dept 1	Dept 2	Dept 3	Dept 4	Dept 5	Dept 6	Dept 7	Dept 8
Dept 1		30						
Dept 2	10		15	10				
Dept 3		5		12	2		15	
Dept 4		6			10	10		
Dept 5				8		8	10	12
Dept 6					2		30	
Dept 7						13		2
Dept 8				10	6		15	

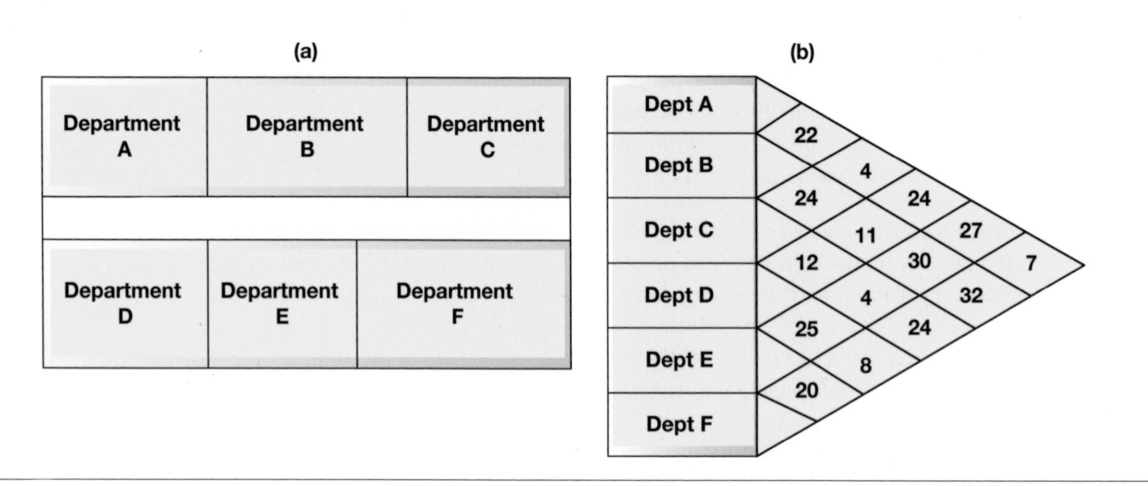

Figure 7.19 (a) the current layout of the educational kits producer, and (b) the current interdepartmental flow of parts and products (in pallet loads)

'This new range will totally replace our existing products, and although I believe our existing layout is fine for the current product range, I think that we will need to reconfigure our layout when we make the transition to the new product range' (COO of the company). The estimate for the flow between the departments when the new product range is introduced is shown in Figure 7.20.

(a) Is the COO right in thinking that the current layout is right for the current product range?
(b) Assuming that the estimate of future interdepartmental flow is correct, how would you rearrange the factory?

6 A computer games developer is moving into new offices. The new office has a floor space of approximately 300 square metres in the form 20 metres by 15 metres. The company has six departments, as identified in Figure 7.21. This also shows the approximate area required by each department and the degree of closeness required between each department.

7 The operations manager of a specialist company assembling seabed monitors that record pollution levels had a dilemma. *'At the moment, we are producing around 40 seabed monitoring stations per year, using what is basically a fixed-position layout. However, as volume increases over the next few*

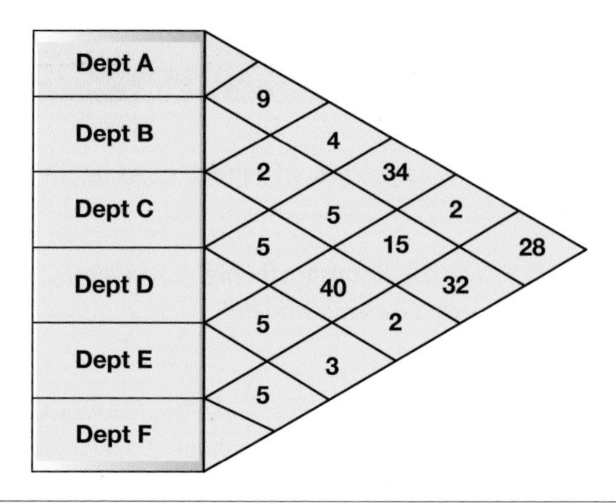

Figure 7.20 Estimated interdepartmental flow when the new product range is introduced

Department	Approximate area (sq. mtrs)
Administration	35
Social media	35
Planning	45
Developers	80
IT and software	60
Lounge and coffee area	45

Code	Degree of closeness
A	Essential
E	Very important
I	Important
O	Neutral
U	Unimportant
X	Not desirable

Figure 7.21 The required areas and closeness for the six departments of the computer games developer

years, we could move over to using a cell layout with two cells, one for use assembling monitors for tropical waters and one for colder waters. We know that the fixed costs associated with our current layout are €20,000 per year and the variable cost is €380 per unit. Using a cell layout, we think that the fixed cost per year will be €35,000 per year, but it could be as high as €40,000 or as low as €30,000 per year. We are clearer on the variable cost, which we are fairly sure will be €60 per unit'. How much would the company's volume need to increase to be certain that the cell layout would be less costly?

8 Normally, patients undergoing surgery remain stationary with surgeons and other theatre staff performing their tasks around the patient. A surgeon frustrated by spending time drinking tea while patients were prepared for surgery redesigned the process so he moves continually between two theatres. While he is operating on a patient in one theatre, his anaesthetist colleagues are preparing a patient for surgery in another theatre. After finishing with the first patient, the surgeon 'scrubs up', moves to the second operating theatre and begins the surgery on the second patient. While he is doing this, the first patient is moved out of the first operating theatre and the third patient is prepared. What are the advantages and disadvantages of this layout?

9 Re-read the 'Operations in practice' example 'Rolls-Royce factory is designed on environmental principles'. Why is designing such a factory particularly important for a company like Rolls-Royce?

10 Why do tech companies such as Google design particularly exciting offices?

Selected further reading

This is a relatively technical chapter and, as you would expect, most books on the subject are technical. Here are a few of the more accessible.

Boxall, P. and Winterton, J. (2018) Which conditions foster high-involvement work processes? A synthesis of the literature and agenda for research, *Economic and Industrial Democracy*, 39 (1), 27–47.
Academic, but comprehensive and insightful.

Gillen, N. (2019) *Future Office: Next-generation workplace design*, RIBA Publishing, London.
Takes an interdisciplinary approach, but essentially an architectural perspective.

Plunkett, D. and Reid, O. (2014) *Detail in Contemporary Office Design* (Detailing for Interior Design), Laurence King, London.
An interior designer's take on how offices can look.

Rosenbaum, M.S. and Massiah, C. (2011) An expanded servicescape perspective, *Journal of Service Management*, 22 (4), 471–490.
Academic but a good review of the research literature.

Saval, N. (2015) *Cubed: A Secret History of the Workplace*, Anchor Books, New York, NY.
An interesting history, and unexpectedly funny.

Stephens, M.P. (2019) *Manufacturing Facilities Design and Material Handling*, 6th edn, Purdue University Press, West Lafayette, IN.
Exactly what it says, thorough.

Van Meel, J., Martens, Y. and van Ree, H.J. (2010) *Planning Office Spaces: A Practical Guide for Managers and Designers*, Laurence King, London.
Exactly what the title says. A practical guide that includes both the 'flow' and the aesthetic aspects of office design.

Notes on chapter

1. The information on which this example is based is taken from: Urry, J. (2017) Inside Ducati: MCN walk around the Bologna factory, *Motorcycle News*, 21 September; Hickey, S. (2014) Death of the desk: the architects shaping offices of the future, *Guardian*, 14 September; Segran, E. (2015) Designing a happier office on the super cheap, Fast Company, 30 March.
2. The information on which this example is based is taken from: Booth, R. (2017) Francis Crick Institute's £700m building too noisy to concentrate, *Guardian*, 21 November.
3. Koontz, C. (2005) Retail interior layout for libraries, Information Today, Inc., January/February, http://www.infotoday.com/mls/jan05/koontz.shtml (accessed September 2021).
4. Schein, E.M. (1999) *The Corporate Culture Survival Guide: Sense and Nonsense About Culture Change*, Jossey-Bass, San Francisco, CA.
5. Palvalin, M. (2019) What matters for knowledge work productivity? *Employee Relations*, 41 (1), 209–27, https://doi.org/10.1108/ER-04-2017-0091 (accessed September 2021).
6. Economist (2013) Montessori management: the backlash against running firms like progressive schools has begun, *Economist* print edition, 7 September.
7. Waber B., Magnolfi, J. and Lindsay, G. (2014) Workspaces that move people, *Harvard Business Review*, October.
8. The information on which this example is based is taken from: Economist (2019) Future of the workplace: redesigning the corporate office, *Economist* print edition, 28 September; Economist (2019) Why open-plan offices get a bad rap, *Economist* print edition, 24 October; Waber, B., Magnolfi, J. and Lindsay, G. (2014) Workspaces that move people, *Harvard Business Review*, October.

9. The idea of 'servicescapes was originally explored by Bitner, M.J. (1992) Servicescapes: the impact of physical surroundings on customers and employees, *Journal of Marketing*, 56 (2), 57–71.
10. The information on which this example is based is taken from: Urbanist Architecture (2020) Virtual reality in architecture: visit your home before it's been built with VR, 12 April, https://urbanistarchitecture.co.uk/urbanist-4d-reality-virtual-reality-technology-in-architectural-design/ and Revizto.com (accessed September 2021).
11. The information on which this example is based is taken from: Rolls-Royce (2020) Birds, bees, roses and trees all thriving at the home of Rolls-Royce, Rolls-Royce Media Information, Goodwood, 2 July; Burstein, L. (2015) An inside look at the Rolls-Royce assembly plant in Goodwood, Robb Report, 23 October, https://robbreport.com/motors/cars/inside-look-rolls-royce-assembly-plant-goodwood-229474/ (accessed September 2021); Rolls-Royce (2017) Home of Rolls-Royce motor cars, press release, Rolls-Royce Media Information, Goodwood, 7 April.
12. UN Environment Programme and International Energy Agency (2017): Towards a zero-emission, efficient, and resilient buildings and construction sector, Global Status Report 2017.

8 Process technology

KEY QUESTIONS

8.1 What is process technology and why is it getting more important?

8.2 How can one understand the potential of new process technology?

8.3 How can new process technologies be evaluated?

8.4 How are new process technologies developed and implemented?

INTRODUCTION

There is a lot of new process technology around. There can be few, if any, operations that have not been affected by the advances in process technology. And all indications are that the pace of technological development is not slowing down, in fact in many ways it is speeding up. This has important implications for operations managers because all operations use some kind of process technology, whether it is a simple cloud computing service or the most complex and sophisticated artificial intelligence-driven automated factory. But this chapter is not particularly about specific technologies; there are too many of them, and they are changing too fast. Rather it is concerned with the questions that operations managers will need to ask, whatever the technology they are dealing with. All operations managers need to understand, in broad terms, what emerging technologies can do, how they do it, what advantages the technology can give, and what constraints it might impose on the operation. Figure 8.1 shows where the issues covered in this chapter relate to the overall model of operations management activities.

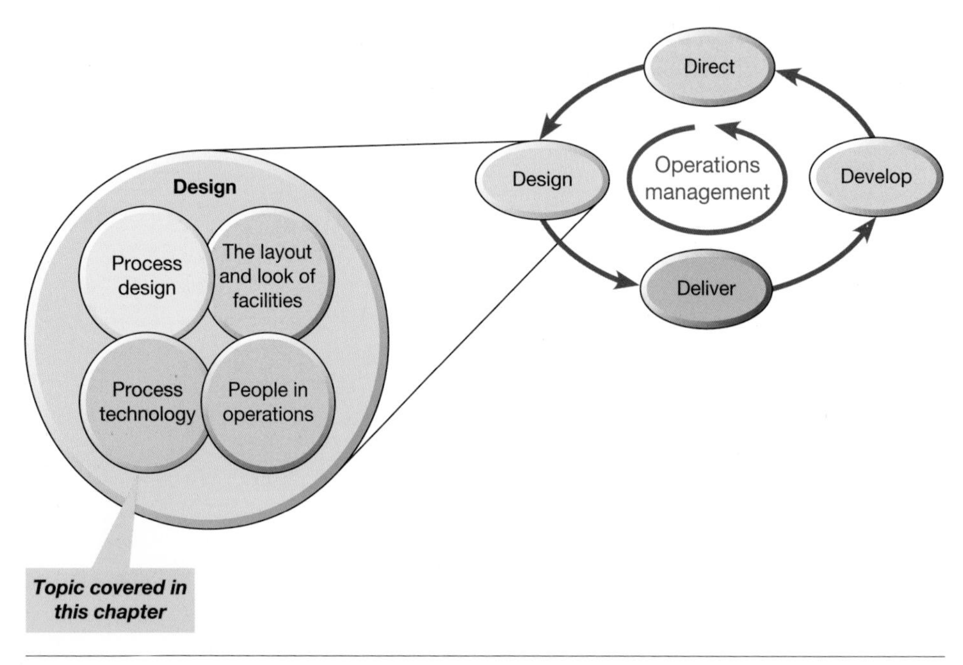

Figure 8.1 This chapter examines process technology

8.1 What is process technology and why is it getting more important?

The idea of harnessing technology to make operations more effective is not new. For example, the regular use of some form of automation to replace human work activities has been happening for at least the past 300 years. What *is* new is the sheer scope, sophistication and combination of technologies that are being deployed or developed to be part of operations activities in almost all parts of the economy. This has important implications, not only for how technology will be applied, but also for how operations are organised to make the most of new technologies' emerging capabilities. Even more important will be the rate and extent of technology change that operations managers will have to cope with. How operations managers deal with process technology is now one of the most important decisions that will shape the capabilities of their operations.

Process technology versus product technology

We take the word 'technology' to mean the use of scientific knowledge for real-world purposes, and we shall focus upon *process* technology as distinct from *product* or *service* technology. In manufacturing operations, it is a relatively simple matter to separate the two. For example, the **product technology** of a computer is embodied in its hardware and software. But the process technology that manufactured the computer is the technology that made and assembled all the different components. In service operations, it can be more difficult. For example, theme parks such as Walt Disney World use flight simulator technologies in some of their rides. These are large rooms mounted on a moveable hydraulic platform that, when combined with wide-screen projection, give a realistic experience of, say, space flight. But is it product/service or process technology? It clearly processes Disney's customers, yet the technology is also part of the product – the customers' experience. Product/service and process technologies are, in effect, the same thing. The formal definition of process technology that we shall use here is the machines, equipment and devices that *create* and/or *deliver* products and services. Process technologies range from milking machines to exam marking software, from body scanners to bread ovens, from mobile phones to milling machines. In fact, process technology is pervasive in all types of operations and has a very significant effect on quality, speed, dependability, flexibility and cost. That is why we devote a whole chapter to it.

Even when technology seems peripheral to the actual creation of goods and services, it can play a key role in *facilitating* the direct transformation of inputs to an operation. For example, the IT that runs planning and control activities and stock control systems can be used to help managers and operators control and improve their processes. This type of technology is called **indirect process technology**. It is becoming increasingly important. Some businesses spend more on the computer systems that control their processes than they do on the direct process technology that acts on their material, information or customers.

Operations principle

Process technology is the machines, equipment and devices that create and/or deliver products and services.

It is worth noting, however, that the distinction between 'product/service' technology and 'process' technology can depend on context. What is one business's product/service technology is another's process technology. For example, the product/service technology that software firms embed in their planning and control systems is their customers' (indirect) process technology. Fanuc, the Japanese robot manufacturer, has a factory in Oshino, Japan, where its own industrial robots themselves produce industrial robots, supervised by a staff of only four workers per shift. The robots could be classed as *both* product and process technology.

What is 'new' in new technologies?

Why is new process technology becoming so important? One can argue that it is mainly for two reasons. The first is that most new process technologies have a greater capability than what they are replacing; in other words, new process technologies are capable of doing things that older technologies could not do or do as well. Second, these increased capabilities have a greater scope of application;

they can be applied in sectors of the economy, and in types of operation, where process technology used to be far less important.

New technologies often have increased capabilities

Even technologies, such as robots, that have become commonplace in many operations are becoming cheaper, more proficient and more adaptable. Algorithms used, for example, by banks to carry out routine transactions, or by parcel delivery companies to plan routes, can outperform human decision-making. When combined with artificial intelligence (AI) they can perform and/or support activities, such as medical diagnostics, previously requiring expert human judgement. Some technologies are also becoming less expensive. Over a 20-year period, the cost of industrial robots has dropped by half, while the cost of labour in developed economies has risen over 100 per cent. Not surprising, then, that labour substitution accounts for some of the adoption of new technologies. But do not assume that it is always the main driver. Other performance benefits can be even more important, as we shall discuss later.

OPERATIONS IN PRACTICE

Go figure, or not[1]

What can artificial intelligence (AI) do and what can't it do? Well, the boundary is changing all the time. But some things are easier than others. One significant event was a five-game match, played between arguably the best professional Go player (called Lee Sedol) and AlphaGo, a computer Go program developed by Google DeepMind. The computer won the contest by four games to one. Significant because, although seemingly simple, Go is a far more complex game than chess, and AI developers had become obsessed with mastering the game. The size of a Go board means that the number of games that can be played on it is colossal: probably around 10^{170}, which is almost 100 orders of magnitude greater than the number of atoms in the observable universe (estimated to be around 10^{80}). Before AlphaGo was developed, the best Go programs were little better than a skilled amateur. The breakthrough of AlphaGo was to combine some of the same ideas as the older programs with new approaches that focused on how the computer could develop its own 'instinct' about the best moves to play. It uses a technique that its makers have called 'deep learning' that allows the computer to develop an understanding of the instinctive rules of the game that experienced players can understand but cannot fully explain. The excitement over AlphaGo's capabilities was partly based on the idea that such AI is a 'general purpose' technology (like electricity, it is capable of affecting entire economies) and is the basis of any application that requires pattern recognition.

Yet, although AI techniques are powerful, they are still limited, and often difficult to adopt (see the section on Moravec's paradox). They struggle with the types of cognitive abilities that underlie human reasoning. They have been compared to an artificial 'idiot savant' that can shine at well-bounded tasks, but can get things wrong when faced with the unexpected. In other words, they lack that difficult to define, but nevertheless important quality, 'common sense'.[2] AI can 'learn' in its basic sense, but it does require very large data sets to learn from. The sheer number and subtleties of possible human interactions, for example, can make AI difficult to apply. Most humans can staff a customer-support helpline, but few could play Go at grandmaster level. However, a game of Go has only two possible results (one wins or loses) both of which can be easily classified, and the game's underlying rules are relatively simple and clear. AI is well suited to such well-defined problems. Building a customer-service chatbot can be much harder. Every customer interaction can end with many different outcomes. Humans can more readily cope with such subtleties because they are more capable of 'top-down' reasoning about the way the world works. This helps to guide them under conditions where 'bottom-up' signals from their senses are unclear or inadequate. AI finds this difficult. Although it is capable within defined limits, it can find even small changes problematic.

Moravec's paradox

The enigma of AI struggling with tasks that most humans find easy is known as Moravec's paradox. It was articulated by Hans Moravec, at Carnegie Mellon University. He observed that it is comparatively easy to make computers exhibit adult-level performance such as tricky logical problems, yet difficult or impossible to give them the skills of a one-year-old in tasks that require perception and mobility skills. In other words, with AI, the complex is easy, and the easy is complex. However, the skills that are normally defined as 'easy', such as those humans learn instinctively, are in reality the result of billions of years of evolution. We are 'tuned' to find them easy. We find them easy not because we have knowingly broken them down into logical steps or articulated all the required computations, yet this is what would be necessary to teach an AI system to do them. As we indicate in the 'Operations in practice' example, 'Go figure, or not', the issue is whether AI can ever master such 'easy' tasks.

New technologies can increasingly be applied in all types of operation

There used to be a simple division between those operations (usually manufacturing) that used a lot of process technology, and those (usually service) that used little or none. This is no longer true. Now there are very few, if any, types of enterprise that are not actively using some kind of technology to support their operations processes. For example, in an airport, we check in and scan our passports using automated machines, progress through security with our bags (and sometimes ourselves) having been subjected to (semi-) automatic scanning and obtain access to the gate by scanning our boarding card or phone image. Even when on the flight, the pilots may actively steer aircraft only for a few minutes before the autopilot takes over for the rest of the journey. But even relatively low-volume, high-variety professional services such as legal and medical services can benefit from new and value-adding technologies.

How do we view new technologies?

Sometimes it is difficult to separate the reality of a new technology from the publicity and speculation that surrounds it, especially when its potential is not yet fully understood. One attempt to illustrate how perceptions of a technology's usefulness develop over time is the 'Gartner Hype Cycle' created by Gartner, the information technology research and consultancy company. It has five sequential (but sometimes overlapping) stages (see Figure 8.2):

- *Stage 1 – 'technology trigger'*: The early stages of a technology; it probably exists in a theoretical or prototype stage (which has aroused media interest), but there are no working practical demonstrations.
- *Stage 2 – 'peak of inflated expectations'*: The technology has developed to the point where it is implemented by some more adventurous 'early adopter' operations. There is press coverage describing both successful and unsuccessful experiences.
- *Stage 3 – 'trough of disillusionment'*: The difficulties of using the technology in practical situations start to demonstrate its shortcomings. This results in something of a backlash, leading to disappointment and disillusionment with the technology.
- *Stage 4 – 'slope of enlightenment'*: Problems with the technology are slowly solved and its potential becomes more realistically understood. It is adopted by an increasing number of operations that learn how to implement it in their context.
- *Stage 5 – 'plateau of productivity'*: The technology, in its developed form, becomes widely adopted, probably with technical standards being shared by users and suppliers.

Operations principle

New process technologies can have increased capabilities and greater scope of application.

Process technology and transformed resources

One common method of distinguishing between different types of process technology is by what the technology actually processes – materials, information or customers. We used this distinction in Chapter 1 when we discussed inputs to operations and processes.

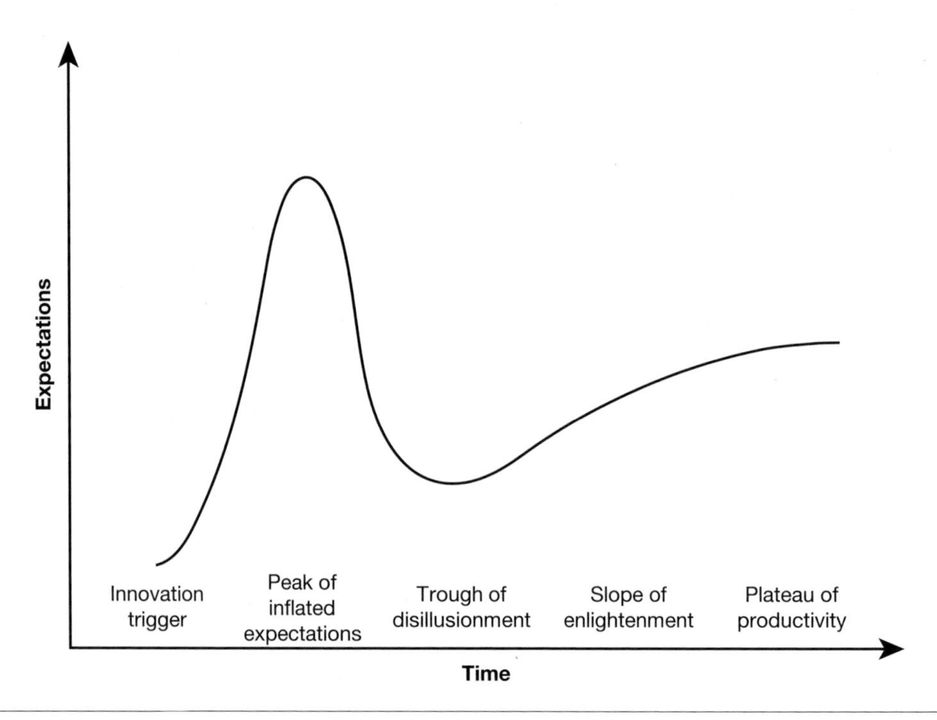

Figure 8.2 The Gartner Hype Cycle
Source: Used with permission from Gartner.

Material-processing technologies

These include any technology that shapes, transports, stores or in any way changes physical objects. It includes the machines and equipment found in **manufacturing operations** (robots, 3D printing, **computer integrated manufacturing** systems and so on), but also includes trucks, conveyors, packing machines, warehousing systems and even retail display units. In manufacturing operations, technological advances have meant that the ways in which materials are processed have been automated compared with the assembly of parts. Although far more automated than once it was, this presents more technical challenges.

OPERATIONS IN PRACTICE

Technology or people? The future of jobs[3]

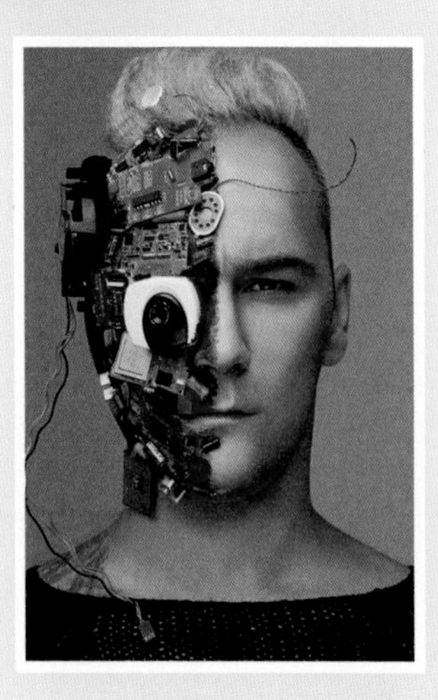

Technological advances have always had an impact on the types of jobs that are in demand by businesses, and by extension, the types of jobs that are eliminated. Much routine work has been usurped by the 'the robot and the spreadsheet'. It is the type of work that is more difficult to break down into a set of standardised elements that is less prone to being displaced by technology. Obvious examples include the type of tasks that involve decision making based on judgement and insight, teaching small children, diagnosing complex medical conditions and so on. However, the future may hold a less certain future for such jobs. As the convenience of data collection and analysis becomes more sophisticated, and process knowledge increases, it becomes easier to break more types of work down into routine constituents, which allows them to be automated. The extent to which such automation will take hold is disputed. Carl Benedikt Frey and Michael Osborne, of the University of Oxford, maintain that the range of jobs that are likely to be automated is far higher than many assume, especially traditionally white-collar jobs such as accountancy, legal work, technical writing and (even) teaching. It is not simply that technology is getting cleverer; in addition, it can exploit the capability to access far more data. Frey and Osborne even go so far as to estimate the probability that technology will mean job losses for certain jobs in the next two decades (bravely, because such forecasting is notoriously difficult). Among jobs most at risk are telemarketers (0.99, where 1.0 = certainty), accountants and auditors (0.94), retail salespersons (0.92), technical writers (0.89) and retail estate agents (0.86). Those jobs least likely to be replaced include actors (0.37), firefighters (0.17), editors (0.06), chemical engineers (0.02), athletic trainers (0.007) and dentists (0.004). Yet another study by the OECD (a group of relatively rich countries) claims these forecasts are too gloomy, and fewer people's jobs are likely to be destroyed by artificial intelligence and robots than has been suggested. However, many people will face a future in which their jobs may change significantly as technology affects the way processes are designed.

Information-processing technology

Information-processing technology, or just **information technology (IT)**, is the most common single type of technology within operations, and includes any device that collects, manipulates, stores or distributes information. Initially, it was the use of internet-based technology that had the most obvious impact on operations – especially those that are concerned with buying and selling activity. Its advantage was that it increased both reach (the number of customers/suppliers who could be reached and the number of items they could be presented with) and richness (the amount of detail that could be provided concerning both the items on sale and customers'/suppliers' behaviour). Subsequently, other types of information-processing technologies came to provide opportunities for process innovation, particularly those involving some form of analytical capability, such as **algorithmic decision-making**, artificial intelligence (AI) and data mining, those involving communication or connectivity, such as **blockchain**, and those capable of processing visual information, such as **augmented reality (AR)**.

Customer-processing technology

Increasingly the human element of service is being reduced with customer-processing technology being used to give an acceptable level of service while significantly reducing costs. There are three types of customer-processing technologies. The first category includes active interaction technology such as vehicles, online shopping, fitness equipment and self-checkout stations. In all of these, customers themselves are using the technology to create the service. By contrast, aircraft, mass transport systems, moving walkways, elevators, cinemas, 'fitness' monitors and most theme park rides are passive interactive technologies; they 'process' (and sometimes control) customers (or aspects of a customer) in some way, but do not expect the customer to take a direct part in the interaction. Some customer-processing technology is 'aware' of customers but not the other way around: for example, security monitoring, or face-recognition technologies in shopping malls or at national border customs areas. The objective of these 'hidden technologies' is to track customers' movements or transactions in an unobtrusive way.

Integrating technologies

Of course, some technologies process more than one type of resource, and/or are combinations of other technologies. These technologies are called 'integrating technologies'. For example, electronic point of sale (EPOS) technology integrates scanning and information technologies and processes shoppers, products and information. Perhaps the most-discussed of the more recent integrating technologies is 'Industry 4.0'. This is the term for the automation and integration of manufacturing technologies. It is explained in more detail later.

How should operations managers manage process technology?

The management of process technology has always been important to operations managers because they are involved in its selection, installation and management. But how do operations managers decide on the best way of enabling their use, especially in circumstances where such technologies have not previously been appropriate? They should be able to do three things:

- First, they need to understand the technology to the extent that they are able to articulate what it should be able to do; not in the sense that they need to be experts in whatever constitutes the core science of the technology, but enough to understand its implications.
- Second, they should be able to evaluate alternative technologies, particularly as they affect the operations they manage, and share in the decisions of which technology to choose.
- Third, they must be able to develop, plan and implement the technology so that it can reach its full potential in contributing to the performance of the operation.

Operations principle

Operations managers need to be able to understand, evaluate and manage new process technologies.

8.2 How can one understand the potential of new process technology?

The first responsibility of operations managers is to gain an understanding of what a process technology can do. But 'understanding process technology' does not (necessarily) mean knowing the details of the science and engineering embedded in the technology. It means knowing enough about the principles behind the technology to be comfortable in evaluating some technical information, capable of dealing with experts in the technology and confident enough to ask relevant questions.

Operations principle

Operations managers should understand enough about process technology to evaluate alternatives.

The four key questions

In particular the following four key questions can help operations managers to grasp the essentials of the technology:

1. What does the technology do that is different from other similar technologies?
2. How does it do it? That is, what particular characteristics of the technology are used to perform its function?
3. What benefits does using the technology give to the operation?
4. What constraints or risks does using the technology place on the operation?

For example, look at the worked example on QB House to think through these four key questions.

Emerging technologies – understand their primary capabilities

The four questions listed above are universal, in the sense that they can help to understand the implications for operations management of any new or emerging technology. But operations managers are not immune from more general public perceptions of the importance of new technologies, which is why it is worthwhile considering how technologies are seen.

Worked example

QB House[4]

One day, in Japan, Kuniyoshi Konishi became so frustrated at having to wait to get his hair cut and then paying over 3,000 yen for the privilege, he decided that there must be a better way to offer this kind of service. *'Why not'*, he said, *'create a no-frills barber shop where the customer could get a haircut in ten minutes at a cost of 1,000 yen (€7)?'* He realised that a combination of technology and process design could speed up the basic task of cutting hair. His chain of barbers is called QB House. Its barbers never handle cash. Each shop has a ticket vending machine that accepts 1,000 yen bills (and gives no change) and issues a ticket that the customer gives the barber in exchange for the haircut. Second, QB House does not take reservations, so no receptionist is needed, or anyone to schedule appointments. Third, QB House developed a system indicating how long customers will have to wait. Electronic sensors under each seat in the waiting area track how many customers are waiting. Green lights displayed outside the shop indicate that there is no waiting, yellow lights indicate a wait of about 5 minutes and red lights indicate that the wait may be around 15 minutes. This system can also keep track of how long it takes for each customer to be served. Fourth, QB has done away with the traditional Japanese practice of shampooing its customers after the haircut to remove any loose hairs. Instead, the barbers use QB House's own 'air wash' system where a vacuum cleaner hose is pulled down from the ceiling and used to vacuum the customer's hair clean. The QB House system proved so popular that its shops spread through many other south-east Asian countries. Each year almost 4,000,000 customers experience QB House's 10-minute haircuts.

Analysis

- *What does the technology do?* – Signals availability of servers, so managing customers' expectations. Avoids hairdressers having to handle cash. Speeds service by substituting 'air wash' for traditional shampoo.
- *How does it do it?* – Uses simple sensors in seats, ticket dispenser and air wash blowers.
- *What benefits does it give?* – Faster service with predictable wait time (dependable service) and lower costs, therefore less expensive prices.
- *What constraints or risks does it impose?* – Risk of customer perception of quality of service. It is not an 'indulgent' service. It is a basic, but value, service in which customers need to know what to expect and how to use it.

Classifying technologies by their primary capabilities

Any in-depth understanding of a technology calls for a knowledge of what, in general terms, it actually does. In other words, what is its 'primary capability? What is it better at than the technology that it replaces? Figure 8.3 shows some technologies that, at the time of writing, were new(ish). This is certainly not a fully comprehensive list. Some technologies are specialist, in the sense that their application is limited to one type of operation. Other technologies are (again, at the time of writing) still very much in their development stage. The intention is not to provide a comprehensive survey of technologies – that could be expanded into a whole book – nor is it to delve into technical details. Rather it is to demonstrate how operations managers have to look 'behind' the technology in order to start to understand what it is intended to do. The figure positions the technologies relative to five 'primary capabilities':

Operations principle

The primary capability of process technologies can be one or more of: thinking/reasoning, seeing/sensing, communicating/connecting, moving physical objects and materials processing.

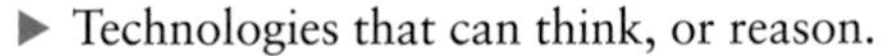

- Technologies that can think, or reason.
- Technologies that can see, or sense.
- Technologies that can communicate, or connect.
- Technologies that can move physical objects.
- Technologies that can process materials.

Technologies that can think, or reason

The best-known class of technology that attempts to replicate (and even surpass) human thinking is artificial intelligence (AI). This is usually taken to mean a computer-based capability that accentuates the creation of an intelligence that reacts and works like humans. It is the underlying technology that supports several other learning activities such as speech recognition, planning and problem solving. Since computers were developed, people have worked with and been augmented by them. AI challenges

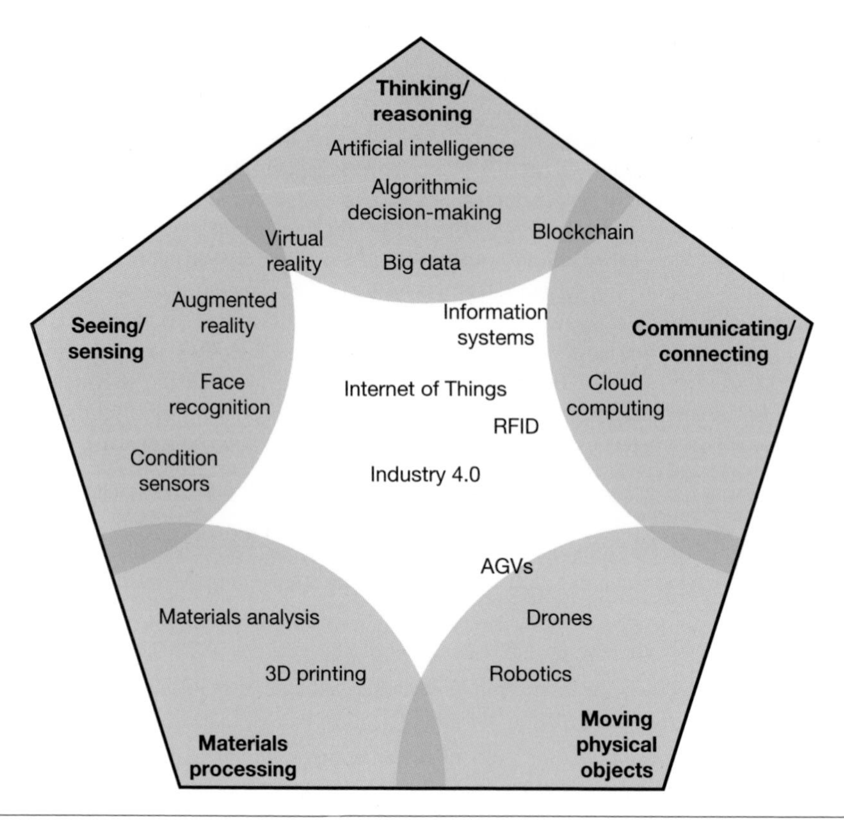

Figure 8.3 New and emerging technologies positioned relative to five 'primary capabilities': thinking/reasoning, seeing/sensing, communicating/connecting, moving physical objects, processing materials

this relationship as computers improve their capabilities to take more control. Somewhat less sophisticated, but more widely used (currently) in operations management is algorithmic decision-making. An algorithm is a predefined sequence of instructions, or rules. Many of the models used in this text are algorithms that could be incorporated into decision-making routines. Algorithmic decision-making may be combined with large data sets (often called 'big data'). *Big data* is a large volume of both structured and unstructured data whose analysis can reveal hidden patterns, correlations and other insights.

Technologies that can see, or sense

Some technologies exploit their ability to manipulate computer-generated or computer augmented visual information. For example, *augmented reality* technologies show an enhanced version of reality where live views of physical real-world environments are augmented with overlaid computer-generated images, thus supplementing one's perception of reality. *Virtual reality* goes further by using entirely computer-generated simulations, with which humans can interact in a seemingly real manner using special helmets and gloves fitted with sensors. Although both augmented and virtual reality technologies are used in entertainment operations, they are also valuable in surgery training, maintenance planning and process design. Both of these involve people using technology to view objects. *Face recognition* is the reverse of this. It uses still or video images of a scene to identify or verify one or more individuals, using a stored database of faces so that those people can be identified for (say) automatic charging for a service, or for security or advertising purposes. *Condition sensor* technologies are more intimate. They sense characteristics of people (e.g. fitness monitors), or materials (e.g. quality control sensors).

Technologies that can communicate, or connect

Arguably, the most significant capability that is increasingly built into many process technologies is the ability to network, connect or communicate with other elements in an operations process. For example, *cloud computing* applications allow dispersed groups of people to collaborate virtually, using shared information (in real time) and shared storage of information. *Blockchain* technology also relies on connected networks, but rather than 'share' it uses distributed databases in such a way that they maintain a shared list of records (called blocks), where each encrypted block of code contains the history of every block that came before it, with each transaction 'timestamped'. There is transparency within the network, but no single point where records could be hacked or corrupted. At a more technical level, communication between physical objects has been made significantly more effective through the use of *RFID* technologies. These devices use radio waves to automatically identify objects, collect data about them, and communicate it into *information systems* (integrated sets of components for collecting, storing and processing data, and for providing information and knowledge).

Technologies that can move physical objects

Robots are often credited with almost human-like abilities, but in fact are used primarily for handling materials, such as loading and unloading work pieces onto a machine, for processing where a tool is gripped by the robot, and for assembly where the robot places parts together. Some robots have some limited sensory feedback through vision control and touch control. A close relation of a robot is an *automated guided vehicle* (AGV). This is a materials-handling system that uses automated vehicles that are programmed to move between different stations (usually in a manufacturing or warehouse setting) without a driver. On a more airborne level, aerial *drones,* both guided and autonomous, are increasingly used in industrial applications. Non-military uses include safety and quality inspection, filming and journalism, search and rescue, precision agriculture and short-haul delivery.

Technologies that can process materials

The various developments in materials-processing technology tend to be too specialised and technical to grab popular attention. Nevertheless, plenty of new (or newish) processes are having a sometimes-significant effect on the economics and practice of materials-processing operations. Everything from miniaturisation, to the use of lasers, to the creation of products with complex shapes and multifunctional materials, has opened novel processing opportunities. One technology that has received publicity is *3D printing,* also known as additive manufacturing. A 3D printer produces a three-dimensional object by laying down layer upon layer of material until the final form is obtained. But 3D printing is not a new technology as such. Since the 1990s it has been used to make prototype products quickly and cheaply prior to full production. Now it is increasingly used for finished products for real customers. Importantly, because the technology is 'additive' it reduces waste significantly. Sometimes as much as 90 per cent of material is wasted in machining some aerospace parts, for example.

Technologies with more than one primary capability

Some of the technologies described above have more than one primary capability, even if one of them is dominant. Virtual reality, for example, is a visual technology, but could not work without a relatively powerful thinking/reasoning capability. Automatic guided vehicles are concerned primarily with moving physical objects, but can frequently communicate, 'see' where they are going, and reason to work out alternative routes. Other important technologies combine even more primary capabilities. The *Internet of Things* (IoT) exploits the potential of RFID technology with its sensors and actuators, connects them using wireless networks and allows information systems and physical networks to merge. SAP, the developer of enterprise resource systems, describes the Internet of Things as: 'A world where physical objects are seamlessly integrated into the information network, and where the physical objects can become active participants in business processes. Services are available to interact with these "smart objects" over the Internet, query and change their state and any information associated with them, taking into account security and privacy issues'.[5]

Industry 4.0

Perhaps the most developed of these technologies that combine several primary capabilities has become known as *Industry 4.0.* The name derives from the contention that there have been four industrial revolutions: first, mechanisation through water and steam power; second, mass production and assembly lines powered by electricity; third, computerisation and automation; and finally, fourth, smart factories combining digital, virtual and physical systems. The name Industry 4.0 was first used publicly in 2011 as 'Industrie 4.0' by a group of businesspeople, political representatives and academics, meeting as an initiative to improve German manufacturing competitiveness. The German federal government formed a working group that published a vision for Industry 4.0, which they saw as the cyber-physical systems that comprise smart machines, storage systems and production facilities capable of autonomously exchanging information, triggering actions and controlling each other independently.

OPERATIONS IN PRACTICE

Love it or hate it, Marmite's energy recycling technology[6]

For those readers who live in regions of the world where Marmite is not a big seller, it is a nutritious savoury spread that contains vitamins, can be eaten in a sandwich, on toast, or used as a cooking ingredient. It is not to everyone's taste, which is why it is advertised with the line '. . . you either love it or hate it'. But behind the clever advertising, Marmite, which is part of Unilever, the large food company, is a pioneer in recycling the leftovers from its production process into energy at the factory where it is made. The factory is in the United Kingdom and every year around 18,000 tonnes of solidified Marmite deposit is left adhering to the surfaces of the machines and handling equipment that are used to produce the product. For years this residue was cleaned off and then either flushed into the sewerage system or sent to landfill sites. Then Unilever installed an anaerobic digester. This is a composter that contains microbes that feed on the waste. As they do this, they release methane, which is burned in a boiler that is connected to a generator that produces power. The system also captures the waste heat that comes through the exhaust and helps heat the factory's water system (see Figure 8.4).

But this example is just one from Unilever's 'Sustainable Living Plan' published by the company every year, detailing the progress it is making globally and nationally towards meeting its sustainable targets. It publishes its performance

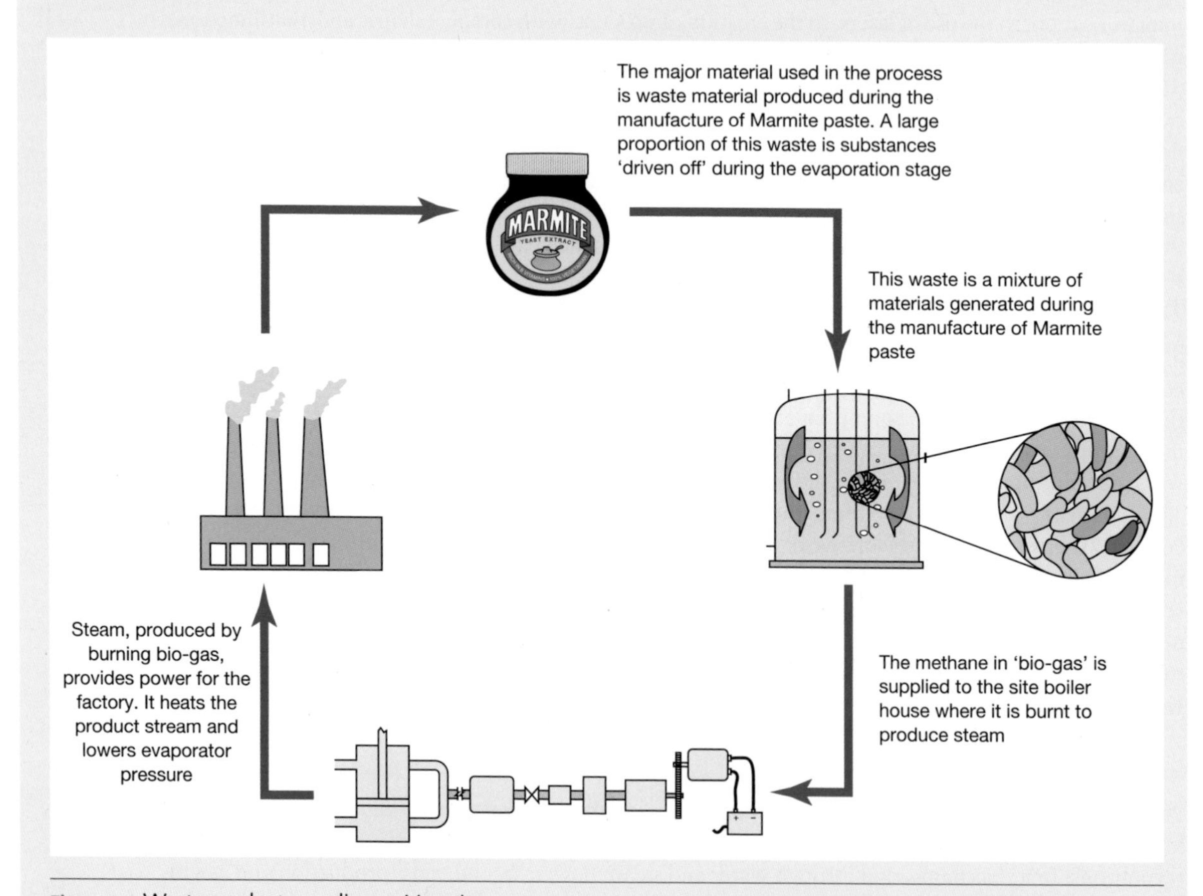

The major material used in the process is waste material produced during the manufacture of Marmite paste. A large proportion of this waste is substances 'driven off' during the evaporation stage
MARMITE
YEAST EXTRACT
This waste is a mixture of materials generated during the manufacture of Marmite paste
The methane in 'bio-gas' is supplied to the site boiler house where it is burnt to produce steam
Steam, produced by burning bio-gas, provides power for the factory. It heats the product stream and lowers evaporator pressure

Figure 8.4 Waste product recycling at Marmite

in three categories. The first is 'areas where we are making genuinely good progress'. These include sustainable sourcing, nutrition and eco-efficiency (including the Marmite project). The second category is 'areas where we have had to consider carefully how to reach our targets but are now ready to scale up', such as a programme to increase the recycling rates of aerosols, encouraging more local councils to collect aerosols kerbside. 'However', the report admitted, 'we have more to do, working in partnership with industry, Government and NGOs to help to increase recycling and recovery rates'. The third category is 'areas where we are finding it difficult to make progress and will need to work with others to find solutions'. This includes targets that require consumer behaviour change, such as encouraging people to eat foods with lower salt levels or reducing the use of heated water in showering and washing clothes.

8.3 How can new process technologies be evaluated?

The most common technology-related decision for operations managers is likely to be whether to adopt an alternative technology to whatever is currently being used. It is an important decision because process technology can have a significant effect on the operation's capability, and no one wants to change expensive technologies too frequently. Yet, with the emergence of so many new process technologies with sometimes ambiguous capabilities, the evaluation process becomes both more difficult and more important. Added to this are new types of technological risk – security, obsolescence, implementation problems and the tendency of some organisations to get carried away with new technology for its own sake. Here we use three sets of criteria for evaluation:

Operations principle

Process technologies can be evaluated in terms of their fit with process tasks, their effect on performance and their financial impact.

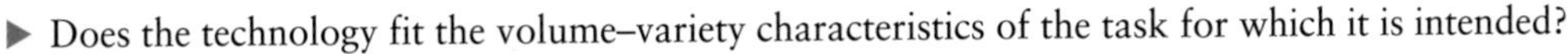

- Does the technology fit the volume–variety characteristics of the task for which it is intended?
- What aspects of the operation's performance does the technology improve?
- Does the technology give an acceptable financial return?

Does the process technology fit the volume–variety characteristics of the task?

Different process technologies will be appropriate for different types of operations, not just because they process different transformed resources, but also because they do so at different levels of volume and variety. High–variety–low–volume processes generally require process technology that is *general purpose,* because it can perform the wide range of processing activities that high variety demands. High–volume–low–variety processes can use technology that is more *dedicated* to its narrower range of processing requirements. Within the spectrum from general purpose to dedicated process technologies, three dimensions in particular tend to vary with volume and variety:

- Its degree of 'automation'
- The capacity of the technology to process work: that is, its 'scale' or 'scalability'.
- The extent to which it is integrated with other technologies: that is, its degree of 'coupling' or 'connectivity'.

The degree of automation of the technology

To some extent, all technology needs human intervention. It may be minimal, for example the periodic maintenance interventions in a petrochemical refinery. Conversely, the person who operates the technology may be the entire 'brains' of the process, for example the surgeon using keyhole surgery techniques. The ratio of technological to human effort it employs is sometimes called the capital intensity of the process technology. Generally, processes that have high variety and low volume will employ process technology with lower degrees of automation than those with higher volume and lower variety. For example, investment banks trade in highly complex and sophisticated financial 'derivatives', often customised to the needs of individual clients, and each may be worth millions of dollars. The back office of the bank has to process these deals to make sure that payments are made on time, documents are exchanged and so on. Much of this processing will be done using relatively general-purpose technology such as spreadsheets. Skilled back-office staff are making the decisions rather than the technology. Contrast this with higher-volume, low-variety products, such as straightforward equity (stock) trades. Most of these products are simple and straightforward and are processed in very high volumes of several thousand per day by 'automated' technology.

The scale/scalability of the technology

There is usually some discretion as to the scale of individual units of technology. For example, the duplicating department of a large office complex may decide to invest in a single, very large, fast copier, or alternatively in several smaller, slower copiers distributed around the operation's various processes. An airline may purchase one or two wide-bodied aircraft or a larger number of smaller aircraft. The advantage of large-scale technologies is that they can usually process items more cheaply than small-scale technologies, but usually need high volume and can cope only with low variety. By contrast, the virtues of smaller-scale technology are the nimbleness and flexibility that is suited to high-variety, lower-volume processing. For example, four small machines can between them produce four different products simultaneously (albeit slowly), whereas a single large machine with four times the output can produce only one product at a time (albeit faster). Small-scale technologies are also more robust. Suppose the choice is between three small machines and one larger one. In the first case, if one machine breaks down, one-third of the capacity is lost, but in the second, capacity is reduced to zero. The advantages of large-scale technologies are similar to those of large-capacity increments discussed in Chapter 5.

The equivalent to scale for some types of information-processing technology is *scalability*. By scalability we mean the ability to shift to a different level of useful capacity quickly and cost-effectively. Scalability is similar to absolute scale in so much as it is influenced by the same volume–variety characteristics. IT scalability relies on consistent IT platform architecture and the high process standardisation that is usually associated with high-volume and low-variety operations.

The coupling/connectivity of the technology

Coupling means the linking together of separate activities within a single piece of process technology to form an interconnected processing system. Tight coupling usually gives fast process throughput. For example, in an automated manufacturing system, products flow quickly without delays between stages, and inventory will be lower – it can't accumulate when there are no 'gaps' between activities. Tight coupling also means that flow is simple and predictable, making it easier to keep track of parts when they pass through fewer stages, or information when it is automatically distributed to all parts of an information network. However, closely coupled technology can be both expensive (each connection may require capital costs) and vulnerable (a failure in one part of an interconnected system can affect the whole system). The fully integrated manufacturing system constrains parts to flow in a predetermined manner, making it difficult to accommodate products with very different processing requirements. So, coupling is generally more suited to relatively low variety and high volume. Higher-variety processing generally requires a more open and unconstrained level of coupling because different products and services will require a wider range of processing activities.

> **Operations principle**
> Process technology in high-volume, low-variety processes is relatively automated, large-scale and closely coupled when compared to that in low-volume, high-variety processes.

New technology is changing the diagonal

Figure 8.5 illustrates these three dimensions of process technology and the implied 'diagonal' between low volume–high variety, and high volume–low variety. It also shows how some developments in technology have, to some extent, overcome the implied trade-off between flexibility and cost. Specifically, digitisation and the vastly increased computing power embedded in many new technologies has made it easier to achieve lower costs while not having to sacrifice the ability to provide variety. For example, the IT systems, databases and algorithms behind internet banking services allow customers to access a very wide variety of customised services, while retaining or enhancing

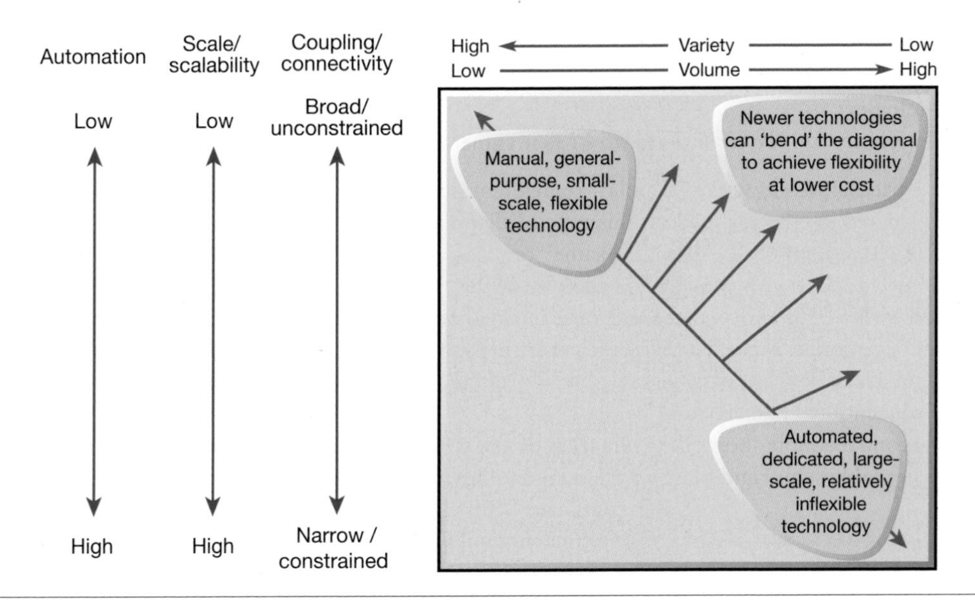

Figure 8.5 Different process technologies are appropriate for different volume–variety combinations, but some newer technologies can achieve both flexibility and low cost

back-office efficiency. Similarly, so-called 3D knitting machines can produces an entire garment including the arms, collars and other parts that would normally be produced separately and sewn on, using a single machine using a single thread. Not only does this allow a wide variety of garments to be produced at relatively low cost, it reduces the waste of material. The effect of this is to 'bend' what has hitherto been accepted as the 'natural' diagonal.

How does the technology improve the operation's performance?

In Chapter 2, we identified the five operations performance objectives on which an operation or process can be judged. So, a sensible starting point for evaluating the impact of any process technology on an operation is to assess how it affects its quality, speed, dependability, flexibility and cost performance. However, two refinements are necessary to the normal list of performance objectives. First, given that some process technologies can perform totally novel tasks (for example, 3D printing can create shapes and use materials in a novel manner), it is worth splitting the criterion of 'quality' into 'specification quality' (what can the technology do?) and 'conformance quality' (can it do it in an error-free manner?). Similarly, when considering the flexibility of technology, it is worth distinguishing between response flexibility (how easy is it to switch between tasks?) and range flexibility (how many different tasks can it perform?). In addition, again given the increased capabilities of some new process technologies, other criteria are worth including in any evaluation.

These assessment questions can be viewed as a starting point. The criteria should be adjusted depending on the operation for which the technology is intended. They are:

- *What can the technology do?* Is it capable of doing something (or some things) that the previous technology could not do?
- *How well can the technology do things?* Is it capable of doing things in an error-free manner?
- *How fast can the technology do things?* Is it capable of doing things more rapidly?
- *How reliably can the technology do things?* Is it capable of doing things with greater dependability?
- *How flexibly (response) can the technology do things?* Is it capable of switching easily between tasks?
- *What range of things can the technology do?* How many different tasks can it perform?
- *How sustainable is the technology?* Does it have a positive environmental impact? (For example, see the 'Operations in practice' example on Marmite's energy recycling technology.)
- *Where can the technology do it?* Can it perform its tasks in alternative locations? (For example, is it portable?)

- *How safely can the technology do it?* Can it perform its tasks without harming people?
- *How connectedly can the technology do it?* Can it communicate or connect with other technologies?
- *How securely can the technology do it?* Is it vulnerable to interference or hacking?

These criteria are shown on a 'polar' representation (see Chapter 2) in Figure 8.6. A technology being evaluated can be mapped on this type of diagram. In this representation, the evaluations are classed as 'worse than' (red), 'about the same' (orange), and 'better than' (green) compared to what the technology is replacing. It shows two examples of this. The first examines the QB House technology, described earlier. This shows that QB House has two advantages over conventional haircutting – it is (usually) faster and it is cheaper. However, there are fewer locations ('where can it be done?') and the waiting time may be marginally less reliable. On all other criteria, it performs at more or less the same level as conventional hairdressers.

The second example shows the evaluation of a 3D printer to be used to produce prototype designs of biodegradable semi-rigid plastic packaging moulds. The performance of the proposed 3D printer is compared with the current method of producing the moulds using conventional machining techniques. It indicates that the new 3D technology would be superior in almost all respects, particularly in terms of its speed, flexibility, range, sustainability (because it wastes far less material) and its ability to connect with other technologies (particularly, in this case, the design system). However, this connectivity would make it less difficult to steal and replicate designs.

Operations principle

Process technology needs to be evaluated on a range of criteria that include its impact on the performance of the operation in which the technology will be used.

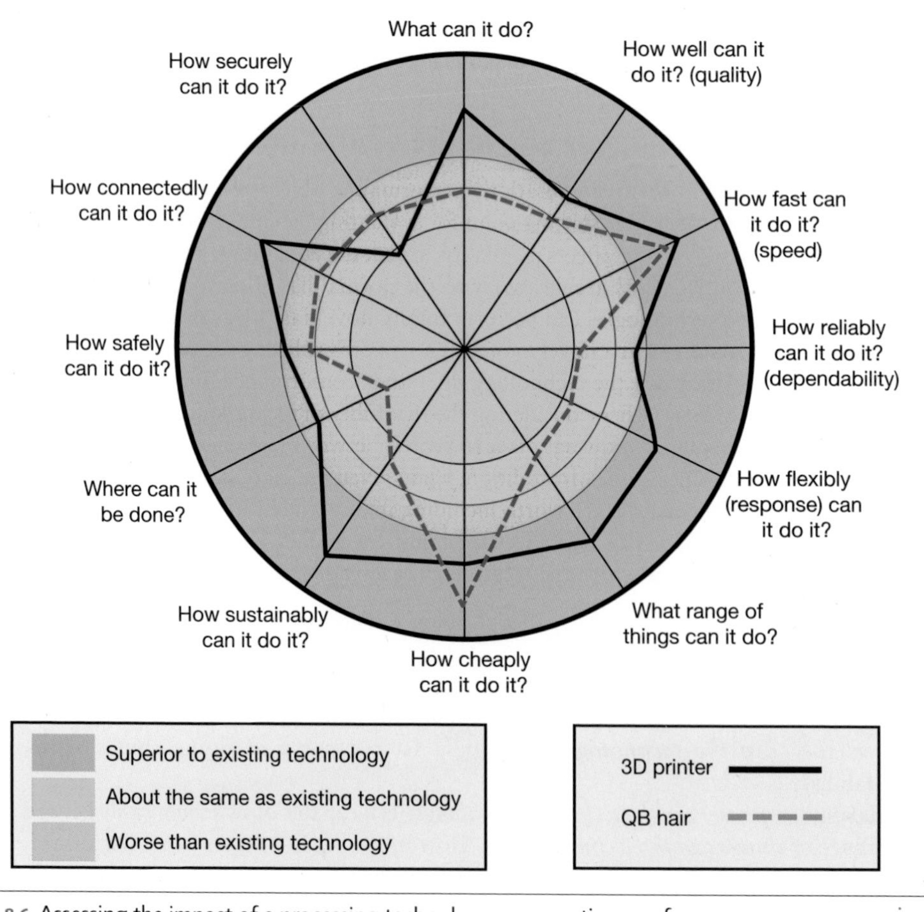

Figure 8.6 Assessing the impact of a processing technology on operations performance

Does the technology give an acceptable financial return?

Assessing the financial value of investing in process technology is in itself a specialised subject. And while it is not the purpose of this text to delve into the details of financial analysis, it is important to highlight one important issue that is central to financial evaluation: while the benefits of investing in new technology can be spread over many years into the future, the costs associated with investing in the technology usually occur up front. So we have to consider the time value of money. Simply, this means that receiving €1,000 now is better than receiving €1,000 in a year's time. Receiving €1,000 now enables us to invest the money so that it will be worth more than the €1,000 we receive in a year's time. Alternatively, reversing the logic, we can ask ourselves how much would have to be invested now to receive €1,000 in one year's time. This amount (lower than €1,000) is called the net present value of receiving €1,000 in one year's time.

For example, suppose current interest rates are 10 per cent per annum; then the amount we would have to invest to receive €1,000 in one year's time is:

$$€1000 \times \frac{1}{(1.10)} = €909.10$$

So the present value of €1,000 in one year's time, *discounted for the fact that we do not have it immediately*, is €909.10. In two years' time, the amount we would have to invest to receive €1,000 is:

$$€1000 \times \frac{1}{(1.10)} \times \frac{1}{(1.10)} = €1000 \times \frac{1}{(1.10)^2} = €826.50$$

The rate of interest assumed (10 per cent in our case) is known as the discount rate. More generally, the present value of €x in n years' time, at a discount rate of r per cent, is:

$$\frac{x}{(1 + r/100)}^{n}$$

Worked example

Blankston warehouse

Blankston warehouse stores and distributes spare parts. It is considering investing in a new 'retrieval and packing' system, which converts sales orders into 'retrieval lists' and uses materials-handling equipment to automatically pick up the goods from its shelves and bring them to the packing area. The capital cost of purchasing and installing the new technology can be spread over three years, and from the first year of its effective operation, overall operations cost savings will be made. Combining the cash that the company will have to spend and the savings that it will make, the cash flow year by year is shown in Table 8.1.

However, these cash flows have to be discounted in order to assess their 'present value'. Here the company is using a discount rate of 10 per cent. This is also shown in Table 8.1. The effective life of this technology is assumed to be six years:

The total cash flow (sum of all the cash flows) = €1.38 million

However, the net present value (NPV) = €816 500

This is considered to be acceptable by the company.

Calculating discount rates, although perfectly possible, can be cumbersome. As an alternative, tables are usually used, such as the one in Table 8.2.

So now the net present value is:

$$NPV = DF \times FV$$

Table 8.1 Cash flows for the warehouse process technology

Year	0	1	2	3	4	5	6	7
Cash flow (€000)	−300	30	50	400	400	400	400	0
Present value (discounted at 10%)	−300	27.27	41.3	300.53	273.21	248.37	225.79	0

Table 8.2 Present value of €1 to be paid in future

Years	3.0%	4.0%	5.0%	6.0%	7.0%	8.0%	9.0%	10.0%
1	€0.970	€0.962	€0.952	€0.943	€0.935	€0.926	€0.918	€0.909
2	€0.942	€0.925	€0.907	€0.890	€0.873	€0.857	€0.842	€0.827
3	€0.915	€0.889	€0.864	€0.840	€0.816	€0.794	€0.772	€0.751
4	€0.888	€0.855	€0.823	€0.792	€0.763	€0.735	€0.708	€0.683
5	€0.862	€0.822	€0.784	€0.747	€0.713	€0.681	€0.650	€0.621
6	€0.837	€0.790	€0.746	€0.705	€0.666	€0.630	€0.596	€0.565
7	€0.813	€0.760	€0.711	€0.665	€0.623	€0.584	€0.547	€0.513
8	€0.789	€0.731	€0.677	€0.627	€0.582	€0.540	€0.502	€0.467
9	€0.766	€0.703	€0.645	€0.592	€0.544	€0.500	€0.460	€0.424
10	€0.744	€0.676	€0.614	€0.558	€0.508	€0.463	€0.422	€0.386
11	€0.722	€0.650	€0.585	€0.527	€0.475	€0.429	€0.388	€0.351
12	€0.701	€0.626	€0.557	€0.497	€0.444	€0.397	€0.356	€0.319
13	€0.681	€0.601	€0.530	€0.469	€0.415	€0.368	€0.326	€0.290
14	€0.661	€0.578	€0.505	€0.442	€0.388	€0.341	€0.299	€0.263
15	€0.642	€0.555	€0.481	€0.417	€0.362	€0.315	€0.275	€0.239
16	€0.623	€0.534	€0.458	€0.394	€0.339	€0.292	€0.252	€0.218
17	€0.605	€0.513	€0.436	€0.371	€0.317	€0.270	€0.231	€0.198
18	€0.587	€0.494	€0.416	€0.350	€0.296	€0.250	€0.212	€0.180
19	€0.570	€0.475	€0.396	€0.331	€0.277	€0.232	€0.195	€0.164
20	€0.554	€0.456	€0.377	€0.312	€0.258	€0.215	€0.179	€0.149

where

DF = the discount factor from Table 8.2

FV = future value

To use the table, find the vertical column and locate the appropriate discount rate (as a percentage). Then find the horizontal row corresponding to the number of years it will take to receive the payment. Where the column and the row intersect is the present value of €1. You can multiply this value by the expected future value, in order to find its present value.

Worked example

Best-health clinic

Best-health clinic is considering purchasing a new analysis system. The net cash flows from the new analysis system are as follows.

Year 1: −€10,000 (outflow of cash)
Year 2: €3,000
Year 3: €3,500
Year 4: €3,500
Year 5: €3,000

Assuming that the real discount rate for the clinic is 9 per cent, using the net present value table (Table 8.3), demonstrate whether the new system would at least cover its costs. Table 8.3 shows the calculations. It shows that, because the net present value of the cash flow is positive, the new system will cover its costs, and will be (just) profitable for the clinic.

Table 8.3 Present value calculations for the clinic

Year	Cash flow		Table factor		Present value
1	(€10,000)	×	1.000	=	(€10,000.00)
2	€3,000	×	0.917	=	€2,752.29
3	€3,500	×	0.842	=	€2,945.88
4	€3,500	×	0.772	=	€2,702.64
5	€3,000	×	0.708	=	€2,125.28
			Net present value	=	€526.09

OPERATIONS IN PRACTICE

Bionic duckweed[7]

Trying to predict the future of technology is never straightforward. Both technology development and technology adoption can sometimes be quicker and sometimes slower than is predicted. Blindly ignoring the possibility of changes in technology is never a sensible strategy. But neither is avoiding taking decisions on technology because one is continually waiting for the next technological breakthrough. It may sound strange, but the phrase 'bionic duckweed' has become to mean avoiding committing to a currently available technology because a better one might come along in the future. The term was first used by a journalist and railway expert called Roger Ford when giving evidence to a UK parliamentary committee. He was criticising a decision to delay investing in the electrification of the UK railway network because of the suggestion that '*we might have fuel-cell-power trains using hydrogen developed from bionic duckweed in 15 years' time*'. This, he said, was being used as a justification for delay on the grounds that it would be wasting resources to have the network electrified now. In other words, don't invest today – there will be bionic duckweed tomorrow. As one commentator put it, '*My own computer . . . . has broken down several times in the first two years of use. I am tempted to buy something new and start again. And yet I keep patching it up and plodding on. Why? Duckweed. The longer I can keep it going, the better and cheaper the replacement will be*'.

8.4 How are new process technologies developed and implemented?

Developing and implementing process technology means organising all the activities involved in making the technology work as intended. No matter how potentially beneficial and sophisticated the technology, it remains only a prospective benefit until it has been implemented successfully. Yet implementation is very context dependent. The way one implements any technology will very much depend on its specific nature, the changes implied by the technology and the organisational conditions that apply during its implementation. In the remainder of this chapter, we look at four particularly important issues that affect technology implementation: the way technology is planned over the long term, the idea of resource and process 'distance', the need to consider customer acceptability and the idea that if anything can go wrong, it will.

Technology planning in the long term – technology roadmapping

Operations managers are involved with the development of process technologies in consultation with other parts of the firm and in the context of some kind of formal planning process. A technology roadmap (TRM) is an approach that provides a structure that coordinates this consultation. Motorola originally developed the approach in the 1970s so that it could support the development of its products and their supporting technologies. A TRM is essentially a process that supports technology development by facilitating collaboration between the various activities that contribute to technology strategy. It allows technology managers to define their firm's technological evolution in advance by planning the timing and relationships between the various elements that are involved in technology planning. For example, these 'elements' could include the business goals of the company, market developments or specific events, the component products and services that constitute related offerings, product/service and process technologies, the underlying capabilities that these technologies represent, and so on. Figure 8.7 shows an example of a technology road map for the development of products/services, technologies and processes for a facilities management service.

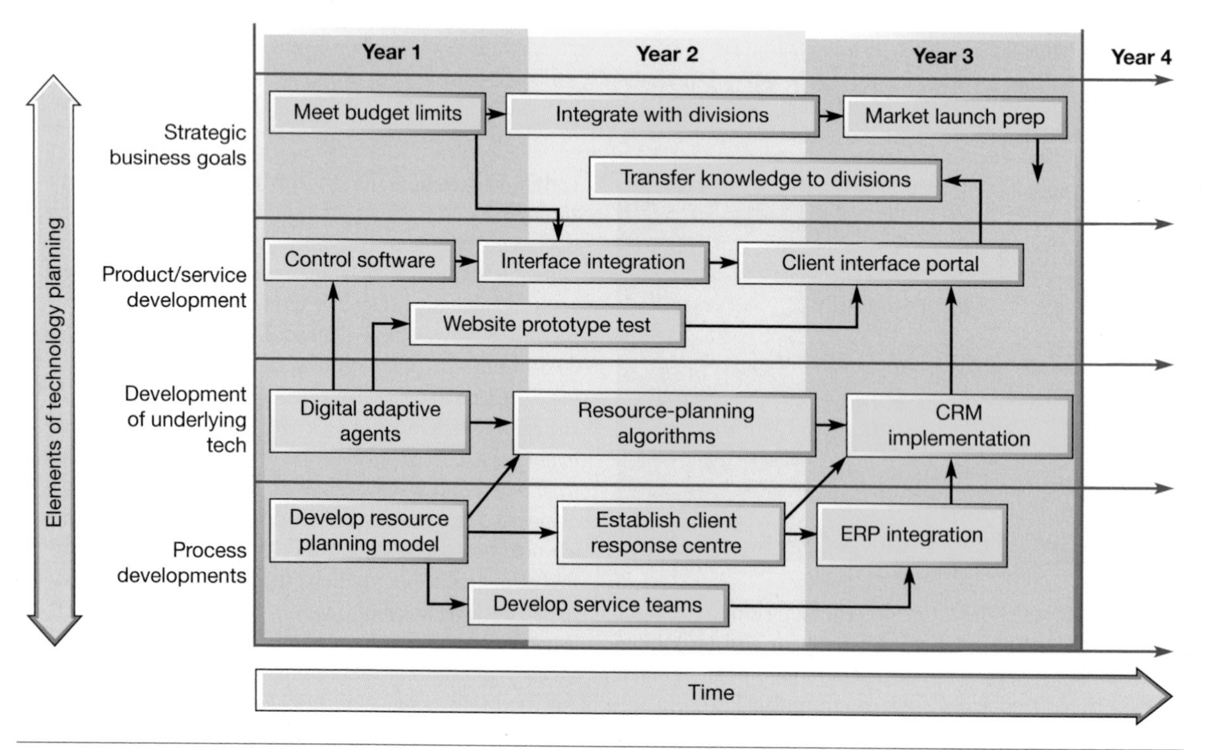

Figure 8.7 Simplified example of a technology roadmap (TRM) for the development of products/services, technologies and processes for a facilities management service

OPERATIONS IN PRACTICE

'Wrong-shaped' parcels post a problem for UK Mail[8]

It was intended to be an investment in a 'transforming technology'. When UK Mail unveiled its £20 million investment in its new fully-automated sorting facility in Coventry in the Midlands of the United Kingdom, it expected that it would give it an edge over its competitors. Guy Buswell, then the Chief Executive of UK Mail, said that the new hub would play a crucial role in the company's network. Aided by its new technology, the company, which competes with Royal Mail for parcel delivery business, had been expected to make healthy pre-tax profits. But that was before the new technology was brought into operation. In practice, the new state-of-the-art sorting equipment struggled to cope with the volume of 'irregular-shaped parcels' it was expected to process. This resulted in the operation having to divert a larger than expected proportion of its sorting to a manual process, which incurred extra operating costs. The setback was *'clearly very disappointing'*, admitted Guy Buswell, when the company announced its second profit warning in four months. The problems also hit the company's share price. The following year UK Mail announced that it would be acquired by Deutsche Post DHL Group.

The benefits of TRM are mainly associated with the way they bring together the significant stakeholders involved in technology strategy and various (and often differing) perspectives they have. It tackles some fundamental questions, such as why do we need to develop our technology? Where do we want to go with our technological capabilities? How far away are we from that objective? In what order should we do things? By when should development goals be reached? Yet TRMs do not offer any solutions to any firm's technological strategic options; in fact, they need not offer options or alternative technology trajectories. They are essentially a narrative description of how a set of interrelated developments should (rather than will) progress. Because of this they have been criticised as encouraging over-optimistic projections of the future. Nevertheless, they do provide, at the very least, a plan against which technology strategy can be assessed.

Resource and process 'distance'

The degree of difficulty in the implementation of process technology will depend on the degree of novelty of the new technology resources and the changes required in the operation's processes. The less that the new technology resources are understood (influenced perhaps by the degree of innovation), the greater their 'distance' from the current technology resource base of the operation. Similarly, the extent to which an implementation requires an operation to modify its existing processes, the greater the 'process distance'. The greater the resource and process distance, the more difficult any implementation is likely to be. This is because such distance makes it difficult to adopt a systematic approach to analysing change and learning from mistakes. Those implementations that involve relatively little process or resource 'distance' provide an ideal opportunity for organisational learning. As in any classic scientific experiment, the more variables that are held constant, the more confidence you have in determining cause and effect. Conversely, in an implementation where the resource and process 'distance' means that nearly everything is 'up for grabs', it becomes difficult to know what has worked and what has not. More importantly, it becomes difficult to know why something has or has not worked.[9] This idea is illustrated in Figure 8.8.

Operations principle

The difficulty of process technology implementation depends on its degree of novelty and the changes required in the operation's processes.

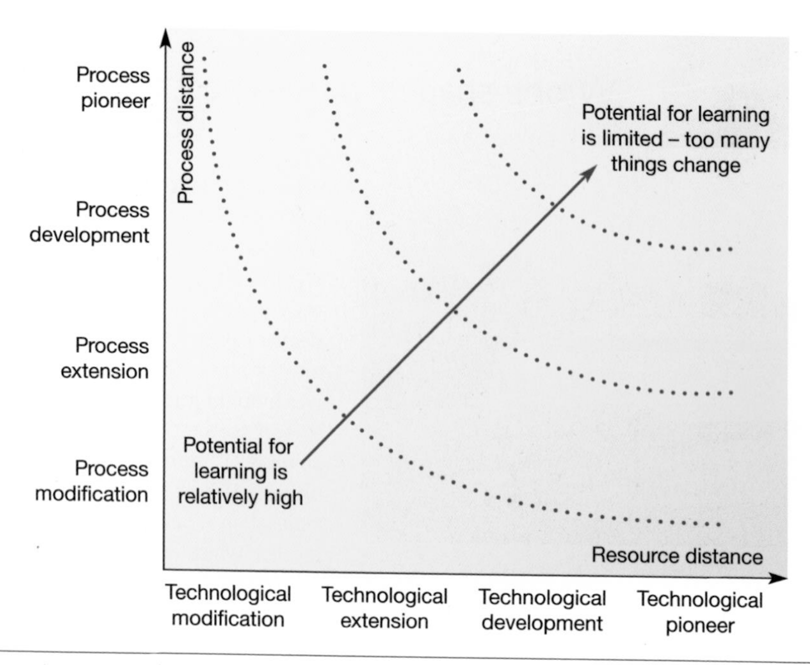

Figure 8.8 Learning potential depends on both technological resource and process 'distance'

Customer acceptability

When an operation's customers interact with its process technology it is essential to consider the customer interaction when evaluating it. If customers are to have direct contact with technology, they must have some idea of how to operate it. Where customers have an active interaction with technology, the limitations of their understanding of the technology can be the main constraint on its use. For example, even some domestic technologies such as smart TVs cannot be used to their full potential by some owners. Other customer-driven technologies can face the same problem, with the important addition that if customers cannot use technologies such as internet banking, there are serious commercial consequences for a bank's customer service. Staff in manufacturing operations may require several years of training before they are given control of the technology they operate. Service operations may not have the same opportunity for customer training.

Walley and Amin[10] suggest that the ability of the operation to train its customers in the use of its technology depends on three factors: complexity, repetition and the variety of tasks performed by the customer. If services are complex, higher levels of 'training' maybe needed: for example, the technologies in theme parks and fast-food outlets rely on customers copying the behaviour of others. Frequency of use is important because the payback for the 'investment' in training will be greater if the customer uses the technology frequently. Also, customers may, over time, forget how to use the technology, but regular repetition will reinforce the training. Finally, training will be easier if the customer is presented with a low variety of tasks. For example, vending machines tend to concentrate on one category of product, so that the sequence of tasks required to operate the technology remains consistent. In other cases, the technology may not be trusted by customers because it is technology and not a person. Sometimes we prefer to put ourselves in the care of a person, even if their performance is inferior to a technology. For example, the use of robot technologies in surgery has distinct advantages over conventional surgery, but in spite of the fact that the surgeon is in control, it is viewed with suspicion by some patients and doctors. When robot surgeons operate without any direct human control, rather than simply mirroring the movement of human surgeons, resistance is likely to be even greater. If, during the early adoption of a technology, there is some accident involving people, the resulting publicity can increase 'customer' resistance. For example, see the 'Operations in practice' example, 'Rampaging robots'.

OPERATIONS IN PRACTICE

Rampaging robots[11]

It is not a big problem (at least not at the moment), but it could become one as robot technologies start to mix directly with customers. Robots can be dangerous, and not just in a highly automated factory environment. (Factory robots can be dangerous. In 2015 a factory worker at a Volkswagen factory was picked up and killed by a robot. He was installing it when he was lifted up by the robotic arm and crushed against a metal plate, suffering fatal chest injuries.) It is the introduction of robotic technologies into the customer environment that could give rise to new areas of reputational risk for companies. For example, in 2016, a robot that was intended to guard against shoplifters accidentally ran over a 16-month-old boy at a shopping centre in Palo Alto, California – ironically, a town famous for high-tech industries. The 130 kg robot, which looks like R2-D2 from *Star Wars*, apparently did not sense that the child had fallen in its path and failed to stop before they collided. According to the boy's mother, '*The robot hit my son's head and he fell down — facing down — on the floor, and the robot did not stop and it kept moving forward*'.

It is an issue that was causing concern (or discussion) decades ago, before robots existed. The author and visionary Isaac Asimov devised his 'Three Laws of Robotics' to protect humans:

1 Don't hurt a human being, or through inaction, allow a human being to be hurt.
2 A robot must obey the orders a human gives it unless those orders would result in a human being harmed.
3 A robot must protect its own existence as long as it does not conflict with the first two laws.

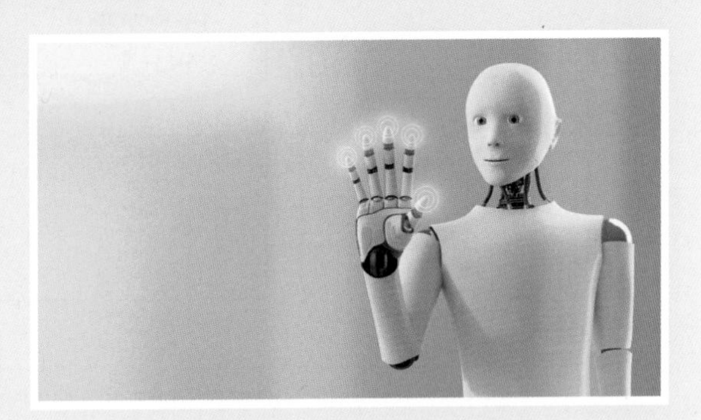

The robot's makers, Knightscope, said the incident was 'absolutely horrifying' and that the company would apologise directly to the family. It also pointed out that its fleet of similar robots had covered 25,000 miles on patrol duty and there had never been an incident like this before. Nevertheless, the shopping centre said it would take the robot out of service temporarily.

Other problems that have been raised by companies fearing legal liability and reputational risk include domestic devices like robot vacuum cleaners hurting pets or humans. A South Korean woman was sleeping on the floor when her robot vacuum 'ate' her hair. Also some 'automated' services could lead to customers confusing what's real and what isn't, resulting in customers revealing more than they intended. For example, 'Invisible Boyfriend' is a service that, for a monthly fee, sends 'pretend' romantic texts and voicemails to your phone – but not all customers realise it is not fully automated, and that there are human operators involved.

Anticipating implementation problems

The implementation of any process technology will need to account for the 'adjustment' issues that almost always occur when making any organisational change. By adjustment issues we mean the losses that could be incurred before the improvement is functioning as intended. But estimating the nature and extent of any implementation issues is notoriously difficult. This is particularly true because more often than not, Murphy's law seems to prevail. This law is usually stated as, 'if anything can go wrong, it will'. This effect has been identified empirically in a range of operations, especially when new types of process technology are involved. Discussing technology-related change specifically (although the ideas apply to almost any implementation), Bruce Chew of the Massachusetts Institute of Technology[12] argues that adjustment 'costs' stem from unforeseen mismatches between the new technology's capabilities and needs and the existing operation. New technology rarely behaves as planned and as changes are made their impact ripples throughout the organisation.

Figure 8.9 is an example of what Chew calls a Murphy curve. It shows a typical pattern of performance reduction (in this case, quality) as a new process technology is introduced. It is recognised that

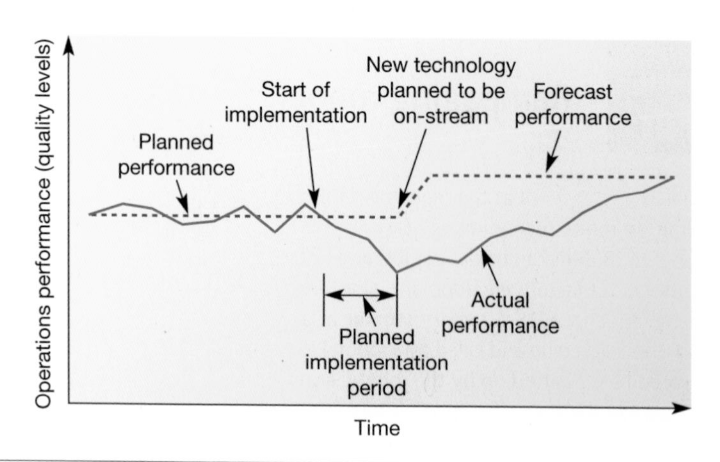

Figure 8.9 The reduction in performance during and after the implementation of a new process reflects 'adjustment costs'

implementation may take some time; therefore, allowances are made for the length and cost of a 'ramp-up' period. However, as the operation prepares for the implementation, the distraction causes performance actually to deteriorate. Even after the start of the implementation this downward trend continues and it is only weeks, indeed maybe months, later that the old performance level is reached. The area of the dip indicates the magnitude of the adjustment costs, and therefore the level of vulnerability faced by the operation.

Responsible operations

In every chapter, under the heading of 'Responsible operations', we summarise how the particular topic covered in the chapter touches upon important social, ethical and environmental issues.

New process technology is never a neutral issue. From the Luddites who reacted violently against the introduction of new mechanical looms in the wool industry to recent privacy campaigners pressing for greater control of online data, there have always been periods of 'disruptive' technological change, and during such periods there have always been corresponding anxieties. The typical response to new technologies is to highlight these anxieties: the human body cannot long survive the speed of trains, television will prevent us from going out for entertainment and so on. Yes, some concerns are real. We have already examined one of the most significant issues in the 'Operations in practice' example of 'Technology or people? The future of jobs', but this is certainly not the only issue that operations managers face. In fact, one of the implications of the increased prevalence and power of new technologies is the widespread extent of their impact. The diagram in Figure 8.10 is by no means comprehensive, but it does indicate the range of ethical concerns connected to how technologies are used.

Process technology affects suppliers. As information technologies become more integrated with suppliers' systems, whatever changes are made by one operation will need coordinating with its suppliers. So, in addition to considering the internal implications of, say, how the *Internet of Things* (IoT) can be prone to cyberattacks, on one's own operation, it becomes morally necessary to consider others in the supply network, including suppliers.

Process technology affects staff. Very obviously, if their jobs are replaced by technology, but also in many other ways. One of the most contentious issues concerns the potential of newer technologies to monitor staff performance (explored in the next chapter). An increasing concern is how people are affected when they have to make decisions in conjunction with technology.

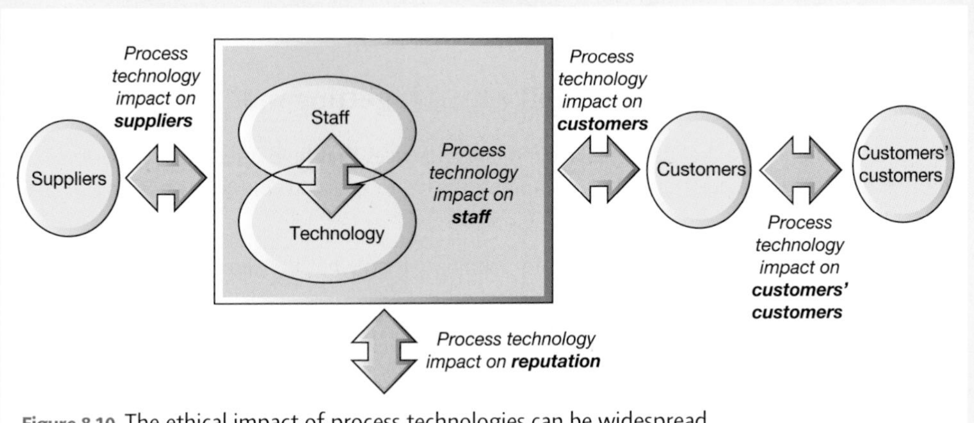

Figure 8.10 The ethical impact of process technologies can be widespread

For example, will clinicians who rely on artificial intelligence-driven diagnostics become complacent? Will lawyers who rely on 'big data' systems to draw up contracts lose their skills, or miss important issues?

Process technology affects customers. Increasingly the preferred (and often, only) method of communicating between an enterprise and its customers is through technology. The internet has become the default medium of information exchange. It is this pivotal role of information and data in customers' (in the broadest sense of the word) lives that raises obvious, and well-debated, issues of data privacy. Some critics of businesses that capture such data argue that it is repackaged as 'prediction products', which are then sold in 'behavioural futures markets'. In effect, this makes all of us the raw material of these products, which are then sold to other companies in closed business-to-business markets.[13] Another, much-debated, issue is the extent to which social media companies have a responsibility to suppress the spread of fake news on their platforms.

Process technology affects customers' customers. When a business's products or services are themselves process technologies, there are ethical issues in how much responsibility one should have for how one's customers use the technology. For example, a payment systems company could integrate an algorithm in its systems that predicts the likelihood of a person defaulting on their payments. This may be intended to be used in conjunction with other information, but how can the developing company ensure that its customers are using the algorithm as intended? A more obvious example is that of face-recognition systems. There are some relatively non-controversial applications of such technology (security on one's laptop), but there are also less straightforward ones. Should a face-recognition system developer supply a customer who they suspect might use it unethically? Should they supply authoritarian states? Should they supply law enforcement agencies before their technology has proved to be unbiased? Or at all?

Process technology affects reputations. The examples of ethical issues related to technology that are cited above are by no means comprehensive. There are many that we have not mentioned, and many that have not yet emerged. All of them have the potential to impact on an organisation's reputation. Arguably there are only two choices facing operations managers. Either avoid using any type of technology that may be reputationally risky, or try to manage the risk. If the latter, it requires that ethical issues are recognised and debated, and a high degree of transparency be encouraged.

Summary answers to key questions

8.1 What is process technology and why is it getting more important?

- Process technologies are the machines, equipment or devices that help operations to create or deliver products and services. Indirect process technology helps to facilitate the direct creation of products and services.
- Most new process technologies have a greater capability than what they are replacing, and many new technologies can be applied in all types of operations.
- One common method of distinguishing between different types of process technology is by what the technology actually processes – materials, information or customers.

8.2 How can one understand the potential of new process technology?

- Operations managers do not need to know the technical details of all technologies, but they do need to know the answers to four key questions: What does it do? How does it do it? What advantages does it give? What constraints does it impose?
- It can be difficult to separate the reality of a new technology from the publicity and speculation that surrounds it. The 'Gartner Hype Cycle' attempts to illustrate this using five sequential stages.
- An understanding of a technology calls for a knowledge of its 'primary capability'. That is, what is it better at than the technology it replaces? For example, its ability to think or reason, see or sense, communicate or connect, move physical objects, or process materials.

8.3 How can new process technologies be evaluated?

- All technologies should be appropriate for the volume–variety characteristics of the task for which they are intended.
- All technologies should be evaluated by assessing the impact that the process technology will have on the operation's performance objectives (quality, speed, dependability, flexibility and cost) and other operational factors.
- All technologies should be evaluated financially. This usually involves the use of some of the more common evaluation approaches, such as net present value (NPV).

8.4 How are new process technologies developed and implemented?

- Implementing process technology means organising all the activities involved in making the technology work as intended.
- A technology roadmap (TRM) is an approach that provides a structure that attempts to assure the alignment of developments (and investments) in technology, possible future market needs, and the new development of associated operations capabilities.
- The resource and process 'distance' implied by the technology implementation will indicate the degree of difficulty.
- Customer acceptability may be a barrier to implementation in customer-processing technologies.
- It is necessary to allow for the adjustment costs of implementation.

CASE STUDY Logaltel Logistics

This case was co-authored with Vaggelis Giannikas, at the University of Bath School of Management

It was two months since Thalia had joined Logaltel, a third-party logistics (3PL) provider owning and operating five large warehouse bases across the United Kingdom (each hosting three to four warehouses). As the new Chief Technology Officer, she had been asked to make a recommendation to the executive board about the best way to invest in new warehouse technology. Investment that potentially could fundamentally change both the operations and the image of the company. *'It was made clear to me that, while the senior team very much supported investing in new technology, it did not want to take any unnecessary risks to the ongoing service provided to our clients'* (Thalia).

Logaltel delivered inventory management services for its clients, who operated in two types of market. Around 80 per cent of the company's revenue came from the clients that served business-to-business (B2B) markets. These clients required Logaltel to receive, store and ship their products stored on pallets. The other 20 per cent (which was growing really fast) came from those same business clients that also served business-to-consumer (B2C) markets. Instead of dealing in pallet loads, this service required Logaltel to pick (assemble) relatively small orders (sometimes containing just a few items or boxes) that needed to be shipped directly to end customers. This rapidly expanding B2C business had resulted in most local warehouse managers assigning separate areas of each warehouse to process B2C orders, although the majority of warehousing space was still assigned to serving B2B operations. The increased complexity of B2C operations had also prompted the development of an upgrade to the firm's warehouse management system (WMS). This is the software system responsible for planning the receiving, storage, picking and shipping operations, the first phase of which had been introduced very recently. It had boosted the performance of both B2B and B2C operations but had been a long and expensive development and implementation process.

New warehouse technology

Under consideration were various innovative technologies that had the potential to automate Logaltel's currently predominantly manual operation for picking goods that used a combination of forklifts, picking carts and staff to pick the items of an order. After a review of available technologies, Thalia shortlisted her two preferred technologies, both involving types of automated guided vehicles (AGVs). One technology was appropriate for B2B operations and used driverless forklifts that could store, pick and move pallets around a warehouse. These could easily operate in the existing warehousing setting, requiring only minor modifications to a warehouse as they interfaced very well with common material-handling systems (storage shelves, pallets, etc.) often found in the industry.

The second technology was more suitable for automating B2C operations. This mobile robotic fulfilment system enabled workers to avoid the time-consuming and costly activity of walking around the warehouse collecting individual items for customer orders. Instead, workers could stand in a predefined area while AGVs carried shelves full of items to them. Although offering significant efficiencies, this technology would require a complete restructuring of B2C warehouse areas because it fundamentally changed the way items were stored and transported. Moreover, there were limitations in the size of items that could be stored on the specialised shelves the AGVs moved around, meaning that traditional operations had still to be used alongside automated ones for the technology to be practically applicable.

Both technologies had the potential to significantly reduce labour costs and improve productivity. According to Logaltel's Chief Financial Officer, both would give a financial return well within the firm's minimum requirements based on expected growth projections. The issue for Thalia was whether to recommend adopting both technologies immediately, or start with implementing just one of them, and if so, which one? Moreover, should implementation start with a pilot project at one site or be adopted throughout the network of warehouses? In an attempt to summarise the different financial factors at play, Thalia's team had helped her put together a table (Table 8.4) with some key information.

In addition, Thalia had spent a lot of her time talking to different stakeholders in the company, trying to get their views on the topic.

Table 8.4 Selected information on the two shortlisted technologies

Criterion		B2B clients		B2C clients	
		Current – manual operations	Proposed technology – driverless forklifts	Current – manual operations	Proposed technology – mobile robotic fulfilment systems
Forecast growth in orders within three years	Worst case	8%		15%	
	Average	10%		35%	
	Best case	12%		45%	
Payback period of new technology investment per growth scenario	Worst case	22 months		30 months	
	Average	18 months		12 months	
	Best case	14 months		10 months	
Scalability		Would require purchasing extra traditional forklifts and hiring more people	Requires purchasing driverless forklifts and integration with new WMS	Would require hiring more people	Requires purchasing extra robots and integration with new WMS
Picking/shipping accuracy		Acceptable	At least as good	Unacceptable	Difficult to estimate, probably significantly better
Warehouse modification needed		None	Minor	Minor	Probably very significant mods needed
Disruption during transition period		None	Little	Minor	Difficult to estimate, would need careful implementation
Compatibility with new WMS		Compatible	Compatible	Some compatibility issues	Difficult to know, could probably be made compatible

The warehouse operations manager

Jamal, a Senior Operations Manager with more than two decades of experience, had his own reservations about investing in the B2C part of the warehouse. *'I have been in the company long enough to know what has always been the thing that actually got us going. For years, this has been corporate clients that ship pallets in and out of the warehouse. I can definitely see a certain rise in e-commerce but how sure are we this is going to last?'*. Moreover, Jamal was reluctant to rely heavily on any automation solution based on his past experience with new technologies: *'All this shiny new stuff, they are simply the new kids on the block. With small exceptions, most of these technologies have failed us time and again. Let alone the health and safety issues they bring with them. It is not that straightforward to have people operating side by side with robots. I say, let's wait for others to use it first. They can prove the technology's value, then we can simply copy what they do. Its's not that I'm against us adopting new tech, but I am sure even better systems will be available in the near future'.*

The business developer

Martha, the firm's Business Development Manager painted a very different picture. She stressed the 'new' of the new technology as she was certain it would help business and offer a positive return on investment at the end of the day. *'This kind of automation is good not only for attracting new customers but also retaining existing ones. Showing around prospective clients in a facility with state-of-the-art equipment often speaks to our ability to offer superior quality services. For existing ones, demonstrating our appetite for innovation shows them that we are always looking for ways to improve our operations and drive down costs that will have an impact on the offering we can provide them. I mean, the financials clearly indicate there is a strong case for adopting both technologies'*. Martha also emphasised the importance of looking at what Logaltel's competition does: *'We can't really avoid adopting these technologies, considering the general market trend. Particularly for the high-growth B2C services, we are already late on this. So, the question is really whether we want to be pioneers and significantly benefit from it, or late adopters behaving like copycats'.*

The union representative

'Believe me, we have this company's best interests in mind when we say we are against the development of these automation technologies', said Vimal, representing his colleagues at the warehouse. *'Not only have these technologies yet to prove themselves but they also fundamentally question the role of our warehouse workers and the value they bring to this company. We are proud to have created an environment that makes people want to stay at Logaltel and give their best selves to the company. Replace that with robots and the company's culture goes into the bin. There is no way you can replace people's creativity and flexibility with automation, and we all know Logaltel operates in a business environment where both are highly valued. I can't see why such a big investment needs to go forward without any testing or evidence it works. I have to admit, I don't know much about the carbon footprint of these technologies, but it seems like charging those robots all the time would require way too much energy. It makes me wonder how this speaks to our sustainability agenda'.*

Thalia knew that getting the union on her side on this topic would be very challenging and was very worried that without the support of warehouse staff, any new technology would be difficult to make work effectively.

With all this in mind and trying to ensure different stakeholders' views were taken into consideration, Thalia was struggling to make up her mind. On top of that she was seriously concerned about the future of a company she wanted to build a career in. On one hand, adopting these new technologies could be the first step towards creating one of the first truly 'dark warehouses' in the UK, a prospect that sounded really promising in her head. On the other hand, the type of business it might be in 10 years down the line considering its 3PL nature could be significantly different to the one it is now. What if the technology is not capable of dealing with new business requirements then? Would it all go to waste, taking her reputation with it?

QUESTIONS

1 What are Thalia's options?

2 How would you evaluate these different options?

3 What do you think her recommendation to the board should be?

Problems and applications

All chapters have 'Problems and applications' questions that will help you practise analysing operations. They can be answered by reading the chapter. Model answers for the first two questions can be found on the companion website for this text.

1 It is a new job, as yet without a formal title, but one commentator has called it being a 'robot wrangler'. They even proposed a possible job advert: 'Wranglers wanted for growing fleets of robots. Your responsibilities will include evaluating robot performance, providing real-time analysis and support for problems. You must be analytical, detail-oriented, friendly – and ready to walk. No advanced degree required'. Elisabeth Reynolds at the Massachusetts Institute of Technology also sees a future for people overseeing robots. *'We use that term "autonomous" a lot when we think about robots, but in fact very few robots are purely autonomous'*, she says.[14] Actual job adverts for this type of job use terms like technicians, monitors, handlers and operations specialists. Journalists have described the role as anything from robot chauffeurs to robot babysitters. Why would such a job be necessary? Isn't the role of new technologies to replace humans?

2 Modern aircraft fly on automatic pilot for most of the time. Most people are blissfully unaware that when an aircraft lands in mist or fog, it is a computer that is landing it. When autopilots can do something better than a human pilot, it makes sense to use autopilots. They can take control of the plane during the long and (for the pilot) monotonous part of the flight between take-off and landing. They can also make landings, especially when visibility is poor because of fog or light conditions. In fact, automatic landings when visibility is poor are safer than when the pilot is in control. On some

flights, the autopilot is switched on within seconds of the aircraft wheels leaving the ground and then remains in charge throughout the flight and the landing. As yet, commercial flights do not take off automatically, mainly because it would require airports and airlines to invest in extra guidance equipment, which would be expensive to develop and install. Also, take-off is technically more complex than landing. Yet some in the airline industry believe that technology could be developed to the point where commercial flights can do without a pilot on the aircraft entirely. If it was developed, what would be the problems and benefits associated with introducing this type of technology?

3 The 'robot milkmaid' can milk between 60 and 100 cows a day. Computer-controlled gates activated by transmitters around the cows' necks allow the cows to enter. The machine then checks their health, connects them to the milking machine and feeds them while they are being milked. If illness is detected in any cow, or if the machine for some reason fails to connect the milking cups to the cow after five attempts, automatic gates divert the cow into a special pen where the farmer can inspect it later. Finally, the machine ushers the cows out of the system. It also self-cleans periodically and can detect and reject any impure milk. Rather than herding all the cows in a 'batch' to the milking machine twice a day, the system relies on the cows being able to find their own way to the machine. Once they have been shown the way to the machine a few times, they go there of their own volition. The cows may make the journey to the machine three or more times per day.

   (a) What advantages do you think this technology gives?
   (b) Do you think the cows mind?
   (c) There is some anecdotal evidence that farmers still go to watch the process. Why do you think this is?

4 The Boeing 737 MAX was grounded in 2019 following two crashes. A Lion Air 737 MAX crashed killing 189 people; a few months later, a second, operated by Ethiopian Airlines, crashed, leaving no survivors. To blame was a new flight-control feature, Maneuvering Characteristics Augmentation System (MCAS). It was designed to prevent the aeroplane's nose from getting dangerously high by automatically lowering it, but under certain conditions, the MCAS lowered the nose so much that pilots found it difficult to maintain control. What does this tell us about the problems with automating 'people-processing' technologies?

5 Process technology can impact all of the operations performance objectives (quality, speed, dependability, flexibility and cost). Think through, and identify, how process technology could affect these performance objectives in the airline industry.

6 There have been a number of changes in medical process technology that have had a huge impact on the way healthcare operations manage themselves. In particular, **telemedicine** has challenged one of the most fundamental assumptions of medical treatment – that medical staff need to be physically present to examine and diagnose a patient. No longer; web-connected devices are now able to monitor an individual's health-related data and communicate the information to healthcare professionals located anywhere in the world. Medical staff are alerted to changing conditions as they occur, and provided with a status report of a person's health so that the appropriate care can be administered. Telemedicine generally refers to the use of communications and information technologies for the delivery of clinical care. Formally, telemedicine is the ability to provide interactive healthcare utilising modern technology and telecommunications. It allows patients to virtually 'visit' with doctors, sometimes live using video links, sometimes automatically in the case of an emergency, sometimes where patient data are stored and sent to doctors for diagnosis and follow-up treatment at a later time. What do you think are the implications of telemedicine for how healthcare operations can be managed?

7 What are known as 'care robots' are being used in Japanese nursing homes to interact with people for social and therapeutic purposes, including dementia patients. What value do you think such technology is adding? And why Japan?

8 Artificial intelligence (AI) is starting to be used for military applications such as making decisions in fighter planes. What do you see as the practical and ethical implications of this?

9 Of all sectors of the economy, robots were particularly slow to be adopted in farming, and agriculture generally. Why do you think that was?

10 Re-read the 'Operations in practice' example on 'Bionic duckweed'. Under what circumstances would the criticism of 'always waiting for the next breakthrough stopping you from taking action now' be misplaced?

Selected further reading

Arthur, W.B. (2010) ***The Nature of Technology: What It Is and How It Evolves*, Penguin, Harmondsworth.**
Popular science in a way, but very interesting on how technologies evolve.

Boden, M.A. (2018) ***Artificial Intelligence: A Very Short Introduction*, Oxford University Press, Oxford.**
A very accessible examination of the philosophical and technological challenges raised by Artificial Intelligence.

Brynjolfsson, E. and Mcafee, A. (2014) ***The Second Machine Age: Work, Progress, and Prosperity in a Time of Brilliant Technologies*, W.W. Norton, New York, NY.**
This is one of the most influential recent books on how technology will change our lives.

Chew, W.B., Leonard-Barton, D. and Bohn, R.E. (1991) Beating Murphy's Law, ***MIT Sloan Management Review*, 32 (3) (Spring), 5–16.**
One of the few articles that treats the issue of why everything seems to go wrong when any new technology is introduced. Insightful.

Christensen, C.M. (2016) ***The Innovator's Dilemma: When New Technologies Cause Great Firms to Fail*, Harvard Business Review Press, Harvard, MA.**
The latest published version of a classic.

Tapscott, D. and Tapscott, A. (2016) ***Blockchain Revolution: How the Technology Behind Bitcoin and Other Cryptocurrencies is Changing the World*, Portfolio Penguin, London.**
A readable book outlining the uses and implications of Blockchain, but broad and generic rather than detailed.

Tegmark, M. (2018) ***Life 3.0: Being Human in the Age of Artificial Intelligence*, Penguin, London.**
An intelligent but accessible treatment.

Notes on chapter

1. The information on which this example is based is taken from: Economist (2020) Businesses are finding AI hard to adopt, *Economist Technology Quarterly*, 11 June; Economist (2016) Artificial intelligence and Go: Showdown, *Economist*, 12 March; Koch, C. (2016) How the computer beat the Go master, *Scientific American*, 19 March.
2. Economist (2020) Artificial intelligence and its limits, *Economist* Technology Quarterly, 13 June.
3. For further information of this subject, see: OECD (2018) Automation, skills use and training, OECD Social, Employment and Migration Working Papers, OECD, ISSN: 1815199X; Economist (2014) The future of jobs: the onrushing wave, *Economist*, 18 January; Economist (2013) Schumpeter: the age of smart machines, *Economist*, 25 May; Finkelstein, D. (2013) Machines are becoming cheaper than labour, *The Times*, 6 November; Groom, B. (2014) Automation and the threat to jobs, *Financial Times*, 26 January; Frey, C.B. and Osborne, M.A. (2013) The future of employment: how susceptible are jobs to computerisation? Oxford Martin School Working Paper, 1 September; Brynjolfsson, E. and Mcafee, A. (2014) *The Second Machine Age: Work, Progress, and Prosperity in a Time of Brilliant Technologies*, W.W. Norton, New York.

4. The information on which this example is based is taken from: QB House website, http://www.qbhouse.com (accessed September 2021).
5. SAP IOT Definition: SAP Research, https://insights.sap.com/what-is-iot-internet-of-things/ (accessed September 2021).
6. The information on which this example is based is taken from: West, K. (2011) Turn up the heat with Marmite, *The Sunday Times*, 2 October, https://www.thetimes.co.uk/article/turn-up-the-heat-with-marmite-vz8d87qx253 (accessed September 2021).
7. The information on which this example is based is taken from: Harford, T. (2020) Why tech isn't always the answer – the perils of bionic duckweed, *Financial Times*, 30 October.
8. The information on which this example is based is taken from: Walsh, D. (2015) Irregular parcels put UK Mail out of shape, *The Times*, 8 August; UK Mail website https://www.ukmail.com.
9. Dosi, G., Teece, D. and Winter, S.G. (1992) Towards a theory of corporate coherence, in Dosi, G., Giametti, R. and Toninelli, P.A. (eds) *Technology and Enterprise in a Historical Perspective*, Oxford University Press, Oxford.
10. Walley, P. and Amin, V. (1994) Automation in a customer contact environment, *International Journal of Operations and Production Management*, 14 (5), 86–100.
11. The information on which this example is based is taken from: Deng, B. (2016) Security robot runs over toddler at shopping centre, *The Times*, 15 July; Times Leader (2016) They, robots, *The Times*, 1 January; Hall, A. (2015) Factory robot grabs worker and kills him, *The Times*, 3 July.
12. Chew, W.B., Leonard-Barton, D. and Bohn, R.E. (1991) Beating Murphy's Law, *MIT Sloan Management Review*, 32 (3) (Spring), 5–16.
13. Zuboff, S. (2019) *The Age of Surveillance Capitalism: The Fight For a Human Future at the New Frontier of Power*, Profile Books, London.
14. Sherman, N. (2018) Wanted: robot wrangler, no experience required, BBC News, 21 March, https://www.bbc.co.uk/news/business-43259903 (accessed September 2021).

9 People in operations

KEY QUESTIONS

9.1 Why are people so important in operations management?

9.2 How can the operations function be organised?

9.3 How do we go about designing jobs?

9.4 How are work times allocated?

INTRODUCTION

Operations management is often presented as a subject, the main focus of which is technology, systems, procedures and facilities – in other words, the non-human parts of the organisation. This is not true, of course. On the contrary, the manner in which an organisation's human resources are managed has a profound impact on the effectiveness of its operations function. In this chapter we look especially at the elements of human resource management that are traditionally seen as being directly within the sphere of operations management. These are strategic and cultural issues, organisation design, job design and the allocation of 'work times' to operations activities. The more detailed (and traditional) aspects of these last two elements are discussed further in the supplement on work study at the end of this chapter. Figure 9.1 shows how this chapter fits into the overall model of operations activities.

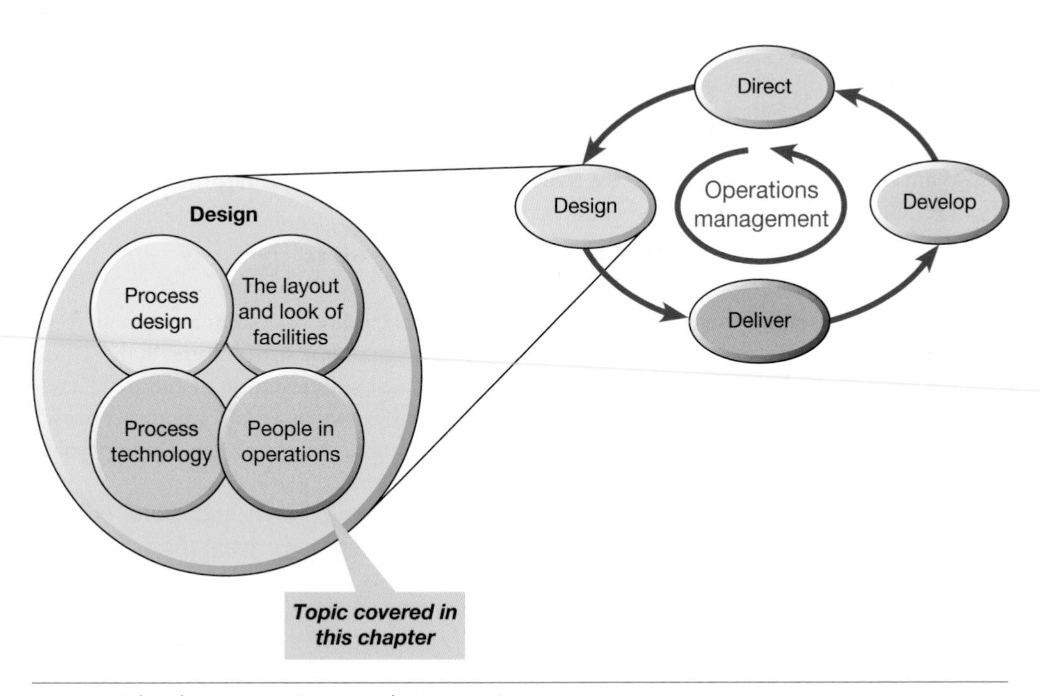

Figure 9.1 This chapter examines people in operations

9.1 Why are people so important in operations management?

To say that an organisation's human resources are its greatest asset is something of a cliché. Yet it is worth reminding ourselves of the importance of the abilities, attitudes and culture of the people who make up the operations function. It is, after all, where most 'human resources' are to be found. It follows that it is operations managers who are most involved in the leadership, development and organisation of people. But the influence of operations management on the organisation's staff is not limited to the topics that are covered in this chapter. Almost everything discussed in this text has a 'people' dimension. Yet, in some chapters, the human perspective is particularly important. In addition to this chapter, Chapters 15, 16 and 17, for example, are concerned largely with how the contribution of the operation's staff can be harnessed. In essence, the issues covered in this chapter define how people go about their working lives. It positions their expectations of what is required of them, and it influences their perceptions of how they contribute to the organisation. It defines their activities in relation to their work colleagues and it channels the flows of communication between different parts of the operation. But, of most importance, it helps to develop the culture of the organisation – its shared values, beliefs and assumptions.

Operations principle

Human resources aspects are especially important in the operations function, where most 'human resources' are to be found.

Operations culture

What do we mean by culture in the context of the operations function? There is a wealth of academic and popular literature that treats the concept of organisational culture, but no single authoritative definition has emerged. Nevertheless, most of us know roughly what is meant by 'culture' in an organisation. It is what it feels like to be part of it. What is assumed about how things get done rather than what is necessarily formally articulated. It is, in the words of one well-known writer on the subject, 'the way we do things around here'.[1] But the idea of 'organisational' culture can also apply to a single function like the operations function. In fact, there is considerable interest among researchers and practitioners in overcoming the cultural differences between different functions that can sometimes lead to what has been called 'cultural fragmentation'. Even though there may be elements of an organisation's culture that are shared across all parts of the enterprise, different functions are very likely to have their own subcultures.

The iceberg metaphor

A common metaphor for understanding organisational culture is to compare it to an iceberg.[2] Most of an iceberg's mass is under the surface of the ocean. And so it is with culture. In the 'culture iceberg' the physical artefacts are the outward or surface physical manifestation of culture. These are things like premises, the physical elements of a service environment (think about servicescapes from Chapter 7), office chairs, computer equipment, staff uniforms and so on. These physical artefacts express some aspects of culture, both to customers and service workers.

The middle layer of the iceberg represents the intangible beliefs and values of the operation. Such intangibles often cascade from the beliefs and values of the organisation's founder or leaders and are passed on to staff through formal instruments such as mission statements or strategy documents, but also by example, performance management systems and in particular reward systems.

The deepest levels of the iceberg in the depths of the ocean are the underlying values, assumptions, beliefs and expectations of the organisation (these often correspond with the values of the founder or leader). This deep level is where one finds the most fervently held values and assumptions. Changing these underlying values assumptions, beliefs and expectations is particularly difficult. Moreover, even if they can be changed, one risks anxiety and defensive behaviours on the part of those being required to change. It is argued that the physical artefacts of the service as well as the intangible activities and

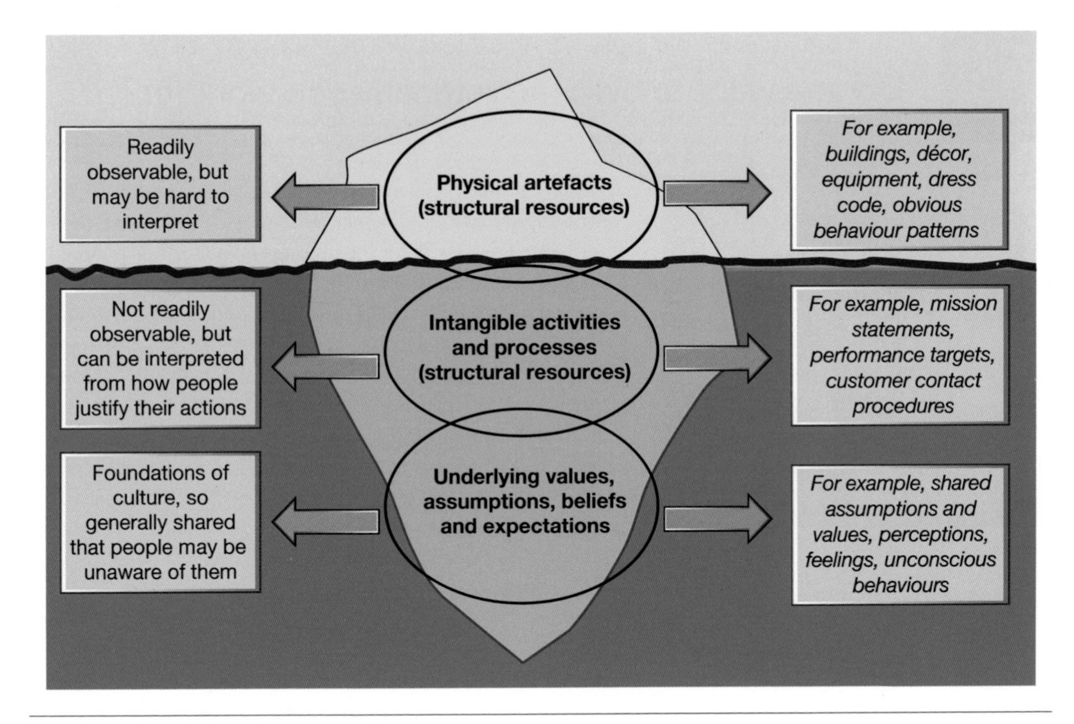

Figure 9.2 The iceberg metaphor of culture

routines cannot be adequately explained unless one delves down into the basic assumptions at the base of the iceberg. The iceberg metaphor is illustrated in Figure 9.2.

Believe, know and behave

Culture is difficult to explain. As was said of one organisation with a particularly strong culture (a university as it happens), 'From the outside looking in, you can't understand it. From the inside looking out, you can't explain it'. As far as the operations function is concerned, it is best summed up by what the operations team *believes*, what they *know* and how they *behave*. It is these three elements of operations culture, belief, knowledge and behaviour, which provide the foundations for how it contributes to the business and how capable it is to improve over time.

- ***What operations believe*** – By 'operations belief', we mean what the people within the operations function accept as self-evident. For example, do operations believe that they have a responsibility to fully understand all other functions' strategies and their implications for operations? Do they develop capabilities within their operations resources and processes that offer a unique and long-lasting strategic advantage?
- ***What operations should know*** – What should the operations team know? Obviously, they should understand the underlying principles that govern how operations and processes work. Only with a thorough understanding of the objectives, concepts, tools and techniques of operations management will the operations function ever contribute fully to the success of any business.
- ***How operations should behave*** – The way operations managers should behave is not fundamentally different from any effective manager. The popular and academic literature has for decades been full of 'key behaviours' for effective leadership, and they don't seem to have changed much for years. 'Don't micromanage your team, empower them while still being available for advice'. 'Be a coach to your team'. 'Be clear and results-oriented but help the team to see how they can achieve them'. 'Have a clear vision and strategy'. 'Always communicate, both ways – and listen to your team'. 'Support open discussion and listen to the team's concerns'. All of these might be obvious, but they make good managerial sense.

OPERATIONS IN PRACTICE

Do you want to own the company you work for?[3]

Torchbox, is a web design and development company based in the United Kingdom in Oxfordshire, Bristol and Cambridge. Employing around 70 people, it provides 'high-quality, cost-effective, and ethical solutions for clients who come primarily, but not exclusively, from the charity, non-governmental organisations and public sectors. The company was founded after Tom Dyson, now the Technical Director, met Olly Willans, now the Creative Director, while working at OneWorld.org, an unusual and progressive web design business, which made some of the first websites for organisations like Oxfam and Christian Aid. Tom and Olly quickly realised that their shared passion, of wanting to drive positive change, warranted a go at starting a digital agency. Thanks to its creativity and technical success, the firm grew to attract a portfolio of respected non-profit clients. By the late 2000s, they were working for the kinds of clients they had dreamed of when they started: Greenpeace, the International Red Cross, WWF. They also had managed to recruit a group of talented, enthusiastic staff and long-term associates, not only in their offices, but also further afield including the United States, Hungary, the Philippines, Greece and India. One particular success, said Tom Dyson, was when the Royal College of Art, asked them to build a new content management system. They called it Wagtail, after the sweet little birds that congregate on the lawn outside their Oxfordshire office. Within a year or two, Wagtail was much more famous than Torchbox.

In the later years of the 2010s, Olly and Tom (now both in their 40s) started thinking about what might be next for Torchbox and themselves. The standard 'exit' for business owners is what is called a trade sale – selling your company to a bigger one, for a cash fee and a clause where the current owners work for the business for a specified period. In the United Kingdom there are very few independently owned digital agencies the size of Torchbox, so they would have been an attractive acquisition for a larger company. Yet, neither Tom nor Olly could imagine a large company such as an advertising agency, or a big tech business, making a credible commitment to maintaining Torchbox's ethical focus, which was the primary reason that most of their staff had joined the company. Their solution was, in their words, not to sell up, but to sell down. On Employee Ownership Day in 2019, they handed over the ownership of Torchbox to its employees. The agreement was that the new owners (the employees) would pay the original shareholders, not from their own money but from the company's future profits, over the next four or five years. Said Tom, *'Olly and I started Torchbox, but its successes have all been down to the wonderful, creative, thoughtful people who work here. We're very happy that they are the new owners. We'll continue to push Torchbox forward, doing the jobs we love until then, and perhaps much longer, if they'll let us. They've elected a board of employee trustees who will oversee our work, and who have ultimate control over the business. When they've paid off the original sellers, they can decide to distribute all the profits between themselves, or give it to charity, or invest in something new, or work four-day weeks, or whatever they decide.'*

Marketing Manager, Lisa Ballam, recognised that the new ownership implied some significant changes for her and all employees. *'There isn't a rule book (yet) for becoming employee owned, or a guide to the best approach for moving forward; so we're still finding our feet, testing and learning, to discover what works best for us. It's quite a cultural shift to go from having two people (plus a senior management team) in charge, to having a business collectively owned. Making decisions effectively and decisively is vital in a fast-moving business environment, so our company is still run day-to-day by our senior management team, but our elected board of employee trustees are now consulted on all important strategic decisions. Employees are also able to share opinions and commit to initiatives that matter to them and work towards shared common goals through elected voice groups'.*

Operations are socio-technical systems

The idea that operations management is primarily concerned with 'technical' issues is as widespread as it is false. Any practising operations manager will confirm both the importance of people-related issues and the high proportion of their time devoted to them. Effective operations management demands both 'technical' and human understanding. One well-established model for describing this idea is that of

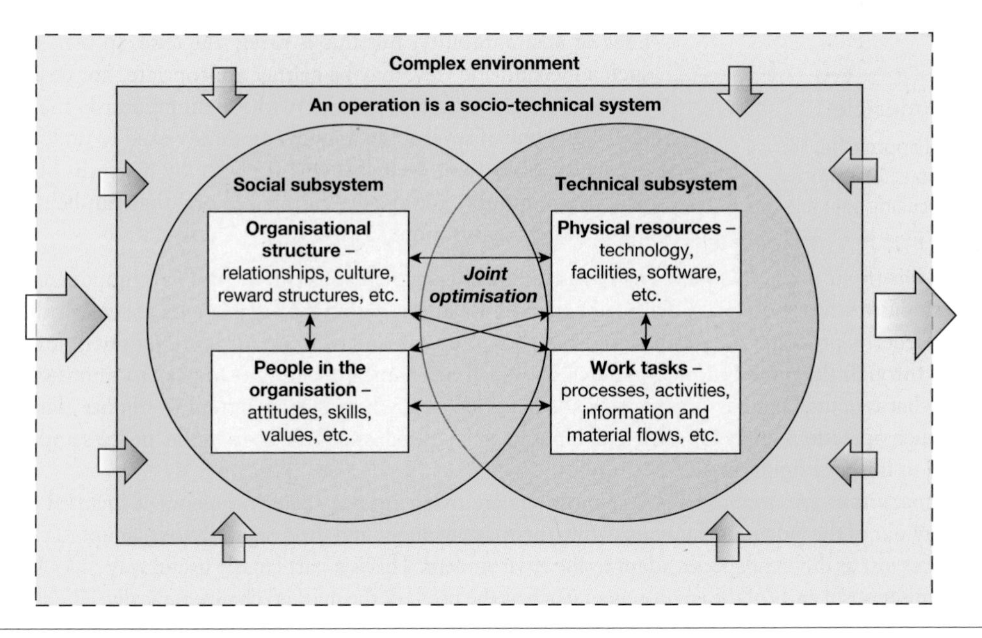

Figure 9.3 The socio-technical approach requires that social and technical aspects of the operation are given consideration, as is the relationship between them

socio-technical systems (STS). The socio-technical systems approach is a way of thinking about complex organisations, such as most operations, that acknowledges the interaction between people and technology within an organisation, and between both and the complex environment in which they operate. The general form of socio-technical systems is illustrated in Figure 9.3. What is often termed the 'social' subsystem includes the formal and informal organisational structures within an operation, as illustrated by reporting relationships, lines of responsibility and reward structures, as well as the characteristics of the people within the operation such as their attitudes, skills and values. The 'technical' subsystem is defined in a wider sense than is normal in operations management. It includes the technology that is used to transform inputs into outputs, but also embraces software, knowledge, facilities layout, processes and the flow of transformed resources through the operation.

Crucially, a socio-technical approach requires consideration of all aspects of both the social and technical subsystems in order to achieve 'joint optimisation'. The term was first used by researchers at the Tavistock Institute in London. It was noted that potential innovative technical changes would often fail because the implications for the people involved (the social subsystem) were not sufficiently considered. Priority, they said, should therefore be given to systems design that properly considers the complex relationships between technology, the organisation, people, business processes and systems supporting these processes. The core principle of the socio-technical approach is that improvements to operations and their systems can only be fully achieved if both the social and technical elements are treated as interdependent, because changes to one subsystem will necessitate changes to the other.

9.2 How can the operations function be organised?

The issues of 'how the operations function should be organised relative to the rest of the enterprise' is clearly one that is broader than the operations function itself. It is a decision for the whole firm, yet it is important for operations because it defines the function's internal position and relationships. But first, it is useful to look at how 'organisations' can be described.

Perspectives on organisations

How we illustrate organisations says much about our underlying assumptions of what an 'organisation' is and how it is supposed to work. For example, the illustration of an organisation as a conventional 'organogram' implies that organisations are neat and controllable with unambiguous

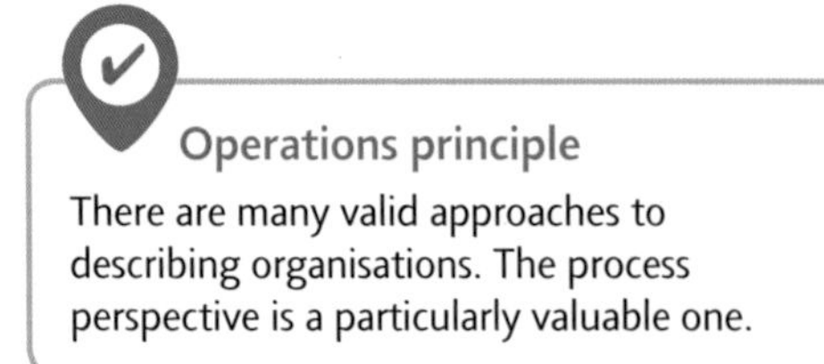
Operations principle
There are many valid approaches to describing organisations. The process perspective is a particularly valuable one.

lines of accountability. But this is rarely the case. In fact, taking such a mechanistic view may be neither appropriate, nor desirable. Seeing an organisation as though it was unambiguously machine-like is just one of several metaphors commonly used to understand organisations. One well-known analysis by Gareth Morgan proposes a number of 'images' or 'metaphors' that can be used to understand organisations, as follows:[4]

- ***Organisations are machines*** – The resources within organisations can be seen as 'components' in a mechanism whose purpose is clearly understood. Relations within the organisation are clearly defined and orderly, processes and procedures that should occur usually do occur, and the flow of information through the organisation is predictable. Such mechanical metaphors appear to impose clarity on what is actually messy organisational behaviour. But, where it is important to impose clarity (as in much operations analysis), such a metaphor can be useful, and is the basis of the 'process approach' used in this and similar texts.
- ***Organisations are organisms*** – Organisations are living entities. Their behaviour is dictated by the behaviour of the individual humans within them. Individuals, and their organisations, adapt to circumstances just as different species adapt to the environment. This is a particularly useful way of looking at organisations if parts of the environment (such as the needs of the market) change radically. The survival of the organisation depends on its ability to exhibit enough flexibility to respond to its environment.
- ***Organisations are brains*** – Like brains, organisations process information and make decisions. They balance conflicting criteria, weigh up risks and decide when an outcome is acceptable. They are also capable of learning, changing their model of the world in the light of experience. This emphasis on decision making, accumulating experience and learning from that experience is important in understanding organisations. They consist of conflicting groups where power and control are key issues.
- ***Organisations are cultures*** – An organisation's culture is usually taken to mean its shared values, ideology, pattern of thinking and day-to-day ritual. Different organisations will have different cultures stemming from their circumstances and their history. A major strength of seeing organisations as cultures is that it draws attention to their shared 'enactment of reality'. Looking for the symbols and shared realities within an organisation allows us to see beyond what the organisation says about itself.
- ***Organisations are political systems*** – Organisations, like communities, are governed. The system of government is rarely democratic, but nor is it usually a dictatorship. Within the mechanisms of government in an organisation are usually ways of understanding alternative philosophies, ways of seeking consensus (or at least reconciliation) and sometimes ways of legitimising opposition. Individuals and groups seek to pursue their aims through the detailed politics of the organisation. They form alliances, accommodate power relationships and manage conflict. Such a view is useful in helping organisations to legitimise politics as an inevitable aspect of organisational life.

Forms of organisation structure

There are many different ways of defining 'organisation structure'; here it is seen as the way in which tasks and responsibilities are divided into distinct groupings, and how the responsibility and coordination relationships between the groupings are defined. Most organisation designs attempt to divide an organisation into discrete parts that are given some degree of authority to make decisions within their part of the organisation. All but the very smallest of organisations need to delegate decision making in this way; it allows specialisation so decisions can be taken by the most appropriate people. The main issue is what dimension of specialisation should be used when grouping parts of the organisation together. There are three basic approaches to this:

- Group resources together according to their *functional purpose* – for example, sales, marketing, operations, research and development, finance, etc.
- Group resources together by the *characteristics of the resources themselves* – for example, by clustering similar technologies together (extrusion technology, rolling, casting, etc.). Alternatively, it may be done by clustering similar skills together (audit, mergers and acquisitions, tax, etc.). It may also be done according to the resources required for particular products or services (chilled food, frozen food, canned food, etc.).

- Group resources together by the *markets* that the resources are intended to serve – for example, by location (North America, South America, Europe and Middle East, South East Asia, etc.), or by the type of customer (small firms, large national firms, large multinational firms, etc.).

There are an almost infinite number of possible organisational structures. However, some pure types of organisation have emerged that are useful in illustrating different approaches to organisational design:

- ***The U-form organisation*** – The unitary form, or U-form, organisation clusters its resources primarily by their functional purpose. Figure 9.4(a) shows a typical U-form organisation with a pyramid management structure, each level reporting to the managerial level above. Such structures can emphasise process efficiency above customer service and the ability to adapt to changing markets. But the U-form keeps together expertise and can promote the creation and sharing of technical knowledge. The problem then with the U-form organisation is not so much the development of capabilities, but the flexibility of their deployment.
- ***The M-form organisation*** – This form of organisational structure emerged because the functionally based structure of the U-form was cumbersome when companies became large, often with complex markets. It groups together either the resources needed for each product or service group, or alternatively, those needed to serve a particular geographical market, in separate divisions. The separate functions may be distributed throughout the different divisions (see Figure 9.4(b)), which can reduce economies of scale and operating efficiency. But it does allow each individual division to focus on the specific needs of its markets.

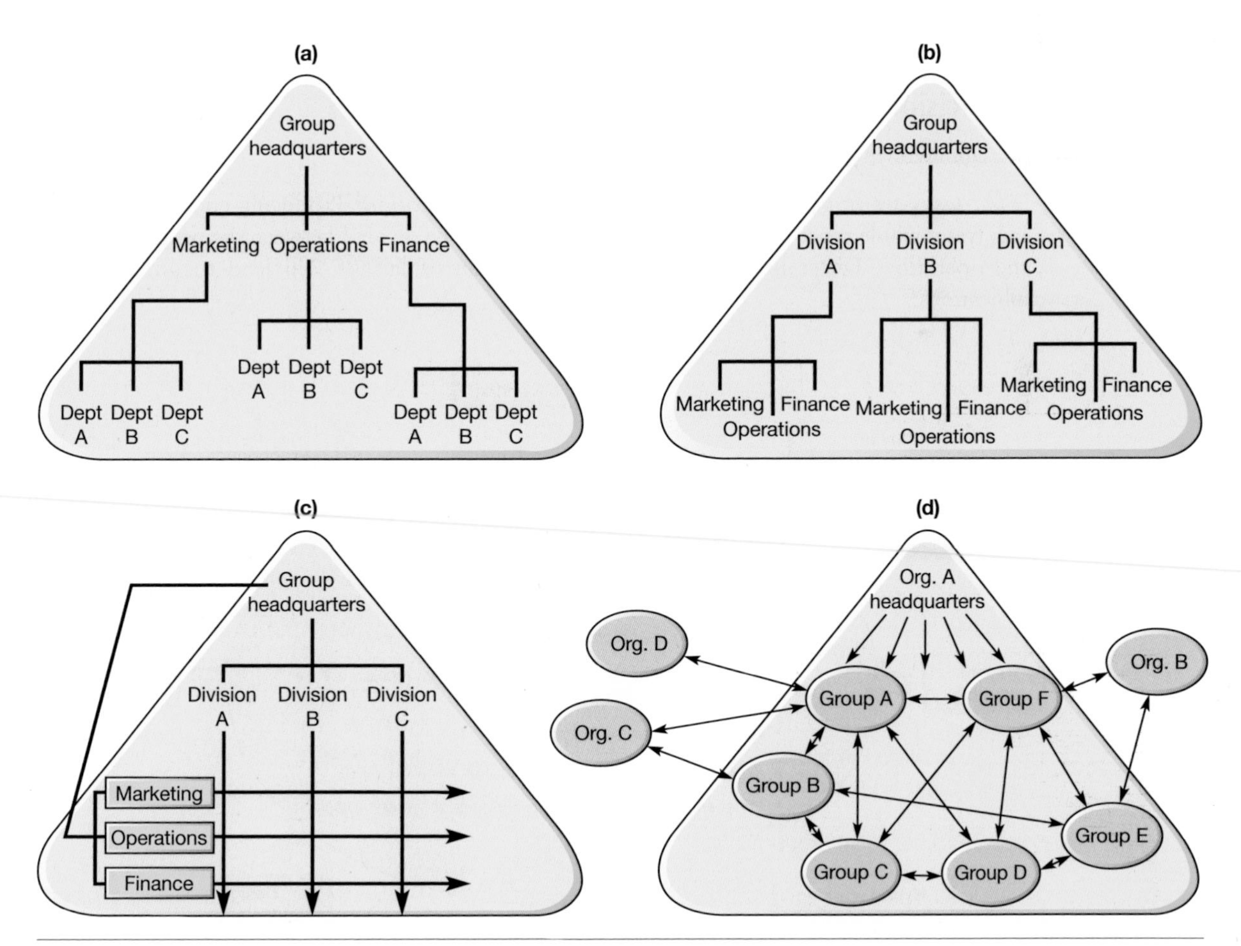

Figure 9.4 (a) U-form organisations give prominence to functional grouping of resources. (b) The M-form separates the organisation's resources into separate divisions. (c) Matrix form structures the organisation's resources so that they have two (or more) levels of responsibility. (d) N-form organisations form loose networks internally between groups of resources and externally with other organisations

- *Matrix forms* – Matrix structures are a hybrid, usually combining the M-form with the U-form. In effect, the organisation has two different structures simultaneously (see Figure 9.4(c)). In a matrix structure each resource cluster has at least two lines of authority, for example both to the division and to the functional groups. While a matrix organisation ensures the representation of all interests within the company, it can be complex and sometimes confusing.
- *The N-form organisation* – The 'N' in N-form stands for 'network'. In N-form organisations, resources are clustered into groups as in other organisational forms, but with more delegation of responsibility for the strategic management of those resources. N-forms have relatively little hierarchical reporting and control. Each cluster of resources is linked to the others to form a network, with the relative strength of the relationships between clusters changing over time, depending on circumstances (see Figure 9.4(d)). Senior management set broad goals and attempt to develop a unifying culture but do not 'command and control' to the same extent as in other organisation forms.

9.3 How do we go about designing jobs?

Job design is concerned with how we structure each individual's jobs, the team to which they belong (if any), their workplace and their interface with the technology they use. In this section we deal with what is usually considered to be the central people-related responsibility of operations managers – job design. It is a huge topic and we can only deal with some of the influences on, and approaches to, job design. The influences on job design that we deal with here are illustrated in Figure 9.5.

The decisions of job design

Job design involves a number of separate yet related elements:

- *What tasks are to be allocated to each person in the operation?* Producing goods and services involves a whole range of different tasks, which need to be divided between the people who staff the operation. Different approaches to the division of labour will lead to different task allocations.

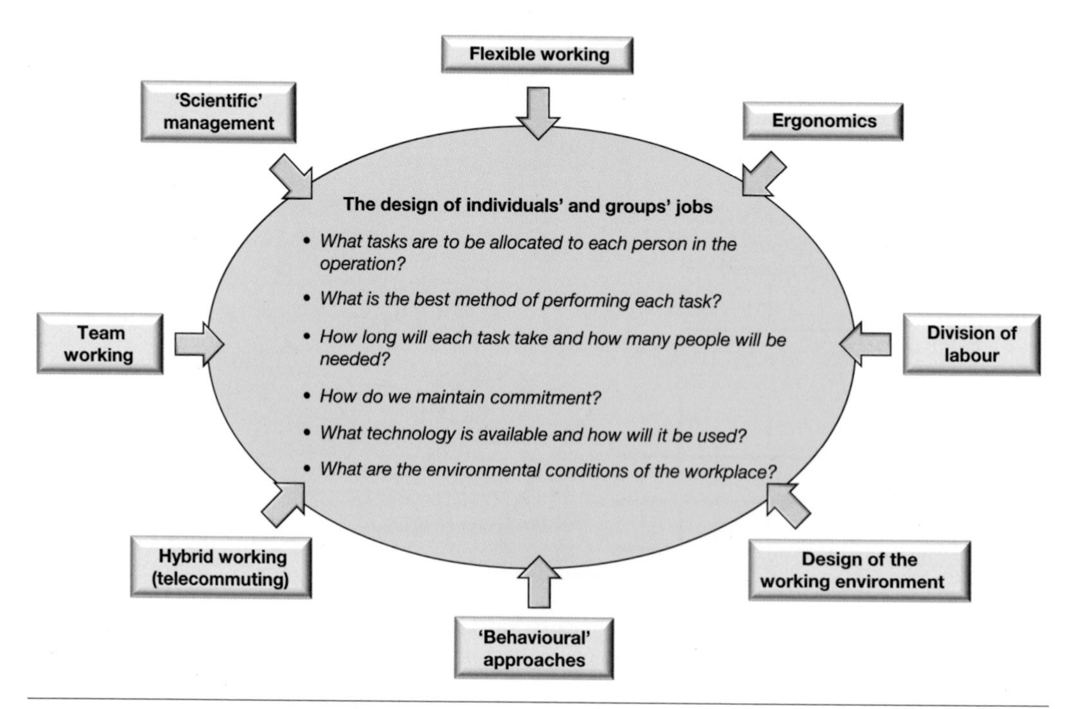

Figure 9.5 Some of the influences on job design

- *What is the best method of performing each job?* Every job should have an approved (or best) method of completion. And although there are different ideas of what is 'best', it is generally the most efficient method but that fits the task and does not unduly interfere with other tasks.
- *How long will it take and how many people will be needed?* Work measurement helps us calculate the time required to do a job, and therefore how many people will be needed.
- *How do we maintain commitment?* Understanding how we can encourage people and maintain job commitment is, arguably, the most important of the issues in job design. This is why behavioural approaches, including empowerment, teamwork and flexible working are at the core of job design.
- *What technology is available and how will it be used?* Many operational tasks require the use of technology. Not only does the technology need to be appropriately designed, but so does the interface between the people and the hardware.
- *What are the environmental conditions of the workplace?* The conditions under which jobs are performed will have a significant impact on people's effectiveness. Although often considered a part of job design, we treat it separately in this chapter.

Task allocation – the division of labour

Any operation must decide on the balance between using specialists or generalists. This idea is related to the division of labour – dividing the total task down into smaller parts, each of which is accomplished by a single person or team. It was first formalised as a concept by the economist Adam Smith in his *Wealth of Nations* in 1776. Perhaps the epitome of the division of labour is the assembly line, where products move along a single path and are built up by operators continually repeating a single task. This is the predominant model of job design in most mass-produced products and in some mass-produced services (fast food, for example). There are some *real advantages* in division-of-labour principles:

- *It promotes faster learning.* It is obviously easier to learn how to do a relatively short and simple task than a long and complex one. This means that new members of staff can be quickly trained and assigned to their tasks.
- *Automation becomes easier.* Dividing a total task into small parts raises the possibility of automating some of those small tasks. Substituting technology for labour is considerably easier for short and simple tasks than for long and complex ones.
- *Reduced non-productive work.* This is probably the most important benefit of division of labour. In large, complex tasks the proportion of time spent picking up tools and materials, putting them down again and generally finding, positioning and searching can be very high indeed. For example, one person assembling a whole motor car engine would take two or three hours and involve much searching for parts, positioning, and so on. Around half the person's time would be spent on these reaching, positioning, finding tasks (called non-productive elements of work). Now consider how a motor car engine is actually made in practice. The total job is probably divided into 20 or 30 separate stages, each staffed by a person who carries out only a proportion of the total. Specialist equipment and materials-handling devices can be devised to help them carry out their job more efficiently. Furthermore, there is relatively little finding, positioning and reaching involved in this simplified task. Non-productive work can be considerably reduced, perhaps to under 10 per cent, which would be very significant to the costs of the operation.

However, there are also serious drawbacks to highly divided jobs:

- *Monotony.* The shorter the task, the more often operators will need to repeat it. Repeating the same task, for example every 30 seconds, eight hours a day and five days a week, can hardly be called a fulfilling job. As well as any ethical objections, there are other, more obviously practical, objections to jobs that induce such boredom. These include the increased likelihood of absenteeism and staff turnover, the increased likelihood of error and even the deliberate sabotage of the job.
- *Physical injury.* The continued repetition of a very narrow range of movements can, in extreme cases, lead to physical injury. The over-use of some parts of the body (especially the arms, hands

and wrists) can result in pain and a reduction in physical capability. This is sometimes called repetitive strain injury (RSI).

- *Low flexibility*. Dividing a task up into many small parts often gives the job design a rigidity that is difficult to change under changing circumstances. For example, if an assembly line has been designed to make one particular product but then has to change to manufacture a quite different product, the whole line will need redesigning. This will probably involve changing every operator's set of tasks, which can be a long and difficult procedure.
- *Poor robustness*. Highly divided jobs imply materials (or information) passing between several stages. If one of these stages is not working correctly, for example because some equipment is faulty, the whole operation is affected. On the other hand, if each person is performing the whole of the job, any problems will only affect that one person's output.

Operations principle

There are both positive and negative effects of the division of labour, but it is still a significant factor in job design.

Designing job methods – scientific management

The term 'scientific management' became established in 1911 with the publication of the book of the same name by Frederick Taylor (this whole approach to job design is sometimes referred to, pejoratively, as *Taylorism*). In this work he identified what he saw as the basic tenets of scientific management:[5]

- All aspects of work should be investigated on a scientific basis to establish the laws, rules and formulae governing the best methods of working.
- Such an investigative approach to the study of work is necessary to establish what constitutes a 'fair day's work'.
- Workers should be selected, trained and developed methodically to perform their tasks.
- Managers should act as the planners of the work (analysing jobs and standardising the best method of doing the job) while workers should be responsible for carrying out the jobs to the standards laid down.
- Cooperation should be achieved between management and workers based on the 'maximum prosperity' of both.

The important thing to remember about scientific management is that it is not particularly 'scientific' as such, although it certainly does take an 'investigative' and, within limits, a rationalist approach to improving operations. Perhaps a better term for it would be 'systematic management'. It gave birth to two separate, but related, fields of study: method study, which determines the methods and activities to be included in jobs; and work measurement, which is concerned with measuring the time that should be taken for performing jobs. Together, these two fields are often referred to as work study and are explained in detail in the supplement to this chapter. Both still have an influence over how many operations managers approach job design. However, the approach is controversial – see the 'Responsible operations' section later in the chapter.

Designing the human interface – ergonomic workplace design

Ergonomics is concerned primarily with the physiological aspects of job design. Physiology is about the way the body functions. It involves two aspects: first, how a person interfaces with his or her immediate working area; and second, how people react to environmental conditions. We will examine the second aspect of ergonomics later in this chapter. Ergonomics is sometimes referred to as human factors engineering or just 'human factors'. Both aspects are linked by two common ideas:

- There must be a fit between people and the jobs they do. To achieve this fit there are only two alternatives. Either the job can be made to fit the people who are doing it or, alternatively, the people can be made (or perhaps less radically, recruited) to fit the job. Ergonomics addresses the former alternative.
- It is important to take a 'scientific' approach to job design: for example, collecting data to indicate how people react under different job design conditions and trying to find the best set of conditions for comfort and performance.

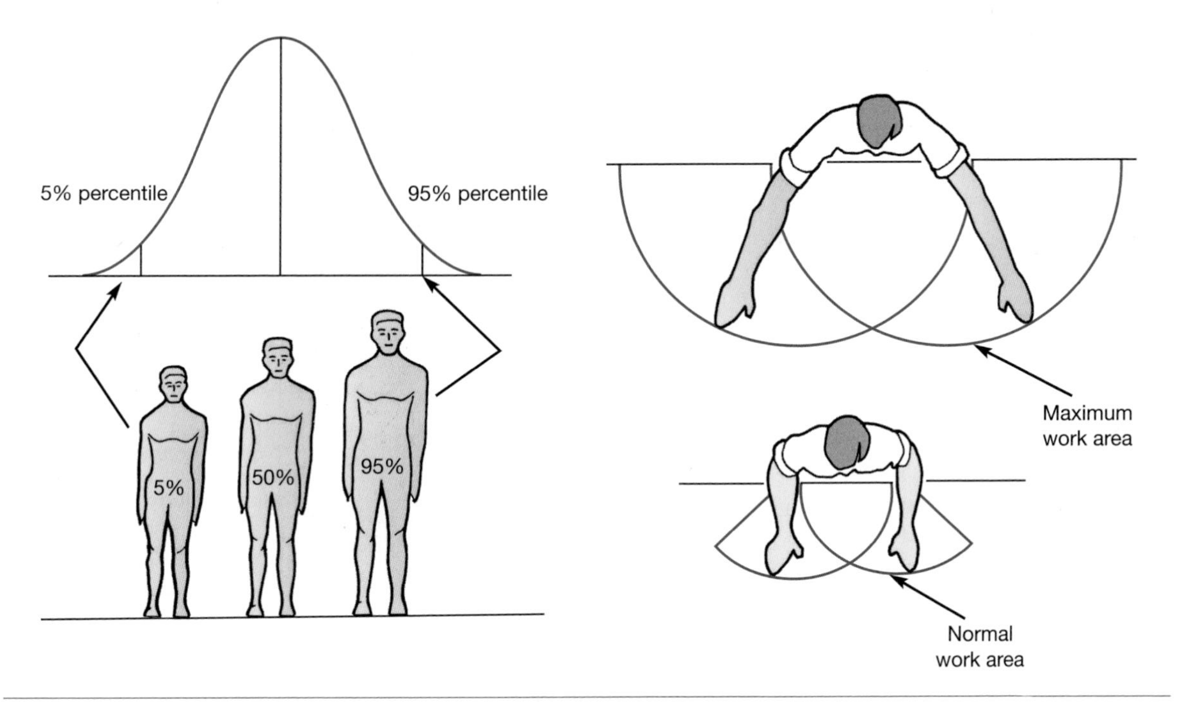

Figure 9.6 The use of anthropometric data in job design

Anthropometric aspects

Many ergonomic improvements are primarily concerned with what are called the anthropometric aspects of jobs – that is, the aspects related to people's size, shape and other physical abilities. The design of an assembly task, for example, should be governed partly by the size and strength of the operators who do the job. The data that ergonomists use when doing this are called anthropometric data. Because we all vary in our size and capabilities, ergonomists are particularly interested in our range of capabilities, which is why anthropometric data are usually expressed in percentile terms. Figure 9.6 illustrates this idea. This shows the idea of size (in this case, height) variation. Only 5 per cent of the population are smaller than the person on the extreme left (5th percentile), whereas 95 per cent of the population are smaller than the person on the extreme right (95th percentile). When this principle is applied to other dimensions of the body, for example arm length, it can be used to design work areas. Figure 9.6 also shows the normal and maximum work areas derived from anthropometric data. It would be inadvisable, for example, to place frequently used components or tools outside the maximum work area derived from the 5th percentile dimensions of human reach.

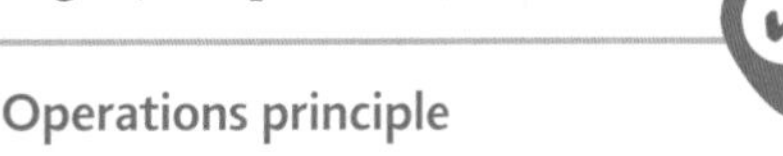

Operations principle

Ergonomic considerations in job design can prevent excessive physical strain and increase efficiency.

OPERATIONS IN PRACTICE

Exoskeleton devices take the strain[6]

Exoskeletons are not a new idea; since the 1960s they have been suggested for the enhancement of people's natural physical capabilities, usually for medical or military purposes. In the natural world, they are defined as 'a hard outer layer that covers, supports, and protects the body of an invertebrate animal such as an insect or crustacean'. But, in the context of physical work, it is usually taken to refer to a powered exoskeleton – a wearable powered mobile device that assists human movement or positioning to allow increased strength and/or endurance. Long the subject of science fiction, they are starting to be tested in industrial conditions. For example, typical of any automotive assembly lines, Ford requires its assembly employees to adopt positions that involve reaching overhead for long

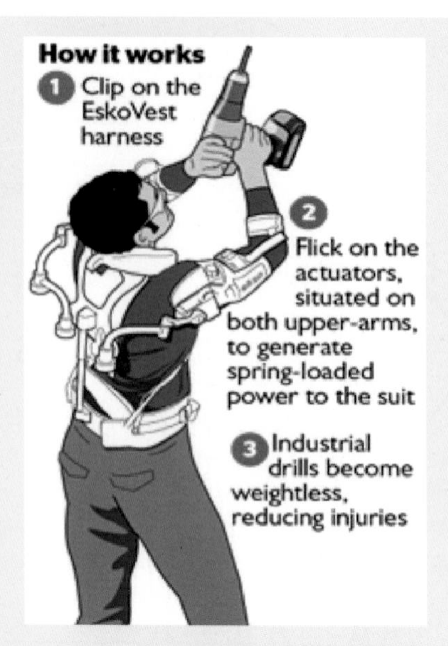

periods. This can result in aching back, neck and shoulders as well as general fatigue. This is why the company has partnered with California-based Ekso Bionics to trial an upper-body exoskeleton known as the EksoVest, which elevates and supports the arms. According to Marty Smets, an ergonomics expert at Ford who works on human systems and virtual manufacturing, Ford has been working on wearable robotics solutions since 2011, not to give their operatives superhuman strength, but to prevent injury. *'Right now, we're just using upper body supports, but we do have interest in other systems. Our goal right now is to just figure out how to integrate exoskeletons in our plants. Once we get them into our plants we can begin to replicate and figure out what the sweet spots are for application.'*

How it works
1 Clip on the EskoVest harness
2 Flick on the actuators, situated on both upper-arms, to generate spring-loaded power to the suit
3 Industrial drills become weightless, reducing injuries

Designing for job commitment – behavioural approaches to job design

Jobs that are designed purely on division of labour, scientific management or even purely ergonomic principles can alienate the people performing them. Job design should also take into account the desire of individuals to fulfil their needs for self-esteem and personal development. This is where motivation theory and its contribution to the behavioural approach to job design is important. This achieves two important objectives of job design. First, it provides jobs that have an intrinsically higher quality of working life – an ethically desirable end in itself. Second, because of the higher levels of motivation it engenders, it is instrumental in achieving better performance for the operation, in terms of both the quality and the quantity of output. This approach to job design involves two conceptual steps: first, exploring how the various characteristics of the job affect people's motivation; and second, exploring how individuals' motivation towards the job affects their performance at that job.

Typical of the models that underlie this approach to job design is that by Hackman and Oldham[7] shown in Figure 9.7. Here a number of 'techniques' of job design are recommended in order to affect particular core 'characteristics' of the job. These core characteristics of the job are held to influence various positive 'mental states' towards the job. In turn, these are assumed to give certain performance outcomes. In Figure 9.7 some of the 'techniques' (which Hackman and Oldham originally called 'implementing concepts') need a little further explanation:

- Combining tasks means increasing the number of activities allocated to individuals.
- Forming natural work units means putting together activities that make a coherent whole.
- Establishing client relationships means that staff make contact with their internal customers directly.
- Vertical loading means including 'indirect' activities (such as maintenance).
- Opening feedback channels means that internal customers feed back perceptions directly.

Hackman and Oldham also indicate how these techniques of job design shape the core characteristics of the resulting job, and further, how the core characteristics influence people's 'mental states'. Mental states are the attitude of individuals towards their jobs: specifically, how meaningful they find the job, how much responsibility and control they feel they have over the way the job is done, and how

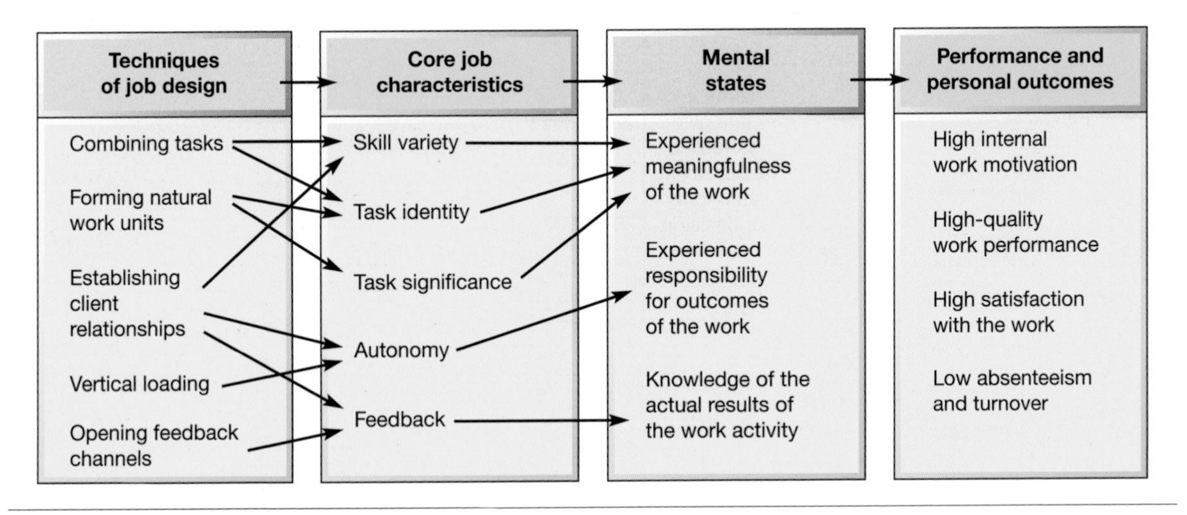

Figure 9.7 A typical 'behavioural' job design model

much they understand about the results of their efforts. All of these mental states influence people's performance at their job in terms of their motivation, quality of work, satisfaction with their work, turnover and absenteeism.

Job rotation

If increasing the number of related tasks in the job is constrained in some way, for example by the technology of the process, one approach may be to encourage job rotation. This means moving individuals periodically between different sets of tasks to provide some variety in their activities. When successful, job rotation can increase skill flexibility and make a small contribution to reducing monotony. However, it is not viewed as universally beneficial either by management (because it can disrupt the smooth flow of work) or by the people performing the jobs (because it can interfere with their rhythm of work).

Job enlargement

The most obvious method of achieving at least some of the objectives of behavioural job design is by allocating a larger number of tasks to individuals. If these extra tasks are broadly of the same type as those in the original job, the change is called job enlargement. This may not involve more demanding or fulfilling tasks, but it may provide a more complete and therefore slightly more meaningful job. If nothing else, people performing an enlarged job will not repeat themselves as often, which could make the job marginally less monotonous. So, for example, suppose that the manufacture of a product has traditionally been split up on an assembly-line basis into 10 equal and sequential jobs. If that job is then redesigned so as to form two parallel assembly lines of five people, the output from the system as a whole would be maintained but each operator would have twice the number of tasks to perform. This is job enlargement. Operators repeat themselves less frequently and presumably the variety of tasks is greater, although no further responsibility or autonomy is necessarily given to each operator.

Job enrichment

Job enrichment means not only increasing the number of tasks, but also allocating extra tasks that involve more decision making, greater autonomy and greater control over the job. For example, the extra tasks could include maintenance, planning and control, or monitoring quality levels. The effect is both to reduce repetition in the job and to increase autonomy and personal development. So, in the assembly-line example, each operator, as well as being allocated a job that is twice as long as that previously performed, could also be allocated responsibility for carrying out routine maintenance and such tasks as record-keeping and managing the supply of materials. Figure 9.8 illustrates the

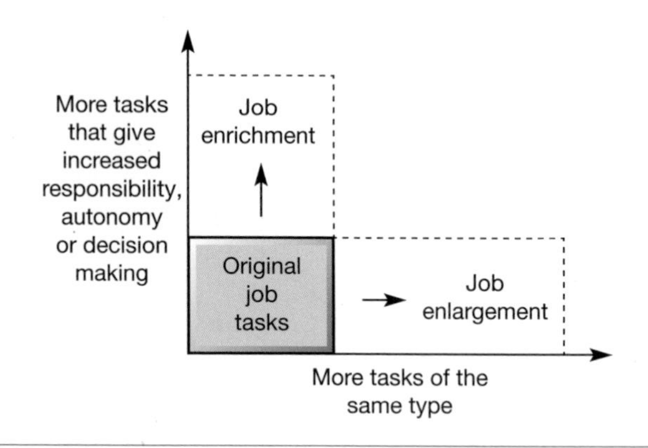

Figure 9.8 Job enlargement and job enrichment

difference between what are called horizontal and vertical changes. Broadly, horizontal changes are those that extend the variety of *similar* tasks assigned to a particular job. Vertical job changes are those that add responsibilities, decision making or autonomy to the job. Job enlargement implies movement only in the horizontal scale, whereas job enrichment certainly implies movement on the vertical scale and perhaps on both scales.

OPERATIONS IN PRACTICE

Michelin calls it 'responsabilisation'[8]

This is how Édouard Michelin (then a director of Michelin) put it, back in 1928, *'The spirit of empowerment has always been one of our values and is part of the Michelin Group's "genetic code". One of our principles is to give responsibility to the person who carries out a given task because* [they] *know a lot about it'*. It is an approach to empowerment that survives today. In French, Michelin calls it 'responsabilisation', which roughly translates into English as a mixture of empowerment and accountability. It is an initiative that is part of the Group's efforts to streamline its organisational structures, increase responsiveness and efficiency, and encourage faster decision-making. Team empowerment is seen as an essential part of this aim. *'Not only does it foster initiative and dialogue'*, says Michelin, *'it also enables decisions to be made close to operations and customers. Our Empowering Organisations are being developed throughout the Group, building trust-based relationships that encourage all employees to take part in our transformation. We are empowering front-line teams to organise themselves and find the right solutions to meet given objective in a framework defined by Management. In this way, managers are resuming their role as advisors, who develop people's capabilities and train their teams'*. In essence, 'responsabilisation' means shifting more operational responsibility to those people who work on the factory floor. This involves learning new skills: how to work effectively in teams, how to structure projects, how to manage conflicts and how to communicate in non-confrontational ways. Rather than issue direct instructions, team leaders act as coaches or, if any conflict arises, as referees. Workers in the teams allocate responsibility between themselves for tasks such as production scheduling, safety procedures, quality control and so on. After the company introduced the idea of greater worker autonomy in one of its factories, staff there said that they felt happier and more productive as a result. The company felt that it was such a success that it extended the practice to six factories in Europe and North America.

Empowerment

Empowerment is an extension of the *autonomy* job characteristic prominent in the behavioural approach to job design. However, it is usually taken to mean more than autonomy. Whereas autonomy means giving staff the *ability* to change how they do their jobs, empowerment means giving staff the *authority* to make changes to the job itself, as well as how it is performed. This can be designed into jobs to different degrees. At a minimum, staff could be asked to contribute their suggestions for how the operation might be improved. Going further, staff could be empowered to redesign their jobs. Further still, staff could be included in the strategic direction and performance of the whole organisation. The *benefits* of empowerment are generally seen as providing fast responses to customer needs (including dissatisfied customers), employees feeling better about their jobs and interacting with customers with more enthusiasm, promoting 'word-of-mouth' advertising and customer retention. However, there are *costs* associated with empowerment, including higher selection and training costs, perceived inequity of service and the possibility of poor decisions being made by employees.

OPERATIONS IN PRACTICE

Hybrid working divides opinions[9]

The practice of spending at least some of the working week doing so from home had been common for some jobs, particularly when it became possible to connect seamlessly with one's company IT systems (called 'remote working', 'telecommuting', 'flexible working', 'working from home', etc.). Yet when the worldwide COVID-19 pandemic forced many people to abandon their regular offices, the practice became widespread. It was controversial because some regarded working from home as an excuse for working less hard and others pointed out (with some justification) that productive and creative communication with co-workers would be reduced. In fact, years before, Marissa Mayer, the then boss of Yahoo, ruled that employees of the company could no longer work from home, but must come into the office to work. It was a move that prompted a debate about how much freedom employees should be given to decide how, when and where they should do their jobs. Perhaps most surprising was that Ms Mayer's decision seemed to go against the trend, especially in high-tech companies, to allow and even encourage a degree of working from home. Surveys had shown that home-based working in some industries, especially information systems, engineering and science, was both increasing and popular. Also, given that many of these technology firms produced the hardware and software that make working from home possible, it seemed only sensible to let their employees use them.

Yahoo's view was echoed by Goldman Sachs, the investment bank. Its Chief executive, David Solomon, said that although working from home had become the norm for billions of people during the pandemic, it should be seen as a temporary thing. An innovative and collaborative business like that could not countenance it becoming the new normal. Rather it was an aberration that it would correct as quickly as possible. The arguments against hybrid working usually focus on the lack of opportunity for the serendipitous meetings that can promote the energy and 'buzz' that gives birth to new ideas. Moreover, some people find working from home can be isolating, with vastly reduced social contact.

Yet, other companies, including banks, such as HSBC and Lloyds, took the opportunity to cut their office space (in HSBC's case, by 40 per cent) around the world in anticipation of more hybrid working. Similarly, PwC, the multinational professional services partnership, believed that the pandemic fundamentally changed attitudes to work for ever, ending 'presenteeism' (the assumption that being present at the office was the same as doing productive work). However, Kevin Ellis, its Chairman, did admit that he missed the atmosphere and the culture of the office, foreseeing a future of three or four days a week in the office. But changes to working practice were not confined to 'office' work. For example, the pandemic interrupted a planned project to redesign Danone Mexico's new environmentally friendlier bottling process for its mineral water. Because of travel bans, technical experts could not travel to Mexico to install new technology. Instead, they trained local technicians via Zoom video calls. It did take longer than originally planned (two weeks to prepare and implement rather than four days) but cost less than sending European technicians would have done. More importantly, the method will be repeated. Danone's then Chief Executive, Emmanuel Faber, said the change in its tactic indicated a company-wide shift in approach towards a more flexible workforce.

Team working

A development in job design that is closely linked to the empowerment concept is that of team-based work organisation (sometimes called self-managed work teams). This is where staff, often with overlapping skills, collectively perform a defined task and have a high degree of discretion over how they actually perform the task. The team would typically control such things as task allocation between members, scheduling work, quality measurement and improvement, and sometimes the hiring of staff. To some extent most work has always been a group-based activity. The concept of teamwork, however, is more prescriptive and assumes a shared set of objectives and responsibilities. Groups are described as teams when the virtues of working together are being emphasised, such as the ability to make use of the various skills within the team. Teams may also be used to compensate for other organisational changes such as the move towards flatter organisational structures. When organisations have fewer managerial levels, each manager will have a wider span of activities to control. Teams that are capable of autonomous decision-making have a clear advantage in these circumstances. The benefits of teamwork can be summarised as:

- improving productivity through enhanced motivation and flexibility;
- improving quality and encouraging innovation;
- increasing satisfaction by allowing individuals to contribute more effectively;
- making it easier to implement technological changes in the workplace because teams are willing to share the challenges this brings.

Flexible working

The nature of most jobs has changed significantly over the last 25 years. New technologies, more dynamic marketplaces, more demanding customers and a changed understanding of how individuals can contribute to competitive success have all had their impact. Also changing is our understanding of how home life, work and social life need to be balanced. Alternative forms of organisation and alternative attitudes to work are being sought, which allow, and encourage, a degree of flexibility in working practice that matches the need for flexibility in the marketplace. From an operations management perspective, three aspects of flexible working are significant: skills flexibility, time flexibility and location flexibility.

- ***Skills flexibility*** – A flexible workforce that can move across several different jobs could be deployed (or deploy themselves) in whatever activity is in demand at the time. In the short term, staff at a supermarket may be moved from warehouse activities to shelf replenishment in the store to the checkout, depending on what is needed at the time. In the longer-term sense, multi-skilling means being able to migrate individuals from one skill set to another as longer-term demand trends become obvious. So, for example, an engineer who at one time maintained complex equipment by visiting the sites where such equipment was installed may now perform most of their activities by using remote computer diagnostics and 'helpline' assistance. The implication of job flexibility is that a greater emphasis must be placed on training, learning and knowledge management. Defining what knowledge and experience are required to perform particular tasks and translating these into training activities are clearly prerequisites for effective multi-skilling.

Critical commentary

Teamwork is not only difficult to implement successfully, but it can also place undue stress on the individuals who form the teams. Some teams are formed because more radical solutions, such as total reorganisation, are being avoided. Teams cannot compensate for badly designed organisational processes; nor can they substitute for management's responsibility to define how decisions should be made. Often teams are asked to make decisions but are given insufficient responsibility to carry them out. In other cases, teams may provide results but at a price. The Swedish car maker Volvo introduced self-governing teams in the 1970s and 1980s, which improved motivation and morale but eventually proved prohibitively expensive.[10] Perhaps most seriously, teamwork is criticised for substituting one sort of pressure for another. Although teams may be autonomous, this does not mean they are stress-free. Top-down managerial control is often replaced by excessive peer pressure, which is in some ways more insidious.

- ***Time flexibility*** – Not every individual wants to work full-time. Many people, often because of family responsibilities, only want to work for part of their time, sometimes only during specific parts of the day or week (because of childcare responsibilities, etc.). Likewise, employers may not require the same number of staff at all times. They may, for example, need extra staff only at periods of heavy demand. Bringing both the supply of staff and the demand for their work together is the objective of 'flexible time' or **flexi-time working** systems. These may define a *core* working time for each individual member of staff and allow other times to be accumulated flexibly. Other schemes include **annual hours** schemes, one solution to the capacity management issue described in Chapter 11.
- ***Location flexibility*** – The service sector in most economies in the Global North now accounts for between 70 and 80 per cent of all employment. Even within the manufacturing sector, the proportion of people with indirect jobs (those not directly engaged in making products) has also increased significantly. One result of all this is that the number of jobs that are not 'location-specific' has increased. Location-specific means that a job must take place in one fixed location. So, a shop worker must work in a shop and an assembly line worker must work on the assembly line. But many jobs could be performed at any location where there are communication links to the rest of the organisation. The realisation of this has given rise to what is known as hybrid working, flexible working', 'home working', telecommuting, **teleworking**, and mobile working. In other words, working from home, at least part of the time. See the 'Operations in practice' example, 'Hybrid working divides opinions'.

Critical commentary

There is always a big difference between what is technically possible and what is organisationally feasible. Hybrid working/telecommuting does have its problems, especially those types of telecommuting that deny individuals the chance to meet with colleagues often face difficulties. Problems can include the following:

- *Lack of socialisation* – offices are social places where people can adopt the culture of an organisation as well as learn from each other. It is naïve to think that all knowledge can be codified and learned formally at a distance.
- *Effectiveness of communication* – a large part of the essential communication we have with our colleagues is unplanned and face to face. It happens on 'chance meet' occasions, yet it is

important in spreading contextual information as well as establishing specific pieces of information necessary to the job.

- *Problem-solving* – it is still often more efficient and effective to ask a colleague informally for help in resolving problems than to formally frame a request using communications technology.
- *It is lonely* – isolation among mobile or home workers is a real problem. For many of us, the workplace provides the main focus for social interaction. A computer screen is no substitute.

Work-related stress

The idea that there is a link between job design and work stress is not new. Even some of the early scientific management pioneers accepted that working arrangements should not result in conditions that promoted stress. Now it is generally accepted that stress can seriously undermine the quality of people's working lives and, in turn, the effectiveness of the workplace. Here stress is defined as 'the adverse reaction people have to excessive pressures or other types of demand placed on them'.[11] In addition to the obvious ethical reasons for avoiding work-related stress, there are also business-related benefits, such as the following:

- Staff feel happier at work; their quality of working life is improved and they perform better.
- Introducing improvements is easier when 'stress' is managed effectively.
- Employment relations: problems can be resolved more easily.
- Attendance levels increase and sickness absence reduces.

Table 9.1 illustrates some of the causes of stress at work and what operations managers can do about it.

Table 9.1 Causes of stress at work and what can be done about it

Causes of stress	What can be done about it
Staff can become overloaded if they cannot cope with the amount of work or type of work they are asked to do	Change the way the job is designed, assess training needs and whether it is possible for employees to work more flexible hours
Staff can feel disaffected and perform poorly if they have no control or say over how and when they do their work	Actively involve staff in decision making, the contribution made by teams, and how reviewing performance can help identify strengths and weaknesses
Staff feel unsupported: levels of sick absence often rise if employees feel they cannot talk to managers about issues that are troubling them	Give staff the opportunity to talk about the issues causing stress, be sympathetic and keep them informed
A failure to build relationships based on good behaviour and trust can lead to problems related to discipline, grievances and bullying	Check the organisation's policies for handling grievances, unsatisfactory performance, poor attendance and misconduct, and for tackling bullying and harassment
Staff will feel anxious about their work and the organisation if they don't know their role and what is expected of them	Review the induction process, work out an accurate job description and maintain a close link between individual targets and organisational goals
Change can lead to huge uncertainty and insecurity	Plan ahead so change is not unexpected. Consult with employees so they have a real input, and work together to solve problems

Work–life balance

A number of factors have made it increasingly difficult to separate work life from personal life. First is the general decline in the number of people working in operations where working times are very clearly delineated (usually those that employ routine, high-volume processes) towards operations where activities (and processes) are less formal and/or defined – and so are working times. Second, there is less distinction between what are clearly 'work' technologies (laptops, mobile devices, etc.) and personal devices, meaning that it is difficult to remain 'unconnected' from work emails, phone calls, etc. Third, as more people work from home, the discipline to set limited working hours is not always easy. Finally, some organisational cultures confuse 'working longer' with 'working better'. Ensuring that there is an appropriate split between work and personal life is usually taken to mean that work should not interfere unreasonably with family obligations and personal interests (although what exactly constitutes 'unreasonable' can be a cause of some disagreement). From an organisation's perspective, the case for addressing work–life balance is usually made by stressing the following benefits:

- Employee retention is improved – staff who feel overloaded are more likely to seek alternative employment.
- Reputation – the promotion of a healthy work–life balance will develop a reputation that will help companies to attract more able staff.
- Without balance, staff 'burn-out' will eventually lead to a higher incidence of physical or mental health problems – an ethical as well as an economic (increased absenteeism) problem.
- Higher levels of staff performance – it is usually assumed that stressed or overloaded staff are less effective at getting work done, but this is disputed. Although some studies show that staff with a healthy work–life balance work more effectively, others say that there is 'insufficient evidence to support the notion that work–life practices enhance performance by means of reduced work–life conflict'.[12]

Many of the mechanisms for promoting better work–life balance are the measures that are described in this part of the chapter, such as various forms of flexible working, home working, job sharing, on-site child care and so on.

OPERATIONS IN PRACTICE

The stress of high customer contact jobs[13]

Those jobs that are on the front line of dealing directly with customers (particularly a lot of customers, all the time, of all different types) can be particularly stressful. Not all customers will be reasonable, patient, courteous or even sane. The people who have these high customer contact roles need support, training and perhaps a special aptitude. And there is plenty of advice for staff who have to deal with customers who are angry because they feel that the level of service they have received is inadequate. Such advice usually includes such things as: acknowledge the (perceived) problem, try to put yourself in the position of the complainer, get the all facts straight, and try to rectify the problem. It isn't easy, but if complaints can be resolved to the satisfaction of the customer, there can be significant benefits. Some surveys indicate that 90 per cent of customers whose complaints are resolved are happy to use the service again, and may even go on to become advocates for the service. Nevertheless, maintaining tolerance and politeness in the face of some particularly difficult customers can be more than, even experienced, staff can bear. That certainly was the case with Steven Slater, formerly an air steward on the US airline JetBlue. He was working on a New York flight when he had to arbitrate after one passenger began arguing with another passenger about space in the overhead luggage compartment during boarding. The first passenger swore at Mr Slater and pulled down the compartment door on his head. Later, when the plane landed, they seemingly refused to follow Mr Slater's request to remain in their seat and got up to take their bag from the overhead locker while the plane was still taxiing. Again, the passenger allegedly swore at Mr Slater. It was then that his patience ran out in a particularly dramatic fashion. He went to the intercom and broadcast to the whole plane: *'To the passenger who just called me a motherf*****: F*** you. I've been in this business for 28 years and I've had it.'* He then collected his hand luggage (and two beers from the trolley), opened the cabin door, activated the inflatable chute, announced, *'to those of you who have shown dignity and respect for 20 years, have a great ride'* and slid out of the (fortunately stationary) plane on to the runway. As a way to give up your job, it's not recommended. He was later arrested and charged with criminal mischief and reckless endangerment.

How should the working environment be designed?

The aspect of ergonomics that we examined earlier was concerned with how a person interfaces with the physical aspects of their immediate working area, such as its dimensions. But the subject also examines how people interface with their working environment. By this we mean the temperature, lighting, noise environment, and so on. It will obviously influence the way jobs are performed. Working conditions that are too hot or too cold, insufficiently illuminated or glaringly bright, excessively noisy or irritatingly silent will all influence the way jobs are carried out. Many of these issues are often covered by occupational health and safety legislation, which controls environmental conditions in workplaces throughout the world. A thorough understanding of this aspect of ergonomics is necessary to work within the guidelines of such legislation.

Operations principle
Designing working environments is an important part of job design.

Working temperature

Predicting the reactions of individuals to working temperature is not straightforward. Individuals vary in the way their performance and comfort vary with temperature. Furthermore, most of us judging 'temperature' will also be influenced by other factors such as humidity and air movement. Nevertheless, some general points regarding working temperatures provide guidance to job designers:

- Comfortable temperature range will depend on the type of work being carried out; lighter work requires higher temperatures than heavier work.
- The effectiveness of people at performing vigilance tasks reduces at temperatures above about 29°C; the equivalent temperature for people performing light manual tasks is a little lower.
- The chances of accidents occurring increase at temperatures that are above or below the comfortable range for the work involved.

OPERATIONS IN PRACTICE

Music while you work?[14]

Background music at work is not new. It has been used in the workplace for centuries. As far back as the industrial revolution orchestras and singers would be hired occasionally to perform for workers in the quieter factories. Later, in the 1940s, the BBC launched a radio programme called *Music While You Work*. Broadcasting twice a day, it was made especially for factory workers. Artists who were booked for the show were told to 'play material with an upbeat rhythm that would keep the workers' attention', in the belief that it would improve productivity. But playing music at work is not always free. In the United Kingdom, for example, the law requires businesses that play any recorded music in public to get licenses from the Performing Right Society (PRS), which collects fees and pays royalties to composers and their publishers. Listening to a device through headphones, however, is free. But does music help or hinder?

Some bodies definitely think that it helps. Musicworks (which is an organisation supported by the PRS, so it is not exactly independent) cites studies that show that music in the workplace promotes a positive mood, can build team

spirit, improves alertness and can reduce the number of workplace accidents. It can also, they say, cut the number of sick days and increase workplace productivity. One study by Teresa Lesiuk at the University of Miami found IT specialists who listened to music completed tasks more quickly and came up with better ideas than those who didn't. But not everyone is convinced. *'If people need a high level of concentration, it could be a distraction'*, says Dr Carolyn Axtell, at the Institute of Work Psychology. *'When people choose to listen, there can be positive effects – it can be relaxing and help manage other distractions such as noise. But when it's imposed, they can find it annoying and stressful'*. However, individuals can differ in their reaction to music and problems occur when colleagues clash. *'You can look away if you don't want to see something, but you can't close your ears'*, she says.

In another study, researchers at London University studied the apparently common practice of surgeons playing music in the operating theatre (playlists ranged from gentle classical music through heavy metal, to electronic dance music). Patients didn't complain, being anaesthetised, but other members of the surgical team were not always happy. Music could damage communication in a surgical team, preventing team members from hearing instructions. Even worse, when sound levels are uneven and a new track blasts out unexpectedly, or when a surgeon turns up the volume when their favourite song comes on, other team members can be disturbed. But notwithstanding the sometimes conflicting findings from researchers, some themes do emerge:

- How 'immersive' a task is makes a difference when evaluating music's effectiveness in increasing productive output. 'Immersive' refers to the variability and creative demand of the task. Creating an entirely original piece of work from scratch that demands a lot of creativity is 'immersive'. Performing more routine tasks such as answering emails is not. When the task is routine, clearly defined and repetitive, music is probably useful for most people.
- Music affects your mood. Apparently, it isn't the background noise of the music itself, but rather the improved mood that your favourite music creates that is the reason for the increase in productivity. In one study, information technology specialists who listened to music completed their tasks more quickly and came up with better ideas than those who didn't, because the music improved their mood.
- In open-plan offices where background chatter can be too much for some people to handle, headphones can help some people.
- Music does not help learning. It has a negative effect on absorbing and retaining new information, because it demands too much of your attention.
- Listening to music with lyrics, especially interesting and/or new lyrics, detracts from performing immersive tasks. Listening to lyrics activates the language centre of your brain, so trying to perform other language-related tasks is particularly difficult.

(Full disclosure: most of this text was written while listening to music.)

Illumination levels

The intensity of lighting required to perform any job satisfactorily will depend on the nature of the job. Some jobs that involve extremely delicate and precise movement, surgery for example, require very high levels of illumination. Other, less delicate jobs do not require such high levels.

Noise levels

The damaging effects of excessive noise levels are perhaps easier to understand than some other environmental factors. Noise-induced hearing loss is a well-documented consequence of working environments where noise is not kept below safe limits. When considering noise levels, bear in mind that the recommended (and often legal) maximum noise level to which people can be subjected over the working day is 90 decibels (dB) in the United Kingdom (although in some parts of the world the legal level is lower than this). Also bear in mind that the decibels unit of noise is based on a logarithmic scale, which means that noise intensity doubles about every 3 dB. In addition to the damaging effects of high levels of noise, intermittent and high-frequency noise can also affect work performance at far lower levels, especially on tasks requiring attention and judgement.

Responsible operations

In every chapter, under the heading of 'Responsible operations', we summarise how the particular topic covered in the chapter touches upon important social, ethical and environmental issues.

Although we have included (the somewhat mis-termed) 'scientific' management earlier, as just one among other influences on job design, some would argue that operations management is still excessively influenced by its ideas. Obviously, Taylor (1856–1915), the originator of 'scientific' management, was a product of his times, yet even in his lifetime, criticisms of the scientific management approach were being voiced. In a submission to the United States Commission on Industrial Relations, scientific management was described as:[15]

- Being in 'spirit and essence a cunningly devised speeding up and sweating system'.
- Intensifying the 'modern tendency towards specialisation of the work and the task'.
- Condemning 'the worker to a monotonous routine'.
- Putting 'into the hands of employers information that may be used to the detriment of workers'.
- Tending to 'transfer to the management all the traditional knowledge, the judgement and skills of workers'.
- Greatly intensifying 'unnecessary managerial dictation and discipline'.
- Tending to 'emphasise quantity of product at the expense of quality'.

Some of these criticisms could be seen as reflecting the nature of operations management, with its focus on productivity, specialisation, standardisation and the use of IT-based planning and control systems. All these could take responsibility and agency away from the staff working in the operation, and some very modern operations have been criticised for working practices not too far removed from Taylor's ideas. Look at the staff in the fulfilment centres central to the success of online retailing, whose performance is strictly measured and controlled. Or look at the staff in theme parks whose behaviour and personal appearance is defined and restricted; one can see the tension between conventional operations objectives and the requirements of a more inclusive and civilised approach to the role of humans working within many operations. In particular, two themes evident in the criticism of scientific management-influenced job design do warrant closer attention. The first is that inevitably, it results in, at best standardised, or at worst highly divided jobs, and thus reinforces the negative effects of excessive division of labour mentioned previously. Second, scientific management formalises the separation of the judgemental, planning and skilled tasks, which are done by 'management', from the routine, standardised and low-skill tasks, which are left for 'workers'. Such a separation, at the very least, deprives the majority of staff of an opportunity to contribute in a meaningful way to their jobs (and, incidentally, deprives the organisation of their contribution). Both of these themes lead to the same point: that the jobs designed under 'scientific' management principles lead to low motivation among staff, frustration at the lack of control over their work, and alienation from the job.

9.4 How are work times allocated?

Without some estimate of how long it takes to complete an activity, it would not be possible to know how much work to allocate to teams or individuals, to know when a task will be completed, to know how much it costs, to know if work is progressing according to schedule, and many other vital pieces of information that are needed to manage any operation. Without some estimate of work times, operations managers are 'flying blind'. At the same time, it does not need much thought before it becomes clear that measuring work times must be difficult to do with any degree of accuracy, or confidence. The time you take to do any task will depend on how skilled you are at the task, how much experience you have, how energetic or motivated you are, whether you have the appropriate tools, what the environmental conditions are, how tired you are, and so on. So, at best, any 'measurement' of how long a task will, or should, take will be an estimate. It will be our 'best guess' of how much time to allow for the task. That is why we call this process of estimating work times, 'work time allocation'. We are allocating a time for completing a task because we need to do so for many important operations management decisions. For example, work times are needed for:

- planning how much work a process can perform (its capacity);
- deciding how many staff are needed to complete tasks;
- scheduling individual tasks to specific people;
- balancing work allocation in processes (see Chapter 6);
- costing the labour content of a product or service;
- estimating the efficiency or productivity of staff and/or processes;
- calculating bonus payments (less important than it was at one time).

Notwithstanding the weak theoretical basis of work measurement, understanding the relationship between work and time is clearly an important part of job design. The advantage of structured and systematic work measurement is that it gives a common currency for the evaluation and comparison of all types of work. So, if work time allocation is important, how should it be done? In fact, there is a long-standing body of knowledge and experience in this area. This is generally referred to as 'work measurement', although as we have said, 'measurement' could be regarded as indicating a somewhat spurious degree of accuracy. Formally, work measurement is defined as 'the process of establishing the time for a qualified worker, at a defined level of performance, to carry out a specified job'. Although this is not a precise definition, generally it is agreed that a *specified job* is one for which specifications have been established to define most aspects of the job. A *qualified worker* is 'one who is accepted as having the necessary physical attributes, intelligence, skill, education and knowledge to perform the task to satisfactory standards of safety, quality and quantity'. **Standard performance** is 'the rate of output which qualified workers will achieve without over-exertion as an average over the working day provided they are motivated to apply themselves to their work'.

The techniques of work measurement

At one time, work measurement was firmly associated with an image of the 'efficiency expert', 'time and motion' man or 'rate fixer', who wandered around factories with a stopwatch, looking to save a few cents or pennies. And although that idea of work measurement has (almost) died out, the use of a stopwatch to establish a **basic time** for a job is still relevant, and used in a technique called 'time study'. Time study and the general topic of work measurement is treated in the supplement to this chapter.

OPERATIONS IN PRACTICE: Technology and surveillance at work[16]

Monitoring and analysing how people work is not new. Work study has always been used to increase productivity by examining and evaluating methods of work. But whereas traditionally observing how people go about their jobs has been performed 'up front' and been obvious to whoever is being examined, increasingly electronic surveillance technology is being used to track how we do our jobs. It is claimed to be a lot more effective, but a lot more controversial. Surveillance technologies range from simply requiring workers to tap in and out of their workplace (a method that has been used for over a century) to employee identification badges with integrated biometric measuring potential. These can track staff location, movements, interactions, the durations of any conversations, even the tone of voice being used in the conversation. Other technologies monitor whether you're at your desk, how often you are interrupted, what emails and phone calls are being made and even what speech patterns are being used. All of this is in support of increasing productivity and devising new work methods, say its supporters. Not so, say its detractors. There are dangers inherent in such surveillance, especially as technology creates more opportunities for companies to monitor their employees in unprecedented ways. Certainly, not everyone likes to be monitored. One study[17] found (unsurprisingly) that the more people felt their privacy to be violated at work, the more dissatisfied they were. Also, when they believe the surveillance to be unnecessary or too intense, they are more likely to find ways to subvert, sabotage or trick the surveillance systems. Objections can be more formalised. For example, when journalists at *The Telegraph* (a UK newspaper) discovered that a tracking device had been added to their desks by senior management, their union objected and the sensors (which monitored body heat, indicating when employees were at their desk and how often they moved around) were removed. *'The right to be consulted on new procedures governing such data is enshrined in law'*, said the union's assistant general secretary. '[We] *will resist Big Brother-style surveillance in the newsroom'*. Nevertheless, surveillance can be deemed necessary. In 2018 the UK Food Standards Agency announced that closed-circuit television will be installed at all 900 meat-cutting plants it monitored under plans to improve hygiene and reduce the risk of food poisoning. It had discovered that several plants had breached food safety rules by altering use-by dates on meat or dropping meat on the floor and putting it back on to the production line.

As well as time study, there are other work measurement techniques in use. They include the following:

- *Synthesis from elemental data* is a work measurement technique for building up the time for a job at a defined level of performance by totalling element times obtained previously from studies of other jobs containing the elements concerned or from synthetic data.
- *Predetermined motion–time systems* (PMTS) are a work measurement technique whereby times established for basic human motions (classified according to the nature of the motion and the conditions under which it is made) are used to build up the time for a job at a defined level of performance.
- *Analytical estimating* is a work measurement technique that is a development of estimating whereby the time required to carry out the elements of a job at a defined level of performance is estimated from knowledge and experience of the elements concerned.
- *Activity sampling* is a technique in which a large number of instantaneous observations are made over a period of time of a group of machines, processes or workers. Each observation records what is happening at that instant and the percentage of observations recorded for a particular activity or delay is a measure of the percentage of time during which that activity or delay occurs.

Critical commentary

The criticisms aimed at work measurement are many and various. Among the most common are the following:

- All the ideas on which the concept of a standard time is based are impossible to define precisely. How can one possibly give clarity to the definition of qualified workers, or specified jobs, or especially a defined level of performance?
- Even if one attempts to follow these definitions, all that results is an excessively rigid job definition. Most modern jobs require some element of flexibility, which is difficult to achieve alongside rigidly defined jobs.
- Using stopwatches to time human beings is both degrading and (usually) counterproductive. At best it is intrusive, at worst it makes people into 'objects for study'.
- The rating procedure implicit in time study is subjective and usually arbitrary. It has no basis other than the opinion of the person carrying out the study.
- Time study, especially, is very easy to manipulate. It is possible for employers to 'work back' from a time that is 'required' to achieve a particular cost. Also, experienced staff can 'put on an act' to fool the person recording the times.

Summary answers to key questions

9.1 Why are people so important in operations management?

- People are any organisation's greatest asset. Often, most people are to be found in the operations function.
- The importance of people and social issues is reflected in the socio-technical approach.

9.2 How can the operations function be organised?

- One can take various perspectives on organisations. How we illustrate organisations says much about our underlying assumptions of what an 'organisation' is. For example, organisations can be described as machines, organisms, brains, cultures or political systems.
- There are an almost infinite number of possible organisational structures. Most are blends of two or more 'pure types', such as the U-form, the M-form, matrix forms and the N-form.

9.3 How do we go about designing jobs?

- There are many influences on how jobs are designed. These include: the division of labour; scientific management; method study; work measurement; ergonomics; and behavioural approaches, such as job rotation, job enlargement and job enrichment, empowerment, team working and flexible working (including hybrid working, also known as 'telecommuting').

9.4 How are work times allocated?

- The best-known method is time study, but there are other work measurement techniques including synthesis from elemental data, predetermined motion–time systems (PMTS), analytical estimating and activity sampling.

CASE STUDY Grace faces (three) problems

Grace Whelan, Managing Partner of McPherson Charles, was puzzled. Three of her most successful teams seemed to be facing similar problems with their staff, even though each team had very different tasks, processes and types of staff. Every year the firm surveyed its entire staff in order to gauge their views, levels of satisfaction with their jobs and development needs. It was the results from the latest survey that surprised Grace. *'The results of the survey are really unanticipated. Only last year everything seemed fine. Now staff morale has evidently slumped in all three teams. Yet the partners who lead all of these teams are first class. Outstanding lawyers and good leaders'.*

McPherson Charles, based in Bristol in the west of England, had grown rapidly to be one of the biggest law firms in the region, with 21 partners and around 400 staff. Three years previously the firm had reorganised into 15 teams, each headed by a 'lead partner' and specialising in practising one type of law. It had proved to be a good organisational structure, which encouraged each team to organise themselves appropriately for the type of clients that they dealt with. In particular, three teams had flourished under this structure, 'family law', 'property' and 'litigation'. Now it was these very teams whose staff were showing signs of dissatisfaction.

Before the results of the survey were published to all staff, Grace knew that she would need to have worked out some kind of response to the issues raised. She decided to go and see each of the lead partners in the three teams. The first person she decided to talk to was Simon Reece, who led the family law team. Before doing so she explained what his team did.

Family law

'They are called the "family law" team but basically what they do is to help people through the trauma of divorce, separation and break up. Their biggest "high-value" clients come to them because of word-of-mouth recommendation. Last year they had almost a hundred of these "high-value" clients and they all valued the personal touch that they were able to give them, getting to know them well and spending time with them to understand the often "hidden" aspects of their case. Of course, not all their clients are the super-rich. About a third of the annual family law income comes from about 750 relatively routine divorce and counselling cases'.

Simon was blunt about the declining levels of staff satisfaction in his team. *'The problem is that working with the "high-value" clients is just more fun and more rewarding than the routine "bread-and-butter" work. So my people who do that kind of work, usually the more experienced ones, don't want to take on the routine stuff. With "high-value" cases you have to be able to untangle the personal issues from the business ones. Interviewing these clients cannot be rushed. They tend to be wealthy people with complex assets. We will often have to drop everything and go off half-way round the world to meet and discuss their situation. There are no standard procedures, every client is different, and everyone has to be treated as an individual. So we have a team of individuals who rise to the challenge each time and give great service. By contrast, the routine work is a lot less interesting, yet sometimes very harrowing. The more junior staff who tend to take on the routine cases can sometimes feel themselves to be "second-class citizens". Many of them would like to get more experience with the complex high-value work, but I can't take the risk of giving them that degree of responsibility, the work is too valuable. Also, frankly, the senior people who deal with the high-value work don't want to give up their more glamorous work. I have been trying to make sure that everyone in my team who wants to has a mix of interesting and routine work over the year. It's the only way to develop them in the long term. You have to encourage them to exercise and develop their professional judgement. They are empowered to deal with any issues themselves or call on one of the more senior members of the team for advice if appropriate. It is important to give this kind of responsibility to them so that they see themselves as part of a team. But there are still tensions between senior and more junior staff. We are thinking about adopting an open-plan office arrangement centred around our specialist library of family case law, to try and encourage more cooperation'.*

Litigation

Grace was less concerned about the litigation team, led by Hazel Lewis. *'The litigation team has been our best success story. The have grown far faster than any other part of the firm, and a lot of that is down to Hazel. She provides a key service for our commercial client base. Their primary work consists of handling bulk collections of debt. The group has 17 clients of which 5 provide 85 per cent of total volume. They work closely with the accounts departments of the client companies and have developed a semi-automatic approach to debt collection. It's a great service that Hazel has largely automated'.*

Hazel had led the litigation team since it had been set up four years ago. As well as being the partner in charge of litigation, unusually she and her assistant were the only qualified lawyers in the team. *'Our problems in the litigation team are not really because of any internal tensions or disputes. Broadly, our people are happy with what they do and how they are supervised. The issue is just that we are so different from the rest of the firm. Apart from myself and Raymond* [her assistant] *everyone else in the team are either technicians who look after and develop the systems that we use, or people who have worked in processing or call centres, before they came to us. And between us we have developed a smart operation here. Our staff input data received from their clients into the system, from that point everything progresses through a pre-defined process, letters are produced, queries responded to and eventually debts collected, ultimately through court proceedings if necessary. Work tends to come in batches from clients, and varies according to the time of year and client sales activities. At the moment things are fairly steady: we had almost 900 new cases to deal with last week. The details of each case are sent over by the client; our people input the data onto our screens and set up a standard diary system for sending letters out. Some people respond quickly to the first letter and often the case is closed within a week or so; other people ignore letters and eventually we initiate court proceedings. We know exactly what is required for court dealings and have a pretty good process to make sure all the right documentation is available on the day. Our problem is that the rest of the firm does not see us as being "proper lawyers", and they are right, we're not. But it does get difficult for our people, being looked down upon all the time. Our salary structure is different, our bonus scheme is different, and how we measure performance is different. But there is a solution. Because we have expanded so much, we need more space than is available in this building. I think that we should think about moving the litigation team. There is a great location out by the airport that could be expanded in the future if needed. There is really no reason for us to be located with the other teams'.*

Property

The 'property' team is one of the largest parts of the firm and is established in the local market with an excellent reputation for being fast, friendly and giving value for money. Most of their work is 'domestic', acting for individuals buying or selling their home, or their second home. Each client is allocated to a solicitor who becomes their main point of contact. But, given that they can have up to a hundred domestic clients a week, most of the work is actually carried out by the rest of the team of 'paralegal' staff (staff with qualifications less than a fully qualified lawyer) behind the scenes.

Kate Hutchinson, who led the property team, was proud of the process she and her team had set up. *'There is a relatively standard process to domestic property sales and purchases and we think that we are pretty efficient at managing these standard jobs. Our process has four stages, one dealing with land registry searches, one liaising with banks who are providing the mortgage finance, one to make sure surveys are completed and one section that finalises the whole process to completion. We believe that this degree of specialisation can help us achieve the efficiencies that are becoming important, as the market gets more competitive. Our particular problem is that increasingly we are also getting more complex "special" jobs. These are things like "volume re-mortgage" arrangements and rather complex "one-off" jobs, where a mortgage lender transfers a complex set of loan assets to another lender. These "special" jobs are always more complex than the domestic work and they are not popular with our staff. They don't always fit easily into our standard process, and they disrupt the routine of working. For example, sometimes there are occasions when fast completion is particularly important and that can throw us a bit'.*

Grace was more worried about the property team than Kate appeared to be. The firm had recently formed partnerships with two large speculative builders, which dealt in special 'plot sales' that would also be classed as non-standard 'specials' by Kate. Grace knew that all these 'specials' did involve a lot of work and could occupy several members of the team for a time. But they were an important source of revenue. Currently the team was dealing with up to 25 'specials' each week, and this would certainly increase. Grace suspected that Kate was mistaken to try to follow the same process with them as the normal domestic jobs. Maybe trying to do different things on the same process was the cause of the dissatisfaction in the team?

QUESTIONS

1. **What are the problems among the staff of each of the three teams?**
2. **What are the individual 'services' offered by each of the three teams?**
3. **How would you describe each team's process in terms of the jobs of its staff?**
4. **What do you think each team leader should be doing to try to overcome their team's problems?**

Problems and applications

All chapters have 'Problems and applications' questions that will help you practise analysing operations. They can be answered by reading the chapter. Model answers for the first two questions can be found on the companion website for this text.

1 Using technology for staff surveillance is clearly a controversial issue. Re-read the 'Operations in practice' example that discusses using technology for surveillance.

   (a) Draw up a list of possible positives and negatives that could result from staff surveillance.
   (b) Which industries do you think might be the most eager to trial staff surveillance?

2 Operations managers can have a profound influence on how organisations implement their **human resource strategy** (the overall long-term approach to ensuring that an organisation's human resources provide a strategic advantage). One authority on human resource (HR) strategy (Dave Ulrich, the University of Michigan) proposes four elements to the HR activity:[18]

   - Being 'strategic partner' to the business – aligning HR and business strategy: 'organisational diagnosis', manpower planning, environmental monitoring, etc.
   - Administering HR procedures and processes – running the organisation's HR processes and 'shared services': payroll, appraisal, selection and recruitment, communication, etc.
   - Being an 'employee champion' – listening and responding to employees: 'providing resources to employees', conciliation, career advice, grievance procedures, etc.
   - Being a 'change agent' – managing transformation and change: 'ensuring capacity for change', management development, performance appraisal, organisation development, etc.

   What do you think is the relevance of these HR roles to operations managers?

3 *(It is recommended that you look at the supplement to this chapter before answering this question.)* A hotel has two wings, an east wing and a west wing. Each wing has four 'room-service housekeepers' working 7-hour shifts to service the rooms each day. The east wing has 40 standard rooms, 12 deluxe rooms and 5 suites. The west wing has 50 standard rooms and 10 deluxe rooms. The standard times for servicing rooms are as follows: standard rooms 20 standard minutes (SMs), deluxe rooms 25 SMs and suites 40 SMs. In addition, an allowance of 5 SMs per room is given for any miscellaneous jobs such as collecting extra items for the room or dealing with customer requests. The east-wing housekeepers believe that they have the more demanding job. Are they right?

4 In the example above, one of the housekeepers in the west wing wants to job share with their partner, each working 3 hours per day. Their colleagues have agreed to support them and will guarantee to service all the rooms in the west wing to the same standard each day. Can this arrangement succeed without excessive work for the other three housekeepers?

5 *(This question is based on an original case study by Dr Ran Bhamra, Loughborough University.)* Service Adhesives Ltd produces specialist adhesives. It has always been profitable, but there had been a slowdown in the company's profits. Several improvement initiatives had attempted to reverse the company's declining position, but none had fully taken hold. Some senior management put this down to staff having 'below-average' skills and motivation, and being reluctant to change. Staff turnover was high, and the company had started employing short-term contract labour to cope with fluctuating orders. There had been some tension between temporary and permanent employees. The company organised a visit to one of its customers, called (bizarrely) 'Happy Products'. *'It was like entering another world. Their plant was cleaner, the flow of materials seemed smoother, their staff*

seemed purposeful, it seemed efficient, and everybody worked as a team. I'm sure that team-based approach could be implemented just as successfully in our plant' (CEO, Service Adhesives).

The Happy Products operation made diapers (nappies) and healthcare products and was organised into three product areas, each staffed by five operators. One operator was a team leader responsible for 'first line management'. A second operator was a specially trained health and safety representative. A third was a trained quality representative who also liaised with the quality department. A fourth operator was a trained maintenance engineer, while a fifth was a non-specialist 'floating' operator. This meant that most day-to-day problems could be dealt with immediately, so production output, product quality and line efficiency were controlled exceptionally well. Team members derived great satisfaction from playing a key part in the success of the organisation. Teams were also involved in determining annual performance targets for their specific areas. Service Adhesives decided to adopt a team-based work organisation. However, it realised that it lacked the organisational 'cohesiveness' that Happy Products had. Traditionally, it had prided itself on its hierarchical organisation structure, with five layers of management from the plant director to the shop-floor operatives. The chain of command was strictly enforced by operating procedures enshrined in the comprehensive quality assurance system.

(a) Service Adhesives Ltd currently employs some people on short-term contracts. How could this affect its proposed team-based working structure?
(b) In moving from a traditional to a team-based work structure, what sort of formal (e.g. roles and procedures) and informal (e.g. social groups and communication) barriers is Service Adhesive likely to encounter?
(c) Senior management of Service Adhesives thought that the main reason for ineffective improvement initiatives in the past was the apparent lack of cohesion among the organisation's human resources. Could a team-based work organisation be the answer to their organisational difficulties?
(d) Employee empowerment is a key element of team-based working. What difficulties could Service Adhesives face in implementing empowerment?

6 At W.L. Gore (that makes the high-performance Gore-Tex fabrics) few in the company have any formal job titles or job descriptions. There are no managers, only leaders and associates, people are paid 'according to their contribution' and staff help to determine each other's pay. Its skilled staff (called 'associates') develop, manufacture and sell a range of innovative products. Associates are hired for general work areas rather than specific jobs, and with the guidance of their 'sponsors' (not bosses) and as they develop experience, they commit to projects that match their skills. Teams organise around opportunities as they arise, with associates committing to the projects that they have chosen to work on, rather than having tasks delegated to them. Project teams are small, focused, multi-disciplined, and foster strong relationships between team members. Personal initiative is encouraged, as is 'hands-on' innovation. There are no traditional organisational charts, no chains of command, no predetermined channels of communication. Instead, team members communicate directly with each other and are accountable to the other members of their team. Groups are led by whoever is the most appropriate person at each stage of a project. Leaders are not appointed by senior management; they 'emerge' naturally by demonstrating special knowledge, skill or experience that advances a business objective. Everyone's performance is assessed using a peer-level rating system. The explicit aim of the company's culture is to 'combine freedom with cooperation and autonomy with synergy'.

(a) How is W.L. Gore different from most international corporations?
(b) Why is Gore's way of working particularly appropriate for how it competes in its markets?

7 Among the first large organisation to take flexible working seriously in Europe was Lloyds TSB Banking Group (now called TSB Bank). It adopted flexible working because it was sensitive to the social and economic changes that were affecting both customers and staff. There were benefits of adopting work patterns that reflected its staff's needs and yet still offered quality of service to customers. Recruiting and keeping talented people meant understanding and implementing the

right balance between staff's individual needs, the business's requirement to control its costs, and the customers' expectation of excellent service. A survey of employees' views showed that one of their main concerns was trying to balance the job with outside commitments, such as family and leisure. So, the Group introduced its flexible working policy, called 'Work Options'. It allowed staff to request a different working pattern from the conventional working day. Sometimes this simply involved starting and finishing earlier or later each day, while maintaining the same weekly hours. This could benefit the business. Varying staff's work patterns could mean staffing is more closely aligned with actual customer demand. Job sharing is also used. It suits two staff, who may not want full-time employment and the business can have two people's combined experience, skills and creativity. Job-sharing staff can also be more productive than full-time colleagues. Another form of flexible working is 'compressed working', where staff work a standard one or two weeks within a shorter timescale, for example by working some longer days a week, then taking extra time off to compensate.

(a) What seem to be the main advantages and potential disadvantages of flexible working for staff, the company and customers?
(b) How can a firm such as Lloyds try and overcome any clashes between the requirements of staff, the business and customers?

8 Re-read the 'Operations in practice' example on Torchbox selling itself to its employees. What advantages does such a move have (a) for the employees, (b) for the original owners?

9 Many staff in retail operations spend most of their time interacting with customers, not all of whom are always polite. In fact, some can be abusive. How should such operations balance their responsibility to their customers with their responsibility to their staff?

10 Some high customer contact jobs (such as police forces) have started requiring staff to wear 'body cameras' that record interactions with the public. What do you see as the advantages and disadvantages of doing this?

Selected further reading

Argyris, C. (1998) Empowerment: the emperor's new clothes, *Harvard Business Review*, May–June.
A critical but fascinating view of empowerment.

Bock, L. (2015) *Work Rules!: Insights from Inside Google that Will Transform How You Live and Lead*, John Murray, London.
With an agenda far wider than this chapter, it is nevertheless an absorbing book that gives an insight into an absorbing firm.

Buchanan, D. and Huczynski, A. (2019) *Organizational Behaviour*, 10th edn, Pearson, Harlow.
One of the most popular and best-established books in its field – for good reason.

Dul, J. and Weerdmeester, B. (2008) *Ergonomics for Beginners: A Quick Reference Guide*, 3rd edn, CRC Press, Boca Raton, FL.
Good, practical guidance on the removal from the workplace of physical and mental stresses caused by poor job or environmental design.

Hackman, R.J. and Oldham, G. (1980) *Work Redesign*, Addison-Wesley, Reading, MA.
Somewhat dated but, in its time, ground breaking and certainly hugely influential.

Herzberg, F. (1987) One more time: how do you motivate employees? (with retrospective commentary), *Harvard Business Review*, 65 (5), 109–120.
An interesting look back by one of the most influential figures in the behavioural approach to job design school.

Mullins, L. (2016) ***Management and Organisational Behaviour*****, 11th edn, Pearson, Harlow.**
Another classic, with broad coverage.

Schwartz, J., Riss, S. and Fishburne, T. (2021) ***Work Disrupted: Opportunity, Resilience, and Growth in the Accelerated Future of Work*****, Wiley, Hoboken, NJ.**
Forward-looking and speculative. How consultants see the future of work for us all.

Shorrock, S. (ed.) (2016) ***Human Factors and Ergonomics in Practice*****, CRC Press, Boca Raton, FL.**
An edited book, but with lots of examples of the real practice of human factors and ergonomics.

Notes on chapter

1. Schein, E.H. (1999) *The Corporate Culture Survival Guide: Sense and Nonsense About Culture Change*, Jossey-Bass, San Francisco, CA.
2. Schein, E.H. (1992) *Organizational Culture and Leadership*, 2nd edn, Jossey-Bass, San Francisco, CA.
3. The information on which this example is based is taken from: an interview with Tom Dyson, and the Torchbox website, http://www.torchbox.com (accessed September 2021).
4. Morgan describes these and other metaphors in Morgan, G. (1986) *Images of Organization*, Sage, Thousand Oaks, CA.
5. Hoxie, R.F. (1915) *Scientific Management and Labor*, D. Appleton and company, New York, NY. Or Taylor's original work, first published in 1911: Taylor F.W. (2005) *The Principles of Scientific Management*, 1st World Library – Literary Society.
6. The information on which this example is based is taken from: Byers, D. (2017) Bionic suits to make tools feel weightless, *The Times*, 24 July; Coxworth, B. (2017) Exoskeleton helps Ford workers reach up, New Atlas, 13 November, https://newatlas.com/ford-eksovest/52166/ (accessed September 2021); Goode, L. (2017) Are exoskeletons the future of physical labor? The Verge, 5 December, https://www.theverge.com/2017/12/5/16726004/verge-next-level-season-two-industrial-exoskeletons-ford-ekso-suitx (accessed September 2021).
7. Hackman, J.R., Oldham, G., Janson, R. and Purdy, K. (1975) A new strategy for job enrichment, *California Management Review*, (17) 4, 57–71.
8. The information on which this example is based is taken from: Hill, A. (2017) Power to the workers: Michelin's great experiment, *Financial Times*, 11 May; Hill, A. (2017) Michelin chief Jean-Dominique Senard devolves power to workers, *Financial Times*, 14 May; Michelin (2017) 2016 Annual Report.
9. The information on which this example is based is taken from: Nixey, C. (2020) Death of the office, *Economist 1843 Magazine*, 29 April; Economist (2020) Countering the tyranny of the clock, *Economist* print edition, 17 October; Treanor, J. (2021) Has Goldman's DJ just pulled the plug on WFH? *The Sunday Times*, 28 February; Hill, A. (2020) Future of work: how managers are harnessing employees' hidden skills, *Financial Times*, 1 September.
10. Berggren, C. (1992) *The Volvo Experience: Alternatives to Lean Production in the Swedish Auto Industry*, ILR Press.
11. The Health and Safety Executive (HSE) of the UK Government.
12. Beauregard, T. A. and Henry, L.C. (2009) Making the link between work-life balance practices and organizational performance, *Human Resource Management Review*, 19 (1), 9–22.
13. The information on which this example is based is taken from: Bone, J., Robertson, D. and Pavia, W. (2010) Plane rumpus puts focus on crews' growing revolution in the air, *The Times*, 11 August.
14. The information on which this example is based is taken from: Jones, A. (2015) The riff: dangers of music at work, *Financial Times*, 5 August; Ciotti, G. (2014) How music affects your productivity, *Fast Company*, 11 July; BBC (2013) Does music in the workplace help or hinder?, *Magazine Monitor*, 9 September.
15. Hoxie R.F. (1916) *Scientific Management and Labor*, published originally by The United States Commission on Industrial Relations, Scientific Management; and in the *Monthly Review of the U.S. Bureau of Labor Statistics*, 2 (1) (January 1916), 28–38.
16. The information on which this example is based is taken from: Derousseau, R. (2017) The tech that tracks your movements at work, BBC Worklife, 14 June, https://www.bbc.com/worklife/article/20170613-the-tech-that-tracks-your-movements-at-work (accessed September 2021); Solon, O. (2017) Big Brother isn't just watching: workplace surveillance can track your every move, *Guardian*, 6 November; Staats, B.R., Dai, H., Hofmann, D. and Milkman, K.L. (2016) Motivating process compliance through individual electronic monitoring: an empirical examination of hand hygiene in healthcare, *Management Science*, 63 (5), 1563–85; Webster, B. (2018) CCTV to monitor hygiene in meat factories, *The Times*, 3 March.
17. Samaranayake, V. and Gamage, C. (2012) Employee perception towards electronic monitoring at work place and its impact on job satisfaction of software professionals in Sri Lanka, *Telematics and Informatics*, 29 (2), 233–44.
18. Best explained in Ulrich, D. (1996) *Human Resource Champions: The Next Agenda for Adding Value and Delivering Results*, Harvard Business Review Press.

Supplement to Chapter 9
Work study

INTRODUCTION

A tale is told of Frank Gilbreth (the founder of method study) addressing a scientific conference with a paper entitled 'The best way to get dressed in the morning'. In his presentation, he rather bemused the scientific audience by analysing the 'best' way of buttoning up one's waistcoat in the morning. Among his conclusions was that waistcoats should always be buttoned from the bottom upwards (to make it easier to straighten his tie in the same motion; buttoning from the top downwards requires the hands to be raised again). Think of this example if you want to understand scientific management and **method study** in particular. First of all, he is quite right. Method study and the other techniques of scientific management may often be without any intellectual or scientific validation, but by and large they work in their own terms. Second, Gilbreth reached his conclusion by a systematic and critical analysis of what motions were necessary to do the job. Again, these are characteristics of scientific management – detailed analysis and painstakingly systematic examination. Third (and possibly most important), the results are relatively trivial. A great deal of effort was put into reaching a conclusion that was unlikely to have any earth-shattering consequences. Indeed, one of the criticisms of scientific management, as developed in the early part of the twentieth century, is that it concentrated on relatively limited, and sometimes trivial, objectives.

The responsibility for its application, however, has moved away from specialist 'time and motion' staff to the employees who can use such principles to improve what they do and how they do it. Further, some of the methods and techniques of scientific management, as opposed to its philosophy (especially those that come under the general heading of 'method study'), can in practice prove useful in critically re-examining job designs. It is the practicality of these techniques that possibly explains why they are still influential in job design almost a century after their inception.

Method study in job design

Method study is a systematic approach to finding the best method. There are six steps:

1 Select the work to be studied.
2 Record all the relevant facts of the present method.
3 Examine those facts critically and in sequence.
4 Develop the most practical, economic and effective method.
5 Install the new method.
6 Maintain the method by periodically checking it in use.

Step 1 – Selecting the work to be studied

Most operations have many hundreds and possibly thousands of discrete jobs and activities that could be subjected to study. The first stage in method study is to select those jobs to be studied that will give the most return on the investment of the time spent studying them. This means it is unlikely that it will be worth studying activities which, for example, may soon be discontinued or are only performed occasionally. On the other hand, the types of job that should be studied as a matter of priority are those which, for example, seem to offer the greatest scope for improvement, or which are causing bottlenecks, delays or problems in the operation.

Step 2 – Recording the present method

There are many different recording techniques used in method study. Most of them:

- record the sequence of activities in the job;
- record the time interrelationship of the activities in the job; or
- record the path of movement of some part of the job.

Perhaps the most commonly used recording technique in method study is process mapping, which was discussed in Chapter 6. Note that here we are recording the present method of doing the job. It may seem strange to devote so much time and effort to recording what is currently happening when, after all, the objective of method study is to devise a better method. The rationale for this is, first of all, that recording the present method can give a far greater insight into the job itself, and this can lead to new ways of doing it. Second, recording the present method is a good starting point from which to evaluate it critically and therefore improve it. In this last point the assumption is that it is easier to improve the method by starting from the current method and then criticising it in detail, than by starting with a 'blank sheet of paper'.

Step 3 – Examining the facts

This is probably the most important stage in method study and the idea here is to examine the current method thoroughly and critically. This is often done by using the so-called 'questioning technique'. This technique attempts to detect weaknesses in the rationale for existing methods so that alternative methods can be developed (see Table 9.2). The approach may appear somewhat detailed and tedious, yet it is fundamental to the method study philosophy – everything must be critically examined. Understanding the natural tendency to be less than rigorous at this stage, some organisations use pro-forma questionnaires, asking each of these questions and leaving space for formal replies and/or justifications, which the job designer is required to complete.

Table 9.2 The method study questioning technique

Broad question	Detailed question
The purpose of each activity (questions the fundamental need for the element)	What is done? Why is it done? What else could be done? What should be done?
The place in which each element is done (may suggest a combination of certain activities or operations)	Where is it done? Why is it done there? Where else could it be done? Where should it be done?
The sequence in which the elements are done (may suggest a change in the sequence of the activity)	When is it done? Why is it done then? When should it be done?
The person who does the activity (may suggest a combination and/or change in responsibility or sequence)	Who does it? Why does that person do it? Who else could do it? Who should do it?
The means by which each activity is done (may suggest new methods)	How is it done? Why is it done in that way? How else could it be done? How should it be done?

Table 9.3 The principles of motion economy

Broad principle	How to do it
Use the human body the way it works best	▶ Work should be arranged so that a natural rhythm can become automatic ▶ Motion of the body should be simultaneous and symmetrical if possible ▶ The full capabilities of the human body should be employed ▶ Arms and hands as weights are subject to the physical laws and energy should be conserved ▶ Tasks should be simplified
Arrange the workplace to assist performance	▶ There should be a defined place for all equipment and materials ▶ Equipment, materials and controls should be located close to the point of use ▶ Equipment, materials and controls should be located to permit the best sequence and path of motions ▶ The workplace should be fitted both to the tasks and to human capabilities
Use technology to reduce human effort	▶ Work should be presented precisely where needed ▶ Guides should assist in positioning the work without close operator attention ▶ Controls and foot-operated devices can relieve the hands of work ▶ Mechanical devices can multiply human abilities ▶ Mechanical systems should be fitted to human use

Step 4 – Developing a new method

The previous critical examination of current methods has by this stage probably indicated some changes and improvements. This step involves taking these ideas further in an attempt to:

- eliminate parts of the activity altogether;
- combine elements together;
- change the sequence of events so as to improve the efficiency of the job; or
- simplify the activity to reduce the work content.

A useful aid during this process is a checklist such as the revised principles of motion economy. Table 9.3 illustrates these.

Steps 5 and 6 – Install the new method and regularly maintain it

The method study approach to the installation of new work practices concentrates largely on 'project-managing' the installation process. It also emphasises the need to monitor regularly the effectiveness of job designs after they have been installed.

Work measurement in job design

Basic times

Terminology is important in work measurement. When a *qualified worker* is working on a *specified job* at *standard performance*, the time they take to perform the job is called the basic time for the job. Basic times are useful because they are the 'building blocks' of time estimation. With the basic times for a range of different tasks, an operations manager can construct a time estimate for any longer activity that is made up of the tasks. The best-known technique for establishing basic times is probably time study.

Time study

Time study is 'a work measurement technique for recording the times and rate of working for the elements of a specified job, carried out under specified conditions, and for analysing the data so as to obtain the time necessary for the carrying out of the job at a defined level of performance'. The technique takes three steps to derive the basic times for the elements of the job:

- observing and measuring the time taken to perform each element of the job;
- adjusting, or 'normalising', each observed time;
- averaging the adjusted times to derive the basic time for the element.

Step 1 – Observing, measuring and rating

A job is observed through several cycles. Each time an element is performed, it is timed using a stopwatch. Simultaneously with the observation of time, a rating of the perceived performance of the person doing the job is recorded. Rating is 'the process of assessing the worker's rate of working relative to the observer's concept of the rate corresponding to standard performance. The observer may take into account, separately or in combination, one or more factors necessary to carrying out the job, such as speed of movement, effort, dexterity, consistency, etc.'. There are several ways of recording the observer's rating. The most common is on a scale that uses a rating of 100 to represent standard performance. If an observer rates a particular observation of the time to perform an element at 100, the time observed is the actual time that anyone working at standard performance would take.

Step 2 – Adjusting the observed times

The adjustment to normalise the observed time is:

$$\frac{\text{observed rating}}{\text{standard rating}}$$

where standard rating is 100 on the common rating scale we are using here. For example, if the observed time is 0.71 minutes and the observed rating is 90, then:

$$\text{Basic time} = \frac{0.71 \times 90}{100} = 0.64 \text{ mins}$$

Step 3 – Averaging the basic times

In spite of the adjustments made to the observed times through the rating mechanism, each separately calculated basic time will not be the same. This is not necessarily a function of inaccurate rating, or even the vagueness of the rating procedure itself; it is a natural phenomenon of the time taken to perform tasks. No human activity can be repeated in *exactly* the same time on every occasion.

Standard times

The standard time for a job is an extension of the basic time and has a different use. Whereas the basic time for a job is a piece of information that can be used as the first step in estimating the time to perform a job under a wide range of conditions, standard time refers to the time *allowed* for the job under specific circumstances. This is because standard time includes allowances, which reflect the rest and relaxation allowed because of the conditions under which the job is performed. So the standard time for each element consists principally of two parts, the basic time (the time taken by a qualified worker, doing a specified job at standard performance) and an allowance (this is added to the basic time to allow for rest, relaxation and personal needs).

Allowances

Allowances are additions to the basic time, intended to provide the worker with the opportunity to recover from the physiological and psychological effects of carrying out specified work under specified conditions and to allow for personal needs. The amount of the allowance will depend on the nature of the job. The way in which relaxation allowance is calculated, and the exact allowances given for each of the factors that determine the extent of the allowance, varies between different organisations. Table 9.4 illustrates the allowance table used by one company, which manufactures domestic appliances. Every job has an allowance of 10 per cent. The table shows the further percentage allowances to be applied to each element of the job. In addition, other allowances may be applied for such things as unexpected contingencies, synchronisation with other jobs, unusual working conditions and so on.

Figure 9.9 shows how average basic times for each element in the job are combined with allowances (low in this example) for each element to build up the standard time for the whole job.

Table 9.4 An allowances table used by a domestic appliance manufacturer

Allowance factors	Example	Allowance (%)
Energy needed		
Negligible	none	0
Very light	0–3 kg	3
Light	3–10 kg	5
Medium	10–20 kg	10
Heavy	20–30 kg	15
Very heavy	Above 30 kg	15–30
Posture required		
Normal	Sitting	0
Erect	Standing	2
Continuously erect	Standing for long periods	3
Lying	On side, face or back	4
Difficult	Crouching, etc.	4–10
Visual fatigue		
Nearly continuous attention		2
Continuous attention with varying focus		3
Continuous attention with fixed focus		5
Temperature		
Very low	Below 0°C	over 10
Low	0–12°C	0–10
Normal	12–23°C	0
High	23–30°C	0–10
Very high	Above 30°C	over 10
Atmospheric conditions		
Good	Well ventilated	0
Fair	Stuffy/smelly	2
Poor	Dusty/needs filter	2–7
Bad	Needs respirator	7–12

Job Pack 20 × pt # 73/2A　　Location Packing Dept.　　Observer FWT

Element		Observation					Observation					Average basic time	Allowances	Element standard time
		1	2	3	4	5	6	7	8	9	10			
Make box	Observed time	0.71	0.71	0.71	0.69	0.75	0.68	0.70	0.70	0.70	0.68			
	Rating	90	90	90	90	80	90	90	90	90	90			
	Basic time	0.64	0.64	0.63	0.62	0.60	0.61	0.63	0.65	0.63	0.61	0.626	10%	0.689
Pack × 20	Observed time	1.30	1.32	1.25	1.33	1.33	1.28	1.32	1.32	1.30	1.30			
	Rating	90	90	100	90	90	90	90	90	90	90			
	Basic time	1.17	1.19	1.25	1.20	1.20	1.15	1.19	1.19	1.17	1.17	1.168	12%	1.308
Seal and secure	Observed time	0.53	0.55	0.55	0.56	0.53	0.53	0.53	0.55	0.49	0.51			
	Rating	90	90	90	90	90	90	90	90	100	100			
	Basic time	0.48	0.50	0.50	0.50	0.48	0.48	0.48	0.50	0.49	0.51	0.495	10%	0.545
Assemble outer	Observed time	1.12	1.21	1.20	1.25	1.41	1.27	1.27	1.15	1.20	1.23			
fix and label	Rating	100	90	90	90	90	90	90	100	90	90			
	Basic time	1.12	1.09	1.08	1.13	1.27	1.14	1.14	1.15	1.08	1.21	1.138	12%	1.275

Raw standard time		3.817
Allowances for total job	5%	0.191
Standard time for job		4.01 SM

Figure 9.9 Time study of a packing task – standard time for the whole task calculated

Worked example

The Monrovian Embassy

Two work teams in the Monrovian Embassy have been allocated the task of processing visa applications. Team A processes applications from Europe, Africa and the Middle East. Team B processes applications from North and South America, Asia and Australasia. Team A has chosen to organise itself in such a way that each of its three team members processes an application from start to finish. The four members of Team B have chosen to split themselves into two sub-teams. Two open the letters and carry out the checks for a criminal record (no one who has been convicted of any crime other than a motoring offence can enter Monrovia), while the other two team members check for financial security (only people with more than Monrovian $1,000 may enter the country). The head of consular affairs is keen to find out if one of these methods of organising the teams is more efficient than the other. The problem is that the mix of applications differs region by region. Team A typically processes around two business applications to every one tourist application. Team B processes around one business application to every two tourist applications.

A study revealed the following data:

Average standard time to process a business visa = 63 standard minutes
Average time to process a tourist visa = 55 standard minutes

Average weekly output from Team A is:

85.2 business visas
39.5 tourist visas

Average weekly output from Team B is:

53.5 business visas
100.7 tourist visas

All team members work a 40-hour week.

The efficiency of each team can be calculated by comparing the actual output in standard minutes and the time worked in minutes.

So Team A processes:

$$(85.2 \times 63) + (39.5 \times 55) = 7{,}540.1 \text{ standard minutes of work}$$

$$\text{in } 3 \times 40 \times 60 \text{ minutes} = 7{,}200 \text{ minutes}$$

$$\text{So its efficiency} = \frac{7540.1}{7200} \times 100 = 104.72\%$$

Team B processes:

$$(53.5 \times 63) + (100.7 \times 55) = 8{,}909 \text{ standard minutes of work}$$

$$\text{in } 4 \times 40 \times 60 \text{ minutes} = 9{,}600 \text{ minutes}$$

$$\text{So its efficiency} = \frac{8909}{9600} \times 100 = 92.8\%$$

The initial evidence therefore seems to suggest that the way Team A has organised itself is more efficient.

PART THREE

Deliver

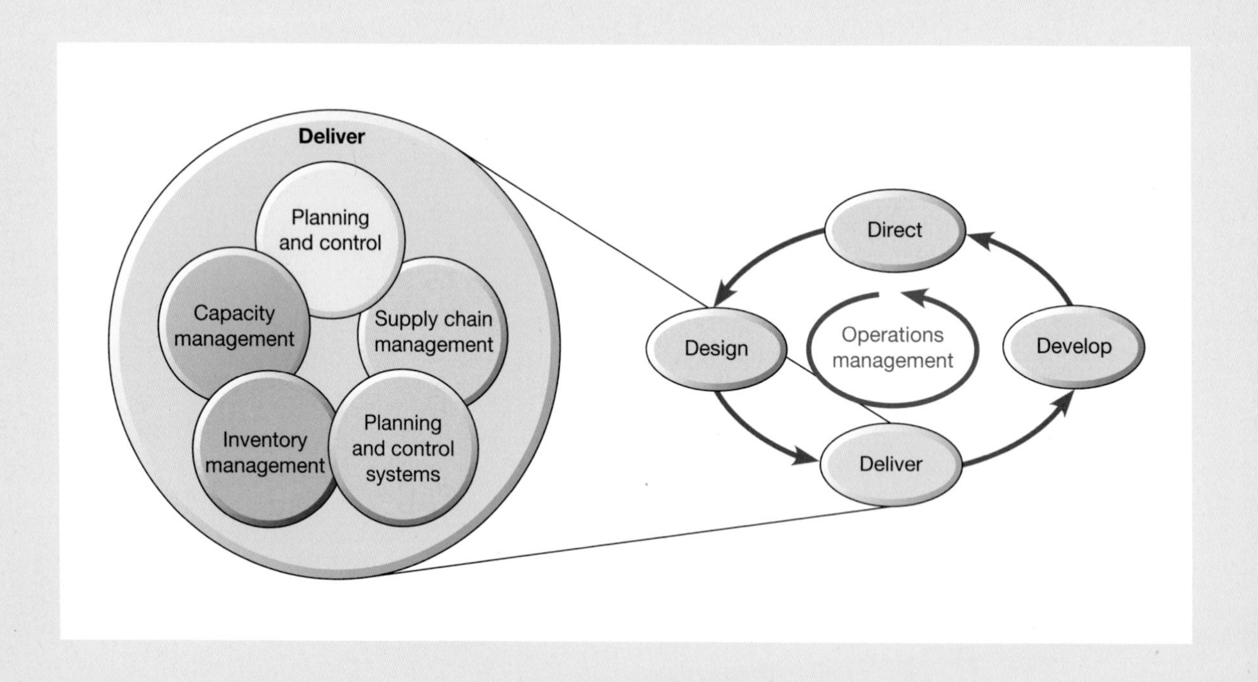

All the activities involved in the design of an operation should have provided the nature and shape of the transforming resources that are capable of satisfying customers' demands. Products and services then have to be created and delivered to customers. This is done by planning and controlling the activities of the transforming resources on a day-to-day basis to ensure the appropriate supply of products and services to meet the requirements of the market. This part of the text will look at five different aspects of planning and controlling the delivery of products and services as they flow through processes, operations and supply networks. The chapters in this part are:

- **Chapter 10 Planning and control**

  This examines how operations organise the delivery of their products and services on an ongoing basis so that customers' demands are satisfied.

- **Chapter 11 Capacity management**

  This explains how operations need to decide how to vary their capacity (if at all) as demand for their products and services fluctuates.

- **Chapter 12 Supply chain management**

  This describes how operations relate to each other in the context of a wider network of suppliers and customers, and how these relationships can be managed.

- **Chapter 13 Inventory management**

  This looks at how transformed resources accumulate as inventories as they flow through processes, operations or supply networks.

- **Chapter 14 Planning and control systems**

  This describes how systems are needed to manage the very large amounts of information required to plan and control operations, and how enterprise resources planning (ERP) is used to do this.

10 Planning and control

KEY QUESTIONS

10.1 What is planning and control?

10.2 How do supply and demand affect planning and control?

10.3 What is 'loading'?

10.4 What is 'sequencing'?

10.5 What is 'scheduling'?

10.6 What is 'monitoring and control'?

INTRODUCTION

The design of an operation determines the resources with which it creates its services and products, but then the operation has to deliver those services and products on an ongoing basis. And central to an operation's ability to deliver is the way it plans its activities and controls them so that customers' demands are satisfied. This chapter introduces and provides an overview of some of the principles and methods of planning and control. Later chapters in this part of the text develop some specific issues that are vital to an operation delivering its services and products. These issues start with managing capacity and move through managing inventory, providing an overview of supply chain management and looking at how planning and control systems, particularly enterprise resources planning (ERP), manage the information that ensures effective delivery. Figure 10.1 shows where this topic fits into the activities of operations management.

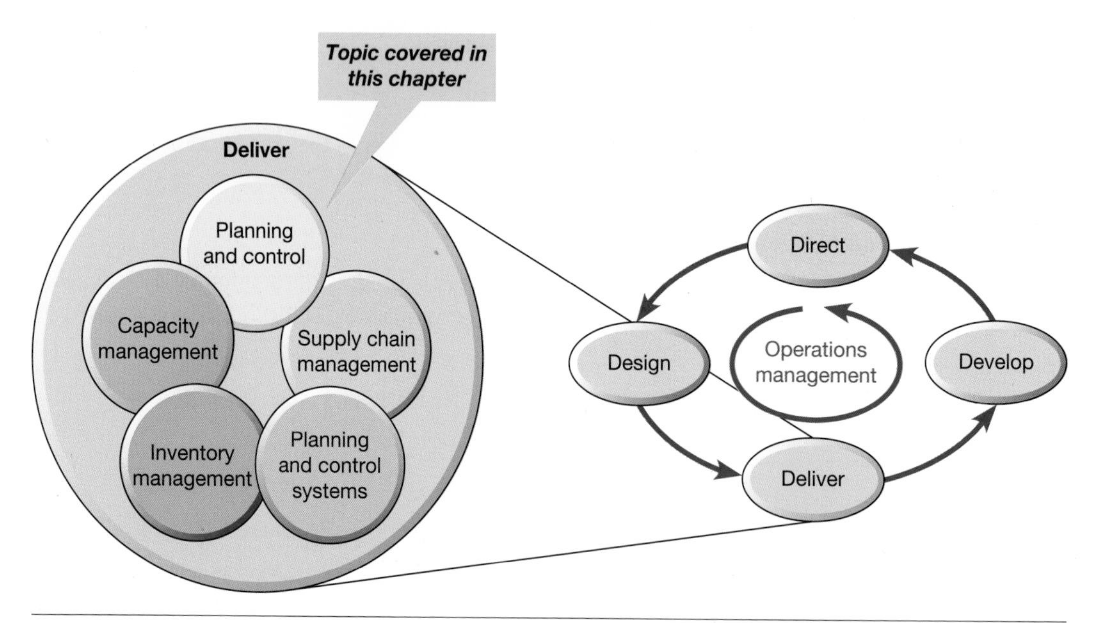

Figure 10.1 This chapter examines planning and control

10.1 What is planning and control?

Planning and control is concerned with the activities that attempt to reconcile the demands of the market and the ability of the operation's resources to deliver. It provides the systems, procedures and decisions that bring different aspects of supply and demand together. Consider, for example, the way in which routine surgery is organised in a hospital. When a patient arrives and is admitted to the hospital, much of the planning for the surgery will already have happened. The operating theatre will have been reserved, and the doctors and nurses who staff the operating theatre will have been provided with all the information regarding the patient's condition. Appropriate pre-operative and post-operative care will have been organised. All this will involve staff and facilities in different parts of the hospital, all of whom must have been given the same information and their activities coordinated. Soon after the patient arrives, they will be checked to make sure that the condition is as expected (in much the same way as material is inspected on arrival in a factory). Blood, if required, will be cross-matched and reserved, and any medication will be made ready (in the same way that all the different materials are brought together in a factory). Any last-minute changes may require some degree of re-planning. For example, if the patient shows unexpected symptoms, observation may be necessary before the surgery can take place. Not only will this affect the patient's own treatment, but other patients' treatment may also have to be rescheduled (in the same way as machines will need rescheduling if a job is delayed in a factory). At least some of these activities will be evident to the hospital's customers and are necessary for the planning and control of its resources.

Operations principle

Customers' perceptions of an operation will partially be shaped by its planning and control system.

OPERATIONS IN PRACTICE

Operations control at Air France[1]

'In many ways a major airline can be viewed as one large planning problem which is usually approached as many independent, smaller (but still difficult) planning problems. The list of things which need planning seems endless: crews, reservation agents, luggage, flights, through trips, maintenance, gates, inventory, equipment purchases. Each planning problem has its own considerations, its own complexities, its own set of time horizons, its own objectives, but all are interrelated' (Rikard Monet, Air France).

Air France has 80 flight planners working 24-hour shifts in their flight planning office at Roissy, Charles de Gaulle. Their job is to establish the optimum flight routes, anticipate any problems such as weather changes, and minimise fuel consumption. Overall, the goals of the flight planning activity are first, and most important, safety, followed by economy and passenger comfort. Increasingly powerful computer programs process the mountain of data necessary to plan the flights, but in the end many decisions still rely on human judgement. Even the most sophisticated **expert systems** only serve as support for the flight planners. Planning Air France's schedule is a massive job that includes the following:

- *Frequency* – for each airport how many separate services should the airline provide?
- *Fleet assignment* – which type of plane should be used on each leg of a flight?
- *Banks* – at any airline hub where passengers arrive and may transfer to other flights to continue their journey, airlines like to organise flights into 'banks' of several planes, which arrive close together, pause to let passengers change planes, and all depart close together.
- *Block times* – a block time is the elapsed time between a plane leaving the departure gate at an airport and arriving at its gate in the arrival airport. The longer the

allowed block time the more likely a plane will keep to schedule, even if it suffers minor delays, but the fewer flights can be scheduled.

- *Planned maintenance* – any schedule must allow time for planes to have time at a maintenance base.
- *Crew planning* – pilot and cabin crew must be scheduled to allocate pilots to fly planes on which they are licensed and to keep within the maximum 'on duty' allowances.
- *Gate plotting* – if many planes are on the ground at the same time there may be problems in loading and unloading them simultaneously.
- *Recovery* – many things can cause deviations from any plan in the airline industry. Allowances must be built in that allow for recovery.

For flights within and between Air France's 12 geographic zones, the planners construct a flight plan that will form the basis of the actual flight only a few hours later. All planning documents need to be ready for the flight crew, who arrive two hours before the scheduled departure time. Being responsible for passenger safety and comfort, the captain always has the final say and, when satisfied, co-signs the flight plan together with the planning officer.

The difference between planning and control

Notice that we have chosen to treat 'planning and control' together. This is because the division between 'planning' and 'control' is not clear, either in theory or in practice. However, there are some general features that help to distinguish between the two. Planning is a formalisation of what is intended to happen at some time in the future. But a plan does not guarantee that an event will actually happen. Rather it is a statement of intention. Although plans are based on expectations, during their implementation things do not always happen as expected. Customers change their minds about what they want and when they want it. Suppliers may not always deliver on time, process technology may fail, or staff may be absent through illness. Control is the process of coping with these types of change. It may mean that plans need to be redrawn in the short term. It may also mean that an 'intervention' will need to be made in the operation to bring it back 'on track' – for example, finding a new supplier who can deliver quickly, getting process technology up and running again, or moving staff from another part of the operation to cover for the absentees. Control activities make the adjustments that allow the operation to achieve the objectives that the plan has set, even when the assumptions on which the plan was based do not hold true.

Operations principle

Planning and control are separate but closely related activities.

Long-, medium- and short-term planning and control

The nature of planning and control activities changes over time. In the very long term, operations managers make plans concerning what they intend to do, what resources they need, and what objectives they hope to achieve. The emphasis is on planning rather than control, because there is little to control as such. They will use forecasts of likely demand described in aggregated terms. For example, a hospital will make plans for '2,000 patients' without necessarily going into the details of the individual needs of those 2,000 patients. Similarly, the hospital might plan to have 1,000 nurses and 200 doctors but again without deciding on the specific attributes of the staff. Operations managers will focus mainly on volume and financial targets.

Medium-term planning and control is more detailed. It looks ahead to assess the overall demand, which the operation must meet in a partially disaggregated manner. By this time, for example, the hospital must distinguish between different types of demand. The number of patients coming as accident and emergency cases will need to be distinguished from those requiring routine operations. Similarly, different categories of staff will have been identified and broad staffing levels in each category set. Just as important, contingencies will have been put in place that allow for slight deviations from the plans. These contingencies will act as 'reserve' resources and make planning and control easier in the short term.

In short-term planning and control, many of the resources will have been set and it will be difficult to make large changes. However, short-term interventions are possible if things are not going to plan. By this time, demand will be assessed on a totally disaggregated basis, with all types of surgical

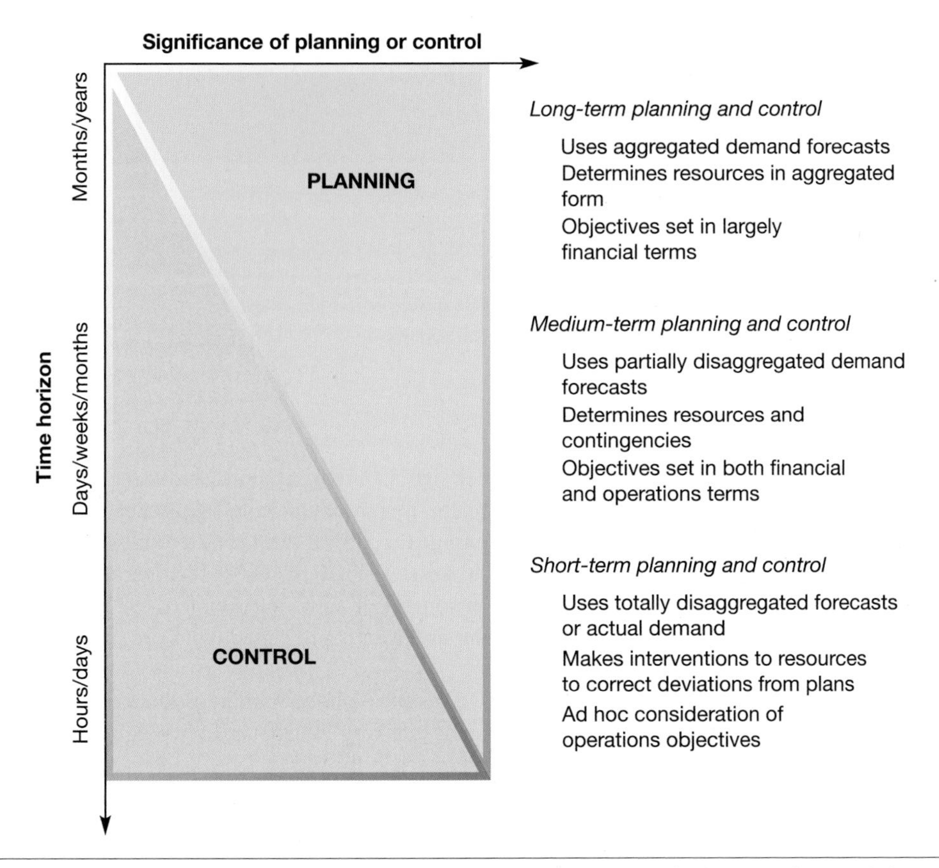

Figure 10.2 The balance between planning and control activities changes in the long, medium and short terms

procedures treated as individual activities. More importantly, individual patients will have been identified by name, and specific time slots booked for their treatment. In making short-term interventions and changes to the plan, operations managers will be attempting to balance the quality, speed, dependability, flexibility and costs of their operation dynamically on an ad hoc basis. It is unlikely that they will have the time to carry out detailed calculations of the effects of their short-term planning and control decisions on all these objectives, but a general understanding of priorities will form the background to their decision making. Figure 10.2 shows how the control aspects of planning and control increase in significance closer to the date of the event.

The volume–variety effect on planning and control

As we have found previously, the volume and variety characteristics of an operation will have an effect on its planning and control activities. Operations that produce a high variety of services or products in relatively low volume will have customers with different requirements and use different processes from operations that create standardised services or products in high volume (see Table 10.1).

Take two contrasting operations – an architects' practice and an electricity utility. The architects' high variety of customised services means they cannot produce designs in advance of customers requesting them. Because of this, the time it will take to finally deliver their services to customers will be relatively slow. Customers will understand this but will expect to be consulted extensively as to their needs. The details and requirements of each job will emerge only as each individual building is designed to the client's requirements, so planning occurs on a relatively short-term basis. The individual decisions that are taken in the planning process will usually concern the timing of activities and events – for example, when a design is to be delivered, when building should start, when each

Table 10.1 The volume–variety effects on planning and control

Volume	Variety	Customer responsiveness	Planning horizon	Major planning decision	Control decisions	Robustness
Low	High	Slow	Short	Timing	Detailed	High
↓	↓	↓	↓	↓	↓	↓
High	Low	Fast	Long	Volume	Aggregated	Low

individual architect will be needed to work on the design. Control decisions also will be at a relatively detailed level. A small delay in fixing one part of the design could have significant implications in many other parts of the job. For an architect, planning and control cannot be a totally routine matter; projects need managing on an individual basis. However, the robustness of the operation (that is, its vulnerability to serious disruption if one part of the operation fails) will be relatively high. There are probably plenty of other things to get on with if an architect is prevented from progressing one part of the job.

The electricity utility, on the other hand, is very different. Volume is high, production is continuous and variety is non-existent. Customers expect instant 'delivery' whenever they plug in an appliance. The planning horizon in electricity generation can be very long. Major decisions regarding the capacity of power stations are made years in advance. Even the fluctuations in demand over a typical day can be forecast in advance. Popular television programmes can affect minute-by-minute demand and these are scheduled weeks or months ahead. The weather, which also affects demand, is more uncertain, but can to some extent be predicted. Individual planning decisions made by the electricity utility are not concerned with the timing, but rather the volume of output. Control decisions will concern aggregated measures of output such as the total kilowatts of electricity generated, because the product is more or less homogeneous. However, the robustness of the operation is very low because, if a generator fails, the operation's capability of supplying electricity from that part of the operation also fails.

Operations principle

The volume–variety characteristics of an operation will affect its planning and control activities.

10.2 How do supply and demand affect planning and control?

If planning and control is the process of reconciling demand with supply, then the nature of the decisions taken to plan and control an operation will depend on both the nature of demand and the nature of supply in that operation. In this section, we examine some differences in demand and supply, which can affect the way in which operations managers plan and control their activities.

Uncertainty in supply and demand

If the future was perfectly predictable, planning would be straightforward – and control would not be needed. But the future is not perfectly predictable, and uncertainty makes planning and control more difficult. Sometimes the supply of inputs to an operation may be uncertain. Planned activities may take longer than expected. Similarly, demand may be unpredictable. A fast-food outlet inside a shopping centre does not know how many people will arrive, when they will arrive and what they will order. It may be possible to predict certain patterns, such as an increase in demand over the lunch periods, but a sudden rainstorm that drives shoppers indoors into the centre could significantly and unpredictably increase short term demand. Both supply and demand uncertainty make planning and control more difficult, but a combination of supply and demand uncertainty is particularly difficult.

Dependent and independent demand

Some operations can predict demand with relative certainty because demand for their services or products is dependent upon some other factor which is known. This is known as dependent demand. For example, the demand for tyres in a car factory is not a totally random variable. The process of demand forecasting is relatively straightforward. It will consist of examining the manufacturing schedules in the car plant and deriving the demand for tyres from these. If 600 cars are to be manufactured on a particular day, then it is simple to calculate that 3,000 tyres will be demanded by the car plant (each car has five tyres) – demand is dependent on a known factor, the number of cars to be manufactured. Because of this, the tyres can be ordered from the tyre manufacturer to a delivery schedule that is closely related to the demand for tyres from the plant (as in Figure 10.3). In fact, the demand for every part of the car plant will be derived from the assembly schedule for the finished cars.

Other operations will act in a dependent-demand manner because of the nature of the service or product they provide. For example, a custom-made dressmaker will not buy fabric and make up dresses in many different sizes just in case someone comes along and wants to buy one. Nor will a high-class restaurant begin to cook food just in case a customer arrives and requests it. In both these cases, a combination of risk and the perishability of the product or service prevent the operation from starting to create the goods or services until it has a firm order. Planning and control in dependent-demand situations is largely concerned with how the operation should respond when demand has occurred.

> **Operations principle**
> Planning and control systems should be able to cope with uncertainty in demand.

By contrast, some operations are subject to independent demand. They need to supply future demand without knowing exactly what that demand will be; or in the terminology of planning and control, they do not have firm 'forward visibility' of customer orders. For example, the Ace Tyre

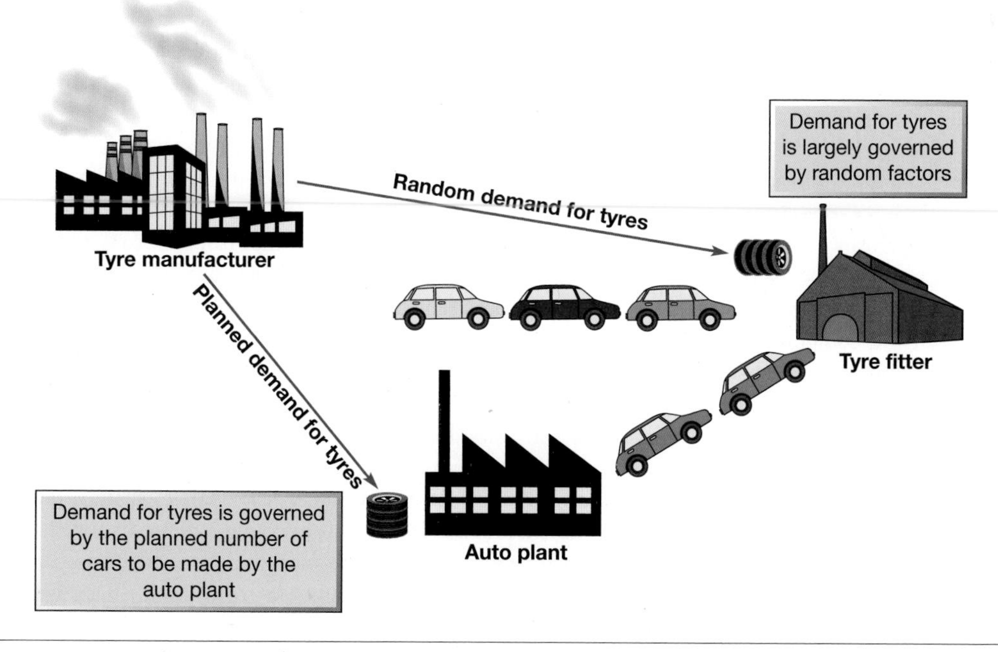

Figure 10.3 Dependent demand is derived from the demand for something else; independent demand is more random

Company, which operates a drive-in tyre replacement service, will need to manage a stock of tyres. In that sense it is exactly the same task that faced the manager of tyre stocks in the car plant. However, demand is very different for Ace Tyres. It cannot predict either the volume or the specific needs of customers. It must make decisions on how many and what types of tyres to stock, based on demand forecasts and in the light of the risks it is prepared to run of being out of stock. This is the nature of independent-demand planning and control. It makes 'best guesses' concerning future demand, attempts to put the resources in place that can satisfy this demand, and attempts to respond quickly if actual demand does not match the forecast. Inventory planning and control, treated in Chapter 13, is typical of independent-demand planning and control.

> **Operations principle**
> Planning and control systems should distinguish between dependent and independent demand.

Responding to demand

It is clear then that the nature of planning and control in any operation will depend on how it responds to demand, which is in turn related to the types of services or products it produces. For example, an advertising agency will start the process of planning and controlling the creation of an advertising campaign only when the customer (or client as the agency will refer to them) confirms the contract with the agency. The creative 'design' of the advertisements will be based on a 'brief' from the client. Only after the design is approved are the appropriate resources (director, scriptwriters, actors, production company, etc.) contracted. The actual shooting of the advertisement and post-production (editing, putting in the special effects, etc.) then goes ahead, after which the finished advertisements are 'delivered' through television slots. This is shown in Figure 10.4 as a 'Design, resource, create and deliver to order' operation.

Other operations might be sufficiently confident of the nature of demand, if not its exact details, to keep 'in stock' most of the resources it requires to satisfy its customers. Certainly, it will keep its transforming resources, if not its transformed resources. However, it will still make the actual service or product only when it receives a firm customer order. For example, a website designer will have most of its resources (graphic designers, software developers, specialist development software, etc.) in place, but must still design, create and deliver the website after it understands its customer's requirements. (See the 'Operations in practice' example of Torchbox in Chapter 9.) This is shown in Figure 10.4 as a 'Design, create and deliver to order' operation.

Some operations offer relatively standard services or products, but do not create them until the customer has chosen which particular service or product to have. So a house builder that has standard designs might choose to build each house only when a customer places a firm order. Because the design of the house is relatively standard, suppliers of materials will have been identified, even if the building operation does not keep the items in stock itself. This is shown in Figure 10.4 as a 'Create and deliver to order' operation. In manufacturing it would be called a '**Make to order**' operation.

Some operations have services or products that are so predictable that they can start to 'create' them before specific customer orders arrive. Possibly the best-known example of this is Dell Computers, where customers can 'specify' their computer by selecting between various components through the company's website. These components will have already been created (usually by suppliers) but are assembled to a specific customer order. This is shown in Figure 10.4 as a 'Partially create and deliver to order' operation. In manufacturing it would be called an 'Assemble to order' operation.

When an operation's services or products are standardised, there is the potential to create them entirely before demand is known. Almost all domestic products, for example, are 'Created to stock', or '**Make to stock**' (shown in Figure 10.4), from which they are delivered to customers. Taking this evolving logic to its conclusion, some operations require their customers to collect their own services or products. This is the 'Collect/choose from stock' illustration in Figure 10.4. IKEA and most high-street retail operations are like this.

> **Operations principle**
> The planning and control activity will vary depending on how much work is done before demand is known.

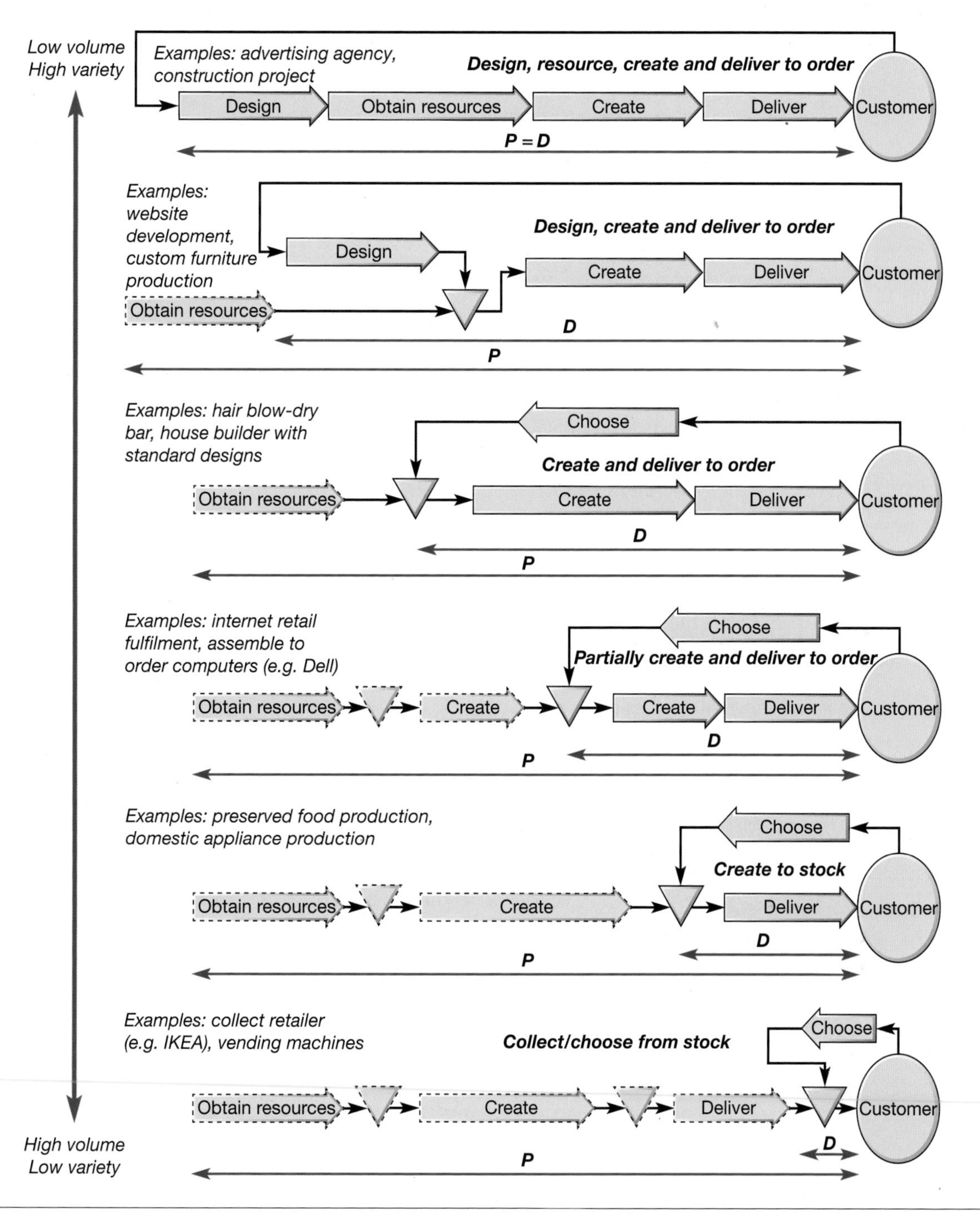

Figure 10.4 The *P:D* ratio of an operation indicates how long the customer has to wait for the service or product as compared with the total time needed to carry out all the activities to make the service or product available to the customer

One point to note in the operations illustrated in Figure 10.4 is that there is a relationship between how operations respond to demand and their volume–variety characteristics. It is easy to see that 'Design, resource, create and deliver to order' operations are intended for low-volume and high-variety businesses. By definition, designing different services or products will result in high variety, and performing each activity for each customer would be too cumbersome for a high-volume business. Conversely, 'Create to stock' or 'Collect/choose from stock' clearly rely on standardised services or products.

P:D ratios

Another way of characterising the graduation between 'Design, resource, create and deliver to order' and 'Collect/choose from stock' planning and control is by using a *P:D* ratio. This contrasts the total length of time customers have to wait between asking for the service or product and receiving it, called the demand time, *D*, and the total throughput time from start to finish, *P*. Throughput time is how long the operation takes to design the service or product (if it is customised), obtain the resources, create and deliver it.

P and *D* times depend on the operation

P and *D* are illustrated for each type of operation in Figure 10.4. Generally, the ratio of *P* to *D* gets larger as operations move from 'Design, resource, create and deliver to order' to 'Collect/choose from stock'. In other words, as one moves down this spectrum towards the 'Create to stock' and 'Collect/choose from stock' end, the operation has anticipated customer demand and already created the services and products even though it has no guarantee that the anticipated demand will really happen. This is a particularly important point for the planning and control activity. The larger the *P:D* ratio, the more speculative the operation's planning and control activities will be. In its extreme form, the 'Collect/choose from stock' operation, such as a high-street retailer, has taken a gamble by designing, resourcing, creating and delivering (or more likely, paying someone else to do so) products to its shops before it has any certainty that customers will want them. Contrast this with a 'Design, resource, create and deliver to order' operation as in the advertising agency mentioned earlier. Here, *D* is the same as *P* and speculation regarding the volume of demand in the short term is eliminated because everything happens in response to a firm order. So by reducing their *P:D* ratio, operations reduce their degree of speculative activity and also reduce their dependence on forecasting (although bad forecasting will lead to other problems).

But do not assume that when the *P:D* ratio approaches 1, all uncertainty is eliminated. The volume of demand (in terms of the number of customer 'orders') may be known, but not the time taken to perform each 'order'. Take the advertising agency again: during each stage of the process, from design to delivery, it is common to have to seek the customer's approval and/or feedback many times. Moreover, there will almost certainly be some recycling back through stages as modifications are made. And, in a similar way to how simultaneous development works in new service and product design (see Chapter 4), a stage can be started before the previous one has been completed. So, for example, the video shoot director will have started prior to the artwork design being completed. This is illustrated in Figure 10.5. So here, it is the timings that are uncertain.

Operations principle

The *P:D* ratio of an operation contrasts how long customers have to wait for a service or product with its total throughput time.

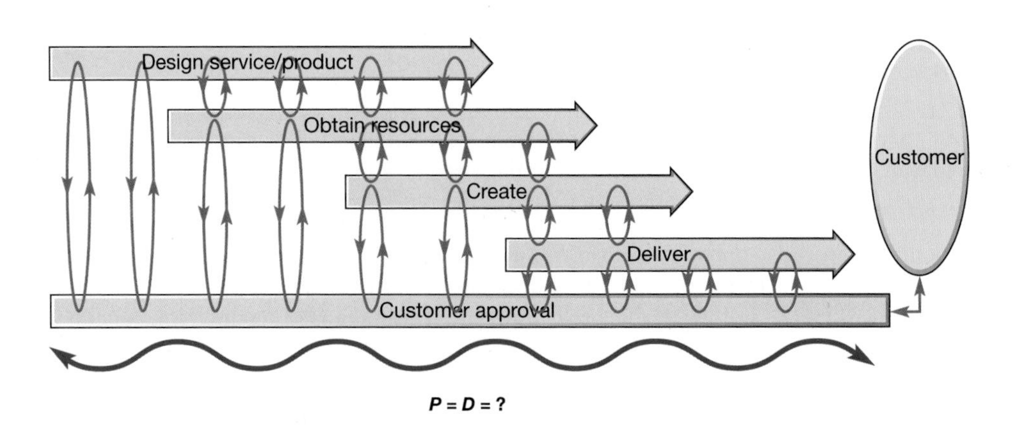

Figure 10.5 The relationship between stages in some 'Design, resource, create and deliver to order' operations, such as an advertising agency, can be complex with frequent consultation and unpredictable recycling

Sales and operations planning (S&OP)

One of the problems with traditional operations planning and control is that, although several functions were often routinely involved in the process, each function could have a very different set of objectives. For example, marketing could be interested in maximising revenues and ensuring continuity of delivery to customers. Operations are likely to be under pressure to minimise cost (perhaps achieved through relatively long and stable operating levels). Finance will be interested in reducing working capital and inventory, and also reducing fixed costs. And so on. Yet these, and other functions such as engineering or human resources management, are all impacted by operations planning decisions and are probably involved in their own planning processes that partly depend on the output from the operations planning process. Sales and operations planning (S&OP) was first promoted as an important element of planning as manufacturing resource planning (see the supplement to Chapter 14) became a commonly used process. Early manufacturing resource planning implementations were often made less effective by the system being driven by unachievable plans. This is the dilemma that S&OP is intended to address. It is a planning process that attempts to ensure that all tactical plans are aligned across the business's various functions and with the company's longer-term strategic plans.

It is a formal business process that looks over a period of 18 to 24 months ahead. In other words, it is not a purely short-term process. In fact, S&OP developed as an attempt to integrate short- and longer-term planning, as well as integrating the planning activities of key functions. It is an aggregated process that does not deal with detailed activities, but rather focuses on the overall (often aggregated) volume of output. Generally, it is a process that happens monthly, and tends to take place at a higher level, involving more senior management than traditional operations planning. S&OP also goes by many names. It can be called integrated business planning, integrated business management, integrated performance management, rolling business planning and regional business management, to name a few. It has also been noted[2] that some organisations continue to use the phrase 'S&OP', although they may mean something quite different.

The activities of planning and control

Planning and control requires the reconciliation of supply and demand in terms of volumes, timing and quality. In this chapter we will focus on an overview of the activities that plan and control volume and timing (most of this part of the text is concerned with these issues). There are four overlapping activities: loading, sequencing, scheduling, and monitoring and control (see Figure 10.6). Some caution is needed when using these terms. Different organisations may use them in different ways, and even textbooks in the area adopt different definitions. For example, some authorities term what we have called planning and control as 'operations scheduling'. However, the terminology of planning and control is less important than understanding the basic ideas described in the remainder of this chapter.

Operations principle

Planning and control activities include loading, sequencing, scheduling, and monitoring and control.

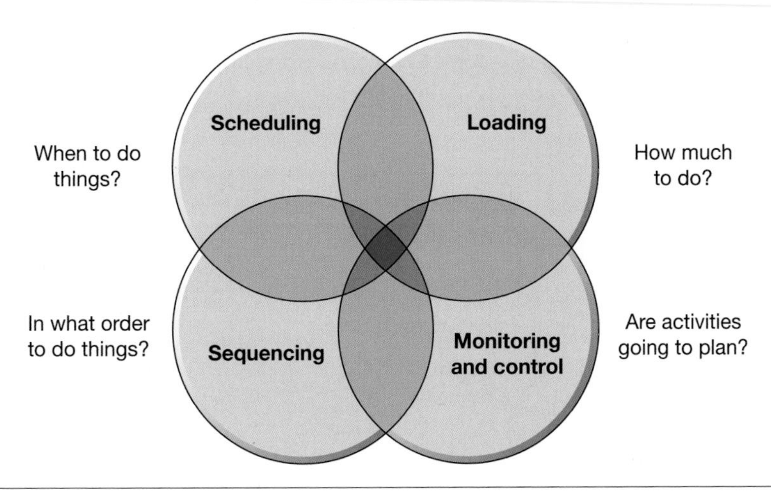

Figure 10.6 Planning and control activities

10.3 What is 'loading'?

Loading is the amount of work that is allocated to a work centre. For example, a machine on the shop floor of a manufacturing business is available, in theory, 168 hours a week. However, this does not necessarily mean that 168 hours of work can be loaded onto that machine. Figure 10.7 shows what erodes this available time. For some periods the machine cannot be worked: for example, it may not be available on statutory holidays and weekends. Therefore, the load put onto the machine must take this into account. Of the time that the machine is available for work, other losses further reduce the available time. For example, time may be lost while changing over from making one component to another. If the machine breaks down, it will not be available. If machine reliability data are available, these must also be taken into account. Sometimes the machine may be waiting for parts to arrive or be 'idling' for some other reason. Other losses could include an allowance for the machine being run below its optimum speed (for example, because it has not been maintained properly) and an allowance for the 'quality losses' or defects that the machine may produce. Of course, many of these losses should be small or non-existent in a well-managed operation. However, the **valuable operating time** available for productive working, even in the best operations, can be significantly below the maximum time available. This idea is taken further in Chapter 11 when we discuss the measurement of capacity.

> **Operations principle**
> For any given level of demand, a planning and control system should be able to indicate the implications for the loading on any part of the operation.

Finite and infinite loading

Finite loading is an approach that only allocates work to a work centre (a person, a machine, or perhaps a group of people or machines) up to a set limit. This limit is the estimate of capacity for the work centre (based on the times available for loading). Work over and above this capacity is not accepted. Figure 10.8 first shows how the load on the work centres is not allowed to exceed the capacity limit. Finite loading is particularly relevant for operations where:

- *it is possible to limit the load* – for example, it is possible to run an appointment system for a general medical practice or a hairdresser;
- *it is necessary to limit the load* – for example, for safety reasons only a finite number of people and weight of luggage are allowed on aircraft;
- *the cost of limiting the load is not prohibitive* – for example, the cost of maintaining a finite order book at a specialist sports car manufacturer does not adversely affect demand, and may even enhance it.

Infinite loading is an approach to loading work that does not limit accepting work, but instead tries to cope with it. The second diagram in Figure 10.8 illustrates this loading pattern where capacity constraints have not been used to limit loading so the work is completed earlier. Infinite loading is relevant for operations where:

- *it is not possible to limit the load* – for example, an accident and emergency department in a hospital should not turn away arrivals needing attention;

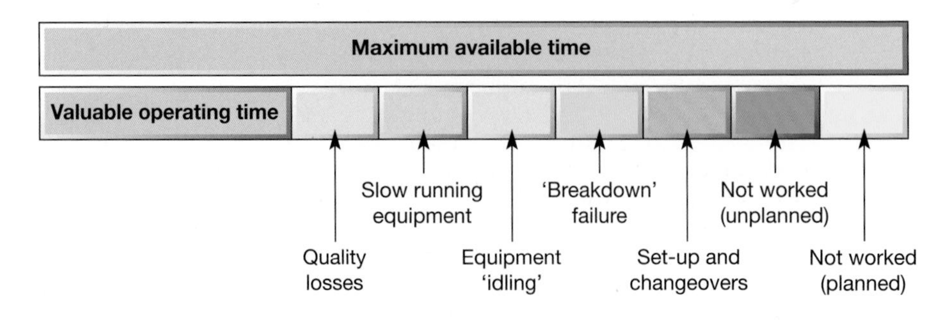

Figure 10.7 The reduction in the valuable operating time available

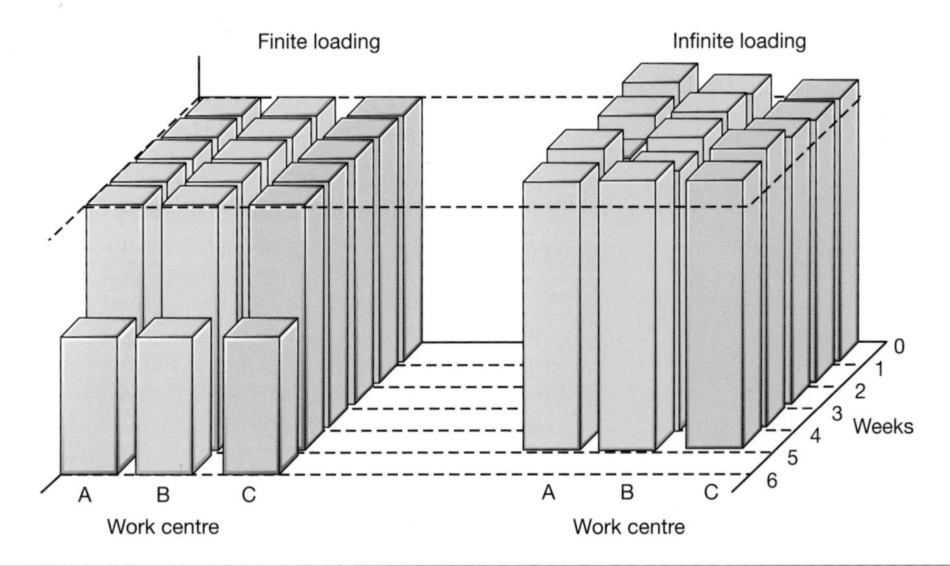

Figure 10.8 Finite and infinite loading of jobs on three work centres A, B and C. Finite loading limits the loading on each centre to its capacity, even if it means that jobs will be late. Infinite loading allows the loading on each centre to exceed its capacity to ensure that jobs will not be late

- *it is not necessary to limit the load* – for example, fast-food outlets are designed to flex capacity up and down to cope with varying arrival rates of customers. During busy periods, customers accept that they must queue for some time before being served. Unless this is extreme, the customers might not go elsewhere;
- *the cost of limiting the load is prohibitive* – for example, if a retail bank turned away customers at the door because a set number were inside, customers would feel less than happy with the service.

In complex planning and control activities where there are multiple stages, each with different capacities and with a varying mix arriving at the facilities, such as a machine shop in an engineering company, the constraints imposed by finite loading make loading calculations complex and not worth the considerable computational power that would be needed.

10.4 What is 'sequencing'?

Whether the approach to loading is finite or infinite, when work arrives, decisions must be taken on the order in which the work will be tackled. This activity is termed sequencing. The priorities given to work in an operation are often determined by a predefined set of rules, some of which are relatively complex. Some of these are summarised below.

Physical constraints

The physical nature of the inputs being processed may determine the priority of work. For example, in an operation using paints or dyes, lighter shades will be sequenced before darker shades. On completion of each batch, the colour is slightly darkened for the next batch. This is because darkness of colour can only be added to and not removed from the colour mix. Sometimes the mix of work arriving at a part of an operation may determine the priority given to jobs. For example, when fabric is cut to a required size and shape in garment manufacture, the surplus fabric would be wasted if not used for another product. Therefore, jobs that fit together physically may be scheduled together to reduce waste. The sequencing issue described in the 'Operations in practice' example 'Can airline passengers be sequenced?' is of this type.

OPERATIONS IN PRACTICE

Can airline passengers be sequenced?[3]

Like many before him, Dr Jason Steffen, a professional astrophysicist, was frustrated by the time it took to load him and his fellow passengers onto the aircraft. He decided to devise a way to make the experience a little less tedious. So, for a while, he neglected his usual work of examining extra-solar planets, dark matter and cosmology, and experimentally tested a faster method of boarding aircraft. He found that, by changing the sequence in which passengers are loaded onto the aircraft, airlines could potentially save both time and money. Using a computer simulation and the arithmetic techniques routinely used in his day-to-day job, he was able to find what seemed to be a superior sequencing method. His simulations showed that the most common way of boarding passengers was the least efficient. This is called the 'block method' where blocks of seats are called for boarding, starting from the back. Previously, other experts in the airline industry had suggested boarding those in window seats first followed by middle and aisle seats. This is called the Wilma method. But according to Dr Steffen's simulations, two things slow down the boarding process. The first is that passengers may be required to wait in the aisle while those ahead of them store their luggage before they can take their seat. The second is that passengers already seated in aisle or middle seats frequently have to rise and move into the aisle to let others take seats nearer the window. So Dr Steffen suggested a variant of the Wilma method that minimised the first type of disturbance and eliminated the second. He suggested boarding in alternate rows, progressing from the rear forward, window seats first. Using this approach (now called the Steffen method), first, the window seats for every other row on one side of the plane are boarded. Next, alternate rows of window seats on the opposite side are boarded. Then, the window seats in the skipped rows are filled in on each side. The procedure then repeats with the middle seats and the aisles (see Figure 10.9).

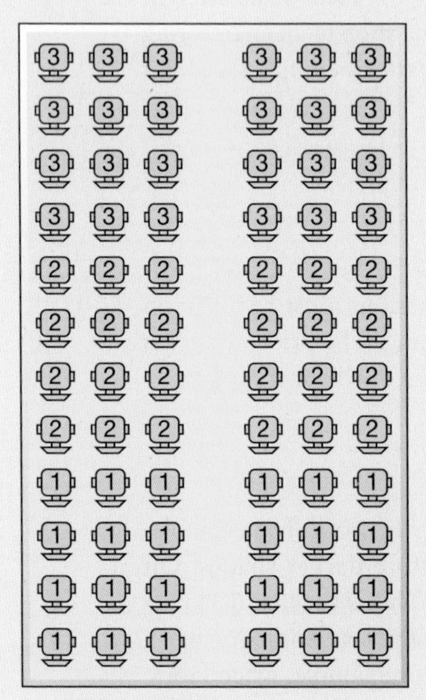

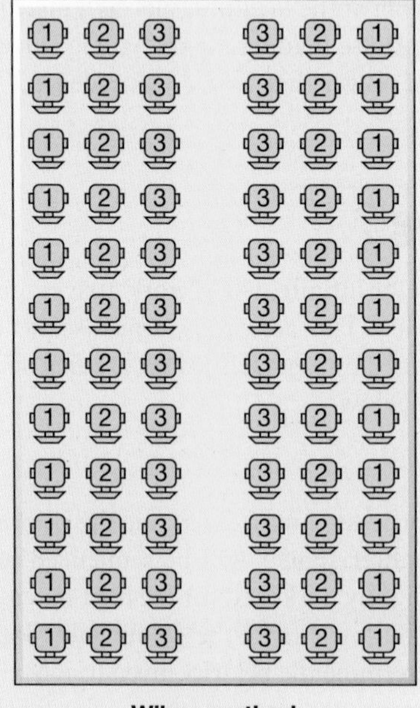

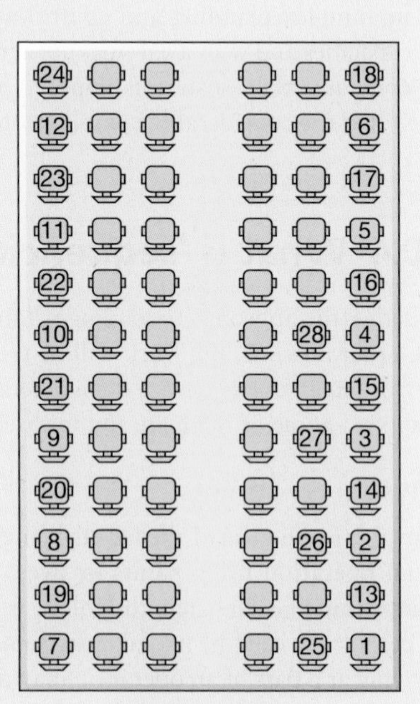

Figure 10.9 The best way to sequence passengers onto an aircraft

Later, the effectiveness of the various approaches were tested using a mock-up of a Boeing 757 aircraft and 72 luggage-carrying volunteers. Five different scenarios were tested: 'block' boarding in groups of rows from back to front, one by one from back to front, the 'Wilma method', the Steffen method, and completely random boarding. In all cases, parent–child pairs were allowed to board first. It was assumed that families were likely to want to stay together. As Dr Steffen had predicted, the conventional block approach came out as the slowest, with the strict back-to-front approach not much better. Completely random boarding (unallocated seating), which is used by several low-cost airlines, fared much better, most probably because it randomly avoids space conflicts. The times for fully boarding the 72 passengers using each method were as follows: 'block' boarding – 6:54 minutes, back-to-front – 6:11 minutes, random boarding – 4:44 minutes, Wilma method – 4:13 minutes, Steffen method – 3:36 minutes.

The big question is, 'would passengers really be prepared to be sequenced in this way as they queue to board the aircraft?' Some airlines argue that directing passengers on to a plane is a little like herding cats.

Customer priority

Operations will sometimes use customer priority sequencing, which allows an important or aggrieved customer, or item, to be 'processed' prior to others, irrespective of the order of arrival of the customer or item. This approach is typically used by operations whose customer base is skewed, containing a mass of small customers and a few large, very important customers. Some banks, for example, give priority to important customers. Similarly, in hotels, complaining customers will be treated as a priority because their complaint may have an adverse effect on the perceptions of other customers. More seriously, the emergency services often have to use their judgement in prioritising the urgency of requests for service. For example, Figure 10.10 shows a typical triage system used in hospitals to prioritise patients (see the 'Operations in practice' example 'The trials of triage'). However, customer priority sequencing, although giving a high level of service to some customers, may erode the service given to many others. This may lower the overall performance of the operation if work flows are disrupted to accommodate important customers.

Due date (DD)

Prioritising by due date means that work is sequenced according to when it is 'due' for delivery, irrespective of the size of each job or the importance of each customer. For example, a support service, such as a printing unit, will often ask when copies are required, and then sequence the work according to that due date. Due date sequencing usually improves the delivery dependability and average delivery speed. It may not provide optimal productivity, as a more efficient sequencing of work may reduce total costs. However, it can be flexible when new, urgent work arrives at the work centre.

1	Immediate resuscitation	Patient in need of immediate treatment for preservation of life
2	Very urgent	Seriously ill or injured patients whose lives are not in immediate danger
3	Urgent	Patients with serious problems, but apparently stable conditions
4	Standard	Standard cases without immediate danger or distress
5	Non-urgent	Patients whose conditions are not true accidents or emergencies

Figure 10.10 A triage prioritisation scale

OPERATIONS IN PRACTICE

The trials of triage[4]

Dominique-Jean Larrey was a military surgeon in the French army during Napoleon's campaigns. When treating battlefield casualties, Larrey had to decide (regardless of their military rank) which soldiers needed medical attention most urgently. To help in what was a difficult decision, both medically and ethically, he developed the concept of distinguishing between urgent and non-urgent patients. This became known as 'triage', from the French trier (to separate out). It is still used today to sequence patients waiting for limited medical resources such as staff, beds, intensive care and ventilators. The main difference is that, in Napoleonic times, many injuries meant certain death. Now, technological progress means that most can be treated – if the resources are available. For example, hospital accident and emergency departments have limited staff, beds and equipment, and patients arrive at random. It is then the job of the hospital's reception and medical staff rapidly to devise the priority of patients that meets most of the necessary criteria. Patients with very serious injuries, or presenting symptoms of a serious illness, generally need to be attended to urgently. Patients in some discomfort, but whose injuries or illnesses are not life-threatening, will have to wait until the urgent cases are treated. Routine non-urgent cases will have the lowest priority. In many circumstances, these patients will have to wait many hours, if the hospital is busy. They may even be turned away if the hospital is too busy with more important cases.

Triage has always presented moral dilemmas and trade-offs. This was especially exposed during the COVID-19 pandemic. Doctors were faced with the most appalling decisions about how to allocate scarce resources. Under extreme resource shortage, it may not be the patients who are the most sick that are given priority access to resources. In fact, beyond a certain point, a severely sick patient may, under some circumstances, be less likely to receive treatment. It can be argued that, especially if treatment is likely to be distressing to the patient, and a person is inescapably going to die, it is simply cruel to add to their affliction by, for example, putting them on a ventilator. Moreover, the equipment could be better used for a patient who is less ill but more likely to survive. Making such trade-offs takes its toll on medical staff. During the pandemic there were reports of doctors weeping in hospital corridors because of the choices they had to make. Although most professionals agreed that resources were allocated to the patients who had the greatest chances of successful treatment, and who had the greatest life expectancy, it still meant having to take some brutal decisions that literally meant life or death. Doctors reported that it helped if the criteria and decision framework for distinguishing between patients were decided in advance, and patients and their families were carefully and sensitively informed. It may also be better for someone other than front-line doctors to make difficult decisions. Some states in the US have triage officers or committees that make such decisions, with front-line doctors free to appeal a decision if they think it is mistaken.

Last in, first out (LIFO)

Last in, first out (LIFO) is a method of sequencing usually selected for practical reasons. For example, unloading an elevator is more convenient on a LIFO basis, as there is only one entrance and exit. However, it is not an equitable approach. Patients at hospital clinics may be infuriated if they see newly arrived patients examined first.

First in, first out (FIFO)

Some operations serve customers in exactly the sequence they arrive in. This is called first in, first out sequencing (FIFO), or sometimes 'first come, first served' (FCFS). For example, UK passport offices receive mail, and sort it according to the day when it arrived. They work through the mail, opening it in sequence, and process the passport applications in order of arrival. Queues for tickets at theme parks may be designed so that one long queue snakes around the lobby area until the row of counters is reached. When customers reach the front of the queue, they are served at the next free counter.

Longest operation time (LOT)

Operations may feel obliged to sequence their longest jobs first, called longest operation time sequencing. This has the advantage of occupying work centres for long periods. By contrast, relatively small jobs progressing through an operation will take up time at each work centre because of the need to change over from one job to the next. However, although longest operation time sequencing keeps utilisation high, this rule does not take into account delivery speed, reliability or flexibility. Indeed, it may work directly against these performance objectives.

Shortest operation time first (SOT)

Most operations at some stage become cash constrained. In these situations, the sequencing rules may be adjusted to tackle short jobs first; this is called shortest operation time sequencing. These jobs can then be invoiced, and payment received to ease cash-flow problems. Larger jobs that take more time will not enable the business to invoice as quickly. This has an effect of improving delivery performance, if the unit of measurement of delivery is jobs. However, it may adversely affect total productivity and can damage service to larger customers.

Judging sequencing rules

All five performance objectives, or some variant of them, could be used to judge the effectiveness of sequencing rules. However, the objectives of dependability, speed and cost are particularly important. So, for example, the following performance objectives are often used:

- meeting 'due date' promised to customer (dependability);
- minimising the time the job spends in the process, also known as 'flow time' (speed);
- minimising work-in-progress inventory (an element of cost);
- minimising idle time of work centres (another element of cost).

Worked example

Steve Smith, website designer

Steve Smith is a website designer in a business school. Returning from his annual vacation (he finished all outstanding jobs before he left), five design jobs are given to him upon arrival at work. He gives them the codes A to E. Steven has to decide in which sequence to undertake the jobs. He wants both to minimise the average time the jobs are tied up in his office and, if possible, to meet the deadlines (delivery times) allocated to each job.

His first thought is to do the jobs in the order they were given to him, i.e. first in, first out (FIFO):

Sequencing rule – first in first out (FIFO)

Sequence of jobs	Process time (days)	Start time	Finish time	Due date	Lateness (days)
A	5	0	5	6	0
B	3	5	8	5	3
C	6	8	14	8	6
D	2	14	16	7	9
E	1	16	17	3	14
Total time in process		60	**Total lateness**		32
Average time in process (total/5)		12	**Average lateness (total/5)**		6.4

Alarmed by the average lateness, he tries the due date (DD) rule:

Sequencing rule – due date (DD)

Sequence of jobs	Process time (days)	Start time	Finish time	Due date	Lateness (days)
E	1	0	1	3	0
B	3	1	4	5	0
A	5	4	9	6	3
D	2	9	11	7	4
C	6	11	17	8	9
Total time in process		42	**Total lateness**		16
Average time in process (total/5)		8.4	**Average lateness (total/5)**		3.2

Better! But Steve tries out the shortest operation time (SOT) rule:

Sequencing rule – shortest operation time (SOT)

Sequence of jobs	Process time (days)	Start time	Finish time	Due date	Lateness (days)
E	1	0	1	3	0
D	2	1	3	7	0
B	3	3	6	5	1
A	5	6	11	6	5
C	6	11	17	8	9
Total time in process		38	**Total lateness**		16
Average time in process (total/5)		7.6	**Average lateness (total/5)**		3.2

This gives the same degree of average lateness but with a lower average time in the process. Steve decides to use the SOT rule.

Comparing the three sequencing rules described in the worked example, together with the two other sequencing rules described earlier and applied to the same problem, gives the results summarised in Table 10.2. The SOT rule results in both the best average time in process and the best (or least bad) in terms of average lateness. Although different rules will perform differently depending on the circumstances of the sequencing problem, in practice the SOT rule generally performs well.

Table 10.2 Comparison of five sequencing decision rules

Rule	Average time in process (days)	Average lateness (days)
FIFO	12	6.4
DD	8.4	3.2
SOT	7.6	3.2
LIFO	8.4	3.8
LOT	12.8	7.4

10.5 What is 'scheduling'?

Having determined the sequence that work is to be tackled in, some operations require a detailed timetable showing at what time or date jobs should start and when they should end – this is scheduling. Schedules are familiar statements of volume and timing in many consumer environments. For example, a bus schedule shows that more buses are put on routes at more frequent intervals during rush-hour periods. The bus schedule shows the time each bus is due to arrive at each stage of the route. Schedules of work are used in operations where some planning is required to ensure that customer demand is met. Other operations, such as rapid-response service operations where customers arrive in an unplanned way, cannot schedule the operation in a short-term sense. They can respond only at the time demand is placed upon them.

OPERATIONS IN PRACTICE

Sequencing and scheduling at London's Heathrow airport[5]

Heathrow is the United Kingdom's busiest airport, and the busiest two-runway airport in the world, welcoming around 1,300 combined take-offs and landings each day. Landing around 650 aircraft in a day, air traffic controllers have one of the most complex sequencing jobs to perform as they decide which aircraft to call down next from their waiting areas (known as 'stacks') to land on one of the two runways. Many airports use a sequencing policy based on 'first come, first served'. However, this does not always give the best airport performance, where performance is assessed by such measures as runway utilisation, total aircraft throughput, passenger throughput and passenger waiting time. For very busy airports such as Heathrow, a more sophisticated sequencing approach is needed. For most of the time at Heathrow, one runway is used solely for take-offs and the other solely for landings (known as a 'segregated' operating mode). However, at particularly busy times, both runways can be used for landings. Safety considerations are, of course, paramount in deciding on an appropriate sequence. There must be a minimum time and distance between aircraft when they take off or land. This is because of what is known as the 'wake vortex' – turbulence that is caused by the 'lift' component of flight (without which the aircraft could not fly). Lift is caused by the pressure difference between the upper and lower surfaces of the wing. Wake vortices can result in turbulent conditions if an aircraft follows too close to the previous one, which passengers would find uncomfortable, and possibly distressing. It could even cause possible damage to the following aircraft. The magnitude of a wake vortex depends on the size of the aircraft, large aircraft causing more air turbulence. So, following a large aircraft means leaving a (relatively) long time before another aircraft can land. Conversely, a light aircraft generates little air turbulence and therefore only a (relatively) small time delay is needed before other aircraft can land. In other words, the sequence in which aircraft are called to land will determine the total time taken to complete landing. But, in addition to deciding the sequence in which aircraft will land, controllers must also construct a schedule that determines a landing time for each aircraft. This schedule should:

- allow sufficient time for an aircraft to fly safely from its current position in the stack to the runway so that it will land at the appropriate position in the sequence;
- make sure that no aircraft run low on fuel while waiting to land;
- ensure that aircraft do not land too close together.

To make the situation more difficult, weather can also complicate things. Aircraft have to take off and land against the wind, so the landing direction depends on the prevailing wind (which can change). This is why meteorological experts are constantly monitoring weather conditions prevailing at 30,000 feet. Also, the airport must try to minimise the noise nuisance caused to local communities, which means that no landings are allowed before 04.30 with a maximum of 16 flights before 06.00, preferably the quietest planes.

The complexity of scheduling

The scheduling activity is one of the most complex tasks in operations management. First, schedulers must deal with several different types of resource simultaneously. Machines will have different capabilities and capacities; staff will have different skills. More importantly, the number of possible schedules increases rapidly as the number of activities and processes increases. For example, suppose one machine has five different jobs to process. Any of the five jobs could be processed first and, following that, any one of the remaining four jobs, and so on. This means that there are:

$$5 \times 4 \times 3 \times 2 = 120 \text{ different schedules possible}$$

In other words, for n jobs there are $n!$ (factorial n) different ways of scheduling the jobs through a single process. But when there are (say) two machines, there is no reason why the sequence on machine 1 would be the same as the sequence on machine 2. If we consider the two sequencing tasks to be independent of each other, for two machines there would be:

$$120 \times 120 = 14{,}400 \text{ possible schedules of the two machines and five jobs}$$

So a general formula can be devised to calculate the number of possible schedules in any given situation, as follows:

$$\text{Number of possible schedules} = (n!)^m$$

where n is the number of jobs and m is the number of machines.

In practical terms, this means that there are often many millions of feasible schedules, even for relatively small scheduling tasks. This is why scheduling rarely attempts to provide an 'optimal' solution but rather satisfies itself with an 'acceptable' feasible one.

Forward and backward scheduling

Forward scheduling involves starting work as soon as it arrives. Backward scheduling involves starting jobs at the last possible moment to prevent them from being late. For example, assume that it takes six hours for a contract laundry to wash, dry and press a batch of workwear. If the work is collected at 8.00 am and is due to be picked up at 4.00 pm, there are more than six hours available to do it. Table 10.3 shows the different start times of each job, depending on whether they are forward or backward scheduled.

The choice of backward or forward scheduling depends largely upon the circumstances. Table 10.4 lists some advantages and disadvantages of the two approaches.

> **Operations principle**
> An operation's planning and control system should allow for the effects of alternative schedules to be assessed.

Gantt charts

One crude but simple method of scheduling is by use of the Gantt chart. This is a simple device, which represents time as a bar, or channel, on a chart. The start and finish times for activities can be indicated on the chart and sometimes the actual progress of the job is also indicated. The advantage of Gantt charts is that they provide a simple visual representation both of what should be happening and of what actually is happening in the operation. Furthermore, they can be used to 'test out' alternative schedules. It is a relatively simple task to represent alternative schedules (even if it is a far from simple task to find a schedule that fits all the resources satisfactorily). Figure 10.11 illustrates a Gantt

Table 10.3 The effects of forward and backward scheduling

Task	Duration	Start time (backwards)	Start time (forwards)
Press	1 hour	3.00 pm	1.00 pm
Dry	2 hours	1.00 pm	11.00 am
Wash	3 hours	10.00 am	8.00 am

Table 10.4 Advantages of forward and backward scheduling

Advantages of forward scheduling	Advantages of backward scheduling
High labour utilisation – workers always start work to keep busy	Lower material costs – materials are not used until they have to be, therefore delaying added value until the last moment
Flexible – the time slack in the system work to be loaded	Less exposed to risk in case of schedule change by the customer
	Tends to focus the operation on customer due dates

chart for a specialist software developer. It indicates the progress of several jobs as they are expected to move through five stages of the process. Gantt charts are not an optimising tool, they merely facilitate the development of alternative schedules by communicating them effectively.

Scheduling work patterns

Where the dominant resource in an operation is its staff, then the schedule of work times effectively determines the capacity of the operation itself. The main task of scheduling, therefore, is to make sure that sufficient numbers of people are working at any point in time to provide a capacity appropriate for the level of demand at that point in time. This is often called staff rostering. Operations such as call centres, postal delivery, policing, holiday couriers, retail shops and hospitals will all need to schedule the working hours of their staff with demand in mind. This is a direct consequence of these operations having relatively high 'visibility' (we introduced this idea in Chapter 1). Such operations cannot store their outputs in inventories and so must respond directly to customer demand. For example, Figure 10.12 shows the scheduling of shifts for a small technical 'hot line' support service for a small software company. It gives advice to customers on their technical problems. Its service times are 04:00 hours to 20:00 hours on Monday, 04:00 hours to 22:00 hours Tuesday to Friday, 06:00 hours to 22:00 hours on Saturday, and 10:00 hours to 20:00 hours on Sunday. Demand is heaviest on Tuesday to Thursday, starts to decrease on Friday, is low over the weekend and starts to increase again on Monday.

The scheduling task for this kind of problem can be considered over different timescales, two of which are shown in Figure 10.12. During the day, working hours need to be agreed with individual staff members. During the week, days off need to be agreed. During the year, vacations, training periods and other blocks of time where staff are unavailable need to be agreed. All this has to be scheduled such that:

- capacity matches demand;
- the length of each shift is neither excessively long nor too short to be attractive to staff;
- working at unsocial hours is minimised;

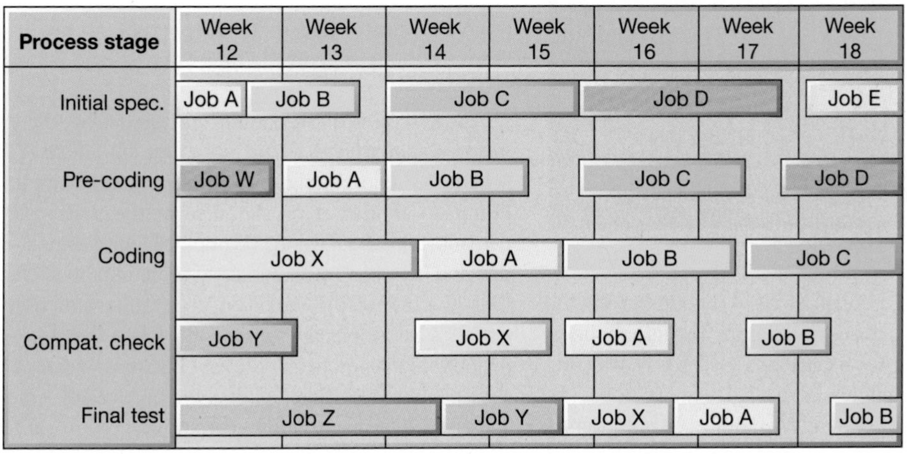

Figure 10.11 Gantt chart showing the schedule for jobs at each process stage

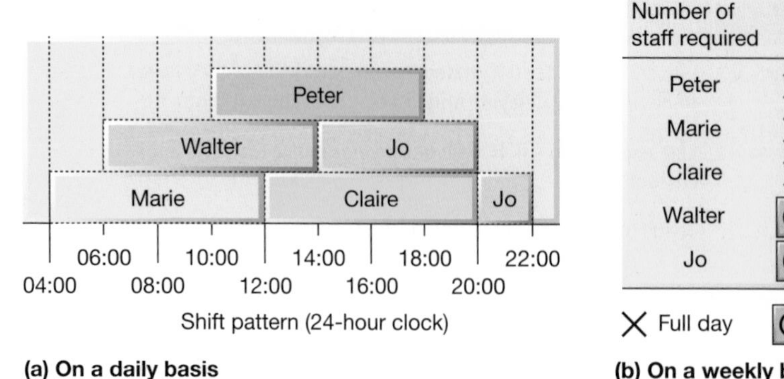

(a) On a daily basis

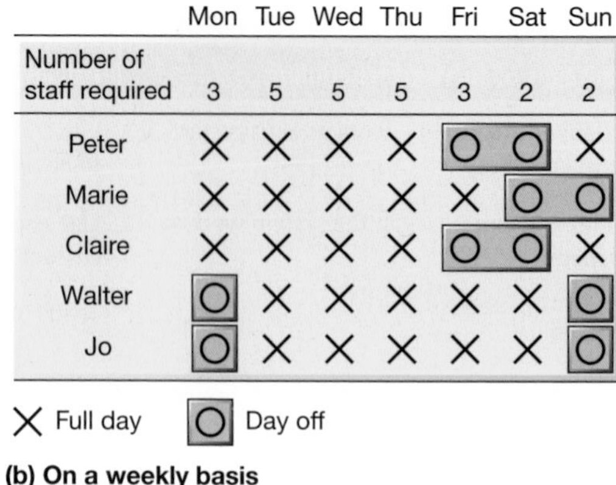

Mon Tue Wed Thu Fri Sat Sun
Number of staff required 3 5 5 5 3 2 2
Peter
Marie
Claire
Walter
Jo
Full day
Day off

(b) On a weekly basis

Figure 10.12 Shift scheduling in a support service for a small software company

- days off match agreed staff conditions – in this example, staff prefer two consecutive days off every week;
- vacation and other 'time-off' blocks are accommodated;
- sufficient flexibility is built into the schedule to cover for unexpected changes in supply (staff illness) and demand (surge in customer calls).

Scheduling staff times is one of the most complex of scheduling problems. In the relatively simple example shown in Figure 10.12 we have assumed that all staff have the same level and type of skill. In very large operations with many types of skills to schedule and uncertain demand (for example, a large hospital) the scheduling problem becomes extremely complex. Some mathematical techniques are available but most scheduling of this type is, in practice, solved using heuristics (rules of thumb), some of which are incorporated into commercially available software packages.

OPERATIONS IN PRACTICE

Ryanair cancels flights after 'staff scheduling' errors[6]

Ryanair, the largest European airline, is used to making the headlines for both good and bad reasons, but its announcement of multiple flight cancellations, potentially affecting up to 285,000 passengers, was not its finest hour. The cause, according to Ryanair Marketing Officer Kenny Jacobs, was that *'We have messed up in the planning of pilot holidays and we're working hard to fix that'*. It was unable to fix a pilot schedule that accommodated staff holidays, but left enough pilots to staff scheduled flights. Ryanair said that actually less than 2 per cent of its flights would be cancelled, and that it would not stop it hitting its annual punctuality target of 90 per cent. Yet this did not stop passengers complaining about the resulting uncertainty.

The root cause of the damaging staff shortage were two factors – annual holidays and flight time limitations (FTLs). Scheduling pilots' holidays was always problematic for European airlines, partly because of the seasonality of demand. Short-haul flying in Europe is highly seasonal, with demand between Easter and early September much higher than the rest of the year. This is an important time for airlines such as Ryanair, because it is when they make the majority of their profits. Because of this, most airlines prefer their pilots to take their holidays in one block of a month, somewhere between September and March. To complete the remainder of their holiday allocation, they can arrange to take 'ad hoc' days as and when the airline's plans permit. Complicating the scheduling task are FTLs. These stipulate the maximum hours pilots can work as 100 hours in any 28 days, 900 hours in a calendar year or 1,000 hours in any rolling 12-month period. Staff schedulers have to make sure their schedules conform to these limits. These rules are unbreakable. Ryanair's problem came when the Irish Aviation Authority told the airline that it had to change its calendar for the calculation of pilots' hours and leave from its previous April to March calendar, to one using the calendar year to keep it in line with the rules adopted by European regulators. Ryanair said the change meant it had to allocate annual leave to pilots (at relatively short notice) in September and October. Kenny Jacobs said, *'Most of the cancellations were due to a backlog of staff leave which has seen large numbers of the airline's staff book holidays towards the end of the year'*. The airline tried asking pilots to 'sell back' their annual leave. However, this was not always possible within the FTL rules, and some pilots were reluctant to change their holiday plans at short notice.

Theory of constraints (TOC)

An important concept, closely related to scheduling, that recognises the importance of planning to known capacity constraints, is the theory of constraints (TOC). It focuses scheduling effort on the bottleneck parts of the operation. By identifying the location of constraints, working to remove them, and then looking for the next constraint, an operation is always focusing on the part that critically determines the pace of output. The approach that uses this idea is called optimised production technology (OPT). Its development and the marketing of it as a proprietary software product were originated by Eliyahu Goldratt.[7] It helps to schedule production systems to the pace dictated by the most heavily loaded resources, that is, bottlenecks. If the rate of activity in any part of the system exceeds that of the bottleneck, then items are being produced that cannot be used. If the rate of working falls below the pace at the bottleneck, then the entire system is underutilised. The 'principles' underlying OPT demonstrate this focus on bottlenecks.

OPT principles

1. Balance flow, not capacity. It is more important to reduce throughput time rather than achieving a notional capacity balance between stages or processes.
2. The level of utilisation of a non-bottleneck is determined by some other constraint in the system, not by its own capacity. This applies to stages in a process, processes in an operation, and operations in a supply network.
3. Utilisation and activation of a resource are not the same. According to the TOC, a resource is being *utilised* only if it contributes to the entire process or operation creating more output. A process or stage can be *activated* in the sense that it is working, but it may only be creating stock or performing other non-value-added activity.
4. An hour lost (not used) at a bottleneck is an hour lost forever out of the entire system. The bottleneck limits the output from the entire process or operation, therefore the underutilisation of a bottleneck affects the entire process or operation.
5. An hour saved at a non-bottleneck is a mirage. Non-bottlenecks have spare capacity anyway. Why bother making them even less utilised?
6. Bottlenecks govern both throughput and inventory in the system. If bottlenecks govern flow, then they govern throughput time, which in turn governs inventory.
7. You do not have to transfer batches in the same quantities as you produce them. Flow will probably be improved by dividing large production batches into smaller ones for moving through a process.
8. The size of the process batch should be variable, not fixed. The circumstances that control batch size may vary between different products. (See discussion of the EBQ model in Chapter 13.)
9. Fluctuations in connected and sequence-dependent processes add to each other rather than averaging out. So, if two parallel processes or stages are capable of a particular average output rate, in parallel, they will never be able to produce the same average output rate.
10. Schedules should be established by looking at all constraints simultaneously. Because of bottlenecks and constraints within complex systems, it is difficult to work out schedules according to a simple system of rules. Rather, all constraints need to be considered together.

10.6 What is 'monitoring and control'?

Having created a plan for the operation through loading, sequencing and scheduling, each part of the operation has to be monitored to ensure that planned activities are indeed happening. Any deviation from the plans can then be rectified through some kind of intervention in the operation, which itself will probably involve some re-planning. Figure 10.13 illustrates a simple view of control. The output from a work centre is monitored and compared with the plan, which indicates what the work centre is supposed to be doing. Deviations from this plan are taken into account through a re-planning activity and the necessary interventions made to the work centre, which will (hopefully) ensure that the new plan is carried out. Eventually, however, some further deviation from planned activity will be detected and the cycle is repeated.

> **Operations principle**
>
> A planning and control system should be able to detect deviations from plans within a timescale that allows an appropriate response.

Push and pull control

One element of control, then, is periodic intervention into the activities of the operation. An important decision is how this intervention takes place. The key distinction is between intervention signals, which push work through the processes within the operation, and those which pull work only when it is required. In a **push system of control**, activities are scheduled by means of a central system and completed in line with central instructions, such as an MRP system (see Chapter 14). Each work centre pushes out work without considering whether the succeeding work centre can make use of it. Work centres are coordinated by means of the central operations planning and control system. In practice, however, there are many reasons why actual conditions differ from those planned. As a consequence, idle time, inventory and queues often characterise push systems. By contrast, in a **pull system of control**, the pace and specification of what is done are set by the 'customer' workstation, which 'pulls' work from the preceding (supplier) workstation. The customer acts as the only 'trigger' for movement. If a request is not passed back from the customer to the supplier, the supplier cannot produce anything or move any materials. A request from a customer not only triggers production at the supplying stage, but also prompts the supplying stage to request a further delivery from its own suppliers. In this way, demand is transmitted back through the stages from the original point of demand by the original customer.

The inventory consequences of push and pull

Understanding the differing principles of push and pull is important because they have different effects in terms of their propensities to accumulate inventory in the operation. Pull systems are far less likely to result in inventory build-up and are therefore favoured by lean operations (see Chapter 16). To understand why this is so, consider an analogy: the 'gravity' analogy is illustrated

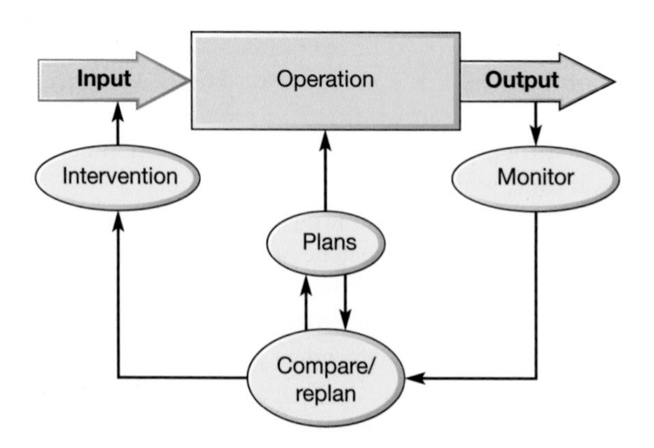

Figure 10.13 A simple model of control

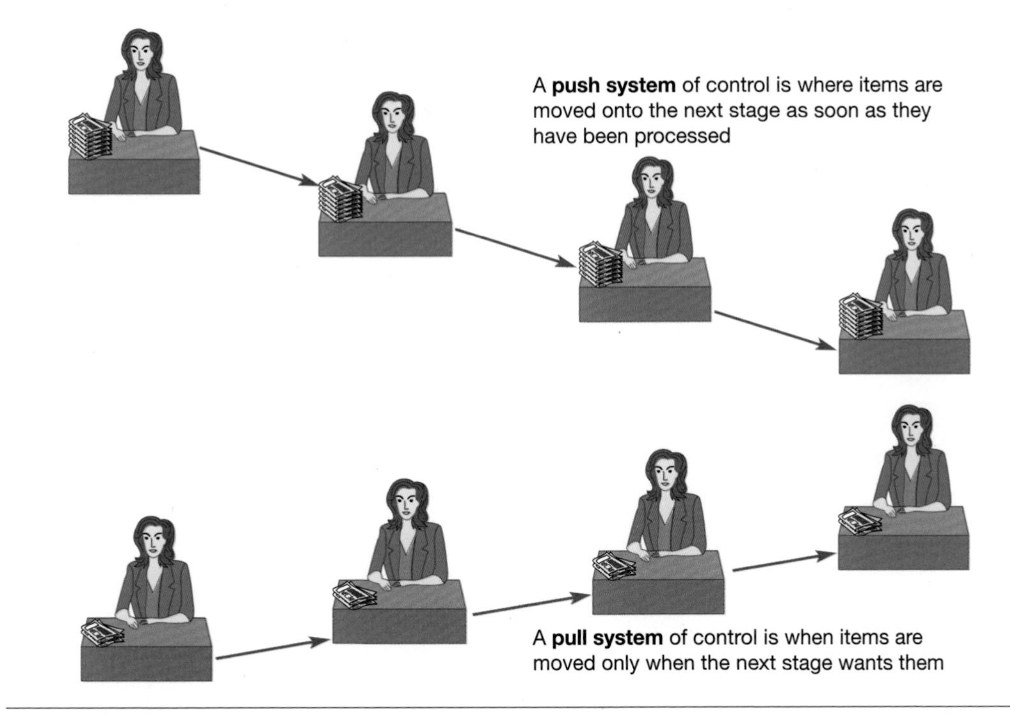

A **push system** of control is where items are moved onto the next stage as soon as they have been processed
A **pull system** of control is when items are moved only when the next stage wants them

Figure 10.14 Push versus pull: the gravity analogy

in Figure 10.14. Here a push system is represented by an operation, each stage of which is on a lower level than the previous stage. When items are processed by each stage, gravity pushes them down the slope to the next stage. Any delay or variability in processing time at that stage will result in the items accumulating as inventory. In the pull system, items cannot naturally flow uphill, so they can only progress if the next stage along deliberately pulls them forward. Under these circumstances, inventory cannot accumulate as easily.

Operations principle
Pull control reduces the build-up of inventory between processes or stages.

Drum, buffer, rope

The drum, buffer, rope concept comes from the theory of constraints (TOC) described earlier. It is an idea that helps to decide exactly where in a process control should occur. Most do not have the same amount of work loaded onto each separate work centre (that is, they are not perfectly balanced). This means there is likely to be a part of the process that is acting as a bottleneck on the work flowing through the process. TOC argues that the bottleneck in the process should be the control point of the whole process. It is called the drum because it sets the 'beat' for the rest of the process to follow. Because it does not have sufficient capacity, a bottleneck is (or should be) working all the time. Therefore, it is sensible to keep a buffer of inventory in front of it to make sure that it always has something to work on. Because it constrains the output of the whole process, any time lost at the bottleneck will affect the output from the whole process, so it is not worthwhile for the parts of the process before the bottleneck to work to their full capacity. All they would do is produce work that would accumulate further along in the process up to the point where the bottleneck is constraining the flow. Therefore, some form of communication between the bottleneck and the input to the process is needed to make sure that activities before the bottleneck do not overproduce. This is called the rope (see Figure 10.15).

Operations principle
The constraints of bottleneck processes and activities should be a major input to the planning and control activity.

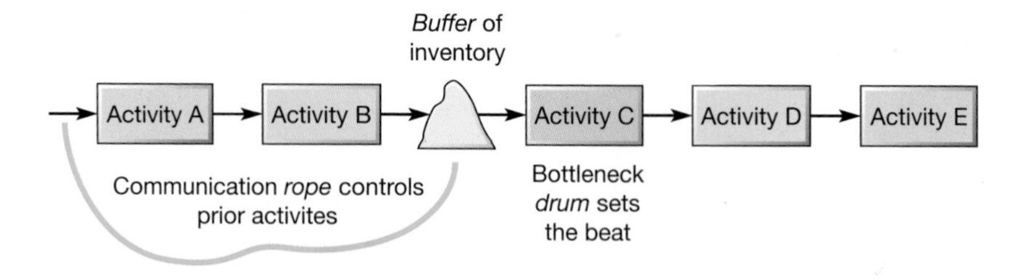

Figure 10.15 The drum, buffer, rope concept

Critical commentary

Most of the perspectives on control taken in this chapter are simplifications of a far messier reality. They are based on models used to understand mechanical systems such as car engines. But anyone who has worked in real organisations knows that organisations are not machines. They are social systems, full of complex and ambiguous interactions. Simple models such as these assume that operations objectives are always clear and agreed, yet organisations are political entities where different and often conflicting objectives compete. Local government operations, for example, are overtly political. Furthermore, the outputs from operations are not always easily measured. A university may be able to measure the number and qualifications of its students, for example, but it cannot measure the full impact of its education on their future happiness. Also, even if it is possible to work out an appropriate intervention to bring an operation back into 'control', most operations cannot perfectly predict what effect the intervention will have. Even the largest of burger bar chains does not know exactly how a new shift allocation system will affect performance. Also, some operations never do the same thing more than once anyway. Most of the work done by construction operations are one-offs. If every output is different, how can 'controllers' ever know what is supposed to happen? Their plans themselves are mere speculation.

Controlling operations is not always routine

The simple monitoring control model in Figure 10.13 helps us to understand the basic functions of the monitoring and control activity. But, as the critical commentary box says, it is a simplification. Some simple routine processes may approximate to it, but many other operations do not. In fact, some of the specific criticisms cited in the critical commentary box provide a useful set of questions, which can be used to assess the degree of difficulty associated with control of any operation. In particular:

- Is there consensus over what the operation's objectives should be?
- Are the effects of interventions into the operation predictable?
- Are the operation's activities largely repetitive?

Starting with the first question, are strategic objectives clear and unambiguous? Many operations are just too complex to articulate every aspect of their objectives in detail. Nor is there always agreement on what the objectives should be. In social care organisations, for example, some managers are charged with protecting vulnerable members of society, others with ensuring that public money is not wasted, and others may be required to protect the independence of professional staff. A further assumption in the simplified control model is that there is some reasonable knowledge of how to bring about the desired outcome. That is, when a decision is made, one can predict its effects with a reasonable degree of confidence. It is assumed that interventions intended to bring a process back under control will indeed have the intended effect. Yet, this implies that the relationships between the intervention and the resulting consequence within the process are predictable, which in turn assumes that the degree of process knowledge is high. For example, if an organisation decides to relocate in order to be more convenient for its customers, it may or may not prove to be correct. Customers may react in a manner that was not predicted. In fact, many operations decisions are taken where the cause–effect relationship is only partly understood. The final assumption about control is that control interventions are made in a repetitive way and occur frequently (for example, checking on a process, hourly or daily). This means that the operation has the opportunity to learn how its interventions affect the process, which considerably facilitates control. However, some control situations are non-repetitive, offering less opportunity for learning.

Figure 10.16 illustrates how these questions can form a 'decision tree'-type model that indicates how the nature of operations control may be influenced.[8] Operational control is relatively straightforward: objectives are unambiguous, the effects of interventions are known, and activities are repetitive. This type of control can be done using predetermined conventions and rules. There are, however, still some challenges to successful routine control. It needs operational discipline to make sure that control procedures are systematically implemented. The main point though is that any divergence from the conditions necessary for routine control implies a different type of control.

Operations principle

Planning and control is not always routine, especially when objectives are ambiguous, the effects of interventions into the operation are not predictable and activities not repetitive.

Expert control

If objectives are unambiguous, yet the effects of interventions relatively well understood, but the activity is not repetitive (for example, installing or upgrading software or IT systems) control can be delegated to an 'expert'; someone for whom such activities are repetitive because they have built their knowledge on previous experience elsewhere. Making a success of expert control requires that such experts exist and can be 'acquired' by the firm. It also requires that the expert takes advantage of the control knowledge already present in the firm and integrates his or her 'expert' knowledge with the support that potentially exists internally. Both of these place a stress on the need to 'network', in terms of both acquiring expertise and then integrating that expertise into the organisation.

Trial-and-error control

If strategic objectives are relatively unambiguous, but effects of interventions not known, yet the activity is repetitive, the operation can gain knowledge of how to control successfully through its own failures. In other words, although simple prescriptions may not be available in the early stages of making control interventions, the organisation can learn how to do it through experience. For example,

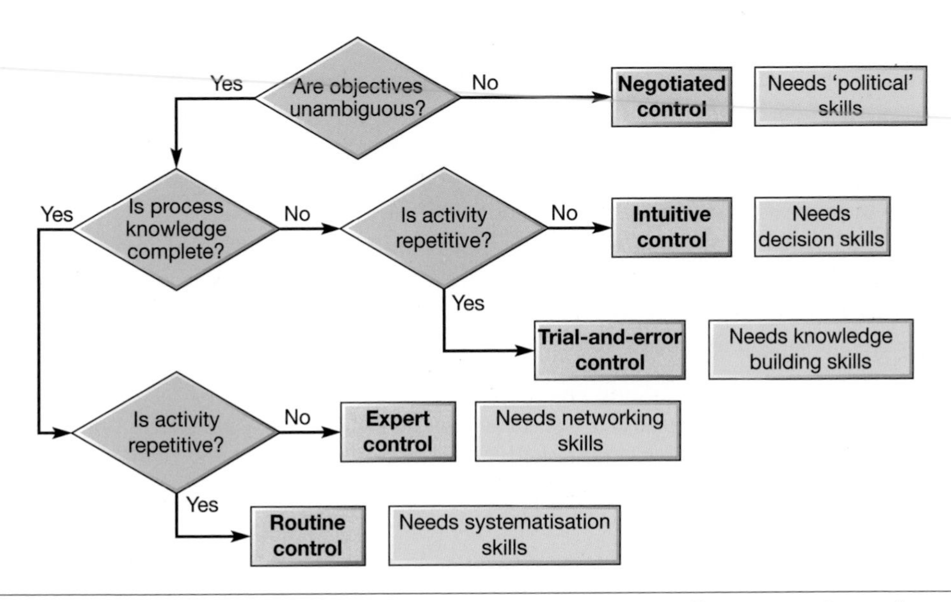

Figure 10.16 Control is not always routine, different circumstances require different types of control

if a fast-food chain is opening new stores into new markets, it may not be sure how best to arrange the openings at first. But if the launch is the first of several, the objective must be not only to make a success of each launch, but equally (or more) important, it must learn from each experience. It is these knowledge-building skills that will ultimately determine the effectiveness of trial-and-error control.

Intuitive control

If objectives are relatively unambiguous (so it is clear what the operation is trying to do), but effects of control interventions not known, and nor are they repetitive, learning by trial and error is not possible. Here control becomes more of an art than a science. And in these circumstances control must be based on the management team using its intuition to make control decisions. Many strategic operations processes fall into this category – for example, setting up a strategic supply partnership (see Chapter 12). Objectives are clear (jointly survive in the long term, make an acceptable return, and so on) but not only are control interventions not repetitive and their effects not fully understood, sometimes the supplier's interests may be in conflict with yours. Yet, simply stating that 'intuition' is needed in these circumstances is not particularly helpful. Instinct and feelings are, of course, valuable attributes in any management team, but they are the result, at least partly, of understanding how best to organise their shared understanding, knowledge and decision-making skills. It requires thorough decision analysis, not to 'mechanistically' make the decision, but to frame it so that connections can be made, consequences understood and insights gained.

Negotiated control

The most difficult circumstance for strategic control is when objectives are ambiguous. This type of control involves reducing ambiguity in some way by making objectives less uncertain. Sometimes this is done simply by senior managers 'pronouncing' or arbitrarily deciding what objectives should be, irrespective of opposing views. For example, controlling the activities of a childcare service can involve very different views among the professional social workers making day-to-day decisions. Often a negotiated settlement may be sought, which then can become an unambiguous objective. Alternatively, outside experts could be used, either to help with the negotiations, or to remove control decisions from those with conflicting views. But, even within the framework of negotiation, there is almost always a political element when ambiguities in objectives exist. Negotiation processes will be, to some extent, dependent on power structures.

Responsible operations

In every chapter, under the heading of 'Responsible operations', we summarise how the particular topic covered in the chapter touches upon important social, ethical and environmental issues.

Planning and control in general, and particularly scheduling, can be complex, even when the resources or jobs being scheduled are inanimate. Schedules may be well put together or not, but at least the things being scheduled are not in a position to complain. Staff scheduling is different. Ask people to work at inconvenient times, or work for too-long periods of time, or when high levels of demand overload their ability to do a good job, and they may not only complain, they will also become dissatisfied, they may even resign. Poor staff scheduling can have serious consequences for employees and the performance of their operation. We briefly described staff scheduling earlier in this chapter, but it is worth expanding on why it poses particular problems. There are a number of factors:

- The obvious issue is that demand on the operation may not match the times when staff want to work. For example, many high customer contact operations, such as restaurants or call centres, experience high demand at lunchtime when their staff also want lunch.
- Part-time staff may not be always available at exactly the same time every day or every week.
- Full-time staff should reasonably expect schedules to accommodate short-term leave requests and longer-term holiday arrangements.

- Staff can be unexpectedly unavailable because of sickness.
- The increasing popularity of flexible working arrangements means that staff scheduling is becoming more complex.

One review of staff scheduling indicated that people in some occupations are increasingly required to work difficult schedules, sometimes with irregular hours, little notice of schedule changes, and inadequate partial consultation.[9] Yet, staff scheduling, technically complex though it is, needs to accommodate not only the needs of employees, but also the human consequences of the resulting schedules. Volatile or badly planned schedules can result in a feeling of pressure and anxiety for employees, affecting both their effectiveness and well-being. What do people want from a staff schedule?

- *Flexibility* – Schedules that allow individual flexibility are commonly associated with reduced absenteeism and turnover, as are compressed work schedules where staff work longer each day, but fewer days per week. Presumably this is because it gives give larger periods of free time.
- *Stability* – Most people want their hours to be roughly the same each week, helping them to organise their time effectively. Schedules based on a fixed, or reasonably stable, pattern, are generally popular because they are less disruptive to employees' personal lives.
- *Notification* – Most people appreciate the opportunity to plan their personal lives more than a few days in advance. While they may understand that circumstances change, providing as much notice of changes as possible is usually welcomed.

Note how, in describing what staff 'generally' want from staff schedules, we qualify our points in terms of 'most people', or 'commonly'. In other words, there are some staff who do not share these preferences. This raises the possibility, suggested by some authorities, that work schedules should be aligned with the needs, desires and personalities of individual employees. In other words, a 'one size fits all' approach to staff scheduling is neither equitable nor effective. Schedules should perhaps be customised to meet the needs and desires of individual employees. Creating more 'idiosyncratic' schedules may even help to find a balance between the need to provide enough staff to meet demand while also meeting the preferences of employees.

Summary answers to key questions

10.1 What is planning and control?

- Planning and control is the reconciliation of the potential of the operation to supply products and services, and the demands of its customers on the operation. It is the set of day-to-day activities that run the operation on an ongoing basis.
- A plan is a formalisation of what is intended to happen at some time in the future. Control is the process of coping with changes to the plan and the operation to which it relates. Although planning and control are theoretically separable, they are usually treated together.
- The balance between planning and control changes over time. Planning dominates in the long term and is usually done on an aggregated basis. At the other extreme, in the short term, control usually operates within the resource constraints of the operation but makes interventions into the operation to cope with short-term changes in circumstances.

10.2 How do supply and demand affect planning and control?

- The degree of uncertainty in demand affects the balance between planning and control. The greater the uncertainty, the more difficult it is to plan, and greater emphasis must be placed on control.

- This idea of uncertainty is linked with the concepts of dependent and independent demand. Dependent demand is relatively predictable because it is dependent on some known factor. Independent demand is less predictable because it depends on the chances of the market or customer behaviour.
- The different ways of supplying demand can be characterised by differences in the *P:D* ratio of the operation. The *P:D* ratio is the ratio of total throughput time of services or products to demand time.
- The volume and variety characteristics of an operation will have an effect on its planning and control activities.

10.3 What is 'loading'?

- Loading is the amount of work that is allocated to a work centre. It dictates the amount of work that is allocated to each part of the operation.
- Finite loading is an approach that only allocates work to a work centre up to a set limit. Infinite loading is an approach to loading work that does not limit accepting work, but instead tries to cope with it.

10.4 What is 'sequencing'?

- Sequencing decides the order in which work is tackled within the operation. It determines the priorities given to work in an operation, usually determined by some predefined set of rules, known as sequencing rules.
- Sequencing rules include factors such as the physical constraints of items, customer priority and due dates.

10.5 What is 'scheduling'?

- Scheduling determines the detailed timetable of activities, and when activities are started and finished. It can be a complex endeavour because the number of possible schedules increases rapidly as the number of activities and processes increases.
- The two main approaches are forward and backward scheduling. Forward scheduling involves starting work as soon as it arrives. Backward scheduling involves starting jobs at the last possible moment to prevent them from being late.

10.6 What is 'monitoring and control'?

- Monitoring and control involves detecting what is happening in the operation, replanning if necessary, and intervening to impose new plans.
- Two important types are 'pull' and 'push' control. Pull control is a system whereby demand is triggered by requests from a work centre's (internal) customer. Push control is a centralised system whereby control (and sometimes planning) decisions are issued to work centres, which are then required to perform the task and supply the next workstation. In manufacturing, 'pull' schedules generally have far lower inventory levels than 'push' schedules.
- The ease with which control can be maintained varies between operations.

CASE STUDY Audall Auto Servicing

It had been ten years since Dan Audall had founded Audall auto servicing as an independent vehicle servicing and repair business. Previously he had been the manager of the servicing department of a 'premium' car dealership, the experience of which had convinced him that there was demand in the area for servicing one of the 'rival' makes of car. '*We were continually getting requests from owners to service or repair their vehicles, partly because we had a good reputation, but mainly because there were no local dealerships who could do that kind of work. Owners had to use small independents or travel a long distance to get to the nearest dealership. I persuaded the car company that I could provide appropriate service for their vehicles without taking significant business away from their other franchised service centres. It was a gamble, but they backed me*'.

That was ten years ago, and Dan's gamble had paid off. Audall Auto Servicing had grown to the point where he had invested in a modern service centre close to his first location. The outline plans for the new centre are shown in Figure 10.17. It had five servicing bays, parts store, a car wash and a customer waiting area. '*Although the new building is the same nominal capacity as the old one, it gives us more room for the technicians to move about, and the customer waiting area is*

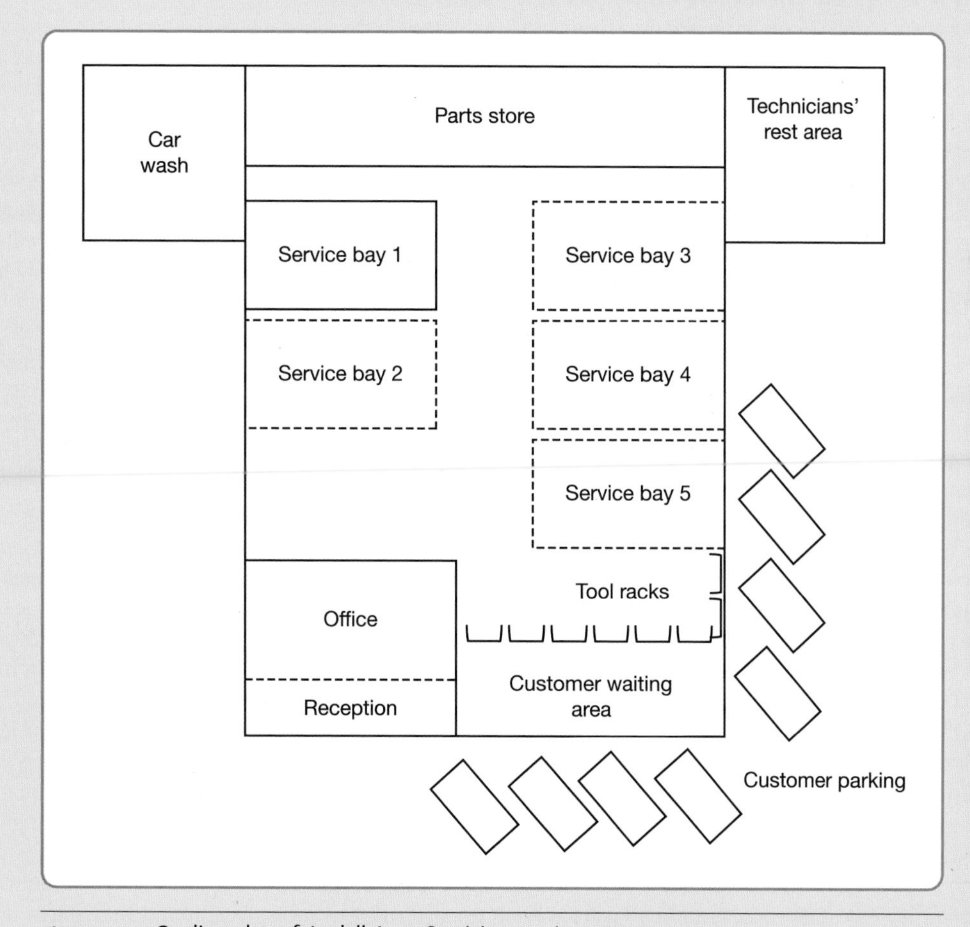

Figure 10.17 Outline plan of Audall Auto Servicing service centre

a distinct improvement in the service that we can offer customers. There is also room next to this building that we could use for expansion, although I would prefer to wait a couple of years before committing to this investment. Before then, I think that we could do more business with our existing capacity' (Dan Audall).

Dan's conviction that the operation could do more business with its current capacity was based on performance figures covering the first two months in the new building, showing that the servicing bays were, on average, only 83 per cent utilised. This was a figure that Dan believed could be improved, as did Diya Chopra, his office manager. However, Diya thought that the room for improvement would be limited. *'No week is ever perfectly predictable. There are just too many uncertainties. Even a minor service (usually every 6,000 miles or 10,000 km) can throw up problems that can take two or three times the time that we have allowed. Major services (usually every 12,000 miles or 18,000 km) are even less predictable. With these "standard" services, customers often want us to keep the vehicle until it's repaired rather than make another booking. But the most unpredictable are what we call the "short-term" repairs where a customer wants their vehicle "up and running" as soon as possible. We call these jobs "short-term" not because they take little time, but because they are usually booked in at short notice. We have to give good service to our (minor and major) servicing customers, but there is some flexibility in planning these jobs. At the other extreme, short-term emergency repair work for customers has to be fitted into our schedule as quickly as possible. If someone is desperate to have their car repaired at very short notice, we sometimes ask them to drop their car in as early as they can and pick it up as late as possible. This gives us the maximum amount of time to fit it into the schedule. There are a number of service options open to customers. We can book short jobs in for a fixed time and do it while they wait. Most commonly, we ask the customer to leave the car with us and collect it later'*.

The technicians

The company employed five technicians, two trainee technicians, three part-time valeting assistants (who cleaned customers' cars), two part-time receptionists, Diya, and two office assistants who reported to her. Each technician worked in their own service bay, with the trainees assisting them as required. Two of the technicians had worked for Dan since the company was founded. They were the most experienced and tended to be allocated a mixture of major servicing jobs and 'short-term' repairs that might require more advanced diagnostic work. One of the other technicians was only recently qualified and was usually allocated the more routine jobs, such as minor servicing and MOT checks (the UK Government requires vehicles over a certain age to be tested every year; these are known as 'Ministry of Transport' or MOT tests). The remaining two technicians were allocated a mixture of work. *'We are going to have to reconsider how we allocate jobs in the near future. The more experienced technicians will always do more of the repair and major servicing than the more junior technicians, but we can't keep on giving them all the interesting work. The more junior people will get frustrated if they only do routine work. Also, we have tended to keep the senior technicians lightly loaded so that they can be free for the unpredictable short-term repairs. This means that, if demand is lighter than usual, they are less heavily loaded that the others, However, they are quite good at helping the others out, if this happens'* (Diya Chopra).

Scheduling the service bays

Most days the service centre has to deal with 15–30 jobs, taking from half an hour up to a whole day, or very occasionally even longer. Most jobs have a time allowance. A minor service is usually allowed one and a half hours, a major service around twice as long, and MOT tests will take half an hour. Short-term repairs could take anything between half an hour and all day. Diya would make an estimate of how long the job would take, often in consultation with one of the senior technicians, but some jobs would be difficult to estimate. *'Some jobs are easier than others to estimate. One of the guys will say something like, "Sounds like a cam belt, should take a couple of hours". At other times they will say, "Goodness knows* (or words to that effect)*, sorry, don't know how long it will take". That is why I have to leave times free in the week's schedule'*.

Figure 10.18 shows a typical schedule for the service centre at the beginning of the week, before many 'short-term' jobs have been allocated. At this stage, the schedule is purely nominal. For example, this particular week, only one 'short-term' job had been programmed into the schedule by early Monday morning, but within a couple of hours the unused space on the schedule would fill up. Often within an hour of starting work on a Monday morning, the schedule would have changed. Early in the week, this is usually because a standard service has taken longer than expected because a problem has been found. As Diya explained, *'Every day we have to cope with the unexpected. A technician may find that extra work is needed, customers may want extra work doing, and technicians are sometimes ill, which reduces our capacity. Occasionally parts may not be available so we have to arrange with the customer for the vehicle to be rebooked for a later time. Every week up to ten or twelve customers just don't turn up. We automatically text them the day before, but even so they still forget to bring their car in so we have to rebook them in at a later time. We can cope with most of these uncertainties because our technicians can be flexible and most are also willing to work overtime when needed. The important thing is important to manage customers' expectations. If there is a chance that the vehicle may not be ready for them, it shouldn't come as a surprise when they try to collect it'*.

Even with some flexibility from the technicians, as the week progresses, short-term repairs are increasingly likely to disrupt the schedule. The actual schedules were recorded on a computer-based scheduling system that the centre had been using for several years. Diya found the system useful, but

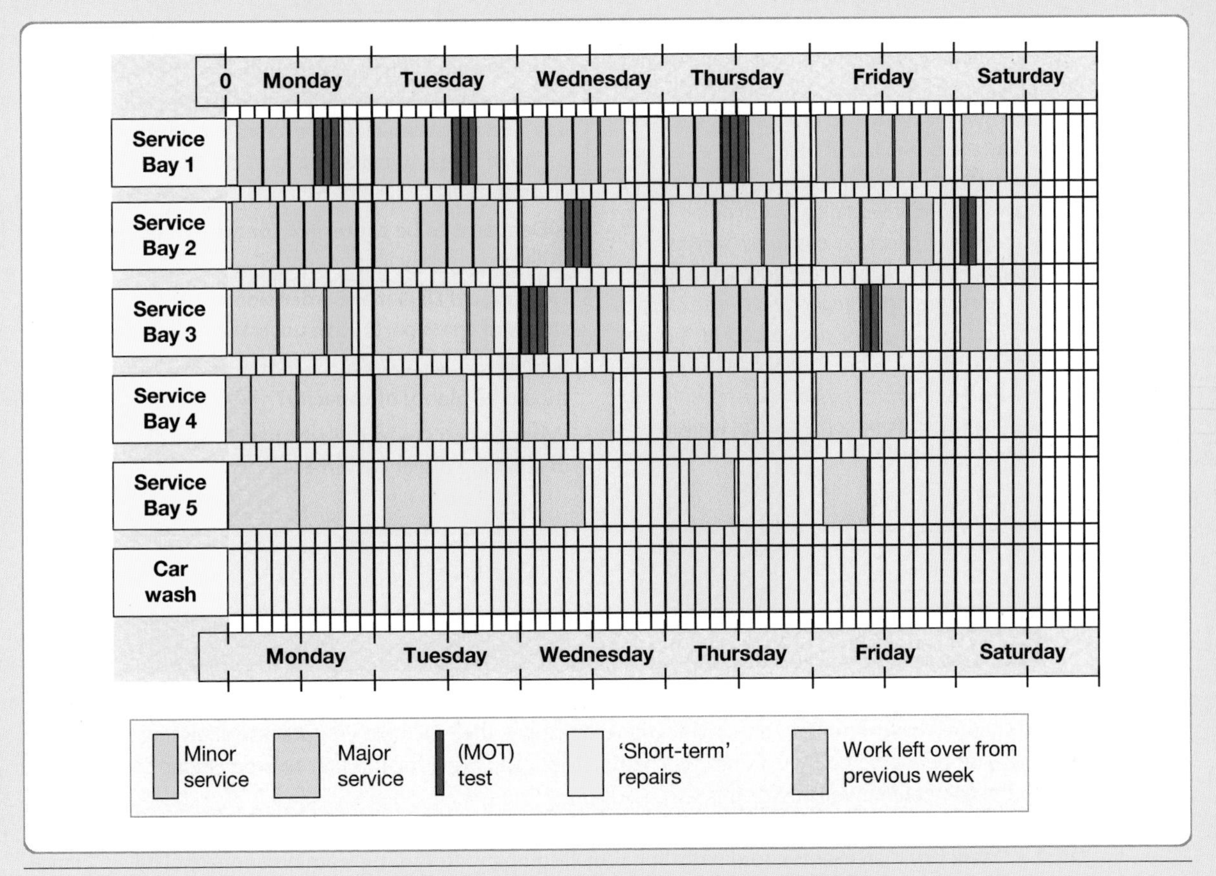

Figure 10.18 Audall Auto Servicing Schedule for the upcoming week (as of Monday 08.00)

limited. *'We enter all jobs into the scheduling system. On screen, it shows the total capacity we have day by day, all the jobs that are booked in, the amount of free capacity still available and so on. We use this to see when we have the capacity to book a customer in, and then enter all the customer's details. The car maker has issued "standard times" for all the major jobs. However, you have to modify these standard times a bit to take account of circumstances. That is where the senior technicians' experience comes in. Of course, the system does not really make any decisions as to which jobs have priority, nor can it automatically reschedule when things change. We make all the decisions. It's really a convenient system for keeping track of what's happening. It also works out overtime payments, issues invoices and calculates workshop utilisation figures'.*

The car wash

An unexpected bottleneck in the workshop was the car wash facility. Diya explained, *'We normally process around twenty to twenty-five vehicles a day. Valeting each of them takes between 10 minutes and 20 minutes. There should be plenty of capacity, but very often vehicles are waiting for up to an hour before they can be cleaned. This means that customers can be waiting for longer than they expect, and longer than we have promised them. Dan firmly believes that we should never return a car that has not been fully valeted, but I know that, under pressure, the valeting staff sometimes rush the cleaning. I don't think customers mind too much'.*

Spare parts

Parts stocks could be an important factor in keeping to schedule. Diya thought that their spare parts stock control was pretty good, but could be better. *'We keep all the most commonly used parts in stock, but if a repair needs a part that is not in stock, we can usually get it from our parts distributors within a day. Every evening our planning system prints out the jobs that we should be doing the next day and the parts that are likely to be needed for each job. This allows the parts staff to pick out the parts for each job so that the technicians can receive them first thing the next morning without any delay. The problem is that because we normally get a part within a day, we can become complacent. If a part is not available for any reason, it can delay a repair, which disrupts our schedule and can upset the customer. Maybe we should keep some "emergency stocks" of parts that we have had problems getting quickly, even if they are not often needed. It would tie up cash, but would help to protect the schedule'.*

Direct customer reservations

Dan and Diya had been investigating the costs and benefits of investing in a new computer-based scheduling system. There were several systems on the market, specifically designed for scheduling automobile workshops. Many of them were intended for larger workshops than Audall's, but some incorporated an interface that would allow customers to book their vehicle into the workshop in a specific time slot. Dan was intrigued by the possibility of being able to do this but was also hesitant. *'It would mean a very different way of working. I'm not sure whether it would be worth doing, especially in the way it would impact on customer service. For example, at the moment Diya can judge the degree of urgency of each job and talk to the customer, managing their expectations where appropriate. She is particularly good at making decisions that reconcile the needs of the customers and our resources in some way. For Diya, this involves attempting to maximise the utilisation of her workshop resources while keeping customers satisfied.*

QUESTIONS

1 The schedule in Figure 10.18 indicates that (a) the more predictable jobs tend to be loaded earlier in the day, and (b) technicians are allocated different types of job. If these two assumptions are true, do they seem sensible?

2 Is Dan right to be concerned that the workshop's utilisation is only 83 per cent?

3 How should Diya make a decision about keeping stocks of spare parts with uncertain delivery?

4 Why is the car wash such a bottleneck when there should be plenty of capacity?

5 What do you see as the advantages and disadvantages of allowing direct customer reservations?

Problems and applications

All chapters have 'Problems and applications' questions that will help you practise analysing operations. They can be answered by reading the chapter. Model answers for the first two questions can be found on the companion website for this text.

1 Mark Key is an events coordinator for a small company. Returning from his annual holiday in France, he is given six events to plan. He gives them the codes A–F. He needs to decide upon the sequence in which to plan the events and wants to minimise the average time the jobs are tied up in the office and, if possible, meet the deadlines allocated. The six jobs are detailed in Table 10.5. Determine a sequence based on using (a) the FIFO rule, (b) the due date rule, (c) the shortest operation time rule. (d) Which of these sequences gives the most efficient solution and which gives the least lateness?

Table 10.5 The six jobs that Mark has to sequence

Sequence of jobs	Process time (days)	Due date
A	4	12
B	3	5
C	1	7
D	2	9
E	2	15
F	5	8

2 It is week 35 of a busy year at Ashby Architects and Jo Ashby is facing a big problem. Both her two junior partners have been diagnosed with a serious illness contracted on a trip to scope out a prospective job in Lichtenstein. So Jo has to step in and complete the outstanding jobs that were being worked on by the two juniors. The outstanding jobs are shown in Table 10.6.

Table 10.6 Outstanding jobs that Jo will need to complete.

Job	Due date (week)	Weeks of work remaining
Ashthorpe lavatory block	40	2.0
Bugwitch bus shelters	48	5.0
Crudstone plc HQ	51	3.0
Dredge sewage works	52	8.0

Jo has heard that a sequencing rule called the critical ratio (CR) will give efficient results. The priority of jobs using the CR rule is defined by an index computed as follows:

$$CR = \frac{\text{Time remaining}}{\text{Workdays remaining}} = \frac{\text{Due date} - \text{Today's date}}{\text{Workdays remaining}}$$

Using this rule, in what priority should Jo give the jobs?

3 It takes 6 hours for a contract laundry to wash, dry and press (in that order) a batch of overalls. It takes 3 hours to wash the batch, 2 hours to dry it and 1 hour to press it. Usually, each day's batch is collected and ready for processing at 8.00 am and needs to be picked up at 4.00 pm. The two people who work in the laundry have different approaches to how they schedule the work. One schedules 'forward'; forward scheduling involves starting work as soon as it arrives. The other schedules 'backwards'; backward scheduling involves starting jobs at the last possible moment that will prevent them from being late.

(a) Draw up a schedule indicating the start and finish time for each activity (wash, dry and press) for both forward and backward approaches.
(b) What do you think are the advantages and disadvantages of these two approaches?

4 Read the following descriptions of two cinemas:

Kinepolis in Brussels is one the largest cinema complexes in the world, with 28 screens, a total of 8,000 seats, and four showings of each film every day. It is equipped with the latest projection technology. All the film performances are scheduled to start at the same times every day: 4.00 pm, 6.00 pm, 8.00 pm and 10.30 pm. Most customers arrive in the 30 minutes before the start of the film. Each of the 18 ticket desks has a networked terminal and a ticket printer. For each customer, a screen code is entered to identify and confirm seat availability for the requested film. Then the number of seats required is entered and the tickets are printed, though these do not allocate specific seat positions. The operator then takes payment by cash or credit card and issues the tickets. This takes an average of 19.5 seconds, and a further 5 seconds is needed for the next customer to move forward. An average transaction involves the sale of approximately 1.7 tickets.

The UCI cinema in Birmingham has eight screens. The cinema incorporates many 'state-of-the-art' features, including the high-quality THX sound system, fully computerised ticketing and a video games arcade off the main hall. In total the eight screens can seat 1,840 people; the capacity (seating) of each screen varies, so the cinema management can allocate the more popular films to the larger screens and use the smaller screens for the less popular films. The starting times of the eight films at UCI are usually staggered by 10 minutes, with the most popular film in each category (children's, drama, comedy, etc.) being scheduled to run first. Because the films are of different durations, and since the manager must try to maximise the utilisation of the seating, the scheduling task is complex. Ticket staff are continually aware of the remaining capacity of each 'screen' through their terminals. There are up to four ticket desks open at any one time. The target time per overall transaction is 20 seconds. The average number of ticket sales per transaction is 1.8. All tickets indicate specific seat positions, and these are allocated on a first come, first served basis.

(a) What are the main differences between the two cinemas from the perspectives of their operations managers?
(b) What are the advantages and disadvantages of the two different methods of scheduling the films onto the screens?
(c) Find out the running times and classification of eight popular films. Try to schedule these onto the UCI Birmingham screens, taking account of what popularity you might expect at different times. Allow 20 minutes for emptying, cleaning, and admitting the next audience, and 15 minutes for advertising before the start of the film.

5 Think through the following three brief examples. What type of control (according to Figure 10.16) do you think they warrant?

The Games Delivery Authority (GDA) was a public body responsible for developing and building the new venues and infrastructure for the 'International Games' and their use after the event. The GDA appointed a consortium responsible for the overall programme's quality, delivery and cost, in addition to health and safety, sustainability, equality and diversity targets. The Games Park was a large construction programme spreading across five separate local government areas, including transport developments, retail areas and local regeneration projects. Sustainability was central to the development. *'"Sustainability" was ingrained into our thinking – from the way we planned, built and worked, to the way we play, socialise and travel'*. To ensure they stuck to commitments, the GDA set up an independent body to monitor the project. All potential contractors tendering for parts of the project were aware that a major underlying objective of the Games initiative was regeneration. The Games site was to be built on highly industrialised and contaminated land.

The supermarket's new logistics boss was blunt in his assessment of its radical supply chain implementation. *'Our rivals have watched in utter disbelief'*, he said. *'Competitors looked on in amazement as we poured millions into implementing new IT systems and replaced 21 depots with a handful of giant automated "fulfilment factories".'*

'In hindsight, the heavy reliance on automation was a big mistake, especially for fast moving goods', said the company's CIO. *'When a conventional facility goes wrong, you have lots of options. You have flexibility to deal with issues. When an automated "fulfilment factory" goes wrong, frankly, you're buggered'*. Most damning was the way that the supermarket pressed on with the implementation of the automated facilities before proving that the concept worked at the first major site. *'I'd have at least proved that one of them worked before building the other three'*, he said. *'Basically, the whole company was committed to doing too much, too fast, trying to implement a seven-year strategy in a three-year timescale'*.

'It's impossible to overemphasise just how important this launch is to our future', said the CEO. *'We have been losing market share for seven quarters straight. However, we have very high hopes for the new XC10 unit'*. And most of the firm's top management team agreed with her. Clearly the market had been maturing for some time now, and was undoubtedly getting more difficult. New product launches from competitors had been eroding market share. Yet competitors' products, at best, simply matched the firm's offerings in all benchmark tests. *'Unless someone comes up with a totally new technology, which is very unlikely, it will be a matter of making marginal improvements in product performance and combining this with well-targeted and coordinated marketing. Fortunately, we are good at both of these. We know this technology, and we know these markets. We are also clear what role the new XC10 should play. It needs to consolidate our market position as the leader in this field, reduce the slide in market share by half, and re-establish our customers' faith in us. Margins, at least in the short term, are less important'*.

6 Re-read the 'Operations in practice' example 'Operations control at Air France'. How do the planning and control tasks compare to those in a motor servicing garage?

7 Re-read the 'Operations in practice' example 'Can airline passengers be sequenced?'. What problems could airlines face if they attempt to implement the Steffen method?

8 What might a Gantt chart for the mass production of chicken salad sandwiches look like?

9 If you had to make a case against the use of triage in sequencing the treatment of patients at an accident and emergency department in a hospital, what points would you make?

10 How does scheduling staff shift patterns differ from scheduling truck maintenance?

Selected further reading

Chapman, S.N. (2005) *Fundamentals of Production Planning and Control*, Pearson, Harlow.
A detailed textbook, intended for those studying the topic in depth.

Goldratt, E.Y. and Cox, J. (1984) *The Goal*, North River Press, Croton-on-Hudson, NY.
Don't read this if you like good novels, but do read it if you want an enjoyable way of understanding some of the complexities of scheduling. It particularly applies to the drum, buffer, rope concept described in this chapter.

Jacobs, F.R., Berry, W.L., Whybark, D.C. and Vollmann, T.W. (2010) *Manufacturing Planning & Control Systems for Supply Chain Management*, 6th edn, McGraw Hill Higher Education, New York, NY.
The latest version of the 'bible' of manufacturing planning and control.

Kehoe, D.F. and Boughton N.J. (2001) New paradigms in planning and control across manufacturing supply chains – the utilisation of internet technologies, *International Journal of Operations & Production Management*, 21 (5/6), 582–93.
Academic, but interesting.

Pinedo, M.L. (2016) *Scheduling: Theory, Algorithms, and Systems*, 5th edn, Springer, New York, NY.
A very technical, but well-established text.

Notes on chapter

1. The information on which this example is based is taken from: Caswell, M. (2020) Air France to operate 50 per cent of schedules during November and December, *Business Traveller*, 28 September; Farman, J. (1999) 'Les Coulisses du Vol', Air France, talk presented by Richard E. Stone, NorthWest Airlines at the IMA Industrial Problems Seminar, 1998.
2. Coldrick, A., Ling, D. and Turner, C. (2003) Evolution of sales and operations planning: from production planning to integrated decision making, StrataBridge Working Paper.
3. The information on which this example is based is taken from: Barro, J. (2019) Here's why airplane boarding got so ridiculous, *New York Magazine Intelligencer*, 9 May, https://nymag.com/intelligencer/2019/05/heres-why-airplane-boarding-got-so-ridiculous.html (accessed September 2021); The Economist (2011) Please be seated: a faster way of boarding planes could save time and money, *Economist* print edition, 3 September.
4. The information on which this example is based is taken from: Economist (2020) Triage under trial: the tough ethical decisions doctors face with covid-19, *Economist* print edition, 2 April; Jones, C. (2020) What a career in intensive care nursing has taught me about triage, *Financial Times*, 6 February.
5. The information on which this example is based is taken from: Heathrow website, https://www.heathrow.com (accessed September 2021). For a technical explanation of the aircraft landing algorithm, see Cecen, R.K., Cetek, C. and Kaya, O. (2020) Aircraft sequencing and scheduling in TMAs under wind direction uncertainties, *The Aeronautical Journal*, 124 (1282), 1896–912; and Beasley, J.E., Sonander, J. and Havelock, P. (2001) Scheduling aircraft landings at London Heathrow using a population heuristic, *Journal of the Operational Research Society*, 52, 483–93.
6. The information on which this example is based is taken from: Calder, S. (2017) Ryanair cancellations: the truth behind why 2,000 flights are due to be scrapped, *Independent*, 19 September.
7. Goldratt, E.Y. and Cox, J. (1984) *The Goal*, North River Press.
8. Based on an original model described in Hofstede, G. (1981) Management control of public and not-for-profit activities, *Accounting, Organizations and Society*, 6 (3), 193–211.
9. Bolino, M.C., Kelemen, T.K. and Matthews, S.H. (2020) Rethinking work schedules? Consider these 4 questions, *Harvard Business Review*, 6 July.

11 Capacity management

KEY QUESTIONS

11.1 What is capacity management?

11.2 How is demand measured?

11.3 How is capacity measured?

11.4 How is the demand side managed?

11.5 How is the supply side managed?

11.6 How can operations understand the consequences of their capacity management decisions?

INTRODUCTION

Capacity management is the activity of understanding the nature of an operation's demand and supply and coping with mismatches between them. It involves an attempt to forecast demand and measure the ability to supply products and services, followed by the selection of appropriate demand-side and supply-side responses based on performance objectives and long-term outlook. In doing this, operations managers must be able to understand and reconcile two competing requirements. On the one hand is the importance of maintaining customer satisfaction by delivering services and products to customers reasonably quickly. On the other is the need for operations (and their extended supply networks) to maintain efficiency by minimising the costs of excess capacity. This is why capacity management is so important – it has an impact on both revenue and costs, and therefore profitability (or the general effectiveness of service delivery in not-for-profit operations). In this chapter, we look at these competing tensions at an **aggregated** level. At this level, managers essentially bundle together different products and services to gain a general view of demand and capacity. For example, a hotel might think of demand and capacity in terms of 'room nights per month'; this ignores the number of guests in each room and their individual requirements, but it is a good first approximation. Figure 11.1 shows where this chapter fits in the structure of the text. At the end of the chapter is a supplement on queuing for those wishing to go into more detail on this important sub-topic of capacity management.

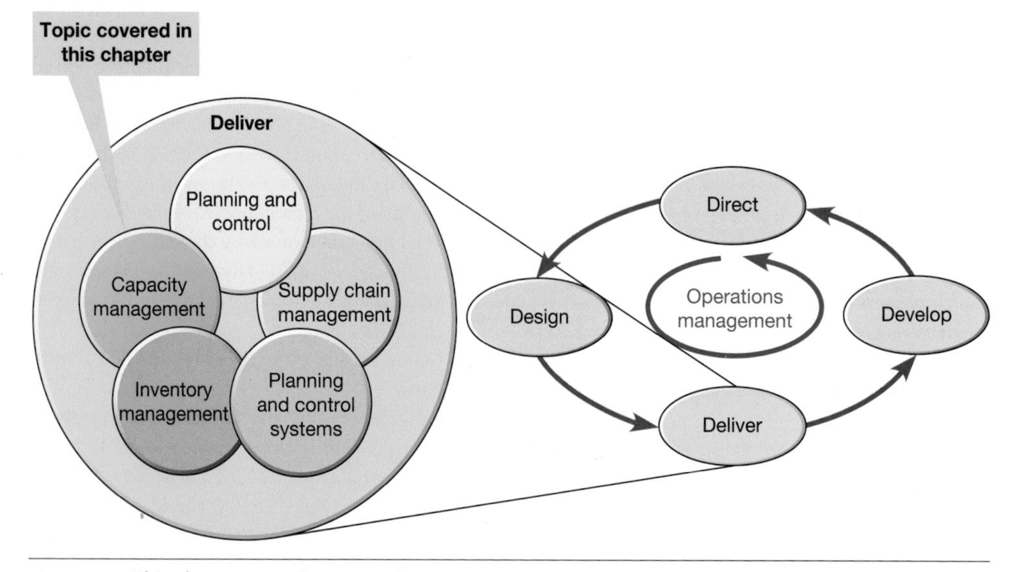

Figure 11.1 This chapter examines capacity management

11.1 What is capacity management?

Capacity management is concerned with understanding the nature of demand and supply (capacity) and attempting to reduce mismatches between them in a way that reconciles the competing demands of customer satisfaction and resource efficiency. These decisions are also made within the constraints of the operation, the ability of its suppliers to supply, the availability of staff and so on. As such, each level of capacity decision is made with the constraints of a higher level. In the other direction, short-term decisions provide important feedback for planning over longer-term time horizons. This interaction effect between different time horizons is illustrated in Figure 11.2.

In Chapter 5, we examined long-term capacity decisions as they relate to the structure and scope of operations. In Chapter 10, we examined the more short-term capacity decisions around allocation, sequence and resourcing of tasks. In this chapter, we focus more on the medium-term aspects of capacity management, where decisions are being made largely within the constraints of the physical capacity set by the operation's long-term capacity strategy. Medium-term capacity management usually involves assessing demand forecasts with a time horizon of between 2 and 18 months, during which time planned output can be varied, for example, by changing the number of hours that resources are used. In practice, however, few forecasts are accurate, and most operations also need to respond to changes in demand that occur over an even shorter timescale – termed short-term capacity management. For example, hotels and restaurants have unexpected and apparently random changes in demand from night to night, but also know from experience that certain days are, on average, busier than others.

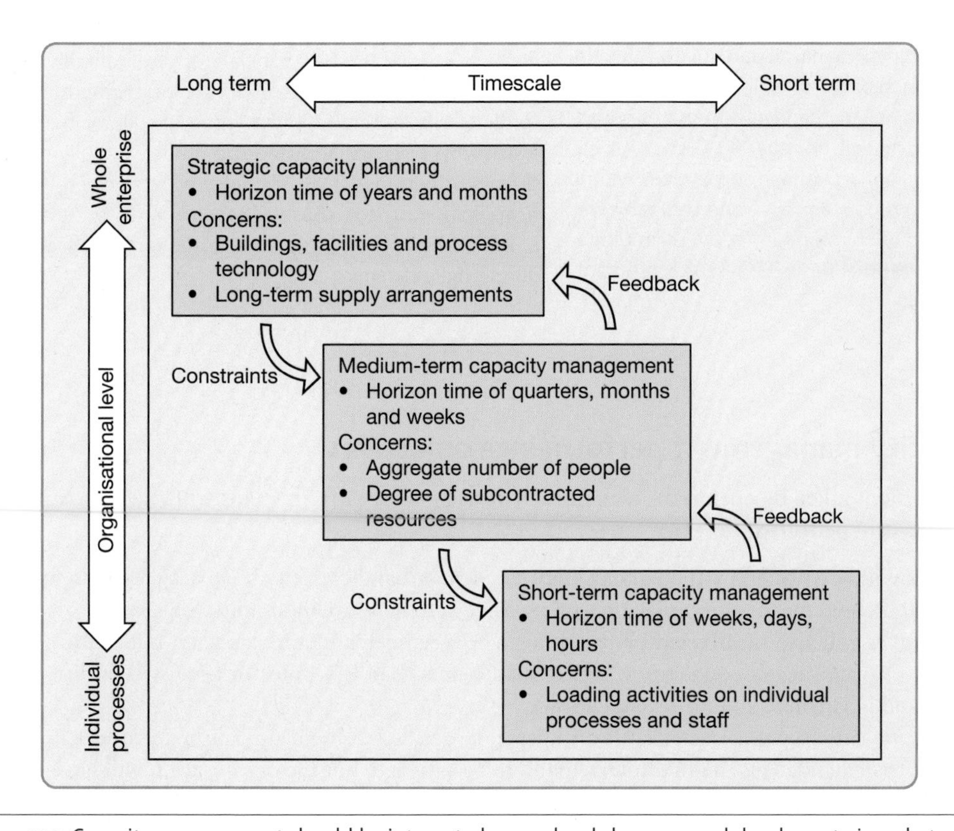

Figure 11.2 Capacity management should be integrated across levels because each level constrains what can be done in the level below and can provide feedback for the level above

OPERATIONS IN PRACTICE

3M's COVID-19 surge capacity[1]

During the COVID-19 pandemic, demand for protective gear soared, and none more so than N95 respirator masks that were used to keep front-line healthcare workers safe. These masks (also known as FFP2 masks in the European Union) used a fine mesh of synthetic polymer fibres to filter at least 95 per cent of airborne particulates, thereby limiting the transmission of the infection.

In the early months of the pandemic, mask manufacturers such as Honeywell, Kimberley-Clark, Ambu, Medicom, Teleflex, Shanghai Dasheng and Moldex-Metric all looked to ramp up production. But one company was arguably a step ahead of its competitors. 3M is a multinational operating in industrial business, safety equipment, healthcare, electronics and energy, and consumer goods. The firm had used failures in the past to prepare itself for major surges in demand for its N95 masks. During the SARS outbreak in 2002–3, 3M had realised it did not have sufficient operational flexibility to respond to sudden jumps in demand. Therefore, the company decided to invest in **surge capacity** – assembly lines that would remain dormant until required – throughout its network of factories across the globe. This would enable rapid ramping up of capacity in the face of future crises and their related demand spikes. Later emergencies, such as the H1N1 flu pandemic in 2009, the Ebola outbreak in West Africa between 2013 and 2016, and one of the worst bush-fire seasons in Australian history in 2019, allowed 3M to refine its capacity-flexing approaches further.

As the sheer scale of the COVID-19 pandemic became evident, 3M shifted many of its employees to overtime working and then turned on its surge capacity. In just two months, the firm had doubled its global production of N95 masks to 100 million per month. To achieve this, the firm not only needed its spare capacity of physical infrastructure, it also relied on a large proportion of its employees to continue working through the crises, in spite of the new restrictions (e.g. social distancing) that were placed on them. Furthermore, 3M's suppliers were involved in the firm's planning for these kinds of 'X-factor' events, enabling the entire supply network to surge its capacity during the pandemic.

Capacity management performance objectives

The decisions taken by operations managers in devising their capacity plans will affect several different aspects of performance:

- *Costs* will be affected by the balance between demand and capacity. Capacity levels in excess of demand could mean underutilisation of capacity and therefore high unit-cost.
- *Revenues* will also be affected by the balance between demand and capacity, but in the opposite way. Capacity levels equal to or higher than demand at any point in time will ensure that all demand is satisfied, and no revenue lost.
- *Working capital* will be affected if an operation decides to build up finished product inventory prior to demand. This might allow demand to be satisfied, but the organisation will have to fund the inventory until it can be sold.
- *Quality* of services might be affected by a capacity plan that involves large fluctuations in capacity levels, by hiring temporary staff for example. The new staff and the disruption to the routine working of the operation could increase the probability of errors being made.
- *Speed* of response to customer demand could be enhanced either by the deliberate provision of surplus capacity to avoid queuing, or through the build-up of product inventories.

- *Dependability* of supply will also be affected by how close demand levels are to capacity. The closer demand gets to the operation's capacity ceiling, the less able it is to cope with any unexpected disruptions.
- *Flexibility*, especially volume flexibility, will be enhanced by surplus capacity. If demand and capacity are in balance, the operation will not be able to respond to any unexpected increase in demand.

A framework for capacity management

There are a series of activities involved in capacity management, which are shown in Figure 11.3. The most common first step on the demand side of the 'equation' is to measure (forecast) demand for services and products over different time periods. This involves selecting from a range of qualitative (panel, Delphi and scenario planning) and quantitative (time series and causal models) tools to support more accurate prediction of demand. The second step is typically on the supply side of the framework and involves measuring the capacity to deliver services and products. Here, the impacts of mix, time frame and output specification should be considered. The third step is to consider if and how to manage demand using demand management and yield management techniques. The fourth step is to manage the supply side by determining the appropriate level of average capacity and then deciding whether to either keep this constant (level capacity plan) or to adjust capacity in line with changing demand patterns (chase capacity plan). Finally, operations managers must understand the consequences of different capacity management decisions on both the demand side and supply side of the framework.

11.2 How is demand measured?

The first task of capacity management is to understand the patterns of demand for products and services over different time frames (hourly, daily, weekly, monthly, annually, etc). Importantly, knowing that demand may be rising or falling, while a useful start, is not enough in itself. Knowing the *rate of change* is often vital for business planning. For example, a firm of lawyers may have to decide the point at which, in its growing business, it will have to take on another partner. Hiring a new partner could take months, so it needs to be able to forecast when it expects to reach that point and then when it needs to start its recruitment drive.

Operations principle
Understanding patterns of demand for products and services is critical for successful capacity management.

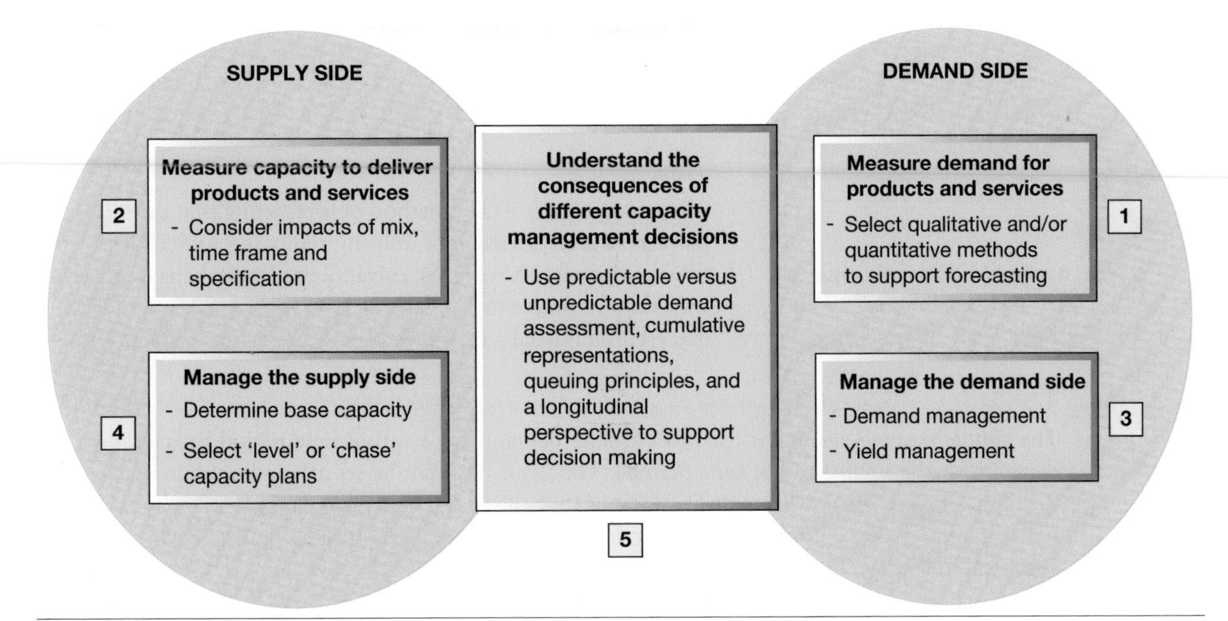

Figure 11.3 Capacity management framework

Qualitative approaches to forecasting

Managers sometimes use qualitative methods based on judgement and past experience to forecast demand. Three of the most popular methods are the panel approach, the Delphi method and scenario planning.

Panel approach

Just as panels of sports pundits gather to speculate about likely outcomes, so too do politicians, business leaders, stock market analysts, banks and airlines. The panel acts like a focus group allowing everyone to talk openly. Although there is the great advantage of several brains being better than one, it can be difficult to reach a consensus, or sometimes the views of the loudest or highest status may emerge (the bandwagon effect). Although more reliable than one person's views, the panel approach still has the weakness that everybody, even the experts, can get it wrong.

Delphi method

Perhaps the best-known approach to generating forecasts using experts is the Delphi method. This is a more formal method, which attempts to reduce the influences from procedures of face-to-face meetings. It employs a survey of experts, where replies are analysed and anonymous summaries are sent back to all experts. The experts are then asked to reconsider their original forecasts in the light of the replies and arguments put forward by the other experts. This process is repeated several times to conclude either with a consensus or at least a narrower range of decisions. One refinement of this approach is to allocate weights to the individuals and their suggestions, based on, for example, their experience, their past success in forecasting, and other people's views of their abilities. The obvious problems associated with this method include constructing an appropriate questionnaire and selecting an appropriate panel of experts.

Scenario planning

One method for dealing with situations of even greater uncertainty is scenario planning. This is usually applied to long-range forecasting, again using a panel. The panel members are usually asked to devise a range of future scenarios. Each scenario can then be discussed, with inherent risks considered. Unlike the Delphi method, scenario planning is not necessarily concerned with arriving at a consensus, but looking at a range of options and putting plans in place to try to avoid the ones that are least desired and taking action to follow the most desired.

Quantitative approaches to forecasting

Managers sometimes prefer to use quantitative methods to forecast demand. Two key approaches are time series analysis and causal modelling techniques. Time series examines the pattern of past behaviours to forecast future behaviour. Causal modelling is an approach that describes and evaluates the cause–effect relationships between key variables.

Operations principle

Time series methods of forecasting use past patterns of demand to make predictions.

Time series analysis

Times series analysis is a method of forecasting that examines the pattern of time series data and, by removing underlying variations with assignable causes, extrapolates future behaviour. Here, we examine simple moving average, simple exponential smoothing, trend-adjusted exponential smoothing and seasonal models.

Simple moving-average forecasting

The *simple moving-average* is used to estimate demand for a future time period by averaging the demand for the *n* most recent time periods. The value of *n* can be set at any level but is usually in the range of 3 to 7. So, for example, if *n* is set at four, the next period's demand is forecast by taking the moving average of the previous four periods' actual demand. Thus, if the forecast demand for week *t* is F_t and the actual demand for week *t* is A_t, then:

$$F_t = \frac{A_{t-1} + A_{t-2} + A_{t-3} + A_{t-4}}{4}$$

Simple exponential smoothing

The main disadvantage of moving averages is that they do not use data from beyond n periods in forecasting. The *exponential smoothing approach* forecasts demand in the next period by taking into account the actual demand in the current time period and the forecast that was previously made. It does so according to the following formula:

$$F_t = \alpha(A_{t-1}) + (1 - \alpha)F_{t-1}$$

where

F_t = new forecast
A_{t-1} = previous period's *actual demand*
F_{t-1} = previous period's forecast demand
α = smoothing constant.

The smoothing constant α is, in effect, the weight that is given to the last (and therefore assumed to be most important) piece of information available to the forecaster. However, the other expression in the formula includes the forecast for the current period, which included the previous period's actual demand, and so on. In this way, all previous data has an (albeit diminishing) effect on the next forecast.

Worked example

Forecasting demand at Eurospeed using an exponential smoothing method

Table 11.1 shows the data for Eurospeed's parcels forecasts using an exponential smoothing method, where $\alpha = 0.2$. For example, the forecast for week 35 is:

$$F_{35} = (0.2 \times 67.0) + (0.8 \times 68.3) = 68.04$$

Table 11.1 Exponentially smoothed forecast calculated with smoothing constant $\alpha = 0.2$

Week (t)	Actual demand (thousands) (A)	Forecast ($F_t = \alpha A_{t-1} + (1 - \alpha) F_{t-1}$) ($\alpha = 0.2$)
20	63.3	60.00
21	62.5	60.66
22	67.8	60.03
23	66.0	61.58
24	67.2	62.83
25	69.9	63.70
26	65.6	64.94
27	71.1	65.07
28	68.8	66.28
29	68.4	66.78
30	70.3	67.12
31	72.5	67.75
32	66.7	68.70
33	68.3	68.30
34	67.0	68.30
35		**68.04**

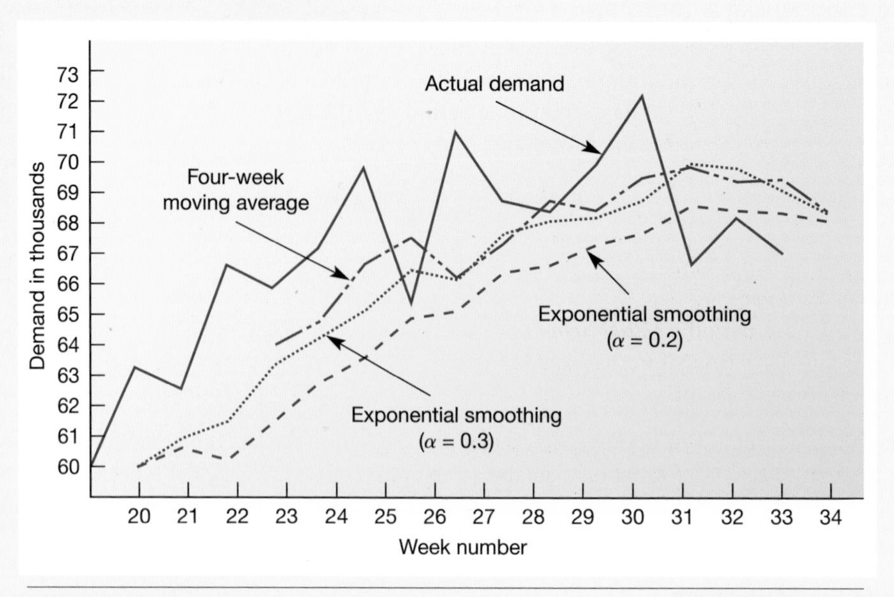

Figure 11.4 A comparison of a moving-average forecast and exponential smoothing with the smoothing constant $\alpha = 0.2$ and 0.3

The value of α governs the balance between the *responsiveness* of the forecasts to changes in demand, and the *stability* of the forecasts. The closer α is to 0, the more forecasts will be dampened by previous forecasts (not very sensitive but stable). Figure 11.4 shows the Eurospeed volume data plotted for a four-week moving average, exponential smoothing with $\alpha = 0.2$ and exponential smoothing with $\alpha = 0.3$.

Trend-adjusted exponential smoothing

The main disadvantage of simple exponential smoothing is that it assumes a stable underlying average. When there is a trend in the average, exponentially smoothed forecasts lag behind the changes in underlying demand. While higher smoothing constants (>0.5) help to reduce forecast errors, there may still be a lag if the average is systematically changing. Therefore, it is possible to include a trend within exponentially smoothed forecasts to improve accuracy. The new formula is:

$$\text{FIT}_t = F_t + T_t$$

where

FIT_t = forecast including trend
F_t = exponentially smoothed forecast
T_t = exponentially smoothed trend.

For a *trend-adjusted forecast*, we must smooth both the average (F_t) and the trend (T_t). The smoothing constant is shown with the α symbol for the average and the β symbol for the trend. To arrive at the forecast including trend (FIT_t), we must compute the two parts of the equation:

$$F_t = \alpha(A_{t-1}) + (1 - \alpha)(F_{t-1} + T_{t-1})$$
$$T_t = \beta(F_t - F_{t-1}) + (1 - \beta)T_{t-1}$$

where

F_t = exponentially smoothed forecast for period t
T_t = exponentially smoothed trend for period t
A_t = actual demand for period t
α = smoothing constant for the average
β = smoothing constant for the trend.

Seasonality in forecasting

Most organisations experience seasonal patterns in their demand. Sometimes the causes of seasonality are climatic (holidays), sometimes festive (gift purchases), sometimes financial (tax processing), or social, or political. For most of us, we typically think of seasonality in annual terms. However, in forecasting the term is used to describe any regularly repeating changes in demand (quarterly, monthly, weekly, daily or hourly). For example, utility companies may experience larger annual seasonality, but will also face seasonal patterns over the week and across the day. A popular technique for incorporating seasonality in forecasting is the *multiplicative seasonal model*, which involves the following five steps (for simplicity, here we assume that there is no other trend in the data, apart from seasonality):

Operations principle

In demand forecasting, 'seasonality' refers to any repeating pattern of demand – annual, quarterly, monthly, weekly, daily, or even hourly.

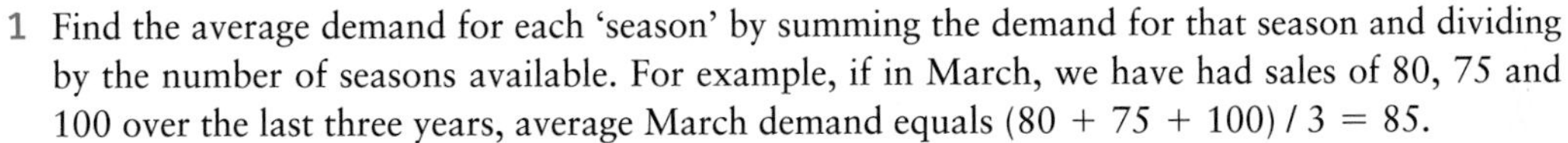

1. Find the average demand for each 'season' by summing the demand for that season and dividing by the number of seasons available. For example, if in March, we have had sales of 80, 75 and 100 over the last three years, average March demand equals (80 + 75 + 100) / 3 = 85.
2. Calculate average demand over all 'seasons' by dividing total average demand by the number of seasons. For example, if total average annual demand is 1320 and there are 12 seasons (months), average demand equals 1320 / 12 = 110.
3. Compute seasonal index by dividing average season demand (step 1) over average demand (step 2). For example, March seasonal index equals 85 / 110 = 0.77.
4. Estimate the next time period's (in this case, annual) total demand using one or more of the qualitative or quantitative methods described in this section.
5. Divide this estimate by the number of seasons (in this case, 12 months) and multiply by the seasonal index to provide a seasonal forecast.

Worked example

Seasonal forecasting for Matsuyama Consulting

Matsuyama Consulting, based in Japan, expects to have an annual demand for 7,500 hours of supply chain strategy consulting in 2023. Using the multiplicative seasonal model, we can forecast demand for June, July and August of that year (Table 11.2):

$$\text{June 2023 forecast} = (7{,}500 / 12) \times 1.10 = 687.50$$

$$\text{July 2023 forecast} = (7{,}500 / 12) \times 1.32 = 825.00$$

$$\text{August 2023 forecast} = (7{,}500 / 12) \times 0.79 = 493.75$$

Table 11.2 Seasonal forecast for Matsuyama Consulting

Month	2020	2021	2022	Ave demand 2020–22	Ave monthly demand	Seasonal index
Jan	450	475	475	466.67	570.14	0.82
Feb	500	500	550	516.67	570.14	0.91
Mar	625	600	575	600.00	570.14	1.05
Apr	600	600	650	616.67	570.14	1.08
May	550	600	600	583.33	570.14	1.02
Jun	600	625	650	625.00	570.14	1.10
Jul	700	750	800	750.00	570.14	1.32
Aug	450	400	500	450.00	570.14	0.79
Sep	500	450	450	466.67	570.14	0.82
Oct	550	500	525	525.00	570.14	0.92
Nov	650	600	650	633.33	570.14	1.11
Dec	600	600	625	608.33	570.14	1.07
Total average annual demand				6841.67		

Causal models

Causal models often employ complex techniques to understand the strength of relationships between the network of variables and the impact they have on each other. 'Simple regression' models try to determine the 'best fit' expression between two variables. For example, suppose an ice-cream company is trying to forecast its future sales. After examining previous demand, it can see that the main influence on demand at the factory is the average temperature of the previous week. To understand this relationship, the company plots demand against the previous week's temperatures. This is shown in Figure 11.5. Using this graph, the company can make a reasonable prediction of demand, once the average temperature is known, provided that the other conditions prevailing in the market are reasonably stable.

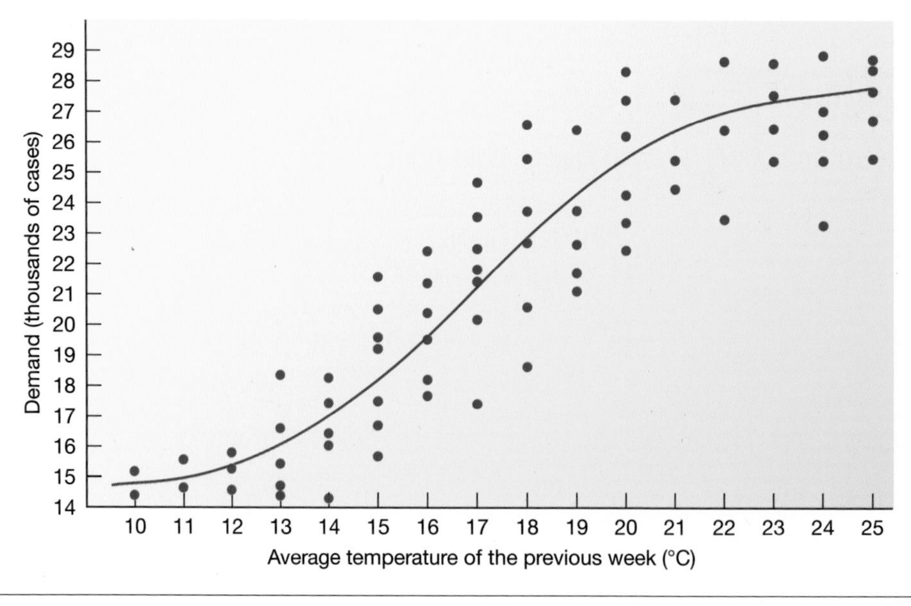

Figure 11.5 Regression line showing the relationship between the previous weeks' average temperature and demand

If they are not, then these other factors that have an influence on demand will need to be included in a 'multiple regression' model. These more complex networks comprise many variables and relationships, each with their own set of assumptions and limitations. While developing such models and assessing the importance of each of the factors is beyond the scope of this text, many techniques are available to help managers undertake this more complex modelling and also feed back data into the model to further refine and develop it, in particular structural equation modelling.

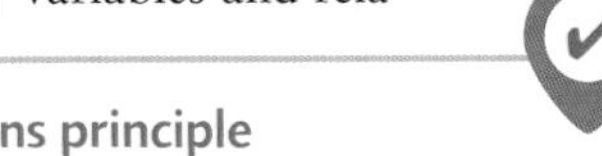

Operations principle

Causal models make predictions by examining the impact that one or more variables have on demand.

OPERATIONS IN PRACTICE — How artificial intelligence helps with demand forecasting[2]

Of the many potential applications of artificial intelligence (AI), arguably one of the most useful is its use in demand forecasting. In particular, advances in machine learning improved the availability and reliability of forecasts as well as the speed with which they can be produced. Like traditional methods of forecasting, AI-based systems begin with a set of assumptions and rules, which, when applied to available data, produce a forecast. The 'deep learning' of AI methods can learn and adapt these rules using huge amounts of historic data. However, machine learning models suffer from one major limitation – they work far more effectively when input data is similar to the data upon which they were 'trained'. In the face of sudden unexpected rises or falls in demand, predictive algorithms struggle to reflect the reality on the ground, because the system has not been 'trained' for any particularly large changes. In other words, they become unreliable when the past is incapable of giving helpful insights into the future.

For example, the COVID-19 pandemic had impacts on almost all facets of business life and one of the most dramatic areas was the effect on demand patterns. Around the world, firms manufacturing face masks, hand sanitisers, cleaning products, toilet paper, home and garden equipment, medical supplies, canned and frozen goods, games and home fitness apparatus saw dramatic increases in demand over short time periods. Likewise, the demand soared for services such as food delivery, logistics, online video calling and social media platforms. Other firms experienced massive drops in their demand – those connected with the travel, oil and gas, hospitality and leisure, and auto industries were particularly affected. The consequence on demand forecasting was significant. Many firms experienced an almost overnight shift from relatively stable and predictable demand variation – the easiest pattern for an operations manager to deal with – to much more volatile and less predictable demand variation – the most challenging pattern to deal with. This meant that much of the historical data that would normally be essential in developing high-quality demand forecasts were instead wildly inaccurate. For those firms that had invested in machine learning models to support demand forecasting, artificial intelligence (AI) algorithms were simply incapable of dealing with the radical shifts in 'normal' human behaviour during the crises.

This fundamental limitation meant that many firms required significant manual intervention to override automatically generated capacity plans and left some commentators questioning the suitability of machine learning models (and indeed other methods relying on historical data), especially in more dynamic business contexts. However, the counter argument was that the nature of AI is a 'learning technology'. With the right kind of human intervention, machine learning models could be 're-trained' to take account of more dynamic data sets. Some experts suggested that one approach would be for AI training to incorporate much longer time frames and, in doing so, include other 'black swan' events over the past century. In doing so, they argued, machine learning models could be improved significantly.

Making forecasts useful for operations managers

There are three key ways to assess the usefulness of a demand forecast from an operations manager's perspective – its level of accuracy, its ability to indicate relative uncertainty, and its expression in terms that are useful for capacity management.

It is as accurate as possible

The process of capacity management is hugely aided if forecasts are as accurate as possible because, whereas demand can change instantaneously, there is usually a lag between deciding to change capacity and the change taking effect. It is possible to assess the relative accuracy of demand forecasts by calculating *forecast error* and three methods are popular for this:

$$\textbf{mean absolute deviation, MAD} = \frac{\sum |E_t|}{n}$$

$$\textbf{mean squared error, MSE} = \frac{\sum E_t^2}{n}$$

$$\textbf{mean absolute per cent error, MAPE} = \frac{\sum [(E_t / A_t)\ 100]}{n}$$

where

n = number of forecast periods.

Worked example

Calculating forecast error at Sinh Restaurant

Table 11.3 shows the forecast and actual customers served each week in Sinh (Lion), a small restaurant located in Kolkata, India.

Table 11.3 Calculating forecast errors

(t)	Actual sales (A_t)	Forecast (F_t)	Absolute error (E_t)	Error² (E_t^2)	Absolute per cent error
1	2500	2250	250	62500	10.00
2	2600	2200	400	160000	15.38
3	2580	2900	320	102400	12.40
4	2700	3000	300	90000	11.11
5	2250	3100	850	722500	37.78
6	2600	2450	150	22500	5.77

We can calculate the MAD, MSE and MAPE as follows:

$$\text{MAD} = \frac{\sum |E_t|}{n} = \frac{2{,}270}{6} = 378.33$$

$$\text{MSE} = \frac{\sum E_t^2}{n} = \frac{1{,}159{,}900}{6} = 193{,}316.67$$

$$\text{MAPE} = \frac{\sum [(E_t / A_t)\ 100]}{n} = \frac{92.45}{6} = 15.41$$

It gives an indication of relative uncertainty

Perhaps most importantly, good forecasts give an indication of relative uncertainty. Decisions to operate extra hours and recruit extra staff are usually based on forecast levels of demand, which could in practice differ considerably from actual demand, leading to unnecessary costs or unsatisfactory customer service. For example, a forecast of demand levels in a supermarket may show initially slow business that builds up to a lunch time rush. After this, demand slows, only to build up again for the early evening rush, and it finally falls again at the end of trading. The supermarket manager can use this forecast to adjust checkout capacity throughout the day, for example. However, while this may be an accurate average demand forecast, no single day will conform exactly to predicted patterns. Of equal importance is an estimate of how much actual demand could differ from the average. This can be found by examining demand statistics to build up a distribution of demand at each point in the day. The importance of this is that the manager now has an understanding of when it will be important to have reserve staff, perhaps filling shelves, but on call to staff the checkouts should demand warrant it.

It is expressed in terms that are useful for capacity management

If forecasts are expressed only in money terms and give no indication of the demands that will be placed on an operation's capacity, they will need to be translated into realistic expectations of demand, expressed in the same units as the capacity (for example, staff, machines, space, etc.). Nor should forecasts be expressed in money terms, such as sales, when those sales are themselves a consequence of capacity planning. For example, some retail operations use sales forecasts to allocate staff hours throughout the day. Yet sales will also be a function of staff allocation. Better to use forecasts of 'traffic', the number of customers who potentially could want serving if there is sufficient staff to serve them.

11.3 How is capacity measured?

The second task of capacity management is to understand the nature of capacity (i.e. the ability to supply). The capacity of an operation is the *maximum level of value-added activity over a period of time* that the process can achieve under normal operating conditions. Critically, this definition reflects the scale of capacity but more importantly, its *processing capabilities*. Suppose a pharmaceutical manufacturer invests in a new 1,000-litre capacity reactor or a property company purchases a 500-vehicle capacity car park. This information gives you a good sense of the *scale* of capacity but it is far from a useful measure of capacity for an operations manager. Instead, the pharmaceutical company will be concerned with the level of output (i.e. the processing capability) that can be achieved from the 1000-litre reactor vessel. If a batch of standard products can be produced every hour, the planned processing capacity could be as high as 24,000 litres per day. If the reaction takes four hours, and two hours are used for cleaning between batches, the vessel may only produce 4,000 litres per day. Similarly, the car park may be fully occupied by office workers during the working day, 'processing' only 500 cars per day. Alternatively, it may be used for shoppers staying on average only one hour, and theatre-goers occupying spaces for three hours, in the evening. The processing capability would then be up to 5,000 cars per day.

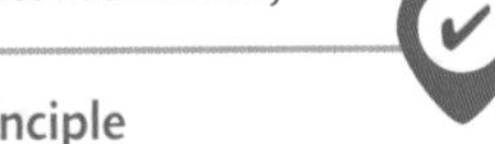

Operations principle

Capacity is the maximum level of value-added activity over a period of time that the process or operation can achieve under normal operating conditions.

OPERATIONS IN PRACTICE

Next-generation signal technology expands railway capacity[3]

In many parts of the world, railway networks are at breaking point, struggling to deal with the demands placed on them daily. At peak times of the day, some commuter lines are near gridlock, and many commentators put the blame on capacity shortages. But recall our earlier definition – the capacity of an operation is the *maximum level of value-added activity over a period of time* that can be achieved under normal operating conditions. Therefore, for railway networks, as with airports, seaports, roads and other transport infrastructure, it's not just about what resources you have (the scale of capacity), it's about what you can do with it (the processing capabilities of capacity). Herein lies the problem – increasing the scale of capacity is often difficult. It can be expensive, very slow and often politically sensitive. Therefore, while they wait for longer-term increments in capacity, railways (as with other operations) often have to make the best of what they have.

One way this can be achieved is through the digitisation of railway signalling systems. Traditional signalling was

developed nearly two centuries ago using an approach called 'block working' whereby only one train can be in a 'block' at any one time. However, the fixed position of signals means that it is impossible to increase capacity without risking safety. In addition, many railways around the world now combine a wide array of different systems to manage trains – in one area there may be advanced software automatically routing train pathways, in another signal workers continue to operate manual wire-systems to control semaphore arms on the side of the track. This makes for very complex, inefficient and often dysfunctional operations networks!

Then things started to change. Replacing the 'working block' method of managing train movements and its associated physical infrastructure of signals and signal boxes, came the European Train Control System (ETCS). This system created a digital 'moving block' by transmitting the actual location of trains via sensors placed close together along the track, in much the same way as aircraft transponders report the position of a plane to air traffic control. Large rail operating centres then used advanced software to calculate the optimum distance between trains, factoring in train speeds, braking distances and stations on the route. The effect was to create a safety buffer zone around each train, while minimising unnecessary distance between trains. When implemented, the ETCS was expected to have a positive effect on safety and general reliability through improved coordination, and also create increases in effective capacity. Conventional railway networks have a mix of high-speed passenger trains and low-speed freight trains, which creates significant problems for the fixed position 'working block' system. The 'moving blocks' used by the ETCS could increase effective *processing* capacity by around 40 per cent.

The system was rolled out on a number of dedicated high-speed lines, such as the Wuhan–Guangzhou route in China and the TGV route in France. However, the bigger tests were expected to come as moving-block technology was implemented across more complex networks with many more legacy systems to replace. These included projects to install ETCS on trans-European 'corridor' routes connecting different EU countries, specific routes or regions in Australia, Hungry, Italy and New Zealand, and across the entire railway networks of Belgium, Denmark, Germany, Israel and the United Kingdom. But even these projects looked small when compared with intended roll-out of ETCS across India's enormous railway network, which moves over 20 million passengers and 3 million tonnes of freight daily. The initial plan was to roll out the technology across the entire 65,000km network with a ₹780 billion ($12 billion) contract over six years. However, cost concerns led to scaling back, with the Indian Railway Board instead deciding instead to undertake a full-scale pilot on a 780 km section of the Delhi to Kolkata route before finalising its future roll-out strategy.

Measuring capacity may sound simple but can in fact be relatively hard to define unambiguously unless the operation is standardised and repetitive. For example, a ride at a theme park might be designed to process batches of 60 people every three minutes – a capacity to convey 1,200 people per hour. In this case, an *output capacity measure* is the most appropriate measure because the output from the operation does not vary in its nature. For many operations, however, the definition of capacity is not so obvious. When a much wider range of outputs places varying demands on the process, for instance, output measures of capacity are less useful. Here *input capacity measures* are frequently used to define capacity. Almost every type of operation could use a mixture of both input and output measures, but in practice, most choose to use one or the other (see Table 11.4).

Operations principle
Any measure of capacity should reflect the ability of an operation or process to supply demand.

The effect of activity mix on capacity measurement

An operation's ability to supply is partly dependent on *what* it is being required to do. For example, a hospital may have a problem in measuring its capacity because the nature of the service varies significantly. Output depends on the mix of activities in which the hospital is engaged and, because

Table 11.4 Input and output capacity measures for different operations

Operation	Input measure of capacity	Output measure of capacity
Hospital	**Beds available**	Number of patients treated per week
Air-conditioner plant	Machine hours available	**Number of units per week**
Theatre	**Number of seats**	Number of customers entertained per week
University	**Number of students**	Students graduated per year
Electricity company	Generator size	**Megawatts of electricity generated**
Retail store	**Sales floor area**	Number of items sold per day
Airline	**Number of seats available**	Number of passengers per week
Brewery	Volume of fermentation tanks	**Litres per week**

(*Note*: The most commonly used measure is shown in bold.)

most hospitals perform many different types of activities, output is difficult (though not impossible!) to predict. Some of the problems caused by variation mix can be partially overcome by using aggregated capacity measures (remember that 'aggregated' means that different products and services are bundled together in order to get a broad view of demand and capacity).

Capacity is a function of service/product mix, duration and product service specification.

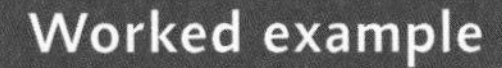

The impact of activity mix on drone production capacity

A small engineering firm based in Stockholm, Sweden, manufactures a range of three commercial drones – *Vortex*, *Elysia* and *Moln* (cloud). The *Vortex* model can be assembled in 2.5 hours, the *Elysia* in 1.5 hours and the *Moln* in 0.75 hours. The firm has 600 staff hours available for assembly per week. Assuming that the demand mix for the *Vortex*, *Elysia* and *Moln* drones are in a ratio of 2:2:4, the time needed to assemble $2 + 2 + 4 = 8$ units is:

$$(2 \times 2.5) + (2 \times 1.5) + (4 \times 0.75) = 11 \text{ hours}$$

The number of units assembled per week is therefore:

$$\frac{600}{11} \times 8 = 436.4 \text{ units}$$

If the activity mix changed to a ratio of 3:2:3, the time needed to assemble $3+2+3 = 8$ units is:

$$(3 \times 2.5) + (2 \times 1.5) + (3 \times 0.75) = 12.75 \text{ hours}$$

Now the number of units assembled per week (i.e. the operation's new capacity) is:

$$\frac{600}{12.75} \times 8 = 376.5 \text{ units}$$

The effect of time frame on capacity measurement

The level of activity and output that may be achievable over short periods of time is not the same as the capacity that is sustainable on a regular basis. For example, a tax return processing office, during its peak periods at the end (or beginning) of the financial year, may be capable of processing 120,000

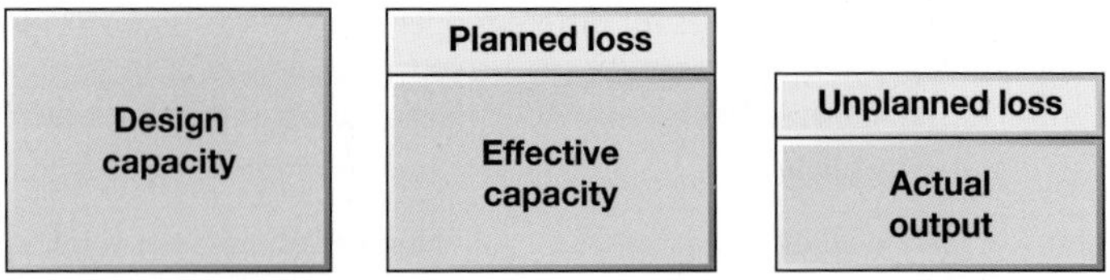

Design capacity

Planned loss

Effective capacity

Unplanned loss

Actual output

Figure 11.6 Design capacity, effective capacity and actual output

applications a week. It does this by extending the working hours of its staff, discouraging its staff from taking vacations during this period, avoiding any potential disruption to its IT systems, and maybe just by working hard and intensively. Nevertheless, staff do need vacations, they cannot work long hours continually, and eventually the information system will have to be upgraded. As such, when measuring capacity, operations managers should consider three different measures of capacity as shown in Figure 11.6. Design capacity is the theoretical capacity of an operation that its technical designers had in mind when they commissioned it. Effective capacity is the capacity of an operation after planned losses are accounted for. Finally, *actual output* is the capacity of an operation after both planned and unplanned losses are accounted for. For example, quality problems, machine breakdowns, absenteeism and other avoidable problems all take their toll. This offers two measures of capacity performance:

Operations principle

Useable capacity is rarely equal to theoretical or 'design' capacity.

$$\text{Utilisation} = \frac{\text{Actual output}}{\text{Design capacity}}$$

$$\text{Efficiency} = \frac{\text{Actual output}}{\text{Effective capacity}}$$

Capacity 'leakage'

This reduction in capacity, caused by both predictable and unpredictable losses, is sometimes called 'capacity leakage' and one popular method of assessing this leakage is the overall equipment effectiveness (OEE) measure that is calculated as follows:

$$\text{OEE} = a \times p \times q$$

where a is the availability of a process, p is the performance or speed of a process and q is the quality of product or services that the process creates. OEE works on the assumption that some capacity leakage occurs, causing reduced availability. For example, availability can be lost through time losses such as set-up and changeover losses (when equipment, or people in a service context, are being prepared for the next activity), and breakdown failures (when the machine is being repaired, or in a service context, where employees are being trained, or absent). Some capacity is lost through speed losses such as when equipment is idling (for example, when it is temporarily waiting for work from another process) and when equipment is being run below its optimum work rate. In a service context, the same principle can be seen when individuals are not working at an optimum rate, for example, call centre employees in the quiet period after the winter holiday season. Finally, not everything processed by an operation will be error free, so some capacity is lost as a result of inspection activities, rework or complaint handling.

For processes to operate effectively, they need to achieve high levels of performance against all three dimensions – availability, performance (speed) and quality. Viewed in isolation, these individual metrics are important indicators of performance, but they do not give a complete picture of the process's *overall* effectiveness. In contrast, combining all three dimensions, as in OEE, gives a more accurate reflection of the valuable operating time as a percentage of the capacity something was designed to have. Figure 11.7 gives an illustration of OEE applied to assess the capacity of a client support service team in a small software company.

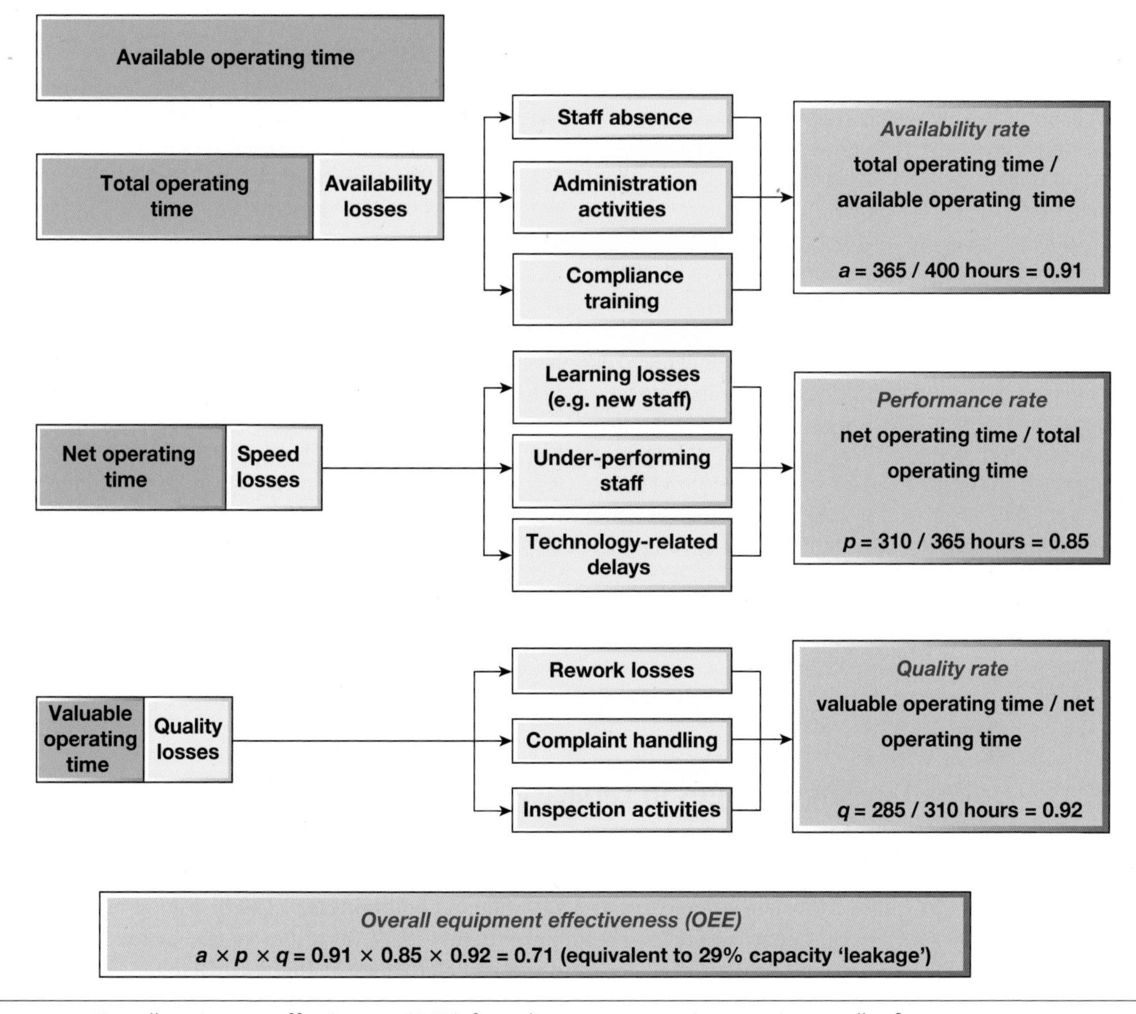

Figure 11.7 Overall equipment effectiveness (OEE) for a client support service team in a small software company

OPERATIONS IN PRACTICE

Mass transport systems have limited options in coping with demand fluctuation[4]

Anyone who commutes regularly in a large urban area knows the frustration of congestion at peak travelling times. Frustrating for passengers, certainly, but also frustrating for the operations managers who have to cope with, sometimes widely variable, demand. The mass transport systems on which most large cities depend are faced with one of the most difficult capacity management problems. Their transportation facilities are prone to congestion because of three characteristics of travel – demand varies very significantly over time, supply is relatively fixed over long periods of time, and their service is not storable. The London Underground is a good illustration of this. It is the oldest and busiest mass rapid transport (MRT) system in the world. But from the perspective of Transport for London (TfL), who operate the system, an equally pressing issue is how to cope with the fluctuation in demand during each day. In most parts of the London Underground, the morning peak time lasts from 07:30 to 09:30 and the evening peak from around 16:40 to 18:30. The network is largely closed overnight during the week, with some exceptions for all-night services on Friday and Saturday nights and special occasions.

However, there is some evidence that demand is 'spreading out' as passengers deliberately try to avoid travelling in peak periods. This is a pattern reflected in other mass transport systems. Singapore's Land Transport Authority (LTA) launched an incentive called Travel Smart Journeys (TSJ), aimed at distributing peak-hour demand more evenly by rewarding commuters along congested areas to consider alternative transport modes or routes. Some commentators see such shifting patterns, particularly the increase in off-peak travel, as reflecting the ways in which people are changing the way they live and work. In London, the shift to more people working from home and the number of self-employed people increased much faster than the growth in overall employment in the city. To encourage this trend some MRT systems adopt differential pricing (off-peak lower fares, etc.). The other main difficulty for most MRT systems is the more-or-less fixed capacity of their networks. More and better trains would help but are expensive. On London Underground there is an ongoing programme of commissioning new trains, some of which allow passengers to move through the train to spread out more efficiently. But there are few options to flex capacity. Essential maintenance is carried out during the nightly closures, and more substantial track maintenance is occasionally done during weekends when demand is lower. Similarly, cleaning work can be performed at night. London Underground has nearly 10,000 night workers. It also helps if passengers can be encouraged to board and exit trains as quickly as possible. The newer trains can help this. They are not much faster between stations, but they do draw passengers off the platform faster and this helps the system to stay fluid at peak times.

Understanding changes in capacity

While many operations are most concerned with dealing with changes in demand, some operations also have to cope with variation in *capacity* (if it is defined as 'the ability to supply'). For example, Figure 11.8 shows the demand and capacity variation of two businesses. The first is a domestic appliance repair service. Capacity is relatively stable with only small variations caused by the field service operatives preferring to take their vacations at particular times of the year. Demand, by contrast, fluctuates more significantly, with peak demand being approximately twice the level of the low point in demand. The second business is a food manufacturer producing frozen spinach. The demand for this product is relatively constant throughout the year but the capacity of the business varies significantly. During the growing and harvesting season capacity to supply is high, but it falls off almost to zero for part of the year. Yet although these examples are different, the essence of the capacity management activity is essentially similar – both are dealing with *gaps* between supply and demand.

Operations principle

Capacity management decisions should reflect both predictable and unpredictable variations in capacity and demand.

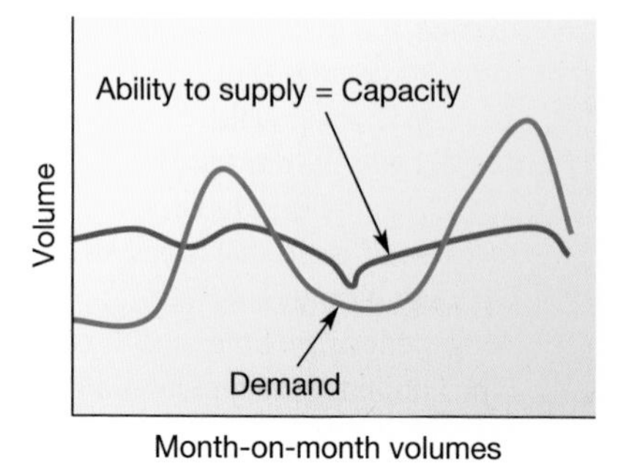

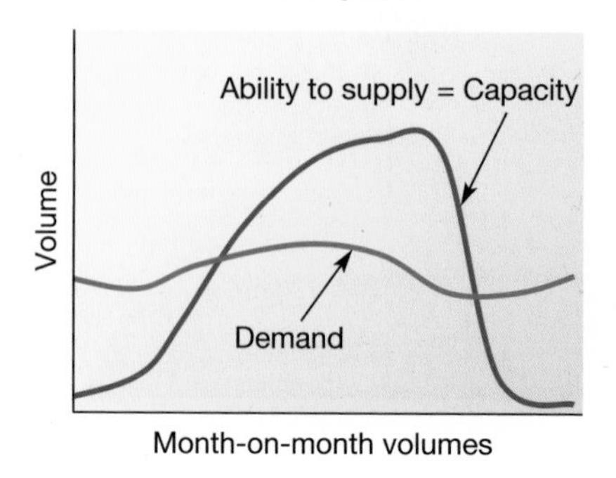

Figure 11.8 Volatility in demand versus volatility in capacity

11.4 How is the demand side managed?

Earlier in the chapter, we discussed the value of improved forecasting in helping operations managers know what demand for services and products to expect. Demand patterns clearly have a big influence on the way operations function and therefore many organisations will seek to influence them in some way. Referred to as demand management, this involves changing the pattern of demand to bring it closer to available capacity. Figure 11.9 illustrates how this achieved – either by stimulating off-peak demand or by constraining peak demand. There are a number of methods used to manage demand:

Operations principle

Demand management involves changing the pattern of demand by stimulating off-peak demand or constraining peak demand.

- *Price differentials* – adjusting price to reflect demand (see the 'Operations in practice' example on surge pricing). For example, skiing and camping holidays are cheapest at the beginning and end of the season and are particularly expensive during school vacations.
- *Scheduling promotion* – varying the degree of market stimulation through promotion and advertising in order to encourage demand during normally low periods. For example, turkey growers in the United Kingdom and the United States make vigorous attempts to promote their products at times other than Christmas and Thanksgiving.
- *Constraining customer access* – customers may only be allowed access to the operation's products or services at particular times. For example, reservation and appointment systems in various settings.
- *Service differentials* – allowing service levels to reflect demand (implicitly or explicitly) by letting service deteriorate in periods of high demand and increase in periods of low demand.
- *Creating alternative products or services* – developing services or products aimed at filling capacity in quiet periods. For example, most universities fill their accommodation and lecture theatres with conferences and company meetings during vacations. Ski resorts may provide organised mountain activity holidays in the summer, and garden tractor companies may make snow movers in the autumn and winter.

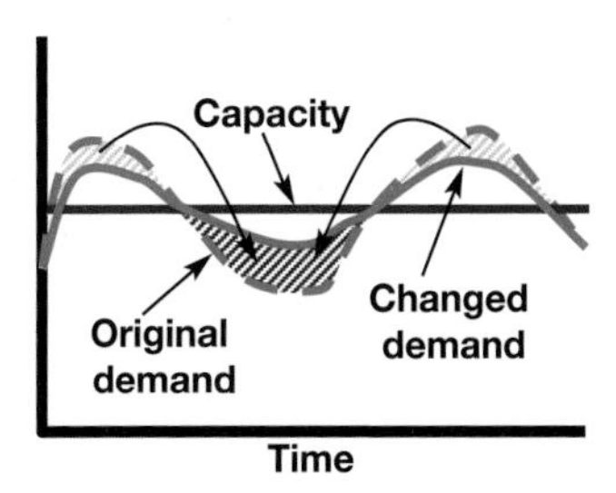

Figure 11.9 Demand management plan

OPERATIONS IN PRACTICE

Surge pricing helps manage demand for taxis and art galleries[5]

Surge (or dynamic) pricing is a demand management technique that relies on frequent adjustments in price to influence supply and (especially) demand. For example, some electricity suppliers charge different rates for energy depending on when it is consumed. Similarly, in countries with road charging, tolls are set at higher levels during peak times in an effort to keep traffic flowing. But perhaps the best-known example of surge pricing is the algorithm used by the cab-hailing app Uber. During times of excessive demand or inadequate supply, when the number of people wanting a ride exceeds the number of available drivers, Uber applies a multiplier to increase its normal fares based on the scarcity of available drivers. The company says that it does this to make sure those who need a ride can get one. Moreover, surge pricing helps ensure that pick-ups are available quickly and reliably. For its drivers, surge pricing means higher fares and a steady stream of ride requests.

Even leading arts venues such as the Royal Academy of Arts, the Barbican and the National Portrait Gallery in London introduced surge pricing, by charging higher prices at popular times. The justification used by arts venues is usually that the extra income helps them to subsidise admissions for less popular exhibitions. It also helps to smooth visitor numbers throughout the week. However, some critics dislike the intrusion of such commercial demand management methods into the art world. Even in the unambiguously commercial world of cab hailing, surge pricing can be deeply unpopular with customers. In the press and on social media, customers complain that Uber are taking advantage of them. But some marketing experts say it is, at least partly, a matter of perception, and as well as capping their multiplier, Uber should make the way they calculate it more transparent, limit how often prices are adjusted, communicate the benefits of the technique, and change its name (*certainty pricing* and *priority pricing* have been suggested).

Yield management

In operations that have relatively inflexible capacities, such as airlines and hotels, it is important to use the capacity of the operation for generating revenue to its full potential. One approach used by such operations is called yield management. This is really a collection of methods, some of which we have already discussed, which can be used to ensure that an operation maximises its potential to generate profit. Yield management is especially useful where capacity is relatively fixed; the market can be fairly clearly segmented; the service cannot be stored in any way; the service is sold in advance; and the marginal cost of making a sale is relatively low.

Airlines, for example, fit all these criteria. They adopt a collection of methods to try to maximise the yield (i.e. profit) from their capacity. Over-booking capacity may be used to compensate for passengers who do not show up for the flight. However, if more passengers show up than they expect, the airline will have a number of upset passengers. By studying past data on flight demand, airlines try to balance the risks of over-booking and under-booking. Operations may also use price discounting at quiet times, when demand is unlikely to fill capacity. For example, hotels will typically offer cheaper room rates outside of holiday periods to try to increase naturally lower demand. In addition, many larger chains will sell heavily discounted rooms to third parties who in turn take on the risk (and reward) of finding customers for these rooms.

11.5 How is the supply side managed?

Operations principle

Managing the supply side involves setting base capacity and using 'level' or 'chase' plans to manage the supply of services or products.

We now turn from managing the demand side of the capacity management framework to the supply side. Here, decisions include setting the base capacity level, and then using two key methods of managing supply – level capacity plans, where nominal capacity is kept constant; and chase capacity plans, where capacity is adjusted to 'chase' fluctuations in demand over time.

Setting base capacity

The most common starting point in managing the supply side is to decide the 'base level' of capacity and then adjust it periodically up or down to reflect fluctuations in demand. Three factors are important to consider in setting this base level:

- The operation's performance objectives.
- Perishability of the operation's outputs.
- Variability in demand or supply.

The effect of performance objectives on the base level

Base levels of capacity should be set primarily to reflect an operation's performance objectives (see Figure 11.10). Setting the base level of capacity high above average demand will result in relatively low levels of utilisation of capacity When an operation's fixed costs are high,

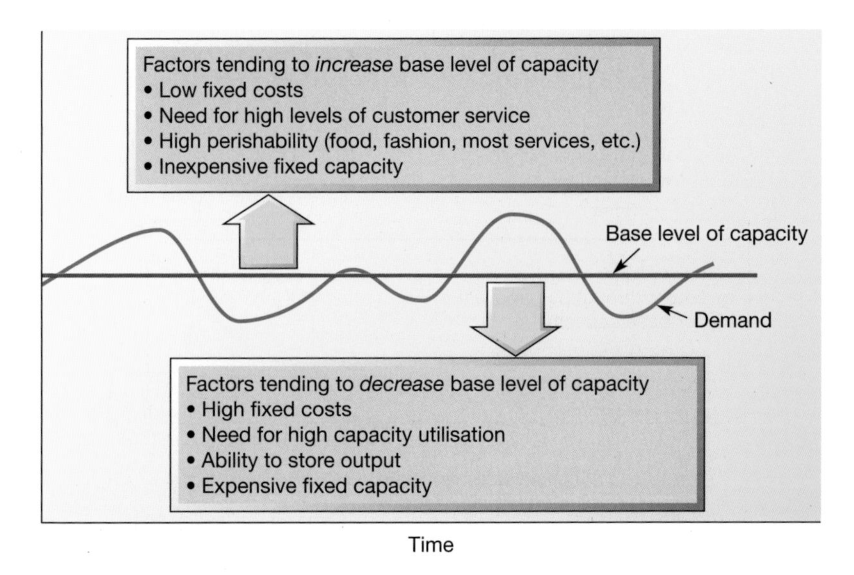

Figure 11.10 Base level of capacity should reflect the relative importance of the operation's performance objectives

underutilisation has significant detrimental effects. Conversely, high base levels of capacity result in a capacity 'cushion' for much of the time, so the ability to flex output to give responsive customer service will be enhanced. When the output from the operation is capable of being stored, there may also be a trade-off between fixed capital and working capital in where base capacity level is set. A high level of base capacity can require considerable investment, while a lower base level would reduce the need for capital investment but may require inventory to be built up to satisfy future demand, thus increasing working capital. For some operations, building up inventory is either risky because products have a short shelf-life (for example, perishable food, high-performance computers, or fashion items) or because the output cannot be stored at all (for example, most services).

The effect of perishability on the base level

When either supply or demand is perishable, base capacity will need to be set at a relatively high level because inputs to the operation or outputs from the operation cannot be stored for long periods. For example, a factory that produces frozen fruit will need sufficient freezing, packing and storage capacity to cope with the rate at which the fruit crop is being harvested during its harvesting season. Similarly, a hotel cannot store its accommodation services. If an individual hotel room remains unoccupied, the ability to sell for that night has 'perished'. In fact, unless a hotel is fully occupied every single night, its capacity is always going to be higher than the average demand for its services.

The effect of demand or supply variability on the base level

Variability, either in demand or capacity will reduce the ability of an operation to process its inputs. The consequences of variability in individual processes were discussed in Chapter 6. As a reminder, the greater the variability in arrival time (demand) or activity time (supply) at a process, the more the process will suffer both high throughput times *and* reduced utilisation. This principle holds true for whole operations, and because long throughput times mean that queues will build up in the operation, high variability also affects inventory levels. The implication of this is that the greater the variability, the more extra capacity will need to be provided to compensate for the reduced utilisation of available capacity. This is illustrated in Figure 11.11.

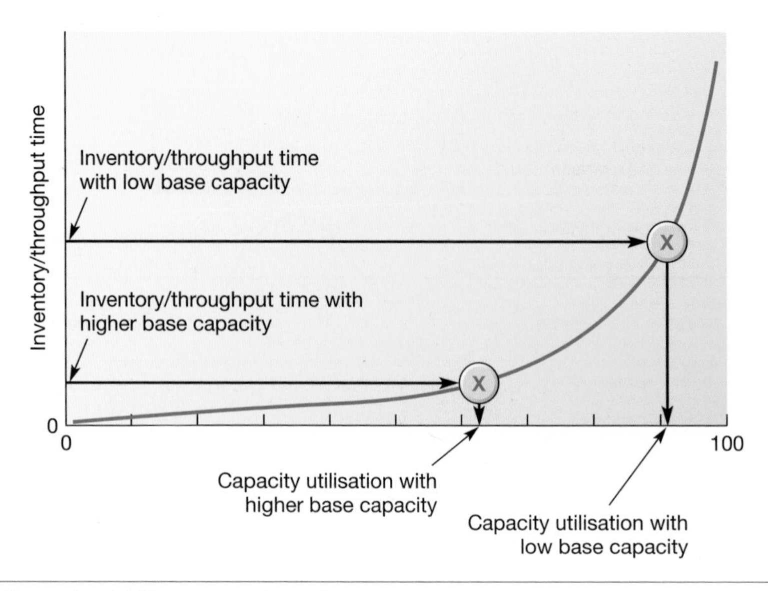

Figure 11.11 Effects of variability on capacity utilisation

Level capacity plan

Once base capacity is set, the first alternative supply-side approach is a 'level capacity plan', where capacity is fixed throughout the planning period regardless of the fluctuations in forecast demand (see Figure 11.12a). Level capacity plans offer stable employment patterns, high process utilisation and often high productivity with low unit costs. Unfortunately, they can also create considerable inventories of materials, customers or information. In addition, level capacity plans are not well suited to 'perishable' products, such as foods and some pharmaceuticals, for products where fashion changes rapidly and unpredictably (for example, fashion garments) or for customised products.

In many service settings, low utilisation effects can make level capacity plans prohibitively expensive but may be considered appropriate where the opportunity costs of individual lost sales are very high; for example, in the high-margin retailing of jewellery and in (real) estate agents. It is also possible to set the capacity somewhat below the forecast peak demand level in order to reduce the degree of underutilisation. However, in the periods where demand is expected to exceed planned capacity, customer service may deteriorate. Customers may have to queue for long periods or may be 'processed' faster and less sensitively.

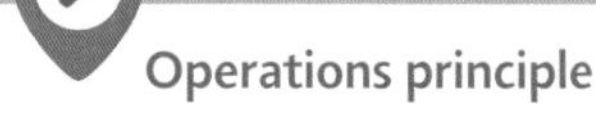

Operations principle

The higher the base level of capacity, the less capacity fluctuation is needed to satisfy demand.

Chase (demand) capacity plan

In contrast to level capacity plans, chase capacity plans attempt to match demand patterns closely by varying levels of capacity (see Figure 11.12b). Chase capacity strategies are much more challenging than level capacity plans, as different numbers of staff, different working hours and even different amounts of equipment may be necessary in each period. For this reason, pure chase plans are unlikely to appeal to operations that manufacture standard, non-perishable products. Also, where manufacturing operations are particularly capital-intensive, this approach would require a high level of physical capacity, much of which would be used only occasionally. A pure chase plan is more usually adopted by operations that are not able to store their output, such as some customer-processing operations or manufacturers of perishable products. It avoids the wasteful provision of excess staff that occurs with a level capacity plan, and yet should satisfy customer demand throughout the planned period. Where output can be stored, the chase demand policy might be

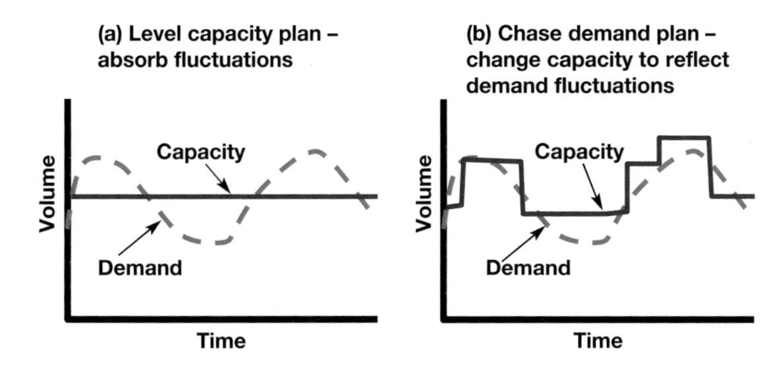

Figure 11.12 (a) 'Level' capacity plan versus (b) 'chase' capacity plan

adopted in order to minimise or eliminate finished goods inventory, especially if the nature of future demand (in terms of volume or mix) is relatively unpredictable. There are a number of different methods for adjusting capacity (see Table 11.5), although they may not all be feasible for all types of operation.

Operations principle

The 'chase' (demand) approach is most useful when output cannot be stored or when demand is both volatile and unpredictable.

Table 11.5 Methods of executing a chase (demand) capacity plan

Method of adjusting capacity	Advantages	Disadvantages
Overtime – staff working longer than their normal working times	Quickest and most convenient	Extra payment and agreement from staff normally necessary. Can reduce productivity over long periods
Annualised hours – staff contracting to work a set number of hours per year rather than a set number of hours per week	Without many of the costs associated with overtime the number of staff available to an organisation can be varied throughout the year to reflect demand	When very large and unexpected fluctuations in demand are possible, all the negotiated annual working time flexibility can be used before the end of the year
Staff scheduling – arranging working times (start and finish times) to vary the aggregate number of staff available for working at any time	Staffing levels can be adjusted to meet demand without changing job responsibilities or hiring in new staff	Providing start and finish (shift) times that both satisfy staff's need for reasonable working times and shift patterns as well as providing appropriate capacity can be difficult
Varying the size of the workforce – hiring extra staff during periods of high demand and laying them off as demand falls, or hire and fire	Reduces basic labour costs quickly	Hiring costs and possible low productivity while new staff go through the learning curve. Lay-offs may result in severance payments and possible loss of morale in the operation and loss of goodwill in the local labour market
Using part-time staff – recruit staff who work for less than the normal working day (at the busiest periods)	Good method of adjusting capacity to meet predictable short-term demand fluctuations	Expensive if the fixed costs of employment for each employee (irrespective of how long they work) are high
Skills flexibility – designing flexibility in job design and job demarcation so that staff can transfer across from less busy parts of the operation	Fast method of reacting to short-term demand fluctuations	Investment in skills training needed and may cause some internal disruption
Subcontracting/outsourcing – buying, renting or sharing capacity or output from other operations	No disruption to the operation	Can be very expensive because of subcontractor's margin and subcontractor may not be as motivated to give same service, or quality. Also a risk of leakage of knowledge
Change output rate – expecting staff (and equipment) to work faster than normal	No need to provide extra resources	Can only be used as a temporary measure, and even then can cause staff dissatisfaction, a reduction in the quality of work, or both

Responsible operations

In every chapter, under the heading of 'Responsible operations', we summarise how the particular topic covered in the chapter touches upon important social, ethical and environmental issues.

The ethics of the gig economy[6]

The gig economy (also referred to as zero-hours contracting) describes the trend of organisations to employ subcontractors on a freelance basis rather than relying on full-time employees. In these settings, an employer does not offer any guarantee of a specific number of hours of work for an individual. Neither is any person working under a zero-hours (or gig) contract obliged to accept those hours when they are offered. From a capacity management perspective, these developments have helped to maintain high levels of customer service even in the face of changeable demand, while simultaneously achieving high levels of utilisation. Operations avoid the fixed costs of employees or facilities when demand drops but could quickly ramp up capacity when demand increases. Such approaches are employed across a wide range of industries, including arts and design, transportation, construction, accommodation, media, education and professional services.

Uber is arguably the most famous of all gig economy companies worldwide. Its technology platform connects those wanting taxi rides, food delivery and transportation of small packages with individuals who want to provide a service. It gives subcontracted drivers considerable flexibility over when and where they work, generating what is described as a perfectly competitive supply market. Uber has also developed other operations in over 600 cities worldwide with extensions to its core ride-sharing service including UberBOAT (a water taxi service), UberMOTO (transportation by motorcycle), UberEats (a meal delivery service) and UberRUSH (a courier package service).

Those in favour of gig contracting highlight that it offers organisations significant flexibility to meet varied demand patterns and gives suppliers the freedom to take on work as and when they choose. In contrast, the idea of fluctuating the workforce to match demand, either by using part-time staff or using zero-hours contracts, or by hiring and firing, is considered unethical by some. It is the responsibility of businesses, they argue, to engage in a set of activities that are capable of sustaining employment at a steady level. In addition, hiring people on a short-term contract, in practice, leads to them being offered poorer conditions of service and leads to a state of permanent anxiety as to whether they will keep their jobs. On a more practical note, in an increasingly global business world where companies often have sites in different locations, those countries that allow hiring and firing are more exposed to 'downsizing' than those where legislation makes this difficult. In addition, for knowledge-oriented organisations, there are significant risks when using contracts that do not tie-in workers in a meaningful way. In other words, talent leaves at the end of the day and a firm can't be sure it will return in the morning!

Such concerns have given rise to varied legal status and conditions for gig working around the world. For example, at the time of writing, they are allowed in Hong Kong, Malaysia, Norway, Singapore and the United States (though not typically referred to as zero-hours contracts), allowed but increasingly regulated in the Netherlands, the United Kingdom and Sweden, and not generally allowed in Austria, Belgium, China, Czech Republic, France, Germany, Italy and Spain. In addition, moves towards greater legal rights for 'gig workers' (such as sick pay, maternity leave and paid holidays) has reduced some of its operational advantages for employers. For example, in a landmark court case brought by James Farrar and Yaseen Aslam, the UK's Supreme Court ruled that Uber drivers were to be treated as workers rather than independent contractors. The decision had major implications for Uber, with its 70,000 UK-based operatives now entitled to seek a minimum wage, holiday pay and pensions, and it is facing ongoing legal challenges in multiple other countries. The legal case also sent ripples through the gig economy more broadly, as companies anticipated a tightening of rules surrounding their use of gig workers in some locations.

11.6 How can operations understand the consequences of their capacity management decisions?

When making capacity management decisions, managers are attempting to balance the need to provide a responsive and customer-oriented service with the need to minimise costs. For this reason, most organisations choose to follow a mixture of the approaches outlined in this chapter. For example, an accounting firm may seek to bring forward some of its peak demand by offering discounts to selected clients (demand management plan). Capacity may also be increased through the use of outsourced suppliers during the busiest months of the year (chase capacity plan). However, some capacity may be constrained (for example, specialist advisory services offered by the firm) and therefore clients may still experience delays during high demand periods (level capacity plan).

Operations principle

Most organisations mix demand-side (demand management and yield management) and supply-side (level and chase plans) capacity management strategies to maximise performance.

Before an operation adopts one or more of the three 'pure' capacity management plans (demand management, level capacity or chase capacity), it should examine the likely consequences of its decisions. Four methods are particularly useful in this assessment:

- factoring in predictable versus unpredictable demand variation;
- using cumulative representations of demand and capacity;
- using queuing principles to make capacity management decisions;
- taking a longitudinal perspective that considers short- and long-term outlooks.

Factoring in predictable versus unpredictable demand variation

When demand is stable and predictable, the life of an operations manager is relatively easy! If demand is changeable, but this change is predictable, capacity adjustments may be needed, but at least they can be planned in advance. With unpredictable variation in demand, if an operation is to react to it at all, it must do so quickly; otherwise the change in capacity will have little effect on the operation's ability to deliver products and services as needed by their customers. Figure 11.13 illustrates how the objective and tasks of capacity management vary depending on the balance between predictable and unpredictable variation.

		Unpredictable variation	
		Low	**High**
Predictable variation	**High**	*Objective* – Adjust planned capacity as efficiently as possible *Capacity management tasks* • Evaluate optimum mix of methods for capacity fluctuation • Work on how to reduce cost of putting plan into effect	*Objective* – Adjust planned capacity as efficiently as possible and enhance capability for further fast adjustments *Capacity management tasks* • Combination of those for predictable and unpredictable variation
	Low	*Objective* – Make sure the base capacity is appropriate *Capacity management tasks* • Seek ways of providing steady capacity effectively	*Objective* – Adjust capacity as fast as possible *Capacity management tasks* • Identify sources of extra capacity and/or uses for surplus capacity • Work on how to adjust capacity and/or uses of capacity quickly

Figure 11.13 The nature of capacity management depends on the mixture of predictable and unpredictable demand and capacity variation

Using cumulative representations of demand and capacity

Figure 11.14 shows a cumulative representation of demand and capacity for a chocolate factory using a level capacity plan that produces at a rate of 14.03 tonnes per productive day.

It shows that although total demand peaks in September, because of the restricted number of available productive days, the peak demand per productive day occurs a month earlier in August. Second, it shows that the fluctuation in demand over the year is significant. The ratio of monthly peak demand to monthly lowest demand is 6.5:1, but the ratio of peak to lowest demand per productive day is 10:1. Demand per productive day is more relevant to operations managers, because productive days represent the time element of capacity.

The most useful consequence of plotting demand and capacity on a cumulative basis is that the feasibility and consequences of a plan can be assessed. Figure 11.14 indicates that the current output plan meets cumulative demand by the end of the year. Up to around day 168, the line representing cumulative production is above that representing cumulative demand. However, by day 198, around 3,025 tonnes have been demanded but only 2,778 tonnes produced. The shortage is therefore 247 tonnes. For any

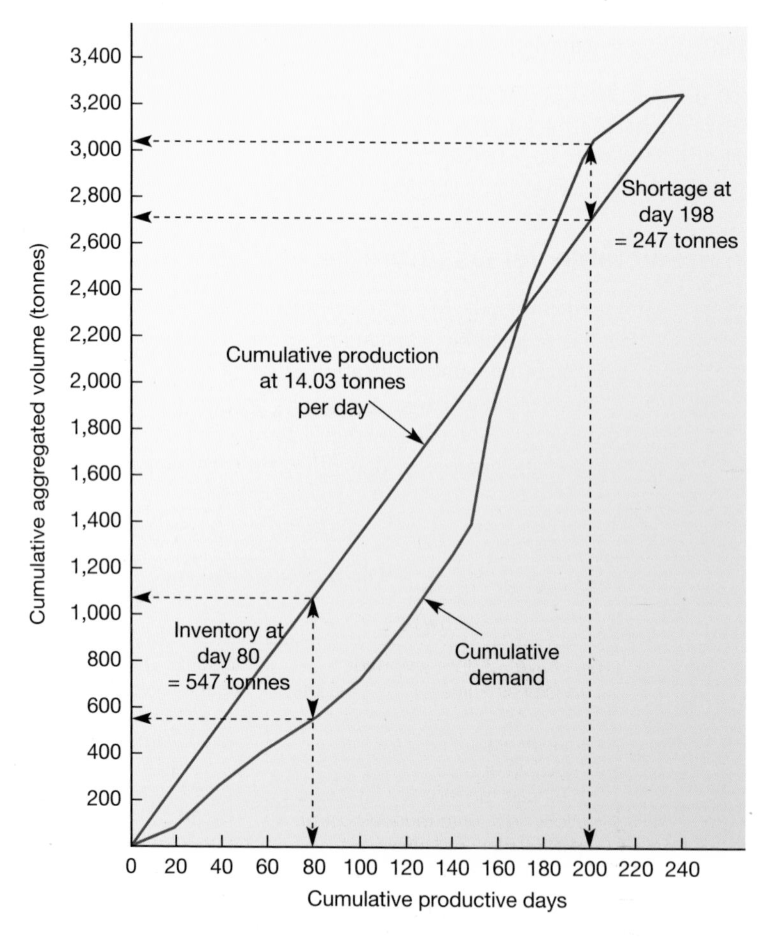

	J	F	M	A	M	J	J	A	S	O	N	D
Demand (tonnes/month)	100	150	175	150	200	300	350	500	650	450	200	100
Productive days	20	18	21	21	22	22	21	10	21	22	21	18
Demand (tonnes/day)	5	8.33	8.33	7.14	9.52	13.64	16.67	50	30.95	20.46	9.52	5.56
Cumulative days	20	38	59	80	102	124	145	155	176	198	219	237
Cumulative demand	100	250	425	575	775	1,075	1,425	1,925	2,575	3,025	3,225	3,325
Cumulative production (tonnes)	281	533	828	1,122	1,431	1,740	2,023	2,175	2,469	2,778	3,073	3,325
Ending inventory (tonnes)	181	283	403	547	656	715	609	250	(106)	(247)	(150)	0

Figure 11.14 A level capacity plan that produces shortages in spite of meeting demand at the end of the year

capacity plan to meet demand as it occurs, its cumulative production line must always lie *above* the cumulative demand line. This makes it a straightforward task to judge the adequacy of a plan, simply by looking at its cumulative representation. An impression of the inventory implications can also be gained from a cumulative representation by judging the area between the cumulative production and demand curves. Figure 11.15 illustrates an adequate level capacity plan for the chocolate manufacturer, together with the costs of carrying inventory. It is assumed that inventory costs £2 per tonne per day to keep in storage. The average inventory each month is taken to be the average of the beginning- and end-of-month inventory levels, and the inventory carrying cost each month is the product of the average inventory, the inventory cost per day per tonne and the number of days in the month.

Operations principle

For any capacity plan to meet demand as it occurs, its cumulative production line must always lie above its cumulative demand line.

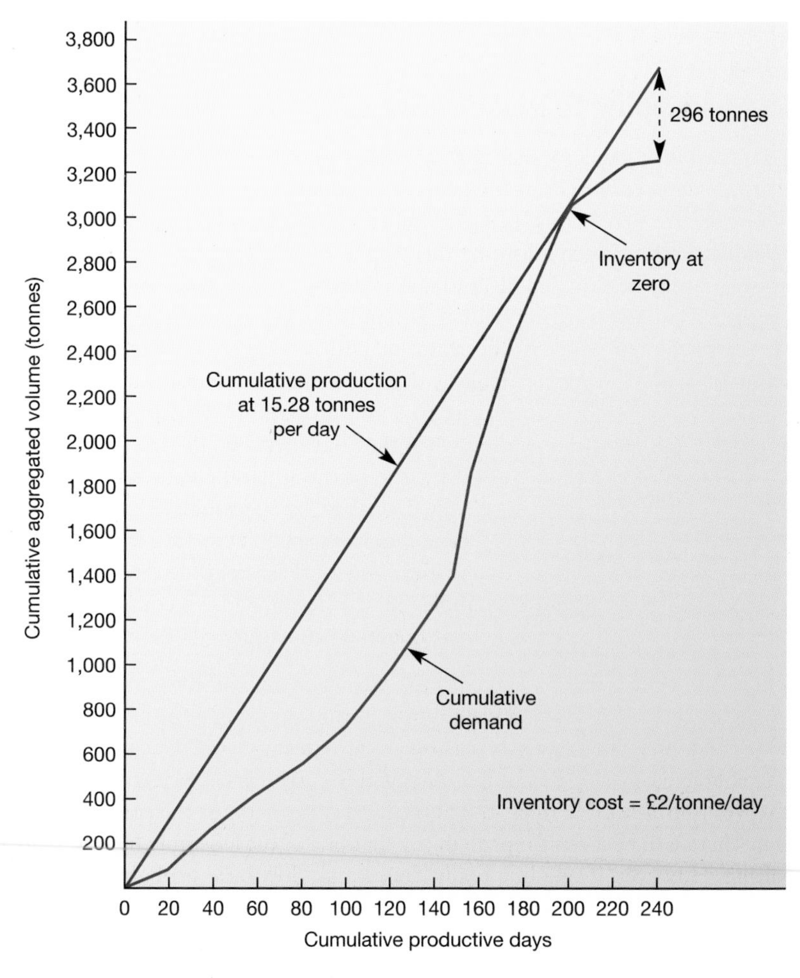

	J	F	M	A	M	J	J	A	S	O	N	D
Demand (tonnes/month)	100	150	175	150	200	300	350	500	650	450	200	100
Productive days	20	18	21	21	22	22	21	10	21	22	21	18
Demand (tonnes/day)	5	8.33	8.33	7.14	9.52	13.64	16.67	50	30.95	20.46	9.52	5.56
Cumulative days	20	38	59	80	102	124	145	155	176	198	219	237
Cumulative demand	100	250	425	575	775	1,075	1,425	1,925	2,575	3,025	3,225	3,325
Cumulative production (tonnes)	306	581	902	1,222	1,559	1,895	2,216	2,368	2,689	3,025	3,346	3,621
Ending inventory (tonnes)	206	331	477	647	784	820	791	443	114	0	121	296
Average inventory (tonnes)	103	270	404	562	716	802	806	617	279	57	61	209
Inventory cost for month (£)	4,120	9,720	16,968	23,604	31,504	35,288	33,852	12,340	11,718	2,508	2,562	7,524

Total inventory cost for year = £191,608

Figure 11.15 A level capacity plan that meets demand at all times during the year

Chase (demand) capacity plans can also be illustrated on a cumulative representation. Rather than the cumulative production line having a constant gradient, it would have a varying gradient representing the production rate at any point in time. If a pure demand chase plan were adopted, the cumulative production line would match the cumulative demand line. The gap between the two lines would be zero and hence inventory (or the queue, if we were taking a service example) would be zero. Although this would eliminate inventory-carrying costs, as we discussed earlier, there would be costs associated with changing capacity levels. Usually, the marginal cost of making a capacity change increases with the size of the change. For example, if the chocolate manufacturer wishes to increase capacity by 5 per cent, this can be achieved by requesting its staff work overtime – a simple, fast and relatively inexpensive option. If the change is 15 per cent, overtime cannot provide sufficient extra capacity and temporary staff will need to be employed – a more expensive solution, which also would take more time. Increases in capacity of above 15 per cent might only be achieved by subcontracting some work out. This would be even more expensive.

Using queuing principles to make capacity management decisions

Cumulative representations of capacity plans are useful where the operation has the ability to store its finished goods as inventory. However, for operations which, by their nature, cannot store their output, such as most service operations, capacity management decisions are best considered using waiting line or queuing theory. When adopting this perspective, operations managers are accepting that while some demand may be satisfied instantly, at other times customers may have to wait. This is particularly true when the arrival of individual demands on an operation are difficult to predict, the time to create a service or product is uncertain, or both. These circumstances make providing adequate capacity at all points in time particularly difficult. Figure 11.16 shows the general form of this capacity issue. Customers arrive according to some probability distribution and wait to be processed (unless part of the operation is idle); when they have reached the front of the queue, they are processed by one of the n parallel 'servers' (their processing time also being described by a probability distribution), after which they leave the operation. There are many examples of this kind of system and some of these are illustrated in Table 11.6. All of these examples can be described by a common set of elements that define their queuing behaviour.

Operations principle

Using queuing principles to make capacity management decisions is useful for operations that cannot store their output.

- ***The source of customers*** – In queue management, 'customers' are not always human. 'Customers' could for example be trucks arriving at a weighbridge, orders arriving to be processed or machines waiting to be serviced, etc. The source of customers for queuing systems can be either *finite* or *infinite*. A finite source has a known number of possible customers. For example, if one maintenance person serves four assembly lines, the number of customers for the maintenance person is known, i.e. four. There will be a certain probability that one of the assembly lines will break down and need repairing. However, if one line really does break down the probability of another line needing repair is reduced because there are now only three lines to break down. So, with a finite source of customers the probability of a customer arriving depends on the number of customers already being serviced. By contrast, an infinite customer source assumes that there is a large number of potential customers so that it is always possible for another customer to arrive no matter how many are being serviced. Most queuing systems that deal with outside markets have infinite, or 'close-to-infinite', customer sources.
- ***Servers*** – A server is the facility that processes the customers in the queue. In any queuing system, there may be any number of servers configured in different ways. In Figure 11.16 servers are configured in parallel, but some may have servers in a series arrangement. For example, on entering a self-service restaurant you may queue to collect a tray and cutlery, move on to the serving area where you queue again to order and collect a meal, move on to a drinks area where you queue once more to order and collect a drink, and then finally queue to pay for the meal. In this case you have passed through four servers (even though the first one was not staffed) in a series arrangement. Of course, many queue systems are complex arrangements of series and parallel connections. There is also likely to be variation in how long it takes to process each customer. Even if customers do not

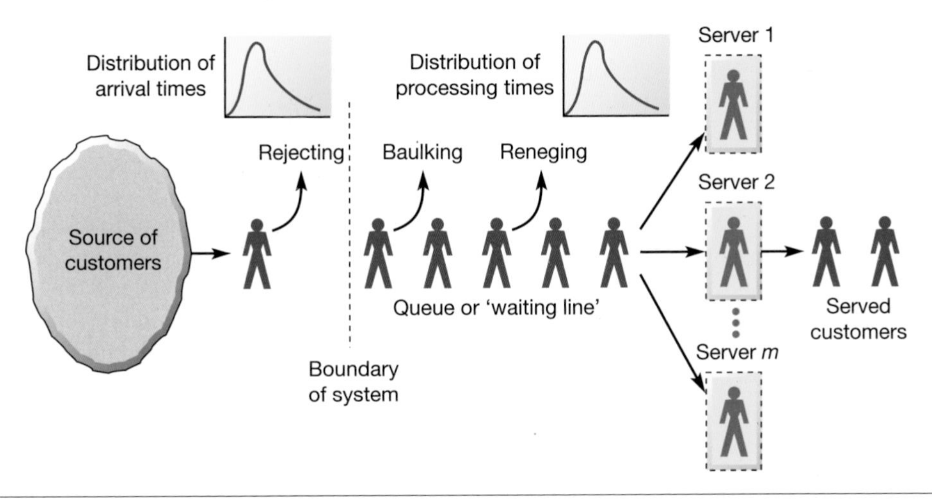

Figure 11.16 Capacity management as a queuing problem

have differing needs, human servers will vary in the time they take to perform repetitive serving tasks. Therefore, processing time, like arrival time, is usually described by a probability distribution.

- ***The arrival rate*** – This is the rate at which customers needing to be served arrive at the server or servers. Rarely do customers arrive at a steady and predictable rate. Usually there is variability in their arrival rate. Because of this it is necessary to describe arrival rates in terms of probability distributions. The important issue here is that, in queuing systems, it is normal that at times no customers will arrive and at other times many will arrive relatively close together.
- ***The queue*** – Customers waiting to be served form the queue or waiting line itself. If there is relatively little limit on how many customers can queue at any time, we can assume that, for all practical purposes, an infinite queue is possible. These queues are not always physical in nature of course – consider, for example, customers waiting for a customised product to be delivered or a patient sitting on a waiting list for six months prior to an operation.
- ***Queue discipline*** – This is the set of rules that determine the order in which customers waiting in the queue are served. Most simple queues, such as those in a shop, use a *first come, first served* queue discipline. The various sequencing rules described in Chapter 10 are examples of different queue disciplines.
- ***Rejecting*** – If the number of customers in a queue is already at the maximum number allowed, then the customer could be rejected by the system. For example, during periods of heavy demand some websites will not allow customers to access part of the site until the demand on its services has declined.
- ***Balking*** – When a customer is a human being with free will (and the ability to get annoyed) they may refuse to join the queue and wait for service if it is judged to be too long. In queuing terms this is called balking.

Table 11.6 Examples of operations that have parallel processors

Operation	Arrivals	Processing capacity
Supermarket	Shoppers	Checkouts
Hospital clinic	Patients	Doctors
Graphic artist	Commissions	Artists
Custom cake decorators	Orders	Cake decorators
Ambulance service	Emergencies	Ambulances with crews
Maintenance department	Breakdowns	Maintenance staff

- *Reneging* – This is similar to balking but here the customer has queued for a certain length of time and then (perhaps being dissatisfied with the rate of progress) leaves the queue and therefore the chance of being served.

Variability effects on queues

The dilemma in managing the capacity of a queuing system is how many servers to have available at any point in time in order to avoid unacceptably long queuing times or unacceptably low utilisation of the servers. Because of the probabilistic arrival and processing times, only rarely will the arrival of customers match the ability of the operation to cope with them. Sometimes, if several customers arrive in quick succession and require longer-than-average processing times, queues will build up in front of the operation. At other times, when customers arrive less frequently than average and also require shorter-than-average processing times, some of the servers in the system will be idle. So even when the *average* capacity (processing capability) of the operation matches the *average* demand (arrival rate) on the system, both queues and idle time will occur.

If the operation has too few servers (that is, capacity is set at too low a level), queues will build up to a level where customers become dissatisfied with the time they are having to wait, although the utilisation level of the servers will be high. Conversely, if too many servers are in place (that is, capacity is set at too high a level), the time that customers can expect to wait will not be long, but the utilisation of the servers will be low. This is why the capacity planning and control problem for this type of operation is often presented as a trade-off between customer waiting time and system utilisation. What is certainly important in making capacity decisions is being able to predict both of these factors for a given queuing system. The supplement to this chapter details some of the simpler mathematical approaches to understanding queue behaviour.

Customer perceptions of queuing

Queues are not generally something we want, but they can be managed to make them more satisfactory to customers. After all, an important aspect of how customers judge the service they receive is how they perceive the time spent queuing. Because of this, the management of queuing systems usually involves attempting to manage customers' perceptions and expectations in some ways. Below are a set of 'principles' that can help in evaluating and improving queues (of course, in cases where the queue itself can't be removed through process improvement).

Operations principle

Customer reactions to having to queue will be influenced by more factors than simply waiting time.

1 Unoccupied time feels longer than occupied time.
2 Pre-process waits feel longer than in-process waits.
3 Anxiety makes the wait seem longer.
4 Uncertain waits feel longer than known, finite waits.
5 Unexplained waits feel longer than explained waits.
6 Unfair waits feel longer than equitable waits.
7 The more valuable the service, the longer customer will 'happily' wait.
8 Solo waiting feels longer than group waiting.
9 Uncomfortable waits feel longer than comfortable waits.
10 New or infrequent users feel they wait longer than frequent users.

These principles have led a number of organisations to experiment with a range of interventions aimed at providing a more comfortable waiting experience for customers and in doing so mitigate the negative effects of queuing. These include the use of music, lighting, scent, art, furnishing and colour; and social elements such as employee visibility, customer interaction, and video games for children. It is also worth noting that in some circumstances there are important *positive* effects of queues. These include the way that some queues can increase the perceived value of the product or service, generate increased demand via shortage perceptions, give time for customer decision-making or mental preparation, or increase levels of positive anticipation.

Taking a longitudinal perspective that considers short- and long-term outlooks

Our emphasis so far has been on the planning aspects of capacity management. In practice, capacity management is a far more dynamic process, which involves controlling and reacting to *actual* demand and *actual* capacity as it occurs. The capacity control process can be seen as a sequence of partially reactive capacity decisions. At the beginning of each period, operations management considers its forecasts of demand, its understanding of current capacity and, if appropriate, how much inventory has been carried forward from the previous period. Based on all this information, it makes plans for the following period's capacity. During the next period, demand might or might not be as forecast and the actual capacity of the operation might or might not turn out as planned. But whatever the actual conditions during that period, at the beginning of the next period the same types of decisions must be made, in the light of the new circumstances.

Operations principle

The learning from managing capacity in practice should be captured and used to refine both demand forecasting and capacity planning.

The success of capacity management is generally measured by some combination of costs, revenue, working capital and customer satisfaction (which goes on to influence revenue). This is influenced by the actual product or service and the capacity available to the operation in any period. However, capacity management is essentially a forward-looking activity. Overriding other considerations of what one or more capacity strategies to adopt is usually the difference between the long- and short-term outlook for the volume of demand. Figure 11.17 illustrates some appropriate capacity management strategies depending on the comparison of long- and short-term outlooks. If long-term outlook for demand is 'good' (in the sense that it is higher than current capacity can cope with) then it is unlikely that even 'poor' (demand less than capacity) short-term demand would cause an operation to make large, or difficult to reverse, cuts in capacity. Conversely if long-term outlook for

Long-term outlook for volume \ Short-term outlook for volume	Decreasing below current capacity	Level with current capacity	Increasing above current capacity
Decreasing below current capacity	Reduce capacity (semi) permanently. For example, reduce staffing levels; reduce supply agreements.	Plan to reduce capacity (semi) permanently. For example, freeze recruitment; modify supply agreements.	Increase capacity temporarily. For example, increase working hours, and/or hire temporary staff; modify supply agreements.
Level with current capacity	Reduce capacity temporarily. For example, reduce staff working hours; modify supply agreements.	Maintain capacity at current level.	Increase capacity temporarily. For example, increase working hours, and/or hire temporary staff; modify supply agreements.
Increasing above current capacity	Reduce capacity temporarily. For example, reduce staff working hours, but plan to recruit; modify supply agreements.	Plan to increase capacity above current level; plan to increase supply agreements.	Increase capacity (semi) permanently. For example, hire staff; increase supply agreements.

Figure 11.17 Capacity management strategies are partly dependent on the long- and short-term outlook for volumes

demand is 'poor' (in the sense that it is lower than current capacity) then it is unlikely that even 'good' (demand more than capacity) short-term demand would cause an operation to take on large, or difficult to reverse, extra capacity.

Summary answers to key questions

11.1 What is capacity management?

- Capacity management is the activity of understanding the nature of demand for services and products, and effectively planning and controlling capacity.
- Capacity decisions are taken across multiple time horizons and each level of capacity decision is made within the constraints of a higher level. In the other direction, short-term decisions provide important feedback for longer-term planning.
- **Long-term capacity management** (or strategy) focuses on introducing or deleting major increments of capacity (see Chapter 5). Short-term capacity decisions focus on allocation, sequence and resourcing of tasks (see Chapter 10). In this chapter, we focus more on the medium-term aspects of capacity management, where decisions are being made largely within the constraints of the physical capacity set by the operation's long-term capacity strategy.
- The process of capacity management includes (1) measuring and forecasting changes in aggregate demand; (2) measuring capacity (ability to supply products and services); (3) managing the demand side; (4) managing the supply side; and (5) understanding the consequences of different capacity management decisions.

11.2 How is demand measured?

- Organisations can attempt to forecast demand using a mix of qualitative methods (panel, Delphi and scenario planning) and quantitative methods (time series and causal models).
- Good demand forecasts should: (1) be as accurate as possible; (2) give a clear indication of relative uncertainty; and (3) be expressed in terms that are useful for capacity management (e.g. units per hour, operatives per month, etc.).
- Operations must find some balance between having better forecasts and being able to cope without perfect forecasts. The resources invested in forecasting should reflect the varying sensitivity to forecast error.

11.3 How is capacity measured?

- The capacity of an operation is the *maximum level of value-added activity over a period of time* that the process can achieve under normal operating conditions.
- Capacity can be measured by the availability of its input resources or by the output that is created.
- The usage of capacity is measured by the factors 'utilisation' and 'efficiency'. A useful measure of capacity leakage is overall equipment effectiveness (OEE).
- Some operations can increase their output by changing the specification of the product or service (although this is more likely to apply to a service).

11.4 How is the demand side managed?

- 'Demand management' involves changing the pattern of demand to bring it closer to available capacity. This is achieved by either stimulating off-peak demand or constraining peak demand.
- A number of methods are used to manage demand, including price differentials, scheduling promotion, constraining customer access, service differentials and creating alternative products or services.
- Yield management is a common method of coping with mismatches when outputs cannot be stored.

11.5 How is the supply side managed?

- Capacity planning often involves setting a base level of capacity and then planning capacity fluctuations around it. The level at which base capacity is set depends on three main factors: the relative importance of the operation's performance objectives, the perishability of the operation's outputs, and the degree of variability in demand or supply.
- 'Level capacity' plans involve no change in capacity and require that the operation absorb demand–capacity mismatches, usually through under- or over-utilisation of its resources, or the use of inventory.
- 'Chase' (demand) plans involve the changing of capacity to track demand as closely as possible. Methods include overtime, annualised hours, staff scheduling, varying workforce size, using part-time staff, increasing skills flexibility, subcontracting, zero hours contracting and changing the output rate.

11.6 How can operations understand the consequences of their capacity management decisions?

- Most organisations choose to follow a mixture of capacity management approaches because single 'pure' approaches do not match their required combination of competitive and operational objectives.
- Understanding of the balance between predictable and unpredictable variation in demand is critical in deciding the most appropriate mix of capacity management strategies.
- Representing demand and output in the form of cumulative representations allows the feasibility of alternative capacity plans to be assessed.
- In many operations, particularly service operations, a queuing approach can be used to explore the consequences of capacity strategies.
- Taking a longitudinal perspective, considering both long-term and short-term outlook for demand, allows further evaluation of alternative capacity management decisions.

CASE STUDY FreshLunch

(This case was co-authored with Vaggelis Giannikas at School of Management, University of Bath)

'Carlos, are you ready to head out then?' Antônia called across the office. *'Too right! After the morning I've had, I could do with the break!'* Carlos laughed, as he grabbed his wallet and sunglasses. As the two headed towards the lift (elevator), they entered into a deep conversation. *'So, what do you fancy Antônia?'* After a short pause to think, Antônia responded, *'Well, we could go to Byōōdē – their red curry is definitively one of the best around and the Pad Thai's pretty good too. Or Pollo Picante? I had the chicken with chimichurri sauce the other day and it was really good'*. As the lift descended from the 32nd floor of their building, the conversation continued. *'If we're looking for something hot Antônia, I guess we could also try the new Indian thali place? Rebecca went there last week and said it was excellent, though she did mention it was pretty slow service. Besides, it's such a nice day, maybe we could have something cold instead? There's the sushi at Kazoku – it's so fresh and there's loads of choice'*. Antônia thought for moment, *'Well I'm absolutely fine not having anything hot, but I'm not in the mood for Sushi today to be honest'*. There were just so many options she reflected, *'How's about FreshLunch instead?'* Carlos smiled as they walked out into bright May sunshine, *'Sounds like a good plan to me!'*

A few blocks away, Sofia had already been serving customers at FreshLunch for an hour and, as usual, things were picking up quickly towards the lunchtime rush. She was a chemist by training and had spent the first six years after her graduation working for a large multinational in a research laboratory based in Norway. But her passion was food and fresh produce. Having completed a part-time executive MBA, Sofia had changed her career direction dramatically and set up as a restaurant owner. Knowing how demanding busy customers were, Sofia had established FreshLunch utilising the techniques she learned from her studies in an attempt to manage her operation effectively. FreshLunch had taken the traditional cafeteria-style approach often found in universities and large hotels, and developed a process that offered quality, variety and speed. The process was simple, involving five sequential steps from order placement to delivery (see Figure 11.18). Having collected any items from the fridge (drinks and sweet treats), the customer first had to choose the base for their meal, a selection between rice and couscous. Next, they would choose their main protein, including chicken, lamb, steak, salmon and grilled vegetables. Then, two sides were selected to accompany the main meal, from a choice of around ten different plates full of vegetables and salads. Finally, dressings and sauces were available before the customer moved to payment at the end of the process. For each step, the customer moved along with their tray, which was passed from one 'assembler' to the next until it was put in a bag and handed to the customer by the cashier.

Since its opening, FreshLunch had always been very busy around lunchtime with long queues created at the counter, some of which extended outside the restaurant itself. Sofia was happy that her hard work over the past three years was paying off. However, she was beginning to appreciate that these long queues were not translating into the profits needed to create a sustainable business. Sofia tried to think again about how some of the things she had learned during her MBA programme might help her tackle the situation. But while she had applied so much to the business and made plenty of improvements, she increasingly felt like she was too close to the problems to see them clearly. *'What I need'*, she reflected, *'is a fresh pair of eyes'*. That evening, scrolling through her social media, Sofia noticed that her friend Zuri had just posted an interesting piece on the challenges of demand forecasting during 'X-factor' events. *'Now why didn't I think of her before?'* she thought. After their time together on the MBA programme, Zuri's small consulting business, which helped its customers analyse and improve their operations, had grown substantially, so she was clearly doing something right. Sofia dialled her number and after a few rings, Zuri answered, *'Sofia! Long time, no chat! How are you and how's things in the restaurant business?'*

Fifteen minutes later, their conversation turned from generally catching up to FreshLunch. *'I find it really hard to predict what my customers are going to choose every day and I often end up having to throw away quite a lot of food. I tried cooking fewer portions but then had a lot of annoyed customers and I can't risk bad reviews!'* Zuri already had a few ideas in mind but decided to ask a few more questions to understand the business better, *'What about the customers? Do*

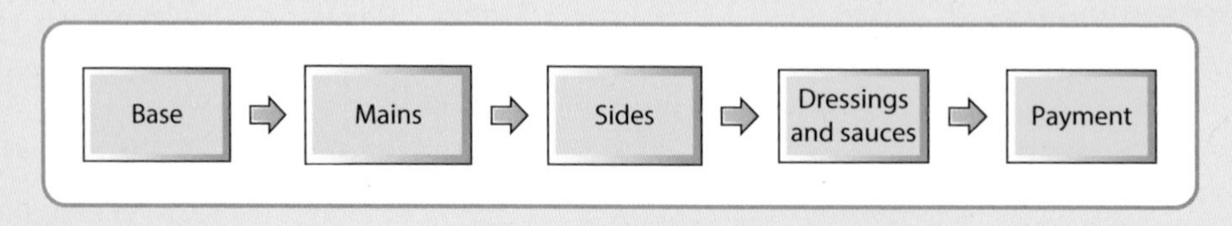

Figure 11.18 FreshLunch process

★★★★★ 4 hours ago

Lovely salads and very friendly staff. We really enjoyed our meal and it's good if you have a mix of meat eaters and vegetarians. We'll be back!

★★★ 6 hours ago

The food is really tasty as long as you don't mind waiting. If you're in a hurry, go somewhere else. FreshLunch need to sort out their queues.

★★★★ 1 day ago

Love the relaxed atmosphere and you never feel rushed making your choices. The food is always really fresh (so the name of the place is spot-on!) and we like how simple the process is.

★★★★★ 4 days ago

I love this place. The food is perfect and simple, and the staff are on it. I try to go about 11.45 so I don't get stuck in the long queues.

★★★ 1 week ago

Pretty good food and reasonable price. Wish they had a bigger selection of mains and sides as I go regularly and am getting a bit bored with the same stuff.

★★★★ 1 week ago

The food is great and I love the roasted cauliflower salad. A bit inconsistent on the steaks and lamb – seems like they get cooked differently every time I order them.

★ 2 weeks ago

Never again! We waited for ages and when we were finally served, they didn't have any chicken and half of the side options weren't available either. Loads of other better options nearby!

★★★★ 3 weeks ago

What a great little find. I had the salmon with baby spinach and a beetroot and feta salad. My colleague had chicken with some lentils and a Greek salad. We both really liked everything. Fast becoming one of our go-to places. Shame they don't serve breakfast, as we're often in work for 6am.

★★★★ 1 month ago

Best place for fresh salads in the area. The slow service isn't ideal, but the food's worth waiting for.

Figure 11.19 Recent customer reviews of FreshLunch

you know if they are happy with what they get?', she said. *'I love how busy we are'*, replied Sofia, *'but some customers have already started posting negative reviews due to the long waits. To be fair, it's normally only 20 minutes but since most people only have an hour for lunch, I completely understand where they're coming from'*. Zuri quickly did a search for FreshLunch and began scrolling through some of the most recent customer reviews (see Figure 11.19).

As Zuri read, Sofia continued, *'Unfortunately, the huge rent is kind of killing me, to be honest. You can imagine how expensive it is to rent even a small place in the centre of London. It's also hard to find good cooks and waiters as FreshLunch can only offer them contracts for 50–80 per cent of their time'*. Zuri leaned back in her chair, *'OK Sofia, let me have a think about this over the weekend and I'll get back to you with my thoughts. It's been great catching up'*. As she took a sip of her drink, she thought about how this could be a good exercise for the new associates in her company.

On Monday morning, it was Zuri's time to pick up the phone and call her friend. *'Sofia, I'd really like to help you with this. We've just hired a small group of young associates and I'd be happy to assign them to work with you. From my perspective, it would give me a chance to see these guys in action and get a feel for how they work as a team, before I set them off on the paid jobs. And for you, it'd be some free consulting – feels like a win-win, right? I can give them a little bit of supervision, but not much, as "ts basically pro bono [without charge]'*. Sofia was delighted, *'Wow Zuri, that would be fantastic! And I'm happy to give them a bit of advice on any client interaction issues that come up during this'*. Zuri knew that could prove extremely helpful. Sofia had always been excellent at giving constructive feedback. *'Good thinking – that would be brilliant. To get this started, can you please send me over some information about FreshLunch? I will ask my associates to get in touch with a data spec first thing tomorrow'*.

Over the rest of the week, Sofia collected the information that Zuri's team had requested for the project. She started with some basic information such as opening times (11.00–15.30) and the daily menu (see Figure 11.20).

FreshLunch

✦ OPEN EVERY WEEKDAY 11.00 – 15.30 ✦

STEP 1:
Choose a base

COUSCOUS
RICE

STEP 2:
Choose a main

CHINESE CHICKEN	£7.95
FLANK STEAK	£8.95
MEDITERRANEAN CHICKEN	£7.95
ORGANIC GRILLED VEGETABLES	£7.45
PERSIAN LAMB	£8.95
THAI SALAMON	£8.95

STEP 3:
Choose two sides

BABY SPINACH
BEETROOT AND FETA SALAD
CHEESE CROQUETTES
CHICKPEAS WITH SPICY DRESSING
GNOCCHI NAPOLITANA
GREEK SALAD
LENTILS, HALLOUMI AND HERBS
OVEN BAKED SWEET POTATO FRIES
PEA AND MINT SALAD
ROASTED CAULIFLOWER AND BROCCOLI SALAD

STEP 4:
Dressings and sauces

GREEK VINAIGRETTE
HONEY MUSTARD
LEMON DRESSING
NONNA'S PESTO
PEPPERCORN SAUCE
WILD GARLIC MAYO
YOGHURT AND MINT

✦ Sweet desserts ✦

CARROT CAKE	£3.45
GIANT MACARON (ALL FLAVOURS)	£3.45
HAZELNUT CHOCOLATE MOUSSE	£3.45
SUMER FRUIT SALAD	£3.45

Dishes may contain allergens. If you have any dietary requirements, please speak to a member of staff.

✦ Beverages ✦

ELDERFLOWER PRESSÉ	£2.95
FRESH APPLE JUICE	£3.95
FRESH ORANGE JUICE	£3.95
GINGER BLITZ SMOOTHIE	£5.45
ICELANDIC STILL WATER	£1.85
MANGO AND CARROT SMOOTHIE	£5.45
RHUBARB PRESSÉ	£2.95
SUMMER FRUIT SMOOTHIE	£5.45

Figure 11.20 FreshLunch daily menu

Note: Some sides changed periodically based on seasonality and popularity

Table 11.7 FreshLunch demand across a 'typical' day

Time slot	Meals	Customers (measured in number of receipts)
11.00–11.30	10	10
11.30–12.00	20	19
12.00–12.30	38	30
12.30–13.00	89	68
13.00–13.30	154	121
13.30–14.00	92	66
14.00–14.30	24	22
14.30–15.00	12	11
15.00–15.30	4	4

She also spent some time putting together information that could be used to analyse the demand patterns for FreshLunch. Luckily, she had recently installed a software package that allowed her to collect and analyse point-of-sale (POS) data. Now Sofia felt she was actually beginning to make some use of it. She remembered one of her professors talking about organisations 'drowning in data' and was starting to appreciate what she meant! To keep it simple, she began with what felt to her like a typical day and broke it down into 30-minute time slots (see Table 11.7). She included information on how many customers typically visited, but also the number of actual meals prepared, as some customers would order more than one meal. In addition, Sofia was asked by the associates to provide some information about daily sales of meals over recent weeks (see Table 11.8).

At Zuri's firm, Bankole Consulting, the associates were looking forward to working on the project. For the group, it was a chance to get stuck into the world of consulting and prove that they were ready to step up to the firm's paying clients. It was also an opportunity to repay some of that faith that Zuri had shown in them when making her hires. After an initial meeting with Zuri, they started analysing the information that Sofia had sent over. They also decided to pay a visit to FreshLunch to get some first-hand experience of the operation. With the approval of Sofia, they behaved as normal customers, queuing for food, ordering and then eating at the bench by the window. During their visit, they drew the layout of the main floor of the restaurant (see Figure 11.21) as this could be useful for their discussions with Zuri. They also looked at the basement area – used for storing ingredients, crockery and utensils – and an area on the upper floor, which functioned as a break room and office space for the shift manager.

During their visit, the team collected data on the size of the queues – something not currently captured by Sofia. There were two parts of the queue: the *assembly queue* was formed

Table 11.8 FreshLunch meals sold over a three-week period

Day	Med chicken	Chinese chicken	Flank steak	Persian lamb	Thai salmon	Grilled veg	TOTAL
Monday	139	44	28	83	105	155	554
Tuesday	83	33	34	66	57	57	330
Wednesday	102	53	44	89	75	80	443
Thursday	80	30	33	64	60	63	330
Friday	133	55	83	139	100	46	556
Monday	134	62	29	84	95	157	561
Tuesday	84	40	30	67	48	63	332
Wednesday	121	36	44	89	76	81	447
Thursday	85	34	34	68	57	60	338
Friday	129	62	84	138	101	45	559
Monday	141	56	30	85	96	158	566
Tuesday	88	39	34	68	51	61	341
Wednesday	104	55	44	90	77	81	451
Thursday	78	36	34	70	61	61	340
Friday	136	56	87	136	102	47	564
							6,712

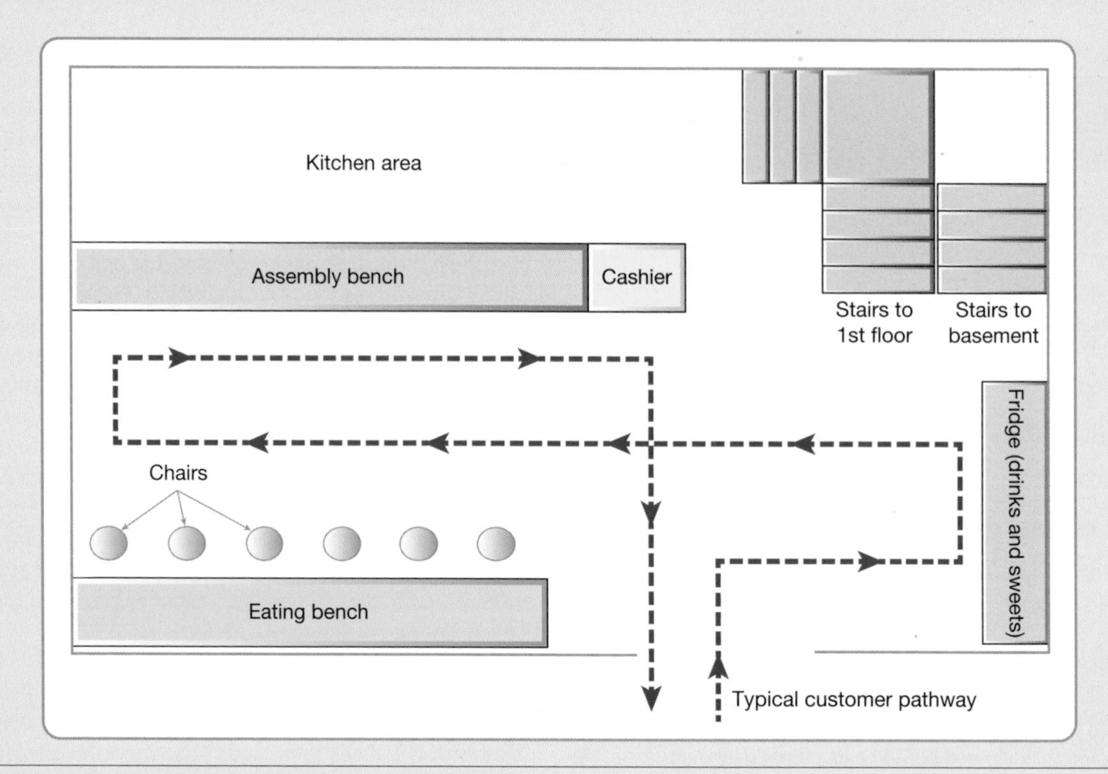

Figure 11.21 FreshLunch restaurant layout

Table 11.9 Queue size during associates visit to FreshLunch

Time	11.30	11.45	12.00	12.15	12.30	12.45	13.00	13.15	13.30	13.45	14.00	14.15	14.30
Queue	0	0	1	3	6	11	15	21	18	8	3	2	0

between the point a customer started giving their order and the point of payment at the cashier. This was always moving relatively quickly as the employees were used to receiving orders and serving the customers efficiently. The second, and more concerning, part of the queue was the one formed by people *waiting* to place their order. The team noted the size of the queue at 15-minute intervals over a 3-hour period (see Table 11.9). Most of the people in the queue spent their time talking to each other, checking their phones or looking at the printed menus to decide what to order. Zuri's associates also noticed that some people left the queue (nine in total during the three busiest periods) and others decided not to join the queue (around 4–5 in the 15-minute intervals when the queue exceeded 10 customers). Finally, the associates talked to Sofia and her employees to gain further insights into FreshLunch.

Selected quotes from staff at FreshLunch

'This 90 minutes during lunch is hell. It is always busy and there is nothing worse than hungry people.'

'The space upstairs is quite nice, but we hardly ever get the time to use it. By the time things calm down, our shift is over.'

'It's nice to get a free meal at the end of the day but then again, we often have so many leftovers, they need to go to the bin. Especially the sides; sometimes we end up having whole trays of unsold food and you can't really tell what is going to sell well on a particular day.'

'The job pays well for the hours I have to work, but I need to have a second job in the evenings to make ends meet.'

'It's pretty boring in the morning and after 14:30.'

QUESTIONS

1. **What do you think are the key issues faced by FreshLunch and what are the underlying reasons for these issues?**
2. **What advice would you give Sofia and how would you prioritise potential improvements?**

Problems and applications

All chapters have 'Problems and applications' questions that will help you practise analysing operations. They can be answered by reading the chapter. Model answers for the first two questions can be found on the companion website for this text.

1 In March, a law firm predicted April demand for 360 client consultations. Actual April demand was 410. Using a smoothing constant chosen by management of $\alpha = 0.20$, what is the forecast for May demand using the exponential smoothing mode?

2 The degree of effort (and cost) to devote to forecasting is often a source of heated debate within organisations. This often comes down to two opposing arguments. One goes something like this. *'It's important for forecasts to be as accurate as possible. We cannot plan operations capacity otherwise'*. The counter-argument is very different. *'Demand will always be uncertain, that is the nature of demand. Get used to it'*. Discuss the relative merits of these two perspectives.

3 A German car manufacturer defines 'utilisation' as the ratio of actual output for a process to its design capacity, where design capacity is the capacity of a process as it is designed to operate. However, it knows that it is rarely possible to achieve this theoretical level of capacity, which is why the company uses a measure that it calls 'effective capacity'. This is the actual capacity of a process, once maintenance, changeover, other stoppages,and loading have been considered. The ratio of actual output for a process to its effective capacity is defined as its 'efficiency'.

The company has a painting line with a design capacity of 100 square metres per minute and the line is operated 24 hours a day, 7 days a week (168 hours). Records for a week show the following lost time in production:

1	Product changeovers (set-ups)	18 hours
2	Regular maintenance	12 hours
3	No work scheduled	6 hours
4	Quality sampling checks	8 hours
5	Shift change times	8 hours
6	Maintenance breakdown	16 hours
7	Quality failure investigation	12 hours
8	Paint stock-outs	6 hours
9	Labour shortages	6 hours
10	Waiting for paint	5 hours
	Total	100 hours

During this week, production was only $100 \times 60 \times (168 - 100) = 408{,}000$ square metres per week. What is the painting line's 'utilisation' and 'efficiency' according to the company's definitions?

4 In a typical 7-day period, the planning department of the pizza company programs its 'Pizzamatic' machine for 148 hours. It knows that changeovers and set-ups take 8 hours and breakdowns average 4 hours each week. Waiting for ingredients to be delivered usually accounts for 6 hours, during which the machine cannot work. When the machine is running, it averages 87 per cent of its design speed. An inspection has revealed that 2 per cent of the pizzas processed by the machine are not up to the company's quality standard. Calculate the OEE of the 'Pizzamatic' machine.

5 Seasonal demand is particularly important to the greetings card industry. Mother's Day, Father's Day, Halloween, Valentine's Day and other occasions have all been promoted as times to send (and buy) appropriately designed cards. Now, some card manufacturers have moved on to 'non-occasion' cards, which can be sent at any time. The cards include those intended to be sent from a parent to a child with messages such as 'Would a hug help?', 'Sorry I made you feel bad' and 'You're perfectly wonderful – it's your room that's a mess'. Other cards deal with more serious adult themes such as friendship ('you're more than a friend, you're just like family') or even alcoholism ('this is hard to say, but I think you're a much neater person when you're not drinking'). Some card companies have founded 'loyalty marketing groups' that 'help companies communicate with their customers at an emotional level'. They promote the use of greetings cards for corporate use, to show that customers and employees are valued.

(a) What seem to be the advantages and disadvantages of these strategies?
(b) What else could card companies do to cope with demand fluctuations?

6 A pizza company has a demand forecast for the next 12 months that is shown in the table below. The current workforce of 100 staff can produce 1,500 cases of pizzas per month.

(a) Prepare a production plan that keeps the output level. How much warehouse space would the company need for this plan?
(b) Prepare a demand chase plan. What implications would this have for staffing levels, assuming that the maximum amount of overtime would result in production levels of only 10 per cent greater than normal working hours?

Pizza demand forecast

Month	Demand (cases per month)
January	600
February	800
March	1,000
April	1,500
May	2,000
June	1,700
July	1,200
August	1,100
September	900
October	2,500
November	3,200
December	900

7 Revisit the 'Operations in practice' example, '3M's COVID-19 surge capacity'. With reference to this and other operations that you are familiar with:

(a) What are the main challenges in long-term demand forecasting and how these might be addressed?
(b) How can longer-term capacity investments retain sufficient flexibility if forecasts turn out to be inaccurate?

8 Why do airlines often 'overbook' (sell more tickets than they have seats available) and what are the risks of doing so?

9 Re-examine the chocolate manufacturer's demand shown in Figure 11.14. Use this data to explore two alternative plans:

Plan 1 – Produce at 8.7 tonnes per day for the first 124 days of the year, then increase capacity to 29 tonnes per day by heavy use of overtime, hiring temporary staff and some subcontracting. Then produce at 29 tonnes per day until day 194, after which reduce capacity back to 8.7 tonnes per day for the rest of the year. The costs of changing capacity by such a large amount (the ratio of peak to normal capacity is 3.33:1) are calculated by the company as follows: Cost of changing from 8.7 tonnes/day to 29 tonnes/day = £110,000. Cost of changing from 29 tonnes/day to 8.7 tonnes/day = £60,000.

Plan 2 – Produce at 12.4 tonnes per day for the first 150 days of the year, then increase capacity to 29 tonnes per day by use of overtime and hiring some temporary staff. After which, produce at 29 tonnes per day until day 190, then reduce capacity back to 12.4 tonnes per day for the rest of the year. The costs of changing capacity in this plan are smaller because the degree of change is smaller (a peak to normal capacity ratio of 2.34:1), and they are calculated by the company as being as follows: Cost of changing from 12.4 tonnes/day to 29 tonnes/day = £35,000. Cost of changing from 29 tonnes/day to 12.4 tonnes/day = £15,000.

10 If you were managing a small farm park that attracted visitors to observe farming exhibits, watch cows being milked, visit a farm shop and café, etc., how would you go about determining the appropriate capacity for the car park that served the visitors?

Selected further reading

Gilliland, M., Tashman, L. and Sglavo, U. (2015) ***Business Forecasting: Practical Problems and Solutions*****, John Wiley & Sons, Hoboken, NJ.**
A collection of papers from focused on forecasting practitioners.

Gunther, N.J. (2007) ***Guerrilla Capacity Planning*****, Springer, New York, NY.**
This book provides a tactical approach for planning capacity in both product-based and service-based contexts. Particularly interesting for those new to the ideas of capacity planning as it covers basic and more advanced demand forecasting techniques as well as 'classic' capacity responses.

Kolassa, S. and Siemsen, E. (2016) ***Demand Forecasting for Managers*****, Business Expert Press, New York, NY.**
This book is aimed at simplifying demand forecasting by avoiding the complex formulas and focusing more on principles and heuristics of forecasting. A very useful guide for those wanting a bit more on forecasting but wary of the maths!

Manas, J. (2014) ***The Resource Management and Capacity Planning Handbook: A Guide to Maximizing the Value of Your Limited People Resources*****, McGraw-Hill Education, New York, NY.**
A practitioners' guide, particularly focused on managed human resource capacity to deliver better performance.

Ord, K., Fildes, R. and Kourentzes, N. (2017) ***Principles of Business Forecasting*****, 2nd edn, Wessex Press, Inc., New York, NY.**
A very detailed textbook covering demand in real depth.

Notes on chapter

1. The information on which this example is based is taken from: Gruley, B. and Clough, R. (2020) How 3M plans to make more than a billion masks by the end of the year, *Bloomburg Businessweek*, 25 March; Technavio (2020) Coronavirus outbreaks boosts the sales of the world's top 10 N95 mask manufacturers, blog, 8 April.
2. The information on which this example is based is taken from: Heaven, W. (2020) Our weird behaviour during the pandemic is messing with AI models, *MIT Technology Review*, 11 May; S&P Global (2020) Industries most and least impacted by COVID-19 from a probability of default perspective, blog, 22 March; McKinsey & Company (2020) COVID-19: Briefing materials: Global health and crisis response, 6 July, https://www.mckinsey.com/~/media/mckinsey/business%20functions/risk/our%20insights/covid%2019%20implications%20for%20business/covid%2019%20july%209/covid-19-facts-and-insights-july-6.pdf (accessed September 2021).
3. The information on which this example is based is taken from: Das, A.K. (2019) Six bidders vie for Indian Railways ETCS Level 2 pilot project, *International Railway Journal*, 7 November; Rail Technology Magazine (2017) Network Rail awards landmark £150m ETCS signalling contract, 20 December; Railway Pro (2018) India to install ETCS Level 2 on its entire broad-gauge network, 8 March; Jha, S. (2018) Modi blocks Indian Railways ETCS plan, *International Railway Journal*, 11 April.
4. The information on which this example is based is taken from: Chong, A. (2019) What will it take for LTA's latest anti-congestion plan to work? *Channel News Asia*, International Edition, 13 May; Economist (2015) Squeezing in: what the London Underground reveals about work in the capital, *Economist* print edition, 23 May.
5. The information on which this example is based is taken from: Gadher, D. (2019) Art-lovers see red at surge pricing, *The Sunday Times*, 18 August; The Economist (2016) A fare shake: jacking up prices may not be the only way to balance supply and demand for taxis, *Economist*, 14 May; Dholakia, U.M. (2015) Everyone hates Uber's surge pricing – here's how to fix it, *Harvard Business Review*, 21 December.
6. The information on which this example is based is taken from: Cornelissen, J. and Cholakova, M. (2019) Profits Uber everything? The gig economy and the morality of category work, *Strategic Organisation*, 23 December; Russon, M. (2021) Uber drivers are workers not self-employed, Supreme Court rules, BBC News, 19 February; O'Brien, S. (2021) Uber's UK drivers to get paid vacation, pensions following Supreme Court ruling, *CNN Business*, 17 March; The New York Times (2018) What will New York do about its Uber problem?, 7 May.

Supplement to Chapter 11 Analytical queuing models

INTRODUCTION

In the main part of Chapter 11, we described how the queuing approach (in the United States it would be called the 'Waiting line approach') can be useful in thinking about capacity, especially in service operations. It is useful because it deals with the issue of variability, both of the arrival of customers (or items) at a process and of how long each customer (or item) takes to process. Where variability is present in a process (as it is in most processes, but particularly in service processes, the capacity required by an operation cannot easily be based on averages but must include the effects of the variation. Unfortunately, many of the formulae that can be used to understand queuing are extremely complicated, especially for complex systems, and are beyond the scope of this text. In fact, computer programs are almost always used now to predict the behaviour of queuing systems. However, studying queuing formulae can illustrate some useful characteristics of the way queuing systems behave.

Notation

Unfortunately, there are several different conventions for the notation used for different aspects of queuing system behaviour. It is always advisable to check the notation used by different authors before using their formulae. We shall use the following notation:

c_a = coefficient of variation of arrival times

c_e = coefficient of variation of process time

m = number of parallel servers at a station

r_a = arrival rate (items per unit time) = $1/t_a$

r_e = processing rate (items per unit time) = m/t_e

t_a = average time between arrival

t_e = mean processing time

u = utilisation of station = r_a/r_e = $(r_a\, t_e)/m$

W = expected waiting time in the system (queue time + processing time)

W_q = expected waiting time in the queue

WIP = average work in progress (number of items) in the queue

WIP_q = expected work in progress (number of times) in the queue

Some of these factors are explained later.

Variability

The concept of variability is central to understanding the behaviour of queues. If there were no variability there would be no need for queues to occur because the capacity of a process could be relatively easily adjusted to match demand. For example, suppose one member of staff (a server) serves customers at a bank counter who always arrive exactly every five minutes (i.e. 12 per hour). Also suppose that every customer takes exactly five minutes to be served, then because:

(a) the arrival rate is $\leq$ processing rate, and
(b) there is no variation,

no customer need ever wait because the next customer will arrive when, or after, the previous customer leaves. That is, $WIP_q = 0$.

Also, in this case, the server is working all the time, again because exactly as one customer leaves the next one is arriving. That is, $u = 1$.

Even with more than one server, the same may apply. For example, if the arrival time at the counter is five minutes (12 per hour) and the processing time for each customer is now always exactly 10 minutes, the counter would need two servers, and because:

(a) arrival rate is $\leq$ processing rate, and
(b) there is no variation,

again, $WIP_q = 0$, and $u = 1$.

Of course, it is convenient (but unusual) if arrival rate/processing rate equals a whole number. When this is not the case (for this simple example with no variation):

$$\text{Utilisation} = \text{processing rate}/(\text{arrival rate multiplied by } m)$$

For example, if arrival rate, r_a = 5 minutes

processing rate, r_e = 8 minutes

number of servers, m = 2

then, utilisation, $u = 8/(5 \times 2) = 0.8$ or 80%

Incorporating variability

The previous examples were not realistic because the assumption of no variation in arrival or processing times very rarely occurs (unless demand is carefully scheduled). We can calculate the average or mean arrival and process times but we also need to take into account the variation around these means. To do that we need to use a probability distribution – Figure 11.22 contrasts two processes with different arrival distributions. The units arriving are shown as people, but they could be jobs arriving at a machine, trucks needing servicing, or any other uncertain event. The top example shows low variation in arrival time where customers arrive in a relatively predictable manner. The bottom

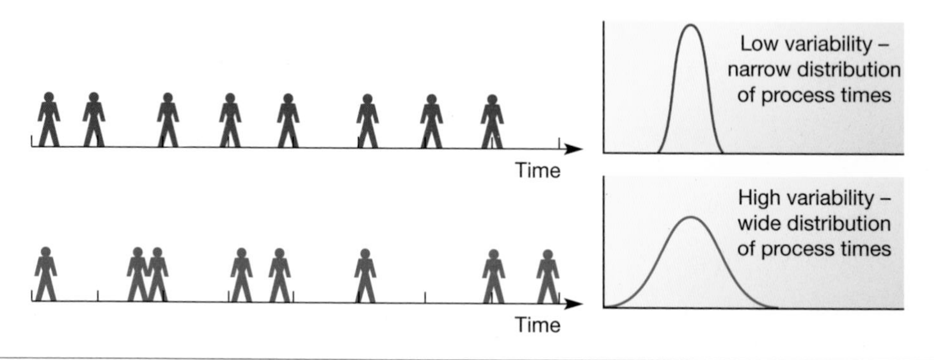

Figure 11.22 Low and high arrival variation

example has the same average number of customer arriving but this time they arrive unpredictably with sometimes long gaps between arrivals and at other times two or three customers arriving close together. Of course, we could do a similar analysis to describe processing times. Again, some would have low variation, some higher variation and others somewhere in-between.

In Figure 11.22, high arrival variation has a distribution with a wider spread (called 'dispersion') than the distribution describing lower variability. Statistically the usual measure for indicating the spread of a distribution is its standard deviation, σ. But variation does not only depend on standard deviation. For example, a distribution of arrival times may have a standard deviation of 2 minutes. This could indicate very little variation when the average arrival time is 60 minutes. But it would mean a very high degree of variation when the average arrival time is 3 minutes. Therefore, to normalise standard deviation, it is divided by the mean of its distribution. This measure is called the coefficient of variation of the distribution. So:

$$c_a = \text{coefficient of variation of arrival times} = \sigma_a/t_a$$

$$c_e = \text{coefficient of variation of processing times} = \sigma_e/t_e$$

Incorporating Little's law

In Chapter 6, we discussed one of the fundamental laws of processes that describes the relationship between the cycle time of a process (how often something emerges from the process), the work in progress within the process and the throughput time of the process (the total time it takes for an item, person or piece of information to move through the whole process, including waiting time). It was called Little's law and it was denoted by the following simple relationship:

$$\text{Throughput time} = \text{Work in progress} \times \text{Cycle time}$$

Therefore:

$$\text{Work in progress} = \text{Throughput time/Cycle time}$$

Or:

$$\text{WIP} = T/C$$

We can make use of Little's law to help understand queuing behaviour. Consider the queue in front of a station:

$$\text{Work in progress in the queue} = \text{the arrival rate at the queue (equivalent to 1/cycle time)} \times \text{waiting time in the queue (equivalent to throughput time)}$$

$$\text{WIP}_q = r_a \times W_q$$

and

$$\text{Waiting time in the whole system} = \text{the waiting time in the queue} + \text{the average process time at the station}$$

$$W = W_q + t_e$$

We will use this relationship later to investigate queuing behaviour.

Types of queuing system

Conventionally, queuing systems are characterised by four parameters, displayed as $A/B/m/b$:

A – the distribution of arrival times (or more properly interarrival times, the elapsed times between arrivals)

B – the distribution of process times

m – the number of servers at each station

b – the maximum number of items (or people) allowed in the system.

The most common distributions used to describe *A* or *B* are either:

(a) the exponential (or Markovian) distribution denoted by M; or
(b) the general (for example normal) distribution denoted by G.

So, for example, an M/G/1/5 queuing system would indicate a system with exponentially distributed arrivals, process times described by a general distribution such as a normal distribution, with one server and a maximum number of items (or people) allowed in the system of 5. This type of notation is called Kendall's notation.

Queuing theory can help us investigate any type of queuing system, but in order to simplify the mathematics, we shall here deal only with the two most common situations. Namely:

M/M/*m* – the exponential arrival and processing times with *m* servers and no maximum limit to the queue;

G/G/*m* – general arrival and processing distributions with *m* servers and no limit to the queue.

And first we will start by looking at the simple case when *m* (number of servers) = 1.

M/M/1 queuing systems

For M/M/1 queuing systems, the formulae are as follows:

$$\text{WIP} = \frac{u}{1 - u}$$

Using Little's law,

$$\text{WIP} = \text{Cycle time} \times \text{Throughput time}$$

$$\text{Throughput time} = \text{WIP/Cycle time}$$

Then:

$$\text{Throughput time} = \frac{u}{1 - u} \times \frac{1}{r_a} = \frac{t_e}{1 - u}$$

and since, throughput time in the queue = total throughput time − average processing time, then:

$$\begin{aligned} W_q &= W - t_e \\ &= \frac{t_e}{1 - u} - t_e \\ &= \frac{t_e - t_e(1 - u)}{1 - u} = \frac{t_e - t_e - ut_e}{1 - u} \\ &= \frac{u}{(1 - u)} t_e \end{aligned}$$

again, using Little's law:

$$\text{WIP}_q = r_a \times W_q = \frac{u}{(1 - u)} t_e r_a$$

and since,

$$u = \frac{r_a}{r_e} = r_a t_e$$

$$r_a = \frac{u}{t_e}$$

then,

$$\text{WIP}_q = \frac{u}{(1-u)} \times t_e \times \frac{u}{t_e}$$

$$= \frac{u^2}{(1-u)}$$

M/M/*m* systems

When there are *m* servers at a station the formula for waiting time in the queue (and therefore all other formulae) needs to be modified. Again, we will not derive these formulae but just state them:

$$W_q = \frac{u^{\sqrt{2(m+1)}-1}}{m(1-u)} t_e$$

From which the other formulae can be derived as before.

Worked example

A bank wishes to decide how many staff to schedule during its lunch period. During this period customers arrive at a rate of 9 per hour and the enquiries that customers have (such as opening new accounts, arranging loans, etc.) take on average 15 minutes to deal with. The bank manager feels that four staff should be on duty during this period but wants to make sure that the customers do not wait more than 3 minutes on average before they are served. The manager has been told that the distributions that describe both arrival and processing times are likely to be exponential. Therefore,

$r_a = 9$ per hour, therefore

$t_a = 6.67$ minutes

$r_e = 4$ per hour, therefore

$t_e = 15$ minutes

The proposed number of servers

$$W_q = \frac{u^{\sqrt{2(m+1)}-1}}{m(1-u)} t_e$$

$$m = 4$$

therefore, the utilisation of the system, $u = 9/(4 \times 4) = 0.5625$.

From the formula for waiting time for a M/M/*m* system,

$$W_q = \frac{0.5625^{\sqrt{10}-1}}{4(1-0.5625)} \times 0.25$$

$$= \frac{0.5625^{2.162}}{1.75} \times 0.25$$

$$= 0.042 \text{ hours}$$

$$= 2.52 \text{ minutes}$$

Therefore, the average waiting time with 4 servers would be 2.52 minutes, which is well within the manager's acceptable waiting tolerance.

G/G/1 systems

While the assumption of exponential arrival and processing times used in M/M/*m* systems above are convenient as far as mathematical derivation. However, in practice, process times in particular are rarely truly exponential. This is why it is important to have some idea of how a G/G/1 and G/G/M queue behaves, where it is assumed that arrivals and processing follow a normal distribution. However, exact mathematical relationships are not possible with such distributions. Therefore, some kind of approximation is needed. The one here is in common use, and although it is not always accurate, it is for practical purposes. For G/G/1 systems the formula for waiting time in the queue is as follows:

$$W_q = \left(\frac{c_a^2 + c_e^2}{2}\right)\left(\frac{u}{(1 - u)}\right) t_e$$

There are two points to make about this equation. The first is that it is exactly the same as the equivalent equation for an M/M/1 system but with a factor to take account of the variability of the arrival and process times. The second is that this formula is sometimes known as the VUT formula because it describes the waiting time in a queue as a function of:

V – the variability in the queuing system

U – the utilisation of the queuing system (that is, demand versus capacity), and

T – the processing times at the station.

In other words, we can reach the intuitive conclusion that queuing time will increase as variability, utilisation or processing time increase.

For G/G/*m* systems

The same modification applies to queuing systems using general equations and *m* servers. The formula for waiting time in the queue is now as follows:

$$W_q = \left(\frac{c_a^2 + c_e^2}{2}\right)\left(\frac{u^{\sqrt{2(m+1)}-1}}{m(1 - u)}\right) t_e$$

Worked example

'I can't understand it. We have worked out our capacity figures and I am sure that one member of staff should be able to cope with the demand. We know that customers arrive at a rate of around 6 per hour and we also know that any trained member of staff can process them at a rate of 8 per hour. So why is the queue so large and the wait so long? Have a look at what is going on there please.'

Sarah knew that it was probably the variation, both in customers arriving and in how long it took each of them to be processed, that was causing the problem. Over a two-day period when she was told that demand was more or less normal, she timed the exact arrival times and processing times of every customer. Her results were as follows:

$$\text{The coefficient of variation, } c_a \text{ of customer arrivals} = 1$$

$$\text{The coefficient of variation, } c_e \text{ of processing time} = 3.5$$

$$\text{The average arrival rate of customers, } r_a = 6 \text{ per hour,}$$

$$\text{therefore, the average interarrival time} = 10 \text{ minutes}$$

$$\text{The average processing rate, } r_e = 8 \text{ per hour}$$

$$\text{therefore, the average processing time} = 7.5 \text{ minutes}$$

$$\text{Therefore, the utilisation of the single server, } u = 6/8 = 0.75$$

Using the waiting time formula for a G/G/1 queuing system:

$$W_q = \left(\frac{1 + 12.25}{2}\right)\left(\frac{0.75}{1 - 0.75}\right)7.5$$

$$= 6.625 \times 3 \times 7.5 = 149.06 \text{ mins}$$

$$= 2.48 \text{ hours}$$

Also because,

$$WIP_q = \text{arrival rate } (r_a) \times \text{waiting time in queue } (W_q)$$

$$WIP_q = 6 \times 2.48 = 14.68$$

So, Sarah had found out that the average wait that customers could expect was 2.48 hours and that there would be an average of 14.68 people in the queue.

'Ok, so I see that it's the very high variation in the processing time that is causing the queue to build up. How about investing in a new computer system that would standardised processing time to a greater degree? I have been taking with our technical people and they reckon that, if we invested in a new system, we could cut the coefficient of variation of processing time down to 1.5. What kind of a different would this make?'

Under these conditions with $c_e = 1.5$:

$$W_q = \left(\frac{1 + 2.25}{2}\right)\left(\frac{0.75}{1 - 0.75}\right)7.5$$

$$= 1.625 \times 3 \times 7.5 = 36.56 \text{ mins}$$

$$= 0.61 \text{ hours}$$

Therefore,

$$WIP_q = 6 \times 0.61 = 3.66$$

In other words, reducing the variation of the process time has reduced average queuing time from 2.48 hours down to 0.61 hours and has reduced the expected number of people in the queue from 14.68 down to 3.66.

12 Supply chain management

KEY QUESTIONS

12.1 What is supply chain management?

12.2 How should supply chains compete?

12.3 How should relationships in supply chains be managed?

12.4 How is the supply side managed?

12.5 How is the demand side managed?

12.6 What are the dynamics of supply chains?

INTRODUCTION

How is it that businesses such as IKEA, JD.com, Amazon, Zara, Spotify, Singapore Airlines, Google, Apple, 7-Eleven Japan and Maersk achieve notable results in highly competitive markets? Partly, it is down to their services and products, but it is also a result of excellence in managing their supply networks. Given the typically large (and increasing) proportion of activities that are outsourced by operations, managing supply networks is a particularly vital activity. In Chapter 5 we explored the structure and scope of operations; by contrast, this chapter is more concerned with how supply chains and networks are subsequently managed. This involves determining key performance objectives for the supply network, deciding on supplier relationships (transactional versus partnership), developing sourcing strategies for different products and services, selecting appropriate suppliers, negotiating the terms of their engagement, managing day-to-day supply, improving suppliers' capabilities over time, and attempting to mitigate supply chain dynamics. Figure 12.1 shows where project management fits in the overall model of operations management.

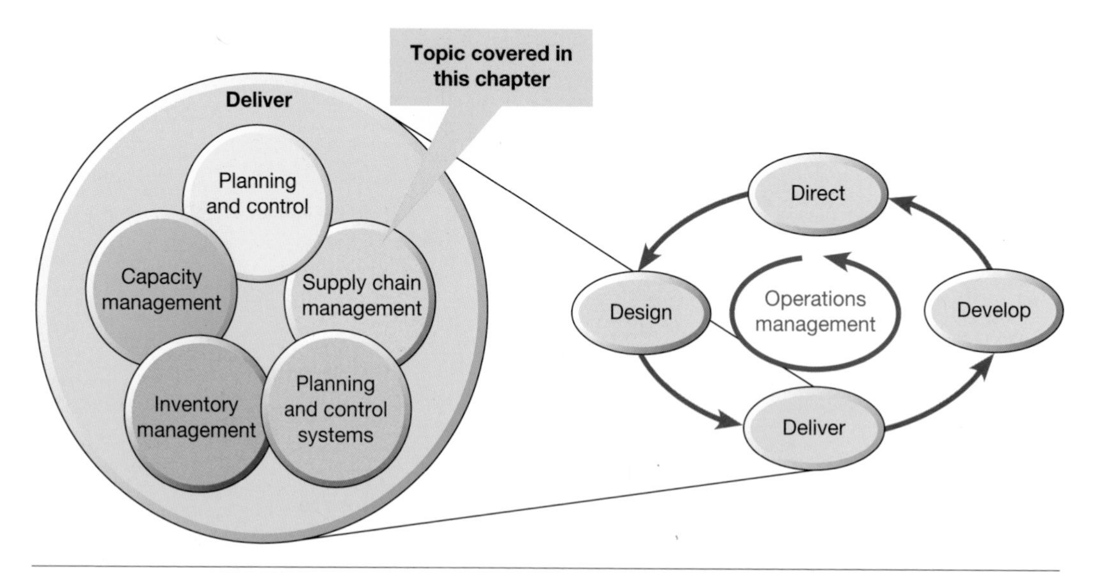

Figure 12.1 This chapter examines supply chain management

12.1 What is supply chain management?

Supply chain management (SCM) is the management of the relationships and flows between the 'string' (or chain) of operations and processes that deliver value in the form of services and products to the ultimate consumer. As illustrated in Figures 12.2 and 12.3, supply *chains* are technically different to supply *networks*. In large supply networks there can be many hundreds of supply chains of linked operations passing through a single operation. Confusingly, the terms supply network and supply chain are often (mistakenly) used interchangeably. It is also worth noting that the 'flows' in supply chains are not restricted to the downstream flow of products and services from suppliers through to customers. Although the most obvious failures in supply chain management occur when downstream product or service flows fail to meet customer requirements, the root causes may be failures in the upstream flows of *information*. As such, supply chain management is as much concerned with managing information flows (upstream and downstream) as it is with managing the flow of products and services.

Supply chain management applies to non-physical flow

Most books, blogs and articles that focus on the challenges of supply chain management continue to focus on 'material transformation' operations – operations that are concerned with the creation, movement, storage or sale of physical products. However, supply chain management applies equally

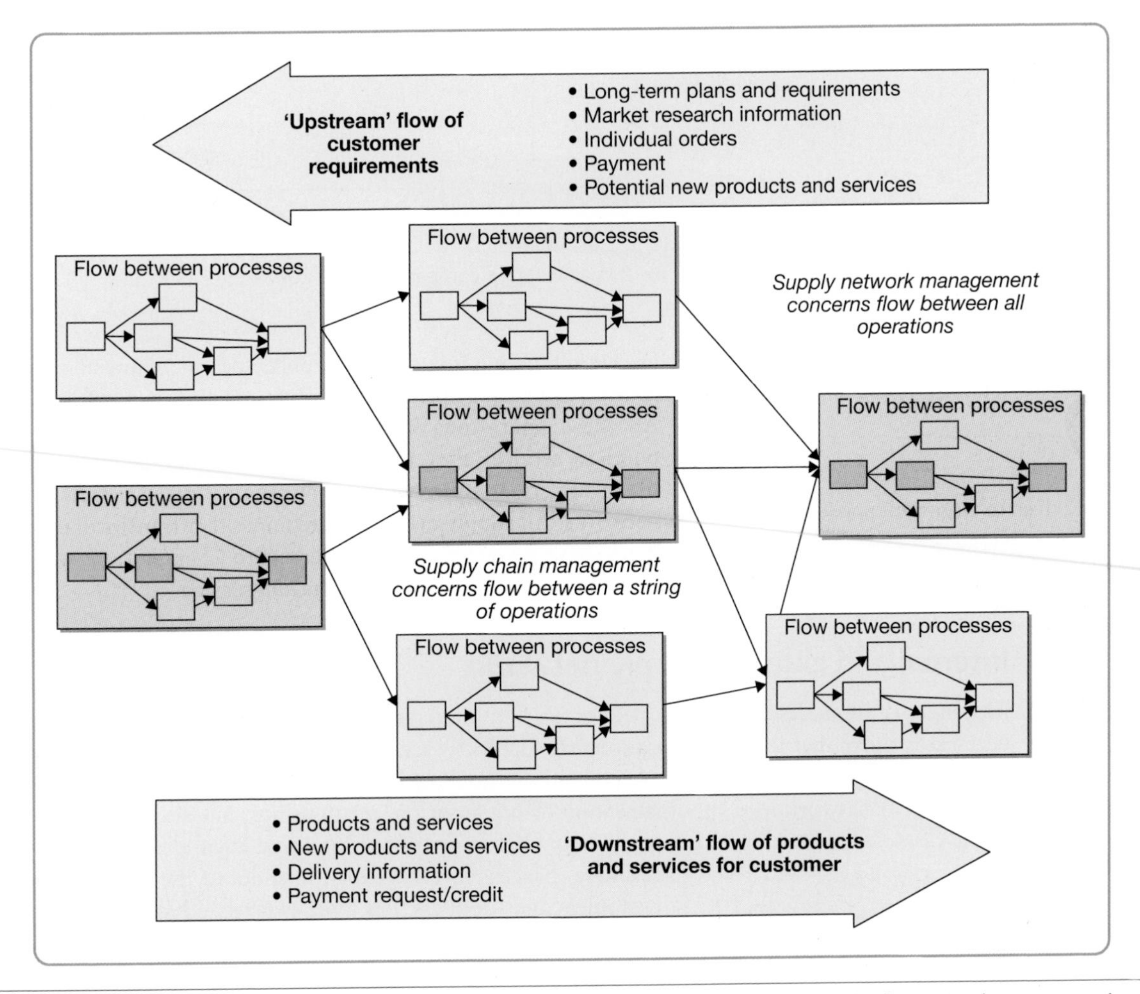

Figure 12.2 Supply chain management is concerned with managing the flow of materials and information between a string of operations that form the strands or 'chains' of a supply network

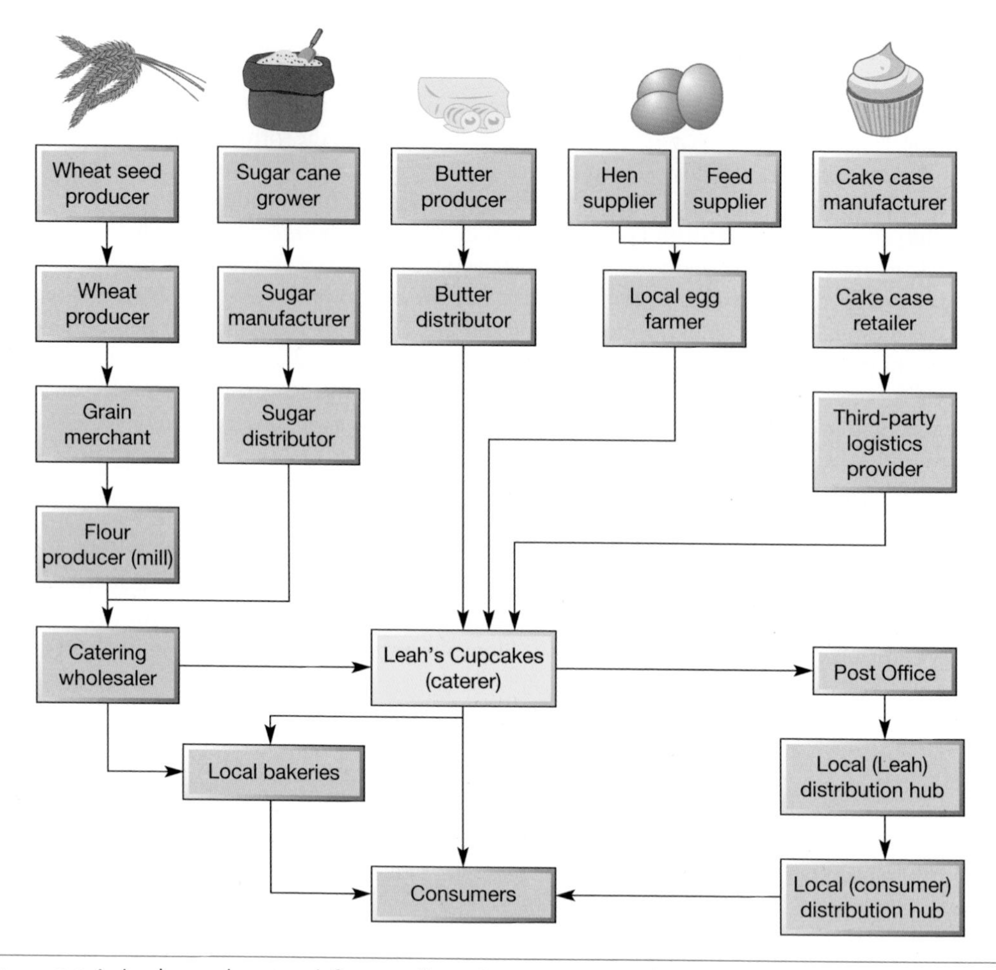

Figure 12.3 A simple supply network for a small catering company (Leah's Cupcakes)

to operations with largely or exclusively intangible inputs and outputs; such as financial services, retail shopping malls, insurance providers, healthcare operations, consultants, universities and so on. All these operations have suppliers and customers, they all purchase services, they all must choose how they get their services to consumers. In other words, they all must manage their supply networks. All supply networks, even ones that transform physical items, have service elements – referring to Chapter 1, most operations supply a mixture of products *and* services.

Operations principle

Supply chain management applies equally to non-physical flows between operations and processes as well as physical flows.

Internal and external supply networks

Although we often describe supply networks as an interconnection of 'organisations', this does not necessarily mean that these 'organisations' are distinctly separate entities belonging to and managed by different owners. In Chapter 1, we pointed out how the idea of networks can be applied, not just at the supply network level of 'organisation-to-organisation' relationships, but also at the 'process-to-process' within-operation level and even at the 'resource-to-resource' process level. We also introduced the idea of internal customers and suppliers. Put these two related ideas together and one can understand how many of the issues that we will be discussing in the context of 'organisation-to-organisation' supply networks can also provide insight for internal 'process-to-process' supply networks.

Operations principle

Supply chain management concepts apply to internal relationships between processes as well as the external relationships between operations.

OPERATIONS IN PRACTICE

Zipline's drone-enabled supply network[1]

When drones started to be deployed in supply chain management, one of their most successful users was not a big logistics provider but Zipline, who built the world's largest autonomous logistics networks, operating at national scale across multiple territories, delivering blood and medical supplies in parts of the Global South. In Rwanda, Zipline helped establish a valuable alternative to road transportation for high-value, low-weight items. Although the country was investing in its transport links, over 80 per cent of its road network remained unsurfaced, making road travel challenging at best, and almost impossible during the country's rainy season. For critical medical supplies, where delivery delays meant significant negative patient outcomes, the problem was acute. Also, medical supplies were hard to forecast or had short shelf lives, ideal for drone delivery, as were orders that dipped into safety stocks of drugs. With a small fleet of autonomous drones, each with carrying capacity of 1.5 kg (equivalent to 3 × 500 ml of blood) and capable of delivering anywhere within a 22,500-square-kilometre service area, Zipline was able to establish a highly effective delivery network for medical supplies, capable of reaching over half of Rwanda's 12 million population.

Given the urgency of many of its orders, the firm focused on reducing the time to get drones airborne. One innovation was moving the GPS circuitry from the drone to its battery, enabling continuous GPS connection, saving precious minutes by establishing a stable signal prior to launch. Modular construction enabled orders to be placed inside the fuselage, which was put on the launcher, after which its wing section and battery module were attached. This made the drone easier to handle and ensured that problems picked up in pre-flight checks could be resolved quickly by switching out a faulty module. The often lengthy pre-flight checking was reduced by using a mobile inspection application and computer vision algorithms. The overall effect was to reduce the time between order receipt and drone launch to just five minutes. After which, a motorised launcher accelerated the drones up to their maximum speed of 100 km per hour in just 0.3 seconds. Not only was speed important, so was *dependability* of service. Having a back-up for each of the critical components on the drone, in case the primary component failed, helped reduce failed deliveries. At its destination, the drone would simply drop its medical supplies box using a parachute made from wax paper and biodegradable tape, removing the need for delivery site infrastructure.

For some clients, Zipline operated a 'cross-docking' solution. Here, a client would prepare its packages which would then be consolidated and sent to Zipline's distribution centres. The firm would then schedule delivery with end recipients at a time that was convenient for them. Other clients preferred to use Zipline as a third-party logistics (3PL) provider. This involved the firm receiving inventory from its clients and holding it in its distribution centres ready to pack and ship rapidly as soon as orders were received. Ultimately, Zipline's aim was to create a supply network that was both efficient (lean) when it could be, but responsive (agile) when it needed to be. Higher costs of delivery using drones were inevitable, but these were partially offset against savings from reduced inventory in the supply network and less product obsolescence. In addition, the benefits in terms of speed, dependability and, most critically, patient health were substantial.

12.2 How should supply chains compete?

Supply chain management shares one common, and central, objective – to satisfy the end customer. All stages in the various chains that form the supply network must eventually include consideration of the final customer, no matter how far an individual operation is from the end customer. When a customer decides to make a purchase, they trigger action back along a whole series of supply chains in the network. Thus, each operation in each chain should be satisfying its own customer, but also making sure that eventually the end customer is also satisfied.

OPERATIONS IN PRACTICE

Supply chain excellence at JD.com – the rise of an e-commerce titan[2]

Alongside Amazon, Alibaba and Suning.com, JD.com is now established as one of the largest and most successful e-commerce companies worldwide. With revenues exceeding 745 billion yuan (over $114 billion), the company offers a range of around 6 million products to over 500 million active customers, predominantly within China (up from 362 million in 2019).

Underpinning JD.com's rapid growth in the last decade is a highly sophisticated data-driven supply network with a global reputation for speed and dependability. With over 1,000 warehouses covering approximately 21 million square metres, JD.com offers warehousing, cold chain delivery, cross-border distribution and inventory analysis to brands from around the world, enabling them to reach JD.com customers even when they have no physical presence in China. However, it is not simply the scale of JD.com's operations that is impressive, it is the significant emphasis it places on technology and innovation within its supply network. For example. within its warehousing, the firm have invested heavily in robotics and automation to organise, pack and distribute up to 100 million packages each day. The company completed construction of one of the world's first fully automated warehouses in Shanghai. It also led in the development of several forms of robotic delivery, including driverless trucks, and investment in the largest drone distribution infrastructure in the world, with 150 hubs aimed at delivering products to more rural areas.

Importantly, the digitisation of JD.com's supply network has generated huge amounts of data – a staggering 30 petabytes every day! The firm has been able to leverage this data in the development of its machine learning and artificial intelligence capabilities, giving it significant insights into both consumer behaviour and product movement within its network. To support the continued development of its 'smart supply chain', JD.com set up a research lab focused on logistics and automation – JD-X. Its role was to focus on the continued development of automated fulfilment, blockchain technology, IoT, image and vision recognition, and deep learning technologies. In addition, significant investments were made in the development of algorithms that predict trends in consumer demand more accurately (for example, using speech recognition and voice fingerprinting technology). In fact, the firm's expertise in supply chain fulfilment became so strong, it led to the creation of JD Logistics, a business group within the parent company offering smart supply chain services to businesses in a wide range of industries. In recognition of its excellence, JD.com was selected as a finalist (alongside the UN World Food Programme, Alibaba, Lenovo, OCP and Memorial Sloan Kettering) for one of the most prestigious awards in advanced analytics and operations research, the INFORMS Franz Edelman Award.

Performance objectives for supply networks

The objectives of supply chain management are like those for individual operations – to deliver services and products to end customers that meet their expectations in terms of quality, speed, dependability, flexibility, cost and sustainability.

Quality

The quality of a service or product when it reaches the customer is a function of the quality performance of every operation in the chain that supplied it. The implication of this is that errors in each stage of the chain can multiply in their effect on end-customer service. For example, if each of seven stages in a supply chain has a 1 per cent error rate, only 93.2 per cent of products or services will be of good quality on reaching the end customer (0.99^7). Therefore, only by every stage taking some responsibility for its own *and its suppliers'* performance, can a supply network achieve high end-customer quality.

Speed

Speed has two meanings in a supply chain context. The first is how fast customers can be served, (the elapsed time between a customer requesting a service and receiving it in full). However, fast customer response can be achieved simply by over-resourcing or overstocking within the supply chain. For example, very large stocks in a retail operation can reduce the chances of stock-out to almost zero, so reducing customer waiting time virtually to zero. Similarly, an accounting firm may be able to respond quickly to customer demand by having a very large number of accountants on standby waiting for demand spikes. An alternative perspective on speed is the time taken for goods and services to *move through the chain*. So, for example, products that move quickly down a supply chain from raw material suppliers through to retailers will spend little time as inventory, because to achieve fast throughput time, material cannot dwell for significant periods as inventory. This in turn reduces the working capital requirements and other inventory costs in the supply chain, so reducing the overall cost of delivering to the end customer. Achieving a balance between speed as responsiveness to customers' demands and speed as fast throughput (although they are not incompatible) will depend on how the supply network is choosing to compete.

Dependability

Dependability in a supply chain context is similar to speed in so much as one can almost guarantee 'on-time' delivery by keeping excessive resources within the chain. However, dependability of throughput time is a much more desirable aim because it reduces uncertainty within the network. If individual operations do not deliver as promised on time, there will be a tendency for customers to over-order, or order early, to provide insurance against late delivery. The same argument applies if there is uncertainty regarding the *quantity* of services or products. Therefore, delivery dependability is often measured as 'on time, in full' in supply chains.

Flexibility

In a supply chain context, flexibility is usually taken to mean the network's ability to cope with changes and disturbances. Very often this is referred to as supply chain agility. The concept of agility includes previously discussed issues such as focusing on the end customer and ensuring fast throughput and responsiveness to customer needs. But, in addition, agile supply chains are sufficiently flexible to cope with changes, either in customer demand or in the supply capabilities of operations within the chain.

Cost

In addition to the costs incurred within each operation to transform its inputs into outputs, the supply network as a whole incurs additional costs that derive from operations doing business with one another. These 'transaction' costs include the costs of finding appropriate suppliers, setting up supplier agreements (e.g. tendering, negotiation and contracting), running ongoing supply (e.g. ordering, expediting, monitoring supply performance, invoicing and payment), dealing with failure, supplier training, and potentially the costs of exiting an unsatisfactory relationship. It is important to consider the transaction costs of trading with a supplier as opposed to simply the price of the service or product. (For more on this issue, see critical commentary later in this chapter on approaches to supplier selection.) Many of the recent developments in supply chain management, such as partnership agreements or reducing the number of suppliers, are an attempt to minimise such transaction costs.

Sustainability

Any organisation that subscribes to sustainability objectives will want to make sure that it purchases its input products and services from suppliers that are similarly responsible. This may involve, for example, buying from local suppliers where possible, sourcing supplies from suppliers with ethical practices, choosing environmentally friendly products and services, improving working conditions within supply chains, or moving away from (environmentally unfriendly) air freight transportation methods.

OPERATIONS IN PRACTICE

Twice around the world for Wimbledon's tennis balls

(Our thanks to Professor Mark Johnson of Warwick Business School for this example)

The Wimbledon 'Grand Slam' tennis tournament is the oldest and arguably the most prestigious tennis tournament in the world. Slazenger, the sports equipment manufacturer, has been the official ball supplier for Wimbledon since 1902. Yet those balls used at Wimbledon, and the materials from which they are made, will have travelled 81,385 kilometres between 11 countries and across four continents before they reach Centre Court. Professor Mark Johnson, of Warwick Business School said: *'It is one of the longest journeys I have seen for a product. On the face of it, travelling more than 80,000 kilometres to make a tennis ball does seem fairly ludicrous, but it just shows the global nature of production, and in the end, this will be the most cost-effective way of making tennis balls. Slazenger are locating production near the primary source of their materials in Bataan in the Philippines, where labour is also relatively low cost'.*

The complex supply chain is illustrated in Figure 12.4. It sees clay shipped from South Carolina in the United States, silica from Greece, magnesium carbonate from Japan, zinc oxide from Thailand, sulphur from South Korea, and rubber from Malaysia to Bataan where the rubber is vulcanised – a chemical process for making the rubber more durable. Wool is then shipped from New Zealand to the United Kingdom, where it is pressed into felt and then flown back to Bataan in the Philippines. Meanwhile, petroleum naphthalene from Zibo in China and glue from Basilan in the Philippines are brought to Bataan where Slazenger manufacture the ball. Finally, the tins, which contain the balls, are shipped in from Indonesia and once the balls have been packaged, they are sent to Wimbledon. *'Slazenger shut down the factory in the UK years ago and moved the equipment to Bataan in the Philippines'*, says Professor Johnson. *'They still get the felt from Stroud [UK], as it requires a bit more technical expertise. Shipping wool from New Zealand to Stroud and then sending the felt back to the Philippines adds a lot of miles, but they obviously want to use the best wool for the Wimbledon balls'.*

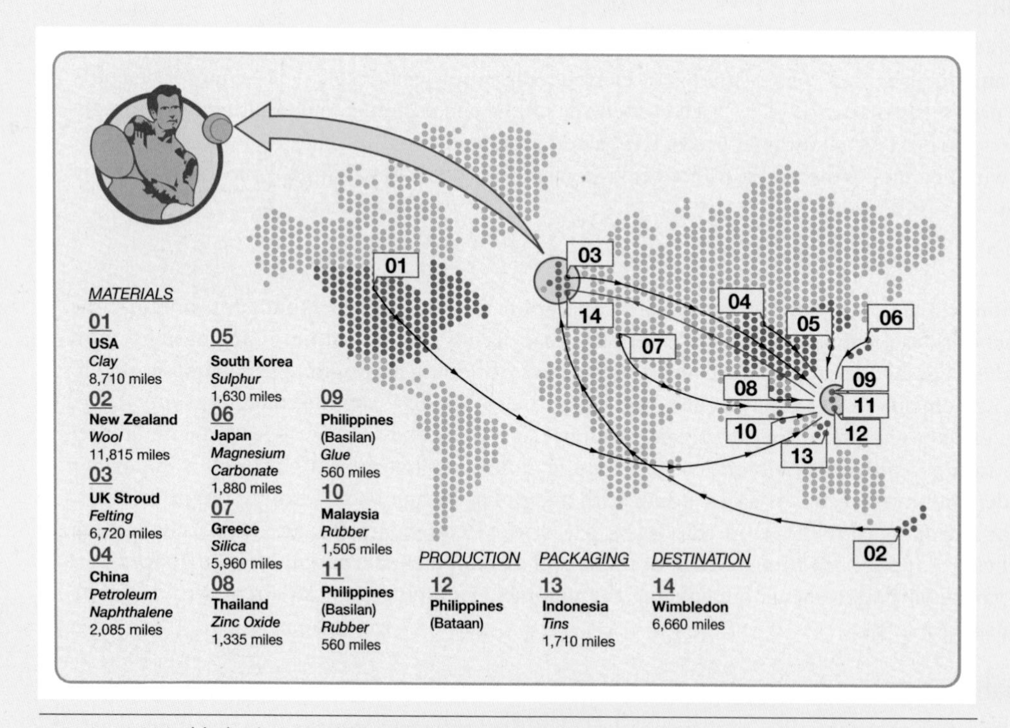

Figure 12.4 Wimbledon's tennis balls travel over 80,000 kilometres in their supply network

Lean versus agile supply networks

As discussed in Chapters 1 and 2, different services or products often exhibit clear differences in how they compete in markets. One popular approach used in supply chain management is to distinguish between services or products that are 'functional' and those that are 'innovative'. Functional services or products have relatively stable and predictable demand, while demand for innovative services or products will be far more uncertain. In addition, the profit margins are typically much higher for innovative, as opposed to functional, offerings. Even within the same company, there may be *both* functional and innovative categories in evidence. For example, some of the work carried out by consultants is very standardised, with just small variations from one client to the next, while other work is highly tailored for each specific project.[3] Hospitals have routine 'standardised' surgical procedures, such as cataract removal, but also must provide very customised emergency post-trauma surgery. Shoe manufacturers may sell classics that change little over time, alongside fashion shoes that last a single season.

Depending on the nature of the product or service, different supply networks are likely to be more suitable (an idea originally proposed by Professor Marshall Fisher).[4] For functional offerings, efficient (or lean) supply chain policies are typically most appropriate. This includes keeping service capacity or inventories low, especially in the downstream parts of the network, to maintain fast throughput. There is also a significant focus on maximising utilisation of all resources in the supply network to minimise costs. Information must flow quickly up and down the chain to maximise the amount of time to adjust schedules efficiently.

In contrast, innovative offerings are more suited to responsive (or agile) supply chain policies. Here, the emphasis is on high service levels and responsive supply to the end customer. The service capacity or inventory in the network will be deployed as close as possible to the customer. In this way, the network can supply even when dramatic changes occur in customer demand. Fast throughput from the upstream parts of the network will still be needed to replenish downstream operations. Figure 12.5

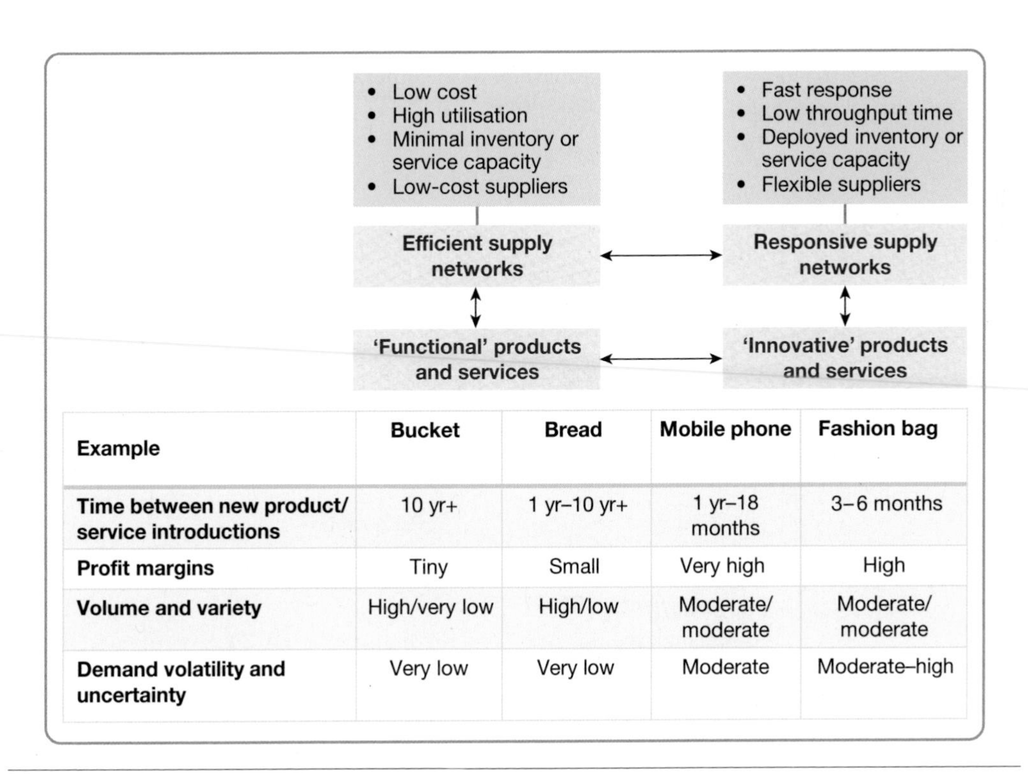

Example	Bucket	Bread	Mobile phone	Fashion bag
Time between new product/ service introductions	10 yr+	1 yr–10 yr+	1 yr–18 months	3–6 months
Profit margins	Tiny	Small	Very high	High
Volume and variety	High/very low	High/low	Moderate/ moderate	Moderate/ moderate
Demand volatility and uncertainty	Very low	Very low	Moderate	Moderate–high

Figure 12.5 Aligning product and service characteristics with supply network design

Operations principle

'Functional' products and services suit efficient (lean) supply chain management; 'innovative' products and services suit responsive (agile) supply chain management.

outlines some of the key characteristics of functional versus innovative offerings. It also illustrates the natural alignment between these characteristics and the type of supply network that is most appropriate. In practice, some firms take a more hybrid approach in operating their supply networks. For example, Inditex, one of the world's largest and most successful fashion retailers, operates a predominantly agile supply network, but still has aspects of lean within some of its activities. We provide more detail on lean in Chapter 16.

12.3 How should relationships in supply chains be managed?

The 'relationship' between operations in a supply network is the basis on which the exchange of services, products, information and money is conducted. Managing supply chains is about managing relationships, because relationships influence the smooth flow between operations and processes. Different forms of relationship will be appropriate in different circumstances. Two dimensions are particularly important – *what* the company chooses to outsource, and *who* it chooses to supply it. In terms of *what* is outsourced, key questions are 'how many activities should be outsourced?' (doing everything in-house at one extreme, to outsourcing everything at the other extreme) and 'how important are the outsourced activities?' (outsourcing only trivial activities at one extreme, to outsourcing even core activities at the other extreme). We discussed these in detail in Chapter 5 when exploring the scope of supply networks.

When dealing with the question of *who* is chosen to supply services and products, again two questions are important, 'how many suppliers will be used by the operation?' (using many suppliers to perform the same set of activities at one extreme, through to using only one supplier for each activity at the other) and 'how close are the relationships?' (from 'arm's length' relationships at one extreme, through to close and intimate relationships at the other). Figure 12.6 illustrates this way of characterising relationships.

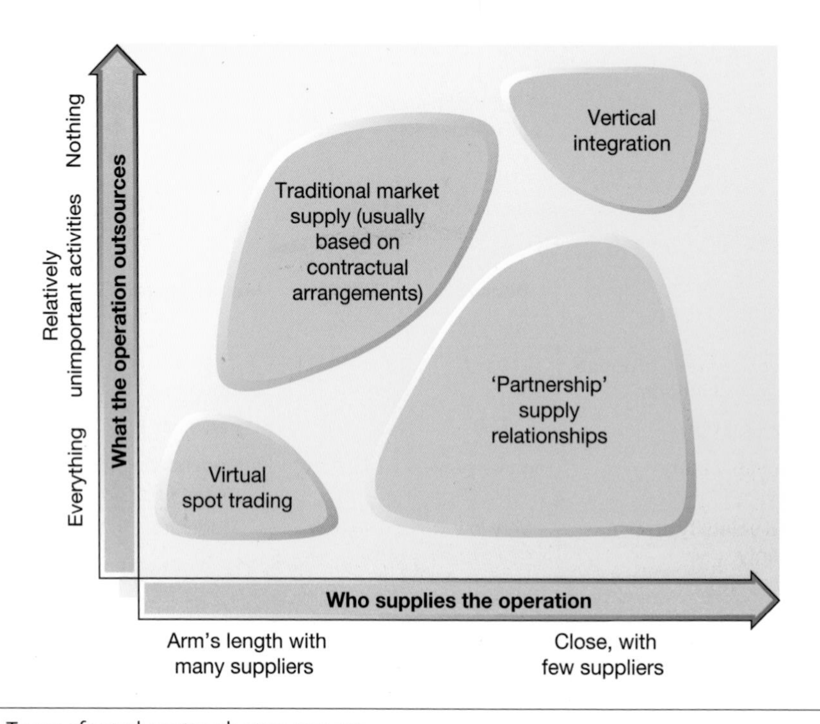

Figure 12.6 Types of supply network arrangement

Table 12.1 Benefits of transactional and partnership supply chain relationships

Transactional relationship benefits	Partnership relationship benefits
▶ Retains competition between alternative suppliers ▶ If demand changes, customer can change the number and type of suppliers; a faster and cheaper alternative to redirecting internal activities ▶ Wider variety of innovation sources (though access may be harder than partnership relationships) ▶ Useful in assessing a supplier before potential move to partnership model ▶ Good for one-off or irregular purchases	▶ Higher levels of loyalty to customers than transactional relationships ▶ Reduced time and effort of frequent re-contracting ▶ Reduced transaction costs (though dependent on the nature of sourcing) ▶ Reduced cost of compliance monitoring ▶ Fewer quality failures and unanticipated failures ▶ Earlier failure identification ▶ Emphasis on joint problem-solving during failure episodes rather than assigning blame ▶ Generation of increased value by leveraging shared competencies

'Transactional' versus 'partnership' relationships

Transactional (or contractual) relationships involve purchasing services and products in a 'pure' market fashion, often seeking the 'best' supplier every time it is necessary to make a purchase. Transactional relationships can be either long- or short-term, but there is no guarantee of anything beyond the immediate contract. They are appropriate when short-term benefits are important. In contrast, partnership (or collaborative) relationships are longer-term and involve a commitment to work together over time to gain mutual advantage. These relationships emphasise cooperation, frequent interaction, information sharing, joint problem-solving and sometimes even profit sharing. True partnerships imply mutual benefit, and often mutual sacrifice. If it is not in the culture of a business to give up some freedom of action, it is very unlikely to ever make a success of partnerships. Table 12.1 outlines some of the key benefits of the two extremes.

It is very unlikely that any business will find it sensible to engage exclusively in one type of relationship or another. Most businesses will have a portfolio of, possibly, widely differing relationships. In addition, there are degrees to which any particular relationship can be managed on a transactional or partnership basis. The real question is: Where, on the spectrum from transactional to partnership, should each relationship be positioned? While there is no simple formula for choosing the 'ideal' form of relationship, there are some important factors that can sway the decision. The most obvious issue will concern how a business intends to compete in its marketplace. If price is the main competitive factor, then the relationship could be determined by which approach offers the highest potential savings. On one hand, market-based contractual relationships could minimise the actual price paid for purchased services, while partnerships could minimise the transaction costs of doing business. If a business is competing primarily on product or service innovation, the type of relationship may depend on where innovation is likely to happen. If innovation depends on close collaboration between supplier and customer, partnership relationships are needed. On the other hand, if suppliers are busily competing to outdo each other in terms of their innovations, and especially if the market is turbulent and fast growing (as with many software and internet-based industries), then it may be preferable to retain the freedom to change suppliers quickly using market mechanisms.

Operations principle
All supply chain relationships can be described by the balance between their 'transactional' and 'partnership' elements.

12.4 How is the supply side managed?

Once a decision has been made to buy services or products (as opposed to doing activities in-house), managers must decide on sourcing strategies for different products and services, select appropriate suppliers, manage ongoing supply and improve suppliers' capabilities over time. These activities are usually the responsibility of the purchasing or procurement function within the business. Purchasing should provide a vital link between the operation itself and its suppliers.

Sourcing strategy

In Chapter 5, we outlined a number of issues concerning the configuration of a supply network. Changing the shape of the supply network may involve reducing the number of suppliers to the operation so as to develop closer relationships, and bypassing or disintermediating operations in the network. Here, we go a bit further by examining four key sourcing approaches – *multiple sourcing, single sourcing, delegated sourcing* and *parallel sourcing*.

Multiple sourcing involves obtaining a product or service component from more than one supplier. It is commonly seen in competitive markets where switching costs are low and performance objectives are primarily focused on price and dependability. Multiple sourcing can help maintain competition in the supply market, reduce supply risk and increase flexibility in the face of supplier failure or changes in customer demand. In addition, some firms like to multi-source to prevent supplier dependence, thus allowing for changes in purchase volumes without the risk of supplier bankruptcy. However, the disadvantage of multiple sourcing is that it becomes hard to encourage supplier commitment and as such limits the opportunity to develop a partnership approach to supply chain management.

Single sourcing involves buying all of one product or service component from a single supplier. Often these represent a high proportion of total spend or are of strategic importance. In other cases, however, firms simply prefer the simplicity (and reduced transaction costs) of single sourcing. Many single-source arrangements have a longer-term focus than multiple-sourcing arrangements and focus on a wider range of performance objectives. However, single-source arrangements can carry an increased risk of lock-in and a reduction in the buyer's bargaining power.

Delegated sourcing involves a tiered approach to managing supplier relationships. This means that one supplier is responsible for delivering a package of services as opposed to an individual service, or an entire sub-assembly as opposed to a single part. This has the advantage of reducing the number of tier-one suppliers significantly while simultaneously allowing a focus on strategic partners. However, delegated sourcing can alter the dynamics of the supply market and risk creating 'mega-suppliers' with significant power in the network.

Parallel sourcing has the aim of providing the advantages of both multiple sourcing and single sourcing simultaneously. It involves having single-source relationships for services for different service packages. If a supplier is deemed unsatisfactory, it is possible to switch to the alternative supplier that currently provides the *same* service but for a *different* service package. The advantage of this sourcing approach is that it maintains competition and allows for switching. However, managing delegated sourcing arrangements is relatively complex.

Supply base reduction

The last 30 years has seen a trend for organisations to reduce their supplier base. This trend, building on the idea of partnership relationships, is partly an acknowledgment that organisations have limited resources and therefore some decide to develop fewer, higher-quality, relationships with key suppliers. Supply base reduction often results in significant reductions in daily running costs, such as the costs of ordering, expediting, supplier visits and various failure interventions. However, there are also potential downsides, notably around increased supply risk with some suppliers more able to act opportunistically. In addition, power dynamics typically shift in favour of suppliers as the buyer becomes increasingly dependent on its remaining suppliers.

Making the sourcing strategy decision

Given that each sourcing strategy has its advantages and disadvantages, a key challenge is to decide which is most suitable. Here, we can explore two key questions – *what is the risk in the supply market?* and *what is the criticality of the service or prooduct to the business?* Considering risk, we can consider the number of alternative suppliers, how easy it is to switch from one supplier to another, exit barriers, and the cost of bringing operations back in-house. For criticality, managers often consider a service or product component's importance in terms of volume purchased, percentage of total purchase cost, or the impact on business growth. By looking at these two dimensions, one can position service or product components broadly in one of four key quadrants – leverage, strategic, non-critical or bottleneck[5] – and select appropriate sourcing strategies. Figure 12.7 shows this for a smartphone manufacturer:

Operations principle

Sourcing strategy is governed partly by the risk in the supply market, and the criticality of the service or product to the business.

- *Non-critical* – The packaging for transportation and display, and the screws that hold the phone components together account for a relatively low proportion of the total cost of the product. In addition, with many alternative suppliers, the supply risk is low. For the non-critical quadrant, multiple-sourcing strategies tend to be most common, though supply base reduction initiatives sometimes see single-supplier arrangements but with short contract terms.
- *Bottleneck* – The power pack for a smartphone is relatively low cost compared to other components that make up the product. However, the limited supply alternatives and relatively high switching costs increase supply risk. For services or products in the bottleneck category, single sourcing is common because of a lack of choice in the supply market. In addition, firms sometimes look to reduce the specificity of their requirements and in doing so, increase the number of supplier options available to them.
- *Leverage* – The touch screen and display, and to a lesser extent camera(s) and speaker, account for a high proportion of the purchasing cost of the smartphone. However, these components are easier to source as there are a relatively large number of available (and reliable) suppliers. Suppliers in the leverage quadrant typically need to be price competitive, given the strong bargaining position that abundant supply gives to the buyer. For leverage services or products, bundling of requirements allows a shift towards delegated sourcing in many cases.

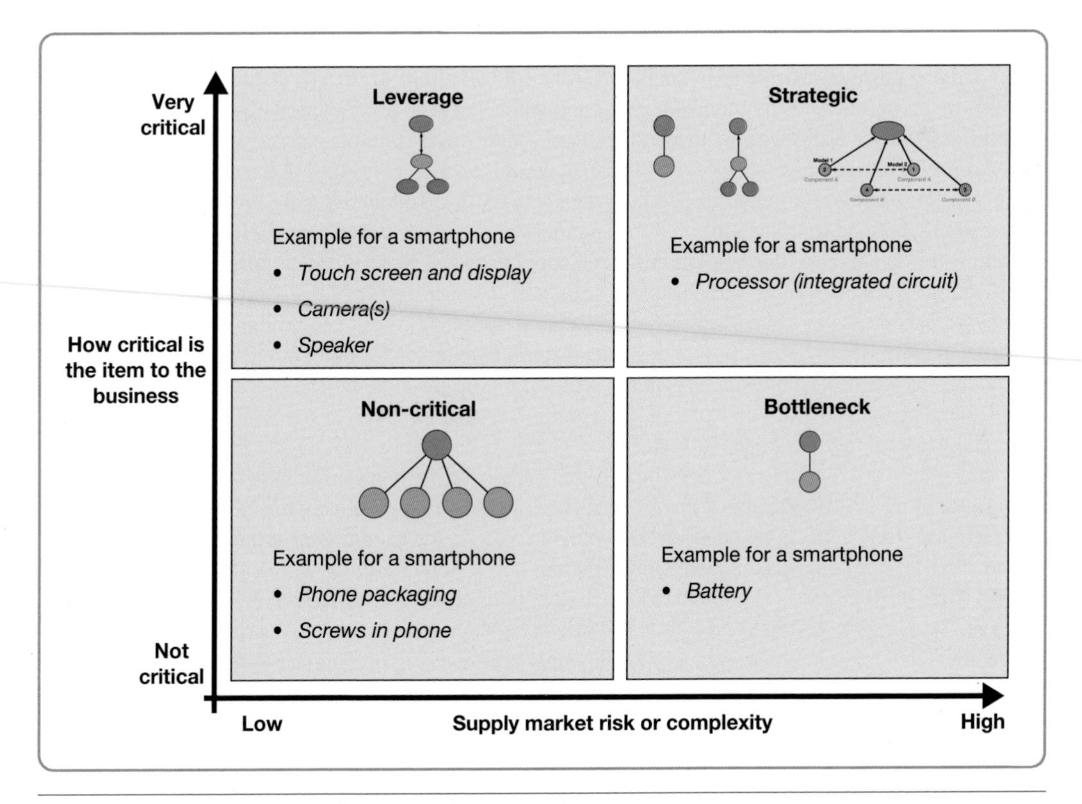

Figure 12.7 Key sourcing groups for a smartphone manufacturer

▶ *Strategic* – Strategic services or products are both complex to acquire and critical to the business, accounting for a significant proportion of total spending. In this example, the processor (the 'brain' of the smartphone) sits in the strategic quadrant. There are relatively few firms capable of supplying components to sufficient quality and so the cost of switching is high. For strategic services or products, single-sourcing approaches remain popular. However, given the associated risks of single sourcing noted earlier, some firms have moved to delegated or parallel sourcing approaches.

OPERATIONS IN PRACTICE

Considering the longer-term effects of COVID-19 on managing supply networks[6]

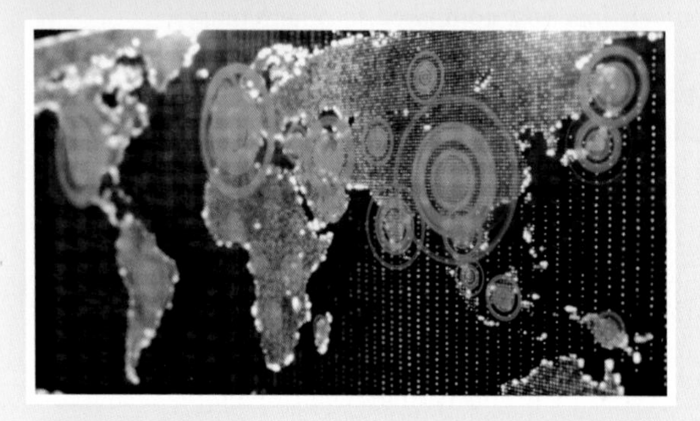

The world economy is now, more than ever, reliant on global supply chains. As such, the COVID-19 pandemic presented a myriad of challenges for operations looking to work across international boundaries. As the pandemic gathered pace and demand patterns became increasingly hard to predict, crisis meetings became commonplace in boardrooms around the world. To buffer against demand uncertainty, many firms increased their inventory and service capacity levels significantly – a major burden in terms of working capital, space and coordination.

Yet it was not simply the demand side that was presenting a challenge. Getting supplies became increasingly problematic as national lockdowns in different regions of the world played havoc with supply chain operations, logistics and distribution. For example, in the food industry many suppliers were simply unable to move their products economically. The cancellation of most international flights meant a lack of shipping capacity for many products that would typically travel in the holds of passenger aircraft. Many service-oriented firms faced reduction in their effective capacity due to absence of employees (a combination of COVID cases and stress-related time off) and radically altered work patterns.

However, there were also big winners in pandemic-affected supply networks. Many online retailers saw demand skyrocket with customers often forced to stay at home or severely restrict travel. While some returned to previous shopping habits as restrictions eased, a large proportion had fundamentally shifted their buying behaviour towards online channels. As a result, significant investments were seen within warehousing and last-mile distribution, especially given customers' increasing expectations regarding the speed of delivery. The pandemic also saw a general shortening of many supply chains, evidenced through the increasing popularity of apps and other platforms connecting suppliers directly with buyers.

Looking to the future, several commentators anticipated that the trend of supply chain shortening would continue given the re-evaluation of the risks posed by extended global networks. Further impetus for the shortening of supply chains was provided by protectionist trade policies that occurred during and after the pandemic. However, other experts cautioned that while these reactions were understandable, in many cases they would not fundamentally improve the resilience of supply networks to future pandemic-type events. In the face of national lockdowns imposed during a major crisis, domestic suppliers were as likely to be affected as overseas suppliers. As such, a firm increasing its proportion of domestic suppliers might see limited reduction in risk exposure. Instead of fundamentally reconfiguring the *location* of suppliers, some commentators argued that attention should be given to the *visibility* and *resilience* of global supply networks. To do this would require significant investment in various integrating technologies such as IoT devices or sensor technology that would provide data on the condition of assets (materials, information and indeed customers) moving through supply chains, operations and processes. In addition, several experts highlighted the value of AI technology to support the real-time tracking of various risk indicators as a way of providing early flagging of potential disruptions within supply networks.

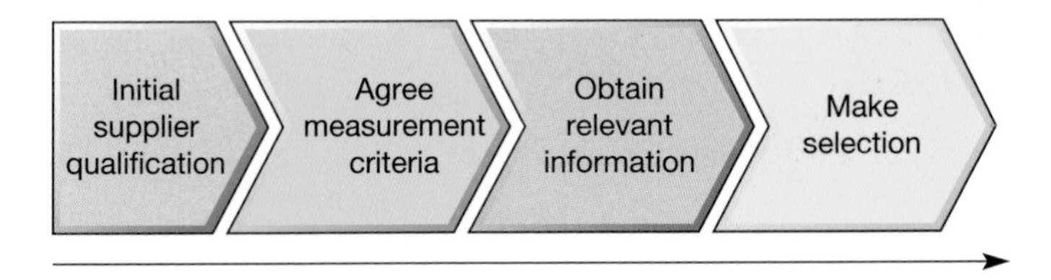

Figure 12.8 The supplier selection process

Supplier selection

In conjunction with deciding on sourcing strategies for different services and products, organisations much select appropriate suppliers. Given the trends of outsourcing, supply base rationalisation, supplier involvement in new product/service development, and longer-term supplier relationships, the selection process is extremely important to the success of organisations. Figure 12.8 outlines the four key steps in supplier selection:

- *Initial qualification* – This is aimed at reducing possible suppliers to a manageable set for subsequent assessment. Pre-qualification criteria vary, but typically focus on establishing a minimum threshold against technical capability (does the supplier have knowledge to supply to the required specification?); operations capability (does the supplier have the process knowledge to ensure consistent, responsive, dependable and reasonable cost supply?); and financial capability (does the supplier have the financial strength to fund the business in the short and long terms?)
- *Agree measurement criteria* – This stage focuses on deciding the relative importance of key performance objectives (quality, speed, dependability, flexibility, cost, sustainability and others). For these objectives, measurable criteria are then needed. For example, for cost a firm might consider unit price, pricing terms (e.g. volume discounts), exchange rate effects and so on.
- *Obtain relevant information* – This stage involves gathering further information on the shortlist of potential suppliers. This may include additional levels of detail in delivery options and cost structure, site visits and tests (e.g. test orders in small quantities) to assess competence prior to potential ramp-up of supply. The amount of time and effort invested in information search is partly influenced by the strategic importance of the purchase and the perceived capability of the supply base. For example, when the supply market is generally considered to be capable, and the strategic importance of the purchase is low, limited information search is appropriate. Conversely, if the firm is buying a service of strategic importance and the supply market is more uncertain, more investment in information search will be necessary. The type of purchase also influences the process. Low levels of information search are needed for *repeat or routine re-buys* (for example, placing an order for a well-established service with an existing supplier). *Modified re-buys*, (for example, where new services are bought from known suppliers, or where existing services are bought from new suppliers) justify moderate information search. Finally, *new buys* (for example, sourcing entirely new services from unknown suppliers) have high levels of uncertainty and require extensive information search.
- *Make selection* – Having arrived at a group of viable alternatives, selection may be supported using various multi-criteria decision-making models. These models aim to provide quantifiable information for each key selection criteria and a weighting of their relative importance to allow for an objective assessment of different suppliers (see worked example and critical commentary below).

Operations principle

Measurement criteria for supplier selection should align with the overall performance objectives of the supply network.

Worked example

Supplier selection for a legal service provider

A commercial bank in Belgium has decided to change its legal services provider. The three legal firms have been evaluated using a 1–10 scale against a set of key criteria. Each of these criteria has a weighting, again from 1–10, signifying its relative importance to the bank in its purchasing decision. Table 12.2 illustrates that, based on the 'weighted score method', Juris Civilis is the preferred supplier. If subsequent negotiations reach a satisfactory conclusion, they would be selected by the bank to provide legal services.

Table 12.2 Weighted supplier selection criteria legal services provider

Selection criteria	Weight	Sullivan & Anderson	Juris Civilis	Altium Legal
Quality of service	10	7 (10 × 7 = 70)	8 (10 × 8 = 80)	6 (10 × 6 = 60)
Range flexibility	8	5 (8 × 5 = 40)	7 (8 × 7 = 56)	6 (8 × 6 = 48)
Volume flexibility	4	8 (4 × 8 = 32)	3 (4 × 3 = 12)	7 (4 × 7 = 28)
Average service speed	7	5 (7 × 5 = 35)	8 (7 × 8 = 56)	9 (7 × 9 = 63)
Record of dependability	9	4 (9 × 4 = 36)	8 (9 × 8 = 72)	7 (9 × 7 = 63)
Sustainability	4	5 (4 × 5 = 20)	6 (4 × 6 = 24)	7 (4 × 7 = 28)
Potential innovation	3	5 (3 × 5 = 15)	7 (3 × 7 = 21)	7 (3 × 7 = 21)
Price of service	6	9 (6 × 9 = 54)	6 (6 × 6 = 36)	5 (6 × 5 = 30)
Total weighted score		**302**	**357**	**341**

Critical commentary[7]

The fundamental approach taken to supplier selection requires careful consideration and is an issue that has been subject to significant academic and practitioner debate over the years. Some organisations continue to favour a *piece-price* approach whereby the lowest unit price of a service or product typically determines the winning supplier. This approach has the advantages of simplicity, minimal data and a clear incentive structure for buyers. However, it ignores other non-price cost elements, risks and aspects of revenue generation. The *total landed cost* (*TLC*) approach addresses some of these limitations by factoring in the costs of logistics, handling, tax, tariffs and trade compliance. However, arguably TLC's treatment of non-price cost elements remains relatively narrow.

One approach that recognises the complexities of cost in sourcing decisions is the *total cost of ownership* (*TCO*) approach, which attempts to quantify the long-term price of a service or product (i.e. its total cost or life cycle cost) as opposed to its short-term price (i.e. its unit price). In addition to the costs considered within TLC, this approach considers various *sunk costs* (e.g. design and development, supplier evaluation, supplier certification and negotiation costs), *overhead costs* (e.g. incoming inspection, supplier monitoring, storage and distribution, and supplier development) and *life cycle costs* (e.g. spare parts, differences in service warranties, product disposal and switching costs in the event of relationship termination). While TCO arguably represents a richer approach to supplier selection relative to the *piece-price* or TLC approaches, it has been criticised for continuing to anchor decision makers on cost as opposed to other performance variables.

An alternative approach, proposed by John Gray, Susan Helper and Beverly Osborn is *total value contribution (TVC)*. The TVC perspective argues that sourcing decisions should not focus on minimising costs but instead should be driven by the maximisation of an organisation's *long-term value*. So, beyond factors required by a TCO analysis, this approach factors three key elements into sourcing decisions – *risk* (e.g. intellectual property loss, brand damage, shortages, disruptions and downtime), *revenue* (e.g. service quality, impact on consumer demand and aspects of triple bottom line performance) and the *value of options* (e.g. potential learning from suppliers, future growth prospects and a supplier's innovative capabilities). The key advantage of adopting a TVC approach, the authors argue, is that it anchors decision making on customer value as opposed to cost. In doing so, it can reduce the risk of selecting a supplier based on what is easily measured as opposed to what really matters. In addition, TVC explicitly considers, and tries to counteract, various cognitive biases and traditional procurement incentives that tend to reinforce excessive cost focus in supplier selection. However, TVC is limited by the difficulty of quantifying different dimensions of value as well the significant time required to collect data on the wide range of selection criteria (note, the TVC process does suggest quantification only on factors for which there is a meaningful difference between potential sources, possibly saving some of this time).

Deciding which approach to use in supplier selection, from the *piece-price* at one extreme and the TVC approach at the other, is likely to be contingent on the importance of non-price factors of the service or product being sourced in driving the long-term value of the organisation.

Negotiating with suppliers during supplier selection

A key part of the selection process is negotiation. The approach to negotiation is naturally affected by initial decisions around performance priorities, supplier type (i.e. traditional versus partnership) and sourcing configuration (e.g., single, multi, delegated or parallel). However, regardless of the general approach, managers who are well-informed on different negotiation techniques are likely to see better outcomes. During negotiations between buyers and suppliers, several tactics are often in evidence, including *emotion*, *logic*, *threat*, *bargaining* and *compromise*. Table 12.3 outlines some important considerations when using each tactic.

While it is essential for negotiators to be skilled at the tactics described in Table 12.3, this 'advocacy approach', whereby a negotiator advocates for one party to gain its most favourable outcome, carries several risks. These include an emphasis on short-term solutions at the expense of potential longer-term gains, personalised conflicts, damage to the buyer–supplier relationship, and increased likelihood of a 'lose–lose' outcome. An alternative approach is 'collaborative negotiation', also called principled negotiation or mutual gains bargaining. This requires negotiators to adopt a mindset that emphasises meeting the needs of both parties, and involves a transparent process, investment in developing personal relationships, using creative brainstorming techniques (growing the pie rather than dividing a 'fixed pie') and favouring longer-term win–win agreements.

Managing ongoing supply

Managing supply relationships is not just a matter of choosing the right suppliers and then leaving them to get on with day-to-day supply. It is also about ensuring that mechanisms are in place that give suppliers the right (and consistent) information and encouragement to maintain smooth supply. Customers may see suppliers as having the responsibility for ensuring appropriate supply 'under any circumstances'. Yet, if a customer and supplier see themselves as 'partners', the free flow of information, and a mutually supportive tolerance of occasional problems, is the best way to ensure smooth supply.

Table 12.3 Common tactics used in buyer-supplier negotiations

	Considerations for using negotiation tactic
EMOTION	▶ Positive and negative emotions are extremely influential in 'framing' the negotiation ▶ Anger is the most common emotion in negotiation, but also the most damaging to potential settlement ▶ Positive emotions support collaborative approaches to negotiation ▶ Use emotions early in negotiation to maximise influence ▶ Be sincere and don't let emotions control you ▶ Use to increase 'perceived value' when bargaining
LOGIC	▶ Ensure your logic is credible and use one powerful argument rather than multiple weaker arguments ▶ Try to get your logic in before the other negotiating party ▶ Use logic to counter emotion-based tactics
THREAT	▶ Avoid threatening where possible and be credible with any threat ▶ If you do use threat, threaten the business rather than the person ▶ Discreet or veiled threats are less risky than explicit threats ▶ Uuse 'if' in any threat to give the opponent room to manoeuvre
BARGAINING	▶ Avoid exposing your position too early in negotiations ▶ Move in small and decreasing steps and extract return for any movement where possible ▶ Don't move too quickly and avoid appearing too keen to move
COMPROMISE	▶ Use compromise after other tactics ▶ Use extreme (but credible) position to maximise value when using compromise ▶ Avoid suggesting compromise too early in a negotiation as it shuts down more creative solutions and favours the other negotiating party

Responsible operations[8]

In every chapter, under the heading of 'Responsible operations', we summarise how the particular topic covered in the chapter touches upon important social, ethical and environmental issues.

Embedding environmental and ethical practices in supply networks

Global supply networks continue to be subject to extensive criticism and controversy when considering the subject of responsibility. Many commentators have argued that organisations have typically failed to do enough to tackle the major disconnect between their corporate social responsibility standards and the business practices of those suppliers operating in their supply chains.

Considering the environment, global supply networks are accused of creating significant harm to the planet. Concerns include the use of fossil fuels, unnecessary waste of materials, energy use, toxic waste, deforestation, water pollution, harm to air quality, loss of biodiversity and long-term damage to ecosystems. In response, many organisations have looked to reduce environmental damage and attempted to make their supply chains more transparent. For example, the IPE Green Supply Chain Map is a real-time tool that shares information on Chinese supplier facilities for a wide range of leading international brands, such as Gap Inc, Samsung, Tesco, Carrefour, Levi's, Puma, Esprit, Adidas, C&A, Inditex, Primark and Nike.

When it comes to ethical behaviour, a major source of concern in recent years has been practices commonly associated with 'indecent work', forced labour or modern slavery. This has resulted in efforts to establish (or reinforce) freedom of association and collective bargaining within some economies. These are seen as an essential bedrock on which worker rights and improved conditions are built. However, efforts to improve the working conditions of those within supply networks is far from straightforward and many global brands have struggled to impose values across their supply networks. For example, when H&M announced that it was going to ban cotton sourced from the Xinjiang region of China (an area responsible for around 20 per cent of the world's cotton supply) because of human rights and forced labour concerns, many applauded the move. Customers in Global North markets, human rights charities, and many of their own employees were pleased to see the firm taking what they saw as a strong ethical stand. However, in China, outcry over the fast-fashion retailer's decision led to H&M being removed from several e-commerce platforms, including JD.com, Tmall and Taobao. In addition, the firm's app was banned from the largest telecoms company in China, Huawei. Other firms, such as Adidas, Nike, Burberry and Calvin Klein, were also affected when they expressed concerns over working conditions in China. Gigabyte Technology Co, the manufacturer of graphics cards, gaming laptops and motherboards, lost over 20 per cent of its share price in a matter of days after claiming its brands were superior to those produced in China in a 'low-cost, low-quality way'. Its comments led to a social media storm and the removal of product listings by several China-based e-commerce platforms, including Suning.com and JD.com.

These incidents, and others like it, highlighted a key tension for many global brands. On one hand, the increasing stakeholder pressure to do more when it comes to environmental or social responsibility; on the other, a fear that increased supply network transparency exposes practices that are hard to ignore, but that by addressing may create very damaging impacts for the business itself. In fact, some experts have argued that the dominant 'buyer-centric approach' to responsibility in supply networks – where the buyer decides standards and then attempts to force adherence to them – is fundamentally flawed. Instead, they contend that a more holistic approach is needed that engages a wide range of stakeholders in seeking to establish norms and control behaviours. In addition, to ensure environmental or ethical interventions are effective, significant tailoring to local contexts is essential, they argue. In other words, one size doesn't appear to fit all when it comes to addressing responsibllity in supply chain management.

Perception differences in supply chain relationships

One of the biggest issues for successful SCM is the mismatch between how customers and suppliers perceive both what is required and how the relationship is performing. Figure 12.9 illustrates the four key gaps that sometimes emerge in supply chain relationships. Taking the *customer perspective* (Scenario A in Figure 12.9) first, you (the customer) have an idea about what you really want from a supplier. While this may not be formalised using a service-level agreement (SLA), it is always hard to capture everything about what is required. There may be a gap between how you as a customer describe what is required and how the supplier interprets it. This is the *requirements perception gap*. Similarly, as a customer, you will have a view on how your supplier is performing. Misalignment between your view and how your supplier believes it is performing is evidence of a *fulfilment perception gap*. Third, the *supplier improvement gap* is the gap between what the customer wants from their supplier and how they perceive the supplier's performance. This will influence the kind of supplier development goals set by the customer. Finally, the *supplier performance gap* is the gap between your supplier's perceptions of your needs and its own assessment of performance. This will indicate how a supplier initially sees itself improving its own performance.

From a *supplier perspective* (Scenario B in Figure 12.9), the same approach can be used to understand *customer* perceptions, both of their requirements and their view of performance. What is less common, but can be equally valuable, is to use this model to examine the question of whether customer

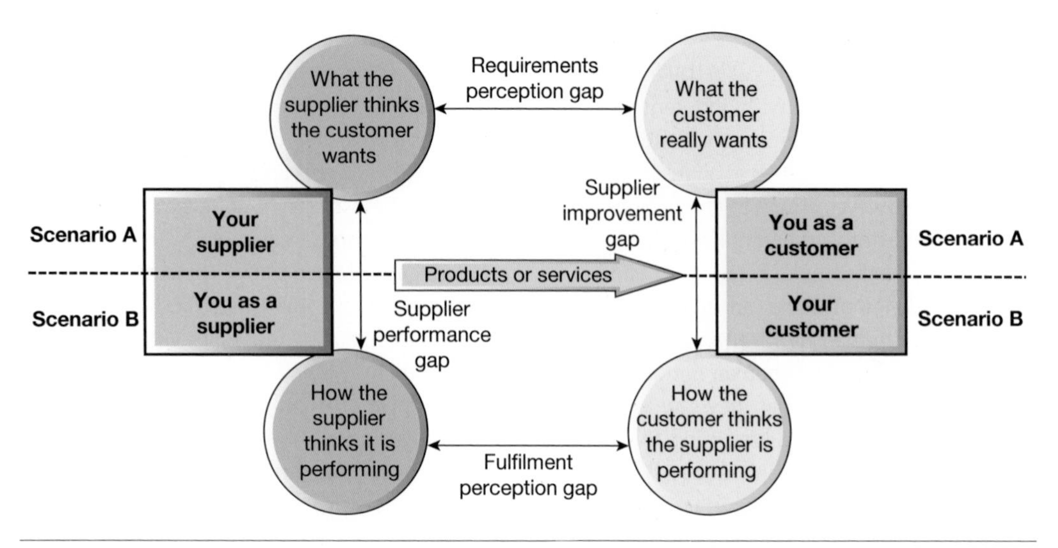

Figure 12.9 Potential perception mismatches in supply chain relationships

Operations principle

Unsatisfactory supply chain relationships can be caused by requirement, fulfilment, supplier improvement and supplier performance gaps.

requirements and perceptions of performance are either accurate or reasonable. For example, customers may be placing demands on suppliers without fully considering their consequences. It may be that slight modifications to the service would not inconvenience the customers and yet would provide significant benefits to the supplier that could then be passed on to the customer. Similarly, a customer may be incompetent at measuring supplier performance, in which case current good service may not be recognised.

Supply chain relationships are multi-tiered configurations

Up to this point, we have largely treated supply chain relationships as if they simply exist between two entire organisations. However, supply chain relationships are multi-tiered configurations, as illustrated in Figure 12.10. As such, a formal contract between two organisations in a supply network must be interpreted by managers in these organisations. In turn, the service itself is then delivered by

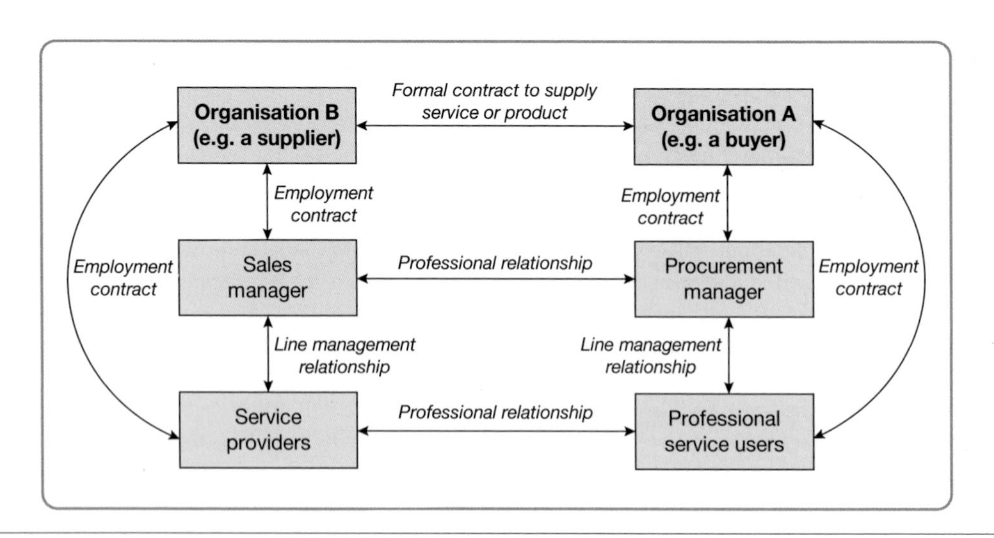

Figure 12.10 Supply chain relationships are multi-tiered configurations

Source: Developed in collaboration with Juri Matinheikki and Katri Kauppi of Aalto University, Finland, and Erik van Raaij of Erasmus University Rotterdam, Netherlands

service personnel and received by service users. At each of these levels, there is essentially a relationship. Ideally, these relationships will be consistent in terms of attitudes, actions and perceptions. But there are often significant differences across these levels of a supply chain relationship. Furthermore, we also see a complex relationship dynamic between the organisation and its employees. Again, disconnects between what the organisation wants its managers and employees to deliver and how the service is actually delivered on the ground can create significant problems in multi-tiered supply chain relationships.

Developing suppliers

Unless a relationship is purely market-based and transactional, it is in a customer's long-term interests to take some responsibility for developing supplier capabilities. Helping a supplier to improve not only enhances the service from the supplier; it may also lead to greater supplier loyalty and long-term commitment. This is why some particularly successful businesses invest in supplier development teams whose responsibility is to help suppliers improve their own operations processes. The process of supplier development can be broken down into four key stages:

1. *Select product or service and supplier for development* – Not all suppliers can be developed for all offerings, but those that provide strategic products or services (in the 'strategic' quadrant of Figure 12.7) are likely candidates. From this quadrant, those with relatively weak current performance but with potential for improvement are likely to be appealing, as are suppliers where switching costs are high.
2. *Form a project team and gain buy-in for supplier development* – The project team should bring together key stakeholders from the buyer, supplier and other relevant parties. We discuss this issue in detail within the project management chapter, but it is worth noting the importance of cooperation from key stakeholders. Gaining 'buy-in' can require significant time, sensitivity and effort in the early stages of the supplier development process, especially if the supplier's management are relatively defensive of development needs.
3. *Agree goals and measures for supplier development* – Like all projects, supplier development requires clear scoping in terms of timelines, costs and key deliverables to ensure that both buyer and supplier agree on what constitutes success. This stage should also involve risk evaluation. For example, if a lack of supplier commitment is identified as a risk, mitigation might include more investment in the internal marketing of potential benefits or refined financial incentives (such as profit-sharing).
4. *Implement, monitor and learn* – It is important to monitor progress and intervene if the project is deviating from its performance targets. Similarly, it is important to celebrate success as this helps to reinforce the value of supplier development to all stakeholders. In addition, learning from supplier development initiatives can be extremely powerful in informing subsequent projects with other suppliers.

Critical commentary

When considering supplier development, we focused on the most common approach, whereby a customer 'imposes' initiatives and performance requirements on its suppliers and sub-suppliers. Arguably, this perspective has two major limitations. First, most suppliers have several important customers who may all place conflicting development demands upon them. This can encourage suppliers to mislead customers over the extent to which they are adopting changes requested, which in turn may lead to trust erosion in trading relationships. Second, critics note that the traditional 'cascade' and 'intervention' approaches necessarily assume that the ideas flowing from the customer are superior to those of other parties in the supply network. However, in practice, many suppliers are equally if not more knowledgeable than their customers. These criticisms have led to some 'mutual' development initiatives whereby both customers and suppliers contribute knowledge and expertise in determining development priorities. Others have gone further, arguing that *network development* is more appropriate than supplier development. Here, the aim is to improve the overall network rather than a single organisation. Network development activities include network-wide technology initiatives (see the 'Operations in practice' example on TradeLens), stakeholder suggestion programmes and supply network councils.

12.5 How is the demand side managed?

Supply chain management is not only concerned with supply-side activities. Managers must also make important decisions related to the demand-side part of the network. Here, we examine two important issues – logistics services and customer relationship management.

Logistics services

Logistics (or distribution) is the activity of moving products from suppliers to their customers. For many firms, logistics is not something they consider, because they either focus solely on non-physical outputs, or the products they do deliver are such a small proportion of their business that they simply use distribution services on an ad hoc basis. However, for other more product-oriented firms, managing logistics is often a critical issue. This is especially the case when the costs of distribution account for a large proportion of their total costs.

Some organisations operate *first-party logistics* (1PL), whereby the logistics activity is an entirely internal process. For example, a supermarket might collect products from a supplier or have its own fleet of vans to deliver to customers. Some firms decide to outsource logistics services over a specific segment of a supply chain. This is known as *second-party logistics* (2PL). For example, the supermarket may hire a maritime shipping company to transport and, if necessary, store products from a specific collection point to a specific destination. *Third-party logistics* (3PL) is when a firm contracts a logistics company to work with other transport companies to manage their logistics operations more fully. It is a broader concept than 2PL and can involve transportation, warehousing, inventory management and even packaging or re-packaging products. *Fourth-party logistics* (4PL) is a yet broader idea. Accenture, the consulting group, originally termed 4PL as 'an integrator that assembles the resources, capabilities, and technology of its own organisation and other organisations to design, build and run comprehensive supply chain solutions'. Finally (almost inevitably), some firms are selling themselves as *fifth-party logistics providers* (5PL), mainly by defining themselves as broadening their scope further to e-business.

Volume, size and value

In selecting methods of transportation for logistics, organisations typically consider the volume, size and value characteristics of their products. For example, a firm distributing low volumes of small and high-value products around the world is much more likely to adopt air freight whereas a firm distributing high volumes of bulky and low-value products across its supply network is more likely to use maritime transportation. Typically, the cost of moving a given product from one location to another is highest for air, then road, rail, water and finally pipeline. In selecting the methods of transportation that are most suitable, costs of different options clearly count. However, the choice may also be influenced by 'vertical trade-offs', benefits that may change over time because of improvements in one method of transportation versus another (for example, the development of a new high-speed rail link or improved fuel economy in shipping). In addition, 'lateral trade-offs' consider the balance between costs of a particular transportation method and possible benefits elsewhere. For example, Zara predominantly adopts (high-cost) air freight for its worldwide logistics, but gains significantly from faster delivery times and ability to hold much lower inventories across its supply network.

OPERATIONS IN PRACTICE

Donkeys – the unsung heroes of supply networks[9]

The last 150 years has witnessed huge changes in the way that products are transported across supply networks. In shipping, 'containerisation' has enabled levels of efficiency across maritime operations previously unthinkable, paving the way for globalisation. The same period saw the Wright brothers make the first successful powered flight in 1903, ultimately leading to the creation of the aviation sector and air freight as a new mode of transportation. Similarly, Karl Benz's 'motorwagen' patent in 1886 eventually led to a new dominant mode of ground-based transportation. Yet, throughout all this turbulence, there remains an unsung hero of supply networks around the world – the humble donkey.

For thousands of years, people have used animals of all shapes and sizes to move their possessions and supplies from one location to another, including camels, dogs, elephants, goats, llamas, oxen, reindeer and sheep to name a few. But by far the most widely used animals in modern supply chains are donkeys (alongside ponies, mules or horses, they are collectively known as 'packhorses'). Around the world, donkeys fulfil a wide range of roles, acting as trucks, ambulances, tractors and school buses. They carry sugar cane in Peru, transport crops in Argentina, move flowers in Jamaica and maize in Malawi, pull snow sledges in the United States, collect rubbish in Mexico, serve as mobile beach bars in Brazil, carry fish in Mauritania, collect water in Senegal, move families in Namibia, pull mobile libraries in Ethiopia, haul building materials in Kenya, lead livestock in Yemen, are ambulances for women in labour in Afghanistan, work in India's brick kilns and plough land in China. In many tourist destinations around the world, donkeys, mules, ponies and horses are used to pull tourist carriages or to give children rides on beaches.

The decision to use a donkey for transportation follows the same fundamental process as any other mode of transport. First, transport options must be identified. These choices are naturally constrained by the terrain over which products need to be moved and the existing supply chain infrastructure. Second, judgements must then be made concerning the relative performance benefits (in terms of quality, speed, dependability, flexibility, cost and sustainability) of available methods. As such, even when an alternative might be *technically* possible, the donkey may still represent the best balance of supply chain performance. This helps explains why, in a world of increasingly advanced technological solutions, there remains an important place for the donkey and its packhorse cousins. Yet, despite their role in helping build civilisation as we know it, packhorses are sadly some of the most mistreated animals on the planet. Cruelty comes in the form of malnutrition, beatings, overloading, poorly fitting packs and long hours in extreme climates. Several animal charities around the world have lobbied to establish codes of conduct for animals involved in transportation. In addition, some charities provide access to cheap or free veterinary treatment, as well as offering advice to owners on how to care for their animals.

Logistics and the Internet of Things (IoT)

Internet-based communication has had a significant impact on logistics and distribution, in particular through the adoption of the 'Internet of Things' (IoT). At its simplest, an IoT is a network of physical objects (such as products, equipment, materials handling devices, trucks) that have electronics, software and sensors implanted in them that can gather and exchange data. Combined with global positioning systems (GPS), it permits instantaneous tracking of trucks, materials and people, allowing logistics companies, warehouses, suppliers and customers to share knowledge of where products are in the network and where they are going next. This permits more effective coordination and creates the potential for cost savings. For example, an important issue for transportation companies is 'back-loading'. When the company is contracted to transport goods from A to B, its vehicles may have to return from B to A empty. Back-loading means finding a potential customer who wants their goods transported from B to A in the right time frame. IoT increases the likelihood that a company can fill their vehicles on both the outward and return journeys. Similarly, IoT enables 'track-and-trace' technologies so package distribution companies can inform and reassure customers that their service is being delivered as promised.

Critical commentary

The use of technology in supply chain management is not universally welcomed and can be viewed as preventing closer partnership relationships. In addition, many firms continue to underestimate the challenge of successfully assimilating such new technologies across their supply networks. As a result, the expected returns from some supply chain improvement initiatives often fail to materialise. In addition, there are concerns that some supply chain technologies have the potential to threaten employment, individual privacy and supply chain security. For example, IoT could raise the potential for being hacked. Yet, any web-based connectivity will always create new vulnerabilities. In one view, we no longer have objects with computers embedded in them, we have computers with objects attached to them. Supply chain security is, arguably, given too little emphasis in the use of IoT.

Customer relationship management (CRM)

There is a story that is often quoted to demonstrate the importance of using information technology to analyse customer information. It goes like this, Walmart, the huge US-based supermarket chain, did an analysis of customers' buying habits and found a statistically significant correlation between purchases of beer and purchases of nappies (diapers), especially on Friday evenings. The reason? Fathers were going to the supermarket to buy nappies for their babies, and because fatherhood restricted their ability to go out for a drink as often, they would also buy beer. Supposedly this led the supermarket to start locating nappies next to the beer in their stores, resulting in increased sales of both!

Whether the story is true or not, it does illustrate the potential of analysing data to understand customers. This is the basis of customer relationship management (CRM). It is a method of learning more about customers' needs and behaviours to develop stronger relationships with them. Although CRM usually depends on information technology, it is misleading to see it as a 'technology'. CRM brings together all the disparate information about customers to gain insight into their behaviour and their value to the business. It helps to sell services more effectively and increase revenues by:

- providing offerings that are more closely aligned to customer needs;
- retaining existing customers and discovering new ones;
- offering better customer service;
- cross selling services more effectively.

CRM helps organisations understand who their customers are and what their value is over a lifetime. It does this by building several steps into its customer interface processes. First the business must determine the needs of its customers and how best to meet those needs. For example, banks may keep track of its customers' age and lifestyle so that it can offer appropriate services like mortgages or pensions to them when they fit their needs. Second, the business must examine all the different ways and parts of the organisation where customer-related information is collected, stored and used. Businesses may interact with customers in different ways and through different people. For example, salespeople, call centres, technical staff, operations and distribution managers may all, at different times, have contact with customers. CRM systems should integrate this data. Third, all customer-related data must be analysed to obtain a holistic view of each customer and identify where service can be improved.

Critical commentary

Despite its name, some critics of CRM argue that the greatest shortcoming is that it is insufficiently concerned with directly helping customers. CRM systems are sold to executives as a way to increase efficiency, force standardised processes and gain better insight into the state of the business. But they rarely address the need to help organisations resolve customer problems, answer customer questions faster or help them solve their own problems. This may explain the trend towards a shift in focus from automating internal front-office functions to streamlining processes such as online customer support.

12.6 What are the dynamics of supply chains?

There are dynamics that exist between firms in supply chains that cause errors, inaccuracies and volatility, and these increase for operations further upstream. This effect is known as the **bullwhip effect**, so called because a small disturbance at one end of the chain causes increasingly large disturbances as it works its way towards the other end. Its main cause is a perfectly understandable and rational desire by the different links in the supply chain to manage their levels of service activity and inventory sensibly. To demonstrate this, examine the production rate and stock levels for the supply chain shown in Table 12.4. This is a four-stage supply chain where an original equipment manufacturer (OEM) is served by three tiers of suppliers. The demand from the OEM's market has been running at a rate of 100 items per period, but in period 2, demand reduces to 95 items per period. All stages in the supply chain work on the principle that they will keep in stock one period's demand. This is a simplification but not a gross one, as many operations gear their inventory levels or service capacity to their demand rate. The column headed 'stock' for each level of supply shows the starting stock at the beginning of the period and the finish stock at the end of the period. At the beginning of period 2, the OEM has 100 units in stock (that being the rate of demand up to period 2). Demand in period 2 is 95 and so the OEM knows that it would need to produce sufficient items to finish up at the end of the period with 95 in stock (this being the new demand rate). To do this, it need only manufacture 90 items; these, together with five items taken out of the starting stock, will supply demand and leave a finished stock of 95 items. The beginning of period 3 finds the OEM with 95 items in stock. Demand is also 95 items and therefore its production rate to maintain a stock level of 95 will be 95 items per period. The original equipment manufacturer now operates at a steady rate of producing 95 items per period. Note, however, that a change in demand of only five items has produced a fluctuation of 10 items in the OEM's production rate.

Carrying this same logic through to the first-tier supplier, at the beginning of period 2, the second-tier supplier has 100 items in stock. The demand that it must supply in period 2 is derived from the production rate of the OEM. This has dropped down to 90 in period 2. The first-tier supplier therefore must produce sufficient to supply the demand of 90 items (or the equivalent) and leave one month's demand (now 90 items) as its finished stock. A production rate of 80 items per month will achieve this. It will therefore start period 3 with an opening stock of 90 items, but the demand from the OEM has now risen to 95 items. It therefore must produce sufficient to fulfil this demand of 95 items and leave 95 items in stock. To do this, it must produce 100 items in period 3. After period 3 the first-tier supplier then resumes a steady state, producing 95 items per month. Note again, however, that the fluctuation has been even greater than that in the OEM's production rate, decreasing to 80 items a period, increasing to 100 items a period, and then achieving a steady rate of 95 items a period. Extending the logic back to the third-tier supplier, it is clear that the further back up the supply chain an operation is placed, the more drastic are the fluctuations.

Operations principle

Demand fluctuations become progressively amplified as their effects work back up the supply chain, a dynamic known as the 'bullwhip effect'.

Table 12.4 Fluctuations of production levels along supply chain in response to small change in end-customer demand (Starting stock (a) + production (b) = finishing stock (c) + demand, that is production in previous tier down (d): see explanation in text. Note all stages in the supply chain keep one period's inventory, c = d.)

Period	Third-tier supplier		Second-tier supplier		First-tier supplier		Original equipment mfr		Demand
	Prodn	Stock	Prodn	Stock	Prodn	Stock	Prodn	Stock	
1	100	100	100	100	100	100	100	100	100
		100		100		100		100	
2	20	100	60	100	80	100	90	100	95
		60		80		90		95	
3	180	60	120	80	100	90	95	95	95
		120		100		95		95	
4	60	120	90	100	95	95	95	95	95
		90		95		95		95	
5	100	90	95	95	95	95	95	95	95
		95		95		95		95	
6	95	95	95	95	95	95	95	95	95
		95		95		95		95	

3 Orders 2 Orders 1 Orders OEM Orders Market

Items Items Items Items

This simple demonstration ignores several other factors that make the fluctuations even more pronounced. These include lack of forecasting, errors in forecasting, quantity discounting (which encourages less-frequent but larger orders), price fluctuations, time lags between the flows of information (order) and flow of materials or service provision (deliveries), variable delivery times and panic ordering in anticipation of shortages or in reaction to them. Figure 12.11 shows the net result of all these effects in supply chains. As we move further away from the end customer, the amplitude and variance in order patterns increases significantly.

Supply chain dynamics have several harmful impacts on those operating in a supply network. These include the costs of outsized facilities and excess inventories to deal with demand spikes that are then often underutilised. For human resources, service capacity typically oscillates between underutilisation and overutilisation, and many firms are forced to hire, fire and re-hire employees as they experience the volatile demand patterns caused by the bullwhip effect. Furthermore, irregular patterns of work cause inefficiencies, delays (with the extra costs of expediting orders) and both customer and staff dissatisfaction. In addition, we have assumed to this point that the end-customer demand is fundamentally *stable* (Figure 12.11a). Sales of services or products can of course be *unstable*, either because of the fundamental nature of demand, promotion activities or panic buying (as seen during the COVID-19 pandemic for some services and products). In such contexts, the bullwhip effect is even more extreme (Figure 12.11b).

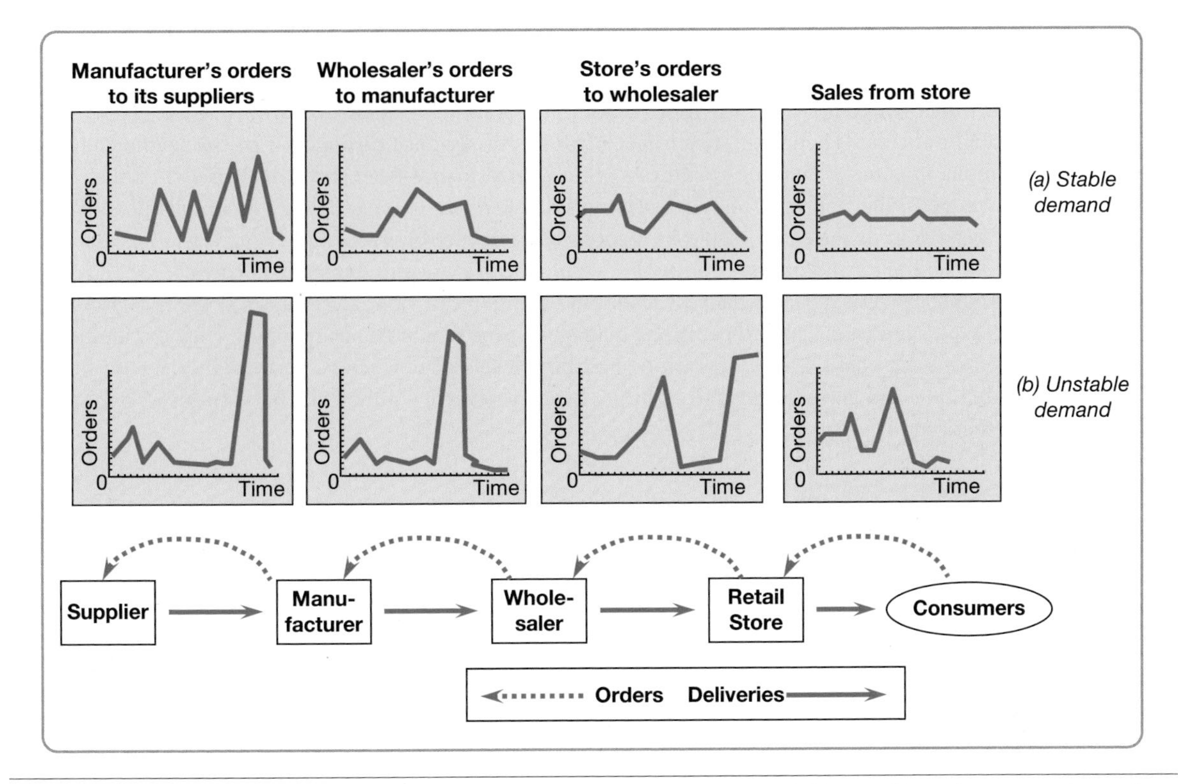

Figure 12.11 Typical supply chain dynamics for (a) stable end-customer demand and (b) unstable end-customer demand

Finally, there is some emerging evidence that the bullwhip effect may be even more pronounced within service supply networks compared to more product-oriented contexts. Factors that may worsen service bullwhips include the lack of accurate data on current rework volumes within a supply chain that would otherwise help to signal the likely formation of new bullwhip effects; the disruption caused by manual rework in many automated service environments; and the lower levels of supply network coordination in many service settings that limit the speed of reaction to emergent bullwhip problems.

Controlling supply chain dynamics

The first step in improving supply chain performance involves attempting to reduce the bullwhip effect. This usually means coordinating the activities of the operations in the chain in several ways:[10]

Channel alignment in supply networks

Channel alignment means the adjustment of scheduling, material movements, stock levels, pricing and other sales strategies to bring all the operations in the chain into line with each other. This goes beyond the provision of information. It means that the systems and methods of planning and control decision-making are harmonised through the chain. For example, even when using the same information, differences in forecasting methods or purchasing practices can lead to fluctuations in orders between operations in the chain. One way of avoiding this is to allow an upstream supplier to manage the inventories of its downstream customer. This is known as vendor-managed inventory (VMI). So, for example, a packaging supplier could take responsibility for the stocks of packaging materials held by a food manufacturing customer. In turn, the food manufacturer takes responsibility for the stocks of its products that are held in its customer's, the supermarket's, warehouses.

Operational efficiency in supply networks

'Operational efficiency' in this context means the efforts that each operation in the chain makes to reduce its own complexity, the cost of doing business with other operations in the chain, and its throughput time. The cumulative effect of this is to simplify throughput in the whole chain. For example, imagine a chain of operations whose performance level is relatively poor: quality defects are frequent, the lead time to order products and services is long, delivery is unreliable and so on. The behaviour of the chain would be a continual sequence of errors and effort wasted in replanning to compensate for the errors. Poor quality would mean extra and unplanned orders being placed, and unreliable delivery and slow delivery lead times would mean high safety stocks or spare service capacity. Just as important, most operations managers' time would be spent coping with the inefficiency. By contrast, a chain whose operations had high levels of operations performance would be more predictable and have faster throughput, both of which would help to minimise supply chain fluctuations.

Information sharing

Accurate demand information is extremely helpful in efforts to control supply chain dynamics, so it makes sense to share demand information, free of distortions, across the supply network. An obvious improvement is to make information on end-customer demand available to upstream operations. Electronic point-of-sale (EPOS) systems used by many retailers attempt to do this. Sales data from checkouts or cash registers is consolidated and transmitted to the warehouses, transportation companies and supplier manufacturing operations that form its supply chain. Similarly, electronic data interchange (EDI) helps to share information and can affect the economic order quantities shipped between operations in the supply chain. The other rapidly developing approach to gaining a trusted overview of supply networks is the use of blockchain technology (see the following section).

> **Operations principle**
> Supply chain dynamics (the bullwhip effect) can be reduced by aligning activities in supply networks, improving operational efficiency and increasing information sharing.

Blockchain technology in supply networks

Many new technologies find applications for which they seem to have been designed – robots in manufacturing processing, face recognition in retail and security operations, and so on. Blockchain (technically, 'distributed ledger') technology is slightly different, having evolved from its first, and best-known, application as the accounting method for the virtual currency Bitcoin to be used in a variety of business domains including supply chain management.

A blockchain is a decentralised, digitised, public ledger (list) of transactions (movements, authorisations, payments, etc.). A 'block' is a record of new transactions. When each block is completed (verified) it is added to the chain, thus creating a chain of blocks, or 'blockchain'. There are five basic principles underlying the technology:[11]

- ***It uses a distributed database*** – All participants on a blockchain have access to the entire database and its complete history. No single participant controls the data or the information. Every participant can verify the records of its transaction partners directly, without an intermediary.
- ***It uses peer-to-peer (P2P) transmission*** – All communication occurs directly between peers (or rather their computers, known as nodes) rather than through a central node. Each node in the network stores and forwards information to all other nodes.
- ***It is transparent and can be used anonymously*** – Every transaction is visible to anyone with access to the system. Each node (user) on a blockchain has a unique 30-plus-character alphanumeric address that identifies it. Users can either choose to remain anonymous, or alternatively provide proof of their identity to others. Transactions occur between blockchain addresses.
- ***Its records are irreversible*** – Once a transaction is entered in the database (and the accounts updated) the records cannot be altered, because they're linked to every transaction record that came before them (which is why it is called a 'chain'). Computational algorithms and approaches

are used to ensure that the recording on the database is permanent, chronologically ordered and available to all others on the network.

- *It uses computational logic* – Because the distributed ledgers are digital, all blockchain transactions can be tied to a known computational logic. This means that participants can set up algorithms and rules that automatically trigger transactions between nodes.

The role of blockchain technology in supply chain management

So why has blockchain gained traction in supply chain management? First, supply networks are often complex and involve many operations. However, for a distributed ledger network, the larger the number of nodes, the more secure it is, so the innate complexity of supply networks actually strengthens the technology. Second, there are a very large number of transactions necessary for the smooth running of most supply networks. As such, blockchain solutions are especially useful if the supply network crosses national boundaries; customs, certification and other documents are required by many regulatory authorities (see the 'Operations in practice' example, 'TradeLens – blockchain revolutionises shipping'). Third, blockchain technology is secure, with its cryptographic techniques making it practically impossible for hackers to alter information. The validation rules of blockchains mean that a potential hacker would need to access more than half of the nodes in the blockchain (which is why it is more secure to have more nodes in the blockchain). Fourth, the security of blockchains means that all parties can trust the provenance (history) of supplies. This is particularly important when dealing with food supplies, luxury goods (that are frequently forged) or ethical goods (avoiding 'blood diamonds', for example). Finally, supply chains are 'threads of operations through supply networks' – the key word being 'networks'. Blockchain technology is 'distributed' – the whole concept is based on how supply chains operate as parts of networks. Figure 12.12 illustrates how blockchains can be used in a supply chain context.

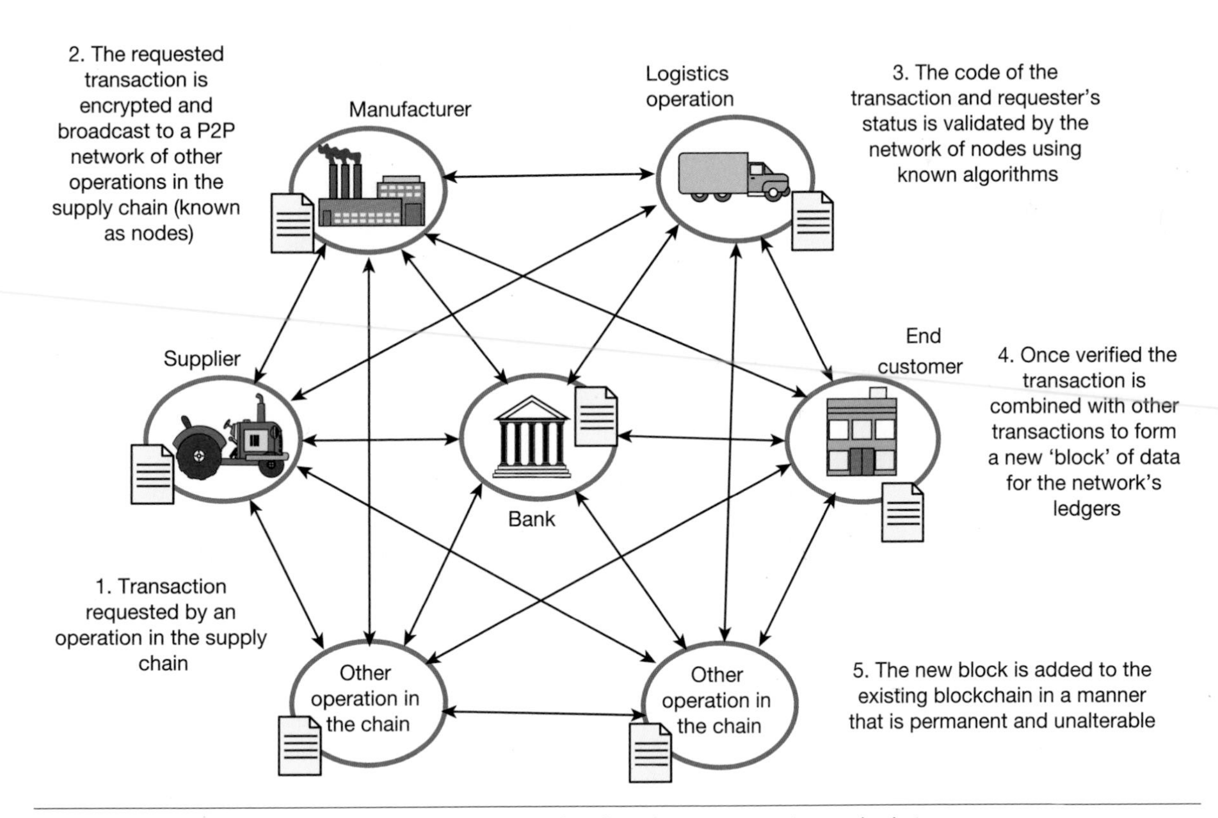

Figure 12.12 How blockchain technology works to record and verify transactions in supply chains

OPERATIONS IN PRACTICE

TradeLens – blockchain revolutionises shipping[12]

Although the technology was still very much in its infancy, the shipping giant Maersk announced in January 2018 that, subject to receipt of regulatory approvals, it was teaming up with IBM to form a joint venture to provide more efficient and secure methods for conducting global trade using blockchain technology. IBM was already recognised as a leading provider of blockchain technology. An early member of Hyperledger, an open-source collaborative effort created to advance cross-industry blockchain technologies, it had worked with other clients to implement blockchain applications. The aim, it said, would be to offer a jointly developed platform, built on open standards, and designed for use by the entire global shipping ecosystem. According to commentators, joining up with Maersk made sense, because of the complexity of its integrated transport, container shipping, ports and logistics operations in 130 countries. More than 80 per cent of the products consumers use daily are carried by the ocean shipping industry. But, traditionally, it had been a document-intensive business. The maximum cost of generating and processing all the required trade documentation to organise many of these shipments was estimated to be as high as one-fifth of the actual physical transportation costs. Blockchain technology was seen as ideally suited to large networks of disparate partners across complex supply chains. Maersk executives said that, with access to a shared, trusted record of transactions, the world's shipping companies would save money and be able to better compete on enhanced services.

The CEO of the joint venture, Michael J. White said, *'Today, a vast amount of resources are wasted due to inefficient and error-prone manual processes. The pilots confirmed our expectations that, across the industry, there is considerable demand for efficiency gains and opportunities coming from streamlining and standardising information flows using digital solutions. Our ambition is to apply these learnings to establish a fully open platform whereby all players in the global supply chain can participate and extract significant value. We look forward to further expanding our ecosystem of partners as we progress toward a global solution'*.

By the end of 2018, Maersk and IBM announced the name of their platform venture – 'TradeLens'. Over the following years, the platform expanded rapidly in the shipping sector as many of the largest ocean carriers and their many ecosystem partners, representing more than 50 per cent of global container cargo, adopted TradeLens. The collaborative nature of the platform was critical in encouraging competitors, including CMA CGM, Hapag-Lloyd, MSC Mediterranean, Namsung and Ocean Network Express (ONE), to engage with TradeLens. For example, when announcing Hapag-Lloyd's adoption of the platform, Martin Gnass, Managing Director IT, said *'Expanding digital collaboration is critical to the evolution of the container shipping industry [...] we can collectively accelerate that transformation to provide greater trust, transparency and collaboration across supply chains and help promote global trade'*.

Summary answers to key questions

12.1 What is supply chain management?

- Supply chain management is the management of relationships and flows between chains of operations and processes. Technically, it is different from supply network management, which looks at all the operations or processes in a network, but the two terms are often used interchangeably.
- Many of the principles of managing external supply chains (flow between operations) are also applicable to internal supply chains (flow between processes and departments).

12.2 How should supply chains compete?

- The central objective of supply chain management is to satisfy the needs of the end customer.
- So, each operation in the chain (and each chain in the supply network) should contribute to whatever mix of quality, speed, dependability, flexibility, cost and sustainability the end customer requires.
- Individual operations failure in any of these objectives can be multiplied throughout the chain. So, although each operation's performance may be adequate, the performance of the whole chain could be poor.
- An important distinction is between lean and agile supply chain networks. Broadly, lean (or efficient) supply networks are appropriate for stable 'functional' services or products, while agile (or responsive) supply networks are more appropriate for more unpredictable services and products.

12.3 How should relationships in supply chains be managed?

- Supply chain relationships can be described on a spectrum from transactional and 'arm's-length' relationships, through to close and long-term partnership relationships.
- The types of relationships adopted may be dictated by the structure of the market itself.

12.4 How is the supply side managed?

- Managing supply-side relationships involves determining sourcing strategy, selecting appropriate suppliers, managing ongoing supply activity, and supplier development.
- Sourcing strategies include multiple sourcing, single sourcing, delegated sourcing and parallel sourcing. Their selection is influenced by the complexity and risk of the supply market and the criticality of a service or product to the business.
- Supplier selection involves trading off different supplier attributes, often using scoring assessment methods.
- Managing ongoing supply chain relationships involves clarifying supply expectations, often using service-level agreements (SLAs).
- Supplier development can benefit both suppliers and customers, especially in partnership relationships. Very often barriers are the mismatches in perception between customers and suppliers.

12.5 How is the demand side managed?

- Logistics is a critical part of supply chain management. Organisations must decide whether to adopt *first-party logistics* (1PL), *second-party logistics* (2PL), *third-party logistics* (3PL), *fourth-party logistics* (4PL) or *fifth-party logistics* (5PL).
- Deciding which transportation method to use for a two-step process. First, viable transport options must be identified. Second, selection is made based on relative performance benefits (in terms of quality, speed, dependability, flexibility, cost and sustainability) of available methods.
- Customer relationship management (CRM) is a method of learning more about customers' needs and behaviours to develop stronger relationships with them. It helps to increase revenues by providing services that more closely align with customer needs.

12.6 What are the dynamics of supply chains?

- Supply chains have a dynamic that is often called the *bullwhip effect*. It means that relatively small changes at the demand end of the chain increasingly amplify into large disturbances as they move upstream.
- Methods to reduce the bullwhip effect include *channel alignment* through standardised planning and control methods that coordinate the whole network; *improved operational efficiency* of each part of the chain to prevent local errors multiplying; and *improved information sharing* across the network.
- Blockchain technology is increasingly used to give a trusted overview of supply chain transactions.

CASE STUDY Big or small? EDF's sourcing dilemma

(This case was co-authored with Jas Kalra, Bartlett School of Construction and Project Management, University College London, and Jens Roehrich and Brian Squire, School of Management, University of Bath)

It was a warm afternoon as Stefano Moretto, Commercial Director of Hinkley Point C (HPC), and Eva Glines, Senior Supply Chain Engagement Manager, stood looking out of their office. Stefano, having recently joined EDF, had been tasked with establishing a supply network for the recently approved HPC project – the first of a number of new nuclear power stations aimed at supporting the United Kingdom (UK) Government's intended ambition of creating a 'clean, safe and affordable' energy future.

EDF and the Hinkley Point C (HPC) project

As with many other projects that the two had worked on, HPC would be complex and EDF would be right at the centre of things. The firm would hold detailed knowledge of the power station's design, maintain codes and standards, contribute to the design process at a strategic level and ensure the execution of the detailed design at an operational level. Over the coming years, EDF and its partners would need to build, test and commission two massive reactors, turbine halls and an electricity substation on the site, located in the South-West of the UK. Managing construction would involve people from many different organisations and disciplines working alongside each other. It would be EDF's responsibility to ensure that all relationships were underpinned by a consistent set of values and behaviours. However, at that moment, Stefano and Eva were preoccupied with another piece of the jigsaw – the site operations services.

Site operations services

Site operations services involved all the services not required for the actual construction of HPC. *People and organisation* services were there to make workers' lives more convenient and pleasant. Examples included catering, hospitality, cleaning, security and transportation. *Space and infrastructure* services were concerned with the physical infrastructure of the site. Examples included running the site's network of permanent and temporary roads safely and securely, as well as building maintenance, heating, lighting, plumbing, fire safety, etc. At the time, EDF expected to spend over £500 million on its site operations services over the period of HPC's construction (in fact, this figure rose to over £1 billion as the project developed). Given the large number of people likely to be on site at any one time, they were going to be vital to the project's success. *'It'll be like a small town'*, said Stefano. *'We're going to have a lot of people to move about, accommodate and feed.'* *'I know'*, said Eva, *'and the clock's ticking! We need to make some decisions on the way we're going to go with this soon.'* The two then sat back down and began to discuss some of the key services needed at the HPC site and its associated development sites in nearby towns:

Catering

The HPC catering contract covered all aspects of catering at the HPC site and associated developments (AD). The scope of supply included: food production and preparation; food and beverage vending across the site; bar operations; management of food waste; chilled and ambient food distribution; cleaning and maintenance of kitchen and food servery areas; hospitality services; and management of all catering-related staff.

Transportation

A transport service provider would be responsible for operating and managing a bus service to transport construction and office workers between the HPC main site, associated development sites, car-parks and selected local towns in the area. Bus services would be operated in the following ways:

- *Park and ride* – Workers would embark at a single point of departure, travel directly to HPC main site, disembark at the site perimeter, pass through security and embark on an internal bus to be dropped off at the contractors' compound.
- *Direct service* – Buses would drop-off/pick-up at a number of different points to deliver workers to the HPC main site. They would then follow the same process as above.
- *Direct secure* – Operated from specific locations, workers would pass through security before embarking on buses that would deliver them directly to their place of work on site. These buses would pass through a 'fast-gate' at the site perimeter, negating the need for internal bus transfer.

- *Internal buses* – The transport provider would provide buses on the internal HPC site circuit to use in conjunction with the *park and ride* and *direct service* offerings.

Accommodation management

EDF would be building new campus accommodation buildings at the HPC site and in Bridgwater, a town close to the site. Once completed, the contracted firm would need to run these facilities. This would include: management of the hotel services for campus sites; day-to-day management of the catering contractor (see above); management of the campus buildings, campus grounds and sports facilities; creating additional sales from hotel services; management of the security of guests while in the accommodation; 24/7 reception services; and management of all campus accommodation staff.

Facilities management

The facilities management contract would provide services to the HPC site and some of the associated developments. The scope of the services required would include: daily operational management of all temporary building facilities; general office services, including reception, porters and drivers, postal services and room booking; daily office cleaning, window cleaning and specialist cleaning (server rooms, etc.); domestic waste removal including confidential and segregated recycling; mechanical, electrical and building fabric maintenance, internal plumbing and drainage, and pest control; and audio-visual (AV) equipment management, maintenance and support.

Infrastructure operations and maintenance

EDF would construct the necessary permanent and temporary road network and other infrastructure for the HPC project both at the main construction site and at the associated development sites. Once constructed, these would need to be operated and maintained to ensure that the sites would be managed in a safe, secure and efficient way. The infrastructure operations and maintenance (O&M) contract would cover this activity.

Big or small?

'*OK, so we're pretty clear on what we need Eva. The next big question is big or small?*' Eva knew what Stefano was referring to – two competing views on the best strategy for sourcing EDF's site operations services. On one side was the argument that EDF should use one large 'generalist' supplier for each of the five main categories identified. These suppliers were typically multinational companies (MNCs) capable of providing a one-stop shop for the complete service solution required in any given contract. It was a tried-and-tested solution for sourcing in projects of this size. On the other side was the view that the firm should instead look to award contracts to local specialist small and medium-sized enterprises (SMEs) where possible. To do this would require breaking down some of the categories into more 'bitesize' contracts or possibly encouraging consortia of local suppliers who could jointly deliver site service requirements (see Table 12.5).

'*Eva, you know I've been working with some of the MNCs for years. They've got proven experience and expertise. Given that you, me and the team have got to set up over 150 tier-1 supply contracts over the next few months, maybe we're better going with what we know?*' Eva thought for a moment and responded: '*I see that argument, but sometimes I don't really rate these [MNC] firms. My experience is often they're hard work in negotiations and it doesn't get a whole lot better once they start providing the services. Also, they have too many other customers to be really concerned about giving us top service quality. At least smaller suppliers are likely to really put the effort in. Besides, I think we're agreed that it's at least worth thinking about whether we could really do something different this time.*'

'*All good points Eva*', said Stefano. He was also very keen to invite local SMEs to take advantage of the opportunities afforded by the HPC project. '*How good would it be to actually make a difference to the region through our supply chain?*', he thought. Still, earlier in the day, a conference call with another manager, discussing a leading global catering supplier intending to bid for the catering contract, had left Stefano unsure. His colleague had argued that it made more sense to let an experienced contractor with global presence help EDF manage the uncertainties associated with these contracts. Then he thought about some of the conversations that he'd had over the last two weeks. '*Another issue I'm thinking about here,*' Stefano continued, '*is what some of the service managers, who will eventually be responsible for the quality of services, have been saying to me. They seem to think that the local suppliers may not have the capability to deliver on this scale. Ultimately, if these services are not performed right, this project won't get off the ground.*'

Eva sighed. Stefano was right that most of the suppliers in the region were indeed small businesses, with no experience of delivering on the scale that would be required by EDF. But in the back of her mind was the feeling that not developing a local supply base would be a missed opportunity. '*I don't really agree with them on this Stefano. To be honest, much of this is just about a fear of doing things differently. Yes, MNCs are the safe and familiar option, but we're not talking about rocket science stuff here! All these services should be possible for local suppliers to deliver, surely?*'

Stefano thought for a moment. '*That's true Eva. Still, local SMEs are going to need a lot of upskilling to align with our needs. Developing a bespoke, local supply base with capabilities up to the quality standard we need will be a lot of effort. Remember, this project is already a very BIG jigsaw – I suspect some of the stakeholders might not think it's a good idea to add yet more pieces! And another thing, it's not just servicing the initial three or four thousand people on site that worries me, it's the ability to scale up to the seven, eight or nine thousand we'll probably have at the height of construction.*'

Table 12.5 Early expressions of interest for site operations service contracts

Catering	Transportation	Accommodation management	Facilities management	Infrastructure operations and maintenance
MNC1 Global, experienced caterer	**MNC2** International transport service provider	**MNC3** Global hotel construction and operations	**MNC4** Global facilities management and utility company	**MNC5** Global civil engineering contractor, highway, road construction, facilities maintenance
SME1 Wholesale ingredients and condiments	**NC1** National bus company	**SME8** Bed and breakfast owner	**SME11** Mechanical and electrical engineers	**SME13** Groundworks, civil engineering, projects
SME2 Dairy and eggs producer	**SME7** Local bus company	**SME9** Independent laundry company	**SME12** Civil engineer and pipeline contractor	**SME14** Temporary traffic management service
SME3 Greengrocer		**SME10** Bricklaying, carpentry, decorating, plastering, rendering and roofing		
SME4 Tea and coffee supplier				
SME5 Baker				
SME6 Butcher				

Eva shook her head. '*The services might be complex to manage, but they're not really capital- or technology-intensive. I think capability development in this area would be a lot easier to achieve than the manufacturing operations. From my conversations with the local chamber of commerce, it's clear that local businesses are very keen to work with us. We just need to develop a process of engaging more with local businesses, as I think quite of few of them wouldn't even think to bid for this work at the moment. I also wonder if we should encourage them to club together for some of these bits of work. It wouldn't just help now, but even more as we look to scale up further down the line.*'

'*Maybe Eva, maybe.*' Stefano quietly reflected that he was glad to be working with someone who was so passionate and shared his desire to do something value-added for the community. Eva's inputs certainly helped, but he remained undecided as to the best route forward. In addition to the issues they had just discussed, there was the broader political and public pressure to create economic and social value for the region. Construction of HPC was controversial and a recent piece of research had revealed that a substantial proportion of local residents were opposed to the project, while the acceptance of others was 'potentially fragile'. So, any good news stories were likely to be appreciated by the firm's leadership. Stefano stood up again. '*Come on Eva, I think we need to get the team together and try and make a decision. Let's go for a coffee first – I think we're in for a long day!*'

QUESTIONS

1 How do the characteristics of different site operations services influence the sourcing decision (MNCs versus local SMEs)?

2 What other factors are affecting the decision?

3 If HPC were to adopt a sourcing strategy with a preference for local SMEs:

(a) How might it engage effectively with local businesses to encourage bids?

(b) How might it effectively configure these sourcing arrangements?

(c) What approaches might it take to supplier capability development?

Note: **All names and quantities mentioned in this case study are fictional.**

Problems and applications

All chapters have 'Problems and applications' questions that will help you practise analysing operations. They can be answered by reading the chapter. Model answers for the first two questions can be found on the companion website for this text.

1 The COO of Super Cycles was considering her sourcing strategy. *'I have two key questions, for each of our outsourced parts: what is the risk in the supply market, and what is the criticality of the product or service to our business?'* Four key outsourced components are shown in table below. What approaches to sourcing these components would you recommend?

Four outsourced components for Super Cycles

Component	Cost (as a proportion of total material cost)	Suppliers	Ease of changing supplier
The inner tubes	3%	Many alternative suppliers	Very easy. Could do it in days
Frame tubing	15%	Only one supplier capable at the moment. Could take a long time to develop a new supplier	Difficult in the short term, possible in the longer term
Carbon-fibre stem and bars	32%	Relatively large number of available suppliers	Relatively easy. Would probably take a few weeks for new contract
'Groupset' gearing system	35%	Few suppliers who are capable of manufacturing these components to sufficient quality	Complex to source. Could switch supplier in the longer term but would pose quality risk

2 A chain of women's apparel retailers had all their products made by Lopez Industries, a small but high-quality garment manufacturer. They worked based on two seasons: Spring/Summer season and Autumn/Winter. *'Sometimes we are left with surplus items because our designers have just got it wrong'*, said the retailer's chief designer. *'It is important that we are able to flex our order quantities from Lopez during the season. Although they are a great supplier in many ways, they can't change their production plans at short notice'.* Lopez Industries was aware of this. *'I know that they are happy with our ability to make even the most complex designs to a high level of quality. I also know that they would like us to be more flexible in changing our volumes and delivery schedules. I admit that we could be more flexible within the season. Partly, we can't do this because we must buy in cloth at the beginning of the season based on the forecast volumes from our customers. Even if we could change our production schedules, we could not get extra deliveries of cloth. We only deal with high-quality and innovative cloth manufacturers who are very large compared to us, so we do not represent much business for them'.* A typical cloth supplier said: *'We compete primarily on quality and innovation. Designing cloth is as much of a fashion business as designing the clothes into which it is made. Our cloth goes to tens of thousands of customers around the world. These vary considerably in their requirements, but presumably all of them value our quality and innovation'.* Use the supply network behaviour (gap) model and its gaps to analyse the relationships between the players in this chain.

3 The example of the bullwhip effect shown in Table 12.4 shows how a simple 5 per cent reduction in demand at the end of supply chain causes fluctuations that increase in severity the further back an operation is placed in the chain.

(a) Using the same logic and the same rules (i.e. all operations keep one period's inventory), what would the effect on the chain be if demand fluctuated period by period between 100 and 95?

That is, period 1 has a demand of 100, period 2 has a demand of 95, period 3 a demand of 100, period 4 a demand of 95 and so on?

(b) What happens if all operations in the supply chain decided to keep only half of the period's demand as inventory?

(c) Find examples of how supply chains try to reduce this bullwhip effect.

4 If you were the owner of a small local retail shop, what criteria would you use to select suppliers for the goods that you wish to stock in your shop? Visit two or three shops that are local to you and ask the owners how they select their suppliers. In what way were their answers different from what you thought they might be?

5 Many companies devise a policy on ethical sourcing, covering such things as workplace standards and business practices, health and safety conditions, human rights, legal systems, child labour, disciplinary practices, wages and benefits, etc.

(a) What do you think motivates a company to draw up a policy of this type?

(b) What other issues would you include in such a supplier selection policy?

6 Airline catering is a tough business. Meals must be of a quality that is appropriate for the class and type of flight, yet the airlines that are their customers are always looking to keep costs as low as possible, menus must change frequently and they must respond promptly to customer feedback. Forecasting passenger numbers is difficult. Suppliers are advised of likely numbers for each flight several days in advance, but the actual minimum number of passengers is only fixed six hours before take-off. Also, flight arrivals can be delayed, upsetting work schedules – even when on time, no more than 40 minutes are allowed before the flight takes off again. Airline caterers usually produce food on, or near, airports, using their own staff. Catering companies' suppliers are also usually airline specialists that are also located near the caterers. A consortium of Northern Foods, a leading food producer (that normally supplies retailers) and DHL won a large contract at Heathrow Airport against the traditional suppliers. DHL was already a large supplier to 'airside' caterers there, with its own premises at the airport. Northern Foods made the food at its existing factories and delivered it to DHL, which assembled it onto airline catering trays and transferred them to the aircraft.

(a) Why would an airline use a catering services company rather than organise its own on-board services?

(b) What are the main operations objectives that a catering services company must achieve to satisfy its customers?

(c) Why is it important for airlines to reduce turn-around time when an aircraft lands?

(d) Why was the Northern Foods–DHL consortium a threat to more traditional catering companies?

Selected further reading

Chopra, S. and Meindl, P. (2015) ***Supply Chain Management: Strategy, Planning, and Operation*****, 5th edn, Pearson, Harlow.**

One of the best of the specialist texts.

Christopher, M. (2016) ***Logistics and Supply Chain Management*****, 5th edn, Pearson, Harlow.**

Updated version of a classic that gives a comprehensive treatment on supply chain management by one of the gurus of the subject.

Voss, C. and Raz, T. (2016) ***Never Split the Difference: Negotiating as if Your Life Depended On It*****, Random House.**

A really interesting book examining a wide range of negotiation settings and building on some of the ideas of mutual gains bargaining and win-win approaches to negotiation.

Vyas, N. (2019) ***Blockchain and the Supply Chain: Concepts, Strategies and Practical Applications*****, Kogan Page, London.**
A practical book examining the potentially transformative effect of blockchain technology on supply networks, plus some very interesting insights on the origins of supply chain management, starting with Egypt in 2500 BC!

Notes on chapter

1. The information on which this example is based is taken from: Banker, S. (2017) Drones deliver life saving supplies in Africa, *Forbes*, 13 October; Stewart, J. (2017) Blood-carrying, life-saving drones take off for Tanzania, *Wired*, 24 August; https://flyzipline.com/how-it-works/ (accessed September 2021).
2. The information on which this example is based is taken from: van Marle, G. (2021) E-commerce giant JD.com applies to spin-off supply chain arm, The Loadstar, 19 February; CIW Team (2021) JD.com annual customers grew 30% to 472 million in 2020, China Internet Watch, 12 March; JD.com 'About us', https://corporate.jd.com/aboutUs (accessed September 2021); JD.com announces first quarter 2021 results (2021), JD.com, press release, https://ir.jd.com/news-releases/news-release-details/jdcom-announces-first-quarter-2021-results (accessed September 2021).
3. For more on operational aspects of consultancies, see Brandon-Jones, A., Lewis, M., Verma, R. and Walsman, M. (2016) Examining the characteristics and managerial challenges of professional services: an empirical study of management consultancy in the travel, tourism, and hospitality sector, *Journal of Operations Management*, (42–43), March, 9–24.
4. Fisher, M.L. (1997) What is the right supply chain for your product?, *Harvard Business Review*, 75 (2).
5. Adapted from Kraljic, P. (1983) Purchasing Must Become Supply Management, *Harvard Business Review*, September.
6. The information on which this example is based is taken from: Evans, J. (2020) Covid-19 crises highlights supply chain vulnerability, *Financial Times*, 28 May; MacDowall, A. (2021) Managing warehousing in a changed world, *Financial Management*, 5 January; Bonadio, B., Huo, Z., Levchenko, A. and Pandalai-Nayar, N. (2020) The role of global supply chains in the COVID-19 pandemic and beyond, Vox EU/CEPR, 25 May; Harapko, S. (2021) How COVID-19 impacted supply chains and what comes next, EY, 18 February.
7. Our thanks to John Gray and Beverly Osborn at The Ohio State University, and Susan Helper at Case Western Reserve University, for their perspectives in the writing of this critical commentary. Further details of the *TVC* approach can be found in Gray, John V., Helper, S. and Osborn, B. (2020) Value first, cost later: total value contribution as a new approach to sourcing decisions, *Journal of Operations Management*, 66 (6), 735–50.
8. The information on which this responsible operations box is based is taken from: Williams, G.A. (2021) China Cancels H&M, *Jing Daily*, 24 March; Indvik, L. (2021) Fashion, Xinjiang and the perils of supply chain transparency, *Financial Times*, 9 April; Danigelis, A. (2018) Supply chain transparency map reduces time, expense for big brands, *Eco-Business*, 1 February; Kuruvilla, S. and Li, C. (2021) Freedom of association and collective bargaining in global supply chains: a research agenda, *Journal of Supply Chain Management*, 57 (2), 43–57; Bloomburg (2021) Gaming gear maker Gigabyte dives after mocking 'Made in China', *Bloomberg News*, 12 May.
9. The information on which this example is based is taken from: author visit to The Donkey Sanctuary, Sidmouth, UK, 2020, https://www.thedonkeysanctuary.org.uk/ (accessed September 2021); Hameed, A., Tariq, M. and Yasin, M.A. (2016) Assessment of welfare of working donkeys and mules using health and behavior parameters, *Journal of Agricultural Science and Food Technology*, 2 (5), 69–74.
10. Our thanks to Stephen Disney at University of Exeter, for his help with this section.
11. Iansiti, M. and Lakhani, K.R. (2017) The truth about blockchain, *Harvard Business Review*, January–February.
12. The information on which this example is based is taken from: del Castillo, M. (2018) Shipping blockchain: Maersk spin-off aims to commercialize trade platform, Coindesk.com, 16 January; Slocum, H. (2018) Maersk and IBM to form joint venture applying blockchain to improve global trade and digitize supply chains, Maersk, press release, 18 January; Maersk (2019) TradeLens blockchain-enabled digital shipping platform continues expansion with addition of major ocean carriers Hapag-Lloyd and Ocean Network Express, Maersk, press release, 2 July.

13 Inventory management

KEY QUESTIONS

INTRODUCTION

Operations managers often have an ambivalent attitude towards inventories. On the one hand, they are costly, sometimes tying up considerable amounts of working capital. They are also risky because items held in stock could deteriorate, become obsolete or just get lost and, furthermore, they take up valuable space in the operation. On the other hand, they provide some security in an uncertain environment that one can deliver items in stock, should customers demand them. This is the dilemma of inventory management: in spite of the cost and the other disadvantages associated with holding stocks, they do facilitate the smoothing of supply and demand. In fact, inventories only exist because supply and demand are not exactly in harmony with each other (see Figure 13.1).

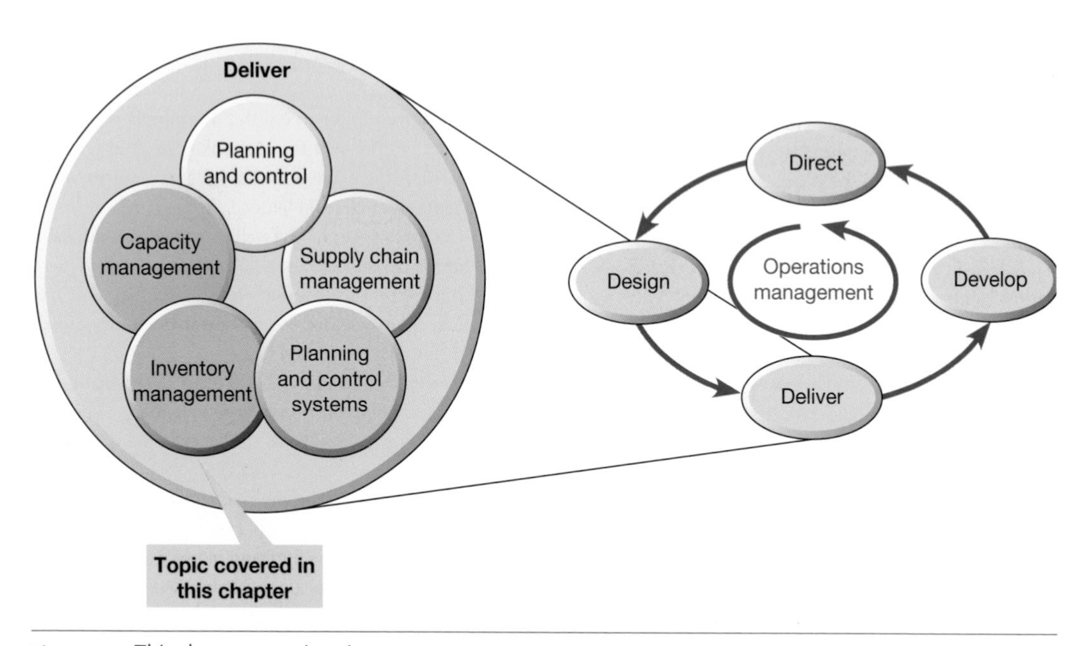

Figure 13.1 This chapter examines inventory management

13.1 What is inventory?

Inventory is a term used to describe the accumulations of materials, customers or information as they flow through processes or networks. Occasionally the term is also used to describe transforming resources, such as rooms in hotels or cars in a vehicle hire firm, but here we use the term for the accumulation of the transformed resources that flow through processes, operations or supply networks. Physical inventory (sometimes called 'stock') is the accumulation of physical materials such as components, parts, finished goods or physical (paper) information records. Queues are accumulations of customers, physical as in a queuing line or people in an airport departure lounge, or waiting for service at the end of phone lines or on-line requests. Databases are stores for accumulations of digital information, such as medical records or insurance details. Managing these accumulations is what we call 'inventory management'. And it's important. Material inventories in a factory can represent a substantial proportion of cash tied up in working capital. Minimising them can release large quantities of cash. However, reducing them too far can lead to customers' orders not being fulfilled. Customers held up in queues for too long can get irritated, angry, and possibly leave, so reducing revenue. Databases are critical for storing digital information and while storage may be inexpensive, maintaining databases may not be.

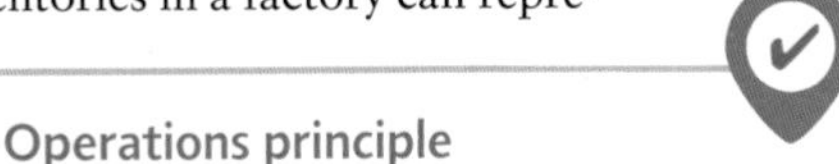

Operations principle

Inventories are accumulations of transformed resources; either physical items (called 'stock'), people (called queues) or information (called databases).

All processes, operations and supply networks have inventories

Most things that flow do so in an uneven way. Rivers flow faster down steep sections or where they are squeezed into a ravine. Over relatively level ground they flow slowly, and form pools or even large lakes where there are natural or human-made barriers blocking their path. It's the same in operations. Passengers in an airport flow from public transport or their vehicles, then have to queue at several points including check-in, security screening and immigration. They then have to wait again (a queue, even if they are sitting) in the departure lounge as they are joined (batched) with other passengers to form a group of several hundred people who are ready to board the aircraft. They are then squeezed down the air bridge as they file in one at a time to board the plane. Likewise, in a tractor assembly plant, stocks of components are brought into the factory in bulk, then stored next to the assembly line ready for use. Finished tractors will also be stored until the transporter comes to take them away in ones or tens to the dealers or directly to the end customer. Similarly, a government tax department collects information about citizens' finances from various sources, including employers, tax returns, information from banks or other investment companies, and stores this in databases until they are checked, sometimes by people, sometimes automatically, to create our tax codes and/or tax bills. In fact, because most operations involve flows of materials, customers and/or information, at some points they are likely to have material and information inventories and queues of customers waiting for goods or services (see Table 13.1).

Inventories are often the result of uneven flows. If there is a difference between the timing or the rate of supply and demand at any point in a process or network then accumulations will occur. A common analogy is the water tank shown in Figure 13.2. If, over time, the rate of supply of water to the tank differs from the rate at which it is demanded, a tank of water (inventory) will be needed to maintain supply. When the rate of supply exceeds the rate of demand, inventory increases; when the rate of demand exceeds the rate of supply, inventory decreases. So, if an operation or process can match supply and demand rates, it will also succeed in reducing its inventory levels. But most organisations must cope with unequal supply and demand, at least at some points in their supply chain.

Table 13.1 Examples of inventory held in processes, operations or supply networks

Process, operation or supply network	'Inventories'		
	Physical inventories	Queues of customers	Information in databases
Hotel	Food items, drinks, toilet items	At check-in and check-out	Customer details, loyalty card holders, catering suppliers
Hospital	Dressings, disposable instruments, blood	Patients on a waiting list, patients in bed waiting for surgery, patients in recovery wards	Patient medical records
Credit card application process	Blank cards, statements	Customers waiting on the phone	Customers' credit and personal information
Computer manufacturer	Components for assembly, packaging materials, finished computers ready for sale	Customers waiting for delivery of their computer	Customers' details, supplier information

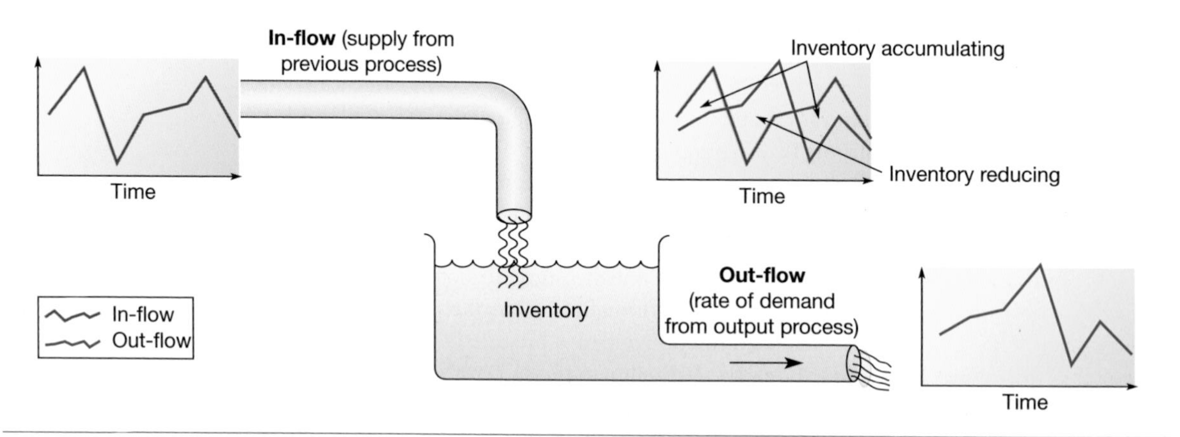

Figure 13.2 Inventory is created to compensate for the differences in timing between supply and demand

Inventoried information can be different

There is a complication when using this 'water flow' analogy to represent flows and accumulations (inventories) of information. Inventories of information can either be stored because of uneven flow, in the same way as materials and people, or stored because the operation needs to use the information to process something in the future. For example, an online retail operation will process each order it receives, and inventories of information may accumulate because of uneven flows. But, in addition, customer details could be permanently stored in a database. This information will then be used, not only for future orders from the same customer, but also for other processes, such as targeting promotional activities. In this case the inventory of information has turned from a transformed resource into a transforming resource, because it is being used to transform other information rather than being transformed itself. So, whereas managing physical material concerns ordering and holding the right amounts of goods or materials to deal with the variations in flow, and managing queues is about the level of resources to deal with demand, a database is the accumulation of information but may not cause an interruption to the flow. Managing databases is about the organisation of the data, its storage, security and retrieval (access and search).

OPERATIONS IN PRACTICE

An inventory of energy[1]

Inventory exists to smooth out the differences over time between supply and demand. And the bigger the gap between supply and demand, the more useful inventory is. But for some industries there is a big problem – they deal in things that cannot be stored very easily. And probably the best illustration of this is the business of generating and supplying energy. First, demand can fluctuate wildly, especially in countries that use large amounts of energy for cooling or heating. Nor can generating capacity be planned purely on the basis of average demand. In electricity generation, aggregated demand and average usage don't count for much when demand can spike with little warning. Second, supply, especially of the most convenient or cleaner forms of energy are not always available at the right time. For example, wind does not blow all the time. Worse than that, it tends to be at its strongest at night, when demand is low. Third, in most countries, regulators require energy firms to preserve a safety margin over total estimated demand to safeguard a reliable supply to citizens. Finally, energy is not easy to store. It would be ideal if energy firms could easily store excess energy, such as that produced by wind turbines at night, for later use at peak times. This so-called 'time shifting' would counteract the irregular supply from 'green' sources such as wind and solar power, which would make them simpler to integrate into the grid. If energy could be stored it would also permit what energy companies call 'peak shaving', using stored energy instead of having to buy more expensive energy on the spot (short-term) market. So how can energy be stored? Batteries can deliver power for short periods but are expensive for storing (or discharging) energy at the high rates (hundreds of megawatts) or the huge quantities (thousands of megawatt hours). However, this is changing. *'You have seen prices fall through the floor'*, said Claire Curry, a senior analyst at Bloomberg New Energy Finance. *'Batteries have been very expensive'*, she says. *'However, because prices are falling so dramatically we are now seeing some cases where the utility or grid operators see value in a battery'*.

The most practical method of energy storage, and the most widely used, is pumped-storage hydropower (PSH). This method harnesses water and gravity to 'store' off-peak power and release it during periods of high demand by using off-peak electricity to pump water from one reservoir up to another higher one. The water is then released back down to the lower reservoir, when power is needed, through a turbine that produces electricity. The drawback to traditional PSH is that it requires two reservoirs at different heights. Which is why, if greener energy is to be stored, new methods need to be developed. One idea is to use wind turbines to pump water from a deep central reservoir out to sea, which is allowed to flow back into the reservoir through turbines that produce electricity. Another is to pump water to raise a piston that sinks back down through a generator. Another is to use modified railway cars on a specially built track that utilise off-peak electricity to get to the top of a hill, and releasing the cars run back down the track so that their motion drives a generator. Other ideas include using compressed air to store the energy, using argon gas to transfer heat between two vast tanks filled with gravel, and storing energy in molten salt. But for whichever method proves the most effective at creating energy inventories, there will be rewards, both in terms of the potential market and in enabling the better use of sustainable energy.

13.2 Why should there be any inventory?

There are plenty of reasons to avoid accumulating inventory where possible. Table 13.2 identifies some of these, particularly those concerned with cost, space, quality and operational/organisational issues.

So why have inventory?

On the face of it, it may seem sensible to have a smooth and even flow of materials, customers and information through operational processes and networks, and thus not have any accumulations. In fact, inventories provide many advantages for both operations and their customers. If a customer has to go

Table 13.2 Some reasons to avoid inventories

	'Inventories'		
	Physical inventories	Queues of customers	Information in databases
Cost	Ties up working capital and there could be high administrative and insurance costs	Primarily time-cost to the customer, i.e. wastes customers' time	Cost of set-up, access, update and maintenance
Space	Requires storage space	Requires areas for waiting, or phone lines for held calls	Requires memory capacity. May require secure and/or special environment
Quality	May deteriorate over time, become damaged or obsolete	May upset customers if they have to wait too long. May lose customers	Data may be corrupted or lost or become obsolete
Operational/ organisational	May hide problems (see Chapter 16 on lean operations)	May put undue pressure on the staff and so quality is compromised for throughput	Databases need constant management; access control, updating and security

to a competitor because a part is out of stock or because they have had to wait too long or because the company insists on collecting all their personal details each time they call, the value of inventories seems undisputable. The task of operations management is to allow inventory to accumulate only when its benefits outweigh its disadvantages. The following are some of the benefits of inventory:

- ***Physical inventory is an insurance against uncertainty*** – Inventory can act as a buffer against unexpected fluctuations in supply and demand. For example, a retail operation can never forecast demand perfectly over the lead time. It will order goods from its suppliers such that there is always a minimum level of inventory to cover against the possibility that demand will be greater than expected during the time taken to deliver the goods. This is safety, or **buffer, inventory**. It can also compensate for the uncertainties in the process of the supply of goods into the store. The same applies with the output inventories, which is why hospitals always have a supply of blood, sutures and bandages for immediate response to accident and emergency patients. Similarly, auto-servicing services, factories and airlines may hold selected critical spare parts inventories so that the most common faults can be repaired without delay. See the 'Operations in practice' example, 'Safety stocks for coffee and COVID'.

Operations principle
Inventory should accumulate ONLY when the advantages of having it outweigh its disadvantages.

OPERATIONS IN PRACTICE
Safety stocks for coffee and COVID[2]

Drastic, inherently unpredictable, interruptions to supply have always been recognised as the justification for safety stocks – at least in theory. Earthquakes in Japan or machinery breakdowns in a Polish-based supplier's factory are the types of rare but inconvenient occurrences for which safety stocks are intended, and some operations will probably keep stocks to protect themselves. The problem is that what seems obvious when disaster strikes can seem like a waste of resources in normal times. For example, partly because of their political neutrality and self-reliance, Switzerland has traditionally kept relatively large stockpiles of food, medicine and animal feed. It is a policy it has maintained since the 1920s. Companies in the Swiss coffee trade, such as Nestlé, are required by law to store large quantities of raw coffee. Together, these safety stocks are enough to keep the Swiss in coffee for three months. But it is expensive, and it is the government that pays. So, the Swiss Federal Office for National Economic Supply decided that it should no longer pay for vast safety stocks of coffee, saying that the drink was not 'vital for life'. What it underestimated was the reaction of the Swiss public, who consume about 9 kg of coffee per person annually (twice as much as Americans). One opinion poll sponsored by Migros (a supermarket chain, which also owns a coffee brand), found that two-thirds of respondents could 'barely imagine a life without coffee'. Faced with the reaction, the Federal Office quickly suspended its decision.

More difficult are the decisions to keep safety stocks when the items concerned are not like coffee, consumed every day by everyone. Take the case of healthcare operations at the start of the COVID-19 pandemic. Countries varied in the amount and type of safety stocks they held at the start of the outbreak. Most health systems had recognised the possibility of a pandemic, even though it was seen as a 'black swan event' – an event that has a very high impact, but a very low probability of happening. And most kept some kind of safety stockpile. But often the safety stocks were designed to cover 'normal' fluctuations or spikes in demand, not the massive increase sparked by COVID-19. In the United Kingdom for example, the stocks of personal protective equipment (PPE) were a maximum of five- or six-weeks' worth of stock either in place or *en route*. Also, what constitutes the type of emergency that can release safety stock? In the United Kingdom, technically, the stockpile could be triggered only when the World Health Organization declared a 'pandemic influenza', yet COVID-19 was not flu. The government body responsible had to intervene (relatively quickly) to order the release of the stockpile. But the items in the safety stockpile were designed for a flu pandemic, and COVID-19 was a different disease with a higher hospitalisation rate. So, it contained the surgical face masks, FFP3 respirator masks, gloves and aprons needed to tackle an influenza outbreak, but not sufficient fluid-repellent gowns and visors that were to prove critical for treating a novel virus like COVID-19, which could survive much longer outside the body. Nor can safety stock items usually be quickly replaced - that's why they are kept as inventory. The panic to buy PPE at the beginning of the pandemic saw governments and private health organisations alike scrambling to replenish their stocks. States and companies were involved in a bidding war. Consignments of PPE were spirited away from airports before they could be sold to rivals. Some shipments were even diverted when the supplier received a higher bid.

- ***Physical inventory can counteract a lack of flexibility*** – Where a wide range of customer options is offered, unless the operation is perfectly flexible, stock will be needed to ensure supply when it is engaged In other activities. This is sometimes called cycle inventory. For example, Figure 13.3 shows the inventory profile of a baker who makes three types of bread. Because of the nature of the mixing and baking process, only one kind of bread can be produced at any time. The baker will have to produce each type of bread in batches large enough to satisfy the demand for each kind of bread between the times when each batch is ready for sale. So, even when demand is steady and predictable, there will always be some inventory to compensate for the intermittent supply of each type of bread.
- ***Physical inventory allows operations to take advantage of short-term opportunities*** – Sometimes opportunities arise that necessitate accumulating inventory, even when there is no immediate demand for it. For example, a supplier may be offering a particularly good deal on selected items for a limited time period, perhaps because they want to reduce their own finished goods

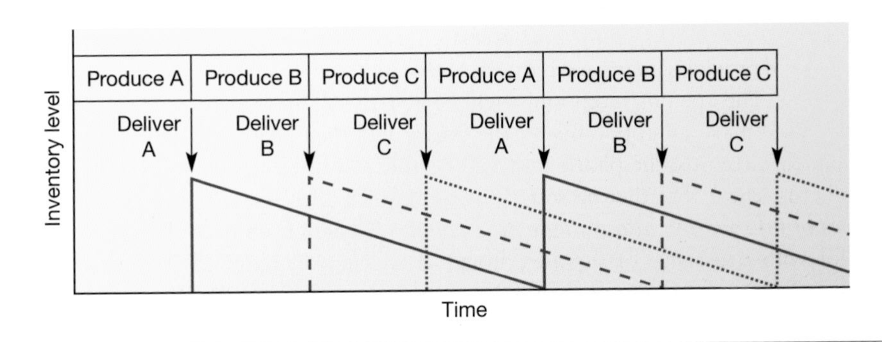

Figure 13.3 Cycle inventory in a bakery

inventories. Under these circumstances a purchasing department may take advantage opportunistically of the short-term price advantage.

- ***Physical inventory can be used to anticipate future demands*** – Medium-term capacity management (covered in Chapter 11) may use inventory to cope with demand capacity. Rather than trying to make a product (such as chocolate) only when it is needed, it is produced throughout the year ahead of demand and put into inventory until it is needed. This type of inventory is called anticipation inventory and is most commonly used when demand fluctuations are large but relatively predictable.
- ***Physical inventory can reduce overall costs*** – Holding relatively large inventories may bring savings that are greater than the cost of holding the inventory. This may be when bulk-buying gets the lowest possible cost of inputs, or when large order quantities reduce both the number of orders placed and the associated costs of administration and material handling. This is the basis of the 'economic order quantity' (EOQ) approach that will be treated later in this chapter.
- ***Physical inventory can increase in value*** – Sometimes the items held as inventory can increase in value and so become an investment. For example, dealers in fine wines are less reluctant to hold inventory than dealers in wine that does not get better with age. (However, it can be argued that keeping fine wines until they are at their peak is really part of the overall process rather than inventory as such). A more obvious example is inventories of money. The many financial processes within most organisations will try to maximise the inventory of cash they hold because it is earning them interest.
- ***Physical inventory fills the processing 'pipeline'*** – 'Pipeline' inventory exists because transformed resources cannot be moved instantaneously between the point of supply and the point of demand. When a retail store places an order, its supplier will 'allocate' the stock to the retail store in its own warehouse, pack it, load it onto its truck, transport it to its destination, and unload it into the retailer's inventory. From the time that stock is allocated (and therefore it is unavailable to any other customer) to the time it becomes available for the retail store, it is pipeline inventory. Especially in geographically dispersed supply networks, pipeline inventory can be substantial.
- ***Queues of customers help balance capacity and demand*** – This is especially useful if the main service resource is expensive, for example doctors, consultants, lawyers or expensive equipment such as CAT scanners. By waiting a short time after their arrival, and creating a queue of customers, the service always has customers to process. This is also helpful where arrival times are less predictable, for example where an appointment system is not used or not possible.
- ***Queues of customers enable prioritisation*** – In cases where resources are fixed and customers are entering the system with different levels of priority, the formation of a queue allows the organisation to serve urgent customers while keeping other less urgent ones waiting. In some circumstances it is not usual to have to wait 3–4 hours for treatment in an accident and emergency ward, where more urgent cases take priority.
- ***Queuing gives customers time to choose*** – Time spent in a queue gives customers time to decide what products/services they require; for example, customers waiting in a fast-food restaurant have time to look at the menu so that when they get to the counter they are ready to make their order without holding up the server.
- ***Queues enable efficient use of resources*** – By allowing queues to form, customers can be batched together to make efficient use of operational resources. For example a queue for an elevator makes better use of its capacity; in an airport, by calling customers to the gate, staff can load the aircraft more efficiently and quickly.
- ***Databases provide efficient multi-level access*** – Databases are a relatively cheap way of storing information and providing many people with access, although there may be restrictions or different levels of access. The doctor's receptionist will be able to call up your records to check your name and address and make an appointment, the doctor will then be able to call up the appointment and the patient's records, the pharmacist will be able to call up the patient's name and prescriptions, and cross check for other prescriptions and known allergies, etc.
- ***Databases of information allow single data capture*** – There is no need to capture data at every transaction with a customer or supplier, though checks may be required.
- ***Databases of information speed the process*** – Amazon, for example, if you agree, stores your delivery address and credit card information so that purchases can be made with a single click, making it fast and easy for the customer.

Table 13.3 Some ways in which physical inventory may be reduced

Reason for holding inventory	Example	How inventory could be reduced
As an insurance against uncertainty	Safety stocks for when demand or supply is not perfectly predictable	▶ Improve demand forecasting ▶ Tighten supply, e.g. through service-level penalties
To counteract a lack of flexibility	Cycle stock to maintain supply when other products are being made.	▶ Increase flexibility of processes, e.g. by reducing changeover times (see Chapter 16) ▶ Using parallel processes producing output simultaneously (see Chapter 6)
To take advantage of relatively short-term opportunities	Suppliers offer 'time-limited' special low-cost offers	▶ Persuade suppliers to adopt 'everyday low prices' (see Chapter 2)
To anticipate future demands	Build up stocks in low-demand periods for use in high-demand periods	▶ Increase volume flexibility by moving towards a 'chase demand' plan (see Chapter 11)
To reduce overall costs	Purchasing a batch of products in order to save delivery and administration costs	▶ Reduce administration costs through purchasing process efficiency gains ▶ Investigate alternative delivery channels that reduce transport costs
To fill the processing 'pipeline'	Items being delivered to customer	▶ Reduce process time between customer request and dispatch of items ▶ Reduce throughput time in the downstream supply chain (see Chapter 12)

Reducing physical inventory

For the remainder of this chapter we will focus on physical inventory, largely because this is what most operations managers assume is meant by the term 'inventory'. Moreover, we assume that the objective of those who manage physical inventories is to reduce the overall level (and/or cost) of inventory while maintaining an acceptable level of customer service. Table 13.3 identifies some of the ways in which physical inventory may be reduced.

The effect of inventory on return on assets

One can summarise the effects on the financial performance of an operation by looking at how some of the factors of inventory management impact on 'return on assets', a key financial performance measure. Figure 13.4 shows some of these factors:

- Inventory governs the operation's ability to supply its customers. The absence of inventory means that customers are not satisfied with the possibility of reduced revenue.

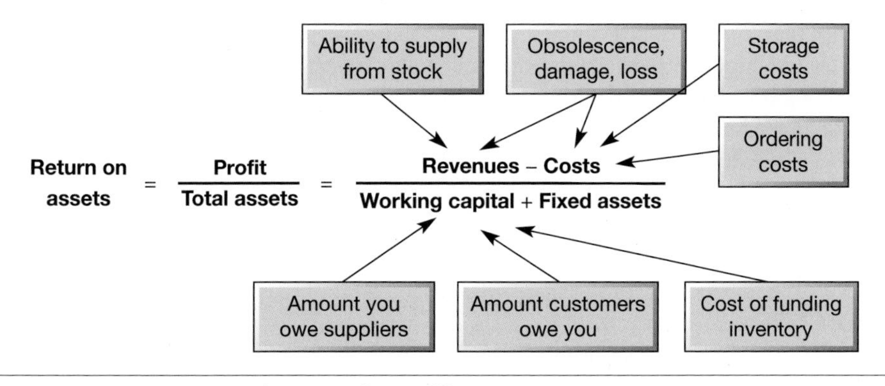

Figure 13.4 Inventory management has a significant effect on return on assets

- Inventory may become obsolete as alternatives become available, or could be damaged, deteriorate or simply get lost. This increases costs (because resources have been wasted) and reduces revenue (because the obsolete, damaged or lost items cannot be sold).
- Inventory incurs storage costs (leasing space, maintaining appropriate conditions, etc.). This could be high if items are hazardous to store (for example, flammable solvents, explosives, chemicals) or difficult to store, requiring special facilities (for example, frozen food).
- Inventory involves administrative and insurance costs. Every time a delivery is ordered, time and costs are incurred.
- Inventory ties up money, in the form of working capital, which is therefore unavailable for other uses, such as reducing borrowings or making investment in productive fixed assets (we shall expand on the idea of working capital later).
- Inventory contracts with suppliers can dictate the timing of when suppliers need to be paid. If they require paying before the operation receives payment from *its* customers (as is normal), the difference between the amount the operation owes suppliers and the amount suppliers owe the operation adds to working capital requirements.

Operations principle

Inventory management can have a significant effect on return on assets.

Day-to-day inventory decisions

Wherever inventory accumulates, operations managers need to manage the day-to-day tasks of managing inventory. Orders will be received from internal or external customers; these will be dispatched and demand will gradually deplete the inventory. Orders will need to be placed for replenishment of the stocks; deliveries will arrive and require storing. In managing the system, operations managers are involved in three major types of decisions:

- *How much to order* – Every time a replenishment order is placed, how big should it be? (Sometimes called the *volume decision*.)
- *When to order* – At what point in time, or at what level of stock, should the replenishment order be placed? (Sometimes called the *timing decision*.)
- *How to control the system* – What procedures and routines should be installed to help make these decisions? Should different priorities be allocated to different stock items? How should stock information be stored?

13.3 How much should be ordered? The volume decision

To illustrate this decision, consider again the example of the food and drinks we keep at our home. In managing this inventory, we implicitly make decisions on *order quantity*, which is how much to purchase at one time. In making this decision we are balancing two sets of costs: the costs associated with going out to purchase the food items and the costs associated with holding the stocks. The option of holding very little or no inventory of food and purchasing each item only when it is needed has the advantage that it requires little money, since purchases are made only when needed. However, it would involve purchasing provisions several times a day, which is inconvenient. At the very opposite extreme, making one journey to the local superstore every few months and purchasing all the provisions we would need until our next visit reduces the time and costs incurred in making the purchase but requires a very large amount of money each time the trip is made – money that could be invested elsewhere. We might also have to invest in extra cupboard units and a very large freezer. Somewhere between these extremes there will lie an ordering strategy that will minimise the total costs and effort involved in the purchase of food.

Inventory costs

The same principles apply in commercial order-quantity decisions as in the domestic situation. In making a decision on how much to purchase, operations managers must try to identify the costs that will be affected by their decision. Earlier we examined how inventory decisions affect some of the important components of return on assets. Here we take a cost perspective and re-examine these components in order to determine which costs go up and which go down as the order quantity increases. In the following list, the first three costs will decrease as order size is increased, whereas the next four generally increase as order size is increased:

1. *Cost of placing the order* – Every time that an order is placed to replenish stock, a number of transactions are needed, which incur costs to the company. These include preparing the order, communicating with suppliers, arranging for delivery, making payment and maintaining internal records of the transaction. Even if we are placing an 'internal order' on part of our own operation, there are still likely to be the same types of transactions concerned with internal administration.
2. *Price discount costs* – Often suppliers offer discounts for large quantities and cost penalties for small orders.
3. *Stock-out costs* – If we misjudge the order-quantity decision and our inventory runs out of stock, there will be lost revenue (opportunity costs) of failing to supply customers. External customers may take their business elsewhere, internal customers will suffer process inefficiencies.
4. *Working capital costs* – After receiving a replenishment order, the supplier will demand payment. Of course, eventually, after we supply our own customers, we in turn will receive payment. However, there will probably be a lag between paying our suppliers and receiving payment from our customers, when we will have to fund the costs of inventory. This is called the *working capital* of inventory. The costs associated with it are the interest we pay the bank for borrowing it, or the opportunity costs of not investing it elsewhere.
5. *Storage costs* – These are the costs associated with physically storing the goods. Renting, heating and lighting the warehouse, as well as insuring the inventory, can be expensive, especially when special conditions are required such as low temperature or high security.
6. *Obsolescence costs* – When we order large quantities, this usually results in stocked items spending a long time stored in inventory. This increases the risk that the items might either become obsolete (in the case of a change in fashion, for example) or deteriorate with age (in the case of most foodstuffs, for example).
7. *Operating inefficiency costs* – According to lean philosophies, high inventory levels prevent us seeing the full extent of problems within the operation. This argument is explored in Chapter 16.

It is worth noting that it may not be the same organisation that incurs the costs. For example, sometimes suppliers agree to hold consignment stock. This means that they deliver large quantities of inventory to their customers to store but will only charge for the goods as and when they are used. In the meantime, they remain the supplier's property so do not have to be financed by the customer, who does, however, provide storage facilities.

Inventory profiles

An inventory profile is a visual representation of the inventory level over time. Figure 13.5 shows a simplified inventory profile for one particular stock item in a retail operation. Every time an order is placed, Q items are ordered. The replenishment order arrives in one batch instantaneously. Demand for the item is then steady and perfectly predictable at a rate of D units per month. When demand has depleted the stock of the items entirely, another order of Q items instantaneously arrives, and so on. Under these circumstances:

$$\text{The average inventory} = \frac{Q}{2} \text{ (because the two shaded areas in Figure 13.5 are equal)}$$

$$\text{The time interval between deliveries} = \frac{Q}{D}$$

$$\text{The frequency of deliveries} = \text{the reciprocal of the time interval} = \frac{D}{Q}$$

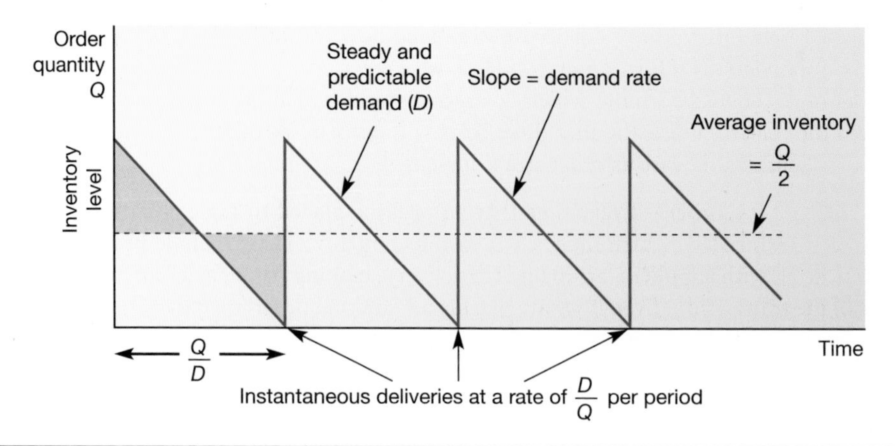

Figure 13.5 Inventory profiles chart the variation in inventory level

The economic order quantity (EOQ) formula

The most common approach to deciding how much of any particular item to order when stock needs replenishing is called the economic order quantity (EOQ) approach. This approach attempts to find the best balance between the advantages and disadvantages of holding stock. For example, Figure 13.6 shows two alternative order-quantity policies for an item. Plan A, represented by the unbroken line, involves ordering in quantities of 400 at a time. Demand in this case is running at 1,000 units per year. Plan B, represented by the dotted line, uses smaller but more frequent replenishment orders. This time only 100 are ordered at a time, with orders being placed four times as often. However, the average inventory for plan B is one-quarter of that for plan A.

To find out whether either of these plans, or some other plan, minimises the total cost of stocking the item, we need some further information, namely the total cost of holding one unit in stock for a period of time (C_h) and the total costs of placing an order (C_o). Generally, holding costs are taken into account by including:

- working capital costs;
- storage costs;
- obsolescence risk costs.

Order costs are calculated by taking into account:

- cost of placing the order (including transportation of items from suppliers if relevant);
- price discount costs.

In this case, the cost of holding stocks is calculated at £1 per item per year and the cost of placing an order is calculated at £20 per order.

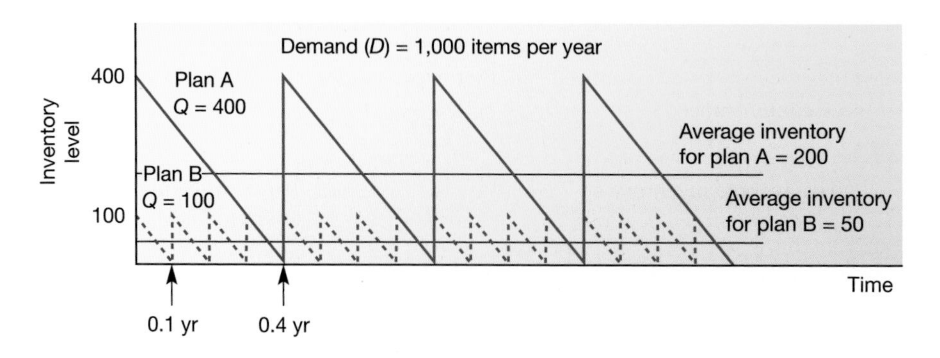

Figure 13.6 Two alternative inventory plans with different order quantities (Q)

We can now calculate total holding costs and ordering costs for any particular ordering plan as follows:

$$\text{Holding costs} = \text{Holding cost/unit} \times \text{Average inventory}$$

$$= C_h \times \frac{Q}{2}$$

$$\text{Ordering costs} = \text{Ordering cost} \times \text{Number of orders per period}$$

$$= C_o \times \frac{D}{Q}$$

So, total cost,

$$C_t = \frac{C_h Q}{2} + \frac{C_o D}{Q}$$

We can now calculate the costs of adopting plans with different order quantities. These are illustrated in Table 13.4. As we would expect with low values of Q, holding costs are low but the costs of placing orders are high because orders have to be placed very frequently. As Q increases, the holding costs increase but the costs of placing orders decrease. Initially the decrease in ordering costs is greater than the increase in holding costs and the total cost falls. After a point, however, the decrease in ordering costs slows, whereas the increase in holding costs remains constant and the total cost starts to increase. In this case the order quantity, Q, which minimises the sum of holding and order costs, is 200. This 'optimum' order quantity is called the *economic order quantity* (*EOQ*). This is illustrated graphically in Figure 13.7.

A more elegant method of finding the EOQ is to derive its general expression. This can be done using simple differential calculus as follows. From before:

$$\text{Total cost} = \text{Holding cost} + \text{Order cost}$$

$$C_t = \frac{C_h Q}{2} + \frac{C_o D}{Q}$$

The rate of change of total cost is given by the first differential of C_t with respect to Q:

$$\frac{dC_t}{dQ} = \frac{C_h}{2} + \frac{C_o D}{Q^2}$$

Table 13.4 Costs of adoption of plans with different order quantities

Demand (D) = 1,000 units per year Order costs (C_o) = £20 per order			Holding costs (C_h) = £1 per item per year		
Order quantity (Q)	*Holding costs* ($0.5Q \times C_h$)	+	*Order costs* [$(D/Q) \times C_o$]	=	*Total costs*
50	25		20 × 20 = 400		425
100	50		10 × 20 = 200		250
150	75		6.7 × 20 = 134		209
200	100		5 × 20 = 100		200*
250	125		4 × 20 = 80		205
300	150		3.3 × 20 = 66		216
350	175		2.9 × 20 = 58		233
400	200		2.5 × 20 = 50		250

* Minimum total cost.

The lowest cost will occur when $\frac{dC_t}{dQ} = 0$, that is:

$$0 = \frac{C_h}{2} + \frac{C_oD}{Q_o^2}$$

where Q_o = the EOQ. Rearranging this expression gives:

$$Q_o = \text{EOQ} = \sqrt{\frac{2C_oD}{C_h}}$$

When using the EOQ:

$$\text{Time between orders} = \frac{\text{EOQ}}{D}$$

$$\text{Order frequency} = \frac{D}{\text{EOQ}} \text{ per period}$$

Sensitivity of the EOQ

Examination of the graphical representation of the total cost curve in Figure 13.7 shows that, although there is a single value of Q, which minimises total costs, any relatively small deviation from the EOQ will not increase total costs significantly. In other words, costs will be near-optimum provided a value of Q that is reasonably close to the EOQ is chosen. Put another way, small errors in estimating either holding costs or order costs will not result in a significant deviation from the EOQ. This is a particularly convenient phenomenon because, in practice, both holding and order costs are not easy to estimate accurately.

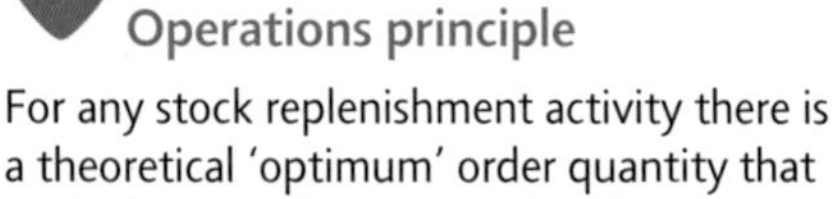

Operations principle

For any stock replenishment activity there is a theoretical 'optimum' order quantity that minimises total inventory-related costs.

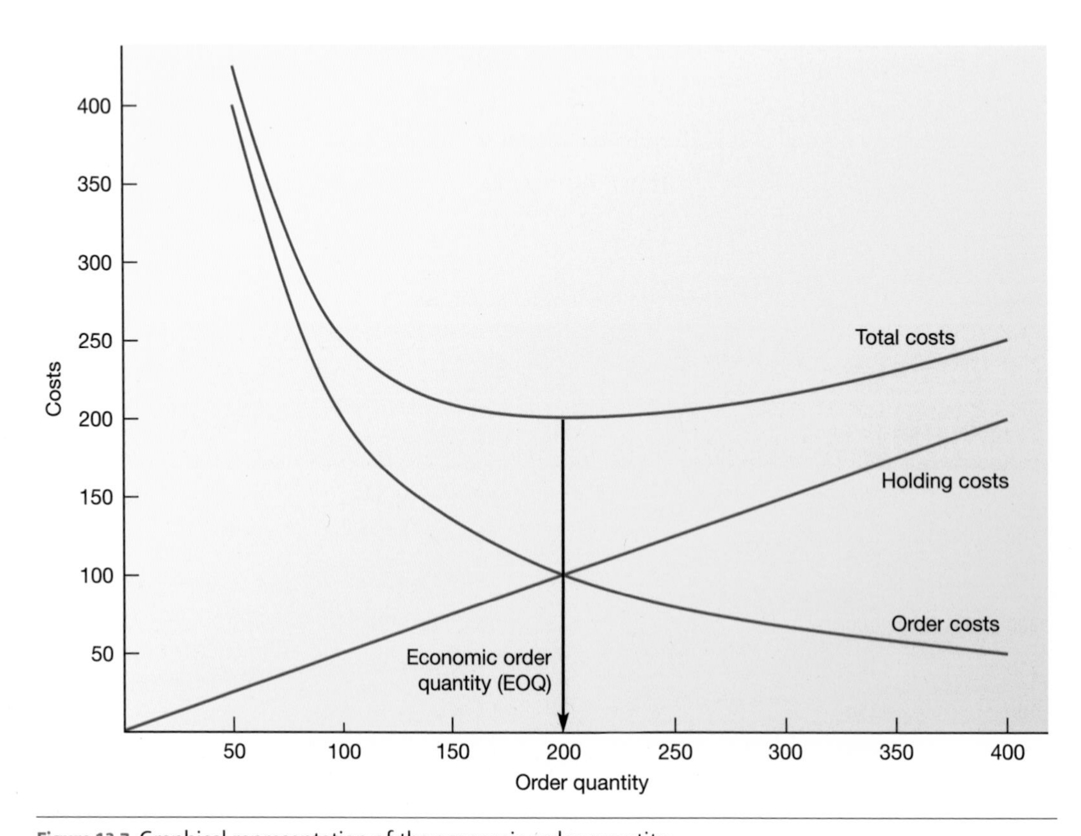

Figure 13.7 Graphical representation of the economic order quantity

Worked example

CVM building materials

CVM building materials obtains its bagged cement from a single supplier. Demand is reasonably constant throughout the year, and last year the company sold 2,000 tonnes of this product. It estimates the costs of placing an order at around £25 each time an order is placed, and calculates that the annual cost of holding inventory is 20 per cent of purchase cost. The company purchases the cement at £60 per tonne. How much should the company order at a time?

$$\text{EOQ for cement} = \sqrt{\frac{2C_oD}{C_h}}$$

$$= \sqrt{\frac{2 \times 25 \times 2{,}000}{0.2 \times 60}}$$

$$= \sqrt{\frac{100{,}000}{12}}$$

$$= 91.287 \text{ tonnes}$$

After calculating the EOQ the operations manager feels that placing an order for 91.287 tonnes *exactly* seems somewhat over-precise. Why not order a convenient 100 tonnes?

Total cost of ordering plan for $Q = 91.287$:

$$= \frac{C_hQ}{2} + \frac{C_oD}{Q}$$

$$= \frac{(0.2 \times 60) \times 91.287}{2} + \frac{25 \times 2{,}000}{91.287}$$

$$= £1{,}095.454$$

Total cost of ordering plan for $Q = 100$:

$$= \frac{(0.2 \times 60) \times 100}{2} + \frac{25 \times 2{,}000}{100}$$

$$= £1{,}100$$

The extra cost of ordering 100 tonnes at a time is £1,100 − £1,095.45 = £4.55. The operations manager therefore should feel confident in using the more convenient order quantity.

Gradual replacement – the economic batch quantity (EBQ) model

Although the simple inventory profile shown in Figure 13.5 made some simplifying assumptions, it is broadly applicable in most situations where each complete replacement order arrives at one point in time. In many cases, however, replenishment occurs over a time period rather than in one lot. A typical example of this is where an internal order is placed for a batch of parts to be produced on a machine. The machine will start to produce the parts and ship them in a more or less continuous stream into inventory, but at the same time demand is continuing to remove parts from the inventory. Provided the rate at which parts are being made and put into the inventory (P) is higher than the rate at which demand is depleting the inventory (D) then the size of the inventory will increase. After the batch has been completed the machine will be reset (to produce some other part), and demand will continue to deplete the inventory level until production of the next batch begins. The resulting profile is shown in Figure 13.8. Such a profile is typical for cycle inventories supplied by batch processes, where items are produced internally and intermittently. For this reason the

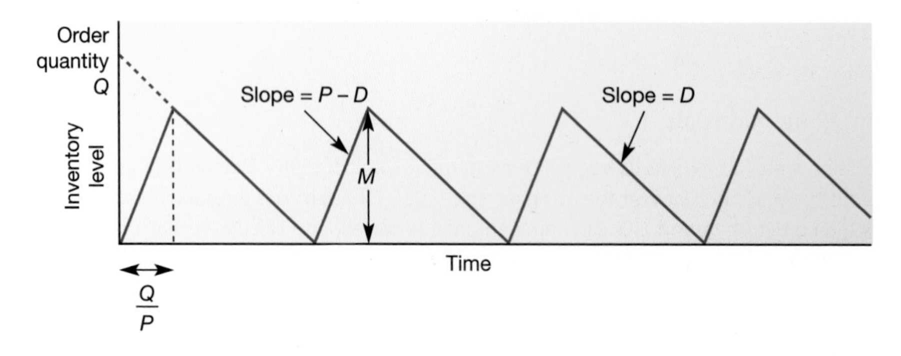

Figure 13.8 Inventory profile for gradual replacement of inventory

minimum-cost batch quantity for this profile is called the economic batch quantity (EBQ). It is also sometimes known as the economic manufacturing quantity (EMQ), or the production order quantity (POQ). It is derived as follows:

$$\text{Maximum stock level} = M$$

$$\text{Slope of inventory build-up} = P - D$$

Also, as is clear from Figure 13.8:

$$\text{Slope of inventory build-up} = M \div \frac{Q}{P}$$

$$= \frac{MP}{Q}$$

So,

$$\frac{MP}{Q} = P - D$$

$$M = \frac{Q(P - D)}{P}$$

$$\text{Average inventory level} = \frac{M}{2}$$

$$= \frac{Q(P - D)}{2P}$$

As before:

$$\text{Total cost} = \text{Holding cost} + \text{Order cost}$$

$$C_t = \frac{C_h Q(P - D)}{2P} + \frac{C_o D}{Q}$$

$$\frac{dC_t}{dQ} = \frac{C_h(P - D)}{2P} - \frac{C_o D}{Q^2}$$

Again, equating to zero and solving Q gives the minimum-cost order quantity EBQ:

$$\text{EBQ} = \sqrt{\frac{2C_o D}{C_h(1 - (D/P))}}$$

Worked example

Clonacola

The manager of Clonacola bottle-filling plant, which bottles soft drinks, needs to decide how long a 'run' of each type of drink to process. Demand for each type of drink is reasonably constant at 80,000 per month (a month has 160 production hours). The bottling lines fill at a rate of 3,000 bottles per hour, but take an hour to clean and reset between different drinks. The cost (of labour and lost production capacity) of each of these changeovers has been calculated at £100 per hour. Stock-holding costs are counted at £0.10 per bottle per month.

$$D = 80{,}000 \text{ per month}$$

$$= 500 \text{ per hour}$$

$$\text{EBQ} = \sqrt{\frac{2C_oD}{C_h(1-(D/P))}}$$

$$= \sqrt{\frac{2 \times 100 \times 80{,}000}{0.10(1-(500/3{,}000))}}$$

$$\text{EBQ} = 13{,}856$$

The staff who operate the lines have devised a method of reducing the changeover time from 1 hour to 30 minutes. How would that change the EBQ?

$$\text{New } C_o = £50$$

$$\text{New EBQ} = \sqrt{\frac{2 \times 50 \times 80{,}000}{0.10(1-(500/3{,}000))}}$$

$$= 9{,}798$$

Critical commentary

The approach to determining order quantity that involves optimising costs of holding stock against costs of ordering stock, typified by the EOQ and EBQ models, has always been subject to criticisms. Originally these concerned the validity of some of the assumptions of the model; more recently they have involved the underlying rationale of the approach itself. The criticisms fall into four broad categories, all of which we shall examine further.

- The assumptions included in the EOQ models are simplistic.
- The real costs of stock in operations are not as assumed in EOQ models.
- The models are really descriptive, and should not be used as prescriptive devices.
- Cost minimisation is not an appropriate objective for inventory management.

Responding to the criticisms of EOQ

In order to keep EOQ-type models relatively straightforward, it was necessary to make assumptions. These concerned such things as the stability of demand, the existence of a fixed and identifiable ordering cost, that the cost of stock-holding can be expressed by a linear function, shortage costs that were identifiable, and so on. While these assumptions are rarely strictly true, most of them can approximate to reality. Furthermore, the shape of the total cost curve has a relatively flat optimum point, which means that small errors will not significantly affect the total cost of a near-optimum order quantity. However, at times the assumptions do pose severe limitations to the models. For example, the assumption of steady demand (or even demand that conforms to some known probability distribution) is untrue for a wide range of the operation's inventory problems. For example, a bookseller might be very happy to adopt an EOQ-type ordering policy for some of its most regular and stable products such as dictionaries and popular reference books. However, the demand patterns for many other books could be highly erratic, dependent on critics' reviews and word-of-mouth recommendations. In such circumstances it is simple inappropriate to use EOQ models.

Cost of stock

Other questions surround some of the assumptions made concerning the nature of stock-related costs. For example, placing an order with a supplier as part of a regular and multi-item order might be relatively inexpensive, whereas asking for a special one-off delivery of an item could prove far more costly. Similarly with stock-holding costs – although many companies make a standard percentage charge on the purchase price of stock items, this might not be appropriate over a wide range of stock-holding levels. The marginal costs of increasing stock-holding levels might be merely the cost of the working capital involved. On the other hand, it might necessitate the construction or lease of a whole new stock-holding facility such as a warehouse. Operations managers using an EOQ-type approach must check that the decisions implied by the use of the formulae do not exceed the boundaries within which the cost assumptions apply. In Chapter 16 we explore the 'lean' approach that sees inventory as being largely negative. However, it is useful at this stage to examine the effect on an EOQ approach of regarding inventory as being more costly than previously believed. Increasing the slope of the holding cost line increases the level of total costs of *any* order quantity, but more significantly, shifts the minimum cost point substantially to the left, in favour of a lower economic order quantity. In other words, the less willing an operation is to hold stock on the grounds of cost, the more it should move towards smaller, more frequent ordering.

Using EOQ models as prescriptions

Perhaps the most fundamental criticism of the EOQ approach again comes from 'lean' philosophies. The EOQ tries to optimise order decisions. Implicitly the costs involved are taken as fixed, in the sense that the task of operations managers is to find out what are the true costs rather than to change them in any way. EOQ is essentially a reactive approach. Some critics would argue that it fails to ask the right question. Rather than asking the EOQ question of 'What is the optimum order quantity?' operations managers should really be asking, 'How can I change the operation in some way so as to reduce the overall level of inventory I need to hold?' The EOQ approach may be a reasonable description of stock-holding costs but should not necessarily be taken as a strict prescription over what decisions to take. For example, many organisations have made considerable efforts to reduce the effective cost of placing an order. Often they have done this by working to reduce changeover times on machines. This means that less time is taken changing over from one product to the other, and therefore less operating capacity is lost, which in turn reduces the cost of the changeover. Under these circumstances, the order cost curve in the EOQ formula reduces and, in turn, reduces the effective economic order quantity. Figure 13.9 shows the EOQ formula represented graphically with increased holding costs (see the previous discussion) and reduced order costs. The net effect of this is to significantly reduce the value of the EOQ.

Should the cost of inventory be minimised?

Many organisations (such as supermarkets and wholesalers) make most of their revenue and profits simply by holding and supplying inventory. Because their main investment is in the inventory, it is critical that they make a good return on this capital, by ensuring that it has the highest possible 'stock turn' (defined later in this chapter) and/or gross profit margin. Alternatively, they may also be concerned to maximise the use of space by seeking to maximise the profit earned per square metre. The EOQ model does not address these objectives. Similarly for products that deteriorate or go out of fashion, the EOQ model can result in excess inventory of slower-moving items. In fact, the EOQ model is rarely used in such organisations, and there is more likely to be a system of periodic review (described later) for regular ordering of replenishment inventory. For example, a typical builders' supply merchant might carry around 50,000 different items of stock (SKUs). However, most of these

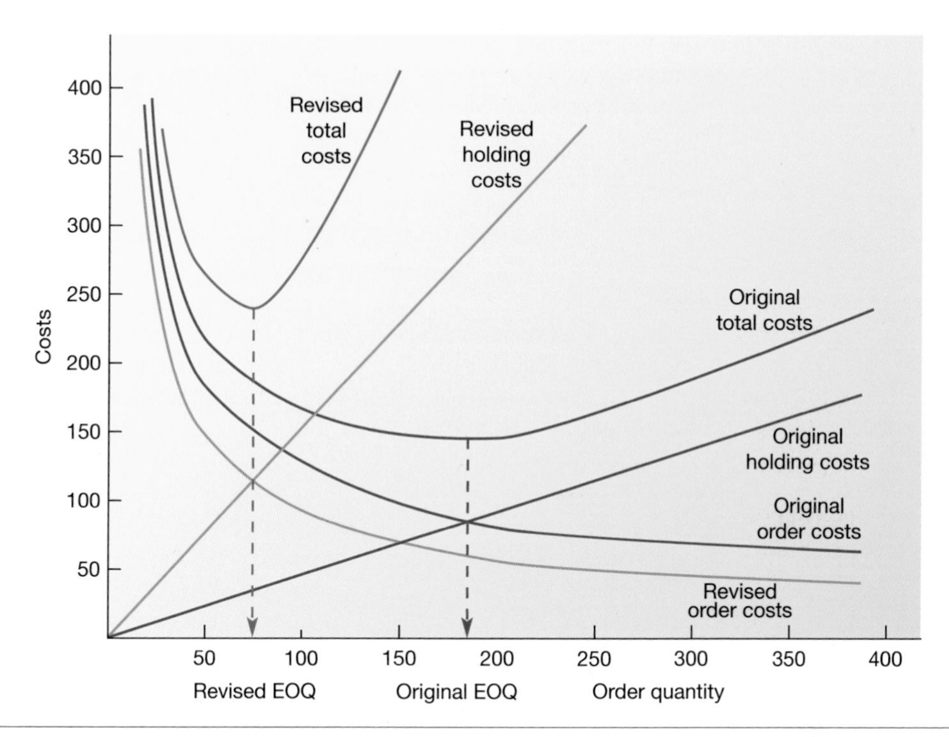

Figure 13.9 If the true costs of stock holding are taken into account, and if the cost of ordering (or changeover) is reduced, the economic order quantity (EOQ) is much smaller

cluster into larger families of items such as paints, sanitary ware or metal fixings. Single orders are placed at regular intervals for all the required replenishments in the supplier's range, and these are then delivered together at one time. For example, if such deliveries are made weekly, then on average, the individual item order quantities will be for only one week's usage. Less popular items, or ones with erratic demand patterns, can be individually ordered at the same time, or (when urgent) can be delivered the next day by carrier.

If customers won't wait – the news vendor (single-period) inventory problem

A special case of the inventory order-quantity decision is when an order quantity is purchased for a specific event or time period, after which the items are unlikely to be sold. A simple example of this is the decision taken by a newspaper vendor of how many newspapers to stock for the day. If the news vendor should run out of papers, customers will either go elsewhere or decide not to buy a paper that day. Newspapers left over at the end of the day are worthless, and demand for the newspapers varies day-by-day. In deciding how many newspapers to carry, the news vendor is in effect balancing the risk and consequence of running out of newspapers against that of having newspapers left over at the end of the day. Retailers and manufacturers of high-class leisure products, such as some books and popular music CDs, face the same problem. For example, a concert promoter needs to decide how many concert T-shirts emblazoned with the logo of the main act to order. The profit on each T-shirt sold at the concert is £5 and any unsold T-shirts are returned to the company that supplies them, but at a loss to the promoter of £3 per T-shirt. Demand is uncertain but is estimated to be between 200 and 1,000. The probabilities of different demand are as follows:

Demand level	200	400	600	800
Probability	0.2	0.3	0.4	0.1

How many T-shirts should the promoter order? Table 13.5 shows the profit that the promoter would make for different order quantities and different levels of demand.

Table 13.5 Pay-off matrix for T-shirt order quantity (profit or loss in £)

Demand level	200	400	600	800
Probability	0.2	0.3	0.4	0.1
Promoter orders 200	1,000	1,000	1,000	1,000
Promoter orders 400	400	2,000	2,000	2,000
Promoter orders 600	−200	1,400	3,000	3,000
Promoter orders 800	−800	800	4,000	4,000

We can now calculate the expected profit that the promoter will make for each order quantity by weighting the outcomes by their probability of occurring.

If the promoter orders 200 T-shirts:

$$\text{Expected profit} = 1{,}000 \times 0.2 + 1{,}000 \times 0.3 + 1{,}000 \times 0.4 + 1{,}000 \times 0.1$$
$$= £1{,}000$$

If the promoter orders 400 T-shirts:

$$\text{Expected profit} = 400 \times 0.2 + 2{,}000 \times 0.3 + 2{,}000 \times 0.4 + 2{,}000 \times 0.1$$
$$= £1{,}680$$

If the promoter orders 600 T-shirts:

$$\text{Expected profit} = -200 \times 0.2 + 1{,}400 \times 0.3 + 3{,}000 \times 0.4 + 3{,}000 \times 0.1$$
$$= £1{,}880$$

If the promoter orders 800 T-shirts:

$$\text{Expected profit} = -800 \times 0.2 + 800 \times 0.3 + 2{,}400 \times 0.4 + 4{,}000 \times 0.1$$
$$= £1{,}440$$

The order quantity that gives the maximum profit is 600 T-shirts, which results in a profit of £1,880.

The importance of this approach lies in the way it takes a probabilistic view of part of the inventory calculation (demand), something we shall use again in this chapter.

OPERATIONS IN PRACTICE

Mr Ruben's bakery[3]

Be careful about treating the news vendor problem on a product-by-product basis. It is a powerful idea, but needs to be seen in context. Take the famous (in New York) City Bakery, in Manhattan. It is run by Maury Rubin, a master baker, who knows the economics of baking fresh products. Ingredients and rent are expensive. It costs Mr. Rubin $2.60 to make a $3.50 croissant. If he makes 100 and sells 70, he earns $245 but his costs are $260, and because all goods are sold within a day (his quality standards mean that he won't sell leftovers), he loses money. Nor can he raise his prices. In his competitive market, he says, shoppers bristle when the cost of baked goods

passes a certain threshold. However, Mr. Ruben has two 'solutions'. First, he can subsidise his croissants by selling higher-margin items such as fancy salads and sandwiches. Second, he uses data to cut waste, by studying sales to detect demand trends, so that he can fine-tune supply. He monitors the weather carefully (demand drops away when it rains) and carefully inspects school calendars so he can reduce the quantities he bakes during school holidays. Each day in the morning, he makes sure that pastries are prepared, but then he checks sales every 60–90 minutes before making the decision to adjust supply or not. Only when the numbers are in do the pastries go into the oven. Having no croissants left by the end of the day is a sign of success.

13.4 When should an order be placed? The timing decision

When we assumed that orders arrived instantaneously and demand was steady and predictable, the decision on when to place a replenishment order was self-evident. An order would be placed as soon as the stock level reached zero. This would arrive instantaneously and prevent any stock-out occurring. If replenishment orders do not arrive instantaneously, but have a lag between the order being placed and it arriving in the inventory, we can calculate the timing of a replacement order as shown in Figure 13.10. The lead time for an order to arrive in this case is two weeks, so the re-order point (ROP) is the point at which stock will fall to zero minus the order lead time. Alternatively, we can define the point in terms of the level that the inventory will have reached when a replenishment order needs to be placed. In this case this occurs at a re-order level (ROL) of 200 items.

However, this assumes that both the demand and the order lead time are perfectly predictable. In most cases, of course, this is not so. Both demand and the order lead time are likely to vary to produce a profile that looks something like that in Figure 13.11. In these circumstances it is necessary to make the replenishment order somewhat earlier than would be the case in a purely deterministic situation. This will result in, on average, some stock still being in the inventory when the replenishment order arrives. This is buffer (safety) stock. The earlier the replenishment order is placed, the higher will be the expected level of safety stock (s) when the replenishment order arrives. But because of the variability of both lead time (t) and demand rate (d), there will sometimes be a higher-than-average level of safety stock and sometimes lower. The main consideration in setting safety stock is not so much the average level of stock when a replenishment order arrives but rather the probability that the stock will not have run out before the replenishment order arrives.

The key statistic in calculating how much safety stock to allow is the probability distribution, which shows the lead-time usage. The lead-time usage distribution is a combination of the distributions that describe lead-time variation and the demand rate during the lead time. If safety stock is set below the

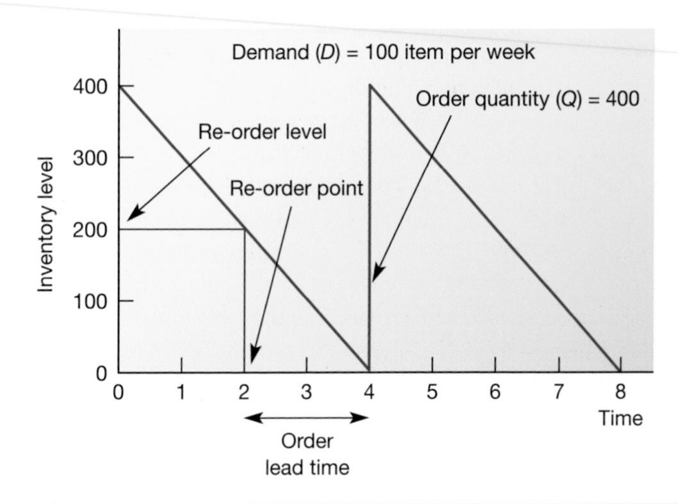

Figure 13.10 Re-order level (ROL) and re-order point (ROP) are derived from the order lead time and demand rate

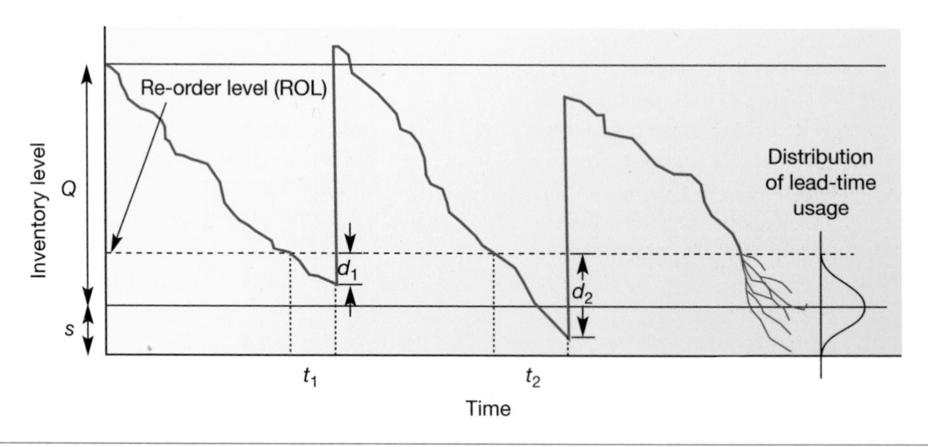

Figure 13.11 Safety stock (s) helps to avoid stock-outs when demand and/or order lead time are uncertain

lower limit of this distribution then there will be shortages every single replenishment cycle. If safety stock is set above the upper limit of the distribution, there is no chance of stock-outs occurring. Usually, safety stock is set to give a predetermined likelihood that stock-outs will not occur. Figure 13.11 shows that, in this case, the first replenishment order arrived after t_1, resulting in a lead-time usage of d_1. The second replenishment order took longer, t_2, and demand rate was also higher, resulting in a lead-time usage of d_2. The third order cycle shows several possible inventory profiles for different conditions of lead-time usage and demand rate.

Operations principle

For any stock replenishment activity, the timing of replenishment should reflect the effects of uncertain lead time and uncertain demand during that lead time.

Worked example

Knacko running shoes

Knacko imports running shoes for sale in its sports shops and it can never be certain of how long, after placing an order, the delivery will take. Examination of previous orders reveals that out of 10 orders: one took one week, two took two weeks, four took three weeks, two took four weeks and one took five weeks. The rate of demand for the shoes also varies between 110 pairs per week and 140 pairs per week. There is a 0.2 probability of the demand rate being either 110 or 140 pairs per week, and a 0.3 chance of demand being either 120 or 130 pairs per week. The company needs to decide when it should place replenishment orders if the probability of a stock-out is to be less than 10 per cent.

Both lead time and the demand rate during the lead time will contribute to the lead-time usage. So the distributions that describe each will need to be combined. Figure 13.12 and Table 13.6 show how this can be done. Taking lead time to be either one, two, three, four or five weeks, and demand rate to be either 110, 120, 130 or 140 pairs per week, and also assuming the two variables to be independent, the distributions can be combined as shown in Table 13.6. Each element in the matrix shows a possible lead-time usage with the probability of its occurrence. So if the lead time is one week and the demand rate is 110 pairs per week, the actual lead-time usage will be $1 \times 110 = 110$ pairs. Since there is a 0.1 chance of the lead time being one week, and a 0.2 chance of demand rate being 110 pairs per week, the probability of both these events occurring is $0.1 \times 0.2 = 0.02$.

We can now classify the possible lead-time usages into histogram form. For example, summing the probabilities of all the lead-time usages that fall within the range 100–199 (all the first column) gives a combined probability of 0.1. Repeating this for subsequent intervals results in Table 13.7.

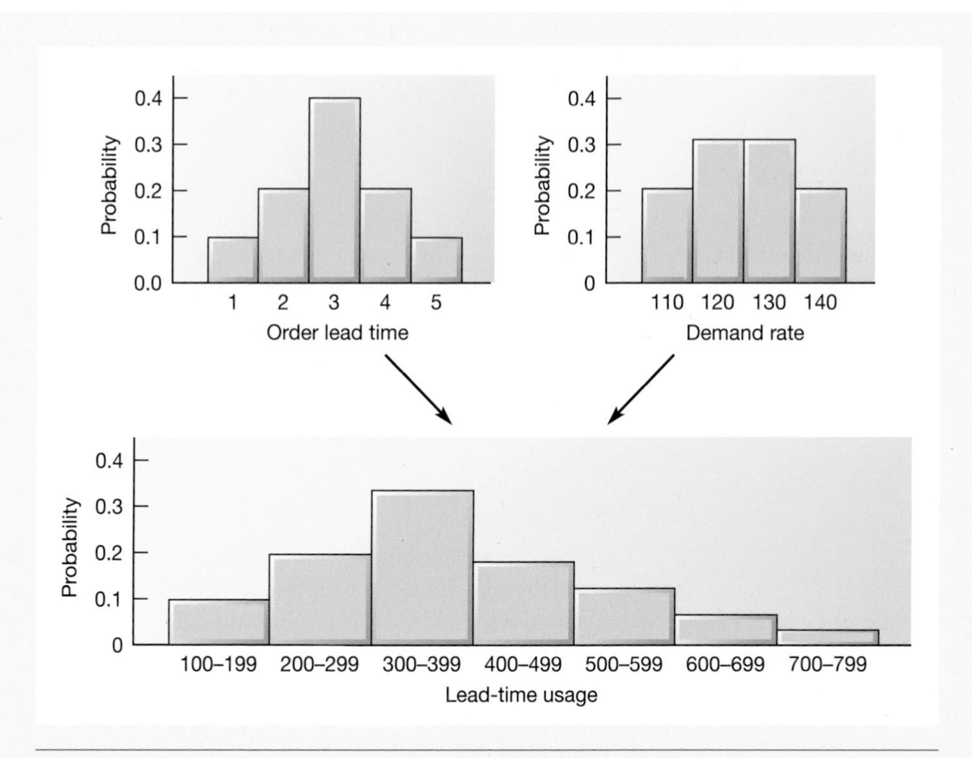

Figure 13.12 The probability distributions for order lead time and demand rate combine to give the lead-time usage distribution

Table 13.6 Matrix of lead-time and demand-rate probabilities

			Lead-time probabilities				
			1	2	3	4	5
			0.1	0.2	0.4	0.2	0.1
	110	0.2	110	220	330	440	550
			(0.02)	(0.04)	(0.08)	(0.04)	(0.02)
Demand-rate probabilities	120	0.3	120	240	360	480	600
			(0.03)	(0.06)	(0.12)	(0.06)	(0.03)
	130	0.3	130	260	390	520	650
			(0.03)	(0.06)	(0.12)	(0.06)	(0.03)
	140	0.2	140	280	420	560	700
			(0.02)	(0.04)	(0.08)	(0.04)	(0.02)

Table 13.7 Combined probabilities

Lead-time usage	100–199	200–299	300–399	400–499	500–599	600–699	700–799
Probability	0.1	0.2	0.32	0.18	0.12	0.06	0.02

Table 13.8 Combined probabilities

Lead-time usage X	100	200	300	400	500	600	700	800
Probability of usage being greater than X	1.0	0.9	0.7	0.38	0.2	0.08	0.02	0

This shows the probability of each possible range of lead-time usage occurring, but it is the cumulative probabilities that are needed to predict the likelihood of stock-out (see Table 13.8).

Setting the re-order level at 600 would mean that there is only a 0.08 chance of usage being greater than available inventory during the lead time, i.e. there is a less than 10 per cent chance of a stock-out occurring.

Continuous and periodic review

The approach we have described to making the replenishment timing decision is often called the continuous review approach. This is because, to make the decision in this way, there must be a process to review the stock level of each item continuously and then place an order when the stock level reaches its re-order level. The virtue of this approach is that, although the timing of orders may be irregular (depending on the variation in demand rate), the order size (Q) is constant and can be set at the optimum economic order quantity. Such continual checking on inventory levels can be time-consuming, especially when there are many stock withdrawals compared with the average level of stock, but in an environment where all inventory records are computerised, this should not be a problem unless the records are inaccurate.

An alternative and far simpler approach, but one that sacrifices the use of a fixed (and therefore possibly optimum) order quantity, is called the periodic review approach. Here, rather than ordering at a predetermined re-order level, the periodic approach orders at a fixed and regular time interval. So the stock level of an item could be found, for example, at the end of every month and a replenishment order placed to bring the stock up to a predetermined level. This level is calculated to cover demand between the replenishment order being placed and the following replenishment order arriving. Figure 13.13 illustrates the parameters for the periodic review approach.

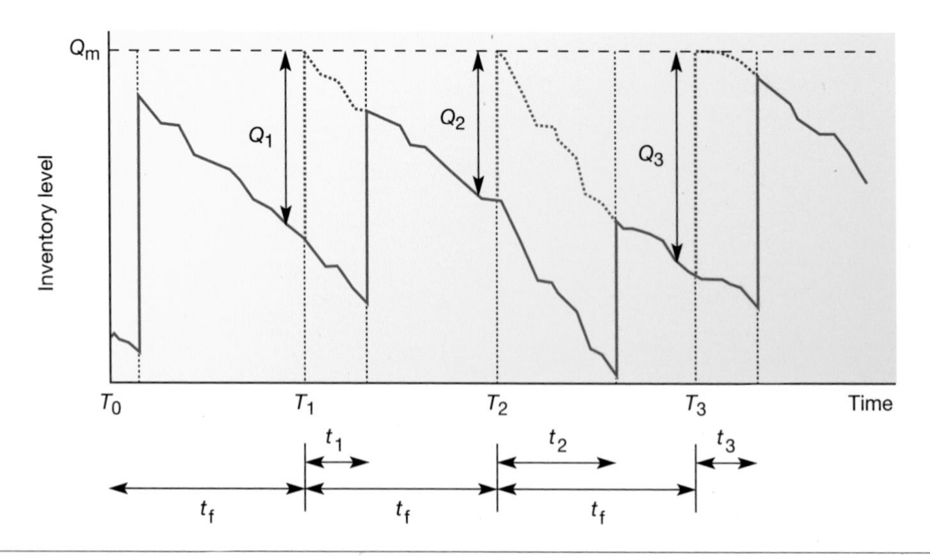

Figure 13.13 A periodic review approach to order timing with probabilistic demand and lead time

At time T_1 in Figure 13.13 the inventory manager would examine the stock level and order sufficient to bring it up to some maximum, Q_m. However, that order of Q_1 items will not arrive until a further time of t_1 has passed, during which demand continues to deplete the stocks. Again, both demand and lead time are uncertain. The Q_1 items will arrive and bring the stock up to some level lower than Q_m (unless there has been no demand during t_1). Demand then continues until T_2, when again an order Q_2 is placed, which is the difference between the current stock at T_2 and Q_m. This order arrives after t_2, by which time demand has depleted the stocks further. Thus the replenishment order placed at T_1 must be able to cover for the demand that occurs until T_2 and t_2. Safety stocks will need to be calculated, in a similar manner to before, based on the distribution of usage over this period.

The time interval

The interval between placing orders, t_1, is usually calculated on a deterministic basis, and derived from the EOQ. So, for example, if the demand for an item is 2,000 per year, the cost of placing an order £25, and the cost of holding stock £0.50 per item per year:

$$\text{EOQ} = \sqrt{\frac{2C_oD}{C_h}} = \sqrt{\frac{2 \times 2{,}000 \times 25}{0.50}} = 447$$

The optimum time interval between orders, t_f, is therefore:

$$t_f = \frac{\text{EOQ}}{D} = \frac{447}{2{,}000}\text{ years}$$
$$= 2.68\text{ months}$$

It may seem paradoxical to calculate the time interval assuming constant demand when demand is, in fact, uncertain. However, uncertainties in both demand and lead time can be allowed for by setting Q_m to allow for the desired probability of stock-out based on usage during the period t_f + lead time.

Two-bin and three-bin systems

Keeping track of inventory levels is especially important in continuous review approaches to re-ordering. A simple and obvious method of indicating when the re-order point has been reached is necessary, especially if there are a large number of items to be monitored. The two- and three-bin systems illustrated in Figure 13.14 are such methods. The simple two-bin system involves storing the

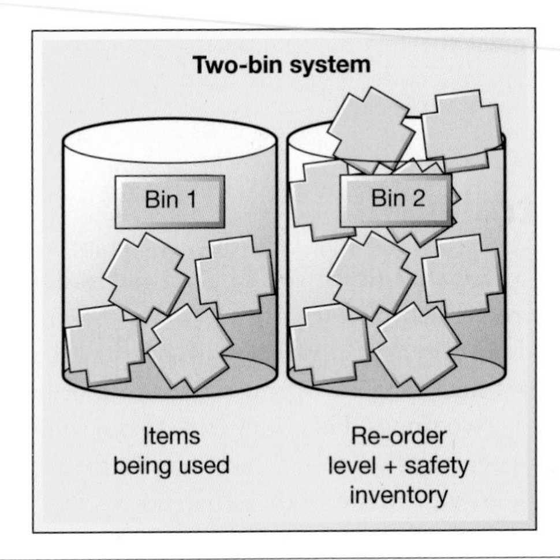

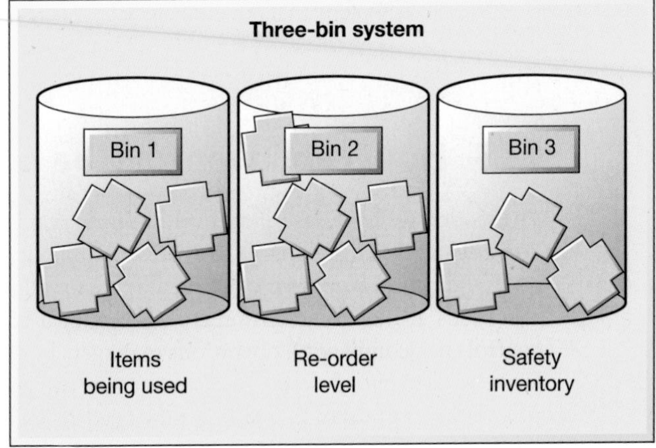

Figure 13.14 The two-bin and three-bin systems of re-ordering

re-order point quantity plus the safety inventory quantity in the second bin and using parts from the first bin. When the first bin empties, that is the signal to order the next re-order quantity. Sometimes the safety inventory is stored in a third bin (the three-bin system), so it is clear when demand is exceeding that which was expected. Different 'bins' are not always necessary to operate this type of system. For example, a common practice in retail operations is to store the second 'bin' quantity upside-down behind or under the first 'bin' quantity. Orders are then placed when the upside-down items are reached.

OPERATIONS IN PRACTICE

Amazon's 'anticipatory inventory'[4]

Could an item be on its way from a stocking point to you even before you think of ordering it? Is it possible for a company accurately to forecast your order and ship it to you before you place it? Of course, forecast accuracy and time to deliver are related. Poor forecasts mean that the wrong items will be stored, which in turn means that delivery will be delayed until the right items are received. But what if a supplier could know what its customers were going to order, even before they do? That is the ambition of Amazon's online retail operation. It filed a patient to protect its system for the technology that hopes to predict what its customers will buy, even before they have clicked the 'order' button. The company, which is the world's largest online retailer, calls its new system 'anticipatory shipping' and perceives it as a way to speed up its delivery times. Amazon's patent application reveals the thinking behind the system. It says that one substantial disadvantage to the virtual storefront model is that in many instances, customers cannot receive their merchandise immediately upon purchase, but must instead wait for product to be shipped to them. The availability of expedited shipping methods from various common carriers may to some extent compensate for the delay in shipment, but often at substantial additional cost that may rival the price paid for the merchandise. Such delays may dissuade customers from buying items from online merchants, particularly if those items are more readily available locally. The approach is reported as using several elements to predict what purchases a person may make. Factors to be taken into account could include, age, income, previously purchased items, searched for items, 'wish lists' and maybe even the time a user's cursor lingers over a product. Armed with this information, Amazon could ship items that are likely to be ordered to the inventory 'hub' nearest to the customer. So, when a customer really does order, the item can be delivered far faster.

13.5 How can inventory be controlled?

The models we have described, even the ones that take a probabilistic view of demand and lead time, are still simplified compared with the complexity of real stock management. Coping with many thousands of stocked items, supplied by many hundreds of different suppliers, with possibly tens of thousands of individual customers, makes for a complex and dynamic operations task. In order to control such complexity, operations managers have to do two things. First, they have to discriminate between different stocked items, so that they can apply a degree of control to each item that is appropriate to its importance. Second, they need to invest in an information-processing system that can cope with their particular set of inventory control circumstances.

Inventory priorities – the ABC system

In any inventory that contains more than one stocked item, some items will be more important to the organisation than others. Some, for example, might have a very high usage rate, so if they ran out many customers would be disappointed. Other items might be of particularly high value, so excessively high inventory levels would be particularly expensive. One common way of discriminating between different stock items is to rank them by the **usage value** (their usage rate multiplied by their individual value). Items with a particularly high usage value are deemed to warrant the most careful control, whereas those with low usage values need not be controlled quite so rigorously. Generally, a relatively small proportion of the total range of items contained in an inventory will account for a large proportion of the total usage value. This phenomenon is known as the **Pareto law** (after the person who described it), sometimes referred to as the 80/20 rule. It is called this because, typically, 80 per cent of an operation's sales are accounted for by only 20 per cent of all stocked item types. The Pareto law is also used elsewhere in operations management (see, for example, Chapter 15). Here the relationship can be used to classify the different types of items kept in an inventory by their usage value. **ABC inventory control** allows inventory managers to concentrate their efforts on controlling the more significant items of stock:

- *Class A items* are those 20 per cent or so of high usage value items that account for around 80 per cent of the total usage value.
- *Class B items* are those of medium usage value, usually the next 30 per cent of items, which often account for around 10 per cent of the total usage value.
- *Class C items* are those low usage value items which, although comprising around 50 per cent of the total types of items stocked, probably only account for around 10 per cent of the total usage value of the operation.

Operations principle

Different inventory management decision rules are needed for different classes of inventory.

Worked example

Selectro electrical wholesaler

Table 13.9 shows all the parts stored by Selectro, an electrical wholesaler. The 20 different items stored vary in terms of both their usage per year and cost per item as shown. However, the wholesaler has ranked the stock items by their usage value per year. The total usage value per year is £5,569,000. From this it is possible to calculate the usage value per year of each item as a percentage of the total usage value, and from that a running cumulative total of the usage value as shown. The wholesaler can then plot the cumulative percentage of all stocked items against the cumulative percentage of their value. So, for example, the part with stock number A/703 is the highest-value part and accounts for 25.14 per cent of the total inventory value. As a part, however, it is only one-twentieth or 5 per cent of the total number of items stocked. This item, together with the next highest-value item (D/012), account for only 10 per cent of the total number of items stocked, yet account for 47.37 per cent of the value of the stock, and so on.

This is shown graphically in Figure 13.15. Here the wholesaler has classified the first four part numbers (20 per cent of the range) as Class A items, and will monitor the usage and ordering of these items very closely and frequently. A few improvements in order quantities or safety stocks for these items could bring significant savings. The next six part numbers, C/375 through to A/138 (30 per cent of the range) are to be treated as Class B items with slightly less effort devoted to their control. All other items are classed as Class C items whose stocking policy is reviewed only occasionally.

Table 13.9 Warehouse items ranked by usage value

Stock no.	Usage (items/year)	Cost (£/item)	Usage value (£000/year)	% of total value	Cumulative % of total value
A/703	700	20.00	1,400	25.14	25.14
D/012	450	2.75	1,238	22.23	47.37
A/135	1,000	0.90	900	16.16	63.53
C/732	95	8.50	808	14.51	78.04
C/375	520	0.54	281	5.05	83.09
A/500	73	2.30	168	3.02	86.11
D/111	520	0.22	114	2.05	88.16
D/231	170	0.65	111	1.99	90.15
E/781	250	0.34	85	1.53	91.68
A/138	250	0.30	75	1.34	93.02
D/175	400	0.14	56	1.01	94.03
E/001	80	0.63	50	0.89	94.92
C/150	230	0.21	48	0.86	95.78
F/030	400	0.12	48	0.86	96.64
D/703	500	0.09	45	0.81	97.45
D/535	50	0.88	44	0.79	98.24
C/541	70	0.57	40	0.71	98.95
A/260	50	0.64	32	0.57	99.52
B/141	50	0.32	16	0.28	99.80
D/021	20	0.50	10	0.20	100.00
Total			5,569	100.00	

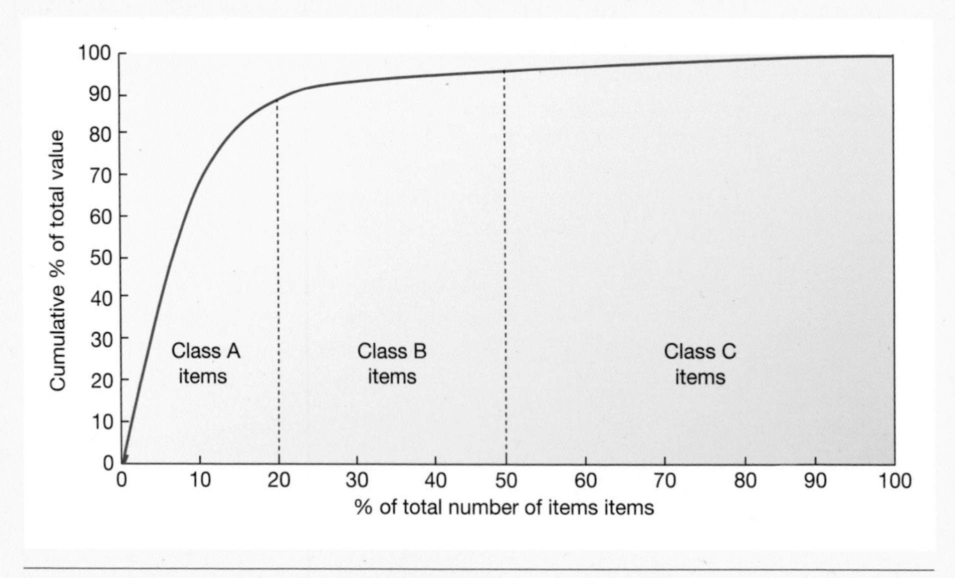

Figure 13.15 Pareto curve for items in a warehouse

Although annual usage and value are the two criteria most commonly used to determine a stock classification system, other criteria might also contribute towards the (higher) classification of an item:

- *Consequence of stock-out* – High priority might be given to those items that would seriously delay or disrupt other operations, or the customers, if they were not in stock.
- *Uncertainty of supply* – Some items, although of low value, might warrant more attention if their supply is erratic or uncertain.
- *High obsolescence or deterioration risk* – Items that could lose their value through obsolescence or deterioration might need extra attention and monitoring.

Some more complex stock classification systems might include these criteria by classifying on an A, B, C basis for each. For example, a part might be classed as A/B/A meaning it is an A category item by value, a class B item by consequence of stock-out and a class A item by obsolescence risk.

Critical commentary

This approach to inventory classification can sometimes be misleading. Many professional inventory managers point out that the Pareto law is often misquoted. It does not say that 80 per cent of the SKUs (stock keeping units) account for only 20 per cent of inventory value. It accounts for 80 per cent of inventory 'usage' or throughput value; in other words, sales value. In fact, it is the slow-moving items (the C category items) that often pose the greatest challenge in inventory management. Often these slow-moving items, although only accounting for 20 per cent of sales, require a large part (typically between one half and two thirds) of the total investment in stock. This is why slow-moving items are a real problem. Moreover, if errors in forecasting or ordering result in excess stock in 'A class' fast-moving items, it is relatively unimportant in the sense that excess stock can be sold quickly. However, excess stock in slow-moving C items will be there a long time. According to some inventory managers, it is the A items that can be left to look after themselves, it is the B and even more the C items that need controlling.

Measuring inventory

In our example of ABC classifications we used the monetary value of the annual usage of each item as a measure of inventory usage. Monetary value can also be used to measure the absolute level of inventory at any point in time. This would involve taking the number of each item in stock, multiplying it by its value (usually the cost of purchasing the item) and summing the value of all the individual items stored. This is a useful measure of the investment that an operation has in its inventories but gives no indication of how large that investment is relative to the total throughput of the operation. To do this we must compare the total number of items in stock against their rate of usage. There are two ways of doing this. The first is to calculate the amount of time the inventory would last, subject to normal demand, if it were not replenished. This is sometimes called the number of weeks' (or days', months', years', etc.) *cover* of the stock. The second method is to calculate how often the stock is used up in a period, usually one year. This is called the stock turn or turnover of stock and is the reciprocal of the stock-cover figure mentioned earlier.

Worked example

Boncorko

Boncorko wine importer holds stocks of three types of wine, Chateau A, Chateau B and Chateau C. Current stock levels are 500 cases of Chateau A, 300 cases of Chateau B and 200 cases of Chateau C. Table 13.10 shows the number of each held in stock, their cost per item and the demand per year for each.

▶

Table 13.10 Stock, cost and demand for three stocked items

Item	Average number in stock	Cost per item (£)	Annual demand
Chateau A	500	3.00	2,000
Chateau B	300	4.00	1,500
Chateau C	200	5.00	1,000

$$\text{The total value of stock} = \sum(\text{average stock level} \times \text{cost per item})$$
$$= (500 \times 3) + (300 \times 4) + (200 \times 5)$$
$$= 3{,}700$$

The amount of *stock cover* provided by each item stocked is as follows (assuming 50 sales weeks per year):

$$\text{Chateau A, stock cover} = \frac{\text{Stock}}{\text{Demand}} = \frac{500}{2{,}000} \times 50 = 12.5 \text{ weeks}$$

$$\text{Chateau B, stock cover} = \frac{\text{Stock}}{\text{Demand}} = \frac{300}{1{,}500} \times 50 = 10 \text{ weeks}$$

$$\text{Chateau C, stock cover} = \frac{\text{Stock}}{\text{Demand}} = \frac{200}{1{,}000} \times 50 = 10 \text{ weeks}$$

The *stock turn* for each item is calculated as follows:

$$\text{Chateau A, stock turn} = \frac{\text{Demand}}{\text{Stock}} = \frac{2{,}000}{500} = 4 \text{ times/year}$$

$$\text{Chateau B, stock turn} = \frac{\text{Demand}}{\text{Stock}} = \frac{1{,}500}{300} = 5 \text{ times/year}$$

$$\text{Chateau C, stock turn} = \frac{\text{Demand}}{\text{Stock}} = \frac{1{,}000}{200} = 5 \text{ times/year}$$

To find the average stock cover or stock turn for the total items in the inventory, the individual item measures can be weighted by their demand levels as a proportion of total demand (4,500). Thus:

$$\text{Average stock cover} = \left(12.5 \times \frac{2{,}000}{4{,}500}\right) + \left(10 \times \frac{1{,}500}{4{,}500}\right) + \left(10 \times \frac{1{,}000}{4{,}000}\right)$$
$$= 11.11$$

$$\text{Average stock turn} = \left(4 \times \frac{2{,}000}{4{,}500}\right) + \left(5 \times \frac{1{,}500}{4{,}500}\right) + \left(5 \times \frac{1{,}000}{4{,}500}\right)$$
$$= 4.56$$

Inventory information systems

Most inventories of any significant size are managed by computerised systems. The many relatively routine calculations involved in stock control lend themselves to computerised support. This is especially so since data capture has been made more convenient through the use of barcode readers and the point-of-sale recording of sales transactions. Many commercial systems of stock control are available, although they tend to share certain common functions.

Updating stock records

Every time a transaction takes place (such as the sale of an item, the movement of an item from a warehouse into a truck, or the delivery of an item into a warehouse) the position, status and possibly value of the stock will have changed. This information must be recorded so that operations managers can determine their current inventory status at any time.

Generating orders

The two major decisions we have described previously, namely how much to order and when to order, can both be made by a computerised stock control system. The first decision, setting the value of how much to order (Q), is likely to be taken only at relatively infrequent intervals. Originally almost all computer systems automatically calculated order quantities by using the EOQ formulae covered earlier. Now more sophisticated algorithms are used, often using probabilistic data and based on examining the marginal return on investing in stock. The system will hold all the information that goes into the ordering algorithm but might periodically check to see if demand or order lead times, or any of the other parameters, have changed significantly and recalculate Q accordingly. The decision on when to order, on the other hand, is a far more routine affair, which computer systems make according to whatever decision rules operations managers have chosen to adopt: either continuous review or periodic review. Furthermore, the systems can automatically generate whatever documentation is required, or even transmit the re-ordering information electronically through an electronic data interchange (EDI) system.

Generating inventory reports

Inventory control systems can generate regular reports of stock value for the different items stored, which can help management monitor its inventory control performance. Similarly, customer service performance, such as the number of stock-outs or the number of incomplete orders, can be regularly monitored. Some reports may be generated on an exception basis. That is, the report is generated only if some performance measure deviates from acceptable limits.

Forecasting

Inventory replenishment decisions should ideally be made with a clear understanding of forecast future demand. The inventory control system can compare actual demand against forecast and adjust the forecast in the light of actual levels of demand. Control systems of this type are treated in more detail in Chapter 14.

OPERATIONS IN PRACTICE

France bans the dumping of unsold stock[5]

The French government became the first country to ban retailers from destroying unsold stock when it announced that the practice would be outlawed for all goods by 2023. Brune Poirson, the ecology minister, said that stock worth €800 million went unsold in France every year, of which only €140 millions' worth was given to charities and the rest was destroyed. It had become a widespread practice because occasionally businesses end up with either too little or too much inventory to serve their markets. Too little inventory will result in reduced customer service. But too much can be even more problematic, particularly for some businesses that trade in high-value, 'brand integrity' goods. What does a business do when demand slows and it can't sell any surplus stock without affecting its brand? Four years before the French government's announcement, Burberry, the upmarket fashion brand, had to defend its decision to destroy £19 millions' worth of its products that it could not sell through its discount outlet stores. At its annual meeting in London, the company said that it was looking to reduce the amount of wasted stock 'every single season', but also said that destroying surplus stock was a common practice among luxury goods companies. The company's outgoing chief executive, said: '*We have a process where we have a sale, then packs go to* [a discount] *outlet . . . There are some raw materials at the end of that process that we do have to destroy because of intellectual property. It's a common practice but it's something we're enormously mindful of. Every single season we look at how we can reduce, and we have reduced it over the years*'.

Burberry is not alone. When sales of Cartier and Montblanc products slowed sharply, partly because of a crackdown on corruption in China, the overstocking by dealers and an uncertain outlook for growth, the Swiss luxury group Richemont that owns the brands bought back stock from some of its Hong Kong dealers. The watches that were bought back were either reallocated to other regions or, in the case of older models that were no longer selling, were dismantled and recycled. With some luxury goods, the tax rules in some countries actively encourage scrapping surplus stock. For example, if a company makes a bottle of perfume, its cost is a relatively tiny amount (the value comes through advertising and the effect it has on public perception). But the tax loss that the company can claim comes from destroying the product is based on its retail price, not production cost. Of course, there are perfectly legitimate reasons for destroying surplus stock. Any business is responsible for protecting its intellectual property and its brand. However, stock destruction as a means of maintaining 'brand integrity' can backfire. After bags of slashed and cut clothing were found outside one of its New York stores, the clothing retailer H&M had to promise that it would stop destroying new, unworn clothing that it could not sell, and would instead donate the garments to charities.

Common problems with inventory systems

Our description of inventory systems has been based on the assumption that operations (a) have a reasonably accurate idea of costs such as holding cost, or order cost, and (b) have accurate information that really does indicate the actual level of stock and sales. But data inaccuracy often poses one of the most significant problems for inventory managers. This is because most computer-based inventory management systems are based on what is called the **perpetual inventory principle**. This is the simple idea that stock records are (or should be) automatically updated every time that items are recorded as having been received into an inventory or taken out of the inventory. So:

Opening stock level + Receipts in − Dispatches out = New stock level

Operations principle

The maintenance of data accuracy is vital for the day-to-day effectiveness of inventory management systems.

Any errors in recording these transactions, and/or in handling the physical inventory can lead to discrepancies between the recorded and actual inventory, and these errors are perpetuated until physical stock checks are made (usually quite infrequently). In practice there are many opportunities for errors to occur, if only because inventory transactions are numerous. This means that it is surprisingly common for the majority of inventory records to be inaccurate. The underlying causes of errors include:

- keying errors – entering the wrong product code;
- quantity errors – a miscount of items put into or taken from stock;
- damaged or deteriorated inventory not recorded as such, or not correctly deleted from the records when it is destroyed;
- the wrong items being taken out of stock, but the records not being corrected when they are returned to stock;
- delays between the transactions being made and the records being updated;
- items stolen from inventory (common in retail environments, but also not unusual in industrial and commercial inventories).

Responsible operations

In every chapter, under the heading of 'Responsible operations', we summarise how the particular topic covered in the chapter touches upon important social, ethical and environmental issues.

The nature of inventory is that it hangs around, and while it hangs around, it can lose value, even to the point where it can only be thrown away. But discarding items on which energy and materials and labour have been expended is at best a waste, and at worst ethically highly questionable. When the stock is upmarket fashion goods it can be reputationally damaging (see the 'Operations in practice' example, 'France bans the dumping of unsold stock', earlier). When it is food that is being wasted, it is somehow far more distasteful. Dumping food when some people don't have enough to eat is seen by many to be ethically indefensible. For food retailers, a major reduction in waste food could come from better forecasting. It can both reduce waste and cut costs. In Japan, which throws away over 6 million tonnes of food waste every year (the highest food waste per capita in Asia) retailers are using sophisticated artificial intelligence and advanced technologies to predict demand more accurately. For one Japanese convenience store chain, Lawson Inc, the disposal of food waste is its largest cost after labour costs, so it started using AI in an attempt to halve food waste at all of its stores. A complicating factor is that Japan's consumers can be extremely fussy, demanding perfectly presented food items with plenty of 'shelf life'. However, in a further attempt to avoid waste, the Japanese e-commerce firm, Kuradashi, started to offer unsold foods at a discount. Kuradashi's founder, Tatsuya Sekito, explained: *'Japanese shoppers tend to be picky, but we attract customers by offering not just a sale but a chance to donate a portion of purchases to a charity, raising awareness about social issues.'*[6]

But food waste is not an issue just for retailers. Further back up the supply chain, food can also be wasted at the farms where it is grown. Sometimes waste happens when mistakes are made in packaging food, sometimes when a retailer backs out of a purchase, and sometimes because of exceptionally good weather producing a glut. When this happens, farmers in the United Kingdom have the option of donating surplus food to charity. One of the most active charities is Fareshare, a national network of charitable food redistributors.[7] They take good quality surplus food from right across the food industry and get it to front-line charities and community groups. Sometimes a farmer, grower or manufacturer has food that they can donate, but do not have the ability to deliver it. Fareshare can act as an orchestrator, ensuring that the donation does not go to waste. For example, gluts of fruit can be sent for freezing or juicing. More problematic is the role of packaging in preventing food waste. Certainly, the design and use of packaging can have an impact on food waste through the supply chain. Yet packaging is considered by some to be environmentally 'unsound', partly because of the increasing awareness of the damage from plastic pollution. Yet plastic packaging could be seen a 'necessary evil' that can reduce food waste, both during processing and in the home. A much-quoted example is that the use of just 1.5 g of plastic film for wrapping a cucumber can extend its shelf life from three days to 14 days.[8]

But, if all else fails, and food cannot be used to provide nutrition, it could still play a useful environmental role. It can provide energy. For example, Sainsbury's, the UK supermarket, send food waste to a food recycling plant where it's converted into gas and fertiliser. The gas is then exported to the national gas grid. The company buys back certified carbon-neutral energy from the same plants for use in its stores, for power and heating. One Sainsbury's superstore in Staffordshire runs entirely on electricity generated from food waste. Food from the company, along with other waste suppliers, is sent to a plant near the store, where it is converted into gas and used to generate electricity on site, which is then supplied directly to the supermarket via a cable.[9]

Summary answers to key questions

13.1 What is inventory?

- Inventory, or stock, is the stored accumulation of the transformed resources in an operation. Sometimes the words 'stock' and 'inventory' are also used to describe transforming resources, but the terms *stock control* and *inventory control* are nearly always used in connection with transformed resources.

- Almost all operations keep some kind of inventory, most usually of materials but also of information and customers (customer inventories are normally called queues).

13.2 Why should there be any inventory?

- Inventory occurs in operations because the timing of supply and the timing of demand do not always match. Inventories are needed, therefore, to smooth the differences between supply and demand.
- There are five main reasons for keeping physical inventory:
  — to cope with random or unexpected interruptions in supply or demand (buffer inventory);
  — to cope with an operation's inability to make all products simultaneously (cycle inventory);
  — to allow different stages of processing to operate at different speeds and with different schedules;
  — to cope with planned fluctuations in supply or demand (anticipation inventory);
  — to cope with transportation delays in the supply network (pipeline inventory).
- Inventory is often a major part of working capital, tying up money that could be used more productively elsewhere.
- If inventory is not used quickly, there is an increasing risk of damage, loss, deterioration or obsolescence.
- Inventory invariably takes up space (for example, in a warehouse), and has to be managed, stored in appropriate conditions, insured and physically handled when transactions occur. It therefore contributes to overhead costs.

13.3 How much should be ordered? The volume decision

- This depends on balancing the costs associated with holding stocks against the costs associated with placing an order. The main stock-holding costs are usually related to working capital, whereas the main order costs are usually associated with the transactions necessary to generate the information to place an order.
- The best-known approach to determining the amount of inventory to order is the economic order quantity (EOQ) formula. The EOQ formula can be adapted to different types of inventory profile using different stock behaviour assumptions.
- The EOQ approach, however, has been subject to a number of criticisms regarding the true cost of holding stock, the real cost of placing an order, and the use of EOQ models as prescriptive devices.
- One approach to this problem, the news vendor problem, includes the effects of probabilistic demand in determining order quantity.

13.4 When should an order be placed? The timing decision

- Partly this depends on the uncertainty of demand. Orders are usually timed to leave a certain level of average safety stock when the order arrives. The level of safety stock is influenced by the variability of both demand and the lead time of supply. These two variables are usually combined into a lead-time usage distribution.
- Using re-order level as a trigger for placing replenishment orders necessitates the continual review of inventory levels. This can be time-consuming and expensive. An alternative approach is to make replenishment orders of varying size but at fixed time periods.

13.5 How can inventory be controlled?

- The key issue here is how managers discriminate between the levels of control they apply to different stock items. The most common way of doing this is by what is known as the ABC classification of stock. This uses the Pareto curve to distinguish between the different values of, or significance placed on, types of stock.
- Inventory is usually managed through sophisticated computer-based information systems, which have a number of functions: the updating of stock records, the generation of orders, and the generation of inventory status reports and demand forecasts. These systems critically depend on maintaining accurate inventory records.

CASE STUDY Supplies4medics.com

Founded almost 20 years ago, supplies4medics.com has become one of Europe's most successful direct mail suppliers of medical hardware and consumables to hospitals, doctors' and dentists' surgeries, clinics, nursing homes and other medical-related organisations. Its physical and online catalogues list just over 4,000 items, categorised by broad applications such as 'hygiene consumables' and 'surgeons' instruments'. Quoting their website:

'We are the pan-European distributors of wholesale medical and safety supplies... We aim to carry everything you might ever need; from nurses' scrubs to medical kits, consumables for operations, first-aid kits, safety products, chemicals, fire-fighting equipment, nurse and physicians' supplies, etc. Everything is at affordable prices – and backed by our very superior customer service and support – supplies4medics is your ideal source for all medical supplies. Orders are normally dispatched same-day, via our European distribution partner, the Brussels Hub of DHL. You should therefore receive your complete order within one week, but you can request next-day delivery if required, for a small extra charge. You can order our printed catalogue on the link at the bottom of this page, or shop on our easy-to-use online store'.

Last year turnover grew by over 25 per cent to about €120 million, a cause for considerable satisfaction in the company. However, profit growth was less spectacular; and market research suggested that customer satisfaction, although generally good, was slowly declining. Most worrying, inventory levels had grown faster than sales revenue, in percentage terms. This was putting a strain on cash flow, requiring the company to borrow more cash to fund the rapid growth planned for the next year. Inventory holding is estimated to be costing around 15 per cent per annum, taking account of the cost of borrowing, insurance and all warehousing overheads.

Pierre Lamouche, the Head of Operations summarised the situation faced by his department: *'As a matter of urgency, we are reviewing our purchasing and inventory management systems! Most of our existing re-order levels (ROL) and re-order quantities (ROQ) were set several years ago, and have never been recalculated. Our focus has been on rapid growth through the introduction of new product lines. For more recently introduced items, the ROQs were based only on forecast sales, which actually can be quite misleading. We estimate that it costs us, on average, €50 to place and administer every purchase order, since most suppliers are still not able to take orders online or by EDI. In the meantime, sales of some products have grown fast, while others have declined. Our average inventory (stock) cover is about 10 weeks, but – amazingly – we still run out of critical items! In fact, on average, we are currently out of stock of about 500 SKUs (stock keeping units) at any time. As you can imagine, our service level is not always satisfactory with this situation. We really need help to conduct a review of our system, so have employed a mature intern from the local business school to review our system. He has first asked my team to provide information on a random, representative sample of 20 items from the full catalogue range'* (see Table 13.11).

Table 13.11 Representative sample of 20 catalogue items

Sample Number	Catalogue reference number*	Sales unit description**	Sales unit cost (€)	Last 12 months' sales (units)	Inventory as at last year end (units)	Re-order quantity (units)
1	11036	Disposable aprons (10pk)	2.40	100	0	10
2	11456	Ear-loop masks (box)	3.60	6,000	1,200	1,000
3	11563	Drill type 164	1.10	220	420	250
4	12054	Incontinence pads large	3.50	35,400	8,500	10,000
5	12372	150 ml syringe	11.30	430	120	100
6	12774	Rectal speculum 3-prong	17.40	65	20	20
7	12979	Pocket organiser blue	7.00	120	160	500
8	13063	Oxygen trauma kit	187.00	40	2	10
9	13236	Zinc oxide tape	1.50	1,260	0	50
10	13454	Dual head stethoscope	6.25	10	16	25
11	13597	Disp. latex catheter	0.60	3,560	12	20
12	13999	Roll-up wheelchair ramp	152.50	12	44	50
13	14068	WashClene tube	1.40	22,500	10,500	8,000
14	14242	Cervical collar	12.00	140	24	20
15	14310	Head wedge	89.00	44	2	10
16	14405	Three-wheel scooter	755.00	14	5	5
17	14456	Neonatal trach. tube	80.40	268	6	100
18	14675	Mouldable strip paste	10.20	1,250	172	100
19	14854	Sequential comp. pump	430.00	430	40	50
20	24943	Toilet safety frame	25.60	560	18	20

* Reference numbers are allocated sequentially as new items are added to catalogue

** All quantities are in sales units (e.g. item, box, case, pack)

QUESTIONS

1 Prepare a spreadsheet-based ABC analysis of Usage Value. Classify as follows:

A items: top 20 per cent of usage value
B items: next 30 per cent of usage value
C items: remaining 50 per cent of usage value

2 Calculate the inventory weeks for each item, for each classification, and for all the items in total. Does this suggest that the Head of Operations's estimate of inventory weeks is correct?

3 If so, what is your estimate of the overall inventory at the end of the base year, and how much might that have increased during the year?

4 Based on the sample, analyse the underlying causes of the availability problem described in the text.

5 Calculate the EOQs for the A items.

6 What recommendations would you give to the company?

Problems and applications

All chapters have problems and application questions that will help you practise analysing operations. They can be answered by reading the chapter. Model answers for the first two questions can be found on the companion website for this text.

1 A supplier makes monthly shipments to 'House & Garden Stores', in average lot sizes of 200 coffee tables. The average demand for these items is 50 tables per week, and the lead time from the supplier 3 weeks. 'House & Garden Stores' must pay for inventory from the moment the supplier ships the products. If it is willing to increase its lot size to 300 units, the supplier will offer a lead time of 1 week. What will be the effect on cycle and pipeline inventories?

2 A local shop has a relatively stable demand for tins of sweetcorn throughout the year, with an annual total of 1,400 tins. The cost of placing an order is estimated at £15 and the annual cost of holding inventory is estimated at 25 per cent of the product's value. The company purchases tins for 20p. How much should the shop order at a time, and what is the total cost of the plan?

3 A fruit canning plant has a single line for three different fruit types. Demand for each type of tin is reasonably constant at 50,000 per month (a month has 160 production hours). The canning process rate is 1,200 per hour, but it takes 2 hours to clean and re-set between different runs. The cost of these changeovers (C_o) is calculated at £250 per hour. Stock-holding is calculated at £0.10 per tin per month. How big should the batch size be?

4 *'Our suppliers often offer better prices if we are willing to buy in larger quantities. This creates a pressure on us to hold higher levels of stock. Therefore, to find the best quantity to order we must compare the advantages of lower prices for purchases and fewer orders with the disadvantages of increased holding costs. This means that calculating total annual inventory-related costs should now not only include holding costs and ordering costs, but also the cost of purchased items themselves'* (Manager, Tufton Bufton Port Importers Inc.). One supplier to Tufton Bufton Port Importers Inc. (TBPI) has introduced quantity discounts to encourage larger order quantities. The discounts are shown below:

Order quantity	Price per bottle
0–100	€15.00
101–250	€13.50
251+	€11.00

TBPI estimates that its annual demand for this particular wine is 1,500 bottles, its ordering costs are €30 per order, and its annual holding costs are 20 per cent of the bottle's price.

(a) How should TBPI go about deciding how many to order?
(b) How many should they order?

5 Most countries have blood collection and distribution services that collect from donors, process the blood by either breaking the blood down into its constituent parts or keeping it whole, and transport the blood from collection centres to hospitals in response to both routine and emergency requests.

(a) What are the factors that constitute inventory holding costs, order costs and stock-out costs in such a blood service?
(b) What makes this particular inventory planning and control example so complex?
(c) How might the efficiency with which a blood service controls its inventory affect its ability to collect blood?

6 Re-read the 'Operations in practice' example, 'An inventory of energy'. It mentions the potential of battery storage of energy, but stresses the cost of this method. What do you think would be the implications for energy distribution if batteries become both cheaper and more effective?

7 Re-read the 'Operations in practice' example, 'Safety stocks for coffee and COVID'. List and comment on the differences between the two examples (coffee and COVID) described.

8 Inventory management systems should not only get order quantities and timing right, they also need to make sure that stocks are in the right place. One airline found this out when a shortage of toilet paper and 'the wrong kind of headphones' delayed a London to Barbados flight for five hours. What could have been the possible consequences of mismanaging stocks in this case?

9 In the colder parts of the world (such as northern Europe and North America) where winter snow and ice can cause huge disruption to everyday life, local governments 'grit' the roads (actually rock salt, a mixture of salt and grit) when they believe weather conditions warrant it. If you were in charge of this operation, what would be the main decisions that you would need to take, and what factors would you take into account in order to make these decisions?

10 Two retail managers who had been at university together, Rosanne and Abeke, were having dinner together. *'It's such a coincidence that we both went into garment retailing'*, said Rosanne, *'but at the moment I'm having real problems getting repenishment stocks from our central warehouse. The margin we make is so small that we can't afford to be out of stock for long, or our profits suffer'.* Abeke had a different problem, she said. *'That isn't much of a problem for me. You sell casual wear with a strong brand, but low margins. Your customers will return later if something is out of stock. I sell fashion garments with far higher margins, but clothes that are essentially an impulse buy. If I fail to sell something because it's not in stock, a lot of potential profit walks out of the door'.* How do you think inventory practice should differ for the operations that the two friends work for?

Selected further reading

Axsäter, S. (2015) *Inventory Control*, 3rd edn, Springer, Heidelberg.
A traditional, but comprehensive textbook that takes an 'operational research' quantitative approach.

Bragg, S.M. (2021) *Inventory Management*, 4th edn, Accounting Tools, Centennial, CO.
A supply chain and financial approach to the subject.

Emmett, S. and Granville, D. (2007) *Excellence in Inventory Management: How to Minimise Costs and Maximise Service*, Cambridge Academic, Shelford.
A practical guide.

Muller, M. (2011) *Essentials of Inventory Management*, 2nd edn, Amacom.
Straightforward treatment.

Relph, G. and Milner, C. (2015) *Inventory Management: Advanced Methods for Managing Inventory within Business Systems*, Kogan Page, London.
An advanced book that covers most topics in the subject, including the 'k-curve' that is not included in this chapter.

Silver, E.A., Pyke, D.F. and Thomas, D.J. (2021) *Inventory and Production Management in Supply Chains*, 4th edn, CRC Press, Boca Raton, FL.
Current and practical. Strong on managing inventory in multiple locations.

Vandeput, N. (2020) *Inventory Optimization: Models and Simulations*, De Gruyter, Berlin.
For those who want a very quantitative treatment of the subject.

Wild, T. (2017) *Best Practice in Inventory Management*, 3rd edn, Routledge, Abingdon.
A straightforward and readable practice-based approach to the subject.

Notes on chapter

1. The information on which this example is based is taken from: Koen, A. and Antunez, P.F. (2020) How heat can be used to store renewable energy, The Conversation, 25 February; Gosden, E. (2017) Power shift brings energy market closer to holy grail, *The Times*, 17 April; The Economist (2012) Energy storage: packing some power, *Economist Technology Quarterly*, 3 March.
2. The information on which this example is based is taken from: The Economist (2019) A nation of have-beans, Defending Switzerland's coffee stockpile, *Economist* print edition, 21 November; Foster, P. and Neville, S. (2020) How poor planning left the UK without enough PPE, *Financial Times*, 1 May; Britt, H. (2020) What is safety stock and how can businesses use it to ensure continuity?, Thomasnet.com, 8 April, https://www.thomasnet.com/insights/what-is-safety-stock/ (accessed September 2021); Anderson, H. (2020) COVID-19: preparing your supply chain in times of crisis, publicissapient.com, 8 April, https://www.publicissapient.com/insights/coronavirus_and_managing_the_supply_chain_amid_a_crisis (accessed September 2021).
3. The information on which this example is based is taken from: The Economist (2015) Croissantonomics: lessons in managing supply and demand for perishable products, *Economist* print edition, 29 August. However, unfortunately Mr Ruben's efforts did not manage to save the bakery, which went out of business in 2019.
4. The information on which this example is based is taken from: Ulanoff, L. (2014) Amazon knows what you want before you buy it, Mashable, 21 January, https://mashable.com/2014/01/21/amazon-anticipatory-shipping-patent/#Ryy4twKmRiqb (accessed September 2021); Duke, S. (2014) He knows what you want — before you even want it, *The Sunday Times*, 2 February; Ahmed, M. (2014) Amazon will know what you want before you do, *The Times*, 27 January; Bernard, Z. (2018) Amazon is spending more and more on shipping out your orders, Business Insider, 13 February, https://www.businessinsider.com/amazons-logistics-costs-are-growing-really-fast-charts-2018-2 (accessed September 2021).
5. The information on which this example is based is taken from: Sage, A. (2019) France to ban luxury brands from dumping unsold stock, *The Times*, 24 September; Leroux, M. (2016) Burberry boss defends stock destruction, *The Times*, 15 July; Atkins, R. (2016) Richemont buys back and destroys stock as sales fall, *Financial Times*, 20 May; Dwyer, J. (2010) A clothing clearance where more than just the prices have been slashed, *New York Times*, 5 January.
6. Based on Kajimoto, T. (2021) Japanese companies go high-tech in the battle against food waste, Reuters, 28 February.
7. Fareshare website, https://fareshare.org.uk (accessed September 2021).
8. Dora, M. and Iacovidou, E. (2019) Why some plastic packaging is necessary to prevent food waste and protect the environment, The Conversation, 7 June.
9. Sainsbury's website, https://www.about.sainsburys.co.uk/sustainability/plan-for-better/our-stories/2017/we-got-the-power (accessed September 2021).

14 Planning and control systems

KEY QUESTIONS

14.1 What are planning and control systems?

14.2 What is enterprise resource planning, and how did it develop into the most common planning and control system?

14.3 How should planning and control systems be implemented?

INTRODUCTION

One of the most important issues in planning and controlling operations is managing the sometimes vast amounts of information generated by the activity. It is not just the operations function that is the author and recipient of this information; almost every other function of a business will be involved. So, it is important that all relevant information that is spread throughout the organisation is brought together, and that, based on this information, appropriate decisions are taken. This is the function of planning and control systems. They bring information together, help to make decisions then inform the relevant parts of the operation about decisions such as when activities should take place, where they should happen, who should be doing them, how much capacity will be needed, and so on. In this chapter we shall look in particular at the dominant form of planning and control system – **enterprise resource planning (ERP)**. It grew out of a set of calculations known as **materials requirements planning (MRP)**, which is described in the supplement to this chapter. Figure 14.1 shows where this topic fits in our overall model of operations activities.

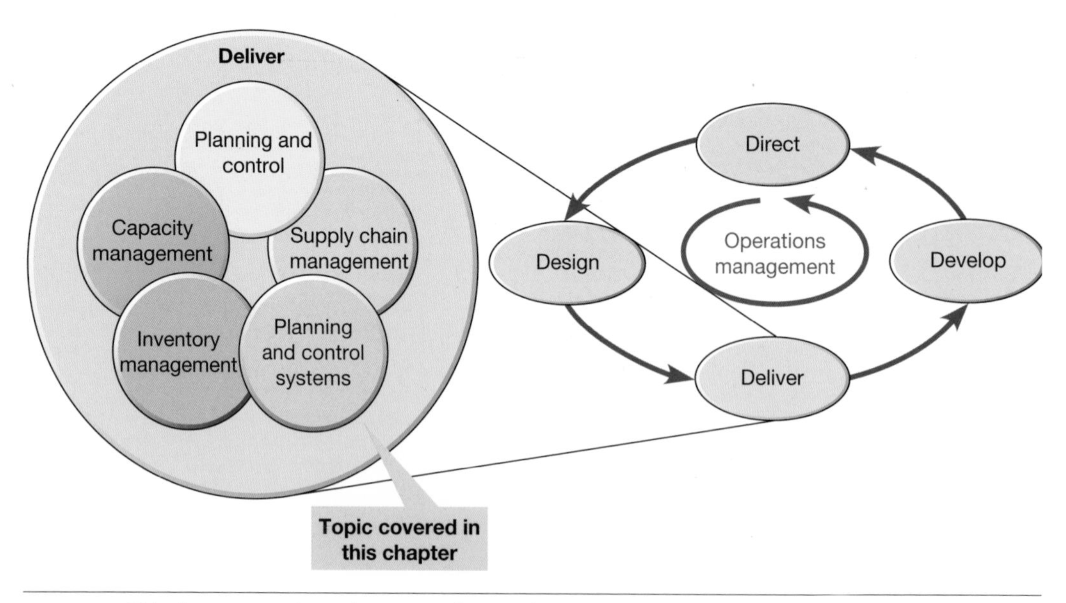

Figure 14.1 This chapter examines planning and control systems

14.1 What are planning and control systems?

In Chapter 10 we described the activity of planning and control as being concerned with managing the ongoing allocation of resources and activities to ensure that the operation's processes are both efficient and reflect customer demand. Planning and control activities are distinct but often overlap. Planning determines what is *intended* to happen at some time in the future. Control is the process of *coping* when things do not happen as intended. Control makes the adjustments that help the operation to achieve the objectives that the plan has set, even when the assumptions on which the plan was based do not hold true.

Planning and control systems

Planning and control systems are the information processing, decision support and execution mechanisms that support the operations planning and control activity. Although planning and control systems can differ, they tend to have a number of common elements. These are: a customer interface that forms a two-way information link between the operation's activities and its customers; a supply interface that does the same thing for the operation's suppliers; a set of overlapping 'core' mechanisms that perform basic tasks such as loading, sequencing, scheduling, and monitoring and control; a decision mechanism involving both operations staff and information systems that makes or confirms planning and control decisions. It is important that all these elements are effective in their own right and work together. Figure 14.2 illustrates these elements. In sophisticated systems they may be extended to include the integration of this core operations resource planning and control task with other functional areas of the firm.

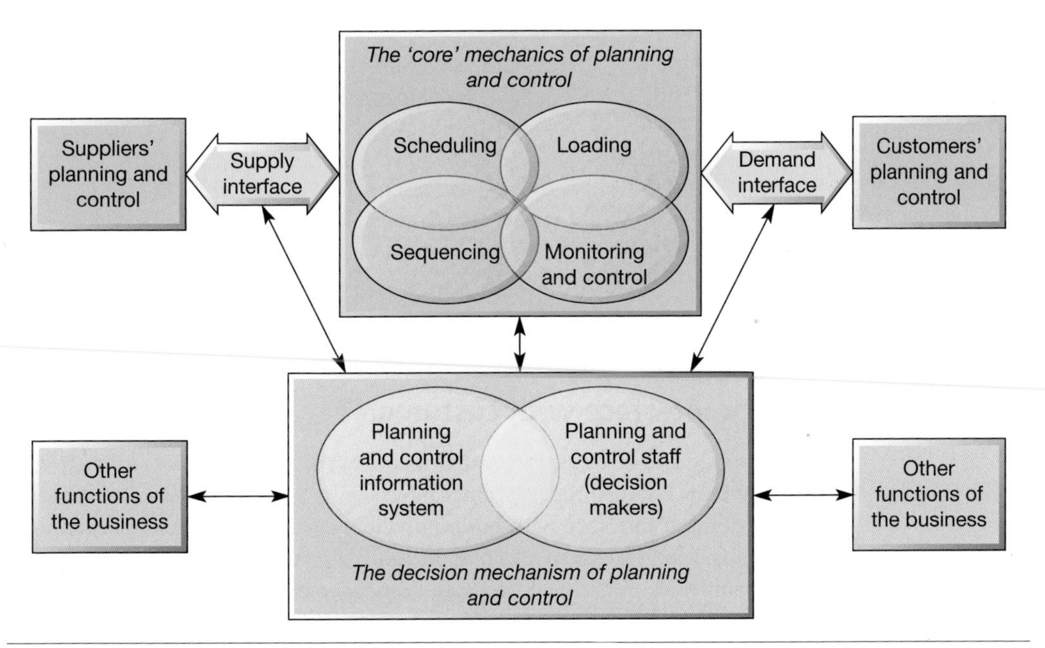

Figure 14.2 Planning and control systems interface with the internal planning and control mechanisms, customers, suppliers and the other functions of the business

OPERATIONS IN PRACTICE

Butcher's Pet Care coordinates its ERP[1]

It may not be a glamorous business, but pet food is certainly big business. It is also competitive, with smaller suppliers battling against the giants like Nestlé. One of the most successful of the smaller European producers of dog food is Butcher's Pet Care, located in the Midlands of the United Kingdom, that takes a positive moral and ethical approach towards the dog food it produces. It also needs to be super-efficient at coordinating its production and distribution if it is to compete with larger rivals. Listen to how Butcher's IT manager, Malcolm Burrows, explains his vision of how its planning and control system helps it do this.

Why implement a new planning and control system?

'There were specific goals that needed to be achieved, as the legacy systems created long processes, and it was an issue to find out what was in the warehouse, etc. A lot of manual planning tasks took place outside of the system, whereas now the planning, the enterprise resource planning (ERP), and the scheduling of material coming in is a lot better.'

What were the benefits of the new system?

'We're definitely getting a better view of what stock we're holding, and a much quicker response in being able to change product fore-ordering. As you can probably guess, within an environment whereby we are supplying to supermarkets and supply chains, there are regularly promotions that affect the manufacturing we produce to, and it's a fairly quick turnaround. So from that point of view, the system does have a core value in that we can respond and meet requirements much quicker and easier.'

What were the challenges in implementing the system?

'It was a very big cultural change for the staff . . . As with any ERP [planning and control, see later] *system, business and process mapping is crucial. The interesting challenges were working out how we needed to change to get the best out of the system, and that we had agreed a timeline for its implementation.'*

How did you train staff to use the system?

'We had a core project team, and they were the "champions" who had to go out and then work within their areas. [The] *IT* [department] *really cannot dictate that;* [users] *need to be able to have that autonomy to say "this is how we want to operate it". We will get involved if there are technical queries, but otherwise the "champions"* [are in charge].*'*

How does the system interface with customers?

The part of the resource planning and control system that manages the way customers interact with the business on a day-to-day basis is called the 'customer interface' or sometimes 'demand management'. This is a set of activities that interface with both individual customers and the market more broadly. Depending on the business, these activities may include customer negotiation, order entry, demand forecasting, order promising, updating customers, keeping customer histories, post-delivery customer service and physical distribution.

Customer interface defines the customer experience

The customer interface is important because it defines the nature of the customer experience. It is the public face of the operation (the 'line of visibility' as it was called in Chapter 6). Therefore, it needs to be managed like any other 'customer-processing' process, where the quality of the service is defined by the gap between customers' expectations and their perceptions of the service they receive. The right-hand side of Figure 14.3 illustrates a typical customer experience of interacting with a planning and control customer interface. The experience itself will start before any customer contact is initiated. Customer expectations will have been influenced by the way the business presents itself through promotional activities, the ease with which channels of communication can be used (for example, design of the website) and so on. The question is, 'does the communication channel give any indication of the kind of service response (for example, how long will we have to wait?) that the customer can expect?' At the first point of contact when an individual customer requests services or products, their request must be understood, delivery possibly negotiated and a delivery promise made. Customers may value feedback as to the progress of their request. At the point of delivery, not only are the products and services handed over to the customer, but there may also be an opportunity to explain the nature of the delivery and gauge customers' reactions. Following the completion of the delivery there may also be some kind of post-delivery action, such as a phone call to confirm that all is well.

Operations principle

Customers' perceptions of an operation will partially be shaped by the customer interface of its planning and control system.

The customer interface acts as a trigger function

Acceptance of an order should prompt the customer interface to trigger the operation's processes. Exactly what is triggered will depend on the nature of the business, as we explained in Chapter 10. Some building and construction companies are willing to build almost any kind of construction and will keep relatively few of their own resources within the business, but rather hire them in when needed. In such a 'resource-to-order' operation the customer interface triggers the task of hiring in the relevant equipment and purchasing the appropriate materials. If the construction company

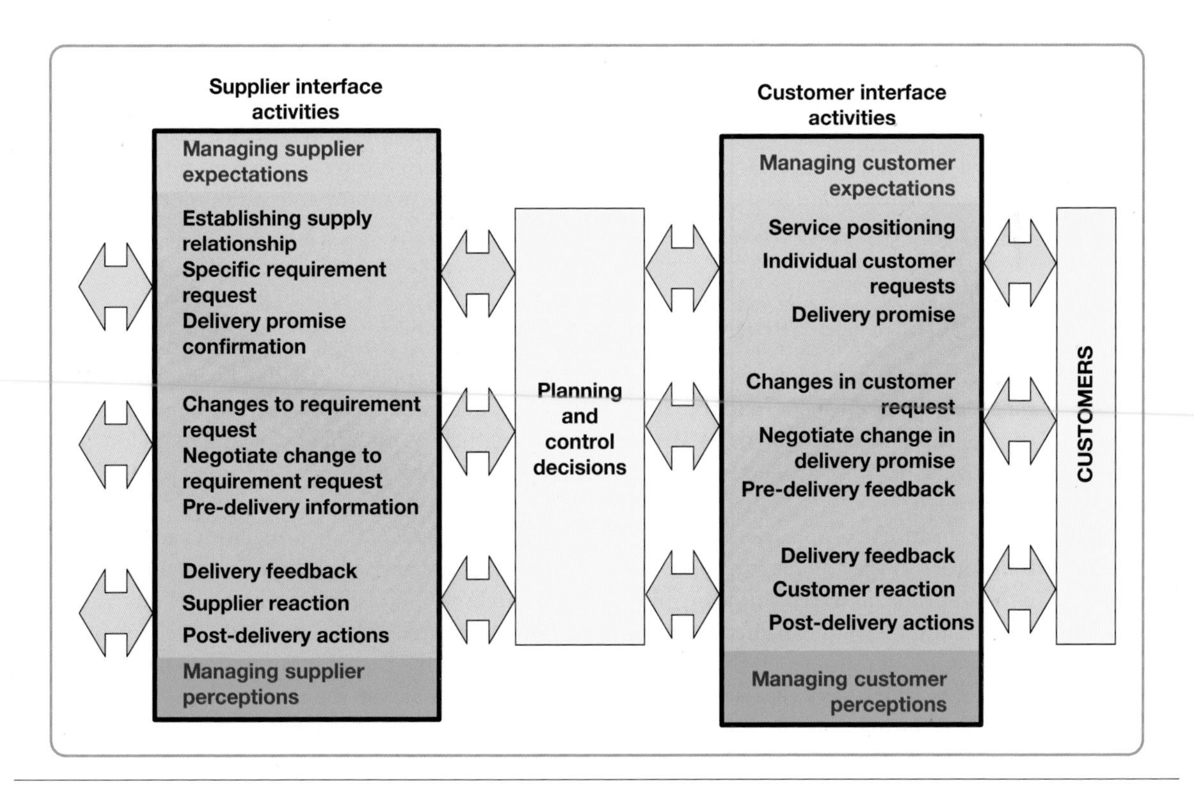

Figure 14.3 The customer and supplier interfaces as 'customer experiences'

confined itself to a narrower range of construction tasks, thereby making the nature of demand slightly more predictable, it would be likely to have its own equipment and labour permanently within the operation. Here, accepting a job would only need to trigger the purchase of the materials to be used in the construction, and the business is a 'produce-to-order' operation. Some construction companies will construct pre-designed standard houses ahead of demand for them. Operations of this type 'produce ahead of order'.

How does the system interface with suppliers?

The supplier interface provides the link between the activities of the operation itself and those of its suppliers. The timing and level of activities within the operation or process will have implications for the supply of products and services to the operation. Suppliers need to be informed so that they can make products and services available when needed. In effect this is the mirror image of the customer interface. The supplier interface is concerned with managing the supplier experience to ensure appropriate supply. It is no less important. Ultimately, customer satisfaction is influenced by supply effectiveness because that in turn influences delivery to customers.

Operations principle

An operation's planning and control system can enhance or inhibit the ability of its suppliers to support delivery effectiveness.

The supplier interface has both a long- and short-term function. It must be able to cope with different types of long-term supplier relationship and also handle individual transactions with suppliers. To do the former it must understand the requirements of all the processes within the operation and also the capabilities of the suppliers. The left-hand side of Figure 14.3 shows a simplified sequence of events in the management of a typical supplier–operation interaction that the supplier interface must facilitate. When the planning and control activity requests supply, the supplier interface must have identified potential suppliers and might also be able to suggest alternative materials or services if necessary. Formal requests for quotations may be sent to potential suppliers if no supply agreement exists. This issue was discussed in Chapter 12 as supplier development. To handle individual transactions, the supplier interface will need to issue formal purchase orders. These may be stand-alone documents or, more likely, electronic orders. They are important because they often form the legal basis of the contractual relationship. Delivery promises will need to be formally confirmed. While waiting for delivery, it may be necessary to negotiate changes in supply and track progress to get early warning of potential changes to delivery.

Hierarchical planning and control[2]

The activity of operations planning and control can be complicated. Demand is usually uncertain, supply can be problematic, and the composition of products and services is often complex, with many components and sub-components. And to add to the difficulty, the cumulative lead times for sourcing components and for production itself are usually longer than customers are prepared to wait (see Chapter 10).

The 'hierarchical approach' to operations planning recognises these difficulties and tries to bring some order to the complexity by dividing up the many interrelated planning and control decisions into sub-problems to reflect the organisational hierarchy. So, decisions at a high level link with decisions at lower levels in an effective manner. Decisions that are made at the higher level will of course impose some constraints on the lower-level decisions. And the execution of detailed decisions at the lower level will provide the necessary feedback so that the quality of higher-level decision-making can be judged. In this way the hierarchical approach separates different kinds of decisions at different levels in the organisation and over different time periods. It allows a degree of stability in the planning process so that relatively complex operations are, to some extent, protected against too many short-term changes. In addition, it gives a certain amount of independence to the planners at different levels. Figure 14.4 illustrates this hierarchical approach. How well this approach works will depend largely

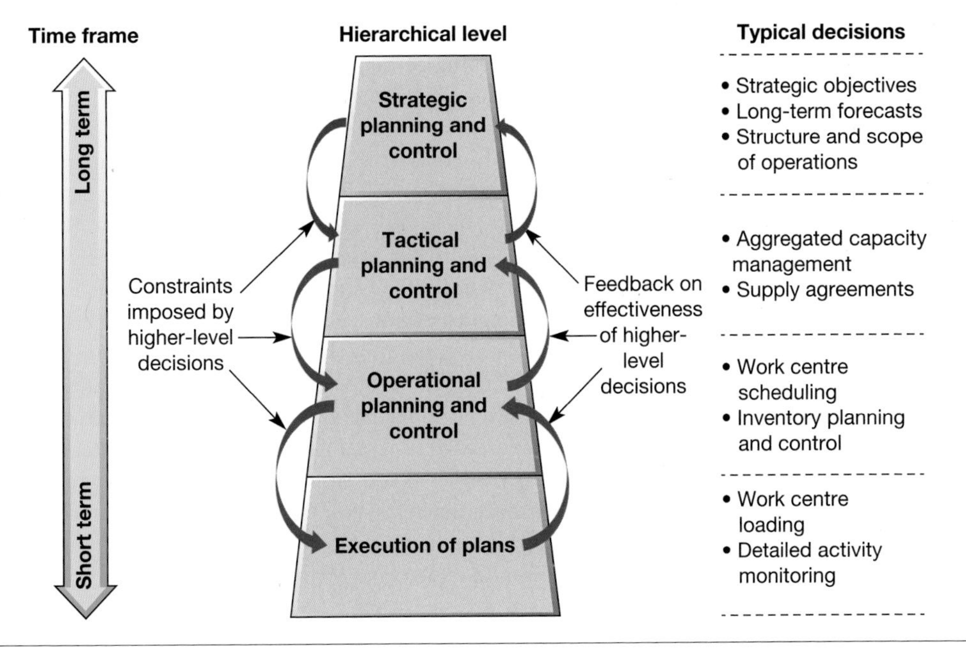

Figure 14.4 The general structure of hierarchical production planning and control

on how effectively and consistently the boundaries between the levels of the hierarchy are managed. Each hierarchical level is likely to have its own set of decision rules and methods with different planning horizons, levels of detail of information and forecasts, scope of the planning activity and managerial authority, all of which can lead to problems in translating the decisions at one level to another.

Operations principle

Hierarchical planning and control systems separate different kinds of decisions at different levels in the organisation and over different time periods.

Critical commentary

Hierarchical planning and control looks to be both rational and straightforward; however, making it work in practice can be problematic. Several questions need to be addressed. How many levels are needed? What should constrain what and how tightly? What should one plan in advance? Does a hierarchical approach reduce the speed of decision making by requiring continual upward referral? How much autonomy and local control should be devolved to lower levels or to distributed production facilities? Is stability achieved by rigidity and at the expense of speed and responsiveness? In addition, data must be accurate, timely and in common formats. Effective transition between the levels also requires a significant degree of managerial discipline.

14.2 What is enterprise resource planning, and how did it develop into the most common planning and control system?

To understand enterprise resource planning (ERP), imagine that you are holding a party in two weeks' time and expect 40 people to attend. As well as drinks, you decide to provide sandwiches and snacks. Some simple calculations and assumptions will give an estimate of the provisions you need. You may already have some food and drink that you could use, so you will take it into account when making your shopping list. You may also want to take into account that you will prepare some of the food in advance and freeze it. In fact, planning a party requires a series of interrelated decisions about the quantity and timing of the *materials* needed. The party also has financial implications. You may have to agree a temporary increase to your credit card limit. Again, this requires some forward planning, and calculations of how much it is going to cost and how much extra credit you require. This is the basis for ERP. It is a process that helps companies make volume and timing calculations (like the party, but on a much larger scale, and with a greater degree of complexity).

So, even for this relatively simple activity, the key to successful planning is how we generate, integrate and organise all the information on which planning and control depends. Of course, in business operations it is more complex than this. Companies usually sell many different products to many hundreds of customers who are likely to vary their demand for the products. This is a bit like organising 200 parties one week, 250 the next and 225 the following week, all for different groups of guests with different requirements who keep changing their minds about what they want to eat and drink. This is what an ERP system does; it automates and integrates core business processes such as customer demand, scheduling operations, ordering items, keeping inventory records and updating financial data. It helps companies 'forward plan' these types of decisions and understand all the implications of any changes to the plan.

Operations principle

ERP systems automate and integrate core business processes.

How did ERP develop?

The (now) large companies that have grown almost exclusively on the basis of providing ERP systems include SAP and Oracle. Yet to understand ERP, it is important to understand the various stages in its development, summarised in Figure 14.5. It started with materials requirements planning (MRP), which became popular during the 1970s, although the planning and control logic that underlies it had, by then, been known for some time. What popularised MRP was the availability of computer power to drive the basic planning and control mathematics. We will deal with MRP in detail in the supplement to this chapter.

Manufacturing resource planning (MRP II) expanded out of MRP during the 1980s. Again, it was a technology innovation that allowed the development. Connected networks, together with increasingly powerful desktop computers, allowed a much higher degree of processing power and communication between different parts of a business. Also, MRP II's extra sophistication allowed the forward modelling of 'what-if' scenarios. The strength of MRP and MRP II always lay in the fact that it could explore the *consequences* of any changes to what an operation was required to do. So, if demand changed, the MRP system would calculate all the 'knock-on' effects and issue instructions accordingly. This same principle also applies to ERP, but on a much wider basis. ERP has been defined as:

> *a complete enterprise wide business solution. The ERP system consists of software support modules such as: marketing and sales, field service, product design and development, production and inventory control, procurement, distribution, industrial facilities management, process design and development, manufacturing, quality, human resources, finance and accounting, and information services. Integration between the modules is stressed without the duplication of information.*[3]

So, ERP systems allow decisions and databases from all parts of the organisation to be integrated so that the consequences of decisions in one part of the organisation are reflected in the planning and control systems of the rest of the organisation (see Figure 14.6). ERP is the equivalent of the organisation's central nervous system, sensing information about the condition of different parts of the business and relaying the information to other parts of the business that need it. The information is updated in real time by those who use it and yet is always available to everyone connected to the ERP system.

Also, the potential of internet-based communication has provided a further boost to ERP development. Many companies have suppliers, customers and other businesses with which they collaborate,

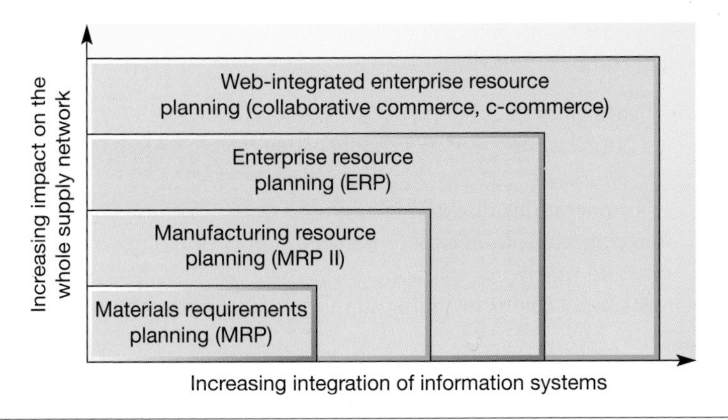

Figure 14.5 The development of ERP

which themselves have ERP-type systems. An obvious development is to allow these systems to communicate. However, the technical, as well as organisational and strategic, consequences of this can be formidable. Nevertheless, many authorities believe that the true value of ERP systems is only fully exploited when such web-integrated ERP (known by some people as 'collaborative commerce', or c-commerce) becomes widely implemented.

The benefits of ERP

ERP is generally seen as having the potential to very significantly improve the performance of many companies in many different sectors. This is partly because of the enhanced visibility that information integration gives, but it is also a function of the discipline that ERP demands. Yet this discipline is itself a 'double-edged' sword. On one hand, it 'sharpens up' the management of every process within an organisation, allowing best practice to be implemented uniformly through the business. No longer will individual idiosyncratic behaviour by one part of a company's operations cause disruption to all other processes. On the other hand, it is the rigidity of this discipline that is both difficult to achieve

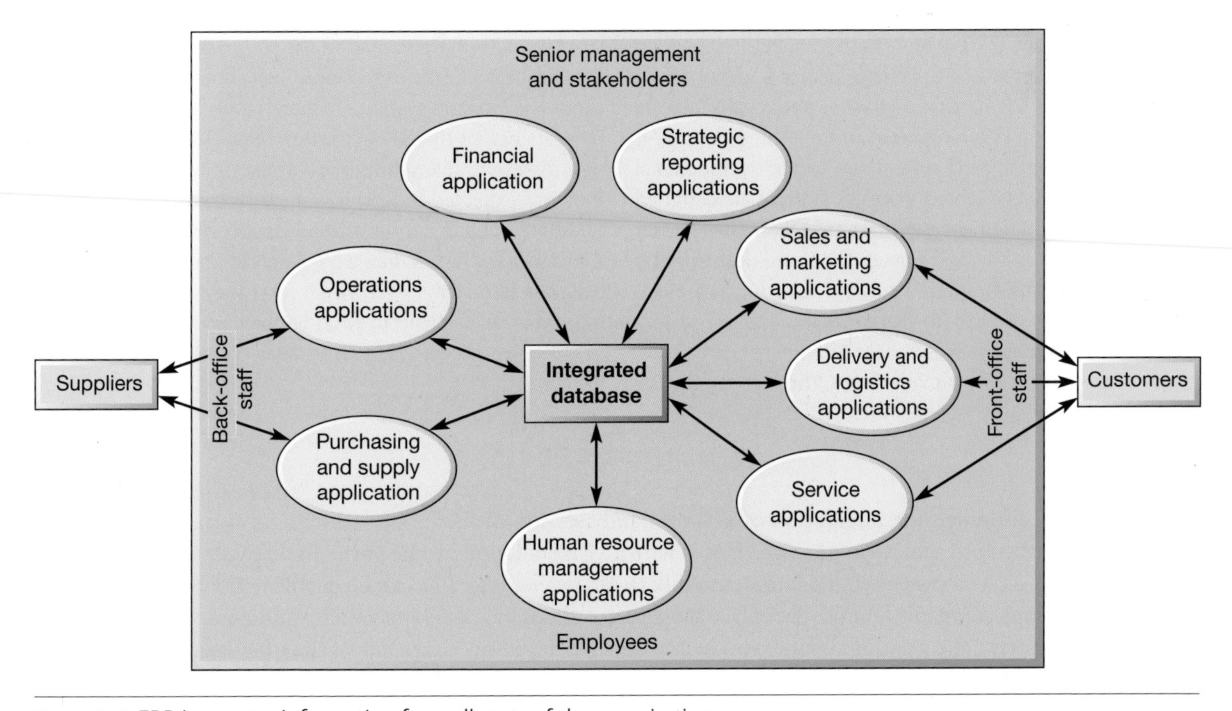

Figure 14.6 ERP integrates information from all parts of the organisation

and (arguably) inappropriate for all parts of the business. Nevertheless, the generally accepted benefits of ERP are usually held to be the following:

- Visibility of what is happening in all parts of the business.
- Discipline of forcing business process-based changes is an effective mechanism for making all parts of the business more efficient.
- A 'sense of control' of operations that will form the basis for continuous improvement.
- A more sophisticated communication with customers, suppliers and other business partners, often giving more accurate and timely information.
- Integration of whole supply chains including suppliers' suppliers and customers' customers.

OPERATIONS IN PRACTICE

The computer never lies – really?[4]

Write a piece of information on a piece of scrap paper, and it carries little authority. Print it in an official-looking report and you are likely to have more faith in it. Let it appear on the screen as part of a core planning system and few would question it. It's a natural response, yet it can also be dangerous. Unquestioning belief in something because it is contained within an IT system can lead to dysfunctional and, at worst, tragic consequences. IT systems are designed by humans, and humans can make mistakes. This is what happened after the UK Post Office and Fujitsu, the Japanese systems company, launched its Horizon project. Described at the time as 'the largest non-military IT system in Europe', its purpose was to help sub-postmasters (independent retailers who provide Post Office services to the public) to keep track of such routine retail operations tasks such as sales transactions and inventory control. From the moment of its implementation there were complaints from sub-postmasters that the system was flawed. The complaints were dismissed by the Post Office as teething difficulties. But when the Horizon system (falsely) indicated that some sub-postmasters had committed a series of thefts, fraudulent activities and false accounting, the Post Office's management chose to believe the IT system and aggressively prosecuted over 700 sub-postmasters. Over a 14-year period, many were convicted for false accounting and theft, and went to prison. Others were ruined financially and felt shamed in their communities. Tragically, one former sub-postmaster killed himself after he was falsely accused of stealing £60,000. Others died with the false convictions still hanging over them. Nine years from its implementation the story was publicised by the magazine *Computer Weekly*, which raised questions about the Horizon system's failures. However, it took over 20 years before the UK's Court of Appeal ruled in favour of a representative group of sub-postmasters, a decision that was expected to lead to almost all convictions being quashed and substantial compensation for those falsely convicted. The lesson drawn by many commentators is that believing IT systems, even when what they indicate seems questionable, can lead to what has been described as 'one of the most grotesque miscarriages of justice in British history'.[5]

Although the integration of several databases lies at the heart of ERP's power, it is nonetheless difficult to achieve in practice. This is why ERP installation can be particularly expensive. Attempting to get new systems and databases to talk to old (legacy) systems can be problematic. Not surprisingly, many companies choose to replace most, if not all, of their existing systems simultaneously. Common systems and relational databases help to ensure the smooth transfer of data between different parts of the organisation. In addition to the integration of systems, ERP usually includes other features that make it a powerful planning and control tool.

ERP changes the way companies do business

Arguably the most significant issue in many companies' decision to buy an off-the-shelf ERP system is its compatibility with the company's current business processes and practices. The advice that is emerging from the companies that have adopted ERP is that it is extremely important to make sure that their current way of doing business will fit (or can be changed to fit) with a standard ERP package. In fact, one of the most common reasons for companies to decide not to install ERP is that they cannot reconcile the assumptions in the software of the ERP system with their core business processes. If, as most businesses find, their current processes do not fit, they could do one of two things. They could change their processes to fit the ERP package. Alternatively, they could modify the software within the ERP package to fit their processes. Both of these options involve costs and risks. Changing business practices that are working well will involve reorganisation costs as well introducing the potential for errors to creep into the processes. Adapting the software will both slow down the project and introduce potentially dangerous software 'bugs' into the system. It would also make it difficult to upgrade the software later on.

> **Operations principle**
>
> ERP systems are only fully effective if the way a business organises its processes is aligned with the underlying assumptions of its ERP system.

OPERATIONS IN PRACTICE — The ERP for a chicken salad sandwich

Pre-packed sandwiches are generally produced in factories during the afternoon, evening and night time for delivery during the evening and the night time, and the morning of the following day. But that is only one half of the story. The other half concerns how a sandwich company manages the *quantity* of ingredients to order, the quantity of sandwiches to be made and the chain of implications for the whole company. Almost all sandwich companies use an ERP system that has at its core an MRP II package. The MRP II system has the two normal basic drivers of, first, a continually updated sales forecast, and second, a product structure database. In this case the **product structure** and/or **bill of materials** is the 'recipe' for the sandwich, and within the company this database is called the 'Recipe Management System'. The 'recipe' for a chicken sandwich (its bill of materials) is shown in Table 14.1.

Figure 14.7 shows the ERP system used by one sandwich company. Orders are received from customers electronically through the EDI system. These orders are then checked through a 'validation system' that checks the order against current product codes and expected quantities to make sure that the customer has not made any mistakes, such as forgetting to order some products. After validation the orders are transferred through the central database to the MRP II system that performs the main requirements breakdown. Based on these requirements and forecasted requirements for the next few days, orders are placed to the company's suppliers for raw materials and packaging. Simultaneously, confirmation is sent to customers, accounts are updated, staffing schedules are finalised for the next two weeks (on a rolling basis), customers are invoiced, and all this information is made available both to the customers' own ERP systems and the transportation company's planning system.

Table 14.1 Bill of materials for a chicken salad sandwich

FUNCTION: MBIL		*MULTI-LEVEL BILL INQUIRY*					
PARENT: BTE80058 *RV:* *PLNR: LOU*		*UM:EA*	*DESC:* *RUNLT:* *PLN POL: N*		*HE CHICKEN SALAD TRY 0* *FIXED LT: 0 DRWG: WA1882*		*LA*
LEVEL *1 . . . 5* *. . . 10*	*PT USE*	*SEQN*	*COMPONENT*	*C T*	*PARTIAL DESCRIPTION*	*QTY*	*UM*
1	PACK	010	FTE80045	P	H.E. CHICKENS	9	EA
2	ASSY	010	MBR–0032	P	BREAD HARVESTE	2	SL
3	HRPR	010	RBR–0023	N	BREAD HARVESTE	.4545455	EA
2	ASSY	020	RDY–0001	N	SPREAD BUTTER	.006	KG
2	ASSY	030	RMA–0028	N	MAYONNAISE MYB	.01	KG
2	ASSY	040	MFP–0016	P	CHICKEN FRESH	.045	KG
3	HRPR	010	RFP–0008	N	CHICKEN FRESH	1	KG
	ASSY	050	MVF–0063	P	TOMATO SLICE 4	3	SL
3	ALTI	010	RVF–0026	P	TOMATOES PRE–S	.007	KG
4	HRPR	010	RVF–0018	N	TOMATOES	1	KG
2	ASSY	060	MVF–0059	P	CUCUMBER SLICE	2	SL
3	ALTI	010	RVF–0027	P	CUCUMBER SLICE	.004	KG
4	TRAN	010	RVF–0017	N	CUCUMBER	1	KG
2	ASSY	070	MVF–0073	P	LETTUCE COS SL	.02	KG
3	HRPR	010	RVF–0015	N	LETTUCE COS	1	KG
2	ASSY	080	RPA–0070	N	WEBB BASE GREY	.00744	KG
2	ASSY	090	RPA–0071	N	WEBB TOP WHITE	.0116	KG
2	ASSY	100	RLA–0194	N	LABEL SW H	1	EA
2	ASSY	110	RLA–0110	N	STICKER NE	1	EA
1	PACK	010	RPA–0259	N	SOT LABEL	1	EA
1	PACK	030	RPA–0170	N	TRAY GREEN	1	EA

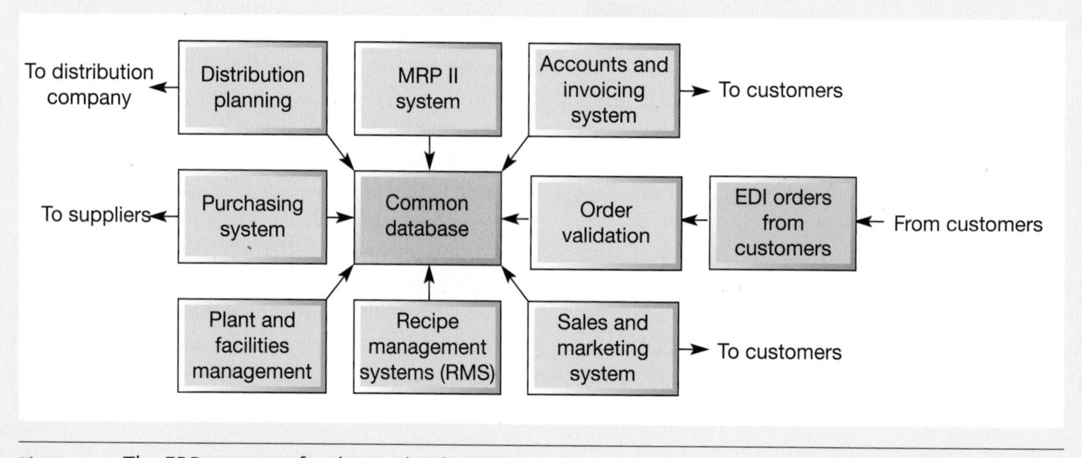

Figure 14.7 The ERP structure for the sandwich company

Web-integrated ERP

Perhaps the most important justification for embarking on ERP is the potential it gives the organisation to link up with the outside world. For example, it is much easier for an operation to move into internet-based trading if it can integrate its external internet systems into its internal ERP systems. However, as has been pointed out by some critics of the ERP software companies, ERP vendors were not prepared for the impact of e-commerce and had not made sufficient allowance in their products for the need to interface with internet-based communication channels. The result of this has been that whereas the internal complexity of ERP systems was designed only to be intelligible to systems experts, the internet has meant that customers and suppliers (who are non-experts) are demanding access to the same information. So, important pieces of information such as the status of orders, whether products are in stock, the progress of invoicing, etc., need to be available, via the ERP system, on a company's website.

One problem is that different types of external company often need different types of information. Customers need to check the progress of their orders and invoicing, whereas suppliers and other partners want access to the details of operations planning and control. Not only that, but they want access all the time. The internet is always there, but web-integrated ERP systems are often complex and need periodic maintenance. This can mean that every time the ERP system is taken offline for routine maintenance or other changes, the website also goes offline. To combat this, some companies configure their ERP and e-commerce links in such a way that they can be decoupled, so that ERP can be periodically shut down without affecting the company's web presence.

Supply chain ERP

The step beyond integrating internal ERP systems with immediate customers and suppliers is to integrate all the ERP and similar systems along a supply chain. Of course, this can never be straightforward and is often exceptionally complicated. Not only do different ERP systems have to communicate together, they have to integrate with other types of systems. For example, sales and marketing functions often use systems such as customer relationship management (CRM) that manage the complexities of customer requirements, promises and transactions. Getting ERP and CRM systems to work together is itself often difficult. Sometimes the information from ERP systems has to be translated into a form that CRM and other e-commerce applications are able to understand. Nevertheless, such web-integrated ERP, or c-commerce (collaborative commerce), applications are emerging and starting to make an impact on the way companies do business. Although implementing supply chain ERP is a formidable task, the benefits are potentially great. The costs of communicating between supply chain partners could be dramatically reduced and the potential for avoiding errors as information and products move between partners in the supply chain is significant. Yet as a final warning note, it is as well to remember that although integration can bring all the benefits of increased transparency in a supply chain, it may also transmit systems failure. If the ERP system of one operation within a supply chain fails for some reason, it may block the effective operation of the whole integrated information system throughout the chain.

Operations principle

The effectiveness of ERP systems depends partly on suppliers' and customers' ERP systems.

14.3 How should planning and control systems be implemented?

By their nature, planning and control systems are designed to address the problems of information having to be obtained from different parts of a business. It is not surprising then that any planning and control system will be complex and difficult to get right. Implementing this type of system will necessarily involve crossing organisational boundaries and integrating internal processes that cover many, if not all, functional areas of a business. Building a single system that simultaneously satisfies the requirements of operations managers, marketing and sales managers, finance managers and everyone else in the organisation is never going to be easy. It is likely that each function will have its own set of processes and a well-understood system that has been designed for its specific needs. Moving everyone onto a single, integrated system that runs off a single database is going to be potentially very unpopular. Change is not always popular, and ERP asks almost everyone to change how they do their jobs.

The particular challenges of information technology (IT) implementation

Surprisingly, given the ubiquity of IT systems such as planning and control systems, the cost effectiveness of investment in IT is not altogether straightforward. Generally, there is a positive connection between investment in IT and increased operations effectiveness, even if the benefits can vary widely. As one authority put it, 'there's no bank where companies can deposit IT investment and withdraw an "average" return . . . [A] strategy of blindly investing in IT and expecting productivity to automatically rise is sure to fail'.[6] Moreover, there is a high failure rate for IT projects (often cited as between 35 per cent and 75 per cent, although the definition of 'failure' is debated). Yet there is extensive agreement that the most common reasons for failure are connected in some way with managerial, implementation or organisational factors. And of these managerial, implementation or organisational factors, one of the main issues is the degree of alignment and integration between a firm's overall IT strategy and the general strategy of the firm.

Of course, different kinds of IT pose different kinds of challenge. The impact of some IT is limited to a defined and (relatively) limited part of the operation. This type of IT is sometimes called 'function IT' because it facilitates a single function or task.[7] Examples include computer-aided design (CAD), spreadsheets and simple decision support systems. The organisational challenges for this type of technology can usually be treated separately from the technology itself. Put another way, function IT can be adopted with or without any changes to other organisational structures. Yet this does not mean that no organisational, cultural or development challenges will be faced. Often the effectiveness of the technology can be enhanced by appropriate changes to other aspects of the operation. By contrast, 'enterprise IT' extends across much of, or even the entire organisation. Because of this, enterprise IT will need potentially extensive changes to the organisation. And the most common (and problematic) enterprise IT systems are ERP systems. The third IT category is 'network IT'. Network IT facilitates exchanges between people and groups inside and/or outside the organisation. However, it does not necessarily predefine how these exchanges should work. For example, simple email is a network IT. It has brought significant changes to how operations and supply networks function, but the changes are not imposed by the technology itself; rather they emerge over time as people gain experience of using the technology. The challenge with this type of technology is to learn how to exploit its emergent potential. This is the challenge for many operations as they extend their ERP systems to encompass the whole, or even a part, of their supply chain.

OPERATIONS IN PRACTICE

It's not that easy[8]

ERP systems have been called the nerve centre of any operation, and like actual nervous systems, they are difficult to deal with and they can cause severe pain when they go wrong. And ERP implementation can go wrong, even when undertaken by experienced professionals. Look at these examples.

Lidl

Like many large companies, the German supermarket chain, Lidl, had put up with its in-house inventory management system that, after being in place for many years, was starting to creak. It commissioned SAP, the expert enterprise software firm, to install a totally new system. However, during the implementation process, it became clear that there was a clash between how Lidl preferred to account for inventory (at purchase price) and how most retailers do it (at the retail price they sell the goods for). SAP's system was set up for the latter. Because Lidl didn't want to change their accounting practice, SAP had to try to customise its system. This resulted in a series of implementation problems. What would have been a complex project was not helped by staff turnover at Lidl's IT department. In 2018, seven years after starting, and after spending nearly €500 million, the project was scrapped.

Woolworth's Australia

Another retailer, this time in Australia, also had significant ERP implementation problems. Woolworths is Australia's largest supermarket chain (not to be confused with the now defunct UK operation). Operating almost 1,000 stores across Australia, it employs 115,000 staff in its stores and supply networks. Its new ERP system was intended to modernise the company's planning and control efforts, but when, after six years of planning, it went live with the new system, problems emerged almost immediately. The most obvious symptom that something was wrong was the empty shelves in many of the company's stores. Apparently, a malfunction in the new system prevented the company from placing orders with its (many) suppliers. The investigation into what went wrong showed that one of the main problems was that during the ERP implementation process insufficient attention was given to understanding and documenting the processes that were actually used by staff in their day-to-day running of the business. Much of the detail of these processes was in the heads of staff, rather than formally recorded. So, when individuals left the company, they took key pieces of information with them. In effect, the ERP implementation failed because of a loss of corporate memory.

Oriola Finland

Oriola Finland is a Finnish health and well-being company with a strong position in the Swedish and Finnish healthcare markets. In Sweden, Oriola owns Kronans Apotek, the third-largest pharmacy chain in the country. It is a major distributor of health and well-being products that employs just under 3,000 people. But its ability to deliver its pharmaceuticals was severely compromised when a major ERP system upgrade went wrong. Oriola Finland deliver thousands of medications to pharmacists around the country, including insulin, cancer medications and anti-psychotics, so any disturbance in the supply chain doesn't just cause lost sales, it can damage people's health. This was a particularly serious disruption, given that 46 per cent of all drugs sold in Finland were supplied by Oriola, and switching to an alternative distributor would have been difficult to do quickly. The problem was partly one of failure to sufficiently anticipate the disruption that any IT system change can cause. It is believed that Oriola did not anticipate any major supply disruption as the switchover happened, yet its ordering system was inoperative for days. The lesson is that, when transitioning to a new ERP, it is usually best to always plan for the worst. Eventually, Oriola did manage to sort out the problems with its new ERP, and also hired extra staff to process the backlog of orders. But the incident cost the company millions of euros and caused damage to its reputation for supply reliability.

Waste Management

ERP implementation failures can sometimes end up in the law courts. Waste Management, Inc. is the leading provider of waste and environmental services in North America. When it announced that it was suing its ERP supplier, SAP, over the failure of an ERP implementation, it said that it was seeking the recovery of more than $100 million in project expenses as well as the savings and benefits that the SAP software was promised to deliver. It said that SAP promised that the software could be fully implemented throughout all of Waste Management within 18 months, and that its software was an 'out-of-the-box' solution that would meet Waste Management's needs without any customisation or enhancements. However, according to Waste Management, the SAP implementation team discovered significant 'gaps' between the software's functionality and Waste Management's business requirements. Waste Management had discovered that these gaps were already known to the product development team in Germany even before the SLA (service-level agreement) was signed. But members of SAP's implementation team had reportedly blamed Waste Management for the functional gaps and had submitted change orders requiring that Waste Management pay for fixing them. Five years after the complaint, the dispute was settled when SAP made a one-time cash payment to Waste Management.

Implementation critical success factors

One of the key issues in ERP implementation is what critical success factors (CSFs) should be managed to increase the chances of a successful implementation. In this case, CSFs are those things that the organisation must 'get right' in order for the ERP system to work effectively. Much of the research in this area has been summarised by Finney and Corbett,[9] who distinguish between the broad organisation-wide, or strategic, factors, and the more project specific, or tactical, factors. These are shown in Table 14.2.

Of course, some of these CSFs could be appropriate for any kind of complex implementation, whether of an ERP system, or some other major change to an operation. But that is the point. ERP implementation certainly has some specific technical requirements, but good ERP implementation practice is very similar to other complicated and sensitive implementation. Again, what is different about ERP is that it is enterprise-wide, so implementation should always be considered on an enterprise-wide level. Therefore, there will at all times be many different stakeholders to consider, each with their own concerns. That is why implementing an ERP system is always going to be an exercise in change management. Only if the anxieties of all relevant groups are addressed effectively will the prospect of achieving superior system performance be high.

At a purely practical level, many consultants who have had to live through the difficulties of implementing ERP have summarised their experiences. The following list of likely problems with an ERP implementation is typical (and really does reflect reality):[10]

- The total cost is likely to be underestimated.
- The time and effort to implement it is likely to be underestimated.

Table 14.2 Strategic and tactical critical success factors (CSF) related to successful ERP implementation.

Strategic critical success factors	Tactical critical success factors
▶ Top management commitment and support – strong and committed leadership at the top management level is essential to the success of an ERP implementation	▶ Balanced team – the need for an implementation team that spans the organisation, as well as one that possesses a balance of business and IT skills
▶ Visioning and planning – articulating a business vision to the organisation, identifying clear goals and objectives, and providing a clear link between business goals and systems strategy	▶ Project team – there is a critical need to put in place a solid core implementation team that comprises the organisation's 'best and brightest' individuals. These individuals should have a proven reputation and there should be a commitment to 'release' these individuals to the project on a full-time basis
▶ Project champion – the individual should possess strong leadership skills as well as business, technical and personal managerial competences	▶ Communication plan – planned communication among various functions and organisational levels (specifically between business and IT personnel) is important to ensure that open communication occurs within the entire organisation, as well as with suppliers and customers
▶ Implementation strategy and timeframe – implement the ERP under a time-phased approach	▶ Project cost planning and management – it is important to know up front exactly what the implementation costs will be and dedicate the necessary budget
▶ Project management – the ongoing management of the implementation plan	▶ IT infrastructure – it is critical to assess the IT readiness of the organisation, including the architecture and skills. Infrastructure might need to be upgraded or revamped
▶ Change management – this concept refers to the need for the implementation team to formally prepare a change management programme and be conscious of the need to consider the implications of such a project. One key task is to build user acceptance of the project and a positive employee attitude. This might be accomplished through education about the benefits and need for an ERP system. Part of this building of user acceptance should also involve securing the support of opinion leaders throughout the organisation. There is also a need for the team leader to negotiate effectively between various political turfs. Some authorities also stress that in planning the ERP project, it must be looked upon as a change management initiative, not an IT initiative	▶ Selection of ERP – the selection of an appropriate ERP package that matches the business's processes ▶ Consultant selection and relationship – some authorities advocate the need to include an ERP consultant as part of the implementation team ▶ Training and job redesign – training is a critical aspect of an implementation. It is also necessary to consider the impact of the change on the nature of work and the specific job descriptions ▶ Troubleshooting/crisis management – it is important to be flexible in ERP implementations and to learn from unforeseen circumstances, as well as be prepared to handle unexpected crisis situations. Troubleshooting skills will be an ongoing requirement of the implementation process

Source: Based on Finney, S. and Corbett, M. (2007) ERP implementation: a compilation and analysis of critical success factors, *Business Process Management Journal*, 13 (3), 329–47

- ▶ The resourcing from both the business and the IT function is likely to be higher than anticipated.
- ▶ The level of outside expertise required will be more than anticipated.
- ▶ The changes required to business processes will be greater than expected.
- ▶ Controlling the scope of the project will be more difficult than expected.
- ▶ There will never be enough training.
- ▶ The need for change management is not likely to be recognised until it is too late, and the changes required to corporate culture are likely to be grossly underestimated. (This is the single biggest failure point for ERP implementations.)

Responsible operations

In every chapter, under the heading of 'Responsible operations', we summarise how the particular topic covered in the chapter touches upon social, ethical and environmental issues.

The very purpose of ERP systems is the source of the ethical issues faced by operations managers implementing such systems – they integrate the information needs and interests of a broad range of stakeholders. This leads to two sets of issues in particular. The first relates to how an operation balances the (sometimes conflicting) interests of stakeholder groups during the design and implementation of ERP systems. The second set of issues concerns the nature of stakeholders' access to the huge amounts of information held within ERP systems.

ERP links departments within operations, and when extended to a firm's supply network, different operations (such as suppliers, business partners and customers). It streamlines information flow, allowing greater cooperation between resources and operations. However, different stakeholders are likely to have different interests and priorities. Information flows are not neutral. They can confer power to whoever holds the information. Within an operation, how does one balance information flows to, for example, employee representatives, when that information could imply job losses, against the ethical imperative for openness and transparency? Similarly with external entities, such as suppliers. How should information flows to suppliers be designed when that information might be interpreted as something that could lead to commercial disadvantage?

Regarding access to information when an ERP system is operational, accessibility protocols must make sure that information stored within the ERP system about employees, customers and business partners is only accessible to those who have the right to see and use it. Appropriate security must be built-in to the system that precludes unsanctioned access. Especially with the increasing use of mobile devices, hacking, snooping and other types of unsanctioned access to data could become a concern. In fact, in many regions the privacy and accuracy of information is enshrined in law. Certainly, there is an ethical responsibility to protect information. Customers especially give their information to an operation on condition that it will be adequately guarded.

Summary answers to key questions

14.1 What are planning and control systems?

- ▶ Planning and control systems are the information processing, decision support and execution mechanisms that support the operations planning and control activity.

- Planning and control systems can take various forms, but usually have some common elements such as customer and supplier interfaces, an information system, a set of decision rules, and functions to schedule, sequence, load and monitor operations activities.
- Hierarchical planning and control systems separate different kinds of decisions at different levels in the organisation and over different time periods

14.2 What is enterprise resource planning, and how did it develop into the most common planning and control system?

- ERP is an enterprise-wide information system that integrates all the information from many functions, which is needed for planning and controlling operations activities. This integration around a common database allows for transparency.
- It often requires very considerable investment in the software itself, as well as its implementation. More significantly, it often requires a company's processes to be changed to bring them in line with the assumptions built into the ERP software.
- ERP can be seen as the latest development from the original planning and control approach known as MRP.
- Although ERP is becoming increasingly competent at the integration of internal systems and databases, there is the even more significant potential of integration with other organisations' ERP (and equivalent) systems.
- In particular, the use of internet-based communication between customers, suppliers and other partners in the supply chain has opened up the possibility of wider integration.

14.3 How should planning and control systems be implemented?

- Because planning and control systems are designed to address problems of information fragmentation, implementation will be complex and cross organisational boundaries.
- There are a number of critical success factors (CSFs) that the organisation must 'get right' in order for the ERP system to work effectively. Some of these are broad, organisation-wide, or strategic, factors. Others are more project-specific, or tactical, factors.

CASE STUDY Psycho Sports Ltd

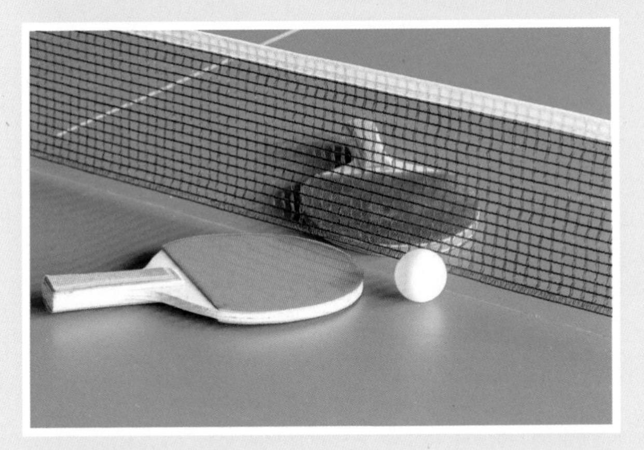

Peter Townsend knew that he would have to make some decisions pretty soon. His sports goods manufacturing business, Psycho Sports, had grown so rapidly over the last two years that he would soon have to install some systematic procedures and routines to manage the business. His biggest problem was in manufacturing control. He had started making specialist high-quality table tennis bats but now made a wide range of sports products, including tennis balls, darts and protective equipment for various games. Furthermore, his customers, once limited to specialist sports shops, now included some of the major sports retail chains.

'We really do have to get control of our manufacturing. I keep getting told that we need what seems to be called an MRP system. I wasn't sure what this meant and so I have bought a specialist production control book from our local bookshop and read all about MRP principles. I must admit, these academics seem to delight in making simple things complicated. And there is so much jargon associated with the technique, I feel more confused now than I did before.

'Perhaps the best way forward is for me to take a very simple example from my own production unit and see whether I can work things out manually. If I can follow the process through on paper then I will be far better equipped to decide what kind of computer-based system we should get, if any!'

Peter decided to take as his example one of his new products: a table tennis bat marketed under the name of the 'high-resolution' bat, but known within the manufacturing unit more prosaically as part number 5654. Figure 14.8 gives the product structure for this table tennis bat, showing it made up of two main assemblies: a handle assembly and a face assembly. In order to bring the two main assemblies together to form the finished bat, various fixings are required, such as nails, connectors, etc.

The gross requirements for this particular bat are shown below. The bat is not due to be launched until Week 13 (it is now Week 1), and sales forecasts have been made for the first 23 weeks of sales:

Weeks 13–21 inclusive, 100 per week
Weeks 22–29 inclusive, 150 per week
Weeks 30–35 inclusive, 200 per week.

Peter also managed to obtain information on the current inventory levels of each of the parts that made up the finished bat, together with cost data and lead times. He was surprised,

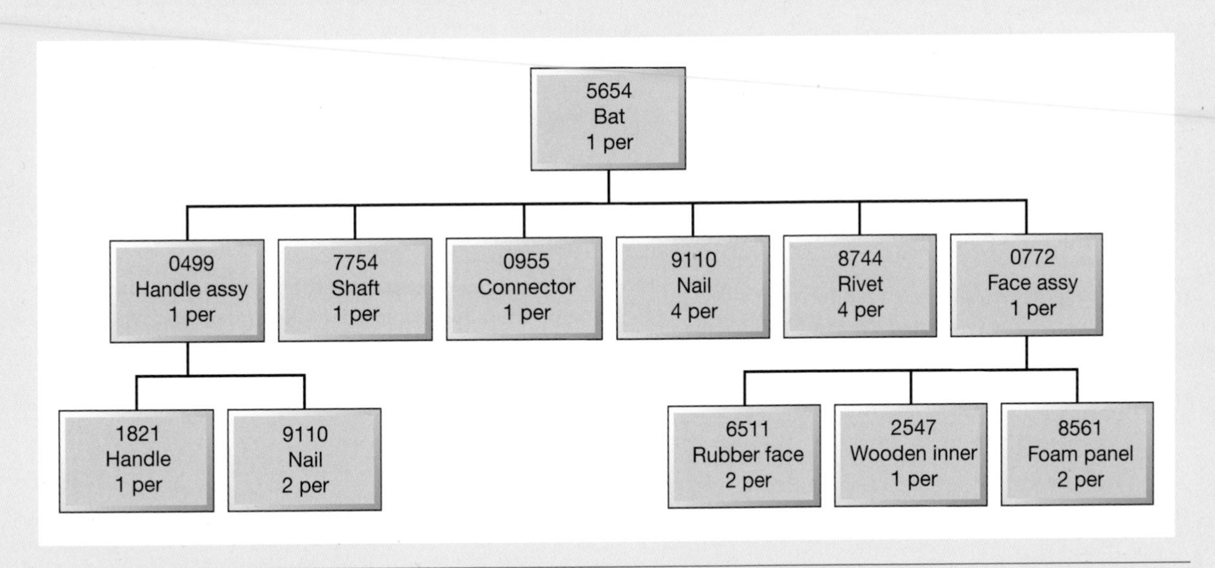

Figure 14.8 Product structure for bat 5654

Table 14.3 Inventory, cost and lead-time information for parts

Part no.	Description	Inventory	EQ	LT	Std cost
5645	Bat	0	500	2	12.00
0499	Handle assy	0	400	3	4.00
7754	Shaft	15	1,000	5	1.00
0955	Connector	350	5,000	4	0.02
9110	Nail	120	5,000	4	0.01
8744	Rivet	3,540	5,000	4	0.01
0772	Face assy	0	250	4	5.00
1821	Handle	0	500	4	2.00
6511	Rubber face	0	2,000	10	0.50
2547	Wooden inner	10	300	7	1.50
8561	Foam panel	0	1,000	8	0.50

LT = lead time for ordering (in weeks); EQ = economic quantity for ordering; Std cost = standard cost in £.

however, how long it took him to obtain this information. *'It has taken me nearly two days to get hold of all the information I need. Different people held it, nowhere was it conveniently put together, and sometimes it was not even written down. To get the inventory data, I actually had to go down to the stores and count how many parts were in the boxes'.* The data Peter collected were as shown in Table 14.3.

Peter set himself six exercises, which he knew he would have to master if he was to understand fully the basics of MRP.

Exercise 1

Draw up:

(a) the single-level bill of materials for each level of assembly;
(b) a complete indented bill of materials for all levels of assembly.

Exercise 2

(a) Create the materials requirements planning records for each part and sub-assembly in the bat.
(b) List any problems that the completed MRP records identify.
(c) What alternatives are there that the company could take to solve any problems? What are their relative merits?

Exercise 3

Based on the first two exercises, create another set of MRP records, this time allowing one week's safety lead time for each item: that is, ensuring the items are in stock the week prior to when they are required.

Exercise 4

Over the time period of the exercise, what effect would the imposition of a safety lead time have on average inventory value?

Exercise 5

If we decided that our first task was to reduce inventory costs by 15 per cent, what action would we recommend? What are the implications of our action?

Exercise 6

How might production in our business be smoothed?

QUESTIONS

1 Why did Peter have such problems getting to the relevant information?

2 Perform all the exercises that Peter set for himself. Do you think he should now fully understand MRP?

Problems and applications

All chapters have 'Problems and applications' questions that will help you practise analysing operations. They can be answered by reading the chapter. Model answers for the first two questions can be found on the companion website for this text.

1 Rolls-Royce is one of the world's largest manufacturers of gas turbines. Their complex products typically have around 25,000 parts and hundreds of sub-assemblies, and their production is equally complex, with over 600 external suppliers. This makes planning a complex task, which is why Rolls-Royce was one of the earliest users of ERP to help with the task. Up to that point, the company had developed its own software, which had become increasingly expensive. It was also risky because customised and complex software could be difficult to update and often could not exchange or share data. So, the company decided to implement a standard ERP system from SAP. Because it was a 'commercial' off-the-shelf system it would force the company to adopt a standardised approach. Also, it would fully integrate all the company's systems, and updates would be made available by SAP. Finally, it would be able to use a single database, whose modules included product information, resource information, inventory, external suppliers, order processing information and external sales. Yet the company knew that many ERP implementations had been expensive disasters and was determined to ensure that this did not happen in Rolls-Royce. The project was too important. It was the largest single element within its strategic plan. So, it had a core technical team that led the design of the systems, and a large implementation team that was spread around the businesses. The result was a significant reduction in inventory, improved customer service, and substantially improved business information and controls.[11] What decisions did Rolls-Royce take in adopting its ERP system?

2 SAP is a large European software company selling ERP systems. It is well known for developing a network of 'business partners' to develop new products, sell its 'solutions', implement them into customers' operations, provide service, educate end users, and several other activities. If you were managing SAP's relationship with its partners, how would you ensure their long-term collaboration?

3 Re-read the 'Operations in practice' example, 'The ERP for a chicken salad sandwich'.
The company on which this example is based found it difficult to implement its ERP system. '*It was a far bigger job than we thought*, according to the company's operations director, '*We had to change the way we organised our processes so that they would fit in with the ERP system that we bought. But that was relatively easy compared to making sure that the system integrated with our customers', suppliers' and distributors' systems. Because some of these companies were also implementing new systems at the time, it was like trying to hit a moving target*. Why do you think that integrating an ERP system with those of suppliers and customers is so difficult?

4 Re-read the 'Operations in practice' example, 'It's not that easy'. Why did things go wrong, specifically with the relationship between SAP and Waste Management?

5 (It is advised that you read the supplement to this chapter before attempting this question.) Your company has developed a simple, but amazingly effective mango peeler. It is constructed from a blade and a supergrip handle that has a top piece and a bottom piece. The assembled mango peeler is packed in a simple recycled card pack. All the parts simply clip together and are bought in from suppliers, which can deliver the parts within one week of orders being placed. Given enough parts, your company can produce products within a day of firm orders being placed. Initial forecasts indicate that demand will be around 500 items per week. You and your suppliers all work 5-day weeks.

(a) Draw the component structure for the product.
(b) Draw up an MRP table (similar to Figure 14.12 in the supplement to this chapter) assuming that the economic order quantity (EOQ) for all parts is 500.
(c) Develop a schedule indicating when and how many of each component should be ordered (your scheduler tells you that, actually, the EOQ for all parts is 1,500).

6 A lunch kiosk serves two meals every day: veggie fritters and mushroom stroganoff, the recipes for which are as follows:

Veggie fritters (serves 10) – Prepare the 'veggie mix' by grating 500 g of carrots, 500 g of courgettes (zucchini), and chopping 300 g of mushrooms, 100 g of onions and 50 g of parsley. Prepare the batter by beating together 4 eggs, 500 g of flour and 500 ml of cream. Combine the veggie mix with the batter and fry as small disks of approximately 10 cm in 100 ml of oil. Keep warm and serve.

Mushroom stroganoff (serves 10) – Gently fry 400 g of finely chopped onions and 10 g of crushed garlic in 20 ml of oil. When cooled, mix with 1,000 ml of cream and gently heat until reduced slightly to make the 'cream base'. Fry 2,000 g of mushrooms in 100 ml of oil until soft. Mix the mushrooms with the cream base and cover with 50 g of chopped parsley.

(a) Draw the component structures for these two products.
(b) If the kiosk sells 50 portions of veggie fritters and 30 portions of mushroom stroganoff every day, how much of each ingredient should it order every day?

7 The nature of the product structure is closely related to the design of the product. This is reflected in the product structure shape. The shape is partly determined by the number of components and parts used at each level – the more that are used, the wider the shape. There are some recognised typical shapes of product structure – 'A', 'T', 'V' and 'X'. In A-shape structures, a product only has a limited product range to offer the customer. However, because there is little variety, the volumes of standardised production can give some economies of scale. A 'T'-shape product structure is typical of operations that have a small number of raw materials and a relatively standard process, but which produce a very wide range of highly customised end products. The 'V'-shape product structure is where a small number of raw materials are used to create a wide range of products and by-products. An X structure is when an operation has standardised designs of a small number of standard modules, to which a wide range of features and options can be added, giving a wide range of finished products. What structures would you use to describe the following?

(a) The game described in the supplement to this chapter.
(b) A manufacturer producing personal name and address labels.
(c) A petrochemical supplier, blending a few raw materials into a larger number of products.
(d) A kitchen unit manufacturer making standardised bodies to which a wide range of doors and fittings can be added.

8 Using a web search, find information on three different ERP suppliers' products. Compare and contrast, ideally using a tabular presentation:

(a) the main modules offered;
(b) the extent to which customisation is claimed to be possible;
(c) the apparent advantages and disadvantages of the systems.

9 Based on web searches, identify two examples of 'successful' ERP implementation, one from manufacturing and the other from a service or government organisation. Summarise the claimed benefits that are stated as having been achieved in each case. If available, highlight the underlying conditions and/or reasons for success.

10 Using a cookery book, choose three similar, fairly complex recipe items such as layered and decorated gateaux (cakes), or desserts. For each, construct the indented bill of materials and identify all the different materials, sub-assemblies and final products with one set of part numbers (i.e. no duplication). Using the times given in the recipes (or your own estimates), construct a table of lead times (e.g. in minutes or hours) for each stage of production and for procurement of the ingredients. Using these examples (and a bit of your own imagination!), show how this information could be used with an MRP system to plan and control the batch production processes within a small cake or dessert factory making thousands of each product every week. Show part of the MRP records and calculations that would be involved.

Selected further reading

Akhtar, J. (2016) ***Production Planning and Control with SAP ERP*****, 2nd edn, SAP Press/Rheinwerk, Boston, MA.**
A good practical treatment.

Atkinson, R. (2013) ***Enterprise Resource Planning (ERP) The Great Gamble: An Executive's Guide to Understanding an ERP Project*****, Xlibris, Bloomington, IN.**
A basic book. Don't look for great depth, but it is a good introduction.

Bradford, M. (2020) ***Modern ERP: Select, Implement, And Use Today's Advanced Business Systems*****, 4th edn, Dr Marianne Bradford, Raleigh, NC.**
A good solid class text.

Davenport, T.H. (1998) Putting the enterprise into the enterprise system, ***Harvard Business Review*****, July–August.**
Covers some of the more managerial and strategic aspects of ERP.

Jacobs, F.R., Berry, W.L., Whybark, D.C. and Vollmann, T.W. (2010) ***Manufacturing Planning & Control Systems for Supply Chain Management*****, 6th edn, McGraw Hill Higher Education, New York, NY.**
The latest version of the 'bible' of manufacturing planning and control. Explains the 'workings' of MRP and ERP in detail.

Koch, C. and Wailgum, T. (2007) ERP definition and solutions, www.cio.com.
CIO.com has some really useful articles, of which this is one of the most thought provoking.

MacCarthy, B.L. (2006) Organizational, systems and human issues in production planning, scheduling and control, in Hermann, J. (ed.) ***Handbook of Production Scheduling*****, International Series in Operations Research and Management Science, Springer, New York, NY.**
This is an academic paper, but don't be put off. It's a good and sensible overview of the topic by one of the best authorities in the area.

Srivastava, D. and Batra, A. (2010) ***ERP Systems*****, I K International Publishing House, New Delhi.**
An in-depth study of ERP systems and their benefits, including implementation.

Turbit, N. (2005) ERP Implementation – The Traps, The Project Perfect White Paper Collection, www.projectperfect.com.au
Practical (and true).

Notes on chapter

1. The information on which this example is based is taken from: Allan, K. (2009) Butcher's Pet Care relies on IT that can co-ordinate its ERP, *Engineering & Technology Magazine*, 21 July.
2. For a good explanation of this and similar issues, see: MacCarthy, B.L. (2006) Organizational, systems and human issues in production planning, scheduling and control, in Hermann, J. (ed.) *Handbook of Production Scheduling*, International Series in Operations Research and Management Science, Springer, New York, NY.
3. Wight, O. (1984) *Manufacturing Resource Planning: MRP II*, Oliver Wight Ltd.
4. The information on which this example is based is taken from: Ellson, A. (2021) Post Office scandal: hundreds could claim compensation after convictions quashed, *The Times*, 24 April; Dixon, H. (2021) Call to prosecute Post Office bosses over 'biggest miscarriage in British legal history', *The Telegraph*, 23 April.
5. Thornhill, J. (2021) Post Office scandal exposes the dangers of automated injustice, *Financial Times*, 29 April.

6. Brynjolfsson, E. (1994) Technology's true payoff, *Information Week*, October.
7. This categorisation is described in McAfee, A. (2007) Managing in the Information Age, Harvard Business School, Teaching note 5-608-011.
8. The information on which this example is based is taken from: Fruhlinger, J., Wailgum, T. and Sayer, P. (2020) 16 famous ERP disasters, dustups and disappointments, CIO.com, 20 March, https://www.cio.com/article/2429865/enterprise-resource-planning-10-famous-erp-disasters-dustups-and-disappointments.html (accessed September 2021); Novacura (2019) 4 ERP implementation failures with valuable lessons, The Novacura Flow blog, 19 February, https://www2.novacura.com/blog/why-do-erp-implementations-fail (accessed September 2021); Kanaracus, C. (2008) Waste Management sues SAP over ERP implementation, InfoWorld, 27 March.
9. Based on a review of the research in this area by Finney, S. and Corbett, M. (2007) ERP implementation: a compilation and analysis of critical success factors, *Business Process Management Journal*, 13 (3), 329–47.
10. Turbit, N. (2005) ERP Implementation – The Traps, The Project Perfect White Paper Collection, www.projectperfect.com.au (accessed September 2021).
11. Communication with Julian Goulder, Director, Logistics Processes and IT, Rolls Royce.

Supplement to Chapter 14

Materials requirements planning (MRP)

INTRODUCTION

Materials requirements planning (MRP) is an approach to calculating how many parts or materials of particular types are required and what times they are required. This requires data files which, when the MRP program is run, can be checked and updated. Figure 14.9 shows how these files relate to each other. The first inputs to materials requirements planning are customer orders and forecast demand. MRP performs its calculations based on the combination of these two parts of future demand. All other requirements are derived from, and dependent on, this demand information.

Master production schedule

The master production schedule (MPS) forms the main input to materials requirements planning and contains a statement of the volume and timing of the end products to be made. It drives all the production and supply activities that eventually will come together to form the end products. It is the basis for the planning and utilisation of labour and equipment, and it determines the provisioning of materials and cash. The MPS should include all sources of demand, such as spare parts, internal production promises, etc. For example, if a manufacturer of earth excavators plans an exhibition of its products and allows a project team to raid the stores so that it can build two pristine examples to be exhibited, this is likely to leave the factory short of parts. MPS can also be used in service organisations. For example, in a hospital operating theatre there is a master schedule that contains a statement of which operations are planned and when. This can be used to provision materials for the operations, such as the sterile instruments, blood and dressings. It may also govern the scheduling of staff for operations.

The master production schedule record

Master production schedules are time-phased records of each end product, which contain a statement of demand and currently available stock of each finished item. Using this information, the available inventory is projected ahead in time. When there is insufficient inventory to satisfy forward demand, order quantities are entered on the master schedule line. Table 14.4 is a simplified example of part of

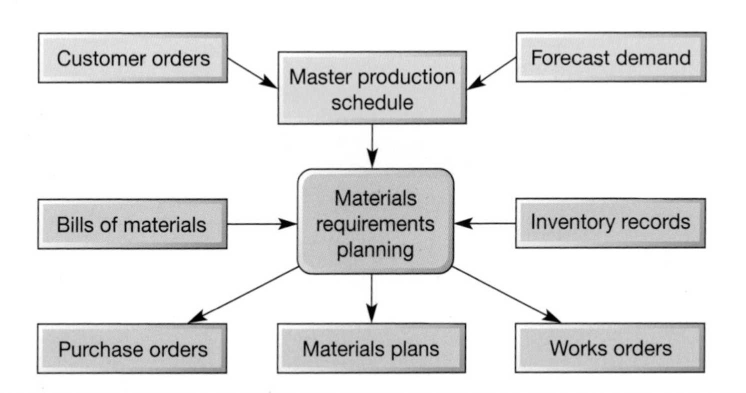

Figure 14.9 Materials requirements planning (MRP) schematic

Table 14.4 Example of a master production schedule

		Week number								
		1	2	3	4	5	6	7	8	9
Demand		10	10	10	10	15	15	15	20	20
Available		20	10	0	0	0	0	0	0	0
MPS		0	0	10	10	15	15	15	20	20
On hand	30									

a master production schedule for one item. In the first row the known sales orders and any forecast are combined to form 'Demand'. The second row, 'Available', shows how much inventory of this item is expected to be in stock at the end of each weekly period. The opening inventory balance, 'On hand', is shown separately at the bottom of the record. The third row is the master production schedule, or MPS; this shows how many finished items need to be completed and available in each week to satisfy demand.

Chase or level master production schedules

In the example in Table 14.4, the MPS increases as demand increases and aims to keep available inventory at 0. The master production schedule is 'chasing' demand (see Chapter 11) and so adjusting the provision of resources. An alternative 'levelled' MPS for this situation is shown in Table 14.5. Level scheduling involves averaging the amount required to be completed to smooth out peaks and troughs; it generates more inventory than the previous MPS.

Available to promise (ATP)

The master production schedule provides the information to the sales function on what can be promised to customers and when delivery can be promised. The sales function can load known sales orders against the master production schedule and keep track of what is available to promise (ATP) (see Table 14.6). The ATP line in the master production schedule shows the maximum that is still available in any one week, against which sales orders can be loaded.

The bill of materials (BOM)

From the master schedule, MRP calculates the required volume and timing of assemblies, subassemblies and materials. To do this, it needs information on what parts are required for each product. This is called the 'bill of materials'. Initially it is simplest to think about these as a product structure. The

Table 14.5 Example of a 'level' master production schedule

		Week number								
		1	2	3	4	5	6	7	8	9
Demand		10	10	10	10	15	15	15	20	20
Available		31	32	33	34	30	26	22	13	4
MPS		11	11	11	11	11	11	11	11	11
On hand	30									

Table 14.6 Example of a level master production schedule including available to promise

		Week number								
		1	2	3	4	5	6	7	8	9
Demand		10	10	10	10	15	15	15	20	20
Sales orders		10	10	10	8	4				
Available		31	32	33	34	30	26	22	13	4
ATP		31	1	1	3	7	11	11	11	11
MPS		11	11	11	11	11	11	11	11	11
On hand	30									

product structure in Figure 14.10 is a simplified structure showing the parts required to make a simple board game. Different 'levels of assembly' are shown with the finished product (the boxed game) at level 0, the parts and sub-assemblies that go into the boxed game at level 1, the parts that go into the sub-assemblies at level 2 and so on.

A more convenient form of the product structure is the 'indented bill of materials'. Table 14.7 shows the whole indented bill of materials for the board game. The term 'indented' refers to the indentation of the level of assembly, shown in the left-hand column. Multiples of some parts are required; this means that MRP has to know the required number of each part to be able to multiply up the requirements. Also, the same part (for example, the TV label, part number 10062) may be used in different parts of the product structure. This means that MRP has to cope with this commonality of parts and, at some stage, aggregate the requirements to check how many labels in total are required.

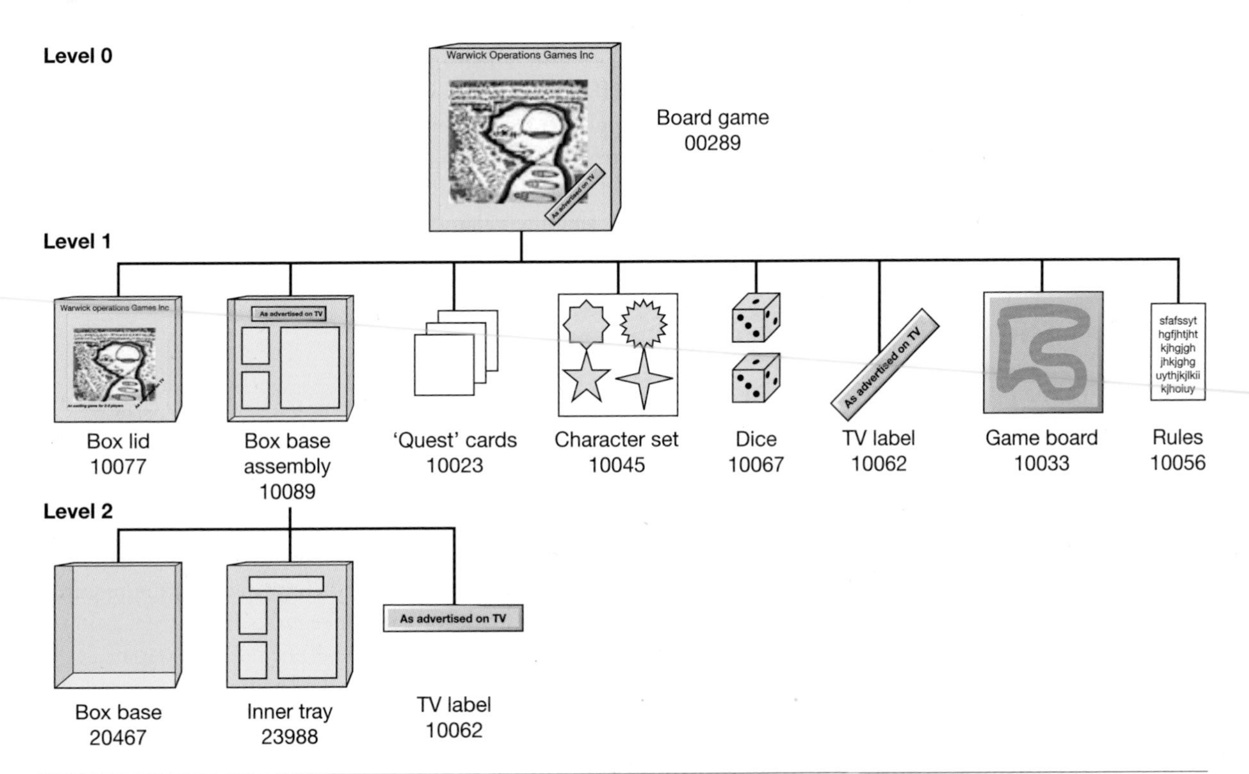

Figure 14.10 Product structure for a simple board game

Table 14.7 Indented bill of materials for board game

Part number: 00289 Description: Board game Level: 0			
Level	**Part number**	**Description**	**Quantity**
0	00289	Board game	1
. 1	10077	Box lid	1
. 1	10089	Box base assy	1
. . 2	20467	Box base	1
. . 2	10062	TV label	1
. . 2	23988	Inner tray	1
. 1	10023	Quest cards set	1
. 1	10045	Character set	1
. 1	10067	Die	2
. 1	10062	TV label	1
. 1	10033	Game board	1
. 1	10056	Rules booklet	1

Inventory records

MRP calculations need to recognise that some required items may already be in stock. So, it is necessary, starting at level 0 of each bill, to check how much inventory is available of each finished product, sub-assembly and component, and then to calculate what is termed the 'net' requirements: that is, the extra requirements needed to supplement the inventory so that demand can be met. This requires that three main inventory records are kept: the item master file, which contains the unique standard identification code for each part or component, the transaction file, which keeps a record of receipts into stock, issues from stock and a running balance, and the location file, which identifies where inventory is located.

The MRP netting process

The information needs of MRP are important, but they are not the 'heart' of the MRP procedure. At its core, MRP is a systematic process of taking this planning information and calculating the volume and timing requirements that will satisfy demand. The most important element of this is the MRP netting process. Figure 14.11 illustrates the process that MRP performs to calculate the volumes of materials required. The master production schedule is 'exploded', examining the implications of the schedule through the bill of materials, checking how many sub-assemblies and parts are required. Before moving down the bill of materials to the next level, MRP checks how many of the required parts are already available in stock. It then generates 'works orders', or requests, for the net requirements of items. These form the schedule, which is again exploded through the bill of materials at the next level down. This process continues until the bottom level of the bill of materials is reached.

Back-scheduling

In addition to calculating the volume of materials required, MRP also considers when each of these parts is required: that is, the timing and scheduling of materials. It does this by a process called back-scheduling, which takes into account the lead time (the time allowed for completion of each

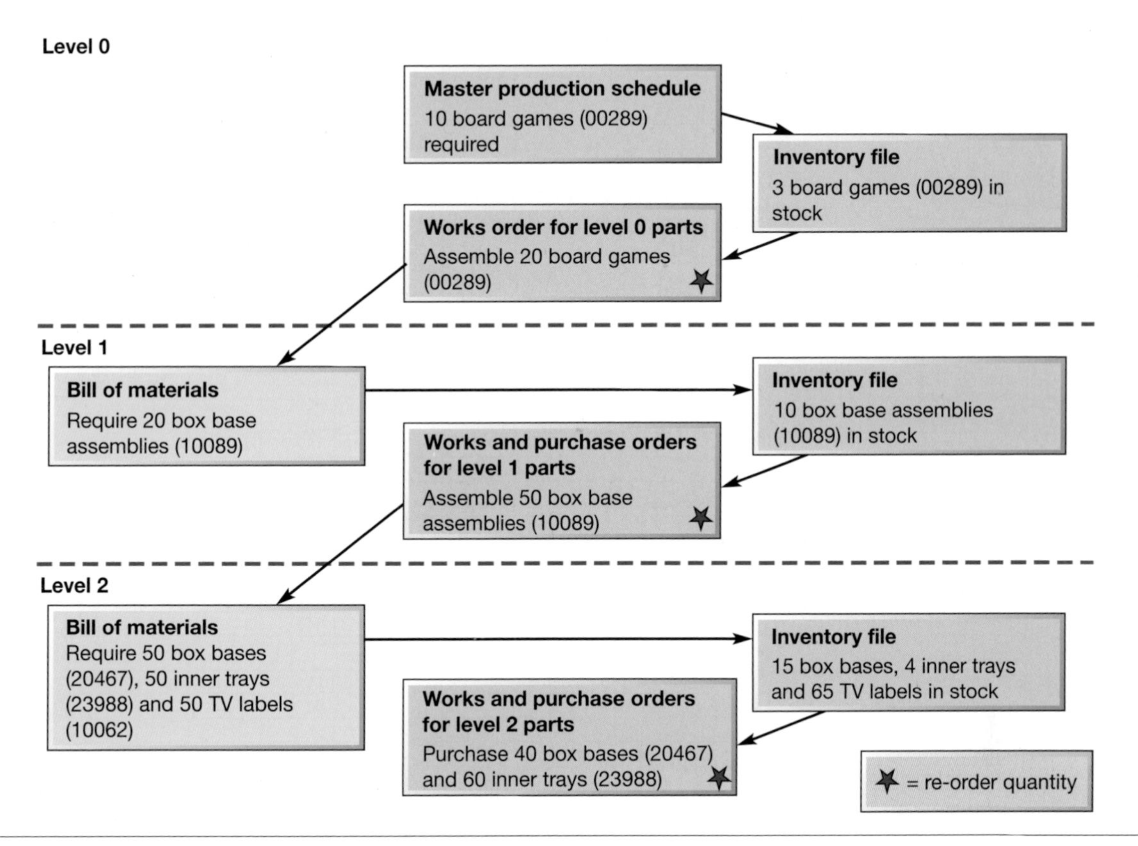

Figure 14.11 The MRP netting calculations for the simple board game

stage of the process) at every level of assembly. Again, using the example of the board game, assume that 10 board games are required to be finished by a notional planning day, which we will term day 20. To determine when we need to start work on all the parts that make up the game, we need to know all the lead times that are stored in MRP files for each part (see Table14.8). Using the lead-time information, the programme is worked backwards to determine the tasks that have to be performed and the purchase orders that have to be placed. Given the lead times and inventory levels shown in Table 14.8, the MRP records shown in Figure 14.12 can be derived.

Table 14.8 Back-scheduling of requirements in MRP

Part no.	Description	Inventory on-hand day 0	Lead time (days)	Re-order quantity
00289	Board game	3	2	20
10077	Box lid	4	8	25
10089	Box base assy	10	4	50
20467	Box base	15	12	40
23988	Inner tray	4	14	60
10062	TV label	65	8	100
10023	Quest cards set	4	3	50
10045	Character set	46	3	50
10067	Die	22	5	80
10033	Game board	8	15	50
10056	Rules booklet	0	3	80

00289: Treasure Hunt game Assembly lead time = 2 Re-order quantity = 20

Day Number:	0	1	2	3	4	5	6	7	8	9	10	11	12	13	14	15	16	17	18	19	20
Requirements Gross																					10
Scheduled Receipts																					
On-hand Inventory	3	3	3	3	3	3	3	3	3	3	3	3	3	3	3	3	3	3	3	3	13
Planned Order Release																			20		

10077: Box lid Purchase lead time = 8 Re-order quantity = 25

Day Number:	0	1	2	3	4	5	6	7	8	9	10	11	12	13	14	15	16	17	18	19	20
Requirements Gross																			20		
Scheduled Receipts																					
On-hand Inventory	4	4	4	4	4	4	4	4	4	4	4	4	4	4	4	4	4	4	9	9	9
Planned Order Release											25										

10089: Box base assembly Assembly lead time = 4 Re-order quantity = 50

Day Number:	0	1	2	3	4	5	6	7	8	9	10	11	12	13	14	15	16	17	18	19	20
Requirements Gross																			20		
Scheduled Receipts																					
On-hand Inventory	10	10	10	10	10	10	10	10	10	10	10	10	10	10	10	10	10	10	40	40	40
Planned Order Release															50						

20467: Box base Purchase lead time = 12 Re-order quantity = 40

Day Number:	0	1	2	3	4	5	6	7	8	9	10	11	12	13	14	15	16	17	18	19	20
Requirements Gross															50						
Scheduled Receipts																					
On-hand Inventory	15	15	15	15	15	15	15	15	15	15	15	15	15	15	5	5	5	5	5	5	5
Planned Order Release			40																		

23988: Inner tray Purchase lead time = 14 Re-order quantity = 60

Day Number:	0	1	2	3	4	5	6	7	8	9	10	11	12	13	14	15	16	17	18	19	20
Requirements Gross															50						
Scheduled Receipts																					
On-hand Inventory	4	4	4	4	4	4	4	4	4	4	4	4	4	4	14	14	14	14	14	14	14
Planned Order Release	60																				

10062: TV label Purchase lead time = 8 Re-order quantity = 100

Day Number:	0	1	2	3	4	5	6	7	8	9	10	11	12	13	14	15	16	17	18	19	20
Requirements Gross															50				20		
Scheduled Receipts																					
On-hand Inventory	65	65	65	65	65	65	65	65	65	65	65	65	65	65	15	15	14	15	95	95	95
Planned Order Release											100										

10023: Quest card set Purchase lead time = 3 Re-order quantity = 50

Day Number:	0	1	2	3	4	5	6	7	8	9	10	11	12	13	14	15	16	17	18	19	20
Requirements Gross																			20		
Scheduled Receipts																					
On-hand Inventory	4	4	4	4	4	4	4	4	4	4	4	4	4	4	4	4	4	4	34	34	34
Planned Order Release																50					

10045: Character set Purchase lead time = 3 Re-order quantity = 50

Day Number:	0	1	2	3	4	5	6	7	8	9	10	11	12	13	14	15	16	17	18	19	20
Requirements Gross																			20		
Scheduled Receipts																					
On-hand Inventory	46	46	46	46	46	46	46	46	46	46	46	46	46	46	46	46	46	46	26	26	26
Planned Order Release																					

10067: Die Purchase lead time = 5 Re-order quantity = 80

Day Number:	0	1	2	3	4	5	6	7	8	9	10	11	12	13	14	15	16	17	18	19	20
Requirements Gross																			40		
Scheduled Receipts																					
On-hand Inventory	3	3	3	3	3	3	3	3	3	3	3	3	3	3	3	3	3	3	3	3	13
Planned Order Release														80							

10033: Game board Purchase lead time = 15 Re-order quantity = 50

Day Number:	0	1	2	3	4	5	6	7	8	9	10	11	12	13	14	15	16	17	18	19	20
Requirements Gross																			20		
Scheduled Receipts																					
On-hand Inventory	8	8	8	8	8	8	8	8	8	8	8	8	8	8	8	8	8	8	38	38	38
Planned Order Release				50																	

10056: Rules booklet Purchase lead time = 3 Re-order quantity = 80

Day Number:	0	1	2	3	4	5	6	7	8	9	10	11	12	13	14	15	16	17	18	19	20
Requirements Gross																			20		
Scheduled Receipts																					
On-hand Inventory	0	0	0	0	0	0	0	0	0	0	0	0	0	0	0	0	0	0	60	60	60
Planned Order Release																80					

Figure 14.12 Extract from the MRP records for the simple board game (lead times indicated by arrows)

MRP capacity checks

The MRP process needs a feedback loop to check whether a plan was achievable and whether it has actually been achieved. Closing this planning loop in MRP systems involves checking production plans against available capacity and, if the proposed plans are not achievable at any level, revising them. All but the simplest MRP systems are now closed-loop systems. They use three planning routines to check production plans against the operation's resources at three levels:

- Resource requirements plans (RRPs) involve looking forward in the long term to predict the requirements for large structural parts of the operation, such as the numbers, locations and sizes of new plants.
- Rough-cut capacity plans (RCCPs) are used in the medium to short term, to check the master production schedules against known capacity bottlenecks, in case capacity constraints are broken. The feedback loop at this level checks the MPS and key resources only.
- Capacity requirements plans (CRPs) look at the day-to-day effect that the works orders issued from the MRP have on the loading of individual process stages.

Summary of supplement

- MRP stands for materials requirements planning, which is a dependent demand system that calculates materials requirements and production plans to satisfy known and forecast sales orders. It helps to make volume and timing calculations, based on an idea of what will be necessary to supply demand in the future.
- MRP works from a master production schedule, which summarises the volume and timing of end products or services. Using the logic of the bill of materials (BOM) and inventory records, the production schedule is 'exploded' (called the MRP netting process) to determine how many sub-assemblies and parts are required, and when they are required.
- Closed-loop MRP systems contain feedback loops that ensure that checks are made against capacity to see if plans are feasible.
- MRP II systems are a development of MRP. They integrate many processes that are related to MRP, but that are located outside the operation's function.

PART FOUR

Development

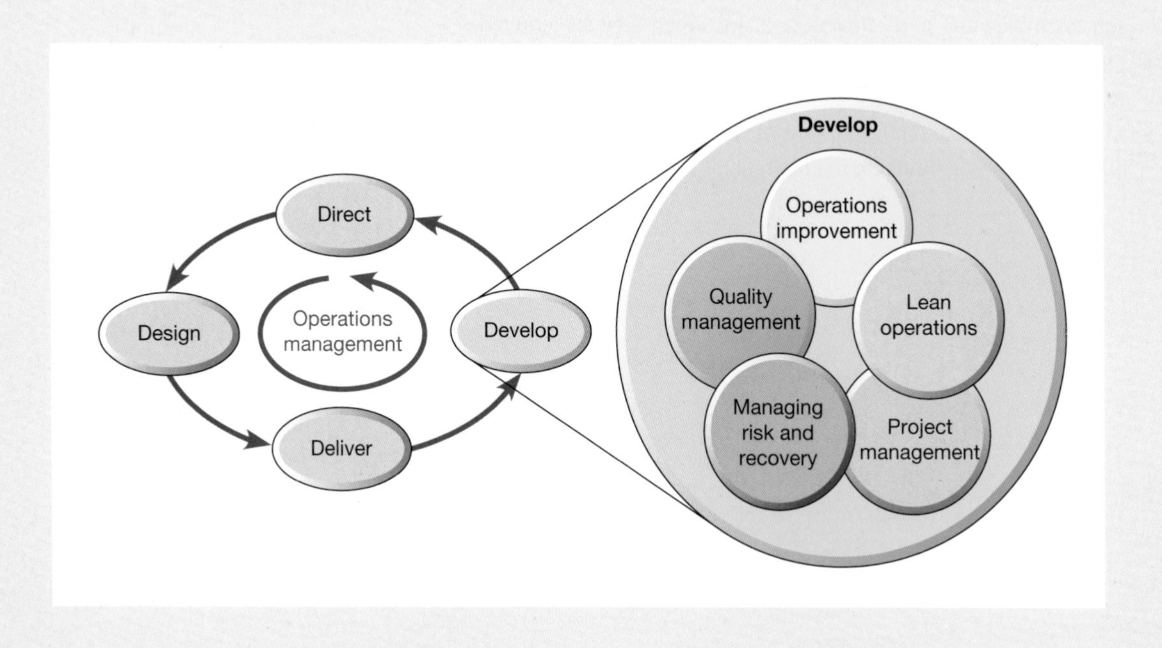

Even when an operation's direction is set, its design finalised and its deliveries planned and controlled, the operations manager's task is not finished. Even the best operation will need to improve and develop, partly because customers' expectations are likely to be rising, and partly because the operation's competitors will also be improving. This part of the text looks at five key issues for operations development. The chapters in this part are:

- **Chapter 15 Operations improvement**

  This examines how managers can make their operation perform better through the use of the many elements of new (and not so new) improvement approaches.

- **Chapter 16 Lean operations**

  This looks at an approach to operations management that started as a planning and control concept (called 'just-in-time'), but is now seen primarily as an approach to improvement.

- **Chapter 17 Quality management**

  This identifies some of the ideas of quality management and how they can be used to facilitate improvement.

- **Chapter 18 Managing risk and recovery**

  This examines how operations managers can reduce the risk of things going wrong and how they can recover when they do.

- **Chapter 19 Project management**

  This looks at how operations managers can project manage (among other things) improvement activities to organise the changes that improvement inevitably requires.

15 Operations improvement

KEY QUESTIONS

15.1 Why is improvement so important in operations management?

15.2 What are the key elements of operations improvement?

15.3 What are the broad approaches to improvement?

15.4 What techniques can be used for improvement?

15.5 How can the improvement process be managed?

INTRODUCTION

Improvement means to make something better. And all operations, no matter how well managed, are capable of being better. Of course, in one sense, all of operations management is concerned with doing things better, but there are some issues that relate specifically to the activity of improvement itself. These are the issues that we deal with in the next five chapters. Yet, at one time improvement was not central to operations managers, who were expected simply to 'run the operation', 'keep the show on the road' and 'maintain current performance'. No longer. Now the emphasis has shifted markedly towards making improvement one of the main responsibilities of operations managers. Moreover, the study of improvement as a specific activity has attracted significant attention. Some of this attention focuses on specific techniques and prescriptions, while some looks at the underlying philosophy of improvement. Both aspects are covered in this chapter. Figure 15.1 shows where this topic fits into our overall model of operations management.

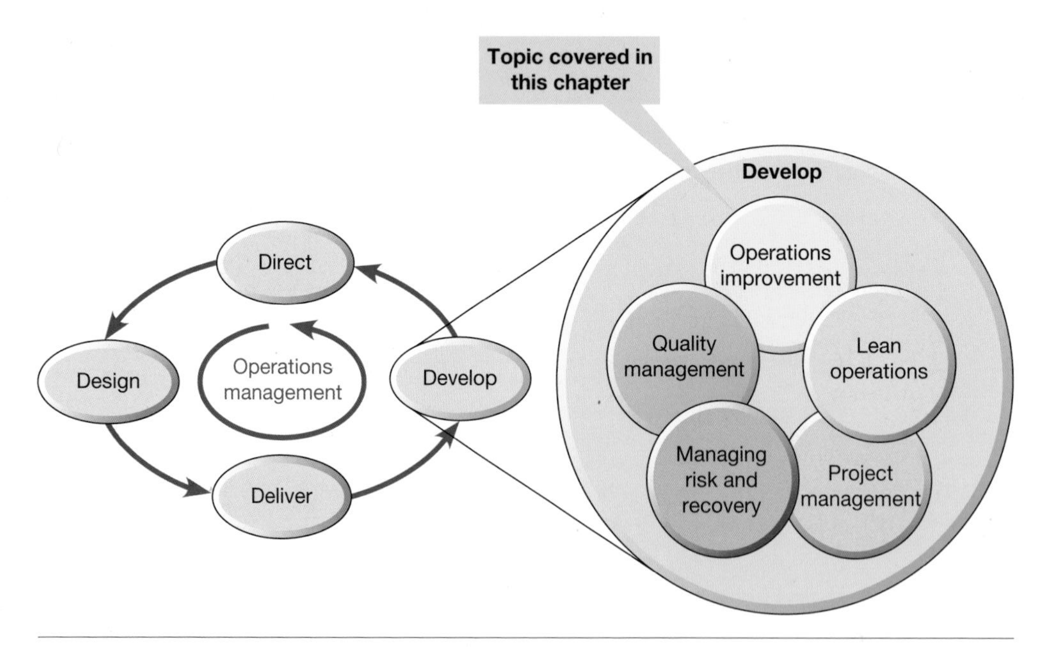

Figure 15.1 This chapter examines operations improvement

15.1 Why is improvement so important in operations management?

Why is operations improvement so important? Well, who doesn't want to get better? And businesses are (or should be) just the same as people – they generally want to get better. Not just for the sake of their own excellence, although that may be one factor, but mainly because improving operations performance has such an impact on what any organisation is there to do. Emergency services want to reach distressed people faster and treat them better because by doing so they are fulfilling their role more effectively. Package delivery businesses want to deliver more reliably, at lower cost and reducing emissions because it means happier customers, higher profits and less pollution. Development charities want to target their aid and campaign for improvement in human conditions as wisely and efficiently as possible because more money will find its way to beneficiaries rather than be wasted or consumed in administration. Not surprising then that the whole emphasis of operations management has shifted towards emphasising improvement. Operations managers are judged not only on how they meet their ongoing responsibilities of producing products and services to acceptable levels of quality, speed, dependability, flexibility and cost, but also on how they improve the performance of the operations function overall.

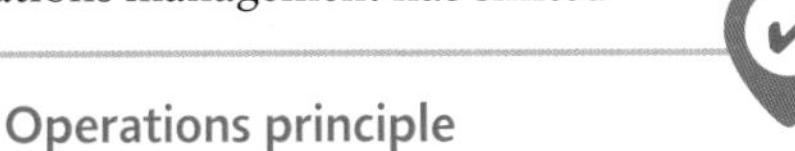

Operations principle

Performance improvement is the ultimate objective of operations management.

Why the focus on improvement?

Various reasons have been suggested to explain the shift towards a focus on improvement in professional operations managers' activities:

- There is a perceived increase in the intensity of competitive pressures (or 'value for money' in not-for-profit or public sector operations). In fact, economists argue about whether markets are really getting more competitive. As far as improvement is concerned it doesn't matter; there is a *perception* of increased competitive pressure, and certainly the owners of operations (shareholders or governments) are less likely to tolerate poor returns or value for money.
- The nature of world trade is changing. Emerging economies are becoming important as both producers and consumers. This has introduced cost pressures in countries with relatively expensive labour and infrastructure costs; it has introduced new challenges for global companies, such as managing complex supply chains; and it has accelerated demand for resources (materials, food, energy) pushing up (or destabilising) prices for these commodities.
- New technology has introduced opportunities to both improve operations practice and disrupt existing markets.
- The interest in operations improvement has resulted in the development of many new ideas and approaches to improving operations. The more ways there are to improve operations, the more operations will be improved.
- The scope of operations management has widened from a subject associated largely with manufacturing to one that embraces all types of enterprises and processes in all functions of the enterprise. Because of this extended scope, operations managers have seen how they can learn from each other.

The Red Queen effect

The scientist Leigh Van Valen was looking to describe a discovery that he had made while studying marine fossils. He had established that, no matter how long a family of animals had already existed, the probability that the family will become extinct is unaffected. In other words, the struggle for survival never gets easier. However well a species fits with its environment, it can never relax. The analogy that Van Valen drew came from *Through the Looking Glass*, by Lewis Carroll. In the book, Alice had encountered living chess pieces and, in particular, the **Red Queen**. '*"Well, in our country"*, said Alice, still panting a little, *"you'd generally get to somewhere else – if you ran very fast for a long* *time, as we've been doing". "A slow sort of country!"* said the Queen. *"Now, here, you see, it takes all the running you can do, to keep in the same place. If you want to get somewhere else, you must run at least twice as fast as that!"*'[1]

In many respects this is like business. Improvements and innovations may be imitated or countered by competitors. For example, in the automotive sector, the quality of most firms' products is very significantly better than it was two decades ago. This reflects the improvement in those firms' operations processes. Yet their relative competitive position, in many cases, has not changed. Those firms that have improved their competitive position have improved their operations performance *more than* competitors. Where improvement has simply matched competitors, survival has been the main benefit. The implications for operations improvement are clear. It is even more important, especially when competitors are actively improving their operations.

An important distinction in the approach taken by individual operations is that between radical or 'breakthrough' improvement, on one hand, and continuous or 'incremental' improvement on the other.

Radical or breakthrough change

Radical or breakthrough improvement (or 'innovation'-based improvement as it is sometimes called) is a philosophy that assumes that the main vehicle of improvement is major and dramatic change in the way the operation works. The introduction of a new, more efficient machine in a factory, the total redesign of a computer-based hotel reservation system, and the introduction of an improved degree programme at a university are all examples of breakthrough improvement. The impact of these improvements is relatively sudden, abrupt and represents a step change in practice (and hopefully performance). Such improvements are rarely inexpensive, usually calling for high investment of capital, often disrupting the ongoing workings of the operation, and frequently involving changes in the product/service or process technology. The bold line in Figure 15.2(a) illustrates the pattern of performance with two breakthrough improvements. The improvement pattern illustrated by the dotted line in Figure 15.2(a) is regarded by some as being more representative of what really occurs when operations rely on pure breakthrough improvement. Breakthrough improvement places a high value on creative solutions. It encourages free thinking and individualism. It is a radical philosophy in so much as it fosters an approach to improvement that does not accept many constraints on what is possible. 'Starting with a clean sheet of paper', 'going back to first principles' and 'completely rethinking the system' are all typical breakthrough improvement principles.

Operations principle
Performance improvement sometimes requires radical or 'breakthrough' change.

Continuous or incremental improvement (kaizen)

Continuous improvement, as the name implies, adopts an approach to improving performance that assumes many small incremental improvement steps. For example, modifying the way a product is fixed to a machine to reduce changeover time, simplifying the question sequence when taking a hotel reservation, and rescheduling the assignment completion dates on a university course so as to smooth the students' workload are all examples of incremental improvements. While there is no guarantee that such small steps towards better performance will be followed by other steps, the whole philosophy of continuous improvement attempts to ensure that they will be. Continuous improvement is not concerned with promoting small improvements per se. It does view small improvements, however, as having one significant advantage over large ones – they can be followed relatively painlessly by other small improvements (see Figure 15.2(b)). Continuous improvement is also known as kaizen. Kaizen is a Japanese word, the definition of which is given by Masaaki Imai[2] (who has been one of the main proponents of continuous improvement) as follows: 'Kaizen means improvement. Moreover, it means improvement in personal life, home life, social life and work life. When applied to the workplace, kaizen means continuing improvement involving everyone – managers and workers alike'.

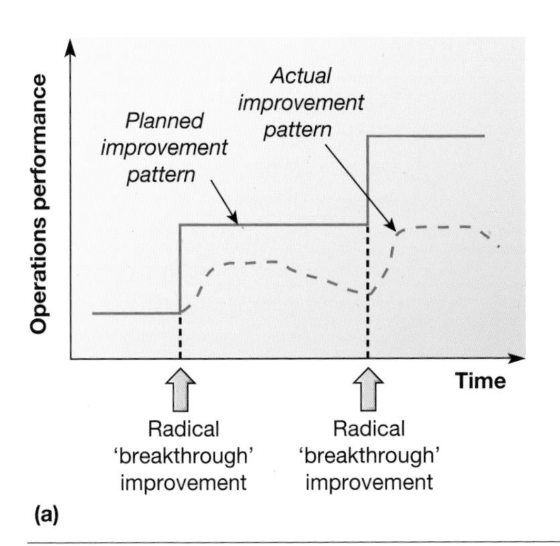

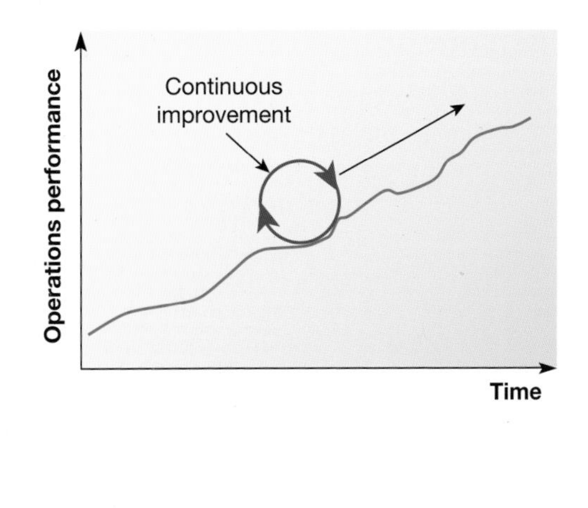

Figure 15.2 (a) Radical 'breakthrough' (planned and actual) and (b) continuous improvement

In continuous improvement, it is not the *rate* of improvement that is important; it is the *momentum* of improvement. It does not matter if successive improvements are small; what does matter is that every month (or week, or quarter, or whatever period is appropriate) some kind of improvement has actually taken place.

Operations principle
Performance improvement almost always benefits from continuous improvement.

OPERATIONS IN PRACTICE Kaizen at Amazon[3]

At one point in Amazon's development of its fulfilment centres, it believed that most issues could be solved with technology. It learned that engaging its front-line staff in continuous improvement could be at least equally effective. For example, the company had been attempting to automate a large part of its fulfilment centres. However, the automation was devised for books, but it did not work well on the other goods that Amazon was introducing, such as shoes. So, when the shoebox reached the mechanism in the automated system that was intended to bring the shoes to the packing line, they were hurled out of the box. Given their experience, Amazon adopted an 'autonomation' approach, using people for complex tasks and using machines to support them. In another kaizen improvement, the time to scan products being put onto shelves in one of the fulfilment centres, called the 'stow line', was taking longer than anticipated. Each person on the line had a trolley full of products to stow on the shelves, and a scanner. The products and the corresponding shelf number had to be scanned so that the computer knew where each product was located. The standard time target for this task was 20 minutes per trolley. But when one of the company's senior managers tried to perform this task, it took him 45 minutes. One of the reasons was that he had to scan some things four times before the scanner recognised them. It was obvious that, at least in part, rather than simply being incompetent, his performance was affected by an abnormality – the bad performance of the scanner. After analysing all the deviations from expected performance that had been reported by the staff, and looking for their root causes, it was found that managers were unclear as to how battery life affected scanner performance. In fact, there were several hours of low productivity because of the low battery charge at the end of each scanner's operating time, and there was no satisfactory process in place to check and reload the scanner batteries. The root-cause analysis helped the company to put a new process in place to monitor and load the scanners to avoid low-charge periods.

Exploitation or exploration

A closely related distinction to that between continuous and breakthrough improvement is the one that management theorists draw between what they call 'exploitation' versus 'exploration'. Exploitation is the activity of enhancing processes (and products) that already exist within a firm. The focus of exploitation is on creating efficiencies rather than radically changing resources or processes. Its emphasis is on tight control of the improvement process, standardising processes, clear organisational structures and organisational stability. The benefits from exploitation tend to be relatively immediate, incremental and predictable. They also are likely to be better understood by the firm and fit into its existing strategic framework. Exploration, by contrast, is concerned with the exploration of new possibilities. It is associated with searching for and recognising new mindsets and ways of doing things. It involves experimentation, taking risks, simulation of possible consequences, flexibility and innovation. The benefits from exploration are principally long term but can be relatively difficult to predict. Moreover, any benefits or discoveries that might come may be so different from what the firm is familiar with that it may not find it easy to take advantage of them.

Organisational 'ambidexterity'

It is clear that the organisational skills and capabilities needed to be successful at exploitation are likely to be very different from those that are needed for the radical exploration of new ideas. Indeed, the two views of improvement may actively conflict. A focus on thoroughly exploring for totally novel choices may consume managerial time, effort and the financial resources that would otherwise be used for refining existing ways of doing things, reducing the effectiveness of improving existing processes. Conversely, if existing processes are improved over time, there may be less motivation to experiment with new ideas. So, although both exploitation and exploration can be beneficial, they may compete both for resources and for management attention. This is where the concept of 'organisational ambidexterity' becomes important. Organisational ambidexterity means the ability of a firm to both exploit and explore as they seek to improve; to be able to compete in mature markets where efficiency is important, by improving existing resources and processes, while also competing in new technologies and/or markets where novelty, innovation and experimentation are required.

Operations principle

Organisational ambidexterity is the ability to both exploit existing capabilities and explore new ones as they seek to improve.

The structure of improvement ideas

There have been hundreds of ideas relating to operations improvement that have been proposed over the last few decades. To understand how these ideas relate to each other it is important to distinguish between four aspects of improvement:

- The elements contained within improvement approaches – these are the fundamental ideas of what improves operations. They are the 'building blocks' of improvement.
- The broad approaches to improvement – these are the underlying methodologies or sets of beliefs that form a coherent philosophy and shape how improvement should be accomplished. Some improvement approaches/methodologies have been used for over a century (for example, some work study approaches, see Chapter 9), others are relatively recent (for example, Six Sigma, explained later). But do not think that approaches to improvement are different in all respects; there are many elements that are common to several approaches.
- The improvement ***techniques*** – there are many 'step-by-step' techniques, methods and tools that can be used to help find improved ways of doing things; some of these use quantitative modelling and others are more qualitative.
- The ***management*** of improvement – how the process of improvement is managed is as important, if not more important, than understanding the elements and approaches to improvement. The improvement activity must be organised, resourced and generally controlled for it to be effective at actually achieving demonstrable improvement.

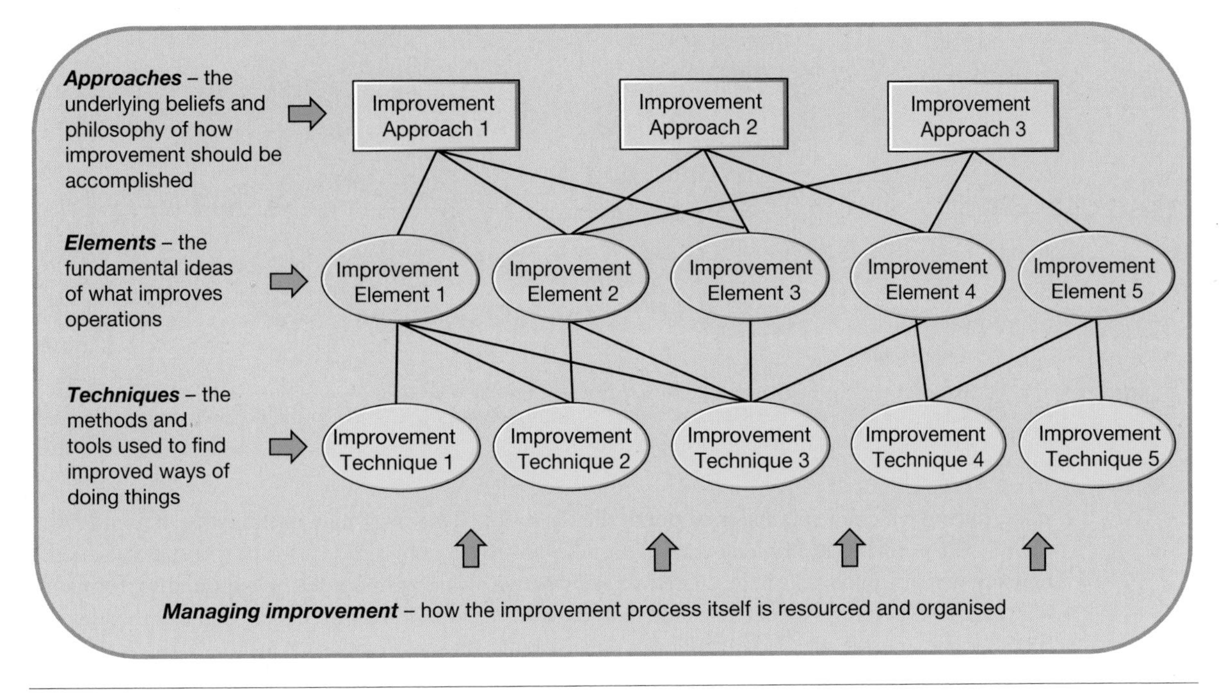

Figure 15.3 How the four aspects of improvement – approaches, elements, techniques and management – relate

The rest of this chapter will treat each of these aspects of improvement. The best way to understand improvement is to deal with the elements contained within improvement approaches first, then see how they come together to form broad approaches to improvement, and then examine some typical improvement techniques, before looking briefly at how operations improvement can be managed. Figure 15.3 illustrates the structure of the four aspects of improvement.

15.2 What are the key elements of operations improvement?

The elements of improvement are the individual fundamental ideas of improvement. Think of these elements of improvement as the building blocks of the various improvement approaches that we shall look at later. Here we explain some, but not all (there are lots), of the more common elements in use today.

> **Operations principle**
> The various approaches to improvement draw from a common group of elements.

Improvement cycles

An important element within some improvement approaches is the use of a literally never-ending process of repeatedly questioning and re-questioning the detailed working of a process or activity. This repeated and cyclical questioning is usually summarised by the idea of the improvement cycle, of which there are many, but two are widely used models – the PDCA (or PDSA) cycle (sometimes called the Deming Cycle, named after the famous quality 'guru', W.E. Deming) and the DMAIC (pronounced de-make) cycle, made popular by the Six Sigma approach (see later).

The PDCA (or PDSA) cycle

The PDCA cycle model is shown in Figure 15.4(a). It starts with the P (for plan) stage, which involves an examination of the current method or the problem area being studied. This involves collecting and analysing data so as to formulate a plan of action that is intended to improve performance. Once a plan for improvement has been agreed, the next step is the D (for do) stage. This is the implementation

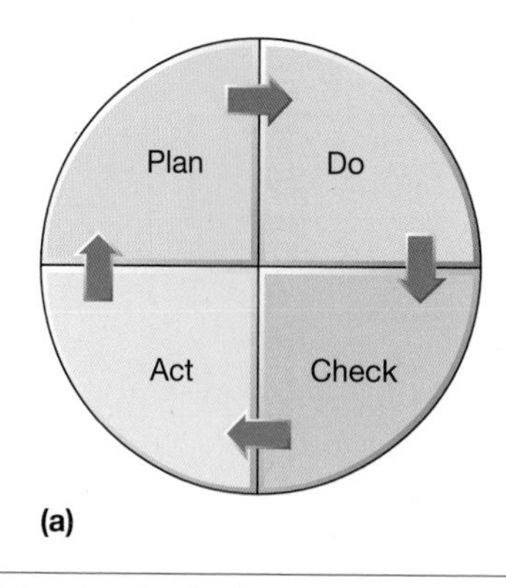

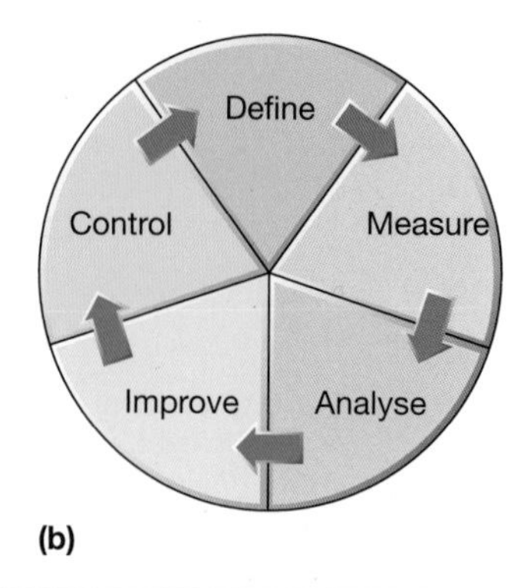

Figure 15.4 (a) The plan-do-check-act, or 'Deming' improvement cycle, and (b) the define-measure-analyse-improve-control, or DMAIC Six Sigma improvement cycle

stage, during which the plan is tried out in the operation. This stage may itself involve a mini-PDCA cycle as the problems of implementation are resolved. Next comes the C (for check) stage where the new implemented solution is evaluated to see whether it has resulted in the expected performance improvement. Some versions of this idea use the term 'study' instead of 'check' and call the idea the 'PDSA' cycle, but the idea is basically the same. Finally, at least for this cycle, comes the A (for act) stage. During this stage, the change is consolidated or standardised if it has been successful. Alternatively, if the change has not been successful, the lessons learned from the 'trial' are formalised before the cycle starts again. You may also find this cycle called the Deming cycle, Deming wheel or Shewhart cycle.

OPERATIONS IN PRACTICE

Disco balls and rice lead to innovative improvement[4]

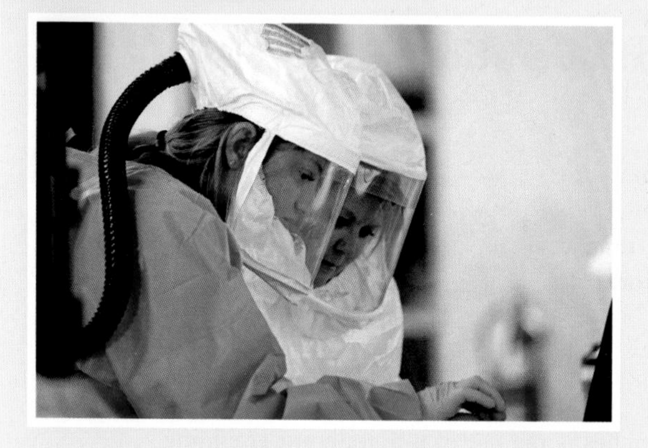

Operations improvement means solving problems, and problem-solving benefits greatly from the creativity of a (preferably multidisciplinary) team. This was demonstrated by the Surrey and Sussex Healthcare Trust in the United Kingdom, when they were faced with an increased need to decontaminate 'personal protection' respirator hoods used to protect medical staff (and patients) during sensitive medical procedures. Respirator hoods allow medical staff who are exposed to a high 'viral load' to breathe purified air. (Complete with face shield and neck cape, respirator hoods are shaped like motorcycle helmets.) However, after they have been used, the hoods need to be decontaminated, sometimes by exposing them to ultraviolet (UV) light. Devising a process to do this in a speedy and effective manner was the problem facing the hospital's improvement team. One member of the team explains. *'We brought together a diverse team from various medical specialisms and based our discussions around the PDCA/PDSA improvement cycle. We had to make sure that all hoods were exposed evenly to sufficient UV light to decontaminate them. First, we experimented with putting the hoods in devices like "shop cages", with the hoods hanging off broomsticks. It kind of worked, but we had to keep going into the UV room to stop the process and rotate the hoods, which obviously wasn't the most efficient use of time. We carried out direct observations of the process, timing different set-ups and variations. We brought in colleagues from infection control, electrical medical engineering, and estate and facilities management to brainstorm ideas that would improve the process. We finally came up with the novel idea of using disco ball motors (usually used in dance clubs) attached to the ceiling with chains. The hoods would hang on the disco motors and automatically rotate. We also borrowed crash mats*

from the physiotherapy department in case the hoods fell. They are too expensive to risk damage'.

However, when the team monitored the process over time, they found that the UV light was not always reaching inside the hood. They decided it would be necessary to tip the hood at a 45-degree angle to ensure light penetration. They experimented by hanging various gadgets on the front of the hood before finally settling on inserting an s-shaped hook through a hole on the front of the hood and hanging bags filled with rice to tip the front of the hood to the required angle. The team's checks using sensors confirmed that the new decontamination process had achieved 100 per cent detection rate. In addition, the PDCA process had prompted further improvement outside the original scope of the problem. *'We devised a simple system for numbering the hoods and allocating them a "place card". When the hoods are issued the place card is filled in with the user's name and where in the department they are located. At a glance I know, for example, that hood 29 is with Dr T from the Intensive Care Unit and that number 27 is in theatres and so on. It's simple and quick. We don't have to consult complicated spreadsheets'.*

The DMAIC cycle

The DMAIC cycle is in some ways more intuitively obvious than the PDCA cycle in so much as it follows a more 'experimental' approach (Figure 15.4(b)). The DMAIC cycle starts with (D) defining the problem or problems, partly to understand the scope of what needs to be done and partly to define exactly the requirements of the process improvement. Often at this stage a formal goal or target for the improvement is set. After definition comes (M) the measurement stage. This stage involves validating the problem to make sure that it really is a problem worth solving, using data to refine the problem and measuring exactly what is happening. Once these measurements have been established, they can be (A) analysed. The analysis stage is sometimes seen as an opportunity to develop hypotheses as to what the root causes of the problem really are. Such hypotheses are validated (or not) by the analysis and the main root causes of the problem identified. Once the causes of the problem are identified, work can begin on (I) improving the process. Ideas are developed to remove the root causes of problems, solutions are tested and those solutions that seem to work are implemented, formalised and results measured. The improved process needs then to be continually monitored and (C) controlled to check that the improved level of performance is sustaining. After this point, the cycle starts again and defines the problems that are preventing further improvement. Remember though, it is the last point about both cycles that is the most important – the cycle starts again. It is only by accepting that in a continuous improvement philosophy these cycles quite literally never stop and that improvement becomes part of every person's job.

A process perspective

Even if some improvement approaches do not explicitly or formally include the idea that taking a process perspective should be central to operations improvement, almost all do so implicitly. This has two major advantages. First, it means that improvement can be focused on what actually happens rather than which part of the organisation has responsibility for what happens. In other words, if improvement is not reflected in the process of creating products and services, then it is not really improvement as such. Second, as we have mentioned before, all parts of the business manage processes. This is what we call operations as activity rather than operations as a function. So, if improvement is described in terms of how processes can be made more effective, those messages will have relevance for all the other functions of the business in addition to the operations function.

End-to-end processes

Some improvement approaches take the process perspective further and prescribe exactly how processes should be organised. One of the more radical prescriptions of business process reengineering (BPR, see later), for example, is the idea that operations should be organised around the total process that adds value for customers, rather than the functions or activities that perform the various stages of the value-adding activity. We have already pointed out the difference between conventional processes within a specialist function, and an **end-to-end business process** in Chapter 1. Identified

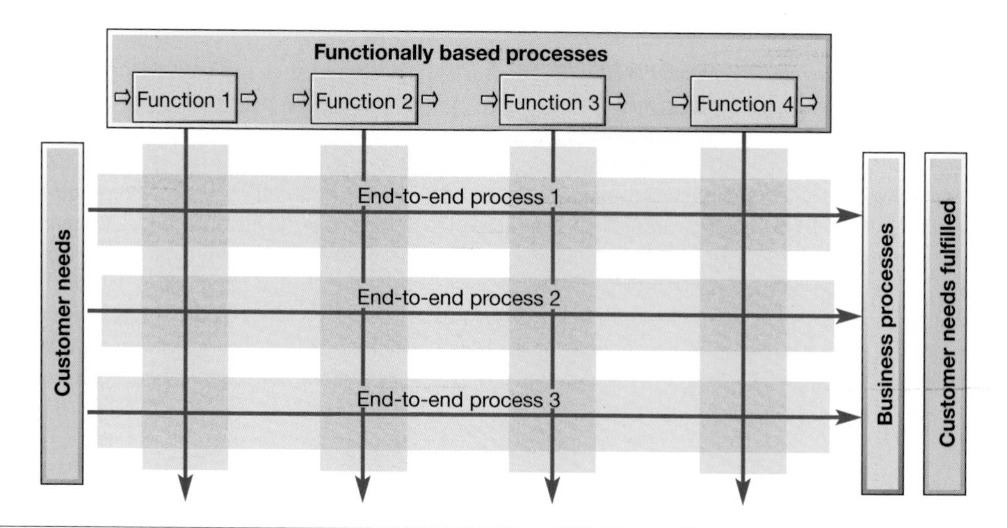

Figure 15.5 End-to-end processes focus directly on customer requirements and often cut across conventional organisational boundaries

customer needs are entirely fulfilled by an 'end-to-end' business process, which often cuts across conventional organisational boundaries. Figure 15.5 illustrates this idea.

Evidence-based problem solving

There has been a resurgence of the use of quantitative techniques in improvement approaches. Six Sigma (see later) promotes systematic use of (preferably quantitative) evidence. Yet Six Sigma is not the first of the improvement approaches to use quantitative methods (some of the TQM gurus promoted statistical process control, for example) although it has done a lot to emphasise the use of quantitative evidence. In fact, much of the considerable training required by Six Sigma consultants is devoted to mastering quantitative analytical techniques. However, the statistical methods used in improvement activities do not always reflect conventional academic statistical knowledge as such. They emphasise observational methods of collecting data and the use of experimentation to examine hypothesis. Techniques include graphical methods, analysis of variance and two-level factorial experiment design. Underlying the use of these techniques is an emphasis on the scientific method, responding only to hard evidence, and using statistical software to facilitate analysis.

OPERATIONS IN PRACTICE

The checklist manifesto[5]

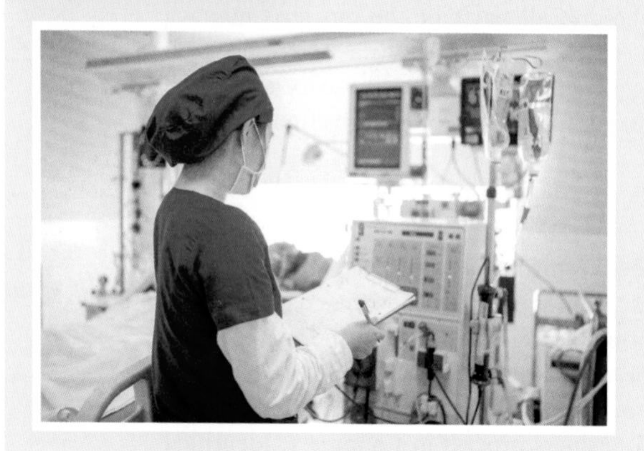

Improvement methodologies are often associated with repetitive operations. Performing the same task repeatedly means that there are plenty of opportunities to 'get it right'. The whole idea behind continuous improvement derives from this simple idea. By contrast, operations that have to perform more difficult activities, especially those that call for expert judgement and diagnostic ability must call for equally complex improvement approaches, no? Well, no, according to Atul Gawande, a doctor at the prestigious Johns Hopkins Hospital. Mr Gawande thinks that the very opposite is true. Although medicine is advancing at an astounding rate and medical journals produce learned papers adding the results of advanced research to an ever-expanding pool of knowledge, medics are less good at the basics. Surgeons carry out over 200 major operations a year, unfortunately not all of them successful, but the medical profession overall does not always have a reliable method for learning from its mistakes. Atul Gawande's idea is that his and similar 'knowledge-based' professions are in danger of sinking under the weight of facts. Scientists are accumulating more and more information and professions are fragmenting into ever narrower specialisms. Mr Gawande tells the story of Peter Pronovost, a specialist in critical care at Johns Hopkins Hospital, who tried to reduce the number of patients who were becoming infected on account of the use of intravenous central lines. There are five steps that medical teams can take to reduce the chances of contracting such infections. Initially Pronovost simply asked nurses to observe whether doctors took the five steps. What they found was that, at least one-third of the time, they missed one or more of the steps. So nurses were authorised to stop doctors who had missed out any of the steps and, as a matter of course, ask whether existing intravenous central lines should be reviewed. As a result of applying these simple checklist-style rules, the ten-day line-infection rates went down from 11 per cent to zero. In one hospital, it was calculated that, over a year, this simple method had prevented 43 infections, 8 deaths and saved about $2 million. Using the same checklist approach the hospital identified and applied the method to other activities. For example, a check in which nurses asked patients about their pain levels led to untreated pain reducing from 41 per cent to 3 per cent. Similarly, the simple checklist method helped the average length of patient stay in intensive care to fall by half. When Pronovost's approach was adopted by other hospitals, within 18 months 1,500 lives and $175 million had been saved.

Mr Gawande describes checklists used in this way as a 'cognitive net' – a mechanism that can help prevent experienced people from making errors due to flawed memory and attention, and ensure that teams work together. Simple checklists are common in other professions. Civil engineers use them to make certain that complicated structures are assembled on schedule. Chefs use them to make sure that food is prepared exactly to the customer's taste. Airlines use them to make sure that pilots take off safely and also to learn from, now relatively rare, crashes. Indeed, Mr Gawande is happy to acknowledge that checklists are not a new idea. He tells the story of the prototype of the Boeing B17 Flying Fortress that crashed after take-off on its trial flight in 1935. Most experts said that the bomber was 'too complex to fly'. Facing bankruptcy, Boeing investigated and discovered that, confronted with four engines rather than two, the pilot forgot to release a vital locking mechanism. But Boeing created a pilot's checklist, in which the fundamental actions for the stages of flying were made a mandated part of the pilot's job. In the following years, B17s flew almost 2 million miles without a single accident. Even for pilots, many of whom are rugged individualists, says Mr Gawande, it is usually the application of routine procedures that saves planes when things go wrong, rather than 'hero-pilotry' so fêted by the media. It is discipline rather than brilliance that preserves life. In fact, it is discipline that leaves room for brilliance to flourish.

Customer-centricity

There is little point in improvement unless it meets the requirements of the customers. However, in most improvement approaches, meeting the expectations of customers means more than this. It involves the whole organisation in understanding the central importance of customers to its success and even to its survival. Customers are seen not as being external to the organisation but as the most important part of it. However, the idea of being customer-centric does not mean that customers must be provided with everything that they want. Although 'What's good for customers' may frequently be the same as 'What's good for the business', it is not always. Operations managers are always having to strike a balance between what customers would like and what the operation can afford (or wants) to do.

Voice of the customer (VOC)

The 'voice of the customer' (VOC) is an idea that is closely related to the idea of customer centricity. The term means capturing a customer's requirements, expectations, perceptions and preferences in some depth. Sometimes a VOC exercise is done as part of new service and product development as part of quality function deployment (QFD), which was explained in Chapter 4. Sometimes it is part of a more general improvement activity. There are several ways to do this, but they usually involve using market research to derive a comprehensive set of customer requirements, which is ordered into a hierarchical structure, often prioritised to indicate the relative importance of different aspects of operations performance.

Systems and procedures

Improvement is not something that happens simply by getting everyone to 'think improvement'. Some type of system that supports the improvement effort may be needed. An improvement system (sometimes called a 'quality system') is the processes and resources for implementing improvement. It specifies the organisational responsibilities for improvement as well as the procedures and processes that support improvement activities.

Reduce process variation

Processes change over time, as does their performance. Some aspect of process performance (usually an important one) is measured periodically (either as a single measurement or as a small sample of measurements). These are then plotted on a simple time scale. This has several advantages. The first is to check that the performance of the process is acceptable (capable). They can also be used to check if process performance is changing over time, and to check on the extent of the variation in process performance. In Chapter 17 we illustrate how random variation in the performance of any process can obscure what is really happening within the process. So a potentially useful method of identifying improvement opportunities is to try to identify the sources of random variation in process performance.

Synchronised flow

Synchronised flow means that items in a process, operation or supply network flow smoothly and with even velocity from start to finish. This is a function of how inventory accumulates within the operation. Whether inventory is accumulated to smooth differences between demand and supply, or as a contingency against unexpected delays, or simply to batch for purposes of processing or movement, it all means that flow becomes asynchronous. It waits as inventory rather than progressing smoothly on. Once this state of perfect synchronisation of flow has been achieved, it becomes easier to expose any irregularities of flow, which may be the symptoms of more deep-rooted underlying problems.

Emphasise education/training

Several improvement approaches stress the idea that structured training and organisation of improvement should be central to improvement. Not only should the techniques of improvement be fully understood by everyone engaged in the improvement process, the business and organisational context of improvement should also be understood. After all, how can one improve without knowing what kind of improvement would best benefit the organisation and its customers? Furthermore, education and training have an important part to play in motivating all staff towards seeing improvement as a worthwhile activity. Some improvement approaches in particular place great emphasis on formal education. Six Sigma for example (see later) and its proponents often mandate a minimum level of training (measured in hours) that they deem necessary before improvement projects should be undertaken.

Perfection is the goal

Almost all organisation-wide improvement programmes will have some kind of goal or target that the improvement effort should achieve. And while targets can be set in many ways, some improvement authorities hold that measuring process performance against an absolute target encourages improvement. By an 'absolute target' one literally means the theoretical level of perfection, for example, zero errors, instant delivery, delivery absolutely when promised, infinite flexibility, zero waste, etc. Of course, such perfection may never be achievable. That is not the point. What is important is that current performance can be calibrated against this target of perfection to indicate how much more improvement is possible. Improving (for example) delivery accuracy by 5 per cent may seem good until it is realised that only an improvement of 30 per cent would eliminate all late deliveries.

Waste identification

All improvement approaches aspire to eliminate waste. In fact, any improvement implies that some waste has been eliminated, where waste is any activity that does not add value. But the identification and elimination of waste is sometimes a central feature. For example, as we will discuss in Chapter 16, it is arguably the most significant part of the lean philosophy.

Include everybody

Harnessing the skills and enthusiasm of every person and all parts of the organisation seems an obvious principle of improvement. The phrase 'quality at source' is sometimes used, stressing the impact that every individual has on improvement. The contribution of all individuals in the organisation may go beyond understanding their contribution to 'not making mistakes'. Individuals are expected to bring something positive to improving the way they perform their jobs. The principles of 'empowerment' are frequently cited as supporting this aspect of improvement. When Japanese improvement practices first began to migrate in the late 1970s, this idea seemed even more radical. Yet now it is generally accepted that individual creativity and effort from all staff represents a valuable source of development. However, not all improvement approaches have adopted this idea. Some authorities believe that a small number of internal improvement consultants or specialists offer a better method of organising improvement. However, these two ideas are not incompatible. Even with improvement specialists used to lead improvement efforts, the staff who operate the process can still be used as a valuable source of information and improvement ideas.

Develop internal customer–supplier relationships

One of the best ways to ensure that external customers are satisfied is to establish the idea that every part of the organisation contributes to external customer satisfaction by satisfying its own internal customers. This idea is explored further in Chapter 17, as is the related concept of service-level agreements (SLAs). It means stressing that each process in an operation has a responsibility to manage these internal customer–supplier relationships. They do this primarily by defining as clearly as possible what their own and their customers' *requirements* are. In effect this means defining what constitutes 'error-free' service – the quality, speed, dependability and flexibility required by internal customers.

15.3 What are the broad approaches to improvement?

By the broad approaches to improvement, we mean the underlying sets of beliefs that form a coherent philosophy that shapes how improvement should be accomplished. But do not think that approaches to improvement are different in all respects; there are many elements that are common to several approaches. Some of these approaches have been, or will be, described in other chapters. For example, both lean operations (Chapter 16) and TQM (in Chapter 17) are discussed in some detail. So, in this section we will only briefly examine TQM and lean operations, specifically from an improvement perspective and, so that we can demonstrate how these approaches overlap, we add two further approaches – business process reengineering (BPR) and Six Sigma. One can think about approaches to improvement as being alternative lenses through which to view operations improvement (see Figure 15.6).

Operations principle

There is no single universal approach to improvement. Rather, there are many alternative approaches.

Total quality management as an improvement approach

Total quality management was one of the earliest management 'fashions'. Its peak of popularity was in the late 1980s and early 1990s. As such it has suffered from something of a backlash. Yet the general precepts and principles that constitute TQM are still hugely relevant. Few, if any, managers have not heard of TQM and its impact on improvement. Indeed, TQM has come to be seen as an approach to

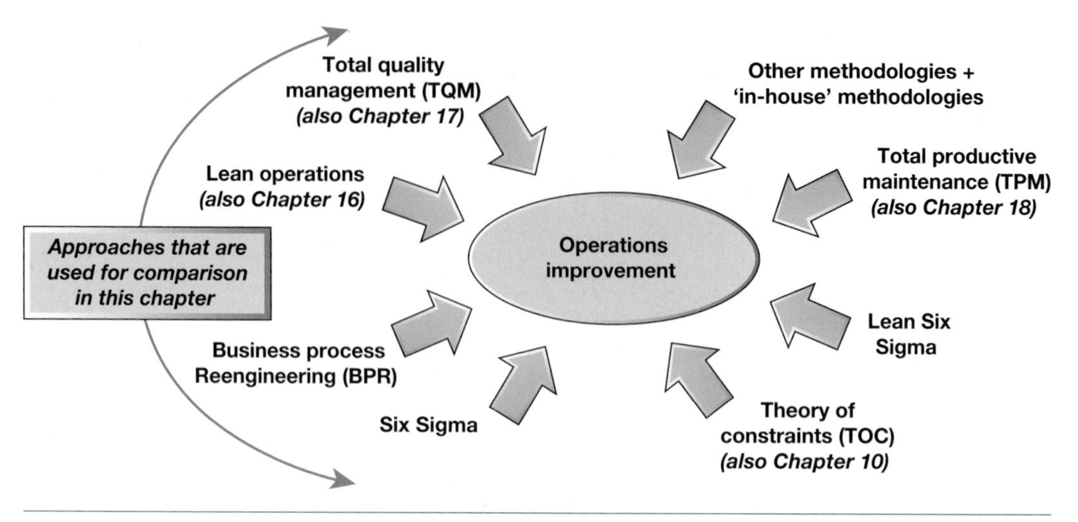

Figure 15.6 There are many 'approaches'/'methodologies' that can be used to form the basis of operations improvement, four of which are compared in this chapter to illustrate how most approaches share some elements

the way operations and processes should be managed and improved generally. Even if TQM is not the label given to an improvement initiative, many of its elements will almost certainly have become routine. It is best thought of as a philosophy for how to approach improvement. This philosophy, above everything, stresses the 'total' of TQM. It is an approach that puts quality (and indeed improvement generally) at the heart of everything that is done by an operation. As a reminder, this totality can be summarised by the way TQM lays stress on the following elements (see Chapter 17):

- Meeting the needs and expectations of customers.
- Improvement covers all parts of the organisation (and should be group-based).
- Improvement includes every person in the organisation (and success is recognised).
- Including all costs of quality.
- Getting things 'right first time', i.e. designing-in quality rather than inspecting it in.
- Developing the systems and procedures that support improvement.

Lean as an improvement approach

The idea of 'lean' spread beyond its Japanese roots and became fashionable in the West at about the same time as TQM. And although its popularity has not declined to the same extent as TQM, several decades of experience have diminished the excitement once associated with the approach. But, unlike TQM, it was seen initially as an approach to be used exclusively in manufacturing and primarily for planning and control. Now, lean is widely applied as an improvement approach in all types of operations. The lean approach aims to meet demand instantaneously, with perfect quality and no waste (see Chapter 16). The key elements of lean when used as an improvement approach are as follows:

- waste elimination;
- customer-centricity;
- internal customer–supplier relationships;
- perfection is the goal;
- synchronised flow;
- reduce variation;
- include all people.

Some organisations, especially now that lean is being applied more widely in service operations, view waste elimination as the most important of all the elements of the lean approach. In fact, they sometimes see the lean approach as consisting almost exclusively of waste elimination. What they fail to realise is that effective waste elimination is best achieved through changes in behaviour. It is the behavioural change brought about through synchronised flow and customer triggering that provides the window onto exposing and eliminating waste.

Business process reengineering (BPR)

The idea of business process reengineering originated in the early 1990s when Michael Hammer proposed that rather than using technology to automate work, it would be better applied to doing away with the need for the work in the first place ('don't automate, obliterate'). In doing this he was warning against establishing non-value-added work within an information technology system where it would be even more difficult to identify and eliminate. All work, he said, should be examined for whether it adds value for the customer and if not, processes should be redesigned to eliminate it. In doing this, BPR was echoing similar objectives in both scientific management and, more recently, lean approaches. But BPR, unlike those two earlier approaches, advocated radical changes rather than incremental changes to processes. Shortly after Hammer's article, other authors developed the ideas, again the majority of them stressing the importance of a radical approach to elimination of non-value-added work.[6]

BPR has been defined as:[7] '. . . .the fundamental rethinking and radical redesign of business processes to achieve dramatic improvements in critical, contemporary measures of performance, such as cost, quality, service and speed'. But there is far more to it than that. In fact, BPR was a blend of several ideas, which had been current in operations management for some time. Lean concepts, process flow charting, critical examination in method study, operations network management and customer-focused operations all contribute to the BPR concept. It was the potential of information technologies to enable the fundamental redesign of processes, however, which acted as the catalyst in bringing these ideas together. It was the information technology that allowed radical process redesign even if many of the methods used to achieve the redesign had been explored before. The main principles of BPR can be summarised in the following points:

- Rethink business processes in a cross-functional manner that organises work around the natural flow of information (or materials or customers).
- Strive for dramatic improvements in performance by radically rethinking and redesigning the process.
- Have those who use the output from a process, perform the process. Check to see if all internal customers can be their own supplier rather than depending on another function in the business to supply them (which takes longer and separates out the stages in the process).
- Put decision points where the work is performed. Do not separate those who do the work from those who control and manage the work.

Worked example

Torvill's Total Trading

Torvill's Total Trading decided to reorganise (or reengineer) around business processes. Figure 15.7(a) shows its original organisation. The company purchases consumer goods from several suppliers, stores them, and sells them on to retail outlets. At the heart of the operation is the warehouse, which receives the goods, stores them, and packs and dispatches them when they are required by customers. Orders for more stock are placed by purchasing, which also takes charge of materials planning and stock control. Purchasing buy the goods based on a forecast, which is prepared by marketing, which takes advice from the sales department, which is processing customers' orders. When a customer does place an order, it is the sales department's job to instruct the warehouse to pack and dispatch the order and tell the finance department to invoice the customer for the goods. So, traditionally, five departments have between them organised the flow of materials and information within the total operation. But at each interface between the departments there is the possibility of errors and miscommunication arising. Furthermore, *who is responsible for looking after the customer's needs?* Currently, three separate departments all have dealings with the customer. Similarly, *who is responsible for liaising with suppliers?* This time, two departments have contact with suppliers.

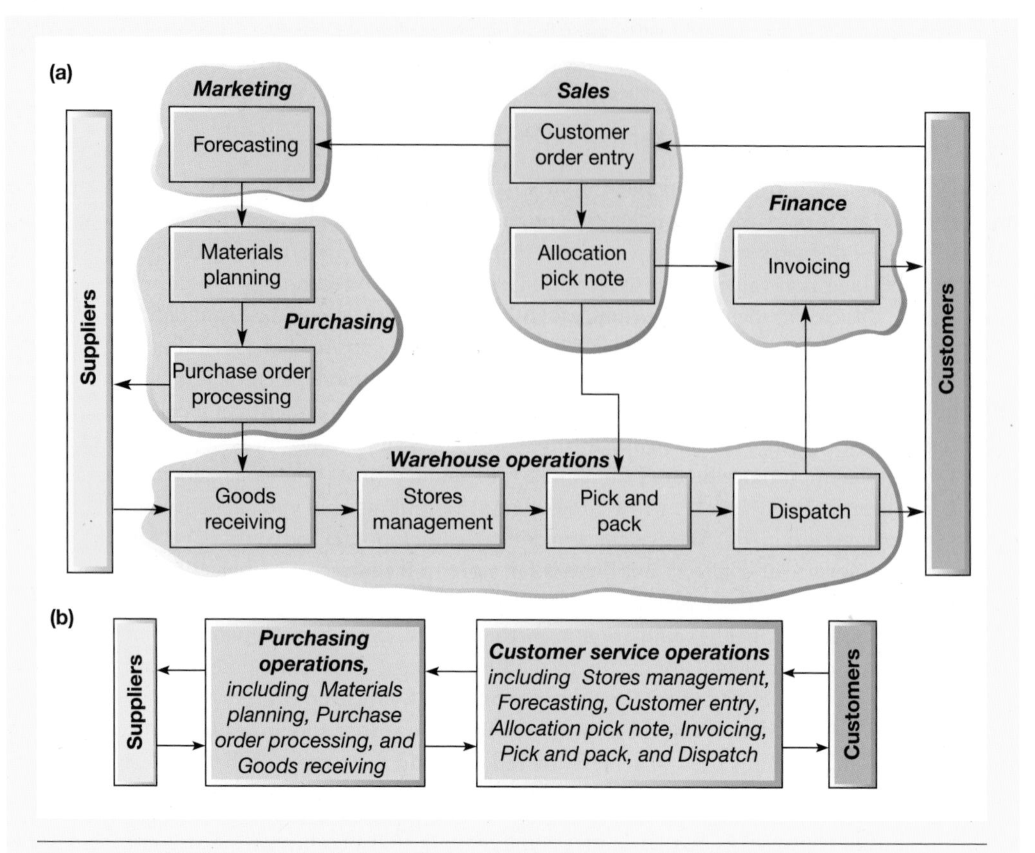

Figure 15.7 (a) Before and (b) after reengineering a consumer goods trading company

Eventually the company reorganised around two essential business processes. The first process (called purchasing operations) dealt with everything concerning relationships with suppliers. It was this process's focused and unambiguous responsibility to develop good working relationships with suppliers. The other business process (called customer service operations) had total responsibility for satisfying customers' needs. This included speaking 'with one voice' to the customer.

Critical commentary

BPR has aroused considerable controversy, mainly because BPR sometimes looks only at work activities rather than at the people who perform the work. Because of this, people become 'cogs in a machine'. Many of these critics equate BPR with the much earlier principles of scientific management, pejoratively known as 'Taylorism'. Generally, these critics mean that BPR is overly harsh in the way it views human resources. Certainly, there is evidence that BPR is often accompanied by a significant reduction in staff. Studies at the time when BPR was at its peak often revealed that the majority of BPR projects could reduce staff levels by over 20 per cent. Often BPR was viewed as merely an excuse for getting rid of staff. Companies that wished to 'downsize' were using BPR as the pretext, putting the short-term interests of the shareholders of the company above either their longer-term interests or the interests of the company's employees. Moreover, a combination of radical redesign together with downsizing could mean that the essential core of experience was lost from the operation. This leaves it vulnerable to any marked turbulence since it no longer possesses the knowledge and experience of how to cope with unexpected changes.

Six Sigma

The Six Sigma approach was first popularised by Motorola, the electronics and communications systems company. When it set its quality objective as 'total customer satisfaction' in the 1980s, it started to explore what the slogan would mean to its operations processes. They decided that true customer satisfaction would be achieved only when its products were delivered when promised, with no defects, with no early-life failures and when the product did not fail excessively in service. To achieve this, Motorola focused initially on removing manufacturing defects. However, it soon came to realise that many problems were caused by latent defects, hidden within the design of its products. These may not show initially but eventually could cause failure in the field. The only way to eliminate these defects was to make sure that design specifications were tight (i.e. narrow tolerances) and its processes very capable.

Motorola's Six Sigma quality concept was so named because it required the natural variation of processes ($\pm$ 3 standard deviations) should be half their specification range. In other words, the specification range of any part of a product or service should be $\pm$ 6 the standard deviation of the process (see Chapter 17). The Greek letter sigma (σ) is often used to indicate the standard deviation of a process, hence the Six Sigma label. Figure 15.8 illustrates the effect of progressively narrowing process variation on the number of defects produced by the process, in terms of defects per million. The defects per million measure is used within the Six Sigma approach to emphasise the drive towards a virtually zero-defect objective.[8] Now the definition of Six Sigma has widened to well beyond this rather narrow statistical perspective. General Electric (GE), who were probably the best-known of the early adopters of Six Sigma, regarded it as a disciplined methodology of defining, measuring, analysing, improving and controlling the quality in every one of the company's products, processes and transactions; the ultimate goal of which was to virtually eliminate all defects. So, now Six Sigma should be seen as a broad improvement concept rather than a simple examination of process variation, even though this is still an important part of process control, learning and improvement.

Measuring performance

The Six Sigma approach uses several related measures to assess the performance of operations processes:

- *A defect* – a failure to meet customer required performance (defining performance measures from a customer's perspective is an important part of the Six Sigma approach).
- *A defect unit or item* – any unit of output that contains a defect (i.e. only units of output with no defects are not defective, defective units will have one or more than one defects).
- *A defect opportunity* – the number of different ways a unit of output can fail to meet customer requirements (simple products or services will have few defect opportunities, but very complex products or services may have hundreds of different ways of being defective).
- *Proportion defective* – the percentage or fraction of units that have one or more defect.
- *Process yield* – the percentage or fraction of total units produced by a process that are defect free (i.e. 1 – proportion defective).

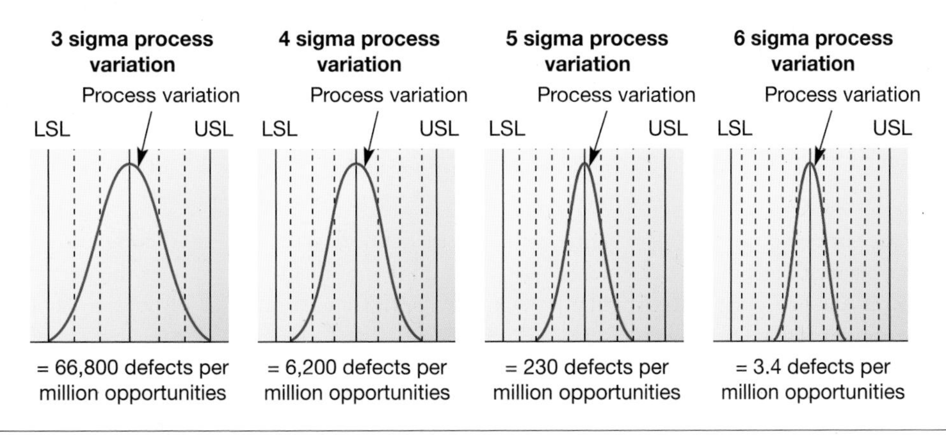

Figure 15.8 Process variation and its impact on process defects per million

- *Defect per unit (DPU)* – the average number of defects on a unit of output (the number of defects divided by the number of items produced).
- *Defects per opportunity* – the proportion or percentage of defects divided by the total number of defect opportunities (the number of defects divided by the number of items produced × the number of opportunities per item).
- *Defects per million opportunities (DPMO)* – exactly what it says – the number of defects that the process will produce if there were one million opportunities to do so.
- *The Sigma measurement* – derived from the DPMO and is the number of standard deviations of the process variability that will fit within the customer specification limits.

Worked example

Transient Insurance

Transient Insurance checks details of insurance claims and arranges for customers to be paid. It samples 300 claims at random at the end of the process. They find that 51 claims had one or more defects and there were 74 defects in total. Four types of error were observed: coding errors, policy conditions errors, liability errors and notification errors.

$$\text{Proportion defective} = \frac{\text{Number of defects}}{\text{Number of units processed}}$$

$$= \frac{51}{300} = 0.17 \text{ (17\% defective)}$$

$$\text{Yield} = 1 - \text{proportion of defectives}$$

$$= 1 - 0.17 = 0.83 \text{ or (83\% yield)}$$

$$\text{Defects per unit} = \frac{\text{Number of defects}}{\text{Number of units processed}}$$

$$= \frac{74}{300} = 0.247 \text{ (or 24.7) DPU}$$

$$\text{Defects per opportunity} = \frac{\text{Number of defects}}{\text{Number of units produced} \times \text{Number of opportunities}}$$

$$= \frac{74}{300 \times 4} = 0.062 \text{ DPO}$$

$$\text{Defects per million opportunities} = \text{DPO} \times 10^6$$

$$= 62{,}000 \text{ DPMO}$$

Although the scope of Six Sigma is disputed, elements frequently associated with Six Sigma include the following:

- *Customer-driven objectives* – Six Sigma is sometimes defined as 'the process of comparing process outputs against customer requirements'. It uses a number of measures to assess the performance of operations processes. In particular, it expresses performance in terms of defects per million opportunities (DPMO).
- *Use of evidence* – Although Six Sigma is not the first of the new approaches to operations to use statistical methods, it has done a lot to emphasise the use of quantitative evidence.
- *Structured improvement cycle* – The structured improvement cycle used in Six Sigma is the DMAIC cycle.

- *Process capability and control* – Not surprisingly, given its origins, process capability and control is important within the Six Sigma approach.
- *Process design* – Latterly, Six Sigma proponents also include process design in the collection of elements that define the Six Sigma approach.
- *Structured training and organisation of improvement* – The Six Sigma approach holds that improvement initiatives can be successful only if significant resources and training are devoted to their management.

The 'martial arts' analogy

The terms that have become associated with Six Sigma experts (and denote their level of expertise) are Master Black Belt, Black Belt and Green Belt. Master Black Belts are experts in the use of Six Sigma tools and techniques, as well as how such techniques can be used and implemented. Master Black Belts are seen primarily as teachers, who can not only guide improvement projects, but also coach and mentor Black Belts and Green Belts who are closer to the day-to-day improvement activity. They are expected to have the quantitative analytical skills to help with Six Sigma techniques and also the organisational and interpersonal skills to teach and mentor. Given their responsibilities, it is expected that Master Black Belts are employed full time on their improvement activities. Black Belts can take a direct hand in organising improvement teams. Like Master Black Belts, Black Belts are expected to develop their quantitative analytical skills and also act as coaches for Green Belts. Black Belts are dedicated full time to improvement, and although opinions vary on how many Black Belts should be employed in an operation, some organisations recommend one Black Belt for every 100 employees. Green Belts work within improvement teams, possibly as team leaders. They have significant amounts of training, although less than Black Belts. Green Belts are not full-time positions; they have normal day-to-day process responsibilities but are expected to spend at least 20 per cent of their time on improvement projects.

OPERATIONS IN PRACTICE

Six Sigma at Wipro[9]

There are many companies that have benefited from Six Sigma-based improvement, but few have gone on to be able to sell the expertise that they gathered from applying it to themselves. Wipro is one of these. Wipro is a global information technology, consulting and outsourcing company with over 200,000 employees serving over 1,100 clients in six continents. It provides a range of business services from 'business process outsourcing' (doing processing for other firms) to 'software development', and from 'information technology consulting' to 'cloud computing'. (Surprisingly for a global IT services giant, Wipro was actually started in 1945 in India as a vegetable oil company.) Wipro also has one of the most developed Six Sigma programmes in the IT and consulting industries, especially in its software development activities where key challenges included reducing the data transfer time within the process, reducing the risk of failures and errors, and avoiding interruption due to network downtime. For Wipro, Six Sigma simply means a measure of quality that strives for near perfection. It means:

- Having products and services that meet global standards.
- Ensuring robust processes within the organisation.

- Consistently meeting and exceeding customer expectations.
- Establishing a quality culture throughout the business.

Individual Six Sigma projects were selected on the basis of their probability of success and were completed relatively quickly. This gave Wipro the opportunity to assess the success and learn from any problems that had occurred. Projects were identified on the basis of the problem areas under each of the critical business processes that could adversely impact business performance. Because Wipro took a customer-focused definition of quality, Six Sigma implementation was measured in terms of progress towards what the customer finds important (and what the customer pays for). This involved improving performance through a precise quantitative understanding of the customer's requirements. Wipro say that its adoption of Six Sigma has been an unquestionable success, whether in terms of customer satisfaction, improvement in internal performance, or in the improvement of shareholder value.

However, as the pioneers of Six Sigma in India, Wipro's implementation of the process was not without difficulties and, they stress, opportunities for learning from them. To begin with, it took time to build the required support from the higher-level managers, and to restructure the organisation to provide the infrastructure and training to establish confidence in the process. In particular, the first year of deployment was extremely difficult. Resourcing the stream of Six Sigma projects was problematic, partly because each project required different levels and types of resources. Also, the company learned not to underestimate the amount of training that would be required. To build a team of professionals and train them for various stages of six sigma was a difficult job. (In fact, this motivated Wipro to start its own consultancy that could train its own people.) Nevertheless, regular and timely reviews of each project proved particularly important in ensuring the success of a project and Wipro had to develop a team of experts for this purpose.

Critical commentary

One common criticism of Six Sigma is that it does not offer anything that was not available before. Its emphasis on improvement cycles comes from TQM, its emphasis on reducing variability comes from statistical process control, its use of experimentation and data analysis is simply good quantitative analysis. The only contribution that Six Sigma has made, argue its critics, is using the rather gimmicky martial arts analogy of Black Belt, etc., to indicate a level of expertise in Six Sigma methods. All Six Sigma has done is package pre-existing elements together in order for consultants to be able to sell it to gullible chief executives. In fact, it's difficult to deny some of these points. Maybe the real issue is whether it is really a criticism. If bringing these elements together really does form an effective problem-solving approach, why is this is a problem? Six Sigma is also accused of being too hierarchical in the way it structures its various levels of involvement in the improvement. It is also expensive. Devoting such large amounts of training and time to improvement is a significant investment, especially for small companies. Nevertheless, Six Sigma proponents argue that the improvement activity is generally neglected in most operations and if it is to be taken seriously, it deserves the significant investment implied by the Six Sigma approach. Furthermore, they argue, if operated well, Six Sigma improvement projects run by experienced practitioners can save far more than their cost. There are also technical criticisms of Six Sigma. Most notably, in purely statistical terms the normal distribution, which is used extensively in Six Sigma analysis, does not actually represent most process behaviour. Other technical criticisms (that are not really the subject of this text) imply that aiming for the very low levels of defects per million opportunities, as recommended by Six Sigma proponents, is far too onerous.

Differences and similarities

In this chapter we have chosen to explain four improvement approaches very briefly. It could have been more. Enterprise resource planning (ERP, see Chapter 14), total productive maintenance (TPM, see Chapter 18), Lean Sigma (a combination of lean and Six Sigma) and others could have been added. But these four constitute a representative sample of the most used approaches. Nor do we have the space to describe them fully. But there are clearly some common elements between some of these approaches that we have described. Yet there are also differences between them in that each approach includes a different set of elements and therefore a different emphasis, and these differences need to be understood. For example, one important difference relates to whether the approaches emphasise a gradual, continuous approach to change, or whether they recommend a more radical 'breakthrough' change. Another difference concerns the aim of the approach. What is the balance between whether the approach emphasises *what* changes should be made or *how* changes should be made? Some approaches have a firm view of what is the best way to organise the operation's processes and resources. Other approaches hold no view on what an operation should do but rather concentrate on how the management of an operation should decide what to do. Indeed, we can position each of the elements and the approaches that include them.

This is illustrated in Figure 15.9. The approaches differ in the extent that they prescribe appropriate operations practice. BPR, for example, is very clear in what it is recommending. Namely, that all processes should be organised on an end-to-end basis. Its focus is *what* should happen rather than *how* it should happen. To a slightly lesser extent, lean is the same. It has a definite list of things that processes should or should not be – waste should be eliminated, inventory should be reduced, technology should be flexible, and so on. Contrast this with both Six Sigma and TQM that focus to a far greater extent on *how* operations should be improved. Six Sigma has relatively little to say about what is good or bad in the way operations resources are organised (apart from it emphasising the negative effects of process variation). Its concern is largely the way improvements should be made – using evidence, using quantitative analysis, using the DMAIC cycle and so on. They also differ in terms of whether they emphasise gradual or rapid change. BPR is explicit in its radical nature. By contrast, TQM and lean both incorporate ideas of continuous improvement. Six Sigma is relatively neutral on this issue and can be used for small or very large changes.

Operations principle

There is significant overlap between the various approaches to improvement in terms of the improvement elements they contain.

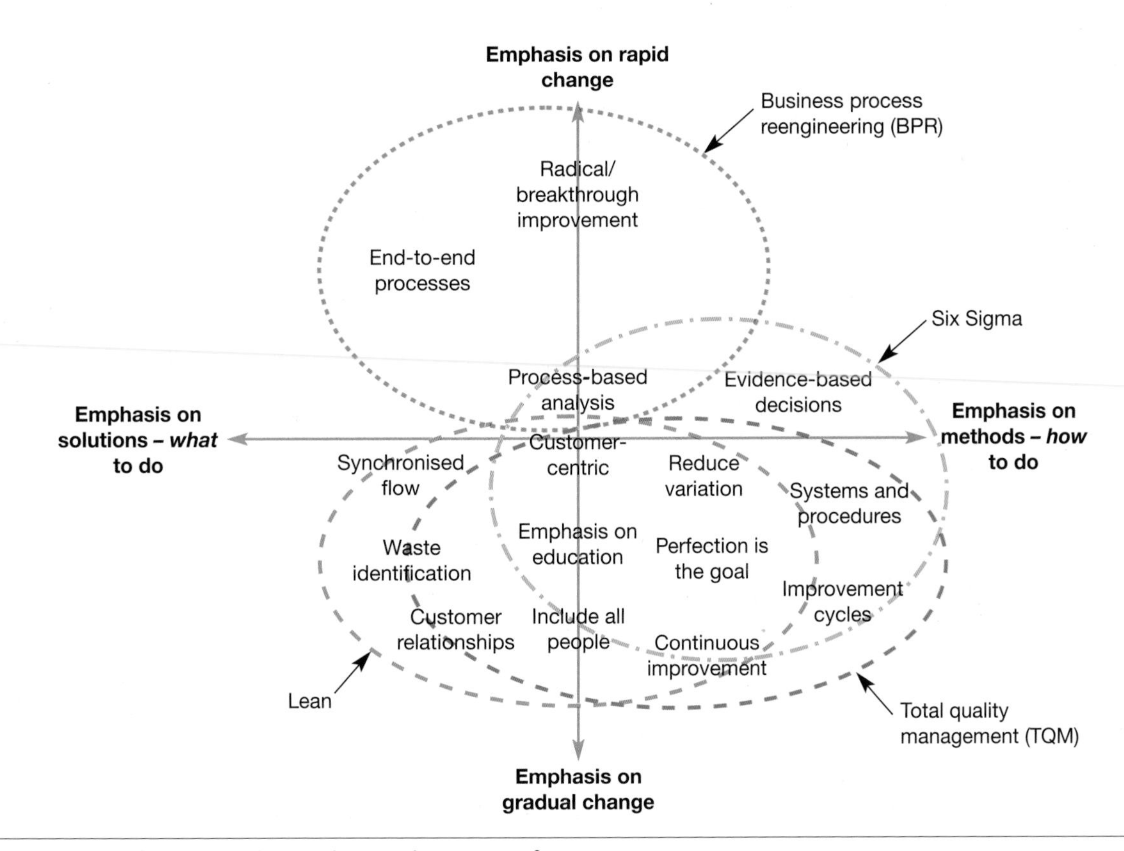

Figure 15.9 The four approaches on the two dimensions of improvement

Lean Sigma[10]

As if to emphasise the shared elements of the various approaches to operations improvement, some organisations are blending two or more approaches to form hybrids that try to combine their best characteristics. The best known of these is **Lean Sigma** (also called Lean Six Sigma or Six Sigma Lean). As its name suggests, Lean Six Sigma is a combination of lean methods and Six Sigma concepts. It attempts to build on the experience, methods and tools that have emerged from the several decades of operational improvement and implementation using lean and Six Sigma approaches separately. Lean Sigma includes the waste reduction, fast throughput time and impact of lean with the data-driven rigour and variation control of Six Sigma. Some organisations also include other elements from other approaches. For example, the continuous improvement and error-free quality orientation of TQM is frequently included in the concept.

15.4 What techniques can be used for improvement?

Improvement techniques are the 'step-by-step' methods and tools that can be used to help find improved ways of doing things. Some of these use quantitative modelling and others are more qualitative. All the techniques described in this text and its supplements can be regarded as 'improvement' techniques. However, some techniques are particularly useful for improving operations and processes generally. Here we select some techniques that either have not been described elsewhere or need to be reintroduced, in their role of helping operations improvement particularly.

> **Operations principle**
> Improvement is facilitated by relatively simple analytical techniques.

Scatter diagrams

Scatter diagrams provide a quick and simple method of identifying whether there is evidence of a connection between two sets of data: for example, the time at which you set off for work every morning and how long the journey to work takes. Plotting each journey on a graph, which has departure time on one axis and journey time on the other, could give an indication of whether departure time and journey time are related, and if so, how. Scatter diagrams can be treated in a far more sophisticated manner by quantifying how strong the relationship between the sets of data is. But however sophisticated the approach, this type of graph only identifies the existence of a relationship, not necessarily the existence of a cause–effect relationship. If the scatter diagram shows a very strong connection between the sets of data, it is important evidence of a cause–effect relationship, but not proof positive. It could be coincidence!

Worked example

Kaston Pyral Services Ltd (A)

Kaston Pyral Services Ltd (KPS) installs and maintains environmental control, heating and air-conditioning systems. It has set up an improvement team to suggest ways in which it might improve its levels of customer service. The improvement team had completed its first customer satisfaction survey. The survey asked customers to score the service they received from KPS in several ways. For example, it asked customers to score services on a scale of 1–10 on promptness, friendliness, level of advice, etc. Scores were then summed to give a 'total satisfaction score' for each customer – the higher the score, the greater the satisfaction. The spread of satisfaction scores puzzled the team and they considered what factors might be causing such differences in the way their customers viewed them. Two factors were put forward to explain the differences:

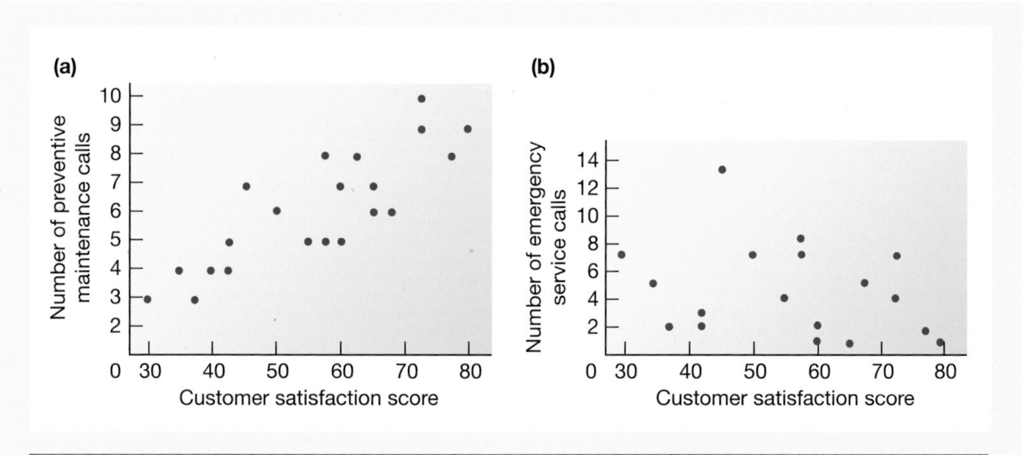

Figure 15.10 Scatter diagrams for customer satisfaction versus (a) number of preventive maintenance calls and (b) number of emergency service calls

(a) the number of times in the past year the customer had received a preventive maintenance visit;
(b) the number of times the customer had called for emergency service.

All this data was collected and plotted on scatter diagrams as shown in Figure 15.10. It shows that there seems to be a clear relationship between a customer's satisfaction score and the number of times the customer was visited for regular servicing. The scatter diagram in Figure 15.10(b) is less clear. Although all customers who had very high satisfaction scores had made very few emergency calls, so had some customers with low satisfaction scores. As a result of this analysis, the team decided to survey customers' views on its emergency service.

Process maps (flow charts)

Process maps (sometimes called flow charts in this context) can be used to give a detailed understanding prior to improvement. They were described in Chapter 6 and are widely used in improvement activities. The act of recording each stage in the process quickly shows up poorly organised flows. Process maps can also clarify improvement opportunities and shed further light on the internal mechanics or workings of an operation. Finally, and probably most importantly, they highlight problem areas where no procedure exists to cope with a particular set of circumstances.

Worked example

Kaston Pyral Services Ltd (B)

As part of its improvement programme, the team at KPS is concerned that customers are not being served well when they phone in with minor queries over the operation of their heating systems. These queries are not usually concerned with serious problems, but often concern minor irritations, which can be equally damaging to the customers' perception of KPS's service. Figure 15.11 shows the process map for this type of customer query. The team found the map illuminating. The procedure had never been formally laid out in this way before, and it showed up several areas where information was not being recorded or some other opportunity for improvement existed.

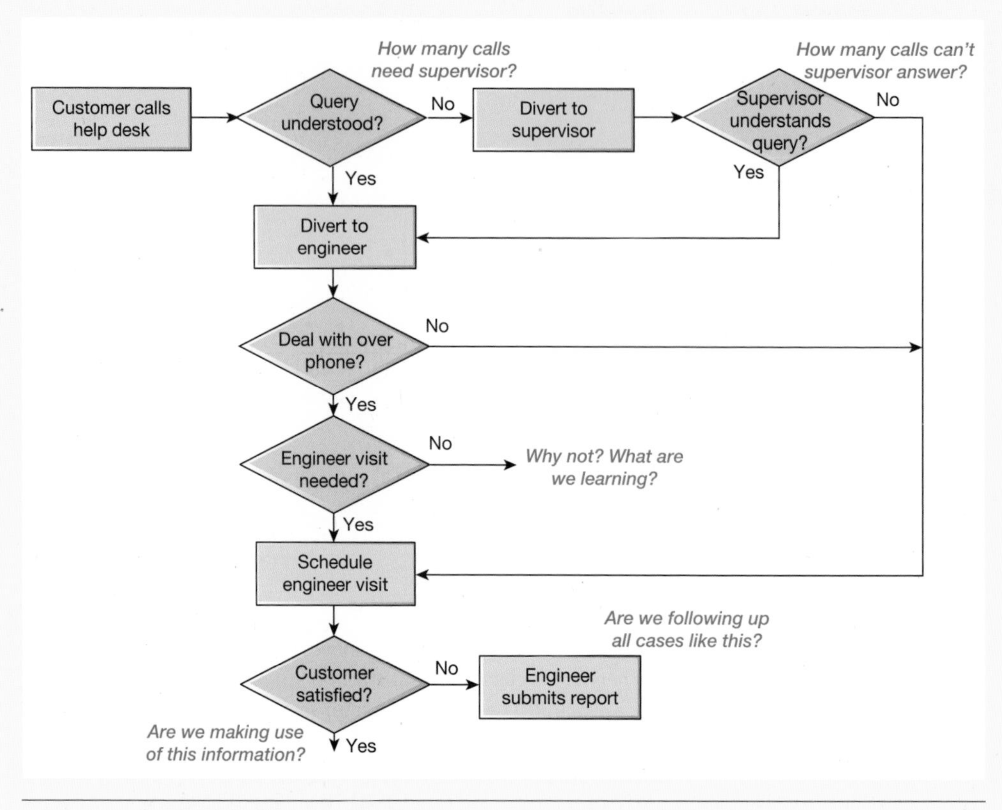

Figure 15.11 Process map for customer query

Cause–effect diagrams

Cause–effect diagrams are a particularly effective method of helping to search for the root causes of problems. They do this by asking what, when, where, how and why questions, but also add some possible 'answers' in an explicit way. They can also be used to identify areas where further data is needed. Cause–effect diagrams (which are also known as Ishikawa diagrams) have become extensively used in improvement programmes. This is because they provide a way of structuring group brainstorming sessions. Often the structure involves identifying possible causes under the (rather old-fashioned) headings of: machinery, manpower, materials, methods and money. Yet in practice, any categorisation that comprehensively covers all relevant possible causes could be used.

Worked example

Kaston Pyral Services Ltd (C)

The improvement team at KPS was working on a particular area that was proving a problem. Whenever service engineers were called out to perform emergency servicing for a customer, they took with them the spares and equipment that they thought would be necessary to repair the system. Although engineers could never be sure exactly what materials and equipment they would need for a job, they could guess what was likely to be needed and take a range of spares and equipment that would cover most eventualities. Too often, however, the engineers would find that they needed a spare that they had not brought with them. The cause–effect diagram for this particular problem, as drawn by the team, is shown in Figure 15.12.

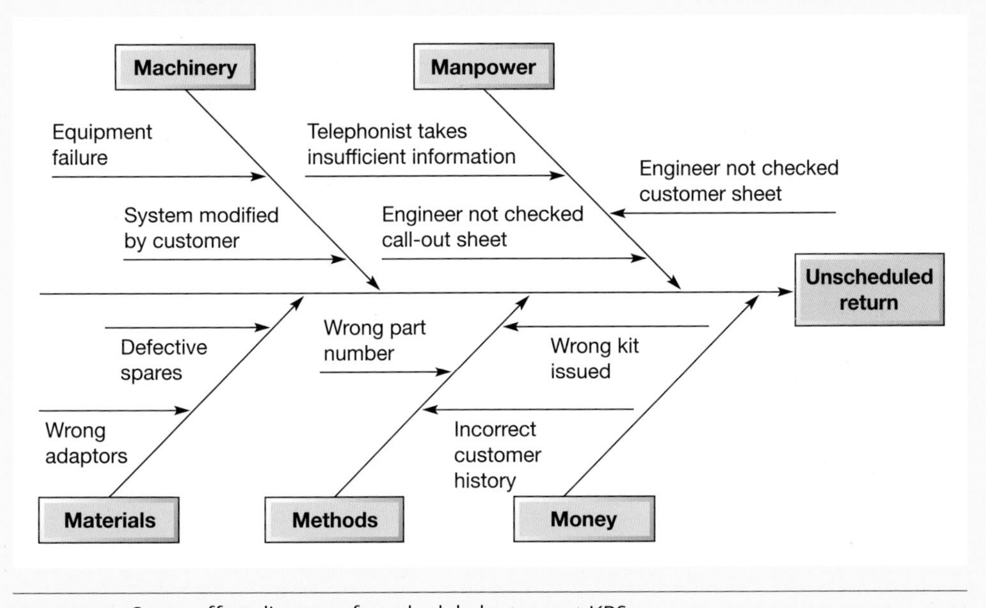

Figure 15.12 Cause–effect diagram of unscheduled returns at KPS

Pareto curves

In any improvement process, it is worthwhile distinguishing between what is important and what is less so. The purpose of the Pareto curve (that was first introduced in Chapter 13) is to distinguish between the 'vital few' issues and the 'trivial many'. It is a relatively straightforward technique, which involves arranging items of information on the types of problems or causes of problems into their order of importance (usually measured by 'frequency of occurrence'). This can be used to highlight areas where further decision making will be useful. Pareto analysis is based on the phenomenon of relatively few causes explaining the majority of effects. For example, most revenue for any company is likely to come from relatively few of the company's customers. Similarly, relatively few of a doctor's patients will probably occupy most of his or her time.

Worked example

Kaston Pyral Services Ltd (D)

The KPS improvement team that was investigating unscheduled returns from emergency servicing (the issue that was described in the cause–effect diagram in Figure 15.12) examined all occasions over the previous 12 months on which an unscheduled return had been made. They categorised the reasons for unscheduled returns as follows:

1 The wrong part had been taken to a job because, although the information that the engineer received was sound, they had incorrectly predicted the nature of the fault.
2 The wrong part had been taken to the job because there was insufficient information given when the call was taken.
3 The wrong part had been taken to the job because the system had been modified in some way not recorded on KPS's records.
4 The wrong part had been taken to the job because the part had been incorrectly issued to the engineer by stores.

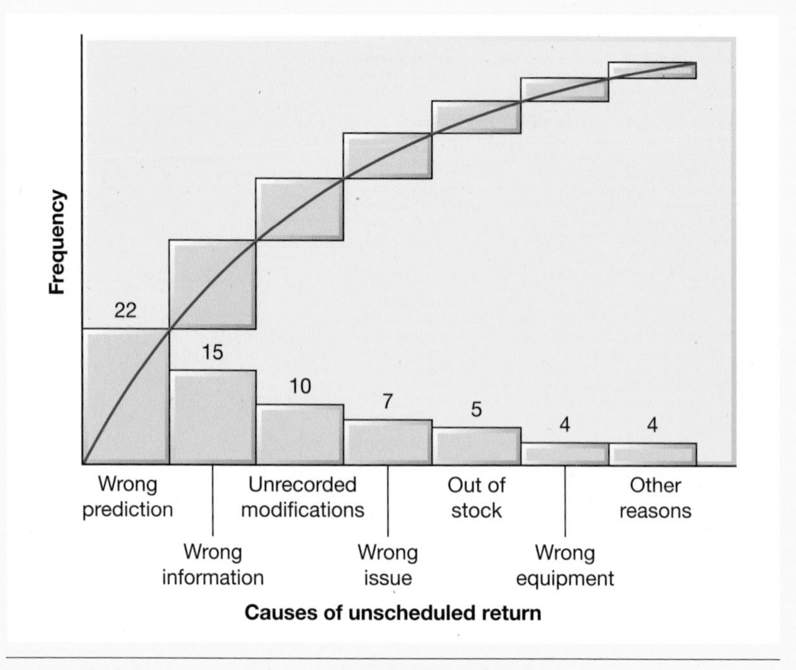

Figure 15.13 Pareto curve for causes of unscheduled returns

5 No part had been taken because the relevant part was out of stock.
6 The wrong equipment had been taken for whatever reason.
7 Any other reason.

The relative frequency of occurrence of these causes is shown in Figure 15.13. About one-third of all unscheduled returns were due to the first category, and more than half the returns were accounted for by the first and second categories together. It was decided that the problem could best be tackled by concentrating on how to get more information to the engineers who would enable them to predict the causes of failure accurately.

Why–why analysis

Why–why analysis starts by stating the problem and asking *why* that problem has occurred. Once the reasons for the problem occurring have been identified, each of the reasons is taken in turn and again the question is asked *why* those reasons have occurred, and so on. This procedure is continued until either a cause seems sufficiently self-contained to be addressed by itself or no more answers to the question 'Why?' can be generated.

Worked example

Kaston Pyral Services Ltd (E)

The major cause of unscheduled returns at KPS was the incorrect prediction of reasons for the customer's system failure. This is stated as the 'problem' in the why–why analysis in Figure 15.14. The question is then asked, why was the failure wrongly predicted? Three answers are proposed: first, that the engineers were not trained correctly; second, that they had insufficient knowledge of the particular product installed in the customer's location; and third, that they had insufficient knowledge of the customer's

of successful practices in the way it manages its process does not mean that adopting those same practices in another context will prove equally successful. It is possible that subtle differences in the resources within a process (such as staff skills or technical capabilities) or the strategic context of an operation (for example, the relative priorities of performance objectives) will be sufficiently different to make the adoption of seemingly successful practices inappropriate.

Improvement as learning

Many of the abilities and behaviours often associated with the successful management of improvement are directly or indirectly related to learning in some way. This is not surprising given that operations improvement implies intervention or change to the operation, and change will be evaluated in terms of whatever improvement occurs. This evaluation adds to our knowledge of how the operation really works, which in turn increases the chances that future interventions will also result in improvement. This idea of an improvement cycle was discussed earlier. What is important is to realise that it is a learning process, and it is crucial that improvement is organised so that it encourages, facilitates and exploits the learning that occurs during improvement. This requires us to recognise that there is a distinction between single- and double-loop learning.

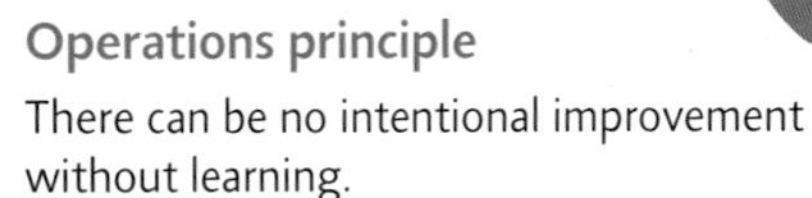

Operations principle

There can be no intentional improvement without learning.

Single- and double-loop learning

Single-loop learning occurs when there is repetitive and predictable link between cause and effect. This is like the idea of 'routine control' that we discussed in Chapter 10. Output characteristics from a process are measured and associated with the input conditions that caused it. Every time an operational error or problem is detected, it is corrected and, in doing so, more is learned about the process. However, this happens without questioning or altering the underlying values and objectives of the process, which may, over time, create an unquestioning inertia that prevents it adapting to a changing environment. Double-loop learning, by contrast, questions the fundamental objectives, service or even the underlying culture of the operation. This kind of learning implies an ability to challenge existing operating assumptions in a fundamental way. It seeks to re-frame competitive assumptions and remain open to any changes in the competitive environment. But being receptive to new opportunities sometimes requires abandoning existing operating routines, which may be difficult to achieve in practice, especially as many operations reward experience and past achievement (rather than potential) at both an individual and group level. Figure 15.15 illustrates single- and double-loop learning.

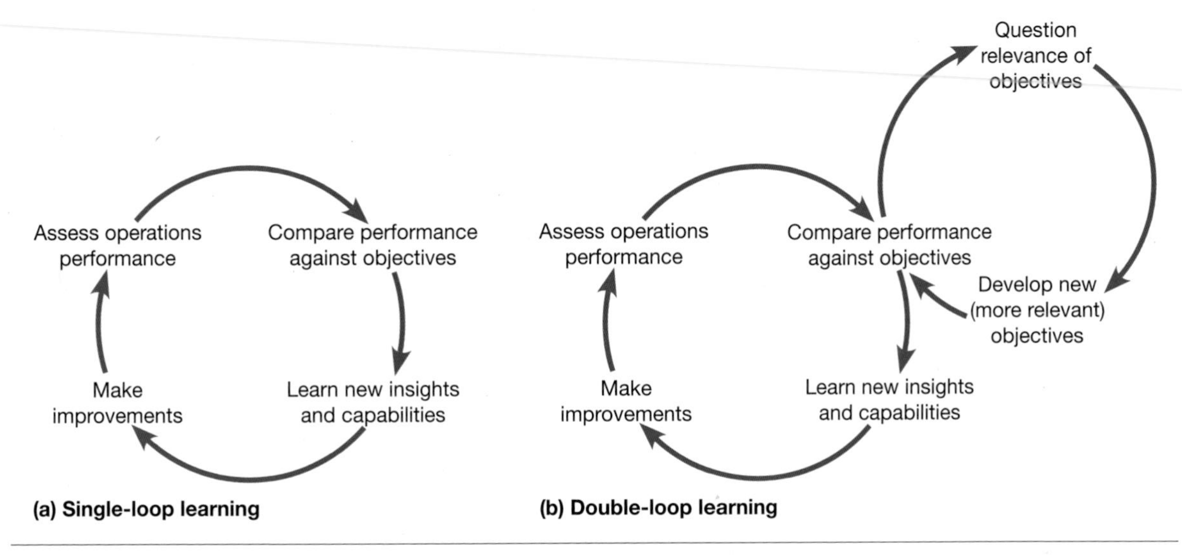

Figure 15.15 Single- and double-loop learning

OPERATIONS IN PRACTICE

Learning from Formula 1[13]

As driving jobs go, there could be no bigger difference than between a Formula 1 racing driver weaving their way through some of the fastest competitors in the world and a supermarket truck driver quietly delivering beans, beer and bacon to distribution centres and stores. But they have more in common than one would suspect. Both Formula 1 and truck drivers want to save fuel, either to reduce pit stops (Formula 1) or keep delivery costs down (heavy goods vehicles). And although grocery deliveries in the suburbs do not seem as thrilling as racing round the track at Monza, the computer-assisted simulation programs developed by the Williams Formula 1 team are being deployed to help Sainsbury's (a British supermarket group) drivers develop the driving skills that could potentially cut its fuel bill by up to 30 per cent. The simulator technology, which allows realistic advanced training to be conducted in a controlled environment, was developed originally for the advanced training of Formula 1 drivers and was developed and extended at the Williams Technology Centre in Qatar. It can now train drivers to a high level of professional driving skills and road safety applications.

Williams F1's Chief Executive, Alex Burns commented, *'Formula 1 is well recognised as an excellent technology incubator. It makes perfect sense to embrace some of the new and emerging technologies that the Williams Technology Centre in Qatar is developing from this incubator to help Sainsbury's mission to reduce its energy consumption and enhance the skills and safety of those supporting its crucial logistics operation'.* Sainsbury's energy-related improvement programmes tackle energy supply (for example, wind, solar and geothermal energy) as well as energy consumption (for example, switching to LED lighting, CO_2 refrigeration, etc.). Learning from Formula 1 will help Sainsbury's to improve further in the field of energy efficiency. Roger Burnley, Sainsbury's retail and logistics director, said, *'We are committed to reducing our environmental impact and as a result, we are often at the very forefront of technological innovation. By partnering with Williams F1, we can take advantage of some of the world's most advanced automotive technology, making our operations even more efficient and taking us a step closer to meeting our CO_2 reduction targets'.*

Knowledge management

Central to the idea of learning how to do things better is the idea of 'knowledge'. In an operations context, 'knowledge' is taken to mean any fact, information or skill that is obtained through direct experience or education. Note how we distinguish between two sources of knowledge – experience (doing things) and education (explaining or describing what experience has taught you for the benefit of other people). Doing something may lead to you knowing more about it but having to articulate it or explain it makes your knowledge more valuable because it can be shared with others. It is this process of formalising experience that distinguishes between 'tacit' knowledge and 'explicit' knowledge:

- Tacit knowledge is knowledge that is in people's heads rather than written or formally articulated or described. An example of tacit knowledge that is often used is the knowledge of how to ride a bicycle. If you can do it, it is easy to understand, but explaining how to do it in precise terms is very difficult.
- Explicit knowledge is that which is set out in definite form. It can be transmitted in formal, organised language. It has been 'codified', arranged into systematic language. It is probably included in manuals, records or process maps. Explicit knowledge can be relatively easily communicated between individuals formally and systematically.

Improvement (at least as far as operations managers are concerned) relies on the continual transformation of experience (tacit knowledge) into a formal, recognised 'better way of doing things' (explicit knowledge). The activity of managing how knowledge is formalised in this way is called 'knowledge management' (often abbreviated to KM).

OPERATIONS IN PRACTICE

Schlumberger's InTouch technology for knowledge management[14]

Schlumberger is a global company working in the oilfield services industry, supplying the latest technology to: 'optimize reservoir performance for customers working in the oil and gas industry'. The company often operates in difficult and challenging environments, so deploying technology to manage its knowledge base is vital for Schlumberger's continued success. It describes knowledge management (KM) as the 'development and deployment of processes and technology to improve organizational performance and reduce costs for Schlumberger and its customers by enabling individuals to capture, share and apply their overall knowledge – in real time'. Or, as the company sometimes puts it more simply, 'apply everywhere what you learn anywhere'. According to Susan Rosenbaum, Schlumberger's Director of Knowledge Management, *'Knowledge is respected as an important asset at Schlumberger. We've had technological solutions internally to capture knowledge since before the term "knowledge management" entered the popular business lexicon. But, while such systems are essential, the key is in how we make use of these tools. It's the sustained interaction between our people that makes the difference'.*

As is normal in KM, technology is important. Schlumberger's proprietary InTouch system is central for knowledge capture and sharing at Schlumberger, which has a direct impact on its customers' experience. The InTouch database, which contains more than 1 million knowledge items and receives 8 million views per year, is typically the first recourse for field engineers experiencing a persistent technical problem. It also comprises a team of 125 dedicated InTouch engineers available to help solve field issues one on one. These specialists, who 'sleep with beepers and cell phones', have at least five years of field experience and are drawn from all of the company's product and domain segments. Their location within the company's research and technology centres gives them immediate access to the scientists and engineers involved in developing the products and services in the first place. Schlumberger also supports internal Eureka technical bulletin boards, many of which log 20 or more discussion threads per week. *'You have field and InTouch engineers interacting through the InTouch system'*, says Rosenbaum. *'But you also have field engineers helping other field engineers on the bulletin boards. InTouch engineers routinely scan these discussion threads to glean information and spot experienced contacts'.* Increasingly, the flow of knowledge is cyclical, making it more robust than ever. *'Field engineers can flag content on the InTouch database that they feel is outdated, to ensure it gets checked'*, says Rosenbaum. *'We're using the power of the people to keep our information up to date'.*

KM is a way of not having to 'reinvent the wheel', and of building on the previous experience. It is also a way of supporting improvement activities because of its potential to combine ideas from all parts of an organisation and its external contacts. KM has two distinct, but connected, functions:

- It *collects* knowledge together, often codifying tacit into explicit knowledge, allowing anyone with access to the knowledge base to search for, and use, the knowledge whenever and wherever it is needed. This requires the building of large information repositories such as databases.
- It *connects* individual staff (who themselves are holders of tacit knowledge) with the formal codified knowledge that has been collected, and to each other. Connecting individuals is particularly important because it is not always possible to completely codify tacit knowledge into explicit knowledge. People need to interact with the tacit knowledge that is embodied in the people who have the understanding derived from direct experience, to gain the insights that may not be obvious in its formal codified form.

Responsible operations

In every chapter, under the heading of 'Responsible operations', we summarise how the particular topic covered in the chapter touches upon important social, ethical and environmental issues.

There was a time when all one expected from an employee was that they turn up on time, perform their tasks 'to specification' and did not undermine or upset their work colleagues. Working for financial reward alone was fine for some people. They simply 'did the job', no more, no less. And of course, there are still jobs like this, employers who expect nothing further. There are even employees who prefer such minimum engagement. However, this attitude, both in employees and employers, is far less common. Increasingly, both now would accept that employees have some responsibility to contribute to improving their (and their colleagues') activities. The question is, 'how much should enterprises expect of their staff over and above what they are technically contracted to do? Can one reasonably expect everyone in an operation to be enthusiastic about generating improvement ideas and implementing them? Or to put it more bluntly, 'should you expect (or even force) people to participate in improvement?' It is a real ethical issue for operations managers because so many approaches to improvement assume that, if given appropriate leadership, everyone will be willing to contribute to improvement and participate in improvement activities.

One could argue that expecting everyone in an organisation to contribute to improvement is perfectly reasonable in a modern workplace (although some would dispute this). But, to what extent? For example, is it reasonable for someone to say, 'I am perfectly happy to contribute improvement ideas and work hard to make them, and my colleagues' ideas, work. However, this does not mean that my employer can force me to participate in specific improvement activities if I don't think that they are effective as far as I'm concerned'? In other words, encouraging or asking people to participate in improvement activities is not the same as making it compulsory – in effect forcing people to participate. But there are a range of defendable positions on this issue, and companies take different ethical views. Some reject the idea of any compulsory participation. Rather they see operations managers as being responsible for creating an environment in which people feel secure and empowered to contribute and participate in improvement. They would argue that no one likes to be 'over-managed' or forced to adopt a particular methodology of improvement. When people feel forced, they may even disrupt all improvement efforts. Why not, they say, just let people work their own approach themselves? Some companies seek to avoid the issue by, in the short term, selecting volunteers from existing employees or, in the long-term, recruiting only people who are willing to participate in a manner the organisation finds acceptable. At the other end of the spectrum, some companies see participation in whatever improvement methodology they favour as a condition of employment. For example, Toyota, one of the most successful promoters of operations improvement, expect all staff to participate in their improvement methodology.

Summary answers to key questions

15.1 Why is improvement so important in operations management?

- Improvement is now seen as the prime responsibility of operations management. Furthermore, all operations management activities are really concerned with improvement in the long term. Also, companies in many industries are having to improve simply to retain their position relative to their competitors. This is sometimes called the 'Red Queen' effect.
- A common distinction is between radical or breakthrough improvement on one hand, and continuous or incremental improvement on the other.
- This distinction is closely associated with the distinction between the exploitation of existing capabilities versus the exploration of new ones. The ability to do both is called 'organisational ambidexterity'.

15.2 What are the key features of operations improvement?

- There are many 'elements' that are the building blocks of improvement approaches. The ones described in this chapter are:
  - — improvement cycles;
  - — a process perspective;
  - — end-to-end processes;
  - — radical change;
  - — evidence-based problem solving;
  - — customer centricity;
  - — systems and procedures;
  - — reducing process variation;
  - — synchronised flow;
  - — emphasise education/training;
  - — perfection is the goal;
  - — waste identification;
  - — including everybody;
  - — developing internal customer–supplier relationships.

15.3 What are the broad approaches to improvement?

- What we have called 'the broad approaches to improvement' are relatively coherent collections of some of the 'elements' of improvement. The four most common are total quality management (TQM), lean, business process reengineering (BPR) and Six Sigma.
- There are differences between these improvement approaches. Each includes a different set of elements and therefore a different emphasis. They can be positioned on two dimensions. The first is whether the approaches emphasise a gradual, continuous approach to change or a more radical 'breakthrough' change. The second is whether the approach emphasises *what* changes should be made or *how* changes should be made.

15.4 What techniques can be used for improvement?

- Many of the techniques described throughout this text could be considered improvement techniques, for example statistical process control (SPC).
- Techniques often seen as 'improvement techniques' include: scatter diagrams, flow charts, cause–effect diagrams, Pareto curves and why–why analysis.

15.5 How can the improvement process be managed?

- Improvement does not just happen by itself. It needs organising, information must be gathered so that improvement is treating the most appropriate issues, responsibility for looking after the improvement effort must be allocated, and resources must be allocated. It must also be linked to the organisation's overall strategy.
- The process of benchmarking is often used as a means of obtaining competitor performance standards.
- An organisation's ability to improve its operations performance depends to a large extent on its 'culture', that is 'the pattern of shared basic assumptions . . . that have worked well enough to be considered valid. . . '. A receptive organisational culture that encourages a constant search for improved ways to do things can encourage improvement.
- Many of the abilities and behaviours related to an improvement culture relate to learning in some way. This involves two types of learning, single- and double-loop learning.

CASE STUDY Sales slump at Splendid Soup Co.

Meghana was only three months into her new role as Vice President for Sales, yet she was already starting to feel frustrated. *'When I was appointed, it was made clear that my main task was to grow market share. Yet it was obvious that the market was changing and would change further. There are now so many competitors in the chilled soup market, it's considerably tougher to retain existing customers or gain new ones. We need to decide how we will position our products given this market shift'.*

Growing sales

As VP for Sales at Splendid Soup Co., Meghana had started to question the company's approach to growing sales. Although still the leading brand of chilled soup, its soup brands had been experiencing declining sales and revenue over the 12-month period prior to Meghana's appointment and had continued its decline. Meghana believed that part of the problem was the company's use of the strategy of 'special offer promotions', where a deal was struck between the distributor and the supermarket chain to promote the soup at a reduced price for a limited period only. This could be highly successful in the short term, with some supermarkets selling the soup faster than it could be produced. Typically, Splendid Soup Co. would respond to this increased demand by ramping up production, bringing in extra supplies of fresh ingredients and putting in place additional workforce capacity. Unfortunately, once the promotion finished, orders would decrease sharply. Both consumers and supermarkets (rightly) predicted that the soup would come 'on offer' again soon, so why pay full price? When this happened, the Splendid Soup Co. factory would be left with surplus (and often out-of-date) materials. On at least one occasion in the previous year, a significant amount of soup in the supply pipeline had to be thrown away or sold at a markdown price.

Meghana saw her challenge as having two aspects. *'Strategically, the real challenge lies with convincing the senior team that the regular cycle of promotion activity actually decreases market share. We can't rely on old tactics any more. Over the years we have sought to stimulate sales through regular promotions alongside a dynamic product offering, bringing new flavours to market every month. Delivering value through our range of exciting flavours has been the cornerstone of our brand, but we have let product variety increase far too much. Currently we produce over 100 different varieties of soup in our factory. I couldn't tell you what they all are, or even if our customers like them. It's an expensive and inefficient way to run a business and we can't produce that level of variety without increasing our prices. The other issue we have is with our sales processes; they verge on the chaotic. Our forecasting, sales information, customer relationship and product inventory processes all need critically examining. In my previous company, the operations people used an approach called Six Sigma to improve their processes. It seemed to be very successful. Certainly, they improved their responsiveness as well as reducing costs'.*

Factory flexibility

Benit, the company's Operations Manager, had expressed strong support for Meghana's views. *'I have been complaining for the last few years about our product proliferation and the frequent introductions of new flavours. Of course, innovation is important to the company, and we have to keep up with food trends, but the complexity of constantly switching between different product lines is further complicated by the need to meet standards of the gluten free, halal, vegetarian and vegan varieties that the company promises to deliver. Avoiding contamination is a persistent challenge for the factory. Recently a small chicken bone was found by a consumer inside a carton of carrot and coriander soup. They (understandably) complained to the supermarket where they bought it. I was involved in the angry exchange with the supermarket, who referred to the incident as a "never event". It was embarrassing for me personally and damaging to our reputation. The other thing I agree with is Meghana's criticism of the chaos caused by promotions. It certainly doesn't make for a stable planning environment'.*

Benit also supported Meghana's view that an improvement initiative was needed but was less enthusiastic about adopting Six Sigma principles. *'I'm sceptical about all these fads. I've heard of Six Sigma of course and I don't have anything against it, but it's not a magic bullet. It won't solve all our problems overnight. We must get some structural problems fixed first. The variety we have to cope with, and the sales turbulence, are complicated by our strange reporting structure'.* Benit was referring here to the quality function of the company reporting directly to the chief executive officer (CEO), rather than to Benit himself. The reporting structure had been prescribed some years previously, when the then

From: CEO's office

To: Senior Management Team

Subject: Soup business

Hi All,

As you will have seen from our 2nd quarter results, to say that the Soup business needs reviewing is something of an understatement.

It seems obvious that we have to quickly come up with some quick ideas on how we can recover both market share and profitability.

I am firmly of the belief that we have to think differently, think smarter and most of all think radically, if we are going to meet this year's budget commitment. We cannot just rely on the other product categories to support the Soup business as we did last financial year. It is clear that, without some drastic action we will again be in a LOSS-MAKING situation in the business that started this firm. As things stand, we cannot carry on supporting Soup with all its overheads and the level of trade and marketing support it needs.

I would like you all to consider what we can do to turn things around in the next few weeks so we can review ideas at the next monthly meeting. There may even be actions that you can take prior to the meeting.

I know that you are all passionate about the brand, of which the Soup business is an integral part.

Gem – As you know I'm on annual leave for the next two weeks. Can I suggest that you call an emergency meeting in my absence to start the discussion.

Sabine

Figure 15.16 Sabin Roche's email to her senior management team

CEO of the company had introduced a total quality management (TQM) initiative. Reporting directly to the CEO was supposed to indicate the importance of quality to the company. However, Benit believed it was now time for a change. *'None of the executives who were involved in deciding the structure are still with the company. Certainly, the quality function has lost interest in TQM as a broad set of improvement principles. They see themselves as inspectors who simply check incoming raw materials and outgoing finished product. Basically, they are "box tickers" rather than innovators. And they aren't particularly good at that. After the "chicken bone" incident, we lost that supermarket's business. I think we could have recovered, but the quality control team showed a lack of attention to the customer's complaints. They simply brushed aside the customer's comments, regarding them as "exaggerated" since it was merely human error that had produced the one small mistake'.*

View from the top

Sabine Roche had been appointed as CEO of Splendid Soup Co. a year earlier after a successful career in confectionery marketing. She was fully aware of the need to reconcile the need for sales growth with the problems it was causing for operations in the factory. Both Meghana and Benit had met with her to explain their views. *'I understand the frustration within the senior management team, but I definitely do not want to impose my views on how to improve the situation. Neither am I keen on importing some idea from outside to find a solution. For improvement to happen we must generate our own solutions and we must do it together. I can't be at the next monthly senior team meeting, which is probably a good thing* [see Sabine's email to the senior team in Figure 15.16]. *It will allow the team to come up with ideas of their own based on their experience'.*

Despite Sabine's hopes, the next meeting did not reach any degree of consensus. Meghana was particularly disappointed. *'I remember the meeting as being very frustrating. I had the marketing team there, and the sales team and the head of strategy. I wanted them to understand my view that promotions cannibalise sales. I put together a convincing PowerPoint presentation encompassing actual sales and cost of sales data. I presented the data in 'at-a glance' tables and graphs to clearly and visually demonstrate my point that these promotions cost a lot of money and do more harm than good! Everybody liked it, heads were nodding in agreement around the table. Five minutes later, the head of strategy announced his big idea: to abandon local suppliers and import chicken from Brazil, cutting costs of producing our most popular chicken soup. How does cheap chicken from Brazil help us build market share? They had entirely missed the point. I still feel that there really has to be a better way to increase market share, halt declining sales and improve our performance generally'.*

QUESTIONS

1 **Summarise what you consider to be the problems facing Splendid Soup Co. Which of these would you consider strategic issues and which operational issues?**

2 **How might the principles, methods and tools associated with Six Sigma, lean, BPR and/or TQM help this company to think differently about keeping sales buoyant and increasing market share?**

3 **Is Benit correct in viewing Six Sigma as simply a 'management fad'?**

4 **If you were Sabine, how would you try to promote improvement and a return to profitability of the company's soup brands?**

Problems and applications

All chapters have 'Problems and applications' questions that will help you practise analysing operations. They can be answered by reading the chapter. Model answers for the first two questions can be found on the companion website for this text.

1 Sophie was sick of her daily commute. '*Why*', she thought, '*should I have to spend so much time in the morning stuck in traffic listening to some babbling halfwit on the radio? We can work flexi-time, after all. Perhaps I should leave the apartment at some other time?*' So resolved, Sophie deliberately varied her time of departure from her usual 8.30 am. Also, being an organised soul, she recorded her time of departure and her journey time each day. Her records are shown in Table 15.1.

   (a) Draw a scatter diagram that will help Sophie decide on the best time to leave her apartment.

   (b) How much time per (5-day) week should she expect to be saved from having to listen to a babbling halfwit?

2 '*Everything we do can be broken down into a process*', said Lucile, COO of an outsourcing business for the 'back-office' functions of a range of companies. '*It maybe more straightforward in a manufacturing business, but the concept of process improvement is just as powerful in service operations. Using this approach our team of Black Belts has achieved 30 per cent productivity improvements in six months. I think Six Sigma is powerful because it is the process of comparing process outputs against customer requirements. To get processes operating at less than 3.4 defects per million opportunities means that you must strive to get closer to perfection and it is the customer that defines the goal. Measuring defects per opportunity means that you can actually compare the process of, say, a human resources process with a billing and collection process*'.

Table 15.1 Sophie's journey times (in minutes)

Day	Leaving time	Journey time	Day	Leaving time	Journey time	Day	Leaving time	Journey time
1	7.15	19	6	8.45	40	11	8.35	46
2	8.15	40	7	8.55	32	12	8.40	45
3	7.30	25	8	7.55	31	13	8.20	47
4	7.20	19	9	7.40	22	14	8.00	34
5	8.40	46	10	8.30	49	15	7.45	27

(a) What are the benefits of being able to compare the number of defects in a human resources process with those of collection or billing?
(b) Why is achieving defects of less than 3.4 per million opportunities seen as important by Lucile?
(c) What do you think are the benefits and problems of training Black Belts and taking them off their present job to run the improvement projects rather than the project being run by a member of the team that has responsibility for actually operating the process?

3 Develop cause–effect diagrams for the following types of problems:
- Staff waiting too long for their calls to be answered by their IT helpdesk.
- Poor food in the company restaurant.
- Poor lecturing from teaching staff at a university.
- Customer complaints that the free plastic toy in their breakfast cereal packet is missing.
- Staff having to wait excessively long periods to gain access to the coffee machine.

4 For over 10 years, a hotel group had been developing self-managed improvement groups within its hotels. At one hotel reception desk, staff were concerned about the amount of time the reception desk was left unattended. To investigate this, the staff began keeping track of the reasons they were spending time away from the desk and how long each absence kept them away. Everyone knew that reception desk staff often had to leave their post to help or give service to a guest. However, no one could agree what was the main cause of absence. Collecting the information was itself not easy because the staff had to keep records without affecting customer service. After three months, the data shown in Table 15.2 was presented in the form of a Pareto curve. It came as a surprise to reception staff and hotel management that making photocopies for guests was the main reason for absence. Fortunately, this was easily remedied by moving the photocopier to a room adjacent to the reception area, enabling staff to keep a check on the reception desk while they were making copies.

(a) Do you think it was wise to spend so much time on examining this particular issue? Isn't it a trivial issue?
(b) Should this information be used to reflect improvement priorities? In other words, was the group correct to give priority to avoiding absence through photocopying, and should its next priority be to look at the time checking files in the back office?

Table 15.2 Reasons for staff time away from the reception desk

Reason for being away from reception desk	Total number of minutes away
Checking files in back office	150
Providing glasses for night drinks	120
Providing extra key cards	90
Providing medication	20
Providing extra stationery	70
Providing misc. items to rooms	65
Providing night drinks	40
Making photocopies	300
Carrying messages to meeting rooms	125
Locking and unlocking meeting rooms	80
Providing extra linen	100

5 A transport services company provided a whole range of services to railway operators. Its reputation for quality was a valuable asset in its increasingly competitive market. '*We are continually looking for innovation in the way we deliver our services because the continuous improvement of our processes is the only way to make our company more efficient*', said the company's CEO. '*We use a defined set of criteria to identify critical processes, each of which is allocated a "process owner" by our quality steering committee. This is helped by the company's "process excellence index" (PEI), which is an indicator of the way a process performs, particularly how it is designed, controlled and improved. The EPI score, which is expressed on a scale of 1 to 100, is calculated by the process owner and registered with the quality department. With this one figure we can measure the cost, reliability and quality of each process so that we can compare performance. If you don't measure, you can't improve. And if you don't measure in the correct way, how can you know where you are? Employee recognition is also important. Our suggestion scheme is designed to encourage staff to submit ideas that are evaluated and rated. No individual suggestion is finally evaluated until it has been fully implemented. Where a team of employees puts ideas forward, the score is divided between them, either equally or according to the wishes of the team itself. These employee policies are supported by the company's training schemes, many of which are designed to ensure all employees are customer-focused*'.

(a) What seem to be the key elements in this company's approach to improvement?
(b) Do you think this approach is appropriate for all operations?

6 Step 1 – As a group, identify a 'high-visibility' operation that you all are familiar with. This could be a type of quick-service restaurant, record stores, public transport systems, libraries, etc.

Step 2 – Once you have identified the broad class of operation, visit a number of them and use your experience as customers to identify:

(a) the main performance factors that are of importance to you as customers; and
(b) how each store rates against the others in terms of their performance on these same factors.

Step 3 – Draw an importance–performance matrix (see Chapter 3) for one of the operations that indicates the priority they should be giving to improving their performance.

Step 4 – Discuss the ways in which such an operation might improve its performance and try to discuss your findings with the staff of the operation.

7 There is an old saying that goes, 'How do you learn to charm a snake? Answer – start with a slow, non-venomous snake. Then a slow, venomous snake. Then try a fast, non-venomous snake. Then a fast, venomous snake'. How could this be applied to operations improvement?

8 The idea of 'organisational ambidexterity', described earlier sounds attractive. What do you think are the barriers to achieving it, and what could be done to overcome them?

9 '*This coffee is terrible*', said Anita, the catering manager. '*We have to look into it before we get even more complaints*'. How might a cause–effect diagram be used to investigate the causes of the coffee being so bad? Draw a diagram that would help Anita.

10 Benchmarking can be used to learn from other types of operation. What could, for example, a supermarket learn, by studying an airport?

Selected further reading

Ahlstrom, J. (2015) ***How to Succeed with Continuous Improvement: A Primer for Becoming the Best in the World*****, McGraw-Hill Professional, New York, NY.**

This is very much a practical guide. Slightly evangelical, but it gets the message over.

Cunningham, P. (2020) ***BASICS: Be Always Sure Inputs Create Success: 12 Lean Six Sigma Tools and Techniques to Reduce the Cost of Quality from the Coal Face Out*****, Routledge, Abdingdon.**

Very much a 'how to do it' book for practitioners. Simple explanations of the key ideas, if somewhat manufacturing biased.

Goldratt, E.M. and Cox, J. (2014), *The Goal: A Process of Ongoing Improvement*, North River Press, Great Barrington, MA.
Updated version of a classic. A graphic novel version is also available.

Hindo, B. (2007) At 3M, a struggle between efficiency and creativity: how CEO George Buckley is managing the yin and yang of discipline and imagination, *Businessweek*, 11 June.
Readable article from the popular business press.

Zu, X., Fredendall, L.D. and Douglas, T.J. (2008) The evolving theory of quality management: the role of Six Sigma, *Journal of Operations Management*, 26 (5), 630–650.
As it says. . .

Notes on chapter

1. Carroll, L. (1871) *Through the Looking Glass*, Penguin Classics, 2008.
2. Imai, M. (1986) *Kaizen: The Key to Japan's Competitive Success*, McGraw-Hill, New York, NY.
3. The information on which this example is based is taken from: Onetto, M. (2014) When Toyota met e-commerce: lean at Amazon, *McKinsey Quarterly*, No 2, 1 February.
4. We are grateful to Sue Jenkins, Director of Kaizen, Sussex Healthcare NHS Trust, for this example.
5. For a full explanation, see: Gawande, A. (2010) *The Checklist Manifesto: How to Get Things Right*, Profile Books, London; Aaronovitch, D. (2010) The Checklist Manifesto: review, *The Times* 23 January.
6. Hammer, M. (1990) Reengineering work: don't automate, obliterate, *Harvard Business Review*, July–August, https://hbr.org/1990/07/reengineering-work-dont-automate-obliterate (accessed September 2021).
7. Davenport, T. (1995) The fad that forgot people, Fast Company, 31 November.
8. Note: These defects per million (DPM) figures assume that the mean and/or SD may vary over the long term so the 3 sigma DPM is actually based on 1.5 sigma and 6 sigma on 4.5 sigma. These distributions are assumed to be 'one-tailed', as the shift is usually one direction.
9. For more information on Wipro's Six Sigma work, see: Harvin, H. (2020) Six Sigma Training & Implementation at Wipro, Henry Harvin.com, blog, 2 March, https://www.henryharvin.com/blog/six-sigma-training-implementation-at-wipro/ (accessed September 2021); Sharma, M., Pandla, K. and Gupta, P. (2014) A case study on Six Sigma at Wipro Technologies: thrust on quality, Working Paper, The Jaipuria Institute of Management; Wipro website, https://www.wipro.com (accessed September 2021).
10. There are many books and publications that explain the benefits of combining lean and Six Sigma. For example, see Byrne, G., Lubowe, D. and Blitz, A. (2007) *Driving Operational Innovation using Lean Six Sigma*, IBM Institute for Business Value; Brue, G. (2005) *Six Sigma for Managers: 24 Lessons to Understand and Apply Six Sigma Principles in Any Organization*, McGraw-Hill Professional Education Series, New York, NY.
11. Shenkar, O. (2010) *Copycats: How Smart Companies Use Imitation to Gain a Strategic Edge*, Harvard Business Press, Boston, MA.
12. The information on which this example is based is taken from: The Economist (2016) The great escape: what other makers can learn from the revival of Triumph motorcycles, *Economist* print edition, 23 January.
13. The information on which this example is based is taken from: Sample, I. (2020) F1 team helps build new UK breathing aid for Covid-19 patients, *Guardian*, 30 March; West, K. (2011) Formula One trains van drivers, *The Sunday Times*, 1 May; Williams F1 Team (2010) Williams in collaboration with Sainsbury's, f1network.net, 12 November, http://www.f1network.net/main/s107/st164086.htm (accessed September 2021).
14. The information on which this example is based is taken from: Schlumberger Press Release (2010) Schlumberger Cited for Knowledge Management, Schlumberger Press Office, 3 December; Deltour, F., Plé, L. and Sargis-Roussel, C. (2013) Eureka! Developing online communities of practice to facilitate knowledge sharing at Schlumberger, IESEG School of Management, LEM, case study 313-122-1.

16 Lean operations

KEY QUESTIONS

16.1 What is lean?

16.2 How does lean consider flow?

16.3 How does lean consider (and reduce) waste?

16.4 How does lean consider improvement?

16.5 How does lean consider the role of people?

16.6 How does lean apply throughout the supply network?

INTRODUCTION

The focus of lean operations is to achieve a smooth flow of materials, information or customers that delivers exactly what customers value (perfect quality), in exact quantities (neither too much nor too little), exactly when needed (not too early nor too late), exactly where required (in the right location) and at the lowest possible cost. Originally developed in manufacturing operations, the idea of lean operations (with its antecedents and derivatives, such as just-in-time and stockless production) initially focused on planning and controlling operations so that they could deliver products and/or services without 'waste'. Its objectives were to create value to the customer by delivering defect-free products and services in the shortest time possible and at the lowest cost to the organisation. Over the years, the concept of lean was increasingly applied in service organisations. Just as important, its emphasis shifted more towards a holistic view of lean as an approach to operations and process improvement. While its role in the planning and control activity remains central to achieving smooth, synchronised flow (without waste), so is its pursuit of achieving perfect quality, in the right quantity, the right place and the right time. For this reason, in practice lean is seen predominantly as an improvement approach. Figure 16.1 places lean in the overall model of operations management.

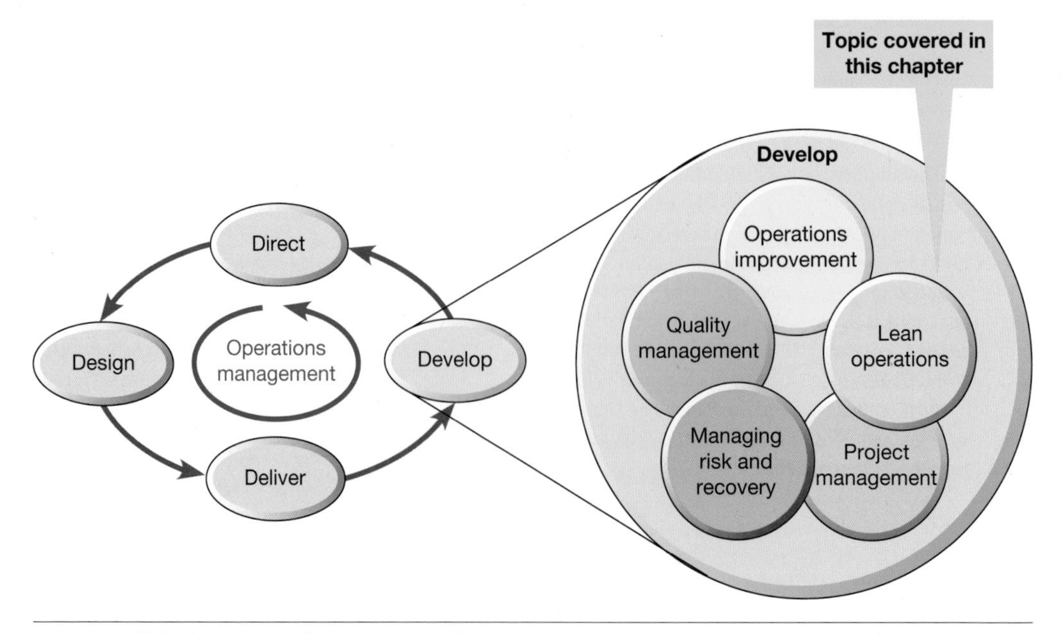

Figure 16.1 This chapter examines lean operations

16.1 What is lean?

There are several approaches one can take to defining 'lean'. The most common, and the approach we took when describing lean in the introduction to this chapter, is to define it by describing what it is trying to achieve. Lean attempts to deliver exactly what customers value (perfect quality), in exact quantities (neither too much nor too little), exactly when needed (not too early nor too late), exactly where required (in the right location) and at the lowest possible cost. But be careful of definitions of lean. They are many and varied, but few are capable of succinctly or totally capturing its essence. Some definitions are too narrow, because they emphasise the tools and techniques used in lean operations, or the individual elements that constitute lean. Others can be too broad or ambiguous in that they fail to provide specific guidance as to how to achieve the objectives listed in our definition. The alternative approach is to identify the role lean plays within the whole range of operations management activities.

Lean is a philosophy, an approach to planning and control, and a set of improvement ideas

Lean can be seen as having three related, but distinct, roles, each of which take a different perspective:

- Lean is a philosophy that places customer value at the heart operations.
- Lean is an approach to planning and controlling flow in operations.
- Lean is a set of ideas to improve operations performance.

Figure 16.2 illustrates this idea.

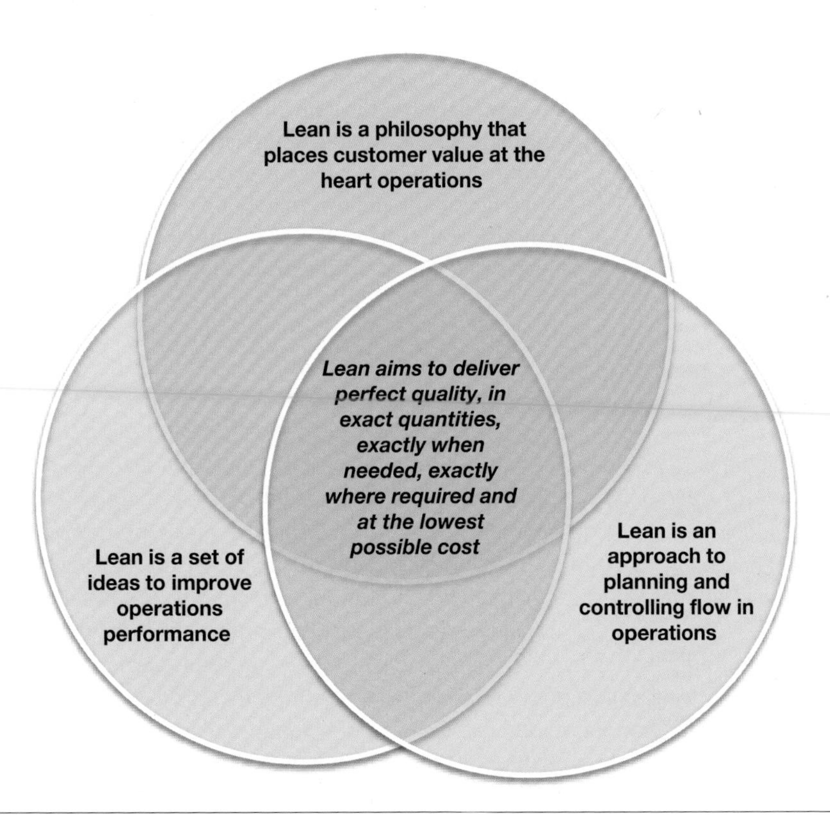

Figure 16.2 Lean can be viewed as having three overlapping roles – lean is a philosophy that places the customer at the heart of operations, lean is an approach to planning and controlling flow in operations, and lean is a set of ideas on how to improve operations performance

Lean is a philosophy

In this chapter we will explain most of the many elements that go together to constitute lean. All these elements are important, but one in particular binds the other elements together. That is that lean is a philosophy that places customer value at the heart of operations. By referring to it as a *philosophy* we mean a set of coherent ideas that form an overarching rationale for how operations should be managed. As we describe the various elements of lean, note how they support each other. They are far more that a disparate set of prescriptions. They form a logical and consistent approach to managing operations generally. Indeed, many of these elements are widely regarded as 'best practice' for operations management. The central objective of the lean philosophy is to gear all activities to adding value for the customer. This leads to the reoccurring theme of lean attempting to eliminate all non-essential, non-value-adding activities.

> **Operations principle**
> Lean is a set of coherent ideas that stress delivery of value from the customers' perspective, which forms an overarching rationale for how operations should be managed.

Lean is an approach to planning and controlling flow

With its roots in manufacturing, lean was originally seen as a method of controlling the movement of transformed resources within an operation to achieve economies of 'flow'. It was, in effect, an approach to synchronising flow through the planning and control of activities within the operation. Then, usually called 'just-in-time', its emphasis was on smooth flow and, through that, the reduction of between-process inventories. We explain this aspect later, when we look at how lean considers flow.

Lean is a set of improvement ideas

The 'engine room' of a lean operation is a collection of improvement tools and techniques that are the means for identifying and eliminating waste (any activity that does not add value from the customers' perspective). The growing popularity of lean, with continuous improvement as one of its core principles and ways of working, helped to shift the focus of operations management generally towards viewing improvement as its main purpose. Moreover, the emphasis that lean places on the contribution of 'people' (meaning the staff who perform the value-adding activities) has also reinforced the degree to which lean is seen as primarily an improvement vehicle. Lean relies on the empowerment of people, skilled in improvement methods, to use their collective knowledge to improve processes in ways that enhance value from the perspective of the customer.

> **Operations principle**
> Improvement tools and techniques associated with lean will not deliver improvement without the engagement of people.

Critical commentary

Some lean proponents would argue that the concepts of lean have a unique influence over the whole subject of operations management. Lean principles have something to tell us about everything in the subject from quality management to inventory management, from job design to product design, not forgetting of course, process improvement. So, why does this text, like most, separate these ideas into a separate chapter? It is an artificial segmentation. Admittedly, it helps to explain lean ideas and knowledge as clearly as possible yet segmenting in this way inevitably means imposing artificial boundaries between various topics. There are some particularly evangelical proponents of the lean philosophy who claim that the underlying ideas of lean have now comprehensively replaced more 'traditional' views of operations management. Rather, they say, lean principles should be the foundation for the whole of operations and process management. And they are right, of course. Nevertheless, the ideas behind lean are both counterintuitive enough and important enough to warrant separate treatment.

The evolution of lean

Although just-in-time principles had been used in many manufacturing operations for years, the term 'lean' was first popularised by John Krafcik to mean 'less of everything'. Krafcik was a researcher on the MIT programme that led to the 1990 publication of the book '*The Machine that Changed the World*'.[1] The book examined the working practices at Toyota Motor Company to understand how the company radically improved its (then) poor reputation for quality. Narrowly avoiding bankruptcy in the 1950s, it became one of the most successful car manufacturers in the world (see the 'Operations in practice' example). Toyota's experience of 'just-in-time' had enabled it to produce products only once a customer had placed an order, thereby producing only what was needed, when it was needed, in the amount that it was needed; an approach that was radically different to the mass production approach of other car manufacturers at the time. The book did much to popularise the idea of lean, and appealed even to those who were not car enthusiasts. As a result, organisations in almost every sector have sought to learn from Toyota's methods in a bid to deliver higher-quality products and services that use less resource yet deliver more value to the customer.

OPERATIONS IN PRACTICE

Toyota: the lean pioneer[2]

Seen as the leading practitioner and the main originator of the lean approach, the Toyota Motor Company has progressively synchronised all its processes simultaneously to give high-quality, fast throughput and exceptional productivity. It has done this by developing a set of practices that has largely shaped what we now call 'lean' but which Toyota calls the Toyota Production System (TPS). Progressively, Toyota has expanded the application of lean principles from its manufacturing into purchasing, finance, logistics and its dealership network.

Arguably, Toyota's strength lies in understanding the differences between the tools used within its operations and the overall philosophy of its production system: continuous improvement. Activities and processes are constantly being challenged and pushed to a higher level of performance, enabling the company to continually innovate and improve. Improvement ideas are ideally simple and low cost, allowing for a high degree of trial and error. While some adopters of lean principles may think they have 'done lean', Toyota simply changes the goal to constantly challenge improvement. As such, we see a key distinction between those that see lean as a specific endpoint to be achieved through the application of a series of improvement tools, and those such as Toyota who treat lean as a philosophy that defines a way of conducting business. This perspective may also explain Toyota's heavy investment in its mentorship programme to support those within its supply network in successfully embedding lean principles. Taiichi Ohno,[3] a chief engineer at Toyota Motor Company, and widely credited with the development of the Toyota Production System, apparently had a saying: 'No problem is a problem'. This sums up the philosophy of lean, that we should always look for ways to improve. Better never stops.

16.2 How does lean consider flow?

Central to the lean philosophy is the pursuit of flow. Here 'flow' means the way transformed resources move between value-added activities without waste. To achieve this, lean operations manage flow in a very different way to traditional approaches of production.

Improving flow through using pull control

The best way to understand how pull control, used in the lean approach, differs to a traditional approach to flow is to contrast the two simple processes in Figure 16.3. The traditional approach assumes that each stage in the process will place its output in an inventory that 'buffers' that stage from the next one downstream in the process. The next stage down will then (eventually) take outputs from the inventory, process them, and pass them through to the next buffer inventory. This is known as a 'push' approach to production where items are processed regardless of whether there is demand. These buffers of inventory 'insulate' each stage from its neighbours, making each stage relatively independent so that if, for example, stage A stops operating for some reason, stage B can continue, at least for a time. The larger the buffer inventory, the greater the degree of insulation between the stages. This insulation has to be paid for in terms of inventory (in production operations) or queues (in service operations), thereby slowing throughput times because products, customers or information will spend time waiting between stages in the process.

The main argument against this traditional approach lies in the very condition it seeks to promote, namely the insulation of the stages from one another. When a problem occurs at one stage, the problem will not immediately be apparent elsewhere in the system. The responsibility for solving the problem will be centred largely on the people within that stage, and the consequences of the problem will hopefully be prevented from spreading to the whole system. However, contrast this with the lean process illustrated in Figure 16.3(b). Here, products, customers or information are processed and then passed directly to the next stage 'just-in-time' for them to be processed further. The aim is to match demand perfectly – neither too much nor too little, only when it is needed. Problems at any stage have a very different effect in such a system. Now if stage A stops processing, stage B will notice immediately and stage C very soon after. Stage A's problem is now quickly exposed to the whole process, which is immediately affected by the problem. This means that the responsibility for solving the problem is no longer confined to the staff at stage A, it is now shared by everyone and it is now too important to be ignored. In other words, by preventing items accumulating between stages, the operation has increased the chances of the intrinsic efficiency of the plant being improved. Traditional approaches seek to encourage efficiency by using inventory

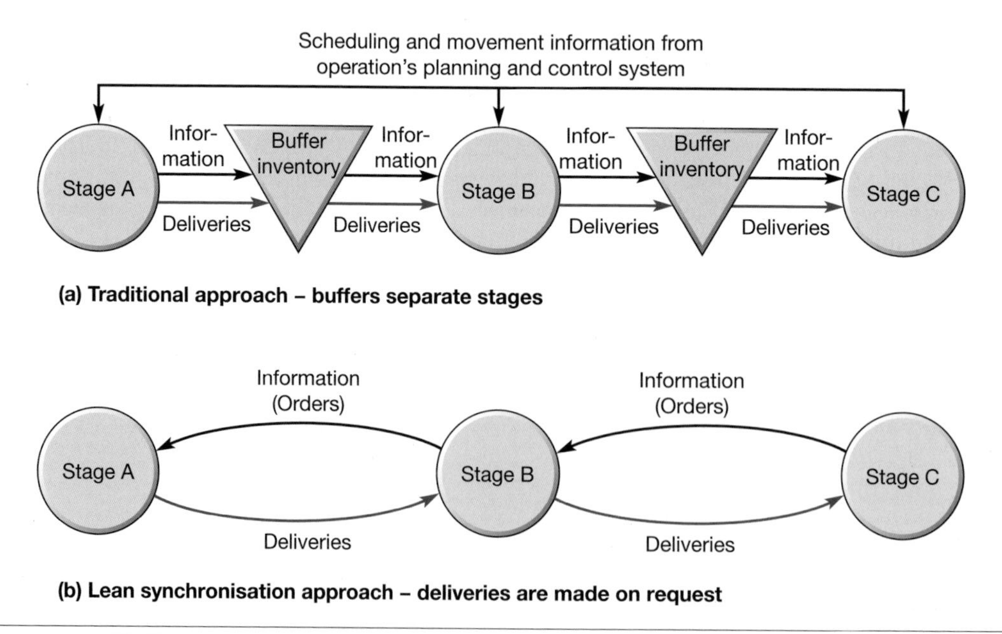

Figure 16.3 (a) Traditional flow versus (b) lean flow between stages

to protect each part of the process from disruption. The lean approach takes the opposite view. Exposure of the system to problems can both make them more evident and change the 'motivation structure' of the whole system towards solving the problems as soon as possible.

Operations principle

Eliminating or reducing buffer inventories exposes operational problems to the whole system, creating a sense of urgency (motivation) to resolve problems as soon as they occur.

Using kanbans to support pull control

The arrangement shown in Figure 16.3(b) is what we described in Chapter 10 as 'pull control'. We contrasted pull control with the more traditional approach used in mass production, where products are 'pushed' into an inventory to buffer against fluctuations in demand. By contrast, pull control matches supply to demand wherever possible without the need for inventory. Consider for example how some fast-food restaurants cook and assemble food and place it in the warm area only when the customer-facing server has sold an item. Production is being triggered only by real customer demand.

The most common method to support pull control in lean operations is the use of '**kanbans**'. Kanbans are simple signalling devices that prevent the accumulation of materials, customer and information inventories. The word 'kanban' is the Japanese for card or signal. It is sometimes called the 'invisible conveyor' that controls the transfer of items between the stages of an operation. In its simplest form, it is a card used by a customer stage to instruct its supplier stage to send more items. In some companies, kanbans remain physical – solid plastic markers or even coloured ping-pong balls; in others, electronic point-of-sales (EPOS) systems generate digital kanbans for further 'production' (delivery of stock). Whichever kind of kanban is being used, the principle is always the same: the receipt of a kanban triggers the movement, production or supply of one unit of product or one standard amount of service activity. If two kanbans are received, two units of work are 'produced', and so on. Some companies use 'kanban squares' – marked spaces on the shop floor to fit one or more work pieces or containers. Only the existence of an empty square triggers production at the stage that supplies the square. Theoretically, kanbans exist because of imperfect information about demand. Ideally, the number of kanbans should be reduced over time as process improvements reduce waste and enhance flow.

OPERATIONS IN PRACTICE — A very simple kanban

In those regions where milk is home-delivered, placing an empty milk bottle on a residential doorstep signals to the delivery person that more milk is needed. The photograph here illustrates the use of a very simple kanban to signal the reorder of essential supplies in a service operation. It shows the inside of a store cupboard in the blood science department of a hospital. The cupboard stores (among other things) caps for blood vials (shown in the picture). What is not shown is that there are no doors on any of the cupboards in this department. During an improvement exercise staff decided to remove the cupboard doors to improve the visibility of the state of essential supplies 'at a glance'. The removal of the cupboard doors also reduced 'waste' arising from the activity of having to open cupboard doors to monitor supplies. (Although opening cupboard doors is a tiresome endeavour, you might not want to do this at home.)

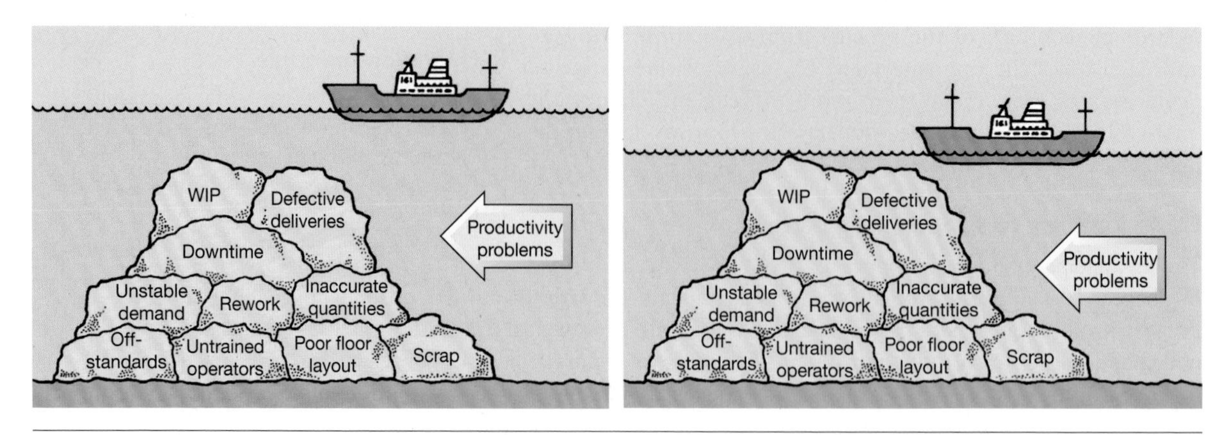

Figure 16.4 Reducing the level of materials, customers or information inventories (the water) allows operations management (the ship) to see the problems in the operation (the rocks) and work to reduce them

Improving flow through reducing inventory

Lean sees accumulations of inventory (both finished products and work-in-progress), as a 'blanket of obscurity' that lies over the system and prevents problems being noticed. The obscuring effects of inventory, be they product, customer or information inventories, are often illustrated diagrammatically, as in Figure 16.4. The many problems of the operation are shown as rocks in a riverbed that cannot be seen because of the depth of the water. The water in this analogy represents the inventory (materials, customers or information) in the operation. Yet even though the rocks cannot be seen, they slow the progress of the river's flow and cause turbulence. Gradually reducing the depth of the water (inventory) exposes the worst of the problems that can be resolved, after which the water is lowered further, exposing more problems, and so on. The same argument also applies to the flow between whole processes, or whole operations. For example, stages A, B and C in Figure 16.3 could be a supplier operation, a manufacturer and a customer's operation, respectively.

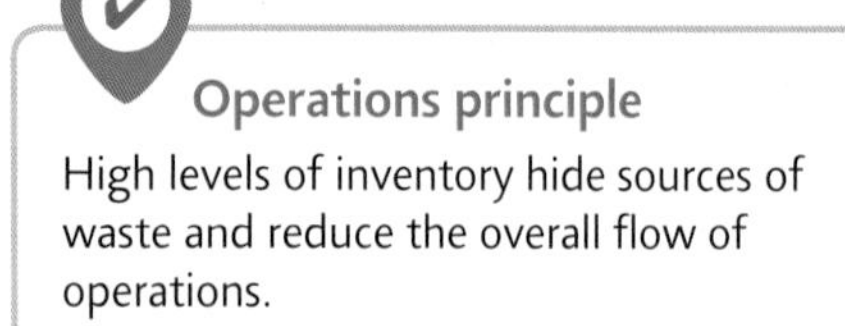

Operations principle

High levels of inventory hide sources of waste and reduce the overall flow of operations.

OPERATIONS IN PRACTICE

The rise of the personal kanban

Kanban principles of reducing work in progress apply as much to 'knowledge' work as they do to operations transforming physical items. Their goal is to reduce work in progress to create the 'mental space' to focus on the important tasks and work quickly and flexibly towards a desired outcome. Excess work in progress often results in getting caught up in a daily round of reactive problem-solving, responding to emails and endless meetings, obscuring problems that affect our ability to complete more value-adding work. Using a kanban whiteboard is becoming an increasingly popular tool that clarifies information on 'stopping starting and starting finishing' activities. For example, a simple whiteboard (or similar) can be divided into three simple sections (see Figure 16.5):

1 To do (activities not yet started, but will need to be started).
2 Doing (activities being worked on right now).
3 Done (activities that have been completed).

Using Post-it notes (or a whiteboard marker pen) one can move tasks from one column to the next. This process can enhance personal productivity since the visualisation of work helps the user to schedule tasks according to priority, focus on fewer tasks, and avoid beginning new

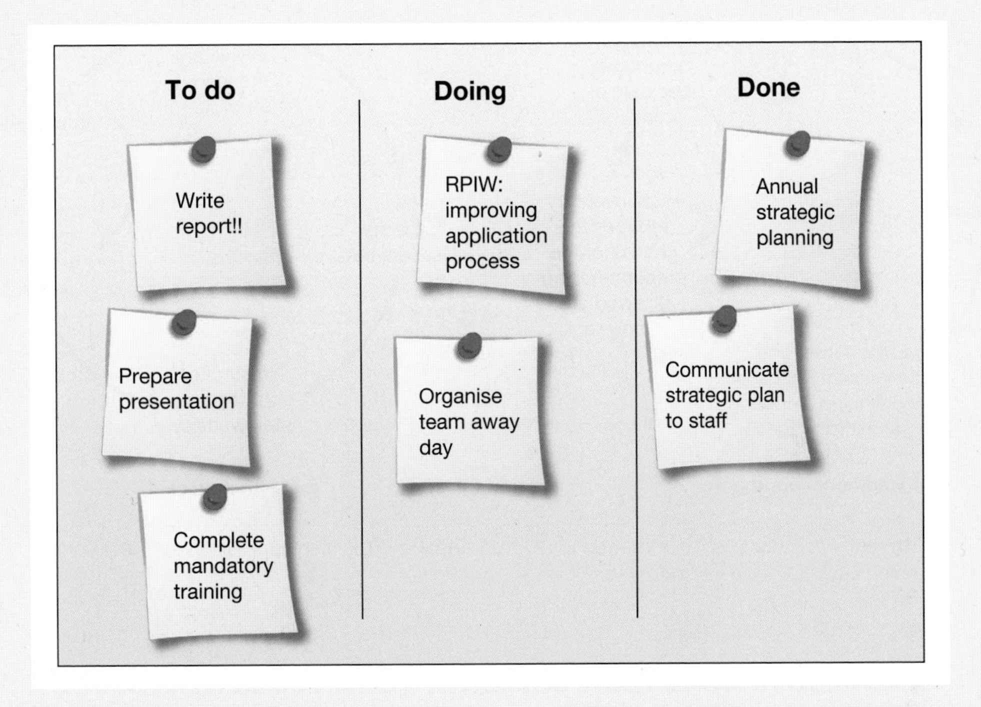

Figure 16.5 A personal Kanban can enhance individual productivity simply by classifying jobs as 'to do', 'doing' and 'done'

items from the 'to-do' column until others have been completed. Given that lean operations emphasise smooth throughput, achieved partly by a reduction in in-process inventory, it is generally regarded as sensible to keep the number of items under 'doing' (in-process inventory) to as few as possible.

Improving flow by decreasing capacity utilisation

Return to the process shown in Figure 16.3. When stoppages occur in the traditional system, the buffers allow each stage to continue working and thus achieve high-capacity utilisation. High-capacity utilisation is central to mass-production approaches, where low unit costs are achieved through large-scale production known as 'economies of scale'. Often extra 'production' goes into buffer inventories or queues of customers. By contrast, with little or no inventory to buffer against stoppages, a production problem becomes immediately apparent, affecting the whole process. This will necessarily lead to lower capacity utilisation, at least in the short term. In organisations that place a high value on the utilisation of capacity this can prove particularly difficult to accept. However, there is no point in producing output just for its own sake. In fact, producing just to keep utilisation high is not only pointless, it is counterproductive, because the extra inventory produced merely serves to make improvements less likely. Further, lean also stresses a reduction in all types of process variability via continuous improvement efforts. As gains from continuous process improvement efforts begin to surface, the improvement path moves towards the point where throughput time is short and capacity utilisation high. It manages to do this because of the reduction in process variability. See Figure 16.6 for the logic of how lean regards capacity utilisation.

Operations principle

Focusing on lean can reduce capacity utilisation in the short term. Continuous improvement efforts should reduce the quantity of defects and reduce process variability over time, which should ultimately return the operation to high levels of capacity utilisation, enhancing value to the customer at a lower production cost.

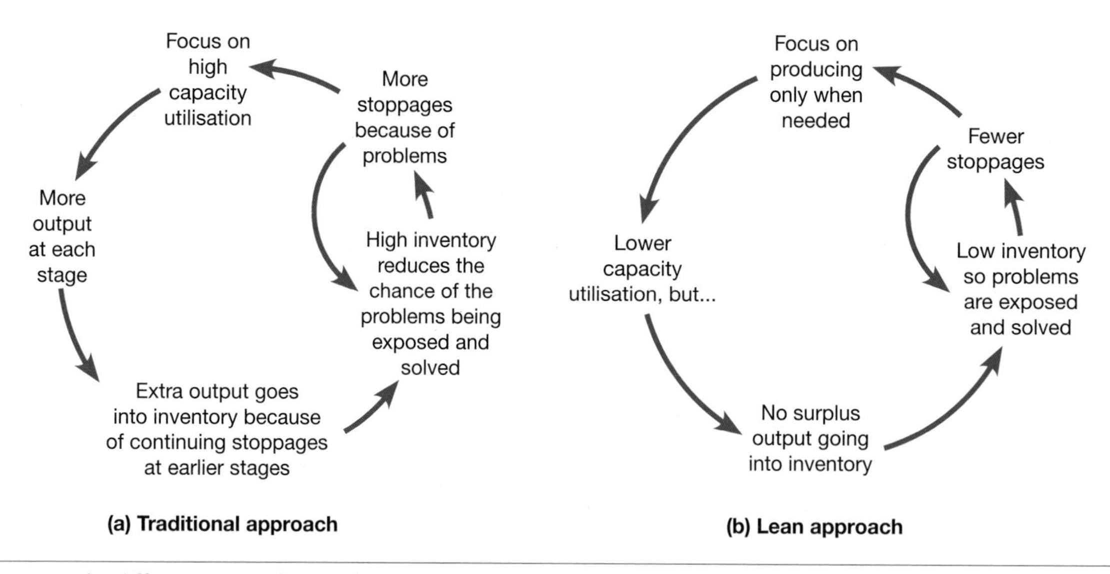

Figure 16.6 The different views of capacity utilisation in (a) traditional and (b) lean approaches to operations

16.3 How does lean consider (and reduce) waste?

Arguably the most significant part of the lean operations is its focus on the elimination of all forms of waste. Waste can be defined as any activity that does not add value from the perspective of the customer and is not therefore an activity that the customer would pay for or benefit from. Hence, what constitutes waste depends on the customer's perception of value. Within many processes, as little as 5 per cent of total throughput time is spent directly adding value. This means that for 95 per cent of its time, an operation is adding cost to the service or product that is not directly delivering value from the perspective of the customer. Such calculations can alert even relatively efficient operations to the enormous waste that lies dormant within all operations. Relatively simple requests, such as applying for a driving licence, may take only a few minutes to actually process, yet take days (or weeks) to be returned.

Causes of waste – muda, mura, muri

The terms muda, mura and muri are Japanese words conveying three causes of waste that should be reduced or eliminated.

Muda

Muda refers to activities in a process that are wasteful because they do not add value to the operation or the customer (see 'Types of waste' below). The main causes of these non-value-adding activities are likely to be poorly communicated objectives (including not understanding the customer's requirements), or the inefficient use of resources, and failing to adopt a systematic approach to the continuous reduction of waste. The implication of this is that, for an activity to be effective, it must be properly recorded and communicated to whoever is performing it. Standard work and visual management are effective tools for communicating objectives and ensuring efficient use of resources.

Mura

Mura means 'lack of consistency' or unevenness that is often caused by variations in customer demand or variations in the way that a process is executed. Mura often leads to the overburdening of some resources while others remain idle or wait. Customer needs can vary, in terms of what they want, how much they want and when they want it. However, processes usually find it more convenient to change what they do relatively infrequently, because every change implies some kind of cost. That is why hospitals schedule specialist clinics only at particular times, and why machines often make a batch of similar products together. Yet responding to customer demands exactly and instantaneously requires a high degree of process flexibility. Again, standard work is central to reducing variations in the way a process is executed while set-up reduction and level scheduling can smooth the production process and allow a close matching of capacity to demand.

Symptoms of inadequate process flexibility include the following:

- ***Large batches*** – sending batches of materials, customers, or information through a process inevitably increases inventory as the batch moves through the whole process.
- ***Delays between activities*** – the longer the time (and the cost) of changing over from one activity to another, the more difficult it is to synchronise flow to match customer demand instantaneously.
- ***More variation in activity mix than in customer demand*** – if the mix of activities in different time periods varies more than customer demand varies, then some 'batching' of activities must be taking place.

Muri

Muri means absurd or unreasonable. It is based on the idea that unnecessary or unreasonable requirements put on a process and its resources will result in poor outcomes. Reducing muri requires smoothing demand (mura). In other words, waste can be caused by failing to carry out basic operations planning tasks such as prioritising activities (sequencing) and understanding the time (scheduling) and resources (loading) needed to perform activities. Avoiding muri requires understanding the nature of demand (stable or unpredictable) in order to calculate takt time. Takt time refers to the required rate of daily production to satisfy customer demand. Knowing takt time means the operation can ensure capacity is aligned to demand.

These three causes of waste are obviously related. When a process is inconsistent (mura), it can lead to the overburdening of equipment and people (muri), which, in turn, will cause all kinds of non-value-adding activities (muda).

Types of waste

Muda, mura and muri are three interrelated causes of waste. Now we turn to types of waste (focusing on types of 'muda'), which apply to all types of operation and which form the core of lean philosophy. Seven types of waste are identified:

1. ***Over-production*** – this is when an operation produces more than is required by the customer. Making too much, too early or 'just in case' can lead to product obsolescence or large quantities of inventory that obscure problems, as discussed in earlier sections. A pull system should only produce output in line with actual customer demand. (Over-production is considered the most serious type of waste, according to Toyota.)
2. ***Waiting*** – any kind of 'waiting' impedes the flow of the product/service to the end customer. Most people have better things to do with their time than wait for a resource (a machine or a service professional, for example) to become available. Any time when products, customers or information wait as inventory or queues, is time when no value is being added.
3. ***Transport*** – concerns the movement of products, customers or information from one place to another that does not add any value to the customer. Moving items or customers around the operation, together with double and triple handling, does not add value. Layout changes that bring processes closer together, and improvements in transport methods and workplace organisation can all reduce waste. Layout changes can also dramatically reduce the physical movement of staff, leaving them less tired at the end of the working day.
4. ***Overprocessing*** – this is where more work is done to a product/process than is required by the customer. Overprocessing means spending time and money on tasks that are not valued by the

customer. A common example of overprocessing is when a customer is required to provide the same information for multiple similar but different documents.

5. ***Inventory*** – presents a capital expenditure that has not yet produced any income; regardless of type (product, customer, information) all inventories should become a target for elimination. However, it is only by tackling the causes of inventory or queues, such as irregular flow (mura), that it can be reduced. The ability of inventories to mask production problems and impede flow have already been discussed in this chapter.
6. ***Motion*** – refers to the motion of the workers and equipment; excess motion wastes time and can cause injury/damage. Excess motion includes having to search for materials, tools or equipment due to poor organisation.
7. ***Defect/rework*** – when defects occur, extra costs and delays ensue. Lean implies exact levels of quality. If there is variability in quality levels, then customers will not consider themselves as being adequately supplied. Symptoms of poor variability and therefore a high propensity for defects include poor reliability of equipment or staff. Unreliable equipment or staff usually indicates a lack of conformance in quality levels. It also means that there will be irregularity in supplying customers. Either way, it impedes flow. Similarly, defective products or services (waste caused by poor quality) is significant in most operations. Errors in the service or product cause both customers and processes to waste time until they are corrected.

An eighth waste?

Many lean practitioners would argue there's an *eighth* waste – the misuse of people. Misaligning individual talents, skills, creativity and knowledge relative to task generates a waste of available knowledge and skills. Examples include having skilled workers completing unskilled tasks or limiting decision-making authority to a high level within a business.

How improving layout design reduces waste

The smooth flow of materials, information and people in the operation is a central idea of lean. Long process routes provide opportunities for delay and inventory build-up, add no value to the customer, and slow down throughput time. So, the first contribution any operation can make to enhance flow is to reconsider the basic layout of its processes. Primarily, reconfiguring the layout of a process to aid lean involves moving it down the 'natural diagonal' of process design that was discussed in Chapter 6. Broadly speaking, this means moving from functional layouts towards cell-based layouts, or from cell-based layouts towards line layouts. Either way, it is necessary to move towards a layout that brings more systematisation and control to the process flow. At a more detailed level, typical layout techniques include placing workstations close together so that inventories of products or customers simply have no space to build up, and arranging workstations in such a way that all those who contribute to a common activity are in sight of each other and can provide mutual help. For example, at the Virginia Mason Medical Center in Seattle, a leading proponent of lean in healthcare, many of the waiting rooms have been significantly reduced in their capacity or removed entirely. This forces a focus on the flow of the whole process because patients have literally nowhere to be 'stored'. By contrast, some hospitals have multiple waiting zones or 'sub-waits' where patients arrive for clinics in batches and progress from one waiting room to another, producing the very antithesis of smooth flow.

How improving process flexibility reduces waste

Responding exactly and instantaneously to customer demand implies that operations resources need to be sufficiently flexible to change both what they do and how much they do of it without incurring high cost or long delays. In fact, flexible processes can significantly enhance smooth flow.

OPERATIONS IN PRACTICE

Waste reduction in airline maintenance[4]

Aircraft maintenance is important. Planes have a distressing tendency to fall out of the sky unless they are checked, repaired and generally maintained regularly. So, the overriding objective of the operations that maintain aircraft must be the quality of maintenance activities. But it is not the only objective. Improving maintenance turnaround time can reduce the number of aircraft an airline needs to own, because they are not out of action for as long. Also, the more efficient the maintenance process, the more profitable is the activity and the more likely a major airline with established maintenance operations can create additional revenue streams by doing maintenance for other airlines. Figure 16.7 shows the path taken by maintenance staff before and after lean analysis. The objectives of the lean analysis were to preserve, or even improve, quality levels while at the same time improving the cost of maintaining airframes and increasing the availability of airframes by reducing turnaround time.

The lean analysis focused on identifying waste in the maintenance process. Two findings emerged from this. First, the sequence of activities on the airframe itself was being set by the tasks identified in the technical manuals supplied by the engine, body, control system and other suppliers. No one had considered all the individual activities together to work out a sequence that would save maintenance staff time and effort. The overall sequence of activities was defined and allocated with 'standard work' for the preparation of tools, materials and equipment. Second, maintenance staff would often be waiting until the airframe became available. Yet some of the preparatory work and set-ups did not need to be done while the airframe was present. Therefore, why not get maintenance staff to do these tasks when they would otherwise be waiting, before the airframe became available? The result of these changes was a substantial improvement in cost and availability. In addition, work preparation was conducted in a more rigorous and standardised manner and maintenance staff were more motivated because many minor frustrations and barriers to their efficient working were removed.

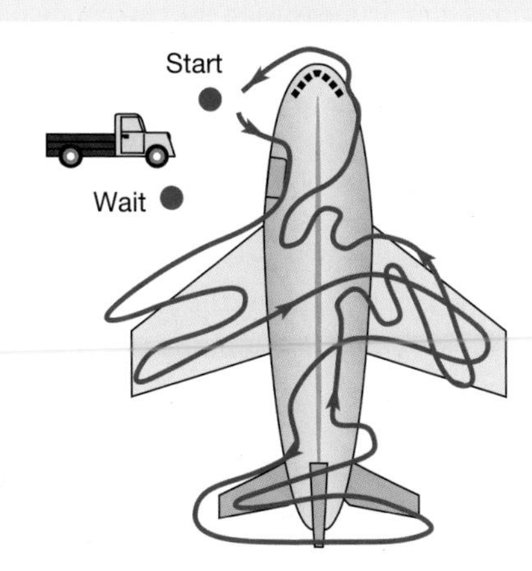

Before:
- Maintenance staff follow the steps as detailed in the technical documentation
- The overall sequence of tasks is not optimised
- Preparation work and set-ups included as part of the task

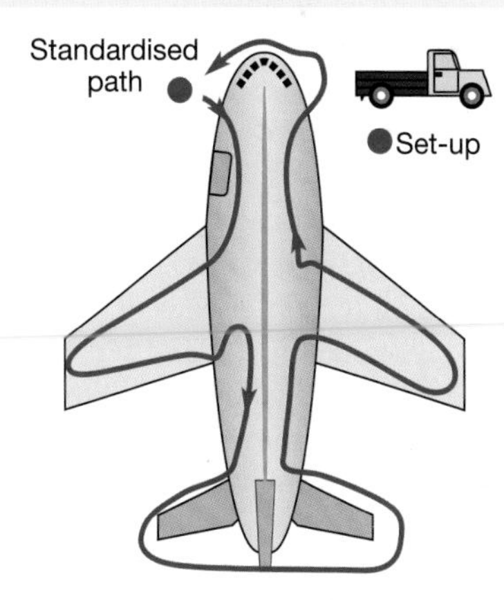

After:
- The overall sequence of tasks is defined and allocated to minimise non-value-added activity
- Preparation work and set-ups may be done ahead of time to minimise aircraft contact time
- Increased productivity and reduced aircraft waiting time

Figure 16.7 Aircraft maintenance procedures subject to waste reduction analysis

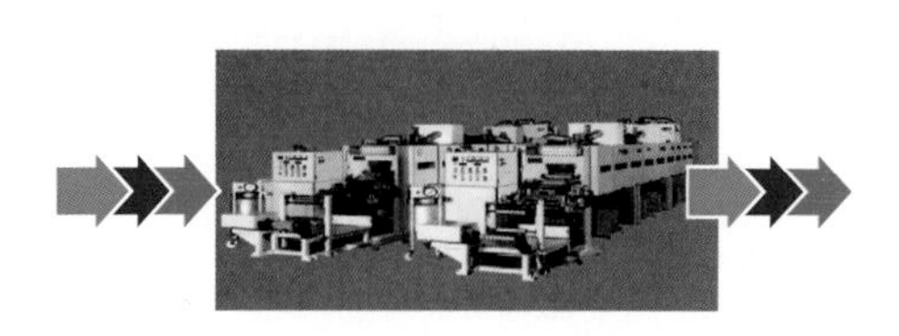

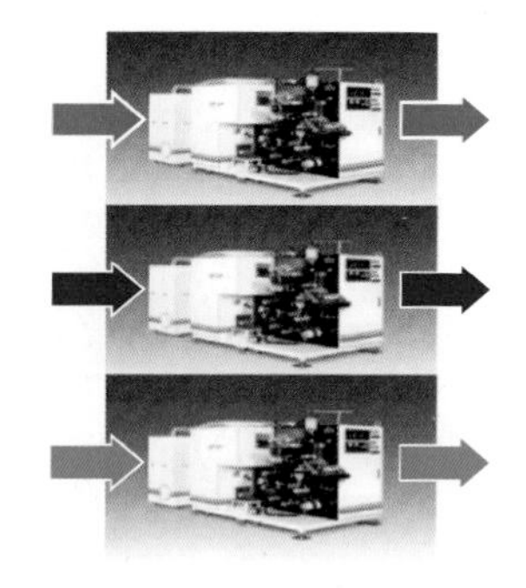

Figure 16.8 Using several small machines rather than one large one allows simultaneous processing, is more robust and is more flexible

Using small-scale simple process technology

Lean often involves moving towards smaller-scale process technology to reduce fluctuations in flow volume. For example, in Figure 16.8 one large machine produces a batch of A, followed by a batch of B, followed by a batch of C. However, if three smaller machines are used, they can each produce A, B or C simultaneously. The system is also more robust. If one large machine breaks down, the whole system ceases to operate. If one of the three smaller machines breaks down, it is still operating at two-thirds effectiveness. Small machines are also easily moved, so that layout flexibility is enhanced, and the risks of making errors in investment decisions are reduced. However, investment in capacity may increase in total because parallel facilities are needed, so utilisation may be lower (see the earlier arguments).

Reducing changeover times (set-up reduction)

Responding to demand only when it is needed usually requires a degree of flexibility in processes, both to cope with unexpected demand and to allow quick changeover between one activity and another. Changeover time can often be reduced significantly – compare, for example, the time it takes you to change a car tyre, with the sub-3 seconds taken by a Formula 1 team. Set-up reduction can be achieved by a variety of methods such as the following:

- ***Measure and analyse changeover activities*** – Sometimes simply measuring the current changeover times, recording them and analysing exactly what activities are performed can help to improve changeover times.
- ***Separate external and internal activities*** – 'External' activities are simply the activities that can be carried out while the process is continuing. For example, processes could be getting ready for the next customer or job while waiting for it to arrive (see the example of aircraft maintenance described earlier). 'Internal' activities are those that cannot be carried out while the process is going on (e.g. interviewing the customer while completing a service request for the previous customer). By identifying and separating internal and external activities, the intention is to do as much as possible while the step/process is continuing.
- ***Convert internal to external activities*** – The other common approach to changeover time reduction is to convert work that was previously performed during the changeover to work that is performed outside the changeover period. There are three major methods of achieving the transfer of internal set-up work to external work:
  - — Pre-prepare activities or equipment instead of having to do it during changeover periods.
  - — Speed up any required changes of equipment, information or staff, for example by using simple devices.
  - — Practise changeover routines – not surprisingly, the constant practice of changeover routines and the associated learning curve effect tends to reduce changeover times.

Implementing standard work

Standard work refers to the step-by-step documentation of the most efficient way of performing a process; the document should explain why a process step is completed in a certain way. Far from making a process rigid, standard work serves as a baseline for further improvement efforts. When process efficiency (and quality) is improved upon, then standard work is updated so that it should always present the best-known way of completing a process. Without standard work, gains from process improvement will not be sustained. Standard work is an important accompaniment to set-up reduction as it helps to ensure the process is performed the same way, every time. Standard work is also effective in facilitating flexible working among members of staff, enabling them to quickly assimilate the knowledge required to perform tasks in areas where they don't normally work.

Operations principle

Standard work ensures processes are performed consistently each time without variation. Standard work should represent the best-known way to complete a process and it should also serve as a baseline for continuous improvement.

Eliminating waste through minimising variability

One of the biggest causes of variability is a variation in the quality of items. This is why a discussion of lean should always include an evaluation of how quality conformance is ensured within processes, referred to as 'mura' earlier. In particular, the principles of statistical process control (SPC) can be used to understand quality variability. Chapter 17 and its supplement on SPC examines this subject, so in this section we shall focus on other causes of variability. The first of these is variability in the mix of items moving through processes, operations or supply networks.

OPERATIONS IN PRACTICE

Rapid changeover for Boeing and Airbus[5]

Fast changeovers are particularly important for airlines because they can't make money from aircraft that are sitting idle on the ground. It is called 'running the aircraft hot' in the industry. For many airlines, the biggest barrier to running hot is that their markets are not large enough to justify passenger flights during the day and night. So, in order to avoid aircraft being idle overnight, they must be used in some other way. That was the motive behind Boeing's 737 'Quick Change' (QC) aircraft. With this aircraft, airlines have the flexibility to use it for passenger flights during the day and, with less than a one-hour changeover (set-up) time, use it as a cargo aeroplane throughout the night. Boeing engineers designed frames that hold entire rows of seats that could smoothly glide on and off the aircraft, allowing twelve seats to be rolled into place at once. When used for cargo, the seats are simply rolled out and replaced by special cargo containers designed to fit the curve of the fuselage and prevent damage to the interior. Before reinstalling the seats, the sidewalls are thoroughly cleaned so that, once the seats are in place, passengers cannot tell the difference between a QC aircraft and a normal 737.

The rapid changeover concept has also been developed by Airbus. It announced that in future it intends to introduce new 'living area' options on its A330 family of aircraft, positioned in the cargo hold. There will be a variety of different pod types, including family sleeping areas, meeting rooms, a children's play area and gym space. Each pod has been designed to be taken in and out of the cargo holds quickly (as with Boeing's 'Quick Change') to allow rapid changeover in capacity use depending on the needs of the market and route at any given time.

Levelling product or service schedules (Heijunka)

Levelled scheduling (or heijunka, pronounced hi-june-kuh) means keeping the mix and volume of flow between stages at an even rate over time to reduce mura (unevenness) and muri (overburden). The move from conventional to levelled scheduling is illustrated in Figure 16.9. Conventionally, if a mix of items (different styles of shoes for example) were required in a time period (usually a month), a batch size would be calculated for each item and the batches produced in some sequence. Figure 16.9(a) shows three items that are produced in a 20-day time period in an operation:

Quantity of Shoe type A required = 3000
Quantity of Shoe type B required = 1000
Quantity of Shoe type C required = 1000

Batch size of Shoe A = 600
Batch size of Shoe B = 200
Batch size of Shoe C = 200

Starting at day 1, the unit commences producing Shoe A. During day 3, the batch of 600 As is finished and dispatched to the next stage. The batch of Shoe B is started but is not finished until day 4. The remainder of day 4 is spent making the batch of Shoe C and both batches are dispatched at the end of that day. The cycle then repeats itself. The consequence of using large batches is, first, that relatively large amounts of inventory accumulate within and between the units, and second, that most days are different from one another in terms of what they are expected to produce (in more complex circumstances, no two days would be the same).

Now suppose that the flexibility of the unit could be increased (through the use of smaller machines and set-up reduction for example) to the point where the batch sizes for the items are a quarter of their previous levels without loss of capacity (see Figure 16.9(b)):

Batch size of Shoe A = 150
Batch size of Shoe B = 50
Batch size of Shoe C = 50

A batch of each item can now be completed in a single day, at the end of which the three batches are dispatched to their next stage. Smaller batches of inventory are moving between each stage, which

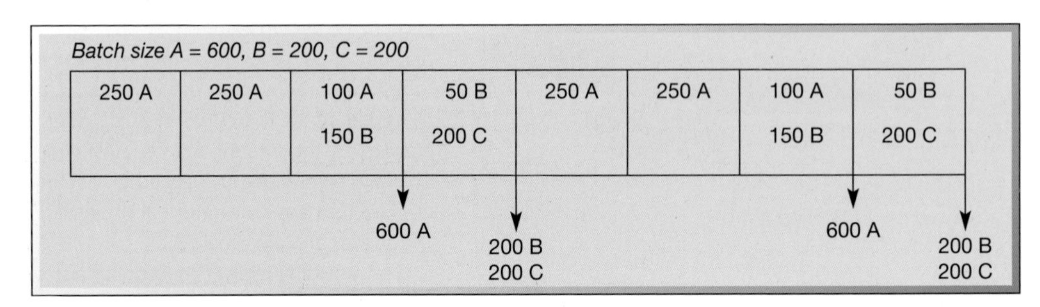

(a) Scheduling in large batches

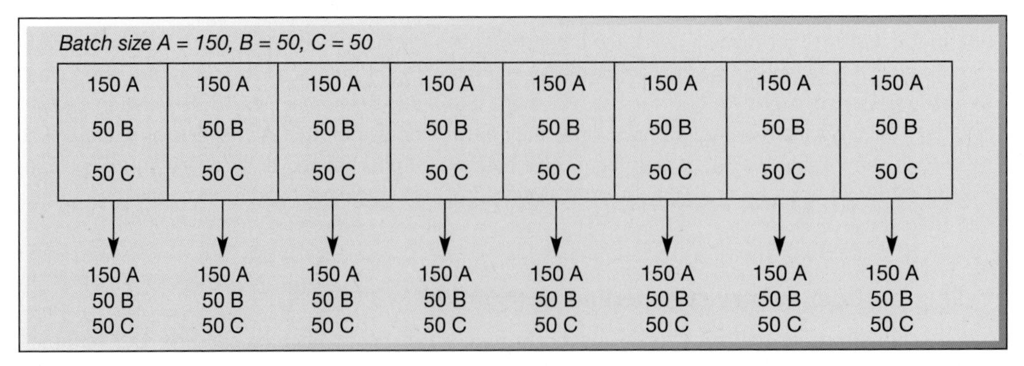

(b) Levelled scheduling

Figure 16.9 Levelled scheduling equalises the mix of products/service delivered each day

will reduce the overall level of work in progress in the operation. Just as significant, however, is the effect on the regularity and rhythm of production at the unit. Now every day in the month is the same in terms of what needs to be processed. This makes planning and control of each stage in the operation much easier. For example, if on day 1 of the month the daily batch of Shoe A was finished by 11.00 am, and all the batches were successfully completed in the day, then the following day the unit will know that, if it again completes all the As by 11.00 am, it is on schedule. When every day is different, the simple question 'Are we on schedule to complete our processing today?' requires some investigation before it can be answered.

Critical commentary

Of the criticisms of heijunka, two are particularly relevant. The first is that it is difficult to implement heijunka when demand is direct and highly uncertain. This is especially true of emergency services such as fire, police and healthcare services. Yet even emergency services experience demand that is, to some extent, predictable (consider the increased police presence required on Friday nights in a busy city centre). But, when unpredictability persists, 'spare' capacity should be available to respond quickly when a customer's 'emergency' presents itself. The second criticism is that, although the goal of heijunka is to reduce or eliminate variation, there are good reasons for grouping activities together in batches. It is argued that it minimises the inefficiencies that always occur when changing over from one activity to another. Expecting processes to be sufficiently flexible to change as often as demanded by heijunka is unrealistic. Operations implementing lean philosophy and practices are unlikely to achieve heijunka in the early stages of implementation, since it depends on embedding all the other practices designed to enhance flow (for example, eliminate waste and reduce variation). Further, heijunka depends upon understanding demand, and smoothing demand where possible. Therefore, heijunka should be considered an aspirational goal for a production system, and one that organisations may move closer to over time.

Levelling delivery schedules

A similar concept to levelled scheduling can be applied to many transportation processes (illustrated in Figure 16.10). For example, a chain of convenience stores may need to make deliveries of all the different types of products it sells every week. Traditionally it may have dispatched a truck loaded with one particular product around all its stores so that each store received the appropriate amount of the product that would last them for one week. This is equivalent to the large batches discussed in the previous example. An alternative would be to dispatch smaller quantities of all products in a single truck more frequently. Then, each store would receive smaller deliveries more frequently, inventory levels would be lower and the system could respond to trends in demand more readily because more deliveries mean more opportunity to change the quantity delivered to a store.

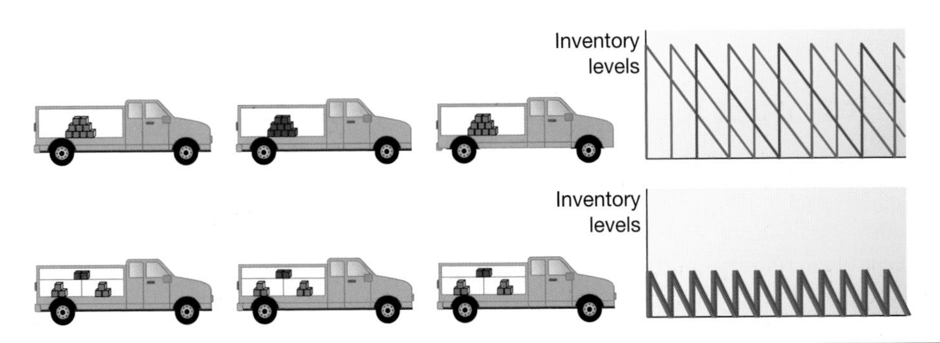

Figure 16.10 Delivering smaller quantities more often can reduce inventory levels

OPERATIONS IN PRACTICE

Jamie's 'lean' cooking[6]

Most people do not have the time to devote long hours to cooking. This might be why the celebrity chef Jamie Oliver has written a book, *Jamie's 30-Minute Meals*, whose philosophy is that cooking a delicious dinner should be as quick and cheap as a takeaway. The book presents 50 ready-made menus with three to four courses per menu designed to take no more than 30 minutes to prepare. To achieve this performance Jamie has, perhaps inadvertently, applied the principles and methods of lean to the everyday activity of cooking.

Let's imagine that your family is coming over for dinner and you want to surprise them with a new Indian multi-course meal with chicken, rice, salad on the side and of course a dessert. Traditionally, you would search and look up four different recipes, one for each dish. Because all recipes come from different places, you need to figure out the quantity of food to buy, doing the maths in the case of shared ingredients across the dishes, how to allocate pots, pans and other equipment to the different ingredients, and most importantly, you need to figure out in what order to prepare things, especially if you want all your dishes ready at the same time. Jamie's approach significantly reduces this complexity by ensuring dishes are prepared right when the next step in the process needs it, regardless of which dish it is. In other words, dishes are not cooked in sequence, one after another, but they are prepared and completed simultaneously.

If we identify all the tasks related to preparing the salad (e.g. chopping the vegetables) with the letter A, cooking the rice (e.g. boiling) with letter B, cooking the chicken with letter C, and finally making the dessert with the letter D, then in the traditional way of cooking our task scheduling would look something like AAAA BBBBBBB CCCCCCC DDDD. This results in batching, waiting time and causing dishes to be ready before the dinner is supposed to be served. Conversely, Jamie Oliver's 30-minute cooking involves scheduling tasks in a sequence like ABACDACBADCBABDC, where single tasks related to different dishes flow smoothly from one to the next: the chef chops a salad ingredient (A), then boils the rice (B), then chops more salad ingredients (A) while the chicken (C) is being roasted in the oven and a part of the dessert (D) is being prepared. This way, all dishes are ready at the same time, just in time, and nothing is prepared before it has to be, avoiding any form of waste. Such a levelled approach to scheduling is called heijunka.

In addition, Jamie's lean cooking builds on reduced set-up times. At the beginning of each recipe, the equipment needed to prepare the menu is presented under the headline 'To Start'. Other necessary preparations, such as heating the oven, are also specified. Having all equipment ready from the start saves time in the process, and is, according to Jamie, a prerequisite for getting done in 30 minutes. The use of simple equipment that is suitable for many different purposes also makes the process quicker as changeovers are minimised. Finally, the recipe is an example of 'standard work', detailing each process step, often including a simple explanation of why we perform the work this way. The rationale is to make the most out of the time available, eliminating the 'faffing around' in cooking (non-value-added activity in OM language) and leaving only what is strictly 'good, fast cooking', without compromising on quality.

16.4 How does lean consider improvement?

Lean objectives are often expressed as ideals, for example: 'to meet demand instantaneously with perfect quality and no waste'. While any operation's current performance may be far removed from such ideals, a fundamental lean belief is that it is possible to get closer to them over time. This is why the concept of continuous improvement is such an important part of the lean philosophy. If its aims are set in terms of ideals that individual organisations may never fully achieve, then the emphasis must be on the way in which an organisation moves closer to the ideal state. As we explained in Chapter 15, the Japanese word for continuous improvement is kaizen, and it is a key part of the lean philosophy.

The Rapid Process Improvement Workshop (RPIW)

The Rapid Process Improvement Workshop (RPIW) is a common vehicle for leveraging improvement at a process level that is often associated with a lean approach to improvement. RPIWs (sometimes called kaizen events) are often three-to-five-day workshops that bring together a diverse team of employees to collectively examine a process with the purpose of thinking about how that process could be improved in line with lean principles. The RPIW team is usually made up of a cross-section of employees whose daily work may 'touch' the focal process (the one that is being examined), but they may not know how their work contributes to the overall process or how their activity impacts on that of another. Hence when a process is collectively mapped by RPIW participants it is not uncommon to hear members say that they never realised what was happening in other parts of the process. They had only ever understood their own part of the process.

Often (but not always) an RPIW is facilitated by one or more individuals that form part of a centralised improvement team, skilled in the use of process improvement methods. If resource allows, the improvement team conducts a detailed analysis of performance data prior to the RPIW and that data is shared with the RPIW team on the first morning. Using data in the form of a run chart illustrating product defects, throughput times and customer satisfaction scores, for example, can be an effective mechanism for engendering a collective acknowledgement that the process is underperforming. Classical change theorists such as John Kotter and Kurt Lewin, for example, have long maintained the importance of acknowledging the existence of a problem as a vital first step for any process improvement intervention.[7]

Across the three-to-five days of the RPIW, team members move from acknowledgement of a problem and a collective desire for change, to understanding the problem using techniques such as 'go-see', process maps and value stream mapping (see later), to thinking about the desired 'to-be' state of a process, where non-value-adding activity is eliminated to enhance flow. At this conjecture, usually day two of the workshop, tools and techniques such as cause–effect diagrams (see Chapter 15) are used to understand the 'root cause' of a problem and prioritise improvements before the team conducts small rapid experiments using repetitive cycles of PDCA (again, see Chapter 15) to test ideas for process changes. The selection of people for the RPIW is also important. An effective team could include someone who has authority to lead change, someone who has 'social capital' (i.e. is connected to others who can support a process change) to bolster the acceptability of process changes, as well as people with appropriate 'technical' knowledge. An RPIW team should also incorporate a 'process owner' who agrees to take forward actions from the event and monitor progress from the changes.

Encouraging improvement by 'stopping the line'

Named after a Japanese lantern, the '**Andon Cord**' is used to stop activities when a defect of some sort is detected. Pulling the cord 'stops the line'. It represents a visual (and sometimes audible) cue that clearly signals the presence and location of a problem. In operations adopting this approach, staff have 'line-stop authority', in other words, they have an obligation to alert others if a problem occurs. Pulling the Andon Cord stops the production line, forcing immediate attention on addressing a problem. Although this may seem to reduce the efficiency of the line (interrupting flow), the idea is that this loss of efficiency in the short term is less than the accumulated losses of allowing defects to continue in the process. In some cases, unless problems are tackled immediately, they may never be corrected. Online retailer Amazon has adopted the concept of the Andon Cord in response to customer complaints (see the 'Operations in practice' example, 'Autonomy at Amazon').

OPERATION IN PRACTICE

Autonomy at Amazon[8]

Amazon is a strong believer in delegating responsibility to those on the front line of its service centres. Every day, service agents at Amazon receive calls from customers who are unhappy with some aspect of the product delivered to them. Employees dealing with these complaints are now empowered to make judgements on the extent to which such complaints may be systemic. In cases where they suspect it's a repetitive defect, service agents can 'stop the line' for a particular product ('pulling the Andon Cord'). This involves taking the product off the website while the problem is fully investigated. According to Amazon, the improved visibility of the system has eliminated tens of thousands of defects a year and has also given service agents a strong sense of being able to deal effectively with customer complaints. Now an agent can not only refund the individual customer, they are also able to tell the customer that others won't receive products until the problem has been properly investigated. Amazon claims that around 98 per cent of the times when the product is pulled in this way, there really is a systemic problem, highlighting the value of giving its service agents autonomy to make decisions as to when and when not to stop the line.

Gemba walks – the principle of go-see

Going to the '**gemba**' (also called 'genba') means to go to the place where the work actually takes place; only then can a true appreciation of the process be gained. Organisations implementing lean use the idea of a 'gemba walk', where managers regularly visit the place where the job is done to understand the work from the perspective of those who perform the work. The gemba walk enhances the visibility of operations, fostering greater understanding of how work gets done, and a greater 'connectedness' between management and the 'shop floor'. The guiding maxim of the gemba walk is 'those who do the work know best how to improve the work'.

> **Operations principle**
> There is no substitute for seeing the way processes actually operate in practice.

Critical commentary

Much of our explanation of lean ideas could be seen as 'technical', in the sense that it deals with issues of how to manage operations resources at a tangible and specific level. However, many advocates of the lean approach would argue that the important lesson from successful lean implementation is that it should be embedded in organisational culture. In other words, lean should become 'the way we do things around here'. However, changing culture enough to impact on the success of lean is not straightforward, taking (arguably) years, and few organisations have fully integrated lean into 'the way we do things around here'. Why is this? First, many organisations see lean primarily as a way of cutting cost (by doing more with less). And, while lean should deliver a more efficient use of resources, it does so through seeking small incremental improvements. Efficiency savings will accumulate, but in the short term at least, savings will most likely be small. Companies can lose interest and seek other ways to deliver cost savings faster. Second, some organisations using Rapid Process Improvement Workshops may not place sufficient emphasis on their continuous nature. While the RPIW can deliver substantial improvements, the gains need to be followed by other gains, possibly by improving other connected processes. These two points lead to the third critical point – lean is a journey, not a destination. Admittedly, this is something of a management cliché; however, organisations wishing to develop lean operations must see it as a long-term commitment. Some lean proponents claim that the first five years of lean implementation involve developing the improvement capability of employees, developing an understanding of lean and aligning organisational strategy with improvement goals.

Value stream mapping for understanding flow and identifying sources of waste

Improving flow requires understanding how materials, information and people flow through a process. Value stream mapping is a simple but effective approach that records not only the direct activities of creating products and services, but also the 'indirect' information systems that support the direct process. It is called 'value stream' mapping because it focuses on distinguishing between value-adding and non-value-adding activities. It is similar to process mapping (see Chapter 6) but different in three ways:

- It uses a broader range of information than most process maps, incorporating calculations of cycle times, waiting times and process efficiency.
- It is usually at a higher level (5–10 activities) than most process maps.
- It often has a wider scope and is capable of adopting a supply network perspective.

A value stream refers to the entire set of activities that encompass the transformation of materials, information and customers from beginning to end. Value stream mapping is seen by many practitioners as a starting point to help recognise waste and identify its causes. It is a four-step technique that identifies waste and suggests ways in which activities can be streamlined. First, it involves identifying the value stream (the process, operation or supply chain) to map. Second, it involves physically mapping a process, then above it mapping the information flow that enables the process to occur. This is the so-called 'current state' map. Third, problems are diagnosed and changes suggested, making a future state map that represents the improved process, operation or supply chain. Finally, the changes are implemented. Figure 16.11 shows a value stream map for an industrial air-conditioning installation service. The service process itself is broken down into five relatively large stages and various items of data for each stage are marked on the chart. The type of data collected here does vary, but all types of value stream map compare the total throughput time with the amount of value-added time within the larger process. In this case, only 8 of the 258 hours of the process are value-adding.

Keeping things simple – the 5S technique

The 5S terminology comes originally from Japan, and although the translation into English is approximate, they are generally taken to represent the following:

- **Sort** (*seiri*). Eliminate what is not needed and keep what is needed.
- **Straighten** (*seiton*). Position things in such a way that they can be easily reached whenever they are needed.
- **Shine** (*seiso*). Keep things clean and tidy; no refuse or dirt in the work area.
- **Standardise** (*seiketsu*). Maintain cleanliness and order – perpetual neatness.
- **Sustain** (*shitsuke*). Develop a commitment and pride in keeping to standards.

5S can be thought of as a simple housekeeping methodology to organise work areas, which focuses on visual order, organisation, cleanliness and standardisation. It helps to eliminate all types of waste relating to uncertainty, waiting, searching for relevant information, creating variation and so on. By eliminating what is unnecessary, and making everything clear and predictable, clutter is reduced, needed items are always in the same place and work is made easier and faster.

Adopting visual management

Visual management is one of the lean techniques designed to make the current and planned state of the operation or process transparent to everyone, so that anyone (whether working in the process or not) can very quickly see what is going on. It usually employs some kind of visual sign, such as a notice

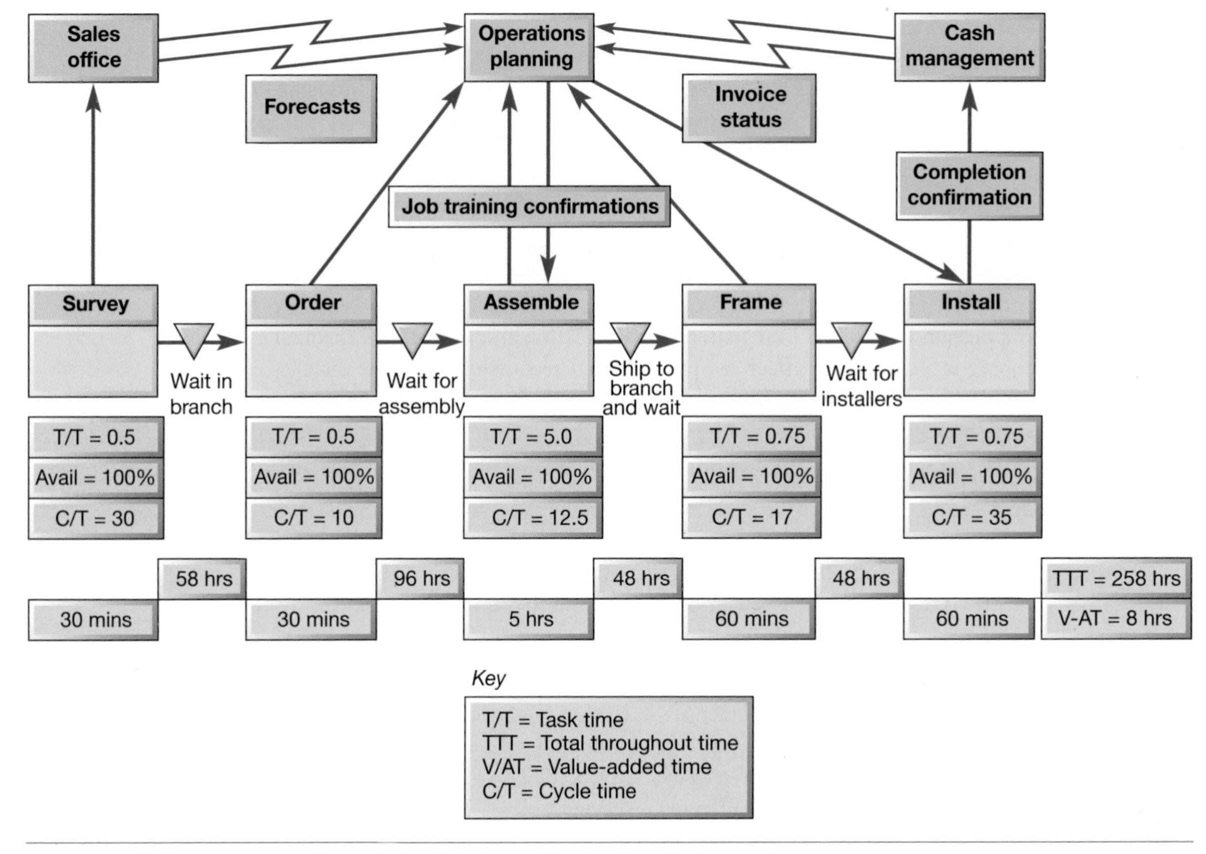

Figure 16.11 Value stream map for an industrial air-conditioning installation service

board, computer screen, or simply lights or other signals, which convey what is happening. Although a seemingly trivial and usually simple device, visual management has several benefits. It:

- acts as a common focus for team meetings;
- demonstrates methods for safe and effective working practice;
- communicates to everyone how performance is being judged;
- assesses at a glance the current status of the operation;
- increases understanding of tasks and work priorities;
- judges your and others' performance;
- identifies the flow of work, what has been and is being done;
- identifies when something is not going to plan;
- shows what agreed standards should be;
- provides real-time feedback on performance to everyone involved;
- reduces the reliance on formal meetings.

OPERATIONS IN PRACTICE — Visual management at KONKEPT

The finance operation of KONKEPT, the online toy retailer based in Singapore, was having problems. Service levels were low and complaints high, as the office attempted to deal with payments from customers, invoices from suppliers and requests for information from its distribution centre; all while demand was increasing. It was agreed that the office's processes were chaotic and poorly managed, with little understanding of priorities or how each member of staff was contributing. To remedy this state of affairs, the manager responsible for the office first tried to bring clarity to the process by defining individual and team roles, and started establishing visual management. The staff

mapped processes and set performance objectives collectively. These objectives were shown on a large board placed so everyone in the office could see it. At the end of each day, process supervisors updated the board with each process's performance for the day. Also indicated on the board were visual representations of various improvement projects being carried out by the teams. Every morning, staff gathered in what was called 'the morning huddle' to discuss the previous day's performance, identify how it could be improved, review the progress of ongoing improvement projects, and plan for the upcoming day's work. For the staff at KONKEPT, the experience illustrated the three main functions of visual management:

- to act as a communication mechanism;
- to encourage commitment to agreed goals;
- to facilitate cooperation between team members.

Adopting total productive maintenance (TPM)

Total productive maintenance aims to eliminate the variability in operations processes caused by the effect of breakdowns. This is achieved by involving everyone in the search for maintenance improvements. Staff are encouraged to assume ownership of their equipment and to undertake routine maintenance and simple repair tasks. These principles apply equally to service operations. For example, at a car wash, service employees regularly maintain their power hoses to prevent unnecessary downtime, while university employees may be encouraged to regularly 'clean' email inboxes, delete old files on their computers, and update software with the aim of maintaining system availability speed and protecting from viruses. By doing so, maintenance specialists can then be freed to develop higher-order skills for improved maintenance systems. TPM is discussed in more detail in Chapter 18 on risk and recovery.

16.5 How does lean consider the role of people?

As we have implied in earlier discussions, for lean to be effective throughout the organisation, it requires serious attention to people-related issues. This applies at all levels in the organisation. At an operation-wide level it implies cultural change (where lean becomes 'the way we do things around here'). At a process or group level it means accepting the need for team-based improvement events, physically visiting processes as they happen, as in gemba walks, using techniques such as value stream mapping and applying visual management. At an individual level, it means training staff in lean principles and techniques, and encouraging a high degree of personal responsibility, engagement and 'ownership' of the job. All are indicative of the importance of the role of people in lean operations.

Basic working practices

Similarly, what lean proponents refer to as 'basic working practices' are sometimes used to encourage the 'involvement of everyone' and 'respect for people'. These practices include the following:

- ***Discipline*** – Work standards that are critical for the safety of staff, the environment and quality must be followed by everyone all the time.
- ***Flexibility*** – It should be possible to expand responsibilities to the extent of people's capabilities. This applies as equally to managers as it does to shop-floor personnel. Barriers to flexibility, such as grading structures and restrictive practices, should be removed.
- ***Equality*** – Unfair and divisive personnel policies should be discarded. Many companies implement the egalitarian message through to company uniforms, consistent pay structures that do not differentiate between full-time staff and hourly-rated staff, and open-plan offices.
- ***Autonomy*** – Delegate responsibility to people involved in direct activities so that management's task becomes one of supporting processes. Delegation includes giving staff the responsibility for

stopping processes in the event of problems, scheduling work, gathering performance-monitoring data and general problem-solving.

- *Development of personnel* – Over time, the aim is to create more company members who can support the rigours of being competitive.
- *Quality of working life (QWL)* – This may include, for example, involvement in decision making, security of employment, enjoyment and working area facilities.
- *Creativity* – This is one of the indispensable elements of motivation. Creativity in this context means not just doing a job, but also improving how it is done and building the improvement into the process.
- *Total people involvement* – Staff take on more responsibility to use their abilities to the benefit of the company as a whole. They are expected to participate in activities such as the selection of new recruits, dealing directly with suppliers and customers over schedules, quality issues and delivery information, spending improvement budgets, and planning and reviewing work done each day through communication meetings.

OPERATIONS IN PRACTICE

Respect![9]

There is a style of popular TV show (*Gordon Ramsay's Kitchen Nightmares* is a UK example) that shows how the heat of the kitchen, combined with the critical eye of the customer, can flame the temper of an angry chef, abusing employees into submission. By contrast, New York restaurateur and Chief Executive of Union Square Hospitality Group Danny Meyer is quoted as saying *'business, like life, is how you make people feel'*. While the concept of being nice to customers is widely understood in the hospitality industry, Meyer goes further to suggest that customers can taste the uncivil behaviours of employees in the food that is served. And what applies to commercial kitchens is also an elementary, yet often overlooked, aspect of lean operations – 'respect for people'. Far from being a glib reminder about the importance of being nice, many successful businesses contend respectful behaviours (between a company's employees and between employees and their customers) are intrinsically related to performance. This is because the effects of 'incivility' are far reaching and almost always negative: productivity suffers, quality suffers, innovation suffers and talented individuals leave their organisations. Promoting civility may save customers from bad-tasting food, but for some complex environments, such as airlines and hospitals, incivility can lead to more catastrophic process errors. In these high-reliability contexts, promoting respectful behaviours can literally save lives.

Linking respectful behaviours to patient safety, Virginia Mason Medical Center, a hospital based in Seattle, sought to develop a culture of respectful behaviours among its 5,500 employees. But teaching people to be nice is not something that can be accomplished quickly. First, the hospital needed to define and understand what 'respectful behaviours' really meant in the context of the operations environment, and then how to enact them, even under pressure. As part of the exploration of civility, the hospital engaged hundreds of employees (and patients) in an initiative to share their experiences of what they regarded as civil and uncivil behaviours. It then used these as a basis for defining its top ten 'foundational behaviours' (see Figure 16.12). To help establish these behaviours, a series of mandatory training workshops were designed that involved a theatre troupe re-enacting disrespectful interactions based on employees' reported experiences across a broad range of job categories, positions and experiences. Subsequent discussions explored how to de-escalate emotions so that more respectful behaviours could become established. Toolkits were also developed to help staff practice respectful behaviours successfully in the event of conflict, and all staff were required to choose two of the ten foundation behaviours for self-development each year. Since the start of the initiative, Virginia Mason senior management report significant improvements in the numbers of staff who feel more able to speak up if they see something that may negatively affect patient care, thereby enhancing patient safety.

Respect *for* People

THE VIRGINIA MASON EXPERIENCE: PATIENTS & FAMILIES, TEAM MEMBERS, COMMUNITY

Our Foundational Behaviors

We all have a role in sustaining a community where everyone feels valued, included and respected.

1 | Be a team player

Working together collaboratively creates an environment where everyone feels engaged. Ask others how you can be helpful. If issues come up, trust that people mean well, and share timely, specific and caring feedback with each other.

2 | Listen to understand

Listening well shows people that you are giving them your full attention. Ask questions if you don't understand what others are saying or how they feel. Be open and curious about ideas that are different from yours. Patience helps – interrupting may leave others feeling not heard.

3 | Share information

Sharing the information people need helps them feel prepared and included. As you do so, make room in the conversation for others to speak. Notice if you have a strong preference for or against something, and be open to other ways of looking at the situation.

4 | Keep your promises

Following through on commitments as soon as possible builds trust and lets others know you care. If you aren't able to keep your word, let others know right away.

5 | Speak up

Speaking up creates a safe environment for patients and team members. Enhance physical and emotional safety by sharing observations and concerns, listening and taking action when needed. Use "I" or "we" when sharing feedback; saying "you" may make others feel defensive.

6 | Connect with others

Smiling and making a personal connection help people feel comfortable interacting. Honoring differences and being kind build trust and a sense of safety. Engaging with others helps them feel included.

7 | Walk in their shoes

Seeking to understand various points-of-view and experiences can help patients, their families and team members feel valued. People may think or act in ways that are unfamiliar to you, and these are opportunities to learn from them. Consider how your actions affect others.

8 | Be encouraging

Giving encouragement shows you care about others' well-being. Notice and celebrate people's growth, effort and contributions whenever you can to inspire them and those around them. Vary your approach with each person to match the way they like to be treated.

9 | Express gratitude

Sharing a heartfelt, timely "thank you" can make others feel appreciated. Be sure to include everyone involved. Ask others how they like to receive thanks – publicly, in-person or privately with a note or via the team member Applause system.

10 | Grow and develop

Committing to personal development can help you gain new skills, knowledge and confidence. Sharing your expertise can help others grow, too. Seek and receive feedback openly to enhance your self-awareness and abilities.

Figure 16.12 Respectful behaviours at Virginia Mason

16.6 How does lean apply throughout the supply network?

Although most of the concepts and techniques discussed in this chapter are devoted to the management of stages within processes and processes within an operation, the same principles can apply to the whole supply network. In this context, the stages in a process are the whole businesses, operations or processes between which products flow. And as any business starts to approach lean it will eventually come up against the constraints imposed by the lack of synchronisation of the other operations in its supply network. So, achieving further gains must involve trying to spread lean practice outwards to its partners. Ensuring entire supply networks are lean is clearly a far more demanding task than doing the same within a single process. The nature of the interaction between whole operations is far more complex than between individual stages within a process. A far more complex mix of products and services is likely to be being provided and the whole network is likely to be subject to a less predictable set of potentially disruptive events. To make a supply network lean means more than making each operation in the network lean. Rather one needs to apply the lean philosophy to the supply chain as a whole, something that is extremely challenging in practice.

Essentially, the principles of lean are the same for a supply network as they are for a process. Fast throughput throughout is still valuable and will save cost. Lower levels of inventory will still make it easier to achieve lean. Waste is just as evident (and even larger) at the level of the supply network and reducing waste is still a worthwhile task. Streamlined flow, exact matching of supply and demand, enhanced flexibility and minimising variability are all still tasks that will benefit the whole network. The principles of pull control can work between whole operations in the same way as they can between stages within a single process. In fact, the principles and the techniques of lean are essentially the same no matter what level of analysis is being used. In addition, because lean is being implemented on a larger scale, the benefits will also be proportionally greater.

> **Operations principle**
> The advantages of lean apply at the level of the process, the operation and the supply network.

One of the weaknesses of lean is that it is difficult to achieve when conditions are subject to unexpected disturbance. This is especially a problem with applying lean principles in the context of the whole supply network. Whereas unexpected fluctuations and disturbances do occur within operations, local management has a reasonable degree of control that it can exert in order to reduce them. Outside the operation, within the supply network, it is far more difficult. Nevertheless, it is generally held that, although the task is more difficult and although it may take longer to achieve, the aim of lean is just as valuable for the supply network as a whole as it is for an individual operation.

Critical commentary

It can be argued that lean principles are sometimes taken to an extreme. When lean ideas first started to have an impact on operations practice, some authorities advocated the reduction of between-process inventories to zero. While in the long term this provides the ultimate in motivation for operations managers to ensure the efficiency and reliability of each process stage, it does not admit the possibility of some processes always being intrinsically less than totally reliable. An alternative view is to allow inventories (albeit small ones) around process stages with higher-than-average uncertainty. This at least allows some protection for the rest of the system. The same ideas apply to lean delivery between factories. Severe disruption to supply chains, such as from the effects of the Japanese tsunami in 2011 and COVID-19 in 2020, caused many overseas factories to close for a time because of a shortage of key parts.

Responsible operations

In every chapter, under the heading of 'Responsible operations', we summarise how the particular topic covered in the chapter touches upon important social, ethical and environmental issues.

There are two elements of lean operations that both bind the concept together and where responsible operations managers need to think carefully about how they treat them. The first is the sometimes relentless focus on waste elimination, and the other is how the 'respect for people' concept is practised.

Waste elimination is clearly central to lean, but it can be taken to extremes. There can be important and necessary elements to any job that are not immediately obvious to those who don't do the job. A naïve or over-enthusiastic approach to identifying and eliminating waste can remove elements that play a non-obvious but important role. Just as important, an obsession with waste elimination can leave operating staff feeling pressured and stressed. For example, periods spent not working (downtime) would generally be regarded as 'waste'. But, even in relatively routine jobs, such periods could be necessary for respite and reflection. Removing them could simply add to a sense of being pressurised. An excessive focus on waste reduction can clash with the other important aspect of lean philosophy – respect for people.

While advocates of lean highlight the importance of the 'respect for people' pillar, few seek to explain what it is or why it's important. Perhaps it's because the concept is fundamentally awkward given the anticipated response: '*of course you must be nice to people*'. What is perhaps less obvious is the cost of incivility on measures of productivity and quality of work output. Bullying and general rudeness towards others are surprisingly common. According to Professors Porath and Pearson[10] writing in the *Harvard Business Review*, 98 per cent of workers reported experiencing uncivil behaviour, with over half stating they were treated rudely at least once every week. The research identified the following consequences of incivility:

- 48 per cent intentionally decreased their work effort.
- 38 per cent intentionally decreased the quality of their work.
- 66 per cent said that their performance declined.
- 78 per cent said that their commitment to the organisation declined.
- 12 per cent said that they left their job because of the uncivil treatment.
- 25 per cent admitted to taking out their frustration on customers.

Further, the effects of incivility are not limited to those on the receiving end of rudeness but also negatively affect the productivity of those who observe the poor behaviour. And if anyone is still not convinced of the importance of 'respect for people', in July 2018 an emergency physician gave a presentation at an international lean healthcare conference, with the no-nonsense title: 'Assholes kill people'.

Summary answers to key questions

16.1 What is lean?

- Lean attempts to deliver exactly what customers value (perfect quality), in exact quantities (neither too much nor too little), exactly when needed (not too early nor too late), exactly where required (in the right location) and at the lowest possible cost.
- Lean can be seen as having three distinct, roles, each of which take a different perspective.
  - — Lean is a philosophy that places customer value at the heart operations.
  - — Lean is an approach to planning and controlling flow in operations.
  - — Lean is a set of ideas to improve operations performance.

16.2 How does lean consider flow?

- Lean uses 'pull control' to manage flow of products and services as opposed to the more conventional 'push control'. The most common method to support pull control is the use of 'kanbans', which are simple signalling devices that prevent the accumulation of materials, customer and information inventories.
- Lean advises the removal of inventory to expose operations problems. It places less emphasis on maximising capacity utilisation given there is little value in producing products or processing customers if the next stage in the process is not ready to receive them.

16.3 How does lean consider (and reduce) waste?

- Lean identifies three causes of waste – muda (presence of activities that add no value to the customer), mura (processes that lack consistency) and muri (unnecessary or unreasonable requests placed on the process).
- Lean identifies seven types of muda that, together, form impediments to flow. They are waste from over-production, waiting time, transportation, over-processing, inventory, motion and defects (rework). An eighth waste is also frequently cited: failing to access and nurture the talent and knowledge of an organisation's people.
- Other methods of removing waste include examining the layout of the operation, improving process flexibility and minimising variability.

16.4 How does lean consider improvement?

- Continuous improvement forms the core of lean operations, but objectives are often expressed as ideals.
- The Rapid Process Improvement Workshop (RPIW) is a common vehicle for leveraging improvement at a process level.
- Methods for effecting improvement include halting the process when a defect is discovered, using value-stream maps, using the 5S technique and using visual management.

16.5 How does lean consider the role of people?

- Lean is as much about people as it is about methods and tools. Cultural change is an important goal of lean implementation, with particular emphasis on the involvement of all staff in driving improvement on an ongoing basis (kaizen) so that lean becomes 'the way we do things around here'.
- Lean requires operations to pay attention to 'respectful' behaviours. The role of people, and respectful behaviours, represent a central but frequently overlooked aspect of lean operations.

16.6 How does lean apply throughout the supply network?

- Most of the concepts and techniques of lean, although usually described as applying to individual processes and operations, also apply to the whole supply network.

CASE STUDY St Bridget's Hospital: seven years of lean[11]

When the decision was taken to introduce lean to Saint Bridget's hospital seven years ago, the stated intention was to enhance productivity, while at the same time improving the quality of patient care. St Bridget's is one of the main hospitals in the Götenborg area of Sweden. Run by a private company, St Bridget's is a little different from any other Swedish hospital: to its patients, treatment is free following payment of a minimal charge that is universal in Sweden. In recent years, St Bridget's has developed a reputation for being at the forefront of lean implementation in a hospital. Although initially sceptical about the applicability of lean to a complex service organisation, executives at the hospital (including the chief executive) now speak regularly at healthcare conferences about the experiences of implementing lean at St Bridget's.

'I never thought lean implementation would deliver the kind of benefits that it has', said Denize Ahlgren, who became Chief Executive five years ago. *'We initially adopted lean methods in the hope that we could reduce our costs and if I'm really honest, we did hope to do "more with less". But despite several RPIWs taking place, and our staff being almost evangelical about the difference the RPIWs had made to their work, it was hard to evidence tangible cost savings. After a while we realised that financial gains through continuous improvement methods like lean rely on an accumulation of many small gains over several years before it becomes visible at an organisational level, particularly in a complex organisation like a hospital. Further, finding ways to quantify those gains and directly attribute them to the implementation of lean methods was really challenging. As an organisation we took a leap of faith: that improving value to our patients, through the continuous elimination of waste, would deliver better and safer care that ultimately costs less'*.

After seven years of implementing lean methods, St Bridget's has developed a more mature understanding of lean as a philosophy. Chief Operating Officer Lars Andersson reflects on what he calls St Bridget's 'lean journey': *'When we started, we saw lean as very much a set of tools for identifying and eliminating waste. Given our pressures around finance and the need to see more patients with the same (or less) resource, staff were increasingly seeing lean as a cost-cutting exercise. Subsequently, doctors were resistant to engaging with lean, expressing their contempt of efforts to standardise their work and reduce medical resources to save money'*. Denize recalled one incident where a cardiac surgeon began yelling abuse at a young nurse because they had only laid out one set of surgical instruments, rather than the three sets he usually had (just in case any of them broke). *'It wasn't even the nurse's fault'*, said Denize, *'the configuration of the surgical tray had been changed as a result of a RPIW where a decision had been taken to reduce the time it took to set up the room and help theatres start patients' operations on time. The change also reduced the running costs of theatres since all instruments needed to be sterilised after surgery, whether they were used or not. Ultimately, reducing set-up times allowed us to increase the number of operations we were able to perform in one day – which means more patients getting timely treatment'*. Reflecting on the incident, Denize highlights the importance of engaging people in improvement. *'No one likes to have improvement "done" to them. Surgeons are particularly sensitive to change, especially if they feel managers are taking decisions without consulting their expertise or understanding why a process is set-up that way'*, she said. *'This incident certainly exemplified the cultural problems we were having as an organisation at the time'*. While the surgeon had good reasons for rejecting the change (surgical instruments can break, and therefore having an additional set just in case was considered essential), the behaviour directed at the young nurse was not acceptable. The surgeon was required to complete a course on respectful behaviour but sadly the young nurse took a period of absence before leaving St Bridget's and nursing for good.

Anders Karlsson is the head of a team of improvement facilitators at St Bridget's. He describes the last seven years as a learning process, through which they have developed a more mature understanding of lean operations. *'We now recognise that lean has to be more than a set of tools to streamline processes, we must engage, empower and connect our staff to work together and collectively make St Bridget's the best place for staff to work and deliver the best care to our patients. We now have a significant number of staff trained in lean methods or having participated in RPIWs. When we started, we did small RPIWs that were often very successful, but on reflection those areas where improvement took place were disconnected from other parts of the organisation and they were never going to deliver the organsiational impact the leadership team aspired to. Today though, everyone in this organisation understands the importance of continuous improvement. Not everyone has completed lean training, but I'd like to think that everyone knows someone who has been trained in lean methods and they can work together to improve together. Doctors, nurses, anaesthestists, even our support staff, regularly collaborate and share improvement ideas. Improvement is no longer considered an "add on" or something that's nice to do if we have time, it's now an integral part of our everyday work'*.

A different style of leadership

Denize is keen to point out that while stable leadership is important for successful lean implementation in any organisation, St Bridget's found adopting a different leadership 'style' was also important. *'Perhaps one of the most difficult aspects of lean implementation is the need for a different kind of leadership. Most of us become leaders because we have demonstrated that we are good at problem solving and fire-fighting. In my career I've seen many a leader banging their fists on the table, demanding staff fix problems today! I've also seen professional staff on the receiving end of this treatment reduced to tears and hiding from their manager in the toilets! Lean teaches us that the people who do the work, know the work, are best placed to improve the work. One of the most important lessons I share when I'm talking at conferences is that leaders, whether they are the chief executive, or a world elite heart surgeon, or a ward manager, or head of the hospital canteen. . . leaders must empower staff to lead change at the front line. We must become "problem framers" rather than problem solvers. By framing problems, we give permission to the people who know the work to improve the work'.*

A clear set of values: The St Bridget's Way

In the years prior to Denize's appointment as CEO, St Bridget's had experienced several changes of leadership. With each new leader came a new solution for improving performance. Staff were frustrated and reported they were tired of managers claiming allegiance to the latest 'management fad'. St Bridget's annual staff survey showed staff were not happy working at the hospital, and bullying was reportedly commonplace; further, patient survey data revealed many patients would not recommend the hospital to family and friends. *'When I became CEO I knew there was a lot of work to do to turn the organisation around. Making it a better place for our staff to work and to deliver the highest-quality, safe care to patients was my number one priority'.* Immediately upon appointment Denize set out to listen to and consult with all 4,500 employees across the whole organisation. She explains: *'We used a crowdsourcing platform to "gather the wisdom of the crowds". We asked all staff to tell us what it feels like to work here and what they feel the expected behaviours and values of St Bridget's hospital should be. The exercise was very successful, almost all employees expressed their opinions directly to the leadership via the platform. From this exercise, leadership were able to clearly articulate a set of organisational values that all staff could subscribe to'.* Denize explains her belief that complex organsiations require simply stated values and objectives. *'That way you can orient all of that towards achieving the same shared goals'.*

At St Bridget's these three organisational values can be found proudly displayed on walls throughout the hospital:

1 To provide the best possible care (no avoidable harm, and no waste).

2 Joy and pride in work (high staff morale, respectful behaviours and empowered to lead improvement).

3 The patient is at the centre of everything we do.

'I expect every member of our organisation to know what these values are', said Denize. *'They represent our True North. Everything we do in this organisation, including all improvement activity, should align to our values. We call this "The St Bridget's Way", and it underpins everything we do'.*

Involving everyone

Getting the senior doctors to buy into lean as *the way we do things around here* has been very successful at St Bridget's, explains the Operations Director, Ingrid Karlsson. *'It's taken seven years but they have all done the lean training, they've all been through the process, they're all converts, you know. Take Par Gustafson for example. He's a very mature radiologist, he came here about three years ago from another hospital. Very, very experienced but was uncomfortable about some of the lean methods and associated language (gemba and kaizen for example). He came to me and said "you know, this language, it's all a bit funny". We put him on the lean training course and now, he's an evangelist! He's completely converted. You know, he kind of just gets it'.*

Highlighting the important role of lean principles and methods at St Bridget's, Ingrid recalls a recent RPIW to shorten the diagnostics time for patients with suspected cancer. Ingrid explains: *'We've got four state-of-the-art head scan machines and two very expensive and very powerful body scanners. So, we scan a lot of people, like at most hospitals. Last week we brought 25 people together in a room, radiographers, stenographers, surgeons, all working on this together, and porters, because it's really important to get a cross-section of staff, as we often have very different perspectives. Anyway, we were looking at streamlining the cancer pathway, but as an aside people started saying "Wow, we scan a huge number of patients every day, seven days a week, and we do a full blood test on every patient that gets scanned. Why are we doing that?" Nobody actually knew – we just did it because that's what we've always done! Now 75 per cent of patients that get scanned don't get a blood test. That is massive. It means 75 per cent of patients get a better experience because they're not getting a needle in their arm, they're not waiting. There's pressure taken off the radiology team, there's pressure taken off the pathology team and it's a much better patient experience'.*

It works, it makes things better for the patients

As more parts of the hospital became convinced of the effectiveness of the lean approach, the improvements to patient flow and quality started to accumulate. Some of the first improvements were relatively simple, such as a change of signage (to stop patients getting lost). As part of their lean training, staff are required to implement lean methods such as 5S and visual management. The benefits from these simple tools can be very surprising. In pharmacy for example, it was discovered that 30 per cent of medicines were out of date.

Clearing those medicines not only improved patient safety but also led staff to think about better (more visual) ways of managing stock to avoid running out of some medicines and over-stocking others. This simple task also led to a review of pharmaceutical supplies in collaboration with staff in procurement. *'We were lucky some of the guys working in procurement had lean training and they really understood how lean could improve the performance of their department'*, said Ingrid. *'We estimate savings of over 1 million kr as a result of an ad hoc collaboration between Ranjiv in pharmaceuticals and the guys in procurement – and we didn't even need to run an RPIW!'*

Better never stops!

Denize Ahlgren feels the hospital has come a very long way on its lean journey, but she says it still has a long way to go. *'We have made some impressive gains as an organisation, however, we must keep on improving. Our financial situation is slowly improving but we can't be complacent. Since I took over as CEO I have been very careful not to link our improvement work to saving money. We have recently introduced a waste reduction programme (WRP) where we ask staff to document the waste they have removed through improvement work. This allows us to capture financial savings in ways that align to the goals of the organisation. WRP has been incredibly successful, people are really keen to remove waste. The irony here is that staff want to take it a step further and quantify that waste in terms of cost savings! They are proud to tell us how they have saved money and how much they have saved!'*

'But there's always more we can do', says Denize. *'I recall a few years back how one of our clinicians, Fredrik Olsen, Chief Physician at St Bridget's lower-back pain clinic, thought that his clinic could benefit from a more radical approach. "We need to go to the next level" he said. The whole of Toyota's philosophy is concerned with smooth synchronous flow, yet we haven't fully got our heads round that here. I know that we are reluctant to talk about 'inventories' of patients, but that is exactly what waiting rooms are. They are 'stocks' of people, and we use them in exactly the same way as pre-lean manufacturers did – to buffer against short-term mismatches between supply and demand. What we should be doing is tackling the root causes of the mismatch. Waiting rooms are stopping us from moving towards smooth, value-added, flow for our patients'.*

Fredrik had proposed scrapping the current waiting room for the lower-back pain clinic and replacing it with two extra consulting rooms to add to the two existing consulting rooms. Patients would be given appointments for specific times rather than being asked to arrive 'on the hour' (effectively in batches) as at present. A nurse would take the patients' details and perform some preliminary tests, after which they would call in the specialist physician. Staffing levels during clinic times would be controlled by a nurse, who would also monitor patient arrival, direct them to consulting rooms and arrange any follow-up appointments (for MRI scans, for example).

'I'm still not sure about Fredrik's proposal', admits Denize. *'It seems as though it might be a step too far. Doctors are expensive resources, we can't give patients individual appointment times because if they turn up late, then the doctor has to wait. Patients expect to wait until a doctor can see them and they are happy to do so, so I'm not sure what benefits would result from the proposal. We've now added a TV screen in all of the waiting rooms so that patients can be occupied while they wait. We can't afford to equip two new consulting rooms if they are not going to be fully utilised'.*

QUESTIONS

1. **Denize admits that the hospital initially implemented lean because it needed to save money but now feels that was a mistake. Do you agree that it's unwise to link lean operations with saving money? Outline the arguments for and against the implementation of lean to save money.**
2. **Using examples from the case study, explain what factors have enabled the successful implementation of lean at St Bridget's over the last seven years.**
3. **Denize still cannot see the benefits of Fredrik's proposal. What do you think they might be? Consider the following:**
   - **Are the benefits of scrapping the waiting room in the clinic worth the underutilisation of the four consulting rooms?**
   - **How could Fredrik convince Denize that having extra capacity is a good idea?**
4. **In the chapter, we discuss the central role of people in lean operations. What does this mean in practice, and what examples can you see within the case study that highlight the importance of understanding both the social and technical aspects of lean? (Tip: look at the example of the surgeon who was upset that his surgical instrument trays had been reduced.)**

Problems and applications

All chapters have 'Problems and applications' questions that will help you practise analysing operations. They can be answered by reading the chapter. Model answers for the first two questions can be found on the companion website for this text.

1 Define the concept of lean as it would apply to a hospital.

2 The Zucchero mail-order clothing company in Milan receives order forms, types in the customer details, checks the information provided from the customers and that the products are in stock, confirms payment and processes the order. During an average eight-hour day, 150 orders are processed. Generally, 225 orders are waiting to be processed or 'in progress'. It takes 20 minutes for all activities required to process an order. What is the throughput efficiency of the process?

3 Consider this record of an ordinary flight. '*Breakfast was a little rushed but left the house at 6.15. Had to return a few minutes later, forgot my passport. Managed to find it and leave (again) by 6.30. Arrived at the airport 7.00, dropped Angela off with bags at terminal and went to the long-stay car park. Eventually found a parking space after 10 minutes. Waited 8 minutes for the courtesy bus. Six-minute journey back to the terminal, we start queuing at the check-in counters by 7.24. Twenty-minute wait. Eventually get to check-in and find that we have been allocated seats at different ends of the plane. Staff helpful but takes 8 minutes to sort it out. Wait in queue for security checks for 10 minutes. Security decide I look suspicious and search bags for 3 minutes. Waiting in lounge by 8.05. Spend 1 hour and 5 minutes in lounge reading computer magazine and looking at small plastic souvenirs. Hurrah, flight is called 9.10, takes 2 minutes to rush to the gate and queue for further 5 minutes at gate. Through the gate and on to air bridge, there is continuous queue going onto plane, takes 4 minutes but finally in seats by 9.21. Wait for plane to fill up with other passengers for 14 minutes. Plane starts to taxi to runway at 9.35. Plane queues to take off for 10 minutes. Plane takes off 9.45. Smooth flight to Amsterdam, 55 minutes. Stacked in queue of planes waiting to land for 10 minutes. Touch down at Schiphol Airport 10.50. Taxi to terminal and wait 15 minutes to disembark. Disembark at 11.05 and walk to luggage collection (calling at lavatory on way), arrive luggage collection 11.15. Wait for luggage 8 minutes. Through customs (not searched by Netherlands security who decide I look trustworthy) and to taxi rank by 11.26. Wait for taxi 4 minutes. Into taxi by 11.30, 30 minutes' ride into Amsterdam. Arrive at hotel 12.00*'.

   (a) Analyse the journey in terms of value-added time (actually going somewhere) and non-value-added time (the time spent queuing etc.).
   (b) Visit the websites of two or three airlines and examine their business-class and first-class services to look for ideas that reduce the non-value-added time for customers who are willing to pay the premium.
   (c) Next time you go on a journey, time each part of the journey and perform a similar analysis.

4 An insurance underwriting process consists of the following separate stages.

Stage	Processing time per application (minutes)	Average work in progress before the stage
Data entry	30	50
Retrieve client details	5	1,500
Risk assessment	18	100
Inspection	15	50
Policy assessment	20	100
Dispatch proposal	10	100

What is the value-added percentage for the process? (Hint – use Little's law to work out how long applications have to wait at each stage before they are processed. Little's law is covered in Chapter 6.)

5 Examine the marking process of an assignment you are currently working on. What is the typical elapsed time between handing the assignment in and receiving it back with comments? How much of this elapsed time do you think is value-added time?

6 A production process is required to produce 980 of product X, 560 of product Y and 280 of product Z in a 4-week period. If the process works 7 hours per day and 5 days per week, devise a production schedule per hour that would meet this demand. (Tip: refer to the section on Levelling product or service schedules (Heijunka).)

7 Examine the value-added versus non-value-added times for some other services. For example, posting a letter (the elapsed time is between posting the letter in the box and it being delivered to the recipient). How much of this elapsed time do you think is value-added time? How might you reduce time that adds no value (NVA) from the perspective of the customer?

8 One physician, in an attempt to emphasise the need for civility, used the slogan, 'Assholes kill people'. What did they mean by this?

9 Re-examine the 'Operations in practice' example, 'The rise of the personal kanban'. Make your own 'kanban list', as described in the example.

10 How might a creative business such as an advertising agency or film studio adopt lean principles?

Selected further reading

Bicheno, J. and Holweg, M. (2016) ***The Lean Toolbox: The Essential Guide to Lean Transformation*****, 5th edn, PICSIE Books, Buckingham.**
A practical guide from two of the European authorities on all matters lean.

Mann, D. (2017) ***Creating a Lean Culture: Tools to Sustain Lean Conversions*****, 3rd edn, Productivity Press, New York.**
Treats the behavioural side of lean.

Modig, N. and Ahlstrom, P. (2012) ***This is Lean: Resolving the Efficiency Paradox*****, Rheologica Publishing, Stockholm.**
This book provides a very practical guide to what lean is and its application in a variety of sectors. Not only does the book demonstrate a clear understanding of how the various aspects of lean come together, it does it in a very readable way.

Womack, J.P., Jones, D.T. and Roos, D. (2007) ***The Machine that Changed the World*****, Simon & Schuster, London.**
One of the most influential books on operations management practice of the last fifty years. Firmly rooted in the automotive sector but did much to establish lean/JIT.

Womack, J.P. and Jones, D.T. (2003) ***Lean Thinking: Banish Waste and Create Wealth in Your Corporation*****, Free Press, New York, NY.**
Some of the lessons from The Machine that Changed the World *but applied in a broader context.*

Notes on chapter

1. The latest edition (at the time of writing) is Womack, J.P., Jones, D.T. and Roos, D. (2007) *The Machine that Changed the World*, Simon and Schuster, London.
2. The information on which this example is based is taken from: Toyota website, https://global.toyota/en/company/ (accessed September 2021).
3. Ohno, T. and Bodek, N. (2019) *Toyota Production System: Beyond Large-scale Production,* Productivity Press, New York, NY.
4. The information on which this example is based is taken from: Corbett, S. (2004) Applying lean in offices, hospitals, planes, and trains, presentation at The Lean Service Summit, Amsterdam, 24 June.
5. The information on which this example is based is taken from: Burgess, M. (2018) Airbus is going to start putting beds in airplane cargo holds, *Wired*, 11 April, http://www.wired.co.uk/article/airbus-sleeping-pods-naps-cargo-hold-zodiac-330 (accessed September 2021).
6. Example written and supplied by Janina Aarts and Mattia Bianchi, Department of Management and Organization, Stockholm School of Economics.
7. For a discussion of these ideas, see, for example, Burnes, B. (2004) Kurt Lewin and the planned approach to change: a re-appraisal, *Journal of Management Studies*, 41 (6), 977–1002.
8. The information on which this example is based is taken from: Onetto, M. (2014) When Toyota met e-commerce: lean at Amazon, *McKinsey Quarterly*, No. 2; Liker, J. (2021) *The Toyota Way: 14 Management Principles from the World's Greatest Manufacturer*, 2nd edn, McGraw Hill, New York, NY.
9. Some of the information on which this example is based is taken from: Porath, C. and Pearson, C. (2013) The price of incivility, *Harvard Business Review*, 91 (1–2), 115–21; Chafetz, L.A., Forsythe, A.M., Kirby, N., Blackmore, C.C. and Kaplan, G.S. (2020) Building a culture of respect for people, *NEJM Catalyst Innovations in Care Delivery*, 1 (6).
10. Porath, C. and Pearson, C. (2013) The price of incivility, *Harvard Business Review*, 91 (1–2), 115–21.
11. This case is based on the work of several real hospitals, in Scandinavia and the rest of the world, that have used the concepts of lean operations to improve their performance. However, all names and places are fictional and no connection to any specific hospital is intended.

17 Quality management

KEY QUESTIONS

17.1 What is quality and why is it so important?

17.2 What steps lead towards conformance to specification?

17.3 What is total quality management (TQM)?

INTRODUCTION

Quality management has always been an important part of operations management, but its position and role within the subject have changed. At one time it was seen largely as an essential, but 'routine', activity that prevented errors having an impact on customers (and would have been located unambiguously in the 'Deliver' section of this text). And that function is still there. But increasingly quality management is viewed as also having a part to play in how operations improve. Quality management can contribute to improvement by making the changes to operations processes that lead to better outcomes for customers. In fact, in most organisations, quality management is one of the main drivers of improvement. It is also the only one of the five 'operations performance objectives' to have its own dedicated chapter in this text. Partly this is because of this central role of 'quality' in improvement. But it is also because in many organisations a separate function is devoted exclusively to the management of quality. Figure 17.1 shows where quality management fits into the model of operations activities.

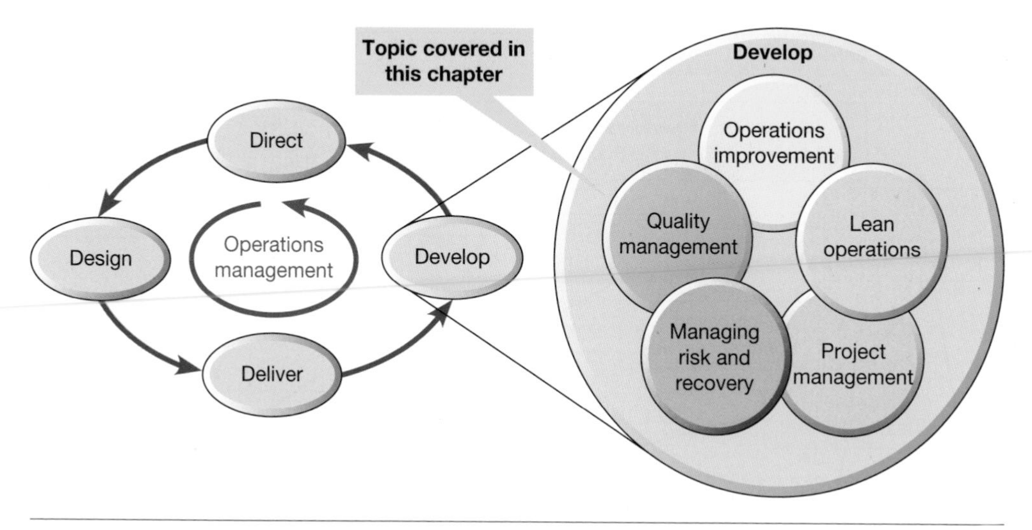

Figure 17.1 This chapter examines quality management

17.1 What is quality and why is it so important?

It is worth revisiting some of the arguments that were presented in Chapter 2 regarding the benefits of high levels of quality. It will help to explain why quality is seen as being so important by most operations. Figure 17.2 illustrates the various ways in which quality improvements can affect other aspects of operations performance. Revenues can be increased by better sales and enhanced prices in the market. At the same time, costs can be brought down by improved efficiencies, productivity and the use of capital. So, a key task of the operations function must be to ensure that it provides quality goods and services, to both its internal and external customers.

The operation's view of quality

There are many definitions of quality. Here we define it as *consistent conformance to customers' expectations*. The use of the word 'conformance' implies that there is a need to meet a clear specification. 'Consistent' implies that conformance to specification is not an ad hoc event but that the service or product meets the specification because quality requirements are used to design and run the processes that produce services or products. The use of 'customers' expectations' recognises that the service or product must take the views of customers into account, which may be influenced by price. Also note the use of the word 'expectations' in this definition, rather than needs or wants.

> **Operations principle**
> Quality is the consistent conformance to customers' expectations.

Customers' view of quality

Past experiences, individual knowledge and history will all shape customers' expectations. Furthermore, customers may each *perceive* a service or product in different ways. One person may perceive a long-haul flight as an exciting part of a holiday; the person in the next seat may see it as a necessary chore to get to a business meeting. So quality needs to be understood from a customer's point of view because, to the customer, the quality of a particular service or product is whatever they perceive it to be. Also, customers may be unable to judge the 'technical' specification of the service

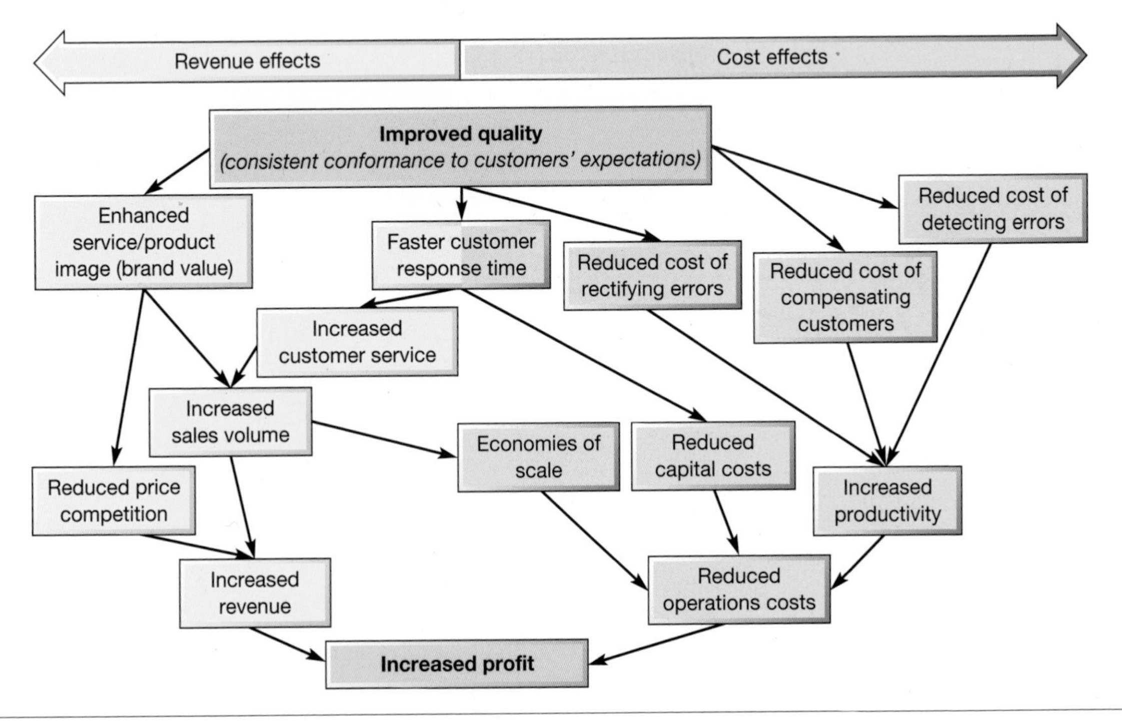

Figure 17.2 Higher quality has a beneficial effect on both revenues and costs

or product and so use surrogate measures as a basis for their perception of quality. For example, a customer may find it difficult to judge the technical quality of dental treatment, except insofar as their teeth do not give any more trouble. The customer may therefore perceive quality in terms the demeanour of the dentist and technician, and how they were treated.

Operations principle
Quality is multifaceted and its individual elements differ for different operations.

OPERATIONS IN PRACTICE — Quality at two operations: Victorinox and Four Seasons

Victorinox and the Swiss Army Knife[1]

The famous Swiss Army Knife is made by the Victorinox Company in its factory in the Swiss town of Ibach. The company has numerous letters from its customers testifying to their product's quality and durability. For example, 'I was installing a new piece of equipment in a sewage treatment plant . . . The knife slipped out of my hand and fell into the aeration tank . . . that is extremely corrosive to metals. Four years later, I received a small parcel with a note from the supervisor of the plant. They had emptied the aeration tank and found my knife . . . it was in astonishingly good condition. I can assure you that very few products could have survived treatment like this, the components would have dissolved or simply disappeared'.

Today, the Victorinox factory assembles 27,000 knives a day. More than 450 steps are required in its manufacture. But a major threat to sales that has been growing is the appearance on the market of fake 'Swiss Army' knives, made mostly in China. Their defence against these fakes is quality says CEO Carl Elsener. '*We have exhausted all legal means for the brand protection of our popular products. Our best means of protection is quality which remains unsurpassed and speaks louder than words*'. It is the 'Victorinox quality control system' that is at the heart of this defence.

Receiving inspection ensures that incoming materials conform to quality specifications. The Victorinox laboratory guarantees that only steel and plastic that comply with its rigorous quality standards are used. Metallurgical inspection is also used by polishing samples, casting them in plastic and etching with an acid. This allows faults in materials to be easily detected. The laboratory also has an 'edge retention test', using special equipment to test the ability of material to retain its edge during a series of cutting tests. During the production of the knives, process control is employed at all stages of the production process, and is the responsibility of the company's employees, who use it to maintain, implement and improve the quality of products. They are also responsible for following the company's quality procedures and for continuous, measurable improvement. At the end of the production process, the 'Final Inspection Department' are responsible for ensuring that all products conform to requirements. Any non-conforming products are isolated and identified. Non-conforming parts are repaired or replaced at the repair department.

Four Seasons Canary Wharf[2]

Four Seasons Hotels are famed for their quality of service, winning countless awards. From its inception the group has had the same guiding principle, 'to make the quality of our service our competitive advantage'. The company has what it calls its Golden Rule: 'Do to others (guests and staff) as you would wish others to do to you'. It guides the whole organisation's approach to quality. '*Quality service is our distinguishing edge and the company continues to evolve in that direction. We are always looking for better, more creative and innovative ways of serving our guests*', says Michael Purtill, the General Manager of the Four Seasons Hotel Canary Wharf in London. '*We have recently refined all of our operating standards across the company enabling us to further enhance the personalised, intuitive service that all our guests receive. All employees are empowered to use their creativity and judgement in delivering exceptional service and making their own decisions to enhance our guests' stay. For example, one morning an employee noticed that a guest had a flat tyre on their car and decided of his own accord to change it for them, which was very much appreciated by the guest.*

'The golden rule means that we treat our employees with dignity, respect and appreciation. This approach encourages them to be equally sensitive to our guests' needs and offer sincere and genuine service that exceeds expectations. Just recently one of our employees accompanied a guest to the hospital and stayed there with him for entire afternoon. He wanted to ensure that the guest wasn't alone and was given the medical attention he needed. The following day that same employee took the initiative to return to the hospital (even though it was his day off) to visit and made sure that that guest's family in America was kept informed about his progress.

'At Four Seasons, we believe that our greatest asset and strength are our people. We pay a great deal of attention to selecting the right people with an attitude that takes great pride in delivering exceptional service. We know that motivated and happy employees are essential to our service culture and are committed to developing our employees to their highest potential. Our extensive training programmes and career development plans are designed with care and attention to support the individual needs of our employees as well as operational and business demands. In conjunction with traditional classroom-based learning, we offer tailor-made web-based learning featuring exceptional quality courses for all levels of employee. Career wise, the sky is the limit, and our goal is to build lifelong, international careers with Four Seasons.

'Our objective is to exceed guest expectations, and feedback from our guests and our employees is an invaluable barometer of our performance. We have created an in-house database that is used to record all guest feedback (whether positive or negative). We also use an online guest survey and guest comment cards, which are all personally responded to and analysed to identify any potential service gaps.

'We continue to focus on delivering individual personalised experiences and our Guest History database remains vital in helping us to achieve this. All preferences and specific comments about service experience are logged on the database. Every comment and every preference is discussed and planned for, for every guest, for every visit. It is our culture that sets Four Seasons apart; the drive to deliver the best service in the industry that keeps our guests returning again and again'.

Reconciling the operation's and the customer's views of quality

The operation's view of quality is concerned with trying to meet customer expectations. The customer's view of quality is what they *perceive* the service or product to be. To create a unified view, quality can be defined as the degree of fit between customers' expectations and customer perception of the service or product.[3] Using this idea allows us to see the customers' view of quality of (and, therefore, satisfaction with) the service or product as the result of the customers comparing their expectations of the service or product with their perception of how it performs. If the service or product experience was better than expected, then the customer is satisfied, and quality is perceived to be high. If the service or product was less than their expectations, then quality is low, and the customer may be dissatisfied. If the service or product matches expectations, then the perceived quality of the service or product is seen to be acceptable. These relationships are summarised in Figure 17.3.

Both customers' expectations and perceptions are influenced by several factors, some of which cannot be controlled by the operation and some of which can be managed. Figure 17.4 shows some factors that will influence the gap between expectations and perceptions. This model of quality can help us understand how operations can manage quality and identify some of the problems in so doing. The bottom part of the diagram represents the operation's 'domain' of quality and the top part the customer's 'domain'. These two domains meet in the actual service or product, which is provided by the organisation and experienced by the customer. Within the operation's domain, management is responsible for designing the service or product and providing a specification of the quality to which the service or product has to be created. Within the customer's domain, their expectations are shaped by such factors as previous experiences with the particular service or product, the marketing image provided by the organisation and word-of-mouth information from other users. These expectations are internalised as a set of quality characteristics.

Operations principle

Perceived quality is governed by the magnitude and direction of the gap between customers' expectations and their perceptions of a product or service.

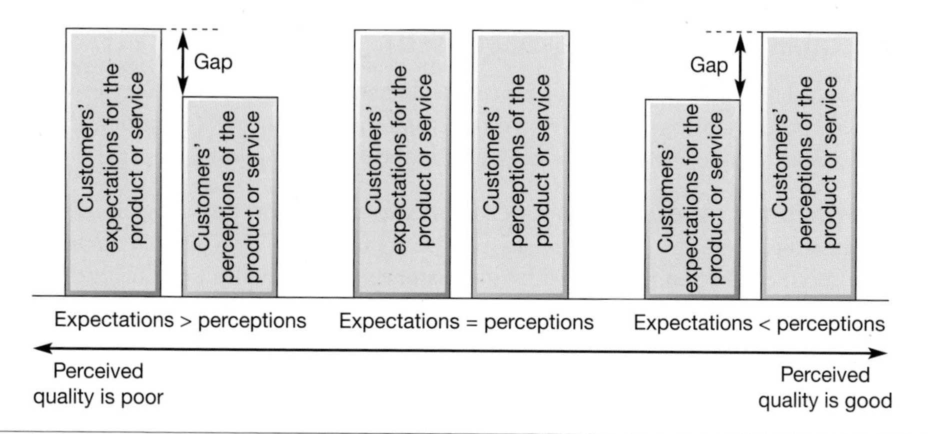

Figure 17.3 Perceived quality is governed by the magnitude and direction of the gap between customers' expectations and their perceptions of the service or product

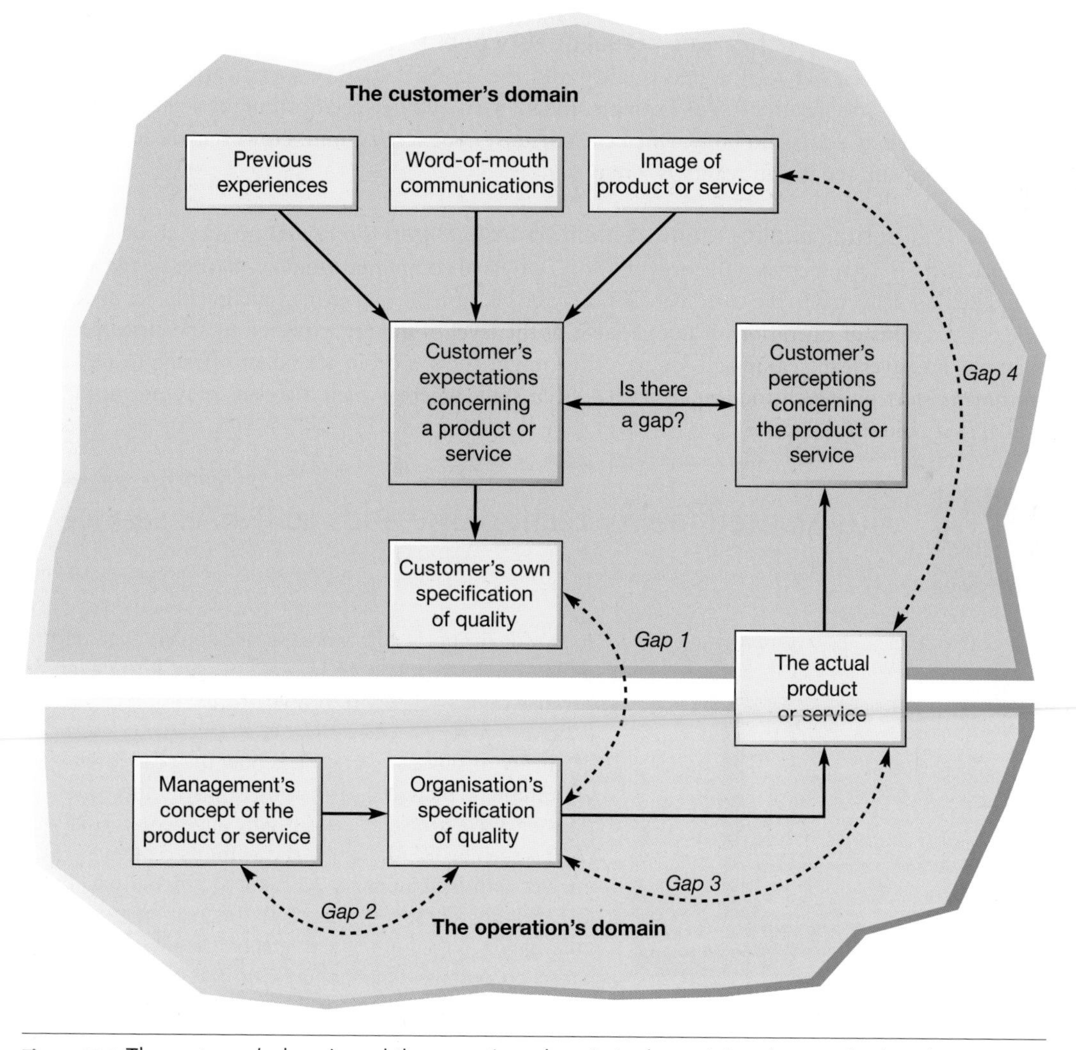

Figure 17.4 The customer's domain and the operations domain in determining the perceived quality, showing how the gap between customers' expectations and their perception of a service or product could be explained by one or more gaps elsewhere in the model

Source: Adapted from Parasuraman, A. *et al.* (1985) A conceptual model of service quality and implications for future research', *Journal of Marketing*, 49, Fall.

How can quality problems be diagnosed?

Figure 17.4 also shows how quality problems can be diagnosed. If the perceived quality gap is such that customers' perceptions of the service or product fail to match their expectations of it, then the reason (or reasons) must lie in other gaps elsewhere in the model as follows.

Gap 1: The customer's specification–operation's specification gap Perceived quality could be poor because there may be a mismatch between the organisation's own internal quality specification and the specification that is expected by the customer. For example, a car may be designed to need servicing every 10,000 kilometres but the customer may expect 15,000-kilometre service intervals.

Gap 2: The concept–specification gap Perceived quality could be poor because there is a mismatch between the service or product concept (see Chapter 5) and the way the organisation has specified quality internally. For example, the concept of a car might have been for an inexpensive, energy-efficient means of transportation, but the inclusion of a climate-control system may have both added to its cost and made it less energy-efficient.

Gap 3: The quality specification–actual quality gap Perceived quality could be poor because there is a mismatch between actual quality and the internal quality specification (often called 'conformance to specification'). For example, the internal quality specification for a car may be that the gap between its doors and body, when closed, must not exceed 7 mm. However, because of inadequate equipment, the gap in reality is 9 mm.

Gap 4: The actual quality–communicated image gap Perceived quality could be poor because there is a gap between the organisation's external communications or market image and the actual quality delivered to the customer. This may be because the marketing function has set unachievable expectations or operations is not capable of the level of quality expected by the customer. For example, an advertising campaign for an airline might show a cabin attendant offering to replace a customer's shirt on which food or drink has been spilt, whereas such a service may not in fact be available should this happen.

OPERATIONS IN PRACTICE

Augmented reality technology adds to IKEA's service quality[4]

A technological revolution swept through retailing operations when online shopping began to take an increasing share of customers' spending. But so-called 'bricks and mortar' retailers – those with a physical presence on the high street or in out-of-town sites – are using technology to provide the quality of service that their online rivals find difficult to match. It is an important issue for retailers because quality of service is a key factor in promoting customer loyalty. And, to customers, an important element of quality of service in retailing is how they can interact with staff and products, particularly to answer their questions or 'solve problems'. This is why augmented reality (AR) is seen by IKEA as an ideal way for its customers to interact with their products. AR is defined by Gartner, the research and advisory company, as the 'real-time use of information in the form of text, graphics, audio and other virtual enhancements integrated with real-world objects. It is this 'real world' element that differentiates AR from virtual reality'.

The objective is to let customers visualise IKEA's furniture to get a realistic impression of how it would look in their home. They do this by using the 'IKEA Place' on their smartphones, an app that allows customers to view 3D representations (from a wide range of angles) of products before deciding which one they want. The app then directs them to the IKEA site to finalise their purchase. Using the app lets customers to make a 'reliable buying' decision, says Michael Valdsgaard, who is in charge of digital transformation at Inter IKEA, the holding company for IKEA. *'Most people postpone a purchase of a new sofa because they're not comfortable making the decision if they aren't sure the colour is going to match* [the rest of the room] *or it fits the style'*, he said. *'Now, we can give them* [those answers] *in their hands, while letting them have fun with home furnishing for free and with no effort. The most important thing for us is that we're not a tech company,* [but] *in order to sell furniture, we have to understand technology and try to move in the direction it's moving. The first* [augmented reality] *experience we had was more like a picture. You could put in a 3D object, but you couldn't really move around it or trust the size of it'.* But the later versions achieved 98 per cent accuracy, with true-to-life representations of the texture, fabric, lighting and shadows.

'Quality', 'quality of service' and 'quality of experience'

The definition of quality that we use here (the *consistent conformance to customers' expectations*) is useful because it can be used to describe 'quality' for either physical products or intangible services, or any offering that combines both tangible and intangible elements. However, not all authorities or individual operations view 'quality' in this way, which can lead to some confusion. For example, the term 'quality of service' (QoS) is often used to describe how an operation serves its customers by combining what we have called 'quality' with some or all of 'speed', dependability' and 'flexibility'. So in Chapter 2 we described the 'quality' of a supermarket as including such factors as the quality of the goods stocked, the cleanliness of the facilities and the courtesy of its staff. But, in assessing its QoS, the supermarket would probably want to include other factors such as the speed of service, the predictability of opening times, stockouts, the range of goods available and so on.

The limitation of QoS is that it may not capture the overall satisfaction with the service as perceived by the users. More useful, claim some providers of service, is to try to assess 'quality of experience' (QoE). Quality of experience is the overall acceptability of the service, *as perceived subjectively by the end user.* More formally it has been defined as 'the degree of delight or annoyance of the user of an application or service. It results from the fulfillment of his or her expectations with respect to the utility and/or enjoyment of the application or service in the light of the user's personality and current state'. QoE is clearly related to, but differs from, QoS in that it expresses, and focuses on, user satisfaction *both objectively and subjectively.*[5] QoS generally includes the aspects of a service that are under the control of the operation creating the service, whereas QoE involves both the aspects governed by the operation and those that are a function of the individual customer and the context in which the service is consumed. Figure 17.5 illustrates the relationship between these ideas, and Table 17.1 shows some typical factors that could be included in assessing the 'quality', 'quality of service' and 'quality of experience' of a supermarket operation and an online education service.

The QoE concept originated in, and has found its most extensive application in, telecommunications operations, information technology (IT) and consumer electronics. Yet its underlying principles have a far wider application. The QoE idea can be applied to any consumer-related business or service where the end user of a service or product could assess its quality subjectively and is partly dependent on the context in which it is consumed. But the dependence of this judgement on the individual subjectivity of the user, and on the context of its consumption (which is beyond the influence of the operation) is both a strength and a weakness for the concept of QoE. An obvious strength is that it focuses operations on the richness of how their offerings are experienced by users. A practical weakness is that it is a difficult idea to operationalise. Subjective metrics of QoE are difficult to design, expensive and time-consuming.

Service guarantees

One method of formalising quality standards from a customer's viewpoint is called a 'service guarantee'. A service guarantee is a promise to recompense the customer for service that fails to meet a defined quality level. It is a way of ensuring quality standards, and of overcoming customers' potential doubts regarding a service. It provides a way of encouraging and rewarding customers who report problems,

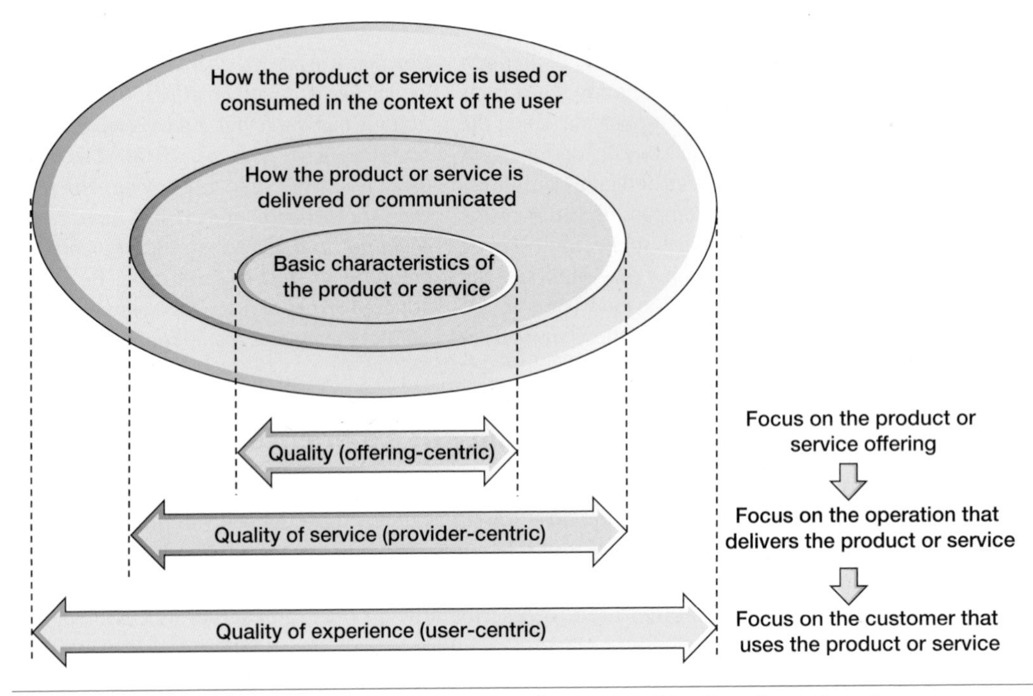

Figure 17.5 The relationship between 'quality', 'quality of service' and 'quality of experience'

Table 17.1 Examples of typical factors that could be included in assessing the 'quality', 'quality of service' and 'quality of experience' of a supermarket operation and an online education service

Operation	Quality	Quality of service (QoS) 'Quality' plus the following . . .	Quality of experience (QoE) 'Quality' plus 'Quality of service' plus the following . . .
Supermarket	Quality of goods Cleanliness Staff courtesy	Speed of service Stockouts Predictability of service (hours of opening) Range of goods stocked	Open when I want to use the service? Perception of speed of service Is what I want available? Nature of other users of the service
Online education service	Quality/accuracy of the lesson content Quality of production values	Various technical measures of network performance (such as throughput, packet loss, delay and jitter)	Relevance of content to me How the content works on the device I am using (display fidelity, transport/stalling quality, etc.) How easy is it for me to navigate the content?

so that the operation is made aware of them and can attempt to rectify them. A good guarantee should be meaningful, in the sense that it is based on customers' expectations. It should be easy to understand and explain exactly what level and type of quality is being promised and what the operation will do if it's not met (including what the customer should expect to receive in compensation). It should include a clear 'easy to invoke' mechanism for customers to trigger the guarantee, and appropriate training and empowerment, so that employees can cope when a guarantee is invoked by a customer.

The sandcone theory

An endorsement of the importance of quality as a driver of improvement generally comes from what is known as the 'sandcone theory'.[6] It comes from the idea that there is a generic 'best' sequence of improvement. It is called the sandcone theory because the sand is analogous to management effort and

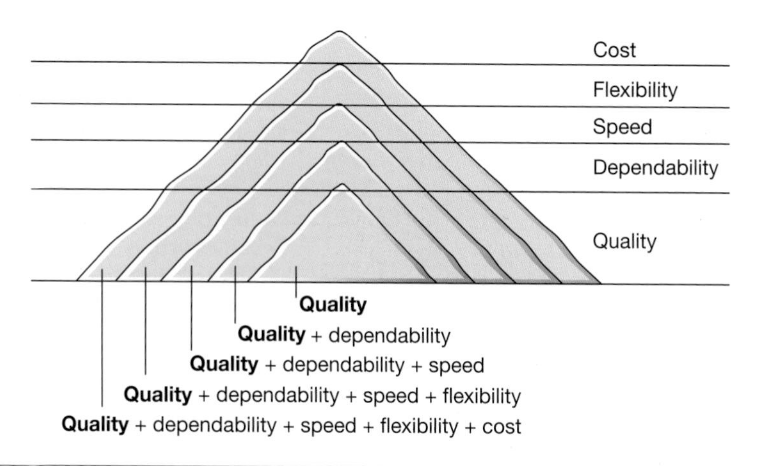

Figure 17.6 The sandcone model of improvement; cost reduction relies on a cumulative foundation of improvement in the other performance objectives

resources. Building a stable sandcone needs a stable foundation of quality, claims the theory, upon which one can build layers of dependability, speed, flexibility and cost (see Figure 17.6). Building up improvement is thus a cumulative process, not a sequential one. Moving on to the second priority for improvement does not mean dropping the first, and so on. According to the sandcone theory, the first priority should be *quality*, since this is a precondition to all lasting improvement. Only when the operation has reached a minimally acceptable level in quality should it then tackle the next issue, that of internal *dependability*. Importantly though, moving on to include dependability in the improvement process will actually require further improvement in quality. Once a critical level of dependability is reached, enough to provide some stability to the operation, the next stage is to improve the *speed* of internal throughput, but again only while continuing to improve quality and dependability further. Soon it will become evident that the most effective way to improve speed is through improvements in response *flexibility*, which is changing things within the operation faster. Again, including flexibility in the improvement process should not divert attention from continuing to work further on quality, dependability and speed. Only now, according to the sandcone theory, should *cost* be tackled head on.

OPERATIONS IN PRACTICE

Virgin Atlantic offers a service guarantee for aviophobes[7]

According to market research firm YouGov, nearly one in six people have a fear of flying. This is clearly an issue for airlines, which do not want a sixth of their potential market reluctant to use their services. That's why the airline Virgin Atlantic made an offer to its customers that if they booked a flight on one particular day, they would will be offered a free place on its 'Flying Without Fear' course. Better still, if they were not cured in time for their trip, the money they spent on their plane ticket would be returned. Shai Weiss, Chief Commercial Officer at Virgin Atlantic, said: *'We want everyone to be able to say "screw it, let's do it" and try something different, fly somewhere new. Hopefully by guaranteeing to cure people of one of the main things holding them back, we can inspire Britain to choose something more positive. Nothing should hold anyone back from seizing the day'.* The Virgin Atlantic 'Flying Without Fear' programme has become, according to the airline, the leading course in the industry, having helped 2,000–3,000 people every year to overcome their fear of flying. They say that they 'want to reassure you that you are not alone in your fear of flying, millions of people suffer similar anxieties. This course has been designed to help you conquer your fears of flying. Our aim is for you to take that holiday or business trip and actually enjoy it. Fear of flying is a phobia that many never, ever face. Now that you are here, we are the best people to help you to get rid of your fear'. People from 4 years old up to 87 (oldest so far) with fears ranging from mild anxiety to complete terror have all been helped by the programme and are now flying. Virgin says that is has a 98 per cent success rate. The programme runs courses more than 20 times a year, and claims to 'help you to learn new ways to think about flying'. And the guarantee? The airline states: 'If you can provide sufficient evidence that your fear of flying is not cured, we will provide a full refund for the flight purchased. This will be determined by our professionals who administer the Flying Without Fear programme'.

Conformance to specification

Conformance to specification means providing a service or producing a product to its design specification. It is usually seen as the most important contribution that operations management can make to the customer's perception of quality. We shall examine how it can be achieved in the remainder of this chapter by describing quality management as six sequential steps.

17.2 What steps lead towards conformance to specification?

Achieving conformance to specification requires the following steps:

Step 1 Define the quality characteristics of the service or product.

Step 2 Decide how to measure each quality characteristic.

Step 3 Set quality standards for each quality characteristic.

Step 4 Control quality against those standards.

Step 5 Find and correct causes of poor quality.

Step 6 Continue to make improvements.

Step 1 – Define the quality characteristics

Much of the 'quality' of a service or product will have been specified in its design and can be summarised by a set of quality characteristics. Table 17.2 shows a list of the quality characteristics that are generally useful. Also, many services have several elements, each with their own quality characteristics, and to understand the quality characteristics of the whole service it is necessary to understand the individual characteristics within and between each element of the whole service.

Step 2 – Decide how to measure each characteristic

These characteristics must be defined in such a way as to enable them to be measured and then controlled. This involves taking a very general quality characteristic such as 'appearance' and breaking it down, as far as one can, into its constituent elements. 'Appearance' is difficult to measure as such, but 'colour match', 'surface finish' and 'number of visible scratches' are all capable of being described in a more objective manner. They may even be quantifiable. Other quality characteristics pose more difficulty. The 'courtesy' of airline staff, for example, has no objective quantified measure. Yet operations with high customer contact, such as airlines, place a great deal of importance on the need to ensure courtesy in their staff. In cases like this, the operation will have to attempt to measure customer *perceptions* of courtesy.

Table 17.2 Quality characteristics for a car, a bank loan and an air journey

Quality characteristic	Car (Material transformation process	Bank loan (Information transformation process)	Air journey (Customer transformation process)
Functionality – how well the service or product does its job	Speed, acceleration, fuel consumption, ride quality, road-holding, etc.	Interest rate, terms and conditions	Safety and duration of journey, onboard meals and drinks, car and hotel booking services
Appearance – the sensory characteristics of the service or product: its aesthetic appeal, look, feel, etc.	Aesthetics, shape, finish, door gaps, etc.	Aesthetics of information, website, etc.	Decor and cleanliness of aircraft, lounges and crew
Reliability – the consistency of the product's or service's performance over time	Mean time to failure	Keeping promises (implicit and explicit)	Keeping to the published flight times
Durability – the total useful life of the service or product	Useful life (with repair)	Stability of terms and conditions	Keeping up with trends in the industry
Recovery – the ease with which problems with the service or product can be resolved	Ease of repair	Resolution of service failures	Resolution of service failures
Contact – the nature of the person-to-person contact that might take place	Knowledge and courtesy of sales staff	Knowledge and courtesy of branch and call centre staff	Knowledge, courtesy and sensitivity of airline staff

Variables and attributes

The measures used by operations to describe quality characteristics are of two types: variables and attributes. Variable measures are those that can be measured on a continuously variable scale (for example, length, diameter, weight or time). Attributes are those that are assessed by judgement and are dichotomous, i.e. have two states (for example, right or wrong, works or does not work, looks OK or not OK). Table 17.3 categorises some of the measures that might be used for the quality characteristics of the car and the airline journey.

Table 17.3 Variable and attribute measures for quality characteristics

Quality characteristic	Automobile		Airline journey	
	Variable	Attribute	Variable	Attribute
Functionality	Acceleration and braking characteristics from test bed	Is the ride quality satisfactory?	Number of journeys that actually arrived at the destination (i.e. didn't crash!)	Was the food acceptable?
Appearance	Number of blemishes visible on car	Is the colour to specification?	Number of seats not cleaned satisfactorily	Is the crew dressed smartly?
Reliability	Average time between faults	Is the reliability satisfactory?	Proportion of journeys that arrived on time	Were there any complaints?
Durability	Life of the car	Is the useful life as predicted?	Number of times service innovations lagged competitors	Generally, is the airline updating its services in a satisfactory manner?
Recovery	Time from fault discovered to fault repaired	Is the serviceability of the car acceptable?	Proportion of service failures resolved satisfactorily	Do customers feel that staff deal satisfactorily with complaints?
Contact	Level of help provided by sales staff (1 to 5 scale)	Did customers feel well served (yes or no)?	The extent to which customers feel well treated by staff (1 to 5 scale)	Did customers feel that the staff were helpful (yes or no)?

Step 3 – Set quality standards

When operations managers have identified how any quality characteristic can be measured, they need a quality standard against which it can be checked; otherwise they will not know whether it indicates good or bad performance. The quality standard is that level of quality which defines the boundary between acceptable and unacceptable. Such standards may well be constrained by operational factors such as the state of technology in the factory and the cost limits of making the product. At the same time, however, they need to be appropriate to the expectations of customers. But quality judgements can be difficult. If one airline passenger out of every 10,000 complains about the food, is that good because 9,999 passengers out of 10,000 are satisfied? Or is it bad because, if one passenger complains, there must be others who, although dissatisfied, did not bother to complain? And if that level of complaint is similar to other airlines, should it regard its quality as satisfactory?

Step 4 – Control quality against those standards

After setting up appropriate standards the operation will then need to check that the products or services conform to those standards, doing things right, first time, every time. This involves three decisions:

1. Where in the operation should it check that it is conforming to standards?
2. Should it check every service or product, or take a sample?
3. How should the checks be performed?

Where should the checks take place?

At the start of the process, incoming resources may be inspected to make sure that they are to the correct specification. For example, a car manufacturer will check that components are of the right specification. A university will screen applicants to try to ensure that they have a high chance of getting through the programme. During the process, checks may take place before a particularly costly process, prior to 'difficult to check', immediately after a process with a high defective rate, before potential damage or distress might be caused, and so on. Checks may also take place after the process itself to ensure that customers do not experience non-conformance.

Check every product and service or take a sample?

While it might seem ideal to check every single service or product, a sample may be more practical for a number of reasons:

- It might be dangerous to inspect everything. A doctor, for example, checks just a small sample of blood rather than taking all of a patient's blood! The characteristics of this sample are taken to represent those of the rest of the patient's blood.
- Checking everything might destroy the product or interfere with the service. Not every light bulb is checked for how long it lasts; it would destroy every bulb. Waiters do not check that customers are enjoying the meal every 30 seconds.
- Checking everything can be time-consuming and costly. It may not be feasible to check all output from a high-volume machine or to check the feelings of every bus commuter every day.

Also, 100 per cent checking may not guarantee that all defects will be identified. Sometimes it is intrinsically difficult. For example, although a doctor may undertake the correct testing procedure, they may not necessarily diagnose a (real) disease. Nor is it easy to notice everything. For example, try counting the number of 'e's on this page. Count them again and see if you get the same score.

Type I and type II errors

Although it reduces checking time, using a sample to make a decision about quality does have its own inherent problems. Like any decision activity, we may get the decision wrong. Take the example of a pedestrian waiting to cross a street. They have two main decisions: whether to continue waiting or to cross. If there is a satisfactory break in the traffic and the pedestrian crosses then a correct decision has been made. Similarly, if that person continues to wait because the traffic is too dense then they have again made a correct decision. There are two types of incorrect decisions or errors, however.

Table 17.4 Type I and type II errors for a pedestrian crossing the road

	Road conditions	
Decision	**Unsafe**	**Safe**
Cross	Type I error	Correct decision
Wait	Correct decision	Type II error

One incorrect decision would be if they decide to cross when there is not an adequate break in the traffic, resulting in an accident – this is referred to as a type I error. Another incorrect decision would occur if they decide not to cross, even though there was an adequate gap in the traffic – this is called a type II error. In crossing the road, therefore, there are four outcomes, which are summarised in Table 17.4.

OPERATIONS IN PRACTICE

Testing cars (close) to destruction[8]

Away from the public eye, at Millbrook Proving Ground, one of Europe's leading independent technology centres for the design, engineering, test and development of automotive and propulsion systems, they treat cars really badly. But all in a good cause. It is where vehicle manufacturers send their new vehicles to be tested, so that any glitches, from irritating rattles to more serious safety problems, can be exposed and corrected before the product reaches the market. The site, in Bedfordshire in the United Kingdom, is hidden away behind security fences and high embankments to discourage car paparazzi taking pictures of new models as they are put through their paces. Vehicle manufacturers also test their new models out on public roads, usually with stick-on panels to disguise them, but for repeatable, carefully measured conditions, a facility like the Millbrook Proving Ground is needed. The site has been called 'an automotive time machine', where a gleaming new model drives in, and about 20 weeks later it drives out (if it can) having been exposed to the equivalent of 10 years of severe weather and wear and tear comparable to being driven around 160,000 miles. During this time, it will have been driven on straight and twisty roads, up and down hills, slowly and very fast, through salt-water baths (to accelerate rusting) and along gravel roads that damage its paintwork. But that's not all. It will be roasted at high temperatures, frozen at down to arctic conditions, and drenched in water to expose any leaks. Also, it will be subjected to the infamous 'Belgian Pavé'. This is a mile-long track made from blocks of paving with rough sections and random depressions. The suspension takes such a beating that after five laps on the track vehicles need to be dowsed in a water trough to cool their shock absorbers down. And during all this wrecking treatment engineers periodically examine the vehicles for signs of wear or damage. This allows carmakers to fine-tune their designs or manufacturing processes to avoid failures that would be expensive and reputationally damaging if they occurred after product launch.

Type I errors are those that occur when a decision was made to do something and the situation did not warrant it. Type II errors are those that occur when nothing was done, yet a decision to do something should have been taken as the situation did indeed warrant it. For example, if a school's inspector checks the work of a sample of 20 out of 1,000 pupils and all 20 of the pupils in the sample have failed, the inspector might draw the conclusion that all the pupils have failed. In fact, the sample just happened to contain 20 out of the 50 students who had failed the course. The inspector, by assuming a high fail rate, would be making a type I error. Alternatively, if the inspector checked 20 pieces of work, all of which were of a high standard, they might conclude that all the pupils' work was good despite having been given, or having chosen, the only pieces of good work in the whole school. This would be a type II error. Although these situations are not likely, they are possible. Therefore any sampling procedure has to be aware of these risks.

How should the checks be performed?

In practice most operations will use some form of sampling to check the quality of their services or products. The most common approach for checking the quality of a sample service or product so as to make inferences about all the output from an operation is called statistical process control (SPC). SPC is concerned with sampling the process during the production of the goods or the delivery of service. Based on this sample, decisions are made as to whether the process is 'in control': that is, operating as it should be. A key aspect of SPC is that it looks at the variability in the performance of processes to check whether the process is operating as it should do (known as the process being 'in control'). In fact variability (or more specifically, reducing variability) is one of the most important objectives of quality improvement. SPC is explained in detail in the supplement to this chapter.

Steps 5 and 6 – Find and correct causes of poor quality and continue to make improvements

The final two steps in our list of quality management activities are, in some ways, the most important, yet also the most difficult. Getting to the root causes of quality problems requires an understanding of improvement techniques, some of which were described in Chapter 15, but it also requires an understanding of the range of possible root causes. Some of these will be errors in whatever technology is used in the operation (see the 'Operations in practice' example, 'Coin counting calculations') but will also include human failures.

OPERATIONS IN PRACTICE

Coin counting calculations[9]

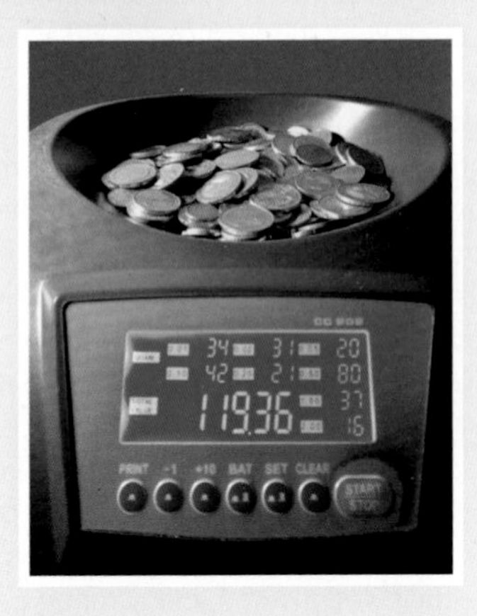

The rise in the use of contactless payment cards was always going to undermine the usefulness of physical money, a trend that the COVID-19 pandemic accelerated. Even before that, the British Treasury had, for the first time in almost half a century, ordered the Royal Mint (that makes coins) to stop producing any low-denomination coins. But what did people do with all their not-totally-unwanted coins? Most retail banks are reluctant to accept a large amount of coins unless they are counted and bagged in standard bags. However, Metro Bank, unlike any other bank in the United Kingdom, have free coin counters, called 'Magic Money Machines'. Yet, like all technology, coin-counting machines exhibit a certain amount of variation. An investigation showed that this variation can mean that the stated worth of the coins put into the machine can range from being perfectly accurate to the very last penny, through to as much as 19 per cent inaccurate. A journalist laboriously sorted and counted by hand £600 worth of coins – 14,500 coins in total – and divided them into bags of coins that were worth exactly £100. He then visited Metro Bank branches in central and west London to see if the machines would count the coins as accurately as he had. In fact, the Magic Money Machines came out of the test reasonably well. Most were accurate, with a margin of error less than 1 per cent, which is pretty good. Even the machine that was 19 per cent out erred in the journalist's favour. In total, he ended up with a net profit of about £30 (the money was donated to Metro Bank's charity partner). Yet even small errors in such machines can add up in absolute terms when one considers that £22.5 million was processed by Metro Bank coin machines in the year prior to the investigation. But not all coin counters have proved as accurate as Metro Bank's. In the United States, TD Bank had to abandon its 'Penny Arcade' coin-counting machines after widespread complaints that the devices were short-changing customers. An investigation had concluded that Penny Arcades in five locations inaccurately counted $300 packets of coins. And in no location did the counting error favour the customer.[10]

The root causes of quality-related human failure

Figure 17.7 illustrates how quality-related human failures can be classified. The first distinction is between 'errors' (mistakes in judgement, where a person should have done something different) and 'violations (acts that are clearly contrary to defined operating procedure), both of which can be subdivided as shown in the figure. Preventing each of these three categories of human error requires different measures:

- *Action errors* – can be reduced through careful job design, consistency of information provided, intuitive interfaces with technology, checklist and reminders.
- *Thinking errors* – can be reduced through planning and appropriate training for all likely situations, the provision of information and diagnostic techniques, and the promotion of regular opportunities to learn from the sharing of experiences.
- *Violations* – can be reduced through working on cultural issues in the case of routine violations, with the promotion of an organisational culture that does not tolerate non-compliance, the protection of 'whistle-blowers', and possibly closer supervision. Situational violations are best tackled by changing working conditions and the elimination of elements of the job that could lead to non-compliance. Exceptional violations can be reduced through a better awareness and understanding of the risks and consequences of choices.

Nevertheless, there is an aspect of quality management that has been particularly important in shaping how quality is improved and the improvement activity made self-sustaining. This is total quality management (TQM). The remainder of the main body of this chapter is devoted to TQM.

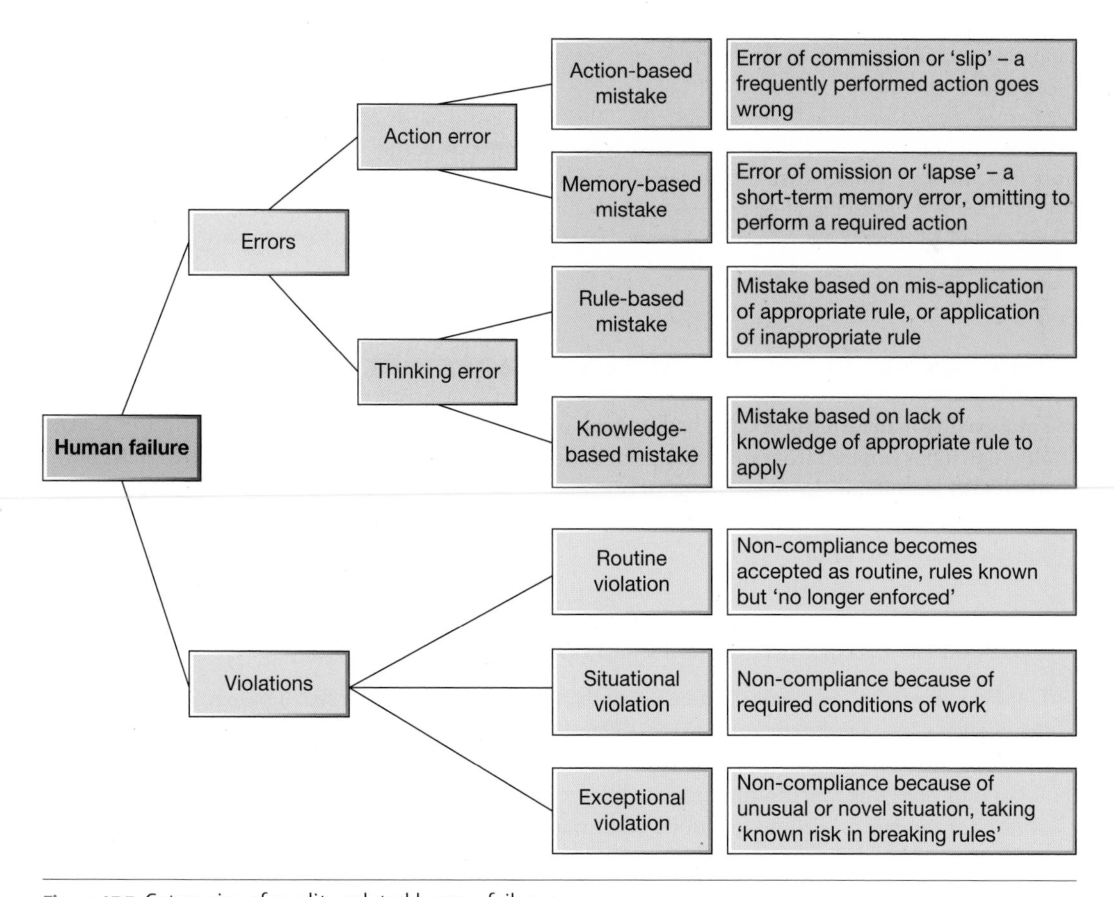

Figure 17.7 Categories of quality-related human failure

OPERATIONS IN PRACTICE

Keyboard errors – autofill and 'fat fingers'[11]

People make mistakes, especially at keyboards. Two types of mistakes have proved both embarrassing and costly. The embarrassing one (which can also be costly) is the use of the autofill function on email and search applications. Fill in the first few letters, and the app does the rest for you. Unless it gets it wrong. Before the United Kingdom voted to leave the European Union, and when the topic was a politically and economically sensitive issue, the Bank of England's head of press mistakenly sent an email to the media revealing that officials were quietly researching the impact of Britain's exit from the EU. There were questions over whether the incident would cost him his job. Following the blunder, in a tightening of security regulations, Bank of England employees were prevented from using the autofill. Instead, staff were asked to write the full name of the recipient of their emails. It did not help productivity, but it reduced the possibility of an email getting to the wrong recipient.

More costly (which can also be embarrassing) is what has become known as 'fat finger syndrome'. For example, feeling sleepy one day, a German bank worker briefly fell asleep on his keyboard when processing a €64 debit (withdrawal) from a pensioner's account, repeatedly pressing the number 2. The result was that the pensioner's account had €222 million withdrawn from it instead of the intended €64. Fortunately, the bank spotted the error before too much damage was done (and before the account-holder noticed). More seriously, the supervisor who should have checked his junior colleague's work was sacked for failing to notice the blunder (unfairly, a German labour tribunal later ruled). Fat-finger trading mistakes are not uncommon. For example, the Swiss bank UBS mistakenly ordered 3 trillion yen (instead of 30 million yen) of bonds in a Japanese video games firm. In another example, a Japanese trader tried to sell one share of a recruitment company at 610,000 yen per share. But he accidentally sold 610,000 shares at one yen each, despite this being 41 times the number of shares available. Unlike the German example, the error was not noticed and the Tokyo Stock Exchange processed the order. It resulted in Mizuho Securities losing 27 billion yen. The head of the Exchange later resigned.

17.3 What is total quality management (TQM)?

Total quality management (TQM) had its peak of popularity in the late 1980s and early 1990s. And although it has suffered from something of a backlash, the general precepts and principles that constitute TQM are still the dominant mode of organising operations improvement. The approach we take here is to stress the importance of the 'total' in total quality management and how it can guide the agenda for improvement.

TQM as an extension of previous practice

TQM can be viewed as a logical extension of the way in which quality-related practice has progressed (see Figure 17.8). Originally quality was achieved by inspection – screening out defects before they were noticed by customers. The quality control (QC) concept developed a more systematic approach to not only detecting, but also treating quality problems. Quality assurance (QA) widened the responsibility for quality to include functions other than direct operations. It also made increasing use of more sophisticated statistical quality techniques. TQM included much of what went before but developed its own distinctive themes to make quality both more strategic and more widespread in the organisation. We will use some of these themes to describe how TQM represents a clear shift from traditional approaches to quality.

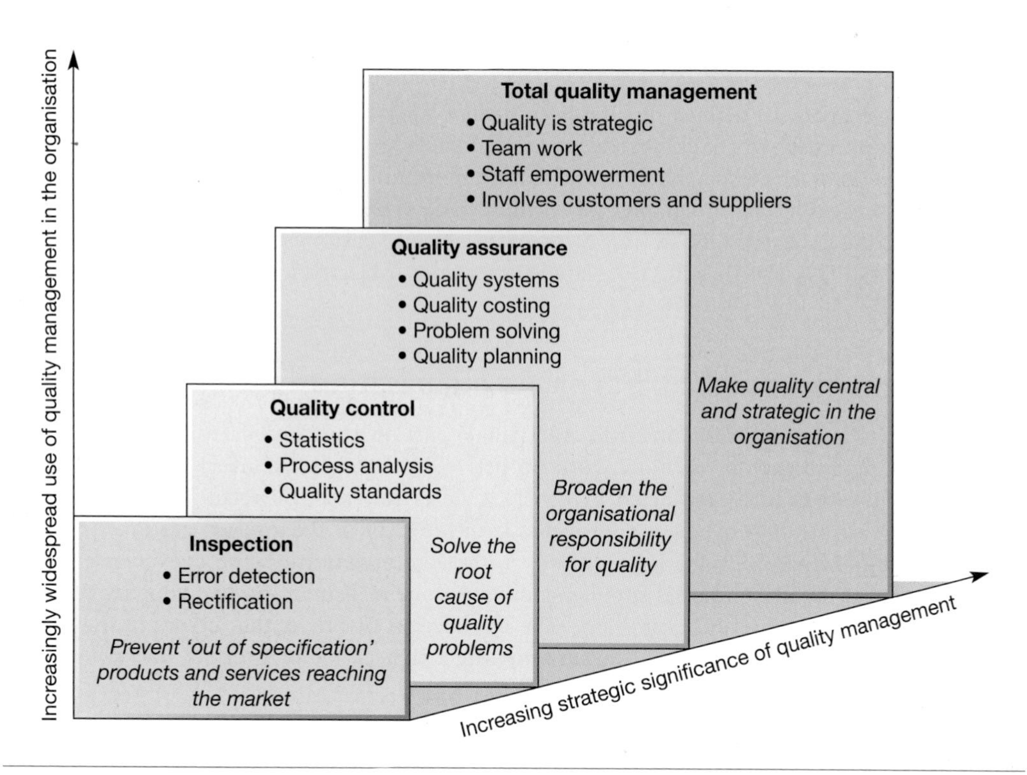

Figure 17.8 TQM as an extension of previous views of quality

The meaning of TQM?

TQM is 'an effective system for integrating the quality development, quality maintenance and quality improvement efforts of the various groups in an organisation so as to enable production and service at the most economical levels which allow for full customer satisfaction'.[12] However, it was the Japanese who first made the concept work on a wide scale, and subsequently popularised the approach and the term 'TQM'. It was then developed further by several so-called 'quality gurus'. Each 'guru' stressed a different set of issues, from which emerged the TQM approach. It is best thought of as a philosophy of how to approach quality improvement. This philosophy, above everything, stresses the 'total' of TQM. It is an approach that puts quality at the heart of everything that is done by an operation and includes all activities within an operation. This totality can be summarised by the way TQM lays particular stress on the following:

- meeting the needs and expectations of customers;
- covering all parts of the organisation;
- including every person in the organisation;
- examining all costs that are related to quality, especially failure costs and getting things 'right first time';
- developing the systems and procedures that support quality and improvement;
- developing a continuous process of improvement (this was treated in Chapter 16).

Not surprisingly, several researchers have tried to establish how much of a relationship there is between adopting total quality management and the performance of the organisation. One of the best-known studies[13] found that there was a positive relationship between the extent to which companies implement TQM and their overall performance. But it found that managers should implement TQM as a whole set of ideas rather than simply picking a few techniques to implement. The same study also suggests that where TQM does not prove successful in improving performance, the problems could be the result of poor implementation rather than the TQM practices themselves and that a serious commitment on the part of top management to TQM is a prerequisite for success.

TQM means meeting the needs and expectations of customers

Earlier in this chapter we defined quality as consistent conformance to customers' expectations. Therefore, any approach to quality management must necessarily include the customer perspective. In TQM, this customer perspective is particularly important. It may be referred to as 'customer centricity' (discussed briefly in Chapter 15) or the 'voice of the customer'. Whatever it is called, TQM stresses the importance of starting with an insight into customer needs, wants, perceptions and preferences. This can then be translated into quality objectives and used to drive quality improvement.

TQM means covering all parts of the organisation

For an organisation to be truly effective, every single part of it, each department, each activity, and each person and each level, must work properly together, because every person and every activity affects and in turn is affected by others. One of the most powerful concepts that has emerged from various improvement approaches is the concept of the internal customer/supplier. This is recognition that everyone is a customer within the organisation and consumes goods or services provided by other internal suppliers, and everyone is also an internal supplier of goods and services for other internal customers. The implication of this is that errors in the service provided within an organisation will eventually affect the service or product that reaches the external customer.

Service-level agreements

Some organisations bring a degree of formality to the internal customer concept by encouraging (or requiring) different parts of the operation to agree service-level agreements (SLAs) with each other. SLAs are formal definitions of the dimensions of service and the relationship between two parts of an organisation. The types of issues which would be covered by such an agreement could include response times, the range of services, dependability of service supply, and so on. Boundaries of responsibility and appropriate performance measures could also be agreed. For example, an SLA between an information systems support unit and a research unit in the laboratories of a large company could define such performance measures as:

- the types of information network services that may be provided as 'standard';
- the range of special information services that may be available at different periods of the day;
- the minimum 'up time', i.e. the proportion of time the system will be available at different periods of the day;
- the maximum response time and average response time to get the system fully operational should it fail;
- the maximum response time to provide 'special' services, and so on.

Operations principle

An appreciation of, involvement in, and commitment to quality should permeate the entire organisation.

Critical commentary

While some see the strength of SLAs as the degree of formality they bring to customer–supplier relationships, there are also some clear drawbacks. The first is that the 'pseudo-contractual' nature of the formal relationship can work against building partnerships (see Chapter 12). This is especially true if the SLA includes penalties for deviation from service standards. Indeed, the effect can sometimes be to inhibit rather than encourage joint improvement. The second (and related) problem is that SLAs, again because of their formal documented nature, tend to emphasise the 'hard' and measurable aspects of performance rather than the 'softer' but often more important aspects. So, a telephone may be answered within four rings, but how the caller is treated, in terms of 'friendliness', may be far more important.

TQM means including every person in the organisation

Every person in the organisation has the potential to contribute to quality and TQM was among the first approach to stress the centrality of harnessing everyone's potential contribution to quality. There is scope for creativity and innovation even in relatively routine activities, claim TQM proponents. The shift in attitude that is needed to view employees as the most valuable intellectual and creative resource that the organisation possesses can still prove difficult for some organisations. Yet most advanced organisations do recognise that quality problems are almost always the results of human error.

TQM means all costs of quality are considered

The costs of controlling quality may not be small, whether the responsibility lies with each individual or a dedicated quality control department. It is therefore necessary to examine all the costs and benefits associated with quality (in fact 'cost of quality' is usually taken to refer to both costs and benefits of quality). These costs of quality are usually categorised as **prevention costs**, **appraisal costs**, **internal failure costs** and **external failure costs**.

- ***Prevention costs*** are those costs incurred in trying to prevent problems, failures and errors from occurring in the first place. They include such things as:
  - — identifying potential problems and putting the process right before poor quality occurs;
  - — designing and improving the design of products and services and processes to reduce quality problems;
  - — training and development of personnel in the best way to perform their jobs;
  - — process control through SPC.
- ***Appraisal costs*** are those costs associated with controlling quality to check to see if problems or errors have occurred during and after the creation of the service or product. They might include such things as:
  - — the setting up of statistical **acceptance sampling** plans;
  - — the time and effort required to inspect inputs, processes and outputs;
  - — obtaining processing inspection and test data;
  - — investigating quality problems and providing quality reports;
  - — conducting customer surveys and quality audits.
- ***Internal failure costs*** are failure costs associated with errors that are dealt with inside the operation. These costs might include such things as:
  - — the cost of scrapped parts and material;
  - — reworked parts and materials;
  - — the lost production time as a result of coping with errors;
  - — lack of concentration due to time spent on troubleshooting rather than improvement.
- ***External failure costs*** are those that are associated with an error going out of the operation to a customer. These costs include such things as:
  - — loss of customer goodwill affecting future business;
  - — aggrieved customers who may take up time;
  - — litigation (or payments to avoid litigation);
  - — guarantee and warranty costs;
  - — the cost to the company of providing excessive capability (too much coffee in the pack or too much information to a client).

The relationship between quality costs

In traditional quality management it was assumed that failure costs reduce as the money spent on appraisal and prevention increases. Furthermore, it was assumed that there is an *optimum* amount of quality effort to be applied in any situation, which minimises the total costs of quality. The argument is that there must be a point beyond which diminishing returns set in – that is, the cost of improving quality gets larger than the benefits which it brings. Figure 17.9(a) sums up this idea. As quality effort is increased, the costs of providing the effort – through extra quality controllers, inspection procedures and so on – increases proportionally. At the same time, however, the cost of errors, faulty products and so on decreases because there are fewer of them. However, TQM proponents believe that this logic is flawed. First, it implies that failure and poor quality are acceptable. Why, TQM proponents argue, should any operation accept the *inevitability* of errors? Some occupations seem to be able to accept a zero-defect standard. No one accepts that pilots are allowed to crash a certain proportion of their aircraft, or that nurses will drop a certain proportion of the babies they deliver. Second, it assumes that costs are known and measurable. In fact, putting realistic figures to the cost of quality is not a straightforward matter. Third, it is argued that failure costs in the traditional model are greatly underestimated. In particular, all the management time wasted by failures and the loss of concentration it causes are rarely accounted for. Fourth, it implies that prevention costs are inevitably high because it involves expensive inspection. But why should quality not be an integral part of everyone's work rather than employing extra people to inspect? Finally, the 'optimum-quality level' approach, by accepting compromise, does little to challenge operations managers and staff to find ways of improving quality. Put these corrections into the optimum-quality effort calculation and the picture looks very different (see Figure 17.9(b)). If there is an 'optimum', it is a lot further to the right, in the direction of putting more effort (but not necessarily cost) into quality.

Operations principle
Effective investment in preventing quality errors can significantly reduce appraisal and failure costs.

The TQM quality cost model

TQM rejects the optimum-quality level concept and strives to reduce all known and unknown failure costs by preventing errors and failure taking place. Rather than looking for 'optimum' levels of quality effort, TQM stresses the relative balance between different types of quality cost. Of the four cost categories, two (costs of prevention and costs of appraisal) are open to managerial influence, while the other two (internal costs of failure and external costs of failure) show the consequences of changes in the first two. So, rather than placing most emphasis on appraisal (so that 'bad products and service don't get through to the customer'), TQM emphasises prevention (to stop errors happening in the first place). That is because the more effort that is put into error prevention, the more internal and external

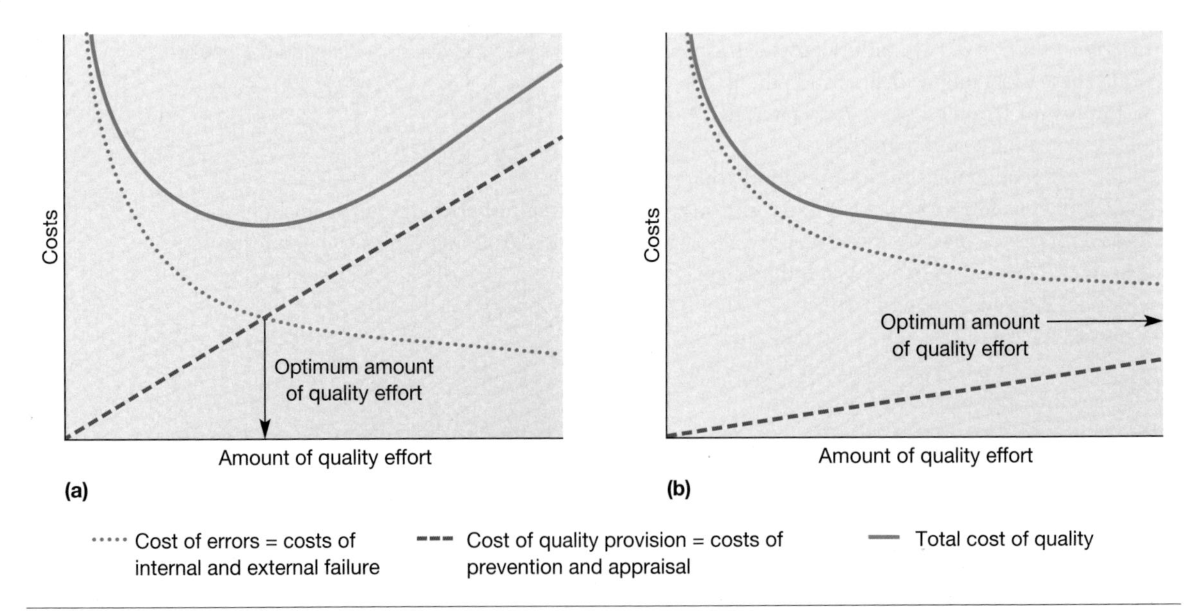

Figure 17.9 (a) The traditional cost of quality model, and (b) the traditional cost of quality model with adjustments to reflect TQM criticisms

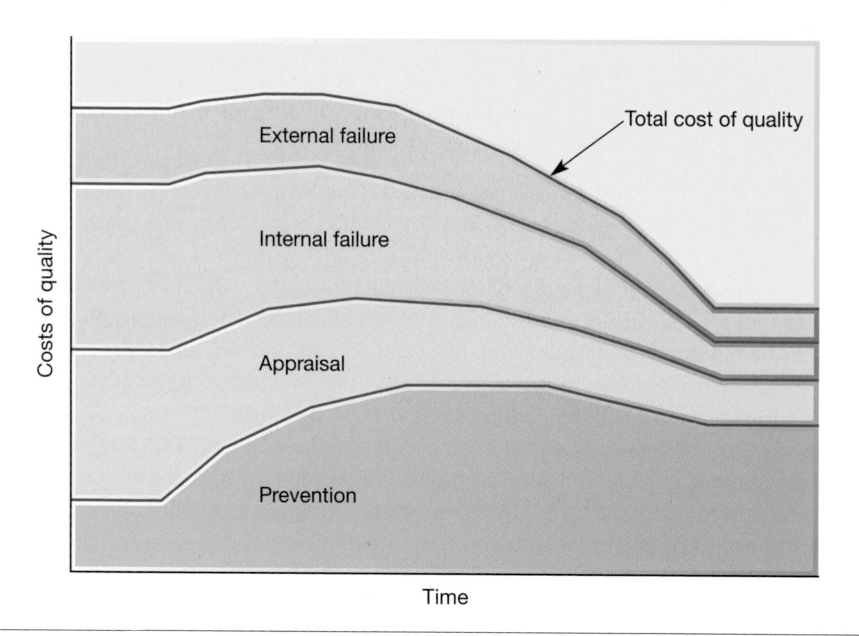

Figure 17.10 Increasing the effort spent on preventing errors occurring in the first place brings a more than equivalent reduction in other cost categories

failure costs are reduced. Then, once confidence has been firmly established, appraisal costs can be reduced. Eventually even prevention costs can be stepped down in absolute terms, though prevention remains a significant cost in relative terms. Figure 17.10 illustrates this idea. Total quality costs may rise initially as investment in some aspects of prevention – mainly training – is increased. However, a reduction in total costs can quickly follow.

Getting things 'right first time'

Accepting the relationships between categories of quality cost as illustrated in Figure 17.10 has a particularly important implication for how quality is managed. It shifts the emphasis from *reactive* (waiting for something to happen) to *proactive* (doing something before anything happens). This change in the view of quality costs has come about with a movement from an inspect-in (appraisal-driven) approach to a design-in (getting it right first time) approach.

TQM means developing the systems and procedures that support quality and improvement

The emphasis on highly formalised systems and procedures to support TQM has declined, yet one aspect is still active for many companies. This is the adoption of the ISO 9000 standard. And although ISO 9000 can be regarded as a stand-alone issue, it is very closely associated with TQM.

The ISO 9000 approach

The ISO 9000 series is a family of standards compiled by the International Organization for Standardization (ISO). The standards symbolise a consensus on good quality management practices, systems, and related supporting standards. The standards apply to any organisation, regardless of size and sector. Organisations using the standards can be certified, the purpose of which was originally to provide an assurance to customers that products or services had been produced to meet their requirements. The best way to do this, it was argued, was to define the procedures, standards and characteristics of the management control system that governs the operation, building quality into an operation's processes. This requires operations to define and record core processes and sub-processes (in a manner very similar to the 'hierarchy of processes' principle that was outlined in Chapter 1). In addition, processes are documented using the process mapping approach that was described in Chapter 6. It also stresses four other principles:

- Quality management should be customer focused. Customer satisfaction should be measured through surveys and focus groups, and improvement against customer standards should be documented.
- Quality performance should be measured. In particular, measures should relate both to processes that create products and services and to customer satisfaction with those products and services. Furthermore, measured data should be analysed in order to understand processes.
- Quality management should be improvement driven. Improvement must be demonstrated in both process performance and customer satisfaction.
- Top management must demonstrate their commitment to maintaining and continually improving management systems. This commitment should include communicating the importance of meeting customer and other requirements, establishing a quality policy and quality objectives, conducting management reviews to ensure the adherence to quality policies, and ensuring the availability of the necessary resources to maintain quality systems.

Quality awards

Various bodies have sought to stimulate improvement through establishing 'quality' awards. The three best-known awards are the Deming Prize, the Malcolm Baldrige National Quality Award and the European Quality Award.

The Deming Prize

The Deming Prize was instituted by the Union of Japanese Scientists and Engineers in 1951 and is awarded to those companies, initially in Japan, but is now open to overseas companies, which have successfully applied 'company-wide quality control' based upon statistical quality control. There are 10 major assessment categories: policy and objectives, organisation and its operation, education and its extension, assembling and disseminating of information, analysis, standardisation, control, quality assurance, effects and future plans. The applicants are required to submit a detailed description of quality practices. This is a significant activity in itself and some companies claim a great deal of benefit from having done so.

Critical commentary

Notwithstanding its widespread adoption (and its revision to take into account some of its perceived failing), ISO 9000 is not seen as beneficial by all authorities and is still subject to some specific criticisms. These include the following:

- The continued use of standards and procedures encourages 'management by manual' and over-systematised decision making.
- The whole process of documenting processes, writing procedures, training staff and conducting internal audits is expensive and time consuming.
- Similarly, the time and cost of achieving and maintaining ISO 9000 registration are excessive.
- It is too formulaic. It encourages operations to substitute a 'recipe' for a more customised and creative approach to managing operations improvement.

OPERATIONS IN PRACTICE

Quality systems only work if you stick to them[14]

When passengers on the Hakata to Tokyo express, one of Japan's famous bullet trains, noticed a burning smell and an unusual sound, they were ordered off the train. The cause turned out to be cracks in the chassis and it marked the latest episode in a long line of quality scandals that had rocked the country, jeopardising the image of 'Japanese quality'. The previous months had seen public admissions by some of Japan's most prestigious names – including Kobe Steel, Mitsubishi Materials, Nissan Motor and Subaru – that their quality tests had been falsified or the results had been fabricated, all to sell products of a lower quality than officially stated. Quality systems had been in place, but often ignored. Quality records had been doctored on materials that had been shipped to make a wide range of products, including the Boeing 787 Dreamliner, nuclear plants and space rockets. It was the shock announcement from Kobe Steel that focused the world's attention on the problem. It confessed that 'improper conduct' had led to the falsification of data relating to 19,300 tonnes of aluminium sheets and poles, 19,400 aluminium components, 2,200 tonnes of copper products and an unspecified amount of iron powder that was supplied to over 200 customers. All these items had been falsely certified as conforming to specifications concerning properties such as tensile strength. Kobe admitted that for up to 10 years its employees had falsified quality checks on tens of thousands of tonnes of metal products, including the aluminium used by Boeing to make the parts that held the 787 together. However, Boeing made it clear that it had been conducting comprehensive inspections and analysis of affected shipments since it was told about Kobe Steel's data falsification. And, notwithstanding the falsification, no deaths or accidents seem to have resulted from the under-specification products.

Nevertheless, Kobe Steel demoted three executives from the aluminium and copper divisions, who had been aware of the data tampering. Two executives had apparently known of the falsification problems for eight years. The company said that they were relieved from their duties and reassigned to lower-ranking roles. Also, the government-backed Japan Industrial certification was revoked at one of its factories owing to 'improper quality management'. But it was the nation-wide inquest as to why so many problems had occurred in so many Japanese companies that was followed closely by quality professionals. Some commentators claimed that it was because of increased pressure to produce profits. When Toray Industries, the chemical company, disclosed data falsification on tyre material, Akihiro Nikkaku, its President, Chief Executive and Chief Operating Officer, blamed the *'pressure to meet productivity targets'*. Other observers pointed to Japanese corporate culture and the reluctance of middle managers to bring quality mistakes to their superiors' attention. Yet others say that the reason that so many scandals have emerged relatively recently is that, among younger employees, revealing bad practice has become more acceptable. Moreover, social media provided a forum and an environment for whistleblowing and airing such grievances, which previously did not exist.

The Malcolm Baldrige National Quality Award

In the early 1980s the American Productivity and Quality Center recommended that an annual prize, similar to the Deming Prize, should be awarded in America. The purpose of the award was to stimulate American companies to improve quality and productivity, to recognise achievements, to establish criteria for a wider quality effort and to provide guidance on quality improvement. The main examination categories are: leadership, information and analysis, strategic quality planning, human resource utilisation, quality assurance of products and services, quality results and customer satisfaction. The process, like that of the Deming Prize, includes a detailed application and site visits.

The EFQM Excellence Model

Originally, the European Foundation for Quality Management (EFQM) founded the European Quality Award, since when the importance of quality excellence has become far more accepted. According to the EFQM, 'Whilst there are numerous management tools and techniques commonly used, the EFQM Excellence Model provides an holistic view of the organisation and it can be used to determine how these different methods fit together and complement each other. The Model . . . [is] . . . an overarching framework for developing sustainable excellence. Excellent organisations achieve and sustain outstanding levels of performance that meet or exceed the expectations of all their stakeholders. The EFQM Excellence Model allows people to understand the cause-and-effect relationships between what their organisation does and the results it achieves'.[15]

The model is based on the idea that it is important to understand the cause-and-effect relationships between what an organisation does (what it terms 'the Enablers') and its results. The EFQM Excellence Model is shown in Figure 17.11. There are five enablers:

- ***Leadership*** – that looks to the future, acts as a role model for values and ethics, inspires trust, is flexible, enables anticipation and so can react in a timely manner.
- ***Strategy*** – that implements the organisation's mission and vision by developing and deploying a stakeholder-focused strategy.
- ***People*** – organisations should value their people, creating a culture that allows mutually beneficial achievement of both organisational and personal goals, develops the capabilities of people, promotes fairness and equality, cares for, communicates, rewards and recognises people, in a way that motivates and builds commitment.
- ***Partnership and resources*** – organisations should plan and manage external partnerships, suppliers and internal resources in order to support strategy and policies and the effective operation of processes.
- ***Processes, products and services*** – organisations should design, manage and improve processes to ensure value for customers and other stakeholders.

Results are assessed using four criteria:

- Customer results – meeting or exceeding the needs and expectations of customers.
- People results – meeting or exceeding the needs and expectations of employees.
- Society results – achieving and sustaining results that meet or exceed the needs and expectations of the relevant stakeholders within society.
- Business results – achieving and sustaining results that meet or exceed the needs and expectations of business stakeholders.

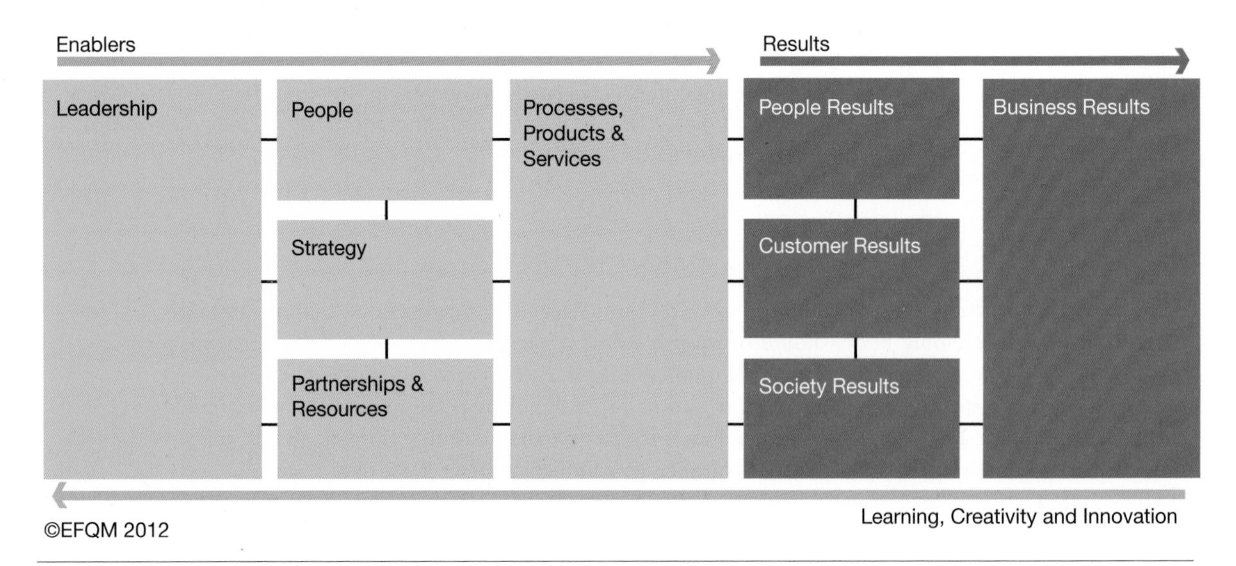

Figure 17.11 The EFQM Excellence Model

Source: Reproduced with the permission of the EFQM

Critical commentary

The similarity of **ISO 14000** to the quality procedures of ISO 9000 is a bit of a giveaway. ISO 14000 can contain all the problems of ISO 9000 (management by manual, obsession with procedures rather than results, a major expense to implement it and, at its worst, the formalisation of what was bad practice in the first place). But ISO 14000 also has some further problems. The main one is that it can become a 'badge for the smug'. It can be seen as 'all there is to do to be a good environmentally sensitive company'. At least with quality standards like ISO 9000 there are real customers continually reminding the business that quality does matter. Pressures to improve environmental standards are far more diffuse. Customers are not likely to be as energetic in forcing good environmental standards on suppliers as they are in forcing the good quality standards from which they benefit directly. Instead of this type of procedure-based system, surely the only way to influence a practice that has an effect at a societal level is through society's normal mechanism – legal regulation. If quality suffers, individuals suffer and have the sanction of not purchasing goods and services again from the offending company. With bad environmental management, however, we all suffer. Because of this, the only workable way to ensure environmentally sensitive business policies is by insisting that our governments protect us. Legislation, therefore, is the only safe way forward.

Green reporting and ISO 14000

Until recently, relatively few companies around the world provided information on their environmental practices and performance. Now environmental reporting is increasingly common. Another emerging issue has been the introduction of the ISO 14000 standard. It has a three-section environmental management system, which covers initial planning, implementation and objective assessment. Although it has had some impact, it is largely limited to Europe.

ISO 14000 makes a number of specific requirements, including the following:

- a commitment by top-level management to environmental management;
- the development and communication of an environmental policy;
- the establishment of relevant legal and regulatory requirements;
- the setting of environmental objectives and targets;
- the establishment and updating of a specific environmental programme, or programmes, geared to achieving the objectives and targets;
- the implementation of supporting systems such as training, operational control and emergency planning;
- regular monitoring and measurement of all operational activities;
- a full audit procedure to review the working and suitability of the system.

Responsible operations

In every chapter, under the heading of 'Responsible operations', we summarise how the particular topic covered in the chapter touches upon important social, ethical and environmental issues.

Some aspects of quality management, or at least the results of quality management, can be considered to contribute directly to responsible operations. Products and services that conform to their clearly communicated specifications clearly benefit the customers who receive them. Similarly, the aim of understanding (and usually minimising) variation in a product's characteristics, benefits both customers and the operation. For example, go round any supermarket and look at all the products that are packaged, bottled or otherwise 'filled' into containers. Bottled drinks, detergent, bags of vegetables, cans of paint; they are all put in their containers in the manufacturing ▶

operations that produce them. And this filling or packing process is, in most countries, governed by strict government regulations. When a package claims to have a certain amount of product, customers have a right to expect that it really does include that amount; otherwise, they are paying for something that they are not getting. In many regions, the law mandates that the average weight must be greater than the declared weight on the container, with the average weight being determined by sampling. But the technology used to fill packages is not always totally consistent. There is always some degree of variation in the amount 'dispensed'. So, if packers or fillers want to conform to legal weights and measures stipulations on minimum fill levels, they must build a margin of safety into filling levels in order to overcome the variation of the filling technology. Quality management's focus on reducing such variation, allows customers to get what they pay for and businesses to avoid what is known as 'giveaway' or 'over-fill'.

As is sometimes noted, it is quality management's (and particularly TQM's) focus on providing value for 'the customer' that is one of its great strengths. In all the quality models covered in this chapter, such as ISO 9000 and the EFQM model, there is a focus on a range of external stakeholders. Yet it is the interests of customers that is often regarded as being paramount, even to the practical exclusion of other stakeholders. Partly this reflects the idea of quality as 'fitness for use' (by the customers). Partly it reflects shorter-term commercial considerations. Yet, as with many aspects of operations management, while not displacing the important role of customer satisfaction in all quality-related decisions, it is worth reiterating the role of other stakeholders. All organisations need to be explicit in how quality management has an impact on society broadly, employees and other stakeholders, as well as on the satisfaction of customers.

Summary answers to key questions

17.1 What is quality and why is it so important?

- The definition of quality used in this text is 'consistent conformance to customers' expectations'. It is important because it has a significant impact on profitability.
- At a broad level, quality is best modelled as the gap between customers' expectations concerning the service or product and their perceptions concerning the service or product.
- Modelling quality this way will allow the development of a diagnostic tool that is based around the perception–expectation gap. Such a gap may be explained by four other gaps:
  - — the gap between a customer's specification and the operation's specification;
  - — the gap between the service or product concept and the way the organisation has specified it;
  - — the gap between the way quality has been specified and the actual delivered quality;
  - — the gap between the actual delivered quality and the way the service or product has been described to the customer.
- Some operations that produce primarily intangible services often use the term 'quality of service' to include elements of speed, dependably and flexibility. Also, increasingly the term 'quality of experience' is used to denote a more user-centric view of quality.
- The 'sandcone' theory of improvement holds that it is generally better to start with improving quality rather than other performance objectives, but then to keep improving quality even as other performance objectives are pursued.

17.2 What steps lead towards conformance to specification?

- There are six steps:
  - define quality characteristics;
  - decide how to measure each of the quality characteristics;
  - set quality standards for each characteristic;
  - control quality against these standards;
  - find and correct the causes of poor quality;
  - continue to make improvements.
- Most quality planning and control involves sampling the operations performance in some way. Sampling can give rise to erroneous judgements, which are classed as either type I or type II errors. Type I errors involve making corrections where none are needed. Type II errors involve not making corrections where they are in fact needed.

17.3 What is total quality management (TQM)?

- TQM is 'an effective system for integrating the quality development, quality maintenance and quality improvement efforts of the various groups in an organisation so as to enable production and service at the most economical levels that allow for full customer satisfaction'.
- It is best thought of as a philosophy that stresses the 'total' of TQM and puts quality at the heart of everything that is done by an operation.
- Total in TQM means the following:
  - meeting the needs and expectations of customers;
  - covering all parts of the organisation;
  - including every person in the organisation;
  - examining all costs that are related to quality, and getting things 'right first time';
  - developing the systems and procedures that support quality and improvement, potentially including 'quality awards'.

CASE STUDY Rapposcience Labs

'There is no doubt that it was a disaster for the laboratory. It was the first time that a client had withdrawn from a contract so soon, and it was our fault entirely. It was also a disaster for Vincent [De Smet]. *I feel sorry for him. I had known him for years. He was a good guy with seemingly unlimited energy and a host of good ideas. But in the end he had to go'.* (Petra Reemer, Chief Scientist, Rapposcience Labs)

Petra Reemer was talking about her predecessor, Vincent De Smet, who was in charge of the Laboratories (simply known internally as 'the Lab') when one of its larger clients, MGQ Services, an extraction services firm, had exercised their right to withdraw from a commercial contract with Rapposcience for 'persistent and significant failure to comply with testing and analytical performance'. This came as a shock to the Lab because, although it was aware that its performance had not been entirely satisfactory, MGQ had not formally complained about the Lab's performance. MGQ's withdrawal not only a created a hole in the Lab's revenue projections, it also attracted enough negative publicity in the industry for the Lab's private equity owners, Brighthorpe Holdings, to replace Vincent De Smet with Petra Reemer. With a background in analytical and industrial forensic testing, Petra started the job of rescuing the Lab's reputation.

Rapposcience Labs

Rapposcience Labs was located at Beveren near Antwerp in Belgium. In the past, it had been one of the most reputable labs for analysing mineral deposit, soil, and mixed inert and biological samples for a number of clients, mainly from extraction (mining), oil and gas, and public environmental agencies. It employed 47 staff, almost all with a science or technical background, the majority in testing and analysis roles, together with some in administrative and sales roles. Up until the MGQ 'disaster', Brighthorpe had adopted a 'hands off' policy towards how the Lab was run. That changed after De Smet's replacement, and Petra Reemer had been given the clear message that she must turn Rapposcience around, or its future would be bleak. *'We lost the MGQ contract in February. Ironically, the previous 12 months had brought in record levels of business for the Lab. Yet it was business won by undercutting rivals on price. In fact, with hindsight, it is obvious that we had been running at a marginal loss all that year. I arrived in March, and I have spent the last month doing my best to reassure our remaining clients that they can still trust us to deliver a timely and trustworthy service. Unfortunately a couple of contracts were up for renewal at that time and, regretfully, we lost them. We are now running at what looks like a sustained loss for the first time in our history'* (Petra Reemer).

The Rapposcience laboratory process

The laboratory divided its activities into four phases of what it called its 'testing cycle'. These were: Pre-contract, Field operations, Analytics and Post-analytics. Table 17.5 summarises these phases.

Pre-contract occurred at the start of the contract and involved agreeing with the client the exact specification of the service to be provided. This usually included the range of sample specifications, how they would be delivered to the lab, the nature of the report that would be prepared, and the contracted performance in terms of analytical accuracy (that

Table 17.5 The testing cycle

Phase of testing cycle			
Pre-contract	**Field operations**	**Analytics**	**Post-analytics**
▶ Sample specification ▶ Delivery to lab agreed ▶ Report outline	▶ Sampling protocols (including training pack) ▶ Containing and recording ▶ Couriering	▶ Sample preparation ▶ Pre-analysis treatment ▶ Analysis	▶ Report generated ▶ Data recording

indicates the veracity of the analysis), precision (that indicates the reproducibility of the analysis), and the timeliness of the report. Laboratory errors had a reported frequency of between 0.012 per cent and 0.6 per cent. Although not large in itself, errors can have huge impact on clients' decision-making as 60–70 per cent of their operational and investment decisions were made on the basis of laboratory tests.

Field operations was the responsibility of the client, but the Lab often supplied the containers used for the samples, and instructions for taking and packaging the sample. Some clients also insisted on more detailed sampling protocols for their field technicians, including training packs.

The analytics phase included all the testing within the Lab itself. This would vary depending on the nature of the tests and the procedures specified in the contract. Generally though, all testing followed three stages; sample preparation, pre-analysis treatment and analysis (see Figure 17.12).

One of the first modifications to the process came when Vincent had decided to split the sample into two parts before it was tested. There was almost always sufficient material to be able to do this, and the advantage was that, if the testing proved inconclusive, or some performance indicators were outside the permitted range, the tests could be repeated. Performance indicators demonstrated whether the analytical process was behaving as planned, if it had revealed a statistical anomaly that required investigation, or when a test had failed. Most contracts specified a particular confidence level for the results (usually 99.5 per cent), but any small error or contamination in the testing procedure could reduce the confidence level. If this happened, the 'back up' sample could be tested. However, this almost certainly meant that the Lab would not be able to meet its promised report delivery time.

The post-analytics phase consisted of preparing the results of the analysis for the client. This was usually a simple report describing the composition of the sample, but some clients also required a more detailed comparative report where sample data were compared with previous sample readings. Even if such comparative reporting was not required, the Lab recorded all sample data.

Initiatives during the De Smet period

Petra Reemer was not unsympathetic to what Vincent De Smet had been trying to do at Rapposcience. Not only had Vincent tried to introduce some worthwhile reforms to the Lab's operating procedures, he was labouring under pressure to increase the profitability of the operation. *'I think that Vincent had been trying to increase the volume of business while keeping staffing levels the same. Presumably he figured that increased revenue with costs held down would equal healthy profitability. He also complicated things by introducing a number of initiatives; all at more or less the same time'.*

One of Vincent's initiatives had been his decision to split the sample into two parts before it was tested. He did this as a 'failsafe' in case there were problems during the analysis phase and the tests had to be repeated. The response of the Lab's technicians to this move had been mixed. Some felt that it was a sensible move that reduced the chances of recording a 'failed through insufficient material' result. Although this did not happen often, it was at best embarrassing to the Lab, and at worst extremely irritating for the client. Others felt that, because there was the possibility of re-testing a sample, there was a tendency to take less care and 'adopt testing shortcuts' because the consequences of testing errors were less serious.

Another of Vincent's innovations had been the introduction of limited statistical process control (SPC). Although the Lab had always recorded measures of its analytical performance, it had not formally examined its analytics process performance in any systematic manner. It was the MGQ contract that Vincent won (and lost) that prompted the Lab to take the potential of SPC seriously. During the pre-contract phase, the Lab had insisted on its use during all testing on its samples, together with periodic SPC summaries being submitted. Vincent had invested in a 'smart laboratory' IT system that was advertised as being able to automate the data management and statistical processes in the Lab. However, almost a year after its partial introduction, the consensus in the Lab was that it had not been a success. *'It was just too sophisticated for us'*, said Petra Reemer, *'we were trying to run before we could walk'*.

The final initiative instituted during Vincent's time as Chief Scientist was an enhanced set of reporting protocols. *'It wasn't a bad idea actually'*, admitted Petra Reemer, *'we already prepared more extensive reports for some clients, so we had the expertise to interpret their test results and advise them on their sampling processes and how they might interpret results. In other words, we have expertise that can add real value for our clients, so why not use it to enhance our quality of service? The problem when Vincent introduced the idea was that he tried to push it as a sales promotion tool. Clients were inclined to dismiss the potential of enhanced reporting because they thought that we were simply trying to get more money out of them'.*

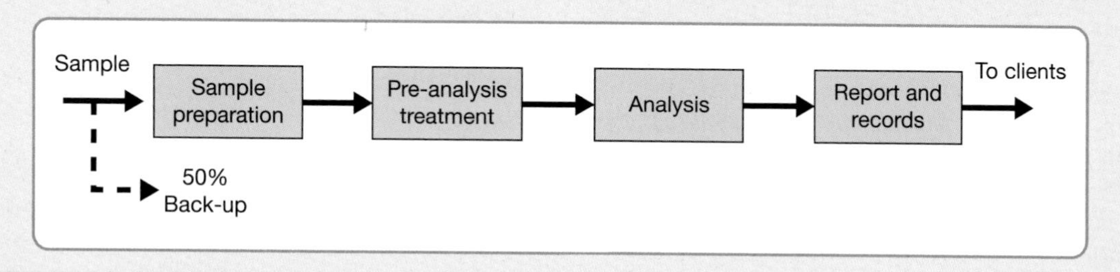

Figure 17.12 The testing phases of Rapposcience Labs laboratory process

Getting back to basics

Petra had taken over from Vincent in March. After three or four weeks talking with all the staff in the Lab, she felt she was ready to shape her plans for the Lab's future. She was convinced that the Lab had to understand what really mattered to clients and then do everything to improve its performance in a way that would have an impact on the quality of service they were providing. Unfortunately, she was also facing pressure from Brighthorpe, the Lab's owners, to cut costs. *'I persuaded them to give me time to restore our reputation. We would find it difficult to do that if we were shedding staff at the same time. Not only would it send the wrong message to the market, it would make it difficult to improve the way we do things. Having said that, we decided to not replace any staff who left the Lab of their own volition. We also delayed any non-essential expenditure. The main objective was to survive long enough to get back to the basics of how we could serve clients better.'*

Her first action was to look at how SPC had been used in the Lab, since it had been introduced. She talked with the Chief Field Engineer at MGQ, who had approved the initial contract that the lab had lost, and who had also insisted on them using SPC. What he said gave Petra much to think about. *'I kind of knew that, when we insisted on Rapposcience using SPC that they really didn't understand what it was all about. They were simply doing it because it was what the client wanted.*

'Their culture said, "If the samples are returned as the specification, then it's OK, if not, then as long as it doesn't happen too often, well that's OK also". They just didn't get that, by seeing their process charts, it enabled us to see more or less exactly what was happening right inside their processes. I take some of the blame on myself. I should have made sure that they fully understood why we were so keen for them to use SPC. It was for them to help themselves by improving their process performance. It wasn't just a whim on our part' (Chief Field Engineer, MGQ).

The first thing Petra did was to hold a series of meetings, first with the supervisors in each department, then with everyone in each department. She was mainly listening to their experiences of using the SPC system that Vincent had imposed, but her secondary motive was to try to judge how much they understood about the fundamentals of SPC. The answer seemed to be 'not a lot'. They were all used to using quite sophisticated statistics within their testing procedures, but not for controlling the performance of the processes themselves. Petra reflected on this. *'I guess it's because the statistics that our technicians use every day are essentially static. They deal with the probability of certain elements or contaminants being present in a single sample. SPC deals with dynamic probabilities – time series in effect – that show whether process behaviour is changing. However, the positive outcome from these meetings was that staff had little problem understanding the basic concepts of SPC, when they were explained. They were not frightened by the maths'.*

Petra realised that, in fact, the biggest problem was attitudinal. *'We had been working for a year with the attitude that testing productivity was paramount. Don't waste time. Get as many tests done as possible every day. It took time to move to an attitude that stressed error-free testing. What was the point of carrying on with testing when the processes themselves were "out of control"? They would only have to be repeated, wasting everyone's time. It may be counter-intuitive, but being slow but methodical, and checking the process regularly, can actually increase effective productivity'.* With the agreement of her staff, Petra devised a set of 'check rules'. These were reference values for all the major procedures in the sample preparation, pre-analysis and analysis stages, which indicated that test results at any stage, although within the limits that indicated a reliable result, were close to those limits. If results violated these 'check rules', the test would be suspended and the sample investigated before it was allowed to progress. Petra had three reasons for instituting the 'check rules'. First, it prevented effort being wasted on samples that could be compromised. Second, it stressed the importance of trying to investigate the root causes of any problems with the process. Third, it emphasised the importance of the Lab's processes in determining its quality of service to customers, and therefore to its profitability and survival.

The 'root cause' programme

By September the Lab's process performance had improved to the point where the number of samples that failed the reliability test had almost halved, and the number of late reports had fallen by over a third. But Petra believed that further improvements were possible. *'The most significant change is in the Lab's culture. Before, staff were simply going through the motions. They were not deliberately being careless, but they were not really digging beneath what they were doing, they were not building their process knowledge. If asked, they would tell you what they were doing rather than why they were doing it. Now there is genuine curiosity about how testing procedures could be made better'.*

Petra wanted to use the staff's new-found interest in the process to make further improvements through what she called the 'root cause' initiative. As the name implies, this was a push to discover what was causing problems in testing. The data collected from those occasions where the check rules had been invoked provided valuable information, which was further supplemented by individual investigations by 'root cause teams' in each department. Petra, with the support of supervisors in each department, had encouraged the formation of these teams, but not made them compulsory. However, most staff elected to become 'root cause team' members.

By the end of October, Petra was in a position to consolidate all the data on the root causes of all the occasions when an error of some sort had occurred in the Lab's processes. This included any defect from ordering tests to reporting and interpretation of the results. Table 17.6 shows the root causes.

What was interesting to Petra was the dominance of errors with a root cause outside the Lab. The data indicated that more than half of all errors were outside the scope of the Lab's responsibility. *'This shouldn't lead us into any form of*

Table 17.6 Root cause by phase of the testing process

Phase of testing	Sample preparation (62% of total errors)	Pre-analysis (19% of total errors)	Analysis (15% of total errors)	Report and record (4% of total errors)
Types of root cause	Mislabelled sample (F) (24%) Badly contained (F) (18%) Preparation error (8%) Request error (F) (6%) Insufficient material (F) (3%) Damaged sample (F) (3%)	Reagent error (6%) Contamination (5%) Spillage (5%) Process violation (3%)	Calibration error (6%) Process violation (4%) In-test calculation (3%) Contamination (2%)	Reading error) (2%) Interpretation error) (1%) Missing data) (1%)
(F) = Root cause in the field (client's responsibility)				

complacency. We can still do a lot to tackle the errors in the phases of the process for which we are clearly responsible. Basic laboratory procedure, such as choosing the correct reagent, violating process rules or allowing contamination, should not be happening. Also, I suspect that we are actually committing more "errors" in the "report and record" phase than it seems. Errors in testing are more obvious, but reporting is not always right or wrong. There are probably opportunities to enhance our service to clients that we are missing. You could class them as just as much of an "error" as a contaminated sample'.

QUESTIONS

1 In hindsight, what were Vincent's mistakes in running the Lab?

2 How did Petra's approach differ, and why was it more successful?

3 Is a 'missed opportunity' in the report and record stage as much of an error as a contaminated sample, as Petra suggests?

4 What do you suggest that Petra does next to improve process quality further?

Problems and applications

All chapters have 'Problems and applications' questions that will help you practise analysing operations. They can be answered by reading the chapter. Model answers for the first two questions can be found on the companion website for this text.

(Read the supplement on statistical process control, before attempting problems 7 and 8.)

1 Human error is a significant source of quality problems. Think through the times that you have (with hindsight) made an error and answer the following questions:

(a) How do you think that human error causes quality problems?
(b) What could one do to minimise human error?

2 The owner of a small wedding photography business realises that the market is changing. *'I used to take a few photos during the wedding ceremony and then formal group shots outside. It rarely took more than two hours. Around 30 photos would go in a standard wedding album. You had to get the photos right, that was really the only thing I was judged on. Now it's different. I spend all day at a wedding, and sometimes late into the evening. You're almost like another guest. All the guests at the wedding now are important. You have to get the best photos while being as discreet as possible. Clients judge you on both the pictures and the way you interact with everyone on the day. The product has changed too. Clients receive a memory stick with around 500 photos and a choice of 10 albums. I also offer photo books with greater customisation. I can offer albums with items such as invitations, confetti and menus; and individual paintings created from photographs. Obviously I*

would have to outsource the paintings. Wedding guests can order photos and related products online. My anxiety is that advertising this service at the wedding will be seen as being too commercial. We have a high level of demand in summer, with weekends booked up two years in advance. I may take on additional photographers during busy periods, but the best ones are busy themselves. Also, the business is about client relations and that's hard to replicate. I often offer clients advice on such things as locations, bands, caterers and florists. Wedding planning is clearly an area that could be profitable to the business. Another option is to move beyond weddings into other areas, such as school photos, birthdays, celebrations, or studio work'.

(a) How has the business changed over time?
(b) What do you think are the key quality challenges facing the business?
(c) What do you think should be done to ensure the business maintains quality levels in the future?

3 Ryanair is the best-known budget airline in Europe, focusing on a popular routes and very low operating costs. For years, the airline's policy on customer service was clear. It aimed to give the lowest air fare available, and a safe and on-time flight. It did not promise to give anything more. Nor was it apologetic about its lack of customer service. If a plane was cancelled, it did not offer overnight accommodation or give a voucher for a restaurant. However, bad publicity eventually prompted a limited rethink by the company. After a drop in its hitherto rapid profit growth, shareholder concern and a loss of what reputation it had for service, the company said it would reform its culture to reform things that unnecessarily annoyed customers, such as fines for small luggage size transgressions and a fee for issuing boarding passes at the airport rather than printing it out at home. Yet Ryanair insisted that these charges were not money-spinning schemes. They were designed to encourage the operational efficiency that kept fares low, in fact fewer than ten passengers a day had to pay for forgotten boarding passes. What does this example tell us about the trade-off between service quality and cost?

4 Understanding type I and type II errors is essential for surgeons' quality planning. For example, in the case of appendicitis, removal of the appendix is necessary because of the risk of its bursting, causing potentially fatal poisoning of the blood. The surgical procedure is relatively simple but there is always a small risk with any invasive surgery. It is also expensive; in the United States around $4,500 per operation. Unfortunately, appendicitis is difficult to diagnose: diagnosis is only 10 per cent accurate. However, a new technique claims to be able to identify 100 per cent of true appendicitis cases prior to surgery. The new technique costs less than $250, which means that one single avoided surgery pays for around 20 tests.

(a) How does this new test change the likelihood of type I and type II errors?
(b) Why is this important?

5 'Tea and Sympathy' (not a made-up name) was a British restaurant and café in the heart of New York's West Village. It became fashionable not only with expatriate Brits but also with native New Yorkers, who were willing to queue to get in. One reason it become famous was for the unusual nature of its service. *'Everyone is treated in the same way'*, said Nicky Perry, one of the two ex-Londoners who ran it. *'We have a firm policy that we don't take any shit'.* This robust attitude to the treatment of customers is reinforced by 'Nicky's Rules', which are printed on the menu:

- Be pleasant to the waitresses – remember Tea and Sympathy girls are always right.
- You will have to wait outside the restaurant until your entire party is present: no exceptions.
- Occasionally, you may be asked to change tables so that we can accommodate all of you.
- If we don't need the table you may stay all day, but if people are waiting it's time to naff off.
- These rules are strictly enforced. Any argument will incur Nicky's wrath. You have been warned.

Nicky's Rules were strictly enforced. If customers objected, they were thrown out. Nicky said that she has had to train 'her girls' to toughen up. *'I've taught them that when people cross the line they can tear their throats out as far as I'm concerned. What we've discovered over the years is that if you are really sweet, people see it as a weakness'.* People were thrown out about twice a week and yet customers still queued for the food (and of course the service).

(a) Why do you think 'Nicky's Rules' help to make the Tea and Sympathy operation more efficient?
(b) The restaurant's approach to quality of service seems very different to most restaurants. Why do you think it seems to work here?

6 Look again at the 'Operations in practice' example that includes a description of the Four Seasons Canary Wharf ('Quality at two operations: Victorinox and Four Seasons').

(a) The company has what it calls its Golden Rule: 'Do to others (guests and staff) as you would wish others to do to you'. Why is this important in ensuring high-quality service?
(b) What do you think the hotel's guests expect from their stay?
(c) How do staff using their own initiative contribute to quality?

7 An animal park in Amsterdam has decided to sample 50 visitors each day (n) to see how many visitors are from overseas. The data below are for the last seven days. If it decided to continue recording these data and plot them on a control chart for attributes, what should the upper and lower control limits be?

Day	Number of overseas visitors
1	7
2	8
3	12
4	5
5	5
6	4
7	8

8 The manager of a sweet shop decides to sample batches of sweets to check that the weight is reasonably consistent. She takes nine samples, each with 10 bags. The data below show the average mean weight for each sample and the weight range. What control limits would a control chart for variables use?

Sample	Weight average in grams	Range (R)
1	10	1.50
2	8	2.00
3	9	3.00
4	9	2.50
5	8	1.50
6	9	1.00
7	11	2.00
8	14	2.50
9	12	2.00

9 Courts in some countries impose curfew restrictions on convicted felons, enforced by attaching 'tags' to their legs so that their movements can be detected. A private security firm sacked two members of staff who tagged a man's false leg allowing him to remove it and break the court-imposed curfew. Was it right to sack these employees? What could the firm have done to prevent this mistake?

10 Is variation on output always a bad thing? Under what circumstances would variation actually add value?

Selected further reading

ASQ Quality Press (2010) *Seven Basic Quality Tools*, ASQ Quality Press, Milwaukee, WI.
Very much a 'how to do it' handbook.

Dale, B.G. (2016) *Managing Quality*, 6th edn, Wiley-Blackwell, Oxford.
The latest version of a long-established, comprehensive and authoritative text.

Kiran, D.R. (2016) *Total Quality Management: Key Concepts and Case Studies*, Butterworth-Heinemann, Oxford.
A good blend of basic principles and examples.

Oakland, J.S., Oakland, R.J. and Turner, M.A. (2020) *Total Quality Management and Operational Excellence: Text with Cases*, 5th edn, Routledge, Abingdon.
Latest version of the classic text from one of the founders of TQM in Europe.

Sartor, M. and Orzes, G. (2019) *Quality Management: Tools, Methods and Standards*, Emerald Publishing, Bingley.
A wide-ranging set of summaries of the key topics in quality management.

Tricker, R. (2019) *Quality Management Systems: A Practical Guide to Standards Implementation*, Routledge, Abingdon.
As its name implies, very much a quality systems approach including the ISO standards.

Notes on chapter

1. For more information, see: Vitaliev, V. (2009) The much-loved knife, *Engineering and Technology*, 4 (13), 58–61.
2. Interview with Michael Purtill, the General Manager of the Four Seasons Hotel Canary Wharf in London. We are grateful for Michael's cooperation (and for the great quality of service at his hotel!).
3. Berry, L.L. and Parasuraman, A. (1991) *Marketing Services: Competing Through Quality*, Free Press, New York, NY.
4. The information on which this example is based is taken from: Pardes, A. (2017) Ikea's new app flaunts what you'll love most about AR, *Wired*, 20 September; Joseph, S. (2017) How Ikea is using augmented reality, Digiday UK, 4 October; Sha, D.Y. and Lai, G.-L. (2012) Exploring the intention of customers to use innovative digital content information technology, IEEE International Conference on Industrial Engineering and Engineering Management (IEEM) December, pp. 1065–9; Augmented Reality (AR), Gartner Glossary, https://www.gartner.com/en/information-technology/glossary/augmented-reality-ar (accessed September 2021).
5. Brunnström, K. *et al.* (2013). Qualinet White Paper on Definitions of Quality of Experience, Qualinet.
6. Ferdows, K. and de Meyer, A. (1990) Lasting improvement in manufacturing perfromance: in search of a new theory, *Journal of Operations Management*, 9 (2), 169–84.
7. The information on which this example is based is taken from: Millington, A. (2018) Virgin Atlantic is offering a full refund on flights booked today if it can't cure a passenger's fear of flying, businessinsider.com, 9 January; Edwards, J. (2018) Why you should book a flight today if you've got a fear of flying, *Cosmopolitan*, 9 January.
8. The information on which this example is based is taken from: Markillie, P. (2011) They trash cars, don't they?, *Intelligent Life Magazine*, Summer.
9. The information on which this example is based is taken from: Walne, T. (2019) Want to exchange a jar of coins for notes?, This is Money, 24 August, https://www.thisismoney.co.uk/money/betterbanking/article-7390729/Want-notes-not-coins-Banks-dont-care-cash-investigation-finds.html (accessed September 2021); Schubber, K. (2016) The Metro Bank coin caper, *Financial Times*, 2 June.
10. As reported in Morgan, R. (2016) TD Bank dumps its faulty coin-counting machines, *New York Post*, 19 May.
11. The information on which this example is based is taken from: Giugliano, F. (2015) Bank of England moves to stamp out fat finger errors, *Financial Times*, 14 June; The Economist (2013) Overtired, and overdrawn, *Economist* print pdition, 15 June; Wilson, H. (2014) Fat fingered trader sets Tokyo alarms ringing, *The Times*, 2 October.
12. Feigenbaum, A.V. (1986) *Total Quality Control*, McGraw Hill, New York, NY.

13. Kaynak, H. (2003) The relationship between total quality management practices and their effects on firm performance, *Journal of Operations Management*, 21 (4), 405–35.
14. For further information, see for example: Wells, P, and Lewis, L. (2018) Japan Inc: a corporate culture on trial after scandals, *Financial Times*, 3 January; Parry, R.L. (2017) Japan's failed corporate culture at root of Kobe Steel scandal, *The Times*, 16 October; Economist (2017) Kobe Steel admits falsifying data on 20,000 tonnes of metal, *Economist* print edition, 12 October; Wells, P. (2017) Kobe Steel demotes three executives over cheating scandal, *Financial Times*, 21 December; Wells, P. and Terazono, E. (2017) Five questions on Kobe Steel and quality controls, *Financial Times*, 11 October.
15. See the EFQM website, http://www.knowledge-base.efqm.org/the-efqm-excellence-model (accessed September 2021).

Supplement to Chapter 17
Statistical process control (SPC)

INTRODUCTION

Statistical process control (SPC) is concerned with checking a service or product during its creation. If there is reason to believe that there is a problem with the process, then it can be stopped and the problem can be identified and rectified. For example, an international airport may regularly ask a sample of customers if the cleanliness of its restaurants is satisfactory. If an unacceptable number of customers in one sample are found to be unhappy, airport managers may have to consider improving its procedures. Similarly, a vehicle manufacturer will periodically check whether a sample of door panels conforms to its standards so it knows whether the machinery that produces them is performing correctly.

Control charts

The value of SPC is not just to make checks of a single sample but to monitor the quality over a period of time. It does this by using control charts, to see if the process seems to be performing as it should, or alternatively if it is 'out of control'. If the process does seem to be going out of control, then steps can be taken *before* there is a problem. Actually, most operations chart their quality performance in some way. Figure 17.13, or something like it, could be found in almost any operation. The chart could, for example, represent the percentage of customers in a sample of 1,000 who, each month, were dissatisfied with the restaurant's cleanliness. While the amount of dissatisfaction may be acceptably small, management should be concerned that it has been steadily increasing over time and may wish to investigate why this is so. In this case, the control chart is plotting an attribute measure of quality (satisfied or not). Looking for trends is an important use of control charts. If the trend suggests the process is getting steadily worse, then it will be worth investigating the process. If the trend is

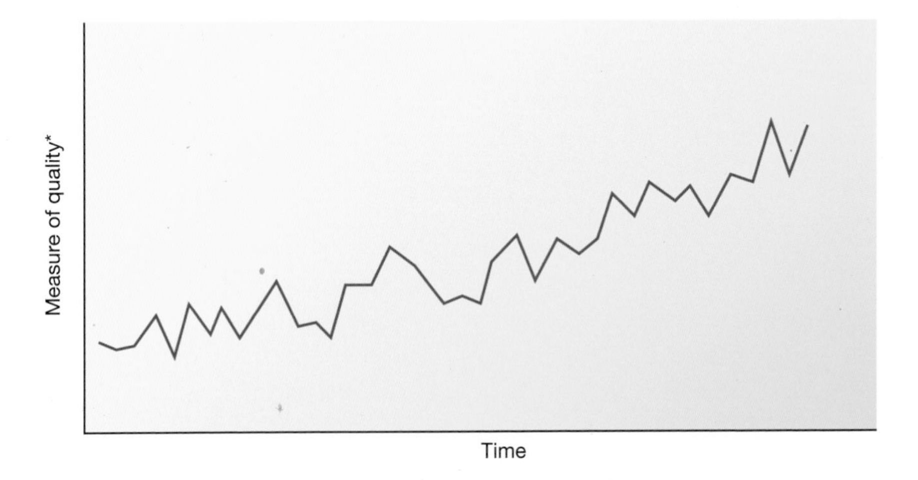

Figure 17.13 Charting trends in quality measures

steadily improving, it may still be worthy of investigation to try to identify what is happening that is making the process better. This information might then be shared with other parts of the organisation or, on the other hand, the process might be stopped as the cause could be adding unnecessary expense to the operation.

Variation in process quality

Common causes

The processes charted in Figure 17.13 showed an upwards trend. But the trend was neither steady nor smooth, it varied, sometimes up, sometimes down. All processes vary to some extent. No machine will give precisely the same result each time it is used. People perform tasks slightly differently each time. Given this, it is not surprising that the measure of quality will also vary. Variations that derive from these *common causes* can never be entirely eliminated (although they can be reduced). For example, if a machine is filling boxes with rice, it will not place *exactly* the same weight of rice in every box it fills. When the filling machine is in a stable condition (that is, no exceptional factors are influencing its behaviour) each box could be weighed and a histogram of the weights could be built up. Figure 17.14 shows how the histogram might develop. The first boxes weighed could lie anywhere within the natural variation of the process but are more likely to be close to the average weight (see Figure 17.14(a)). As more boxes are weighed, they clearly show the tendency to be close to the process average (see Figure 17.14(b) and (c)). After many boxes have been weighed, they form a smoother distribution (Figure 17.14(d)), which can be drawn as a histogram (Figure 17.14(e)), which will approximate to the underlying process variation distribution (Figure 17.14(f)).

Usually this type of variation can be described by a normal distribution with 99.7 per cent of the variation lying within ± 3 standard deviations. In this case the weight of rice in the boxes is described by a distribution with a mean of 206 grams and a standard deviation of 2 grams. The obvious question for any operations manager would be: 'Is this variation in the process performance acceptable?' The answer will depend on the acceptable range of weights that can be tolerated by the operation. This range is called the specification range. If the weight of rice in the box is too small then the organisation might infringe labelling regulations; if it is too large, the organisation is 'giving away' too much of its product for free.

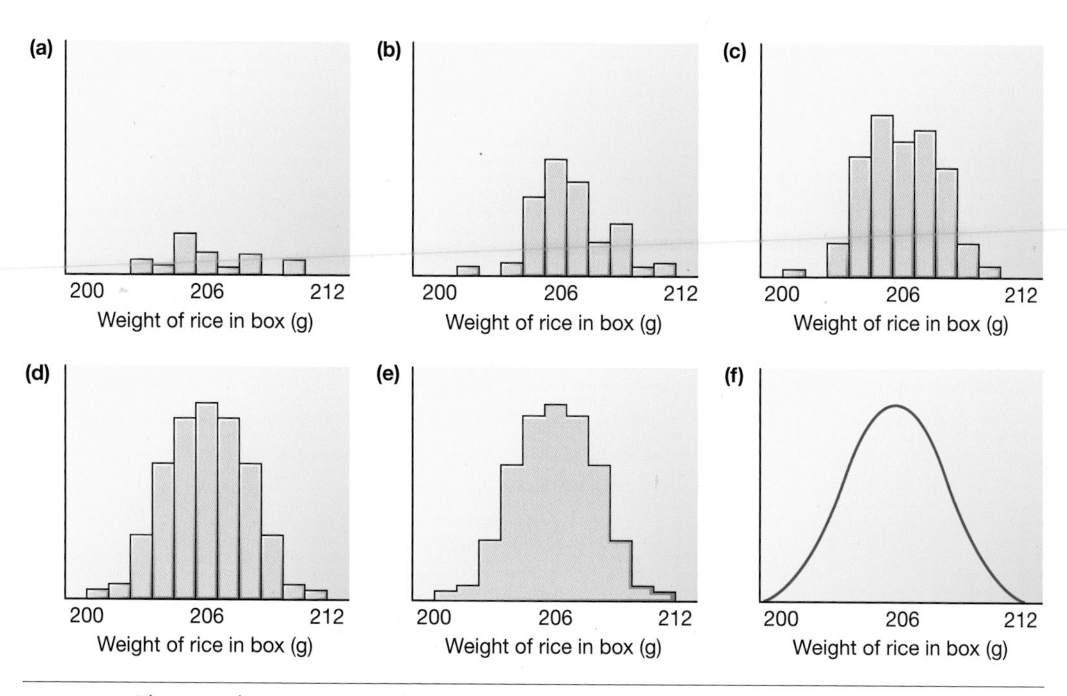

Figure 17.14 The natural variation in the filling process can be described by a normal distribution

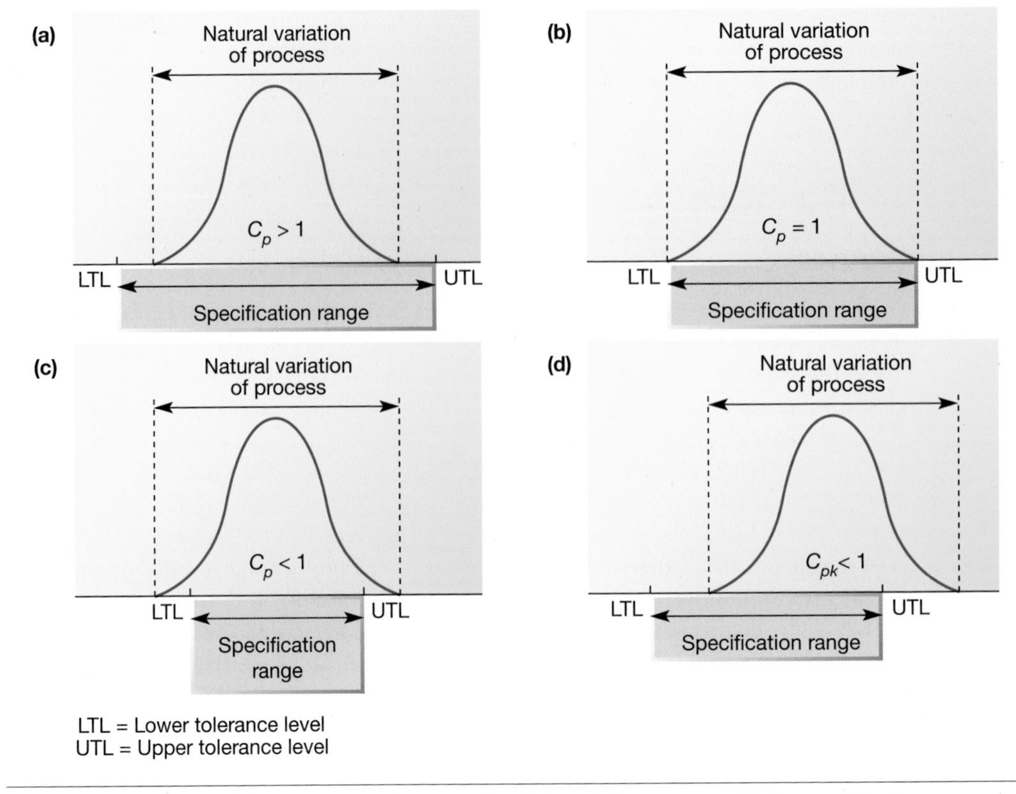

Figure 17.15 Process capability compares the natural variation of the process with the specification range that is required

Process capability

Process capability is a measure of the acceptability of the variation of the process. The simplest measure of capability (C_p) is given by the ratio of the specification range to the 'natural' variation of the process (i.e. ± 3 standard deviations):

$$C_p = \frac{\text{UTL} - \text{LTL}}{6s}$$

where

UTL = the upper tolerance limit

LTL = the lower tolerance limit

s = the standard deviation of the process variability.

Generally, if the C_p of a process is greater than 1, it is taken to indicate that the process is 'capable', and a C_p of less than 1 indicates that the process is not 'capable', assuming that the distribution is normal (see Figure 17.15(a), (b) and (c)).

The simple C_p measure assumes that the average of the process variation is at the mid-point of the specification range. Often the process average is offset from the specification range, however (see Fig. 17.15(d)). In such cases, *one-sided* capability indices are required to understand the capability of the process:

$$\text{Upper one-sided index } C_{pu} = \frac{\text{UTL} - \text{X}}{3s}$$

$$\text{Lower one-sided index } C_{pl} = \frac{\text{X} - \text{LTL}}{3s}$$

where X = the process average.

Sometimes only the lower of the two one-sided indices for a process is used to indicate its capability (C_{pk}):

$$C_{pk} = \min(C_{pu}, C_{pl})$$

Worked example

Boxes of rice

In the case of the process filling boxes of rice, described previously, process capability can be calculated as follows:

$$\text{Specification range} = 214 - 198 = 16 \text{ g}$$

$$\text{Natural variation of process} = 6 \times \text{standard deviation}$$

$$= 6 \times 2 = 12 \text{ g}$$

$$C_p = \text{process capability}$$

$$= \frac{\text{UTL} - \text{LTL}}{6s}$$

$$= \frac{214 - 198}{6 \times 2} = \frac{16}{12}$$

$$= 1.333$$

If the natural variation of the filling process changed to have a process average of 210 grams but the standard deviation of the process remained at 2 grams:

$$C_{pu} = \frac{214 - 210}{3 \times 2} = \frac{4}{6} = 0.666$$

$$C_{pl} = \frac{210 - 198}{3 \times 2} = \frac{12}{6} = 2.0$$

$$C_{pk} = \min(0.666, 2.0)$$

$$= 0.666$$

Assignable causes of variation

Not all variation in processes is the result of common causes. There may be something wrong with the process that is assignable to a particular and preventable cause. Machinery may have worn or been set up badly. An untrained person may not be following prescribed procedures. The causes of such variation are called *assignable causes*. The question is whether the results from any particular sample, when plotted on the control chart, simply represent the variation due to common causes or due to some specific and correctable *assignable* cause. Figure 17.16, for example, shows the control chart for the average impact resistance of samples of door panels taken over time. Like any process the results vary, but the last three points seem to be lower than usual. So, is this a natural (common cause) variation, or the symptom of some more serious (assignable) cause?

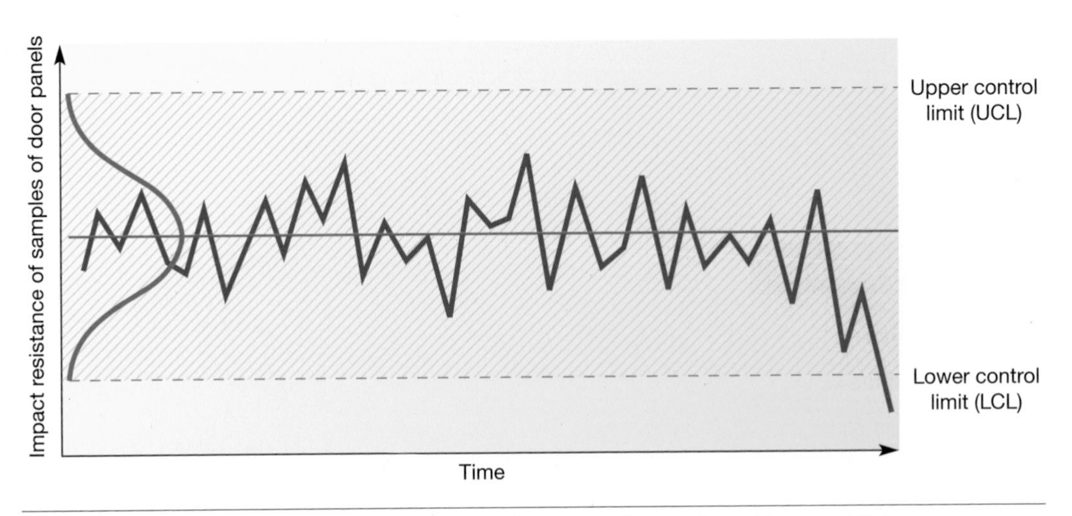

Figure 17.16 Control chart for the impact resistance of door panels, together with control limits

To help make this decision, control limits can be added to the control chart (the red dotted lines), which indicate the expected extent of 'common-cause' variation. If any points lie outside these control limits (the shaded zone) then the process can be deemed out of control in the sense that variation is likely to be due to assignable causes. These control limits could be set intuitively by examining past variation during a period when the process was thought to be free of any variation that could be due to assignable causes. But control limits can also be set in a more statistically revealing manner. For example, if the process that tests door panels had been measured to determine the normal distribution that represents its common-cause variation, then control limits can be based on this distribution. Figure 17.16 also shows how control limits can be added; here put at ± 3 standard deviations (of the population of sample means) away from the mean of sample averages. It shows that the probability of the final point on the chart being influenced by an assignable cause is very high indeed. When the process is exhibiting behaviour that is outside its normal 'common-cause' range, it is said to be 'out of control'. Yet there is a small but finite chance that the (seemingly out of limits) point is just one of the rare but natural results at the tail of the distribution that describes perfectly normal behaviour. Stopping the process under these circumstances would represent a type I error because the process is actually in control. Alternatively, ignoring a result that in reality is due to an assignable cause is a type II error (see Table 17.7).

Control limits are usually set at three standard deviations either side of the population mean. This would mean that there is only a 0.3 per cent chance of any sample mean falling outside these limits by chance causes (that is, a chance of a type I error of 0.3 per cent). The control limits may be set at any distance from the population mean, but the closer the limits are to the population mean, the higher the likelihood of investigating and trying to rectify a process that is actually problem-free. If the control limits are set at two standard deviations, the chance of a type I error increases to about 5 per cent. If the limits are set at one standard deviation, then the chance of a type I error increases to 32 per cent. When the control limits are placed at ± 3 standard deviations away from the mean of

Table 17.7 Type I and type II errors in SPC

	Actual process state	
Decision	In control	Out of control
Stop process	Type I error	Correct decision
Leave alone	Correct decision	Type II error

the distribution, which describes 'normal' variation in the process, they are called the upper control limit (UCL) and lower control limit (LCL).

Critical commentary

When its originators first described SPC more than half a century ago, the key issue was only to decide whether a process was 'in control' or not. Now, we expect SPC to reflect common sense as well as statistical elegance and promote continuous operations improvement. This is why two (related) criticisms have been levelled at the traditional approach to SPC. The first is that SPC seems to assume that any values of process performance that lie within the control limits are equally acceptable, while any values outside the limits are not. However, surely a value close to the process average or 'target' value will be more acceptable than one only just within the control limits. For example, a service engineer arriving only 1 minute late is a far better 'performance' than one arriving 59 minutes late, even if the control limits are 'quoted time ± one hour'. Also, arriving 59 minutes late would be almost as bad as 61 minutes late! Second, a process always within its control limits may not be deteriorating, but is it improving. So rather than seeing control limits as fixed, it would be better to view them as a reflection of how the process is being improved. We should expect any improving process to have progressively narrowing control limits.

Why variability is a bad thing

Assignable variation is a signal that something has changed in the process, which therefore must be investigated. But normal variation is itself a problem because it masks any changes in process behaviour. Figure 17.17 shows the performance of two processes, both of which are subjected to a change in their process behaviour at the same time. The process on the left has such a wide natural variation that it is not immediately apparent that any change has taken place. Eventually it will become apparent because the likelihood of process performance violating the lower (in this case) control limit has increased, but this may take some time. By contrast, the process on the right has a far narrower band of natural variation. Because of this, the same change in average performance is more easily noticed (both visually and statistically). So, the narrower the natural variation of a process, the more obvious are changes in the behaviour of that process. And the more obvious are process changes, the easier it is to understand how and why the process is behaving in a particular way. Accepting any variation in any process is, to some degree, admitting to ignorance of how that process works.

Operations principle

High levels of variation reduce the ability to detect changes in process performance.

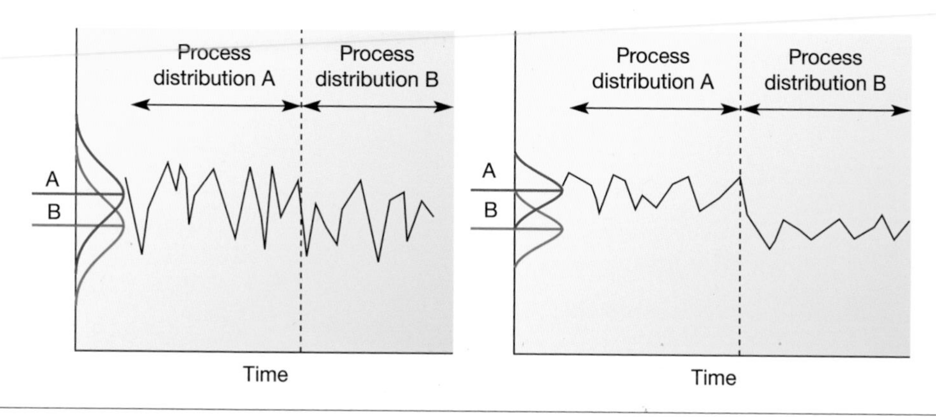

Figure 17.17 Low process variation allows changes in process performance to be readily detected

Control charts for attributes

Attributes have only two states – 'right' or 'wrong', for example – so the statistic calculated is the proportion of wrongs (p) in a sample. (This statistic follows a binomial distribution.) Control charts using p are called 'p-charts'. In calculating the limits, the population mean ($\bar{p}$) – the actual, normal or expected proportion of 'defectives', or wrongs, to rights – may not be known. Who knows, for example, the actual number of city commuters who are dissatisfied with their journey time? In such cases the population mean can be estimated from the average of the proportion of 'defectives' ($\bar{p}$), from m samples each of n items, where m should be at least 30 and n should be at least 100:

$$\bar{p} = \frac{p^1 + p^2 + p^3 \ldots p^n}{m}$$

One standard deviation can then be estimated from:

$$\sqrt{\frac{\bar{p}(1-\bar{p})}{n}}$$

The upper and lower control limits can then be set as:

$$\text{UCL} = \bar{p} + 3 \text{ standard deviations}$$

$$\text{LCL} = \bar{p} - 3 \text{ standard deviations}$$

Of course, the LCL cannot be negative, so when it is calculated to be so it should be rounded up to zero.

Worked example

Aphex Credit

Aphex Credit deals with many hundreds of thousands of transactions every week. One of its measures of the quality of service it gives its customers is the dependability with which it mails customers' monthly accounts. The quality standard it sets itself is that accounts should be mailed within two days of the 'nominal post date' that is specified to the customer. Every week the company samples 1,000 customer accounts and records the percentage that was not mailed within the standard time. When the process is working normally, only 2 per cent of accounts are mailed outside the specified period, that is, 2 per cent are 'defective'.

Control limits for the process can be calculated as follows:

$$\text{Mean proportion defective, } \bar{p} = 0.02$$

$$\text{Sample size } n = 1000$$

$$\text{Standard deviation } s = \sqrt{\frac{\bar{p}(1-\bar{p})}{n}}$$

$$= \sqrt{\frac{0.02(0.98)}{1000}}$$

$$= 0.0044$$

With the control limits at $\bar{p} \pm 3s$:

$$\text{Upper control limit (UCL)} = 0.02 + 3(0.0044) = 0.0332$$

$$= 3.32\%$$

$$\text{and lower control limit (LCL)} = 0.02 - 3(0.0044) = 0.0068$$

$$= 0.68\%$$

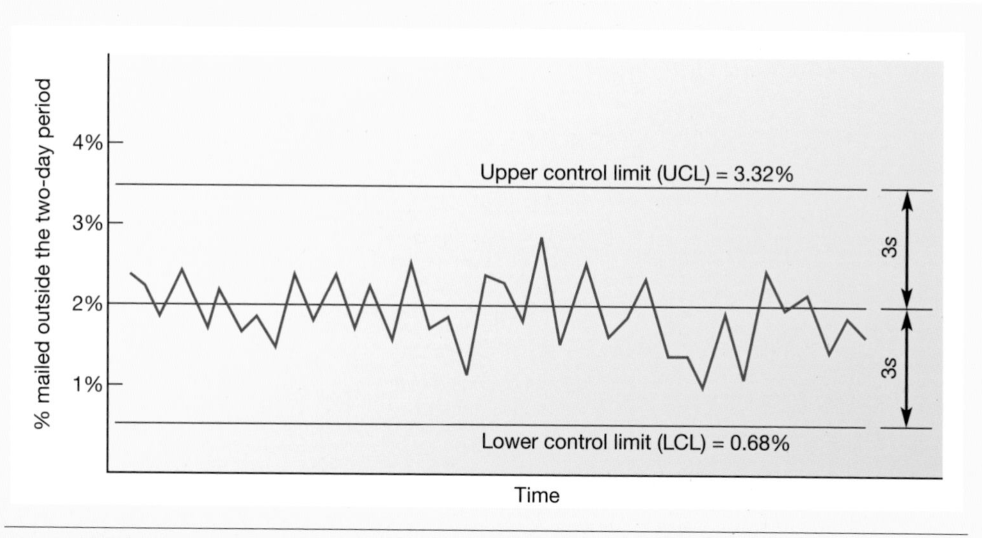

Figure 17.18 Control chart for the percentage of customer accounts that are mailed outside their two-day period

Figure 17.18 shows the company's control chart for this measure of quality over the last few weeks, together with the calculated control limits. It also shows that the process is in control. Sometimes it is more convenient to plot the actual number of defects (c) rather than the proportion (or percentage) of defectives, on what is known as a c-chart. This is very similar to the p-chart but the sample size must be constant and the process mean and control limits are calculated using the following formulae:

$$\text{Process mean } \bar{c} = \frac{c_1 + c_2 + c_3 \ldots c_m}{m}$$

$$\text{Control limits} = \bar{c} \pm 3\sqrt{c}$$

where

c = number of defects

m = number of samples.

Control chart for variables

The most commonly used type of control chart employed to control variables is the $\overline{X}$–R *chart*. In fact, this is really two charts in one. One chart is used to control the sample average or mean ($\overline{X}$). The other is used to control the variation within the sample by measuring the range (R). The range is used because it is simpler to calculate than the standard deviation of the sample.

The means ($\overline{X}$) chart can pick up changes in the average output from the process being charted. Changes in the means chart would suggest that the process is drifting generally away from its supposed process average, although the variability inherent in the process may not have changed (see Figure 17.19).

The range (R) chart plots the range of each sample, that is, the difference between the largest and the smallest measurement in the samples. Monitoring sample range gives an indication of whether the variability of the process is changing, even when the process average remains constant (see Figure 17.19).

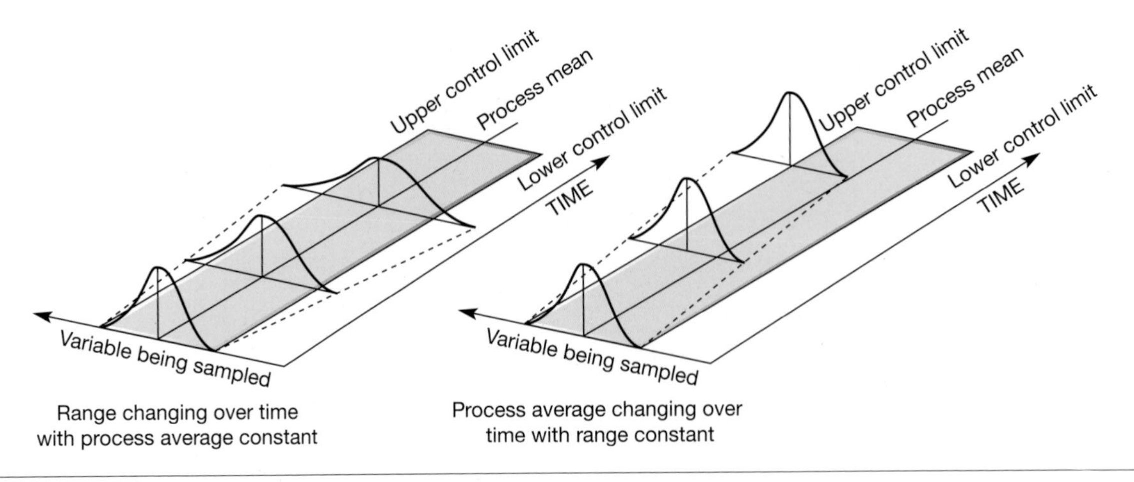

Figure 17.19 The process mean or the process range (or both) can change over time

Control limits for variables control chart

As with attributes control charts, a statistical description of how the process operates under normal conditions (when there are no assignable causes) can be used to calculate control limits. The first task in calculating the control limits is to estimate the grand average or population mean ($\overline{\overline{X}}$) and average range ($\overline{R}$) using m samples each of sample size n.

The population mean is estimated from the average of a large number (m) of sample means:

$$\overline{\overline{X}} = \frac{\overline{X}_1 + \overline{X}_2 + \ \dots \ \overline{X}_m}{m}$$

The average range is estimated from the ranges of the large number of samples:

$$\overline{R} = \frac{R_1 + R_2 + \ \dots \ R_m}{m}$$

The control limits for the sample means chart are:

$$\text{Upper control limit (UCL)} = \overline{\overline{X}} + A_2\overline{R}$$

$$\text{Lower control limit (LCL)} = \overline{\overline{X}} - A_2\overline{R}$$

The control limits for the range charts are:

$$\text{Upper control limit (UCL)} = D_4\overline{R}$$

$$\text{Lower control limit (LCL)} = D_3\overline{R}$$

The factors A_2, D_3 and D_4 vary with sample size and are shown in Table 17.8.

The LCL for the means chart may be negative (for example, temperature or profit may be less than zero) but it may not be negative for a range chart (or the smallest measurement in the sample would be larger than the largest). If the calculation indicates a negative LCL for a range chart then the LCL should be set to zero.

Table 17.8 Factors for the calculation of control limits

Sample size n	A_2	D_3	D_4
2	1.880	0	3.267
3	1.023	0	2.575
4	0.729	0	2.282
5	0.577	0	2.115
6	0.483	0	2.004
7	0.419	0.076	1.924
8	0.373	0.136	1.864
9	0.337	0.184	1.816
10	0.308	0.223	1.777
12	0.266	0.284	1.716
14	0.235	0.329	1.671
16	0.212	0.364	1.636
18	0.194	0.392	1.608
20	0.180	0.414	1.586
22	0.167	0.434	1.566
24	0.157	0.452	1.548

Worked example

GAM

GAM (Groupe As Maquillage) is a contract cosmetics company, based in France but with plants around Europe, which manufactures and packs cosmetics and perfumes for other companies. One of its plants, in Ireland, operates a filling line that automatically fills plastic bottles with skin cream and seals the bottles with a screw-top cap. The tightness with which the screw-top cap is fixed is an important part of the quality of the filling-line process. If the cap is screwed on too tightly, there is a danger that it will crack; if screwed on too loosely it might come loose when packed. Either outcome could cause leakage of the product during its journey between the factory and the customer. The Irish plant had received some complaints of product leakage, which it suspected was caused by inconsistent fixing of the screw-top caps on its filling line. The 'tightness' of the screw tops could be measured by a simple test device, which recorded the amount of turning force (torque) that was required to unfasten the tops. The company decided to take samples of the bottles coming out of the filling-line process, test them for their unfastening torque and plot the results on a control chart. Several samples of four bottles were taken during a period when the process was regarded as being in control. The following data were calculated from this exercise:

$$\text{The grand average of all samples } \overline{\overline{X}} = 812 \text{ g/cm}^3$$

$$\text{The average range of the sample } \overline{R} = 6 \text{ g/cm}^3$$

Control limits for the means ($\overline{X}$) chart were calculated as follows:

$$\text{UCL} = \overline{\overline{X}} + A_2\overline{R}$$

$$= 812 + (A_2 \times 6)$$

From Table 17.6, we know, for a sample size of four, $A_2 = 0.729$. Thus:

$$\text{UCL} = 812 + (0.729 \times 6)$$

$$= 816.37$$

$$\text{LCL} = \overline{\overline{X}} - (A_2\overline{R})$$

$$= 812 - (0.729 \times 6)$$

$$= 807.63$$

Control limits for the range chart (R) were calculated as follows:

$$\text{UCL} = D_4 \times \overline{R}$$

$$= 2.282 \times 6$$

$$= 13.69$$

$$\text{LCL} = D_3\overline{R}$$

$$= 0 \times 6$$

$$= 0$$

After calculating these averages and limits for the control chart, the company regularly took samples of four bottles during production, recorded the measurements and plotted them as shown in Figure 17.20. The control chart revealed that only with difficulty could the process average be kept in control. Occasional operator interventions were required. Also, the process range was moving towards (and once breaking) the upper control limit. The process seemed to be becoming more variable. After investigation it was discovered that, because of faulty maintenance of the line, skin cream was occasionally contaminating the torque head (the part of the line that fitted the cap). This resulted in erratic tightening of the caps.

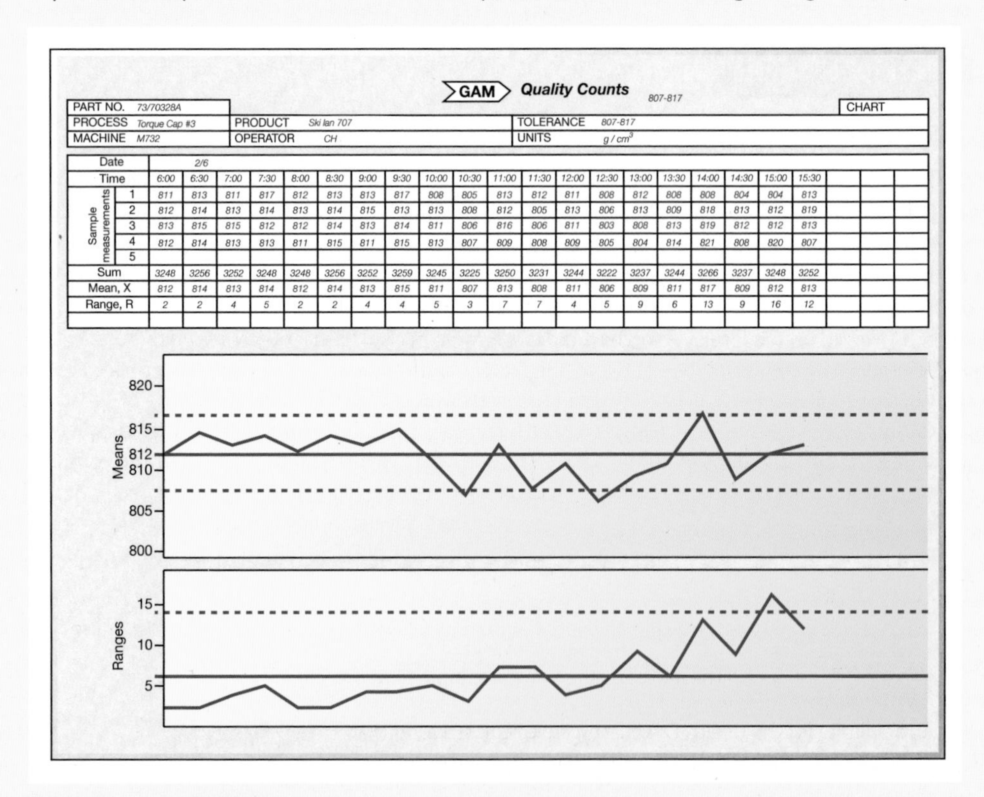

GAM Quality Counts 807-817

PART NO. 73/70328A | CHART
PROCESS Torque Cap #3 | PRODUCT Ski lan 707 | TOLERANCE 807-817
MACHINE M732 | OPERATOR CH | UNITS g / cm³

Date: 2/6

Time	6:00	6:30	7:00	7:30	8:00	8:30	9:00	9:30	10:00	10:30
Sample measurements 1	811	813	811	817	812	813	813	817	808	805
2	812	814	813	814	813	814	815	813	813	808
3	813	815	815	812	812	814	813	814	811	806
4	812	814	813	813	811	815	811	815	813	807
5										
Sum	3248	3256	3252	3248	3248	3256	3252	3259	3245	3225
Mean, X	812	814	813	814	812	814	813	815	811	807
Range, R	2	2	4	5	2	2	4	4	5	3

Time	11:00	11:30	12:00	12:30	13:00	13:30	14:00	14:30	15:00	15:30
Sample measurements 1	813	812	811	808	812	808	808	804	804	813
2	812	805	813	806	813	809	818	813	812	819
3	816	806	811	803	808	813	819	812	812	813
4	809	808	809	805	804	814	821	808	820	807
5										
Sum	3250	3231	3244	3222	3237	3244	3266	3237	3248	3252
Mean, X	813	808	811	806	809	811	817	809	812	813
Range, R	7	7	4	5	9	6	13	9	16	12

Figure 17.20 The completed control form for GAM's torque machine showing the mean ($\overline{X}$) and range ($\overline{R}$) charts

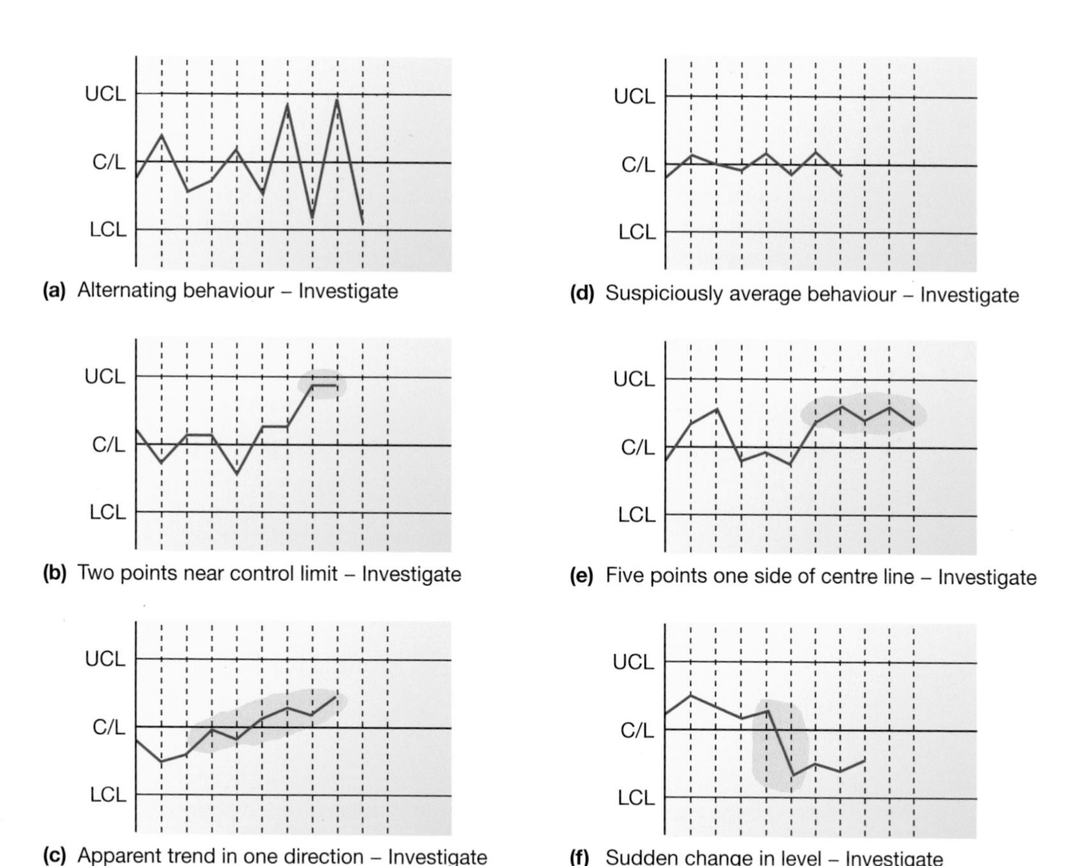

Figure 17.21 In addition to points falling outside the control limits, other unlikely sequences of points should be investigated

Interpreting control charts

Plots on a control chart that fall outside control limits are an obvious reason for believing that the process might be out of control, and therefore for investigating the process. This is not the only clue that could be revealed by a control chart, however. Figure 17.21 shows some other patterns that could be interpreted as behaviour sufficiently unusual to warrant investigation.

Process control, learning and knowledge

The role of process control, and SPC in particular, has changed. Increasingly, it is seen not just as a convenient method of keeping processes in control, but also as an activity that is fundamental to the acquisition of competitive advantage. This is a remarkable shift in the status of SPC. Traditionally it was seen as one of the most *operational*, immediate and 'hands-on' operations management techniques. Yet it is now being connected with an operation's *strategic* capabilities. This is how the logic of the argument goes:

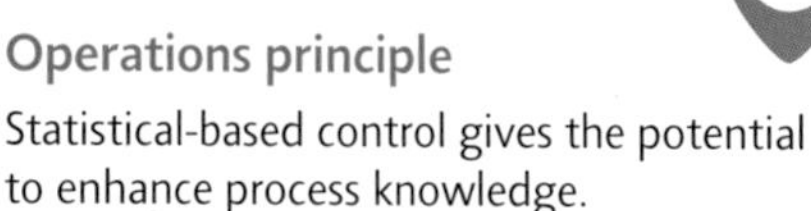

Operations principle
Statistical-based control gives the potential to enhance process knowledge.

1 SPC is based on the idea that process variability indicates whether a process is in control or not.
2 Processes are brought into *control* and improved by progressively reducing process variability. This involves eliminating the assignable causes of variation.
3 One cannot eliminate assignable causes of variation without gaining a better understanding of how the process operates. This involves *learning* about the process, where its nature is revealed at an increasingly detailed level.

4 This learning means that process knowledge is enhanced, which in turn means that operations managers are able to predict how the process will perform under different circumstances. It also means that the process has a greater capability to carry out its tasks at a higher level of performance.
5 This increased *process capability* is particularly difficult for competitors to copy. It cannot be bought 'off-the-shelf'. It comes only from time and effort being invested in controlling operations processes. Therefore, process capability leads to strategic advantage.

In this way, process control leads to learning, which enhances process knowledge and builds difficult-to-imitate process capability.

Summary of supplement

- Statistical process control (SPC) involves using control charts to track the performance of one or more quality characteristics in the operation. The power of control charting lies in its ability to set control limits derived from the statistics of the natural variation of processes. These control limits are often set at ± 3 standard deviations of the natural variation of the process samples.
- Control charts can be used for either attributes or variables. An attribute is a quality characteristic that has two states (for example, right or wrong). A variable is one that can be measured on a continuously variable scale.
- Process control charts allow operations managers to distinguish between the 'normal' variation inherent in any process and the variations that could be caused by the process going out of control.

Selected further reading

Woodall, W.H. (2000) Controversies and contradictions in statistical process control, Paper presented at the *Journal of Quality Technology* session at the 44th Annual Fall Technical Conference of the Chemical and Process Industries Division and Statistics Division of the American Society for Quality and the Section on Physical & Engineering Sciences of the American Statistical Association in Minneapolis, Minnesota, 12–13 October, 2000.
Academic but interesting.

18 Managing risk and recovery

KEY QUESTIONS

18.1 What is risk management?

18.2 How can operations assess the potential causes and consequences of failure?

18.3 How can failures be prevented?

18.4 How can operations mitigate the effects of failure?

18.5 How can operations recover from the effects of failure?

INTRODUCTION

No matter how much effort is put into improvement, all operations will face risk. Some risks emerge from poor operations management practice, such as poor technology maintenance. Some risks come from the operation's supply network, for example relying on unreliable suppliers. Other risks come from broader environmental forces, such as political unrest and environmental disasters. An increasingly important source of risk comes from failures in cybersecurity. In the face of such risks, a 'resilient' operation is one that can identify likely sources of risk, prevent failures occurring, minimise their effects and learn how to recover from them. In a world where the sources and the consequences of risk are becoming increasingly difficult to handle, managing risk is a vital task. Often the problem is that paring down costs, cutting inventories and striving for higher levels of capacity utilisation can all result in higher exposure (see the 'Responsible operations' section at the end of the chapter). In this chapter, we examine both the dramatic and the more routine risks that can prevent operations working as they should. Figure 18.1 shows how this chapter fits into the operation's improvement activities.

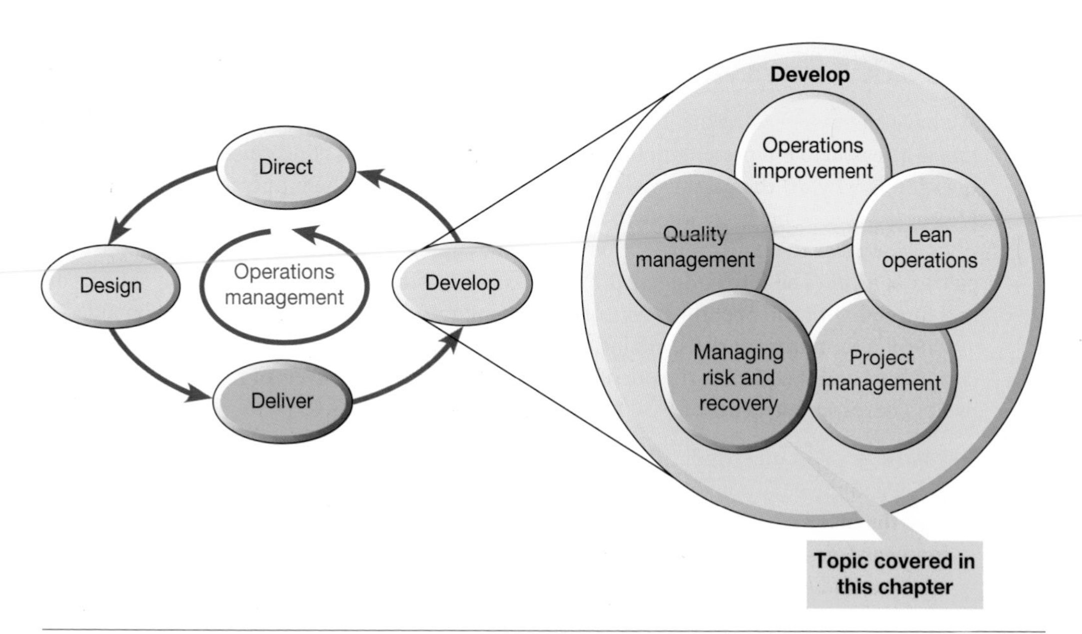

Figure 18.1 This chapter examines managing risk and recovery

18.1 What is risk management?

Risk management is about identifying things that could go wrong, stopping them going wrong, reducing the consequences when things do go wrong, and recovering after things have gone wrong. Things will always go wrong sometimes, but accepting this is not the same thing as tolerating or ignoring it. Although operations managers do generally attempt to minimise both the likelihood and the effect of things going wrong, the methods of coping with it will depend on how serious its consequences are, and how likely it is to occur. At the lower end of the scale, the whole area of quality management is concerned with identifying and reducing every small error in the creation and delivery of products and services. At the other end of the scale, server failure can seriously affect customers, which is why system reliability is such an important measure of performance for IT service providers. Some events, although much less likely (often called 'black swan' events), are so serious in terms of negative consequences that we class them as disasters.

> **Operations principle**
> Failure will always occur in operations. Recognising this does not imply accepting or ignoring it.

This chapter is concerned with all types of risks other than those with relatively minor consequences, as illustrated in Figure 18.2. Some things 'going wrong' are irritating, but relatively unimportant, especially those close to the bottom left-hand corner of the matrix in Figure 18.2. Other risks, especially those close to the top right-hand corner of matrix, are normally avoided as part of operations strategy, because embracing such risks would clearly be foolish. In-between these two extremes is where most operations-related risks occur.

Identify, prevent, mitigate, recover

Managing risk and recovery is relevant to all organisations and generally involves four sets of activities. The first is concerned with understanding what failures could potentially occur in the operation and assessing their seriousness. The second task is to examine ways of preventing failures occurring. The third is to minimise the negative consequences of failure – called failure or risk 'mitigation'. The final task is to devise plans and procedures that will help the operation to recover from failures when they do occur. The remainder of this chapter deals with these four tasks (see Figure 18.3).

> **Operations principle**
> Resilience is governed by the effectiveness of failure prevention, mitigation and recovery.

The VUCA framework

We have characterised risk as being related to 'things going wrong'. While simplistic in some ways, this is a reasonable representation of what risk management is concerned with. However, a further question is why do things 'go wrong'? A framework that can help operations to answer this is the idea of 'VUCA'. This is an acronym for volatility, uncertainty, complexity and ambiguity. They can be considered as the underlying reasons for 'things going wrong', or at least making operations difficult to control. The term originated in the US military to describe the turbulence, unpredictability and instability of the operating environment being faced by armed forces. It was intended to reflect a change in the nature of risks, not just an increase in magnitude. It was then adopted by management consultants and theorists to denote what they saw as increasing levels of risk in the business environment. These four elements of VUCA are illustrated in Figure 18.4:

- *Volatility* – Volatility indicates the speed, nature or magnitude of change in conditions that can happen in the environment in which an operation is working. High levels of volatility mean short time fluctuations in demand, often seen as general turbulence. Managing in highly volatile environments means establishing a vision that accepts surplus (even seemingly redundant) levels of resource and being prepared to use them when needed.
- *Uncertainty* – Uncertainty is the extent to which the future can be confidently predicted. The more uncertain the operating environment is, the less pattern or repetition there is in environmental

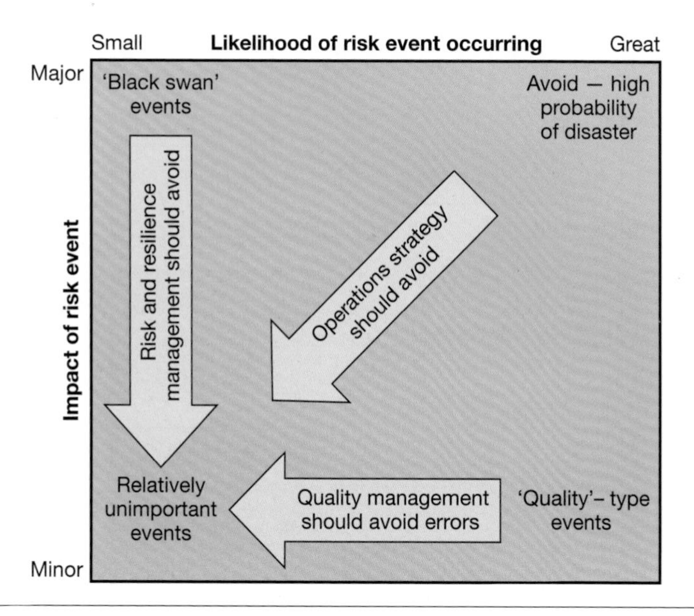

Figure 18.2 How failure is managed depends on its likelihood of occurrence and the negative consequences of failure

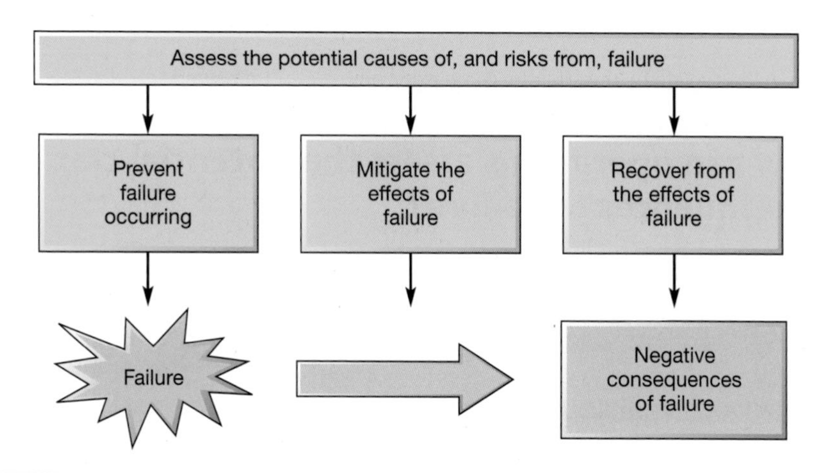

Figure 18.3 Risk management involves failure prevention, mitigating the negative consequences of failure, and failure recovery

conditions, and therefore the more difficult it is to forecast. Managing in highly uncertain environments means increasing the understanding of conditions through collecting, modelling and sharing information.

- ***Complexity*** – Complexity indicates the number of different factors that must be considered and our knowledge of the relationships between them. The more things there are involved in a situation, the more they differ from each other, and the more they are interconnected, the more complex is the environment. Managing in highly complex environments means developing expertise to establish more clarity about the relationship between factors in the environment.
- ***Ambiguity*** – Ambiguity indicates a lack of clarity about exactly what something means. Many different interpretations of available data are considered equally valid. Ambiguity can be the result of misleading, inaccurate or conflicting information. It inhibits the interpretation of information so that it is difficult to draw conclusions. Managing in highly ambiguous environments means developing the agility to overcome the vagueness in ideas and terminology through experimental learning.

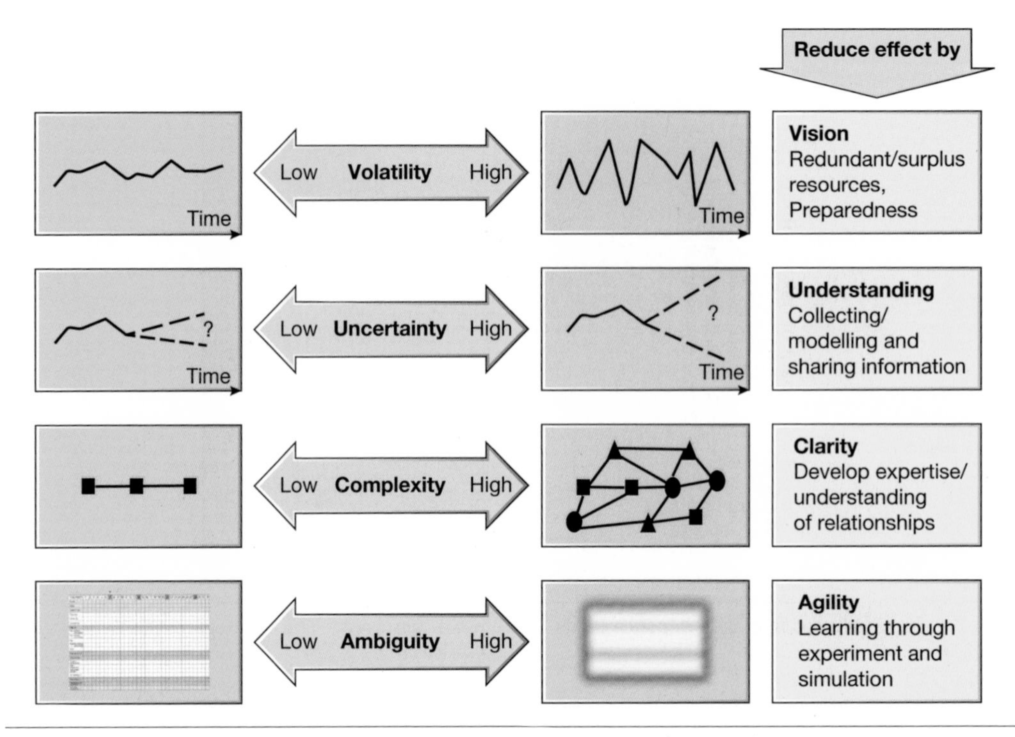

Figure 18.4 The VUCA (volatility, uncertainty, complexity, ambiguity) framework

18.2 How can operations assess the potential causes and consequences of failure?

This task is very much helped by referring to some kind of categorisation of risk. There are many ways of doing this. Here, we use a categorisation based on understanding the potential sources of risk. After this, one can try to assess the likelihood of each type of risk occurring.

Identify the potential sources of risk

This means assessing where failure might occur and what the consequences of failure might be. Often it is a 'failure to understand failure' that results in unacceptable risk. Each potential cause of failure needs to be assessed in terms of how likely it is to occur and the impact it may have. Only then can measures be taken to prevent or minimise the effect of the more important potential failures. The classic approach to assessing potential failures is to inspect and audit operations activities. The vast majority of failures are caused by something that could have been avoided. So, as a minimum starting point, a simple checklist of failure causes is useful. Failure sources are classified as: failures of supply; failures within the operation; 'environmental' risks; failures deriving from the design of products and services; and failures deriving from customer failures.

> **Operations principle**
> A 'failure to understand failure' can be the root cause of a lack of resilience.

Supply failure

Supply failure is any failure in the timing or quality of goods and services delivered into an operation: for example, suppliers delivering the wrong or faulty components, outsourced call centres suffering a telecoms failure, disruption to power supplies and so on. The more an operation relies on suppliers for materials or services, the more it is at risk from failure caused by missing or sub-standard inputs. It is an important source of failure because of the increasing dependence on outsourced activities in many industries, and the emphasis on keeping supply chains 'lean' in order to cut costs. The rise of global sourcing means that parts are shipped around the world, exposing them to risk. Microchips manufactured in Taiwan could be assembled to printed circuit boards in Shanghai that are then assembled into a computer in Ireland and sold in the United States. Perhaps most significantly there tends to be far less inventory in supply chains that could buffer interruptions to supply.

Human failures

The discussion in the previous chapter on quality management is relevant here. We distinguished between two types of human failure – errors and violations. 'Errors' are mistakes in judgement, where a person should have done something different: for example, if the manager of a sports stadium fails to anticipate dangerous crowding during a championship event. 'Violations' are acts that are clearly contrary to defined operating procedure. For example, if a maintenance engineer fails to clean a filter in the prescribed manner, it is eventually likely to cause failure. Catastrophic failures are often caused by a combination of errors and violations.

OPERATIONS IN PRACTICE

Time since last fatal crash . . . 12 years[1]

Improving operations performance is often fundamental to the survival of an organisation. Sometimes, it can be fundamental to the survival of its customers. Take the case of the airline industry. Three incidents during 1994 and 1996 in the United States led to more than 500 passenger deaths, each one grabbing headlines around the world. If this increase in fatal crashes had continued, by 2015 there would have been a major fatal crash every week, costing thousands of lives. That is, of course, assuming people were still willing to fly. Yet the idea of being able to eradicate all possible process errors connected to the operation of maintaining and flying an aircraft, seemed, frankly, to be impossible. Yet thanks to a rigorous and continuous effort to learn from the documented process errors of pilots, mechanics and air-traffic controllers, by 2021 the US domestic airline industry has celebrated 12 years without a fatal crash.

'Data will set you free'

Advances in technology certainly contributed to enhanced reliability, including improvements in cockpit automation that provided more reliable safeguards against pilot errors. But the greatest progress came from a relentless quest to persuade staff to divulge their mistakes. Airlines instructed their people to self-report errors to an incident-reporting system. Those who reported were thanked for doing so. (By contrast, anyone not reporting an incident could lose their job.) The incident reporting revealed common pilot errors, such as veering from assigned altitudes due to distractions or failing to properly position wing flaps and other flight-control surfaces for take-offs. Solutions could be simple and without cost. For example, the practice of 'calling and pointing' was introduced, where flight crew physically pointed to cockpit computers (which control altitude changes) while both pilots verbally double-checked that the correct information had been entered.

Spreading the message throughout the industry, safety official Nick Sabatini was known to reassure audiences at conferences *'data will set you free'*. It was a rallying call to the industry that routine downloading and interrogation of incident data will enable airlines to prevent errors that could potentially lead to fatal crashes. Yet this depends on people feeling safe to divulge mistakes. In parts of the world where reporting of incidents has remained voluntary, improvements have not kept pace with US carriers.

Organisational failure

Organisational failure is usually taken to mean failures of procedures and processes, and failures that derive from a business's organisational structure and culture. This is a huge potential source of failure and includes almost all operations and process management. In particular, failure in the design of processes (such as bottlenecks causing system overloading) and failures in the resourcing of processes (such as insufficient capacity being provided at peak times) need to be investigated.

Technology/facilities failures

By 'technology and facilities' we mean all the IT systems, machines, equipment and buildings of an operation. All are liable to failure, or breakdown. The failure may be only partial: for example, a machine that has an intermittent fault. Alternatively, it can be what we normally regard as a 'breakdown' – a total and sudden cessation of operation. Either way, its effects could bring a large part of the operation to a halt. For example, a computer failure in a supermarket chain could paralyse several large stores until it is fixed.

Cybersecurity

Any advance in processes or technology creates risks. No real advance comes without risk, threats and even danger. A specific type of technological failure is a failure of an operation's technology leading to exposure to cyber-risk. With the increased reliance on internet-based communication in all types of businesses, it has become a major risk factor. Since the original purpose of the internet was not commercial, it is not designed to handle secure transactions. So, there is a trade-off between providing wider access through the internet, and the security concerns it generates. Three developments have amplified cybersecurity concerns. First, increased connectivity means that everyone has the potential to 'see' everyone else. Organisations want to make enterprise systems and information more available to internal employees, business partners and customers. Second, there is reduced 'perimeter' security as more people working from home or through mobile communications. Hackers can exploit lower levels of security in home computers to burrow into corporate networks. Third, it takes time to discover all possible sources of risk, especially as new technologies are introduced.

Most authorities on cybersecurity stress that, stripped of its technological terminology, cyber-risk is just another sort of risk, which should be treated using the same identify, prevent, mitigate, recover framework that we are using here in this chapter.

OPERATIONS IN PRACTICE

Volkswagen and the 'dieselgate' scandal[2]

Some organisational failures can border on criminality. What became known as 'Dieselgate' started out as a scandal that affected only Volkswagen, Germany's largest car company. But it eventually grew until it became a global issue involving many auto industry players. It started (or became evident) when the US Environmental Protection Agency (EPA) found that Volkswagen (VW) had been installing a piece of software onto computers in its cars that falsified emissions data on its vehicles with diesel engines. The software (a so-called 'defeat device') could recognise when a car was being tested so that it could retune the engine's performance to limit nitrogen oxide emissions. After any test, when the car returned to normal road conditions, the level of such emissions increased sharply. An estimated 11 million cars worldwide were fitted with the device. VW's American boss did admit that they had 'totally screwed up', and the group's then chief executive said that VW had 'broken the trust of our customers and the public' (he later resigned because of the scandal). Over a year after the news broke, the US Department of Justice announced that Volkswagen was to pay $4.3 billion under a plea deal with US authorities. This was in addition to a $15 billion civil settlement with car owners and environmental authorities in America, where VW had agreed to buy back some of the affected vehicles. Five years after the scandal broke, the German Federal Court of Justice ruled that Volkswagen car owners were entitled to damages because of the emissions scandal. They said that owners could return their cars and receive the price paid minus a share for using the car in the intervening period.

Product/service design failures

In its design stage, a product or service might look fine on paper; only when it has to cope with real circumstances might inadequacies become evident. One only has to look at the number of product recalls (of cars) or service failures (of banks) to understand that design failures are far from uncommon.

Customer failures

Not all failures are (directly) caused by the operation or by its suppliers. Customers may 'fail' in that they misuse products and services. For example, an IT system might have been well designed, yet the user could treat it in a way that causes it to fail. Customers are not 'always right'; they can be inattentive and incompetent. However, merely complaining about customers is unlikely to reduce the chances of this type of failure occurring. Most organisations will accept that they have a responsibility to educate and train customers, and to design their products and services so as to minimise the chances of failure.

Environmental disruption

Environmental disruption includes all the causes of failure that lie outside of an operation's direct influence. Typically, such disasters include political upheaval, trade wars, weather-related disruption, fire, corporate crime, theft, fraud, sabotage, terrorism, and the contamination of products or processes. This source of potential failure has risen to near the top of many firms' agendas due to a series of major events since the 1990s.

Post-failure analysis

Although sources of failure can often be identified in advance of their occurrence, it is also valuable to use previous failures to learn about sources of potential risk. This activity is called post-failure analysis. It is used to uncover the root cause of failures. This includes such activities as the following:

- *Accident investigation* – Large-scale national disasters such as oil tanker spillages and aeroplane accidents are investigated using specifically trained staff.
- *Failure traceability* – Some businesses adopt traceability procedures to ensure that all their failures (such as contaminated food products) are traceable. Any failures can be traced back to the process that produced them, the components from which they were produced, or the suppliers who provided them.
- *Complaint analysis* – Complaints (and compliments) are a valuable source for detecting the root causes of failures of customer service. The prime function of complaint analysis involves analysing the number and 'content' of complaints over time to understand better the nature of the failure, as the customer perceives it (see later).
- *Fault-tree analysis* – Here a logical procedure starts with a failure or a potential failure and works backwards to identify all the possible causes and therefore the origins of that failure. Fault-tree analysis is made up of branches connected by two types of nodes: AND nodes and OR nodes. The branches below an AND node all need to occur for the event above the node to occur. Only one of the branches below an OR node needs to occur for the event above the node to occur. Figure 18.5 shows a simple tree identifying the possible reasons for a filter in a heating system not being replaced when it should have been.

> **Operations principle**
> Post-failure analysis is an important part of learning from failure.

Likelihood of failure

The difficulty of estimating the chance of a failure occurring varies greatly. Some failures are well understood through a combination of rational causal analysis and historical performance. For example, a mechanical component may fail between 10 and 17 months after its installation in 99 per cent of cases. Other types of failure are far more difficult to predict. The chances of a fire in a supplier's plant are (hopefully) low, but how low? There will be some data concerning fire hazards in this type of plant, but the estimated probability of failure will be subjective.

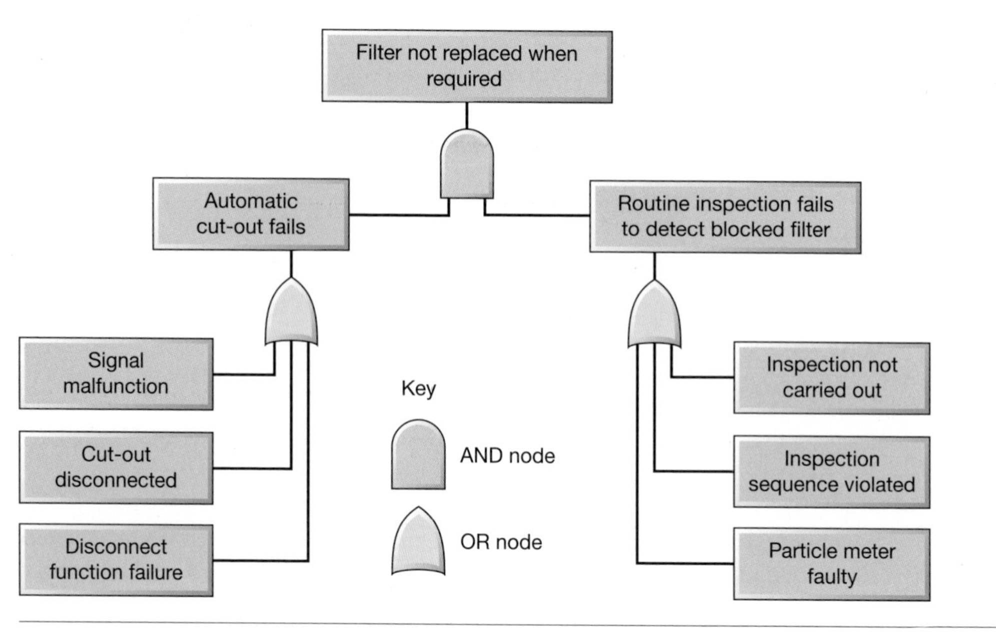

Figure 18.5 Fault-tree analysis for failure to replace filter when required

'Objective' estimates

Estimates of failure based on historical performance can be measured in three main ways:

- ***failure rates*** – how often a failure occurs;
- ***reliability*** – the chances of a failure occurring;
- ***availability*** – the amount of available useful operating time.

Failure rate

Failure rate (FR) is calculated as the number of failures over a period of time. For example, the security of an airport can be measured by the number of security breaches per year, and the failure rate of an engine can be measured in terms of the number of failures divided by its operating time. It can be measured either as a percentage of the total number of products tested or as the number of failures over time:

$$FR = \frac{\text{Number of failures}}{\text{Total number of products tested}} \times 100$$

or

$$FR = \frac{\text{Number of failures}}{\text{Operating time}}$$

Worked example

Component Co.

A batch of 50 electronic components is tested for 2,000 hours. Four of the components fail during the test as follows:

Failure 1 occurred at 12:00 hours

Failure 2 occurred at 14:50 hours

Failure 3 occurred at 17:20 hours

Failure 4 occurred at 19:05 hours

$$\text{Failure rate (as a percentage)} = \frac{\text{Number of failures}}{\text{Number tested}} \times 100 = \frac{4}{50} \times 10 = 8\%$$

The total time of the test = 50 × 2,000 = 100,000 component hours

But:

one component was not operating 2000 − 1,200 = 800 hours

one component was not operating 2000 − 1,450 = 550 hours

one component was not operating 2000 − 1,720 = 280 hours

one component was not operating 2000 − 1,905 = 95 hours

Thus:

Total non-operating time = 1,725 hours

$$\text{Operating time} = \text{Total time} - \text{Non-operating time}$$
$$= 100{,}000 - 1{,}725 = 98{,}275 \text{ hours}$$
$$\text{Failure rate (in time)} = \frac{\text{Number of failures}}{\text{Operating time}} = \frac{4}{98{,}275}$$
$$= 0.000041$$

Bath-tub curves Sometimes failure is a function of time. For example, the probability of an electric lamp failing is relatively high when it is first used, but if it survives this initial stage, it could still fail at any point, and the longer it survives, the more likely its failure becomes. The curve that describes failure probability of this type is called the bath-tub curve. It comprises three distinct stages: the 'infant-mortality' or 'early-life' stage where early failures occur, caused by defective parts or improper use; the 'normal-life' stage when the failure rate is usually low, reasonably constant and caused by normal random factors; and the 'wear-out' stage when the failure rate increases as the part approaches the end of its working life and failure is caused by the ageing and deterioration of parts. Figure 18.6 illustrates three bath-tub curves with slightly different characteristics. Curve A shows a part of the operation that has a high initial infant-mortality failure but then a long, low-failure, normal life followed by the gradually increasing likelihood of failure as it approaches wear-out. Curve B is far less predictable. The distinction between the three stages is less clear, with infant-mortality failure

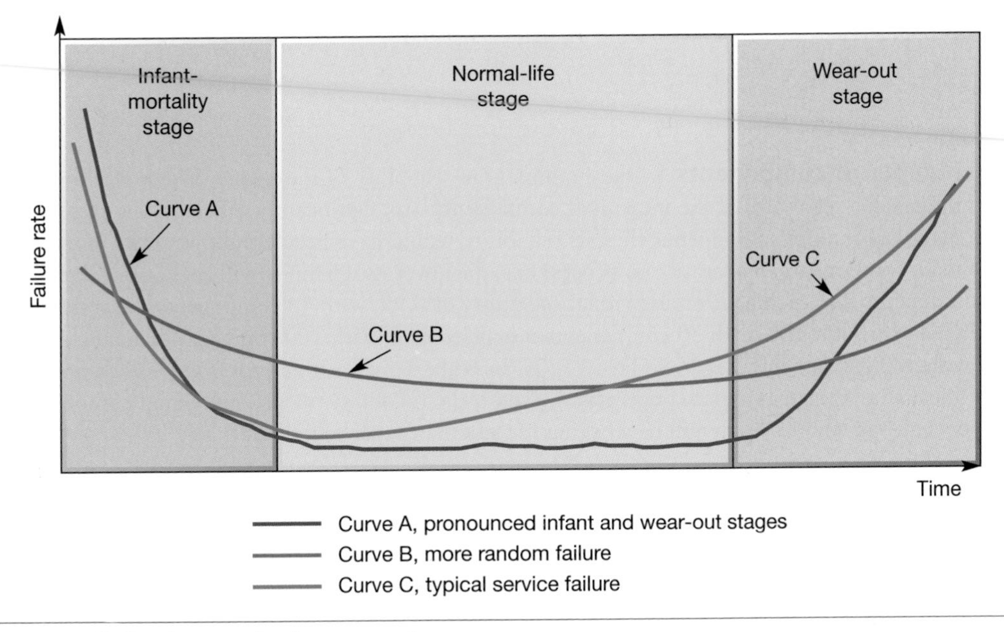

Figure 18.6 Bath-tub curves for three types of processes

subsiding only slowly and a gradually increasing chance of wear-out failure. Failure of the type shown in curve B is far more difficult to manage in a planned manner. The failure of operations that rely more on human resources than on technology, such as some services, can be closer to curve C. They may be less susceptible to component wear-out but more so to staff complacency as the service becomes tedious and repetitive.

Reliability Reliability measures the ability to perform as expected over time. Usually, the importance of any particular failure is determined partly by how interdependent the other parts of the system are. With interdependence, a failure in one component will cause the whole system to fail. So, if an interdependent system has n components, each with their own reliability, $R_1, R_2, \ldots, R_n$, the reliability of the whole system, R_s, is given by:

$$R_s = R_1 \times R_2 \times R_2 \times \ldots R_n$$

where
R_1 = reliability of component 1
R_2 = reliability of component 2
etc.

Worked example

Automated pizza-making

An automated pizza-making machine in a food manufacturer's factory has five major components, with individual reliabilities (the probability of the component not failing) as follows:

Dough mixer	Reliability = 0.95
Dough roller and cutter	Reliability = 0.99
Tomato paste applicator	Reliability = 0.97
Cheese applicator	Reliability = 0.90
Oven	Reliability = 0.98

If one of these parts of the production system fails, the whole system will stop working. Thus the reliability of the whole system is:

$$R_s = 0.95 \times 0.99 \times 0.97 \times 0.90 \times 0.98$$
$$= 0.805$$

The number of components In the example, the reliability of the whole system was only 0.8, even though the reliability of the individual components was significantly higher. If the system had been made up of more components, then its reliability would have been even lower. The more interdependent components an operation or process has, the lower its reliability will be. For one composed of components that each have an individual reliability of 0.99, with 10 components the system reliability will shrink to 0.9, with 50 components it is below 0.8, with 100 components it is below 0.4, and with 400 components it is down below 0.05. In other words, with a process of 400 components (not unusual in a large automated operation), even if the reliability of each individual component is 99 per cent, the whole system will be working for less than 5 per cent of its time.

Mean time between failures An alternative (and common) measure of failure is the mean time between failures (MTBF) of a component or system. MTBF is the reciprocal of failure rate (in time). Thus:

$$\text{MTBF} = \frac{\text{Operating hours}}{\text{Number of failures}}$$

Worked example

Automated pizza-making (*continued*)

In the previous worked example that was concerned with electronic components, the failure rate (in time) of the electronic components was 0.000041. For that component:

$$\text{MTBF} = \frac{1}{0.000041} = 24{,}390.24 \text{ hours}$$

That is, a failure can be expected once every 24,390.24 hours on average.

Availability Availability is the degree to which the operation is ready to work. An operation is not available if it has either failed or is being repaired following failure. There are several different ways of measuring it depending on how many of the reasons for not operating are included. Lack of availability because of planned maintenance or changeovers could be included, for example. However, when 'availability' is being used to indicate the operating time excluding the consequence of failure, it is calculated as follows:

$$\text{Availability } (A) = \frac{\text{MTBF}}{\text{MTBF} + \text{MTTR}}$$

where

MTBF = the mean time between failures of the operation

MTTR = the mean time to repair, which is the average time taken to repair the operation, from the time it fails to the time it is operational again.

Worked example

Poster Company

Poster Company designs and produces display posters for exhibitions and competes largely on the basis of its speedy delivery. One particular piece of equipment that the company uses is causing some problems. This is its large-platform colour laser printer. Currently, the mean time between failures of the printer is 70 hours and its mean time to repair is six hours. Thus:

$$\text{Availability} = \frac{70}{70 + 6} = 0.92$$

The company has discussed its problem with the supplier of the printer, which has offered two alternative service deals. One option would be to buy some preventive maintenance, which would be carried out each weekend (see later for a full description of preventive maintenance). This would raise the MTBF of the printer to 90 hours. The other option would be to subscribe to a faster repair service, which would reduce the MTTR to 4 hours. Both options would cost the same amount. Which would give the company the higher availability?

With MTBF increased to 90 hours:

$$\text{Availability} = \frac{90}{90 + 6} = 0.938$$

With MTTR reduced to 4 hours:

$$\text{Availability} = \frac{70}{70 + 4} = 0.946$$

Availability would be greater if the company took the deal that offered the faster repair time.

'Subjective' estimates Failure assessment, even for subjective risks, is increasingly a formal exercise that is carried out using standard frameworks, often prompted by health and safety, environmental or other regulatory reasons. These frameworks are similar to the formal quality inspection methods associated with quality standards like ISO 9000 (see Chapter 17) that often implicitly assume unbiased objectivity. However, individual attitudes to risk are complex and subject to a wide variety of influences. In fact, many studies have demonstrated that people are generally very poor at making risk-related judgements (think about national lotteries.) But, although people do not always make rational decisions, this does not mean abandoning the attempt.

> **Operations principle**
> Subjective estimates of failure probability are better than no estimates at all.

Critical commentary

The idea that failure can be detected through in-process inspection is increasingly seen as only partially true. Although inspecting for failures is an obvious first step in detecting them, it is not even close to being 100 per cent reliable. Accumulated evidence from research and practical examples consistently indicates that people, even when assisted by technology, are not good at detecting failure and errors. This applies even when special attention is being given to inspection. For example, airport security was significantly strengthened after the 9/11 terrorist attack, yet there is no such thing as 100 per cent security; in a test one in ten lethal weapons that were entered into airports' security systems were not detected. No one is advocating abandoning inspection as a failure detection mechanism. Rather it is seen as one of a range of methods of preventing failure.

Health and safety

One of the paradoxes of operations management is that occupational health and safety (OHS) management is such a large part of many practitioners' work, yet has attracted relatively little academic attention in the OM field. Not only is it a subject of significant practitioner interest, it is clearly important, not only for the potential improvement to operations performance that excellent OHS practices can bring, but also for its clear ethical benefits. Occupational accidents and diseases have very significant adverse consequences. Staff are injured, absent from work and possibly retired early, facilities are damaged, the quantity and quality of output declines, all of which has a negative impact on an operation's performance and reputation. It has been estimated that, globally, this type of incident results in nearly 2.3 million deaths every year and incurs over $2.8 trillion of costs.[3]

In many countries, there has been growing acceptance that the adoption of systematic OHS management systems is necessary to ensure safe and productive work environments. Such systems often stress the importance of adopting the systematic identification of potential hazards, assessment and control of risks, evaluation and periodic review of risk measures, in fact, many of the issues covered in this chapter. However, operations managers have been known to complain about the amount of time and bureaucracy involved in OHS. However, health and safety management does not have to be complicated, costly or time-consuming. In fact, much of what is usually required by legislation is good risk-management practice.

> **Operations principle**
> Ensuring the health and safety of all concerned with their operation's activities is a fundamental responsibility for operations managers.

Failure mode and effect analysis

Having identified potential sources of failure (either in advance of an event or through post-failure analysis) and having then examined the likelihood of these failures occurring through some combination of objective and subjective analysis, managers can move to assigning relative priorities to risk.

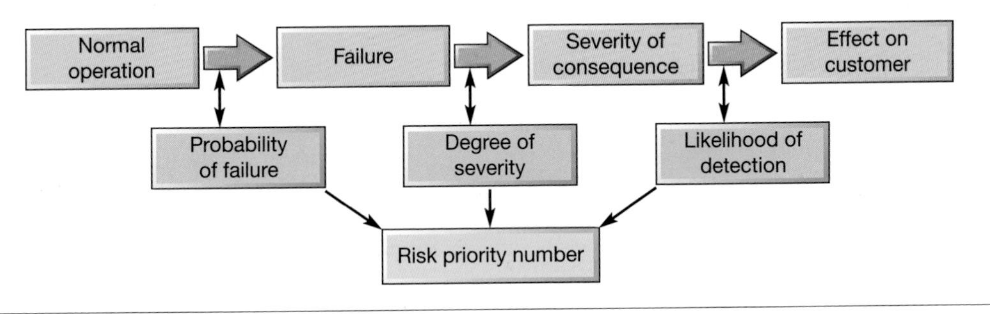

Figure 18.7 Procedure for failure modes effects analysis (FMEA)

The most well-known approach for doing this is failure mode and effect analysis (FMEA). Its objective is to identify the factors that are critical to various types of failure as a means of identifying failures before they happen. It does this by providing a 'checklist' procedure built around three key questions for each possible cause of failure:

- What is the likelihood that failure will occur?
- What would the consequence of the failure be?
- How likely is such a failure to be detected before it affects the customer?

Based on a quantitative evaluation of these three questions, a risk priority number (RPN) is calculated for each potential cause of failure. Corrective actions, aimed at preventing failure, are then applied to those causes whose RPN indicates that they warrant priority (see Figure 18.7).

Worked example

Transport Co.

An FMEA exercise at Transport Co. has identified three failure modes associated with the failure of 'goods arriving damaged' at the point of delivery:

- Goods not secured (failure mode 1).
- Goods incorrectly secured (failure mode 2).
- Goods incorrectly loaded (failure mode 3).

The improvement group that is investigating the failures allocates scores for the probability of the failure mode occurring, the severity of each failure mode, and the likelihood that they will be detected, using the rating scales shown in Table 18.1, as follows:

Probability of occurrence	
Failure mode 1	5
Failure mode 2	8
Failure mode 3	7
Severity of failure	
Failure mode 1	6
Failure mode 2	4
Failure mode 3	4
Probability of detection	
Failure mode 1	2
Failure mode 2	6
Failure mode 3	7

The RPN of each failure mode is calculated:

Failure mode 1 (goods not secured) $5 \times 6 \times 2 = 60$

Failure mode 2 (goods incorrectly secured) $8 \times 4 \times 5 = 160$

Failure mode 3 (goods incorrectly loaded) $7 \times 4 \times 7 = 196$

Priority is therefore given to failure mode 3 (goods incorrectly loaded) when attempting to eliminate the failure.

Table 18.1 Rating scales for FMEA

A. Occurrence of failure		
Description	*Rating*	*Possible failure occurrence*
Remote probability of occurrence It would be unreasonable to expect failure to occur	1	0
Low probability of occurrence Generally associated with activities similar to previous ones with a relatively low number of failures	2	1:20,000
	3	1:10,000
Moderate probability of occurrence Generally associated with activities similar to previous ones which have resulted in occasional failures	4	1:2000
	5	1:1000
	6	1:200
High probability of occurrence Generally associated with activities similar to ones which have traditionally caused problems	7	1:100
	8	1:20
Very high probability of occurrence Near certainty that major failures will occur	9	1:10
	10	1:2

B. Severity of failure	
Description	*Rating*
Minor severity A very minor failure which would have no noticeable effect on system performance	1
Low severity A minor failure causing only slight customer annoyance	2
	3
Moderate severity A failure which would cause some customer dissatisfaction, discomfort or annoyance, or would cause noticeable deterioration in performance	4
	5
	6
High severity A failure which would engender a high degree of customer dissatisfaction	7
	8
Very high severity A failure which would affect safety	9
Catastrophic A failure which may cause damage to property, serious injury or death	10

C. Detection of failure		
Description	*Rating*	*Probability of detection*
Remote probability that the defect will reach the customer (It is unlikely that such a defect would pass through inspection, test or assembly)	1	0 to 5%
Low probability that the defect will reach the customer	2	6 to 15%
	3	16 to 25%
Moderate probability that the defect will reach the customer	4	26 to 35%
	5	36 to 45%
	6	46 to 55%
High probability that the defect will reach the customer	7	56 to 65%
	8	66 to 75%
Very high probability that the defect will reach the customer	9	76 to 85%
	10	86 to 100%

18.3 How can failures be prevented?

Failure prevention is an important responsibility for operations managers. The obvious way to do this is to systematically examine any processes involved and 'design out' any failure points. Many of the approaches used in Chapter 4 on product/service innovation, Chapter 6 on process design and Chapter 17 on quality management can be used to do this. In this section, we will look at three further approaches to reducing risk by trying to prevent failure: building redundancy into a process, 'fail-safeing' some of the activities in the process, and maintaining the physical facilities in the process.

Redundancy

Building in **redundancy** to an operation means having back-up systems or components in case of failure. It can be expensive and is generally used when the breakdown could have a critical impact. Redundancy means doubling or even tripling some parts of a process or system in case one component fails. Nuclear power stations, spacecraft and hospitals all have auxiliary systems in case of an emergency. Some organisations also have 'back-up' staff held in reserve in case someone does not turn up for work. Spacecraft have several back-up computers on board that will not only monitor the main computer but also act as a back-up in case of failure.

One response to the threat of large failures, such as terrorist activity, has been a rise in the number of companies (known as '**business continuity**' providers) offering 'replacement office' operations, fully equipped with normal internet and telephone communications links, and often with access to a company's current management information. Should a customer's main operation be affected by a disaster, business can continue in the replacement facility within days or even hours.

Operations principle
Redundancy is an important failure prevention method, especially when the consequences of failure could be serious.

The effect of redundancy can be calculated by the sum of the reliability of the original process component and the likelihood that the back-up component will both be needed and be working:

$$R_{a+b} = R_a + (R_b \times P\,(\text{failure}))$$

where

R_{a+b} = reliability of component a with its back-up component b

R_a = reliability of a alone

R_b = reliability of back-up component b

P (failure) = the probability that component a will fail and therefore component b will be needed.

Redundancy is often used for servers, where system availability is particularly important. In this context, the industry uses three main types of redundancy:

- ***Hot standby*** – where both primary and secondary (back-up) systems run simultaneously. The data are copied to the secondary server in real time so that both systems contain identical information.
- ***Warm standby*** – where the secondary system runs in the background to the primary system. Data are copied to the secondary server at regular intervals, so there are times when both servers do not contain exactly the same data.
- ***Cold standby*** – where the secondary system is only called upon when the primary system fails. The secondary system receives scheduled data back-ups, but less frequently than a warm standby, so cold standby is used mainly for non-critical applications.

Fail-safeing

The concept of fail-safeing has emerged since the introduction of Japanese methods of operations improvement. Called poka-yoke in Japan (from *yokeru* (to prevent) and *poka* (inadvertent errors)), the idea is based on the principle that human mistakes are to some extent inevitable. What is important is to prevent them becoming defects. Poka-yokes are simple (preferably inexpensive) devices or systems, which are incorporated into a process to prevent inadvertent mistakes by those providing service as well as customers receiving a service. Examples of poka-yokes include:

- trays used in hospitals with indentations shaped to each item needed for a surgical procedure – any item not back in place at the end of the procedure might have been left in the patient;
- checklists that have to be filled in, either in preparation for, or on completion of, an activity, such as a maintenance checklist for a plane during turnaround;
- gauges placed on machines through which a part has to pass in order to be loaded onto, or taken off, the machine – an incorrect size or orientation stops the process;
- the locks on aircraft lavatory doors, which must be turned to switch the light on;
- beepers on ATMs to ensure that customers remove their cards or in cars to remind drivers to take their keys with them;
- limit switches on machines, which allow the machine to operate only if the part is positioned correctly;
- height bars on amusement rides to ensure that customers do not exceed size limitations.

Operations principle

Simple methods of fail-safeing can often be the most cost effective.

OPERATIONS IN PRACTICE

Darktrace uses AI to guard against cyberattacks[4]

As cyber criminals deploy increasingly sophisticated forms of attack, so the industry selling (hopefully) equally sophisticated protection services has grown. One of the most prominent among these is Darktrace, based in Cambridge in the United Kingdom. It was founded by mathematicians from the University of Cambridge and government cyber-intelligence experts in the United Kingdom and the United States. The company is well-established in the application of artificial intelligence (AI) to cyber defence. One great advantage of AI is that it offers the possibility of keeping pace with the ever-evolving nature of cyber threats. Powered by what it describes as unsupervised machine learning, the company's AI is modelled on the human immune system, learning 'on the job', from the data and activity that it observes. AI and machine learning could potentially have a number of advantages over purely human surveillance. A skilled human could look for suspicious patterns, analyse them, devise ways to mitigate the threat, then inform the rest of the business, but it takes time. An analyst might need to spend anything between half an hour and half a day investigating one single suspicious security incident. Darktrace says that its AI cybersecurity solution accelerates this process, conducting continuous investigations in the background to normal operating at a pace and scale beyond the capabilities of a human analyst. Not only that, but AI-driven cybersecurity can handle specialist investigations into a large number of parallel threads concurrently and instantly communicate its findings. However, some cybersecurity commentators believe that, although AI offers the possibility of vastly improved security systems, like any innovative software development, its effectiveness can sometimes be overstated. One problem with such systems is their tendency to show 'false positives', that is, reporting a possible cybersecurity breach when there is actually no threat. Yet, even experts suggesting excessive hype admit that there is certainly a role for AI in cybersecurity, if only because it is particularly good at dealing with vast amounts of information and trying to understand what is normal and what's anomalous.

Maintenance

Maintenance is how organisations try to avoid failure by taking care of their physical facilities. It is an important part of most operations' activities, particularly in operations dominated by their physical facilities such as power stations, hotels, airlines and petrochemical refineries. The benefits of effective maintenance include enhanced safety, increased reliability, higher quality (badly maintained equipment is more likely to cause errors), lower operating costs (because regularly serviced process technology is more efficient), a longer lifespan for process technology, and higher 'end value' (because well-maintained facilities are generally easier to dispose of into the second-hand market).

Operations principle
Maintenance is how organisations try to avoid failure by taking care of their physical facilities.

The three basic approaches to maintenance

In practice, an organisation's maintenance activities will consist of some combination of the three basic approaches to the care of its physical facilities. These are run to breakdown (RTB), preventive maintenance (PM) and condition-based maintenance (CBM).

- **Run-to-breakdown maintenance (RTB)** – As its name implies, this involves allowing the facilities to continue operating until they fail. Maintenance work is performed only after failure has taken place. For example, the televisions, bathroom equipment and telephones in a hotel's guest rooms will probably be repaired or replaced only when they fail. The hotel will keep some spare parts and the staff available to make any repairs when needed. Failure in these circumstances is neither catastrophic (although perhaps irritating to the guest) nor so frequent as to make regular checking of the facilities appropriate.
- **Preventive maintenance (PM)** – This attempts to eliminate or reduce the chances of failure by servicing (cleaning, lubricating, replacing and checking) the facilities at pre-planned intervals. For example, the engines of passenger aircraft are checked, cleaned and calibrated according to a regular schedule after a set number of flying hours. Taking aircraft away from their regular duties for preventive maintenance is clearly an expensive option for any airline. The consequences of failure while in service are considerably more serious, however. The principle is also applied to facilities with less catastrophic consequences of failure. The regular cleaning and lubricating of machines, even the periodic painting of a building, could be considered preventive maintenance.
- **Condition-based maintenance (CBM)** – This attempts to perform maintenance only when the facilities require it. For example, continuous process equipment, such as that used in coating photographic paper, is run for long periods in order to achieve the high utilisation necessary for cost-effective production. Stopping the machine to change, say, a bearing when it is not strictly necessary to do so would take it out of action for long periods and reduce its utilisation. Here condition-based maintenance might involve continuously monitoring the vibrations, for example, or some other characteristic of the line.

How much maintenance?

The balance between preventive and breakdown maintenance is usually set to minimise the total cost of breakdown. Infrequent preventive maintenance will cost little to provide but will result in a high likelihood (and therefore cost) of breakdown maintenance. Conversely, very frequent preventive maintenance will be expensive to provide but will reduce the cost of having to provide breakdown maintenance (see Figure 18.8(a)). The total cost of maintenance appears to minimise at an 'optimum' level of preventive maintenance. However, the cost of providing preventive maintenance may not increase quite so steeply as indicated in Figure 18.8(a). The curve assumes that it is carried out by a separate set of people (skilled maintenance staff) from the 'operators' of the facilities. Furthermore, every time preventive maintenance takes place, the facilities cannot be used productively. This is why the slope of the curve increases, because the maintenance episodes start to interfere with the normal working of the operation. But in many operations some preventive maintenance can be performed by the operators themselves (which reduces the cost of providing it) and at times that are convenient for the operation (which minimises the disruption to the operation). The cost of breakdowns could also be higher than is indicated in Figure 18.8(a). Unplanned breakdowns may do more than necessitate a repair and stop the operation; they can take away stability from the operation, which prevents it being able to improve itself. Put these two ideas together and the minimising total curve and maintenance cost curve look more like Figure 18.8(b). The emphasis is shifted more towards the use of preventive maintenance than run-to-breakdown maintenance.

Total productive maintenance

Total productive maintenance (TPM) is 'the productive maintenance carried out by all employees through small group activities', where productive maintenance is 'maintenance management which recognizes the importance of reliability, maintenance and economic efficiency in plant design'.[5] In Japan, where TPM originated, it is seen as a natural extension in the evolution from run-to-breakdown to preventive maintenance. TPM adopts some of the team-working and empowerment principles discussed in Chapter 9, as well as a continuous improvement approach to failure prevention as discussed in Chapter 16. It also sees maintenance as an organisation-wide issue, to which staff can contribute in some way. It is analogous to the total quality management approach discussed in Chapter 17.

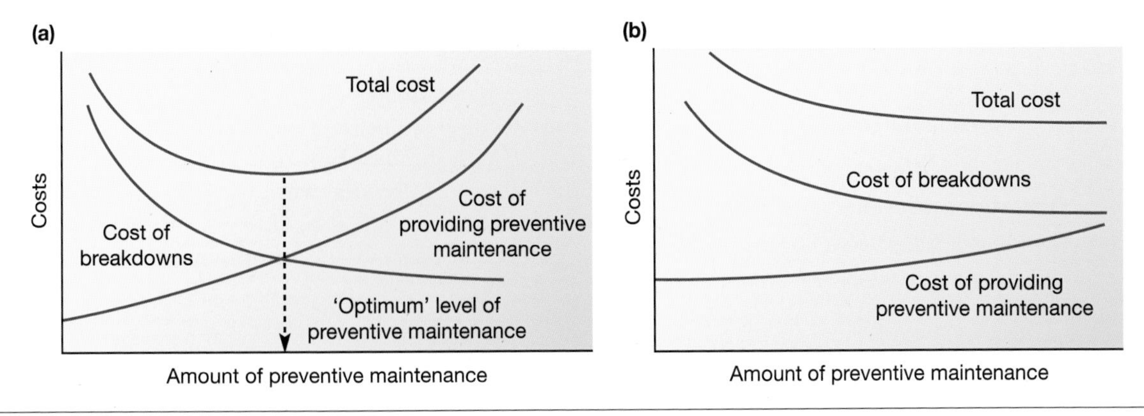

Figure 18.8 Two views of maintenance costs. (a) One model of the costs associated with preventive maintenance shows an optimum level of maintenance effort. (b) If routine preventive maintenance tasks are carried out by operators and if the real cost of breakdowns is considered, the 'optimum' level of preventive maintenance shifts towards higher levels

The five goals of TPM

TPM aims to establish good maintenance practice in operations through the pursuit of 'the five goals of TPM':

1. Improve equipment effectiveness by examining all the losses that occur.
2. Achieve autonomous maintenance by allowing staff to take responsibility for some of the maintenance tasks and for the improvement of maintenance performance.
3. Plan maintenance with a fully worked-out approach to all maintenance activities.
4. Train all staff in relevant maintenance skills so that both maintenance and operating staff have all the skills to carry out their roles.
5. Achieve early equipment management by 'maintenance prevention' (MP), which involves considering failure causes and the maintainability of equipment during its design, manufacture, installation and commissioning.

Critical commentary

Being human, managers often respond to the perception of risk rather than its reality. For example, Table 18.2 shows the cost of each life saved by investment in various road and rail transportation safety (in other words, failure prevention) investments. The table shows that investing in improving road safety is very much more effective than investing in rail safety. And while no one is arguing for abandoning efforts on rail safety, it is noted by some that actual investment reflects more the public perception of rail deaths (low) compared with road deaths (very high).

Table 18.2 The cost per life saved of various safety (failure prevention) investments

Safety investment	Cost per life (€M)
Advanced train protection system	30
Train protection warning systems	7.5
Implementing recommended guidelines on rail safety	4.7
Implementing recommended guidelines on road safety	1.6
Local authority spending on road safety	0.15

18.4 How can operations mitigate the effects of failure?

Even when a failure has occurred, its impact on the customer can, in many cases, be minimised through mitigation actions. Failure (or risk) mitigation means isolating a failure from its negative consequences. It is an admission that not all failures can be avoided. Mitigation can be vital when used in conjunction with prevention to reduce overall risk. One way of thinking about mitigation is as a series of decisions under conditions of uncertainty.

Operations principle

Failure (or risk) mitigation means isolating a failure from its negative consequences.

Failure mitigation actions

The nature of the action taken to mitigate failure will obviously depend on the nature of the risk. In most industries technical experts have established a classification of failure mitigation actions that are appropriate for the types of failure likely to be suffered. So, for example, in agriculture, government agencies and industry bodies have published mitigation strategies for such risks as the outbreak of crop disease, contagious animal infections and so on. Although these classifications tend to be industry specific, the following generic categorisation gives a flavour of the types of mitigation actions that may be generally applicable:

- ***Mitigation planning*** – This is the activity of ensuring that all possible failure circumstances have been identified and the appropriate mitigation actions identified. It is the overarching activity that encompasses all subsequent mitigation actions and may be described in the form of a decision tree or guide rules. It is worth noting that mitigation planning, as well as being an overarching action, also provides mitigation action in its own right. For example, if mitigation planning has identified appropriate training, job design, emergency procedures and so on, then the financial liability of a business for any losses should a failure occur will be reduced. Certainly, businesses that have not planned adequately for failures will be more liable in law for any subsequent losses.
- ***Economic mitigation*** – This includes actions such as insurance against losses from failure, spreading the financial consequences of failure, and 'hedging' against failure. Insurance is the best known of these actions and is widely adopted, although ensuring appropriate insurance and effective claims management is a specialised skill in itself. Hedging often takes the form of financial instruments: for example, a business may purchase a financial 'hedge' against the price risk of a vital raw material deviating significantly from a set price.
- ***Containment (spatial)*** – This means stopping the failure physically spreading to affect other parts of an internal or external supply network. Preventing contaminated food from spreading through the supply chain, for example, will depend on real-time information systems that provide traceability data.
- ***Containment (temporal)*** – This means containing the spread of a failure over time. It particularly applies when information about a failure or potential failure needs to be transmitted without undue delay. For example, systems that give advance warning of hazardous weather such as snowstorms must transmit such information to local agencies such as the police and road-clearing organisations in time for them to stop the problem causing excessive disruption.
- ***Loss reduction*** – This covers any action that reduces the catastrophic consequences of failure by removing the resources that are likely to suffer those consequences. For example, the road signs that indicate evacuation routes in the event of severe weather, or the fire drills that train employees in how to escape in the event of an emergency, may not reduce all the consequences of failure, but can help in reducing loss of life or injury.
- ***Substitution*** – This means compensating for failure by providing other resources that can substitute for those rendered less effective by the failure. It is a little like the concept of redundancy that was described earlier but does not always imply excess resources if a failure has not occurred. For example, in a construction project, the risk of encountering unexpected geological problems may be mitigated by the existence of a separate work plan that is invoked only if such problems are found.

18.5 How can operations recover from the effects of failure?

Failure recovery is the set of actions that are taken to reduce the impact of failure once the customer has experienced its negative effects. Recovery needs to be planned and procedures put in place that can discover when failures have occurred, guide appropriate action to keep everyone informed, capture the lessons learned from the failure, and plan to absorb lessons into any future recovery. All types of operation can benefit from well-planned recovery. For example, a construction company whose mechanical digger breaks down can have plans in place to arrange a replacement from a hire company. The breakdown might be disruptive, but not as much as it might have been if the operations manager had not worked out what to do. Recovery procedures will also shape customers' perceptions of failure.

Even where the customer sees a failure, it may not necessarily lead to dissatisfaction. Indeed, in many situations, customers may well accept that things do go wrong. If there is a metre of snow on the train lines, or if the restaurant is particularly popular, we may accept that the product or service does not work. It is not necessarily the failure itself that leads to dissatisfaction but often the organisation's response to the breakdown. While mistakes may be inevitable, dissatisfied customers are not. A failure may even be turned into a positive experience. A good recovery can turn angry, frustrated customers into loyal ones.

Operations principle

Successful failure recovery can yield more benefits than if the failure had not occurred.

OPERATIONS IN PRACTICE

Pressing the passenger panic button[6]

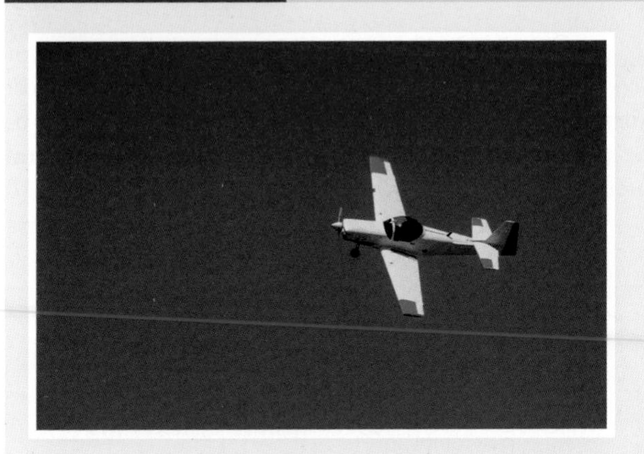

It is every nervous air traveller's nightmare – what if the pilot (or both pilots, if there are two) becomes incapacitated? Such fear is not unwarranted. It is a real, although fortunately rare, danger with light aircraft. One Australian report identified 15 cases in a five-year period of the pilots of small aircraft being incapacitated. If there is an instructor on the ground to give an instant flying lesson to someone on board a plane, the plane and its temporary pilot may be talked down successfully. But it is a very difficult task and can all too easily result in what is known euphemistically in aviation circles as a 'collision with terrain'. Keeping an aircraft flying straight and level while finding an appropriate bearing to arrive at a suitable airfield, and then landing safely, takes a miracle. Which is why Garmin, the American multinational technology company, best known for its satellite-based navigation systems, has, in effect, developed a 'panic button' for passengers who find themselves is such a dangerous position. The company, that also make electronic control systems for aircraft, developed Autoland, a panic button that switches control of the plane to its flight computers, in a similar way to engaging the autopilots used in commercial aircraft. But, in addition, Garmin's system also transmits an emergency radio code alert to air-traffic control and other planes in the area. It will also analyse other factors, such as weather conditions and the amount of available fuel before selecting an appropriate airfield to divert to. When the plane arrives at that airport, the system controls the descent and lands just as a human pilot would. Once on the ground, it even automatically applies the brakes, bringing the plane to a halt, and turns the engine off. During what must be a terrifying experience, passengers are kept informed about what is happening by way of both messages on a screen and voice announcements. They are also warned not to touch the controls, but to sit back and fasten their seat belts.

The complaint value chain

The complaint value chain, shown in Figure 18.9, helps us to visualise the potential value of good recovery at different stages. In Figure 18.9(a) an operation provides service to 5,000 customers, but 20 per cent experience some form of failure. Of these 1,000 customers, 40 per cent decide not to complain, perhaps because it seems like more trouble than it's worth or because the complaint processes are too convoluted. Evidence suggests that around 80 per cent of these non-complainers will switch to an alternative service provider. (Of course, the precise switching percentage will depend on the number of alternatives in the market and the ease of switching.) Another group of the 1,000 customers who experienced a failure do decide to complain, in this case 60 per cent. Some will be satisfied (in this case, 75 per cent) and others will not be (in this case, 25 per cent). Dissatisfied complainers will generally leave the organisation (in this case, 80 per cent), while satisfied complainers will tend to remain loyal (in this case, 80 per cent). So, assuming these percentages are correct, for every 5,000 customers processed by this particular service operation, 530 will switch.

Now let us assume that the operations manager decides to invest in small improvements to all stages in the complaint value chain. In Figure 18.9(b) the company has reduced its failures from 20 to 18 per cent (still very poor of course!), and has encouraged more customers who experienced a failure to come forward and complain. So the percentage complaining has risen from 60 per cent to 70 per cent. It has also made sure that a higher proportion (in this case, up from 75 to 83 per cent) of those who do make the effort to complain are satisfied. The end result is that the number of lost customers falls from 530 to 406. Assuming that the extra 124 customers retained have a value that is equal to, or more than, the costs of improvements, the organisation is making a good investment in its recovery and prevention efforts. What is important to understand here is how a relatively small improvement across the failure and complaint process can have such a significant impact on customer loyalty and switching.

Operations principle

The complaint value chain helps to visualise the potential value of good recovery at different stages.

Failure planning

Organisations need to design appropriate responses to failure that are suitably aligned with the cost and the inconvenience caused by the failure to their customers. Such recovery processes need to be carried out either by empowered front-line staff or by trained personnel who are available to deal with recovery in a way that does not interfere with day-to-day service activities. Figure 18.10

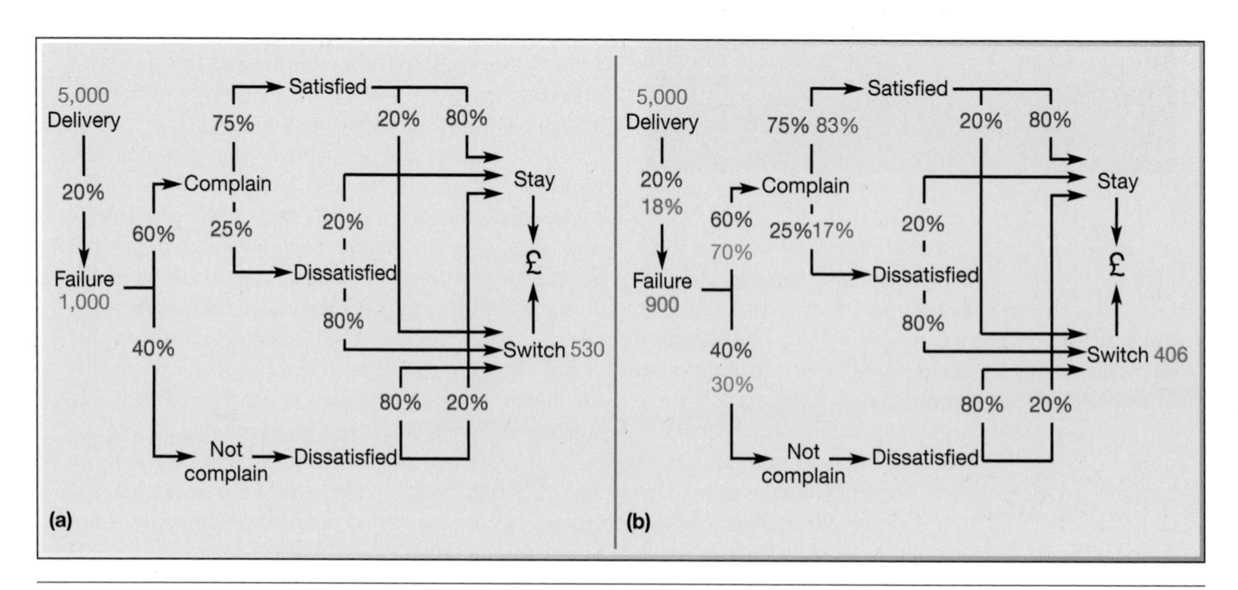

Figure 18.9 Complaint value chain: (a) initial value chain and (b) with small improvements to each step

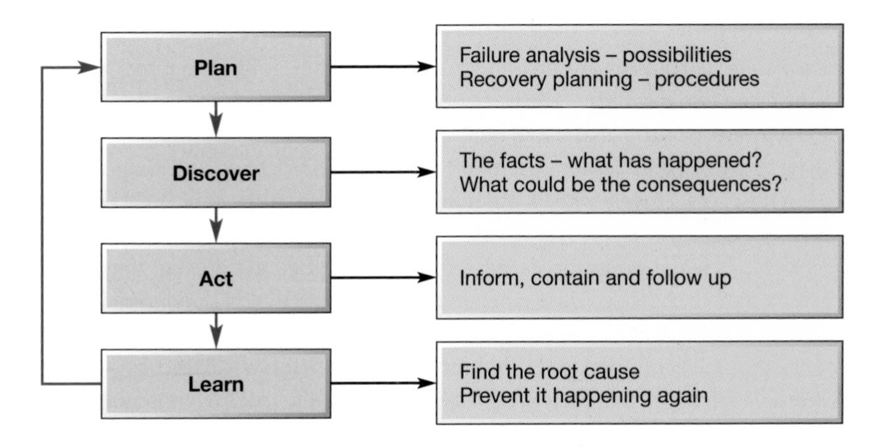

Figure 18.10 Recovery sequence for minimising the impact from failure

illustrates a typical recovery sequence. It is often represented by stage models, one of which is represented in Figure 18.10. We shall follow it through from the point where failure is recognised.

- ***Discover*** – The first thing any manager needs to do when faced with a failure is to discover its exact nature. Three important pieces of information are needed: first of all, what exactly has happened; second, who will be affected by the failure; and, third, why did the failure occur? This last point is not intended to be a detailed inquest into the causes of failure (that comes later) but it is often necessary to know something of the causes of failure in case it is necessary in order to determine what action to take.
- ***Act*** – The discover stage might take only minutes or even seconds, depending on the severity of the failure. If the failure is a severe one with important consequences, we need to move on to doing something about it quickly. This means carrying out three actions, the first two of which could be carried out in reverse order, depending on the urgency of the situation. First, tell the significant people involved what you are proposing to do about the failure. In service operations this is especially important where customers need to be kept informed, both for their peace of mind and to demonstrate that something is being done. Second, the effects of the failure need to be contained in order to stop the consequences spreading and causing further failures. The precise containment actions will depend on the nature of the failure. Third, there needs to be some kind of follow-up to make sure that the containment actions really have contained the failure.
- ***Learn*** – As discussed earlier in this chapter, the benefits of failure in providing learning opportunities should not be underestimated. In failure planning, learning involves revisiting the failure to find out its root cause and then engineering out the causes of the failure so that it will not happen again. This is the key stage for much failure planning.
- ***Plan*** – Learning the lessons from a failure is not the end of the procedure. Operations managers need formally to incorporate the lessons into their future reactions to failures. This is often done by working through 'in theory' how they would react to failures in the future. Specifically, this involves first identifying all the possible failures that might occur (in a similar way to the FMEA approach). Second, it means formally defining the procedures that the organisation should follow in the case of each type of identified failure.

Responsible operations

In every chapter, under the heading of 'Responsible operations', we summarise how the particular topic covered in the chapter touches upon important social, ethical and environmental issues.

Even a cursory glance though this chapter should indicate that there are very many different things that can go wrong with operations. But underlying many of the examples is a central issue – there is an efficiency versus resilience trade-off. In other words, striving for efficient working in normal (steady state) circumstances can undermine an operation's ability to respond effectively to disruptions. Here are two disasters – 10 years apart, but both devastating.

First, the Deepwater Horizon disaster. An explosion and fire ripped through Transocean's Deepwater Horizon drilling rig, working under contract for BP oil company's Macondo well in the Gulf of Mexico. Eleven men lost their lives, 17 were injured, and more than 3 million barrels of oil leaked into the ocean.[7] A pulse of gas shot up, buckling the drill pipe. The emergency valve designed to cap the well in case of an accident, the 'blowout protector', failed, and the gas reached the drill rig, triggering the fatal explosion and prompting the leak.[8] The incident drew attention to the risks of drilling for oil in one of the most difficult, but ecologically rich, parts of the world. Drilling for oil in deep offshore waters, like the Gulf of Mexico, is inherently dangerous, but conditions on this rig were particularly concerning. An official US investigation found that there were many lapses in safety that had contributed to the disaster. BP were found guilty of 'gross negligence' the US court ruled, and found that several decisions leading to the disaster were 'primarily driven by a desire to save time and money, rather than ensuring that the well was secure'. In the years after the Deepwater Horizon crisis, BP also had to cut costs, in part to pay for its legal fees and for the clean-up bills that eventually exceeded $60bn.

Second, the COVID-19 pandemic. In late December 2019, a pneumonia of unknown cause was detected in Wuhan, China. It was reported to the World Health Organization Country Office. The outbreak was declared a Public Health Emergency of International Concern on 30 January 2020.[9] Researchers believed that the coronavirus began in bats, then jumped to an intermediary species that passed it to people.[10] The virus could be spread among humans by way of respiratory droplets within 2 metres or more. Its pneumonia-like symptoms include fever and coughing, causing death in some cases. Its economic symptoms were just as severe. Economic activity dropped drastically around the world. Its impact was variously described as the worst since the Second World War, the great depression of the 1920s and 30s, or the Black Death of 1346–1353 (take your pick). Yet it was undeniably serious for most businesses and governments. As well as hundreds of thousands of people dying, complex but fragile supply chains were disrupted, governments locked down whole industries to prevent the virus spreading, individual organisations closed their locations to preserve staff safety, millions had to work from home, and demand for most services and products shrank. It was this simultaneous impact on supply, operations processes and demand that caused such devastating economic impact, scarring economies for years afterwards.

Why were these two disasters so serious? First the differences. The Deepwater Horizon disaster was largely a function of the internal actions (and mistakes) by the company and its contractors. The economic devastation spread along with the COVID-19 virus was an external event that came upon most organisations with little or no warning. Yet the negative outcomes of both events were partly a result of the efficiency–resilience trade-off – although in different ways. Clearly the cost pressures in the Deepwater Horizon well were excessive. No operation should let its drive for efficiency to override reasonable safety concerns. Yet, no operations activity is entirely risk-free. The impact of COVID-19 could have been mitigated, but only for some businesses, and only to some extent. More robust supply arrangements, high in-process and finished goods stocks, and experience with flexible working arrangements could have helped some, but not all, operations. However, such strategies, which pay off liberally in the event of even the worst case, are usually prohibitively expensive in normal times.

Summary answers to key questions

18.1 What is risk management?

- Risk management is about things going wrong and what operations can do to stop things going wrong. Or, more formally, 'the process that aims to help organisations understand, evaluate and take action on all their risks with a view to increasing the probability of their success and reducing the likelihood of failure'.

- It consists of four broad activities:
  - Understanding what failures could occur.
  - Preventing failures occurring.
  - Minimising the negative consequences of failure (called risk 'mitigation').
  - Recovering from failures when they do occur.

18.2 How can operations assess the potential causes and consequences of failure?

- There are several causes of failure, including design failures, facilities failure, staff failure, supplier failure, cyber failure, customer failure and environmental disruption.
- There are three ways of measuring failure. 'Failure rates' indicate how often a failure is likely to occur. 'Reliability' measures the chances of a failure occurring. 'Availability' is the amount of available and useful operating time left after taking account of failures.
- Failure over time is often represented as a failure curve. The most common form of this is the so-called 'bath-tub curve', which shows the chances of failure as being greater at the beginning and end of the life of a system or part of a system.
- Failure analysis mechanisms include accident investigation, product liability, complaint analysis, critical incident analysis, and failure mode and effect analysis (FMEA).

18.3 How can failures be prevented?

- There are four major methods of improving reliability: designing out the fail points in the operation, building redundancy into the operation, 'fail-safeing' some of the activities of the operation, and maintenance of the physical facilities in the operation.
- Maintenance is the most common way operations attempt to improve their reliability, with three broad approaches. The first is running all facilities until they break down and then repairing them, the second is regularly maintaining the facilities even if they have not broken down, and the third is to monitor facilities closely to try to predict when breakdown might occur.
- Total productive maintenance, where all employees carry out maintenance in small groups, is a particularly useful approach to managing maintenance.

18.4 How can operations mitigate the effects of failure?

- Risk, or failure, mitigation means isolating a failure from its negative consequences.
- Risk mitigation actions include:
  - mitigation planning;
  - economic mitigation;
  - containment (spatial and temporal);
  - loss reduction;
  - substitution.

18.5 How can operations recover from the effects of failure?

- Recovery can be enhanced by a systematic approach to discovering what has happened to cause failure, acting to inform, contain and follow up the consequences of failure, learning to find the root cause of the failure and preventing it taking place again, and planning to avoid the failure occurring in the future.

CASE STUDY Slagelse Industrial Services (SIS)

Slagelse Industrial Services (SIS) had become one of Europe's most respected die casters of zinc, aluminium and magnesium parts for hundreds of companies in many industries, especially automotive and defence. The company cast and engineered precision components by combining the most modern production technologies with precise tooling and craftsmanship. SIS began life as a classic family firm, run by Erik Paulsen, who opened a small manufacturing and die-casting business in his hometown of Slagelse, a town in eastern Denmark, about 100 km south west of Copenhagen. He had successfully leveraged his skills and passion for craftsmanship over many years while serving a variety of different industrial and agricultural customers. His son, Anders, had spent nearly 10 years working as a production engineer for a large automotive parts supplier in the United Kingdom, but eventually returned to Slagelse to take over the family firm. Exploiting his experience in mass manufacturing, Anders spent years building the firm into a larger-scale industrial component manufacturer but retained his father's commitment to quality and customer service. After 20 years he sold the firm to a UK-owned industrial conglomerate and within 10 years it had doubled in size again, and now employed in the region of 600 people and had a turnover approaching £200 million. Throughout this period the firm had continued to target its products into niche industrial markets where its emphasis upon product quality and dependability meant it was vulnerable to price and cost pressures. However, in 2009, in the midst of difficult economic times and widespread industrial restructuring, it had been encouraged to bid for higher-volume, lower-margin work. This process was not very successful but eventually culminated in a tender for the design and production of a core metallic element of a child's toy (a 'transforming' robot).

Interestingly, the client firm, Alden Toys, was also a major customer for other businesses owned by SIS's corporate parent. It was adopting a preferred supplier policy and intended to have only one or two purchase points for specific elements in its global toy business. It had a high degree of trust in the parent organisation and on visiting the SIS site was impressed by the firm's depth of experience and commitment to quality. In 2010, it selected SIS to complete the design and begin trial production.

'Some of us were really excited by the prospect . . . but you have to be a little worried when volumes are much greater than anything you've done before. I guess the risk seemed okay because in the basic process steps, in the type of product if you like, we were making something that felt very similar to what we'd been doing for many years' (SIS Operations Manager).

'Well obviously we didn't know anything about the toy market but then again we didn't really know all that much about the auto industry or the defence sector or any of our traditional customers before we started serving them. Our key competitive advantage, our capabilities, call it what you will, they are all about keeping the customer happy, about meeting and sometimes exceeding specification' (SIS Marketing Director).

The designers had received an outline product specification from Alden Toys during the bid process and some further technical detail afterwards. Upon receipt of this final brief, a team of engineers and managers confirmed that the product could and would be manufactured using an up-scaled version of current production processes. The key operational challenge appeared to be accessing sufficient (but not too much) capacity. Fortunately, for a variety of reasons, the parent company was very supportive of the project and promised to underwrite any sensible capital expenditure plans. Although this opinion of the nature of the production challenge was widely accepted throughout the firm (and shared by Alden Toys' and SIS's parent group), it was left to one specific senior engineer to actually sign both the final bid and technical completion documentation. By early 2011, the firm had begun a trial period of full volume production. Unfortunately, as would become clear later, during this design validation process SIS had effectively sanctioned a production method that would prove to be entirely inappropriate for the toy market but it was not until 12 months later that any indication of problems began to emerge.

Throughout both North America and Europe, individual customers began to claim that their children had been 'poisoned' while playing with the end product. The threat of litigation was quickly levelled at Alden Toys and the whole issue rapidly became a 'full-blown' child health scare. A range of pressure groups and legal damage specialists supported and acted to aggregate the individual claims. Although similar accusations had been made before, the litigants and their supporters focused in on the recent changes made to the production process at SIS and particular the role of Alden Toys in managing its suppliers. '. . . *It's all very well claiming that you trust your suppliers but you simply cannot have the same level of control over another firm in another country. I am afraid that this all comes down to simple economics, that Alden Toys put its profits before children's health. Talk about trust . . . . parents trusted this firm to look out for them and their families and have every right to be angry that board-room greed was more important!'* (legal spokesperson for US litigants when being interviewed on a UK TV consumer rights show).

Under intense media pressure, Alden Toys rapidly convened a high-profile investigation into the source of the contamination. It quickly revealed that an 'unauthorised' chemical had been employed in an apparently trivial metal cleaning and preparation element of the SIS production process. Although when interviewed by the US media, the parent firm's legal director emphasised there was *'no causal link established or any admission of liability by either party'*, Alden Toys immediately withdrew its order and began to signal an intent to bring legal action against SIS and its parent. This action brought an immediate end to production in this part of the operation and the inspection (and subsequent official and legal visits) had a crippling impact upon the productivity of the whole site. The competitive impact of the failure was extremely significant. After over a year of production, the new product accounted for more than one-third (39 per cent) of the factory's output. In addition to major cash-flow implications, the various investigations took up lots of managerial time and the reputation of the firm was seriously affected. As the site operations manager explained, even their traditional customers expressed concerns. *'It's amazing but people we had been supplying for thirty or forty years were calling me up and asking "[Manager's name] what's going on?" and that they were worried about what all this might mean for them . . . these are completely different markets!'*

QUESTIONS

1 What operational risks did SIS face when deciding to become a strategic supplier for Alden Toys?

2 What control problems did it encounter in implementing this strategy (pre- and post-investigation)?

Problems and applications

All chapters have 'Problems and applications' questions that will help you practise analysing operations. They can be answered by reading the chapter. Model answers for the first two questions can be found on the companion website for this text.

1 Although rare, air crashes do happen. Predominantly, the reason for this is human failure such as pilot fatigue. One kind of accident, which is known as 'controlled flight into terrain', where the aircraft appears to be under control and yet still flies into the ground, has a chance of happening less than once in 2 million flights. To occur, a whole chain of minor failures must happen. The pilot at the controls has to be flying at the wrong altitude (one chance in a thousand). The co-pilot would have to fail to cross-check the altitude (one chance in a hundred). The air traffic controllers would have to miss the fact that the plane was at the wrong altitude (one-in-ten chance). Finally, the pilot would have to ignore the ground proximity warning alarm in the aircraft (which can be prone to give false alarms, a one-in-two chance).

   (a) What are your views on the quoted probabilities of each failure described above occurring?
   (b) How would you try to prevent these failures occurring?
   (c) If the probability of each failure occurring could be reduced by a half, what would be the effect on the likelihood of this type of crash occurring?

2 Wyco is a leading international retailer selling clothing and accessories, with stores throughout the world. The countries from which it sources its products include Sri Lanka, Bangladesh, India and Vietnam. It was shocked when a British newspaper reported that an unauthorised subcontractor had used child workers to make some of its products at a factory in Delhi. In response Wyco immediately issued a statement.

   'Earlier this week, the company was informed about an allegation of child labour at a facility in India that was working on one product for Wyco. An investigation was immediately launched. The

company noted that a very small portion of a particular order placed with one of its vendors was apparently subcontracted to an unauthorised subcontractor without the company's knowledge or approval. This is in direct violation of the company's agreement with the vendor under its Code of Vendor Conduct'.

Wyco's CEO said, *'We strictly prohibit the use of child labour. This is a non-negotiable for us. Wyco has a history of addressing challenges like this head-on. Wyco ceased business with 20 factories due to code violations. We have 90 people located around the world whose job is to ensure compliance with our Code of Vendor Conduct. As soon as we were alerted, we stopped the work order and prevented the product from being sold in stores. While such violations are extremely rare, we have called an urgent meeting with our suppliers to reinforce our policies. Wyco has one of the industry's most comprehensive programmes in place to fight for workers' rights. We will continue to work with stakeholder organisations in an effort to end the use of child labour'.*

(a) 'Being an ethical company isn't enough any more. These days, leading brands are judged by the company they keep'. What does this mean for Wyco?
(b) When Wyco found itself with this supply chain problem, did it respond in the right way?

3 An Airbus A320 would not turn left no matter what the pilot tried. Eventually they made an emergency landing. Fortunately, no one was hurt. The cause of the near-disaster was that engineers had forgotten to reactivate four of the five spoilers on the right wing that help the plane to turn. The investigation blamed 'a chain of human errors', by the engineers and by the pilots who had failed to notice the problem before take-off. The A320 is a 'fly-by-wire' aircraft where computer-controlled electrical impulses activate the hydraulically powered spoilers and surfaces, which control the movement of the plane. When the aircraft went for repair to a damaged flap, the engineers had put the spoilers into 'maintenance mode' to block them off from the controls. They had then forgotten to reactivate them. According to the official report, the engineers were not guilty of *'simple acts of neglect or ignorance. Their approach implied that they believed there were benefits to the organisation if they could successfully deliver the aircraft on time. With more complex aircraft, it is no longer possible for maintenance staff to understand all the consequences of any deviation. The avoidance of future accidents with high-technology aircraft depends on total compliance. If a check has been carried out numerous times without any fault being present, it is human nature to anticipate no fault when next the check is carried out'*.

(a) Why should fly-by-wire aircraft pose a more complex maintenance problem than conventional aircraft, which have a physical link between the control and the flaps?
(b) If you were the accident investigator, what questions would you want to ask in order to understand why this failure occurred?

4 A light bulb in the men's lavatories of a firm finally burnt out when it was over 70 years old. It had survived bombs dropped in the Second World War that had devastated buildings in neighbouring streets, shaking buildings in the whole area but leaving the bulb intact and working. It was not even affected by the punk rock bands that played at an adjacent venue and caused residents to complain. When the bulb did eventually fail, the firm had it mounted on a stand and gave it a place of honour. Does this example invalidate the use of failure data in estimating component life?

5 An automated sandwich-making machine in a food manufacturer's factory has six major components, with individual reliabilities as shown in Table 18.3.

(a) What is the reliability of the whole system?
(b) If it is decided that the wrapper in the automated sandwich-making machine is too unreliable and a second wrapper is needed, which will come into action if the first one fails, what will happen to the reliability of the machine?

6 Every time we enter an elevator, we are trusting our lives to the people who designed, made and maintain it. Without effective maintenance, elevators would literally be death traps. Otis, the elevator company, has its 'Otis Maintenance Management System' (OMMS), a programme that takes into account its clients' elevators' maintenance needs. Maintenance procedures are determined by each

Table 18.3 Individual reliabilities of major components

Component	Reliability
Bread slicer	0.97
Butter applicator	0.96
Salad filler	0.94
Meat filler	0.92
Top slice of bread applicator	0.96
Wrapper	0.91

elevator's individual pattern of use, such as frequency of trips, loads carried and conditions of use. Otis also monitors the life cycle characteristics of all its elevators' components. This information on wear and failure is made available to its customers and is used to update maintenance schedules. When an elevator has a problem, a technician can be on their way to a customer within minutes, 24 hours a day, 7 days a week. The service can get the elevators back in service within two and half hours on average. Otis monitors, collects, records, analyses and communicates hundreds of different system functions. If it detects a problem, it calls out a technician. *'Around-the-clock response is important,'* say Otis, *'because problems, don't keep office hours. The remote sensing system identifies most potential problems before they occur'.*

(a) What could be the effects of failure in elevator systems? How does this explain the maintenance service that Otis offers its customers?
(b) What approach(es) to maintenance are implied by the services that Otis offers?
(c) How would you convince potential customers for these services that they are worthwhile?

7 Carlsberg, the brewing company, learned of its crisis late one Friday afternoon. Something appeared to have gone wrong with the 'widget' (the device which gives some canned beer its creamy characteristic) in one of its cans of beer. One customer had found a piece of plastic in his mouth. He had complained to an environmental health official, who had contacted Carlsberg. The company's pre-planned crisis management procedure immediately swung into action. A crisis control group of 12 members, with experts on insurance, legal affairs, quality control and public relations, took control. This group had everyone's telephone number, so any relevant person in the company could be contacted. It also had a control room at one of the company's sites. By the Tuesday they had issued a press release, set up a hotline and taken out national advertising to announce the recall decision. Even though the problem had originated in only one of its six brands, the company decided to recall all of them (a total of 1 million cans) and all production using the suspect widget was halted.

(a) What seem to be the essential elements of this successful recovery from failure?
(b) How do the advantages and disadvantages of deciding whether or not to recall products in a case such as this depend on the likelihood of another potential failure being out there in the market?
(c) Relate this issue to the concept of type I and type II errors dealt with in Chapter 17.

8 What risks does a technology such as the 'Internet of Things' (IoT) pose to all elements of a supply network?

9 How might climate change affect how operations managers view risk management?

10 *'Your feedback is very important to us'*, is something that businesses often tell their customers.

(a) Is this always true?
(b) Why do they do it?

Selected further reading

Hillson, D. (ed.) (2020) ***The Risk Management Handbook: A Practical Guide to Managing the Multiple Dimensions of Risk*****, Kogan Page, London.**
Very much a practical guide, with some interesting ideas about emerging risks.

Hodson, C.J. (2019) ***Cyber Risk Management: Prioritize Threats, Identify Vulnerabilities and Apply Controls*****, Kogan Page, London.**
An obviously specialised, but important look at how cyber risks can cause financial, operational and reputational damage.

Hopkin, P. (2018) ***Fundamentals of Risk Management: Understanding, Evaluating and Implementing Effective Risk Management*****, 5th edn, Kogan Page, London.**
A comprehensive introduction to risk with good coverage of many core frameworks.

Hubbard, D.W. (2020) ***The Failure of Risk Management: Why It's Broken and How to Fix It*****, 2nd edn, Wiley, Hoboken, NJ.**
Provocative and challenging.

Waters, D. (2012) ***Supply Chain Risk Management: Vulnerability and Resilience in Logistics*****, 2nd edn, Kogan Page, London.**
Provides a very detailed and practical guide to considering risks within operations and supply chains.

Notes on chapter

1. The information on which this example is based is taken from: Miller, J. (2021) Lessons from twelve years in pursuit of zero, Gemba Academy, 10 May, https://blog.gembaacademy.com/2021/05/10/lessons-from-twelve-years-in-pursuit-of-zero/ (accessed September 2021).
2. The information on which this example is based is taken from: Amelang, S. and Wehrmann, B. (2020) 'Dieselgate' – a timeline of the car emissions fraud scandal in Germany, Factsheet, Clean Energy Wire, 25 May, https://www.cleanenergywire.org/factsheets/dieselgate-timeline-car-emissions-fraud-scandal-germany (accessed September 2021); Tovey, A. (2017) VW attacked by MPs over failure to release findings of 'dieselgate' investigation, *The Telegraph*, 22 March.
3. Takala, J., Hämäläinen, P., Saarela, K.L., Loke, Y.Y., Manickam, K., Tan, W.J., Heng, P., Tjong, C., Guan, K.L., Lim, S.Y.E. and Gan, S.L. (2014) Global estimates of the burden of injury and illness at work in 2012, *Journal of Occupational and Environmental Hygiene*, 11 (5), 326–37.
4. The information on which this example is based is taken from: Darktrace website, https://www.darktrace.com/en/ (accessed September 2021); Walker, M. (2020) Darktrace: an AI cybersecurity platform that serves as the immune system for enterprise business data by fighting off threats, Credit Card News, 3 February, https://www.cardrates.com/news/darktrace-is-an-ai-based-enterprise-immune-system/ (accessed September 2021); Ismail, N. (2019) Darktrace unveils the Cyber AI Analyst: a faster response to threats, *Information Age*, 4 September; Ross, A. (2019) ML and AI in cyber security: real opportunities overshadowed by hype, *Information Age*, 7 March.
5. Nahajima, S. (1988) *Total Productive Maintenance*, Productivity Press, New York, NY.
6. The information on which this example is based is taken from: The Economist (2019) An emergency landing system that passengers can activate, *Economist* print edition, 28 November.
7. Smithsonian Ocean Portal Team (n.d.) Gulf oil spill, https://ocean.si.edu/conservation/pollution/gulf-oil-spill (accessed September 2021).
8. Borunda, A, (2020) We still don't know the full impacts of the BP oil spill, 10 years later, *National Geographic*, 20 April.
9. World Health Organization Europe (2020) 2019-nCoV outbreak is an emergency of international concern, https://www.euro.who.int/en/health-topics/health-emergencies/international-health-regulations/news/news/2020/2/2019-ncov-outbreak-is-an-emergency-of-international-concern (accessed September 2021).
10. Hill, A. (2020) Covid-19 lays bare managers' efficiency obsession, *Financial Times*, 20 April.

19 Project management

KEY QUESTIONS

19.1 What are projects?

19.2 What is project management?

19.3 How is the project environment understood?

19.4 How are projects defined?

19.5 How are projects planned?

19.6 How are projects controlled and learned from?

INTRODUCTION

Recent years have seen a meaningful increase in the proportion of operations managers' time that is spent working on discrete projects as opposed to 'steady-state' activities – a trend sometimes referred to as '**projectification**'. Yet, despite the increase in project-based activity for operations and their supply networks, many projects are only partially successful. In this chapter, we examine how projects of all shapes and sizes can be executed more successfully. To do this, managers must first understand the innate characteristics of the project as well as the implications of differences between projects. Second, they must appreciate the vital role of effective project management in influencing the success (or failure!) of projects and recognise the key responsibilities and skills of those tasked with running projects. Third, they must understand the environment within which their project is being undertaken and determine how best to manage project stakeholders. Fourth, they must effectively define projects while balancing competing performance objectives of quality, time and cost. Fifth, they must plan projects to help determine the cost and duration of the project and the level of resources that will be required. Finally, they must effectively control projects through their life cycle and ensure that learning between projects is maximised. Figure 19.1 shows where project management fits in the overall model of operations management.

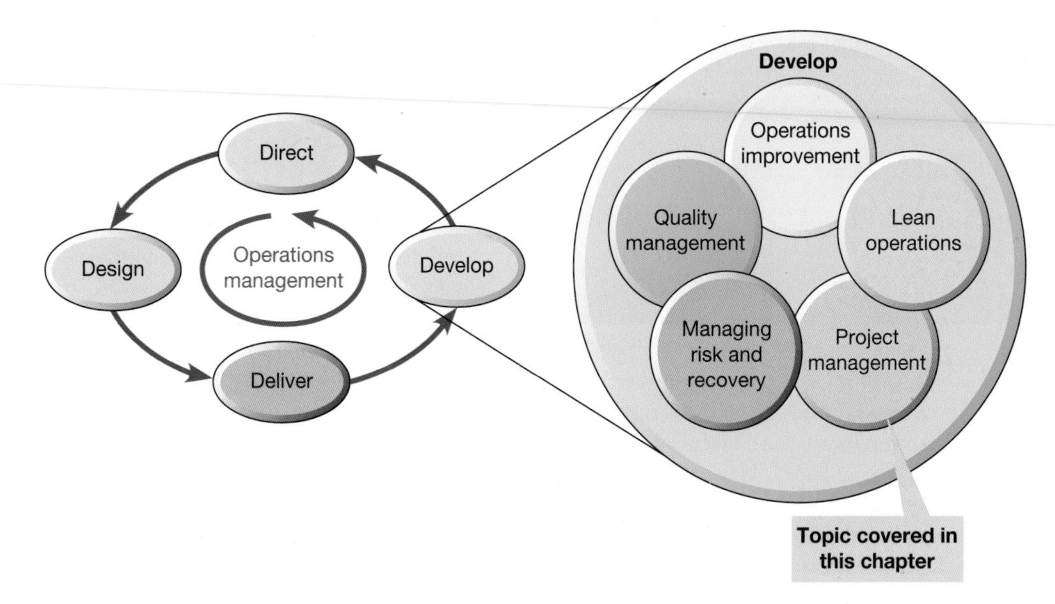

Figure 19.1 This chapter examines project management

19.1 What are projects?

A project is a set of activities that must be completed to deliver a specific goal within a set time frame, using a defined group of resources. Technically, many small-scale operations management endeavours, taking minutes or hours, conform to this definition of a project. However, in this chapter we will focus mainly on relatively larger-scale projects, lasting months, years or even several decades. Although all projects have a defined goal, some of these will be part of a larger purpose. So, most operations improvement (even continuous improvement) can be seen as a series of overlapping 'mini-projects' that cumulatively contribute to a never-ending development effort. Similarly, research and development (R&D) projects sometimes have a specific application in mind, but often, if the research is more 'blue-sky', they don't.

It is also worth pointing out the distinction between *projects*, programmes and *portfolios*. A programme can be seen as a broad effort encompassing several projects connected to a common purpose. An example is NASA's programme of projects with several missions ultimately intended to bring back rock samples, to look for signs of life on ancient Mars. As such, *programme management* will overlay and integrate the individual projects within a programme. Portfolios are bundles of projects grouped together for management convenience. Unlike programmes, the projects in a portfolio are less connected in terms of shared goal, but they do typically have common *resources*. An example would be a company's portfolio of new product and service development (NPSD) projects.

Common features of projects

Projects share several common features – they are:

- temporary – while some projects last hours and others many years, they all have a defined start and end point;
- dedicated to completing a specific goal within key time, cost and quality requirements;
- outcomes that are typically unique or at least highly customised;
- based on many non-routine and complex tasks.

It is these features that generate high levels of risk and uncertainty in the management of projects, and why changed specifications (quality), severe delays (time), cost escalation (cost) and major disputes between key stakeholders are commonplace.

OPERATIONS IN PRACTICE: 'For the benefit of all' – NASA's highs and lows[1]

Space remains one of the most challenging contexts for carrying out successful projects, and since humankind first ventured into space, we have witnessed many project failures. In January 1986, NASA's (National Aeronautics and Space Administration) space shuttle *Challenger* exploded shortly after take-off, killing all on board. The failure was traced to a faulty seal, which ruptured and caused the liquid hydrogen fuel to explode. The NOAA weather satellite was NASA's first mission after a suspension of 32 months to allow investigation of the disaster. Things did not go well – just 71 seconds into its flight, the launch rocket was struck by lightning. With its first stage rockets disabled, ground control destroyed the rocket to minimise the risk of it falling back to Earth.

Project failures, however, have often enabled future successes for NASA. Take the Apollo 6 mission, the final non-manned test for the Saturn V rocket, in April 1968. Shortly after launching, the rocket experienced 'pogo oscillations' (variations in thrust levels caused by changing fuel rates), two engines shut down prematurely in the second stage burn, and the third stage rocket failed to reignite. Yet, in July 1969, the same (improved!) rocket successfully delivered the Apollo Lunar Module into space on the manned Apollo 11 mission that delivered Neil Armstrong and Edwin ('Buzz') Aldrin safely onto the surface of the Moon.

On other occasions, problems have occurred, but a combination of technical capabilities and inventiveness of the key stakeholders have prevented project failure. For example, the *Apollo 13* mission in 1970 was intended to be the third to land on the Moon. However, following a failure in the service module's oxygen tank two days into the mission, the new project objective became the safe return of the crew back to Earth. This involved a move from the service module to the lunar module as a form of 'lifeboat', followed by various improvisations to convert a craft that was originally designed to support two men on the lunar surface for two days, to one able to support three men in space for four days. Fortunately, the revised plan worked and while the mission failed in its primary objectives, it was very successful in its revised scope. In another example, the Hubble Space Telescope had many issues post launch, including problems focusing the telescope due to errors made by European and US scientists in translating the units of measurement! However, the project team found ways to overcome the various failures and the telescope now provides some of the most detailed images of deep space – as well as a significantly expanded understanding of our (tiny) place in the universe.

Some space projects have gone far better than expected, leading to change in scope due to their success. For example, the Cassini mission (a collaboration involving NASA, the European Space Agency and the Italian Space Agency) began with the objective to reach Saturn, some 750 million miles away. On its way, Cassini took photographs of our solar system, including flybys of Earth, Venus and Jupiter – the photos of Jupiter being the most detailed ever of the planet. Cassini also confirmed Einstein's Theory of General Relatively while travelling to Saturn. On arrival, Cassini successfully deployed the Huygens probe, which began returning data to Earth from Saturn's largest moon, Titan. Meanwhile, Cassini continued to collect detailed data and images on the planet and its other moons. The mission, originally expected to last four years, was first extended by two years (the Cassini Equinox Mission) and then again for a further seven years (The Cassini Solstice Mission) as the spacecraft continued to function effectively. It wasn't until 2017, nine years after the planned project end, that Cassini was finally 'de-orbited' to burn up in Saturn's atmosphere, though not before it had completed a number of high-risk passes within Saturn's inner rings to maximise its total scientific contribution.

It's not just NASA and its collaborators that have looked to build on previous experiences to improve their projects. The recent generation of commercial space travel and exploration companies, such as SpaceX, Sierra Nevada Corp, Boeing, Northrop Grumman Innovation Systems, Blue Origin and Virgin Galactic, have all sought to gain insights from established space agencies. This not only offers the benefit of (hopefully) avoiding mistakes that have been made in the past, but has also enabled significant reductions in the costs of development, testing and operations. For example, SpaceX has made major breakthroughs in fuel, engines and, most valuably, in increasing the proportion of its rocket and launch vehicles that it can recover and reuse. The result is that the cost of launching a kilogram of material into space has fallen to (just!) $2,720 for the SpaceX Falcon 9, the rocket used to successfully deliver NASA astronauts to the International Space Station in 2020. This compares to an eyewatering $54,500 per kilogram when the NASA space shuttle programme was operational between 1981 and 2011. In addition to commercial space companies, several countries have demonstrated their increasing aspirations in extra-terrestrial exploration. For example, in 2021 the United Arab Emirates became just the fifth country to successfully send a probe into orbit around Mars. Later that year, China successfully landed a six-wheeled robot, Zhurong, on the surface of the planet, only the second country to successfully complete such a mission.

Differentiating between projects

So far, we have described the *commonalities* of projects – temporary activities, with specific and highly customised goals, within time, cost and quality requirements, usually involving many non-routine and complex tasks. However, it is also critical to understand *differences* between projects. These differences play a critical role in subsequent project management challenges. Here, we focus on differences in the level of *innovation*, *time pressure* and *complexity* of the project being managed. Figure 19.2 illustrates the profiles of four different projects using this form of differentiation and the implications of these differences.

Operations principle

The challenge of managing a project is contingent on its level of innovation, time pressure and complexity.

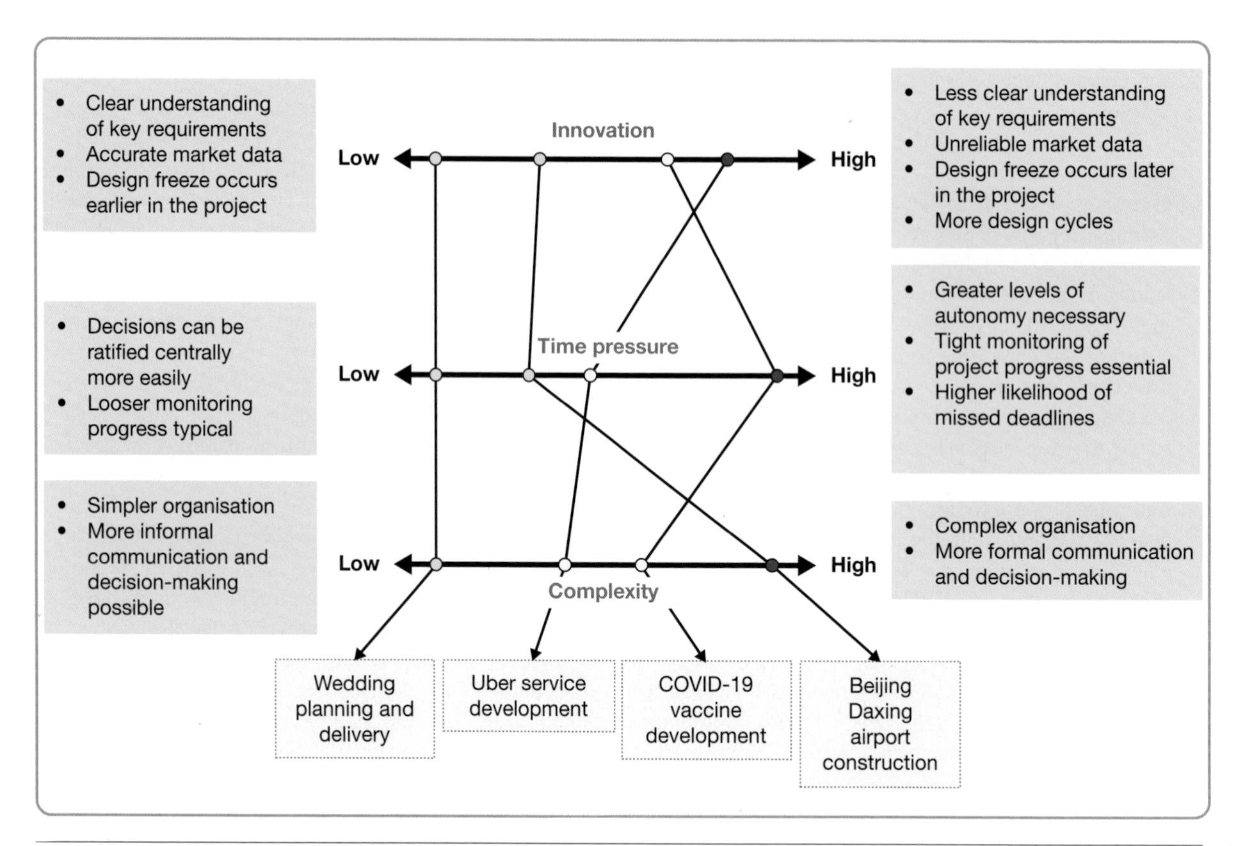

Figure 19.2 Differentiating projects based on level of innovation, time pressure and complexity, and the implications of these differences

Level of project innovation

The first way to differentiate projects is to consider their relative level of innovation. For projects, innovation may entail the delivery of new services or products, the incorporation of new technologies, the development of new routes to market and the transformation of organisational processes, for example. Incremental projects typically involve relatively modest levels of innovation, building upon existing knowledge and/or resources where existing routines are not fundamentally changed. By contrast, radical projects exhibit high levels of novelty that require completely new knowledge and/or resources, and often make existing routines obsolete. Examples of these more innovative projects include the development and launch of Uber (the multinational ride-hailing service) and Airbnb; the rapid emergence of sensor technology in many aspects of farming; the first iPhone, which paved the way for the modern smartphone market; and the development of Netflix, with its huge impact on the home entertainment sector.

Level of project time pressure

The second way to differentiate projects is to consider the relative level of time pressure that they face. It's important to remember that time pressure is not about speed – some projects have urgency but last for many years, others are not urgent but last a few weeks. Some projects face low levels of time pressure where the specific time frame is not deemed critical by the project stakeholders. Many public works and internal projects fall into this category. Some projects face moderate levels of time pressure where completion on time is important for competitive advantage and leadership. Many business-related projects, such as new service or product development, fall into this category. Finally, some projects face high levels of time pressure where there is a specific window of opportunity and any delay can mean project failure. For example, in May 1961, President John F. Kennedy delivered a speech to the US Congress in which he stated, '*I believe that this nation should commit itself to achieving the goal, before this decade is out, of landing a man on the moon and returning him safely to the earth*'. In doing so, he set a time frame that was to be critical to the ambitions of the moon-landing project. Other examples of projects facing high levels of time pressure are those created in response to specific crises such as the Ebola or COVID-19 pandemics, wars or natural disasters.

OPERATIONS IN PRACTICE

When every minute counts in a project – unblocking the Suez Canal[2]

The Suez Canal is a 193 km artificial waterway in Egypt that connects the Red Sea to the Mediterranean Sea. Originally opened in 1869, the canal offers a more direct route between Asia and Europe than the longer journey around Africa. The Suez Canal is one of the busiest trading routes in the world, traversed by over 19,000 vessels every year and accounting for around 12 per cent of global trade. Oil tankers, bulk carriers and container ships together transport well over 1 billion tonnes of cargo every year.

At 07.40 local time on 23 March 2021, the *Ever Given*, loaded with 20,000 containers travelling from Malaysia to the Netherlands, ran aground in the Suez Canal. The crash happened in one of the single-lane stretches and because the vessel was longer than the width of the canal, she became wedged, blocking all traffic north and south.

With the world looking on, the project to free the *Ever Given* began in earnest. The same day, seven tugboats were sent to try to free the ship but to no avail. The following day, a larger tugboat made its way to the stricken vessel and two dredgers began clearing sand and mud from around its bow. Within two days, 156 ships were waiting on either side of the *Ever Given*. By 26 March, this had grown to 237 ships, and multiple carriers, including CMA-CGM, Maersk and MSC, took the decision to reroute vessels around the southern tip of Africa (an increase in travel time of between 7 and 15 days). By 27 March, there were 14 tugs working to free the *Ever Given*, and on 28 March, a specialist tug from the Netherlands and a dredger from Cyprus were sent to support the time-critical project. Finally, on 29 March, six days after running aground and with 367 ships now stranded, the *Ever Given* was finally freed, to the great relief of all stakeholders.

The consequences of the *Ever Given* incident were significant. *Lloyds List* estimated the costs to the world economy were around £9.6 billion per day during the Suez Canal blockage – that's $6.7 million per minute! The owners of the *Ever Given* faced a claim for nearly $1 billion in damages and the ship was impounded in Suez waters until a settlement was reached. Meanwhile, many international supply chains were thrown into disarray because of delayed shipping and rerouting, with many retailers reporting significant disruptions to their operations. The backlog of ships wating to get through the canal following its re-opening lasted several days, extending the disruption yet further. In fact, many experts argued it was almost impossible to accurately assess the full costs of the Suez Canal blockage to businesses worldwide.

The event also highlighted the significant dependence on key trade routes such as the Suez Canal in the international supply of goods and some of their potential vulnerabilities. In the debate that followed, bad weather (30 mph winds and a sandstorm) and crew error were both blamed for the incident. However, other critics suggested that the Suez Canal Authority needed to re-evaluate their regulations on the maximum dimensions of vessels allowed to use the canal (as an ultra-large container vessel, the *Ever Given* had recently qualified as a 'Suezmax'), to reduce the risks of similar disruptions in the future.

Level of project complexity

The third way to differentiate projects is to consider their relative levels of complexity. Some projects exhibit low levels of complexity, often self-contained and with relatively small numbers of key stakeholders. Examples may include planning a wedding, creating an online sales platform, developing a new MBA operations and process management course or writing a new book (!). Other projects face greater levels of complexity, often combining a set of sub-elements and involving many more stakeholders. Examples include constructing a new research and development facility, developing a new portfolio of post-graduate education within a university, or organising a large-scale music festival.

While the sub-elements of the project have a common goal, the added complexity creates significantly higher coordination and integration challenges. Finally, some projects must deal with extremely high levels of complexity, coordinating several major projects to deliver against a common goal. A good example of this kind of project is China's South-to-North water diversion project, a multi-decade infrastructural mega-project expected to be completed in 2050 at a cost of over $70 billion.

19.2 What is project management?

Project management is the activity of defining, planning, controlling and learning from projects. The key stages in this process include:

Stage 1 Understanding the project environment – internal and external factors that may influence the project.

Stage 2 Defining the project – setting the objectives, scope and strategy for the project.

Stage 3 Project planning – deciding how the project will be executed.

Stage 4 Technical execution – performing the technical aspects of the project.

Stage 5 Project control – ensuring that the project is carried out according to plan.

Stage 6 Learning – reviewing project performance to improve future projects.

In the following sections of this chapter, we deal with each of these stages in turn, except for Stage 4. The technical execution of the project is determined by the specific technicalities of individual projects, so is beyond the remit of the chapter. While this 'life cycle' perspective is useful and allows us to consider projects in a sequential manner, it is important to understand that project management is essentially an *iterative* process. Problems or changes that become evident in project control, for example, may require re-planning and may even cause modifications to the original project definition. Going beyond the 'life cycle perspective' described above, project management is also concerned with effectively balancing quality/deliverables, time and cost objectives within the so-called 'iron triangle' (of quality, time and cost). Finally, from an organisational perspective, project management involves managing these life cycles and performance objectives across multiple functions within an organisation.

Operations principle
Project management is the activity of defining, planning, controlling and learning from projects.

OPERATIONS IN PRACTICE — McCormick's AI spice project[3]

Artificial intelligence (AI) is increasingly playing a transformative role in many aspects of business operations. Now we are witnessing its use in new product development projects within the food sector. McCormick, the largest spice company in the world, teamed up with IBM research in February 2019 to develop an AI system aimed at developing new flavour combinations. The collaboration leverages IBM's expertise in machine learning and its proprietary IBM Research AI for Product Composition to sift through data on thousands of ingredients, sales (both its own and within the sector), consumer taste trends, consumer testing information, and hundreds of thousands of existing seasoning mixes, to suggest potential new formulas. The system can also advise on possible substitutes for raw ingredients, relative level of novelty (based on the 'distance' between a flavour combination and its nearest neighbour) and likely human response.

The company has already launched its first products leveraging its new AI system – 'McCormick One' is a range of seasoning for simple one-dish recipes, including Tuscan Chicken, New Orleans Sausage and Bourbon Pork Tenderloin. The new AI system is in sharp contrast to McCormick's traditional approach to new product and service development (NPSD) projects, involving a large team of chefs, nutritionists, food scientists, chemists and chemical engineers building on 'seed formulas' to develop new flavour combinations. The company believe that the AI system can help in its attempts to develop more innovative spice mixes, partly by avoiding cultural biases that may be inherent within its human development team. For example, the system suggested that adding cumin to a pizza seasoning would improve the taste. Such a move had never been considered by McCormick's food scientists, but their subsequent consumer testing backed up the idea.

Not only does McCormick believe that its AI system can help generate more novel flavours, with a lower probability of market rejection, it also saves costs in product development and shortens project time frames by up to two thirds. Some of the time is saved in the rapid creation of many different possible formulas, followed by automated filtering to create a shortlist of potential products for further human evaluation. Further time savings occur in the consumer testing phase of the projects, with feedback fed directly into the system, analysed, and then integrated into flavour revisions. In the highly competitive world of spices and flavourings, such time and cost savings offer substantial commercial benefits.

Project managers and their skill sets

In order to coordinate the efforts of many people in different parts of the organisation (and often outside it as well), all projects need a project manager. The project manager is the person accountable for project delivery and they have several key responsibilities (see Figure 19.3). The project manager organises the project team, with the responsibility, if not always the authority, to run the project on a day-to-day basis.

Project management is a very demanding role, requiring a diverse set of skills including technical project management knowledge, interpersonal skills and leadership ability. Very often, project managers must motivate staff who not only report into managers other than themselves, but also divide their time between several different projects. In addition, they must pay attention to details without losing sight of the big picture, establish an open, communicative environment while remaining wedded to project objectives, and have an ability to remain optimistic while planning for the worst. Such challenges have led to the increasing professionalisation of project management over the last 20 years, with many more of those leading projects now holding professional qualifications.

Operations principle
The activity of project management requires interpersonal as well as technical skills.

Forming project teams and assigning responsibilities

A key role of the project manager is to form a project team and assign responsibilities for key tasks. In forming a team, it is important to consider the diversity of members to ensure that strengths and potential shortcomings of team members are balanced. For example, project teams need individuals that are naturally organised, with the capability to take ideas and make them work in practice. However, too many of this type of team member can limit flexibility and creativity within a project – a major limitation, particularly when working on projects with high levels of innovation. Conversely, projects with too many individuals that are predominantly free-thinking and creative often run into problems because such individuals are often less interested in the 'devil in the detail' of the project.

Once the project team is formed, the project manager needs to assign responsibility for all activities in the project (both to those within the project team and to third parties). A structured way to do this is in the form of a responsibility matrix. In its simplest form, this will simply identify who is responsible for each key activity in the project timeline. In some cases, responsibility matrices will show not only the responsible party but also others who are expected to provide support for the activity. The RACI matrix has emerged as a popular method of visualising responsibility, identifying those who are *responsible*, *accountable*, to be *consulted* and to be *informed*. Table 19.1 provides an example of a RACI matrix for a consulting project focused on new market testing in Chennai, India.

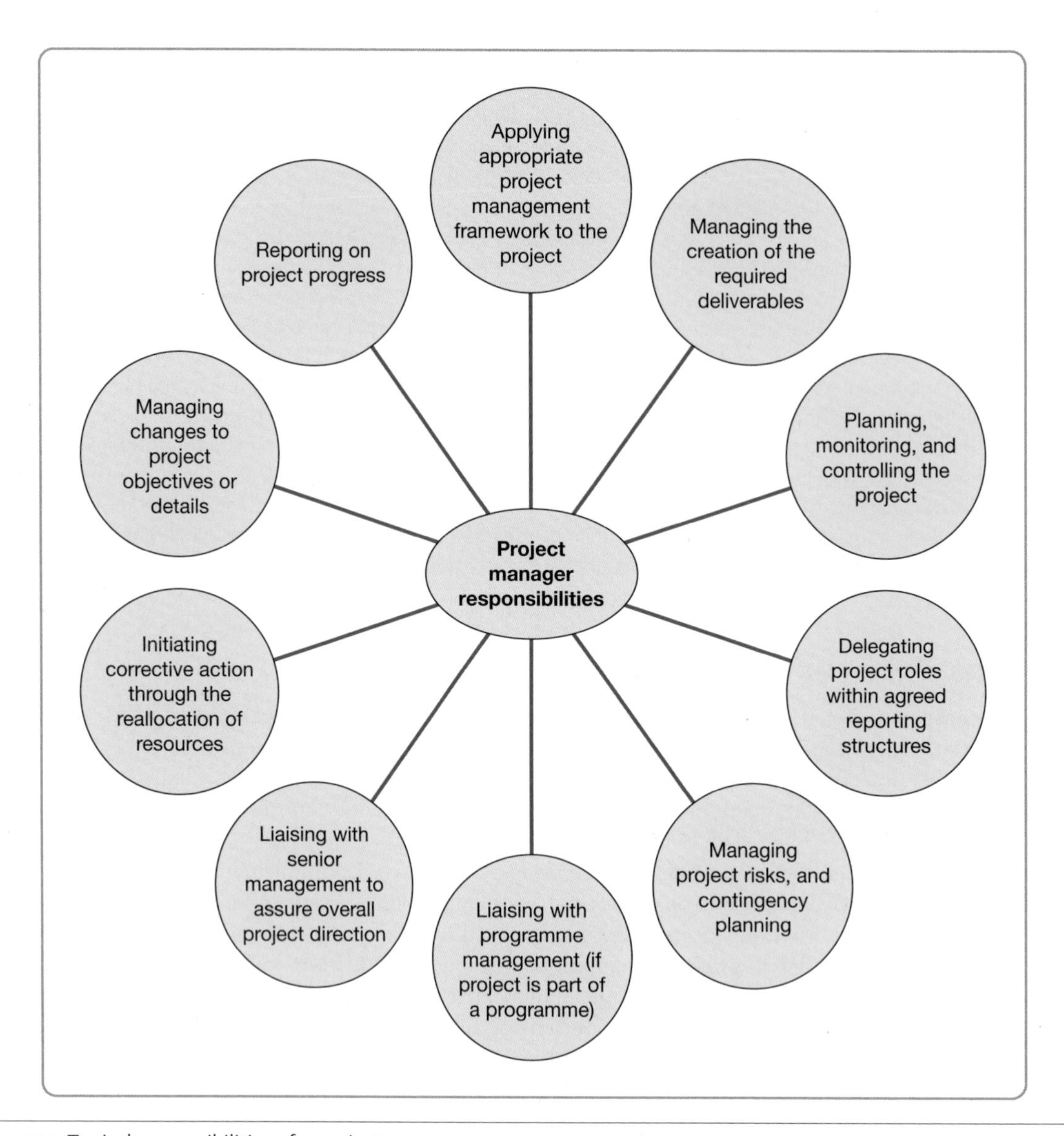

Figure 19.3 Typical responsibilities of a project manager

Table 19.1 RACI matrix for consulting project focused on new market testing in Chennai, India

Deliverable or task	Ekta (Project Sponsor)	Jayesh (Project Manager)	Ritika (Technical lead)	Punya (Analyst)	Shorya (Analyst)	Ashwin (Analyst	Shivani (Client)
Phase 1 (scoping)							
Client kick-off meeting	A	R					C
Needs analysis	I	A	R				C
Analyst contracting	I	A	R				
Client review meeting	A	R	C				C
Phase 2 (data collection)							
Market research (focus group)		I	A	R			C
Market research (survey)		I	A		R		C
Market research (secondary data)		I	A			R	C

Deliverable or task	Ekta (Project Sponsor)	Jayesh (Project Manager)	Ritika (Technical lead)	Punya (Analyst)	Shorya (Analyst)	Ashwin (Analyst	Shivani (Client)
Phase 3 (analysis and report)							
Data analysis (focus group)	I	A	C	R			
Data analysis (survey)	I	A	C		R		
Data analysis (secondary data)	I	A	C			R	
Report (first draft)	I	A	R	C	C	C	
Client presentation	A	R	C				I
Report (final)	A	R	C				C
Project closure	R	C	I				A

R = responsible; A = accountable; C = consult; I = inform

19.3 How is the project environment understood?

Projects do not exist in a vacuum. As such, it is vital that the project team understands the key characteristics of the environment within which their project is being undertaken, and they identify the individuals, groups or entities that have an interest in the project process or outcome. They must then decide how to engage with different stakeholders and how best to manage their competing needs. The project environment comprises all the factors that may affect the project during its life. Figure 19.4 illustrates four key aspects of the project environment.

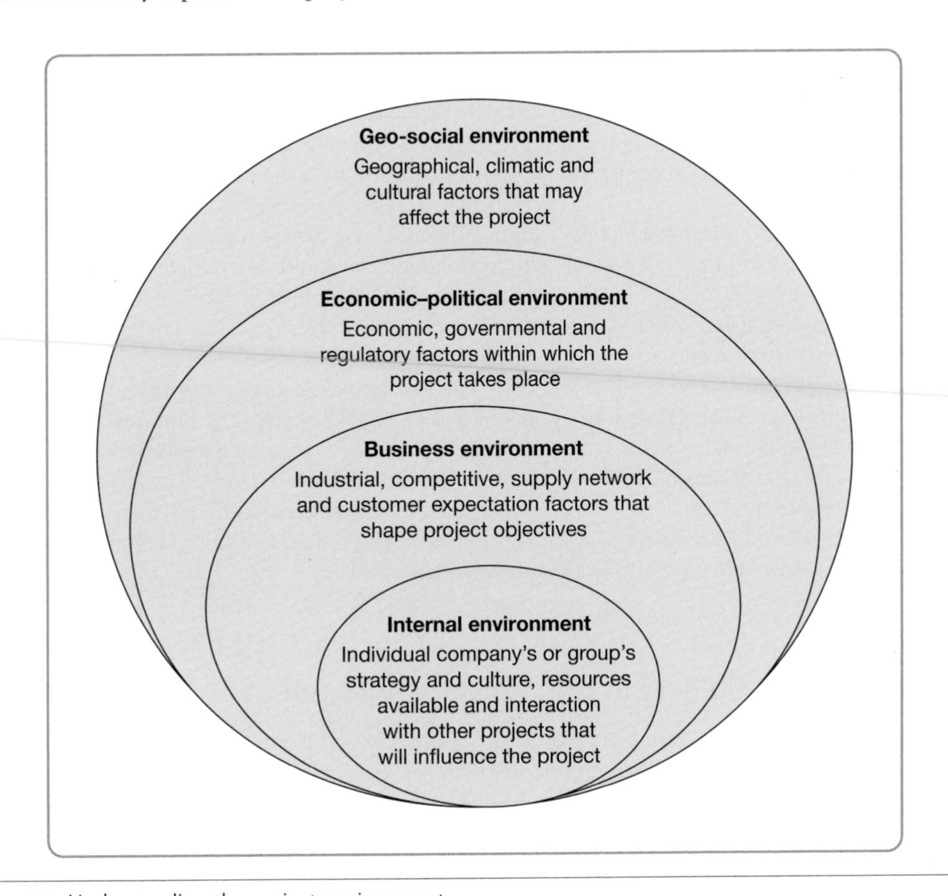

Figure 19.4 Understanding the project environment

OPERATIONS IN PRACTICE

Berlin Brandenburg Airport opens at last[4]

Originally intended to replace the German capital's three ageing airports, Berlin Brandenburg Airport is a major source of embarrassment for a country renowned for delivering things on time and on budget. The infrastructure project, one of the country's largest for decades, was due to open in 2011, with an expected 27 million passengers per year. However, after several major delays and failures, the airport finally received its licence to operate in May 2020 and didn't start operating until late 2020. It's not only the timing of the project that went badly wrong either – the initial budget of €2.83 billion more than doubled to over €7 billion. In fact, it is estimated that it cost around €20 million per month just to run the empty terminal building prior to its opening, plus €13 million per month in lost rental income.

What went so badly wrong? First, the growing popularity of Berlin as a destination meant that the original demand forecasts, made in 2006, were too low. Revised estimates (pre-COVID-19) indicated that the airport now needed to be able to handle somewhere between 30–35 million passengers per year (Dubai International, the world's busiest airport, handles around 90 million passengers). This led to investment in additional terminal space (especially around security, check-in and luggage reclaim), calls for a third runway to be developed and a request by Hatmut Mehdorn, the industrial troubleshooter brought in to save the failing project, to keep open one of the old airports intended for closure, to deal with the excess demand. Other problems included the airport's fire safety systems – an innovative solution that, in the event of fire, pumps smoke under the terminal building rather than through the roof, but which failed to gain regulatory approval for a number of years. Over 1,000 automatic doors in the terminal building had to be reengineered to ensure that they would close properly in an emergency. Other additional costs included additional parking, check-in counters and aircraft gates, rebuilding the airport's entrance hall, extending luggage facilities, and other cost overruns caused by cracking concrete in car parks, re-fitting of pipes and cables, missing conveyor belts, and problems with fire safety walls between the train station and terminal building. To add to these problems, the decision by airport bosses to cancel the contracts of the original consortium of architects and engineering firms, led to significant re-work in planning, as many of the documents and construction expertise became inaccessible.

Then, when it looked as if the project's troubles might finally be over, the coronavirus (COVID-19) pandemic caused a sudden and extreme drop in air travel. Brandenburg Airport faced a new battle to regain old business and capture new clients in a sector that was in crisis. The need for increased capacity, a critical issue during construction, was now more questionable as Berlin's two main airports, Tegel and Schönefeld, experienced 65 per cent reductions in passenger numbers and 17 per cent reductions in cargo at the height of pandemic. While such dramatic reductions in flight statistics were evident in airports around the world, Brandenburg Airport faced a particularly difficult challenge, given its less-established profile and its troubled history. It faced the strong prospect of significant reductions in business from critical intended customers in the low-cost carrier segment, such as easyJet and Ryanair. In addition, one of its large network carriers, Lufthansa, expected to focus its revival efforts on its main hubs in Munich and Frankfurt, instead of its secondary hub in Berlin. Further blows came with the termination of long-haul routes to New York and Philadelphia by Delta Airlines and American Airlines.

The role of stakeholders in the project environment

Once managers have understood the fundamental characteristics of a project, they must consider the stakeholders or agents who are likely to interact with the project, and who could play a critical role in its success or failure. All projects will have stakeholders – complex projects will have many. They are likely to have different views on a project's objectives that may conflict with other stakeholders. Internal stakeholders include the client, the project sponsor, the project team, functional managers, contractors and project support. External stakeholders (i.e. those outside of the core project, rather than outside of the organisation) include end users, suppliers, competitors, lobby groups, shareholders, government agencies and employees. Managing stakeholders can be a subtle and delicate task, requiring significant social and, sometimes, political skills. It is based on three basic activities – identifying stakeholders, understanding their different perspectives and managing stakeholders.

Operations principle
All projects have stakeholders with different interests and priorities.

Identifying project stakeholders

Think of all the individuals, groups or entities who affect or are affected by your work, who have influence or power over it, or have an interest in its successful or unsuccessful conclusion. Figure 19.5 illustrates a stakeholder map for a technology platform project in the third sector (also known as the not-for-profit sector or charity sector) aimed at matching charities with funding opportunities. Even if one decides not to attempt to manage every identified stakeholder, the process of stakeholder mapping is still useful because it gets those working on a project to see the variety of competing forces at play in many projects.

Understanding project stakeholders

Once all stakeholders have been identified, it is important to understand their different perspectives on the project. Some key questions that can help to understand project stakeholders include:

- What financial or emotional interest do they have in the outcome of the project? Is it positive or negative?
- What motivates them most of all?
- What information do they need?
- What is the best way of communicating with them?

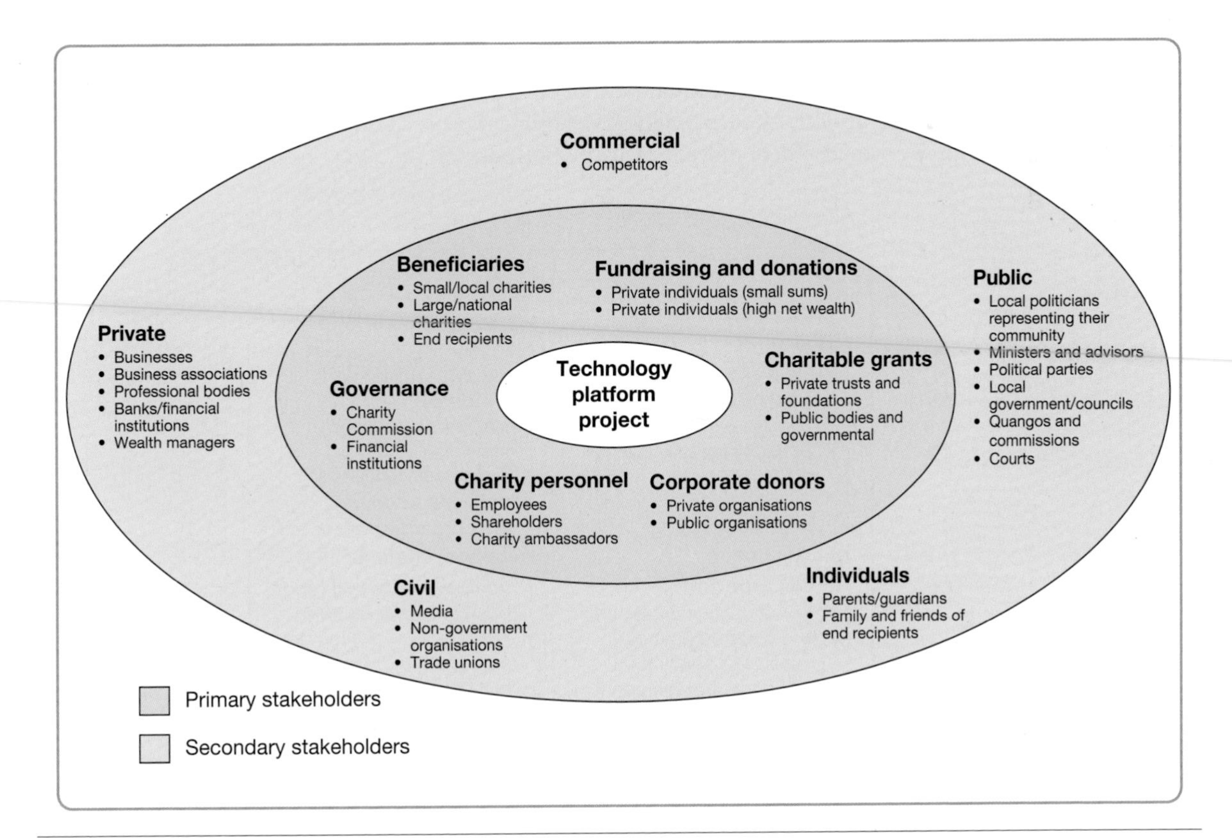

Figure 19.5 Stakeholder mapping for a third sector (not-for-profit) technology platform project

- What is their current opinion of the project?
- Who influences their opinions? Do some of these influencers therefore become important stakeholders in their own right?
- If a stakeholder is not likely to be positive, what will win them around to support the project?
- If you don't think you will be able to win them around, how will opposition be managed?

In seeking to understand these questions, stakeholder consultation becomes a critical activity. Consultation can provide valuable insights and experiences, can improve the legitimacy and buy-in for decisions, can help support relationships with key stakeholders, and can be critical in reducing potential opposition to the project. Figure 19.6 illustrates several key considerations for effective consultation with project stakeholders, considering timing, design, engagement and post-consultation.

Managing stakeholders

Having identified stakeholders and understood their different perspectives on the project, the next step is to decide how best to manage different stakeholders. A popular approach is to differentiate based on stakeholder power and influence. Stakeholders who have the power to exercise a major influence over the project should never be ignored. At the very least, the nature of their interest, and their motivation, should be well understood. But not all stakeholders who have the power to exercise influence over a project will be interested in doing so, and not everyone who is interested in the project has the power to influence it. The power–interest grid, shown in Figure 19.7, classifies stakeholders simply in terms of these two dimensions. Although there will be graduations between them, the two dimensions are useful in providing an indication of how stakeholders can be managed.

High-power and interested groups must be fully engaged, with the greatest efforts made to satisfy them. High-power and less-interested groups require enough effort to keep them satisfied, but not so much that they become bored or irritated with the message. Low-power and interested groups need to be kept adequately informed, with checks to ensure that no major issues are arising. These groups may be very helpful with the detail of the project. Low-power and less-interested groups need monitoring, though without excessive communication. It is also worth noting that stakeholders can *move* positions during a project. As such, ongoing engagement plays a critical role in influencing the ways in which project stakeholders move 'towards' your position or 'away' from it. Therefore, regular

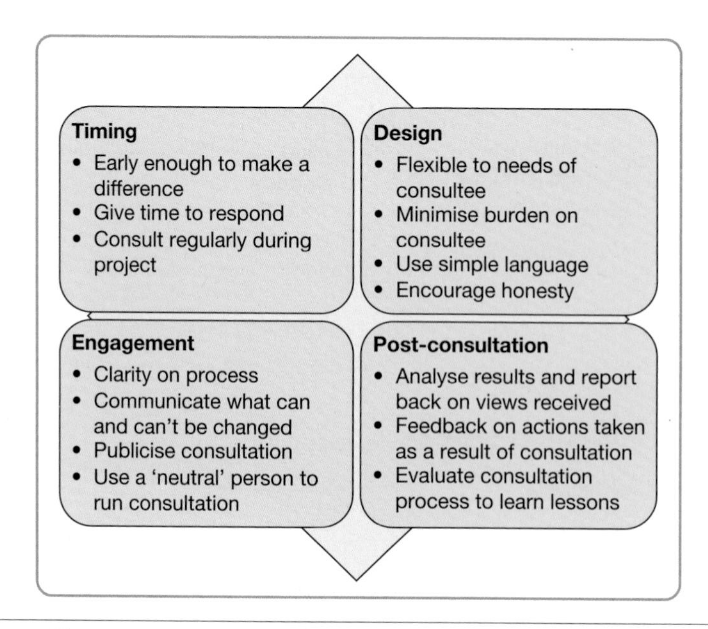

Figure 19.6 Ensuring effective consultation with project stakeholders

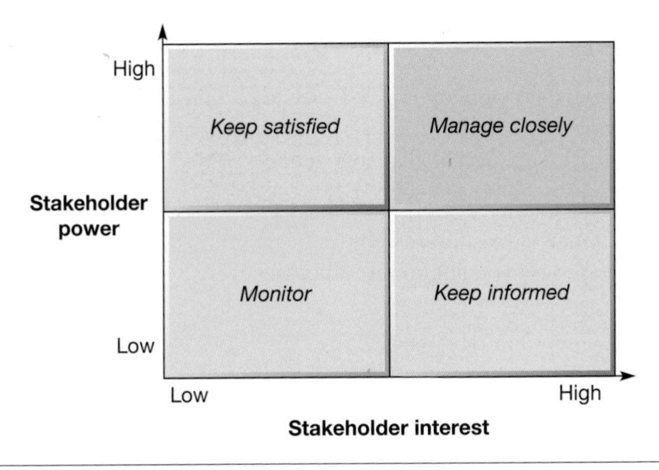

Figure 19.7 Managing project stakeholders based on power and interest

'health checks' with key stakeholders are advisable in projects (especially for projects that are completed over longer time frames).

Responsible operations

In every chapter, under the heading of 'Responsible operations', we summarise how the particular topic covered in the chapter touches upon important social, ethical and environmental issues.

While the importance of understanding and managing stakeholders is increasingly acknowledged in project management, there remain different perspectives on the fundamental role of both stakeholders and stakeholder management. Some argue that stakeholder satisfaction should *not* be seen as a goal in and of itself, but rather stakeholders should be managed only when they have a direct impact on the project's outcomes. From this perspective, stakeholder management is predominantly a *practical* consideration. So, to minimise objections and problems in a project, it makes sense to identify and consult a wide range of stakeholders. Communicating with stakeholders early and frequently can ensure that they fully understand the project and its potential benefits. Furthermore, project managers can use the opinions of powerful stakeholders to shape project scope and in doing so simultaneously improve its perceived quality and gain more resources.

In contrast, others take a much broader view of the core role of stakeholders in a project context. Those adopting this perspective argue that there is an ethical responsibility of those leading projects to look beyond simply satisfying the needs of stockholders. Instead, all stakeholders have 'ownership' of the project and maximising their welfare, alongside stockholders, is therefore a goal in and of itself. Adopting this 'stakeholder perspective' means putting those who have an effect on or are affected by a project at the centre of decision making. Arguably, it provides significant value in driving not only financial success for projects but also in offering benefits to all parties within the broader project context and to society as a whole.

Critics of the stakeholder perspective argue that it risks complicating managerial practice and fundamentally challenges the idea of the corporate objective function. Inevitably, stakeholder requirements are heterogeneous, dynamic, transitory, contradictory and often ambiguous. Yet, project resources are usually finite and difficult to change. Hence, there is an inevitable clash between stakeholder requirements on one hand and project resources on the other. Further, critics note that adopting a stakeholder perspective provides an easy excuse for poor project management decisions, as any decision can be presented retrospectively as an attempt to respond to stakeholder needs.

19.4 How are projects defined?

Before starting the complex task of planning and executing a project, it is essential to be clear about exactly what the project is – its definition. This is not always straightforward, especially for projects with many stakeholders. Three different elements define a project:

- its objectives – the end state that the project is trying to achieve;
- its scope – the exact range of the responsibilities taken on by the project;
- its strategy – how the project is going to meet its objectives.

Project objectives

Objectives help to provide a definition of the project's end point and can therefore be used to monitor progress and identify when success has been achieved. They can be judged in terms of the five performance objectives – quality, speed, dependability, flexibility and cost. However, flexibility is regarded as a 'given' in most projects that, by definition, are to some extent one-offs, and speed and dependability are typically compressed to one composite objective – 'time'. This results in what is known as the 'iron triangle of project management' – quality, time and cost. Although one objective might be particularly important, the other objectives can never be totally forgotten. As illustrated in Figure 19.8, as projects seek improved performance in one dimension, there is likely to be reduced performance in one or both remaining performance dimensions.

Operations principle
Different projects will place different levels of emphasis on cost, time and quality objectives.

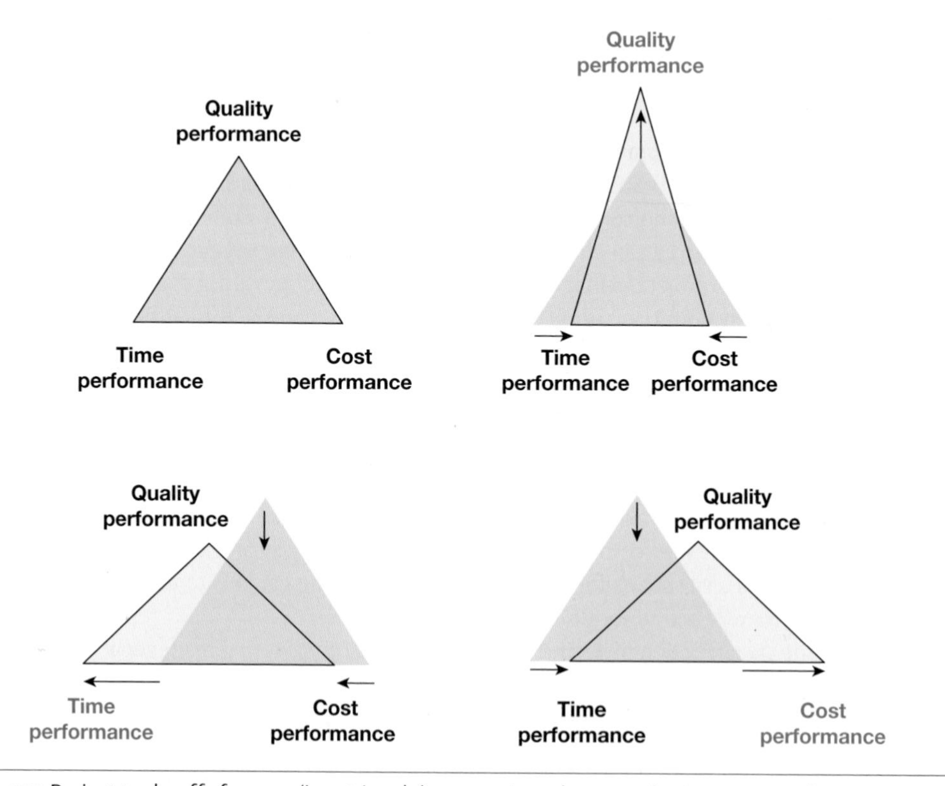

Figure 19.8 Project trade-offs from an 'iron triangle' perspective – how emphasis on one performance dimension impacts other performance dimensions

Critical commentary

When exploring project definition, we noted the of trade-offs between quality, time, and cost – the so-called iron triangle. So, for example, to speed up a project, we can expect increased costs (i.e. cost performance worsens) and possible reductions in deliverables (i.e. quality performance worsens). Although many find the iron triangle perspective useful in making trade-offs explicit during the scoping of projects, others argue that it creates a constrained mindset. When problems present themselves, they are not addressed creatively, but instead simply lead to requests for additional budget or time. In addition, some project scopes are simply unfeasible regardless of expanding time and budget envelopes. In other cases, adding budget and resource can sometimes actually *slow* activities given increased coordination and communication complexity effects.

Good objectives are those which are clear, measurable and, preferably, quantifiable. Clarifying objectives involves breaking down project objectives into three categories – the purpose, the end results and the success criteria. For example, a project that is expressed in general terms as 'improve the budgeting process' could be broken down into:

- Purpose – to allow budgets to be agreed and confirmed prior the annual financial meeting.
- End result – a report that identifies the causes of budget delay, and which recommends new budgeting processes and systems.
- Success criteria – the report should be completed by 30 June, meet all departments' needs and enable integrated and dependable delivery of agreed budget statements. Cost of the recommendations should not exceed $200,000.

Project scope

Project scoping is a boundary-setting exercise that attempts to define the dividing line between what each part of the project will do and what it won't do. Project scoping is critical and failure to scope appropriately or constantly changing scopes are one of the key reasons projects fail. Defining scope is particularly important when part of a project is being outsourced. A supplier's scope of supply will identify the legal boundaries within which the work must be done. Sometimes the scope of the project is articulated in a formal 'project specification'. This is the written, pictorial and graphical information used to define the output, and the accompanying terms and conditions. The project scope will also outline limits or exclusions to the project. This is critical, because perceptions of project success or failure often originate from the extent to which deliverables, limits and exclusions have been clearly stated and understood by all parties during the scoping phase.

OPERATIONS IN PRACTICE

The risk of changing project scope – sinking the *Vasa*

This example was written and kindly supplied by Professor Mattia Bianchi, Stockholm School of Economics.

Project specification changes, alongside poor communication and simple bad luck have always had the ability to bring down even the most high-profile projects. In 1628, the *Vasa*, the most magnificent warship ever built for the Royal Swedish Navy, was launched in front of an excited crowd. It had sailed less than few thousand metres during its maiden voyage in the waters of the Stockholm harbour when, suddenly, after a gun salute was shot in celebration, the *Vasa* heeled over. As water gushed in through the gun

ports, the ship vanished beneath the surface, killing 53 of the 150 passengers. Shocked officials were left questioning how such a disaster could happen.

Yet, as a project, the story of the *Vasa* displayed many of the signs of potential failure. When her construction began in 1625, the *Vasa* was designed as a small traditional warship, like many others previously built by the experienced shipbuilder Henrik Hybertsson. Soon after, the Swedish King, Gustav II Adolphus, at that time fighting the Polish Navy in the Baltic Sea, started ordering a series of changes to the shape and the size of the warship, making its design much longer and bigger than originally envisaged. In addition, the King's spies informed him that the Danes had started building warships with two gun decks, instead of the customary one. This would give them a great advantage in terms of superior firepower from a longer distance. From the battlefront, the King ordered the addition of a second gun deck to the *Vasa*. The message caused consternation when it reached the shipbuilder several months later, but they attempted to comply with the change even though it caused wasteful reworking and complex patching up. Yet more pressure was put on the project when a major storm destroyed ten of the King's ships, making the commissioning of the *Vasa* even more urgent. Then, as a final piece of bad luck (especially for him) the shipbuilder, Hybertsson, died. Nevertheless, just before the ship's completion, a Navy representative, Admiral Fleming, conducted a stability test to assess the seaworthiness of the ship. Notwithstanding the strong signals of instability, the *Vasa* was launched on its maiden voyage – with disastrous results for the King, for the Swedish Navy and the project. The example highlights the major risks of interventions in projects to (radically) change their scope. In this case, not only was *Vasa*'s specification changed, but the schedule was compressed, creating a high risk of project failure.

Project strategy

The third part of a project's definition is the project strategy, which defines, in a general rather than a specific way, how the project is going to meets its objectives. It does this in two ways; by defining the phases of the project, and by setting **milestones**, and/or 'stage gates'. Milestones are important events during the project's life. Stage gates are the decision points that allow the project to move on to its next phase. A stage gate often launches further activities and therefore commits the project to additional costs, etc. Milestone is a more passive term, which may herald the review of a part-complete project or mark the completion of a stage, but does not necessarily have more significance than a measure of achievement or completeness. At this stage, the actual dates for each milestone are not necessarily determined.

However, identifying the significant project milestones and stage gates is very helpful in supporting discussions with key stakeholders and in clarifying boundaries between project phases.

19.5 How are projects planned?

All projects, even the smallest, need some degree of planning. The planning process fulfils four distinct purposes:

- It determines the cost and duration of the project. This enables major decisions to be made, including the decision whether to actually go ahead with the project.
- It determines the level of resources that will be needed.
- It helps to allocate work and to monitor progress. Planning must include the identification of who is responsible for what.
- It helps to assess the impact of any changes to the project.

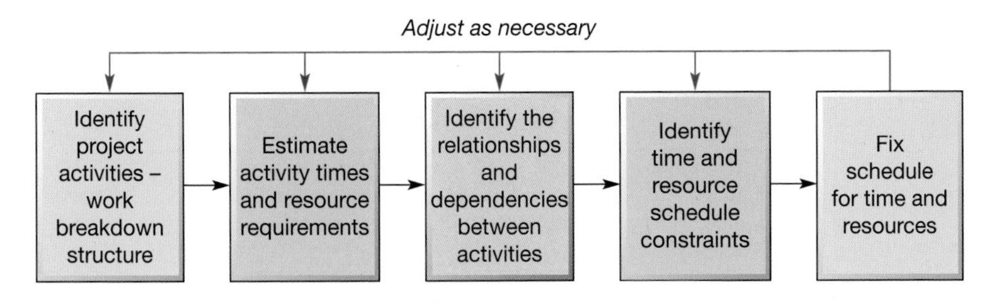

Figure 19.9 Stages in the planning process

Planning is not a one-off process. It may need repeating several times during the project's life as circumstances change. Nor is re-planning a sign of project failure or mismanagement. As discussed earlier, projects can and should be differentiated based on their characteristics – in our case, we have examined the level of innovation, pace and complexity. When managing particularly difficult projects, it is therefore a normal occurrence to revise plans as the project progresses. Figure 19.9 shows the five steps involved in the project planning process.

Operations principle

Project planning is essential for all types of projects, but especially those with higher levels of innovation, pace or complexity.

Identify project activities – work breakdown structure

Some projects are too complex to be planned and controlled effectively unless they are first broken down into manageable portions. This is achieved by structuring the project into a 'family tree' that specifies major tasks or sub-projects. These in turn are divided up into smaller tasks until a defined, manageable series of tasks, called a *work package*, is arrived at. Each work package can be allocated its own objectives in terms of time, cost and quality. Typically, work packages do not exceed 10 days, should be independent from each other, should belong to one sub-deliverable, and should constantly be monitored. The output from this is called the work breakdown structure (WBS). The WBS gives clarity and definition to the project planning process and provides a framework for building up information for reporting purposes.

Example project

As a simple example to illustrate the application of each stage of the planning process, let us examine the following domestic project. The project definition is:

- *purpose* – to make breakfast in bed;
- *end result* – breakfast in bed of boiled egg, toast, and orange juice;
- *success criteria* – plan uses minimum staff resources and time, and product is high quality (egg freshly boiled, warm toast, etc.);
- *scope* – project starts in kitchen at 6.00 am and finishes in bedroom; needs one operator and normal kitchen equipment.

The work breakdown structure is based on the above definition and can be constructed as shown in Figure 19.10.

Estimate activity times and resource requirements

The second stage in project planning is to identify the time and resource requirements of the work packages. Without some idea of how long each part of a project will take and how many resources it will need, it is impossible to define what should be happening at any time during the execution of the project. Estimates are just that, however – a systematic best guess, not a perfect forecast

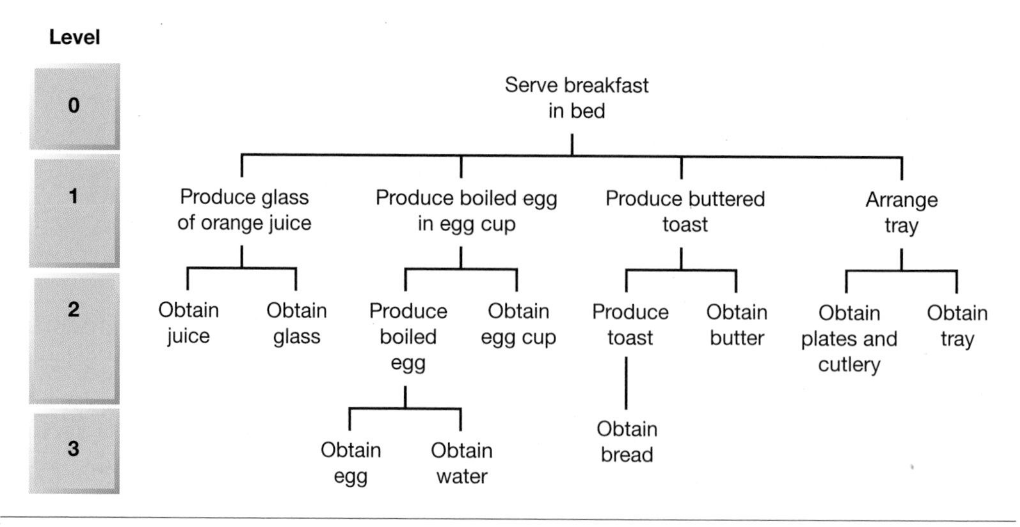

Figure 19.10 A work breakdown structure for a simple domestic project

of reality. Estimates are rarely perfect, but they can be made with some idea of how accurate they might be.

Example project

Returning to our very simple example 'breakfast-in-bed' project, the activities were identified, and times estimated as in Table 19.2. While some of the estimates may appear generous, they consider the time of day and the state of the operator.

There are two approaches typically taken in estimating time or resource needs for a project. Top-down estimates look at the project as a whole and typically use an *analogy approach* (for example, estimating the time of an NPSD project, based on previous similar projects), a *ratio approach* (for example, estimating the cost of building a new house using a cost per square metre calculation), or a *consensus approach* (where a group of experts discuss the project to form a best estimate). Top-down methods are most commonly adopted when very precise estimations are not required or not possible (for example, for highly uncertain projects). Bottom-up approaches to estimation focus on breaking down the project into smaller parts and then estimating the time or resource requirements for each of these parts. With bottom-up estimation, project managers are typically relying on those who will actually be doing the work to come up with an accurate estimate.

Table 19.2 Time and resources estimates for a 'breakfast-in-bed' project

Activity	Effort (person-min)	Duration (min)
Butter toast	1	1
Pour orange juice	1	1
Boil egg	0	4
Slice bread	1	1
Fill pan with water	1	1
Bring water to boil	0	3
Toast bread	0	2
Take loaded tray to bedroom	1	1
Fetch tray, plates, cutlery	1	1

Critical commentary

When project managers talk of 'estimates', they are really talking about guessing. Planning a project happens in advance of the project itself. Therefore, no one really knows how long each activity will take or what it will cost. Of course, some kind of guess is needed for planning purposes. However, some project managers believe that too much faith is put in time and cost estimates. The really important questions, they claim, are how long something *could* take without delaying the whole project and how much something *could* cost without it harming the project's viability. Also, if a single most likely estimate is unreliable, then using three, as one does for probabilistic estimates (see PERT, later), is merely over-analysing what is highly dubious data in the first place. It is also important for project managers to be aware of the likely biases that they and their team may be affected by when developing estimates for time and costs. For example, project planners may be affected by:

- *Anchoring bias* – An over-reliance on an initial piece of information that 'anchors' subsequent judgements (for example, the anchoring effect of an initial cost estimate for a project).
- *Bandwagon effects* – A form of groupthink where individuals believe something because others do (for example, making a similar time estimate to those of other members of the project team).
- *Recency bias* – An over-reliance on more recent forms of information relative to older forms of information (for example, estimates of project supplier risk dominated by experiences from the last completed project).
- *Confirmation bias* – The tendency to search for and select pieces of information that confirm, rather than refute, a given position (for example, purposively selecting examples of activity time completions from previous projects that support a time estimate being made for a new project).

Identify the relationships and dependencies between activities

The third stage of project planning is to understand the interactions between different project work packages. Some activities will, by necessity, need to be executed in a particular order. For example, in the construction of a house, the foundations must be prepared before the walls are built, which in turn must be completed before the roof is put in place. These activities have a dependent or series relationship. Other activities do not have any such dependence on each other. The rear garden of the house could probably be prepared totally independently of the garage being built. These two activities have an independent or parallel relationship.

Gantt charts

Project planning is greatly aided using techniques that help to handle time, resource and relationship complexity. The simplest of these techniques is the *Gantt chart* (or bar chart) that we introduced in Chapter 10. Figure 19.11 shows a Gantt chart for the activities that form the sales system interface project. The bars indicate the start, duration and finish time for each activity. Gantt charts have excellent visual impact, are easy to understand, and they are useful for communicating project status with stakeholders.

Network analysis

As project complexity increases, it becomes more necessary to clearly identify the relationships between activities and to show the logical sequence in which activities must take place. This is commonly done using the **critical path method (CPM)** to clarify the relationships between activities diagrammatically. Though there are alternative methods of carrying out critical path analysis, by far the most common, and the one used in most project management software packages, is the 'activity

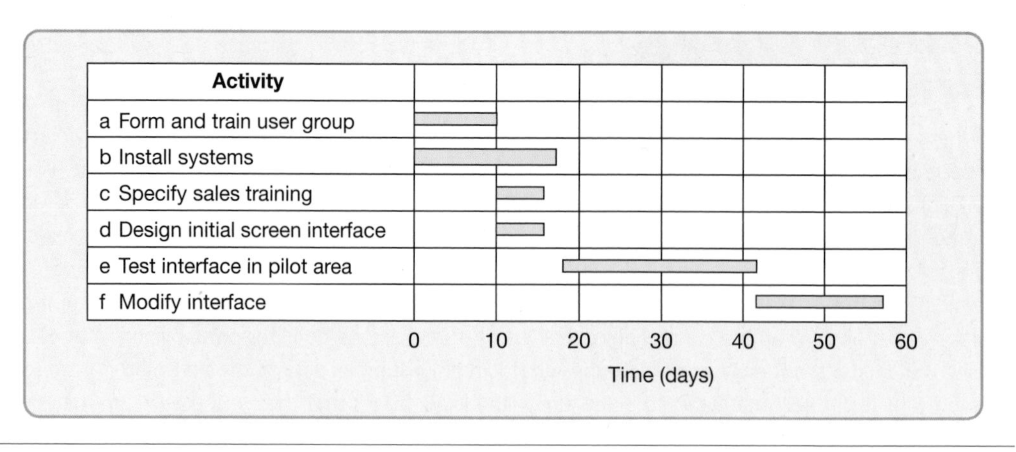

Figure 19.11 Gantt chart for the project to design an information interface for a new sales knowledge management system in an insurance company

on node' (AoN) method. For example, Table 19.3 shows the activities, time estimates, precedence relationships and resources needed (in terms of the number of IT developers) for one phase of a new sales knowledge management system that is being installed in an insurance company.

Figure 19.12 shows the critical path analysis for this project. Activities are drawn as boxes, and arrows are used to define the relationships between them. In the centre of each box is the description of the activity (in this case, 'Activity a', 'Activity b' and so on). Above the description is the duration (D) of the activity (or work package), the earliest start time (EST) and earliest finish time (EFT). Below the description is the latest start time (LST), the latest finish time (LFT), and the 'float' (F) (the number of extra days that the activity could take without slowing down the overall project). The diagram shows that there are several chains of events that must be completed before the project can be considered as finished. In this case, activity chains a–c–f, a–d–e–f and b–e–f must all be completed before the project can be considered as finished. The longest (in duration) of these chains of activities is called the 'critical path' because it represents the shortest time in which the project can be finished, and therefore dictates the project timing. In this case b–e–f is the longest path and the earliest the project can finish is after 57 days.

Activities that lie on the critical path will have the same earliest and latest start times, and earliest and latest finish times. That is why these activities are critical. Non-critical activities, however, have some flexibility as to when they start and finish. This flexibility is quantified into a figure that is known either as 'float' or 'slack'. So, activity c, for example, is only of 5 days duration and it can start any time after day 10 (when activity a is completed) and must finish any time before day 42 (when activities a, b, c, and d are completed). Its 'float' is therefore (42 − 10) − 5 = 27 days (i.e. latest finish time minus earliest start time minus activity duration). Obviously, activities on the critical path have no float; any change or delay in these activities would immediately affect the whole project.

Table 19.3 Time, resource and relationships for the sales system interface design project

Code	Activity	Immediate predecessor(s)	Duration (days)	Resources (developers)
a	Form and train user group	none	10	3
b	Install systems	none	17	5
c	Specify sales training	a	5	2
d	Design initial screen interface	a	5	3
e	Test interface in pilot area	b, d	25	2
f	Modify interface	c, e	15	3

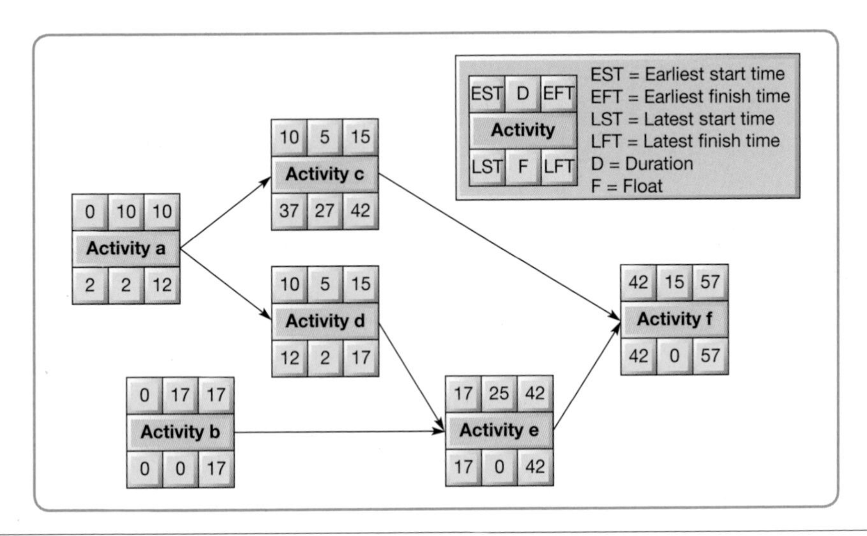

Figure 19.12 Critical path analysis for the project to design an information interface for a new sales knowledge management system in an insurance company

In addition to a critical path (or network) diagram, the idea of float or slack can be shown diagrammatically on a Gantt chart, as in Figure 19.13. Here, the Gantt chart for the project has been revisited, but this time the time available to perform each activity (the duration between the earliest start time and the latest finish time for the activity) has been shown.

Identify time and resource schedule constraints

Once estimates have been made of the time and effort involved in each activity, and their dependencies identified, it is possible to compare project requirements with the available resources. The finite nature of critical resources – such as staff with special skills – means that they should be considered in the planning process. This often has the effect of highlighting the need for more detailed re-planning.

The logic that governs project relationships, as shown in the critical path analysis (or network diagram), is primarily derived from the technical details, but the availability of resources may also impose its own constraints, which can materially affect the relationships between activities. For example, specialist staff may not have the available time to carry out two tasks simultaneously even if the critical analysis has identified that two activities can *technically* run in parallel.

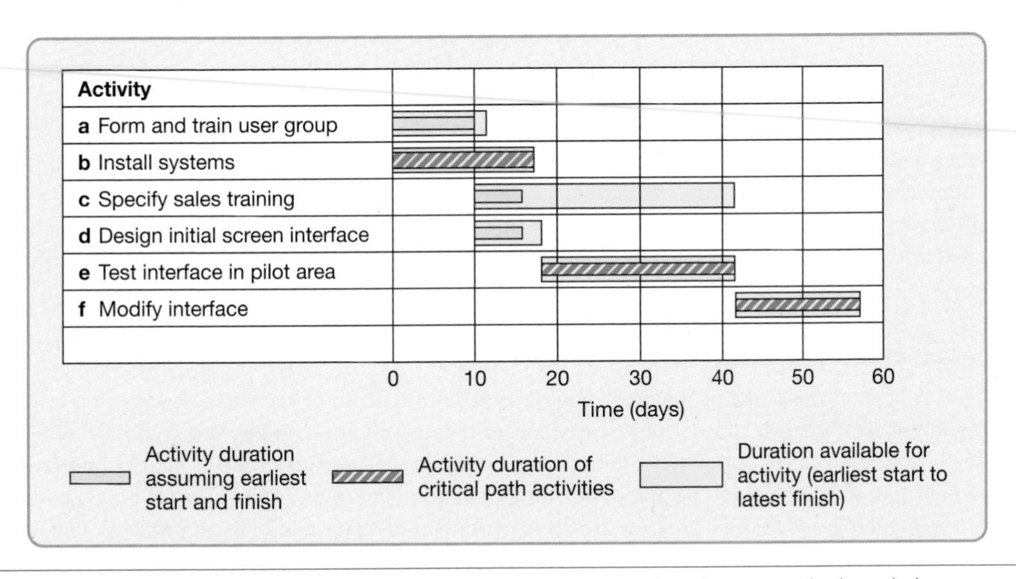

Figure 19.13 Gantt chart for the project to design an information interface for a new sales knowledge management system in an insurance company with latest and earliest start and finish times indicated

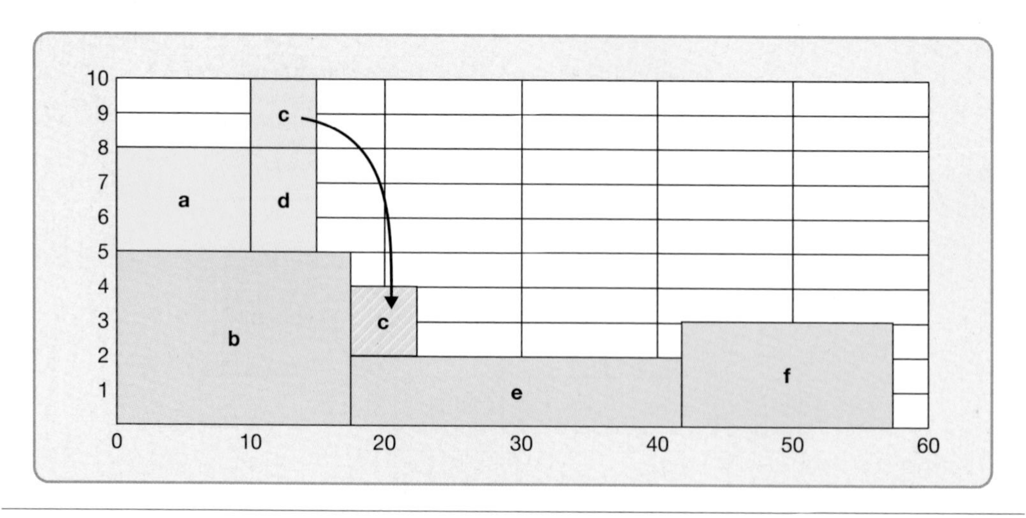

Figure 19.14 Resource profiles for the sales knowledge system interface design, assuming that all activities are started as soon as possible, and assuming that the float in activity c is used to smooth the resource profile

Returning to the sales system interface design project, Figure 19.14 shows the resource profile under two different assumptions. The critical path activities (b–e–f) form the initial basis of the project's resource profile. These activities have no float and can only take place as shown. However, activities a, c and d are not on the critical path, so project managers have some flexibility as to when these activities occur, and therefore when the resources associated with these activities will be required. From Figure 19.14, if one schedules all activities to start as soon as possible, the resource profile peaks between days 10 and 15 when 10 IT development staff are required. However, if the project managers exploit the float for activity c and delays its start until after activity b has been completed (day 17), the number of IT developers required by the project does not exceed 8. In this way, float can be used to smooth resource requirements or make the project fit resource constraints.

Fix schedule for time and resources

Project planners should ideally have several alternatives to choose from. The one that best fits project objectives can then be chosen or developed. While it can be challenging to examine several alternative schedules, especially in very large or very uncertain projects, computer-based software packages such as Bitrix24, Trello, 2-Plan PMS, Asana, MS Project and Producteev make critical path optimisation more feasible. The rather tedious computation necessary in network planning can be performed relatively easily by project planning models. All they need are the basic relationships between activities together with timing and resource requirements for each activity. Earliest and latest event times, float and other characteristics of a network can then be presented, often in the form of a Gantt chart. More significantly, the speed of computation allows for frequent updates to project plans. Similarly, if updated information is both accurate and frequent, such software can provide effective project control data.

Program evaluation and review technique (PERT)

While it is beyond the scope of this text to enter much more detail of the various ways that critical path analysis can be made more sophisticated, programme evaluation and review technique (PERT) is worth noting given its popularity among practising project managers. PERT, as it is universally known, originated in the planning and controlling of major defence projects in the US Navy, with its most spectacular gains in the highly uncertain environment of space and defence projects. The technique recognises that activity durations and costs in project management are not deterministic (fixed), but instead we can use a probability curve to describe the estimate. The natural tendency of some people is to produce optimistic estimates, but these will have a relatively low probability of being correct because they represent the time or cost if everything goes very well. Most likely estimates have the highest probability of proving correct. Finally, pessimistic estimates assume that almost everything that could go wrong does go wrong. As shown in Figure 19.15, it is assumed that these time estimates are consistent with a beta probability distribution, where the mean and variance of the distribution can be estimated as follows:

Operations principle

Probabilistic activity time estimates facilitate the assessment of a project being completed on time.

$$t_e = \frac{t_o + 4t_l + t_p}{6}$$

where

t_e = the expected time for the activity

t_o = the optimistic time for the activity

t_l = the most likely time for the activity

t_p = the pessimistic time for the activity.

The variance of the distribution (V) can be calculated as follows:

$$V = \frac{(t_p - t_o)^2}{6^2} = \frac{(t_p - t_o)^2}{36}$$

The time distribution of any path through a network will have a mean, which is the sum of the means of the activities that make up the path, and a variance, which is a sum of their variances. In Figure 19.15:

$$\text{The mean of the first activity (a)} = \frac{2 + (4 \times 3) + 5}{6} = 3.17$$

$$\text{The variance of the first activity (a)} = \frac{(5 - 2)^2}{36} = 0.25$$

$$\text{The mean of the second activity} = \text{(b)} \frac{3 + (4 \times 4) + 7}{6} = 4.33$$

$$\text{The variance of the second activity (b)} = \frac{(7 - 3)^2}{36} = 0.44$$

$$\text{The mean of the network distribution} = 3.17 + 4.33 = 7.5$$

$$\text{The variance of the network distribution} = 0.25 + 0.44 = 0.69$$

The advantage of this extra information is that we can examine the 'riskiness' of each path through a network, as well as its duration. Given the increased attention on risk management within project management since the turn of the century, this is essential. For example, Figure 19.16 shows a simple two-path network. The top path is the critical one; the distribution of its duration has a mean of 14.5 with a variance of 0.22. The distribution of the non-critical path has a lower mean of 12.66 but a far higher variance of 2.11. The implication of this is that there is a chance that the non-critical path could in reality be critical. Although we will not discuss the probability calculations here, it is possible to determine the probability of any sub-critical path turning out to be critical when the project actually takes place. However, on a practical level, even if the probability calculations are judged not to be worth the effort involved, it is useful to be able to make an approximate assessment of the riskiness of each part of a network.

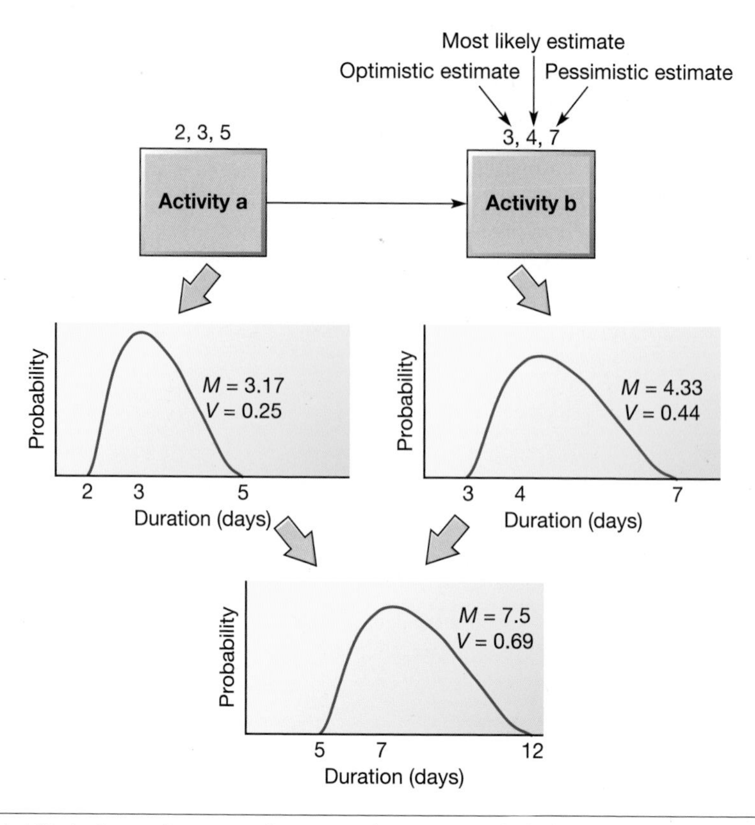

Figure 19.15 Probabilistic time estimates can be summed to give a probabilistic estimate for the whole project

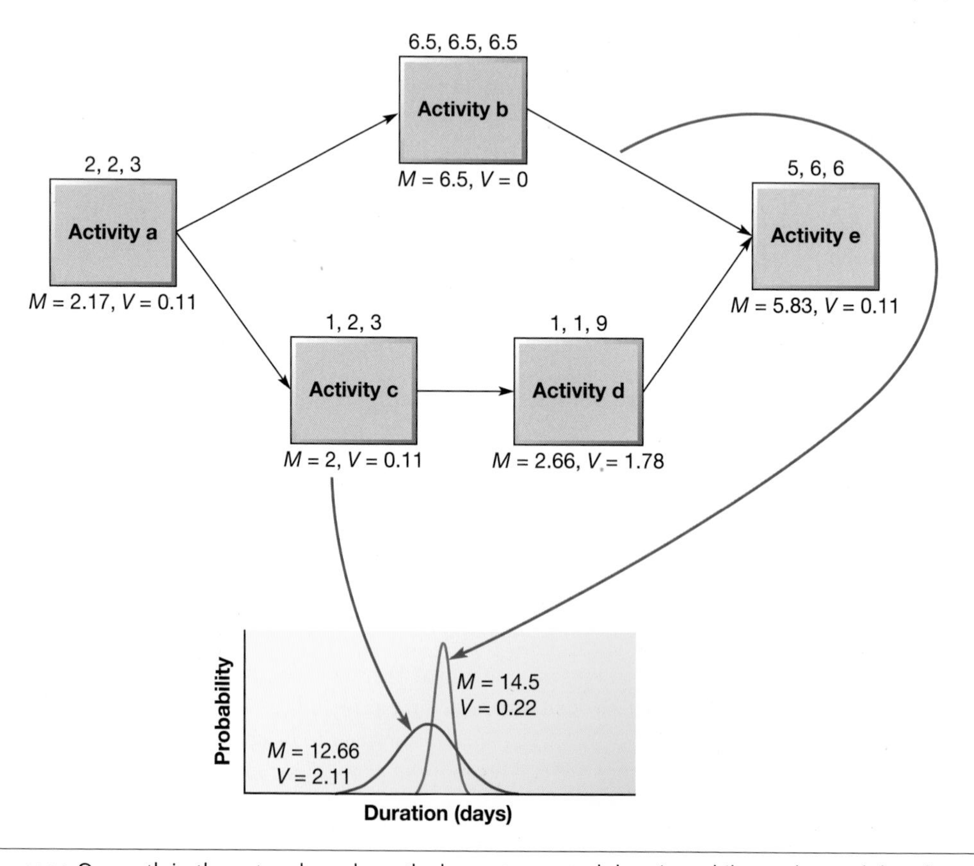

Figure 19.16 One path in the network can have the longest expected duration while another path has the greater variance

19.6 How are projects controlled and learned from?

Understanding the project environment, project definition and project planning stages of project management largely take place before the actual project begins. By contrast, project control and learning deals with activities that take place during the execution of the project and after it ends. It involves five key challenges:

- How to *monitor* the project to check on its progress.
- How to *assess the performance* of the project by comparing monitored observations of the project with the project plan.
- How to *intervene* in the project to make the changes that will bring it back to plan.
- How to *manage matrix tensions* in the project to reconcile the interests of both the project and different organisational functions.
- How to *learn* from the project to improve performance in subsequent projects.

OPERATIONS IN PRACTICE

Ocado's robotics projects[5]

Ocado, the online grocery retailer, remains at the forefront of technology to support its growing operations. While the use of automated warehouse systems is by no means a new phenomenon, the pace of technology adoption has risen sharply in recent years because of both rising labour costs and the availability of better and more cost-effective technologies. Ocado has several projects seeking to leverage new technological opportunities. One is the development of advanced packing robots, capable of handling heavy or hazardous products (to avoid worker injuries) as well as delicate objects, such as fruits, vegetables, salads and eggs. Another recent technology project for Ocado is the development of a humanoid assistant (think of C-3PO from the *Star Wars* movies, but with wheels instead of legs!) aimed at supporting engineers in the maintenance of its product-handling systems. In partnership with the Karlsruhe Institute of Technology in Germany, Ecole Polytechnique Fédérale de Lausanne in Switzerland, University College London in the United Kingdom and Sapienza University in Italy, these robots offer a 'second pair of hands' to engineers, moving tools and materials, and handing them to their human partners as needed. They are also capable of interrupting human actions to offer advice on alternative solutions to common problems. According to Ocado, the aim is to create a fluid and natural interaction between robot and technician within its operations. These two examples point to the continually changing nature of the workplace, as technology is increasingly integrated within many tasks. It also highlights the value (and challenges) of bringing different areas of expertise from geographically dispersed partners to deliver project success.

Project monitoring

Project managers have first to decide what they should be looking for as the project progresses. Common measures include current expenditure to date, supplier price changes, amount of overtime authorised, technical changes to the project, inspection failures, number and length of delays, activities not started on time, missed milestones, etc. Some of these monitored measures affect mainly cost, some mainly time. However, when something affects the quality of the project, there are also time and cost implications. This is because quality problems in project planning and control must usually be solved in a limited amount of time.

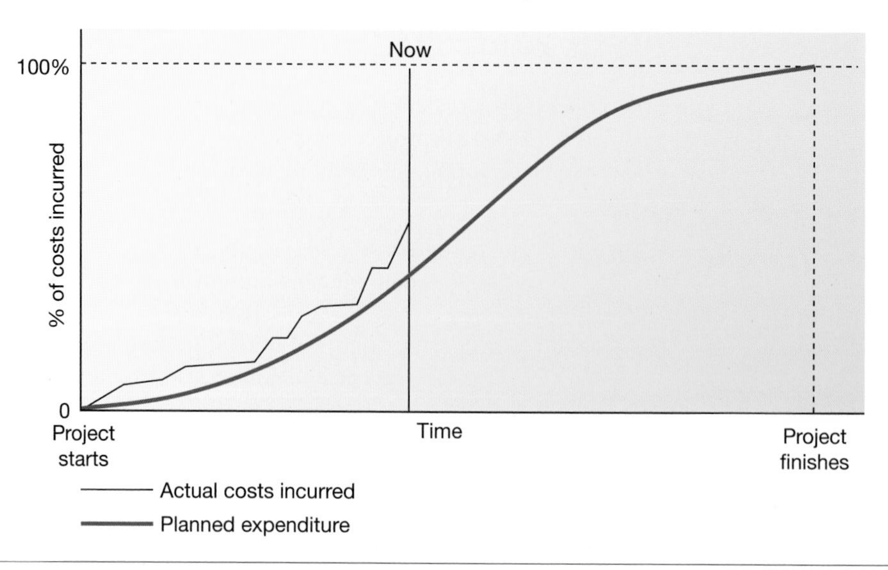

Figure 19.17 Comparing planned and actual expenditure

Assessing project performance

The monitored measures of project performance at any point in time need to be assessed so that the project team can make a judgement concerning overall performance. A typical planned cost profile of a project through its life is shown in Figure 19.17. At the beginning of a project some activities can be started, but most activities will be dependent on others finishing. Eventually, only a few activities will remain to be completed. This pattern of a slow start followed by a faster pace with an eventual tail-off of activity holds true for almost all projects, which is why the rate of total expenditure typically follows an S-shaped pattern, even when the cost curves for the individual activities are linear. It is against this curve that actual costs can be compared to check whether the project's costs are being incurred to plan. Figure 19.17 shows the planned and actual cost figures compared in this way. It shows that the project is incurring costs, on a cumulative basis, ahead of what was planned.

Earned value analysis

Earned value analysis (EVA) is a technique that allows for various comparisons to be made between expected costs and schedules, and the actual performance of a project. Table 19.4 illustrates an EVA for a simple project. This technique is not only useful for determining how well a project is progressing in terms of tasks completed and costs incurred, but it also helps in re-evaluating original budgets and time schedules. In this case (as of week 6 when the EVA was carried out), the project is running 11.4% over budget and 19.5% behind schedule. If things continue in this way for the reminder of the project, it is likely to cost around €117,381 (as opposed to the original budget of €104,000) and be delivered in 12.4 weeks (as opposed to the planned 10 weeks).

Intervention in projects

If the project is obviously out of control in the sense that its costs, quality levels or times are significantly different from those planned, then intervention is almost certainly required. Given the interconnected nature of projects, interventions often require wide consultation. Sometimes intervention is needed even if the project looks to be proceeding according close to plan. For example, the schedule and cost for a project may seem to be 'to plan', but when the project managers project activities and cost into the future, they see that problems are very likely to arise. In this case it is the *trend* of performance that is being used to trigger intervention.

Table 19.4 Earned value analysis (EVA)

Activity	Planned time	Planned budget
1	1 week	€5,500
2	1 week	€8,750
3	1 week	€6,250
4	1 week	€11,000
5	1 week	€15,000
6	1 week	€11,250
7	1 week	€13,750
8	1 week	€9,000
9	1 week	€14,000
10	1 week	€9,500
TOTAL	**10 weeks**	**€104,000**

PROJECT REVIEW – END WEEK 6:

* Work complete: Activities 1–5
* Actual cost (AC) at time of review = €52,500
* Planned value (PV) = sum of weeks 1–6 = €57,750
* Earned Value (EV) = sum of activities completed = €46,500

COST review:

* Cost variance (CV) = EV – AC = €46,500 – €52,500 = (€6,000) (Negative CV, overspend)
* Cost performance index (CPI) = EV/AC = €46,500/€52,500 = 0.886 (CPI < 1 overspend)
* Estimate at completion (EAC) = Budget at completion (BAC) / CPI = €104,000 / 0.886 = €117,381

SCHEDULE review:

* Schedule variance (SV) = EV – PV = €46,500 – €57,750 = (€11,250) (Negative SV, behind schedule)
* Schedule performance index (SPI) = EV / PV = €46,500 / €57,750 = 0.805 (SPI < 1, behind schedule)
* Estimated time to complete (ETC) = Original time estimate / SPI = 10 / 0.805 = 12.4 weeks

Crashing or accelerating activities

One common form of intervention is to 'crash' activities. Crashing is the process of reducing time spans for critical path activities so that the project can be completed in less time. Crashing activities incurs extra costs in terms of overtime working, additional resources or subcontracting work. Figure 19.18 shows an example of crashing a simple network. For each activity, the duration and normal cost are specified, together with the (reduced) duration and (increased) cost of crashing them. Not all activities are capable of being crashed; here activity e cannot be crashed. The critical path is the sequence of activities a, b, c, e. If the total project time is to be reduced, one of the activities *on the critical path* must be crashed. To decide which activity to crash, the 'cost slope' of each is calculated and the activity on the critical path with the lowest cost slope is typically selected. This is activity a, the crashing of which will cost an extra €2,000 and will shorten the project by one week. After this, activity c can be crashed, saving a further two weeks, and costing an extra €5,000. At this point all the activities have become critical and further time savings can only be achieved by crashing two activities in parallel. The shape of the time–cost curve is entirely typical. Initial time savings come relatively inexpensively but further savings are usually more costly.

Operations principle
Only accelerating activities on the critical path(s) will accelerate the whole project.

Managing matrix tensions

In all but the simplest project, project managers usually need to reconcile the interests of both the project itself and the departments contributing resources to the project. When calling on a variety of resources from various departments, projects are operating in a 'matrix management' environment, where projects cut across organisational boundaries and involve staff that are required to report to their own line manager as well as to the project manager. Figure 19.19 illustrates the type of reporting relationship that usually occurs in matrix management structures running multiple projects. A person in department 1, assigned part-time to projects A and B, will be reporting to three different managers,

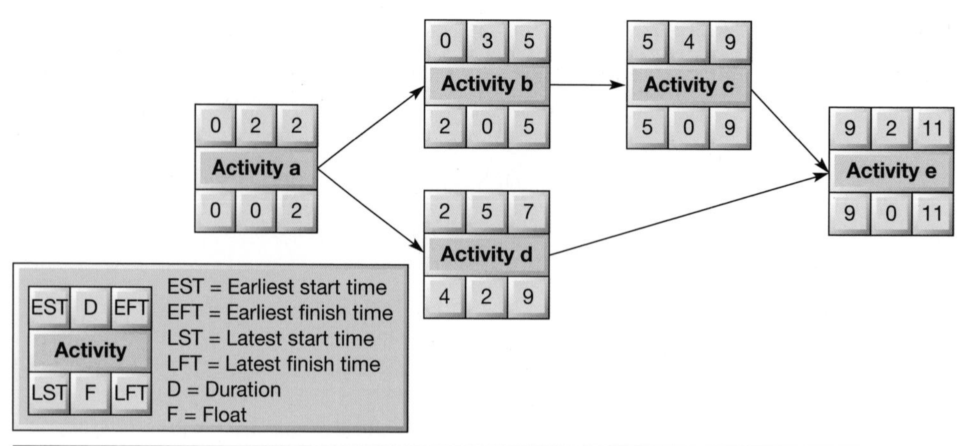

Activity	*Normal*		*Crash*		*Cost slope (€000/week)*
	Cost (€000)	*Time (weeks)*	*Cost (€000)*	*Time (weeks)*	
a	6	2	8	1	2
b	5	3	8	2	3
c	10	4	15	2	2.5
d	5	5	9	4	4
e	7	2	Not possible		–

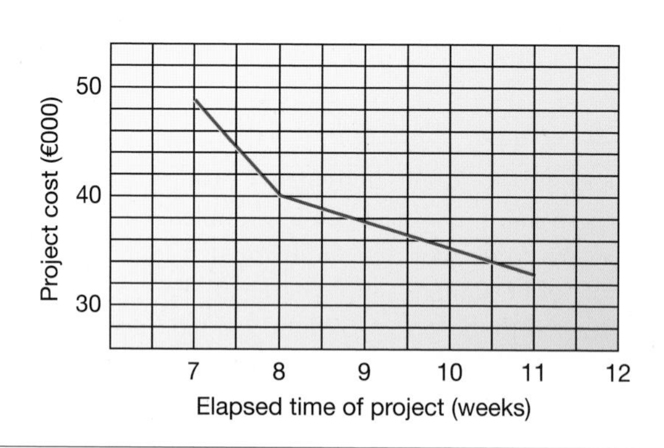

Figure 19.18 Crashing activities to shorten project time becomes progressively more expensive

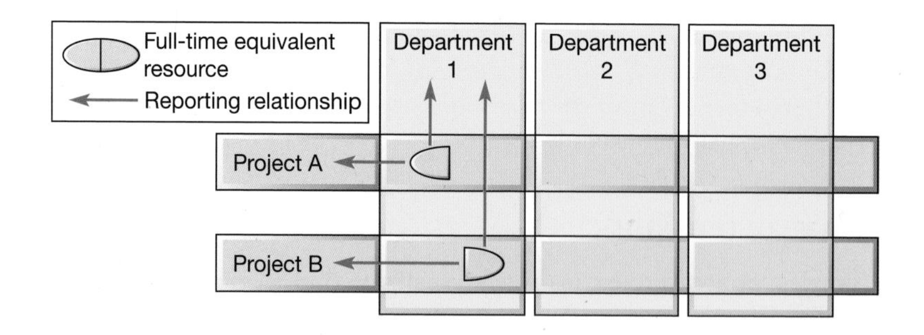

Figure 19.19 Matrix management structures often result in staff reporting to more than one project manager as well as their own department

all of whom will have some degree of authority over their activities. Therefore, matrix management requires a high degree of cooperation and communication between all individuals and departments. Although decision-making authority will formally rest with either the project or departmental manager; most major decisions will need some degree of consensus. To function effectively, matrix management structures should have the following characteristics:

- Effective channels of communication between all managers involved, with relevant departmental managers contributing to project planning and resourcing decisions.
- Formal procedures in place for resolving the management conflicts that do arise.
- Encouragement for project staff to feel committed to their projects as well as to their own department.
- Sufficient time devoted to planning the project and securing the agreement of the line managers to deliver on time and within budget.

Managing project learning

The activity of project management doesn't stop when a project comes to an end – managing the process of project learning is key to future project success. Yet, within most projects, there remains very little formalised learning. This can partly be explained by the key performance objectives for individuals involved in projects – typically focused on the success of an individual project (in terms of quality, time, cost) as opposed to longer-term learning effects and the development of organisational capabilities. As a result, when the project ends, there may be little incentive for stakeholders to spend time reviewing aspects of the project's execution that could have been improved. In addition, when things go wrong, those involved often prefer to move on rather than go back and examine failures. Where organisations have limited formalised learning mechanisms as part of their project processes, there will be little change in the underlying *average* performance of their projects over time (see Figure 19.20 top). However, where organisations place a greater emphasis on formalised learning from one project to another, there is typically an *upwards trend* in project performance, despite variance in project performance remaining (see Figure 19.20 bottom).

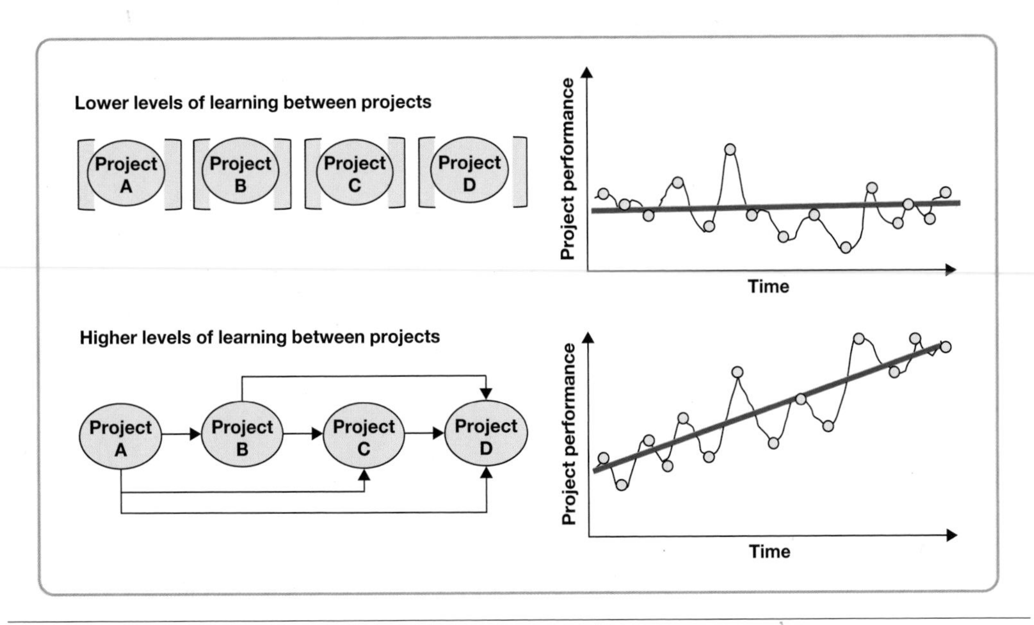

Figure 19.20 Improving project performance over time through learning between projects

Summary answers to key questions

19.1 What are projects?

- A project is a temporary activity aimed at achieving a specific and highly customised goal, within a set time frame, using a defined group of resources.
- Projects involve many non-routine and complex tasks, which means projects are often highly uncertain.
- The process of managing a project involves not only understanding their common characteristics, but also key differences from one project to the next. In this chapter, we focus on differences in the level of *innovation*, *time pressure* and *complexity* of the project being managed.

19.2 What is project management?

- Project management is the activity of understanding the project environment, defining, planning, controlling and learning from projects.
- Beyond the 'life cycle' perspective, project management is also concerned with effectively balancing quality (deliverables), time and cost objectives within the so-called 'iron triangle'.
- From an organisational perspective, project management involves managing these life cycles and performance objectives across multiple functions within an organisation.

19.3 How is the project environment understood?

- The project environment comprises all the factors that may affect the project during its life. These factors include the internal environment, business environment, economic–political environment and geo-social environment.
- Stakeholder management is a key role for project managers. This activity involves identifying stakeholders, understanding their different perspectives, and managing different, and often competing, interests.

19.4 How are projects defined?

- Defining a project involves three related activities – setting project objectives, scoping the project and developing a project strategy.
- Most projects can be defined by the relative importance of three objectives. These are cost – keeping the overall project to its original budget, time – finishing the project by the scheduled finish time, and quality – ensuring that the project outcome is as was originally specified.

19.5 How are projects planned?

- Project planning involves more detailed analysis to help determine the cost and duration of the project and the level of resources that will be required.

- Project planning involves five stages:
  - Identifying the activities within a project.
  - Estimating times and resources for project activities.
  - Identifying the relationships and dependencies between project activities.
  - Identifying time and resource schedule constraints.
  - Fixing the schedule for time and resources.

19.6 How are projects controlled and learned from?

- Project control and learning deals with the management activities that take place during the execution of the project and after its completion. It involves five key challenges:
  - How to *monitor* the project to check on its progress.
  - How to *assess the performance* of the project by comparing monitored observations of the project with the project plan.
  - How to *intervene* in the project to make the changes that will bring it back to plan, including crashing or accelerating activities.
  - How to *manage matrix tensions* in the project to reconcile the interest of both the project and different organisational functions.
  - How to *learn* from the project to improve performance in subsequent projects.

CASE STUDY Kloud BV and Sakura Bank K.K.

(This case was co-authored with Nigel Spinks, Henley Business School, University of Reading)

'Well, that's the bad news!', said Tao, the Managing Director of Kloud BV, a consulting and executive development firm headquartered in Amsterdam, specialising in operations and supply chain improvement. *'The good news is that Chao should be out of hospital in a couple of weeks. It may take a few months before he's fully fit, but it all looks very promising'*. Maria was pleased to hear that things were looking more positive for Chao after his accident. She had only been at the company for six weeks, having taken up a role as a junior project manager, but had already grown to respect and like Chao.

'But', continued Tao, *'that does leave us in a tricky situation. As you know, Chao oversaw the big project with Sakura Bank in Tokyo, which I'm going to look after until he's back at work. He was also just setting up a smaller project for them, training senior managers, which will run out of their facilities in Osaka. I appreciate you're pretty new here, but I'd like you to take on the project management for this one. Chao recommended you, so it seems you've made a very good impression!'* Maria was pleased to hear that Chao, her immediate boss, had a good impression of her. *'Well, I'm very happy to take this on Tao'*, she said, as she quickly looked through the draft proposal for the project that Chao had been developing for Sakura Bank just before his accident (Figure 19.21).

As Maria read through the proposal, she got a clearer idea of what was needed, but she still had several questions. *'There's plenty of information for me here Tao. Still, what constraints do I need to be aware of?'* Tao picked up a notebook from the corner of his desk. *'Good question! I was chatting to Chao earlier today and he mentioned a few things. The client kick-off meeting takes place online next Monday – so that's week 1 on this project. Sakura have already said that ideally, they'd like the residential programme to start in week 6. Do you think that's a realistic time frame? They're also pretty keen that pre-programme activities and the residential programme elements start on Mondays, and that Saturdays and Sundays are non-working days'.* Maria and Tao's discussion then moved to how best to resource the project. Within about 10 minutes, they had identified most of the key players who would be involved:

- **Project sponsor** – Tao (attend online client kick-off meeting and final review with client; will review final report).
- **Project manager** – Maria to replace Chao (run client and kick-off meetings, sign-off trainer contracting and programme design, and do final report and client management).
- **Training lead** – Kavita in Tokyo office (training needs analysis, identify trainers and detailed programme design; on-site lead for residential training).
- **Web design** – Li Wei in Shanghai office (liaise with Kavita and Una)
- **Project support/admin** – Krister (distribute contracts, confirm travel/accommodation/meal bookings, etc., for the meetings/residential programme).
- **Training** – Three external trainers (finalise names once training agenda is completed); prepare materials, support pre-programme online training; one trainer per week for residential training, supported by Kavita as training lead. Most likely three trainers: two days each to develop materials (extra two days internal time to review content, check for overlaps, etc.); three to four days each on online support for pre-programme activities and five days each on residential delivery.
- **Survey** – Una in Shanghai office (design, distribution and analysis of final survey; discuss with Kavita).
- **Invoicing and budget support** – Ruben (track invoicing and do budget close for project).

Maria then turned her attention to an additional note that Chao had made on the key activities in the project, including their time estimates, predecessors and average daily costs (Table 19.5).

Maria thought for a moment. She assumed that Chao had developed his time estimates for each activity based on normal costing but wasn't sure what options there might be to reduce the time of some of these activities. *'Tao, I don't*

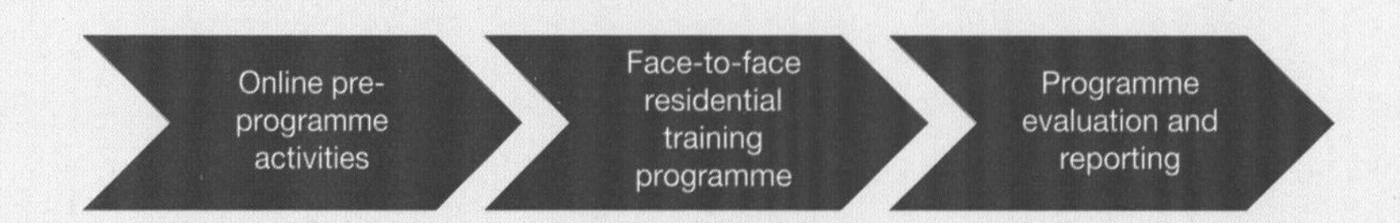

Figure 19.21 Extract from draft proposal for Sakura Bank K.K operations improvement executive development

Table 19.5 Chao's notes on the Sakura Bank K.K. project activities and costs

Programme component	Activity	Optimistic estimate	Likely estimate	Pessimistic estimate	Immediate Predecessors	Daily cost (€)
Programme design	1. Client kick-off meeting (online)	1	1	1	n/a	500
	2. Training needs analysis	3	3	6	1	500
	3. Trainer contracting	2	3	4	2	150
	4. Programme design	4	5	8	2	500
	5. Client review meeting	1	1	1	3, 4	500
	6. Internal kick-off meeting	1	1	1	5	450
	7. Training material creation	6	8	14	6	350
Pre-programme activities	8. Website set-up	5	5	7	6	250
	9. Website go live	1	1	2	7, 8	250
	10. Pre-programme activities	9	10	14	9	350
Residential training	11. Programme administrative arrangements	8	10	12	6	150
	12. Residential training programme	15	15	15	10, 11	1500
Evaluation and reporting	13. Post-course survey	3	4	5	12	150
	14. Final report	2	2	4	13	500
	15. Project closure	2	2	3	14	250

suppose Chao made any notes on possible activity "crashing" did he?' After rummaging around his desk for what seemed like an age, Tao found a bright pink Post-it note hiding under a collection of files, *'Phew, I was starting to think I'd lost this! So, it looks like the training needs analysis could be shortened to two days, but it'll increase the daily cost to €850; the programme design activity can be shortened from five days to four days, but daily costs will increase to €750; for a fixed fee of €4,000, we could get a single more experienced trainer to do the training material creation in four days; and website set-up could be done in three days, but daily costs will increase to €500 per day'.* Maria looked up from her notes, *'OK, that's good to know. Anything else?'* Tao took a sip of water, *'Well, I guess it's important to say that Sakura is an important new client. There's a lot of potential for growth if we can deliver this project and the one I'll be leading effectively! We've heard from a few other firms who've worked with them that they can be quite a challenging client – apparently, they often change their mind on specifications! Oh, and I nearly forgot, to ensure that any project is viable for Kloud, we typically work based on a 20 per cent mark-up between our costs and the price we charge the client. I think that the margin will be pretty tight on this one'.*

Maria left the Managing Director's office and headed for her desk. Tao's final words were ringing in her ears: *'Meet me tomorrow so we can prepare for the kick-off meeting next Monday'*. Sitting down, she looked back over the notes she'd made. Where to begin?

QUESTION

Based on the information you have, develop a project plan for the Sakura Bank K.K. operations improvement training programme, to share with Tao, the Managing Director of Kloud BV. This should include:

- **Project timing: complete a critical path analysis, create a Gantt chart, and consider any uncertainties in time estimates.**
- **Project costing: create a project budget and consider options for 'crashing' activities.**
- **Project resourcing: create a RACI matrix to determine the key responsibilities for those involved in the project.**
- **Project risk: Note any risks you are concerned about and possible mitigation strategies.**

Problems and applications

All chapters have 'Problems and applications' questions that will help you practise analysing operations. They can be answered by reading the chapter. Model answers for the first two questions can be found on the companion website for this text.

1 Revisit the 'Operations in practice' example, 'The risk of changing project scope – sinking the *Vasa*', in this chapter.

(a) Who should be held responsible for this disaster?

(b) What can be learnt from the *Vasa* story for the management of different kinds of modern-day projects?

2 '*Funding comes from a variety of sources; to restore the literally irreplaceable buildings we work on. We try to reconcile historical integrity with commercial viability and rely on the support of volunteers. So, we need to involve all stakeholders all the way through the project*' (Janine Walker, Chief Project Manager, Happy Heritage, a not-for-profit restoration organisation). Her latest project was the restoration of a 200-year-old 'poorhouse' as a visitor attraction, originally built to house the local poor. Janine's team drew up a list of stakeholders and set out to win them over with their enthusiasm for the project. They invited local people to attend meetings, explained the vision and took them to look round the site. Also, before work started, Janine took all the building staff on the same tour of the site as they had taken other groups and the VIPs who provided the funding. '*Involving the builders in the project sparked a real interest in the project and the archaeological history of the site. Often, they would come across something interesting, tell the foreman, who would involve an archaeologist and so preserve an artefact that might otherwise have been destroyed. They took a real interest in their work, they felt involved*'.

(a) Who do you think would be the main stakeholders for this project?

(b) How might not involving them damage the project, and how would involving them benefit the project?

3 Table 19.6 shows the activities, their durations and precedences for designing, writing and installing a bespoke computer database for a commercial bank headquartered in Singapore. Draw a network diagram (activity-on-node) for the project and calculate the fastest time in which the operation might be completed.

Table 19.6 Bespoke computer database project activities

Activity	Duration (weeks)	Activities that must be completed before it can start
1 Contract negotiation	1	–
2 Discussions with main users	2	1
3 Review of current documentation	5	1
4 Review of current systems	6	2
5 Systems analysis (A)	4	3,4
6 Systems analysis (B)	7	5
7 Programming	12	5
8 Testing (prelim)	2	7
9 Existing system review report	1	3,4
10 System proposal report	2	5,9
11 Documentation preparation	19	5,8
12 Implementation	7	7,11
13 System test	3	12
14 Debugging	4	12
15 Manual preparation	5	11

4 The table shows the planned time and budget for a legal consulting project being developed for a client in Copenhagen, Denmark. Complete an earned value analysis (EVA) for the project, based at the end of month 4, given that only activities A, B and C have been completed and spending to date has been €38,250.

Activity	Planned time	Planned budget
A	1 month	€26,000
B	1 month	€10,500
C	1 month	€6,750
D	1 month	€13,000
E	1 month	€9,650
F	1 month	€12,750
G	1 month	€8,750
TOTAL	**7 months**	**€87,400**

5 In the oil industry, project teams are increasingly using virtual reality and visualisation models of offshore structures that allow them to check out not only the original design but any modifications that have to be made during construction.

(a) Why do you think a realistic picture of a completed project helps the process of project management?
(b) Why are such visualisations becoming more important?

6 The idea of the 'critical path' is important in project planning. What different ideas might be meant by the word 'critical'?

7 Re-read the 'Operations in practice' example, '"For the benefit of all" – NASA's highs and lows'. How might an organisation like NASA view the trade-offs implied by the 'iron triangle' perspective?

8 Identify your favourite sporting team (Manchester United, the Toulon rugby team, or if you are not a sporting person, choose any team you have heard of). What kinds of projects do you think they need to manage? For example, merchandising, sponsorship, etc. What do you think are the key issues in making a success of managing each of these different types of projects?

9 Identify a project of which you have been part (for example, moving apartments, a holiday, dramatic production, revision for an examination, etc.).

(a) Who were the stakeholders in this project?
(b) What was the overall project objective (especially in terms of the relative importance of cost, quality and time)?
(c) Were there any resource constraints?
(d) Looking back, how could you have managed the project better?

Selected further reading

There are hundreds of books on project management. They range from the introductory to the very detailed, and from the managerial to the highly mathematical. Here are five general project management books that are worth a look and two journal articles examining more specific aspects of projects:

Cole, R. and Scotcher, E. (2015) ***Brilliant Agile Project Management: A Practical Guide to Using Agile, Scrum and Kanban*****, Pearson, Harlow.**
A practical and modern take on project management.

Davies, A. (2017) ***Projects: A Very Short Introduction*****, Oxford University Press, Oxford.**
A very well-written book on project management, covering key ideas in a very concise way.

Davies, A., Dodgson, M., Gann, D.M. and MacAulay, S.C. (2017) Five rules for managing large, complex projects, ***MIT Sloan Management Review*****, Fall.**
An interesting article focusing on recent research on megaprojects, but giving useful management insights for all large-scale projects.

Kogon, K., Blakemore, S. and Wood, J. (2015) ***Project Management for the Unofficial Project Manager*****, BenBella Books, Dallas, TX.**
A short, easy-to-read book that helps lay a solid foundation in project management and does a good job bringing in the 'people' aspect of projects.

Maylor, H. (2022) ***Project Management*****, 5th edn, Pearson, Harlow.**
A very good introductory text on the subject that takes the reader through the key steps of projects.

Pinto, J.K. (2019) ***Project Management: Achieving Competitive Advantage*****, Global edn, Pearson, Harlow.**
Long-running text with comprehensive coverage of project management.

Whyte, J. (2019) How digital information transforms project delivery models, ***Project Management Journal*****, 50 (2), 177–94.**
An interesting article exploring the implications of digitisation on project management.

Notes on chapter

1. The information on which this example is based is taken from: Howell, E. and Hickock, K. (2020) Apollo 13: the moon-mission that dodged disaster, Space.com, 31 March; Whiting, M. (2018) The legacy of Apollo 6, NASA, 4 April; Taylor Redd, N. (2019) Apollo 11: first men on the Moon, Space.com, 9 May; Overbye, D. (2017) Cassini flies towards a fiery death on Saturn, *New York Times*, 8 September; Rincon, P. (2017) Our Saturn years – Cassini-Huygens' epic journey to the ringed planet, told by the people who made it happen, BBC, 14 September; Cobb, W. (2019) How SpaceX lowered cost and reduced barriers to space, The Conversation, 1 March; Howells, R. (2021) China becomes only the second country to make historic touchdown on Mars, *EuroWeekly News*, 15 May.
2. The information on which this example is based is taken from: George, R. (2021) Wind ...or worse: was pilot error to blame for the Suez blockage?, *Guardian*, 3 April; Magdy, S. (2021) Suez Canal chief: vessel impounded amid financial dispute, AP News, 12 May, https://apnews.com/article/egypt-coronavirus-pandemic-suez-canal-189a9115b7c31cb1dbad5febfb778fa6 (accessed September 2021); Leonard, M. (2021) Timeline: how the Suez Canal blockage unfolded across supply chains, Supply Chain Dive, 6 July, https://www.supplychaindive.com/news/timeline-ever-given-evergreen-blocked-suez-canal-supply-chain/597660/ (accessed September 2021).
3. The information on which this example is based is taken from: Metz, R. (2019) The world's biggest spice company is using AI to find new flavors, CNN Business, 5 February; Wiggers, K. (2019) IBM and McCormick blend new seasonings with AI, Venture Beat, 4 February; Lougee, R (2019) Using AI to develop new flavor experiences, IBM Research Blog, 5 February.
4. The information on which this example is based is taken from: Schuetze, C. (2020) Berlin's newest airport prepares for grand opening. Again, *New York Times*, 29 April; CAPA Centre for Aviation (2020) Berlin Brandenburg Airport's terminal certified for opening – at last, centreforaviation.com, 8 May; L.R.S. (2017) Why Berlin's new airport keeps missing its opening date, *Economist*, 25 January.
5. The information on which this example is based is taken from: Burgess, M. (2018) Ocado's collaborative robot is getting closer to factory work, *Wired*, 11 January; Butler, S. (2018) Ocado to wheel out C3PO-style robot to lend a hand at warehouses, *Guardian*, 11 January.

GLOSSARY

3D printing: also known as additive manufacturing, a 3D printer produces a three-dimensional object by laying down layer upon layer of material until the final form is obtained.

4Vs: acronym for the characteristics of volume, variety, variation and visibility.

5Ss: a simple housekeeping methodology to organise work areas. Originally translated from the Japanese, they are generally taken to mean sort, strengthen, shine, standardise and sustain. The aim is to reduce clutter in the workplace.

ABC inventory control: an approach to inventory control that classes inventory by its usage value and varies the approach to managing it accordingly.

Activity: as used in project management, it is an identifiable and defined task, used together with event activities to form network planning diagrams.

Aggregated planning and control: a term used to indicate medium-term capacity planning that aggregates different products and services together in order to get a broad view of demand and capacity.

Agility: the ability of an operation to respond quickly and at low cost as market requirements change.

Algorithmic decision-making: where decisions are made automatically using a predefined sequence of instructions, or rules.

Allen curve: a relationship that shows a powerful negative correlation between the physical distance between colleagues and their frequency of communication.

Allowances: a term used in work study to indicate the extra time allowed for rest, relaxation and personal needs.

Alpha testing: essentially an *internal* process where the developers or manufacturers (or sometimes an outside agency that they have commissioned) examine the product for errors. Generally, it is also a private process, not open to the market or potential customers.

Andon: a light above a workstation that indicates its state, whether working, waiting for work, broken down, etc.; andon lights may be used to stop the whole line when one station stops.

Annual hours: a type of flexitime working that controls the amount of time worked by individuals on an annual rather than a shorter basis.

Anthropometric data: data that relate to people's size, shape and other physical abilities, used in the design of jobs and physical facilities.

Anticipation inventory: inventory that is accumulated to cope with expected future demand or interruptions in supply.

Appraisal costs: those costs associated with checking, monitoring and controlling quality to see if problems or errors have occurred, an element within quality-related costs.

Approaches to improvement: the underlying sets of beliefs that form a coherent philosophy and shape how improvement should be accomplished.

Artificial intelligence (AI): an area of computer science that emphasises the creation of intelligent machines that work and react like humans.

Attributes of quality: measures of quality that can take one of two states, for example, right or wrong, works or does not work, etc.

Augmented reality: technologies that show an enhanced version of reality where live views of physical real-world environments are augmented with overlaid computer-generated images,

Automated guided vehicle (AGV): a materials handling system that uses automated vehicles programmed to move between different stations without a driver.

B2B: abbreviation of business-to-business operation, meaning those that provide their products or services to other businesses.

B2C: abbreviation of business-to-consumer operation, meaning those that provide their products or services direct to the consumers who (generally) are the ultimate users of the outputs from the operation.

B Corps: an abbreviation for Certified B, a third-party certification administered by the non-profit B Lab, based partly on a company's performance on positive stakeholder impact alongside profit.

Back office: the low-visibility part of an operation.

Backward scheduling: starting jobs at a time when they should be finished exactly when they are due, as opposed to forward scheduling.

Balanced scorecard (BSC): in addition to financial performance, the balanced scorecard also includes assessment of customer satisfaction, internal processes, and innovation and learning.

Barcode: a unique product code that enables a part or product type to be identified when read by a barcode scanner.

Basic time: the time taken to do a job without any extra allowances for recovery.

Basic working practices: principles used to encourage the involvement and respect in lean philosophy.

Batch processes: processes that treat batches of products together, and where each batch has its own process route.

Bath-tub curve: a curve that describes the failure probability of a product, service or process and indicates relatively high probabilities of failure at the beginning and at the end of the life cycle.

Behavioural job design: an approach to job design that considers individuals' desire to fulfil their needs for self-esteem and personal development.

Benchmarking: the activity of comparing methods and/or performance with other processes in order to learn from them and/or assess performance.

Beta testing: when the product or service is released for testing by selected customers. It is an *external* 'pilot test' that takes place in 'real world' (or near real world, because it is still a relatively short trial with a small sample) before commercial production.

Big data: a large volume of both structured and unstructured data whose analysis can reveal hidden patterns, correlations and other insights.

Bill of materials (BOM): a list of the component parts required to make up the total package for a product or service together with information regarding their level in the product or component structure and the quantities of each component required.

Blockchain: a decentralised, digitised, public ledger (list) of transactions (movements, authorisations, payments, etc.), where a 'block' is a record of new transactions. When each block is completed (verified) it is added to the chain, thus creating a chain of blocks, or 'blockchain'.

Blueprinting: a term often used in service design to mean process mapping.

Bottleneck: the capacity-constraining stage in a process; it governs the output of the whole process.

Bottom-up: the influence of operational experience on operations decisions.

Brainstorming: an improvement technique where small groups of people put forward ideas in a creative free-form manner.

Break-even analysis: the technique of comparing revenues and costs at increasing levels of output in order to establish the point at which revenue exceeds cost, that is, the point at which it 'breaks even'.

Breakthrough improvement: an approach to improving operations performance that implies major and dramatic change in the way an operation works; for example, business process reengineering (BPR) is often associated with this type of improvement, also known as innovation-based improvement, contrasted with continuous improvement.

Broad definition of operations: all the activities necessary for the fulfilment of customer requests.

Buffer inventory: an inventory that compensates for unexpected fluctuations in supply and demand; can also be called safety inventory.

Bullwhip effect: the tendency of supply chains to amplify relatively small changes at the demand side of a supply chain such that the disruption at the supply end of the chain is much greater.

Business continuity: the procedures adopted by businesses to mitigate and recover from the effects of major failures.

Business ecosystem: an idea that is closely related to that of co-opetition. Like supply networks, business ecosystems include suppliers and customers. However, they also include stakeholders that may have little or no direct relationship with the main supply network yet interact with it by complementing or contributing significant components of the value proposition for customers.

Business model: the plan that is implemented by a company to generate revenue and make a profit (or fulfil its social objectives if a not-for-profit enterprise).

Business process outsourcing (BPO): the term that is applied to the outsourcing of whole business processes; this need not mean a change in location of the process, sometimes it involves an outside company taking over the management of processes that remain in the same location.

Business process reengineering (BPR): the philosophy that recommends the redesign of processes to fulfil defined external customer needs.

Business processes: processes, often that cut across functional boundaries, which contribute some part to fulfilling customer needs.

Business strategy: the strategic positioning of a business in relation to its customers, markets and competitors, a subset of corporate strategy.

Capacity: the maximum level of value-added activity that an operation, or process, or facility is capable of over a period of time.

Capacity lagging: the strategy of planning capacity levels such that they are always less than or equal to forecast demand.

Capacity leading: the strategy of planning capacity levels such that they are always greater than or equal to forecast demand.

Causal modelling: a quantitative approach to demand forecasting, which describes and evaluates the complex cause–effect relationships between the key variables.

Cause–effect diagrams: also known as Ishikawa diagrams, a technique for searching out the root cause of problems, it is a systematic questioning technique.

Cell layout: locating transforming resources with a common purpose such as processing the same types of product, serving similar types of customers, etc., together in close proximity (a cell).

Chase demand: an approach to medium-term capacity management that attempts to adjust output and/or capacity to reflect fluctuations in demand.

Circular economy: an alternative to the traditional linear economy (or make-use-dispose as it is termed). The idea is to keep products in use for as long as possible, extract the maximum value from them while in use, and then recover and regenerate products and materials at the end of their service life.

Closed-loop economy: *see* 'Circular economy'.

Cluster analysis: a technique used in the design of cell layouts to find which process groups fit naturally together.

Clusters: where similar companies with similar needs locate relatively close to each other in the same geographical area.

Co-creation: where the customer or customers play an important part in the character of the product or service offering.

Combinatorial complexity: the idea that many different ways of processing products and services at many different locations or points in time combine to result in an exceptionally large number of feasible options; the term is often used in facilities layout and scheduling to justify non-optimal solutions (because there are too many options to explore).

Commonality: the degree to which a range of products or services incorporate identical components (also called 'parts commonality').

Community factors: those factors that are influential in the location decision that relate to the social, political and economic environment of the geographical position.

Competitive factors: the factors such as delivery time, product or service specification, price, etc. that define customers' requirements.

Complaint value chain: a model that helps visualise the potential value of good service recovery.

Component structure: *see* 'Product structure'.

Computer-aided design (CAD): the use of computer software to aid in creating and modifying product, service or process drawings.

Computer-integrated manufacturing (CIM): a term used to describe the integration of computer-based monitoring and control of all aspects of a manufacturing process, often using a common database and communicating via some form of computer network.

Concept generation: a stage in the product and service design process that formalises the underlying idea behind a product or service.

Concurrent engineering: *see* 'Simultaneous development'.

Condition-based maintenance (CBM): an approach to maintenance management that monitors the condition of process equipment and performs work on equipment only when it is required.

Content of strategy: the set of specific decisions and actions that shape the strategy.

Continuous improvement: an approach to operations improvement that assumes many, relatively small, incremental improvements in performance, stressing the momentum of improvement rather than the rate of improvement; also known as 'kaizen', often contrasted with breakthrough improvement.

Continuous processes: processes that are high volume and low variety; usually products made on a continuous process are produced in an endless flow, such as petrochemicals or electricity.

Continuous review: an approach to managing inventory that makes inventory-related decisions when inventory reaches a particular level, as opposed to periodic review.

Control: the process of monitoring operations activity and coping with any deviations from the plan; usually involves elements of replanning.

Control charts: the charts used within statistical process control to record process performance.

Control limits: the lines on a control chart used in statistical process control that indicate the extent of natural or common-cause variations; any points lying outside these control limits are deemed to indicate that the process is likely to be out of control.

Co-opetition: an approach to supply networks that defines businesses as being surrounded by suppliers, customers, competitors and complementors.

Core functions: the functions that manage the three core processes of any business: marketing, product/service development and operations.

Corporate social responsibility (CSR): how business takes account of its economic, social and environmental impacts.

Corporate strategy: the strategic positioning of a corporation and the businesses within it.

Crashing: a term used in project management to mean reducing the time spent on critical path activities so as to shorten the whole project.

Create-to-order: *see* 'Make-to-order'.

Critical path: the longest sequence of activities through a project network, it is called the critical path because any delay in any of its activities will delay the whole project.

Critical path method (CPM): a technique of network analysis.

Crowdsourcing: the act of taking an activity traditionally performed by a designated agent and outsourcing it to a large group of people in the form of an open call.

Customer relationship management (CRM): a method of learning more about customers' needs and behaviours by analysing sales information.

Customisation: the variation in product or service design to suit the specific need of individual customers or customer groups.

Cybersecurity: activity that attempts to protect an operation from the failure of an operation's technology leading to exposure to any risk of financial loss, disruption or damage to the reputation of an organisation from some sort of failure of its information technology systems.

Cycle inventory: inventory that occurs when one stage in a process cannot supply all the items it produces simultaneously and so has to build up inventory of one item while it processes the others.

Cycle time: the average time between units of output emerging from a process.

Decision support system (DSS): a management information system that aids or supports managerial decision-making; it may include both databases and sophisticated analytical models.

Delegated sourcing: involves a tiered approach to managing supplier relationships, where one supplier is responsible for delivering an entire sub-assembly as opposed to a single part, or a package of services as opposed to an individual service.

Delivery: the activities that plan and control the transfer of products and services to customers.

Delivery flexibility: the operation's ability to change the timing of the delivery of its services or products.

Delphi method: the best-known approach to generating forecasts using experts is the Delphi method. It employs a survey of experts where replies are analysed, and anonymous summaries are sent back to all experts. The experts are then asked to re-consider their original response in the light of the replies and arguments put forward by the other experts. This process is repeated several times to conclude either with a consensus or at least a narrower range of decisions.

Demand management: an approach to medium-term capacity management that attempts to change or influence demand to fit available capacity.

Demand side: the chains of customers, customers' customers, etc. that receive the products and services produced by an operation.

Dependability: delivering, or making available, products or services when they were promised to the customer.

Dependent demand: demand that is relatively predictable because it is derived from some other known factor.

Design acceptability: the attractiveness to the operation of a process, product or service.

Design capacity: the capacity of a process or facility as it is designed to be, often greater than effective capacity.

Design feasibility: the ability of an operation to produce a process, product or service.

Design funnel: a model that depicts the design process as the progressive reduction of design options from many alternatives down to the final design.

Design screening: the evaluation of alternative designs with the purpose of reducing the number of design options being considered.

Development: a collection of operations activities that improve products, services and processes.

Digital twins: powerful digital 'replicas' that can be used instead of the physical reality of a product. For example, digital twins could monitor and simulate possible future scenarios and predict the need for repairs and other problems before they occur.

Directing: operations activities that create a general understanding of an operation's strategic purpose and performance.

Disaster recovery: this term is used in a similar way to business continuity, but is concerned largely with action plans and procedures for the recovery of critical information technology and systems after a natural or human-induced disaster.

Diseconomies of scale: a term used to describe the extra costs that are incurred in running an operation as it gets larger.

Disintermediation: the emergence of an operation in a supply network that separates two operations that were previously in direct contact.

Division of labour: an approach to job design that involves dividing a task into relatively small parts, each of which is accomplished by a single person.

DMAIC cycle: an increasingly used improvement cycle model, popularised by the Six Sigma approach to operations improvement. DMAIC stands for Defining the problem, Measuring data to refine the problem, Analysing to detect the root causes of the problem really are, Improving the process, and Controlling to check that the improved level of performance is sustaining.

Do or buy: the term applied to the decision on whether to own a process that contributes to a product or service or, alternatively, outsource the activity performed by the process to another operation.

Downstream: the other operations in a supply chain between the operation being considered and the end customer.

Drum, buffer, rope: an approach to operations control that comes from the theory of constraints (TOC) and uses the bottleneck stage in a process to control materials movement.

Dynamic pricing: *see* 'Surge pricing'.

Earned-value control: a method of assessing performance in project management by combining the costs and times achieved in the project with the original plan.

E-business: the use of internet-based technologies either to support existing business processes or to create entirely new business opportunities.

E-commerce: the use of the internet to facilitate buying and selling activities.

Economic batch quantity (EBQ): the number of items to be produced by a machine or process that supposedly minimises the costs associated with production and inventory holding.

Economic bottom line: the part of the triple bottom line that assesses an organisation's economic performance, usually in financial terms.

Economic order quantity (EOQ): the quantity of items to order that supposedly minimises the total cost of inventory management, derived from various EOQ formulae.

Economy of scale: the manner in which the costs of running an operation decrease as it gets larger.

Effective capacity: the useful capacity of a process or operation after maintenance, changeover and other stoppages and loading have been accounted for.

Efficient frontier: the convex line that describes current performance trade-offs between (usually two) measures of operations performance.

EFQM excellence model: a model that identifies the categories of activity that supposedly ensure high levels of quality; now used by many companies to examine their own quality-related procedures.

Elements of improvement: the fundamental ideas of what improves operations.

Emergent strategy: a strategy that is gradually shaped over time and based on experience rather than theoretical positioning.

Emotional mapping: charting how customers' emotions could be engaged (positively and negatively) at each stage of a process.

Empowerment: a term used in job design to indicate increasing the authority given to people to make decisions within the job or changes to the job itself.

End-to-end business processes: processes that totally fulfil a defined external customer need.

Enterprise resource planning (ERP): the integration of all significant resource planning systems in an organisation that, in an operations context, integrates planning and control with the other functions of the business.

Environmental bottom line: the element of the triple bottom line that assesses an organisation's performance in terms of how it affects the natural environment.

Ergonomics: a branch of job design that is primarily concerned with the physiological aspects of job design, with how the human body fits with process facilities and the environment; can also be referred to as human factors, or human factors engineering.

ESG: acronym for environmental, social and governance, a similar idea to corporate social responsibility, but more from an investor's perspective.

European Quality Award (EQA): a quality award organised by the European Foundation for Quality Management (EFQM), based on the EFQM excellence model.

Events: points in time within a project plan; together with activities, they form network planning diagrams.

Expert systems (ES): computer-based problem-solving systems that, to some degree, mimic human problem-solving logic.

External failure costs: those costs that are associated with an error or failure reaching a customer, an element within quality-related costs.

External neutrality: the second stage of Hayes and Wheelwright's four-stage model of operations contribution, where the operations function begins comparing itself with similar companies or organisations in the outside market.

Externally supportive: the final stage of Hayes and Wheelwright's four-stage model of operations contribution where the operations function is the foundation for an organisation's competitive success.

Facilitating products: products that are produced by an operation to support its services.

Fail-safeing: building in, often simple, devices that make it difficult to make the mistakes that could lead to failure; also known by the Japanese term 'poka-yoke'.

Failure analysis: the use of techniques to uncover the root cause of failures; techniques may include accident investigation, complaint analysis, etc.

Failure mode and effect analysis (FMEA): a technique used to identify the product, service or process features that are crucial in determining the effects of failure.

Failure rate: a measure of failure that is defined as the number of failures over a period of time.

Fault-tree analysis: a logical procedure that starts with a failure or potential failure and works backwards to identify its origins.

Finite loading: an approach to planning and control that only allocates work to a work centre up to a set limit (usually its useful capacity).

First-tier: the description applied to suppliers and customers that are in immediate relationships with an operation with no intermediary operations.

Fixed-cost break: the volumes of output at which it is necessary to invest in operations facilities that bear a fixed cost.

Fixed-position layout: locating the position of a product or service such that it remains largely stationary, while transforming resources are moved to and from it.

Flexibility: the degree to which an operation's process can change what it does, how it is doing it, or when it is doing it.

Flexible manufacturing systems (FMS): manufacturing systems that bring together several technologies into a coherent system, such as metal cutting and material handling technologies; usually their activities are controlled by a single governing computer.

Flexi-time working: increasing the possibility of individuals varying the time during which they work.

Focus group: a group of potential product or service users, chosen to be typical of its target market, formed to test their reaction to alternative designs.

Forward scheduling: loading work onto work centres as soon as it is practical to do so, as opposed to backward scheduling.

Four perspectives on operations strategy: categorisation that distinguishes between top-down, bottom-up, outside-in and inside-out perspectives.

Four-stage model of operations contribution: model devised by Hayes and Wheelwright that categorises the degree to which operations management has a positive influence on overall strategy.

Front office: the high-visibility part of an operation.

Functional layout: layout where similar resources or processes are located together (sometimes called process layout).

Functional operations: the idea that every function in an organisation uses resources to produce products and services for (internal) customers; therefore, all functions are, to some extent, operations.

Functional strategy: the overall direction and role of a function within the business; a subset of business strategy.

Gantt chart: a scheduling device that represents time as a bar or channel on which activities can be marked.

Gartner Hype Cycle: an idea created by Gartner, the information technology research and consultancy company that attempts to illustrate how perceptions of a technology's usefulness develop over time.

Gemba: also sometimes called genba, a term used to convey the idea of going to where things actually take place, as a basis for improvement.

Generative design: an approach to exploring alternative designs, involving specifying design goals, parameters and performance requirements, and using software to generate design alternatives.

Gig economy: describes the trend of organisations to employ sub-contractors and freelancers to do a greater proportion of their activities rather than relying on full-time employees. This flexing of capacity is now employed across a wide range of industries, including arts and design, transportation, construction, accommodation, media, ICT, education and professional services.

Globalisation: the extension of operations' supply chains to cover the whole world.

Heijunka: *see* 'Levelled scheduling'.

Heuristics: 'rules of thumb' or simple reasoning short cuts that are developed to provide good but non-optimal solutions, usually to operations decisions that involve combinatorial complexity.

Hierarchy of operations: the idea that all operations processes are made up of smaller operations processes.

High-level process mapping: an aggregated process map that shows broad activities rather than detailed activities (sometimes called an 'outline process map').

House of quality: *see* 'Quality function deployment'.

Hub operation: a common facility in a logistics network through which items or information is routed.

Human factors engineering: an alternative term for ergonomics.

Human resource strategy: the overall long-term approach to ensuring that an organisation's human resources provide a strategic advantage.

Hybrid working: the practice of spending at least some of the working week working from home.

Iceberg metaphor: a metaphor for organisational culture, comparing it to an iceberg, with most of mass under the surface.

Ideas management: the process of collecting innovation ideas from employees, assessing them and, if appropriate, implementing them.

IHIP: acronym for the characteristics of intangibility, heterogeneity, inseparability and perishability.

Immediate supply network: the suppliers and customers that have direct contact with an operation.

Importance–performance matrix: a technique that brings together scores that indicate the relative importance and performance of different competitive factors in order to prioritise them as candidates for improvement.

Improvement cycles: the practice of conceptualising problem-solving as used in performance improvement, in terms of a never-ending cyclical model, for example the PDCA cycle or the DMAIC cycle.

Independent demand: demand that is not obviously or directly dependent on the demand for another product or service.

Indirect process technology: technology that assists in the management of processes rather than directly contributes to the creation of products and services, for example information technology that schedules activities.

Industry 4.0: cyber-physical systems that comprise smart machines, storage systems and production facilities capable of autonomously exchanging information, triggering actions and controlling each other independently.

Infinite loading: an approach to planning and control that allocates work to work centres irrespective of any capacity or other limits.

Information technology (IT): any device, or collection of devices, that collects, manipulates, stores or distributes information, nearly always used to mean computer-based devices.

Infrastructural decisions: the decisions that concern the operation's systems, methods and procedures and shape its overall culture.

Innovation: the act of introducing new ideas to products, services or processes.

Innovation S-curve: the curve that describes the impact of an innovation over time.

Input resources: the transforming and transformed resources that form the input to operations.

Intangible resources: the resources within an operation that are not immediately evident or tangible, such as relationships with suppliers and customers, process knowledge, and new product and service development.

Interactive design: the idea that the design of products and services on one hand, and the processes that create them on the other, should be integrated.

Internal customers: processes or individuals within an operation that are the customers for other internal processes' or individuals' outputs.

Internal failure costs: the costs associated with errors and failures that are dealt with inside an operation but yet cause disruption; an element within quality-related costs.

Internal neutrality: the first stage of Hayes and Wheelwright's four-stage model of operations contribution, where the operations function performance harms the organisation's ability to compete effectively.

Internal suppliers: processes or individuals within an operation that supply products or services to other processes or individuals within the operation.

Internally supportive: the third stage of Hayes and Wheelwright's four-stage model of operations contribution, where the operations functions have typically reached the 'first division' of their markets.

Internet of Things (IoT): the integration of physical objects into an information network where the physical objects become active participants in business processes.

Inventory: also known as stock, the stored accumulation of transformed resources in a process; usually applies to material resources but may also be used for inventories of information; inventories of customers or customers of customers are usually queues.

ISO 9000: a set of worldwide standards that established the requirements for companies' quality management systems; last revised in 2000, there are several sets of standards.

ISO 14000: an international standard that guides environmental management systems and covers initial planning, implementation and objective assessment.

Job design: the way in which we structure the content and environment of individual staff members' jobs within the workplace and the interface with the technology or facilities that they use.

Job enlargement: a term used in job design to indicate increasing the amount of work given to individuals in order to make the job less monotonous.

Job enrichment: a term used in job design to indicate increasing the variety and number of tasks within an individual's job; this may include increased decision-making and autonomy.

Job rotation: the practice of encouraging the movement of individuals between different aspects of a job in order to increase motivation.

Jobbing processes: processes that deal with high variety and low volumes, although there may be some repetition of flow and activities.

Just-in-time (JIT): a method of planning and control and an operations philosophy that aims to meet demand instantaneously with perfect quality and no waste.

Kaizen: Japanese term for continuous improvement.

Kanban: Japanese term for card or signal; it is a simple controlling device that is used to authorise the release of materials in pull control systems such as those used in JIT.

Knowledge management: the management of facts, information and skills acquired through experience or education; the theoretical or practical understanding of a subject.

Lead-time usage: the amount of inventory that will be used between ordering replenishment and the inventory arriving, usually described by a probability distribution to account for uncertainty in demand and lead time.

Lead user: users who are ahead of the majority of the market on a major market trend, and who also have a high incentive to innovate. As these lead users will be familiar both with the positives and negatives of the early versions of products and services, they are a particularly valuable source of potential innovative ideas.

Lean: (also known as lean synchronisation) an approach to operations management that emphasises the continual elimination of waste of all types, often used interchangeably with just-in-time (JIT); it is more an overall philosophy, whereas JIT is usually used to indicate an approach to planning and control that adopts lean principles.

Lean Sigma: a blend of improvement elements from lean and Six Sigma.

Less important factors: competitive factors that are neither order-winning nor qualifying; performance in them does not significantly affect the competitive position of an operation.

Level capacity plan: an approach to medium-term capacity management that attempts to keep output from an operation or its capacity constant, irrespective of demand.

Levelled scheduling: the idea that the mix and volume of activity should even out over time so as to make output routine and regular, sometimes known by the Japanese term 'heijunka'.

Life cycle analysis: a technique that analyses all the production inputs, life cycle use of a product and its final disposal in terms of total energy used and wastes emitted.

Line layout: a more descriptive term for what is technically a product layout.

Line of fit: an alternative name for the 'natural' diagonal of the product process matrix.

Line of visibility: the boundary between the parts of the process visible to the customer and those that are not.

Little's law: the mathematical relationship between throughput time, work-in-progress and cycle time (throughput time = work-in-progress × cycle time).

Loading: the amount of work that is allocated to a work centre.

Location: the geographical position of an operation or process.

Logistics: a term in supply chain management broadly analogous to physical distribution management.

Long-term capacity management: the set of decisions that determine the level of physical capacity of an operation in whatever the operation considers to be long-term; this will vary between industries, but is usually in excess of one year.

Long-thin process: a process designed to have many sequential stages, each performing a relatively small part of the total task; the opposite of short fat process.

Maintenance: the activity of caring for physical facilities so as to avoid or minimise the chance of those facilities failing.

Make-to-order: operations that produce products only when they are demanded by specific customers.

Make-to-stock: operations that produce products prior to their being demanded by specific customers.

Manufacturing resource planning (MRP II): an expansion of materials requirements planning to include greater integration with information in other parts of the organisation and often greater sophistication in scheduling calculations.

Market requirements: the performance objectives that reflect the market position of an operation's products or services, also a perspective on operations strategy.

Mass customisation: the ability to produce products or services in high volume, yet vary their specification to the needs of individual customers or types of customer.

Mass processes: processes that produce goods in high volume and relatively low variety.

Mass services: service processes that have a high number of transactions, often involving limited customisation, for example mass transportation services, call centres, etc.

Master production schedule (MPS): the important schedule that forms the main input to material requirements planning, it contains a statement of the volume and timing of the end products to be made.

Materials requirements planning (MRP): a set of calculations embedded in a system that helps operations make volume and timing calculations for planning and control purposes.

Matrix organisational forms: hybrids of M-form and U-form organisations.

Mean time between failures (MTBF): operating time divided by the number of failures; the reciprocal of failure rate.

Method study: the analytical study of methods of doing jobs with the aim of finding the 'best', or an improved, job method.

M-form organisation: an organisational structure that groups together either its resources needed to produce a product or service group, or those needed to serve a particular geographical area in separate divisions.

Milestones: term used in project management to denote important events at which specific reviews of time, cost and quality can be made.

Mitigation: a term used in risk management to mean isolating a failure from its negative consequences.

Mix flexibility: the operation's ability to produce a wide range of products and services.

Modular design: the use of standardised sub-components of a product or service that can be put together in different ways to create a high degree of variety.

Moravec's paradox: term describing the enigma that even highly sophisticated technologies can find some tasks difficult that most humans find easy.

MRP netting process: the process of calculating net requirements using the master production schedule and the bills of materials.

Muda: all activities in a process that are wasteful because they do not add value to the operation or to the customer.

Multi-skilling: increasing the range of skills of individuals in order to increase motivation and/or improve flexibility.

Multi-sourcing: the practice of obtaining the same type of product, component or service from more than one supplier in order to maintain market bargaining power or continuity of supply.

Mura: a term meaning lack of consistency or unevenness that results in periodic overloading of staff or equipment.

Muri: waste because of unreasonable requirements placed on a process that will result in poor outcomes.

Net promoter score (NPS): a simple scoring formula that subtracts the percentage of customers who would not recommend a service from the percentage of those who would.

Network analysis: overall term for the use of network-based techniques for the analysis and management of projects; includes, for example, critical path method (CPM) and programme evaluation and review technique (PERT).

N-form organisation: networked organisational structures where clusters of resources have delegated responsibility for the strategic management of those resources.

Occupational health and safety (OHS): practices that prevent occupational accidents and diseases or other events that could have adverse consequences for staff.

Omnichannel model: a set of connections, usually in retail networks, that attempts to provide a seamless all-inclusive customer experience by integrating all possible channels so customers can use whichever is the most convenient for them at whatever stage of the transaction.

Offshoring: sourcing products and services from operations that are based outside one's own country or region.

Open sourcing: products or services developed by an open community, including users.

Operating model: a high-level design of the organisation that defines the structure and style that enable it to meet its business objectives.

Operations function: the arrangement of resources that are devoted to the production and delivery of products and services.

Operations management: the activities, decisions and responsibilities of managing the production and delivery of products and services.

Operations managers: the staff of the organisation who have particular responsibility for managing some or all of the resources that compose the operation's function.

Operations resource capabilities: the inherent ability of operations processes and resources; also a perspective on operations strategy.

Operations strategy: the overall direction and contribution of the operation's function with the business; the way in which market requirements and operations resource capabilities are reconciled within the operation.

Optimised production technology (OPT): software and concept originated by Eliyahu Goldratt to exploit his theory of constraints (TOC).

Order fulfilment: all the activities involved in supplying a customer's order, often used in e-retailing but now also used in other types of operation.

Order winners: the competitive factors that directly and significantly contribute to winning business.

Organisational ambidexterity: the ability of an operation to both exploit and explore as it seeks to improve.

Outline process map: *see* 'High-level process mapping'.

Outsourcing: the practice of contracting out work previously done within the operation to a supplier.

Overall equipment effectiveness (OEE): a method of judging the effectiveness of how operations equipment is used.

***P*:*D* ratio:** a ratio that contrasts the total length of time customers have to wait between asking for a product or service and receiving it (*D*) and the total throughput time to produce the product or service (*P*).

Panel approach: a qualitative method of forecasting, using a panel of experts to discuss and agree on likely future demand (or other future events).

Parallel sourcing: having single-source relationships for components or services for different product models or service packages in order to provide the advantages of both multiple sourcing and single sourcing simultaneously.

Pareto law: a general law found to operate in many situations that indicates that 20 per cent of something causes 80 per cent of something else, often used in inventory management (20 per cent of products produce 80 per cent of sales value) and improvement activities (20 per cent of types of problems produce 80 per cent of disruption).

Partnership: a type of relationship in supply chains that encourages relatively enduring cooperative agreements for the joint accomplishment of business goals.

Parts commonality: *see* 'Commonality'.

PDCA (or PDSA) cycle: stands for Plan, Do, Check (Study), Act cycle, perhaps the best known of all improvement cycle models.

Performance management: similar to but broader than performance measurement, attempts to influence decisions behaviour and skills development so that individuals and processes are better equipped to meet objectives.

Performance measurement: the activity of measuring and assessing the various aspects of a process or a whole operation's performance.

Performance objectives: the generic set of performance indicators that can be used to set the objectives or judge the performance of any type of operation; although there are alternative lists proposed by different authorities, the five performance objectives as used in this text are quality, speed, dependability, flexibility and cost.

Performance standards: a defined level of performance against which an operation's actual performance is compared; performance standards can be based on historical performance, some arbitrary target performance, the performance of competitors, etc.

Periodic review: an approach to making inventory decisions that defines points in time for examining inventory levels and then makes decisions accordingly, as opposed to continuous review.

Perpetual inventory principle: a principle used in inventory control that inventory records should be automatically updated every time items are received or taken out of stock.

Pipeline inventory: the inventory that exists because material cannot be transported instantaneously.

Planning: the formalisation of what is intended to happen at some time in the future.

Poka-yoke: Japanese term for fail-safeing.

Polar diagram: a diagram that uses axes, all of which originate from the same central point, to represent different aspects of operations performance.

Polar representation of performance: a method of representing the relative importance of performance objectives for a product or service.

Predetermined motion–time systems (PMTS): a work measurement technique where standard elemental times obtained from published tables are used to construct a time estimate for a whole job.

Preliminary design: the initial design of a product or service that sets out its main components and functions, but does not include many specific details.

Prevention costs: those costs that are incurred in trying to prevent quality problems and errors occurring, an element within quality-related costs.

Preventive maintenance: an approach to maintenance management that performs work on machines or facilities at regular intervals in an attempt to prevent them breaking down.

Principles of motion economy: a checklist used to develop new methods in work study that is intended to eliminate elements of the job, combine elements together, simplify the activity or change the sequence of events so as to improve efficiency.

Process capability: an arithmetic measure of the acceptability of the variation of a process.

Process design: the overall configuration of a process that determines the sequence of activities and the flow of transformed resources between them.

Process distance: the degree of novelty required by a process in the implementation of a new technology.

Process hierarchy: the idea that a network of resources form processes, networks of processes form operations, and networks of operations form supply networks.

Process mapping: describing processes in terms of how the activities within the process relate to each other (may also be called 'process blueprinting' or 'process analysis').

Process mapping symbols: the symbols that are used to classify different types of activity; they usually derive either from scientific management or from information-systems flow-charting.

Process of operations strategy: how operations strategies are put together, often divided into formulation, implementation, monitoring and control.

Process outputs: the mixture of goods and services produced by processes.

Process technology: the machines and devices that create and/or deliver goods and services.

Process types: terms that are used to describe a particular general approach to managing processes; in manufacturing these are generally held to be project, jobbing, batch, mass and continuous processes; in services they are held to be professional services, service shops and mass services.

Process variability: the degree to which activities vary in their time or nature in a process.

Processes: an arrangement of resources that produces some mixture of products and services.

Product layout: locating transforming resources in a sequence defined by the processing needs of a product or service.

Product structure: diagram that shows the constituent component parts of a product or service package and the order in which the component parts are brought together (often called components structure).

Product technology: the embedded technology within a product or service, as distinct from process technology.

Production flow analysis (PFA): a technique that examines product requirements and process grouping simultaneously to allocate tasks and machines to cells in cell layout.

Productivity: the ratio of what is produced by an operation or process to what is required to produce it, that is, the output from the operation divided by the input to the operation.

Product–process matrix: a model derived by Hayes and Wheelwright that demonstrates the natural fit between volume and variety of products and services produced by an operation on one hand, and the process type used to produce products and services on the other.

Product/service flexibility: the operation's ability to introduce new or modified products and services.

Product/service life cycle: a generalised model of the behaviour of both customers and competitors during the life of a product or service; it is generally held to have four stages: introduction, growth, maturity and decline.

Professional services: service processes that are devoted to producing knowledge-based or advice-based services, usually involving high customer contact and high customisation; examples include management consultants, lawyers, architects, etc.

Program evaluation and review technique (PERT): a method of network planning that uses probabilistic time estimates.

Programme: as used in project management, it is generally taken to mean an ongoing process of change comprising individual projects.

Project: a set of activities with a defined start point and a defined end state, which pursue a defined goal using a defined set of resources.

Project manager: the person accountable for project delivery, with several key responsibilities. They organise the project team, with the responsibility, if not always the authority, to run the project on a day-to-day basis. Competent project managers are vital for project success.

Project processes: processes that deal with discrete, usually highly customised, products.

Projectification: a term desrcibing the increasing proportation of individual's time spent working on projects as opposed to 'steady-state' activities.

Prototyping: an initial design of a product or service devised with the aim of further evaluating a design option.

Pull control: a term used in planning and control to indicate that a workstation requests work from the previous station only when it is required, one of the fundamental principles of just-in-time planning and control.

Purchasing: the organisational function, often part of the operations function, that forms contracts with suppliers to buy in materials and services.

Push control: a term used in planning and control to indicate that work is being sent forward to workstations as soon as it is finished on the previous workstation.

Qualified worker: term used in work study to denote a person who is accepted as having the necessary physical attributes, intelligence, skill, education and knowledge to perform the task.

Qualifiers: the competitive factors that have a minimum level of performance (the qualifying level) below which customers are unlikely to consider an operation's performance satisfactory.

Quality: there are many different approaches to defining this. We define it as consistent conformance to customers' expectations.

Quality characteristics: the various elements within the concept of quality, such as functionality, appearance, reliability, durability, recovery, etc.

Quality function deployment (QFD): a technique used to ensure that the eventual design of a product or service actually meets the needs of its customers (sometimes called 'house of quality').

Quality of experience: the overall acceptability of the service, as perceived subjectively by the end user.

Quality-related costs: an attempt to capture the broad cost categories that are affected by, or affect, quality, usually categorised as prevention costs, appraisal costs, internal failure costs and external failure costs.

Quality sampling: the practice of inspecting only a sample of products or services produced rather than every single one.

Quality variables: measures of quality that can be measured on a continuously variable scale, for example length, weight, etc.

Queuing theory: a mathematical approach that models random arrival and processing activities in order to predict the behaviour of queuing systems (also called 'waiting line theory').

Radical improvement: *see* 'Breakthrough improvement'.

Rating: a work study technique that attempts to assess a worker's rate of working relative to the observer's concept of standard performance – controversial and now accepted as being an ambiguous process.

Received variety: the variety that occurs because the process is not designed to prevent it.

Recovery: the activity (usually a predetermined process) of minimising the effects of an operation's failure.

Red Queen effect: the idea that improvement is relative; a certain level of improvement is necessary simply to maintain one's current position against competitors.

Redundancy: the extent to which a process, product or service has systems or components that are used only when other systems or components fail.

Reliability: when applied to operations performance, it can be used interchangeably with 'dependability'; when used as a measure of failure it means the ability of a system, product or service to perform as expected over time; this is usually measured in terms of the probability of it performing as expected over time.

Remainder cell: the cell that has to cope with all the products that do not conveniently fit into other cells.

Re-order level: the level of inventory at which more items are ordered, usually calculated to ensure that inventory does not run out before the next batch of inventory arrives.

Re-order point: the point in time at which more items are ordered, usually calculated to ensure that inventory does not run out before the next batch of inventory arrives.

Repeatability: the extent to which an activity does not vary.

Repetitive strain injury (RSI): damage to the body because of repetition of activities.

Research and development (R&D): the function in the organisation that develops new knowledge and ideas, and operationalises the ideas to form the underlying knowledge on which product, service and process designs are based.

Reshoring: the action of moving business activities that had been moved overseas back to the country from which they were originally relocated (also referred to as 'back-shoring', 'home-shoring' and 'on-shoring').

Resource-based view (RBV): the perspective on strategy that stresses the importance of capabilities (sometimes known as core competences) in determining sustainable competitive advantage.

Resource distance: the degree of novelty required of an operation's resources during the implementation of a new technology or process.

Resource-to-order: operations that buy-in resources and produce only when they are demanded by specific customers.

Respect for people: term used in lean philosophy to promote respectful behaviours, based on the idea that the effects of incivility are far-reaching and almost always negative.

Reverse engineering: the taking apart or deconstruction of a product or service in order to understand how it has been produced (often by a competing organisation).

RFID (radio frequency identification): technologies that use radio waves to automatically identify objects and often collect data about them.

Robotic process automation (RPA): a term for tools that function on the human interface of other computer systems using 'rules-based' decision algorithms.

Robots: automatic manipulators of transformed resources whose movement can be programmed and reprogrammed.

Rostering: a term used in planning and control, usually to indicate staff scheduling; the allocation of working times to individuals so as to adjust the capacity of an operation.

Run-to-breakdown maintenance: an approach to maintenance management that repairs a machine or facility only when it breaks down.

Sales and operations planning (S&OP): a formal business process that looks over a period of 18–24 months ahead in an attempt to integrate short- and longer-term planning, as well as integrating the planning activities of key functions.

SAP: a German company that is the market leader in supplying ERP software, systems and training.

Scenario planning: a method for planning where a range of possible future scenarios and their consequences are discussed.

Scheduling: a term used in planning and control to indicate the detailed timetable of what work should be done, when it should be done and where it should be done.

Scientific management: a school of management theory dating from the early twentieth century; more analytical and systematic than 'scientific' as such, sometimes referred to (pejoratively) as Taylorism, after Frederick Taylor who was influential in founding its principles.

Second-tier: the description applied to suppliers and customers who are separated from the operation only by first-tier suppliers and customers.

Sequencing: the activity within planning and control that decides on the order in which work is to be performed.

Service guarantee: a promise to recompense a customer for service that fails to meet a defined quality level.

Service-level agreements (SLAs): formal definitions of the dimensions and levels of service that should be provided by one process or operation to another.

Service shops: service processes that are positioned between professional services and mass services, usually with medium levels of volume and customisation.

Servicescape: a term used to describe the look and feel of the environment within an operation.

Servitisation: involves (often manufacturing) firms developing the capabilities they need to provide services and solutions that supplement their traditional product offerings.

Set-up reduction: the process of reducing the time taken to change over a process from one activity to the next; also called 'single-minute exchange of dies' (SMED) after its origins in the metal-pressing industry.

Seven types of waste: types of waste identified by Toyota, they are overproduction, waiting time, transport, process waste, inventory, motion and defectives.

Shop-within-a-shop: an operations layout that groups facilities that have a common purpose together; the term was originally used in retail operations but is now sometimes used in other industries, very similar to the idea of a cell layout.

Short-fat processes: processes designed with relatively few sequential stages, each of which performs a relatively large part of the total task; the opposite of long-thin processes.

Simulation: the use of a model of a process, product or service to explore its characteristics before the process, product or service is created.

Simultaneous development: overlapping these stages in the design process so that one stage in the design activity can start before the preceding stage is finished, the intention being to shorten time to market and save design cost (also called 'simultaneous engineering' or 'concurrent engineering').

Single-sourcing: the practice of obtaining all of one type of input product, component or service from a single supplier, as opposed to multi-sourcing.

SIPOC analysis: a method of formalising a process at a relatively general rather than a detailed level, it stands for suppliers, inputs, process, outputs and customers.

Six Sigma: an approach to improvement and quality management that originated in the Motorola Company but that was widely popularised by its adoption by the GE Company in America. Although based on traditional statistical process control, it is now a far broader 'philosophy of improvement' that recommends a particular approach to measuring, improving and managing quality and operations performance generally.

Social bottom line: the element of the triple bottom line that assesses the performance of a business in relation to the people and the society with which it has contact; and/or environmental mission and legal responsibility to respect the interests of workers, the community and the environment as well as shareholders.

Social responsibility: the incorporation of the operation's impact on its stakeholders into operations management decisions.

Socio-technical systems: a way of thinking about complex organisations, such as operations, that stresses the interaction between people and technology.

Speed: the elapsed time between customers requesting products or services and their receiving them.

Stakeholders: the people and groups of people who have an interest in the operation and who may be influenced by, or influence, the operation's activities.

Standard performance: a term used in work measurement to indicate the rate of output that qualified workers will achieve without over-exertion, as an average over the working day provided they are motivated to apply themselves, now generally accepted as a very vague concept.

Standard time: a term used in work measurement indicating the time taken to do a job and including allowances for recovery and relaxation.

Standard work: the step-by-step documentation of the most efficient way of performing a process.

Standardisation: the degree to which processes, products or services are prevented from varying over time.

Statistical process control (SPC): a technique that monitors processes as they produce products or services and attempts to distinguish between normal or natural variation in process performance and unusual or 'assignable' causes of variation.

Stock: alternative term for inventory.

Strategic decisions: those that are widespread in their effect, define the position of the organisation relative to its environment and move the organisation closer to its long-term goals.

Strategic level of operations performance: the five aspects of performance that contribute to the 'economic' aspect of the triple bottom line, cost, revenue, investment, risk and capabilities.

Structural decisions: the strategic decisions that determine the operation's physical shape and configuration, such as those concerned with buildings, capacity, technology, etc.

Sub-contracting: when used in medium-term capacity management, it indicates the temporary use of other operations to perform some tasks, or even produce whole products or services, during times of high demand.

Supply chain: a linkage or strand of operations that provides goods and services through to end customers; within a supply network several supply chains will cross through an individual operation.

Supply chain dynamics: the study of the behaviour of supply chains, especially the level of activity and inventory levels at different points in the chain; its best known finding is the bullwhip effect.

Supply network: the network of supplier and customer operations that have relationships with an operation.

Supply side: the chains of suppliers, suppliers' suppliers, etc. that provide parts, information or services to an operation.

Support functions: the functions that facilitate the working of the core functions, for example accounting and finance, human resources, etc.

Surge capacity: the capacity that an operation can deploy for only a relatively short time, often during unexpectedly high levels of demand.

Surge pricing: surge (or dynamic) pricing is a demand management technique that relies on frequent adjustments in price to influence supply and (especially) demand so that they match each other. For example, some electricity suppliers charge different rates for energy depending on when it is consumed.

Sustainability: the ability of a business to create acceptable profit for its owners as well as minimising the damage to the environment and enhancing the existence of the people with whom it has contact.

Synthesis from elemental data: work measurement technique for building up a time from previously timed elements.

Systemisation: the extent to which standard procedures are made explicit.

Take-back economy: *see* 'Circular economy'.

Takt time: (similar to cycle time) the time between items emerging from a process, usually applied to 'paced' processes.

Tangibility: the main characteristic that distinguishes products (usually tangible) from services (usually intangible).

Telemedicine: the ability to provide interactive healthcare, utilising modern telecommunications technology.

Teleworking: the ability to work from home using telecommunications and/or computer technology.

Theory of constraints (TOC): a philosophy of operations management that focuses attention on capacity constraints or bottleneck parts of an operation; uses software known as 'optimised production technology' (OPT).

Throughput efficiency: the work content needed to produce an item in a process expressed as a percentage of total throughput time.

Throughput time: the time for a unit to move through a process.

Time series analysis: a quantitative approach to forecasting that examines the pattern of past behaviour of a single phenomenon over time, considering reasons for variation in the trend, in order to use the analysis to forecast the phenomenon's future behaviour.

Time study: a term used in work measurement to indicate the process of timing (usually with a stopwatch) and rating jobs; it involves observing times, adjusting or normalising each observed time (rating) and averaging the adjusted times.

Time to market (TTM): the elapsed time taken for the whole design activity, from concept through to market introduction.

Top-down: the influence of the corporate or business strategy on operations decisions.

Total productive maintenance (TPM): an approach to maintenance management that adopts a similar holistic approach to total quality management (TQM).

Total quality management (TQM): a holistic approach to the management of quality that emphasises the role of all parts of an organisation and all people within an organisation to influence and improve quality; heavily influenced by various quality 'gurus', it reached its peak of popularity in the 1980s and 1990s.

Total supply network: all the suppliers and customers who are involved in supply chains that 'pass through' an operation.

Touchpoints: the points in a high-visibility process that are the points of contact between a process and customers.

Trade-off theory: the idea that the improvement in one aspect of operations performance comes at the expense of deterioration in another aspect of performance, now substantially modified to include the possibility that in the long term different aspects of operations performance can be improved simultaneously.

Transactional relationships: purchasing goods and services in a 'pure' market fashion, often seeking the 'best' supplier every time it is necessary to make a purchase.

Transformation process model: a model that describes operations in terms of their input resources, transforming processes, and outputs of goods and services.

Transformed resources: the resources that are treated, transformed or converted in a process, usually a mixture of materials, information and customers.

Transforming resources: the resources that act upon the transformed resources, usually classified as facilities (the buildings, equipment and plant of an operation) and staff (the people who operate, maintain and manage the operation).

Triad: the basic elements of a supply network involving three operations.

Triage: sequencing method that prioritises the most urgent jobs first.

Triple bottom line: (also known as people, plants and profit) the idea that organisations should measure themselves on social and environmental criteria as well as financial ones.

U-form organisation: an organisational structure that clusters its resources primarily by their functional purpose.

Upstream: the operations in a supply chain that are towards the supply side of the operation.

Usage value: a term used in inventory control to indicate the quantity of items used or sold multiplied by their value or price.

Utilisation: the ratio of the actual output from a process or facility to its design capacity.

Valuable operating time: the amount of time at a piece of equipment or work centre that is available for productive working after stoppages and inefficiencies have been accounted for.

Value-added throughput efficiency: the amount of time an item spends in a process having value added to it expressed as a percentage of total throughput time.

Value stream map: a mapping process that aims to understand the flow of material and information through a process or series of processes, it distinguishes between value-added and non-value-added times in the process.

Variation: the degree to which the rate or level of output varies from a process over time, a key characteristic in determining process behaviour.

Variety: the range of different products and services produced by a process, a key characteristic that determines process behaviour.

Vertical integration: the extent to which an operation chooses to own the network of processes that produce a product or service; the term is often associated with the 'do or buy' decision.

Virtual reality: uses entirely computer-generated simulations, with which humans can interact in a seemingly real manner using special helmets and gloves fitted with sensors.

Visibility: the amount of value-added activity that takes place in the presence (in reality or virtually) of the customer, also called 'customer contact'.

Visual management: an approach to making the current and planned state of an operation or process transparent to everyone.

Voice of the customer (VOC): capturing a customer's requirements, expectations and perceptions and using them as improvement targets within an operation.

Volume: the level or rate of output from a process, a key characteristic that determines process behaviour.

Volume flexibility: the operation's ability to change its level of output or activity to produce different quantities or volumes of products and services over time.

VUCA: acronym for volatility, uncertainty, complexity and ambiguity.

Waiting line theory: an alternative term for queuing theory.

Web-integrated ERP: enterprise resource planning that is extended to include the ERP-type systems of other organisations, such as customers and suppliers.

Weighted-score method of location: a technique for comparing the attractiveness of alternative locations that allocates a score to the factors that are significant in the decision, and weights each score by the significance of the factor.

Work breakdown structure: the definition of, and the relationship between, the individual work packages in project management; each work package can be allocated its own objectives that fit in with the overall work breakdown structure.

Work content: the total amount of work required to produce a unit of output, usually measured in standard times.

Work measurement: a branch of work study that is concerned with measuring the time that should be taken for performing jobs.

Work study: the term generally used to encompass method study and work measurement, derived from the scientific management school.

Work-in-progress (WIP): the number of units within a process waiting to be processed further (also called 'work-in-process').

Work–life balance: the imperative to achieve an appropriate split between work and personal life so that work should not interfere unreasonably with family obligations and personal interests.

Workflow: process of design of information-based processes.

Yield management: a collection of methods that can be used to ensure that an operation (usually with a fixed capacity) maximises its potential to generate profit.

Zero defect: the idea that quality management should strive for perfection as its ultimate objective even though in practice this will never be reached.

INDEX

CREDITS

Text:

6 The LEGO Group: Based on the corporate websites of LEGOLAND www.legoland.com; **6 Merlin Entertainment:** Merlin Entertainment https://www.merlinentertainments.biz/ (accessed August 2021); **6 Gizmodo:** Diaz, J. (2008) Exclusive look inside the Lego Factory, Gizmodo, 21 July, http://lego.gizmodo.com/exclusive-look-inside-the-lego-factory-5022769 (accessed August 2021); **10 Médecins Sans Frontières:** The information on which this example is based is taken from www.msf.org and https://blogs.msf.org/about-us (accessed August 2021); **15 Marina Bay Sands:** The information on which this example is based is taken from: the hotel's website, https://www.marinabaysands.com/ (accessed August 2021); **18 Ellen MacArthur Foundation:** The Ellen MacArthur Foundation: Let's build a circular economy, https://ellenmacarthurfoundation.org (accessed August 2021); **18 GreenBiz Group Inc:** Phipps, L. (2018) How Philips became a pioneer of circularity-as-a-service, GreenBiz, 22 August; **18 Koninklijke Philips:** Philips website, The circular imperative, https://www.philips.com/a-w/about/environmental-social-governance/environmental/circular-economy.html (accessed August 2021); **24 Tom Avery:** Based on a personal communication with Tom Avery CEO of Verbier Sky Exclusive; **27 FJÄLLRÄVEN:** The company website, https://www.fjallraven.com/; **27 SGB Media:** Silven, R. (2020) Fjällräven voted most sustainable brand in its industry according to Sweden's Sustainable Brand Index, sgbonline.com, 29 April, https://sgbonline.com/pressrelease/fjallraven-voted-most-sustainable-brand-in-its-industry-according-to-swedens-sustainable-brand-index/ (accessed August 2021); **27 FJÄLLRÄVEN:** Sustainability – Fjällräven SEA. Retrieved from https://fjallravensea.com/pages/sustainability; **27 TYF:** Fjällräven – TYF. Retireved from https://www.tyf.com/pages/fjallraven; **30 BBC:** BBC News (2002) Politicians 'trample over' patient privacy, 1 July, http://news.bbc.co.uk/1/hi/in_depth/health/2002/bma_conference/2077391.stm (accessed August 2021); **42 B Corp:** B Corp website, https://bcorporation.net/about-bcorps (accessed August 2021); **42 Economist:** Economist (2018) Choosing plan B – Danone rethinks the idea of the firm, Business section, Economist print edition, 9 August; **47 Times Newspapers Limited:** Willan, P. (2018) Spread the word: dream job if you're nuts about chocolate, *The Times*, 28 July; **47 Reuters:** Reuters Staff (2018) Ferrero stops production at biggest Nutella plant to assess quality issue, Reuters, 21 February, https://www.reuters.com/article/ferrero-nutella-stop-idUSL-5N20G5NY (accessed August 2021); **47 France24:** france 24 (2019) World's largest Nutella factory reopens after 'quality defect' france24, 25 February, https://www.france24.com/en/20190225-worlds-largest-nutella-factory-reopensafter-quality-defect (accessed August 2021); **47 Times Newspapers Limited:** Sage, A. (2018) Nutella fistfights spread at Intermarché stores across France, *The Times*, 26 January; **49 THE FINANCIAL TIMES LTD:** Palmer, M, (2020) Smart ambulances and wearables offer route to speedier treatments, *Financial Times*, 24 November; **49 London's Air Ambulance Charity:** London's Air Ambulance Service can be found at https://www.londonsairambulance.org.uk (accessed August 2021); **51 Guardian News & Media Limited:** McCurry, J, (2017) Japanese rail company apologises after train leaves 20 seconds early, *Guardian*, 17 November; **51 The Local Europe AB:** The Local (2017) SBB remains most punctual train company in Europe, news@thelocal.ch, 21 March; **61 John Wiley & Sons, Inc:** Skinner, W. (1985) *Manufacturing: The Formidable Competitive Weapon*, John Wiley & Sons, New York, NY; **66 Patrick O'Brien:** Quoted by Patrick O'Brien; **67 Gillian Drakeford:** Gillian Drakeford, IKEA's UK boss; **67, 68 Times Newspapers Limited:** Hipwell, D. (2017) This is no time to sit back and relax – we must deliver, says IKEA's UK boss, *The Times*, 10 February; **67 Ray Gaul:** Quoted by Ray Gaul; **67 Torbjorn Loof:** Quoted by Torbjorn Loof; **67 Jesper Brodin:** Quoted by Jesper Brodin; **68 IKEA:** IKEA website, https://www.inter.ikea.com/ (accessed September 2021); **68 Bloomberg:** Matlack, C. (2018) The tiny Ikea of the future, without meatballs or showroom mazes, *Bloomberg Businessweek*, 10 January; **68 THE FINANCIAL TIMES LTD:** Milne, R. (2018) What will Ikea build next? *Financial Times*, 1 February; **68 Economist:** Economist (2017) Frictionless furnishing: IKEA undertakes some home improvements, *Economist* print edition, 2 November; **68 Sustainable Life Media, Inc.:** Gerschel-Clarke, A. (2016) 'Peak Stuff': why IKEA is shifting towards new business models, Sustainablebrands.com, 17 February;

68 THE FINANCIAL TIMES LTD: Milne, R. (2017) Ikea turns to ecommerce sites in online sales push, *Financial Times*, 9 October; **68 BBC:** Hope, K. (2017) Ikea: why we have a love-hate relationship with the Swedish retailer, BBC News, 17 October; **68 The Telegraph:** Armstrong, A. (2017) Revealed: how after 30 years, Ikea is undergoing a radical overhaul, *The Telegraph*, 15 October; **68 Marc-André Kamel:** Marc-André Kamel of consultants Bain & Company; **74 THE FINANCIAL TIMES LTD:** Braithwaite, T. (2020) How a UK supermarket nourished Silicon Valley's critics, *Financial Times*, 6 November; **74 The Sunday Times:** Chambers, S. (2019) Ocado the disruptor is being disrupted, *The Sunday Times*, 1 December; **75, 190 John Wiley & Sons, Inc:** Hayes, R.H. and Wheelwright, S.C. (1984) *Restoring our Competitive Edge: Competing Through Manufacturing*, John Wiley & Sons, Inc., New York, NY; **82 Kogan Page:** Based on an example from Slack, N. (2017) *The Operations Advantage*, Kogan Page, London; **85 Times Newspapers Limited:** Clark, A. and Ralph, A. (2014) Tesco boss defiantamid 4% plunge in sales, *The Times*, 5 June; **85 THE FINANCIAL TIMES LTD:** Vandevelde, M. (2016) Tesco ditches global ambitions with retreat to UK, *Financial Times*, 21 June; **98 Elsevier:** Based on the work of Carroll, A.B. (1991) The pyramid of social responsibility: towards the moral management of organizational stakeholders, *Business Horizons*, 34 (4) July/August, 39–48; **103 Ray Kroc:** Quoted by Ray Kroc; **104 Berkley Publishing Corp:** Kroc, R.A. (1977) *Grinding it Out: The Making of McDonald's*, St. Martin's Press, New York; **105 THE FINANCIAL TIMES LTD:** Whipp, L. (2015) McDonald's to slim down in home market, *Financial Times*, 18 June; **105 THE FINANCIAL TIMES LTD:** Smith, T. (2015) Where's the beef, *Financial Times*, 22 May; **105 THE FINANCIAL TIMES LTD:** Whipp, L. (2015) McDonald's may struggle to replicate British success, *Financial Times*, 5 May; **105 McDonald's:** McDonald's Annual Report, 2017; **105 Macmillan Publisher:** Kroc, R.A. (1977) Grinding it Out: The Making of McDonald's, St. Martin's Press, New York; **105 Times Newspapers Limited:** Cooper, L. (2015) At McDonald's the burgers have been left on the griddle too long, *The Times*, 24 August; **113 The Economist Newspaper Limited:** The information on which this example is based is taken from: Economist (2018) The invention, slow adoption and near perfection of the zip, *Economist* print edition, 18 December; **123 The Economist Newspaper Limited:** The information on which this example is based is taken from: Economist (2017) One of the world's oldest products faces the digital future, *Economist* print edition, 12 October; **128 BT Group plc :** BT website, How BT innovates, https://www.bt.com/about/innovation/how-bt-innovates (accessed August 2021); **128 BT Group plc:** BT News (2018) BT launches Better World Innovation Challenge for start-ups & SMEs, press release from BT, 8 May; **128 BT Group plc:** BT Group plc Annual Report, Strategic Report, 2019; **128 BT Group plc:** Fransman, M. (2014) *Models of Innovation in Global ICT Firms: The Emerging Global Innovation Ecosystems* (ed. M. Bogdanowicz), JRC Scientific and Policy Reports – EUR 26774 EN. Seville: JRC-IPTS; **132 Productivity Press:** Morgan, J. and Liker, J.K. (2006) *The Toyota Product Development System: Integrating People, Process, and Technology*, Productivity Press, New York, NY; **132 Harvard Business Press:** Sobek II, D.K., Liker, J., and Ward, A.C. (1998) Another look at how Toyota integrates product development, *Harvard Business Review* (July–August); **133 THE FINANCIAL TIMES LTD:** Clegg A (2015) Sustainable innovation: shaped for the circular economy, *Financial Times*, 26 August; **148 Scott Fitzgerald:** Quoted by Scott Fitzgerald; **151 The New York Times Company:** Schuetze, C.F. (2014) Dutch flower auction, long industry's heart, is facing competition, *New York Times*, 16 December; **151 Royal FloraHolland:** Company website, https://www.royalfloraholland.com/en/about-floraholland (accessed August 2021); **154 Industry Dive:** Hernández, A. (2020) Learning from Adidas' Speedfactory blunder, Suppychaindive, 4 February, https://www.supplychaindive.com/news/adidas-speedfactory-blunder-distributed-operations/571678/; **154 Quartz Media, Inc:** Bain, M. (2019) Change of plan, Quartz, 11 November, https://qz.com/1746152/adidas-is-shutting-down-its-speedfactories-in-germany-and-the-us/ (accessed August 2021); **167 THE FINANCIAL TIMES LTD:** Fildes, N. (2017) Vodafone to bring 2,100 call-centre jobs back to UK, *Financial Times*, 13 March; **167 TechTarget:** Flinders, K. (2017) Vodafone brings offshore contact centre work to UK, *Computer Weekly*, 13 March; **169 Fashion United:** Hendriksz, V. (2018) 5 years on: what effect has Rana Plaza had on garment workers lives?, Fashion United, 16 April, https://fashionunited.uk/news/fashion/5-yearson-what-effect-has-rana-plaza-had-on-garment-workerslives/2018041629133; **169 International Labour Organization:** International Labour Organization (n.d.) The Rana Plaza accident and its aftermath, https://www.ilo.org/global/topics/geip/WCMS_614394/lang--en/index.htm (accessed August 2021); **180 Condé Nast:** Zhang, S. (2016) 'How to fit the world's biggest indoor waterfall in an airport', *Wired*, 9 July; **180 Verdict Media Limited:** Airport Technology (2014) Terminal 4, Changi International Airport, https://www.airport-technology.com/projects/terminal-4-changi-international-airport-singapore/ (accessed September 2021); **180 Associated Newspapers Ltd:** Driver, C. (2014) And the winners are . . . Singapore crowned the best airport in the world (and Heathrow scoops top terminal), Mailonline, 28 March, https://www.dailymail.co.uk/travel/article-2591405/Singapore-crowned-best-airportworld-Heathrow-scoops-terminal.html (accessed September 2021); **183 JOURNALISTIC, INC:** Oches, S. (2013) The drive-thru performance study, *QSR magazine*, September; **183 Gannett Satellite Information Network LLC:** Horovitz, A. (2002) Fast Food World says drive-through is the way to go, *USA Today*, 3 April; **183 The New York Times Company:** Richtel, M. (2006) The long-distance journey of a fast-food order, *The New York Times*, 11 April, https://www.nytimes.com/2006/04/11/technology/the-longdistancejourney-of-a-fastfood-order.html (accessed September 2021); **184 Ellen MacArthur Foundation:** The built environment: Achieving a resilient recovery with the circular economy', report by the Ellen MacArthur Foundation, https://www.ellenmacarthurfoundation.org/our-work/activities/covid-19/policy-and-investment-opportunities/the-built-environment (accessed August 2021); **184 Inside Housing:** Wilmore, J, (2019) We take a look around L&G's housing factory, *Inside Housing*, 14 February; **184 Legal & General Group plc:** Legal and General Modular Homes website, https://www.legalandgeneral.com/modular/a-modern-method/; **185 Packaging News:** Qureshi, W. (2020) Ecover relaunches biodegradable detergents in PCR plastic, *Packaging News*, 21 January; **185 Media One Communications Ltd:** Cornwall, S. (2013) Ecover announces world-first in plastic packaging, *Packaging Gazette*, 7 March; **185 Ecover:** Ecover website, http://www.ecover.com (accessed August 2021); **191 Lihua Zhu:** Quoted on Dishang Group's website, www.dishang-group.com (accessed August 2021); **191 Sands Films:** Sutherland, E. (2017) Weihai and mighty, Drapersonline, 16 June; **191 EMAP PUBLISHING LIMITED:** Adapted from www.sandsfilms.co.uk (accessed August 2021); **196 Harvard Business Press:** Shostack, G.L. (1984) Designing services that deliver, *Harvard Business Review*, 62 (1), 133–9; **202 Reach Plc:** Matthews, T. and Trim, L. (2019) London Underground: why it would be better if we stood on both sides of the escalators, MyLondon Local News, 13 August; **202 Evening Standard:** Sleigh, S. (2017) TfL scraps standing only escalators – despite trial being deemed a 'success', *London Evening Standard*, 8 March; **219 Mansueto Ventures LLC:** Segran, E. (2015) Designing a happier office on the super cheap, Fast Company, 30 March; **219 Bauer Media Group:** Urry, J. (2017) Inside Ducati: MCN walk around the Bologna factory, *Motorcycle News*, 21 September; **219 Guardian News & Media Limited:** Hickey, S. (2014) Death of the desk: the architects shaping offices of the future, *Guardian*, 14 September; **222 Guardian News & Media Limited:** Booth, R. (2017) Francis Crick Institute's £700m building too noisy to concentrate, *Guardian*, 21 November; **232 The Economist Newspaper Limited:** Economist (2019) Future of the workplace: redesigning the corporate office, *Economist* print edition, 28 September; **232 The Economist Newspaper Limited:** Economist (2019) Why open-plan offices get a bad rap, *Economist* print edition, 24 October; **232 Harvard Business Publishing:** Waber, B., Magnolfi, J. and Lindsay, G. (2014) Workspaces that move people, *Harvard Business Review*, October; **234 Urbanist Architecture Ltd:** Urbanist Architecture (2020) Virtual reality in architecture: visit your home before it's been built with VR, 12 April, https://urbanistarchitecture.co.uk/urbanist-4d-reality-virtual-reality-technology-in-architectural-design/ (accessed September 2021); **234 Vizerra SA:** Adapted from Revizto.com; **238 Rolls-Royce Motor cars:** Rolls-Royce (2020) Birds, bees, roses and trees all thriving at the home of Rolls-Royce, Rolls-Royce Media Information, Goodwood, 2 July; **238 Robb Report Media, LLC:** Burstein, L. (2015) An inside look at the Rolls-Royce assembly plant in Goodwood, Robb Report, 23 October, https://robbreport.com/motors/cars/inside-look-rolls-royce-assembly-plant-goodwood-229474/ (accessed September 2021); **238 Rolls-Royce Motor cars:** Rolls-Royce (2017) Home of Rolls-Royce motor cars, press release, Rolls-Royce Media Information, Goodwood, 7 April; **252 The Economist Newspaper Limited:** Economist (2020) Businesses are finding AI hard to adopt, *Economist Technology Quarterly*, 11 June; Economist (2016) Artificial intelligence and Go: Showdown, *Economist*, 12 March; **252 Scientific American:** Koch, C. (2016) How the computer beat the Go master, *Scientific American*, 19 March; **254 Gartner:** Used with permission from Gartner.; **257 QB Net co, Ltd:** The information on which this example is based is taken from: QB House website, http://www.qbhouse.com (accessed September 2021); **259 SAP Insights:** SAP IOT Definition: SAP Research, https://insights.sap.com/what-is-iot-internet-of-things/ (accessed September 2021); **261 Times Newspapers Limited:** West, K (2011) Turn up the heat with Marmite, *The Sunday Times*, 2 October, https://www.thetimes.co.uk/article/turn-up-the-heat-with-marmite-vz8d87qx253 (accessed September 2021); **267 THE FINANCIAL TIMES LTD:** Harford, T. (2020) Why tech isn't always the answer – the perils of bionic duckweed, *Financial Times*, 30 October; **269 Times Newspapers Limited:** Based on Walsh, D. (2015) Irregular parcels put UK Mail out of shape, *The Times*, 8 August; **269 UK Mail Limited:** UK Mail website https://www.ukmail.com; **271 Times Newspapers Limited:** Deng, B. (2016) Security robot runs over toddler at shopping centre, *The Times*, 15 July; **271 Times Newspapers Limited:** Times Leader (2016) They, robots, *The Times*, 1 January; **271 Times Newspapers Limited:** Hall, A. (2015) Factory robot grabs worker and kills him, *The Times*,

3 July; **275 Thalia:** Quoted by Thalia, Chief Technology Officer, Logaltel; **276 Jamal:** Quoted by Jamal, a Senior Operations Manager, Logaltel; **276 Martha:** Quoted by Martha, the firm's Business Development Manager, Logaltel; **277, 394 Vaggelis Giannikas:** This case was co-authored with Vaggelis Giannikas, at the University of Bath School of Management; **277 BBC:** Sherman, N. (2018) Wanted: robot wrangler, no experience required, BBC News, 21 March, https://www.bbc.co.uk/news/business-43259903 (accessed September 2021); **282 John Wiley & Sons, Inc:** Schein, E.H. (1999) *The Corporate Culture Survival Guide: Sense and Nonsense About Culture Change*, Jossey-Bass, San Francisco, CA; **284 Torchbox:** Based on an interview with Tom Dyson, and the Torchbox website, www.torchbox.com (accessed September 2021; **292 Times Newspapers Limited:** Byers, D. (2017) Bionic suits to make tools feel weightless, *The Times*, 24 July; **292 New Atlas:** Coxworth, B. (2017) Exoskeleton helps Ford workers reach up, New Atlas, 13 November, https://newatlas.com/ford-eksovest/52166/ (accessed September 2021); **292 Vox Media LLC:** Goode, L. (2017) Are exoskeletons the future of physical labor? The Verge, 5 December, https://www.theverge.com/2017/12/5/16726004/verge-next-level-season-two-industrial-exoskeletons-ford-ekso-suitx (accessed September 2021); **293 University of California Press:** Hackman, J.R., Oldham, G., Janson, R. and Purdy, K. (1975) A new strategy for job enrichment, *California Management Review*, (17) 4, 57–71; **294 THE FINANCIAL TIMES LTD:** Hill, A. (2017) Power to the workers: Michelin's great experiment, *Financial Times*, 11 May; **294 THE FINANCIAL TIMES LTD:** Hill, A. (2017) Michelin chief Jean-Dominique Senard devolves power to workers, *Financial Times*, 14 May; **294 Michelin:** Michelin (2017) 2016 Annual Report; **296 The Economist Newspaper Limited:** Nixey, C. (2020) Death of the office, *Economist 1843 Magazine*, 29 April; **296 The Economist Newspaper Limited:** Economist (2020) Countering the tyranny of the clock, *Economist* print edition, 17 October; **296 Times Newspapers Limited:** Treanor, J. (2021) Has Goldman's DJ just pulled the plug on WFH? *The Sunday Times*, 28 February; **296 THE FINANCIAL TIMES LTD:** Hill, A. (2020) Future of work: how managers are harnessing employees' hidden skills, *Financial Times*, 1 September; **300 Times Newspapers Limited:** Bone, J., Robertson, D. and Pavia, W. (2010) Plane rumpus puts focus on crews' growing revolution in the air, The Times, 11 August; **301 THE FINANCIAL TIMES LTD:** Jones, A. (2015) The riff: dangers of music at work, *Financial Times*, 5 August; **301 Mansueto Ventures LLC:** Ciotti, G. (2014) How music affects your productivity, *Fast Company*, 11 July; **301 BBC:** BBC (2013) Does music in the workplace help or hinder? *Magazine Monitor*, 9 September; **304 BBC:** Derousseau, R. (2017) The tech that tracks your movements at work, BBC Worklife, 14 June, https://www.bbc.com/worklife/article/20170613-the-tech-that-tracksyour-movements-at-work (accessed September 2021); **304 Guardian News & Media Limited:** Solon, O. (2017) Big Brother isn't just watching: workplace surveillance can track your every move, *Guardian*, 6 November; **304 INFORMS:** Staats, B.R., Dai, H., Hofmann, D. and Milkman, K.L. (2016) Motivating process compliance through individual electronic monitoring: an empirical examination of hand hygiene in healthcare, *Management Science*, 63 (5), 1563–85; **304 Times Newspapers Limited:** Webster, B. (2018) CCTV to monitor hygiene in meat factories, *The Times*, 3 March; **306 Grace Whelan:** Grace Whelan, Managing Partner of McPherson Charles; **324 Business Traveller:** Caswell, M. (2020) Air France to operate 50 per cent of schedules during November and December, *Business Traveller*, 28 September; **324 Richard E Stone:** Farman, J. (1999) 'Les Coulisses du Vol', Air France, talk presented by Richard E. Stone, NorthWest Airlines at the IMA Industrial Problems Seminar, 1998; **335 Vox Media LLC:** Barro, J. (2019) Here's why airplane boarding got so ridiculous, *New York Magazine Intelligencer*, 9 May, https://nymag.com/intelligencer/2019/05/heres-why-airplane-boarding-got-so-ridiculous.html (accessed September 2021); **335 The Economist Newspaper Limited:** The Economist (2011) Please be seated: A faster way of boarding planes could save time and money, *Economist* print edition, 3 September; **336 The Economist Newspaper Limited:** Economist (2020) Triage under trial: the tough ethical decisions doctors face with covid-19, *Economist* print edition, 2 April; **336 THE FINANCIAL TIMES LTD:** Jones, C. (2020) What a career in intensive care nursing has taught me about triage, *Financial Times*, 6 February; **339 Heathrow Airport Limited:** Heathrow website, https://www.heathrow.com (accessed September 2021); **339 Cambridge University Press:** For a technical explanation of the aircraft landing algorithm, see Cecen, R.K., Cetek, C. and Kaya, O. (2020) Aircraft sequencing and scheduling in TMAs under wind direction uncertainties, *The Aeronautical Journal*, 124 (1282), 1896–912; **339 Taylor & Francis Group:** Beasley, J.E., Sonander, J. and Havelock, P. (2001) Scheduling aircraft landings at London Heathrow using a population heuristic, *Journal of the Operational Research Society*, 52, 483–93; **343 The Independent:** Calder, S. (2017) Ryanair cancellations: the truth behind why 2,000 flights are due to be scrapped, *Independent*, 19 September; **352 Dan Audall:** Quoted by Dan Audall; **360 Bloomburg L.P:** Gruley, B. and Clough, R. (2020) How 3M plans to make more than a billion masks by the end of the year, *Bloomburg Businessweek*, 25 March;

360 Infiniti Research Limited: Technavio (2020) Coronavirus outbreaks boosts the sales of the world's top N95 mask manufacturers, 8 April; **367 MIT Technology Review:** Heaven, W. (2020) Our weird behaviour during the pandemic is messing with AI models, *MIT Technology Review*, 11 May; **367 S&P Global:** S&P Global (2020) Industries most and least impacted by COVID-19 from a probability of default perspective, 22 March; **367 McKinsey & Company:** McKinsey & Company (2020) COVID-19: Global health and crisis response, 6 July; **370 Simmons-Boardman Publishing:** Das, A.K. (2019) Six bidder vie for Indian Railways ETCS Level 2 pilot project, *International Railway Journal*, 7 November; **370 Cognitive Publishing:** Rail Technology Magazine (2017) Network Rail awards landmark £150m ETCS signalling contract, 20 December; **370 RailwayPRO Communication:** Railway Pro (2018) India to install ETCS Level 2 on its entire broad-gauge network, 8 March; **370 Simmons-Boardman Publishing:** Jha, S. (2018) Modi blocks Indian Railways ETCS plan, *International Railway Journal*, 11 April; **374 Mediacorp Pte Ltd:** Chong, A. (2019) What will it take for LTA's latest anti-congestion plan to work? *Channel News Asia*, International Edition, 13 May; **374 The Economist:** Economist (2015) Squeezing in: what the London Underground reveals about work in the capital, *Economist* Print Edition, 23 May; **376 Times Newspapers Limited:** Gadher, D. (2019) Art-lovers see red at surge pricing, *The Sunday Times*, 18 August; **376 The Economist:** The Economist (2016) A fare shake: jacking up prices may not be the only way to balance supply and demand for taxis, *Economist*, 14 May; **376 Harvard Business Publishing:** Dholakia, U.M. (2015) Everyone hates Uber's surge pricing – here's how to fix It, Harvard Business Review, 21 December; **380 SAGE Publications:** Cornelissen, J. and Cholakova, M. (2019) Profits Uber everything? The gig economy and the morality of category work, *Strategic Organisation*, 23 December; **380 BBC:** Russon, M. (2021) Uber drivers are workers not self-employed, Supreme Court rules, BBC News, 19 February; **380 CNN Business:** O'Brien, S. (2021) Uber's UK drivers to get paid vacation, pensions following Supreme Court ruling, *CNN Business*, 17 March; **380 The New York Times Company:** The New York Times (2018) What will New York do about its Uber problem?, 7 May; **409 Forbes Media LLC:** Banker, S. (2017) Drones deliver life saving supplies in Africa, *Forbes*, 13 October; **409 Condé Nast:** Stewart, J. (2017) Blood-carrying, life-saving drones take off for Tanzania, *Wired*, 24 August; **409 Zipline:** flyzipline.com/how-it-works/ [accessed September 2021]; **410 The Loadstar:** van Marle, G. (2021) E-commerce giant JD.com applies to spin-ff supply chain arm, The Loadstar, 19 February; **410 China Internet Watch:** CIW Team (2021) JD.com annual customers grew 30% to 472 million in 2020, China Internet Watch, 12 March; **410 Jingdong JD.com:** JD.com 'About us' [accessed September 2021], JD.com announces first quarter 2021 results (2021), press release, JD.com [accessed September 2021]; **418 Financial Times Limited:** Evans, J. (2020) Covid-19 crises highlights supply chain vulnerability, *Financial Times*, 28 May; **418 Association of International Certified Professional Accountants:** MacDowall, A. (2021) Managing warehousing in a changed era, *Financial Management*, 5 January; **418 Vox:** Bonadio, B., Huo, Z., Levchenko, A. and Pandalai-Nayar, N. (2020) The role of global supply chains in the COVID-19 pandemic and beyond, 25 May; **418 Ernst & Young Global Limited:** Harapko, S. (2021) How COVID-19 impacted supply chains and what comes next, EY, 18 February; **423 Financial Times Limited:** Williams, G.A. (2021) China Cancels H&M, *Jing Daily*, 24 March; Indvik, L. (2021) Fashion, Xinjiang and the perils of supply chain transparency, *Financial Times*, 9 April; **423 Eco-Business:** Danigelis, A. (2018) Supply chain transparency map reduces time, expense for big brands, Eco-Business, 1 February; **423 John Wiley & Sons, Inc:** Kuruvilla, S. and Li, C. (2021) Freedom of association and collective bargaining in global supply chains: a research agenda, *Journal of Supply Chain Management*, 57 (2), 43–57; **427 The Donkey Sanctuary:** Author visit to The Donkey Sanctuary, Sidmouth, UK, 2020, https://www.thedonkeysanctuary.org.uk/ (accessed September 2021); **427 PEARL Research Journals:** Hameed, A., Tariq, M. and Yasin, M.A. (2016) Assessment of welfare of working donkeys and mules using health and behavior parameters, *Journal of Agricultural Science and Food Technology*, 2 (5), 69–74; **434 CoinDesk:** del Castillo, M. (2018) Shipping blockchain: Maersk spin-off aims to commercialize trade platform, Coindesk.com, 16 January; **434 Maersk:** Slocum, H. (2018) Maersk and IBM to form joint venture applying blockchain to improve global trade and digitize supply chains, press release, Maersk, 18 January; **434 Maersk:** (2019) TradeLens blockchain-enabled digital shipping platform continues expansion with addition of major ocean carriers Hapag-Lloyd and Ocean Network Express, Press release, Maersk, 2 July; **438 Jas Kalra/Jens Roehrich/Brian Squire:** This case was co-authored with Jas Kalra, Bartlett School of Construction and Project Management, University College London, and Jens Roehrich and Brian Squire, School of Management, University of Bath; **445 The Conversation Media Group Ltd:** Koen, A. and Antunez, P.F. (2020) How heat can be used to store renewable energy, The Conversation, 25 February; **445 Times Newspapers Limited:** Gosden, E. (2017) Power shift brings energy market closer to holy grail, *The Times*, 17 April; **445 The Economist:** The Economist (2012) Energy storage: packing some power, *Economist Technology Quarterly*, 3 March; **447 The Economist:** The Economist (2019) A nation of have-beans, Defending Switzerland's coffee stockpile, *Economist* print edition, 21 November; **447 Financial Times Limited:** Foster, P. and Neville, S. (2020) How poor planning left the UK without enough PPE, *Financial Times*, 1 May; **447 Thomas Publishing Company:** Britt, H. (2020) What is safety stock and how can businesses use it to ensure continuity?, Thomasnet.com, 8 April, https://www.thomasnet.com/insights/what-is-safety-stock/ (accessed September 2021); **447 Publicis Sapient:** Anderson, H. (2020) COVID-19: preparing your supply chain in times of crisis, publicissapient.com, 8 April, https://www.publicissapient.com/insights/coronavirus_and_managing_the_supply_chain_amid_a_crisis (accessed September 2021); **461 The Economist:** The information on which this example is based is taken from: The Economist (2015) Croissantonomics – lessons in managing supply and demand for perishable products, *Economist* print Edition, 29 August; **466 Mashable, Inc:** Ulanoff, L. (2014) Amazon knows what you want before you buy it. Mashable, 21 January. Available at: https://mashable.com/2014/01/21/amazon-anticipatory-shipping-patent/#Ryy4twKmRiqb (accessed September 2021); **466 Times Newspapers Limited:** Duke, S. (2014) He knows what you want — before you even want it, *The Sunday Times*, 2 February; **466 Times Newspapers Limited:** Ahmed, M. (2014) Amazon will know what you want before you do, *The Times*, 27 January; **466 Business Insider:** Bernard, Z. (2018) Amazon is spending more and more on shipping out your orders. [online] Business Insider, https://www.businessinsider.com/amazons-logistics-costs-are-growing-really-fast-charts-2018-2 (accessed September 2021); **472 Times Newspapers Limited:** Sage, A. (2019) France to ban luxury brands from dumping unsold stock, *The Times*, 24 September; **472 Financial Times Limited:** Leroux, M. (2016) Burberry boss defends stock destruction, *The Times*, 15 July; Atkins, R. (2016) Richemont buys back and destroys stock as sales fall, *Financial Times*, 20 May; **472 The New York Times Company:** Dwyer, J. (2010) A clothing clearance where more than just the prices have been slashed, *New York Times*, 5 January; **473 Thomson Reuters:** Quoted in, Kajimoto, T. (2021) Japanese companies go high-tech in the battle against food waste, Reuters, 28 February; **482 Ms Keri Allan:** Allan, K. (2009) Butcher's Pet Care relies on IT that can co-ordinate its ERP, *Engineering & Technology Magazine*, 21 July; **486 Oliver Wight Ltd:** Wight, O. (1984) *Manufacturing Resource Planning: MRP II*, Oliver Wight Ltd; **488 The Times:** The information on which this example is based is taken from: Ellson, A. (2021) *The Times*, 24 April; **488 The Telegraph:** Dixon, H. (2021) Call to prosecute Post Office bosses over 'biggest miscarriage in British legal history', *The Telegraph*, 23 April; **492 Informa PLC:** Brynjolfsson, E. (1994) Technology's true payoff, *Information Week*, October; **493 IDG Communications, Inc:** Fruhlinger, J., Wailgum, T. and Sayer, P. (2020) 16 famous ERP disasters, dustups and disappointments, CIO.com, 20 March, https://www.cio.com/article/2429865/enterprise-resource-planning-10-famous-erp-disasters-dustups-and-disappointments.html; **493 Novacura AB:** Novacura (2019) 4 ERP implementation failures with valuable lessons, The Novacura Flow blog, 19 February, https://www2.novacura.com/blog/why-do-erp-implementations-fail; **493 IDG Communications, Inc:** Kanaracus, C. (2008) Waste Management sues SAP over ERP implementation, InfoWorld, 27 March; **494 Emerald Group Publishing Limited:** Based on Finney, S. and Corbett, M. (2007) ERP implementation: a compilation and analysis of critical success factors, *Business Process Management Journal*, 13 (3), 329–47; **495 Project Perfect:** Turbit, N. (2005) ERP Implementation – The Traps, The Project Perfect White Paper Collection, www.projectperfect.com.au; **514 Macmillan Ltd:** Carroll, L. (1871) *Through the Looking Glass*, Penquin Classics, 2008; **515 McKinsey & Company:** The information on which this example is based is taken from: Onetto, M. (2014) When Toyota met e-commerce: lean at Amazon, *McKinsey Quarterly*, No 2, 1 February; **519 Sue Jenkins:** We are grateful to Sue Jenkins, Director of kaizen, Sussex Healthcare NHS Trust, for this example; **521 Metropolitan Books:** Gawande, A. (2010) *The Checklist Manifesto: How to Get Things Right*, Profile Books, London; **521 Times Newspapers Limited:** Aaronovitch, D. (2010) The Checklist Manifesto: review, *The Times*, 23 January; **539 Harvard Business Publishing:** Shenkar, O. (2010) *Copycats: How Smart Companies Use Imitation to Gain a Strategic Edge*, Harvard Business Review Press; **540 The Economist:** The information on which this example is based is taken from: The Economist (2016) The great escape: what other makers can learn from the revival of Triumph motorcycles, *Economist* print edition, 23 January; **542 Times Newspapers Limited:** West, K. (2011) Formula One trains van drivers, *The Sunday Times*, 1 May; **542 Durham Associates Group:** f1network.net, http://www.f1network.net/main/s107/st164086.htm; **543 Schlumberger Limited:** Schlumberger Press Release (2010) Schlumberger Cited for Knowledge Management, Schlumberger Press Office, 3 December; **543 IESEG School of Management:** Deltour, F., Plé, L. and Sargis-Roussel, C. (2013) Eureka! Developing online communities of practice to facilitate knowledge sharing at schlumberger, IESEG School of Management, LEM, case study 313-122-1; **563 McKinsey & Company:** The information on which this example is based is taken from: Corbett, S. (2004) Applying lean in offices, hospitals, planes, and trains, presentation at The Lean Service Summit, Amsterdam, 24 June; **565 Condé Nast:** Burgess, M. (2018) Airbus is going to start putting beds in airplane cargo holds, *Wired*, UK. Retreived from https://www.wired.co.uk/article/airbus-sleeping-pods-naps-cargo-hold-zodiac-330; **568 Janina Aarts/Mattia Bianchi:** Example written and supplied by Janina Aarts and Mattia Bianchi, Department of Management and Organization, Stockholm School of Economics. **570 McKinsey & Company:** The information on which this example is based is taken from: Onetto, M. (2014) When Toyota met e-commerce: lean at Amazon, *McKinsey Quarterly*, No 2; Liker, J. (2021) *The Toyota Way: 14 Management Principles from the World's Greatest Manufacturer*, 2nd edn, McGraw Hill, New York, NY.; **575 Harvard Business Publishing:** Some of the information on which this example is based is taken from: Porath, C. and Pearson, C. (2013) The price of incivility, *Harvard Business Review*, 91 (1–2), pp. 115–121; **575 Virginia Mason Medical Center:** Used with permission from Virginia Mason Institute; **588 The Institution of Engineering and Technology:** Vitaliev, V. (2009) The much-loved knife, *Engineering and Technology Magazine*, 4 (13), 58–61.; **588 Michael Purtill:** Interview with Michael Purtill, the General Manager of the Four Seasons Hotel Canary Wharf in London. We are grateful for Michael's cooperation (and for the great quality of service at his hotel!); **589 SAGE Publications:** Adapted from Parasuraman, A. et al. (1985) 'A Conceptual Model of Service Quality and Implications for Future Research', *Journal of Marketing*, Vol 49, Fall; **591 Condé Nast:** Pardes, A. (2017) ikea's new app flaunts what you'll love most about AR, *Wired*, 20 September; **591 Digiday Media:** Joseph, S. (2017) How Ikea is using augmented reality, Digiday UK, 4 October; **591 Industrial Engineering and Engineering Management:** Sha, D.Y. and Lai, G.-L. (2012) Exploring the intention of customers to use innovative digital content information technology", IEEE International Conference on Industrial Engineering and Engineering Management (IEEM) December, pp. 1065–9; **594 Times Internet Limited:** Millington, A. (2018) Virgin Atlantic is offering a full refund on flights booked today if it can't cure a passenger's fear of flying, uk.businessinsider.com, 9 January; **594 Hearst UK:** Edwards, J. (2018) Why you should book a flight today if you've got a fear of flying, *Cosmopolitan*, 9 January; **597 The Economist:** The information on which this example is based is taken from: Markillie, P. (2011) They trash cars, don't they? *Intelligent life magazine*, Summer; **598 DMG Media:** Walne, T. (2019) Want to exchange a jar of coins for notes?, This is Money, 24 August, https://www.thisismoney.co.uk/money/betterbanking/article-7390729/Want-notes-not-coins-Banks-dont-care-cash-investigation-finds.html; **598 Financial Times Limited:** Schubber, K. (2016) The Metro Bank coin caper, *Financial Times*, 2 June; **598 New York Post:** As reported in Morgan, R. (2016) TD Bank dumps its faulty coin-counting machines, *New York Post*, 19 May; **600 Financial Times Limited:** Giugliano, F. (2015) Bank of England moves to stamp out fat finger errors, *Financial Times*, 14 June; **600 The Economist:** The Economist (2013) Overtired, and overdrawn, *Economist* print edition, 15 June; **600 Times Newspapers Limited:** Wilson, H. (2014) Fat fingered trader sets Tokyo alarms ringing, *The Times*, 2 October; **601 McGraw Hill Education:** Feigenbaum, A.V. (1986) *Total Quality Control*, McGraw Hill, New York; **608 EFQM:** The EFQM Website, www.efqm.org. Reproduced with the permission of the EFQM; **637 Gemba Academy LLC:** The information on which this example is based is taken from: Miller, J. (2021) Lessons from twelve years in pursuit of zero, Gemba Academy, 10 May, https://blog.gembaacademy.com/2021/05/10/lessons-from-twelve-years-in-pursuit-of-zero/; **638 Clean Energy Wire:** Amelang, S. and Wehrmann, B. (2020) 'Dieselgate' – a timeline of the car emissions fraud scandal in Germany, Factsheet, Clean Energy Wire, 25 May, https://www.cleanenergywire.org/factsheets/dieselgate-timeline-car-emissions-fraud-scandal-germany; **638 The Daily Telegraph:** Tovey, A. (2017) VW attacked by MPs over failure to release findings of 'dieselgate' investigation, *The Telegraph*, 22 March; **649 Darktrace:** Darktrace website, https://www.darktrace.com/en/; **649 Digital Brands Inc:** Walker, M. (2020) Darktrace: an AI cybersecurity platform that serves as the immune system for enterprise business data by fighting off threats, Credit Card News, 3 February, https://www.cardrates.com/news/darktrace-is-an-ai-based-enterprise-immune-system/; **649 Bonhill Group:** Ismail, N. (2019) Darktrace unveils the Cyber AI Analyst: a faster response to threats, *Information Age*, 4 September; **649 Bonhill Group:** Ross, A. (2019) ML and AI in cyber security: real opportunities overshadowed by hype, *Information Age*, 7 March; **650 Productivity Press:** Nahajima, S. (1988) *Total productive maintenance*, Productivity Press; New York, NY. **653 The Economist:** The information on which this example is based is taken from: The Economist (2019) An emergency landing system that passengers can activate, *Economist* print edition, 28 November; **665 Future US, Inc:** Howell, E. and Hickock, K. (2020) Apollo 13: The moon-mission that dodged disaster, Space.com, 31 March; **665 NASA:** Whiting, M. (2018) The legacy of Apollo 6, NASA, 4th April; **665 Future US, Inc:** Taylor Redd, N. (2019) Apollo 11: first men on the Moon, Space.com, 9 May; **665 The New York Times Company:** Overbye, D. (2017) Cassini flies towards a fiery death on Saturn, *New York Times*, 8 September; **665 BBC:**

Rincon, P. (2017) Our Saturn years – Cassini-Huygens' epic journey to the ringed planet, told by the people who made it happen, BBC, 14 September; **665 The Conversation Media Group Ltd:** Cobb, W. (2019) How SpaceX lowered cost and reduced barriers to space, The Conversation, 1 March; **665 EuroWeekly News:** Howells, R. (2021) China becomes only the second country to make historic touchdown on Mars, *EuroWeekly News*, 15 May; **667 John F. Kennedy:** Quoted by John F. Kennedy; **667 Guardian News & Media Limited:** George, R. (2021) Wind ...or worse: was pilot error to blame for the Suez blockage?, *Guardian*, 3 April; **667 The Associated Press:** Magdy, S. (2021) Suez Canal chief: vessel impounded amid financial dispute, AP News, 12 May; **667 Industry Dive:** Leonard, M. (2021) Timeline: how the Suez Canal blockage unfolded across supply chains, Supply Chain Dive, 6 July; **669 CNN Business:** Metz, R. (2019) The world's biggest spice company is using AI to find new flavors, CNN Business, 5 February; **669 Venture Beat:** Wiggers, K. (2019) IBM and McCormick blend new seasonings with AI, Venture Beat, 4 February; **669 IBM:** Lougee, R. (2019) Using AI to develop new flavor experiences, IBM Research Blog, 5 February; **672 The New York Times Company:** Schuetze, C. (2020) Berlin's newest airport prepares for grand opening. Again, *New York Times*, 29 April; **672 Informa PLC:** CAPA Centre for Aviation (2020) Berlin Brandenburg Airport's terminal certified for opening – at last, centreforaviation.com, 8 May; **672 The Economist:** L.R.S Berlin (2017) Why Berlin's new airport keeps missing its opening date, *Economist*, 25 January; **678 Mattia Bianchi:** This example was written and kindly supplied by Professor Mattia Bianchi, Stockholm School of Economics; **687 Condé Nast:** Burgess, M. (2018) Ocado's collaborative robot is getting closer to factory work, *Wired*, 11 January; **687 Guardian News & Media Limited:** Butler, S. (2018) Ocado to wheel out C3PO-style robot to lend a hand at warehouses, *Guardian*, 11 January; **696 Nigel Spinks:** This case was co-authored with Nigel Spinks, Henley Business School, University of Reading.

Photo:

5 Alamy Stock Photo: ICP/incamerastock/Alamy Stock Photo; **5 Alamy Stock Photo:** Niels Poulsen DK/Alamy Stock Photo; **9 Shutterstock:** TRMK/Shutterstock; **9 Shutterstock:** Rocketclips, Inc/Shutterstock; **9 Getty Images:** MoMo Productions/Stone/Getty Images; **9 Shutterstock:** Fotos593/Shutterstock; **9 Shutterstock:** Jacob Lund/Shutterstock; **10 Alamy Stock Photo:** Abaca Press/Alamy Stock Photo; **14 Shutterstock:** Vichy Deal/Shutterstock; **18 Getty Images:** Sergey Khakimullin/istock/Getty Images; **24 Getty Images:** Norbert Eisele-Hein/Getty Images; **24 Alamy Stock Photo:** BSTAR IMAGES/Alamy Stock Photo; **27 Shutterstock:** kovop58/Shutterstock; **34 Shutterstock:** VidEst/Shutterstock; **42 Alamy Stock Photo:** Denis Michaliov/ Alamy Stock Photo; **47 Shutterstock:** kamarulzamanganu/Shutterstock; **49 Shutterstock:** Brian Minkoff/ Shutterstock; **50 Alamy Stock Photo:** Malcolm Fairman/Alamy Stock Photo; **53 Shutterstock:** Lutsenko_ Oleksandr/Shutterstock; **55 Alamy Stock Photo:** Paula Solloway/Alamy Stock Photo; **66 Alamy Stock Photo:** Katharine Rose/Alamy Stock Photo; **73 Getty Images:** Chris Ratcliffe/Bloomberg/Getty Images; **81 Getty Images:** DonNichols/istock/Getty Images; **85 Shutterstock:** jax10289/Shutterstock; **88 Shutterstock:** Tero Vesalainen/Shutterstock; **100 Getty Images:** SpVVK/iStock Editorial/Getty Images Plus/Getty images; **103 Alamy Stock Photo:** Alex Segre/Alamy Stock Photo; **112 123RF:** booblgum/123RF; **122 Shutterstock:** Wongtang/Shutterstock; **128 Shutterstock:** metamorworks/Sutterstock; **132 Alamy Stock Photo:** VDWI Automotive/Alamy Stock Photo; **134 Shutterstock:** Jim Holden/Shutterstock; **137 Shutterstock:** LightField Studios/Shutterstock; **148 Shutterstock:** Gabriele Maltinti/Shutterstock; **151 Shutterstock:** Mediagram/ Shutterstock; **153 123RF:** daniel timothy allison/123RF; **159 Shutterstock:** Dmitry Dven/Shutterstock; **166 Getty images:** Melina Mara/The Washington Post/Getty Images; **166 Shutterstock:** 360b/Shutterstock; **168 Shutterstock:** Sk Hasan Ali/Shutterstock; **171 Shutterstock:** Gorodenkoff/Shutterstock; **180 Getty Images:** John Seaton Callahan/Moment Unreleased/Getty Images; **183 Shutterstock:** Aladdin Studio/ Shutterstock; **184 Shutterstock:** Shutterstock; **185 Alamy Stock Photo:** Kristoffer Tripplaar/Alamy Stock Photo; **187 123RF:** Dinis Tolipov/123RF; **187 123RF:** Belchonock/123RF; **187 Getty Images:** LIONEL BONAVENTURE/AFP/Getty Images; **188 Shutterstock:** Supergenijalac/Shutterstock; **188 Shutterstock:** Liunian/Shutterstock; **188 Shutterstock:** Jacob Lund/Shutterstock; **189 123RF:** Lightfieldstudios/123RF; **189 123RF:** Iakov Filimonov/123RF; **191 Shutterstock:** BBC Films/Focus Features/Kobal/Shutterstock; **202 Shutterstock:** I Wei Huang/Shutterstock; **211 123RF:** Julief514/123RF; **219 Getty Images:** Alessia Pierdomenico/Bloomberg/Getty Images; **219 Alamy Stock Photo:** Chris Gascoigne-VIEW/Alamy Stock Photo; **228 Alamy Stock Photo:** Justin Kase z12z/Alamy Stock Photo; **231 Alamy Stock Photo:** Hero Images Inc./Alamy Stock Photo; **234 Alamy Stock Photo:** Cyberstock/Alamy Stock Photo; **237 Shutterstock:** josefkubes/Shutterstock; **242 Getty Images:** RaptTV/The Image Bank/Getty Images; **252 Getty Images:**

Google/Handout/Getty Images News/Getty Images; **254 Shutterstock:** Dmitry Niko/Shutterstock; **267 Getty Images:** Watcharapol_Kun/iStock/Getty Images Plus/Getty Images; **269 Alamy Stock Photo:** MediaWorldImages/Alamy Stock Photo; **271 Getty Images:** Westend61/Getty Images; **275 Getty Images:** Vanit Janthra/iStock/Getty Images Plus/Getty Images; **284 Getty Images:** Maskot/Getty Images; **292 Getty Images:** janiecbros/E+/Getty Images; **294 Alamy Stock Photo:** dpa picture alliance archive/Alamy Stock Photo; **295 Getty Images:** Vadym Pastukh/iStock/Getty Images Plus; **299 Shutterstock:** Lucky Business/Shutterstock; **301 Getty Images:** Luis Alvarez/DigitalVision/Getty Images; **303 Shutterstock:** Africa Studio/Shutterstock; **306 Getty Images:** Thomas Barwick/Stone/Getty Images; **323 Shutterstock:** EQRoy/Shutterstock; **334 Alamy Stock Photo:** Imagedoc/Alamy Stock Photo; **336 Shutterstock:** Shutterstock; **339 Shutterstock:** Adning/Shutterstock; **342 Alamy Stock Photo:** Mimadeo/Alamy Stock Photo; **351 Alamy Stock Photo:** Dmitriy Shironosov/Alamy Stock Photo; **360 Getty images:** dowell/Moment/Getty Images; **367 Alamy Stock Photo:** Chombosan/Alamy Stock Photo; **370 Getty Images:** Naufal MQ/Moment/Getty Images; **373 Getty Images:** Godong/Stone/Getty Images; **375 Getty Images:** Metamorworks/iStock/Getty Images Plus/Getty Images; **392 Shutterstock:** Amritansh Sagar/Shutterstock; **409 Alamy Stock Photo:** Joerg Boethling/Alamy Stock Photo; **410 Getty Images:** Barcroft Media/Getty Images; **418 Getty Images:** MR.Cole_Photographer/Moment/Getty Images; **427 Getty Images:** RavindraJoisa/iStock/Getty Images Plus/Getty Images; **434 Alamy Stock Photo:** MICHAEL KOOREN/Reuters/Alamy Stock Photo; **436 Alamy Stock Photo:** Ben Birchall/PA Images/Alamy Stock Photo; **445 Getty Images:** Yaorusheng/Moment/Getty Images; **446 Shutterstock:** Marian Weyo/Shutterstock; **460 123RF:** Mark Agnor/123RF; **466 Alamy Stock Photo:** Uli Deck/dpa/Alamy Stock Photo; **471 Shutterstock:** yougoigo/Shutterstock; **475 Alamy Stock Photo:** Charlie Neuman/San Diego Union-Tribune/Zuma Press Inc/Alamy Stock Photo; **482 Getty Images:** NickyLloyd/E+/Getty Images; **488 Alamy Stock Photo:** Peter Barritt/Alamy Stock Photo; **489 Shutterstock:** ProKasia/Shutterstock; **492 Shutterstock:** Alex Bascuas/Shutterstock; **497 123RF:** CHARAN RUTTANASUPPHASIRI/123RF; **515 Shutterstock:** Frederic Legrand – COMEO/Shutterstock; **518 Alamy Stock Photo:** BING GUAN/REUTERS/Alamy Stock Photo; **520 Shutterstock:** Hospital man/Shutterstock; **529 Getty Images:** IndiaPictures/Universal Images Group/Getty Images; **540 Shutterstock:** AlivePhoto/Shutterstock; **542 Shutterstock:** Shahjehan/Shutterstock; **543 Alamy Stock Photo:** Robert Evans/Alamy Stock Photo; **546 Shutterstock:** Martin Rettenberger/Shutterstock; **555 Getty Images:** Monty Rakusen/Image Source/Getty Images; **557 Nicola Burgess:** Nicola Burgess; **563 Getty Images:** EXTREME-PHOTOGRAPHER/E+/Getty Images; **565 Alamy Stock Photo:** Aviation Visuals/Alamy Stock Photo; **568 Shutterstock:** Mr Pics/Shutterstock; **570 Alamy Stock Photo:** Rocky Grimes/Alamy Stock Photo; **573 Alamy Stock Photo:** NicoElNino/Alamy Stock Photo; **574 Shutterstock:** VDB Photos/Shutterstock; **587 Getty Images:** Adrian Moser/Bloomberg/Getty Images; **590 Shutterstock:** Zapp2Photo/Shutterstock; **593 Shutterstock:** Song_about_summer/Shutterstock; **597 Getty Images:** fStop Images – Caspar Benson/Brand X Pictures/Getty Images; **598 Shutterstock:** Halawi/Shutterstock; **600 Alamy Stock Photo:** Uwe Umstätter/Westend61 GmbH/Alamy Stock Photo; **607 Getty Images:** RichLegg/E+/Getty Images; **612 Alamy Stock Photo:** LightField Studios/Alamy Stock Photo; **637 Getty images:** Michael Dunning/The Image Bank/Getty Images; **638 Shutterstock:** ssuaphotos/Shutterstock; **649 Getty Images:** Michael S. Williamson/The Washington Post/Getty Images; **653 Alamy Stock Photo:** Ayman Alakhras/Alamy Stock Photo; **658 Shutterstock:** saia/Shutterstock; **664 NASA:** JPL-Caltech/NASA; **667 Alamy Stock Photo:** Suez Canal Authority/dpa picture alliance/Alamy Live News/Alamy Stock Photo; **668 Shutterstock:** monticello/Shutterstock; **672 Shutterstock:** peter jesche/Shutterstock; **678 Getty Images:** Anders Blomqvist/The Image Bank/Getty Images; **687 Shutterstock:** Sorn340 Studio Images/Shutterstock.